Developments in Marine Technology, 11

Practical Design of Ships and Mobile Units

DEVELOPMENTS IN MARINE TECHNOLOGY

Developments in Marine Technology, 11

Practical Design of Ships and Mobile Units

Proceedings of the Seventh International Symposium on
Practical Design of Ships and Mobile Units,
The Hague, The Netherlands, September 1998

edited by

M.C.W. Oosterveld

MARIN - *Maritime Research Institute Netherlands,*
Wageningen, The Netherlands

and

S.G. Tan

MARIN - *Maritime Research Institute Netherlands,*
Wageningen, The Netherlands

1998
ELSEVIER
Amsterdam - Laussanne - New York - Oxford - Shannon - Singapore - Tokyo

ELSEVIER SCIENCE B.V.
Sara Burgerhartstraat 25
P.O. Box 211, 1000 AE Amsterdam, The Netherlands

First edition 1998

Library of Congress Cataloging in Publication Data
A catalog record from the Library of Congress has been applied for.

ISBN: 0 444 82918 0

⊗ The paper used in this publication meets the requirements of ANSI/NISO Z39.48-1992 (Permanence of Paper).

Printed in The Netherlands.

These Proceedings consist of papers presented at the 7th International Symposium on Practical Design of Ships and Mobile Units. The Symposium was held at the Congress Centre in The Hague, The Netherlands, on 20-25 September 1998. The Symposium was organized by:

MARIN	Maritime Research Institute Netherlands
KIvI	Royal Institute of Engineers in The Netherlands
KM	Royal Netherlands Navy
NVTS	Netherlands Association of Maritime Engineers
TNO	Netherlands Organization for Applied Research
TU Delft	Delft University of Technology

These organizations are represented in the Local Organizing Committee.

The Local Organizing Committee organized the Symposium under supervision of the PRADS's Standing Committee. The Symposium benefited from the generous support of a number of Sponsors. These, together with the membership of the committees, are listed in the following.

COMMITTEE OF RECOMMENDATION

Dr. G.J. Wijers, Minister of Economic Affairs of The Netherlands

Mr. M.A. Busker, Chairman Controlling Board MARIN, Chairman Association of Shipyards in The Netherlands (VNSI)

Ir. J.A. Dekker, Chairman Board of Directors of Netherlands Organization for Applied Research (TNO)

Ir. J.M.H. van Engelshoven, President of Royal Institute of Engineers in The Netherlands (KIvI)

Drs. A. Korteland, RA, Chairman of Royal Association of Ship Owners in The Netherlands (KVNR)

Dr. N. de Voogd, Chairman of the Board of Delft University of Technology

Ir. M.J. van der Wal, President of Netherlands Association of Maritime Engineers (NVTS)

PRADS STANDING COMMITTEE

Prof. S. Motora, Honorary Chairman of PRADS, Ship and Ocean Foundation, Tokyo, Japan

Dr. M.W.C. Oosterveld, Chairman PRADS Standing Committee, MARIN, Wageningen, The Netherlands

Ir. S.G. Tan, Secretary PRADS Standing Committee, MARIN, Wageningen, The Netherlands

Dr. L.L. Buxton, University of Newcastle, United Kingdom

Prof. O. Faltinsen, The Norwegian Institute of Technology, Trondheim

Dr.Ing .G. di Filippo, Fincantieri, Trieste, Italy

Prof. H. Kim, Seoul National University, Korea

Prof. J.W. Lee, Inha University, Inchon, Korea

Dr. D. Liu, American Bureau of Shipping, New York, U.S.A.

Prof. H. Maeda, University of Tokyo, Japan

Prof. T. Terndrup Pedersen, Technical University of Denmark, Lyngby, Denmark

Prof. Y.S. Wu, China Ship Scientific Research Center, Wuxi, China

PRADS LOCAL ORGANIZING COMMITTEE

Dr. M.W.C. Oosterveld, Chairman Local Organizing Committee, MARIN, Wageningen

Ir. S.G. Tan, Secretary Local Organizing Committee, MARIN, Wageningen

Prof.Ir. A. Aalbers, Delft University of Technology, Delft

Ir. G.T.M. Janssen, Netherlands Organization for Applied Research (TNO), Delft

Ir. P.J. Keuning, Royal Netherlands Navy, The Hague

Prof.Dr. J.A. Pinkster, Delft University of Technology, Royal Institute of Engineers (KIvI), The Hague

Mr. J. Veltman, Netherlands Association of Maritime Engineers (NVTS), Rotterdam

Prof.Dr. J.H. Vugts, Royal Institute of Engineers, The Hague

SPONSORS

MARIN

Ministry of Economic Affairs of The Netherlands

Municipality of The Hague

TNO

SYMPOSIUM SECRETARIAT

Maritime Research Institute Netherlands
P.O. Box 28, 6700 AA Wageningen, The Netherlands
telephone : +31 317 49 32 19
fax   :   +31 317 49 32 45

PREFACE

These Proceedings contain the papers presented at the 7th International Symposium on Practical Design of Ships and Mobil Units. The Symposium was held at the CONGRESS CENTRE in The Hague, The Netherlands, on 20 - 25 September 1998.

The overall aim of PRADS Conferences is to advance the design of ships and mobile marine structures through the exchange of knowledge and the promotion of discussions on relevant topics in the fields of naval architecture and marine and offshore engineering. Greater international co-operation of this kind can help improve design and production methods and so increase the efficiency, economy and safety of ships and mobile units. Previous symposia have been held in Tokyo ('77 and '83), Seoul ('83 and '95), Trondheim ('87), Varna ('89) and Newcastle ('92).

The main themes of this Symposium are Design Synthesis, Production, Ship Hydromechanics, Ship Structures and Materials and Offshore Engineering.

Proposals for over two hundred papers have been received for PRADS '98 from 25 countries, and 126 have been accepted for presentation at the Conference. Given the high quality of the proposed papers, it has been a difficult task for the Local Organising Committee to make a proper balanced selection.

Some topics which attracted many papers were Design Loads, Design for Ultimate Strength, Impact of Safety and Environment, Grounding and Collision, Resistance and Flow, Seakeeping, Fatigue Considerations and Propulsor and Propulsion Systems. The great current interest in these topics and the high quality of the papers guarantee a successful Conference.

The success of PRADS '98 depends on the great contributions of the participants with a special acknowledgement to the authors.

We as Local Organizing Committee have done our utmost to create the proper atmosphere for an interesting and enjoyable conference.

M.W.C. Oosterveld and S.G. Tan

CONTENTS

DESIGN SYNTHESIS

DESIGN - USE OF PROBABILISTIC METHODS

DESIGN – METHODOLOGY

DESIGN – MISCELLANEOUS

PRODUCTION

SHIP HYDROMECHANICS

HYDROMECHANICS – PROPULSOR AND PROPULSION SYSTEMS, MISCELLANEOUS

SHIP STRUCTURES AND MATERIALS

OFFSHORE ENGINEERING

DESIGN SYNTHESIS

TRA NESS "New Ship Concept In The Framework Of Short Sea Shipping" A european Targeted Research Action: results and exploitation aspects

C. Camisetti

CETENA S.p.A. - Via Savona 2 - 16121 Genova - ITALY

ABSTRACT:

The Targeted Research Action NESS was sponsored by the DGXII EU Commission in order to co-ordinate Brite EuRam II projects within the maritime transport field.

TRA-NESS, activated on the second semester 1994, involved seven projects, twenty-seven partners of eleven European Countries, and the total budget is about 21,5 MECU

This research framework aimed at favouring at most the development of short-sea shipping traffic through suitable exploitation of all available technologies to improve ship efficiency, safety and environment friendliness. Taking into account the restricted field of fast speed transport, and considering the most promising market for fast speed vehicles have mainly followed up the expected innovation.

1. INTRODUCTION

Targeted Research Actions (TRA) are co-ordination structures launched and funded by the European Union under the Industrial and Materials technology programme (Brite EuRam II), aiming at co-ordinating sets of thematically interrelated EU financed research projects (clusters), with the objective of strengthening and broadening the impact of the expected outcomes resulting from the single projects with respect to European wide strategic and industrial priorities.

TRA-NESS, the Targeted Research Action on NEw Ship Concept in the Framework of Short-Sea Shipping, involves 7 projects, 27 partners from 11 European Countries

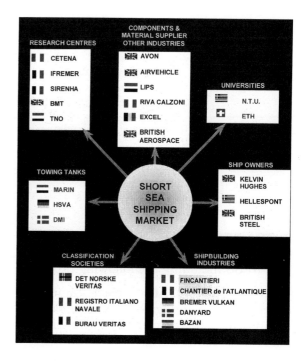

Fig. 1 - TRA-NESS Consortium

2. OBJECTIVES

The overall strategic objective of the TRA-NESS is to favour at most the development of short-sea shipping traffic through suitable exploitation of all available technologies to improve ship efficiency, safety and environment friendliness. The expected innovation has been mainly followed up by taking into account the restricted field of fast speed transport, and considering the most promising market for fast speed vehicles as car ferries traffic.

The research was mainly directed to investigating a new large size Surface Effect Ships (SES) concept for its potential on short-sea shipping transport. However a large part of the research activity results that specific type of ship will have short/medium term exploitation also for more conventional type of ships for passengers and cargo.

Pay load	more than	1000	tons
Displacement	more than	5000	tons
Power	up to	100.000	kW
Speed	more than	50	kn
Cruising range	more than	300 - 500	Miles
Operating range	up to	sea state	5

Target fares			
passengers	0,65 0,95	-	ECU/mile
cars	0,75 1,15	-	ECU/mile

SES target characteristics

3. TRA NESS PROJECTS

TRA NESS including the following projects:

HYDROSES. This project aimed at developing an integrated and advanced theoretical / experimental hydrodynamic procedure for the design of large SES.

MATSTRUTSES. This project is addresses to advanced materials and structural design procedures for large size lightweight SES.

MAINCOMPSES. The main system components for large and fast passenger and cargo SES, such as high performance water jets, fans and seals, are improved in this project. A Ride Control System (RCS) is also developed.

SESLAB. It aims at developing an experimental tool/laboratory for technology assessment at sea and for design of large and fast SES.

SHIP. The ship hull integrity program purposes at developing an advanced stress monitoring system for large ships in operation.

LASERW. The clustered part of this project is addressed at implementing laser beam welding in the shipbuilding sector for high strength structural components.

EUROSLOSH. This project is aiming aims at improving the structural integrity of partially filled tanks by the development of new experimental and numerical analysis procedures for sloshing and impact loads.

TRA NESS is an objective-oriented applied research. It is an interdisciplinary and multi-sectorial research which involving the subjects interested to the Short Sea Shipping market as shown in the figure below:

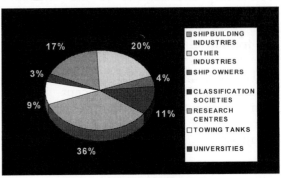

Fig. 2 - The TRA-NESS Partnership

The TRA NESS start from the ex-ante cluster of the four projects relevant to SES (MATSTRUTSES, SESLAB, HYDROSES and MAINCOMPSES) to which three project whose contents and objective were fitting TRA overall objectives were added.

3. RESULTS

SES TARGET VESSEL FEASIBILITY ASSESSMENT

A reference vessel called Target Vessel (T.V.), whose characteristics are shown in the following figure 3, was defined based on market analysis and parametric studies in order to finalise all the research work and to assess ship and main component performances.

Hull forms and ship performances were hydrodynamically optimised in order to reach the following target performance requirements:

Speed (in calm water)	60 kn at 60 000 kW
Vertical Accelerations	Lower than 0.1g RMS in the passenger area at sea state 5 (RCS off)
Speed-loss in waves	30% at sea state 5 (RCS on)

An estimation of the T.V. operability is shown herebelow:

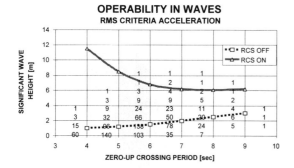

The feasibility assessment of the main system components matching the following target was demonstrated:

Waterjet	Power up to 40 MW Weight/Volume: ~1.4 t/m³ Weight/Power: ~1.75 t/MW Max Flow: ~56 m³/s
Seals	Working pressure 11.000 pa Durability: ~ 1375000 nautical Miles
RCS	Up to 5000 t displacement Integrated control of vent valves, active fins, trim tabs, waterjets
Fans	Max Flow: 1100 ~m³/s Max Pressure: 11,06 kPa Cushion Volume per Fan: ~3150 m³ Suitable for operating with high pressures, large cushion volume, rough sea conditions

Integration of the main components - seals, fans, RCS, water-jet - into the ship system was carried out. Installation aspects, coupling of the water-jet propulsion system with the Main Engines, impact of the sub-systems on the general layout of the ship were investigated with the ultimate scope to ensure optimum ship performance from the point of view of both speed/power and comfort aboard (motion sickness, noise and vibration).

Fig. 3 - Target Vessel Characteristics

Different solutions for hull structure were developed. A structural design in aluminium alloy was carried out and the most relevant components and details were investigated and verified through experimental tests and calculations. The target of a weigh of structure to be the 30% of displacement was achieved.

Three structural composite material solutions were deeply investigated. The detailed design of most critical structural components in advanced material was carried out on the basis of the results of the preliminary strength assessment. Such structural components have been the starting point for a further selection of material and relevant definition of production process and tools.

The feasibility of the structure was demonstrated, even if high cost of raw material may limit the today competitiveness of such a solution.

SHIP COMPONENTS AND SYSTEMS

Waterjet

Analysis on the hydrodynamic, technological and operational aspects of the design of a waterjet propulsion system able to deliver a largerer thrust as compared to the existing ones, necessary by the high power requirements of the Large SES, was carried out.

The hydrodynamic efficiency of the configuration has been studied by CFD calculations, using a free-surface potential-flow code for the study of the best location of the inlet at the aft-body bottom, and a commercial RANSE solver for the selection of the inlet geometry.

The hydrodynamic assessment of the final configuration was performed at wind tunnel on a scaled model. Attention was also given to the problem of pump design (single and double stage) and to fatigue strength and corrosion phenomena.

Seals

Investigation of actual and newly devised front/rear seals geometry and material was performed through specific test-rig experiments and mathematical models. On the one hand, the seals must be resilient to high levels of wear and tear for long periods of time, must endure high values of the air pressure, and must make easy the maintenance/replacement operations.

On the other hand, they must yield the minimum hydrodynamic drag and must produce the minimum air-leakage in still water and rough-sea. The workingness of the innovative configuration has been established on specific model-tests.

Fans

Conceptual design of an improved fan system, supported by specific test-rig experiments, able to feed the cushion with the most appropriate combination of air-flow and pressure in a wide range of conditions, thus ensuring high performances of the ship in her whole operational profile, was performed.

Ride Control System (RCS)

Definition of an innovative Ride Control System (RCS), based on the integrated action of newly-devised Vent Valves (VV) and Inlet Guide Vanes (IGV) fans, able to significantly reduce the wave-induced vertical accelerations and thus the comfort levels on board, through the minimisation of the fluctuations of the air pressure within the seals and the cushion plenum, was performed.

An optimal control strategy was formulated according to the most recent theories of automatic systems. The conceptual design of the RCS was realised by means of a time-domain mathematical simulation model of the ship/seal/fan system, which reliability was assessed on the basis of specific test-rig experiments and seakeeping model-tests.

Stress monitoring system

SHIP generated a stress monitoring system far in advance of the state of the art. Work on Finite Element Analysis (FEA) has been carried far further than was originally planned: this is of central importance for relating the strain at the measurement point on the deck to the stress at the joints at the bottom of the holds, where the most dangerous problems will arise. Work on weather monitoring by navigation radar has proved unsuccessful (with unmodified radar either), but it has been established that the software designed for using the weather data for predictions of stress works correctly, and could be used with a dedicated sea-state monitoring radar.

A commercial spin-off (the SMART system) has been generated, and is currently on market for stress monitoring, although it does not currently incorporate at the present time the sea-state monitoring.

The monitoring of actual stress level in the most loaded structures at sea is particularly important for fast and large new-concept ships. For such ships the uncertainties of the structural behaviour are considerable, while a high operability rate is essential for competitiveness. The goal is that route and speed should not be affected by unnecessary conservative ship operation.

MATERIAL AND PRODUCTION TECHNOLOGIES

A close co-operation to realise a technology transfer by applying new shipbuilding materials and their related production and assembly technology has been established between shipbuilding industries and other industrial fields, particularly aeronautic and civil engineering.

Aluminium alloy

Production and assembly technologies for the T.V. aluminium alloy hull structures have been defined. The problem of welding and assembling massive quantities of thick structures was analysed, welding and assembly procedures were defined and set up, production and assembly tests were performed on two large size full scale components (more than 4x3x4 m).

Composite Structures

Advanced material solutions were investigated for hull and components structures.

Selection and definition of different component concepts, structural typologies, fabrication methods, ship production methods were performed by means of a ranking and elimination process. More than thirty components and joint typologies were analysed. More than ten component fabrication methods and ship production methods were analysed with respect to their production easiness - potential for automation, sensitivity to production and assembly failures, working, workshop and skill requirements - production time, cost for tooling, man-hours, investments waste management and degree of innovation.

A series of preliminary production tests were performed on small scale component and joint samples, to assess both the feasibility of the structural concept and the applicability of the production and assembly processes. The following sandwich panels are investigated:

Balsa/PVC foam /Carbon fibre reinforced epoxy
Balsa core / Carbon fibre reinforced epoxy
PEI foam / PMI foam / Carbon fibre reinf. epoxy
Aluminium honeycomb /Carbon fibre reinf. epoxy
Balsa core / Grass fibre reinforced polyester

T.V. structural design using FRP sandwich panels in carbon fibre reinforced epoxy and balsa or advanced foam core material met strength and stiffness criteria within the weight targeted limits. The cost of such solutions varies according to material cost. Today cost of raw materials is still too high, but development of new carbon fibres (i.e. 50K fibres) and application of innovative core design (i.e. "strong plank") may bring down the cost to a more competitive level. The labour cost may however be lower than that required for metallic materials.

The feasibility of a modern shipyard for the T.V. building able to handle the advanced materials and their related production process was also investigated.

A prefabricated sandwich panel based on aluminium honeycomb core and C.F.R.Plastic laminates, with its four edges closed by aluminium extrusions was also deeply investigated. Basic elements are produced in a separate plant, assembled into modules, and delivered to the shipyard for final assembly. On a technical standpoint, the honeycomb core solution outclasses any other technique and material, but it is much more expensive due to both raw material and production cost.

Laser Welding

The laser welding technology wasn applied to new thin high-strength steel structures fabrication. For welding hull panels a production tools characterised by on line control and seam pilot was tested.

The possibility to weld defect-free subassemblies in the case of accurate cutting and fixating systems was shown. The obtained welding process results faster than conventional methods.

Measured distortion and residual stress are much lower than the typical values induced by arc welding. Strongly reducing the redressing process, the use of laser welding relevantly reduced the total cost of fabrication, particularly for thin plate structures.

DESIGN PROCEDURES AND TOOLS

Design loads definition

Global loads

In principle three approaches are available to define the design loads :

Classification Rules, hydrodynamic numerical tools and experimental tests on a segmented model. For a configuration such as the T.V., for which consolidated experience is virtually fulfil, it is almost compulsory the recourse to direct calculations or model-tests. It is common practice for a ship with unrestricted service to define the design loads based on the long-term extreme value corresponding to a service life of 20 years. As the T.V. is assigned to a restricted service, a different strategy can be adopted. A limiting sea state is prescribed in terms of a significant wave height of 4m beyond which the ship cannot operate. The design loads can thence be defined based on the equivalent short-term extreme value, that is the value which will be exceeded only once on all the encounters of the limiting sea state during the service life of the ship. The procedure to actually determine the design loads is different depending on whether direct calculations or model-tests are performed. As a matter of fact, the reliability of numerical codes are predicting of the hydrodynamic loads on the Target Vessel is still uncertain due to the dominant presence of non-linear effects in the dynamics of lift system. The theoretically devised short-term extreme value must be related to the actual design value via a correlation factor for which determination recourse is to be made to Classification practice. In case of segmented model tests actual measure of the global loads in the limiting sea state is available but, due to the limited duration of the experiments, it is not possible to directly determine the equivalent short-term extreme value, which must be extrapolated by means of a Weibull fitting of the measured peaks. The experience gathered so far allows to conclude that:

- loads are generally higher off-cushion, but survival condition provides lower values than on-cushion, so that design loads should be derived in operational conditions ;

- model-tests are still necessary for a safe definition of the design loads
- Rules and carefully calibrated direct calculations are close to the experiments but not on the conservative side.

Slamming

According to the above procedure, the first step in the definition of the design loads was to establish the limiting sea state up to which the vessel is operational. In the case of the T.V. wet-deck slamming must be considered as the major factor in impairing the operability.

Free-running model-tests carried out within the HYDROSES Project in MARIN high-speed towing-tank showed no slamming evidence up to a significant wave height of 4m, in relatively long waves (about 200m) ; to detect significant wet-deck slamming in these sea conditions it was necessary to switch to off-cushion mode. To be on the safe side, 4m was adopted as the limiting wave height assigning the Target Vessel a DNV service class R_1 in agreement with comparable existing fast ferries. As the design loads for the Target Vessel refer to the on-cushion mode, slamming loads on the wet-deck do not seem to be a primary concern for the dimensioning of the cross-deck structure. However, a deeper insight at the local slamming effects was achieved by means of specific drop-tests at NTUA on a full-scale panel of the wet-deck and slamming pressure measurements during free-running model-tests at the MARIN seakeeping basin.

A good correlation agreement was found between the results of the drop-tests and two-phase hydroelastic calculations with DYTRAN code, indicating that such a code can be a reliable tool in the prediction of wet-deck slamming loads and induced strengths when linked with some

prediction method for the wave-induced vertical motions, such as that specifically developed within the HYDROSES Project.

Sloshing

An extensive analysis of the effect of different physical parameters on the sloshing phenomenon was carried out.

Scale effects and fluid structure interaction was deeply investigated.

Computer Fluid Dynamic (CFD) and experimental technique were improved and the understanding of the effects of sloshing on structures through evaluation of the structural response to sloshing impact as well as of dynamic pressure loads was achieved.

Theoretical design tools

Hydrodynamics

The complex physics of SES was theoretically approached with particular care of the complex interaction effects between lift-system and flexible seals.

This led to the development of the following prediction tools for SES:

Powering (SESFLOW, 3D Rankine-source BEM method for linear steady free-surface potential flow; SESRES, total resistance and steady-running equilibrium accounting for the non-linear interaction between lift-system and flexible seals),

Seakeeping (SESWAVE, 3D Rankine-source BEM method for linear unsteady free-surface potential flow; SESDYN, time-domain simulation method for the wave-induced motions accounting for the non-linear interaction between lift-system and flexible seals motions; SESSLAM, seakeeping post-processor for the evaluation of the local forces induced by wet-deck slamming; SESSTAT, seakeeping post-processor for the short/long-term statistical analysis of time-series),

Manoeuvring (SESMAN, time-domain procedure for the simulation of standard manoeuvres in the horizontal plane).

Such tools were further integrated within a multi-windowing software architecture and implemented with a standard interfacing with a **visualisation package**.

Structures

The general philosophy for a software procedure oriented to structural analysis and design of large SES was defined.

The main steps of the defined procedure was included into modularity software architecture including different software typologies: equation solvers, graphical processors and file manipulators. The main functions that the developed procedure is able to perform are summarised in the following:

– load definition, based upon hydrodynamic analysis ;

– modelling of structure at a suitable degree of detail according to the actual stage of the design process ;

– definition of loading conditions ;

– global static and dynamic analysis, including ultimate strength analysis.

– detailed local static linear and non linear analysis ;

– fatigue analysis, through calculation of the long term stress range distribution for selected structural details and evaluation of their fatigue behaviour on the basis of suitable SN curves using the cumulative damage algorithm according to the Palmgren-Miner law.

System modularity allows easy management of the software, in particular: the use of commercial software (mainly for finite elements), the creation of new software, to be integrated with the existing one, the possibility to manage independently every module by different user, given the necessary input and output.

Experimental design tools

Seslab

A laboratory vessel was conceived and designed in order to allow experimental tests at sea by large-scale models of SES.

Due to the uncertainties related to the scaling effect on the air cushion dynamic simulation the SESLAB is a unique tool to assess the performances of different types and size of large SES at sea.

The SESLAB simulation capability was maximised referring to a market oriented range of vessel typologies and performances, while maintaining a quite simple design in spite of the very complex conception of the ship.

The main SESLAB feature is to perform with only one tool:

- Simulation by a large-scale model of a wide range of SES (length range $65 \div 160$ m; displacement range $500 \div 5000$ t; speed range $40 \div 70$ kn.).
- Measurement of the most important parameters and performances related to the design and the operation of SES.
- Assessment of new and innovative design concepts, both for the platform and for the main components

A cost/benefit analysis was performed showing that utilisation of SESLAB testing facility is economically competitive to a complete set of tests performed at major European Model Basins.

Hydrodynamics

Extensive experimental tests were performed using different facilities (resistance and self-propulsion tests in towing-tank, wind-tunnel tests, PMM tests, test-rig laboratory tests, forced-oscillation tests in depressurised towing-tank, captive and free-running tests in rectilinear and seakeeping tank) and models of different scales (1:25 [6 m length] and 1:35 [3.5 m length]). The scope was to get of a better insight into the aero/hydro-dynamics behaviour of SES (including scale effects) and to provide suitable data for correlation and calibration of physical/mathematical models.

Powering, seakeeping and manoeuvring sea-trials were performed systematically varying the main design parameters on a purposely hired existing SES, for the scope of a better insight into the overall behaviour of SES at sea and of providing suitable data for further correlation and calibration of the physical/mathematical models.

Structures

An extensive testing campaign on materials and structural components was carried out. About 100 materials were investigated for hull and components structure, including high tensile steel, aluminium alloy and composite materials.

Even if a number of applications are related to the SES exercise, the results obtained in this area can be exploited as they are to other more conventional ships.

Fatigue test on samples of the most significant welded joints and on the full scale prototype of a complex structure representing the most critical structural area of a large aluminium alloy structure were carried out. These tests allowed establishing a correlation between the fatigue behaviour of elementary welded joints and those of large complex structures under the most relevant loading

conditions in order to define guidelines for the design and the production of large components with respect to the long-term behaviour.

Two types of tests were performed: Mechanical, and Fire Reaction. Mechanical tests cover both static and fatigue properties of laminates and sandwiches. Within the static program four hundred and twenty tests were performed on five types of laminated and seven types of sandwiches based on hi-tech materials (carbon fibres, PEI/PMI, Light Alloy Honeycomb); tests were addressed to survey tensile, compressive, shear and bending performances. Within the fatigue program fifteen bending tests on two different sandwich typologies were performed. The fire reaction test program comprises more than 60-cone calorimeter test on five different types of sandwiches and ten different types of fire barrier materials. Collected data allowed selecting the most proper materials for the design of the Target Vessel structures as well as provided the information necessary for the design and for the numerical evaluation of the investigated structures.

Full Load Displacement range	FLD	from	77	to	92.2	t
Overall length	LOA				37.0	m
Cushion beam	Bc	from	5.16	to	9.58	m
Cushion pressure range	Pc	from	2.5	to	3.6	kPa
Lift/weight ratio range	L/W	from	0.80	to	0.85	
Propulsion power					1130	Kw
Lift power					195	kW
Electrical power:					75	kVA
Top speed on cushion range (sea state 0)		from	25	to	38	kn
Operational maximum wave height (H1/3)		up to			1.07	m

SESLAB main characteristics:

5. EXPLOITATION ASPECTS

Know-how, tools and products developed within the TRA-NESS are of immediate or potential relevance for exploitation.

Exploitation of technologies developed so far enables the involved industries, and more in general, European shipbuilding and related industries, to increase market competitiveness.

Many technologies were directed to design and construction of large and fast SES car ferry ships. Nevertheless considerable spin-off on more conventional shipbuilding products occurred. Some other technologies were already meant to be applicable to more conventional sea vessels.

The consequence of technology exploitation will be to maintain and possibly increase European market share.

The social impact will then be evident considering that employment level in European shipbuilding industry is currently estimated as 150000 direct/indirect workers.

Although that of fast ferry could be seen as a niche market, the demand of fast ferries had considerably increased in the nineties. In 1997 the ships delivered were 74 and it is expected at least the same trend for the future.

In the same period also demand for higher performances (speed and payload) has increased. At the beginning of the nineties, when TRA-NESS was conceived, fast ferry speeds ranged from 20 to 25 knots, while presently speeds are about 40 knots and dead weight is over 1000 tons.

The T.V. chosen as reference for te TRA-NESS pre-competitive research reaches 55 knots to fulfil future market needs and to compete in the early 21st century with the Japanese SUPER TECHNO LINER in the fast and large SES market.

6 CONCLUSIONS

TRA NESS results show that co-operative product-process oriented research at the pre-competitive and pre-normative level can give significant benefit to the partners, with respect to the effort directly spent.

This research work has given the partners the possibility to efficiently contribute to the overall result, and to establish an effective co-operation basis among a very large number of shipbuilding related industries, research centres and Universities (21 partners of 11 Countries).

7 ACKNOWLEDGEMENTS

The author acknowledges all the Co-ordinators of the TRA-NESS projects for the contribution they provided to the Final Publishable Report::
Mr. L. Sebastiani - HYDROSES,
Mr. G. Puccini - SESLAB&MATSTRUTSES,
Mr. V. Farinetti - MAINCOMPSES,
Mr. M. Dogliani - EUROSLOSH,
Mr. S. Wheeler - SHIP
Messers F.Belid,G. Parames - LASERW.

8 REFERENCES

1. HYDROSES final synthesis publishable Report, CETENA 1996
2. MASTRUTSES final synthesis publishable Report, CETENA 1998
3. MAINCOMPSES final synthesis publishable Report, FINCANTIERI 1997
4. SESLAB final synthesis publishable Report, CETENA 1997
5. EUROSLOSH final synthesis publishable Report, RINA 1996
6. SHIP final synthesis publishable Report, BMT 1997
7. LASERW final synthesis publishable Report, ASTILLEROS ESPAÑOLES 1997
8. TRA-NESS Final Publishable Report, CETENA 1998

Practical Design of Ships and Mobile Units
M.W.C. Oosterveld and S.G. Tan, editors.

Principal trends of container vessels development

W.Chadzyński

Technical University of Szczecin
71-065 Szczecin
Al.Piastów 41

The aim of this paper is to present the principal trends of the development of container vessels throughout the years 1968 - 1988. The analysis of the changes of vessel efficiency indicators such as the utilization of hull inside space and deck area, transport efficiency coefficient etc has been performed. Some methods adopted by the designers to improve the efficiency are reviewed. In the summary the attempt to extrapolate further technical progress directions is undertaken.

1. INTRODUCTION

The introduction of modularized units in sea transportation is connected with the years just after the World War II when the efforts to improve the cargo handling were observed with the following advent of containers used in trade between US main land and their overseas dependencies.The process had started during late fifties and few years later sea container transportation started up. The first container vessels were converted from standard cargo vessels by Matson Navigation Company about 1960 and those were the prototype of that class of the vessels. The unification of containers was fixed up in 1965 when ISO adopted 40 ft and 20 ft standard container size. Those sizes are obligatory nowadays; however the scope of ISO containers includes a number of other accepted sizes and some „non-standard" containers remain in use as well.

In late sixties first purpose-built container vessels started to appear and then from 1969 to 1973 first great boom for container newbuildings was the case with shipbuilding industry.
In that time the significant number of container vessels of 1st, 2nd. and 3rd generation were built.The carrying capability of those vessels was still improved but it appeared soon that the then existing solutions exhausted their possibilities. In pursuit of speed the biggest container vessels became real „fuel eaters" with daily consumption over 150 tonnes per day and even 300 tonnes per day (gas turbine). Still increasing level of general cargo containerisation brought on the need for cheaper transportation means , first of all on the short routes and then the hybridal container/general cargo tweendeckers (without cellguides) appeared in a great number. Those were transformed with the time into a modern type container vessels fitted with cellguides, with the superstructure shifted aft as much as possible. and with still improved carrying capability.

This type of the vessels has been dominating on short and medium routes and has been known as feeder container vessels with a carrying capacity up to 2000 TEU. Big quantity of such vessels were built by the yards throughout the world in overturn of eighties and nineties. The last years introduction of 6000 TEU mammoths on the main routes probably will result in splitting of worlds leading operators and possible releasing of third and fourth generation vessels for secondary routes service. The source says that the potential containerisation of general cargo can reach the level 60 - 65 % of total amount. If this level is reached then the sea container transportation will be potentially in danger of appearing of its first ever annual volume fall. The 3% gap between actual fleet growth and demand may result in the drop of freight rates and further decreasing of the orders volume for new vessels.

2. THE REVIEW OT THE TRENDS IN CONTAINER VESSELS SOLUTION

The annual carrying capability of container vessel can be estimated as in formula

$$P = 2 \cdot k \cdot C \cdot N \qquad (1)$$

where
k - maximum number of containers carried by the vessel
C - cargo space utilisation coefficient
N - the number of voyages per year
The yearly number of voyages

$$A = \frac{365 - M}{T} \qquad (2)$$

where
T - the roundtrips number (including sea voyage, waiting time, container handling time)
M - offhire time (maintenance, repair, delays etc.)

The efficiency of transportation can be estimated using the following factors related to one roundtrip
1. Number of containers carried by the vessel
2. Time spent in the port with particular emphasis on container handling time
3. Time of one roundtrip (vessel speed)
4. Fuel cost
5. Other operating costs (usually related to the vesel size)
Accordingly some general trends can be easily distinguished , all dictated by the economy of transportation reasons:
1. The increasing of overall size of the vessel
2. The combination of the speed, ME output, TEU number allowing to obtain high effectiveness of transportation process.
3. The reduction of port cost
4. The improvement of hold space and deck area utilization within main dimensions given.

To estimate those trends the population of above 100 representative container vessels built in the period 1968 - 1988 years was considered. The population was initially divided into 3 groups i.e
I -   750 - 1500 TEU
II - 1500 - 2600 TEU
III - 2600 - 3700 TEU

It appeared soon that the regularities inside the groups were quite similar for groups I and II, with some deviation within the group III. Finally it was decided that population will be considered as a whole.

2.1. The size of the vessel
The simplest and the most effective way of improving of transport mean productivity is the increasing of it size. This process is occuring for almost every type of cargo vessels and continuing up to the appearing of the factor strongly damping the process. That thesis was confirmed in full extent in the case of great tankers, where ecological danger becomes the damping factor of the vessel size increasing process.
If the case is about container vessels - the first turning point - Panama Canal limitation was achieved very soon at the beginning of its development. It results in very long and narrow hulls limited by Panama breadth limit. To allow further   increase   of TEU number   the Postpanamax idea was created . Occassionally substantial decreasing of water ballast quantity was achieved. The defficiencies of Postpanamax are the high accelerations and resulting forces on the lashing of deck containers, additional torsional moment due to high B/T influencing hull steel weight, and decreased operational flexibility. The Postpanamaxes are the starting point to the new generation of giant container carriers 8000 TEU and over if the necessity of substantial port facilities modernisation does not damp the process. The economy of the transportation with a giant container vessels in terms of one TEU transportation cost is better by abt. 10%, if comparing with 4000 TEU vessel, however the relative port costs are somewhat higher.

2.2. The effectiveness of transportation process
The broadly used term for the vessel efficiency estimation is a combination of a payload, service speed and installed power. For container vessel it may be expressed as transport effectiveness coefficient.

$$TE = \frac{\text{Max TEU number} \cdot \text{service speed}}{\text{Installed ME output}} \qquad (3)$$

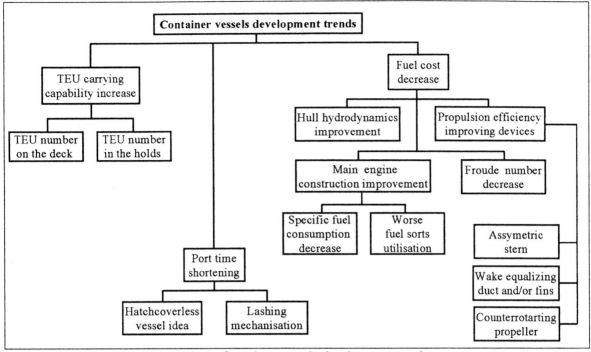

Fig 1. Container vessels development trend

This term is generally useful, since the parameters are readily available for most vessels. The higher TE value, the more efficient is the vessel claimed to be. It must be noted that particular parameters used in this formula are subject to the development because of technical progress throughout the years. The controlling factor in this case is the fuel cost. The decrease of fuel cost may be obtained in the result of hull hydrodynamics improvement, adoption of several devices improving the propulsive efficiency and main engine construction development as shown in Fig.1 The further development of power plant adopted on container vessels will be conditioned also by such factors as the reduction of power plant weight and dimensions, greater flexibility in power plant arrangement,decreasing of harmfullness of exhaust gases. It will not necessarily decrease the fuel consumption because the medium or high speed engines or diesel-electric plant have bigger specific consumption than low speed diesel engines. The general trends observed in the course of the years 1968 - 1988 are identified in Fig 2 showing the changes of TE coefficient and in Fig 3 showing the variation of Froude number.

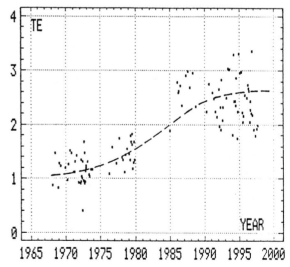

Fig 2. Variation of transport effectiveness coefficient TE with the time

2.3. The port costs

:The reduction of port costs is directly connected with a quantity of particular container manipulation as well as diverse preparatory works. The mechanisation of container lashing

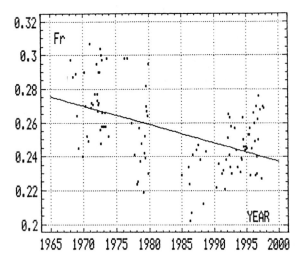

Fig 3. Scatterplot of Froude number

work is one of possible directions of development, but the most spectacular idea in this aspect is a hatchcoverless (open top) vessel concept. This concept allows to avoid:

- time consuming manipulation with lashing equipment during loading and unloading operations,

- hatch cover operation,

This is apparently a promise of substantial reduction of total cost of transportation. More detailed consideration, however, does not confirm such conclusion. The obvious gain resulting from the omitting of hatch covers, lashing equipment, hatch covers and lashing operation cost is balanced by disadvantages of bigger depth of the hull as well as the cellguide construction extended over the deck level.

Financial result of the introducing of open-top vessels if compared with those of hatch cover vessel depends of the vessel size and is more distinct for small vessels and not so clear for large container vessels. At present when hatch cover vessel performance is on the top level the advantages of open-top vessels which are only starting to the competition are not so sharply outlined to convince the owners. The grand scale advent of open - top vessels is probable at the

moment when the replacement of the vessels built at the overturn of eighties and nineties starts.

2.4. The hull space and deck area utilization.

The general solution of cellular container vessel remains the same from early seventies. The engine room and living superstructure of small and medium size vessels is usually stern-located with some exceptions if great slender vessel where the engine room and accommodations are usually semi-stern located. The containers are placed inside vertical cells in the holds and stacked on weather deck and hatch covers. Sometimes the cellguides are extended above weather deck. This arrangement was permanently improved to obtain more and more containers carried, which results in surprisingly great effect regarding vessel carrying capabilities.

The directions of improvement can be presented as follows (see also Fig.5):

1. The increasing of TEU number in the holds realised by:

- increasing of fore part block coefficient, preserving CWL entrance angle as small as possible,

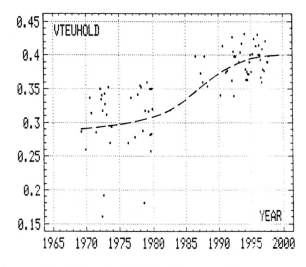

Fig 4. Trend of hull inside space change

17

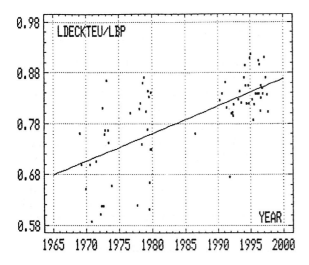

Fig 7. Change of deck length utilization

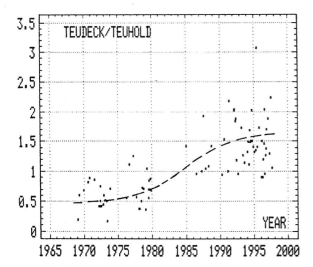

Fig 8. Change of the relation between the number
of containers in the hold and on the deck

- the shortening of forecastle and adoption of
  high wave breaker,
- side location of deck cranes if adopted.

The effectiveness of those steps is much more
efficient which is clearly shown in Fig.6,7,8
where:

- FTEUDECK is a number of deck containers
  multiplicd by the spot area of one container
  related to the vessel area module,

- LDECKTEU is a total length of weather deck
  occupied by the containers related to the
  length bp.
- TEUDECK/TEUHOLD is a relation of deck
  containers number to the hold containers
  number, respectively.

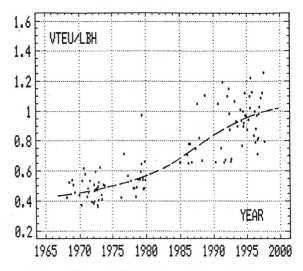

Fig 9. Change of the utilization of container
vessel main dimensions

The total improvement of the utilization of
hold space and deck area throughout the years
1968 - 1988 is quite significant which is clearly
evident in Fig 9. where VTEU/LBH is a total
volume of the deck and hold placed containers
related to the cubic number LBH. The
comparison of main dimensions of two vessels of
approximate same TEU capacity built in 1972 and
in 1996 respectively seems to confirm this
conclusion (Fig.10). It must be noted, however.
that this improvement is substantially better for
the feeder container vessels up to 2000 TEU. The
bigger vessels were more stable as far as main
dimensions utilization and its effectiveness was
improved principally by the increase of vessel
size.

If it is said that the carrying capability of
container vessel is usually based on general
number of containers carried it must be
remembered that this characteristics is closely
composed with the deadweight of the vessel. The
containerised cargo is generally light cargo, so
that the owners are interested in increasing of
TEU number while maintaining of the deadweight

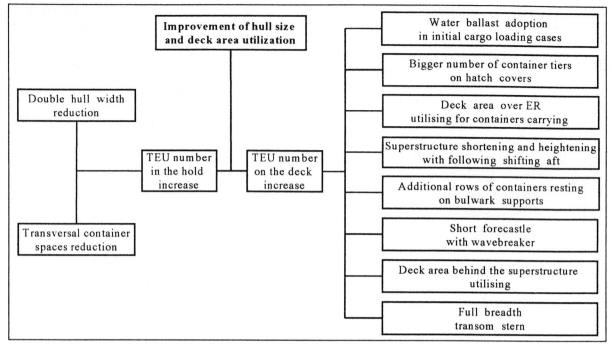

Fig 5. The directions of container vessel capability improvement

- decreasing of cellguides transversal and longitudinal gaps including new type of profiles.

- decreasing double hull width,

- shortening and shifting aft of the engine room,

- adoption of open holds behind the superstructure,

- increasing of hull height.

The effectiveness of all those solutions seems to be rather moderate (excluding some specific cases), which is shown in Fig.4. VTEUHOLD is a total volume of all hold placed containers related to the cubic module LBH.

2. The increasing of deck TEU number realised by:

- adoption of water ballast in DB tanks at the departure loading conditions,

- utilization of full breadth of the vessel by the adoption of bulwark supports for extreme TEU rows.

- shifting aft of the superstructure with the utilizing of weather deck above ER for additional TEU stacks,

- full breadth transom stern utilized for additional TEU stacks on the deck behind the superstructure with the transferring mooring facilities below the deck,

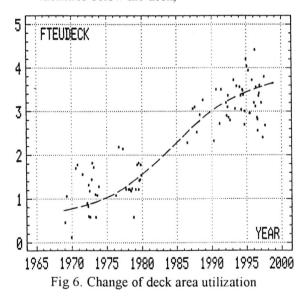

Fig 6. Change of deck area utilization

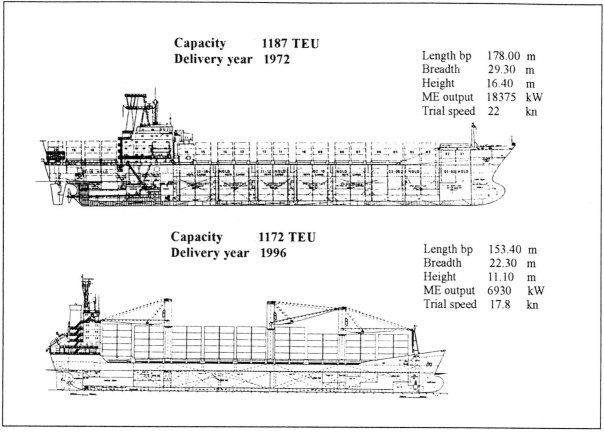

Capacity	1187 TEU		Length bp	178.00	m
Delivery year	1972		Breadth	29.30	m
			Height	16.40	m
			ME output	18375	kW
			Trial speed	22	kn

Capacity	1172 TEU		Length bp	153.40	m
Delivery year	1996		Breadth	22.30	m
			Height	11.10	m
			ME output	6930	kW
			Trial speed	17.8	kn

Fig 10. The comparison of main dimensions of two

vessel of aproximate same capacity built in 1972 on in 1996

within given limits. There are a few typical figures of loaded container weight between 10 and 16 tonnes per TEU. A number of empty containers is usually carried as well as a number of 20 tonnes containers. The general trend of DWT/TEU relation is presented in Fig.11. This relation may be considered as a rough indication of the order of one container weight if the apropriate deduction for the weight of stores and initial water ballast is considered. Some decreasing of DWT/TEU with the passage of time can be seen, which probably results from still better utilisation of hull space and deck area of the vessel.

The technical progress results also in a significant decrease of light ship weight what is shown in Fig 12.

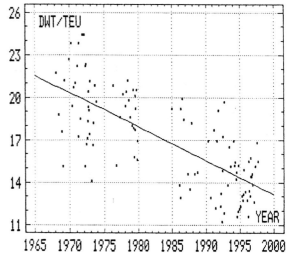

Fig 11. The change of the relation between the deadweight and the number of containers

20

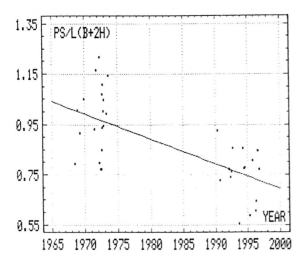

Fig 12. The change of relative light ship weight

3. THE SUMMARY OF TO DATE TRENDS OF CONTAINER VESSELS DEVELOPMENT

The improvement in efficiency of every new technical product including cargo vessels can be represented in a form of relationship between performance obtained and the time. This relation is generally known as S-curve. At the beginning of new vessel type development we can observe the graet activity of the owners in ordering of new vessels. The number of the vessel in service grows very quickly , but their performances are diversified and in average grow very slowly. After the market saturation is achieved the new orders are not occurring so frequently. The yards and owners are gaining the experience and knowledge necessary to make an advance. When new wave of orders comes the average performance of the newbuildings is on the distinctly higher level. Finally the development becomes more and more difficult and further improvement of performances becomes impracticable what is presented on S-curve as the asymptothic limit of the curve. The factors affecting performance of container vessels were presented in Fig.2,4,6,8,9. The factors such as transport effectiveness, vessel hold space and deck area utilization seem to confirm the conclusion that there are some signs of stabilisation in container vessel general solution development. Nowadays the further substantial progress is achieved by the increasing of vessel size. The significant improvement of the performance of particular vessel within given main dimensions will be still possible if new idea of general solution is introduced.

4. THE PROBABLE FURTHER TRENDS OF CONTAINER VESSEL DEVELOPMENT

The analysis of container vessel solution changes over the period 1968 - 1998 allows to indicate the following general directions of improvement:

- moderate increase of hold containers number,
- sharp and still continued increase of deck containers number,
- shortening of time in the port,
- increase of transport efficiency coefficient,

The most promising direction seems to be the increasing of deck container number and the increase of transport efficiency coefficient. The hypothetical directions of progress are possible as follows:

- full utilization of deck area with possible transferring of accommodation to the space below main deck and mooring and anchoring facilities below the poop deck and forecastle deck; special solution for wheelhouse and antennas area is wanted.
- maintaining of vessel depth at the minimum required by LL Convention ,
- -adoption of very slender and very beamy pontoon type hull with the installation of strong cell guide construction directly on weather deck from the transom up to fore wavebreaker,
- decreasing of water ballast capacity,
- increasing of the deck container tiers number up to the limit due to the compression strength of the lowest container in the stack,
- adoption of power plant characterized by small dimensions of particular power sources and big flexibility of engine room arrangement (medium speed diesel engines or diesel electric propulsion)
- possible adoption of progressive slender hull form developed from trimaran form.

REFERENCES

1. H.Poehls. Entwicklungstendenzen der Containerschiffe. Forum TEU 6000+. Hamburg 1995.

2. Jomar Kuvas. Transport Capacity and Economics of Container Ships from a Production Theory Point of View. Transactions of RINA 1974.

3. - . US and World Shipping. Alliances,human factors and flags. Marine/Log June 1966.

4. H.Poehls. Some comparative aspects of modern container ship development. PRADS 1995.

5. L.Nowakowski. The analysis of some technical parameters and prices of container ships (in Polish). CTO - Gdansk 1996

6. W Chadzynski. Container vessels 1990 - 1996, design characteristics, trends, development. II International Conference on Marine Technology ODRA'97. Szczecin 1997.

7. A. Kraus. Container - Transportsysteme der Zukunft. Schiff und Hafen 1/1997.

Practical Design of Ships and Mobile Units
M.W.C. Oosterveld and S.G. Tan, editors.

Hydrodynamic impact on efficiency of inland waterway vessels

A. G. Lyakhovitsky

Scientific Center , State University for Water Communication,
5/7 Dvinskaya str.,198035 St. Petersburg , Russia
Department of Ship Design , State Marine Technical University,
3 Lotsmanskaya str.,190008 St. Petersburg , Russia

It provides the results of systematic research of the energy-saving devices for inland waterway vessels. These devices were based on the theoretical calculations and experimental researches of ship's hydromechanics in shallow water.

1. INTRODUCTION

The shallow water influences on the selection of dimensions , forms of the lines and the propulsion of inland waterway vessels (IWV).The ship limiting draft T must be limited depending on the depth of water h. The speed of ship v can be dependent also on the depth of water. Energy saving and ecology are incompatible with the motion of the ships near critical speed v_c in shallow water

$$v_c = \sqrt{gh} \quad , \qquad (1)$$

where g-gravity acceleration.
The formula (1) is uncomfortable for IWV design. Author suggested the limits between precritical and supercritical zones of ship's motion. The case of deep water was included too [1-3]. These limits were suggested depending on the relative depth of water h/L (L-length of ship) and Froude number

$$F_n = v/\sqrt{gL} \qquad (2)$$

All the vessels may be separated in one of two groups: vessels with precritical speeds (VPS) or vessels with supercritical speeds (VSS).The concrete energy-saving technical proposals are essentially different for each of two groups. A significant reduce of the fuel consumption and ship generated waves influence on small vessels, banks, water creatures may be achieved by improvement of
• hull geometry, including transition from mono-hulls to the multi-hulls;

• ship propulsors, including partially submerged propellers;
• hull-propulsor interaction.

2.VESSELS WITH PRECRITICAL SPEEDS

The increase of loading capacity of VPS on condition of restricted draft causes the increase of form fulness, relative length of cylindrical part of the hull and breath B to draft ratio (B/T). As a result of it the influence of stern hull lines on ship energy losses becomes very strong. The combination of low speeds and above-mentioned hull forms of the VPS will bring to increase the role of viscous resistance. The total resistance coefficient C_T is given

$$C_T = C_F + C_{PV} + C_W \quad , \qquad (3)$$

where C_F - frictional resistance coefficient;
C_{PV} - viscous pressure resistance coefficient or form resistance coefficient;
C_W – wave-making resistance coefficient.
There is a principal change of value of viscous resistance as well as its components correlation C_F and C_{PV} depending of the stern hull lines.
The selection of the stern hull lines IWV depends on the type of ship's propulsor. To evaluate eficiency of the propulsors the term of the maximum hydraulic amidship section A_m is used

$$A_m = BT \qquad (4)$$

Table 1
The results of full-scale measured mils trials

	R143		ST 1305	
	The Tunnel After End	USSR Certificate №1122545	The Tunnel After End	USSR Certificate №1122545
Deadweight capacity , tn	280	280	1300	1300
Draft , m	1,3	1,3	2,9	2,9
Diameter of propellers , m	0,9	0,9	1,7	1,7
Speed on trial , km/h	14,0	14,0	17,6	17,6
Actual power , kw	364	240	845	600
Ratio of actual power : USSR Certificate / The Tunnel After End №1122545	–	0,66	–	0,71

The maximum hydraulic amidship section utilization coefficient β_u is given

$$\beta_u = \mathrm{x}A_0 / A_m \quad , \qquad (5)$$

where x is number of the propulsors; A_0 – hydraulic area of propulsor. Majority of ship's propulsors of IWV (propeller, ducted propeller) have circular hydraulic area. The maximum hydraulic amidship section A_m has right-angled section across the ship. This irreconcilable contradiction leads to two technical solutions:
• the tunnel after end;
• the application of partially submerged propellers.

2.1. The tunnel after end

VPS with tunnel after end usually have slightly immersed transom to prevent air entrainment to the propellers. It practically does not increase hull resistance at designed load draft. But in full load conditions the immersed transom draft becomes much greater , which leads to additional energy losses. In order to lessen this unfavourable effect a special after end was suggested [4-5].
Transition from the traditional tunnel after end to the VPS of new constructive types (USSR Certificate № 1122545) was realized in two river cargo ships R143 and ST 1305. The draft of R143 may be changed from 0,8 to 1,3 m. The draft of ST1305 may be changed from 1,8 to 2,9 m. Both of them were tested with full-scale measurement of fuel consumption and delivered horse power. The results of full-scale measured mils trials is given in Table 1. The total results of the actual power reduction resulted from:

• reduction of the resistance;
• improvement of the hull-propulsor interaction

2.2 The after end and partially submerged propellers

The maximum hydraulic amidship section utilization coefficient β_u (5) is usually done by increasing propeller diameter due to applying the tunnel after end and by multiplying the number of propellers. Another possible way is the use of partially submerged propellers (PSP) with large diameter. The special PSP were designed by F.F.Bolotin. Such propellers were designed for the cargo VPS with displacement 740 tn and small draft 1,3 m. There were designed also the special after end lines with small buttock angles. This special after end ensured the decrease of the resistance very much. The disposition of PSP behind the special after end ensured the increase of hull efficiency. The self-propellered model tests have been carried out with two comparative models in scale model 1:7,321: conventional ship with tunnel after end and river ship of new type (813601) [6]. The comparative characteristics of these models of vessels are given in Table 2.
The application of PSP increased the maximum hydraulic amidship section utilization coefficient β_u (5) in two – three times. Moreover the advantage of propellers over the paddle-wheels was preserved. Transition from traditional type of vessel 813600 to new type of river ship 813601 has made possible to decrease very much the value of viscous resistance and to change its components correlation C_F and C_{PV}. The value of total resistance coefficient and its components are given in Figure 1 for full-scale ship

Table 2
The comparative characteristics of the models

	813600 (The Tunnel After End)	813601 (USSR Certificate №1643309)
Dimensions , m		
Length between perpendiculars	8,47	8,47
Beam on waterline	1,38	1,38
Draft	0,137 – 0,205	0,137 – 0,205
Displacement , m^3		
when T = 0,178 m	1,89	1,89
Diameter of propellers , m	0,137	0,356

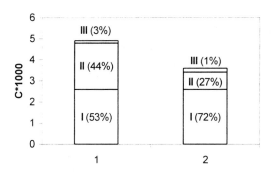

Figure 1. Value and Components Correlation of Resistance Coefficients (Fn = 0,10)
1 – 813600 ; 2 – 813601 ;
I – C_F ; **II** – C_{VP} ; **III** – C_W

when T = 1,5 m by F_n = 0,10. It follows from Figure 1 that the total resistance coefficient decreased by transition from 813600 to 813601 from $4,9 \cdot 10^{-3}$ to $3,6 \cdot 10^{-3}$ (~27%). The quasi – propulsive coefficient η_D by this transition increased more then 30%. Increase of quasi – propulsive coefficient η_D is received with usage of the average comparative experimental data.

The experiments in shallow water showed additional advantages of the model with PSP , for example the improvement of propeller efficiency (quasi - propulsive coefficient η_D)due to the increase of dynamic propeller immersion with decreasing water depth.

Unfortunately, because of some changes in the economic policy of Russia , the river ship of new type 813601 was not built.

3. VESSELS WITH SUPERCRITICAL SPEEDS

The problem of speed growing on the inland waters was connected with practical use of supercritical speeds. These vessels must be design by Froude numbers (2) more than 0,6 (Fn > 0,6) [1]. In Russia of the 60s, the systematic study of the transient regime ship's hydrodynamics was started for creation of, so-called , VSS for internal water-ways [7]. The practical account of the shallow water's influence in designing of VSS is important only near the lower limit of transient regime. After the VSS has overcome the wave barrier , influence of the shallow water becomes less significant. The first ships , built and tested in the natural conditions, have proved this fact. The hydrodynamic studies in the 60s-70s in Russia gave the possibility to substantiate expediency for creation of multi-hull VSS as catamarans and especially trimarans [1,2,7,8,9,10]. These technical solutions have been further developed in the practical ship-building in many countries. The latest years the transient regime of motions is being intensively filled up with constructed and designed vessels of peculiar arrangements. Unfortunately, because of some changes in the economic policy of Russia, intensity of the local research work in this field of the IWV design has declined.

3.1. The wave-making resistance and multi-hull vessels
Total resistance R of the VSS is given

$$R = C_T \frac{\rho V^2}{2} S \ , \qquad (6)$$

where ρ-water dencity, S-wetted surface area.
Within the practical ranges of the principal dimensions, transition from mono-hull to the multi-hulls vessels is always associated with increase in the specific wetted surface area. The specific wetted surface area $\bar{S}$:

$$\bar{S} = S/\nabla^{2/3} \quad , \qquad (7)$$

where ∇ - immersed volume.
The mutual disposition of VSS depending on increase of the specific wetted surface area: mono-hull, three-hull ship with central main hull and small side hulls or vessel with outriggers, catamaran, three-hull ship with even hulls. For example, the specific wetted surface area of three-hull ship with even hulls S_T is given

$$\overline{S_T} = 1,44\overline{S_1} \quad , \qquad (8)$$

where $\overline{S_1}$ -specific wetted surface area of one hull.
One of the many advantages of the trimaran is its low resistance in the transient regime of motion. Interference between the hulls plays the most important role in wave-making resistance generation. These problems were solved by author for three-hull ships [8]. General formula for the wave-making resistance components of the trimaran is given

$$R_W = R_{WMH} + 2R_{WSH} + \Delta R_{WC} + \Delta R_{WT}, \quad (9)$$

where R_{WMH} - wave-making resistance for the main hull of trimaran; R_{WSH} -wave-making resistance for a single trimaran side hull; ΔR_{WS} - a component due to side hulls interference (so-called "catamaran effect"); ΔR_{WT} -a component due to mutual interference of the central and side hulls (so-colled "trimaran effect").
The cases of catamaran, three-hull ship with even hulls and mono-hull were taken as particulars from the general formula (9).
The effect of three-hull configuration on the resistance can be estimated by making use of coefficient K_W , which is equal to the trimaran wave-making resistance coefficient C_{WT} to the wave-making coefficient of the sum of the three individual hulls $C_{W\infty}$

$$K_W = C_{WT}/C_{W\infty} \quad , \qquad (10)$$

The medium and large values of the longitudinal clearens are favorable for any water depth, precritical and supercritical ship's speeds. It is worth to note that increase of longitudinal clearens decreases very much " hump " in wave-making resistance as on deep water as on shallow water.
In Russia of the 70s, the systematic study of the three – hull vessels of different types was begun. The trimaran with main hull with small area of waterline has been studied [10]. The vessel of this type (USSR Certificate №501921(1976)) has been studied later in USA (so – called " O'Neil Hullform") [11]. There have been experimental investigation of the effects of side-hull position on the resistance characteristics of trimarans in USA [12]. The UK is planning to invest the design and built a ship with a trimaran hull form [13]. The british trimaran, built and tested in Thames, have proved success application this type of IWV [14].

3.2. Exposure of coast to the waves

Increase of VSS speed and displacement results in growth in wave generation, which is important for inner water ways. VSS generated waves influence the small vessels, banks, channels, water creatures. The problem of wave impact can be subdivided into two parts:
estimation of wave impact as a function of vessel geometry, speed and channel configuration;
determination of the allowable wave elements from the ecology point of view.
The energy coefficient $\bar{e}$ can be used for comparison of wave impact on environment of different vessel designs [15]

$$\bar{e} = \Delta E/\rho g \nabla L \quad , \qquad (11)$$

where ΔE - part of energy goes to transformation of the free surface.
Energy coefficient is sensitive to hull geometry, number of hulls, dynamic lift devices. Transition from mono-hulls to the multi-hulls is one of the effective ways of reducing wave impact on ecology. Ship generated wave impact on ecology is called ship hydroecology. The amount of energy transmitted from vessel to the surrounding water is indicator of IWV hydroecology.

4. CONCLUSIONS

4.1. We have proposed the general method of approach to the IWV design. This method is proved to conform to the laws of the ship hydrodynamics.

4.2. Two original types VPS are shown. The new type of VPS has a lower fuel consumption and will burn considerably less fuel during its life-time than a less efficient VPS.

4.3. We have noted the prospects for successful application of multi-hulls VSS, especially trimarans, and we have proposed the results of theoretical and experimental investigations of multi-hull vessels.

4.4. We have proposed the evaluation of the IWV wave impact on environment on the basis of the fundamental law of energy conservation. This evaluation may be used for the creation of classification rules of IWV exploitation on the inland waterway.

REFERENCES

1. A.M. Basin, I.O. Velednitsky, A.G. Lyakhovitsky, Hydrodynamics of Ships in Shallow Water, Sudostroenie Publishers, Leningrad, 1976.
2. A.G. Lyakhovitsky, Supercritical Speeds and Multi-Hull Vessels, Int. Symp. on Ship Hydrodynamics, St. Petersburg (1995) 92.
3. A.G. Lyakhovitsky, Energy –Saving Hydromechanics Complexes, J. River Transport, 9 (1992) 16
4. M.V.Begak, S.G.Verchovodov, A.I.Korotkin, A.G. Lyakhovitsky, The Tunnel After End of a River Ship with a Variable Draught, J. Sudostroenie, 9 (1987) 13.
5. A.G. Lyakhovitsky,A.B.Petrov, M.V.Begak, A.I.Korotkin, B.I.Kelim, G.S.Nikitin, After End of a Ship, USSR Certificate № 1122545 (1984).
6. A.G. Lyakhovitsky, V.F.Bavin, F.F.Bolotin, V.M.Zavialov, Yu.M.Sadovnikov, Yu.N.Gorbachov, A.A.Tikhomirov, River Ship, USSR Certificate № 1643309 (1991).
7. A.G. Lyakhovitsky, Influence of the Ship Hydrodynamics on Development on the High-Speed Vessels of the Transient-Regime of Motion, Int. Conf. CRF'96 (1996).
8. A.G. Lyakhovitsky, Theoretical Investigation of the Wave-Making Resistance of the Multi-Hull Vessels, Proc. LIWT, 148 (1974) 14.
9. A.G. Lyakhovitsky, Features of the Trimaran Hydrodynamics and their Consideration in the Design, J. Sudostroenie, 12 (1975) 3.
10. A.G. Lyakhovitsky, Three-Hull Vessel, USSR Certificate № 501921(1976).
11. M.B.Wilson, Ch.Ch.Hsu, Wave Cancellation Multihull Ship Concept, Proc.HPMV'92 (1992) MH 26.
12. B.B. Ackers, H.C. Landen, T,J,Michael, E.R.Miller, J.P.Sodowsky, O.W.Tredennick, J.B.Hadler, An Investigation of the Resistance Characteristics of Powered Trimaran Side-hull Configurations, The Soc. of Naval Arch.and Marine Eng., 13 (1997) 13-1.
13. Plans to Build a British Trimaran Demonstrator Vessel, The Naval Arch., Sept. (1997) 49.
14. C.Fulford, A.Smith, Quest for Quality Influences European Vessel Production, Maritime Reporter/Eng.News, Sept. (1996) 38.
15. A.G. Lyakhovitsky, The problems of River Ships Hydrodynamics, Sudostroenie Publishes, 414 (1985) 50.

1998 Elsevier Science B.V.
Practical Design of Ships and Mobile Units
M.W.C. Oosterveld and S.G. Tan, editors.

Small Waterplane Area Triple Hull (SWATrH) for Mega Yacht Purposes

Dipl.-Ing. Ulrich Heinemann

Heinemann-Yacht-Design
Katschhof 3, D-52062 Aachen, Germany; Via N. Giangi 7, I-47037 Rimini

These pages show the concept of a semisubmersible yacht which will be constructed as a trimaran. The second focus of the concept, optimisation of the shipbuilding production process, leads to a quite unusual design of the superstructure.

1. SEMISUBMERSIBLE-PRINCIPLE

Compared to conventional ships having the same displacement, semisubmersibles have a substantially smaller waterplane (Fig. 1). This smaller waterplane results in a very different motion behaviour of the semisubmersible: the motion amplitudes are much smaller. This principle was already perceived around the last turn of the century. The basic concepts were developed by Nelson (1905).

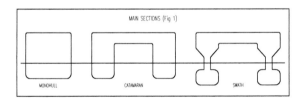

To illustrate the difference in motion behaviour, we take a look at Figure 2. In this [1], a monohull and a SWATH with identical waterline length (146 ft) are compared. Indicated you see the heaving movement's speed over the occurring wave height.

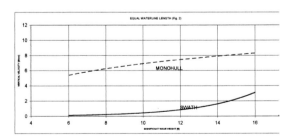

In Figure 3, this comparison is between ships of the same displacement (500 t). In addition, a conventional catamaran has been included. Please note that the significant and not the real wave height is shown.

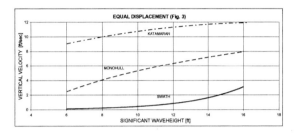

Floating oil riggs, anchored above their oil fields, are designed as semisubmersibles. The great advantage is, that their own movements even in a rough sea are rather small. The benefits are obvious: minor stress on the drilling equipment and the best working conditions for the crew.

This principle was first realised in a sailing ship in the Netherlands in 1971. There, the DUPLUS was built for exploring fossil deposits in extreme depths. Its operational speed was just 8 knots, which, however, seemed sufficient for its purpose. The navy also saw the advantages very soon, as all kinds of missiles could be directed precisely to their target. But for military use, only vessels with a higher speed were and are suitable. Because of that, the US-Navy developed a ship two years later, the KAIMALINO, whose escaping speed reached 25 knots. This ship was designed as a semisubmersible catamaran. Internationally, that type of vessel became known as SWATH (Small Waterplane Area Twin Hull).

In relation to a monohull, semisubmersibles can sail faster in a rising sea. When conventional ships have to slow down quite early in rising wave heights, SWATHs can maintain their operational speeds much longer, until they, too, have to reduce their rate. This is illustrated in Figure 4.

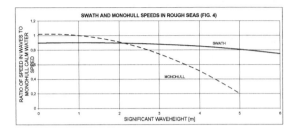

Today, there are about 50 SWATHs around the world, with length from 6 to 260 metres. The applications span from exploration vessels to yachts, ferries, cruisers and battleships.

2. ADVANTAGES AND DISADVANTAGES OF APPLICATION OF THIS PRINCIPLE TO MOTOR YACHTS

There is a large range of advantages of a SWATH compared to a monohull. They apply not only to ships in a general sense but also to motor yachts.

Foremost, their higher seaworthiness has to be mentioned. Comparative tests between a planing vessel (monohull) a displacement craft (monohull) and a SWATH [2] led to a seaworthiness about 15 times higher. A lot less seawater is taken on deck. The occurrence of slamming is neglectable. The operational speed can be maintained almost independently of wave height. Comfort, usability and safety of the crew are, especially in rough weather, definitely higher. Seasickness occurs less often. During these comparative tests with wave heights of seastate 3 to 6, only 5 % of the crew became seasick in a 4-hour-duty on board the SWATH against up to 25 % on the displacement craft.

Manoeuvrability, especially at low speed, is higher than for a monohull. Because of the propulsion units lying usually far from the centre of ship, steering can be done mainly by the engines. A SWATH can turn around on the spot.

With a larger width, a SWATH also has a larger deck area, similar to a catamaran. This enables many design possibilities, that are not available on a monohull. However, a larger width of the ship also requires a wider berth in a harbour and wider passages. Noise level in a SWATH is lower. If the propulsion units are mounted low in the SWATH's floats, the sound producers are more distant from the

crew. The propellers do not send any pulses to the ship's bottom. This also helps to keep the noise level low.

A SWATH can have only one defined operational draught, while a monohull is seaworthy in a certain draught range. A SWATH's stability and its whole sailing behaviour depend sensitively on the floating condition. The installation of water ballast tanks is imperative. However, this also gives the opportunity to adapt the draught to the natural conditions by blowing the tanks empty under certain circumstances, e. g. in a harbour or during inland navigation. Draught is, on principle, slightly greater than for a monohull.

The wetted surface is smaller on a monohull with identical weight, leading to an increase in frictional resistance and, therefore, also in the drive performance of a SWATH. As with rising speed the effect of the wave resistance increases more for the monohull than for the SWATH, this resistance disadvantage only applies to lower speeds.

Because of the general taste in motor yachts, a designer no longer has much room to move. Everyone has his own conception of a motor yacht. Planning vessels and those yachts, that want to look like a planning vessel, are drawn in an aggressive line. Receding lines with single longitudinal edges and a chopped stern as well as a negative sheer are common. Manufacturers, who prefer to address the sailor in their customer, give their products a more shiplike character. Styling elements like screens, diagonally turned forward, and roofed decks are taken from fishing trawlers. Their speed potential, as well, is adapted to their operational use. Without question, many stylistic features have a nautical or technical background, combined with experience. It is very difficult, however, to find new ways in this area, as the designer only has tight creative possibilities. The idea of 'ugly' or 'beautiful' can only be changed over a long period. It took about 60 years from the former limousine boats to the current Miami Beach look. Acceptance of motor catamarans increases in Europe only in small steps. Their introduction follows via the usual lines for monohulls. Nevertheless, a catamaran surely is nothing new to the beholder.

For semisubmersible yachts, no special taste has been generated yet. For the moment, everyone sees such a vessel as something strange and, therefore, as really extraordinary. Friends of science fiction are quicker in liking such ships (illustration without apertures and details, pure shape). There are still many technical and formal possibilities. It is a pleasure to make use of this. Real innovation is still possible in this field. Everything is permitted, that makes sense. The view of the classification societies does not differ much. Although there are classification regulations for SWATH, there are no regulations for this type of application. Sensible and nautical precautionary measures against distress at sea and average (collision bulkhead, etc.) are no contradiction to a semisubmersible.

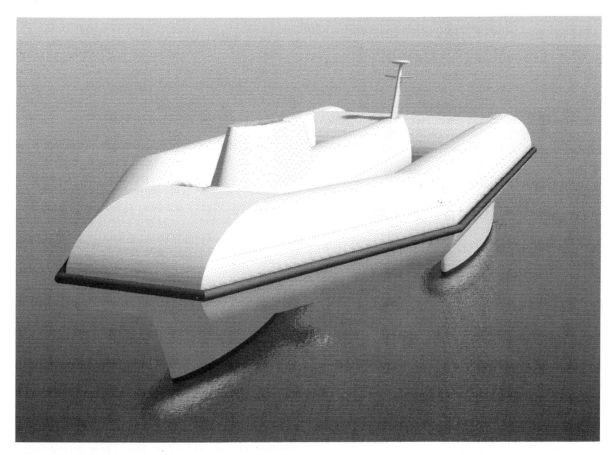

3. DESIGN CONCEPT

The design formulae [3], [4], and [5] resulted in the pre-design ranges or dimensions shown below. These pre-design parameters are based on the SWATHs realised until today. Figure 5 shows length over all (LOA), length between perpendiculars (LPP), beam over all (BOA) and draught (D) of twin hull semisubmersibles dependent on displacement (DSPL).

Shown in Figure 6, you find the engine performance (Pb) prospectively to be installed over the operational speed (Vs).

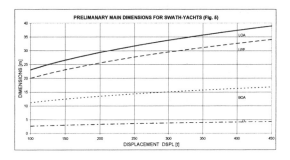

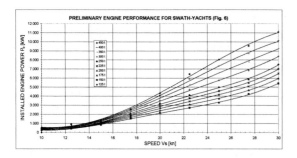

Focus of this design is the optimisation of the shipbuilding production process as well as the range of the vessel. This vessel is meant to be able to cross the Atlantic between Europe and North America for reasonable costs. Minimum range is defined as 2,500 nm, which is the distance between the British Isles and Canada.

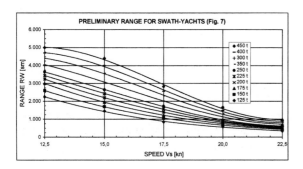

Figure 7 shows the ranges of SWATHs with different displacements as a function of the operational speed (Vs). Therefore, in the beginning the dimensions based on these pre-design statements for a SWATH with 12 kn operational speed can only be the following:

Length over all (LOA)	26.50 m
Length between perpendiculars (LPP)	23.00 m
Beam over all (BOA)	12.50 m
Draught (D)	3.00 m
Displacement (DSPL)	150.00 t
Hull height (DWD)	4.80 m
Clear height (hC)	1.80 m
Diameter of one hull (DH)	1.85 m

For semisubmersibles, SWATH is the state of the art. As the name implies, this is a semisubmersible twin hull. But the semisubmersible principle also applies to ships with more or less than two hulls.

For example, floating intermediate storage tanks of the off-shore industry are designed as monohull semisubmersibles. This concept makes sense for vessels, that require large volumes but not much deck area for superstructures. To get the maximum freedom in designing the living quarters in a motor yacht, this concept is not suitable for application here.

For a SWATH, a designer has to use an even number of drive units. Minimisation of the interference resistances of both parallel struts/hulls makes it necessary to arrange them as far apart as possible. This, in turn, leads to a large width of the vessel.

By arranging the three struts/hulls in the Kelvin's angle, a semisubmersible triple hull, called SWATrH (Small Waterplane Area Triple Hull) by the author, can minimise the occurring interference resistances. In addition, three separate drive units may be installed. This seems to be more economical, as three engines can be bought cheaper than two propulsion units with the same total performance.

The liability of average caused by engine failure is reduced. Each engine can be run separately with the optimum speed, depending on the desired sailing speed. An advantage not to be underestimated in times of rising fuel prices.

The second focus of the concept, optimisation of the shipbuilding production process, leads to a quite unusual design of the superstructure. On the one hand, it must take all stress induced by the struts, and on the other it also has to be very light, and production costs should be low. These prerequisites can be fulfilled best by tubes as engineering elements. They can be produced easily and inexpensively by the metre, and they can take torsional stress even with minimal thickness of the shell.

The midship frame of the yacht is similar to one of an aeroplane. Where a plane has its cargo space, a tubelike superstructure may without difficulty house the technical equipment (piping, wiring, etc.). Installation is facilitated by good accessibility. And the centre of gravity is kept very low, as well.

Above the inner bottom, the interior with all its fittings can be installed, which always have the same radius on the outside. Prefabrication is easy, and the installation can follow later.

An inner ceiling serves as a separation from the area above, that will house, e. g., the ventilation system. Usage of a tube automatically leads to only outside cabins. Thus, natural lighting and ventilation of all rooms is no problem.

The lounge including pantry and bar is situated in the front part of the superstructure. Lying in the junction of the three main tubes, it provides a lot of space and a good view forward as well as sideways. It is the social centre of the ship. From here, you can enter the three longitudinal tubes and, via stairs, the bridge.

The longitudinal tubes accommodate owner (family), crew and guests, respectively. Traditionally, starboard is the owner's side of a ship. Here, he will find enough room for himself and his family in two cabins. A weather deck of his own, separate from the guests, leaves room for private leisure. Port serves as receptacle for guest cabins. The central tube houses the crew cabins and an office for the owner, who can step directly onto the bridge from here. The crew can quickly reach all important places on the ship from their centrally positioned quarters.

A transversal connecting tube forms the rear part of the superstructure. It includes a garage for a van. The car can roll off over a hydraulically retractable gangway of its own. Dinghies and additional aquatics equipment hang from davits between the main tubes. After entering the boats, they can be lowered unhindered.

The wheelhouse is located above the superstructure of the vessel, which, by its arrangement, provides a perfect overview of the nautical situation. It contains not only the control and communication console of the yacht, but a spacious second lounge as well. From here, you can step onto the sundeck. The whole area above the tubes is covered by grating. This results in an enormous resting area, that will also carry a helicopter. Access to the dinghies is given by companionway hatches in the sundeck.

An encircling passage serves on the one hand nautical handling purposes, on the other as connection between the cabins. It can either be closed by vertically adjustable window covers against spray, or be opened to let in the sun. Both outer longitudinal tubes can be reached separately via this passage and gangways.

After finishing the 18th pre-design, the main data have shifted towards the following values, because of several considerations and calculations:

Length over all (LOA)	32.00 m
Length between perpendiculars (LPP)	27.16 m
Beam over all (BOA)	14.10 m
Draught (D)	3.08 m
Displacement (DSPL)	150.00 t
Hull height (DWD)	5.02 m
Clear height (hC)	1.95 m
Diameter of one hull (DH)	2.53 m
Installed engine performance (Pb)	2700 kW
Escaping speed, circa	24 kn

4. DESIGN CONCEPT

As already shown above, tubes will be used as basic elements of the superstructure. Their smeared wall thickness is easily determined in the pre-design [6]. The single tube sections can be manufactured separately without problems, and assembled afterwards on one level.

The struts are clamped between the outer walls of the tubes and the cabin floor; they will be pushed up into the superstructure. This facilitates feeding in stress, as well as missing beams help to keep the room clear.

The hulls will each house a separate propulsion system. Feed and return lines are run through the struts to the superstructure. Every single engine is capable of reaching the cruising speed of 12 knots. All three engines together permit a speed of 25 knots. For voyages across the Atlantic, mainly the front drive will be used, which is located clearly in the frontal area of the vessel. Thus, optimum straight away sailing is given by this arrangement alone, without additional support of a rudder.

The propellers are mounted on the end of the hulls, respectively (not illustrated in the plots in the appendix). Their efficiency is higher than usual because of the almost free flow. The pulsating pressure waves cannot cause resonance vibrations of the ship floor in this kind of vessel. The really horizontal position of the propshafts will, in addition, increase the efficiency of the propellers; and it will affect the vessel's trim less as function of the ship's speed.

To avoid large roll amplitudes, although with very low frequencies, all hulls will be equipped with automatically controlled trimm tabs. The front trimm tabs are topped off, to generate an additional steering momentum by transversal forces. This eliminates the need for conventional rudders, that cause costs and also resistance. Another steering effect comes from usage of both rear propellers, that support the steering by the large lateral distance from each other, especially at crawling speed.

Tanks for fresh water, grey and black water as well as ballast and fuel are mounted in the struts. This keeps the centre of gravity low and the feed lines to the engines very short. Access to all three engine rooms is via central mantubes, that end in the superstructure. Installation of lifts is planned.

The struts are made watertight and separated from the ship by bulkheads and bulkhead hatches. In case of a leak in one of the three hull/strut structure, only very little heel can result. Even sealing a leak during sailing would be possible, if the damaged engine room was pressurised with air. Access when pressurised is also possible, as each mantube has a watertight hatch at both ends.

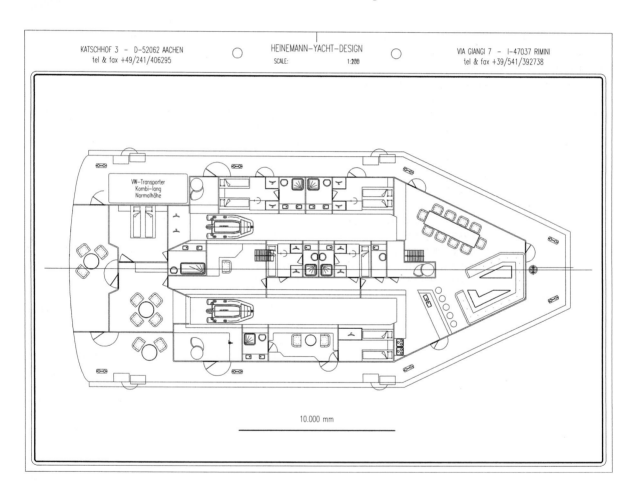

KATSCHHOF 3 – D–52062 AACHEN
tel & fax +49/241/406295

HEINEMANN–YACHT–DESIGN
SCALE: 1:200

VIA GIANGI 7 – I–47037 RIMINI
tel & fax +39/541/392738

VW-Transporter
Kombi-lang
Normalhöhe

10.000 mm

5. EXPECTATIONS

This project is no feasibility exercise, but a concrete order. A short description of the development stage and the underlying thoughts:

- Most components of the ship are produced with CNC-controlled machines. This is a fundamental reason to choose tubes as main components.
- The interior will also be CNC-produced.

- A mannable, free-sailing model (1 : 6.6) is under construction. It will serve to determine, among other parameters, all forces occurring at sea.
- Using the values of these forces, the structure will then be FEA-calculated.
- It is planned to use the vessel commercially as a small passenger ship.

REFERENCES

[1] Developments In SWATH Technology; S.K. Gupta, T.W. Schmidt; Naval Engineers Journal, Mai 1986

[2] Comparative Ship Performance Sea Trials for the U.S. Coast Guard Cutters MELLON and CAPE CORWIN and the U.S. Navy Small Waterplane Area Twin Hull Ship KAIMANILO; D.A. Woolaver, J.B.Peters; DTNSRDC Report 80/037, März 1980

[3] Entwurf der Hauptabmessungen von SWATH-Schiffen; Dr.-Ing. V. Bertram; HANSA 9/92

[4] Leistungsprognose von SWATH-Schiffen in der frühen Entwurfsphase; Dr.-Ing. V. Bertram, J.R. MacGregor; Schiff & Hafen, 10/92

[5] Gewichtsabschätzung von SWATH-Schiffen im Vorentwurfs; Dr.-Ing. V. Bertram, J.R. MacGregor; Schiff & Hafen, 8/93

[6] Zur dynamischen Beanspruchung von SWATH-Schiffen; Richard Gerd Julius Wiefelspütt; Dissertation am Lehrstuhl für Schiffbau, Konstruktion und Statik, RWTH-Aachen, Juli 1993

Practical Design of Ships and Mobile Units
M.W.C. Oosterveld and S.G. Tan, editors.

The Design of a New Concept Sailing Yacht

J.J.Porsius[a], H.Boonstra[b] and J.A.Keuning[c]

[a]Van der Baan & Van Oossanen Naval Architects B.V.,
Costerweg 5, 6702 AA Wageningen, The Netherlands[*]

[b]Delft University of Technology, Section Ship Design
Mekelweg 2, 2628 CD Delft, The Netherlands

[c]Delft University of Technology, Section Shiphydromechanics
Mekelweg 2, 2628 CD Delft, The Netherlands

ABSTRACT

This paper describes several design aspects of a novel type sailing yacht, comprising an unconventional underwater configuration with a bow rudder, a rotating wing mast and a single sail that is operated without any sheets.

The feasibility and critical design aspects of this idea, originated from Van de Stadt Design, was investigated by the department of Marine Technology of the Delft University of Technology. Also, model tests were performed in order to compare the hydrodynamic performance of the bow rudder configuration with a yacht with twin stern rudders.

Although the new design does show advantages in certain conditions, negative aspects such as lack of directional stability, need for continuous adjustment of the sail and the complexity of a sheetless control of the sail, make the feasibility of the concept questionable.

1. INTRODUCTION

A few years ago Van de Stadt Design in Wormerveer (The Netherlands) developed a new design concept, which could best be described as their idea about "the cruising yacht of the future".

The concept aimed to combine maximum (on board living) comfort combined with a reasonable speed potential. The most striking design novelties concern the appendage (keel and rudder) configuration, the replacement of the one or two stern rudders with a single bow rudder and the sail and rig, with a rotating wing mast and no sheets to control the sail, see Figure 1.

The rather unusual rudder configuration was a result of the chosen hull shape, beamy and with flat lines towards the stern. The reasoning behind this and some results of the tanks tests made are handled later in this paper.

A single sail rig was selected, operated without any sheets, increasing the ease of sailing. The absence of stays reduces resistance and disturbance of the flow around the sail.

The sail was fully battened. Sail battens, applied at the full length of the sail, enables the use of roach, creating an elliptical planform. This is a very efficient planform, when considering the aerodynamic performance.

For further aerodynamic improvement over the commonly used rigs, the yacht was designed with a wing mast that can be rotated in every desired position. The sail can be rolled up in the boom, for quick lowering or reefing of the sails.

As the yacht was to be constructed of a sandwich composite with a wooden core, known as 'woodcore', attention was paid at the calculation method of this composite.

This paper will present the advantages and drawbacks of this particular design, by looking at the different aspects separately.

---

[*] Paper is based on MSc. student thesis at DUT

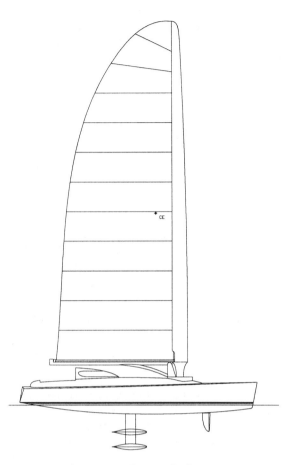

Figure 1 New Concept Design

2. RIG CONSIDERATIONS

2.1. Rotating wingmast

The wing-masted sail is primarily known from the trimarans and catamarans. In fact, the early development of the wing mast in the C-Class catamarans led to a highly efficient rig, within a narrow window of true wind speed. Another effort, described in [1], showed an increase in off-wind drive force of 50%, whereas going-to-windward drag was reduced by 20%.

But, the benefits are only utilised fully if the crew is prepared to adjust their sails to the right shape and the right twist and to trim them to the right angles as they sail. All limits of adjustment must be removed, because these types of rigs are not better than the conventional ones unless they are properly adjusted. The performance is there, but you have to sail more intelligently to get it. *"Be alert, be accurate, or be last."* [1]. This need for adjustment includes the mast too. By nature, wing masts are very stiff in the plane of the sail. Because of this, these masts do not bend sufficiently for adjustment of sail fullness, and so the sail shape cannot be much changed. As a result, a situation with a separation bubble on one side or the other, was the norm in practice. The flow would be 'clean' at one trim angle only. Therefore, in the designing and construction of the mast the correct flexibility must be attained to enjoy the adjustability of the flexible mast and the efficiency of the wing mast.

In heavy weather, the wing mast can put the yacht and its crew in hazardous situations. As it cannot be reefed, a highly efficient high aspect wing is placed in high wind speeds. If it is left feathering in the wind, with little damping of the movement, heavy oscillation could occur. This fluttering could lead to damage and eventually to loss of the mast. If it is stabilised, lift will develop, with the risk of uncontrollable behaviour regarding speed, course, heel, etc.

Another problem may be a situation where the mast is jammed in one position, without any possibility to control it. In this design, the mast was dimensioned in such a way that the yacht does not capsize in the most severe wind condition, with the mast jammed in an unfavourable position.

2.2. Sheetless rig

As a sheeting system was abandoned, alternatives for operating the sail were investigated. The problem of abandoning a sheeting system is counteracting the huge moment induced by the sail force. The solution is found in a balanced rig, like the AeroRig®, see Figure 2. However, since only one sail is used with this design, this was quite impracticable.

An alternative balanced rig was designed, featuring an A-mast, see Figure 3. Some of the benefits are:

- No mast interference, better sail performance
- Balanced rig
- The mast itself can be used as lift generator

Some disadvantages:

- Drag generator
- Its heavy weather performance is unsure

Figure 2 AeroRig®, by permission
of Carbospars Ltd.

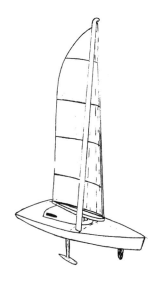

Figure 3 A-mast

A second alternative could lie in operating the boom at the mast by using hydraulic rams to set the desired angle to the mast. An advantage was:

- The rig can be rotated over 360 degrees, enabling gybing over the bow

A disadvantage:

- The system induces a great loading on the mast, which makes the mast design very complex, leading to a heavy mast.

The mast and boom were disconnected in the third alternative, used at this design, which would release the mast from its loading, see Figure 4.

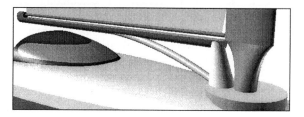

Figure 4 Sheetless boom

The boom is placed on a disk mounted on the cabin roof, which can be rotated over 360 degrees. The disk is fitted with bearings in the deck, immediately connected to a gear wheel. A hydraulic motor drives this wheel. The mast can be set to any desired angle to the boom.

Advantages were:

- 'Clean' mast structure
- Equipment below deck

The complexity and weight of the system were clear disadvantages.

To meet the requirement of the possibility to lower the sail at the boom, a high tensile bar was introduced at the boom, which is connected at the mast and the end of the boom. This bar is used to roll up the sail, after which a protection cover is placed over the sail.

It is clear that the problem caused by the abandoning of sheets of a single sail rig is not easily solved. To counteract the moment, induced by the sail forces, a complex and heavy system is to be fitted on the yacht. It remains to be seen whether this disadvantage is compensated by the increased sailing comfort.

3. SANDWICH CALCULATIONS

The yacht's hull was designed to be made of woodcore sandwich material. This is usually calculated according to the *ABS Guide for Building and Classing Offshore Racing Yachts, 1986*, where the sandwich is treated as a 'common' sandwich, in which the core only contributes in shear strength. For a woodcore sandwich however, this is simplification that could lead to an unnecessary increase in hull weight, as the wooden core is likely to contribute in the strength of the sandwich, both in flexural as in shear strength. The problem is that there is no known method for calculating the required core and skin dimensions of woodcore sandwiches that would give a better result. The section will describe a calculation method for the section modulus of woodcore sandwich panels, in which the core contributes in the strength, based on transformed beam theory.

3.1. Symmetrical Sandwich

In this section the moment of inertia of symmetrical sandwiches, in relation to their neutral axes, is given.

The theory used is that of composite beams, see [3].

The moment of inertia is as follows:

$$I = b_s \cdot \left[\frac{t \cdot (c+t)^2}{2} + \frac{E_{T,c} \cdot c^3}{12 \cdot E_{T,s}} \right] mm^4 / mm,$$

in which

b_s = breath of skins, strip 1 mm

t = thickness of the skins

c = thichness of the core

$E_{T,c}$ = tensile modulus core material

$E_{T,s}$ = tensile modulus skin material

The section modulus is:

$$W = \frac{I}{z_{max}} = \frac{I}{\frac{1}{2}c + t}$$

3.2. Asymmetrical Sandwich

The case gets a little more complicated for asymmetrical sandwiches, built of different materials of different thickness.

The following relationships for the neutral axis and moment of inertia is applicable:

$$z_{axis} = \frac{\frac{1}{2}E_i t_i^2 + E_c c\left(t_i + \frac{1}{2}c\right) + E_o t_o\left(t_i + c + \frac{1}{2}t_o\right)}{E_i t_i + E_c c + E_o t_o}$$

$$I = b_i \left[t_i\left(z_{axis} - \frac{1}{2}t_i\right)^2 + \frac{E_o}{E_i}t_o\left(z_{axis} - \left(t_i + c + \frac{1}{2}t_o\right)\right)^2 \right.$$
$$\left. + \frac{E_c}{E_i}c\left(z_{axis} - \left(t_i + \frac{1}{2}c\right)\right)^2 + \frac{E_c}{12E_i}c^3 \right]$$

Indices:

i = inner skin

o = outer skin

c = core

b_i = breadth of inner skin, strip 1 mm

t = thickness of skins

c = thickness of core

E = tensile modulus

At a height z, with a modulus E, the section modulus becomes:

$$W = \frac{I}{z_{axis} - z} \cdot \frac{E_i}{E}$$

The above method led to a reduction in hull weight of approximately 10%, compared to a hull calculated with the sandwich method, a small percentage of the total weight. However, calculating the hull with transformed beam theory is even more advantageous for lightweight yachts.

4. BOW RUDDER

The philosophy behind the development of such a design concept was based on the following considerations:

In order to be able to obtain a relatively high speed in the running and broad reaching conditions a wide after body with flat and beamy sections, see Figure 5, is considered to be advantageous.

These sections may develop sufficient hydrodynamic lift to be able to support the weight of the craft and so overcome the sharp resistance increase known from ordinary displacement craft at speeds above the "hull speed". In addition this hull geometry with its large and beamy flat-bottomed sections aft has proven to be a very stable platform in running conditions, with or without flying a spinnaker or asymmetrical.

Another important aspect for obtaining high speeds in those conditions is the minimisation of the overall weight of the craft. In order to be able to reduce the weight of the craft and still maintain a sufficiently high transverse stability the metacentric height has to be made as high as reasonably feasible. This allows a minimal ballast weight, which in addition is all concentrated in a bulb at the bottom end of the deep fin keel. A consequence may be the relative large range of "stability" in upside-down position (see also [2]).

The specific shape of the hull lines has so been chosen so that when the ship is heeled to 15 or 20 degrees in the upwind condition, the waterline length is extended and the lines show a almost symmetrical hull shape, see Figure 6, which is considered to be an advantage in those conditions with respect to resistance and side force production. A considerable reduction in the wetted area of the hull due to heeling angle of the yacht is also envisaged, further contributing to a lower overall resistance in the upwind / heeled condition.

So far, the general solution to the problem of the considerable loss of submerged rudder area with these hull shapes when they heel, is found in the application of two rudders both "off centreline" and "with dihedral" instead of the one single rudder at the centreline. This set-up guaranties full downwind control and also in the upwind condition at least one of the rudders is completely submerged without any negative effect of the free water surface disturbance. Also from a redundancy point of view the application of two rudders is beneficial even though they are no longer protected by the (centreline) keel in the case of collision or grounding. The disadvantages of the twin rudder layout obviously lay in the additional resistance arising from the extra appendage and the mechanically more complicated and vulnerable steering device.

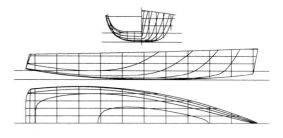

Figure 5 Upright Linesplan

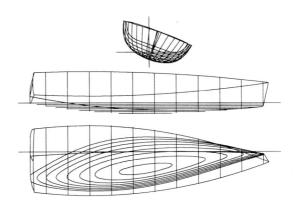

Figure 6 Heeled Linesplan

This led Van de Stadt Design to the idea of the introduction of one single rudder on the centreline near the bow of the yacht in combination with a single keel also on the centreline.

This bow rudder would then no longer be emerged due to the heeling of the yacht so this single rudder would be sufficient. Interest in the bow rudder was also triggered after the successful application during the America's Cup regatta in Perth, 1987.

Without doubt such a "bow" rudder would ask for some skill of the helmsman: in order to let the rudder contribute to the overall side force production of the yacht it should have to generate positively (windward) orientated side force in the stationary condition, which would make a "lee helm" yaw balance of the yacht necessary because the rudder is in front of the keel now. Whether this is acceptable to the "human controller" remains to be seen. In addition the use of a bow rudder also calls for a considerably more aft position of the main foil (the keel), of which the longitudinal position however is strongly dictated by the presence underneath it of the (large amount of) ballast and its position with reference to the centre of buoyancy of the hull.

Serious drawbacks were also envisaged with respect to the course keeping qualities of this bow rudder concept. Much was uncertain about this aspect of the design and available calculation procedures (e.g. [4]) were not considered applicable to the hull and the circumstances under consideration.

Finally the sea keeping behaviour of a design as the one presented here is believed to be advantageous. The large LCB - LCF separation calls for moderate pitch motions in head waves and the relative fine bow shape will prevent a high added resistance and also serious pounding in head waves.

Since a considerable amount of the considerations, which have led to the introduction of the present concept, are related to hydrodynamics, it was decided to carry out an extensive series of model experiments with the two possible variations of the design in order to be able to make a more founded comparison possible.

5. THE MODEL TESTS

The model experiments, which were planned for the two configurations of the design, were intended to make a Velocity Prediction of both concepts possible. To be able to do this the standard tests of the Delft Shiphydromechanics Laboratory for sailing yachts have been carried out. In addition to these tests a simple first assessment test has been carried out with a "free running" model in both configurations to gain some insight in the course keeping qualities.

The tests program consisted of a full upright resistance test from Fn = 0.10 to Fn = 0.70, and a full series of heeled and yawed tests with 0, 10, 20 and 30 degrees of heel and leeway angles ranging from 1 to 10 degrees at least three different forward speeds. The forward speeds selected were made dependent on the heel angle selected and ranged from Fn = 0.25 to Fn = 0.45.

A series of free running tests for the determination of the course keeping capabilities of the two different configurations concluded the tests.

5.1. Upright Resistance

The total, frictional and residuary resistances of the two configurations in the upright condition are presented in Figure 7. The difference in the upright resistance between the two configurations is clearly visible in this graph. This difference in the total

resistance appears to be largely caused by the increase in the residuary resistance of the twin rudder configuration when compared to the single (bow) rudder configuration.

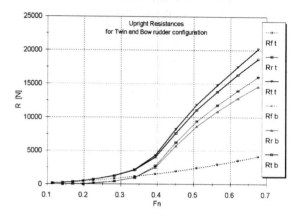

Figure 7 Upright Resistance

5.2. Side Force with Heel and Leeway

In Figure 8 the side force of the yacht in both configurations is presented as a function of the leeway angle for a typical heeling angle of 20 degrees and the different Froude numbers related to the angle of heel such as investigated in the model tests. For the larger leeway angles the side force generation of the hull with the twin rudder configuration is in general somewhat higher, the differences between the two configuration are however small but are consistent over the speed- and heeling angle range investigated. This may be partly explained by the difference in the total lifting generating area of the twin rudder configuration compared with the single rudder configuration.

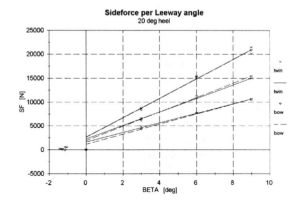

Figure 8 Side force at 20 degrees heel

5.3. Heeled and Induced Resistance

Due to its heeling angle and the side force production, a sailing yacht experiences two types of extra resistance: resistance due to heel and induced resistance due to the lift generated. The heeled resistance is defined as the extra residuary resistance component when the yacht is heeled and with zero side force, whereas the induced resistance is the additional resistance induced by the developed side force.

In Figure 9 the residuary resistance as a function of the generated side force squared is presented for 20 degrees heeling angle and three different Froude numbers respectively.

The lines drawn in these figures are determined by applying a linear least square regression method trough the measurement points obtained from the towing tank data.

In general, the bow rudder configuration generates more induced resistance (i.e. the slope of the resistance curves with respect to the side force squared is steeper) over the entire heel angle and speed range investigated when compared with the twin rudder arrangement.

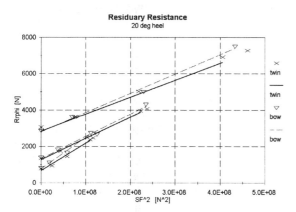

Figure 9 Residuary resistance at 20 degrees heel

5.4. Directional Stability Assessment

Since there were some serious doubts about the positive directional stability of the bow rudder concept is was decided to carry out some additional tests dealing with this problem. Due to the limited time available for such a test it was only possible to carry out some indicative tests which would enlighten the course keeping capabilities of both concepts.

5.4.1. Test procedure

The tests were performed with a more or less free running model in the towing tank. The rudder(s) were put in a zero rudder angle position. The model was free to move transversely. The "tow force" on the model was applied longitudinally in the centre of effort of the sails but at deck level, such as to introduce no serious heeling components.

During these tests the model was brought up to speed (around Fn = 0.25) and once stable in that condition the model was released. If a stable condition persisted, a small disturbance in yaw was supplied and watched if the model tended to return to its original equilibrium condition. The tests have been carried out with both rudder arrangements.

5.4.2. Twin Rudder Configuration

To check the feasibility of this test procedure the test were first carried out with the twin rudder configuration.

The results of these tests came out as were to be expected, knowing that the twin rudder aft configuration is a quite stable configuration. As soon as the model was released, it slowly moved to a stable position a little "off centreline". This small offset of course is necessary to counteract the inevitable side force produced by the hull which is counteracted by the transverse component of the towing force.

Since the model now assumed a stable starting position it was possible to test the course keeping stability by disturbing the model in yaw and sway. After supplying a small disturbance in this direction the resulting motion of the model was clearly very well damped and soon the model came back to its original course and position.

5.4.3. Bow Rudder Configuration

The tests with this appendage layout ended all unsuccessful, i.e. the model immediately started to diverge from its initial course as soon as the run started. Due to its very large excursions in yaw and sway and also due to the limited towing chord length the angle at which the tow force was applied increased very quickly therefore bringing the model to start oscillating fiercely back and forth with ever increasing amplitude.

This combination of large yawing and swaying amplitudes diverged in an uncontrolled motion. The physical restrictions of the towing tank walls necessitated a quick ending to these runs. Change in rudder angles and / or towing force centre of effort

44

did not change this picture dramatically. See Figure 10 for a typical path recording of such a test.

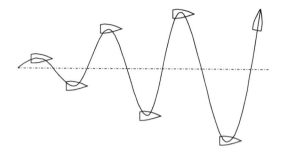

Figure 10 Model path

6. CONCLUSIONS AND RECOMMENDATIONS

The use of the bow rudder doesn't offer a clear improvement over the conventional twin rudder configuration. The tank tests showed more or less equal hydrodynamic performance, except for upright conditions like downwind sailing.
However, one could opt for a maximum of available rudder action when sailing with high speeds at these courses especially. After all, the sail will produce a yawing moment that has to be corrected by the rudder.
The sailing comfort when sailing with a bow rudder is to be questioned; due to the directional instability of the yacht the helmsman has to be alert and give rudder continuously. A feedback control system, which is known from the aviation industry, could offer a solution.

The aerodynamic performance of this particular wingmasted rig stayed insecure. The lack of usable information on this type of mast and sail forced the dimensioning and velocity prediction to be done on assumptions and estimations. It is therefore uncertain whether this cat rig outperforms the sloop rig.
The handling of a wingmasted sail appears to be a specialist's cup of tea. When sailed without adjusting continuously, the performance is not explicitly better than the round-masted sail. The question then arises whether this type of rig is suitable for cruising yachts. After all, an innovation often isn't accepted until it proves to be better.
The abandoning of a main sheet system is questionable. The weight increase, the complexity and the lack of sail controllability of an alternative system are disadvantages that doesn't seem to be compensated by the ease of handling. Especially, when considering that a main sheet can be operated hydraulically too, enabling 'push-button sailing'.

Calculating with a contributing wooden core resulted in a weight decrease of approximately 2%. In this case the gain is therefore not sensational. However, in a market where every weight decrease is welcomed, the racing market for instance, this method could be useful.

REFERENCES

[1] Bethwaite, F., *High Performance Sailing*, International Marine, 1993
[2] Porsius, J.J., H.Boonstra, J.A. Keuning and C.W. van Tongeren, *The Design of a Sailing Yacht with a Bow Rudder*, The Modern Yacht, 1998
[3] Bodig, J. and B.A. Jayne, *Mechanics of Wood*, Van Nostrand Reinhold Company, 1982
[4] Gerritsma, J., *Course Keeping Qualities and Motions in Waves of a Sailing Yacht*, TH Delft, 1968

Practical Design of Ships and Mobile Units
M.W.C. Oosterveld and S.G. Tan, editors.

Enlarged Ship Concept Applied to RORO Cargo/Passenger Vessel

J.M.J. Journée[1], Jakob Pinkster[1] and S.G. Tan[2]

[1]Department of Marine Technology, Delft University of Technology,
Mekelweg 2, 2628 CD Delft, The Netherlands

[2]Research and Development Department, Marin,
P.O. Box 28, 6700 AA Wageningen, The Netherlands

The "Enlarged Ship Concept" (ESC) was successfully applied to a fast semi-planing 26 m. patrol boat by Keuning and Pinkster [1,2]. Their results showed a significant performance improvement both in a technical and economical sense. In order to investigate if ESC may also render a similarly successful design strategy for a RORO/Passenger vessel which is representative for present services in the UK-West Europe route, the underlying study was carried out. The outcome of this study is that some important results are quite the opposite to those of the patrol boat; this is mainly due to the large difference in vessel types and Froude numbers involved. Within a given payload weight, the larger RORO vessels have more cargo carrying capacity in terms of trailers; in other words, the enlarged vessels can carry more trailers if the trailers are not fully laden. Furthermore, the larger vessels are less vulnerable in damaged condition since the lower hold is not used for cargo and can therefore freely be optimally subdivided. Also advantageous is the fact that the draft decreases as the length increases which results in a higher freeboard for the larger vessels. Summarising, it appears that application of ESC to this type of vessel creates more income possibilities for the shipowners and a much safer vessel, but it produces a more expensive ship to buy and exploit.

1. INTRODUCTION

In 1995 Keuning and Pinkster [1] explored the so-called "Enlarged Ship Concept" (ESC) by applying this to a fast 25 knot, semi-planing, 26 m. patrol boat. The Froude number was, based on vessel length, equal to 0.81. The main driver behind this application was the fact that a monohull sailing at high forward speed in head waves may incur unacceptably high vertical accelerations which may hamper the safe operability of the craft. In essence, they improved the seakeeping behaviour and decreased the resistance of the fast patrol vessel by increasing the length in steps of 25% and 50% and so increased also the length to beam ratio, reduced the running trim under speed and improved the general layout of the ship. Their work carried concerned three design concepts, namely a base boat with two enlarged ship configurations. The key to the ESC is that deadweight, i.e. payload, fuel and stores as well as vessel speed remain constant and equal to that of the base boat. The most important results from this study showed, on the one hand, a 68% marked improvement regarding a decrease in vertical acceleration in the wheelhouse in head seas and a 40 % decrease in required propulsion power in calm water at a speed of 25 knots; on the other hand the maximum purchasing price of the largest design alternative was estimated to be only 6% higher than that of the basic 26 m. patrol boat.

In 1997 Keuning and Pinkster [2] presented further research on the ESC topic, extensive model testing related to vessel resistance and motions were carried out and subsequent results were described in detail. This second study confirmed the results of the first study and favoured, once again, the Enlarged Ship Concept. In the meantime, the results from these studies have been applied to a number of new buildings of fast patrol boats in The Netherlands.

Now, the question arises, " Can the ESC also be successfully applied to the common work horse of the seas, the ordinary marine freighter? ".

In the present paper, an attempt was made to answer this question by applying the same ESC to a full time "freight carrying" vessel being a RORO/Passenger Vessel representative for present services in the UK-West Europe route. The base vessel of 157 m. length

46

was lengthened by respectively 25 and 50 per cent, while deadweight and speed remained constant. The consequences with regard to vessel mass, stability and trim, cargo hold configuration, propulsion power, freeboard, net tonnage and building costs were evaluated. On the operability side, seakeeping performance as well as operability were also assessed. Finally costs were determined for the base ship as well as for the two ESC alternatives.

2. THE "BASE SHIP"

The base vessel used for the study was m.v. NORBANK owned by North Sea Ferries and built in 1993 by the Dutch shipyard Van der Giessen-de Noord. This vessel is a well proven design and has been described in more detail in [3]. The vessels main particulars are given in table 1. All design and functional requirements, such as speed, payload, accommodations etc., for the Enlarged Ship Concepts were based on and kept identical to those of this base ship. Relevant design information regarding hull form, stability and trim, masses, building costs etc. of the basic monohull were kindly made especially available to the authors for the work carried out here.

3. THE "ENLARGED SHIP" DESIGNS

To yield the Enlarged Ship Concepts the basic 157.65 m. ship, forthwith designated ESC-0, was enlarged in length only. Two such designs alternatives, ESC-1 and ESC-2, were made, having a length of respectively 197.06 m. and 235.85 m. . The enlarged alternatives are shown in Figure 1 along with the base ship whereas the main design particulars for all designs are given in Table 1.

With regard to engineering of all these alternatives the starting point was relative data related to the base ship. The increase in length was, in both cases, created by inserting a parallel midship section with respective lengths of 25% Lpp and 50% Lpp. In this way the original body plan remained unchanged in both the forward and aft part of all design alternatives; thus keeping the good lines of flow to the propellers and along the bow. Subsequently hydrostatic particulars were computed for the new body plans.

The increase in structural masses of all alternatives was also computed via the original mass data which

was augmented with extra frames and hull plating, taking into account the relevant positions of the centres of gravity of all components of the designs. Also, since an increase in vertical bending moment may be expected to be approximately proportional to the square of the length ratio for the enlarged vessels, i.e. 1.55 and 2.25, an extra allowance has been made for an increase in steel mass of the parallel midship sections of respectively 20% and 45% for the ESC-1 and ESC-2 alternatives. This extra steel, in the form of deckplating, is thought to be placed in the upper deck of the midship section as it is then effectively positioned furthermost from the neutral line and thereby reduce the bending stresses to an equal level of the base ship. The deadweight of the enlarged vessels was placed in such a manner that no trim angles occurred.

The resistance and propulsion calculations were also made for each alternative.

Since the idea behind the Enlarged Ship Concept is equal payload for all possible alternatives, similar main dimensions such as breadth, depth etc., the vessel configuration (i.e. also position of accommodations, wheelhouse etc. with respect to the bow) remains unchanged to that of the basic ship for each design alternative concerned. Consequently, the forward position of the wheelhouse has a distinctive disadvantage with regard to ship motions.

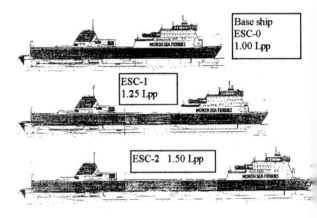

Figure 1. Side elevation of base ship and ESC designs.

4. SHIP RESISTANCE

The still water resistance for all three designs was calculated using the method of Holtrop and Mennen [4] for a range of speeds up to the design speed of 22 knots. This speed corresponds to a Froude number of

0.29 for the base ship. Figure 2-a shows the still water resistance coefficients (C_t) of the three ships, subdivided into frictional (C_f) and residual (C_r) parts. From this figure it appears that, when comparing with the base ship ESC-0 at a speed of 22 knots, ESC-1 has a decrease in resistance coefficient of about 15 per cent while ESC-2 shows 10 per cent decrease only. However, this favourable effect becomes completely lost due to an increase of the wetted surface of the hull with 20 and 45 per cent respectively. As a result of this the total still water resistance will increase with approximately 5 and 30 per cent for ESC-1 and ESC-2, respectively; see Figure 2-b.

An important conclusion regarding still waterresistance is that, when the Enlarged Ship Concept is applied to these ships, there is not a similar profit to be gained as for the fast semi-planing patrol boats from [1] with up to 40%

reduction in still water resistance. This finding may be attributed to the relative low Froude numbers (0.29 for ESC-0) compared to those of the patrol boats (0.81 for the base ship.

Since the vessel resistance is known for m.v. NORBANK (ESC-0), a ratio between actual and computed resistance was determined. This correction coefficient was then applied to the computed resistance of the larger vessels for establishing the required propulsive power.

Since the topic investigated in this paper deals with large seagoing vessels, ship motions are calculated at 20 and 15 knots. When assuming that the still water resistance is proportional to at least the square of the ship speed and using calculated data on added resistance in seaway, a sustained sea speed in rough weather dropped from 22 to 15 knots would expect to be an acceptable average.

Table 1
Main particulars of the base ship and alternative ESC designs

Parameter	Dim.	ESC-0	ESC-1	ESC-2
Loa	m	166.77	206.18	244.97
Lpp	m	157.65	197.06	235.85
Bmld	m	23.40	23.40	23.40
T	m	5.80	4.97	4.50
KB	m	3.26	2.69	2.36
BM	m	9.01	10.25	11.35
KG	m	10.42	10.83	10.87
MG	m	1.85	2.11	2.84
Cb	[-]	0.61	0.64	0.66
Depth to main deck	m	8.60	8.60	8.60
Depth to upperdeck	m	14.40	14.40	14.40
Lightshipweight	t	7417	9126	11176
Deadweight	t	6020	6020	6020
Displacement	t	13437	15146	17196
Speed	kn	22	22	22
Propulsion power	kW	24480	25700	33500
Passengers	no	120	120	120
Lane length upperdeck	m	930	1190	1450
Lane length maindeck	m	910	1170	1430
Lane length hold	m	200	0	0
Trailer capacity @ 40 t	no	156	165	165
Water ballast	t	234	0	0
Gross tonnage	GT	17464	21452	25396
Net tonnage	NT	5239	6436	7619
k_{xx}/B	[-]	0.43	0.43	0.43
k_{yy}/Lpp	[-]	0.29	0.29	0.29
k_{zz}/Lpp	[-]	0.29	0.29	0.29

48

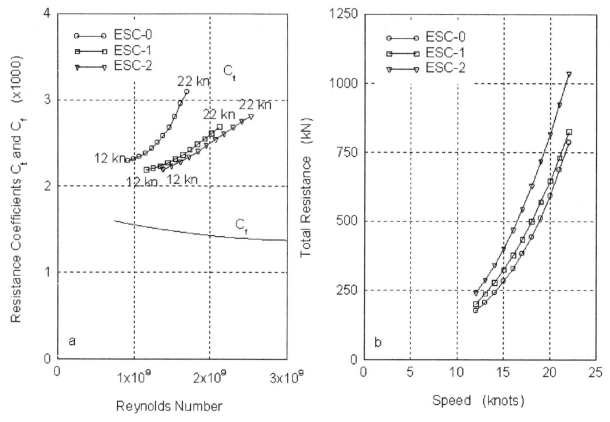

Figure 2. Results of resistance calculations.

5. SHIP MOTIONS

The vessel motions were calculated using the linear strip theory program SEAWAY of the Delft Ship Hydromechanics Laboratory [5]. These calculations were carried out in Beaufort 7 to 12, at wave directions ranging from head to following seas. The energy distribution of the irregular waves in the considered coastal areas was described by uni-directional JONSWAP wave spectra. According to Hasselmann [6], this wave energy distribution is a favourable choice for a fetch limited seaway. A commonly used relationship between period, wave height and Beaufort number was utilised. The long term probability on exceeding a certain sea state was obtained from Global Wave Statistics whereas the limiting criteria of ship motions were obtained from Karppinen [7].

In order to assess the ship's radii of gyration, an analysis has been made of the mass distribution over the length of the various designs.

Figure 3-a shows the vertical significant acceleration amplitude at the bridge in head seas as a function of the Beaufort scale with an acceleration criterion of 0.3 g. At both speeds course can be maintained by ESC-0 in sea states up to Beaufort 8, which will be exceeded during about 2 percent of the year.

However not unexpected, the two enlarged ships ESC-1 and ESC-2 can maintain their course up to Beaufort 9 and 10 respectively. Figure 3-b shows the probability on slamming in head waves, defined by a relative vertical velocity criterion at the bow. Using a slamming criterion of 2 per cent, all ships can maintain their course up to Beaufort 8. The effect of ship size and forward speed on slamming appears to be relatively small.

Figure 3-c shows the horizontal significant acceleration amplitude at the bridge in beam seas as a function of the Beaufort scale with an acceleration criterion of 0.24 g. The effect of forward ship speed is negligible. Course can be maintained by ESC-0 and ESC-1 in sea states up to Beaufort 9, which will be exceeded during less than 1 percent of the year. However, the operability of ESC-2 is limited to

Beaufort 8, which sea state will be exceeded during about 2 per cent of the year.

Figure 3-d shows the significant roll amplitude in beam seas as a function of the Beaufort scale with a roll criterion of 12 degrees. The effect of forward ship speed is negligible. The ships heading can be maintained by ESC-0 and ESC-1 in sea states up to above Beaufort 11, but the operability of ESC-2 is limited to Beaufort 10. However, the probability of occurrence of this sea state is only 0.2 per cent.

From these calculations it was concluded that the overall motional behaviour of ESC-1 is comparable with that of ESC-0. The behaviour of ESC-2 is somewhat better in head seas and somewhat worser in beam seas when compared to the base ship.

The largest impact of all may be found when evaluating bending moments.

Figure 4-a shows the distribution of the vertical bending moment (M_y) in still water over the ship length. Compared to ESC-0, for the enlarged vessels these moments have been increased by approximately 40 and 60 per cent. According to the classical theory of a uniformly loaded elastic beam, simply supported at both ends, the vertical bending moment increases with the square of the length of the beam. When considering a vessel positioned in a longitudinal (quasi static) wave with a length equal to that of the ship and the wave crests at both ends, one can expect a similar increase in vertical bending moments for the enlarged ships (55 and 125 per cent respectively). This simple approach to the problem is confirmed by dynamic calculations of the vertical bending moment (M_y) in head seas (rendering increases of about 50 and 150 per cent respectively); see Figure 4-c.

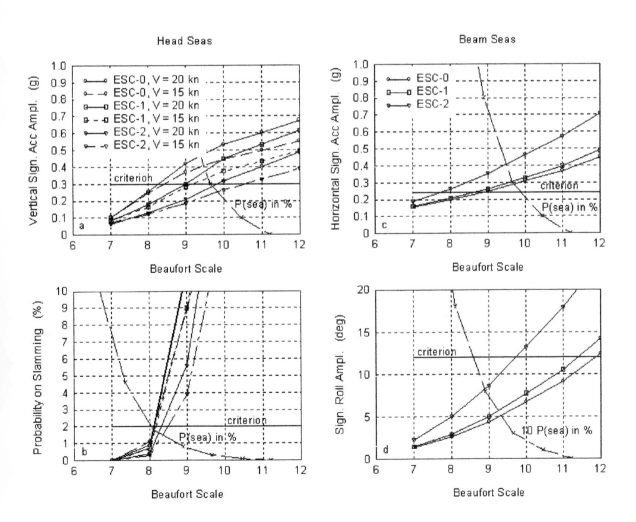

Figure 3. Motional behaviour of ESC's in seaway.

The largest horizontal bending moments (M_z) and torsional moments (M_x) have been found in bow-quartering waves ($\mu=120°$);see figures 4-b and 4-d for the corresponding significant amplitudes. The stresses caused by the torsional moments (M_x) do not play an important role because of the closed character of the midship section. As the lateral bending moments (M_z) are much smaller than the vertical bending moments (M_y) the latter is dominant for this shiptype. Considering similar main frame scantlings for all three designs, the result would be an increase in bending stresses in the outer fibres of the larger vessels in the order of 55 and 125 per cent respectively. To deal with this increase, an increase in scantling mass for the enlarged part of the vessels

has been allowed for about 20 and 45 per cent respectively. Since this extra mass is mainly required in the midship section, the mass distribution is assumed to have the form of a triangle with its base length equal to the length of the enlarged part and its top in the middle of it. This extra mass is distributed as such, in the upper deck of the vessels, having thereby the most optimum effect in reducing bending stresses. As the ends of the enlarged part are the original midship section, here lower scantling dimensions can now be expected due to a distance from amidships. However, this will be, more or less completely, overruled by the effect of an increase of the ship length.

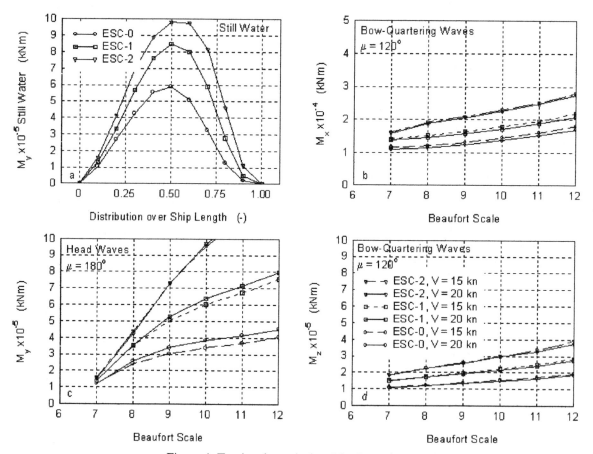

Figure 4. Torsional, vertical and horizontal moments.

6. ECONOMIC EVALUATION

In order to make an economical evaluation the building costs of the different design alternatives were estimated using the original building costs of

the base ship (of which all costs components were known) and correcting this for changes in steel mass of the hull and extra painting costs (i.e. cleaning, preparation and painting) and also for extra machinery costs. The differences in building costs are indexed with regard to the ESC-0 in Table 2. Note the increase in building costs of about 10% for ESC-1 and 28% for ESC-2.

The operational costs of the design alternatives are considered for a scenario of a twenty year economic life of the ship, sailing 18 hours per day at 22 knots, 7 days a week for 48 weeks per year and crewed by 30 persons (3 shifts per 24 hours). The differences in operational costs are indexed with regard to the ESC-0 in Table 2. Note the relatively high increase in operational costs of about 8% for design alternative ESC-1. This increase is even more dramatic in the case of the ESC-2 design alternative (i.e. 18%).

The transport efficiency (TE) - defined, in this particular case as: number of trailers times service speed in m/s over installed power in kW - has been calculated for the three designs. The differences in TE are indexed with regard to ESC-0 in Table 2. When dealing with trailers of 40 ton: an increase in TE of only 1% for ESC-1 is gained, while a decrease of about 20% is calculated for ESC-2. However, when allowing less than 40 ton per trailer and utilising the available trailer space on both D and E decks, the increase of TE becomes 17 and 13 per cent respectively.

Applying the enlarged ship concept to such a RORO vessel as presented in this paper, renders an improvement in concept design with regard to the increase in the transport capacity of non fully laden trailers; the stipulated condition that payload remains constant must still be applied. When allowing fully laden 40 tons trailers of 12.2 m length, the number of trailers transported by the design alternatives are approximatley 6% higher than that of the base ship. This is due to the fact that the larger vessels do not require 234 ton of ballast in the fully loaded condition.

Furthermore, when keeping payload constant, the larger design alternatives have relatively enough space available on both D and E decks for the carriage of homogeneous cargo of respectively 191 trailers of 34.6 ton and 233 trailers of 28.3 ton. This is an increase by respectively 22% and 49% compared to the base ship. Based on a single price per trailer, the earning capacity of the larger alternatives will therefore increase with a similar percentage if, and when, the market has lighter trailers on offer.

If only the D and E decks are utilised for the carriage of trailers, loading and discharging times per trailer will be relatively reduced due to the fact that these decks are more easily accessible than the lower F hold.

Table 2
Results of economical calculations

Index	ESC-0	ESC-1	ESC-2
Building costs	1.00	1.10	1.28
Power at 22 knots	1.00	1.05	1.32
Operational costs	1.00	1.08	1.18
Transport efficiency[1]	1.00	1.01	0.80
Transport efficiency[2]	1.00	1.17	1.13
Trailer capacity[1]	1.00	1.06	1.06
Trailer capacity[2]	1.00	1.22	1.49

[1] 12.2 m. trailers total all in load of 40 tons each
[2] idem with all in load of less than 40 tons each

Although not advocated by the authors, if (the lowest) F deck were included within the cargo carrying capacity, space would be available for yet another 28 and 40 trailers for the alternatives. This would result in the carriage of homogeneous cargo of respectively 219 trailers of 30.1 ton 273 trailers 24.2 ton. This is an increase of 40% and 75% respectively, compared to the base ship. The earning capacity of the alternatives will therefore increase with a similar percentage if, and when, the market has lighter trailers on offer and the price per trailer is independent of the mass carried within.

7. CONCLUSIONS

The following conclusions are drawn with regard to the feasibility of the Enlarged Ship Concept applied to a freight carrying vessel (see also table 2):

- The ESC when applied to such large and relatively moderate Froude number vessels appears, at first glance, to be far less viable than for the fast patrol boat. This is mainly due to the relatively larger increase in building and exploitation costs.

- Heave, pitch and related phenomena on the bridge of such a RORO vessel in waves, although not excessive, are sufficiently reduced by the application of ESC.

- Roll motions on the bridge of such a RORO

vessel in waves are increased by the application of ESC. However this increase is still acceptable with the criteria applied.

- The vertical midship bending moment in rough weather increases largely for the larger design alternatives; in Beaufort 10 the increase is of the same order as the expected increase of the calm water bending moment which is proportional to the square of the ratio between vesssel length and base ship length.

- In the case of the RORO Freighter/Passenger cargo vessels, a definitive advantage of the ESC is the provision of space for the accommodation of lighter cargoes if available which consequently increase the earning capacity and transport efficiency.

- Applying ESC to a RORO vessel renders an improvement in concept design with regard to a significant improvement in survival capability after having suffered the ingress of water into the hull; the condition that the lowest hold remains empty and optimally subdivided for this purpose must be respected.

8. RECOMMENDATIONS

Further optimisation of the enlarged designs of the RORO freighter/passenger ferry may well lead to more promising results and is recommended as follows:

- Optimise the vertical position of the upper deck of the enlarged vessels in order to reduce the vessel mass, while, at the same time, satisfying the requirements regarding allowable stress values due to longitudinal bending moments.

- Optimise the mass of the enlarged vessels by the utilisation of high tensile steel. This will surely reduce the vessel mass while at the same time being able to withstand the higher longitudinal bending stresses.

- Optimise the vessel form with regard to vessel resistance and propulsion. This can be done by optimisation of the longitudinal centre of buoyancy, ships lines, etc.

- Optimise the vessels turn around time by not utilising the F deck for the carriage of trailers.

ACKNOWLEDGEMENT

Although the results and views expressed in this paper are those entirely of the authors, special thanks are due to Shipyard Van der Giessen-de Noord and North Sea Ferries for allowing the authors to use m.v. NORBANK data.

REFERENCES

1. Keuning, J.A. and Pinkster, Jakob, "Optimisation of the seakeeping behaviour of a fast monohull", Fast'95 conference, October 1995.

2. Keuning, J.A. and Pinkster, Jakob, "Further design and seakeeping investigations into the "Enlarged Ship Concept". Fast'97 conference, July 1997.

3. "NORBANK", "A new super freighter for North Sea Ferries, Schip en Werf de Zee, November 1993.

4. Holtrop, J. and Mennen, G.G.J., "A statistical power prediction method", International Shipbuilding Progress, Vol. 25, No. 290, October 1978.

5. Journée, J.M.J., "SEAWAY-Delft, User Manual and Technical Background of Release 4.00", Delft University of Technology, Ship Hydromechanics Laboratory, Report no. 910, 1992.

6. Hasselmann, K., et al., "Measurements of wind-wave growth and swell decay during the Joint North Sea Wave Project (JONSWAP)", Deutches Hydrographisches Institut, Hamburg, 1973.

7. Karppinen, T., "Criteria for seakeeping performance predictions", Technical Research Centre of Finland, Ship Laboratory, Espoo, 1987.

Practical Design of Ships and Mobile Units
M.W.C. Oosterveld and S.G. Tan, editors.

53

Use of non-linear sea-loads simulations in design of ships

L.J.M Adegeest, A. Braathen and R.M. Løseth
Det Norske Veritas, Høvik, Norway.

Abstract

The objective of this paper is to demonstrate an application of a Most Likely Wave approach to ship responses and the use of the nonlinear time-domain program SWAN (DNV-version) to analyze Ultimate Limit State conditions. Two ULS examples are run: a vertical bending moment and a maximum water height on deck. A conditioning of the irregular wave surface is applied such that the linear ULS response occurs at a defined moment in a realistic design sea state. By running SWAN in its nonlinear mode through the same waves, nonlinear corrections to the response in irregular waves are obtained efficiently and accurately.

1. INTRODUCTION

The development of non-linear seakeeping analyses opens new possibilities to determine accurate motions and loads in extreme seas. For extreme loads, non-linear effects are pronounced and have to be considered. At the same time it is expected that direct calculated loads will become more common for use in design as accepted methods become readily available. The implication is that non-linear methods and direct calculations shall be more important in the design phase of ships than they have been in the past.

Most of the traditional sea-keeping analyses are performed in the frequency domain requiring two load cases (real and imaginary parts). A time domain analysis requires a simulation of the sea surface, which enables the estimation of the extreme response. A possible method is to simulate a 'large enough' set of characteristic events having extremes at different points in time at different positions for different responses. By using linear responses in combination with the available wave information, the non-linear analysis can be concentrated on a limited set of characteristic events. A method to define equivalent wave heights or characteristic short time series can then be formulated to reduce the size of the non-linear analysis.

The traditional non-linear wave load analysis for scantling evaluation of ships has been to make corrections on the linear calculated vertical bending moment. Problems are encountered in balancing FE-models when transferring those corrected loads,

pressures and accelerations to a FE-model. By applying a nonlinear 3D forward speed program such as SWAN [1], more consistent and accurate procedures can be followed, such as:

- Consistent combination of non-linear loads components,
- Non-linear load components can be transferred to FE-models for direct scantling evaluations,
- Non-linear simulations improve sea-keeping predictions in general, and load predictions in particular.

SWAN can be run in a linear or non-linear mode. In the nonlinear mode, the hydrostatic and Froude-Krylov pressures are integrated over the exact wetted surface, nonlinear roll damping is used, green water effects are considered and the nonlinear Eulerian motion equations are solved. The radiation and diffraction pressures, including forward speed effects, are computed by linear theory as implemented in the presently used version. This allows nonlinear analyses without dramatically increasing the computational cost.

Computer animations, showing the behaviour of the vessel in waves, can successively be used to demonstrate important physical effects.

2. DESIGN WAVE AND DESIGN SEASTATE

When calculating an Ultimate Limit State (ULS), it is generally necessary to consider non-linear effects.

Regarding vertical hull girder bending moments, non-linear correction factors are often calculated to account for asymmetry in sagging and hogging. To estimate maximum heights of water on deck, it is necessary to simulate in particular wave conditions, which induce large relative motions. All those specific conditions can be determined using the linearly calculated ULS values.

The linear ULS for a particular response is calculated using a set of complex transfer functions $H(\omega)$ in combination with a scatter diagram, a wave spectrum shape $S_{\varsigma\varsigma}(\omega)$, a distribution of heading and speed and a period of operation. By specifying a series of wave frequencies over a large enough interval in the SWAN input, and by running SWAN successively in its linear mode, it is possible to obtain a complete transfer function with just one simulation per heading and speed. Fourier post-processing of the simulated time traces results in the transfer function $H(\omega)$.

Usually the nonlinear ULS is calculated in regular design waves. The amplitude of the regular design wave can be determined by dividing the calculated linear ULS by the peak value of the transfer function. The period of the design wave is set equal to the peak period of the transfer function. In some cases this procedure results in too steep waves and the wave height or the wave period have to be adapted to satisfy a maximum wave steepness of 1/7. However, the regular design wave with maximum steepness 1/7 will never occur repeatedly, and therefore it is interesting to consider an irregular design sea state on the condition that the extreme event will occur. Depending on the type of response, memory effects and phase characteristics, the extreme response does not necessarily coincide with the extreme wave.

Compared with regular design wave analyses, simulations in an irregular design sea state will result in a different inswing into the extreme event. By using a scatter diagram, the irregular design sea state can be found by determining the combination of (H_s, T_z) in which the expected extreme event is maximal according to short-term statistics. Applying the narrow band assumption, the expected extreme single amplitude in a set of n samples can be calculated using the Rayleigh distribution. After calculation of all the transfer functions, the response spectra and spectral properties can be calculated. A summation of Rayleigh-based probabilities of different response

levels over the scatter diagram and weighting the probability distribution of heading and speed finally results in the (linear) ULS. The design sea state in this context is the sea state which most probably gives the largest expected linear extreme value considering the correct exposure duration.

Another way to find the design seastate is by using a contour in the scatter diagram defining the once in the N years combinations of T_z and H_s. This method was published by Winterstein e.a. [2]. Typically, each point on the contour represents a 4-hours condition. The expected largest single response amplitude can be determined in each seastate on the contour in the same way as described above. It should be noted that this value is not necessarily the largest short term extreme value as it is possible that a lower seastate with a higher probability of occurrence results in a larger expected extreme response.

3. MOST LIKELY WAVE THEORY

To limit the use of time-consuming nonlinear simulations to find the maximum response in irregular seas, the basic theory as used to generate Most Likely Wave profiles with conditioned amplitude and frequency has been applied [3,4]. The procedure is as follows.

The first step is the definition of a wave spectrum for which a realisation is to be computed, and to calculate a set of random wave components which represent the specified spectral properties.

The second step is a conditioning of the wave components. Condition is performed such that after realisation of the wave profile, the elevation at $t=0$ and in $x=0$ is equal to a specified amplitude R, the frequency is equal to a user-specified ω^* and the vertical velocity of the wave surface elevation is equal to zero. This realisation is one random profile out of an infinite number of possible realisations satisfying the conditions on spectrum and instantaneous surface elevation.

In order to arrive at the Most Likely Wave profile, the mean of the set of possible random profiles is calculated (Friis Hansen and Nielsen [3]). When choosing the frequency ω^* equal to the mean frequency, the Most Likely Wave profile as derived in [3] reduces to Tromans wave definition [4]. Some remarks have to be made on the applicability of the Most Likely Wave profile in the calculation of extreme responses for ships:

- For nonlinear ship responses, the mean response to a set of random realisations will in general not be equal to the response to the mean, or Most Likely, realisation. Therefore it has to be investigated whether the Most Likely Wave profile or a particular conditioned random wave profile should be used, for example the one with maximum wave height.
- No response memory effects are taken into account. Therefore the method of critical wave episodes was developed by Torhaug [5]. Critical wave episodes were selected based on the occurrence of an extreme linear response calculated in irregular waves.

4. MOST LIKELY EXTREME RESPONSES

In the previous section the Most Likely Wave theory, MLW, is applied to the wave spectrum. The same procedure can also be applied to the response spectrum after which the linear Most Likely Extreme Response can be determined. Using the amplitude and phase information of the frequency response functions, it is possible to derive the underlying irregular wave train.

Definition of waves

In SWAN, the wave surface is defined by a linear superposition of N regular wave components with different amplitudes, frequencies, headings and phases as follows:

$$\zeta(t) = \sum_{j=1}^{N} Z_j \cos(Xk_j \cos\beta_j + Yk_j \sin\beta_j + \varepsilon_j - \omega_j t)$$

where X and Y are the earth-fixed coordinates in which the wave elevation is calculated. k is the deep water wave number, defined as $k = \omega^2/g$. The wave heading β is defined as 180^0 in head waves. Each wave component with frequency ω results in a response component with encounter frequency ω_e. The relation between the wave frequency and encounter frequency for a vessel travelling along the line $Y=0$ is given by:

$$\omega_e = \omega - U\frac{\omega^2}{g}\cos\beta = \omega - Uk\cos\beta$$

where U is the mean forward speed of the vessel. It follows that the wave surface elevation in the origo of the mean ship-fixed system is defined as

$$\zeta(t) = \sum_{j=1}^{N} Z_j \cos(\varepsilon_j - \omega_{e,j}t)$$

Calculation of Most Likely Extreme Responses

After calculating the frequency response function, the response spectrum can be calculated for the design seastate according to $S_{yy}(\omega) = |H(\omega)|^2 S_{\varsigma\varsigma}(\omega)$. Before applying the MLW-procedures to the response spectrum, the spectrum has to be transformed to a frequency of encounter basis as follows:

$$S_{yy}(\omega_e) = S_{yy}(\omega)\left(1 - \frac{2\omega U}{g}\cos\beta\right)^{-1}$$

Now it is possible to generate a random set of amplitudes and phases, or in-phase and out-phase components, of the response at a discrete number of N frequencies, such that:

$$y(t) = \sum_{j=1}^{N}\left(A_j \cos\omega_{e,j}t + B_j \sin\omega_{e,j}t\right)$$

A_j and B_j are independent standard normal distributed variables having zero mean and an expected value for the squares equal to

$$\sigma_A^2 = \sigma_B^2 = E\left[A_j^2\right] = E\left[B_j^2\right] = S_{yy}(\omega_{e,j})\Delta\omega_e$$

Transformation of these Gaussian variables to a phase angle and an amplitude leads to a uniform distribution for the phase angle and a Rayleigh distribution of the amplitude Y_j with

$$E\left[Y_j^2\right] = 2S_{yy}(\omega_{e,j})\Delta\omega_e$$

After determination of the random response components A_j and B_j, the components can be conditioned such that at time $t = 0$, the random response has the prescribed linear ULS amplitude and frequency. In the present study, the mean response frequency $\omega^* = 2\pi/T_0$ was specified. Conditioning of the response components is achieved by applying the theory of the Most Likely Wave model. The output of this conditioning process is a set of modified A_j's and B_j's, A_j^* and B_j^*. Note that both random conditioned sets of response components can be generated as well as the Most Likely Extreme Response, i.e. the set of components giving the mean response.

Derivation of the underlying wave profile

The last step in the procedure is to derive the underlying wave profile, which caused the Most Likely Extreme Response time series according to linear theory. This is achieved using the frequency response functions which were already calculated using SWAN in its linear mode.

For each frequency of encounter, present in the response signal, the amplitude Z_j of the incoming j-th wave component with frequency ω_j is derived according to

$$Z_j^* = \frac{\sqrt{A_j^{*2} + B_j^{*2}}}{|H(\omega_j)|}$$

The phase ε_j of the j-th wave component is derived from

$$\varepsilon_j^* = \arctan(B_j^* / A_j^*) - \varphi_j$$

where φ is the phase angle of the frequency response function at frequency ω. Finally, the underlying conditioned wave profile in the earth-fixed reference frame is defined by:

$$\zeta(t) = \sum_{j=1}^{N} Z_j^* \cos(\varepsilon_j^* - \omega_j t)$$

This wave profile will result in the linear Most Likely Extreme Response with specified amplitude and frequency at time t = 0. By modifying the phases properly, another set of wave components is found which defines the same wave profile in space, but delayed in time such that at time t = t_{amx}, the response obtains the specified magnitude and frequency. This allows a nonlinear inswing into the Most Likely Extreme Response without transient effects, induced at the start of the simulation. The maximum realized wave height and steepness are easily verified against statistically expected values using the wide band distributions given in [8,9].

Nonlinear SWAN calculations

After calculation of the set of wave components inducing a linear conditioned random or Most Likely

Extreme Response, SWAN can be run in its nonlinear mode to calculate the Most Likely Extreme Response including nonlinear effects, i.e. the final ULS value. Linear and nonlinear results can directly be compared in the time domain and nonlinear correction factors are easily derived. A typical simulation length in irregular waves is accordingly reduced to 5 – 10 response cycles. This is comparable with the typical number of response cycles in critical response episodes selected from unconditional randomly generated time traces [5]. A major advantage is believed to be the reduced variability of the estimator of the most likely nonlinear extreme on the condition of the correct frequency choice.

5. EXAMPLES

The first example is the analysis of the ULS vertical bending moment bending condition. Forward speed and bowflare have a considerable impact on the sag/hog ratio as was show in various model test programs, f.e. [6] and therefore, a nonlinear analysis is required. For a 100 meter supply vessel to be applied for well testing,, the linear frequency response functions for vertical bending were calculated first. The ULS vertical midship bending moment was calculated to be 340 MNm (positive in sagging, negative in hogging). The 100-years design seastate was characterized by a JONSWAP spectrum with T_z = 8.2 s, H_s = 10.5 m, a peakedness parameter of 5.0 and a duration of 4 hours.

The calculated Most Likely Extreme Response for sagging (positive bending moment) is shown as a function of time in figure 1. The signal was conditioned such that the linear ULS occurs after 100 seconds of simulation. The nonlinear SWAN result is presented in the same figure. The correction factor for nonlinear sagging was derived to be equal 1.25.

The results of a nonlinear SWAN simulation in one of the many possible conditioned random wave profiles is plotted in figure 2. It can be read that the correction factor for the linear ULS in sagging for this particular condition equals 1.17. Totally, 10 random realizations were run. The realized sagging correction factors varied between 1.09 and 1.27. The mean value was 1.21 with a standard deviation 0.05.

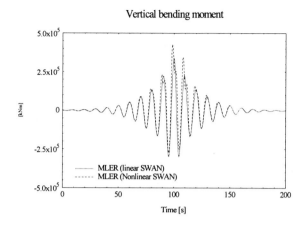

Figure 1. Linear and nonlinear vertical bending moment simulated in the wave profile inducing the Most Likely Extreme Response at t = 100 s.

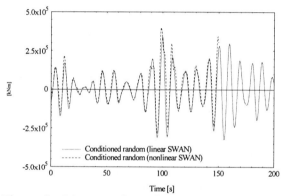

Figure 2. Linear and nonlinear vertical bending moment simulated in a conditioned random wave profile inducing the linear ULS at t = 100 s.

A regular wave ULS analyses resulted in a factor of 1.22. To investigate the correction factor for hogging, the mirrored wave profile has to be used, guaranteeing that the minimum linear vertical bending moment, i.e. maximum hogging, is observed at the specified moment. The wave profiles which induced the MLER and conditioned random response are shown in figure 3. The realized maximum wave steepness was approximately 1/7, which is at the pysical limit for wave breaking.

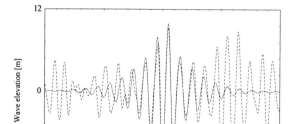

Figure 3. Derived irregular wave profiles after conditioning of the linear vertical bending moment to have an amplitude equal linear ULS and frequency equal mean response frequency at time t = 100 s.

The second example is the calculation of maximum water heights on deck. In SWAN, the water height on deck is calculated using the formulations as presented by Børresen and Tellsgård [7], which are based on Bernoulli's formulation for a falling dam of water. Bow flare effects are accounted for and the influence of water on deck is included in the motion equations. As green water is not a stationary phenomenon, a nonlinear time domain analysis has been performed. The vessel as used in the previous example has been studied. At maximum draught, there was only a free board of three meter at the working deck, and special interest had to be paid to water on deck. After calculation of the relative motion ULS values at different locations along the vessel, the 100-years design seastate for the relative motion at the working deck was characterized by a JONSWAP spectrum with Tz = 6.1 s, Hs = 6.5 m in 150 degrees heading (taking into account a realistic heading distribution).

The Most Likely Extreme Response for the relative motion was determined and the underlying wave profile was derived. Successively, SWAN was run in its nonlinear mode. A snap shot of this simulation is shown in figure 4. The resulting maximum water height on deck was considered as the 100-years ULS water on deck height which could directly be used in the design of the deck structure and in determining the required height of the equipment above the deck.

58

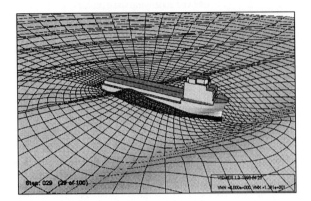

Figure 4 Snap-shot of a nonlinear simulation of ULS water on deck.

6.   CONCLUDING REMARKS

A method for determining the Most Likely Extreme Response has been presented. The method is based on the use of a nonlinear time domain program in combination with a method to condition the irregular wave signal such that the *linear* response ULS will occur at a specified time with a specified frequency. Using DNVs version of SWAN, it was demonstrated that a large scala of ULS problems can be analysed by the engineer and interfacing with FE-models has become simpler because of balanced global loads, pressures and inertia. It was shown that the mean nonlinear correction factor for sagging as observed in conditioned random waves was not equal to the nonlinear correction factor derived from the MLER-simulation. This stresses the advantage of using a well defined Most Likely Extreme approach.

Also effects such as green water on deck could be analyzed efficiently. In combination with modern vizualization techniques, these software tools have become powerful design evaluation tools [10].

More research has to be done on the choice of the characteristic ULS frequency and on conditioning of the wave input regarding maximum realized wave steepnesses. The sensitivity of calculated nonlinear responses to randomness of the input conditioned irregular wave trains has to be studied in more detail as well.

REFERENCES

1. Kring, D., Y.-F. Huang, P. Sclavounos, T. Vada, A. Braathen. Nonlinear ship motions and wave-induced loads by a Rankine method. *In Proc. Of 21ˢᵗ Symp. On Naval Hydrodynamics*, Trondheim, 1996.
2. Winterstein, S.R., T.C. Ude, C.A. Cornell, P.Bjerager, S. Haver. Environmental parameters for extreme response: Inverse form with omission factors. Environmental parameters for extreme response: Inverse form with omission factors, ICOSSAR-93, Innsbruck, 1993.
3. Friis Hansen, P. and L. P. Nielsen. On the new model for the kinematics of large ocean waves. In *OMAE Proceedings, Offshore Technology*, Volume IA, pages 17-24, 1995.
4. Tromans, P.S., Anaturk, A.H.R. and Hagemeijer, P., New model for the kinematics of large amplitude ocean waves applications as a design wave. In *Proc. 1ˢᵗ Int. Offshore Polar Eng. Conf., ISOPE*, pp 64-71, 1991.
5. Torhaug, R. Extreme response of nonlinear ocean structures: Identification of minimal stochastic wave input for the time domain simulation. Ph. D. thesis Stanford University, 1996.
6. Adegeest, L.J.M. Nonlinear hull girder loads in ships, Ph. D. Thesis Delft University of Technology, 1995.
7. Børresen, R. and F. Tellsgård. Non-linear response of vertical motions and loads in regular head waves. DNV rep. 79-1097, 1979.
8. Vinje, T. On the statistical istribution of wave heights in a random seaway. Applied Ocean Research, 11(3):142-152, 1989.
9. Adegeest, L.J.M. Modeling of the probabilistic input for a nonlinear wave-load simulation model, Delft University of Technology, Ship hydromechanics Lab., rep. nR. 933, June 1992.
10. SWAN demo-ULS animations in irregular waves, http://www.dnv.com/

Practical Design of Ships and Mobile Units
M.W.C. Oosterveld and S.G. Tan, editors.

Numerical study of the impact of water on cylindrical shells, considering fluid-structure interactions

M. Arai[a]   and   T. Miyauchi[b]

[a]Department of Naval Architecture and Ocean Engineering, Yokohama National University, 79-5 Tokiwadai, Hodogaya-ku, Yokohama 240-8501, Japan

[b]Sumitomo Heavy Industries, Ltd., 4-7 Uraga-machi, Yokosuka 239, Japan

This paper describes a numerical simulation method for the analysis of water impact on elastic structures. In order to accurately evaluate the structural strength of marine structures under severe water impact, it is necessary to study the characteristics of the dynamic structural response to an impact load that rapidly changes in magnitude during a short period of time. Until now, a simplified two-step method has been widely used for this purpose. In the two-step method, a time series of the hydrodynamic force on a structure is tentatively estimated assuming that the body is rigid, and then a response analysis of the structure is performed by applying the previously estimated force history. However, when the structure is relatively flexible, it is questionable whether the two-step method should be applied because it neglects the coupling effect between the fluid and the structure. In this study, therefore, a numerical simulation method that takes into consideration fluid-structure coupling is developed. The new method consists of two solvers, i.e., a solver that utilizes the computational fluid dynamics (CFD) for the flow field analysis and a modal analysis technique for solving the dynamic structural response. Information such as the fluid pressure and the motion and deformation of structure is exchanged between the two numerical solvers at every cycle of simulation. A series of drop-model experiments was also conducted to verify the accuracy of the developed numerical simulation method. In the experiment, two-dimensional cylindrical shell models made of aluminum plates with different thickness were dropped onto a calm water surface, and the motion of the models and strains on the shell plates were measured.

The following results were obtained from this study:
1. In the case of a lightweight model, the rigid body motion of the dropped model changes remarkably when it impacts the water surface. This deceleration has considerable influence on the pressure field around the model and also on the elastic response of the model structure, i.e., strain time history. The accuracy of the numerical simulation is improved substantially by considering the deceleration effect of the rigid body motion during water entry.
2. Although the predicted elastic response of the two-step method agrees fairly well with the measured response in the case of the cylinder with high rigidity, it fails to predict the elastic response of the cylindrical shell with low rigidity. By carrying out the coupling computation that takes into consideration the elastic deformation of the cylindrical structure, the accuracy of the computed elastic response is improved reasonably and shows the effectiveness of the proposed simulation method.

1. INTRODUCTION

To evaluate the local strength of marine structures such as the bows of ships, the horizontal members of offshore structures, etc., it is necessary to study the characteristics of the dynamic response of structures to an impulsive force that has abrupt changes in magnitude in a short period of time.   Until now, as an analytical solution, it is usual that a series of forces are tentatively calculated by a certain method assuming that the structure is rigid, and then a response analysis of the structure is carried out using the previously obtained force history (We will call this process a "two-step method" in this paper).   In the two-step method, however, the interaction between the structure and the fluid is neglected, so, if the structure is relatively flexible, it is anticipated that the accuracy of the analysis is not sufficient. Accordingly, in this study, a new method was developed by the authors in which the flow field around the structure is calculated by utilizing a computational fluid dynamics (CFD) technique, and details such as the pressure and velocity of the fluid domain, structural deformation, etc., are exchanged at every moment with a dynamic structural response analysis. The accuracy of the proposed numerical method was confirmed by comparing the computed results with data obtained from a model experiment in which an aluminum cylinder was dropped onto a still water surface and dynamic strain responses were measured.

2. NUMERICAL METHOD

2.1 Fluid domain computation
2.1.1 Governing Equations

Assuming the flow to be inviscid and incompressible, the governing equations of two-dimensional flow about the body using the Cartesian coordinate system (i.e., (x,y) coordinate system) are the mass continuity equation:

$$\frac{\partial u}{\partial x} + \frac{\partial v}{\partial y} = 0 \tag{1}$$

and Euler's equations of motion:

$$\frac{\partial u}{\partial t} + u\frac{\partial u}{\partial x} + v\frac{\partial u}{\partial y} = -\frac{1}{\rho}\frac{\partial p}{\partial x} \tag{2}$$

$$\frac{\partial v}{\partial t} + u\frac{\partial v}{\partial x} + v\frac{\partial v}{\partial y} = -\frac{1}{\rho}\frac{\partial p}{\partial y} + g + \dot{V}I \tag{3}$$

where t is time, u and v are the components of the velocity vector in the x and y direction, ρ is the fluid density, p is the pressure, and g is the acceleration due to gravity. Also, $\dot{V}I$ represents the time derivative of the vertical velocity of the center of gravity of the falling body. $\dot{V}I$ is an apparent term and it appears in equation(3) because the falling phenomenon of the body is modeled by using the moving coordinate system fixed to the body [1].

Equations (1)-(3) shown above are the governing equations for the Cartesian coordinate system. In the present numerical method we will rewrite these equations to use them in the body-fitted grid system (i.e., (ξ, η) coordinate system) as shown in Fig.1. For this purpose, we will adopt the volume fluxes U^{ξ} and U^{η} in the ξ and η directions as the primitive variables instead of u and v in equations (1)-(3). The new governing equations for the (ξ, η) coordinate

system are the continuity equation (4) and the momentum equations (5) and (6) [2-5]:

$$\frac{\partial U^{\xi}}{\partial \xi} + \frac{\partial U^{\eta}}{\partial \eta} = 0 \tag{4}$$

$$J\frac{\partial U^{\xi}}{\partial t} + U^{\xi}\frac{\partial U^{\xi}}{\partial \xi} + U^{\eta}\frac{\partial U^{\xi}}{\partial \eta} - U^{\xi}(\boldsymbol{u}\cdot\frac{\partial \boldsymbol{S}^{\xi}}{\partial \xi}) - U^{\eta}(\boldsymbol{u}\cdot\frac{\partial \boldsymbol{S}^{\eta}}{\partial \eta})$$

$$= -(\boldsymbol{S}^{\xi}\cdot\boldsymbol{S}^{\xi})\frac{1}{\rho}\frac{\partial p}{\partial \xi} - (\boldsymbol{S}^{\xi}\cdot\boldsymbol{S}^{\eta})\frac{1}{\rho}\frac{\partial p}{\partial \eta} + J(\boldsymbol{S}^{\xi}\cdot(\boldsymbol{g}+\dot{\boldsymbol{V}}\boldsymbol{I}(t))) \tag{5}$$

$$J\frac{\partial U^{\eta}}{\partial t} + U^{\xi}\frac{\partial U^{\eta}}{\partial \xi} + U^{\eta}\frac{\partial U^{\eta}}{\partial \eta} - U^{\xi}(\boldsymbol{u}\cdot\frac{\partial \boldsymbol{S}^{\eta}}{\partial \xi}) - U^{\eta}(\boldsymbol{u}\cdot\frac{\partial \boldsymbol{S}^{\eta}}{\partial \eta})$$

$$= -(\boldsymbol{S}^{\eta}\cdot\boldsymbol{S}^{\xi})\frac{1}{\rho}\frac{\partial p}{\partial \xi} - (\boldsymbol{S}^{\eta}\cdot\boldsymbol{S}^{\eta})\frac{1}{\rho}\frac{\partial p}{\partial \eta} + J(\boldsymbol{S}^{\eta}\cdot(\boldsymbol{g}+\dot{\boldsymbol{V}}\boldsymbol{I}(t))) \tag{6}$$

Here, $\boldsymbol{g}$ is the acceleration vector due to gravity, and $\dot{\boldsymbol{V}}\boldsymbol{I}$ is the acceleration vector of the center of gravity of the falling body. The volume fluxes U^{ξ} and U^{η} are defined over the faces of the finite difference cell which is used for numerical computation:

$$U^{\xi} = \boldsymbol{S}^{\xi}\cdot\boldsymbol{u}$$
$$U^{\eta} = \boldsymbol{S}^{\eta}\cdot\boldsymbol{u} \tag{7}$$

where

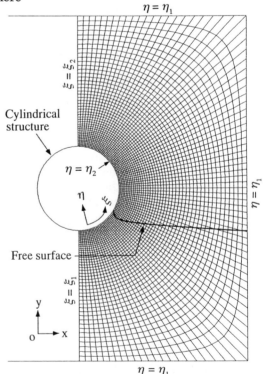

Fig.1 Coordinate systems and an example of finite difference cell arrangement

$$S^\xi = (J\frac{\partial \xi}{\partial x}, J\frac{\partial \xi}{\partial y})$$

$$S^\eta = (J\frac{\partial \eta}{\partial x}, J\frac{\partial \eta}{\partial y}) \tag{8}$$

and

$$J = \begin{vmatrix} \dfrac{\partial x}{\partial \xi} & \dfrac{\partial x}{\partial \eta} \\ \dfrac{\partial y}{\partial \xi} & \dfrac{\partial y}{\partial \eta} \end{vmatrix} \tag{9}$$

S^ξ and S^η in equation(8) are called cell-face vectors [2] which have arc lengths of the cell faces and are oriented in a direction normal to the cell faces. Jacobian J of the transformation has a volume which is equivalent to the cell volume.

In this study, we will solve the governing equations (4)-(6) using a finite difference method [4,5], where fluxes U^ξ, U^η and pressure p are to be solved as the primitive variables. An example of the grid system used for the numerical computation is shown in Fig.1, where only the right-hand side of the fluid domain is modeled, considering the symmetric character of the problem. Also, the locations where the primitive variables are defined are presented in Fig.2.

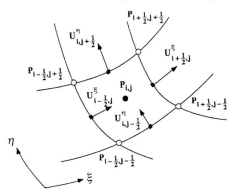

Fig.2 Location of finite difference variables in a typical cell

2.1.2 Boundary Conditions

The boundary conditions necessary to solve the slamming phenomenon of a body dropped onto the water surface are as follows.

Free-surface boundary:

Since spray formation has a significant effect on the hydrodynamic impact pressure, an accurate determination of the free surface deformation is essential. For this purpose, we modified the Volume of Fluid Method (VOF) [6]. Using this method the following kinematic condition is solved for estimating the free surface location.

$$J\frac{\partial F}{\partial t} + U^\xi \frac{\partial F}{\partial \xi} + U^\eta \frac{\partial F}{\partial \eta} = 0 \tag{10}$$

Here, F is a scalar function defined for each cell. F is used to indicate the fraction of the volume of fluid in a cell, i.e., if the cell is fully filled with fluid, F is defined as 1, and if the cell is empty, F is defined as 0. Therefore, the free-surface location is determined to be the place where the F value changes from 1 to 0. On the free surface, the dynamic condition:

$$P_{f.s.} = P_{atm.} \tag{11}$$

is applied, where $P_{f.s.}$ is the pressure at the free surface and $P_{atm.}$ is the atmospheric pressure.

Inflow boundary:

As was mentioned in the explanation of momentum equations, we used the grid system fixed to the falling body. In this way, the downward motion of the body is taken into account by imposing an inflow of fluid with upward velocity VI from the lower boundary of the grid system [1,4].

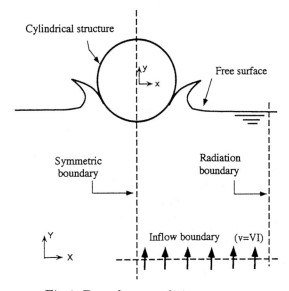

Fig.3 Boundary conditions

Radiation boundary:

At the right end of the physical domain, the

continuous outflow condition is applied:

$$\frac{\partial u}{\partial \xi}\frac{\partial \xi}{\partial x} + \frac{\partial u}{\partial \eta}\frac{\partial \eta}{\partial x} = 0$$

$$\frac{\partial v}{\partial \xi}\frac{\partial \xi}{\partial x} + \frac{\partial v}{\partial \eta}\frac{\partial \eta}{\partial x} = 0 \tag{12}$$

Wall boundary [4]:

In this section, we will show the boundary condition assuming that the wall is rigid. The treatment of a flexible-wall boundary will be explained in Section 2.3. If we assume that the boundary at $\eta = \eta_2$ coincides with the wall of the falling body (see Fig.1), the rigid wall boundary conditions are

$$U^{\eta} = 0 \tag{13}$$

and

$$\frac{\partial p}{\partial \eta} = \frac{\rho}{\boldsymbol{S}^{\eta} \cdot \boldsymbol{S}^{\xi}}\left[U^{\xi}(\boldsymbol{u}\cdot\frac{\partial \boldsymbol{S}^{\eta}}{\partial \xi}) - (\boldsymbol{S}^{\eta}\cdot\boldsymbol{S}^{\xi})\frac{1}{\rho}\frac{\partial p}{\partial \xi} + J\{\boldsymbol{S}^{\eta}\cdot(\boldsymbol{g}+\dot{\boldsymbol{V}}\boldsymbol{I}(t))\}\right] \tag{14}$$

Similar conditions are also used for a symmetric boundary.

2.2 Dynamic structural response analysis of a cylindrical shell

In this study, the dynamic response of a cylindrical shell that drops on the water surface is formulated as a bending vibration problem for a circular ring by assuming that the phenomenon is two-dimensional (Fig.4). Equilibrium equations with regard to the short element $ad\theta$ of the ring are as follows:

$$\frac{\partial T_{\theta}}{a\partial\theta} - \frac{Q_{\theta}}{a} - \mu\frac{\partial^2 v}{\partial t^2} = 0 \tag{15}$$

$$\frac{\partial Q_{\theta}}{a\partial\theta} - \frac{T_{\theta}}{a} - \mu\frac{\partial^2 w}{\partial t^2} + f(\theta,t) = 0 \tag{16}$$

$$-\frac{\partial M_{\theta}}{a\partial\theta} + Q_{\theta} = 0 \tag{17}$$

where $\bar{v}$ is the tangential component of the displacement, $\bar{w}$ is the normal component of the displacement, and μ is the mass per unit length of the element. In the above equations, M_{θ}, Q_{θ}, and T_{θ} are the bending moment, the shearing force, and the axial force at the position θ, respectively. Also, $f(\theta,t)$ is an external force at time t.

The bending moment is expressed as

$$M_{\theta} = -EI(\frac{\partial^2 \bar{w}}{a^2\partial\theta^2} + \frac{\bar{w}}{a^2}) \tag{18}$$

where EI is the flexural rigidity of the ring.

To solve the above equations, we consider that the bending vibration dominates the phenomenon, and we can assume that the elongation of the neutral plane of the ring (dashed line in Fig.4) is sufficiently small [7]:

$$\frac{\partial \bar{v}}{a\partial\theta} - \frac{\bar{w}}{a} = 0 \tag{19}$$

Equations (15)-(19) lead to the following equation of vibration.

$$\frac{EI}{a^4}\left[\frac{\partial^6 \bar{w}}{\partial\theta^6} + 2\frac{\partial^4 \bar{w}}{\partial\theta^4} + \frac{\partial^2 \bar{w}}{\partial\theta^2}\right] - \mu\frac{\partial^2 \bar{w}}{\partial t^2} + \mu\frac{\partial^4 \bar{w}}{\partial\theta^2\partial t^2} = \frac{\partial^2 f(\theta,t)}{\partial\theta^2} \tag{20}$$

Assuming that the $\bar{w}$ in equation(20) can be expressed as an aggregate of the normal mode $W_n(\theta)$ of vibration:

$$\bar{w}(\theta,t) = \sum_{n=1}^{\infty} W_n(\theta)q_n(t) \tag{21}$$

where $q_n(t)$ is the normal coordinate and is a function of time t only.

The normal mode of vibration for a closed circular ring can be expressed as

$$W_n(\theta) = \cos n\theta \tag{22}$$

Expressing $f(\theta,t)$ in equation(20) using the Fourier series and then multiplying equation(20) by $W_n(\theta)$ and integrating it with respect to θ, we obtain

$$m_n\frac{\partial^2 q_n}{\partial t^2} + k_n q_n = F_n \tag{23}$$

This equation can be used to solve for the normal coordinate q_n, where

$$m_n = \mu a\pi(1 + \frac{1}{n^2}) \tag{24}$$

$$k_n = \frac{EI\pi}{a^3}(n^2 - 1)^2 \tag{25}$$

$$F_n = \int_0^{2\pi} f(\theta,t)\cos n\theta \, ad\theta \tag{26}$$

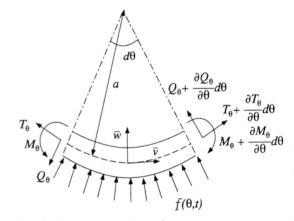

Fig.4 Forces on a ring element

We refer to m_n as the modal mass, k_n as the modal stiffness, and F_n as the generalized force.

The natural frequency of the n-th mode vibration of the circular ring can be estimated by the following equation.

$$\omega_n = \sqrt{\frac{k_n}{m_n}} = \frac{n(n^2-1)}{\sqrt{n^2+1}} \frac{1}{a^2} \sqrt{\frac{EI}{\mu}} \qquad (27)$$

The transient elastic response of the ring is determined by a convolution integral:

$$q_n(t) = \frac{1}{m_n \omega_n} \int_0^t F_n(\tau) \sin \omega_n(t-\tau) d\tau \qquad (28)$$

where the pressure information obtained from the flow field analysis is used as the generalized force F_n.

As $q_n(t)$ can be calculated from equation(28), we can determine the dynamic response of $\overline{w}$ by using equation(21). Furthermore, if we assume that the z coordinate originates from the neutral plane of the ring and that this coordinate is directed in the internal direction of the ring (see Fig.4), and if we also make the non-elongation assumption at the neutral plane, the strain at z can be expressed as [7]:

$$\varepsilon_\theta(z) = -\frac{z}{a}\left(\frac{\partial^2 \overline{w}}{a\partial\theta^2} + \frac{\overline{w}}{a}\right) - \left(\frac{z}{a}\right)^2\left(\frac{\partial^2 \overline{w}}{a\partial\theta^2} + \frac{\overline{w}}{a}\right) \qquad (29)$$

We will compare the strain-time histories calculated from equation(29) with the results of model experiments.

As for the mode shapes expressed by equation(22), we will adopt the three lowest mode shapes shown in Fig.5.

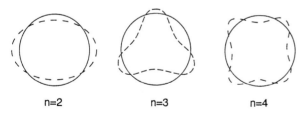

n=2   n=3   n=4

Fig.5 Mode shapes for modal analysis of a cylindrical shell

2.3 Algorithm of fluid-structure interaction computation

In Section 2.1.2, we showed a wall boundary condition assuming that the wall of a structure is rigid. However, in order to handle the elastic response of a cylindrical shell, the influence of the elastic deformation should be fed back to the calculation in the fluid domain. To treat this problem in a numerical computation, we have devised a method which considers the influence of the elastic deformation by adding a source term to the governing equation. In the numerical model, as shown in Fig.6, the location of the wall of the cylindrical shell before deformation is set at $\eta = \eta_2$. Then, the wall deformation δ_η is treated by controlling the source term, which is added in the continuity equation, to express the volume change of the boundary cell adjacent to the wall boundary. Accordingly, a finite difference representation of the continuity equation for the boundary cell (i,j) becomes as follows:

$$\frac{U_{i+1/2,j}^\xi - U_{i-1/2,j}^\xi}{\Delta\xi} + \frac{U_{i,j+1/2}^\eta - U_{i,j-1/2}^\eta}{\Delta\eta} - \frac{\dot{\delta}_\eta|S^\eta|}{\Delta\eta} = 0 \qquad (30)$$

where $\dot{\delta}_\eta$ is the velocity of the elastic deformation of the wall boundary, $|S^\eta|$ is the arc length of the boundary cell at $\eta = \eta_2$, and $\Delta\xi$ and $\Delta\eta$ are the cell widths in the ξ and η directions.

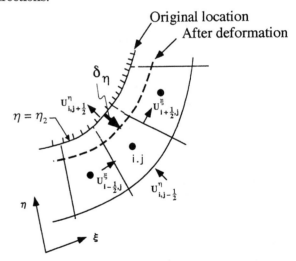

Fig.6 Elastic deformation in boundary cells

It may be understood from the aforementioned explanation that an elastic deformation that exceeds the width of the boundary cell cannot be handled by the present method. In an actual numerical simulation, a cell width such that $\delta_\eta < \Delta\eta/4$ was used in considering the accuracy of the computation. Also, it is assumed that the displacement due to the elastic deformation of the boundary wall occurs only in the η direction.

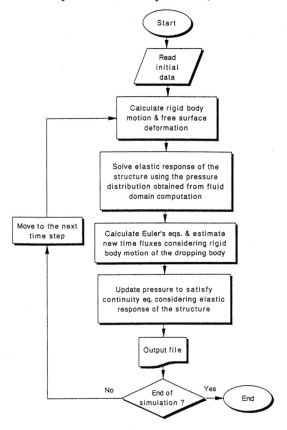

Fig.7 Flow chart for the present simulation method

An outline of the algorithm of the coupling computation is shown in Fig.7. First of all, the instantaneous position of the center of gravity of the dropping body and the free surface location are calculated using the information from the pressure and velocity fields at the previous time step. Then, the elastic response (i.e., displacement and velocity of the wall, strain of the shell, etc.) are calculated using the modal analysis technique. At this time, information regarding the pressure distribution at the wall boundary, which was obtained from the fluid domain computation, is given for the computation of the structural response analysis part as an external-force term in the equation of vibration. Next, the computation advances to the fluid domain analysis. First, the volume flux over the faces of a finite difference cell is predicted by Euler's equations of motion. Then, using the

continuity equation, the iteration process continues until the flux and pressure in the fluid domain are finally settled. The continuity equation(30) that considers the elastic deformation of the wall is used for the cells which are adjacent to the wall boundary. After these procedures, all information regarding the dynamic response of the structure and flow field is determined. These steps are repeated until the simulation time is complete, as determined beforehand.

3. MODEL EXPERIMENT

The cylindrical model that was used for the experiment is shown in Fig.8. The model is made of aluminum plate 3 mm thick, with an outer diameter of 306 mm and a length of 600 mm. The total weight of the model including strain gauges, cables, etc. is 5.2 kg. Strain gauges are arranged to measure the strain in the circumference direction of the cylinder (see Fig.8). The size of the tank used for the model experiment is 44 m long, 2.5 m wide, and 1.8 m deep, and end plates are settled in the tank to create a two-dimensional flow. In the experiment, the model was dropped onto the water surface, keeping the axis horizontal at all times. Measurements were carried out on the motion of the body and the strain that occurred during impact. As for the conditions of the experiments, the distance between the lowest end of the model before dropping and the still water surface was changed. We will refer to this distance as the drop height hereafter. Four experiments with different drop heights were conducted (0.25, 0.5, 0.75, and 1.0 m). Both ends of the model were closed with a rubber membrane with a slight rigidity to prevent water from entering the cylinder.

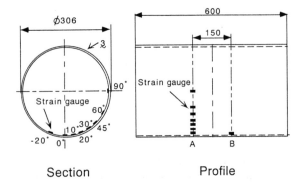

Fig.8 A cylindrical model and strain gauge arrangements

A similar model experiment using a cylindrical shell has been reported by Shibue et al. [8]. They used a steel shell model and measured the dynamic strain response of the shell. Their experimental results supply us with useful information because the rigidity of their model was quite different from that of our aluminum model. Therefore, in the next chapter, we will use not only our experimental data but also Shibue et al's data in evaluating the simulation method shown in this paper. The principal particulars of the steel model quoted from Ref.[8] (we call this Cylinder-1) and those of our aluminum model (we call this Cylinder-2) are compared in Table 1.

Table 1 Principal particulars of cylindrical models and tested conditions

	Cylinder-1 (Shibue et al., 1993)	Cylinder-2 (Yokohama National Univ.)
Material	Steel	Aluminium
Diameter (external)	312.0mm	306.0mm
Plate thickness	5.1mm	3.0mm
Total model weight	23.8kg	5.2kg
Young's modulus	21,000kgf/mm^2	7,500kgf/mm^2
Poisson's ratio	0.30	0.34
Density of material	8.0×10^{-10}kgf·s^2/mm^4	2.76×10^{-10}kgf·s^2/mm^4
Drop height	0.5, 1.0, 1.5m	0.25, 0.5, 0.75, 1.0m

4. COMPARISON OF COMPUTED AND MEASURED RESULTS

4.1 Rigid body motion

The rigid body motion of a dropping cylinder is computed by the following motion equation:

$$(m + m_a)\ddot{Y} = -mg + (f_{imp} + f_b) \quad (31)$$

where Y is the vertical position of the center of gravity of the cylinder in the space-fixed Cartesian coordinate system. As for the force exerted on the cylinder, only buoyancy force (f_b), gravity force (mg), and impact force (f_{imp}) are considered. Buoyancy force is estimated by using the sectional area of the body below the still water surface. Deformation of the water surface adjacent to the body is neglected. Also, the impact force in equation (31) is estimated by

$$f_{imp} = -\frac{dm_a}{dt}\dot{Y} \quad (32)$$

where $m_a = \pi \rho c^2/2$ is the two-dimensional added mass, and c is the wetted half breadth of the partially immersed cylinder.

Figure 9 shows the calculated time-varying

position of the center of gravity of the cylinder. Measured data obtained from an analysis of the videotapes of the model experiment are also shown in the figure. Not only in this example but in all the tested cases, the calculated rigid body motions agreed very well with the measured ones. Thus, it is considered that the motion of water entry of a comparatively lightweight body can be evaluated accurately by the method described in this section. We will apply equations (31) and (32) to the elastic response simulation in the following section in evaluating the motion of the center of gravity of the dropping body.

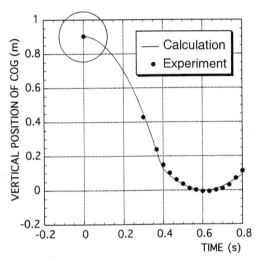

Fig.9 Motion of the cylindrical model onto a still water surface
(Cylinder-2, drop height=0.75m)

4.2 Effect of deceleration and of fluid-structure interactions on the response of the structure

In this section, we will carry out three kinds of numerical simulations and compare the computed and the measured strains.

Cal-A (two-step method, deceleration is considered):

By using the flow analysis described in this paper, the pressure that acts on the cylinder, which is assumed to be a rigid body, is computed tentatively. This pressure information is then given as input to the structural analysis part of the calculation. As for the structural analysis, the dynamic response analysis option of a general purpose FEM (Finite Element Method) solver is

66

utilized. In the calculation, the time change of the entry speed of the cylinder (i.e., the effect of deceleration) is taken into consideration by the method explained in Section 4.1.

Cal-B (coupling calculation, entry speed to be constant):

Dynamic strain is calculated using the fluid-structure coupling computation method described in this paper. However, the entry speed is estimated at the time when the lowest part of the cylinder comes into contact with the water, and this speed is used as a constant vertical velocity throughout the simulation.

Cal-C (coupling calculation, deceleration is considered):

Dynamic strain is calculated using the fluid-structure coupling computation method described in this paper. Also, a time change for the entry speed is considered by the method described in Section 4.1.

In our numerical method, the concept of added mass that has been used in conventional hydro-elastic theories is not used because both the fluid and structural parts are solved simultaneously. Also, in the two-step analysis (Cal-A) we used the FEM solver "ANSYS" to obtain the transient response of the cylindrical structure. According to preliminary simulations, the results of the

FEM analysis and the modal analysis that employed three mode shapes (see Section 2.2) agreed quite well. Therefore, no difference is anticipated between the use of FEM or modal analysis in calculating the structural response of the cylinders. The computed results for Cylinder-1 (relatively rigid and heavy-weight model) and for Cylinder-2 (relatively flexible and lightweight model) are shown as follows together with the results of the model experiments.

Cylinder-1:

Figure 10 shows a comparison between the measured and computed strain-time histories of the inner surface at the bottom of the cylinder. The drop height for this model experiment is 1 m. The computed result of the two-step method (Cal-A) has almost the same amplitude as that of the measured strain. However, the phase of the computed strain does not agree with that of the measured strain. The computed result of Cal-B (constant water entry speed) has a larger response amplitude than the experimental result. This may be caused by the overestimate of the hydrodynamic force exerted on the cylinder after water entry which occurs because the effect of deceleration is neglected in Cal-B. The computed result of Cal-C, in which both the deceleration of the cylinder after water entry and fluid-structure coupling are considered, agrees well with the measured result.

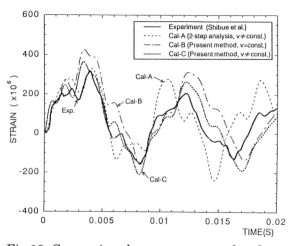

Fig.10 Comparison between computed and measured strain histories (Cylinder-1, drop height=1.0m, strain at $\theta = 0°$ (bottom of the model))

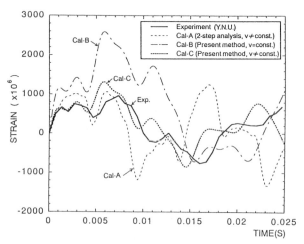

Fig.11 Comparison between computed and measured strain histories (Cylinder-2, drop height=1.0m, strain at $\theta = 0°$ (bottom of the model))

Cylinder-2:

Characteristics of the three kinds of computational methods that were observed in the results for Cylinder-1 appear more clearly in the results for Cylinder-2, which has lower rigidity and weight. In other words, as shown in Fig.11, the computed strains of Cal-A and Cal-B do not agree with that of the model experiment. It is therefore clear that these two computing methods are both insufficient in computing the response of an elastic structure with low rigidity. Accordingly, to accurately evaluate the dynamic response of a structure with low rigidity, it is necessary to consider the deceleration of the structure after water entry and the effect of the fluid-structure interaction. The computed strain for Cal-C, which fully takes into account these two effects, agreed reasonably well with the measured data.

4.3 The relationship between the pressure field and dynamic elastic response

From the discussion in the previous section it is clear that the rigid body motion and the fluid-structure interactions strongly affect the response of the strain of the cylindrical shell. In this section, we will discuss the relationship between the pressure field and the dynamic elastic response of the structure by showing a few more computational results.

Figure 12 shows the pressure fluctuation and free-surface deformation about the cylindrical shell (Cylinder-2) that were obtained from a simulation with a fall height of 0.75 m. In this simulation, both the deceleration of the motion of the cylinder and the fluid-structure interaction are considered. It can be seen from this figure that the pressure field about the cylinder fluctuates in time due to the influence of the elastic deformation of the cylinder. To emphasize the effects of elastic response in the pressure field, the range where the pressure becomes higher than the surrounding region is indicated by the mark "H", and the region where it becomes lower than the surrounding area is indicated by the mark "L". The influence of elastic response is clearly apparent in the petal-like pattern of the pressure contour shown in Fig.12. The computed vertical displacement of the lowest point of the shell is shown in Fig.13. Also, a time history of the strain in the circumference direction at the bottom of the internal face of the shell is shown in Fig.14. As we can determine from Figs.13 and 14, the effect

of the lowest elastic mode (i.e., n=2 mode in Fig.5) is dominant in the displacement response, however, the higher elastic modes (i.e., n=3, 4) strongly affect the response of strain.

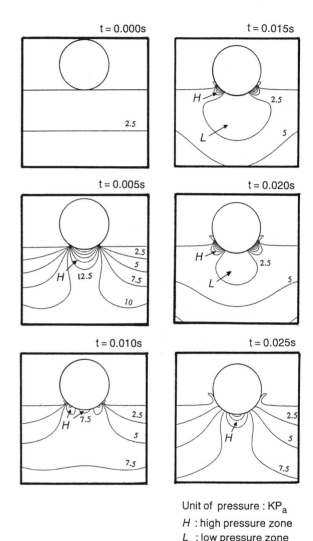

Unit of pressure : KP$_a$
H : high pressure zone
L : low pressure zone

Fig.12 Computed free surface deformation and pressure distribution around a cylinder (Cylinder-2, drop height=0.75m)

68

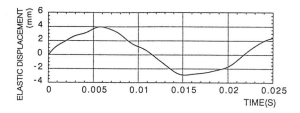

Fig.13 Computed elastic displacement history
(Cylinder-2, drop height=0.75m,
displacement at $\theta = 0°$)

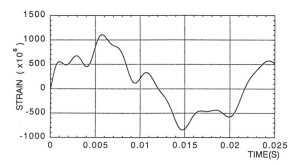

Fig.14 Computed strain history
(Cylinder-2, drop height=0.75m, strain
at $\theta = 0°$)

5. CONCLUSION

1. The motion of a lightweight dropping body changes remarkably during water entry (i.e., a notable deceleration occurs). This motion change affects the pressure field around the body, and the response of the structure changes as well. The numerical simulation method described in this paper can take into account this motion change, and it significantly improves the accuracy of the elastic response computation for the lightweight structure.

2. In the case of a structure with high rigidity, the amplitude of the strain response estimated by the two-step analysis agrees fairy well with the measured response. However, in the case of a structure with low rigidity, neither the amplitude nor the transient pattern of the computed strain agree with the measured data. By using the fluid-structure interaction computation method described in this paper, the accuracy of the elastic response computation for a water-entering structure can be improved.

REFERENCES

1. Arai, M., Matsunaga, K., A Numerical and Experimental Study of Bow Flare Slamming, Journal of The Society of Naval Architects of Japan, Vol.166, pp.343-353(1989) (in Japanese).
2. Rosenfeld,M., Kwak,D., Vinokur,M., A Fractional Solution Method for the Unsteady Incompressible Navier-Stokes Equations in Generalized Coordinate Systems, Journal of Computational Physics, Vol.94, pp.102-137 (1991).
3. Nagahama, M., Nagahama, S., Nekado, Y., Yamamori, T., Hori, T., A 3-Dimensional Analysis of Sloshing by means of Tank Wall Fitted Coordinate System, Journal of The Society of Naval Architects of Japan, Vol.172, pp.487-499(1992) (in Japanese).
4. Arai, M., Cheng, L.Y., Inoue, Y., A Computing Method for the Analysis of Water Impact of Arbitrary Shaped Bodies, Journal of The Society of Naval Architects of Japan, Vol.176, pp.233-240 (1994).
5. Arai, M., Cheng, L.Y., Inoue, Y., A Computing Method for the Analysis of Water Impact of Arbitrary Shaped Bodies (2nd Report) - Application for the Optimization of Horizontal Members of Offshore Structures -, Journal of The Society of Naval Architects of Japan, Vol.177, pp.91-99 (1995).
6. Hirt, C.W. and Nichols, B.D., Volume of Fluid Method (VOF) for Dynamic Free Boundaries, Journal of Computational Physics, No.39, pp.201-225(1981).
7. Kobayashi, S., Theory of Vibration, Maruzen(1994) (in Japanese).
8. Shibue, T., Ito, A., Nakayama, E., Structural Response Analysis of a Cylinder under Water Impact, Journal of The Society of Naval Architects of Japan, Vol.174, pp.479-484(1993) (in Japanese).

Structural response in large twin hull vessels exposed to severe wet deck slamming.

O.D. Økland[a], T. Moan[a] and J.V.Aarsnes[b]

[a] Department of Marine Structures, Norwegian University of Science and Technology.
N-7034 Trondheim, Norway

[b] Advanced Production and Loading AS, P.O. Box 53, N-4801 Arendal, Norway.
Previously with MARINTEK - SINTEF GROUP, N-7034 Trondheim, Norway.

This paper is concerned with the effect of slamming loads against the wet deck of large twin hull vessels with zero forward speed. A numerical model for time domain calculation of slamming loads, and the corresponding structural response, is presented. An experimental investigation has been carried out as well, and measured results are compared with results found by the numerical model. Generally, satisfactory agreement is found between calculated and measured results.

1. INTRODUCTION

Structural weight is an important design parameter for high speed vessels. Optimal balance between safety and weight require rational methods for predicting load effects and structural performance.

Slamming load against the wet deck is one of the design loads that should be considered for a twin hull vessel. For small vessels, slamming loads are important in the design of the local cross structure between the hulls of the vessel. For large crafts, slamming against the wet deck, in combination with continuous wave load, may also lead to large global structural loads.

Kaplan (1991) investigated structural loads on twin hull vessels, caused by continuous wave loads and slamming. A simple momentum consideration was used to determine the slamming load. He considered rigid body modes only, and the structural response was determined by integrating the forces acting on the hull, including the inertia force due vessel accelerations in heave and pitch.

Kvålsvold and Faltinsen (1993) studied slamming loads and corresponding structural response in the local wet deck region of a twin hull. In their investigation a two dimensional boundary value problem was solved in the time domain, to determine the slamming pressure. Hydroelastic

effects due to structural response of the wet deck was included in this solution. It was found that the structural response of the wet deck may influence the slamming load. Effect of structural elasticity has also been studied in later investigations (see e.g. Kvålsvold et al. 1995).

The most important parameter for the magnitude of the wet deck slamming force, is the relative velocity between wave surface and wet deck. For small vessels, the relative velocity is found from wave elevation and rigid body motions. For a large twin hull, the global structural response may also influence the relative velocity. Wu et al. (1995), found that for whipping problems, flexible modes are important in the study of global structural loads.

For extreme cases of wet deck slamming, the ship master will reduce the speed of the ship. This means that the response in severe sea states should be investigated for zero or low forward speed. In the present work, the response of the vessel is described by a normal mode approach. The slamming load is found by solving a boundary value problem in the time domain. Slam induced rigid body motions and hydroelastic effects due to global structural response in the vessel can be accounted for. Longitudinal bending moments and corresponding vertical shear force, for zero speed and head sea, are determined. The numerical model can be extended to handle vessels with forward speed as well.

2. NUMERICAL MODEL

It is assumed that vessel responses are linear for given hydrodynamic loads. For a linear system, the principle of superposition is valid, and time series of response due to different loads can be added to give the total response.

2.1 Equation of motion

Using a Finite Element Method (FEM) approach, the displacements in a structure can be written as a linear combination of it's eigenmodes

$$\mathbf{r}(t) = \sum_{n=1}^{\infty} \mathbf{w}_n q_n(t) \tag{1}$$

where :

$\mathbf{w}_n$ - eigenvector from a free vibration analysis.
$q_n(t)$ - time dependent generalized coordinate.

In general, the first six modes in the above equation are rigid body modes, while the rest are flexible modes. In theory infinitely many flexible modes are needed to describe structural response of the vessel. For practical purposes however, the m first natural modes of the structure will describe the global response with sufficient accuracy.

By introducing Eq. (1) in the equation of motion and pre-multiply with the i'th eigenvector transposed, a generalized equation of motion can be established for the floating vessel.

$$[\mathbf{A}(\omega) + \mathbf{M}]\ddot{\mathbf{q}} + [\mathbf{B}(\omega) + \mathbf{C}]\dot{\mathbf{q}} + [\mathbf{R} + \mathbf{K}]\mathbf{q} = \mathbf{F} \tag{2}$$

Here :

A - hydrodynamic added mass matrix.
M - structural mass matrix.
B - hydrodynamic damping matrix.
C - structural damping matrix.
R - hydrodynamic restoring matrix.
K - structural restoring matrix.
F - generalized exciting force.

The dimension of the matrices is equal to the number of eigenmodes used in the description of vessel deformation.

Fluid motions in an incompressible fluid can be described by a velocity potential ϕ. When linear theory is used, the velocity potential of the fluid surrounding a floating vessel can be written

$$\phi_T = \phi_I + \phi_D + \sum_{r=1}^{m} q_r \phi_r \tag{3}$$

Here :

ϕ_I - velocity potensial of incident wave.
ϕ_D - velocity potensial due to diffracted wave.
ϕ_r - radiation velocity potensial due to unit excitation of mode r.

The velocity potential of an incident wave travelling in the negative x-direction can be written

$$\phi_I(x,z,t) = \frac{\varsigma_a g}{\omega} e^{kz} e^{i(\omega t - kx)} \tag{4}$$

where

ς_a - amplitude of the incident wave
ω - wave frequency
k - wave number

The velocity potentials due to diffraction and radiation are determined by a numerical boundary element method. For slender vessels, the strip theory (Salvesen et al., 1970) is often used. For vessels where three dimensional effects are important, a three dimensional approach (see e.g. Zhao et al., 1988), can be used.

If the origin is located at the free surface, and z is positive upwards, the pressure in the fluid is given by Bernoulli's equation

$$p = -\rho \frac{\partial \phi}{\partial t} - \rho g z - \frac{\rho}{2}\left((\frac{\partial \phi}{\partial x})^2 + (\frac{\partial \phi}{\partial z})^2\right) \tag{5}$$

where ρ is the density of the water, and g is acceleration of gravity.

Generalized force in mode k of our system, due to hydrodynamic pressure on the hull, can be expressed

$$F_k = -\iint_{S_b} \mathbf{u}_k \cdot \mathbf{n} p dS \tag{6}$$

where S_b is the wet body surface, $\mathbf{u}_k$ is the displacement vector of the body surface, and $\mathbf{n}$ is the normal vector of the body surface.

For a harmonic regular motion of the vessel, the radiation potential can be expressed as

$$\phi_R = \sum_{r=1}^{m} \phi_r \bar{q}_r e^{i\omega t} \qquad (7)$$

Further, the quadratic terms in Eq. (5) can be neglected and the generalized linearized force in mode k, due to motion in mode r, may be expressed on the form

$$F_k^R = \sum_{r=1}^{m} \left(\omega^2 A_{kr} - i\omega B_{kr} - R_{kr} \right) \bar{q}_r e^{i\omega t} \qquad (8)$$

In the above equation the coefficients A_{kr}, B_{kr}, and R_{kr} are given by :

$$A_{kr} = \frac{\rho}{\omega^2} \mathrm{Re}\left[\iint_{S_b} \mathbf{u}_k \cdot \mathbf{n}(i\omega\phi_r) dS \right]$$

$$B_{kr} = -\frac{\rho}{i\omega} \mathrm{Im}\left[\iint_{S_b} \mathbf{u}_k \cdot \mathbf{n}(i\omega\phi_r) dS \right]$$

$$R_{kr} = -\rho g \iint_{S_b} \mathbf{u}_k \cdot \mathbf{n} w_r dS$$

where w_r is vertical displacement due to unit excitation of mode r.

The hydrodynamic added mass matrix is generally dependent on the frequency of oscillation. For high frequencies, an asymptotic value is reached. Asymptotic added mass values are used in the determination of response, for whipping induced flexible modes.

The damping present in our system can be divided in two categories.

- Hull damping
- Hydrodynamic damping

Hull damping is zero for rigid body modes, but important for flexible modes. The Rayleigh damping model is often used to describe hull damping. In this model the generalized damping matrix is given as a linear combination of the mass- and stiffness matrices

$$C_{kk} = \alpha_1 M_{kk} + \alpha_2 K_{kk} \qquad (9)$$

where α_1 and α_2 are constant coefficients. Damping in one particular flexible mode k, can be measured in a decay test, as the modal damping ratio v_k. The generalized hull damping matrix can be written

$$C_{kk} = 2 M_{kk} \omega_k v_k \qquad (10)$$

Hydrodynamic damping is dependent of frequency of oscillation. For high frequencies, the generalized hydrodynamic damping matrix B_{jk}, will get close to zero. Investigations of damping (see e.g. Borg 1960), have shown that structural damping will be the dominant source of damping, for the flexible modes of a ship.

In investigations of response due to steady state wave loads, hydrodynamic damping, corresponding to the wave frequency, is the most important source of damping. If response due to impact loads is considered, hydrodynamic damping, corresponding to the natural period of the mode should be used. For flexible modes, one of the above models can be used to describe hull damping.

2.2 Response due to steady state wave load

From Eq. (2), the generalized complex response vector q can be determined for any wave frequency. Real time series of the response for a mode n can be written

$$q_n(t) = |\bar{q}_n| \cos(\omega t + \theta) \qquad (11)$$

where θ is the phase angle between response in mode n and incident wave.

For steady state wave loading, global structural loads on the vessel is determined by integration of forces acting on the vessel. When the response in the different modes have been determined, inertia force due to structural mass can be found, provided a relatively accurate mass distribution is available for the vessel. The response in flexible modes is very small compared to the response in rigid body modes, and can be neglected in the determination of inertia force. Hydrodynamic pressure on the hull has already been established, and by adding inertia force, global shear force and longitudinal bending moment can be determined for any longitudinal position on the vessel.

2.3 Slamming loads

Using the linearized dynamic free-surface condition, the surface elevation can be written

$$\zeta = -\frac{1}{g} \frac{\partial}{\partial t} \left(\phi_I + \phi_D + \phi_R \right) \qquad (12)$$

on z=0.

Wet deck slamming usually occur in the fore part of the vessel. Due to the slender shape of the hulls in the bow region of a high speed vessel, the contributions to wave elevation from diffraction and radiation potentials are small. This means that for zero forward speed, the position of the free surface can be approximated by the wave elevation of the incident wave. For a regular harmonic wave we may write

$$\varsigma(x,t) = \varsigma_a \sin(\omega t - kx) \qquad (13)$$

The vertical position of the wet deck is found by adding the contributions to the vessel response, from the different modes. If it is assumed that shape and motion of the wet deck is independent of the transverse direction, it's vertical position is given by

$$\psi(x,t) = h(x) + \sum_{n=1}^{m} v_n(x)\left[q_n^s(t) + q_n^i(t)\right] \qquad (14)$$

where $v_n(x)$ is vertical displacement of the wet deck for mode n, $h(x)$ is initial vertical distance to wet deck, and superscripts s and i indicate response due to steady state and impact loads respectively.

The response in our system due to impact loads is found from Eq. (2) by a central difference approach, (see e.g. Cheny W and Kincade D 1985). When the position of the wet deck as well as the position of the wave surface is known, we may establish a boundary value problem (BVP) describing the impact between wet deck and wave surface.

A two dimensional fluid flow has been assumed to model the potential set up by an impact against the wet deck. This will be the case if the fluid is trapped between to vertical walls of infinite extension. For a real ship, the validity of this assumption will depend on shape and draft of the hulls, as well as the distance from the impact area to bow or stern of the ship.

By assuming a two dimensional flow, the BVP shown in Figure 1 can be established.

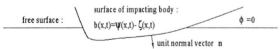

Figure 1. BVP for impact against wet deck.

The surface of the impacting body is given by the position of the wet deck minus the position of the free surface. We have used the free surface condition $\phi=0$, and the boundary condition on the body is given by

$$\frac{\partial \phi}{\partial \mathbf{n}} = -\frac{\partial b(x,t)}{\partial t} n_z \qquad (15)$$

The impact velocity distribution is determined numerically. A central difference approach is used for the contributions from wave elevation and response due to continuous wave loading. Impact induced response is estimated from current and previous time steps. The solution of the above BVP can be found, by a direct numerical method based upon Green's second identity, (see e.g. Kvålsvold 1994). Surface elevation due to the impact velocity potential is neglected. This corresponds to a von Karman (1929) approach, for the wet length.

When the velocity potential on the impacting body has been found, the impact pressure is given from Eq. (5). The last part of this equation is usually small compared to the first part, and has been neglected in our numerical method. The above theory will predict a positive pressure on the wet deck when the vessel is moving out of the water. This pressure is unphysical, and is not supported by experimental investigations. The reason for this error in the theory is our free surface condition, which is a good approximation only for the first stage of the impact. In our approach the pressure is set equal to zero when the relative velocity is negative. For a more thorough discussion of water exit problems, see Greenhow (1988).

Global structural loads due to slamming, can be determined from the excitation of flexible modes. Structural response due to unit excitation of each mode is found during calculation of eigenmodes. When the actual response in the different modes is determined, the global response is found from

$$R^i(t) = \sum_{n=1}^{m} \chi_n q_n(t) \qquad (16)$$

where

χ_n - structural response due to unit excitation of mode no n.

The total response is found by superimposing steady state response and response due to wave impact.

3. EXPERIMENTAL INVESTIGATION

The experimental investigation was carried out in the Ocean basin of MARINTEK in Trondheim. A surface effect ship (SES) operating in the off cushion mode, with zero forward speed was investigated. For an off cushion SES, the influence on ship motions from skirts and seals are small, and to simplify the model a conventional catamaran with small freeboard was used in the investigation.

3.1 Test vessel

The test model was a modified version of the flexible catamaran used earlier by Hermundstad et al (1995). Each hull was divided into three sections, and an aluminum frame was mounted within each section. The rigid sections of the test model were connected by elastic springs. A five degree of freedom force transducer was connected to each of the elastic springs, to measure structural response in the vessel. The structural arrangement of the test model is shown in Figure 2.

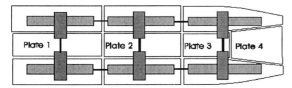

Figure 2. Test model seen from above

The wet deck of the test model consisted of four plates, each connected to the longitudinal aluminum frame by two transverse girders. The suspension system of the wet deck plates is shown in Figure 3.

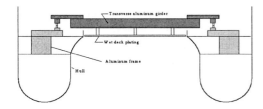

Figure 3. Suspension system for wet deck

The wet deck plates were made not to contribute to the global stiffness of the test model. Flexibility in the connection between transverse girders and aluminum frame is achieved by using connections with a small cross section. These pin connections, are flexible with respect to shear forces and moments, but relatively stiff with respect to axial deformation. Using pin connections, vertical relative

motion between wet deck and hull is prevented. The deck plates are supposed to be stiff, with natural periods well below the natural periods of global flexible modes in the model. Longitudinal stiffeners are used to stiffen the wet deck plates. Plate 1-3 are horizontal at a vertical distance of 0.343m from the base line of the hull. Plate 4 was mounted with a small heel angle with respect to the base line.

Body lines for the test vessel are shown in Figure 4, and main particulars are given in Table 1. A detailed mass distribution for the test model, data about springs and force transducers and geometry of wet deck plates, can be found in Økland (1998).

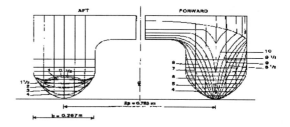

Figure 4. Body lines of test vessel

Table 1
Main particulars for test vessel.

Length over all – L_{OA}	4.10 m
Length between perpendiculars – L_{PP}	3.90 m
Beam at waterline amidships – B_{WL}	1.02 m
Beam of one hull - b_{WL}	0.267 m
Dist. From K to wet deck (parallel part)	0.343 m
Dist. From K to wet deck (front)	0.393 m
Displacement - ∇	303 kg
Draft at aft perpendicular – D_{AP}	0.255 m
Draft at fore perpendicular – D_{FP}	0.255 m
Vertical center of gravity – VCG	0.29 m
Longitudinal center of gravity – LCG	1.61 m
Roll radius of gyration	0.3815 m
Pitch radius of gyration	1.0721 m
Yaw radius of gyration	1.13 m

3.2 Test program and instrumentation

Three categories of tests were carried out :

- Decay tests
- Tests in regular waves
- Tests in irregular waves

Natural frequencies, and damping in the system are determined by decay tests. Decay tests were carried out on the dry test model, and for the test model floating in calm water without ballast.

Wave periods from 0.7 to 2.5 seconds and wave heights in the range 0.02-0.2 m, were used in the regular wave tests. The length of each time series was about 60 seconds. In the irregular tests, the time series had a duration of 4-5 minutes. Motions and structural response were sampled with a frequency of 100 Hz, while parameters related to the slamming force were sampled with 1000 Hz.

The test model was equipped to measure the following load effects :

• Vessel motions and accelerations
• Relative motion between vessel and free surface
• Hydrodynamic pressure on the wet hull
• Slamming force against wet deck plates
• Structural response in connections between rigid sections

Slamming loads were measured by force transducers in the pin connections of the wet deck plates. If the wet deck plates are rigid, the slamming force can be measured directly in the pin connections. Note however that the measured force also include inertia force in the deck plates, due to response of the vessel. This should be accounted for in comparison with experiments.

A detailed description of test program and instrumentation is given in Økland (1998).

4. RESULTS AND DISCUSSION

The theory described in Sec. 2, was implemented in a computer code. The numerical results given under are calculated by this code. The solution of diffraction and radiation potentials used in our code, is based on the theory of the VERES program (Hoff et al, 1995).

4.1 Natural modes and periods

The eigenmodes of the test model are found using the FEM code ABAQUS 5.6. The first symmetric eigenmodes are shown in Figure 5.

The natural frequencies found in the decay tests are shown in Table 2, together with natural frequencies calculated by the FEM-code. Note that the wet decay tests were carried out for a conventional catamaran, not for a SES.

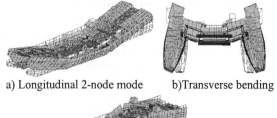

a) Longitudinal 2-node mode   b)Transverse bending

c) Longitudinal 3-node mode

Figure 5. Flexible modes of the test model

Table 2
Natural period of oscillation for test vessel [sec]

Test :		Rigid modes		Flexible modes		
Mode :		No 1	No 2	2-nod	Trv.	3-no
Dry dec.	Exp	-	-	0.170	0.110	0.070
No ballast	Num	-	-	0.163	0.122	0.064
Wet dec.	Exp	0.92	1.0	0.229	0.172	0.083
Cat cond	Num	0.84	0.95	0.215	0.184	0.079
Wet SES (Num)		0.93	1.11	0.232	0.214	0.085

Reasonable agreement is found between measured and calculated results.

4.2 Damping

The modal damping ratios, found in decay tests are shown in Table 3.

Table 3
Modal damping ratios in [%]

	Heave	Pitch	2node	3node
Dry tests	-	-	2.2	1.0
Wet tests	10.0	5.5	2.0	-

The decay tests have been simulated by our theory. Two different approaches have been tried for the damping matrix.

Approach 1 is the standard approach, described in Sec. 2.1. Hydrodynamic damping corresponding to natural frequency are used for rigid body modes. Measured modal damping ratios are used to estimate hull damping. For this particular case, the calculated hydrodynamic damping in flexible modes was small compared to the hull damping measured in the experiments.

Approach 2 is based on modal damping data for the wet system. Here, the generalized damping matrix is calculated from

$$B_{ii} = 2\lambda_i \omega_i (M_{ii} + A_{ii}) \qquad (17)$$

where λ_i and ω_i are damping coefficient and natural frequency from wet decay tests. In this approach, the off-diagonal damping terms for rigid body modes are neglected.

Results for simulated decay in heave are shown in Figure 6. Both approaches show good agreement with measurements. For pitch motion, the damping predicted by approach 1 is too large. This result is most likely caused by a poor description of interaction between hulls in our strip theory model. In the present investigation, interaction between hulls has been neglected when hydrodynamic damping is determined. Full interaction was also tested. In this case the damping was underestimated. For flexible modes both methods predicted damping that agreed with the measured damping. Approach 2 is used for the results presented in the rest of this paper.

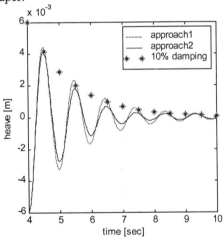

Figure 6. Decay test simulation for heave.

4.3 Transfer functions for motions

The main scope of this paper is to study slamming loads and corresponding structural loads in the vessel. The prediction of slamming loads are however based on responses from steady state wave loads. For the frequencies where slamming occur, pitch was found to be the most important rigid body motion. The comparison between measured, and calculated results in Figure 7, shows that first order pitch motion is well described with our approach.

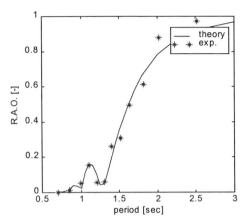

Figure 7. Transfer function for pitch motion

In Figure 8, relative motion in the fore part of the vessel is shown. The agreement between measured and calculated results is fair for periods where slamming is likely to occur. In calculations, maximum relative motion is found for a period of about 1.6 sec. In the measured results however, maximum relative motion seems to occur for wave periods equall to 1.5 sec. The deviation between measured and calculated largest value for relative motion is well below 10 % which is quite good.

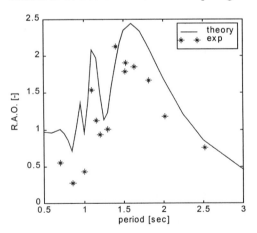

Figure 8 Relative motion for position no. 1

The predicted peak in relative motion for periods about 1.1 sec, corresponds to the natural period in pitch. This peak is observed in experiments as well.

For smaller periods only a few measurements are available. The agreement between calculated and measured results is poor. This range of frequencies is however not so imporant for the determination of slamming loads.

The predicted phase angle was slightly larger than the one found in experiments, for periods in the range 1.5-2.0 sec. This may be a possible reason for the error in estimation of maximum relative motion. Incorrect description of the wave surface due to diffraction, radiation is another explanation.

4.4 Slamming loads

During the experimental investigations all significant slam events were located in the bow region of the vessel. The most severe cases were found in a regular wave test with period 1.5 sec and a wave amplitude of 0.1 m. A time series of measured slamming force against the wet deck for this test is shown in Figure 9. In theory the maximum value of the measured slamming force should be constant during the time series. The varying magnitude of the slams, found in our experiments, is explained by small variations in wave profile and motion of the vessel.

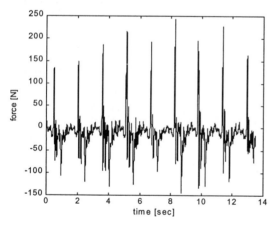

Figure 9. Total slamming force against plate no 3 and 4 as a function of time (T=1.5 sec, H=0.2 m)

In Figure 10, the slamming force have been calculated by three different approaches.

- Vessel motions from steady state loads alone.
- Slam induced rigid body motions added.
- Response in rigid and flexible modes added.

It is clear from this figure that slam induced response in rigid body modes as well as flexible modes are important for the slamming load. Updating for slam induced rigid body response

seems to reduce the slamming force. When flexible modes are included, the slamming force signal varies about the signal found when only rigid modes are included.

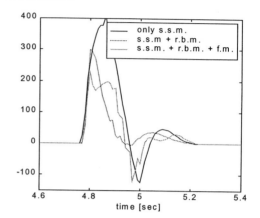

Figure 10. Influence on slamming force from slam induced rigid body motions and hydroelsticity. (s.s.m – steady state motions, r.b.m. – rigid body motions due to impact, f.m. – response in flexible modes due to impact)

In Figure 11, the numerical simulation is compared with the largest and the smallest of the slam events shown in Figure 9.

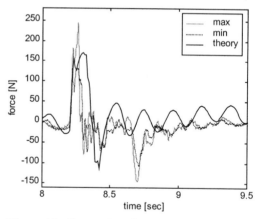

Figure 11. Comparison between theory and exp. For the slamming force (T=1.5sec and H=0.2 m).

It seems that our numerical model, at least to some extent, is capable of describing the physics in this complex problem. The measured slamming force drops to zero, or even negative magnitude, early in the slam. From the above discussion regarding the predicted slamming force, this effect can be explained by excitation of flexible modes. This first drop in the force is not satisfactory described by our model. The inertia force due to excitation of the first flexible mode seems to be over predicted by our theory. Further, the period of oscillation for the first flexible mode seems to be under predicted by our theory. Due to the free surface condition used, our theory is not able to predict slamming force during water exit. In general, our theory will predict a too large force-impulse in the system. This means that the approach is conservative.

As indicated in Figure 9, the slamming force is very sensitive to small changes in wave profile and vessel motions. Figure 12 shows the sensitivity of the slam force to wave elevation. In Figure 13, the sensitive of the slamming force to shift in the pitch phase angle is shown.

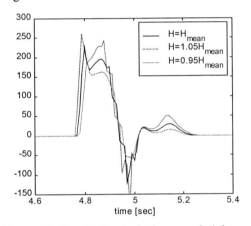

Figure 12. Sensitivity study for wave height.

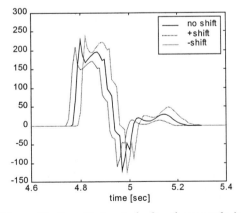

Figure 13. Sensitivity study for phase angle in pitch. Shift of $\pi/20$ rad is used.

4.5 Structural loads

Comparisons between simulated and measured structural loads are given in Figure 14. The results shown are for the springs connecting the front and the mid section. Corresponding results were found for the stern connection.

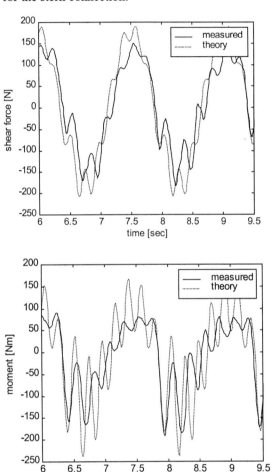

Figure 14 Global structural loads for longitudinal springs in the fore part of the test vessel.

Global structural loads, are dominated by steady state wave loads and the 2-node bending mode. The numerical simulation seems to overestimate the structural response in the 2-node bending mode. This over-prediction is due to the too large impulse predicted by our slamming model.

The period of oscillation for the first flexible mode is larger in experiments than in theory, in particular for the period where the wet deck is in contact with

78

the water. In water entry problems, the impact force is often divided in a part proportional to relative velocity, and one part proportional to acceleration. A possible cause for too short period of oscillation, is a poor description of this "added mass" part of the slam.

In this paper we have focused on response in regular waves. An investigation of extreme structural response in an irregular sea state can be found in Økland and Moan (1998).

5. CONCLUSIONS

A numerical method for investigation of structural loads in large twin hull vessels exposed to severe wet deck slamming has been presented. An experimental investigation has been carried out to validate the numerical results. From comparison between numerical simulations and experimental results, we found :

- Vessel motions are well described by the numerical model.
- The structural response is well described.
- The slamming load model is able to describe important features of wet deck slamming.

Further work should in particular be directed towards improvements of the model predicting the slamming force. Areas that is believed to improve the accuracy of the numerical model are :

- Implementation of 3-D slamming model
- Improve description of surface elevation

The numerical model should also be extended to handle vessels with forward speed, and beam sea.

REFERENCES

Cheny W and Kincade D (1985), "Numerical mathematics and computing." *Brooks/Cole Pblishing Company*, California.

Borg, S. F. (1960), "The Analysis of Ship Structures Subjected to Slamming Loads", *Journal of Ship Research*, Vol4, No. 3, pp. 11-27, 1960

Greenhow M (1988), "Water-entry and –exit of a horizontal circular cylinder." *Applied Ocean Research*, Vol 10 No 4.

Hermunstad OA, Aarsnes JV and Moan T (1995), "Hydroelastic analysis of a flexible catamaran and comparison with experiments." *FAST'95*, pp. 487-500

Hoff JR, Aanesland JV, Holm H and Fathi D (1995), *"VERES -USERS MANUAL."* MMARINTEK Report no 603822.00.01

Kaplan P (1991), "Structural Loads on Advanced MarineVehicles, Including Effects of Slamming." *FAST'91*, Trondheim, Norway, Vol 2, pp 781-795.

Kvålsvold J (1994), "Hydroelastic modelling of wetdeck slamming on multihull vessels" *Dr.Ing. Thesis at Dept of Marine Hydrodynamics - NTH*

Kvålsvold J and Faltinsen OM (1993), "Hydroelastic Modelling of Slamming Against the Wetdeck of a Catamaran." *FAST'93*, pp. 681-697.

Kvålsvold J, Faltinsen OM and Aarsnes JV (1995), "Effect of structural elasticity on slamming against the wetdecks of multihull vessels." *PRADS 95*.

Wu MingKang, Aarsnes JV, Hermundstad O and Moan T. (1997) "A Practical Prediction of Wave induced Structural Response in Ships with Large Amplitude Motion." *Twenty-First Symposium on Naval Hydrodynamics*, Trondheim.

Salvesen, N, Tuck EO, Faltinsen O, "Ship motions and sea loads." *Transactions of the Society of naval Architects and Marine Engineers*, vol 78 pp. 250-287

Troesch AW and Kang CG (1986), "Hydrodynamic Impact Loads on Three-dimensional bodies." *16th Symposium on Naval Hydrodynamics*, U.C. Berkeley, CA.

von Karman T (1929), "The Impact on Seaplane Floats During Landing." *NACA,* TN 321.

Zhao R, Faltinsen O, Krokstad J and Aanesland V (1988), "Wave-Current Interaction Effects on large Volum Structures." Proc. 5th Int. Conf. On Behaviour of offshore Structures.

Økland OD (1998), "Numerical and Experimental Analysis of Twin Hull Vessel exposed to Wet Deck Slamming." *Dr.Ing. Thesis at Department of Marine Structures - NTNU*, Trondheim, (to be published).

Økland OD and Moan T (1998), "Prediction of Slamming Loads and Extreme Structural Response for Large Twin Hull Vessels." *ISOPE'98*, Montreal, Canada

Structural Dynamic Loadings Due to Impact and Whipping

Kenneth Weems[a], Sheguang Zhang[a], Woei-Min Lin[a], James Bennett[b], and Yung-Sup Shin[c]

[a] Science Applications International Corporation, Ship Technology Division
   134 Holiday Court, Suite 318, Annapolis, Maryland 21401 USA
[b] Bath Iron Works Corporation
   46 Church Road, Brunswick, Maine 04011 USA
[c] American Bureau of Shipping, Research and Development Department
   Two World Trade Center, 106th Floor, New York, New York 10048 USA

This paper presents the results of recent efforts to develop a practical computational method that analyzes nonlinear hydrodynamic loads and dynamic responses of a ship in waves. Ships traveling in severe seas often experience impact loads caused by bottom slamming, when the bow emerges and reenters the water; or bow flare impact, when a bow with significant above-the-waterline flare submerges rapidly. The loads generated by impact are of large magnitude but short duration and can induce high-frequency hull vibration usually referred to as "whipping". Whipping responses will amplify the midship global bending moment and accelerate structural fatigue damages. In addition, high pressure concentration in the impact zone will cause local structural damage. These types of loadings are highly nonlinear and have become important considerations in the ship design today.

This paper presents the LAMP System, an integrated computational tool incorporating hydrodynamics, impact, and structural whipping response calculations, that narrows the gap between conceptual and practical design. Computational results are presented to show that impact and whipping related dynamic loadings are important parts of the total structural loadings, and must be included in the analysis for ship structural design, and that the present system can be applied to unconventional hull forms.

1. DESCRIPTION OF THE LAMP SYSTEM

To meet the current need in the design community, the LAMP (Large Amplitude Motion Program) System is being developed as a new multi-level time-domain simulation system for the prediction of the ship motions, wave loads, and structural response. As shown in Figure 1, the LAMP System consists of three closely integrated modules. The first module is for the calculation of ship motions and wave-frequency loads. The second module is for the slamming impact computation. The third module is for computing whipping responses using a non-uniform-section dynamic beam method. In addition, the LAMP System includes an interface to provide loading information for finite element analysis. The principle components of the LAMP System are discussed below.

1.1. Ship Motions and Wave Frequency Loads

LAMP solves the three-dimensional time-domain nonlinear motion and load problems using a

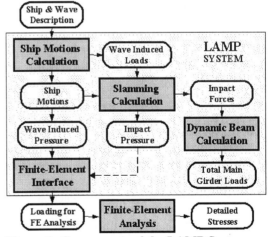

Figure 1: Components of the LAMP System

potential-flow boundary-element method, providing an accurate ship motion prediction and a time history of pressure load over the hull surface. In LAMP, a so-called "body-nonlinear" approach is used (see Lin and Yue, 1990 and 1993; and Lin *et al.*, 1994). In contrast to the linear approach in which the body boundary condition is satisfied on the portion of the hull under the mean water surface, the body-nonlinear approach satisfies the body boundary condition exactly on the portion of the instantaneous body surface below the incident wave. It is assumed that both the radiation and diffraction waves are small compared to the incident wave so that the free surface boundary conditions can be linearized with respect to the incident wave surface. Note that with this formulation, both the body motions and the incident waves can be large relative to the draft of the ship.

Several variations of Lin and Yue's original "body-nonlinear" approach have been developed and are currently available in the LAMP System. LAMP-4 satisfies the free surface boundary condition on the incident wave surface, provides 3-D large-amplitude hydrodynamics, and calculates nonlinear hydrostatic restoring and Froude-Krylov wave forces. LAMP-2 satisfies the free surface boundary condition on the mean wave surface, provides 3-D linear hydrodynamics, and calculates nonlinear hydrostatic restoring and Froude-Krylov wave forces. LAMP-1 differs from LAMP-2 only in that it calculates linear hydrostatic restoring and Froude-Krylov wave forces.

The hydrodynamics problem is solved in the time domain by a 3-D boundary element method using a transient free-surface Green function singularity distribution. In Lin and Yue's original formulation, the transient Green functions are distributed over the hull surface. While effective for most conventional ships, this implementation proved to have significant numerical stability problems for severely non-wall-sided ships, such as the Arsenal Ship discussed later in this paper.

For this reason, a hybrid numerical approach (referred to as the "mixed source formulation" in LAMP) has been implemented which is a combination of the transient Green function (e.g. Lin, et al., 1994) and the Rankine source (e.g. Nakos, Kring and Sclavounos, 1993). In the mixed source formulation, the outer domain is solved with transient Green functions distributed over an arbitrarily shaped matching surface; the inner domain is solved using Rankine sources distributed

over the hull S_b, matching surface S_m, and free surface S_f between the hull and matching surfaces, as shown in Figure 2. The advantage of this formulation is that the Rankine source behaves much better than the transient Green function near the body and free surface juncture and the matching surface can be selected to guarantee good numerical behavior of the transient Green functions. This numerical scheme has resulted in robust motion and load prediction for modern hull forms with non-wall-sided geometry.

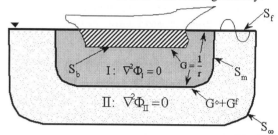

Figure 2: Mixed Source Formulation

In order to calculate the time-domain six-degree-of-freedom coupled motions for any ship heading and speed, LAMP also includes nonlinear models for non-pressure forces including viscous roll damping, propeller thrust, bilge keels, rudder and anti-rolling fins, etc. For oblique seas cases, a PID (Proportional, Integral, and Derivative) course keeping rudder control algorithm and rudder servo model are implemented.

In addition to motion simulations, LAMP calculates the time-domain wave-induced global loads, including the vertical and lateral bending and torsional moments and shear forces, at any cross-section along the length of the ship. Furthermore, at each time step, LAMP calculates the relative motion of the ship and the wave as well as the hydrodynamic pressure distribution over the instantaneous wetted hull surface below the incident wave surface. The relative motion information is used as input for the impact load calculations. The mapped pressure distribution is used to derive input for finite element structural analysis.

1.2. Slamming Calculations for Impact Forces and Pressure

Wave impact loads on the ship can cause high-frequency structural responses. As used here, the term "impact" includes bottom slam, bow flare impact, and stern overhang impact. As impact occurs, the ship structure responds at its structural natural frequency. The total loads at any section of the ship are the sum of the wave frequency loads and the high-frequency whipping loads. Depending on the severity of the impact, the whipping loads can be of the same order of magnitude as the wave-frequency loads. Therefore, in any extreme wave assessment, the effect of impacts must be included in the hydrodynamic load computations.

Most traditional methods for analyzing impact loads rely on semi-empirical force estimates rather than on accurate physics-based prediction of the actual impact pressure distribution. Furthermore, the traditional methods address only head-sea cases with symmetric impact. However, experience indicates that structural failures in oblique seas usually result from asymmetric impact loads. It is important, therefore, that any attempt to resolve the total impact problem include not only the accurate time-domain simulation of the highly nonlinear relative motions in oblique seas, but also the prediction of both the symmetric and asymmetric impact pressures.

In the LAMP System, a post-processor is used for the impact load predictions. It is assumed that the impacts do not affect the global ship motions. The previously computed global ship motions are used to compute relative motion of the ship bow and identify events where impact forces may be significant. The relative ship motion is then used to compute impact loads on 2-D cross sections of the ship for times when bottom slam, bow flare impact, or stern overhang impact occur. The forces from these impact events are then assimilated into an impact force history, which can be used to evaluate whipping loads.

Two levels of impact load computations are currently available: a simple 2-D empirical model for global impact forces and a generalized 2-D Wagner approach for impact forces and pressures. The first level is based on a momentum approach, assuming that the impact force is proportional to the rate of change of added mass of the given cross section. The formula used has been validated against the results obtained by the fully nonlinear boundary element approach of Zhao and Faltinsen (1993). The second approach is a simplified boundary element method based on the fully nonlinear approach of Zhao and Faltinsen. The final results include the vertical and horizontal impact forces as well as the local impact pressure on the hull surface. A second momentum approach, this one using a limited number of 2-D boundary element calculations to compute the sectional added mass properties rather than an empirical model, is currently being implemented (Jore, 1997).

1.3. Dynamic Beam Calculations for Whipping-Associated Loads

Once the sectional impact forces have been computed, the main girder responses are computed using a non-uniform-section dynamic beam method in order to get high-frequency global loads associated with whipping. LAMP uses a one-dimensional finite-element model, in which the ship can be modeled either as a uniform or a variable-mass beam. The total structural bending moment is obtained by combining the wave-frequency and the high-frequency bending moments with proper phasing.

1.4. LAMP's Interface to Structural Finite-Element Codes

The LAMP System calculates the pressure distribution over the instantaneous hull surface below the incident wave surface. The hull pressure information, combined with the acceleration data, can be used for finite element structural analysis. A generic interface between the LAMP System and structural finite element codes has been developed. The interface program reads nodal point coordinates and connectivity information (only surface nodes are needed) used in the finite element code and computes the forces acting on the nodal points. At specified time steps, the interface program writes the nodal point forces and ship acceleration information as outputs for the finite element structural analysis program. Other output from the LAMP/FE interface includes nodal pressure history, sectional main girder loads for FE analysis of partial ship configurations, and external forces by control surfaces, etc. that were modeled in the LAMP simulation but are not included in the pressure distribution. The latter forces must be accounted for so that forces and accelerations are properly balanced in any subsequent structural analysis.

A sample finite element calculation for a container ship in head seas of this interface is shown in Figure 3. The LAMP pressure distribution at one time step of the calculation is shown at the top; the predicted deflection and surface stress from the finite element analysis is shown at the bottom.

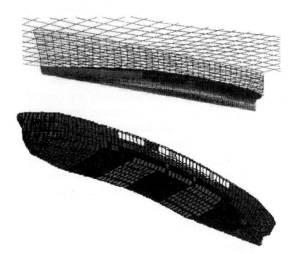

Figure 3: Sample LAMP Pressure and FE Results for a Container Ship in Head Seas

In order to compute detailed local structural responses due to slamming, the local surface pressure distributions due to slamming must also be computed. These impact pressures can be computed as part of 2-D boundary element calculations used to compute the sectional impact loads. LAMP's slamming interface includes procedures for computing the time history of impact pressures at hull surface node points and combining it with the LAMP pressure to get a total pressure history.

A complete slamming pressure distribution could be obtained by computing pressure for a sufficiently large number of sections, which could be added to the LAMP pressure to get a complete combined pressure distribution. Finite element loads, including slamming loads, could be computed using the LAMP/Finite-Element interface described above. However, this approach has not yet been fully implemented in the LAMP System.

2. NUMERICAL EXAMPLES

An extensive validation study of the LAMP ship motion and wave load system is presently ongoing. Selected validation results (including slamming and whipping) for a U.S. Navy AEGIS cruiser are presented here. Sample slamming pressure results are also shown, as are motion and load predictions for an unconventional "tumblehome" configuration analyzed in a recent design effort.

2.1. AEGIS Cruiser in Storm Waves

The U.S. Navy's AEGIS cruiser (CG-47) is a typical modern combatant hull form for which a large amount of model and full-scale motion and load data is available (Hay, *et al.*, 1994), so it has been a principal validation case for the LAMP System. A view of the CG-47 cruiser, with its fine U-shaped bow, sonar dome, and large bow flare, is shown in Figure 4.

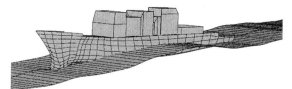

Figure 4: CG-47 AEGIS Cruiser in Waves

Figure 5 shows a comparison of LAMP-1 results and experimental measurements for a two-minute real time record of this ship moving at 10 knots in storm sea conditions. In the top plot, which shows the incident wave elevation at the ship's center of gravity, it can be seen that the maximum wave height in this record reaches about 55 ft (16.8m). The linear hydrodynamic results are shown here in order to illustrate important nonlinear effects.

The remaining three plots show heave and pitch motions and the vertical bending moment at Frame 174 of the model, which is located just forward of the deckhouse. Except for an initial transient period in the calculation, the motion comparisons are excellent. However, comparison for the sectional vertical bending moment is not as good. Although the predicted bending moment compares well with experimental measurement when the wave and ship motions are not large, LAMP-1 under-predicted all the negative (sagging) peaks of the bending moment.

In addition, LAMP-1 did not predict the high-frequency component of the bending moment that appeared in the experimental measurement. These typical results show how linear hydrodynamics are adequate for predicting head seas motion but under-predict peak wave loads, especially near maximum sagging situations.

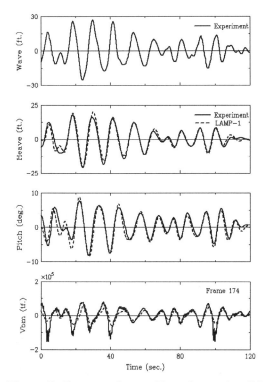

Figure 5: Computation *vs.* Experiment for CG-47 AEGIS Cruiser in Head Storm Seas

To illustrate the importance of nonlinear geometry and hydrodynamics, LAMP-2 results are shown in Figure 6 for the sagging peak at T=100 seconds. These results include the bending moment components at both the wave frequency and high frequency whipping responses due to impact. Near the peak of the sagging moment, not only the magnitude of the moment is comparable, but the frequency and amplitude of the calculated whipping response agree well with the experimental results. Similar results can be obtained using a LAMP-4 computation, indicating that hydrostatic and incident wave effects dominate these vertical loads.

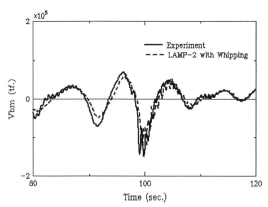

Figure 6: Time History of Vertical Bending Moment Comparison at Frame 174

The same cannot be said for horizontal loads in oblique seas. Similar comparisons have been made for LAMP-2 results and experimental measurements for the CG-47 in 30 degree irregular bow storm seas. As for the head sea case, the predicted heave and pitch compare well with experiments. Roll motions are qualitatively similar but a direct comparison in time is less satisfactory. Similarly, the vertical bending moment compares quite well, but the horizontal (lateral) bending moment does not.

In addition, there are very large lateral high-frequency whipping responses in the experimental data for this naval ship. Large enough, in fact, to suggest that such lateral whipping responses must be considered in the ship's design, especially for naval ships with fine bows and sonar domes. At the time of these calculations, horizontal impact forces could only be roughly approximated, and the lateral whipping responses were significantly under-predicted. An updated impact force calculation currently being implemented, which will not be restricted to vertical velocity only, will allow a more accurate prediction of the lateral slamming forces.

It is important to note that roll, sway, and yaw motions in oblique seas have been shown to be very sensitive to the way in which the ship is steered. Manual rudder control was used for the CG-47 experiments, but unfortunately the rudder action was not recorded. Since the LAMP rudder control system is likely to have resulted in significantly different rudder action, it is very difficult to directly compare the lateral responses. In any event, the development and validation of the LAMP System for oblique sea calculation is currently a top priority of the LAMP development team.

2.2. Slamming Pressure

As described above, LAMP's slamming interface can be used to compute surface pressures due to slamming as well as section impact forces and main girder responses. Figure 8 shows the time history of surface pressure at two node points on the CG-47 during the same storm seas case shown in Figure 5. The node points are located at the bottom of the hull just aft of the sonar dome and further up the side of the hull at the same axial position. A sectional cut through the hull showing the position of the points is inset in the pressure plot.

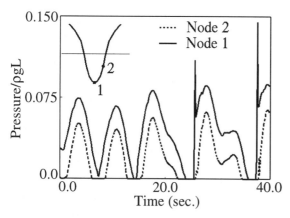

Figure 8: Time History of Surface Pressure at Two Node Points on the CG-47 Hull

The plotted surface non-dimensional pressure includes both the "wave-induced" pressure computed by LAMP, which includes the effects of hydrostatics, the incident wave, forward speed, radiation, and diffraction, and the impact pressure computed by the sectional slamming calculation. Impact pressure is generally small compared to the wave-induced pressure, but it becomes large during the bottom slams at times of 25.2 and 37.2 seconds, as illustrated by the spikes in Figure 8.

2.3. Wave Piercer Hull Form Design

The LAMP System, including the slamming loads calculation, was employed recently for a "wave piercer" high-speed hull form concept under development at Bath Iron Works (General Dynamics/Marine). Because of the unconventional "tumblehome" hull configuration, there was considerable concern about operability and loads in extreme seas and about the ability of current computation techniques to evaluate the ship's seakeeping. A view of the hull is shown in Figure 9.

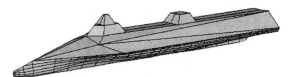

Figure 9: Conceptual "Wave Piercer" Geometry

To address these concerns, a seakeeping study was performed, including both model tests at the U.S. Navy's David Taylor Model Basin and computational analysis with the LAMP System. The conclusion of this study was that the tumblehome configuration had superior seakeeping characteristics *vs.* conventional hulls in both motions and loads and that existing numerical techniques such as LAMP were capable of evaluating the vertical motions of this configuration. The study showed that the wave piercer hull could advance at 32 knots in sea state 7 without major complications.

Figure 10 shows some computational results of this study. The top graph compares the wave-induced vertical bending moments at midships for the wave piercer and for a conventional naval combatant (CG-47) in storm seas. The bottom two plots show the computed slamming-induced bending moment over the same time period. These LAMP results indicate that the maximum non-dimensional wave-induced vertical bending moment for the wave piercer may be only half as large as for the CG-47. In addition, the large slamming-induced vertical bending moment for the CG-47, which is mainly due to bow flare impact, is insignificant for the wave piercer.

So far, only vertical motions and loads have been addressed, and only the vertical motions have been measured experimentally and are available for validation. However, it is hoped that continued development of this hull form will provide an opportunity to further validate both the vertical loads and the horizontal motions and loads in oblique seas for this unconventional hull form.

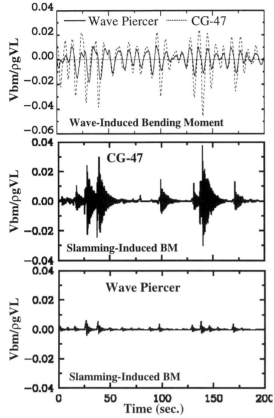

Figure 10: Bending Moments for CG-47 *vs.* Wave Piercer in Head Storm Seas

3. CONCLUSION

The computational method presented here, which integrates 3-D time domain ship-motions calculations with 2-D impact force calculations, provides a practical tool for evaluating wave- and impact-induced sea loads. Results demonstrate the importance of the nonlinear wave effects and impact loads on extreme vertical bending moments and the ability of the LAMP System to predict them. The ongoing validation for oblique sea motions and loads will extend LAMP's applicability for modern ship design, by verifying the prediction of vertical and horizontal loads including whipping effects.

4. ACKNOWLEDGEMENTS

The development of the LAMP System has been supported by DARPA, the U.S. Coast Guard, and the U.S. Navy through an Office of Naval Research project under program manager Dr. Edwin Rood and under the Dynamic Assessment Support System project under Mr. Allen Engle. Other sponsors include the American Bureau of Shipping (ABS) and SAIC. The impact force module has been developed by MARINTEK, Norway. The authors are also grateful to Thomas Treakle and Marian Weems for their assistance in preparing this paper.

5. REFERENCES

Hay, W., J. Bourne, A. Engle, and R. Rubel, "Characteristics of Hydrodynamics Loads Data for a Naval Combatant," *Hydroelasticity in Marine Technology,* Rotterdam, the Netherlands, 1994.

Jore, Arne K., "Numerical Predictions of Slamming Loads on Ships," Siv.ing. Thesis, Norwegian University of Science and Technology, 1997.

Lin, W.M., and D.K.P. Yue, "Numerical Solutions for Large-Amplitude Ship Motions in the Time-Domain," *Proceedings of the Eighteenth Symposium of Naval Hydrodynamics,* The University of Michigan, U.S.A., 1990.

Lin, W.M., and D.K.P. Yue, "Time-Domain Analysis for Floating Bodies in Mild-Slope Waves of Large Amplitude," *Proceedings of the Eighth International Workshop on Water Waves and Floating Bodies,* Newfoundland, Canada, 1993.

Lin, W.M., M.J. Meinhold, N. Salvesen, and D.K.P. Yue, "Large-Amplitude Ship Motions and Wave Loads for Ship Design," *Proceedings of the Twentieth Symposium of Naval Hydrodynamics,* The University of California, U.S.A., 1994.

Nakos, D.E., D. Kring, and P.D. Sclavounos, "Rankine Panel Methods for Transient Free-Surface Flows," *Proceedings 16th Symposium of Naval Hydrodynamics,* Iowa, U.S.A. 1993.

Zhao, R., and O. Faltinsen, "Water Entry of Two-Dimensional Bodies," *Journal of Fluid Mechanics,* vol. 246, pp. 593-612, 1993.

1998 Elsevier Science B.V.
Practical Design of Ships and Mobile Units
M.W.C. Oosterveld and S.G. Tan, editors.

Improved Ship Detail Finite Element Stress Analysis

Neil G. Pegg♣, David Heath♣, Mervyn E. Norwood♦

♣Defence Research Establishment Atlantic, P.O. Box 1012, Dartmouth, Nova Scotia, Canada, B2Y 3Z7
♦Martec Ltd., Suite 400, 1888 Brunswick St., Halifax, Nova Scotia, Canada, B3J 3J8

Abstract

This paper describes an improved approach to finite element modelling of ship structural details for design and in-service assessment. Specialized meshing algorithms have been developed which create finite element meshes of three-dimensional detail structure from predefined templates of the detail boundaries and associated meshing parameters. The ship structural details are divided into classes such as stiffener intersections, brackets, bulkhead-stiffener intersections or cutouts, which would have a common set of meshing parameters. Classes can be identified from lists of standard ship structural details produced by classification societies or for a specific ship type. Parametric stress analyses can be undertaken for these classes in advance to determine suitable meshing parameters which can be stored in a database and used to quickly generate the required detail meshes when required. The overall time to produce a ship detail mesh is reduced from several days using general purpose finite element model generators to a couple of hours.

This improved detail meshing approach is being incorporated into an overall finite element based ship structural analysis program which uses global ship displacement results in a semi-automated top-down analysis of structural details. Specialized spectral sea load codes are used with the global ship model to produce stress spectra or extreme loads in the details for fatigue or ultimate strength analysis. Overall, this approach greatly reduces the time necessary to undertake complex analysis of ship structural details.

Introduction

In their continuous efforts to improve safety and cost efficiency, classification societies and Navies have been developing 'first-principles' rational assessment methods for ship structures [1,2,3,4]. These computer methods are based on numerical modelling of the sea loads (Figure 1), which the vessels are likely to see in different operations, and of the resulting structural behaviour with the finite element method (FEM). Finite element analysis (FEA) is required to obtain reasonable estimates of the hull girder response and of the complex stress patterns occurring in structural details such as cutouts and connections. The rapid advances in computing power have made it possible for an engineer to undertake large scale FEA (10^5 degrees of freedom) in a reasonable time period on inexpensive computer

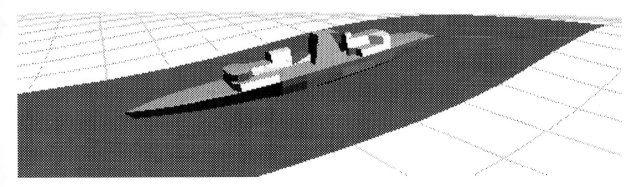

Figure 1: Computer modelling of sea loads and structural response for complete ship

88

hardware. The top-down approach, which uses coarser mesh models of the complete ship hull structure to provide boundary loads to fine mesh models of structural details, has become an accepted approach in many of the rational ship structural analysis tools.

While the computation time for the large finite element (FE) models has become acceptable, there remain some major drawbacks to applying the FEM to routine design and analysis of ship structure. The first is the amount of time required to create the FE models. Even the more advanced general purpose FE mesh generators require significant time and skill to produce models of complex three dimensional shell structure. The second main problem, which is not addressed to any significant extent in this paper, is how to use the results of FE analysis in assessment of the strength and endurance of the structure.

This paper discusses an approach to FE meshing of ship structural details which will significantly decrease the time required to undertake FEA to the point where it can be undertaken in a routine and timely manner. The topic of ship structural detail meshing classes will be discussed followed by a description of a specialized FE mesher that can be used to create FE meshes of the detail classes. Approaches to modelling of damage such as cracks and corrosion and methods of assessing the effects of this damage are then proposed. Several examples of this approach are then given.

Top-Down Ship Structural Detail Analysis

The analysis method used for this work is top-down FEA of the global ship model and the local detail area of interest. The global ship model is a coarse mesh Maestro [5] model of either the entire ship or a longitudinal section. Loads are applied as pressure and inertia forces from three-dimensional sea load codes, and/or as sectional bending moments and shear forces. The detail FE analysis is done with the general purpose code VAST [6]. The detail meshing and interface between the various programs, and of the top-down data, is done by the program MGDSA (Maestro Graphics Detail Stress Analysis). The top-down interface is undertaken by automatically

identifying nodes that are common to the Maestro global model and the detail VAST model. These become master nodes on the boundaries between the two models. Other nodes, which may be created on the boundary of the detail model, are automatically slaved to the master nodes to give the correct translation of displacement boundary conditions between the two models. Of particular value is the facility to automatically create translational and rotational boundary conditions when going from a beam element representation of a stiffener in the global model to a shell element representation in the detail model. In addition to creating the detail meshes and the top-down analysis data, the MGDSA code also has facilities to verify the models and plot displacement and stress results. Current developments of this code include connections to other FE packages such as ANSYS and NASTRAN, an integration of an ultimate strength analysis code and a variety of sea load codes, and a new version of the code using an improved object-oriented data management system.

Ship Structural Detail Classes

The initial developments for detail meshing in MGDSA were based on approaches similar to those used in general purpose FE modellers. While the overall integration of the top-down analysis and development of detail models greatly enhanced the ability to undertake detail FEA of ship structure, the time to produce the detail FE meshes is still often prohibitive. In order to overcome this drawback, an approach to meshing ship structural details is being implemented which will reduce the time to hours instead of days and make routine FEA of details a possibility.

Ships are usually built to a defined set of details which are documented by classification societies, owners (such as Navies), or the builders. Details include stringer and frame intersections, bulkhead and stiffener connections (watertight and non-watertight), penetrations, cutouts, etc. The defined detail may vary in component dimensions (plate thickness, web height, etc.) and possibly have minor variations such as the inclusion of tripping brackets. Detail can thus be defined by a set of 2D patches which are defined by boundary lines (this approach uses interior as well as exterior boundary lines to include cutouts and crack lines) and connected together at common boundaries to form the 3D detail.

The 2D patches are meshed based on a description of the node distribution on the boundary lines consistent with a special purpose FE mesher (discussed below). Parametric stress analysis of the detail classes is undertaken to produce the best predefined set of nodal distributions for meshing the details. The nodal distributions may be defined by functions of the detail geometry as opposed to fixed

values (eg. a function of plate thickness or cutout dimension). Development of the detail classes and the parametric analysis will take considerable time, and is to be done before use in a specific ship analysis. The detail class descriptions and meshing parameters are stored in a database that can be called during a ship analysis to quickly generate the detail FE mesh of interest. As the object-oriented database management system is developed, the specific detail classes will be identified at their locations in the global ship model and include the actual scantling information for that detail. The overall approach to top-down detail meshing is illustrated in the flowchart of Figure 2.

Currently, 30 structural detail classes are being developed for a Canadian frigate. Figure 3 shows a standard longitudinal stringer and frame or beam intersection divided and meshed in modelling planes. In this case there are 8 modelling planes; 2 for each flange (4), 1 for each web (2) and 2 for the plating of the shell. This particular model stops at the frame but could extend through the other side creating an additional number of planes. Each different geometric plane in Figure 3 is bounded by a set of boundary lines. The overall dimensions of the planes and the boundary lines of this class can vary to allow a variety of stiffener dimensions to be accommodated.

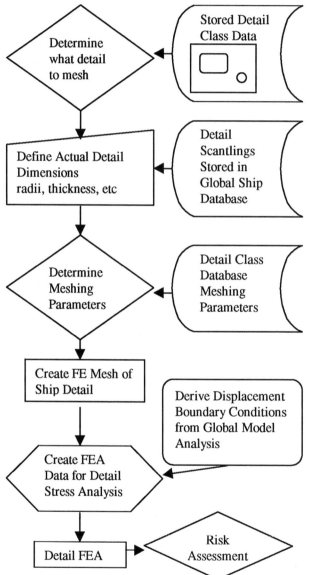

Figure 2: Schematic of top-down analysis with detail class FE mesh data

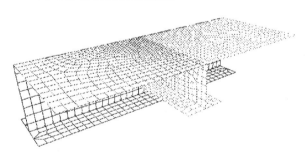

Figure 3: A detail 'class' of a longitudinal/frame intersection showing the modelling planes

Specialized Detail Finite Element Mesher

A meshing algorithm has been developed which uses a paving method to produce a mesh of quad elements on any two-dimensional surface made up of any number of boundary lines. This includes interior (multiple cutouts, crack lines) as well as exterior boundaries. The boundary lines are described by both geometric primitive shapes and node distribution algorithms. Each two-dimensional surface describes one part of a detail class. Figure 4 shows some examples of meshes created by the detail finite element mesher. Work is still underway in testing and improving the algorithm. Special features to control symmetry and limit element size as a function of their location in the mesh domain are being developed and implemented. Assembling the two dimensional planes together produces the three-dimensional models. Compatability between the planes is ensured by using the same boundary line definition at the common boundaries in the meshing process. Curved surfaces are handled by first mapping the curved surface on to a flat surface for meshing and then mapping the element and node distribution back to the curved surface.

Modelling of Structural Damage

The Canadian Department of National Defence's (DND) main interest in developing a first principles computer based approach to analysis of ship structure is to assess the effects of in-service structural damage on ship operational capability. The tool will also be used to assess options for design changes during major refits. This is being done within an overall project entitled ISSMM (Improved Ship Structural Maintenance Management)[7] which is producing computer tools to aid in developing a more efficient repair and maintenance process. The primary questions to be answered are:

⇒ Given the detected damage, does it have to be repaired now or can it wait?
⇒ What operational limits should be placed on the damaged ship to ensure safety?
⇒ What is the most cost effective repair strategy?

In-service damage can consist of corrosion, cracks or deformations from collisions. The tool will also be extended to consider weapons damage. In order to effectively use the ISSMM tool, the modelling and assessment of the detected damage has to be done in a very short timeframe. The database of the global ship and the structural detail classes will be in place to quickly create models of the area of interest and

produce loads for a variety of operating conditions. Additions to the finite element mesher are being

Figure 4: Examples of 2-D meshes created by the specialized detail mesher (a hatch cutout containing a crack propagating upwards and a plate with an arbitrarily placed and sized hole)

developed which will automatically create generic or user defined patterns of corrosion, fatigue cracks or deformation patterns.

The effects of corrosion and deformation will be assessed through changes in stress states over the undamaged structure for the same loads. Assessment of the local ultimate strength of the damaged component will also be undertaken through a semi-automated nonlinear FEA of stiffened panel components (integrated structural unit method). The damaged stiffened panels will be used to produce new load-shortening curves for full global hull cross section ultimate strength analysis.

For crack growth, the complete detail mesher being developed in this program will have the ability to mesh a crack and crack tip located anywhere in the detail of interest (as illustrated in Figure 4). It will be possible for the crack to be progressed according to a particular crack growth law and automatically remeshed to provide new crack tip stress intensity results. Once the detail finite element mesh has been developed and a connection is made to the global model, a series of representative load cases would be run to develop a stress history for crack initiation or crack growth analysis. The Canadian DND is funding development of a series of codes entitled LIFE3D [8] which encompass various crack initiation and crack growth models to be used in conjunction with FEA stress results. The global FE model has also been used to undertake spectral analysis directly by coupling it to a frequency domain sea loads code [9,10]. Stress or stress intensity spectra at the detail of interest resulting from this process will be used in the crack growth or crack initiation models. A question which is being studied at this time, is where to take stress values for crack initiation analysis. The detail class mesh has to be designed to give suitable stress values for the initiation models and material parameters that are being used. A hot-spot stress approach is being investigated which uses models that produce stress in the region of a typical weld-toe location.

Evaluation of the damage results will be undertaken by both deterministic analysis through assessment of the change in safety factor, and probabilistic analysis using limit states for stress, fatigue and ultimate strength.

Examples of Improved Detail Meshing

Large Penetrations in Ship Bottom Structure
During a recent refit of DND's research vessel CFAV QUEST, several large penetrations (the largest being approximately one meter in diameter)

were placed in the midship bottom structure to house sonar instrumentation. MGDSA was used for the analysis of this structural modification [11]. Figure 5 shows the detailed model overlayed on part of the global ship model and the stress results around one of the holes.

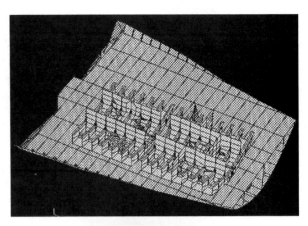

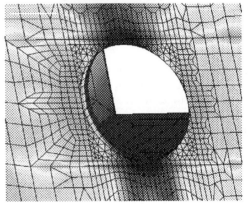

Figure 5: MGDSA detail modelling of large penetrations in ship bottom structure

CPF Main Deck Frame/Stringer Intersection
One of the main types of structure which requires detail meshing is the intersection of transverse frames and longitudinal stiffeners. Figure 6 shows a portion of a global Maestro model of midship bottom and sideshell structure including the frames and some longitudinals. MGDSA has a special feature to automatically convert the MAESTRO strake elements into refined meshes of stiffened panels. This is illustrated in the coarser mesh region of the second model in Figure 6. The specialized

detail mesher was then used to create the mesh of the intersection shown in the central part of the second model in Figure 6. Also included in Figure 6 are the stress results for the detail.

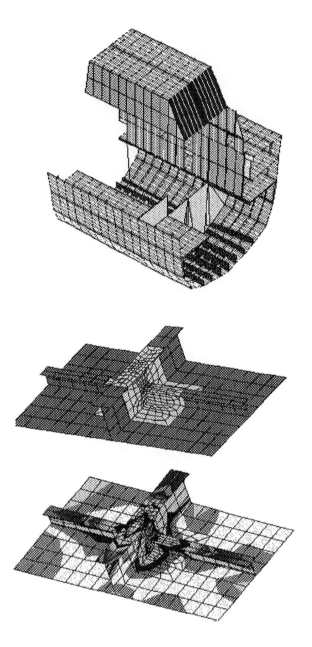

Figure 6: MAESTRO model and MGDSA detail model of a stiffener/frame intersection

Crack Propagation at Main Deck Penetration

A feature of the specialized mesher is its ability to implement a crack into detail structure. The crack line is treated as an internal boundary line in the meshing plane. VAST has special crack-tip elements which are used in the vicinity of the crack tip to give the stress intensity values which are used in fatigue crack growth laws. Figure 7 shows a crack detail created in a longitudinal bulkhead from the Maestro model (the top model of Figure 6), where the crack (indicated by the arrow) is propagating from a deck cutout. The resulting mesh and stress contour plot is shown in Figure 7.

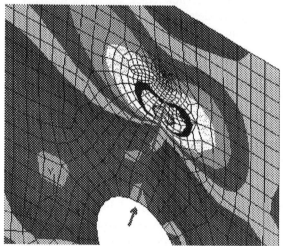

Figure 7: Crack propagating from a bulkhead penetration

Conclusions

At the time of writing, the overall method of detail analysis described in this paper is under development. Most of the components have been developed in an initial state and are being integrated and tested against realistic ship structural analysis scenarios. Many of the components of the proposed method are currently being applied by users of Maestro through the MGDSA program.

The rapid advances in computing power have made it possible to apply extensive numerical computations in routine engineering analysis. The challenge now is to make these methods easily accessible to the users. This requires the development of integrated analysis programs and data management tools and presentation of results to the user in an easily understood and reliable manner. The subjects discussed in this paper are directed towards this end and have demonstrated what can be expected for routine analysis of ship structures in the future.

References

1. 'Veristar, Integrating Design And Analysis And Ship Management', Bureau Veritas program documentation, 1997.
2. 'Safehull, Technical Information', American Bureau of Shipping, 1997.
3. 'Nauticus, The Classification Information Highway', Det Norske Veritas Classification AS, 1995.
4. 'Shipright, Design, Construction And Lifetime Ship Care Procedures', Lloyd's Register Report 95/50/3.11, 1998.
5. MAESTRO - Method for Analysis Evaluation and Structural Optimization, User's Manual - Version 7.0', distributed by Proteus Engineering Ltd, Annapolis, MD, July 1992.
6. VAST - Vibration and Strength Finite Element Code, User's Manual - Version 7.0, Martec Ltd., 1996.
7. Pegg, N.G., Gibson S., "Application of Advanced Analysis Methods to the Life Cycle Management of Ship Structures", Advances in Marine Structures III, DERA, Scotland, 1997.
8. Wallace, J.C., and Chernuka, M.W., 'LIFE3D Three-dimensional Crack-growth Simulation Program Suite User's Manual', Martec Limited, July 1990.
9. Stredulinsky D.C., 'Proposed Method for Calculation of Spectral Response to Random Wave Loading Using Top-Down Finite Element Modelling', DREA Technical Memorandum 97/209, November, 1997.
10. 'Finite Element Based Spectral Analysis Methods for Fatigue Crack Growth and Initiation of a Ship Operating in a Seaway', Martec Limited, September, 1997.
11. Pegg N.G., 'Finite Element Stress Analysis of Midlife Refit Penetrations in QUEST Bottom Structure' DREA Technical Memorandum 97/249, September 1997.

Practical Design of Ships and Mobile Units
M.W.C. Oosterveld and S.G. Tan, editors.

Prediction of the sectional forces and pressures on a free-fall lifeboat during water entry

M. Reaz H. Khondoker

Department of Naval Architecture and Marine Engineering
Bangladesh University of Engineering and Technology
Dhaka-1000, Bangladesh.

The advantages of free-fall lifeboats as lifesaving appliances made them popular very rapidly and such lifeboats are used presently in many maritime vessels and offshore installations. The launch of a free-fall lifeboat from the skid is composed of four phases, viz., (i) sliding or ramp phase, (ii) rotation or restricted fall phase, (iii) free-fall phase and (iv) water entry phase. The forces acting on the lifeboat in the first three phases of its motion are very simple and are less significant in the context of the safety of the structure. However, the boat faces the most important force of hydrodynamic impact during its water entry along with the effects of gravity, buoyancy and drag forces. Among these forces, the hydrodynamic impact is mainly responsible for the high stresses developing in the structure of the boat. Presented in this paper is prediction of the total forces acting on the various sections of the lifeboat during its water entry. The average pressure on a particular section is then computed from the sectional force and the maximum wetted breadth of the section. The results of the study show that the average pressure is maximum at the fore end during first phase of water entry of the lifeboat but at the aft end this is significant during the second phase.

1. INTRODUCTION

Ocean going ships, offshore drilling units and other offshore platforms are required to have some sort of life saving equipment for the emergency evacuation of the occupants on them. Till now conventional types of lifeboats were the most commonly used means for the purpose. However, their use has become doubtful in view of some accidents which has happened recently. This has resulted in revival of the free-fall concept. This concept (free-fall lifeboat) was first advocated about a century back and was rejected on the ground of being inadequate and dangerous. The most important characteristics of these lifeboats are the shorter launching time and minimal possibility of collision with the parent vessel. The advantages of free-fall lifeboats as lifesaving appliances made them popular very rapidly and these are used presently in many maritime vessels and offshore installations.

The basis for evaluation of free-fall concept include the significant loads on the boat during its impact with the water. The hydrodynamic impact of the boat at water entry is a complex problem and the maximum forces are developed very shortly following contact with the surface of the water. These early stages are of utmost importance for the under water trajectory of missiles, for the structural design of the outer skin of ships, seaplane landing, missile and torpedo entry and surface ship and seaplane worthiness, just to mention the other most frequent fields of application. Hydrodynamic principles used in analysing problems of naval ballistics, seaplane impact and wave effect on ships or on stable structures are closely related. With the exception of the important problem of very high velocity impact, the useful theory is the so called momentum theory [1], and the solution of the various problem is reduced to the determination of the variable virtual masses associated with the entering body.

Presented in this paper is prediction of the sectional forces acting on the various sections of lifeboat during its water entry for a typical falling condition. These forces affect directly on the structure of the lifeboat. The average pressure on a particular section is then computed from the sectional force and the maximum wetted breadth of the section.

2. METHODOLOGY

The launch of a free-fall lifeboat from the skid is composed of four phases [2], viz., (i) sliding or ramp phase, (ii) rotation or restricted fall phase, (iii) free-fall phase and (iv) water entry phase. Sliding of the boat begins when it is released and ends when the centre of gravity crosses a point close to the lowest end of the launch skid. The forces acting on the boat during this phase are the gravity force, normal reaction force and the frictional force between the skid and the rail. The rotation phase of the free-fall launch begins as the sliding ends and it continues until the boat is no longer in contact with the launch skid. The forces acting in this phase are the gravity, friction parallel to the launch rail and a force normal to the rail. The free-fall phase of the launch begins at the end of the rotation phase and continues until the boat touches the water surface. The only force acting on the boat at this stage is the gravity force. The water entry of the free-fall lifeboat begins at the end of the free-fall phase.

As has been said earlier, the forces acting on the lifeboat in the first three phases of its motion are very simple and are also less significant for the structure of the boat. However, the boat faces the most important force of hydrodynamic impact during its water entry along with the effects of gravity, buoyancy and drag forces. A detailed derivation of the forces and moments acting on the lifeboat during its water entry phase has been presented by Khondoker [3] and Arai et al [4]. However, the methodology to estimate them is briefly presented in the following.

The volumetric force, the drag force and the force due to momentum transfer in the normal to the longitudinal axis of the boat have been computed using a strip model; the forces per cross-section are calculated using the added mass, relative velocity and acceleration, and they are integrated over the length of the falling lifeboat to obtain the total force. To perform the numerical integration using Newmark-β-method, the boat has been discritized into fourty segments in the longitudinal direction. The forces along the longitudinal axis of the boat has been approximated in total for the entire boat.

The buoyancy force is proportional to the immersed volume of the boat. This volume can be obtained integrating the immersed cross-sectional area along the length of the boat. The buoyancy moment around the centre of gravity of the boat has been computed in similar way. The damping effect of the lifeboat during water entry and exit is incorporated in the form of hydrodynamic drag forces and are calculated using Morison Equations based on the cross-flow principle. The principle assumes that the incident flow can be split into orthogonal components which are independent of each other. The axial drag coefficient depends on both skin friction and end pressure whereas normal drag coefficient is dominated by pressure drag. The drag coefficients for axial and normal flow have been chosen from Hoerner [5].

The hydrodynamic force due to momentum transfer in normal to the longitudinal axis of the boat has been formulated on the basis of a momentum theory [1] and assuming irreversible nature of the impact [6]. The force on an arbitrary cross-section is evaluated from the rate of change of momentum transfer of that section. The rate of change of immersion of a section was taken into account for increasing immersions only. This treatment is based on the considerations of momentum transfer only upon water entry and not during conditions associated with the water exit. The total force due to momentum transfer in the normal direction can be obtained by integrating the sectional forces throughout the length of the lifeboat. However, during water entry the bottom of the front part of the boat first hits the water surface with its axial velocity only. There will be an additional force on the lifeboat due to the hydrodynamic resistance to the axial velocity. The normal component of this force will be influenced by the longitudinal bottom shape of the front part of the boat. This normal force causes rotation of the boat even in presence of axial velocity only. The results of this phenomena is also incorporated in the final equations. Similar approach has been adopted for the momentum transfer along the boat axis but the force has been approximated in total for the entire boat. The average acceleration between the bow and the centre of gravity of the boat has been used along with the axial added mass and axial velocity of the boat to evaluate the rate of change of momentum transfer in the axial direction.

The added mass of a particular section of the lifeboat and its derivatives are functions of the immersion of the keel at that section and are estimated using the flat plate results of von Karman [1] i.e., $\pi\rho c^2$, where c is the instantaneous wetted half width of the section considered. This procedure is followed in order to estimate the added mass and its derivative for the immersion of the section up to its maximum breadth. Considering the flow separation from the boat during water entry, no increase in added mass has been assumed beyond this depth. However, since different cross-sections have different shape and size, input data for those sections have been extracted from the body plan of the lifeboat model. A transverse section is divided into a number of equally distant ordinate and for the immersion between two consecutive ordinates, the wetted half-width is calculated by linear interpolation. The axial added mass for full immersion has been taken on the basis of the added mass for an ellipsoid and the distribution along the axial immersion (throughout the length) is taken from the change of the sectional areas in the fore part of the lifeboat.

3. MATHEMATICAL FORMULATION

The water entry parameters of a free-fall lifeboat along with the coordinate system is shown in Fig. 1.

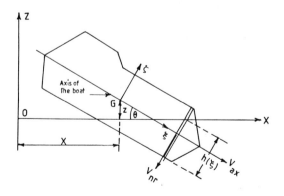

Figure 1. Water entry parameters of a free-fall lifeboat

The fixed global system (x,z) is set with x-axis describing an axis along the still water surface and z-axis corresponding to the vertical axis from the still water surface to the lowest end of the skid. A local coordinate system (ξ, ζ) is set with its origin at the centre of gravity of the lifeboat at any moment and instantaneous lifeboat axis describes the ξ-axis. v_{ax} and v_{nr} are the axial and normal velocities and these can be represented by the following equations:

$$v_{ax} = \dot{x}\cos\theta - \dot{z}\sin\theta \qquad (1)$$
$$v_{nr} = -\dot{x}\sin\theta - \dot{z}\cos\theta + \xi\dot{\theta} \qquad (2)$$

Based on the methodology described in the previous section, the forces on an arbitrary section in the normal direction (ζ) has been evaluated. The hydrodynamic force due to momentum transfer in the normal direction to the axis of the boat on an arbitrary cross-section at position ξ can be given as :

$$F_m(\xi) = \frac{\partial}{\partial t}[m(\xi, h)v_{nr}(\xi)] \qquad (3)$$

$$F_m(\xi) = [\frac{dm}{dt}v_{nr} + m\frac{dv_{nr}}{dt}] \qquad (4)$$

Where dm/dt, the time derivative of the added mass $m(\xi, h)$ for particular section located at a distance ξ from the centre of gravity and having an immersion h is,

$$\frac{dm}{dt} = \frac{\partial m}{\partial h}\frac{dh}{dt} \qquad (5)$$

But dh/dt will be evaluated only when $v_{nr}>0$, which corresponds to conditions of increasing immersion.

The other forces acting on the section are the buoyancy and drag. Their respective contributions can be expressed as:

$$F_b = \rho g A_i(\zeta) \qquad (6)$$
$$F_d = c_d(\zeta)\pi\rho . c_m v_{nr}|v_{nr}| \qquad (7)$$

Here, A_i is the immersed area, c_d is the drag coefficient and c_m is the maximum wetted half breadth of the section considered. Considering all the forces mentioned above and substituting the value of v_{nr} the resultant force of a particular section can be given as:

$$F_{SC} = \rho g A_i (\zeta) + c_d (\zeta) \pi \rho. c_m v_{nr} |v_{nr}|$$

$$-(\dot{x} \sin\theta + \dot{z} \cos\theta) \frac{\partial m}{\partial t} + \dot{\theta} \frac{\partial m}{\partial t} \xi + \ddot{\theta} m \xi$$

$$-(\ddot{x} \sin\theta + \ddot{z} \cos\theta + 2\dot{\theta} v_{ax}) m + v_{ax} \frac{\partial m}{\partial t} \tan\alpha$$

$$(8)$$

The average pressure of the section has been approximated as :

$$P_{sc} = F_{sc}/2c_m \qquad (9)$$

4. RESULTS AND DISCUSSIONS

Numerical simulation has been carried out using a 1.0 m long lifeboat model for the following launching conditions (Fig. 2):

Falling height (H)   = 1.40 m
Falling angle (Θ)   = 30 deg
Skid length ($L_{go'}$)   = 0.80 m &
Guide rail length (L_{ra}) = 0.50 m

The trajectory of the falling lifeboat for the above parameters has been shown in Fig. 2.

As we can see from the figure, during the first phase the boat slides along the skid and is constrained to move only in the axial direction. As it slides off the skid, the couple produced by the gravity force and the reactive force on the launch rail causes it to rotate in clockwise direction. At the end of the rotation phase, the boat leaves the launch skid but it continues to rotate at constant angular velocity until it touches the water surface. As it enters the water, high impact force exerted on the bow of the boat due to change in added mass which is called bow impact. The impact force along with the buoyancy and weight of the boat causes the angular momentum imparted during its rotation phase to be reversed. The rotating motion of the boat in anti-clockwise direction due to the righting moment produced by the fluid forces then leads to the stern impact. At this stage buoyancy and drag forces starts increasing and the impact force reduces until it reaches the maximum immersion (about 0.25m). This is followed by immersion of the longitudinal centre line of the boat into water. Consequently, a large buoyancy force pushes it upward. When it exits the water surface, the boat is almost horizontal with a significant forward velocity and it falls down again. By this time, it has already moved a considerable distance away from the parent vessel.

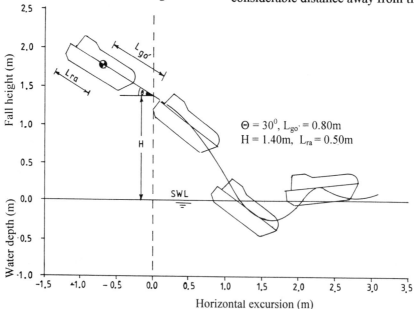

Figure 2. Trajectory of a free-fall lifeboat model (H = 1.40 m, Θ = 30 deg, $L_{go'}$ = 0.80 m, L_{ra} = 0.50 m)

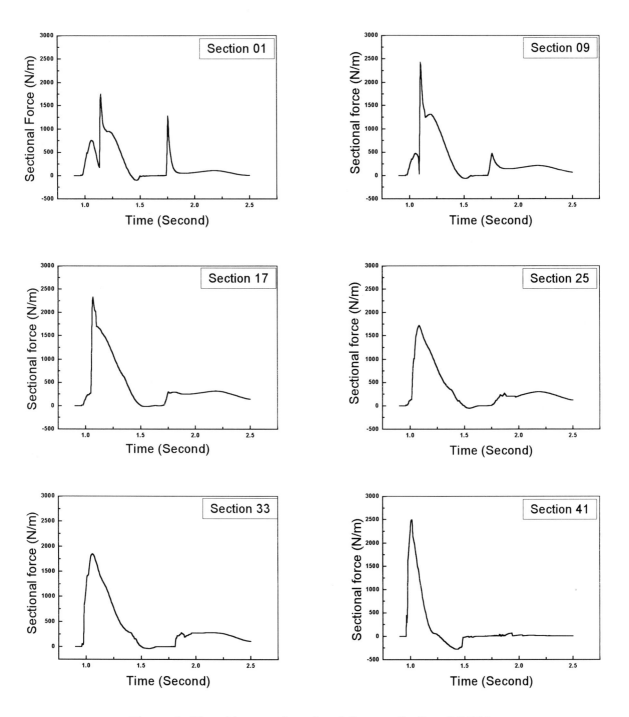

Figure 3. Time history of sectional forces of a free-fall lifeboat

100

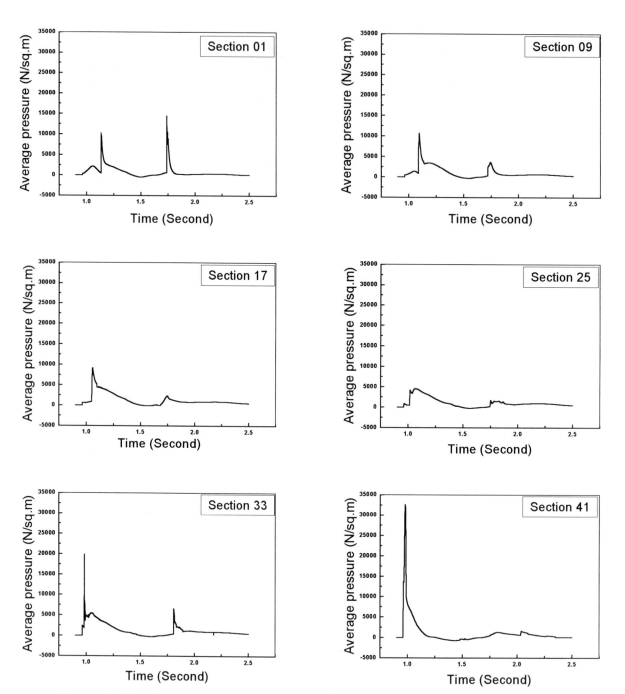

Figure 4. Time history of average pressures in different sections of a free-fall lifeboat

As has been said before, the boat has been discritized into fourty segments and the forces for each section are calculated using the added mass, relative velocity and acceleration. The results are shown for Section 01 (0.0m from the stern), Section 09 (0.2m), Section 17 (0.4m), Section 25 (0.6m), Section 33 (0.8m) and Section 41 (1.0m).

Fig. 3 shows the time history of sectional forces. It is seen that there are two clear peaks of the forces at Section 01 due to bow and stern impacts of the lifeboat and the corresponding magnitude is 750 N/m and 1750 N/m. The third peak is due to the second entry of the lifeboat into the water and the magnitude is very high in this case also (1280 N/m). The time history of sectional force in Section 09 also show similar nature though the magnitude of force during stern impact is higher (2400 N/m) compared to Section 01. The effects due to bow impact and during the second entry are comparatively lower. In Section 17, there is only one peak which is during stern impact and the magnitude is 2300 N/m. The magnitudes of the sectional forces during bow impact is 1700 N/m, 1800 N/m and 2700 N/m respectively in Section 25, Section 33 and Section 41. These sections are also characterized by a single peak occurring during bow impact.

Fig. 4 shows the time history of average pressure in the same sections of the free-fall lifeboat. It is seen that the Section 41 attains the maximum average pressure during the bow impact of the lifeboat and the value is 32600 N/m^2. It is also seen that Section 17 and Section 25 (near midship) experiences minimum pressures though the sectional forces are not minimum in such sections. This is because of the fuller shapes of these sections. The average pressure during the second phase of water entry is maximum at the stern because of the maximum force generated in this section.

5. CONCLUDING REMARKS

Following conclusions can be drawn from the study of the sectional forces and average pressures of a skid launching free-fall lifeboat during water entry.

1. The aft part of the free-fall lifeboat experiences two distinct impacts during the first phase of water entry. They are impact of the bow with the water and the entry or impact of the stern into the water.

2. The sectional force is maximum at the forward end of the boat. However, quite significant peak develops in almost all sections.

3. The average pressure is maximum in the forward part of the boat. The pressure at the stern is also significant and has two peaks during first and second entry respectively.

4. The average pressure at the aft end during second phase of water entry is higher than that of the first phase. But the sectional force is lower in the second phase.

5. The bow of a free-fall lifeboat launching from the skid should be structurally strongest but the stern also should possess sufficient strength.

REFERENCES

1. von Karman, T. , The Impact on Seaplane Floats during Landing, NACA TN 321, (1929).
2. Tasaki, R., Ogawa, A., and Tsukino, Y. : Numerical Simulation and Its Application on the Falling Motion of Free-fall Lifeboats, Journal of The Society of Naval Architects of Japan (SNAJ), Vol. 167 (1990).
3. Khondoker, M.R.H.: Numerical Study of the Behaviour of Free-fall Lifeboat During Water Entry, an unpublished D. Engg. Thesis, Department of Naval Architecture and Ocean Engineering, Yokohama National University, Japan, March (1996).
4. Arai, M., Khondoker M.R.H. and Inoue, Y.: Prediction of the Performance of Free-fall Lifeboat Launching from Skid, Proceedings of the Offshore Mechanics and Arctic Engineering Conference, ASME, Paper No. OMAE-96-430, Florence, Italy (1996).
5. Hoerner, S.F.: Fluid Dynamic Drag, New Jersy, U.S.A., (1958).
6. Boef, W.J.C. : Launch and Impact of Free-fall Lifeboat (Part I and Part II), Ocean Engineering, Vol. 19 (1992).

Practical Design of Ships and Mobile Units
M.W.C. Oosterveld and S.G. Tan, editors.

A Computational Method for Analysis of LNG Vessels with Spherical Tanks

F. Kamsvåg[a], E. Steen[a] and S. Valsgård[a]

[a]Det Norske Veritas AS, Veritasveien 1, 1322 Høvik, Norway

An integrated analysis of the total hull and tank configuration of a spherical tank LNG vessel of Moss-Rosenberg type is presented. The paper gives an outline of the latest development in computer based design analysis in DNV, from hydrodynamic analysis through structural analysis to acceptance criteria. The intention of the analysis was to provide a comprehensive documentation of the structural design of the actual vessel. This means that extended structural analysis were used to put additional steel where it was most needed, and to improve the local fatigue design of selected structural details.

New acceptance criteria for the structural strength of the spheres and cylinders of the containment system were used and are discussed in the paper.

1. INTRODUCTION

The Kvaerner Group of companies developed the spherical LNG tank system in Norway in the early 1970'ies. It has been frequently referred to as the Moss-Rosenberg (MRV) system as the Moss and Rosenberg yards in Norway were the first shipyards where such ships were built. Since then the MRV system has been built in the US (Quincy), Japan (Mitsui, Kawasaki and Mitsubishi), Korea (Hyundai), Germany (HDW) and Finland (Kvaerner-Masa). The MRV system has proved to be a very reliable system with an impressive track record of virtually no off-hire due to the special containment system used. Hence, the MRV system has become one of the most popular systems for transport of LNG.

The "standard" 125 000 m^3 MRV LNG carrier consists of 4-6 spherical tanks built of aluminium (Al-5083-O) with diameters in the range of 35-40 meters and supported by cylindrical skirts via a specially designed equator area. The spheres are unstiffened with shell thickness ranging from 25 mm upwards to some 50 mm and with a thicker equator zone, usually in the order of 160–200 mm. The supporting cylindrical skirts have an unstiffened or vertically stiffened (on the older designs) upper part in aluminium. A specially designed thermal brake is arranged in the transition zone between the upper aluminium part and the lower steel parts of the skirt. The lower steel parts have usually a combination of horizontal ring stiffeners and vertical stiffeners.

An important design consideration with the new system was the "leak-before-failure" principle, which was established in co-operation with US Coast Guard. This entails that if a leak of gas is detected, as the result of a crack through the tank shell, the crack should not become unstable (critical) within 15 days in order to let the vessel reach port and unload her cargo. This became a key design parameter, and in parallel with Kvaerners development, Det Norske Veritas (DNV) carried out material testing of fracture mechanics parameters in order to establish fatigue design criteria for the system. Further, testing of liquid sloshing loads inside the tanks were carried out as well as fairly large-scale laboratory testing on buckling strength of both the spheres and the skirts, ref. [1] and [2]. The testing work was complemented with theoretical developments, hydrodynamic load analyses and Finite Element analyses in order to establish a complete set of design criteria and procedures for these new vessels.

Throughout the 1970-80'ies Finite Element analyses were mainly concentrating on providing the interaction forces between the hull and the tank system. These forces were then applied to axisymmetric models of the tanks in order to determine the stress distribution as basis for allowable stress, buckling and fatigue control.

The objective of the present paper is to outline an analysis procedure for LNG carriers with spherical tanks using the latest development in computer based design analysis. The procedure includes hydrodynamic analysis, structural analysis and the

104

use of current updated acceptance criteria, including stochastic fatigue analysis for the ship hull.

2. WAVE LOAD ANALYSIS

The main aim of the sea keeping and wave load analysis have been to find dimensioning loads for the subsequent structural analysis.

2.1. Wave Load Calculation Procedures

The wave load calculations have been performed using several programs in the SESAM, ref. [3], computer analysis system.

Six different loading conditions have been considered to ensure that the maximum dimensioning loads are found. The dimensioning loads for a LNG vessel with spherical tanks are:
- ❑ vertical wave bending moment (hull/tank)
- ❑ torsion moment (hull/tank cover)
- ❑ inertia forces on the tank system (tanks)

All calculated loads are generally transferred directly from the wave load analysis to the structural analysis. This includes both hydrostatic and hydrodynamic pressures from the sea, internal pressures from cargo/ballast and inertia forces from the structure.

Direct load transfer is performed for zero speed

and linear theory. Effects from forward speed and non-linear effects are included as load factors when applicable.

The calculation procedure is shown in Figure 1.

2.2. Non-linear effects

Non-linear effects from the integration of the wave pressure over the instantaneous position of the hull relative to the waves, such as bottom slamming, bow flare forces and deck wetness, have been determined from a time domain analysis.

The effect is included as correction factors on vertical bending, torsion and acceleration loads for hull and tank systems and is applied for the ultimate strength (ULS) condition. The non-linear effects are assumed to be negligible in the fatigue analysis.

2.2. Forward speed effects

The forward speed in the ultimate strength (ULS) condition is assumed to be low, and will not affect the dimensioning loads.

Forward speed is included in the fatigue analysis as the main contribution to the fatigue is created in moderate sea states. The used forward speed accounts for speed reduction. This speed reduction is based on the exceedance of specific criteria with respect to frequencies of occurrence of slamming, water on deck, added wave resistance and voluntary speed reduction due to heavy weather.

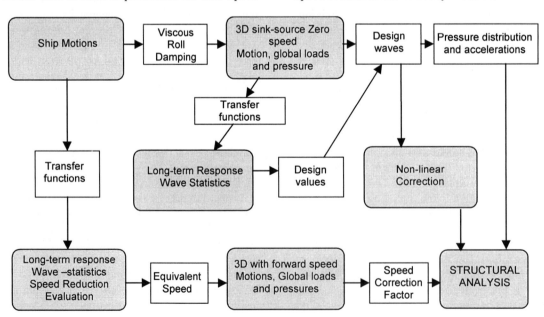

Figure 1. Wave load analysis procedure

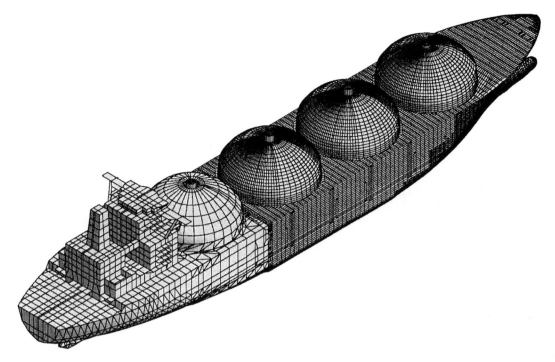

Figure 2. Structural model of the vessel

3. STRUCTURAL ANALYSIS

Several structural analyses are performed in order to ensure that the vessel meets the acceptance criteria stated in the rules. The structural integrity of the vessel (hull and tank system) is checked against yield, buckling and fatigue.

The structural analyses of the hull are performed according to a procedure equivalent to Class Notation CSA-2, ref. [4] as specified in the DNV Rules for Classification of Ships.

The structural analysis of the cargo tank system is performed according to DNV Rules for Classification of Liquefied Gas Carriers and equivalent IMO codes, ref. [5].

3.1. Hull Structural Analysis

A 3-dimensional finite element analysis of the entire hull and tank system, by means of the FEM program SESAM, is performed. The global model extends over the whole ship length assembled by several super-elements. The model is shown in Figure 2. The element model is built up using a combination of shell and beam elements and contains a total of 280.000 degrees of freedom.

The tanks and tank supports are modelled in order to provide a correct tank stiffness and thus assess realistic hull/tank interaction forces.

Hydrostatic and dynamic sea pressures are applied to the global model. The FE-analysis of the global model produces nominal stress responses suitable for checks of the global stress level in the hull girder. The evaluation of wave bending moments, dynamic sea pressure and accelerations is based on direct wave load calculations.

3.2. Hull Ultimate Strength Analysis

The ULS calculations of the hull are performed according to the CSA-2 criteria. This class notation is currently not compulsory for LNG carriers. The CSA-2 procedure is, however, streamlined towards direct calculations as carried out in this project. The CSA-2 requirements are generally stricter than the 1A1 requirements. This leads to increased scantlings in some critical areas and hence to a vessel with higher built-in safety margins.

3.3. Hull Fatigue Analysis

The fatigue analysis follows the direct analysis approach in the draft DNV Classification Note: "Fatigue Assessment of Ship Structures", ref. [6] for a fully stochastic analysis.

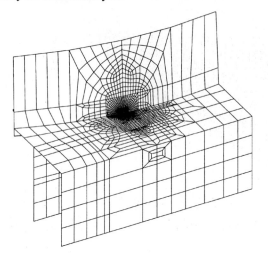

Figure 3. Local model of deck/cover connection

The fatigue calculations are based on environmental data for world-wide trade, and an equal probability of all wave headings is assumed. The fatigue criterion is 25 years in world-wide trade.

Two loading conditions are considered; fully loaded and ballast. It is assumed that the fraction of the total design life in the loaded condition is 0.5 and in the ballast condition 0.4, i.e. the total fraction spent in open sea is 0.9.

Wave load calculations are performed for 8 wave headings and 22 wave periods from 6 to 32 seconds. Infinite water depth is assumed.

The calculated sea pressures, internal tank pressures and accelerations are directly transferred to the global FE model defining 176 load cases (8 headings x 22 periods). The load cases are described by complex unit loads corresponding to a wave amplitude of one meter.

Local stress concentration models are used for the analyses. For these details mesh size is in the order of the plate thickness. Hence, all load effects, global and local, are included in addition to the geometric stress concentration. 8-noded shell elements are used for this purpose. A model of the deck/cover connection is shown in Figure 3.

In the fatigue analysis, the hydrodynamic analysis and the global FE analysis are run for all wave load cases using automatic transfer of the loads to the global FE model. For each load case the deformations of the global FE model are automatically transferred to the local FE model as boundary displacements. In addition the local internal and external hydrodynamic pressures are automatically transferred to the local FE model by the wave load program for details where this is relevant. The fatigue analysis program is then

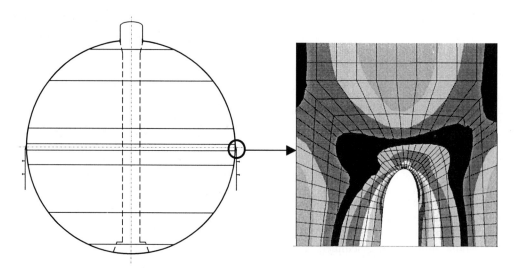

Figure 4. Tank system and close up on equator area

107

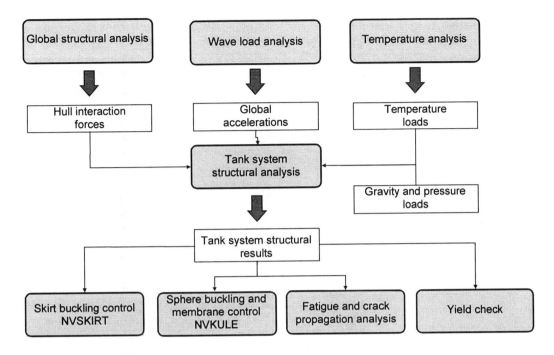

Figure 5. Tank analysis procedure

calculating the part fatigue damage for each cell in the wave scatter diagram based on the principal stresses from the local FE model extrapolated to the hotspot, wave spectrum, wave spreading function, S/N data etc.

Areas of similar design configurations, where low fatigue lives have been experienced in service, have also been pinpointed by a stochastic fatigue analysis carried out for the present design. In such cases, improvements can be achieved with proper weld treatment.

3.4. Tank Ultimate Strength Analysis

Structural analyses and strength assessments are performed for all tank systems. The analyses cover both skirt and sphere.

Tank no. 1 (foremost) is different from the other tanks. Separate structural FE models of Tank no. 1 and Tank no. 2-4 are developed using very fine meshed axisymmetric elements. Both axisymmetric and non-axisymmetric loads may be analysed with this model.

Figure 4 shows the tank system and a close-up on the model in the equator area.

Loads from internal overpressure, gravity, temperature distribution, static and dynamic hull

interaction forces, inertia loads and sloshing forces are included in the analyses.

Both tank structural analyses and strength assessments are performed according to the scheme shown in Figure 5.

The design stresses are based on an envelope stress response in order to be conservative. This means that different load categories (acceleration, hull interaction, sloshing) are combined using their maximum value even if they are calculated for different ship loading conditions. Dynamic stresses are combined using the square root sum of squares method.

The buckling control of the tank spheres and skirts are based on the procedures given in DNV Class Note 30.3, June 1997. The buckling control checks the upper aluminium part, the middle SUS (stainless steel, SS304) part and the lower steel part separately against all relevant buckling modes. Buckling controls are carried out for stresses at centre line and at ship side.

3.5. Tank Fatigue Analysis

Fatigue analyses are performed for the critical welds of the tank system. This covers the tower/sphere connections, the equator area and selected welds on the sphere at centre line and at ship side. Both horizontal and vertical welds are analysed.

FE stress analyses of equator and tower/skirt connections are carried out. Fine meshed solid FE elements have been used in the critical areas for detailed stress evaluations.

Closed form stochastic fatigue analyses are performed for the tower/sphere connection while crack propagation analyses are performed for all selected locations.

The stresses that are used in the fatigue analysis are based on an envelope stress response (see 3.4). This means that phase information is lost and that the dynamic stresses have to be checked both as if membrane stress is dominant at the given location (in phase), and as if bending moment is dominant (out of phase).

4. ACCEPTANCE CRITERIA

4.1. Hull

Since the CSA-2 analyses are based on direct hydrodynamic analyses of the loads, the phase information are kept intact throughout the analyses. The various parts of the vessel are therefore subjected to simultaneously acting multi-axial sets of loads, which calls for acceptance criteria capable of handling multi-axial load situations. The acceptance level for allowable tension stresses is therefore referred to a von Mises equivalent stress equal to 0.85-0.95 of the yield stress. Buckling criteria are given in ref. [7]. The characteristic strength values have been calibrated to a 95% probability of exceedance level using reliability analysis, and the buckling acceptance level has similarly been calibrated to a usage factor of 0.9. More details of this procedure are reported in ref. [8].

4.2. Tank System

The cargo tank system is designed according IMO, ref. [5] and DNV rules, ref. [9]. These rules specify accepted design procedures for type B independent tanks. Thus, based on the directly calculated design stresses the spherical tanks and their skirt support are checked with respect to stress, buckling, fatigue and crack propagation criteria.

The membrane stress and buckling criteria for the sphere are incorporated into the computer program NVKULE, ref. [10]. The buckling criteria for the skirt are incorporated into a similar program called NVSKIRT, ref. [11]. These buckling criteria have recently been improved and made more consistent in order to produce more optimum tank scantlings with respect to safety and weight. Considerable R&D efforts have been used in order to formulate a consistent set of buckling criteria that includes all relevant types of stiffening arrangements in the skirt construction. In particular the criteria for the unstiffened aluminium part, SUS part and steel part between rings have been reformulated in order to carry out more accurate strength assessments. The sphere criteria consider all possible tank fillings and sloshing loads are added for partially filled tanks. Special emphasis is on the circumferential compressive forces generated in the zone just below the equator profile for partially loaded tanks. The new buckling criteria are published as a separate DNV classification note, ref. [12].

The fatigue and crack propagation analyses have been combined with the latest available fatigue and crack propagation parameters for the critical welded connections. These are used together with expected initial crack sizes that can be detected by present inspection methods. Crack propagation criteria include the leak-before-failure concept and detailed assessments of the through thickness crack growth as well as crack growth along the circumference are parts of the design procedure.

5. CONCLUSIONS

A ship hull and tank system analysis is outlined for a LNG carrier with spherical tanks. The analysis is based on direct hydrodynamic calculations with direct load transfer to the global structural model and to fine meshed sub-models used in stochastic fatigue analyses. Separate fine meshed axisymmetric tank models are used in the tank system analyses.

The outlined wave load analysis shows the calculation procedure for a wave load analysis with direct load transfer to the structural models. Effects from forward speed and non-linear effects are accounted for in the analysis.

The referenced analysis is carried out for the complete hull, including the tank system. The hydrodynamic and hydrostatic loads are directly calculated based on the actual loading conditions. Internal and external pressures are, together with inertia loads, automatically transferred from the hydrodynamic program to the global finite element (FE) models.

The acceptance criteria for the hull are according to the CSA-2 (Computational Ship Analysis) Class notation for FE analysis.

Stochastic fatigue analyses are performed for critical areas of the hull. All loads are automatically transferred to local stress concentration models with mesh size in the order of the plate thickness in areas of interest. Areas of similar design configurations, where low fatigue lives have been experienced in service, have also been pinpointed by a stochastic fatigue analysis carried out for the present design.

The global FE analysis interaction forces between hull and tank system are transferred to an axisymmetric FE tank model. This allows for very fine element mesh in interesting parts, as the equator area and the upper and lower tank/dome connections. Detailed stress values found in these areas are used in yield and fatigue/fracture mechanics calculations and tank sphere and skirt supporting structure have been checked against a newly developed set of stress and buckling criteria.

6. ACKNOWLEDGEMENTS

Thanks goes to the team who carried out the analysis i.e. Liv Hovem and Tormod Bøe (wave load analysis), Florus Korbijn (non-linear wave load analysis), Lise Jahren, Rolf Ole Jensen, Hege Bang, Bjørn Jacobsen, Harald Rove and Gunnar Ramstad (FE-models), Erik Ian von Hall (ULS analysis), Torbjørn Lindemark (stochastic fatigue analysis), Agnar Karlsen (fracture mechanics) and Thor Hysing (verification). Valuable discussions with Tryge Tobiesen, Gunnar Rød, Harald Olsen and Mathew Seides on design and approval aspects are highly appreciated.

7. REFERENCES

1. S. Valsgård and G. Foss, "Buckling Research in Det Norske Veritas", Buckling of Shells in Offshore Structures, Granada 1982.
2. J. Odland, "Theoretical and Experimental Buckling Loads of Imperfect Spherical Shell Segments", Journal of Ship Research, Vol. 25, No. 3,Sept. 1981
3. SESAM Technical Description, DNV SESAM AS, Høvik 1991
4. Det Norske Veritas, "Rules for Classification of Ships, Hull Structural Design, Ships with Length 100 Meters and above", Part 3 Ch. 1, Høvik July 1997.
5. International Maritime Organisation (IMO), "International Code for the Construction and Equipment of Ships Carrying Liquefied Gases in Bulk", Resolution MSC.5(48) 1983 Edition, reprinted 1985.
6. Det Norske Veritas, "Fatigue Assessment of Ship Structures", Draft Classification Note 30.7, DNV Report no. 93-0432, Rev. 6, September 1996.
7. Det Norske Veritas, "Buckling Strength Analysis", Classification Note 30.1, 1995
8. Valsgård S., Svensen T.E. and Thorkildsen H., "A Computational method for analysis of container vessels", SNAME Annual Meeting, October 5-6, 1995, Washington D.C.
9. Det Norske Veritas, "Rules for Classification of Ships, Liquefied Gas Carriers", Part 5 Ch. 5, Høvik July 1997.
10. Det Norske Veritas Classification, "NVKULE, User's Manual", DNV Report No. 95-0021, Høvik 1995
11. Det Norske Veritas, "NVSKIRT, User's Manual", DNV Report No. 96-0411, Høvik 1997
12. Det Norske Veritas, "Buckling Criteria of LNG Spherical Cargo Tank Containment System – Skirt and Sphere", Classification Note 30.3, 1997.

Practical Design of Ships and Mobile Units
M.W.C. Oosterveld and S.G. Tan, editors.

The Influence of Adjoining Structures on the Ultimate Strength of Corrugated Bulkheads

Jeom Kee Paik[a] , Anil K. Thayamballi[b] and Sung Geun Kim[a]

[a] Pusan National University, Department of Naval Architecture and Ocean Engineering,
30 Changjeon-Dong, Kumjeong-Ku, Pusan 609-735, Korea

[b] American Bureau of Shipping,
Two World Trade Center, 106th Floor, New York, NY 10048, USA

This paper investigates the influence of adjoining structures on the collapse strength of bulk carrier corrugated bulkheads subject to accidental flooding. A special purpose nonlinear finite element program suitable for direct calculations of strength is developed as part of the study. In the program, the adjoining structures such as lower and upper stools, shedder plates, double bottom and deck structures, are included as options in the automated finite element structural modeling of the corrugated bulkhead. To illustrate application, a hypothetical Capesize bulk carrier bulkhead is analyzed in several different ways to investigate the variation in its ultimate strength and stiffness for different end conditions of the bulkhead. A comparison of the finite element based ultimate strength against predictions based on simplified design formulae is also made.

1. INTRODUCTION

Since the beginning of 1980s to mid 1990s, over 150 bulk carriers have been lost with a loss of more than 1,200 lives[1]. Some (nearly 20) of those vessels apparently disappeared for no known cause. One possible cause of loss of oceangoing vessels is progressive collapse of corrugated bulkheads in a flooded condition, particularly in forward cargo holds when the watertight transverse bulkheads are insufficient to withstand the increased static and dynamic pressures and vertical sectional shear forces. Many studies to reduce bulk carrier casualties have subsequently been undertaken by the International Maritime Organization (IMO), the International Association of Classification Societies (IACS) and also some of the leading classification societies themselves.

In previous practice consistent with the International Load Line Convention, transverse bulkhead locations in ships were normally selected to prevent margin line immersion with a cargo hold accidentally flooded. Traditionally, the design rules given by classification societies have used nominal hydrostatic loads in the intact condition as design loads for the bulkhead. As previously noted, there is a possibility that the collapse of corrugated transverse bulkheads and subsequent progressive flooding may have been a factor in some of the recent bulk carrier losses. In this context, the leading classification societies have now increased their standards and level of safety for the design and long-term maintenance of such bulkheads[2,3].

Most recently, new design requirements have been adopted by IACS and IMO for both existing and new bulk carriers[4]; All new bulk carriers of 150 m length and above and carrying cargoes with a density of 1,000 kg/m³ and above should have sufficient strength to withstand flooding of any one cargo hold, taking into account the dynamic effects resulting from the presence of water in the hold. Existing bulk carriers built before July 1, 1999 and loading bulk cargoes with a density of 1,780 kg/m³ and above should have a transverse watertight bulkhead between the two foremost cargo holds, and the double bottom of the foremost cargo hold, of sufficient strength to withstand similar flooding and related dynamic effects. Cargoes of such densities include iron ore, pig iron, and steel, among others. Existing ships which do not comply with appropriate

112

requirements will presumably have to be reinforced or may have to limit either their loading pattern or move to lighter cargoes such as grain or timber.

Ideally, the strength of the transverse bulkheads in an accidentally flooded condition must be sufficient to avoid collapse resulting in possible progressive flooding. For bulk carrier safety, the set of design lateral pressure load cases for the corrugated transverse bulkheads should then include the condition that one adjacent compartment has been flooded. If the maximum applied bending moment along the span of corrugation becomes larger than the ultimate bending moment the corrugated bulkhead will presumably collapse. The maximum (extreme) bending moment acting on the corrugated bulkheads in a flooded condition would be calculated by classification society rules. And the ultimate strength of the corrugation could be calculated by using the simplified formulas of the IACS and the classification society. The adjoining structures, e.g., lower / upper stools, double bottom and deck structures, may affect the stiffness and strength of the corrugation. In direct calculations, however, the edge conditions of the corrugation should then be carefully accounted for, considering the rigidity of upper / lower stools and other adjacent structures (bottom and deck). For this purpose, the non-linear finite element method could be useful for more accurately taking into account the boundary structure effects on the corrugation which depend on the scantlings and proportions of stools and other adjacent structures.

In a past few years, a wide ranging joint research study on ultimate strength of corrugated bulkheads has been carried out by the American Bureau of Shipping and the Pusan National University[5]. Collapse tests on nine mild steel corrugated bulkhead models under a combination of lateral pressure and axial compressive loads were performed with variation of the corrugation angle and plate thickness[6]. A simple analytical formulation for predicting the ultimate strength of corrugated bulkheads under lateral pressure loads was proposed. Also, a special purpose nonlinear finite element computer program named CORBHD/FEM[7] was developed for efficiently analyzing progressive collapse behavior of corrugated bulkheads subject to lateral pressure loads. The accuracy of the program under known idealized boundary conditions has been verified by comparing with the experimental results.

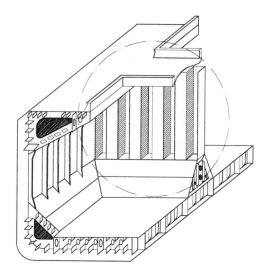

(a) Corrugated transverse bulkhead in the cargo hold of a bulk carrier

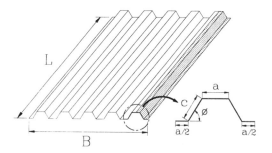

(b) Nomenclature for the corrugated bulkhead structure modeling

(c) A half pitch of corrugation as modeled

Fig. 1 Features of the analysis

The aim of the present study is to numerically investigate the influence of the corrugation end condition due to adjoining structures on the collapse behavior of bulk carrier corrugated bulkheads subject to lateral pressure loads. For this, different types of adjoining structures are included as options in the automated structural modeling using the CORBHD/FEM nonlinear finite element program. As an illustrative example, a hypothetical Capesize bulk carrier corrugated bulkhead is analyzed for various corrugation end conditions. It is seen from the computed results that the ultimate strength of corrugated bulkheads can significantly depend on the rotational restraints posed by the adjoining structures. Based on such computed results, practical end condition assumptions may be eventually recommended, which would avoid the time and cost of explicitly considering the adjoining structures in a nonlinear analysis. The applicability of existing design ultimate strength formulations can also be studied, as will be illustrated. It should be noted that this study deals with the realistic prediction of structural capacities, and not the applied loads themselves.

2. CORBHD/FEM MODELING FOR CORRUGATED BULKHEADS

2.1. Structural modeling

As the representative extent of analysis, a half pitch of corrugation is taken, see Fig. 1. Figure 2 shows the various CORBHD/FEM model types possible for the corrugated bulkhead analysis. The web and the half-flange of the corrugation are modeled by a number of 8 and 4 plate-shell elements, respectively, see Fig. 1.c. In the longitudinal (vertical) direction of the corrugation, a total of over 40 elements are employed. A relatively finer mesh is used at the lower and upper corrugation ends in order to include the deformation of the corrugation at the juncture of the adjoining structure more gradually and precisely.

In the analysis, the lower stool top plate and the upper stool bottom plate may be also included. The corrugated bulkheads normally have shedder plates which may affect their strength and stiffness. The user can select an appropriate option to include those members in the modeling as well.

The bottom and deck structures will impose end rotational restraints to the corrugation behavior which are neither zero nor infinite, and thus the boundary conditions at the junction of corrugation and adjacent structures will never be purely simply supported

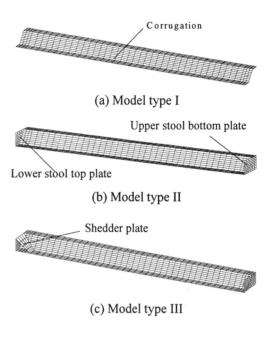

(a) Model type I

(b) Model type II

(c) Model type III

(d) Model type IV

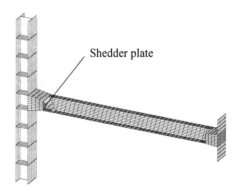

(e) Model type V

Fig. 2 Types of CORBHD/FEM models

or purely clamped which correspond to either zero or infinite rotational restraints, respectively. To take into account these effects properly, the user can directly include adjacent structures, e.g., double bottom and deck structures in the modeling for direct analysis.

Several types of shedder plate arrangements are used in actual corrugated bulkhead construction. In the present version of the program, however, only "usual type" which is most widely used in actual corrugated bulkheads of bulk carriers is considered. Also, the shedder plate is only at lower end of bulkhead. In the program, the shedder plate termination points are decided by the angle of the shedder plates, normally 45 degree.

The adjacent structures, i.e., bottom and deck, are modeled with equivalent plate thickness where the stiffeners are uniformly shared with the parent plate. In the analysis, the adjacent structures are treated as nonlinear elastic, i.e., of nonlinear geometric behavior but linear elastic material behavior.

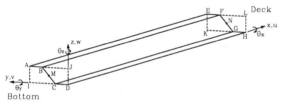

Edges to be restrained	Restrained displacement
DH, AE	$v = \theta_x = \theta_z = 0$
ABCD, EFGH	$v = w = \theta_x = \theta_y = \theta_z = 0$
AI, IC, BJ, JD EK, KG, FL, LH	$v = w = \theta_x = \theta_y = \theta_z = 0$
M	$u = 0$

(a) Both ends are clamped

Edges to be restrained	Restrained displacement
DH, AE	$v = \theta_x = \theta_z = 0$
ABCD	$v = w = \theta_x = \theta_y = \theta_z = 0$
EFGH	$v = w = \theta_x = \theta_z = 0$
AI, IC, BH, JD	$v = w = \theta_x = \theta_y = \theta_z = 0$
EK, KG, FL, LH	$v = w = \theta_x = \theta_z = 0$
M	$u = 0$

(b) Clamped at lower stool end and simply supported at upper stool end

Edges to be restrained	Restrained displacement
DH, AE	$v = \theta_x = \theta_z = 0$
ABCD, EFGH	$v = w = \theta_x = \theta_z = 0$
AI, IC, BJ, JD EK, KG, FL, LH	$v = w = \theta_x = \theta_z = 0$
M	$u = 0$

(c) Both ends are simply supported

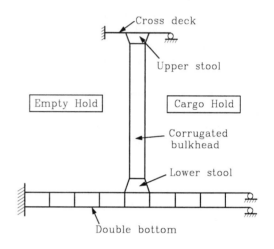

(d) Rotational restraints from the adjoining structures (Clamped at center of cargo hold)

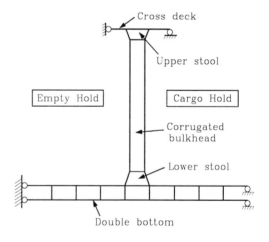

(e) Rotational restraints from the adjoining structures (Simply supported at center of cargo hold)

Fig. 3 Boundary conditions for the corrugated bulkhead models

2.2. Boundary conditions

The boundary conditions of the finite element model of corrugated bulkheads are selected by the user. Three types of the corrugation end conditions, namely both ends simply supported, lower end clamped / upper end simply supported and both ends clamped, are available for the model types I, II and III, see Fig. 3. For the model types IV and V, i.e., with adjoining structures, two types of boundary conditions for the adjacent structures are available in the program as shown in Figs. 3.d and 3.e.

2.3. Load cases

The hydrostatic pressure loads arising from cargo / flooded water in hold are incrementally increased up to the collapse of corrugation, see Fig. 4. Two types of pressure loading pattern are considered, one for pressure loads acting on both of corrugation and inner bottom panel and the other for pressure loads acting on corrugation alone. The water pressure on the outer bottom panel is not specifically considered. Note that the purpose of considering the various load patterns is the realistic prediction of ultimate structural capacity. This study does not directly relate to the magnitude of the loads themselves.

2.4. Initial imperfections

The following form of global initial deflection of the corrugation is used:

$$w_o = A_o \sin\left(\frac{\pi x}{L}\right) \qquad (1)$$

where
A_o = global initial deflection amplitude, taken as 0.0015 L in our study
L = length of corrugation (see Fig. 1.b)

The local initial deflections of the corrugation flange or web are not considered. Also welding induced residual stresses are not taken into consideration in the analysis.

2.5. Rupture criteria

One manner in which potential progressive flooding into an adjacent hold of a bulk carrier can occur is by ductile fracture, i.e., rupture of a corrugated bulkhead and / or its boundaries (including weld seams) after a cargo hold has been flooded.

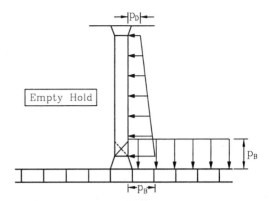

(a) Pressure acting on both of corrugation and inner bottom

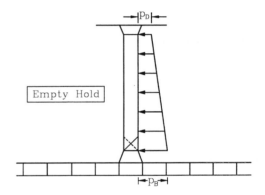

(b) Pressure acting on corrugation alone

Fig. 4 Types of pressure loading patterns available

Therefore, it would be desirable for the structural analyst to get the information related to when and where rupture initiates during the failure of the bulkhead. As the applied pressure loads increase, the CORBHD/FEM program thus checks for the occurrence of the first ductile fracture at the finite element level, considering that the ductile fracture occurs if the total (accumulated) equivalent plastic strain reaches a pre-defined critical rupture strain.

For homogeneous isotropic materials, the increment of equivalent plastic strain for an element can be calculated by

$$d\varepsilon_{eq}^p = \frac{\sigma_x^* \cdot d\varepsilon_x + \sigma_y^* \cdot d\varepsilon_y + \tau_{xy} \cdot d\gamma_{xy}}{\sigma_o} \qquad (2)$$

116

where

$d\varepsilon_{eq}^{p}$ = equivalent plastic strain

$$\sigma_x^* = \frac{2\sigma_x - \sigma_y}{3}, \quad \sigma_y^* = \frac{2\sigma_y - \sigma_x}{3}$$

$d\varepsilon_x$, $d\varepsilon_y$, $d\gamma_{xy}$ = increments of total strains including both elastic and plastic components

σ_x, σ_y, τ_{xy} = average membrane stresses

σ_o = material yield stress

It is assumed that ductile fracture occurs when the total (accumulated) equivalent plastic strain reaches a critical rupture strain which is prescribed by the user in advance, namely

$$\varepsilon_{eq}^{p} \geq \varepsilon_{rcr} \qquad (3)$$

where

ε_{eq}^{p} = total (accumulated) equivalent plastic strain

ε_{rcr} = critical rupture strain

It should be noted that the program will search for only the initiation of rupture in each element and will not account for the effect of any subsequent crack related tearing for the ruptured element.

3. ILLUSTRATIVE EXAMPLES

To verify the accuracy of the CORBHD/FEM program and also to investigate the influence of adjoining structures on the collapse behavior of corrugated bulkheads, the corrugated bulkhead model P90-3 under uniform lateral pressure as tested by the authors[5] was analyzed by using the program.

Three types of structural modeling were considered. Since the original test models are composed of corrugations plus minimal upper / lower stool plates (without adjoining structures), the first type of the calculation model was made to represent the original test model (without adjoining structures) by assuming the simply supported condition at both ends, i.e., no lateral deformation and no rotational restraints. The second and third types of the calculation model were also constructed wherein the original test model is represented with artificial adjoining structures such as those that would be present for the double bottom and deck in a real vessel.

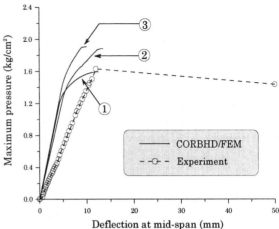

①: Model type II with simply supported condition at both ends
②: Model type V with light adjoining structures
③: Model type V with heavy adjoining structures

Fig. 5 Average pressure load versus deflection at mid-span

In these initial calculations, the plate thickness of the artificial adjoining structures was taken to be either the same as or much larger than that of the corrugation and, as a result the "adjoining structure" rigidity was either small or very large. In the former case it may be expected that due to the weak artificial adjoining structures, while the rotation at the corrugation ends could to some extent be restrained, out-of-plane lateral deformations could occur not only along the span but also at its ends. On the other hand, in the latter case with the strong adjoining structures, out-of-plane lateral deformations as well as the rotation could in some instances be overly restrained at the corrugation ends.

Figure 5 proves these expectations to be true. It is seen from the figure that the numerical solution obtained for the original model (i.e., without adjoining structures) with simply supported edge condition agrees reasonably with the experimental results. There were minor differences in bending stiffness between FEM and experiments due to the fact that in the tests lateral deformations occurred even near the edges while in the analysis model, they could not. The critical rupture strain was assumed to 15%, but no rupture took place until the model reached the ultimate strength.

Figure 5 indicates that adjoining structures such

as deck and bottom structures can potentially affect the stiffness and strength of corrugated bulkheads. If the adjoining structures are light, the lateral deflection at the ends can occur relatively freely, although the rotation at the ends may to some extent be restrained. It should thus be noted that depending on the degree of rigidity of the adjoining structures, the relative accuracy of theoretical and / or numerical procedures which model the corrugated bulkheads assuming idealized (e.g., simply supported) edge conditions will vary compared to reality.

Subsequent to the test model related calculations of Fig. 5, the CORBHD/FEM program was then applied to calculation of the ultimate strength of a transverse corrugated bulkhead of a hypothetical Capesize bulk carrier when subject to the triangular pressure loading pattern in corrugation alone, i.e., with $p_D = 0$. Figure 6 shows examples of related structural deformations at the ultimate limit state. Fig. 7 represents the ongoing relationship between the maximum pressure at the lower end of corrugation and the deflection at the mid-span of corrugation as the analyses proceeded. Based on the results of the above noted calculations for the test model and the Capesize bulkhead (see Fig. 7), the following observations may be made:

1) In a hypothetical case with small rigidity of adjoining structures, lateral deformation could occur at the corrugation ends even though the end rotations are somewhat restrained, the resulting strength and stiffness may in some cases be conceivably smaller than those for the (usual) simply supported edge condition. In general, this would imply that in the direct strength calculations for corrugated bulkheads under lateral pressure, the edge conditions for lateral deformation as well as rotational restraints at both ends should be accounted for correctly.

2) The presence of shedder plates can raise the strength and stiffness of corrugated bulkheads, see Fig. 7. The effects of the shedder plates on the ultimate strength of bulk carrier corrugated bulkheads may thus not be neglected.

3) Due to adjoining structures, i.e., deck and double bottom structures, both the stiffness and the strength of corrugated bulkhead of actual bulk carriers can possibly increase, meaning that rotational restraints due to the adjoining structures are large enough, although the out-of-plane deformation at corrugation ends may be non-zero.

(a) Deformed shape for the type I model

(b) Deformed shape for the type III model with lower end clamped and upper end simply supported

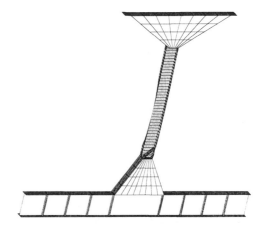

(c) Deformed shape for the type V model

Fig. 6 Deformed shapes for a hypothetical Capesize bulk carrier corrugated bulkhead at the ultimate limit state (under triangular pressure loads acting on corrugation alone)

4) A direct calculation model where both ends of the corrugation are simply supported is one possibility if we wished to avoid modeling the adjoining structures. However, since the lower part of the corrugation is in reality connected to relatively rigid double bottom structures, another possibility may be to assume end conditions which are clamped at the lower end and simply supported at the upper end. To

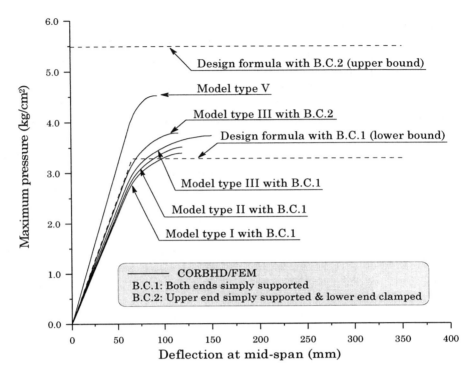

Fig. 7   Maximum pressure load versus deflection at mid-span for a hypothetical Capesize bulk carrier corrugated bulkhead

establish such simplified direct calculation procedures and verify their relative accuracy needs further study, for which the nonlinear finite element program from this study should prove quite useful.

5) In the calculations of Fig. 7, where the critical rupture strain was assumed to be 15%, no ductile fracture appeared in any of the elements until the corrugated bulkhead reached the ultimate strength. The potential for progressive flooding into an adjacent hold is contingent on such ductile fracture, the crack formation related to which apparently appears in the post-ultimate strength regime in this particular case. Because of its specific nature, further study is needed to determine the generality of this particular observation.

4. ULTIMATE STRENGTH DESIGN FORMULATIONS

4.1. Ultimate bending moment formulae

Selected parts of three useful design formulations

for predicting the ultimate bending moment of corrugated bulkheads subject to lateral pressure loads are now noted below. Various application oriented details are omitted, for which the reader should refer to the original sources. Also, it may be noted that the ABS SafeHull formulation has now been largely superseded by the corresponding IACS unified requirements.

ABS SafeHull[2]:

$$M_u = \frac{d}{4}\sigma_o\left(2\,a\,t_f + c\,t_w\right)S_m \times 10^{-3} \quad \text{N-cm} \quad (4)$$

where

t_f, t_w = thickness of the corrugation flange or web

S_m = strength reduction factor

$$= 0.95 - 0.375\left(1.50 - \frac{\sigma_T}{\sigma_o}\right) \le 1.0$$

σ_T = ultimate tensile strength (N/cm²)

Paik, Thayamballi & Chun[6]:

$$M_u = \frac{1}{2}\sigma_{of} \cdot A_f \cdot g + \frac{1}{2}\sigma_{ow} \cdot A_w \cdot \sin\phi \cdot \frac{g^2}{d}$$

$$+ \frac{1}{2}\sigma_{uw} \cdot A_w \cdot \sin\phi \cdot \frac{(d-g)^2}{d}$$

$$+ \frac{1}{2}\sigma_{uf} \cdot A_f \cdot (d-g) \qquad (5)$$

where

A_f = section area of corrugation flange

$= a \cdot t_f$

A_w = cross section of corrugation web

$= c \cdot t_w$

d = vertical height of corrugation web

$= c \cdot \sin\phi$

g = final neutral axis at the ultimate limit state

$$= \frac{d\left[\left(\sigma_{uf} - \sigma_{of}\right)A_f + 2\sigma_{uw}A_w \sin\phi\right]}{2\left(\sigma_{ow} + \sigma_{uw}\right)A_w \sin\phi}$$

ϕ = corrugation angle

σ_{of}, σ_{ow} = yield strengths of corrugation flange, web

σ_{uf}, σ_{uw} = ultimate strengths of corrugation flange, web

IACS[8]:

$$M_u = (0.5 \cdot Z_{le} \cdot \sigma_{a,le} + Z_m \cdot \sigma_{a,m}) \times 10^{-3} \quad \text{kN-m} \qquad (6)$$

where

Z_{le}, Z_m = section modulus of one half pitch corrugation, in cm^3, at the lower end, and midspan of corrugations, respectively

$\sigma_{a,le}, \sigma_{a,m}$ = allowable stresses, in N/mm^2, for the lower end and mid-span of corrugations, respectively

$= \sigma_o$

σ_o = minimum upper yield stress, in N/mm^2, of the material.

4.2. Ultimate pressure load prediction

The above noted formulations provide the ulti-mate bending moments (not the ultimate applied pressure loads) at the ultimate limit state for the corrugation cross section. To predict the corresponding ultimate applied pressure loads it is necessary to get the relationship between the applied pressure loads and the bending moments at the ultimate limit state. In the following, we now derive such relationships using idealized end conditions.

Figure 8 represents a single corrugation that is a proxy for the behavior of a corrugated bulkhead subject to lateral hydrostatic pressure and end moments (e.g., due to rotational restraints of adjoining structures). For simplicity, the corrugation is modeled as a beam with constant cross section (and thus constant bending rigidity). It is assumed that the right end can move freely in the horizontal direction, implying that membrane stresses are not developed.

The lateral hydrostatic pressure distribution with a trapezoidal pattern is assumed to vary linearly between the bottom and deck structures, with the magnitude of pressure per unit length related to one pitch of the corrugation as follows:

$$p = -\frac{p_B - p_D}{L}x + p_B \qquad (7)$$

where

p_B, p_D = lateral pressures at the lower (bottom) or upper (deck) ends of the corrugation

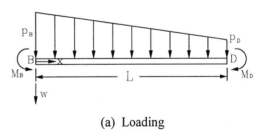

(a) Loading

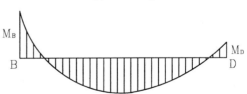

(b) Elastic bending moment distribution

Fig. 8 A single corrugation beam under lateral pressure

The end moments arise from the constraints against angular rotation of the corrugation cross section at the junctures of the bulkhead and the adjoining structure. They thus depend on the torsional rigidity of the adjoining structures.

From the bending moment equilibrium condition, the constraints at the ends of the corrugation can be defined as follows:

at the lower end:

$$\left(\frac{d^2 w}{dx^2}\right)_{x=0} = \frac{C_B}{L}\left(\frac{dw}{dx}\right)_{x=0} \tag{8.a}$$

at the upper end:

$$\left(\frac{d^2 w}{dx^2}\right)_{x=L} = \frac{C_D}{L}\left(\frac{dw}{dx}\right)_{x=L} \tag{8.b}$$

where w is the lateral deflection of the beam, and C_B and C_D are constraint constants at the lower and upper ends of the corrugation, respectively. For simply supported or clamped ends, these two constants will become either zero or infinity, respectively.

The elastic bending moment distribution of the beam is expressed by applying the simple beam theory as:

$$M = EI\frac{d^2 w}{dx^2}$$

$$= M_B - \frac{x}{L}(M_B - M_D) + \frac{p_B}{2}(x^2 - Lx)$$

$$+ \frac{p_B - p_D}{6}\left(Lx - \frac{x^3}{L}\right) \tag{9}$$

where
I = moment of inertia of the beam
E = Young's modulus

Figure 8.b represents the elastic bending moment distribution of the corrugation beam. It is seen from the figure that three extreme values of the bending moments are developed, i.e., at lower end, the upper end and inside the span. By performing the double integration of equation (9) and considering the end

conditions, the lateral deflection may be expressed by

$$w = \frac{p_B + p_D}{24EI}\left(\frac{x^4}{2} - Lx^3 + \frac{L^3 x}{2}\right)$$

$$+ \frac{p_B - p_D}{24EI}\left(-\frac{x^5}{5L} + \frac{x^4}{2} - \frac{Lx^3}{3} + \frac{L^3 x}{30}\right)$$

$$+ \frac{M_B}{EI}\left(-\frac{x^3}{6L} + \frac{x^2}{2} - \frac{Lx}{3}\right) + \frac{M_D}{EI}\left(\frac{x^3}{6L} - \frac{Lx}{6}\right) \tag{10}$$

The end moments at the lower or upper end can be calculated as a function of the constraint coefficients by substituting equation (10) into the equilibrium condition (8) as follows:

$$M_B = \frac{C_B L^2}{120}$$
$$\cdot\left[\frac{(p_B - p_D)(2 - C_D) + (p_B + p_D)(30 - 5C_D)}{12 + 4C_B - 4C_D - C_B C_D}\right] \tag{11.a}$$

$$M_D = \frac{C_D L^2}{120}$$
$$\cdot\left[\frac{(p_B - p_D)(2 + C_B) - (p_B + p_D)(30 + 5C_B)}{12 + 4C_B - 4C_D - C_B C_D}\right] \tag{11.b}$$

The extreme value of the bending moment inside the span will occur at the location where the following condition is satisfied:

$$\frac{dM}{dx} = 0 \tag{12}$$

When both ends and any one point inside the span yield, a collapse hinge mechanism is formed. Depending on the end condition, loading and other details related to the formation of the collapse mechanism will vary. In this study, two types of ideal end conditions, namely 1) both ends simply supported, and 2) the lower end clamped and the upper end simply supported, are now studied in detail.

(1) Condition with both ends simply supported
In this case, the elastic bending moment distribution is represented by Fig. 9.a. The end moments are

zero and the elastic bending moment M along the beam can be obtained by simplifying equation (9) as

$$M = \frac{p_B}{2}(x^2 - Lx) + \frac{p_B - p_D}{6}\left(Lx - \frac{x^3}{L}\right) \quad (13)$$

Also, the maximum bending moment M_{max} is developed at the location given by $x = x_p$ where:

$$x_p = \frac{L}{2} \quad \text{when } p_D = p_B \quad (14.a)$$

$$x_p = \frac{L}{p_B - p_D}\left(p_B - \sqrt{\frac{p_B^2 + p_D^2 + p_D p_B}{3}}\right)$$
$$\text{when } p_B > p_D \quad (14.b)$$

It is considered that the beam collapses if a plastic hinge is formed at any one point inside the span since both ends are already pin joined like a plastic hinge. The in-span plastic hinge is formed when the maximum bending moment reaches the ultimate bending moment of the beam cross section since the corrugation flange / web in compression buckles and the corrugation flange / web in tension yields such that:

$$M_{max} = M_u \quad (15)$$

where M_u = the ultimate bending moment

Substituting equation (13) together with equation (14) (leading to the maximum bending moment) into equation (15), the following expression representing the relationship between the applied pressure loads and the bending moment at the ultimate limit state can be obtained:

$$M_u = \frac{p_{Bu}}{2}(x_p^2 - Lx_p) + \frac{p_{Bu} - p_{Du}}{6}\left(Lx_p - \frac{x_p^3}{L}\right) \quad (16)$$

Further, for the triangular pressure loading pattern, i.e., when $p_D = 0$, the ultimate pressure loads (at the lower end) may be obtained from equation (16) as

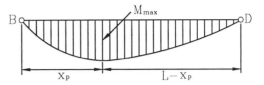

(a) Elastic bending moment distribution

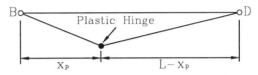

(b) Collapse hinge mechanism

Fig. 9 Moments and collapse mechanism for a corrugation beam with both ends simply supported

$$p_{Bu} = \frac{6L}{x_p(3Lx_p - 2L^2 - x_p^2)} \cdot M_u \quad (17)$$

(2) Condition with lower end clamped and upper end simply supported

A corrugation beam with one end clamped and the other end simply supported condition is now considered, see Fig. 10.

The moment diagram for this case will take the form shown in Fig. 10.a as long as the beam behaves elastically. The elastic bending moment M along the beam can be shown to be obtained by

$$M = \frac{7p_D + 8p_B}{120}(Lx - L^2) + \frac{p_D + 2p_B}{6}Lx$$
$$- \frac{p_B \cdot x^2}{2} + \frac{p_B - p_D}{6L}x^3 \quad (18)$$

The elastic bending moment at the clamped end B will be given by substituting $x = 0$ into equation (18) as follows:

$$M_B = \frac{7p_D + 8p_B}{120}L^2 \quad (19)$$

Also, the maximum elastic bending moment M_e inside the span may be developed at the location

122

where $x = x_e$ such that:

$$x_e = \frac{5L}{8} \qquad \text{when } p_D = p_B \qquad (20.a)$$

$$x_e = \frac{L}{p_B - p_D}\left(p_B - \sqrt{\frac{4p_B^2 + 9p_D^2 + 7p_D p_B}{20}}\right)$$

$$\text{when } p_B > p_D \qquad (20.b)$$

The maximum elastic bending moment M_e inside the span can then be estimated by substituting equation (20) into equation (18). The value of the bending moment at the clamped left end is always greater than that inside the span as long as $p_B \geq p_D$. Therefore, it can be said that a plastic hinge would in this case occur first at the lower clamped end where the largest bending moment develops. The beam can form a plastic hinge mechanism if the clamped end and any one point inside the span yield. However, the location of the plastic hinge inside the span is not so straightforward to determine.

Figure 10.b shows a possible plastic hinge mechanism at the ultimate limit state. The distance from the left end to the plastic hinge inside the span is denoted by x_p which is unknown. Applying the virtual work theorem, the external and internal work(s), denoted by W_E and W_I, respectively, may be estimated by

$$W_E = \int_0^L p_u \, w \, dx$$

$$= \int_0^{x_p} \left(-\frac{p_{Bu} - p_{Du}}{L}x + p_{Bu}\right) \cdot \frac{w_p}{x_p} x \, dx$$

$$+ \int_{x_p}^L \left(-\frac{p_{Bu} - p_{Du}}{L}x + p_{Bu}\right) \cdot \frac{w_p}{L - x_p}(L - x) \, dx$$

$$= \frac{w_p\left\{(L^2 - x_p^2)p_{Du} + (2L^2 - 3Lx_p + x_p^2)p_{Bu}\right\}}{6(L - x_p)}$$

$$(21)$$

$$W_I = M_u(\theta_1 + \theta_3) = M_u(2\theta_1 + \theta_2) \qquad (22)$$

where

p_u = pressure at the ultimate limit state

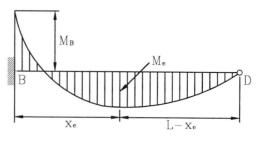

(a) Elastic bending moment distribution

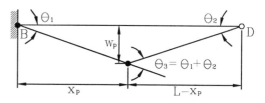

(b) Collapse hinge mechanism

Fig. 10 Moments and collapse mechanism for a corrugated beam with lower end clamped and upper end simply supported

p_{Du}, p_{Bu} = ultimate pressure at the upper or lower end

w_p = deflection at the plastic hinge location

From geometric inter-relationships in Fig. 10.b, the following relationships can be shown:

$$\theta_2 = \frac{x_p}{L - x_p} \cdot \theta_1 \qquad (23)$$

$$w_p = x_p \cdot \theta_1 \qquad (24)$$

Substituting equations (23) and (24) into equations (21) and (22), the external and internal work(s) are then expressed as function of θ_1:

$$W_I = \frac{2L - x_p}{L - x_p} \cdot M_u \theta_1 \qquad (25)$$

$$W_E = \frac{x_p\left\{(L^2 - x_p^2)p_{Du} + (2L^2 - 3Lx_p + x_p^2)p_{Bu}\right\}}{6(L - x_p)}\theta_1 \qquad (26)$$

Upon equating the external work and the internal work, one can then show that

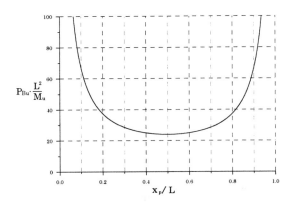

Fig. 11 Ultimate pressure load p_{Bu} varying the location of the plastic hinge inside the span

$$M_u = \frac{x_p \left\{ (L^2 - x_p^2) p_{Du} + (2L^2 - 3Lx_p + x_p^2) p_{Bu} \right\}}{6(2L - x_p)}$$ (27)

When $p_D = 0$, i.e., for a triangular loading pattern, the ultimate pressure load p_{Bu} at the lower end can be estimated by

$$p_{Bu} = \frac{12L - 6x_p}{2L^2 x_p - 3Lx_p^2 + x_p^3} \cdot M_u$$ (28)

Figure 11 shows the variation of the ultimate pressure load as a function of the distance x_p of the plastic hinge location inside the span. The minimum value of the ultimate pressure load which occurs approximately around $x_p = 0.5L$ is as follows:

$$p_{Bu} = 24 \frac{L^2}{M_u}$$ (29)

4.3. Comparison with FEA and discussion

The ultimate pressure loads as obtained above may now be compared with the CORBHD/FEM solutions as shown in Fig. 7. The ultimate pressure loads indicated were obtained by substituting the ultimate bending moments (i.e., equations (4), (5) or (6)) into equation (17), for the case of the both ends simply supported condition, or into equation (29),

for the case of the lower end clamped and the upper end simply supported condition.

It is seen from Fig. 7 that the ultimate pressure loads of the hypothetical bulk carrier corrugated bulkhead, obtained by the CORBHD/FEM program considering the rigidity of real adjoining structures, fall somewhere between the two bounding approximate solutions mentioned above. In fact, the present CORBHD/FEM solutions have been obtained for the load case of Fig. 4.b, i.e., pressure loads acting on corrugation alone. Considering the load case of Fig. 4.a together with the water pressure acting on the outer bottom panel, the ultimate strength can approach the lower bound because the effectiveness on rotational restraints of bottom adjoining structures may decrease. From a strength point of view alone, the condition with both ends simply supported may in some cases be appropriate. Additional needed investigations are continuing in this regard, to better assess the "carry-over" effects from adjoining structure.

5. CONCLUDING REMARKS

As a consequence of several bulk carrier losses during the last decade, the safety assessment of bulk carrier corrugated bulkheads has been of interest, and most recently, bulk carrier corrugated bulkhead requirements have been enhanced by IMO and IACS.

The enhanced requirements are aimed at predicting the ultimate strength of corrugated bulkheads subject to accidental flooding in a relatively simple and reasonably accurate manner, and do suffice for that purpose. In principle, the corrugations have adjoining structures at their ends, which may affect the stiffness and strength of the bulkheads so constructed, and the actual behavior in particular cases could differ from the predictions obtained from design formulae based on idealized end conditions. Hence in any sophisticated direct analysis, the influence of adjoining structures on the collapse strength of corrugated bulkheads will need to be more rigorously accounted for.

The aim of the present study has been to develop and validate related methods to numerically predict the collapse strength characteristics of bulk carrier corrugated bulkheads subject to accidental flooding. As part of the study, a hypothetical Capesize bulk carrier corrugated bulkhead of varying the end condition was analyzed by a special purpose nonlinear finite element program CORBHD/FEM. As would by expected, the adjoining structures do variously affect the strength and stiffness of corrugated bulkheads. For the hypothetical Capesize bulk carrier corrugated bulkhead subject to accidental flooding, limited calculations indicate that ductile rupture may not occur until the ultimate limit state is reached. It is worth noting here that while this study has contributed a valuable tool and several preliminary results, the example calculations have been limited. The generality of the observations noted regarding structural behavior thus remain to be established. Additional studies are on going to more fully investigate the "carry-over" effects from the adjoining structure in a wider range of cases with certainty.

ACKNOWLEDGMENTS

The authors wish to thank the American Bureau of Shipping and the Pusan National University for supporting this joint study. The views expressed in this paper are those of the authors and not necessarily of the institutions they are affiliated with.

REFERENCES

1. RINA, Derbyshire- The search, assessment and survey, The Naval Architect, May 1996, pp. 44-48.

2. ABS, Guide for dynamic based design and evaluation of bulk carrier structures, American Bureau of Shipping, SafeHull Project, March 1995.

3. K. Frystock and J. Spencer, Bulk carrier safety, Marine Structures, Vol. 33, No. 4, October 1996, pp. 309-318.

4. RINA, New bulk carrier rules confirmed by IMO, The Naval Architect, February 1998, pp. 33.

5. J.K. Paik and A.K. Thayamballi, The strength and reliability of transverse bulkheads and hull structure of bulk carriers, Proceedings of International Conference on Design and Operation of Bulk Carriers, The Royal Institution of Naval Architects, London, April 30 & May 1, 1998.

6. J.K. Paik, A.K. Thayamballi and M.S. Chun, Theoretical and experimental study on the ultimate strength of corrugated bulkheads, J. of Ship Research, Vol. 41, No. 4, December 1997, pp. 301-317.

7. J.K. Paik and A.K. Thayamballi, CORBHD/FEM: A computer program for the ultimate strength of bulk carrier corrugated bulkheads, Final Report, Department of Naval Architecture and Ocean Engineering, Pusan National University, Pusan, Korea, May 1997.

8. IACS, Evaluation of scantlings of the transverse watertight corrugated bulkhead between cargo holds Nos. 1 and 2, with cargo hold No. 1 flooded, for existing single side skin bulk carriers, S19, Rev. 1, International Association of Classification Societies, 1997.

Ultimate strength formulation for ship's grillages under combined loadings

S.-R. Cho[a], B.-W. Choi[a] and P. A. Frieze[b]

[a] School of Transportation Systems Engineering, University of Ulsan
Nam-Ulsan P.O. Box 18, Ulsan, 680-749, Korea

[b] PAFA Consulting Engineers,
Hofer House, 185 Uxbridge Road, Hampton TW12 1BN, UK

Grillages are one of the major structural components of various onshore and offshore structures. Generally plates of ship structures are stiffened longitudinally by stiffeners of relatively small size and transversely by girders of larger size. There are many kinds of strength formulations proposed for predicting the ultimate strength of stiffened panels, but few are applicable to grillages.

In this study, a robust ultimate strength formulation has been developed for grillages subjected to combined axial compression, end bending moment and lateral pressure loadings. The so-called the generalised Merchant-Rankine formula is adopted as its basis. For the derivation of the knock-down factors in the formula forty five multi-bay test data have been used exstracted from the open literature and another two hundred and forty four test data have been utilised for the comparison study.

The predictions using the proposed formulation provide improved accuracy compared with other existing approaches. When using the proposed formulation, not only are the scantlings of the plate and longitudinal stiffeners considered but also those of the transverse girders. The numbers of longitudinal stiffeners and transverse girders can be addressed as well. Therefore, with the proposed formulation, optimisation can be performed on all the design variables for grillage structures.

1. INTRODUCTION

Generally most parts of ship structures and the topside decks of offshore structures consist of cross stiffened panels, i.e. grillages. Therefore in designing any efficient marine structures it is essential to employ more accurate strength formulations for grillages instead of those just for panels.

There have been many kinds of strength formulations proposed to predict the ultimate strength of stiffened panels, but few are for grillages. For combined axial compression and lateral pressure loadings, Faulkner[1] proposed a linear interaction equation for beam-column failure and a parabolic interaction one for tripping failure. Hughes[2] suggested a method to predict the failure strength of stiffened panels under combined axial compression and end bending moment. In the method he distinguished the failure modes into three categories, i.e. compression failure of the stiffener, compression failure of the plating and combined failure of the stiffener and plating.

However, the boundary conditions for a single panel of multi-bay grillages are different from those of either simply supported or fixed boundaries, which are frequently assumed in existing methods. The effects of boundary conditions on the ultimate strength of stiffened panels were numerically investigated by Smith[3]. He compared the ultimate strengths of single bay panels having simply supported and fixed ends and of two span models. It was concluded that the flexural rigidity provided by adjacent bays cannot be assumed to be those of either simply supported or fixed in ultimate strength calculations.

For onshore structures various design formulations for stiffened panels have been proposed and adopted in codes and recommendations. Recently Brosowski and Ghavami[4] compared various simple formulations for the design of stiffened plates with the results of tests on seventeen models. In all the formulations they investigated the tripping failure of stiffeners was not explicitly considered. The accuracy and variation of the predictions were quite poor and Murray's method[5] turned out to be the best among them.

There are basically three different approaches to derive ultimate strength formulations for stiffened panels. One is based on the so-called effective width concept. Formulations and procedures proposed in refs 1, 2 and 5-7 are examples of this approach. Beam-column failure strength is predicted using the Perry formula in all except that in ref. 1 The second approach is to derive formulations by regression analysis of test data adopting the plate and stiffener column buckling slenderness parameters as design variables. This type of design equation is proposed mainly for ship structures and examples can be found in refs. 8-10. The third is based on an interaction equation like the Rankine formula. Allen[11] proposed a general interaction formula for stiffened panels considering the interaction of different failure modes. This kind of approach has been successfully applied to ring- stiffened [12] and stringer-stiffened cylinders[13] by the authors

In this paper a procedure is described to derive an ultimate strength formulation for ship's grillages subjected to combined axial compression, end bending moment and lateral pressure. This formulation is based on the generalised Merchant-Rankine formula[14] in which all possible failure modes and their interactions can be considered. The accuracy and consistency of the proposed design equation is discussed. The effects of stiffened panel boundary conditions on ultimate strength are also investigated.

2. RANKINE FORMULAE

In order to derive any versatile design formulation for marine structures, the following phenomena have to be represented therein.

- interaction between yielding and elastic buckling,
- multi-modal elastic buckling and
- effects of combined loadings

Of these, interaction between yielding and elastic buckling can be considered by adopting inelastic buckling formulae[12]. However, the other two phenomena have not been fully accounted for in existing design formulations, which may increase the uncertainty associated with their strength predictions. The generalised Merchant-Rankine formula is of potential use in dealing with these two aspects. At this juncture it seems useful to historically review the Rankine formulae.

In 1866 Rankine[15] proposed an empirical column formula for long struts and pillars as follows:

$$\sigma_u = \frac{f}{1 + \dfrac{4\,l^2}{c\,r^2}} \qquad (1)$$

where σ_u is the failure stress

l is the length of the column

r is the radius of gyration $(= \sqrt{I/A})$ of the cross-section of the column f, c are coefficients depending on the material,

If it is assumed that f and c equal the yield stress of the material and $4\pi^2 E/\sigma_Y$ respectively, eqn (1) can be rewritten as,

$$\frac{\sigma_u}{\sigma_{cr}} + \frac{\sigma_u}{\sigma_Y} = 1 \qquad (2)$$

where

σ_{cr} is the Euler buckling stress of the column, $(\pi^2 E)/(l/r)^2$

σ_Y is the yield stress of the material

The formula given as eqn (2) is the Rankine formula. In order to incorporate other types of structures, where shape imperfections may reduce their elastic buckling stress, the Rankine formula has been modified as follows.

$$\frac{\sigma_u}{\rho\,\sigma_{cr}} + \frac{\sigma_u}{\sigma_Y} = 1 \qquad (3)$$

where ρ is a knock-down factor

In eqn (3) the deteriorating effect of initial shape imperfections on the elastic buckling stress of an ideal structure is assumed to be estimated by the knock-down factor, ρ. The Rankine formula in this modified form indicates the interaction between elastic buckling and yielding is linear. This equation can provide a lower bound for the ultimate strength for some types of structure.

When aiming to predict the mean value of ultimate strength the quadratic interaction between elastic buckling and yielding is more adequate. This quadratic interaction equation, eqn (4), is called the Merchant-Rankine formula[16].

$$\left(\frac{\sigma_u}{\rho\,\sigma_{cr}}\right)^2 + \left(\frac{\sigma_u}{\sigma_Y}\right)^2 = 1 \qquad (4)$$

Allen[16] proposed a modified form of the Rankine formula to consider the local buckling of thin-walled columns as follows:

$$\left(\frac{\sigma_u}{\sigma_Y}\right)^n + \left(\frac{\sigma_u}{\sigma_{cr}}\right)^n + \left(\frac{\sigma_u}{\sigma_{crl}}\right)^n = 1 \qquad (5)$$

where

σ_{crl} is the local elastic buckling stress

n is an imperfection index

Apart from the original intention, eqn (5) can be adopted more generally to take into account different collapse modes under a single loading. However, this equation can over-emphasize the effects of elastic buckling on the ultimate strength.

Recently Odland and Faulkner[14] generalised the Merchant-Rankine formula for 2-dimensional stress states as eqn (6). This formulation has been adopted as the basis of the DNV rules for combined loadings.

$$\left[\frac{\sigma_{xo}}{\rho_x\sigma_{xcr}} + \frac{\sigma_{\theta o}}{\rho_\theta\sigma_{\theta cr}}\right]^2 + \left[\frac{\sqrt{\sigma_x^2 - \sigma_x\sigma_\theta + \sigma_\theta^2}}{\sigma_Y}\right]^2 = 1 \qquad (6)$$

where

σ_x, σ_θ are the axial and average hoop stress respectively

ρ_x, ρ_θ are the axial compression and radial pressure knock-down factors respectively

In eqn (6) the von Mises yield criterion is adopted to monitor the collapse of stocky structures whilst the elastic buckling stress subjected to combined loading is estimated by the linear sum of each component. This equation also accounts for axial tension or internal pressure.

3. NEW FORMULATION

Prior to deriving any ultimate strength formulations for grillages, i.e. multi-bay stiffened plates it is necessary to identify which are their ductile failure modes. The ductile failure modes for grillages can be categolised as follows:

- plate local buckling
- stiffener/plate column buckling between transverse girders
- stiffener tripping
- girder/plate column buckling between longitudinal bulkheads
- girder tripping
- overall grillage buckling;general instability

Among the above the occurrence of the plate local buckling does not imply the failure of the stiffened plates unless the scantling of the stiffeners are extraordinary small, but it may reduce the effective width of the plate. In this paper combined longitudinal axial compression, end bending moment and lateral pressure loadings are considered. Therefore, girder column buckling and its torsional buckling ned not to be considered further because these modes can only occur under transverse compression. General instability causes the collapse of the whole structure and in practice it mainly depends on the scantlings of the girders.

3.1 Basis of the proposed formulation

As mentioned earlier the generalised Merchant-Rankine formula is adopted as the basis of the new ultimate strength formulation. The equation for the new formulation can be written as follows:

$$\left(\frac{\sigma_{xao}}{\rho_c \, \sigma_{ec}} + \frac{\sigma_{xao} + \sigma_{xbso}}{\rho_t \, \sigma_{et}} + \frac{\sigma_{xao}}{\rho_{oa} \, \sigma_{eoax}} \right)^2$$

$$+ \left(\frac{\sigma_{xa} + \sigma_{xb}}{\sigma_Y} \right)^2 = 1 \qquad (7)$$

where

σ_{xa} is applied axial compression stress

σ_{xbs} is bending stress at stiffener flange due to M_{eq},

$\sigma_{xao} = \sigma_{xa}$ for $\sigma_{xa} < 0$,

$\quad = 0$ for $\sigma_{xa} > 0$

$\sigma_{xbso} = \sigma_{xbs}$ for $\sigma_{xbs} < 0$,

$\quad = 0$ for $\sigma_{xbso} > 0$

σ_{ec} is Euler column buckling stress of stiffener including associate plating

$\sigma_Y = (\sigma_{Yp} \, A^p + \sigma_{Ys} \, A^s)/A_{ps}$, mean yield stress

$\sigma_{et} = \frac{1}{I_o}(GJ + \frac{4\pi^2}{L^2} EC_w)$,

elastic tripping stress of stiffener

$\sigma_{xb} = \sigma_Y \times \frac{M_{eq}}{M_P}$, equivalent

bending stress due to end bending moment and lateral pressure

$M_{eq} = M_e + \frac{pbl^2}{16}$

M_e is applied end bending moment

p is applied lateral pressure

$J = (h_{sf} \, t_{sf}^3 + h_{sw} \, t_{sw}^3) / 3$,

St. Venant torsion constant

$I_o = I_w + A_s e_s^2 + I_f$,

moment of inertia of the stiffener

I_w is polar moment of inertia of stiffener web

I_f is polar moment of inertia of stiffener flage

e_s is distance between stiffener centroid (plate excluded) and its toe

$C_w \doteqdot I_f (\frac{h_w + t_f}{2})^2$,

torsional warping constant

$\sigma_{eoax} = \frac{n^2\pi^2 D_y}{a_x B^2} \left[\frac{D_x B^2}{D_y L^2} + \frac{2m^2 D_{xy}}{n^2 D_y} + \frac{m^4 L^2}{n^4 B^2} \right]$,

overall grillage buckling stress

L, B are overall length and breadth (see Fig. 1)

a_x is the average cross-sectional area per unit width of plating and longitudinal stiffeners

D_x, D_y are effective flexural rigidities per unit width of stiffeners with attached plating in longitudinal(x) and transverse(y) directions

D_{xy} is the twisting rigidity per unit width

3.2 Knock-down factors

Having decided the basis of the formulation, the remaining task is to determine the knock-down factors in eqn (7). The knock-down factor for overall grillage buckling, ρ_{eoax}, is assumed to be unity. The other two knock-down factors, ρ_{ec} and ρ_{et}, for stiffener column buckling and tripping respectively, are determined by regression analysis using forty-five multi-bay test data provided in refs. 3, 4 and 18-20. The characteristics of the test data are summarised in Table 1. The parameter ranges are typical of those of ship plates.

Table 1.
Ranges of parameters of test data

ref.	no. of test data	β	λ_c	λ_t	E/σ_Y
3	11	1.42-3.67	0.23-0.75	0.27-0.68	763-827
4	9	0.53-1.31	0.06-0.24	0.11-0.28	689-867
18	7	3.13-3.42	1.24-1.89	0.29-1.09	790
19	12	1.42-1.62	0.33-0.35	0.49-0.53	631-735
20	6	1/87-3.49	0.56-0.70	0.74-0.88	1133-1323
total	45	0.53-3.67	0.06-1.89	0.11-1.09	631-1329

Among the eleven test data from ref. 3 four were tested under combined axial compression and lateral pressure loadings, but the others were under pure axial compression. The slenderness parameters in the table are defined as follows:

β is the plate slenderness, $(b/t)\sqrt{\sigma_Y/E}$

λ_c is the slenderness of the stiffener for column buckling, $\sqrt{\sigma_Y/\sigma_{ec}}$

λ_t is the slenderness of the stiffener for tripping, $\sqrt{\sigma_Y/\sigma_{et}}$

The expressions for the knock-down factors finally determined are as follows:

$$\rho_c = 2.49 \, \beta^{-0.7} \, \lambda_c^{1.2} \, \lambda_t^{0.1},$$

knock-down factor for column buckling of stiffener

$$\rho_t = 2.69 \, \beta^{-0.3} \, \lambda_c^{0.8} \, \lambda_t^{1.1},$$

knock-down factor for tripping of stiffener

$$\rho_{oa} = 1.0,$$

knock-down factor for overall grillage buckling

The detailed procedures to derive the equations for the knock-down factors are described in ref. 21 and details of the test data used in this study are also given therein.

4. DISCUSSION

The newly derived ultimate strength formulation for grillages under combined axial compression, end bending moment and lateral pressure has been substantiated using forty-five grillage(multi-bay stiffened plates) test data. The results are summarised in Table 2. In the table the predictions for single-bay models are also provided: these will be discussed later.

The accuracies of some other existing design formulations are given in Table 3 for the forty-one axial compression test data. Comparing the accuracy of the proposed formulation given in Table 2 with those in Table 3 it can be seen that improvements have been achieved not only for the mean

X_m but also in the COV.

The skewness check of the predictions using the proposed formulation has been performed against various parameters. However, no apparent skewness can be observed, which confirms the robustness of the proposed formulation.

Table 2.
Accuracy of the proposed formulation

boundary condition	loading type	no. of test data	X_m	
			mean	COV
multi-bay models	axial comp.	41	0.990	14.0%
	a. c. + l. p.	4	1.105	6.2%
	total	45	1.000	13.7%
single-bay models – fixed ends	axial comp.	51	1.228	26.4%
	a. c. + l. p.	4	1.005	20.4%
	total	55	1.212	26.5%
single-bay models – simply supports	axial comp.	118	0.920	18.6%
	a. c. + e. b.	71	0.951	20.5%
	total	189	0.932	19.4%

In order to investigate the effects of boundary conditions on the ultimate strength of stiffened panels the predictions using the proposed formulation were calculated for single-bay models having fixed ends and simply supports. The results are summarised in Table 2. As can be seen in the table the predictions are not as good as those for the multi-bay models. In Fig. 2 the predictions for single-bay models are plotted against the stiffener column buckling slenderness ratio, λ_{c}. Apparent skewness can be seen in the figures, which indicates for single-bay panels the possible necessity for a different design formulation from that for multi-bay stiffened plates.

Table 3.
Accuracy of some existing formulations for axial compression test data

formulation	X_m	
	mean	COV
ref. 1	1.016	18.7 %
ref. 6	1.320	39.1 %
ref. 7	1.072	20.8 %
ref. 8	0.976	22.0 %
ref. 9	0.943	18.2 %
ref. 10	0.976	20.9 %

5. CONCLUSIONS

In this study a robust ultimate strength formulation has been developed for grillages subjected to axial compression, end bending moment and lateral pressure adopting the Merchant-Rankine formula as its basis. The accuracy of its predictions improves upon that of other existing approaches. It is found that the boundary conditions for a single panel of multi-bay grillages are different from those of isolated panels with either simply supports or fixed ends. Therefore, it would be appear that one single strength formulation cannot necessarily applied to multi- and single-bay panels for accurate predictions. To improve the accuracy of the proposed formulation is seems necessary to refine the tripping stress for axial compression and for end bending moment and lateral loading. Furthermore in order to make the formulation more versatile it needs to include transverse compression for combined loadings.

REFERENCES
[1] Faulkner, D. 'Strength of welded grillages under combined loads', chap. 22 in <u>Ship structural design concepts</u>, ed. Evans, J.H. Cornell Maritime Press, Cambridge,

1975.

[2] Hughes, O.F. 'Ship structural design: a rationally-based,computer-aided, optimization approach', chap. 14, John Wiley & Sons, New York, 1983.

[3] Smith, C S, 'Compressive strength of welded steel grillages', Trans. RINA vol. 117, pp 325-359, 1975.

[4] Brosowski, B. and Ghavami, K. 'Multi-criteria optimal design of stiffened plates: Part 1. choice of the formula for the buckling load', Thin-Walled Structures, vol. 24, pp 353-369, 1996.

[5] Murray, N W, 'Analysis and design of stiffened plates for collapse load', The Structure Engineer, vol 53, no 3, pp 153-158, 1975.

[6] Carlsen, C A, 'A parametric study of collapse of stiffened plates in compression', The Structural Engineer, Vol 58B, No 2, June 1980, 33-40.

[7] Det Norske Veritas 'Buckling strength analysis', Classification Notes no. 30.1, 1992.

[8] Lin, Y. T. 'Ship longitudinal strength modelling', Ph D Thesis, Dept. of Naval Architecture and Ocean Engineering, Univ. of Glasgow, 1985.

[9] Lee, J. S. 'Reliability analysis of continuous structural systems', Ph D Thesis, Dept. of Naval Architecture and Ocean Eng'g, Univ. of Glasgow, 1989.

[10] Paik, J.K. and Lee, J.M. 'An empirical formulation for predicting the ultimate compressive strength of plates and stiffened plates' Trans. of SNAK, vol 33, no 3, pp 8-21, 1996(in Korean).

[11] Allen, D., Discussion of 'An approximate method for the design of stiffened steel compression panels' by Horne, M.R. and Narayanan. Proc. Inst. Civil Engrs, vol. 61, no. 2 pp 453-455, 1976.

[12] Cho, S-R and Frieze, P.A. 'Strength formulation for ring-stiffened cylinders under combined axial loading and radial pressure', Jour. of Construct. Steel Research, vol. 9, pp 3-34, 1988.

[13] Cho, S-R and Lee, S-B 'New design formulation for predicting the ultimate strength of stringer-stiffened cylinders', Proc. the 9th Asian Tech. Exchange and Advisory meeting on Marine Structures, Hiroshima, pp 337-356, 1995.

[14] Odland, J. and Faulkner, D 'Buckling of curved steel structures – design formulations', in Integrity of offshore structures, eds. Faulkner, D. et al., Applied Science Publisher, London, pp 419-443, 1981.

[15] Rankine, W.J.M. 'Useful rules and tables relating to mensuration, engineering, structures and machines', Charles Griffin and Company, London, pp 210-211, 1866.

[16] Merchant, W. 'The failure load of rigid jointed frameworks as influenced by stability', Structural Engineer, Jour. Inst. of Engg. vol. 32, no. 7, pp 185-190, 1954.

[17] Allen, D. 'Merchant-Rankine approach to member stability', Jour. of Struc. Div., ASCE, vol. 104, no. ST12, 1978.

[18] Bell, A.O., Viner, A.C. and Richardson, W.S. 'Beam and column strength of stiffened plates', Trans. RINA, vol. 105, pp 435-466, 1963.

[19] Dorman, A.P. and Dwight, J.B. 'Test on stiffened compression plates and plate panels', Proc. of the Conf. on Steel Box Girder Bridges, London, Instn of Civil Engineers, pp 63-76, 1973.

[20] Murray, N.W. and Katzer, W. 'The collapse behaviour of stiffened steel plates which have high aspect ratios under axial compression', in Instability and plastic collapse of steel structures, ed. Morris, L.J., Granada, London, pp 432-445, 1983.

[21] Cho, S-R, Choi, B-W and Frieze, P.A. 'Derivation of ultimate strength formulation for grillages subjected to combined axial compression, end bending moment and lateral pressure', Div. of Naval Arch. and Ocean Eng'g, School of Transportation Systems Engineering, Univ. of Ulsan, report no. DNAOE-98-02, 1998(in Korean).

132

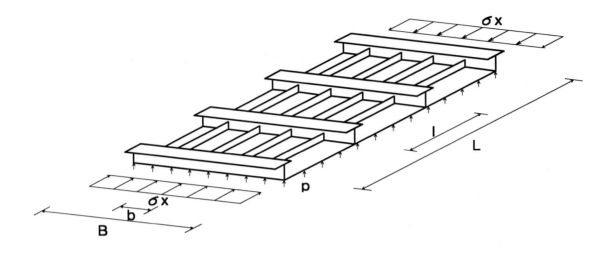

Fig. 1 Graphical notations for grillage

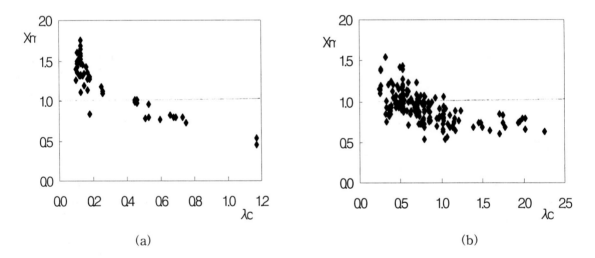

(a)   (b)

Fig. 2 Skewness of the predictions using the proposed formulation for single-bay models
(a) models with fixed ends, (b) models with simply supports

Collision resistance and fatigue strength of New Oiltanker with Advanced Double Hull Structure

J. W. Lee[a], H. Petershagen[b], J. Rörup[b], H. Y. Paik[a], and J. H. Yoon[c]

[a]Inha University, 402-751 Inchon, Korea

[b]Tech. Uni. Hamburg-Harburg, Lämmersieth 90, 22305 Hamburg, Germany

[c]Samsung Heavy Industries, 656-800 Koje Kyungnam, Korea

Based on a new concept of collision energy absorbing system, two types of New Oiltankers with Advanced Double Hull Structure (NOAHS and NOAHS II) are presented through the joint-research program between Inha and Hamburg-Harburg University. A comparative study on collision resistance of these proposed side structures with standard double hull structure of 310 KDWT class VLCC, is carried out. The fatigue investigation of structural detail parts is also included. It contains a comparative fatigue study based on pertinent regulations of two Classification Societies.

1. INTRODUCTION

Collision resistance of oiltanker involves basically two major areas; the structural behavior of the double side hull structure with its components and the rigidity of colliding bow structure. As the installation of soft bow structure is far from the practical application, it seems more attractive to improve the energy absorption capacity of double hull structure. Based on the joint research program on the development of double hull structure of oiltanker between Inha and Hamburg-Harburg University, two types of new oiltankers with advanced double hull structure (NOAHS and NOAHS II) are presented to demonstrate the crashworthiness and a comparative study of these types of new oiltankers is carried out with a standard double hull structure of 310 KDWT VLCC. Finally, according to the pertinent regulation of two Classification Societies, GL and DNV, the fatigue strength of structural detail parts is also investigated.

The joint research program has been carried out under the sponsorship of KOSEF, Korea and DFG, Germany

2. NEW OIL TANKER WITH ADVANCED DOUBLE HULL STRUCTURE (NOAHS & NOAHS II)

The midship tank structure of standard 310 KDWT VLCC is shown in Figure 1, which gives the reference collision resistance against side collision. The main characteristic of standard and new VLCC is listed in Table 1.

Table 1
Characteristic of standard and new VLCC

Width of Double Side	3380mm
Depth of Double Bottom	3000mm
Transverse Spacing	5080mm
Floor Spacing	5080mm
Typical Longi. Stiff. Spacing	900mm
Material of Deck Structure	YP320 HT Steel
Material of Double Side Structure	Mild Steel
Material of Inner Bottom Structure	Mild Steel
Material of Bottom Structure	YP320 HT Steel

134

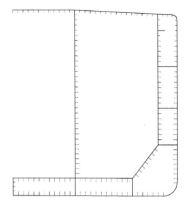

Figure 1. Midship section of
standard VLCC

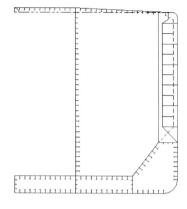

Figure 2. Midship section
of NOAHS

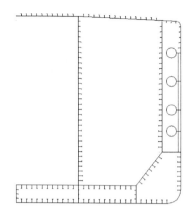

Figure 3. Midship section of
NOAHS II

One type of advanced double hull structure (NOAHS) is composed of crushing and tension plate strips of 11.5mm thickness, which are arranged alternatively in vertical direction as shown in Figure 2. The transverse structure of the double side hull consists of an alternate arrangement of frames on the side shell "side frames" and on the side longitudinal bulkhead "inner frames". A side frame is part of a transverse ring "trans ring", which extends over the whole ship breadth, except the side longitudinal bulkhead. The additional inner frames, which exist only in the double side hull, stiffen the side longitudinal bulkhead. The inner frames are located at a distance of half a frame spacing from the trans rings. The inner frames and the shell frames have a small web depth and large flange width. Five tween-decks in the double side hull act as " collision crushing members and between the shell frames and the inner frames extend four stringers acting as " collision tension spring members", which should act collision energy absorbing system of "NOAHS". The another type of proposed double hull structure is "NOAHS II", which is composed of four large tubes in vertical direction with 1800mm diameter and 40mm thickness, as shown in Figure 3. The numerical simulation for ship collision and crushing damage of the hull

structures are carried out, where the crushing behavior of the plate and tube members in energy absorbing process are analyzed and tested [1-2].

3. EVALUATION OF COLLISION RESISTANCE

3.1 Collision Scenario
The principal dimensions of two ships which are participated in collision scenario, are listed in Table 2.

Table 2
Principle dimensions of two ships in collision scenario

Principle dimension		Struck ship	Striking Ship
Length	[m]	318	264
Breadth	[m]	58	47.8
Depth	[m]	31.25	22.8
Draft	[m]	21.4	14.6
Displacement	[ton]	310,000	150,000

Striking ship collide the center of struck ship's normally. Struck ship's displacement is 310 KDWT and striking ship's one is 150

KDWT Tanker in ballast condition. Observing the object of this study; the development of efficient hull structure in energy absorption, we adopt the orthogonal collision case, as collision scenario. We select two different cases in collision point of the struck ship, NOAHS, as follows.

- Case 1
  : Striking ship collides directly against center of side frame of NOAHS
- Case 2
  : Striking ship collides directly against the midpoint of two side frame of NOAHS

In this case hydrodynamic force is neglected, due to the simplification of structural analysis of developed struck oiltanker. In the analysis, struck ship is modeled by one tank length, and total number of the elements are approx. 16,000. In the striking ship, bulbous bow assumes rigid, rid of all inner elements. Struck ship is assumed in fully loaded condition, move straight to the y-coordinate. Collision speed is set constant value of 10m/s for comparison with Kitamura's result [3]. The orthogonal collision scenario of NOAHS II as a struck ship with 150 KDWT is the similar case as NOAHS's one.

3.2 Contact force

Side collision damages of NOAHS II at bow penetration of 3m are shown in Figure 4. We perform the numerical analysis of the proposed side structure models by using the software MSC/DYTRAN, as a explicit solution method. The contact forces of new proposed Oiltanker, NOAHS of collision scenario case 2 in Figure 5(a) and NOAHS II in Figure 5(b) show, at the initial stage, relatively lower magnitude than the contact forces of standard VLCC and Kitamura's proposed VLCC shown in Figure 5(c) and (d). This phenomenon can be understood due to the flexible stiffness of side shell structure of NOAHS and NOAHS II Oiltanker. However, when the bow penetration approaches to 2m depth, the contact force of standard VLCC and Kitamura VLCC drops dramatically. In contrast to this behavior, the contact forces of NOAHS and NOAHS II increase steady until the rupture initiation of side shell begins at 2.5m for NOAHS and 2m for NOAHS II.

Further more, in case of NOAHS II, the contact force increase quite a lot in magnitude over the bow penetration depth 3m. This behavior is resulted from the

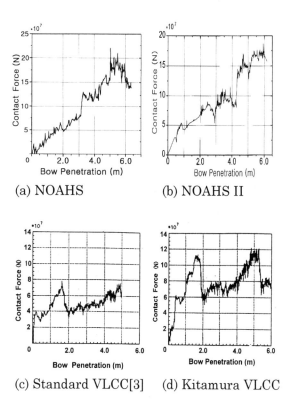

(a) NOAHS   (b) NOAHS II

(c) Standard VLCC[3]   (d) Kitamura VLCC

Figure 5. Contact forces of the struck ships

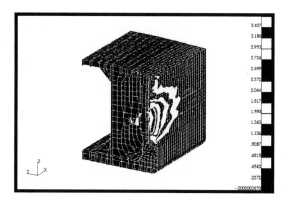

Figure 4. Side view of collision model

effective resistance performances of crushing tubes. At the bow penetration depth about 5m, the contact forces of NOAHS and NOAHS II are still increasing due to the maintaining of rigidity of side structure, while the contact forces in the cases of standard and Kitamura VLCC drop rapidly due to the loss of side structure's rigidity. The maximum value of contact force of NOAHS is greater than that of standard VLCC by approx. 120% and that of Kitamura proposed VLCC by approx. 60%. The maximum penetration depth of 6m for NOAHS and NOAHS II shows the rupture initiation of side BHD, which gives much larger value than that of other two cases. This phenomena is understood due to the global flexible effects of proposed double hull structures.

3.3 The structural absorbing energy

At the initial stages for NOAHS II, the collision energy is absorbed in large quantity by side shell until the penetration depth, 3m. From this depth, the energy absorption of tube member is dominant as shown in Figure 6(b). And the structural members of side shell and crushing plate for NOAHS absorb the major parts of collision energy shown in Figure 6 (a). For the comparison, we show the relationship between the total absorbed energy and bow penetration of four VLCC oiltankers in Figure 7, where the total absorbed energy via bow penetration depth of four cases is shown.

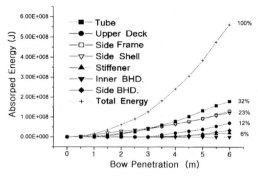

(b) NOAHS II

Figure 6. Energy absorption of structural members of proposed oiltankers

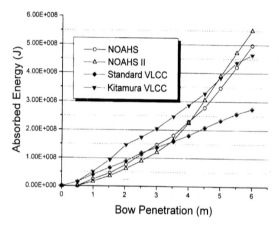

Figure 7. Comparison of absorbed energy

4. FATIGUE STRENGTH

The fatigue investigations were carried out for a preliminary design of the NOAHS-tanker. Contrary to the design of Figure 2 the tension and crushing plate strips have one or two bend angels of 120° degree over the whole length. The first fatigue detail considered is located in the transverse bulkheads at a corrugation close to the side longitudinal bulkhead (inner skin). The dominantly normal stress extends perpendicular to the corrugation and a bending moment is induced. Furthermore a fatigue assessment of the longitudinal end connections has been

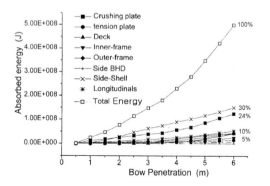

(a) NOAHS

implemented. For VLCC´s the longitudinal end connections are generally considered as a critical fatigue detail. Especially the side shell longitudinals in the region of the load water line are of interest. In the inner skin the longitudinals at the intersections with the transverse bulkheads have only the half stiffener length. As a consequence of this short stiffener length and a soft structure of the girder system the stresses due to the relative deflection between bulkhead and adjacent transverse frame will be increased. The fatigue damage will be estimated for the first asymmetrical side shell stiffener (Z=24.05m above Basis) and for the inner skin stiffener with the greatest relative deflection (Z=19.55m above Basis).

The fatigue damage studies of these details have been carried out to the rules of:
- Germanischer Lloyd [4] and
- Det Norske Veritas [5]

The concepts of the different classes differ in the methods of fatigue life assessment for structural details. The guidelines of the Classification Societies Det Norske Veritas (DNV) and Germanischer Lloyd (GL) are summarised here, partly by quotation of the corresponding rules.

4.1. The Rules of Det Norske Veritas

The fatigue design is based on S-N curves corresponding to test results from smooth specimens without stress concentration and the notch stress range at the structural detail considered. The notch stress range is obtained by multiplication of the nominal stress by K-factors, which are given in [5]. The fatigue life may be calculated based on the S-N fatigue approach under the assumption of linear cumulative damage. Depending on the required accuracy of the fatigue evaluation it may be advisable to divide the design life into a number of time intervals due to different loading conditions and limitations of durability of the corrosion protection. In order to reduce the computational effort, simplified one-slope S-N curves have been derived for typical long-term stress range distributions. Use of the

one-slope S-N curves leads to results on the safe side for calculated fatigue lives exceeding 20 years.

4.2. The Rules of Germanischer Lloyd

Corresponding to their notch effect, welded joints are classified into detail categories considering particulars in geometry and fabrication, including subsequent quality control, and definition of nominal stress. The S-N curves represent sectionwise linear relationships between $\log (\Delta\sigma)$ and $\log (N)$ with a change in slope beyond $5*10^6$ cycles. To account for different influence factors, the design S-N curves have to be corrected for material effect, effect of mean stress, effect of weld shape and influence of importance of the structural element. The stress ranges $\Delta\sigma$ to be expected during the service life of the ship or structural component, respectively, are described by a stress range spectrum. Based on this spectrum and the S-N curve of the detail the fatigue life is calculated under the assumption of linear cumulative damage (Palmgren-Miner rule).

4.3. Fatigue of the Corrugation in the Transverse Bulkhead

The detail considered, see Figure 8, is a cut out in the bulkhead girder for a vertical stiffener with a single-sided support (lug).

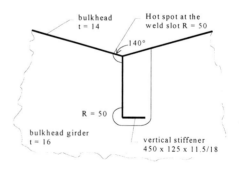

Figure 8. Top view of the bulkhead detail

The stiffener is positioned on the corrugation

of the bulkhead. The assessment is based on the structural stress at the detail considered. The structural stress is estimated by means of a local FE-model, which is loaded at its boundaries with displacements, computed with a global FE-model of the tank hold area. A detail description of this hot spot fatigue assessment is given in [6]. The highest stresses are concentrated at the weld slot of the lug. The results of the fatigue assessment are listed in Table 3.

Table 3
Fatigue results at the bulkhead detail

Classification Society	GL	DNV	
Sailing route	North Atlantic	world wide	
Loading condition	Full	full	Ballast
Fatigue stress range			
[N/mm²]	**1257**	**1600**	**389**
Long term distribution of Weibull shape parameter	$n = 5*10^7$ 1.0	$n = 5*10^7$ 0.9114	
S-N-curve			
Inverse for $n < 5*10^6$	3	3	
Slope: m for $n > 5*10^6$	5	3	
Reference stress range For $n = 2*10^6$ [N/mm²]	115.0	142.2	
Fatigue damage D	**24.1**	**20.8**	

With a fatigue damage of D=20 and more the corrugation in the bulkhead is a very critical detail that has no safety again fatigue cracking in the considered condition.

The high stress range may be explained partly by the reduced size of the finite elements at the hot spot. But definitely the considered detail is very critical and it seems that improvements as a double-sided support for the vertical stiffener can not decrease the fatigue damage below one.

4.4. Fatigue of the Longitudinal End Connections

Figure 9 describes the design of the considered longitudinal end connections. The nominal stress, determined by simple engineering formulae, is the basis of this fatigue assessment. The total stress range is the composition of global stress induced to hull girder bending and the local stress. The local stress is the sign right sum of the stress due to local bending of the longitudinal stiffener between supporting structures and the stress due to relative deflection of the nearest frame to a transverse bulkhead. The relative deflection of the double hull is determined by an FE-model of the tank hold area. A detail description of this fatigue assessment is given in [6].

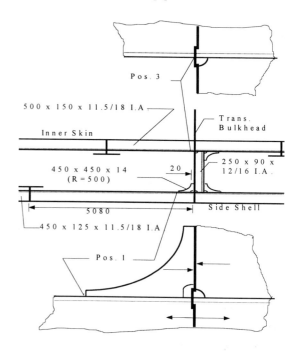

Figure 9: Top view of the stiffener details

Table 4 compares the details of the fatigue assessment. To the rules of GL the relative deflection has a greater effect on the local stress range than to the rules of DNV. For both details this effect leads to a higher total stress range by an assessment to the GL-Rules. Especially at the inner skin stiffener (Pos. 3) the local stress range is mostly induced by the relative deflection. At Pos. 1 a higher fatigue damage results from the DNV-Rules and at Pos. 3 from the GL-Rules.

Table 4
Fatigue results at the longitudinal end connections

		Pos. 1	Pos.1		Pos. 3	Pos. 3	
Classification Society		GL	DNV		GL	DNV	
Loading condition		Full	full	ballast	full	full	ballast
Pressure range	[kN/m²]	63.8	49.4	29.6	37.4	32.2	29.6
Stress range due to local bending		52.0	58.0	34.8	10.3	22.9	12.6
Stress range due to rel. deflection		86.8	26.8	−0.3	183.6	122.8	3.5
Local stress range	*[N/mm²]*	*138.8*	*84.8*	*34.5*	*193.9*	*145.7*	*16.1*
Global stress range due to:							
-vertical hull girder bending		72.5	84.6	96.0	41.1	23.3	46.6
-horizontal hull girder bending		66.8	69.2	47.1	61.0	38.0	43.0
global stress range	*[N/mm²]*	*139.3*	*153.8*	*143.1*	*102.1*	*61.3*	*89.6*
Total stress range (nominal)		*278.1*	*238.6*	*177.6*	*296.0*	*207.0*	*105.7*
Stress concentration		1.0	2.0	2.0	1.0	1.8	1.8
Fatigue stress range	[N/mm²]	**278.1**	**477.2**	**355.2**	**296.0**	**372.6**	**190.3**
Long term stress distribution of		$n = 5*10^7$	$n = 5*10^7$		$n = 5*10^7$	$n = 5*10^7$	
Weibull shape parameter		1.0	0.904	0.878	1.0	0.9114	
S-N-curve							
Inverse slope: m for n < 5*10⁶		3	3		3	3	
for n > 5*10⁶		5	(one slope curve)		5	(one slope curve)	
Reference stress range for n =2*10⁶		88.0	142.2		104.0	142.2	
Fatigue damage D		**0.60**	**0.69**		**0.40**	**0.29**	
Permissible stress range		322.0	474.0	497.7	380.6	467.6	467.6
Stress usage factor		**0.86**	**1.01**	**0.71**	**0.78**	**0.80**	**0.41**

Generally, for identical stress ranges, an assessment to the DNV-Rules results in a slightly higher fatigue damage than to the GL-Rules. But at Pos. 3 the very small stress range in the ballast condition decreases the total fatigue damage to the rules of DNV, while in the simplified assessment to the GL-Rules only the full loaded condition is considered. In respect of the fatigue damage at the considered longitudinal end connections the new tanker design is practicable. The fatigue damage at both details is clearly below one.

REFERENCES

1. J.W.Lee and J.M.Choung, Energy Dissipation and Mean Crushing Strength of Stiffened Plate in Crushing, J. Hydrospace Tech.,SNAK(1996)

2. H.Y.Paik and J.W.Lee, Collision Energy Absorption of Side Hull Construction (in Korean), Proc. The Annual Autumn Conference, SNAK (1997)

3. O.Kitamura, Comprative Study on Collision Resistance of Side Structure, Intern. Conference on Designs and Methodologies for Collision and Grounding Protection of Ships, SNAME & SNAJ (1996)

4. GL (1997): "Rules and Regulations, I - Ship Technology, Part 1: Seagoing Ships, Chapter 1 - Hull Structures" Germanischer Lloyd, Hamburg

5. DNV (1995): "Fatigue Assessment of Ship Structures", Technical Report No.: 93-0432; Det Norske Veritas Classifications

6. J. Rörup; (1998): "Fatigue Strength of Oiltankers with Alternative and Conventional Double Hull Structure"; Tech. Uni. Hamburg-Harburg, Arbeitsbereiche Schiffbau, Report Nr. 591

141

FAILURE CRITERIA FOR SHIP COLLISION AND GROUNDING

L. Zhu [a] [*] and A.G. Atkins [b]

[a] Technical Planning and Development Department, Lloyd's Register of Shipping, UK.

[b] Department of Engineering, The University of Reading, UK.

There is a scarcity of information on failure criteria which can be adopted for ship collision and grounding investigations and design against these accidents. In this paper research work on this important topic is presented with particular reference to application of the Forming Limit Diagram (FLD) and Fracture Forming Limit Diagram (FFLD) to collision and grounding analysis. The work described covers the failure investigation of thin plates dented and torn by hard objects which induce large plastic deformation, necking, fracture and tearing. Both normal and glancing contact have been considered and onset of necking and fracture is predicted successfully. The resulting failure criteria are constructed in terms of fracture toughness and other mechanical properties of the plate material. Future research required to enable this method to be applied to ship collision and grounding failure analysis is highlighted.

1. INTRODUCTION

Over the last 10 years the number of serious ship casualties has remained steady at around 1100 per year[1]. Collision and grounding accounted for nearly one third of the total casualties. These incidents resulted in severe hull damage and pollution of the environment [2].

According to Lloyd's Register of Shipping's World Casualty Statistics[3-4], 367 ships (1.8 million gross tonnage) were reported as total losses during the years 1995-1996

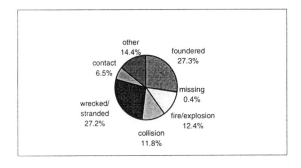

Figure 1 Total losses of ship in gross tonnage during the years 1995-1996.

The breakdown is given in Fig. 1 in which collision and grounding accounts for total losses amounting to 33% in number and 46% in Gross Tonnage (GT).

The major grounding accidents and oil spills became the driving force for new regulations. After the grounding of the Torrey Canyon in 1967, the International Maritime Consultative Organisation (IMCO) adopted the MARPOL Convention in 1973. More than 14 serious pollution accidents during 1973-1977 resulted in the 1978 MARPOL protocol known as MARPOL 73/78. US Congress introduced the Oil Pollution Act 1990 (OPA90) following the grounding of the Exxon Valdez in 1989. IMO also adopted new regulations for new and existing oil tankers in March 1992. Various new designs have been proposed and evaluated to meet these regulations.

There are a number of world wide research projects in the field of ship collision and grounding. Lloyd's Register of Shipping, along with other organisations, has participated in the Tanker Safety Project at MIT since 1991 and the project has now progressed to the second phase which will lead to a calibrated computer program 'DAMAGE' [5]. Other collision and grounding related research activities are reported in refs [6-11].

---

[*] Department of Engineering, The University of Reading from 1990 to1993.

Two main methods are utilised in the simulation of a ship's response to collision and grounding: numerical simulation using the finite element or finite difference technique, and simplified analytical methods.

Numerical simulations are generally used to reproduce the structural response up to onset of failure. Detailed information about stresses and strains can be obtained although the post-failure behaviour is difficult to model. With the rapid development of computer technology, FE is now capable of modelling the response of complex structures. In modelling the failure condition of a plate under a contact force, the simple or equivalent uni-axial rupture strain is often used; this varies with the mesh size used in the FE model. The conventional uni-axial rupture strain criterion is inappropriate without allowance for multi-axial effects since dependence of failure condition on strain path is not catered for.

Simplified analytical methods focus on the force-deformation relationship which is linked to the energy balance of the process. This is an ideal tool for the analysis of rupture and tearing in ship collision and grounding. In using the simplified analytical method it is essential to establish the fracture criteria for the structure under consideration.

It is recognised that the successful simulation of response in either method depends critically on the failure condition adopted as discussed in Volume 3 of [7,8]. There is a general lack of information and research work on calibration with experimental results even for simple structural elements. This is a fundamentally important area in ship collision and grounding research.

Having identified some similarities between plate deformation behaviour in collision and grounding and the metal forming process, it is proposed that Forming Limit Diagrams be used as failure criteria. In this paper, examples of application are given for FLD (necking) in numerical simulations and FFLD in simplified analytical methods.

2. FAILURE CRITERIA

2.1   The Forming Limit Diagram

The production of new alloys and advances in forming processes presented a need for a rapid procedure to assess the forming capacity of a material. The Forming Limit Diagram (FLD) was a milestone in this area. The technique involves printing or etching a grid of small circles on the metal sheet before forming. To define the limit strain, the principal strains (e_1 and e_2) from circles deformed into ellipses within and surrounding necked or fractured zones are measured. These values at a neck or fracture give the "failure" condition. The plot of these measured necking strains is called the Forming Limit Diagram (FLD) and when fracture strains are also included, the plot is called a Fracture Forming Limit Diagram (FFLD).

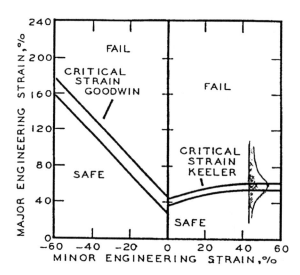

Figure 2 Keeler-Goodwin Diagram for 0.036 in sheet steel

The Keeler-Goodwin FLD shown in Fig. 2 has two limiting strain curves for low-carbon, low strength steel, which was taken collectively from Keeler [12-13] and Goodwin [14] on sheet-formed stampings and deep-drawn stampings. The critical strain band indicates strains for the onset of localised thinning. The FLD is a function of the strain-hardening exponent and the plate thickness [15]. Theoretical aspects of necking prediction have been reviewed by Keeler [16], Ghosh [17], Hosford and Caddell [18], . Ferron and Zeghloul [19] and Atkins [20].

2.2 Use of FLD as a Necking Criterion

There are many factors affecting the the forming limit of metal materials, such as strength level, strain-hardening exponent, strain-rate sensitivity factor, material imperfection, plastic anisotropy and pre-strain. The thickness of sheet metal in relation to the size of mechanical punches (if these are used to deform the material in the construction of FLD's) has an effect; friction between material and punch is also a contributing factor in the process of strain localisation. No attempt has been made to discuss them in detail in this paper.

Various necking FLD's were investigated by Zhu and Atkins [24] for the failure analysis of dented plates. In the tension-tension strain quadrant, Keeler-Goodwin curve can be approximated as

$$\varepsilon_1 = n_t + 0.32\varepsilon_2 \qquad (1)$$

For a plate thickness of 0.914 mm examined by Keeler, n_t had the value of 0.31.

In the tension-compression strain quadrant, localised necking occurs when

$$\varepsilon_1 = n_t + |\varepsilon_2| \qquad (2)$$

where $|\varepsilon_2|$ is the numerical value of the compressive strain [18].

If a plate is stretched biaxially with a known strain path, the failure by necking can be determined using the above relations.

2.3 Use of FFLD as a Fracture Criterion

A metal sheet under in-plane stretching undergoes uniform thickness thinning, then necking and finally fracture. The locus of fracture is represented by the Fracture Forming Limit Diagram in the e_1-e_2 plot.

Based on the work by Atkins and Mai [21-22], the FFLD can be constructed knowing the stress-strain behaviour and the fracture toughness R. For a material that follows $\sigma = \sigma_0 \bar{\varepsilon}^n$, the effective von Mises strain at fracture is given by:

$$\bar{\varepsilon}_f = [\frac{R(n+1)}{\sigma_0 H}]^{\frac{1}{n+1}} \qquad (3)$$

where H is the height of the necked-down region.

Given the strain history of a loading path, $\bar{\varepsilon}_f$ may be decomposed into the in-plane fracture strains $\bar{\varepsilon}_{1f}$ and $\bar{\varepsilon}_{2f}$.

The link between the FFLD and the fracture toughness R was established which makes it possible to predict the FFLD in terms of the independently-measured mechanical properties of the material. Again, with a known strain path, failure by fracture can be determined using the above criteria.

3. FAILURE ANALYSIS OF DENTED PLATE IN NORMAL CONTACT

3.1 Experiment

Static and dynamic denting and the perforation of plates were conducted using a wedge indenter impacting the plate normal to its plane. For the dynamic wedge impact test [23] most specimens exhibited large plastic deformation. The specimens displaying necking, cracking and rupture were analysed [24].

Although the 'sharp' indentors were modelled in numerical simulations as 'line' and 'point' loads for the wedge and cone respectively, in practice the edge or tip was rounded with a small radius.

3.2 Deformation and Failure Process

It was observed that the plate sustained large plastic deformation first, followed by necking and cracking at the two ends of the dent line with subsequent propagation of necking and cracking toward the centre. Afterwards, the plate pushed by the wedge would bend, and tear from the two ends of the dent line in directions perpendicular to the dent line. Bending and tearing happened simultaneously.

A numerical program based on the Variational Finite Difference Method was developed by Zhu [23] and was calibrated against test data for plates under impact and blast loading. The plate large deformation behaviour was studied [25] and an approximate procedure based on the rigid perfectly plastic method was developed [26]. The detailed failure analysis following large plastic deformation given in [24] is reported briefly in this section. The topic of tearing and fracture is discussed in Section 4.

3.3   Necking Analysis of Plates Under Wedge Impact

As the failure condition is strain-path-dependent, it is vitally important to obtain the strain histories of various points along the dent line. Zhu [27] presented detailed information about the dynamic strains of dented plates using an elasto-plastic analysis program[23]. Using the same method, the dynamic strains for the severely damaged plates are calculated.

Defining necking as the failure criterion, the thickness-corrected Keeler-Goodwin Diagram may be used as a failure curve. The necking strains (ε_{xn}, ε_{yn}) and necking time (t_n) can be determined in terms of the strain calculated.

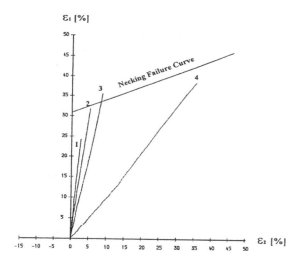

Figure 3 Use of FLD as necking criterion to predict the failure of a dented plate (refer to Fig. 4 for locations)

A typical example is given in Fig. 3 in which the necking failure curve is firstly plotted. On the same diagram, the numerically predicted strain paths [27] of four given locations are also plotted. At any given location, necking will occur when its strain path curve intersects the necking failure curve. For the case examined, necking occurs at location 3 while location 4 does not reach the necking state (Fig. 3).

The predicted necking failure given above is compared with the experimental results shown in Fig. 4, in which the steel plate just reached necking

near the end of the dent line (location 3). Good agreement between the numerical prediction and the experimental results is achieved.

3.4 Discussion

From the above necking prediction of a dented plate, it is seen that the strain path varied with location in the structure. Due to the strain-path-dependent nature of the problem, in a time domain simulation this means that there is a sequence for these points to reach the necking failure.

Providing the impact conditions are known, the response of the plate can be numerically simulated, to produce a complete stress strain history. It must be stated that the strain prediction is only valid before significant necking takes place since the necking may cause complicated changes in stress and strain state prior to fracture. For the simulation of a ruptured specimen, the predicted strain paths following necking can be used for comparative purposes only.

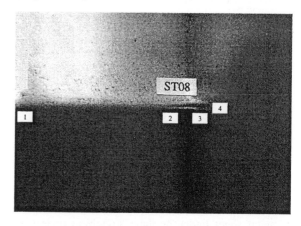

Figure 4 Localised necking of dented plate

It is suggested that, when the failure curve for design is determined, the design curve should be associated with the statistical confidence level for the collected experimental data. A characteristic failure curve below the mean FLD curve should be used to account for the scatter of the test data.

4. TEARING IN GLANCING CONTACT

4.1 Experimental set-up

To simulate a ship hitting a rock when underway the apparatus shown schematically in Fig. 5 was constructed. The indenters are fixed to a carriage running in a rigidly-held, low-friction vertical track attached to the rig holding the test plate. Four series of indenters were examined: balls, cones, pyramids and oblong blocks. Details of test results and analysis are given in [28,29].

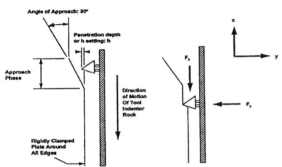

Figure 5 Schematic diagram for experimental arrangement

4.2 Experimental findings

In tests with ball indenter, the stretching deformation is small for small penetration depths, but at increased penetration depths the strain history lines are steeper and the strains greater. The loading lines increase from the instant the indenter contacts the plate to when the indenter reaches the steady-state condition of either just denting or eventually fracturing the plate. When fracture occurs at a sufficiently deep penetration, critical strain levels are reached. A single tear occurs along the direction of motion and is preceded by necking in this direction.

For the cone indenter, the major in-plane strain is also across the direction of motion and the deformation is more localised than for the ball indenters. Figure 6 shows the in-plane strain pair histories in a typical loading path. The strain history measurements are for the very localised stretch region in the distal face around the pointed tip of the indenter. Typical traces of the force and displacement in the x- and y-directions are shown in Figure 7. Loading lines increase from the instant the indenter hits the plate to when fracture occurs.

Deformation without fracture (producing a pointed groove in the direction of motion) occurred only at very low penetration levels.

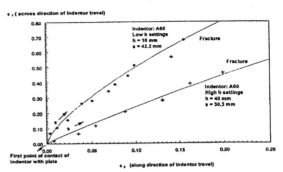

Figure 6 In-plane strain pair histories beneath conical indenters

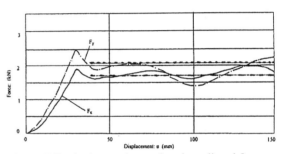

Figure 7 Typical experimental and predicted force-displacement traces for conical indenters

4.3 Energy Balance and Prediction of Force

Once a tear starts, the remote plastic tear field alters because of the change in constraint in the affected area. The critical damage condition for fracture at the running crack tip is likely to remain the same but the remote plastic work changes from that at initiation because the material has now a split. This change in the increments of plastic work done before and after crack initiation has been used to predict transitions from plasticity alone (up to initiation) to subsequent combined plasticity and fracture. That is, before fracture

$$F_{before} \ du_{before} = d\Gamma$$
$$= W_{before}dV_{before} + V_{before} \ dW_{before} \qquad (4)$$

where F is load, u is the load point displacement, Γ is the plastic work, W is the volume-weighted density (non-uniform in the general case) and V is the volume of material undergoing plastic flow. During tearing

$$F_{tear}\, du_{tear} = d\Gamma + Rtda$$
$$= W_{tear}\, dV_{tear} + V_{tear}\, dW_{tear} + Rtda. \qquad (5)$$

where R is the fracture toughness, t is the plate thickness and da is the tear length.

Examples of using the above method are given in [30]. The example below predicts the force-indentation relationship for a plate subjected to glancing contact with balls:

Bulging without Fracture

$$F_{x1} = \cfrac{\cfrac{\sigma_y t^2}{2}(\phi + \omega + \cfrac{\beta}{2}) + \cfrac{4\rho\varepsilon\phi}{t}}{1 - \cfrac{\mu}{\mu \sin\theta + \cos\theta)}}$$

$$ \qquad (6)$$

$$F_{y1} = F_{x1}\tan(\theta - \gamma)$$

Tearing

$$F_{x2} = \cfrac{\cfrac{\sigma_y t^2}{2}(\phi + \omega + \cfrac{\beta}{2}) + Rt}{1 - \cfrac{\mu}{\mu \sin\theta + \cos\theta)}}$$

$$ \qquad (7)$$

$$F_{y2} = F_{x2}\tan(\theta - \gamma)$$

where μ is the coefficient of friction and σ_y is the yield strength; ρ is the radius of the ball, θ is the angle between the normal force and the moving direction of the indenter, γ is the friction angle given by $\gamma = \tan^{-1}\mu$, ϕ is the half contact angle between the ball and the plate, ω is the plate bending angle, β concerns the profile of the bulge ahead of the tear and ε is the effective stretching strain [29].

Good agreement has been achieved between predicted and experimental load-displacement plots for both ball-ended indenters and sharp indenters. A comparison of load-displacement curves for cone indenters is given in Figure 7.

4.4. Discussion
Blunt smooth surfaced indenters produce single cracks in the direction of motion, formed by

stretching over the surface of the indenter. In ductile plate materials, necking will precede cracking. Sharp indenters produce more localised deformations, with the sharp tips or edges of the indenters scoring and cutting the plate.

For blunt smooth-surfaced indenters, the necking and cracking process is very similar to what happens in plate metal pressing and may be described by the "forming limit diagram" (FLD) and particularly by the "fracture forming limit diagram" (FFLD). When cracking occurs by in-plane stretching, the biaxially-dependent fracture strains are those given by the conventional FFLD, even though there are tractions parallel to the surface of the plate in glancing collisions [29].

While the same micromechanics of void growth will still determine the onset of fracture by sharp indenters that score and cut plate, the associated localised stress and strain fields are not as well understood and more research is needed. Propagation of a tear, once initiated, is however easily analysed with equations like (7) for sharp indenters [29]. In this way the extent of damage (length of tear) of a ship with given velocity on hitting one obstacle may be determined. The predictions depend on the coefficient of friction μ, about which there is some uncertainty.

5. CONCLUDING REMARKS

• Necking and fracture failure criteria are proposed based on the FLD and FFLD and have been used in numerical simulation and in simplified analytical methods. The onset of necking and fracture is predicted successfully and the bi-axial stretching effect and the influence of the strain path on the failure condition are taken into account. Simple uniaxial failure strains are not correct in biaxial fields.

• So far, the research work has been performed on thin steel plate. Future work is needed for this method to be applied to the failure analysis of ship plates with defects and welds. FLD's and FFLD's will be determined for such full thickness plates using a 'giant' hydraulic bulger, which is presently being built at Reading University.

REFERENCES

1. D.S. Aldwinckle, Brooking, M.A. and Hart, D.K., 'OPA 90/IMO legislation and the ship emergency response service', Lloyd's Register of Shipping, 1995.
2. J.M. Ferguson,. 'Oil tankers and the environment planning for the future', Lloyd's Register of Shipping, April, 1990.
3. Lloyd's Register of Shipping World Casualty Statistics 1995.
4. Lloyd's Register of Shipping World Casualty Statistics 1996.
5. B C Simonsen, and Wierzbicki, T, 'Theoretical manual on grounding damage of a hull bottom structure, MIT, 1998.
6. T. Kuroiwa, 'Research on structural failure of tankers due to collision and grounding' Proc. Conf. On Prediction Methodology of tanker Structural Failure (ASIS), Tokyo, Japan, 1995.
7. ISSC, Report of Committee V6, Porc. of 12th ISSC, 1994.
8. ISSC, Report of Committee V4, Porc. of 13th ISSC, 1997.
9. H. Ohtsubo, and Wang, G, Aupper-bound solution to the problem of plate tearing, J. of Marine Science and Tech., 1, 46-51, 1995.
10. P T Pederson, 'Collision and grounding mechanics', Proc. Of Ship Safety and Protection of the Environment, WEMT'95, Copenhagen, Denmark, 1995.
11. N. Jones, and Wierzbicki, T.(ed), Structural Crashworthiness and Failure, , 1993.
12. S.P Keeler,. and Backofen, W.A., ' Plastic instability and fracture in sheets stretched over rigid punches', Trans. ASM, 56, 25-48, 1963.
13. S.P Keeler., 'Circular grid system - a valuable aid for evaluating sheet metal formability', SAE paper No. 680092, 1-9, 1968.
14. G.M Goodwin,., 'Application of strain analysis to sheet metal forming problems in the press shop', SAE paper No. 680093, 1-12, 1968.
15. S.P Keeler and Brazier, W.G., 'Relationship between laboratory material characterization and press-shop formability', Micro Alloying 75, Union Carbide, N.Y., 517-30, 1977.
16. S.P Keeler, 'forming limit criteria - sheets', Advances in Deformation Processing, Ed. Burke, J.J and Weiss, V., Plenum Press, New York, 1978.
17. A.K Ghosh, 'Plastic flow properties in relation to localised necking in sheets', Mechanics of Sheet Metal Forming-material behaviour and deformation analysis, Ed: Koistinen, D.P. and Wang, N.M., Plenum Press, NewYork-London, 287-312, 1978.
18. W.F.Hosford, and Caddell, R.M., Metalforming Plasticity -- Mechanics and Metallurgy, Englewood Cliffs, NY:Prentice-Hall, 294-311, 1984.
19. G. Ferron, and Zeghloul, A., 'Strain localisation and fracture in metal sheets and thin-walled structures', Structural Crashworthiness and Failure, edited by Jones, N. and Wierzbicki, T., , 1993.
20. A. G. Atkins Fracture mechanics and metalforming: damage mechanics and the local approach of yesterdayand today in Fracture Research in Retrospect (G. R. Irwin Festschrift, Edited by H.P. Rossmanith), 327-350 Rotterdam: Balkema,1997.
21. A. G. Atkins & Y-W Mai Fracture strains in sheet metalforming and specific essential work of fracture Engng.Fracture Mech. 27, 291-297, 1987.
22. A.G. Atkins & Y-W Mai. Elastic & PlasticFracture, Chapter 4, Chichester: Ellis Horwood, 1985 and 1988.
23. L. Zhu, 'Dynamic inelastic response of ship plates in collision', PhD thesis, University of Glasgow, June 1990.
24. L. Zhu. and A.G. Atkins, 'Failure Analysis of Dented Plates', unpublished research, 1992.
25. L. Zhu and Faulkner, D., 1994, Int. J. Impact Engng, Vol. 15, No.2, 165-178, 1994.
26. L. Zhu., Faulkner, D. and A.G. Atkins, 1994, 'The impact of rectangular plates made from strain-rate sensitive materials', Int. J. Impact Engng, Vol. 15, No.3, 243-255, 1994.
27. L. Zhu. "Stress and strain analysis of plates under wedge impact", Int. J. of Strain Analysis for Engineering Design, Vol. 31, No.1, 1-7, 1996.
28. C. M. Muscat-French. The Tearing of Ships upon Grounding. Report (6 parts) for EPSRC/MTD/MoD., The University of Reading, Department of Engineering, 1996.
29. C M Muscat-Fenech and Atkins, A G., 'Denting & Fracture of Sheet Steel by Blunt & Sharp Obstacles in Glancing Collisions, to be published in Int. J. Impact Engng., 1998.

Practical Design of Ships and Mobile Units
M.W.C. Oosterveld and S.G. Tan, editors.

149

On Ductile Rupture Criteria for Structural Tear in the Case of Ship Collision and Grounding

E. Lehmann[a] and X. Yu[b]

Germanischer Lloyd
Vorsetzen 32, D-20459 Hamburg, Germany

[b]Technical University of Hamburg-Harburg
Lauenbruch Ost 1, D-21079 Hamburg, Germany

In order to determine whether a ship will experience flooding in case of collision and grounding, a method to predict the ductile fracture was proposed. Based on studies within the framework of continuum damage mechanics, a rupture index I_R was introduced. The true stress-strain relationship as well as the stress triaxiality were taken into account. The critical failure load and the location of crack initiation were reliably predicted. Computations based on the nonlinear FEM program MARC showed good agreement between numerical results and experimental values. Compared to another criteria using damage parameters, it is simpler to use the rupture index because it did not require complicated measurement of damage, i. e. one conventional stress-strain diagram from tension test was sufficient.

1. INTRODUCTION

One of the major risks of oil tankers is cargo spillage caused by groundings and collisions. To prevent oil outflow by tanker accidents, new regulations on structural designs, such as OPA'90 and IMO 13F and 13G, were adopted. Since then a lot of researchers have engaged in developing effective measures to reduce the risk of tanker oil pollution. In addition to the new navigation systems to eliminate human error, new structural design to prevent or minimize oil outflow is necessary. In the past decade, a large amount of literature was published on analysing structural behaviour in the event of ship collision and grounding [1].

At accidents like collision or grounding, the hull structures endure extreme large loads and severe deformations including damages such as folding after buckling and tearing after fracture initiation. Under such casualty operations, theories in classic structural mechanics are no more sufficient. Hence new investigations focused on developing new methods to analyse structural behaviour in the case of collision and grounding [2].

Recent experiments with large scale models made substantial contribution to the understanding of structural rupture behaviour.

Compared to the traditional ship structural mechanics, the mechanics of collision and grounding is complicated with large deformation, fracture, dynamic effect, fluid force and contact. Based on experimental and analytical investigations, several useful methods were proposed. Some methods were simple in their mathematics but accurate in prediction of structural energy absorption. The accessment of loads succeeded generally only in terms of *mean force*. Until now, most analytical methods began with assumed rupture modes, i. e. , the location of crack initiation, the path of crack propagation as well as the extent of tearing were predefined. This is questionable considering a structure before it is damaged.

In order to estimate the ability of a ship to resist collision or grounding damage, the energy absorption capacity of structural components was usually determined in correspondence with certain levels of failure. Such a level may be given in terms of a fracture criterion. The widely used criterion, until now, was based on a *mean strain*. However, the use of the mean strain was neither experimentally nor theoretically verified for ship's hull structures. For hull rupture, a uniaxial failure strain was adopted in many analytical approaches and numerical simulations. This treatment could not account for the failure of structures under multi-axial loading which is typical in collision and grounding. Conventional uniaxial rupture strain criterion is inappropriate even for a simple tensile test specimen.

It remains an unsolved problem which determines whether the spillage of liquid cargo may occur. Improved analytical techniques are required to understand the hull rupturing process. Further research is needed for the establishment of a reliable fracture criterion. As emphasized by the ISSC'94 committee V. 6, this is an urgent task in collision and grounding research [3].

In the present paper, a method to predict the ductile fracture was proposed. Based on studies within the framework of continuum damage mechanics, a ruptur index I_R, a key factor determining whether cargo will be spilled, was introduced. Since this criterion relates to intrinsic properties of material and is independent of geometrical configuration and boundary conditions of structures, it is quite general and applicable to different cases. Compared to other criteria using micro-damage parameters, it is simpler to use the present method because it does not require complicated measurement of micro-damage, i. e., one conventional stress-strain diagram from tension test is sufficient.

2. MATERIAL PROPERTIES AT FRACTURE INITIATION

Material properties were obtained through tension test using standard specimens. Theoretically, the engineering stress relates to the initial cross section of specimen. Due to the necking, the actual cross section is smaller, hence the true stress in the necking zone is larger than the corresponding engineering stress. After necking the state of stress is no longer uniform and uniaxial. Such a multi-axial state of stress until the fracture initiation cannot be described in terms of conventional engineering stress.

An the other hand, the fracture elongation relates to a certain gage length L_0. The smaller the gage length is, the larger is the fracture elongation. This dependence on the gage length means that the fracture elongation cannot be used as a material constant to describe the ductile failure. Moreover, it is not a true strain. As shown in Fig. 1, there is a great difference between the true stress-strain relationship and the conventional engineering stress-strain diagram obtained directly from a tension test. The true stress-strain relationship shows the strain-hardening effect which is not negnigible to determine the exact true stress magnitude. Since the ductile rupture of ship structures is resulted by local large strain, the use of a true stress-strain relationship is necessary.

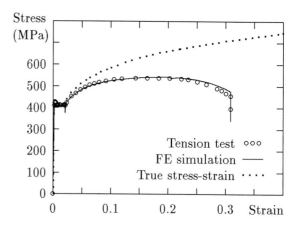

Figure 1: Stress-strain curves for steel St52-3

The failure formation in a tensile specimen could be detected by an ultrasonic technique during the tension test. As schematically illustrated in Fig. 2, a central crack occured over the necked minimal cross section at a certain strain level. Under further tension the crack propagated along the shear band to the free surface and formed the final separation with cup-cone fracture surface.

For such a rupture process, it is more reasonable to consider the true strain, instead of the engineering fracture elongation, over the minimal cross section. The true strain ε_f over this cross section may be approximately obtained using

$$\varepsilon_f = \ln \frac{A_0}{A_f}, \tag{1}$$

where A_0 and A_f are, respectively, the initial

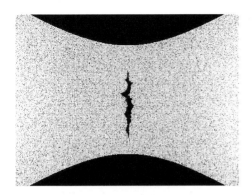

Figure 2: Crack initiation

cross section area and the final area after failure. The magnitude of ε_f is larger than the value of the engineering fracture elongation and can be interpreted as the logarithmic engineering fracture elongation for a zero gage length.

As mentioned the state of stress in the necked region is not uniaxial. The multi-axial state of stress can be well described by the stress triaxiality β, which is here defined as

$$\beta = \frac{3\sigma_m}{\sigma_v}, \tag{2}$$

where σ_m and σ_v are, respectively, the mean stress and the equivalent stress. Thus the following three typical states of stress can be described by their corresponding values of the stress triaxiality:

uniaxial compression: $\beta = -1$;
pure shearing : $\beta = 0$;
uniaxial tension: $\beta = 1$.

It is hardly to realize the ideal state of $\beta = 1$. Even in a smooth tensile specimen, the stress triaxiality at failure lies in the range of $\beta > 1$ since the state of stress in the necked zone is not onedimensional. For the rupture of ship strucutres in the case of collision and grounding, the range for the stress triaxiality is generally $\beta > 1$. The following study was restricted in this range.

Larger magnitudes of stress triaxiality can be obtained by means of notched tensile specimens. Depending on different notch radii R, different stress triaxiality occur in notched specimens under tension. For a round notched specimen with a minimal cross sectional diameter d in the notched

zone, the stress triaxiality may be estimated using Bridgman's approximation:

$$\beta = 1 + 3\ln(1 + \frac{d}{4R}) \tag{3}$$

Equation (3) can also be applied to a smooth tensile bar where R is the contour radius of the necked zone.

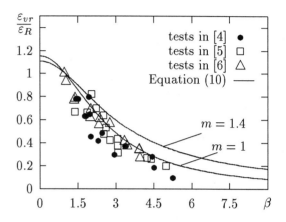

Figure 3: Fracture strain vs. stress triaxiality

A large amount of tensile tests using notched specimens demonstrated that the fracture strain ε_f is highly dependent on the stress triaxiality β. In Fig. 3, for example, some test results in [4]-[6] are plotted. The fracture strain ε_f decreased as the stress triaxiality β increased. Such a dependence of ε_f on β was, unfortunately, not taken into account in many researches on the rupture of ship structures.

3. CONTINUUM DAMAGE MODELS

It is relatively easy to make the prediction for simple tensile specimens that fracture will begin in the necked or notched zone. Ship structures are, however, more complicated. Consequently, it is hardly to predict the first crack initiation under extreme loads using classic theory of ship structural mechanics. Better understanding of their rupture process was obtained through the studies within the framework of the continuum damage mechanics.

Unlike the treatment in classic fracture mechanics dealing with pre-existing macro-crack, it was assumed that the initiation of macro-crack was resulted by the evolution (nucleation, growth and coalescence) of micro-imperfections, i. e. micro-voids or cavities, which were defined as *damages*. For it a continuum damage variable was introduced usually as an internal variable. It is of tensorial nature but reduces to a scalar for isotropic material. In general, the damage depends on the states of stress and strain as well as on other parameters such as strain rate and temperature.

Various damage models for ductile fracture under quasi-static loading were established in the past. The mostly used damage variables are:

the critical void growth ratio R_c, introduced by Rice & Tracey [7];

Gurson's void volume fraction f, modified by Tvergaard & Needleman [8];

the relative intersectional area of microcavities D, introduced by Kachanov & Lemaitre [9-11].

In the follwing, only Lemaitre's model was considered due to its simplicity. The study focused on the isotropic damage.

The ductile plastic damage was defined as

$$D = \frac{\delta S_0}{\delta S}, \qquad 0 \le D \le 1, \tag{4}$$

and its differential constitutive equations was expressed as [11]

$$\dot{D} = \frac{K^2 D}{2ES_0} f(\beta) \varepsilon_v^{2n} \dot{\varepsilon}_v, \tag{5}$$

where K and n are material coefficients. The function

$$f(\beta) = \frac{2}{3}(1+\nu) + \frac{1-2\nu}{3}\beta^2 \tag{6}$$

with Poisson's ratio ν implied the influence of the stress triaxiality.

In the case of proportional loading, the equation for damage evolution was obtained as:

$$D = D_0 \exp[Cf(\beta)(\varepsilon_v^{2n+1} - \varepsilon_{v0}^{2n+1})] \tag{7}$$

For uniaxial tension $\beta = 1$, the critical damage is written in terms of the onedimensional threshold strain ε_0 and the rupture strain ε_R:

$$D_c = D_0 \exp[C(\varepsilon_R^{2n+1} - \varepsilon_0^{2n+1})] \tag{8}$$

During the evolution of damage, the macro-crack is initiated as soon as damage reaches the critical value D_c. The critical damage D_c may be used to predict the ductile rupture of structures. However, this requires a reliable damage measurement which is not easy. Some methods to measure damage evolution are complicated and expensive [9].

Using the notation $m = 2n + 1$ and assuming

$$\varepsilon_{v0}^m = \varepsilon_0^m / f(\beta), \tag{9}$$

the effective rupture strain ε_{vr} for the multi-axial stress and strain states was expressed as a function of the onedimensional rupture strain ε_R and the corresponding stress triaxiality β:

$$\varepsilon_{vr} = \varepsilon_R f^{-m}(\beta) \tag{10}$$

The relationship in Eq. (10) was compared with some tension tests in Fig. 3.

4. RUPTURE INDEX

While the onedimensional rupture strain ε_R in Eq. (10) being constant, the multi-axial rupture strain ε_{vr} is variable, which means that the crack will not certainly occur at the place where the maximal effective strain ε_v exists. As an example, the notched region of a tensile specimen is shown in Fig. 4. The maximal effective strain occured on the edge of the minimal cross section (a), whereas the crack occured at the center (b).

To identify the rupture location in a complex structure, it requires comparing the actual effective strain ε_v to the fracture limit ε_{vr} over the entire structure. By means of FEM calculations it is possible but not convenient since the limit ε_{vr} varies from node to node. For convenience, a *rupture index* (failure factor) I_R, which is generally a plastic micro-damage parameter, is here introduced as:

$$I_R = \varepsilon_v f^m(\beta) \tag{11}$$

The simple form of Eq. (11) may be implemented into any finite element code without additional constitutive model and extra time-consuming algorithm. All the numerical calculations in this paper were carried out using the FE program MARC. The implement of the rupture index I_R

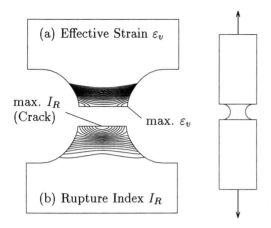

Figure 4: A notched round tensile bar

was realized by means of a user-subroutine. In an FE simulation, the distribution of the rupture index I_R may be, like other parameters such as stresses and strains, plotted by postprocessor. The rupture may be predicted as soon as the condition

$$I_R \geq \varepsilon_R \qquad (12)$$

is fulfilled. The first fracture occurs at the place where the maximal I_R exists. The contours of I_R in Fig. 4(b), e. g. , implied that a crack was initiated at the center of the notched region.

5. DETERMINATION OF ε_R

Most researchers interpreted ε_f in Eq. (1) for smooth specimen as the onedimensional rupture strain ε_R, neglecting the fact that the stress state at failure is no longer uniaxial. More accurate value of ε_R was determined here by the FEM calculation of a tensile specimen. It was only prostulated that a simple tension test had been carried out and a conventional engineering stress-strain diagram was available. The tension test was then simulated using FE model. Important is that the FE model had the same geometries, especially the same gage length L_0, as the real specimen.

As an example, a tension test using flat specimen for the steel St52-3 was simulated with 1/8 FE model as shown in Fig. 5(a). After the tension test, the engineering stress-strain diagram

indicated a fracture elongation of 30% in correspondence of a gage length of $L_0 = 50$ mm. For the entire tension process, the nonlinear large displacement and large strain were considerd. The strain hardening was taken into account using a true stress-strain relationship which was carefully chosen so that the numerical engineering stress-strain results agreed well with the tests as shown in Fig. 1. This required iterative calculations.

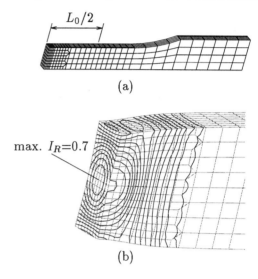

Figure 5: Simulation of a tension test

At the moment that the elongation of the FE model, corresponding to the gage length of $L_0 = 50$ mm, reached 30%, the distribution of the rupture index I_R was illustrated by the contours in Fig. 5(b). The maximal I_R occured at the center of the necked cross section where the first crack initiated. The maximal I_R-value at the failure moment was the searched rupture limit, i. e. $\varepsilon_R = 0.7$, for the steel St52-3. Now the onedimensional rupture strain ε_R, together with the chosen true stress-strain relationship, may be used in analyses of other complex structures consisting of the same material.

6. APPLICATIONS

As applicational examples, two FE simulations were carried out to validate the criterion.

Figure 6 shows a test set-up for the cutting of a longitudinally stiffened cylindric shell. The shell with a diameter of 1.5 m and a thickness of 7 mm

154

was stiffened by twelve flat stiffeners. The width
and the thickness of stiffeners were 140 mm and
7 mm respectively. A thick triangular plate was
machined to a sharp indentor.

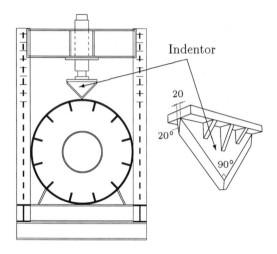

Figure 6: Test set-up

Under quasi-static loading the shell was firstly
dented to large local deflection. The load increased until a critical value of $F_{cr} = 125$ kN
and decreased abruptly when a rupture of the
shell occured suddenly. At this moment an initial
crack with a length of about 55 mm in circumferential direction was observed from the inside
of the shell but the indentor tip had not yet penetrated into the shell. The experimental force-
indentation results are plotted in Fig. 7.

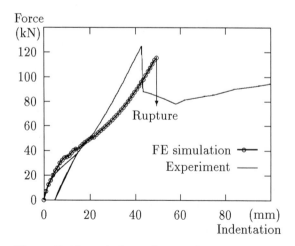

Figure 7: Force-indentation results

Under further loading the shell was penetrated
and torn to a large openning. Meanwhile, the
force remained at a lower level.

The denting process was simulated by an FE
model using shell elements. The indentor was
idealized as rigid contact surface. Figure 8(a)
shows the deformed FE model.

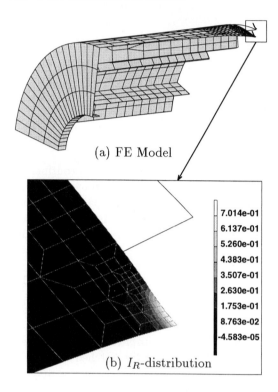

(a) FE Model

(b) I_R-distribution

Figure 8: Deformation und crack location

The material of the shell was the steel St52-3,
whose rupture limit $\varepsilon_R = 0.7$ was already obtained before. The distribution of the rupture
index I_R in the FE model showed that the maxi-
aml I_R-value occured on the inner surface of the
shell, directely underneath the indentor tip. Such
an I_R-distribution in Fig. 8(b) implied a circumferential crack direction. The rupture was predicted when the maximal I_R-value exceeded the
limit of $\varepsilon_R = 0.7$. Thus a failure point in the
numerical force-indentation curve in Fig. 7 was
determined. The numerical result of $F_{cr} = 115.6$
kN correlated well with the experimental result.

As another example, the penetration of a double bottom was calculated. The model test was
reported in [12][13] by the research group of joint

MIT-industry project on tanker safety. Based on the tension test for the steel A366 in [13], the material rupture limit was investigated in the similar way. The tension of a flat specimen was firstly simulated using FE model. The true stress-strain relationship was determined so that the numerical engineering stress-strain curve agreed well with the test diagram. At the failure moment, the maximal I_R-value was obtained in the FE model. Thus a rupture limit of $\varepsilon_R = 0.88$ was determined for the steel A366. Further the true stress-strain relationship and the rupture limit were used in the FE model for the double bottom.

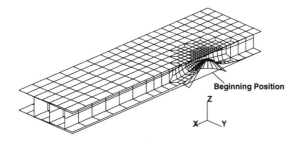

Figure 9: Deformed FE model

Figure 9 shows the deformed FE model. The double bottom was modelled using shell elements, whereas the indentor was idealized as rigid contact surface. Distribution of the rupture index I_R in the FE model in Fig. 10 indicated the first crack location on the formed dome. As soon as the I_R-value exceeded the rupture limit ε_R, a critical rupture force F_{cr} was determined. The corresponding element with the maximal I_R-value was then treated as failed and its stiffness was eliminated. During the further indentation, more elements failed subsequently. In this manner, the crack propagation in the structure was simulated and the decreasing load after the fracture initiation was obtained in the numerical load-indentation curve. It should be noted that only the first element failure was accurately determined, since the "crack" (formed by eliminated elements) in FE model was unrealistic wide so that the sharp crack tip was not precisely modelled. More faithfull FE simulation of the crack propagation is the main objective of our on going research work.

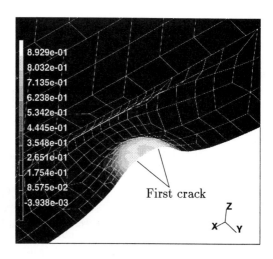

Figure 10: I_R-distribution in FE model

research work.

Figure 11 shows a good agreement between the numerical and experimental force-indentation results until the first fracture initiation.

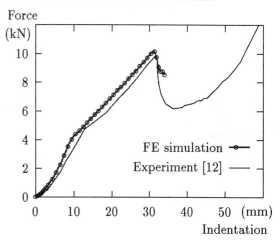

Figure 11: Experimental and numerical results

The examples here demonstrated that the critical rupture loads as well as the fracture locations were reliablly predicted by the present fracture criterion. More successful applications to other structures can be found in [2].

7. CONCLUSIONS

For the ductile failure of structures resulted by large deformation and large strain, an approach

156

to determine the fracture initiation was proposed. Using the rupture index I_R introduced in the present paper, the critical rupture load as well as the crack location can be predicted. For ship structures in the case of collision and grounding, the reliable prediction of the critical rupture load F_{cr} is important to determine whether a cargo spillage may occur.

Unlike the treatment based on the mean strain concept, the stress triaxiality was taken into account in the rupture index. Consequently, the present criterion is valid to the multi-axial state of stress. The rupture index I_R is in principle a plastic damage parameter. Compared to Lemaitre's critical damage D_c, the limit value ε_R is easy to be determined by the FE calculation of a tension test. For it a conventional engineering stress-strain diagram is sufficient. The onedimensional limit strain ε_R is a material constant and independent on structural geometries.

The value of the fracture strain ε_f of a smooth specimen determined using Eq. (1) can be used as a simple approximation for the limit ε_R. However, the strain ε_f, corresponding to a multi-axial stress state $\beta > 1$, is not strictly the onedimensional limit strain. In addition, it is only a mean value for the entire minimal cross section.

It is unreasonable to use directly the classic tensile ultimate σ_u or the engineering fracture elongation to determine the ductile rupture of structures. The engineering fracture elongation depends strongly on the gage length L_0. Hence any account of the fracture elongation without mentioning the corresponding gage length is meaningless.

True stress and true strain may be large at the moment of crack initiation. To determine the exact magnitudes of stress and strain, a true stress-strain relationship must be used. The assumption of ridig-plastic material property is not sufficiently accurate.

The present criterion was successfully used to predict the first fracture initiation of various structures under quasi-static loading. For dynamic process with high speed in which the strain rate plays important role, an appropriate dynamic fracture criterion is needed.

REFERENCES

1. T. Wierzbicki, D. B. Peer, and E. Rady, The anatomy of tanker grounding, Marine Technology, Vol. 30, No. 2, 1993.
2. X. Yu, Structural behaviour with large deformation till crack initiation and with dynamic folding, Doctoral Thesis, University of Hamburg, 1996.
3. N. E. Jeffrey and A. M. Kendrick (eds.), Proceedings of the 12th International Ship and Offshore Structures Congress(ISSC'94),Vol. 3
4. X. M. Kong, Einfluß der Spannungsmehrachsigkeit auf die Schädigung und das Bruch verhalten metallischer Werkstoffe, VDI-Fortschrittberichte, Reihe 18, Nr. 127, VDI-Verlag, 1993.
5. P. J. Bolt, Prediction of ductile failure, Dissertation Technische Universiteit Eindhoven, 1989.
6. W. H. Tai, Plastic damage and structure in mild steels, Eng. Fracture Mech. , Vol. 37, No. 4, 1990.
7. J. R. Rice and D. M. Tracey, On the ductile enlargement of voids in triaxial stress fields, J. Mech. Phys. Solids, Vol. 17, 1969.
8. V. Tvergaard and A. Needleman, Analysis of the cup-cone fracture in a round tensile bar, Acta Metall, Vol. 32, No. 1, 1984.
9. J. Lemaitre and J. Dufailly, Damage measurements, Eng. Fracture Mech. , Vol. 28, Nos. 5/6, 1987.
10. D. Krajcinovic and J. Lemaitre, Continuum Damage Mechanics: theory and applications, Springer-Verlag, 1987.
11. J. L. Chaboche, Continuum damage mechanics: Part II—Damage growth, crack initiation, and crack growth, J. Appl. Mech. Vol. 55, 1988.
12. M. Yahiaoui, M. Bracco, P. Little and K.A. Trauth, Experimental studies on scale models for grounding, Joint MIT-Industry Project on Tanker Safety, Report No. 18, 1994.
13. R. Thunes, Development of analytical models of wedge indentation into unidirectionally stiffened and orthogonally stiffened double hulls, Joint MIT-Industry Project on Tanker Safety, Report No. 21, 1994.

Design of corrugated bulkhead of bulk carrier against accidental flooding load

Hiromu Konishi,[a] Tetsuya Yao,[b] Toshiyuki Shigemi,[c] Ou Kitamura,[d] and Masahiko Fujikubo[b]

[a]Development Department, Nippon Kaiji Kyokai
4-7, Kioi-Cho, Chiyoda-Ku, Tokyo 102-0094, Japan

[b]Department of Naval Architecture & Ocean Engineering, Hiroshima University
1-4-1, Kagamiyama, Higashi-Hiroshma, Hiroshima 739-8527, Japan

[c]Research Center, Nippon Kaiji Kyokai
1-8-3, Ohnodai, Midori-Ku, Chiba 267-0056, Japan

[d]Nagasaki Research & Development Center, Mitsubishi Heavy Industries, Ltd.
5-717-1, Fukahori-Machi, Nagasaki 851-0392, Japan

Elastoplastic large deflection analysis is performed on a 1/2+1/2 holds model of an existing bulk carrier, and the collapse behaviour of a transverse bulkhead of corrugated plating subjected to lateral pressure due to flooding in one of the holds is discussed. Similar analysis is performed also on a half-pitch model of a corrugated bulkhead, and the collapse strength is evaluated. Based on the results of analyses, a simple formula is proposed for the design purpose to estimate the collapse strength of a corrugated bulkhead subjected to accidental flooding load.

1. INTRODUCTION

Due to the occurrence of serious casualties on a large number of bulk carriers at the early 1990's, the International Maritime Organization decided to enforce regulations on the structural strength of existing and new bulk carriers. In this connection, the authors performed series of FEM analyses and considerations on the collapse strength of the corrugated bulkhead exposed to flooding sea water into a cargo hold.

At the beginning, elastoplastic large deflection analysis is performed by the FEM on port side entire structure with 1/2+1/2 holds located forward and aft of the subject transverse bulkhead. Sea water pressure working on the transverse watertight bulkhead and on the double bottom is loaded stepwise to follow the actual loading process.

Then, a half-pitch model of a corrugated bulkhead is generated, and the rationality to use the half-pitch model is examined. The influences of mesh size and the magnitude of load increments are also examined.

After this, half-pitch models are generated for nineteen existing bulk carriers, and a series of collapse analyses was performed applying the FEM to evaluate the collapse strength. Also for this calculation, an actual loading condition at flooding is simulated.

Based on the results of analysis on half-pitch models, a simple formula is derived to evaluate the collapse strength of a corrugated bulkhead applying the Rigid Plastic Mechanism Analysis assuming two plastic hinges. The influence of local buckling at the flange of a corrugated bulkhead as well as the effects of the shedder and the gusset plates are taken into account in the proposed formula. The calculated results by the proposed formula is compared with those by the FEM analyses.

2. 1/2+1/2 HOLDS MODEL ANALYSES

To simulate the actual collapse behaviour of a corrugated bulkhead subjected to lateral pressure due to flooding sea water, 174,000 DWT bulk carrier was considered, and a port side 1/2+1/2 holds model is generated with nearly 67,000 elements. Figure 1 shows the finite element representation of the model. The modelled corrugated bulkhead has shedder plates and gusset plates. The same model is also generated but without gusset plates. Failure of welded connection is considered in the analysis. The pressure is applied always perpendicular to the plate surface.

Sea water pressure working on the transverse watertight bulkhead and on the inner bottom is loaded stepwise to follow the actual flooding process. External sea water pressure and internal heavy ore pressure working on the double bottom plating are also considered.

For the FEM analysis, an explicit finite element code, LS-DYNA3D, specially customised for the collapse simulation of the ship structure is employed. The customisation of the code has been done by the Association for Structural Improvement of the Shipbuilding Industry, Japan in the research project on 'Prediction Methodology of Tanker Structural Failure & Consequential Oil Spil' supported by the Ministry of Transport, Japan. Accuracy of the calculated results has been verified through a number of experiments including full scale collision tests in 1991 and 1997, and large scale grounding tests in 1995, which were carried out as the joint research project of The Netherlands, Germany and Japan.

The solid lines in Fig. 2 represent the water head-lateral deflection relationships obtained by the FEM analysis for the cases with and without gusset plates, respectively. The water head is measured from the lower-end of the corrugated bulkhead. The deflection is non-dimensionalised by the height of a corrugated bulkhead.

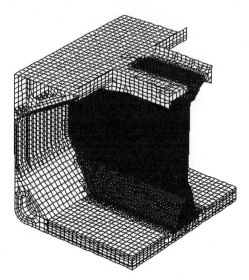

Figure 1. 1/2+1/2 holds model

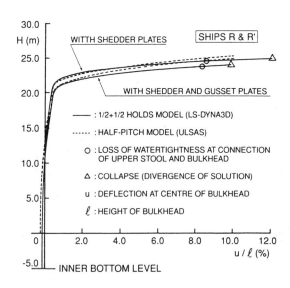

Figure 2. Sea water head-dflection relationship

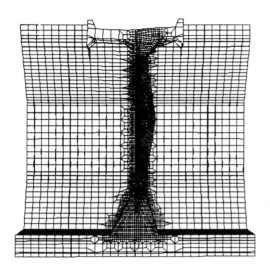

Figure 3. Collapse mode (with shedder plates)

In the case with shedder and gusset plates, overall bending deformation of the bulkhead increases with the increase in flooded sea water head, and local buckling takes place at flanges and webs of the corrugated plates at the compression side of bending near the mid-height part when the water head reaches 20 m above the lower-end of the bulkhead. Soon after this, buckling takes place also at flanges and webs near the lower-end of the bulkhead. Beyond this water head, lateral deflection rapidly increases, and yielded regions spread around the mid-height and the bottom parts of the bulkhead. The upper stool rotates as lateral deflection of the bulkhead increases. At last, plastic hinges are formed near mid-height and bottom parts of the bulkhead.

As the water head goes over 20.5 m, a part of fillet weld connecting the bulkhead and the upper stool begins to fail, but this failure of weld does not propagate. With further increase in the water head, fillet weld connecting the buckhead and the lower stool also starts to fail locally. This failure also does not propagate. After the water head reaches 23.6 m, the plate begins to break at the upper-end of bulkhead, and the watertightness is

partly lost. The solution diverges when the water head approaches to 23.9 m above the lower-end of the bulkhead.

Similar collapse behaviour is observed also in the case with only the shedder plates, but the collapse load is about 1.0 m higher compared to the case with both shedder and gusset plates. The collapse mode for this case is indicated in Fig. 3.

3. HALF-PITCH MODEL ANALYSES

By performing elastoplastic large deflection analysis using a 1/2+1/2 holds model, actual collapse behaviour can be simulated as described above. However, such analysis is time consuming, and is not suitable for series analyses. So, an alternative model, a half-pitch model, is considered, which is shown in Fig. 4 together with boundary conditions. The lower-end is clamped and the upper-end simply supported. These boundary conditions are based on the results of calculation on the 1/2+1/2 holds models.

At first, the bulkheads of the same bulk car-

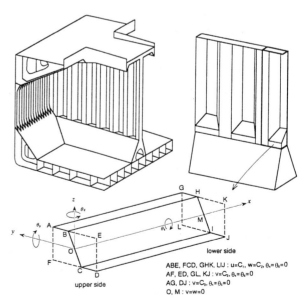

Figure 4. Half-pitch model

160

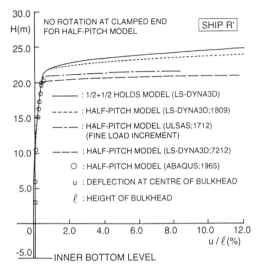

Figure 5. Influence of mesh size and load increments on collapse behaviour

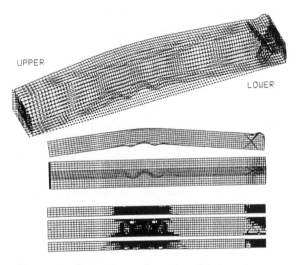

Figure 6. Collapse mode and yielded region

rier in Chapter 2 are analised with the same element size. Forced rotation of 0.002 radian is initially imposed at the lower-end to simulate the effect of ore pressure on the bottom in case of alternate loading. Then, triangular distributed load is given to the corrugated bulkhead until flooding water head reaches the upper-end of the bulkhead, and hereafter the trapezium distributed load on it. The obtained relationships between the sea water head and lateral deflection are plotted by dotted lines in Fig. 2. Comparing the dotted lines with the solid lines, it may be concluded that collapse behaviour of a corrugated bulkhead subjected to flooding load is well simulated by the half-pitch model, and so the collapse strength.

The mesh size in a 1/2+1/2 holds model may be the possible minimum size considering the ability and capacity of a present computer. However, for a halh-pitch model, a finer mesh is possible. So, analysis is performed using half-pitch models dividing original elements into four, and the influence of mesh size is examined. The number of elements of a half-pitch part excluding shedder and gusset plates for the original model is 1809, and for the fine-mesh model 7212. The computer codes LS-DYNA3D and ABAQUS are used. Another model with 1712 elements is also analised using the computer code, ULSAS. For this analysis, small load increments of 1/10 is applied. The results of analyses on the bulkhead with shedder plates are plotted in Fig. 5. The solid line is for the 1/2+1/2 holds model. The dotted, the chain and the broken lines are for half-pitch models with 1809, 1712 and 7212 elements, respectively. For these analyses, forced rotation is not imposed at the clamped end. It is seen that the collapse water level depends on the mesh size as well as the load increments, and the analysis with 1712 elements and small load increments may give satisfactory results for this bulkhead. From the results of finer mesh analysis, it can be said that the strength reserve after the formation of plastic hinges cannot be expected.

For the case with shedder and gusset plates (1712 elements), the collapse mode and the yielded region are shown in Fig. 6. It is seen that local buckling occured at the web and the flange in the compression side of bending near the mid-height and bottom parts of the corrugated bulkhead. The plastic mechanism is formed with two plastic hinges at the bottom and the mid-height

Table 1
Dimensions of corrugated bulkheads for analysis

Ship	Size	a (m)	b (m)	d (m)	ℓ (m)	$t_{f\ell}$ (mm)	$t_{w\ell}$ (mm)	t_{fm} (mm)	t_{wm} (mm)	h_t (m)	t_s (mm)	t_g (mm)	h_g (m)	σ_Y (N/mm²)
A	Handy	0.880	0.863	0.800	11.650	12.5	11.5	-	-	-	11.0	12.5	1.40	235
B	Handy	1.250	1.098	0.950	10.917	20.0	14.0	16.0	12.5	2.60	16.5	20.0	0.45	355
C	Handy	0.970	1.146	0.800	12.780	16.5	16.5	-	-	-	12.0	-	-	315
D	Handy	0.850	0.877	0.800	15.320	22.0	14.0	15.0	12.0	3.05	22.0	-	-	315
E	P'max	0.800	0.881	0.840	11.100	14.5	13.0	-	-	-	11.0	-	-	315
F	P'max	0.900	0.943	0.800	12.890	12.5	12.5	-	-	-	12.5	12.5	1.20	315
G	P'max	0.970	1.044	0.900	12.310	15.5	15.5	-	-	-	14.0	-	-	315
H	P'max	0.938	1.020	0.905	14.810	15.5	14.0	12.5	12.0	3.30	10.5	-	-	355
I	P'max	0.800	0.935	0.840	10.680	14.5	13.0	-	-	-	11.0	-	-	315
J	P'max	0.935	1.000	0.900	14.210	16.5	16.5	13.5	13.5	1.20	16.5	-	-	315
K	Cape	1.150	1.217	1.100	15.000	19.5	17.0	17.0	15.5	3.50	12.0	19.5	0.40	355
L	Cape	1.100	1.077	1.000	15.000	17.5	17.5	-	-	-	14.5	-	-	315
M	Cape	1.200	1.208	1.100	13.200	18.5	18.5	15.0	15.0	6.50	16.0	-	-	315
N	Cape	0.960	0.937	0.885	16.303	22.5	15.5	-	-	-	15.0	-	-	315
O	Cape	1.270	1.350	1.170	16.120	20.0	20.0	17.0	17.0	5.95	13.0	20.0	0.30	315
P	Cape	1.020	1.315	1.100	18.810	25.0	18.5	17.5	14.0	2.00	16.0	25.0	0.50	315
Q	Cape	1.160	1.286	1.055	15.310	21.0	21.0	-	-	-	10.0	-	-	235
R	Cape	1.160	1.159	1.000	14.542	19.0	17.0	16.5	16.0	3.50	12.5	19.0	0.40	355
R'	Cape	1.160	1.159	1.000	14.542	19.0	17.0	16.5	16.0	3.50	12.5	-	-	355

Table 2
Comparison of collapse water heads by the proposed formula and the FEM

Ship	A	B	C	D	E	F	G	H	I	J	K	L	M	N	O	P	Q	R	R'
Proposed	10.3	18.7	10.3	9.3	16.6	9.6	13.5	9.0	16.2	10.0	13.5	11.3	15.2	12.2	8.1	8.1	10.1	12.7	12.1
FEM	10.8	17.4	12.3	10.3	19.2	9.9	15.1	9.3	18.8	10.9	15.0	12.9	16.4	13.9	12.2	9.4	11.5	12.9	14.0

(in m)

parts.

Series of collapse analyses are performed on the corrugated bulkheads of nineteen existing bulk carriers using a half-pitch model with small load increments and relatively fine mesh. The dimensions of individual bulkheads are summarised in Table 1, and the collapse sea water head in Table 2, which is determined as the point that deflection start to increase rapidly. The parameters in Table 1 are explained in the Appendix.

4. SIMPLE METHOD TO EVALUATE COLLAPSE STRENGTH

It was shown in Chapters 2 and 3 that all the corrugated bulkheads collapsed with two plastic hinges located at the bottom and the mid-height parts. This implies that the Rigid Plastic Mecha-

nism Analysis can be applied to evaluate the collapse strength. For several bulkheads in Table 2, analysis is performed with half-pitch models applying uniformly distributed lateral load. The average pressure-deflection relationships under this loading condition are almost the same with those under triangular and then trapezium distrubuted load. Therefore, uniformly distributed load is considered for simplicity when the Rigid Plastic Mechanism Analysis is applied.

Here, a half-pitch of a corrugated bulkhead is modelled as a beam as indicated in Fig. 7. According to the Rigid Plastic Mechanism Analysis, a plastic mechanism is formed when the lateral pressure per unit area reaches:

$$q_f = \frac{8}{C \cdot s_\ell \cdot \ell^2}(0.5 M_{\ell e} + M_m) \qquad (1)$$

where $M_{\ell e}$ and M_m are fully plastic bending mo-

162

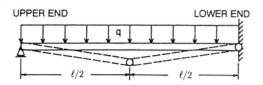

UPPER END   LOWER END

q

$\ell/2$ $\ell/2$

Fig. 7. Assumed plastic mechanism

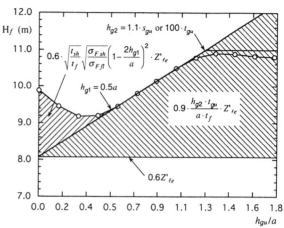

Figure 8. Influences of shedder and gusset plates on collapse waterhead

ment at the lower end and the mid-span of the bulkhead, respectively, and are expressed as:

$$M_{\ell e} = Z_{\ell e} \cdot \sigma_{a,\ell e}, \qquad M_m = Z_m \cdot \sigma_{a,m} \quad (2)$$

Z and σ_a in Eq.(2) are the plastic section modulus and the allowable stress, respectively.

C in Eq.(1) is 1.0 for the exact solution of the Rigid Plastic Mechanism Analysis, but is taken as 1.1 to get better agreement with the FEM results.

The exact location of the plastic hinge is not a mid-span point, but the difference in the collapse load from the exact solution is small. So, the plastic hinge in the span is assumed to form at a mid-span point for simplicity.

The FEM analysis shows that local buckling takes place before plastic hinges have been formed. To account for the influence of local buckling in the flange, the effective width, $a_{ef} = C_e \cdot a$ is used as the width of flange, where C_e is:

$$C_e = \begin{cases} 2.25/\beta - 1.25/\beta^2 & \beta > 1.25 \\ 1.0 & \beta \le 1.25 \end{cases} \quad (3)$$

$\beta = a/t_f \sqrt{\sigma_{Ffl}/E}$ is the slenderness ratio of a flange plate.

Here, at the lower-end of the bulkhead, the web plate is welded to the top plate of the lower stool, but there usually exists no continued member on the backside of the top plate. So, 30% of the web area is considered to be effective, and the web thickness is reduced to 30% of its own thickness when the plastic section modulus, $Z'_{\ell e}$, at the lower-end of the bulkhead is calculated.

In addition to this, at the lower-end of the bulkhead, the influence of shear force has to be considered when the plastic bending moment is

calculated. In the present method, it is approximately accounted by reducing the plastic section modulus to 60% of $Z_{\ell e}'$ at the lower-end of the bulkhead when shedder and gusset plates are not provided.

$$Z_{\ell e(no-s\&g)} = 0.6 Z'_{\ell e} \quad (4)$$

According to the results of the FEM analysis, the collapse water head changes with respect to the gusset height, h_{gu}, as indicated by circles in Fig. 8 when the gusset and the shedder plates are provided at the lower-end of the corrugated bulkhead.

When only the shedder plates are provided, the fully plastic bending moment at the lower-end of the corrugated bulkhead increases nearly to twice because the local buckling is prevented. When gusset plates are also provided, this effect decreases as the gusset height increases, and is almost lost when it becomes nearly a half flange width. However, as the gussets become much more higher, forces which are transmitted from the flange increases. With further increase in the gusset height, larger forces are transmitted to the gusset plates, and the gusset plates become to undergo local buckling. For this reason, the increasing fully plastic bending moment at the lower-end of a bulkhead becomes to show its plateau for higher gusset plates.

To represent such effects of the shedder and the gusset plates, the following contributions are considered.

$$\Delta Z_{\ell e(sh)} = 0.6 \cdot \sqrt{\frac{t_{sh}}{t_f}} \sqrt{\frac{\sigma_{Fsh}}{\sigma_{Ffl}}} \left(1 - \frac{2h_{g1}}{a}\right)^2 \cdot Z'_{\ell e}$$
(5)

$$\Delta Z_{\ell e(gu)} = 0.9 \cdot \frac{h_{g2} \cdot t_{gu}}{a \cdot t_f} \cdot Z'_{\ell e}$$
(6)

In Eq.(5), h_{g1} is equal to the gusset height, h_{gu}, when it is smaller than $a/2$, but is equal to $a/2$ when h_{gu} is greater than $a/2$. h_{g2} in Eq.(6) is equal to the gusset height, h_{gu}, but is equal to the smaller one of $1.1 \cdot s_{gu}$ and $100 \cdot t_{gu}$ when h_{gu} exceeds either of them.

Consequently, the plastic section modulus at the lower end of the corrugated bulkhead is expressed as:

$$Z_{\ell e} = Z_{\ell e(no-s\&g)} + \Delta Z_{\ell e(sh)} + \Delta Z_{\ell e(gu)}$$

$$= 0.3 \cdot \left\{ 2 + 2 \cdot \sqrt{\frac{t_{sh}}{t_f}} \sqrt{\frac{\sigma_{Fsh}}{\sigma_{Ffl}}} \left(1 - \frac{2h_{g1}}{a}\right)^2 \right.$$
$$\left. + \frac{3 \cdot h_{g2} \cdot t_{gu}}{a \cdot t_f} \right\} \cdot Z'_{\ell e}$$
(7)

Explanation of the parameters in Eqs.(1) through (7) is given in the Appendix.

Taking the allowable stresses, $\sigma_{a,\ell e}$ and $\sigma_{a,m}$ as the yielding stress, collapse water heads are calculated for the corrugated bulkheads in Table 1, and the results are summarised in Table 2. Fairly good agreements are observed between the results by the proposed formula and the FEM analyses.

5. CONCLUSIONS

In the present paper, elastoplastic large deflection analyses are performed on the corrugated bulkhead of a bulk carrier subjected to accidental flooding load using $1/2+1/2$ holds models and half-pitch models, and the collapse behaviour and the strength are discussed.

Then a simple design formula is derived to evaluate the collapse water head when the corrugated bulkhead is subjected to flooding load. The formula is based on the Rigid Plastic Mechanism Analysis introducing two plastic hinges at the lower-end and the mid-depth point. The influences of the gusset and the shedder plates are included.

The calculated results are in good agreements with those by the FEM. It can be concluded that the proposed formula gives the accurate collapse water head.

APPENDIX: PARAMETERS IN PROPOSED FORMULA

$Z_{\ell e}$: plastic section modulus at the lower-end considering all factors

$Z'_{\ell e}$: plastic section modulus at the lower-end considering the influence of buckling and non-effectiveness of web

Z_m: plastic section modulus at mid-span considering the influence of buckling

$\sigma_{a,\ell e}$: allowable stress at the lower end

$\sigma_{a,m}$: allowable stress at mid-span

σ_{Fsh}: yielding stress of shedder plate

σ_{Ffl}: yielding stress of flange

s_{gu}: width of gusset plate

t_{gu}: thickness of gusset plate

h_{gu}: height of gusset plate

t_{sh}: thickness of shedder plate

ℓ: height of corrugated bulkhead

a: width of flange

t_f: thickness of flange

s_ℓ: width of half-pitch

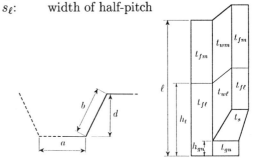

Analysis of the Collision between Rigid Bulb and Side Shell Panel

G. Woisin

Ing.-Büro Gerhard Woisin (Private Marine & Safety Consultant)
Am Haferberg 72, D–21502 Geesthacht (nr. Hamburg), Germany

The inner mechanics of a longitudinally stiffened side shell panel in a collision with a rigid bulb is analysed. The bulb has any parabolic stem contour to what corresponds the *contact line* of both ships. Modelled is the plastic membrane behaviour of the side shell before any fracture takes place or the supporting deep transverses yield. The material is rigid-plastic with proper strain hardening (not with a simply increased constant *working stress*). The collision reaction force can be regarded as composed of three parts due to shear, normal yielding and strain hardening stresses in the panel at the contact line. The two former parts vary mainly with the penetration depth, i.e., they are largely independent of the state of strain that is produced. The smallest third part of the collision reaction force depends, however, in addition strongly on the state of strain in the panel at the contact line. Above all does the with reference to fracture of the panel critical penetration depth depend on it. That part of this study which is conducted on the state of strain in the plastic membrane area on both sides from the contact line (and its extension with the penetration depth also in longitudinal direction), due to lack of space, will be published later. The analytic model applies to the initial penetration phase and agrees well with a static test performed by H. Ito.

1. INTRODUCTION

Whether a protruding bulb is fitted to the bow or not changes the kind of the damages in a ship-ship collision. This concerns particularly the consequences for the laterally struck ship. In one hand is a bulbous bow mostly locally more resistant to being flattened (*harder*) than a bow without a bulb. In the other hand, mainly due to its largely rounded (*softer*) form, does it not as often tend to perforate the other ship's side shell. Above all does a bulb cause mostly damage below the water line to the side shell of the other ship with fracture and following leakage, instead of only above the water line as often happens with a pure raked stem. Therefore has been discussed to ban bulbs with merchant ships or to invent soft bulbs. (Oddly enough, also a further strengthening of the bulb has been proposed to protect the ramming bow from suffering any damage.) Both ideas seem to fail, and that the more as both would increase mainly the safety of the other ships rather than of the own. On the other side does the side shell on both sides of a ship extend too largely to protect it with a *bulb-safe* strengthening by means of additional *collision stringers* at close spacing. It seem to exist options, however, for more intelligent solutions without pure *strengthening* and corresponding additional expense:

a) such a design of the side shell that the penetration depth with the fracture being imminent increases considerably; this would enlarge both, in the side shell absorbed kinetic energy and the collision reaction force before fracture occurs; b) such an optimisation of the form of the (more or less pointed) bulb to maximise the critical penetration depth and with it again the absorbed kinetic energy.

Subject of this study is the *inner mechanics* of a rectangular collision of a *rigid* bulb of a ship's bow into the plane vertical panel of the *elastic-plastic* side shell of another ship. There exist generally mainly three options to approach the problem of the inner collision mechanics:

- physical experiments with complete ships (mostly prohibitive), or with partial structure models on a special test facility;
- numerical models (FEM; still costly), i.e. some kind of mathematically conducted experiments;
- a (simplifying) analytic model.

Because of the large number of parameters and their combinations with ship–ship collisions, even with one design of a ship's side wall, is the third option indicated. Equations of solution in closed form are particularly worthwhile if they, as is mostly the case, explicitly indicate trend and power of influence of the input datas. Exact models rather serve to *validate* such an analytic model.

Previously published analytic models, e.g. [1–3], simplify on the side of the bow the load produced by the bulb as a pointed load, or as one produced by a wedge with a straight vertical edge of a length less than the panel's width, or—in connection with a model that knocks down the side wall structure into longitudinal strips—as a step function. On the side of the laterally struck ship's side wall is applied a rigid-perfectly-plastic material, i.e. the proper strain hardening is replaced by a simply increased (yield) *working stress*. Furthermore are the plastic phenomena of (a), bending at the yield hinge lines, and (b), membrane type distortions, particularly elongation in the longitudinal direction, considered as—with reference to their locations and extensions—being closely interrelated. Often is the plastic membrane area supposed to be same as the (obvious) dented area. This goes then along with the application of full restraint as boundary conditions also in the longitudinal direction of the ship at the ends of the dented area. Either are the yield hinge lines dealt with as developing into more general yield lines with additional and locally concentrated deformation. Or the longitudinal and the transverse elongation is considered as being distributed evenly in plane parts of the plate panel between the yield hinge lines. There do not vary the dimensions of the area with plastic membrane strain with increasing depth of penetration and quantity of collision force, but only by leaps and bounds, namely when deep transverses or stringers, etc., are dented, too.

With the *new analysis* the side shell has longitudinal frames which are supported by web frames in transverse direction. The contact surface with the bulb shall not extend into any decks or deep stringers. This will be typical of what mostly is thought to be the worst case. (*This analysis* may be applied separately and directly also to a longitudinal bulkhead or to an *inner skin* of a cargo hold, as long as the side shell and bulkhead do not act upon each other within the plastic membrane area.)

This analysis simplifies as far as possible and as less as necessary to create a handy but reliable tool for the evaluation of the merits and weak points of different ship designs. Thus are both ships supposed to maintain their positions in the gravity field (i.e., without any heeling, etc.). The bulb is simulated by a parabolic stem contour and the contact surface by a transverse contact line. The material model is rigid-plastic with strain hardening while the strain phase with constant yield stress is ignored. Elastic energy is ignored, as is plastic bending and torsion. The 'membrane work' in the longitudinals is reduced by their rotation and torsion due to the bulb's curvature and also by their 'evasion' due to tipping at the yield hinges. By neglect of these phenomena is the membrane work assessed too large. This is partly compensated by the neglect of torsion and bending work.

The *frame of reference* (x, y, z) has its origin in the point of the first mechanical contact of both ships. It moves with that part of the laterally struck ship which is not deformed. Axis x extends in the struck ship's longitudinal, y in its transverse, and z in the vertical direction. The inner collision mechanics is *symmetric to two planes*: x=0 and z=0. ξ and ζ indicate the longitudinal and the transverse positions and directions *with reference to the surface of the dented plating*. ξ and ζ serve mainly as subscripts with terms of strain and stress. They vary with the initial x, z-coordinates of the material elements of the structure. While the change of the initial x, y-coordinates of any element due to the dent is considered, is that of the z-coordinate neglected. Thus do the longitudinal frames and longitudinal plate strips maintain their vertical position. This assumption should be common with a shallow dent and small strains and is maintained here also for large strains up to fracture and a markedly deep dent.

The *basic idea* is to start from the image that the dent produces a *compact plastic membrane area* (p.m.a.) that extends in the side shell from the contact surface towards both ship's ends and terminates with a certain curved boundary. That area is not generally identical with the dented area but coincides with it only at the maximal width of both at the contact surface. This goes along with the image that, with any longitudinal path from the contact surface to the boundary of the p.m.a., the plastic membrane strain decreases monotonously from a maximum to zero (or any small strain within the yield phase of mild steel). The dimension of the p.m.a. increases with penetration depth δ_0 and collision force F_C in the transverse direction directly with the contact surface. At same time does it increase monotonously in longitudinal direction (as long as the p.m.a. remains within the dented panel). Namely, to balance with a quarter of the p.m.a., a) the increasing surplus of the longitudinal tension forces that arise at the contact area and in the boundary surface, and b) the shear forces along the boundary of the p.m.a., that requires its lengthening as the shear stress should not increase proportional to the penetration depth.

2. BOW PROFILE AND DENT OF SIDE WALL

2.1. Profile of bulb and contact line

The *stem contour* of a bow's bulb is idealised as a power function. The power exponent $p \geq 1$ and constant $c \geq 0$ characterise its curvature, see Fig. 1. At zero time of impact, $t=0$, it is (subscript b for *bow*):

$$y_{b(t=0)}(z) = -c|z|^p, \qquad (1)$$

where the dimension of c is $[c]=L^{1-p}$ with L as the symbol for geometric length (e.g., with $p=2$ it is $[c]=1/L$). $[c]$ is needed in case of a *dimensional check* with calculations with the equations of solution presented in the following. Table 1 contains data for six real ship's bows for which the reaction forces if flattened at a rigid plane obstacle were calculated [4], and for another real bow.

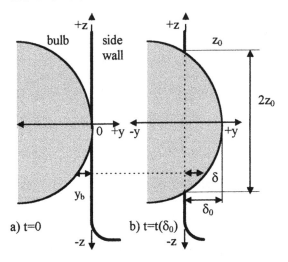

Figure 1. Stem contour of bulb y_b and contact line δ

With $\delta_{0(t)}$ as the maximal depth of the dent in the side wall in transverse direction (in the intersection of planes $x=0$ and $z=0$ at time t) is:

$$y_b(t, z) = \delta_{0(t)} + y_{b(t=0)} = \delta_{0(t)} - c|z|^p. \qquad (2)$$

With z_0 as the half vertical width of the bulb at the stem contour where it intersects the side wall in its imaginary position when the latter is not deformed but overlapping with the bulb, i.e. where $y_b(t,z)=0$, is:

$$\delta_0(t) = c\,z_0{}^p \qquad (3)$$

The profile of the dented side wall's surface in the transverse plane $x=0$ is assumed to coincide with the stem contour in the contact surface, $0 \leq z \leq z_0$:

$$\delta(t, z) = \delta_{0(t)}[1-|z/z_0|^p] = \delta_{0(t)} - c|z|^p \qquad (4a)$$

with, due to eq (3):

$$z_0(t) = [\delta_{0(t)}/c]^{1/p}, \text{ (e.g. with } p=2: z_0=\sqrt{(\delta_0/c))}. \qquad (4b)$$

The contact area between both ships shall be simulated by a *contact line* in the plane $x=0$, what is described by eq (4a)—cf. subsection 2.2.

Table 1: Samples of data for stem contours of bulbs

Type	DW/kt	L_{pp}/m	p	c/m^{1-p}
Coaster	0.5	41.0	3.0	0.5
Pallet-Carrier	1	53.8	3.4	0.45
Tanker	2	69.0	3.5	0.4
General-Cargo	3	78.0	2.0	0.25
Container-Carr.	40	211.5	2.2	0.175
Bulker	150	274	2.5	0.064
Tanker Esso M.	195	272	1.7	0.15

The following convention with exponent p simplifies the writing of some of the equations to come:

$$p_i := p/(ip+1); \; i=1,2,3, \text{ (with exponents: 'pi'} \equiv p_i). \qquad (5)$$

2.2. Assumed surface of the dented panel

Based on general experience with actual collision damages is assumed that the longitudinal frames (and with it the plating in longitudinal direction) keep straight between yield hinges what latter emerge at the boundary of the contact surface with the bulb, here the contact line at $x=0$, and close to the next deep transverses, i.e. at $-x_1$ and $+x_1$.

Vertically reaches the dented part of the side wall to where the stem contour intersects the imaginary position of the side wall, i.e. if not deformed. (A span of the plating from the stem contour to the next longitudinal, that is not deformed, is neglected.) Thus does the contact line extend vertically from $-z_0$ to $+z_0$. The outer surface of the dented side wall (subscript s) is for $-z_0 \leq z \leq +z_0$ and $-x_1 \leq x \leq +x_1$ (Fig.2):

$$y_s(t,x,z)=\delta_{(t)}(1-|x/x_1|)=\delta_{0(t)}\{1-|z/z_0|^p\}(1-|x/x_1|). \qquad (6)$$

For $x=0$ does this agree with eq (2) for the stem's position with the y, z-coordinates, i.e. the contact line.

168

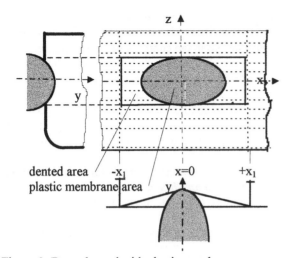

Figure 3. Dented panel with plastic membrane area

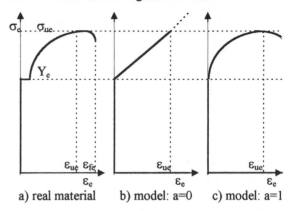

a) real material   b) model: a=0   c) model: a=1

Fig. 3: Real material and applied material model

3. MATERIAL MODEL FOR SHIP SIDE WALL

The ship's steel is modelled with *engineering stress and strain*, σ and ε, as follows, see Fig.3:

$$\sigma_e = Y_e \{1 + H_e\, \varepsilon_e\, [2 - \varepsilon_e/\varepsilon_{ue}]^a\} \qquad (7a)$$

with: Y = lower yield stress;
$H := (\sigma_u - Y)/(Y\varepsilon_u)$, average strain hardening modulus;
ε_u: strain at ultimate stress;
e: subscript: *equivalent* to uni-axial stress state;
a: exponent, either a=0 or a=1, for two different forms of strain hardening.
A yield phase with constant stress is ignored.. A *linear strain hardening* from ε_e=0 to $\varepsilon_e=\varepsilon_{ue}$, a=0, is supposed as basic case:

$$\sigma_e = Y_e (1 + H_e\, \varepsilon_e). \qquad (7b)$$

Yet, the equations of solution become only slightly more complex with a=1 or an unknown 'a'. The model with a=1 concerns the case with $\varepsilon_o > \varepsilon_{u\xi t}$ (index t: for plating), if the longitudinal frames prevent an immediate fracture of the plating at $\varepsilon_0 = \varepsilon_{u\xi t}$.

Due to St.-Venant-Lévy-Mises' law do the three principal strain increments vary proportional to the corresponding three principal stress deviators. This law is not strictly applied here. Rather is the mostly only approximately valid proportional relation of the principal stress deviators and the finite principal strains supposed (Hencky). Moreover, to calculate the equivalent stress and strain are used the engineer-

ing stress and strain rather than, as strictly necessary, the true stress and the logarithmic strain.

For a plane stress state are the ratios of the two principal stresses, $M := \sigma_2/\sigma_1$, and the total principal strains, $L := \varepsilon_2/\varepsilon_1$, interrelated, as well-known, with $L = (1-2M)/(M-2)$. From that follow for the first principal direction (a square root extends as long as the term in brackets that follows the radical sign):

$$Y_1 = Y_e/\sqrt{(1-M+M^2)};\quad H_1 = H_e\sqrt{(1-M+M^2)}/(1-0.5M);$$

and $\varepsilon_1 = \varepsilon_e\,(1-0.5M)/\sqrt{(1-M+M^2)}$. (8)

Similar equations, yet with the ratio of the longitudinal and the transverse normal stress in the plating, $m = \sigma_\zeta/\sigma_\xi$ (instead of M for the principal stresses), are applied in the analysis to the stresses a) in the cross section at the contact line, and b) (with the yield stress condition) at the boundary of the p.m.a..

4. ABOUT THE STATE OF STRAIN

From the balance of the longitudinal forces with a quarter of the p.m.a. for a penetration depth δ_0 is derived its length x_0 and the maximal longitudinal strain at the contact line ε_0. Due to lack of space, this cannot be presented here in detail.

4.1. Elongation s = f (δ) and s_0 = f ($δ_0$)
We need the relationship between the plastic elongation of the p.m.a. in longitudinal direction s_ξ (in the following simply s) and the depth of penetration δ. Both are maximal at z=0: $s = s_0$ and $δ = δ_0$. s (and s_0) is the elongation of the *total* length of the

p.m.a. (not of the half length of it). For a certain s can be derived (by Pythagoras' law):

$$\delta \approx \sqrt{(x_1 s)} \text{ or } s \approx \delta^2/x_1; \ \delta_0 \approx \sqrt{(x_1 s_0)} \text{ or } s_0 \approx \delta_0^2/x_1. \quad (9a,b)$$

From eqs (9a) and (9b) follows also:

$$\delta/\delta_0 \approx \sqrt{(s/s_0)}. \quad (9c)$$

Some radius at the horizontal entrance lines often makes the approximation even better.

4.2. On the longitudinal strain in the p.m.a.

The longitudinal normal plastic strain is at the boundary of the p.m.a., at $-\xi^b$ and $+\xi^b$, zero: $\varepsilon_\xi^b \approx 0$. ε_ξ increases in the p.m.a. with every longitudinal path to a maximum at the contact line: $\max.\varepsilon_\xi = \varepsilon_\xi^c$. To simplify we assume that the distribution of ε_ξ shows along longitudinal paths *geometric affinity* for any couple of (z/z_0)-values. At t=0 is with $|z/z_0| \leq 1$ the longitudinal location of the material at the boundary $-x^b$ and $+x^b$. Therefore is the total plastic elongation between $-\xi^b$ and $+\xi^b$:

$$s(z/z_0) = \{[\xi^b - (-\xi^b)] - [x^b - (-x^b)]\}, \quad (10a)$$

with strain $\varepsilon_\xi^c(z/z_0)$ at the contact line ($\xi=x=0$). With, because of the supposed affinity, same form factor k for the $\varepsilon_\xi(x)$–distribution at every (z/z_0), it is also:

$$s(z/z_0) = k \ \varepsilon_\xi^c x^b, \quad (10b)$$

and its maximum s_0 at z=0 with $x_0 := x^b_{(z=0)}$:

$$s_0 := s_{(t=0)} = k \ \varepsilon_0 x_0. \quad (10c)$$

Therefore is also:

$$s/s_0 = (\varepsilon_\xi^c/\varepsilon_0) (x^b/x_0). \quad (10d)$$

We insert eq (10d) in eq (9c) and get:

$$\delta/\delta_0 \approx \sqrt{\{(\varepsilon_\xi^c/\varepsilon_0)(x^b/x_0)\}}. \quad (11)$$

Furthermore we suppose that the *average* decline of ε_ξ with $|x|$ from the contact line to the boundary is constant in the p.m.a.:

$$\varepsilon_\xi^c/x^b = \varepsilon_0/x_0, \text{ or } \varepsilon_\xi^c/\varepsilon_0 = x^b/x_0. \quad (12)$$

By comparison with eqs (11) and (4a) we receive:

$$\varepsilon_\xi^c/\varepsilon_0 = x^b/x_0 \approx \delta/\delta_0 = 1 - (z/z_0)^p. \quad (13)$$

Thus emerges *geometric affinity of the boundary curve* (of each half of the p.m.a. extending to each side from the contact line) to the profile of the contact line (that again corresponds to the stem contour) —cf. Fig. 3. The affine curves own the same base line from $-z_0$ to $+z_0$. A linear decline of ε_ξ with $|x^b-x|$ gives k=1, what may be used as standard value.

4.3. Factors v_Y and v_H considering two-axial state of stress in the cross section at the contact line

The longitudinal normal stress in the cross section at the contact line varies in the plating with the two-axial state of stress as compared to the uniaxial one (superscript c: *contact line*, is omitted):

$$\sigma_\xi(\varepsilon_\xi) = v \ \sigma_e(\varepsilon_\xi); \ v = f \ (\sigma_\zeta/\sigma_\xi; \ \tau_{\xi\zeta}/\sigma_\xi; \ \varepsilon_\xi/\varepsilon_e) \quad (14a)$$

As v depends also on $\varepsilon_\xi/\varepsilon_e$, it is useful to subdivide σ_ξ into the yield part (Y) and the strain hardening part (H), where v_Y and v_H are independent from $\varepsilon_\xi/\varepsilon_e$:

$$\sigma_\xi = Y_\xi + (Y_\xi H_\xi) \varepsilon_\xi = Y_e v_Y + (Y_e H_e) v_H \varepsilon_\xi; \quad (14b)$$

$$v_Y = [1-m+m^2+3(\tau_{\xi\zeta}/\sigma_\xi)^2]^{-1/2}; \ v_H = (1-0.5m)^{-1}, \quad (15a,b)$$

($m:=\sigma_\zeta/\sigma_\xi$; due to symmetry with plane x=0: $\tau_{\xi\zeta}^c=0$).

In contrast does in the longitudinal frames (represented by the additional plate thickness t*) exist an uniaxial state of stress with longitudinal membrane tension as bending, torsion, normal transverse and shear stresses in it are neglected (indicated by *):

$$\sigma_\xi^*(\varepsilon_\xi) = Y_e + (Y_e H_e) \varepsilon_\xi. \quad (16)$$

In the analysis the factors v_Y and v_H have been used twice:

1) To derive corresponding weighted average factors V_Y and V_H for the change of the integrated longitudinal normal forces in the plating t in the cross section at the contact line. They were applied with the balance of a quarter of the p.m.a. with all longitudinal force components. That influences the half length of the p.m.a. x_0 and the maximal longitudinal strain ε_0.

2) To derive with a given ε_0 but variable m-values the parts F_Y and F_H of the collision force F_C, see section 5, it is useful to introduce different V_{YF} and V_{HF} values: weighted average *effective* factors, that refer to those stress components that act parallel to F_C.

5. DECOMPOSITION OF COLLISION FORCE

The collision reaction force F_c at penetration depth δ_o follows from a free-body diagram with the balance of the forces in the panel at the contact line. The collision reaction force is knocked down into several *parts*: the in-plane shear force (subscript S) and the (for the dented panel) longitudinal normal force (N). The latter is produced by the yield stress (Y) and by the additional stress due to pure strain hardening (H).

Here are considered those components that act parallel to the y-axis and thus balance the collision force. These *effective components* go with the following arithmetic equation:

$$F_C = F_S + F_N = F_S + F_Y + F_H. \tag{17}$$

The two former of the three terms are the major ones and can be determined *largely* as functions of the penetration depth, represented by its maximum δ_o, i.e. without close consideration of the *strain* around the contact line. In contrast, F_H is decisively determined also by the longitudinal strains at the contact line, which quantity is represented by its maximum at $z=0$, ε_o. F_Y and F_H are also influenced by the ratio of the transverse to the longitudinal stress. F_S requires some assumption about the distribution and quantity of the shear *stress* in the cross section at the contact line. To derive the effective components is needed an assumption about the pattern of the dent in the side wall (see subsection 2.2.).

Integration of the partial forces against penetration depth δ_0 gives the terms of the membrane work.

5.1. Force by shear stress $F_S = f(\delta_o)$

Strictly, due to symmetry with $x=0$ and $z=0$, is in both planes the membrane shear stress zero, here: $\tau_{\xi\zeta}^c=0$ (in the following $\tau:=\tau_{\xi\zeta}$; superscript c: *at the contact line*). However, due to reasoning *close* to $x=0$ τ will increase rapidly, and that the more larger (z/z_o) is. Approximately is with $\tau_m := \max.\tau$:

$$\tau^c \approx \tau_m^c (z/z_o). \tag{18}$$

By integration of the effective component of $(\tau^c t)$ ($t=$ thickness of plating) results for all four quarters of the p.m.a. together (the obvious range of integration from $z=0$ to $=z_0$ or from $z/z_0=0$ to $=1$, resp., is from typographical reasons not given with the equations):

$$F_S(\delta_0) = 4\,t\,\tau_m^c \int (z/z_0)[-dy_s/dz]\,dz \tag{19}$$

with due to eq (6): $dy_s/dz = -(\delta_0/z_0)p\,(z/z_0)^{p-1}$; ergo:

$$F_S(\delta_0) = 4\,p_1\,t\,\tau_m^c\,\delta_0. \tag{20}$$

The results of a parameter study with this analysis (due to lack of space not represented here) yield the average of the shear stress in the plating, $\tau_{\xi\zeta}^b$, over the length of the boundary approximately (with $p=2$, $t^*/t=0.5$, $V_Y=V_H=1.15$):

$$\tau_l \approx Y_e\,[0.15 + 0.35(H_e\,\varepsilon_0)^{1/2}], \text{ yet } \tau_l \le 0.5. \tag{21}$$

τ_m^c may be of about same magnitude as τ_l.

5.2. Force by normal yield stress $F_Y = f(\delta_0)$

F_Y is produced in the plating by Y_ξ as part of the total normal stress σ_ξ and in the frames by Y_e. The longitudinal frames are substituted by a continuous additional but separate plate thickness t^*. With the cross surface of a longitudinal frame A_l, distributed over its spacing d, it is: $t^*=A_l/d$. We integrate the components parallel to F_C and get:

$$F_Y(\delta_0, z_0) = 4\,Y_e \int (v_Y\,t + t^*)(\delta/x_1)dz, \tag{22}$$

We insert for $\delta(z)$ the term of eq (4a), introduce V_{YF} as the (*with reference to the component parallel to F_C*) effective average of v_Y in the plating, replace z_0 by the term in eq (4b) and receive:

$$F_Y(\delta_0) = 4\,p_1\,(V_{YF}t + t^*)\,Y_e\,x_1^{-1}\,c^{-1/p}\,\delta_0^{1/p1}; \text{ with:}$$

$$V_{YF}=\int\{1-m+m^2+3(\tau_{\xi\zeta}/\sigma_\xi)^2\}^{-1/2}(\delta/x_1)dz/\int(\delta/x_1)dz, \tag{23}$$

with $m=m^c:=\sigma_\zeta^c/\sigma_\xi^c$. A linear relation should suffice: $m^c \approx m_0\,(1-gz/z_0)$. With consideration of:

a) the complete balance of a quarter of the p.m.a., i.e. including also vertical forces and couples, and

b) how much restraint to the pull-in in vertical direction of the ends of the contact line, what is due to the dent, is given,

can the values m_0 and g be assessed. In the mentioned parameter study with $p=2$, $t^*/t=0.5$ and $V_Y = V_H = 1.15$ was found:

$$m_0 \approx (0.35 - 0.5H_e\varepsilon_0)(1+2.75m^b). \tag{24}$$

E.g., for $m^b=0$ (no restraint in vertical direction at the boundary, i.e. $g=1$) and $H_e=1$ is $m_0 \approx 0.35-0.5\varepsilon_0$.

The influence of τ^c in eq (23) should be small (because τ^c is the less and δ the larger the closer to z=0) and is neglected. For δ applies eq (4a). Numerical integration with Simpson's rule and three ordinates yields:

$$V_{YF} \approx \{(1-m_0+m_0{}^2)^{-1/2}+4(1-0.5^p)(1-m_1+m_1{}^2)^{-1/2}\}/6p_1. \quad (25)$$

E.g., with $m_1=m_0(1-0.5g)$, p=2, $m_0=0.3$, g=1, is $V_{YF}=1.084$; with p=1, $m_0=0.6$, g=1/6, is $V_{YF}=1.151$.

5.3. Force by normal strain hardening $F_H=f(\delta_0,\varepsilon_0)$

With v_H is correspondingly:

$$F_H = 4\,Y_e H_e \varepsilon_0\,(\delta_0/x_1)\int(v_H t+t^*)\{1-(z/z_0)^p\}^2\,dz; \quad (26)$$

or, with V_{HF} as the *effective* weighted average factor and again z_0 replaced by the term of eq (4b):

$$F_H(\delta_0,\varepsilon_0) \approx 8p_1p_2(V_{HF}t+t^*)Y_eH_ex_1^{-1}c^{-1/p}\varepsilon_0\,\delta_0{}^{1/p1}; \text{ with:}$$

$$V_{HF}=\int[1-(z/z_0)^p]^2(1-0.5m^c)^{-1}dz / \int[1-(z/z_0)^p]^2dz. \quad (27)$$

Numerical integration by Simpson's rule yields:

$$V_{HF} \approx \{(1-0.5m_0)^{-1}+4(1-0.5^p)^2(1-0.5m_1)^{-1}\}/(12p_1p_2). \quad (28)$$

With same input data as applied with the examples for V_{YF} is $V_{HF}=1.128$ and 1.404, respectively. A prerequisite to establish the explicit function $F_H=f(\delta_0)$ is that of $\varepsilon_0=f(\delta_0)$—cf. subsection 6.1.

6. VALIDATION OF THE ANALYTIC MODEL

A main problem with the validation of an analytic model on inner collision mechanics by comparison with exact physical or mathematical models is the often met lack of proper boundary conditions for the side wall model. It should be absolutely necessary to make a partial structural model ('test section') so large in length and cross section that the mainly longitudinally produced membrane tensile forces are balanced exclusively *within* the model. The boundary conditions must not change the boundary of the plastic membrane area. Thus it does not suffice that the area of plastic deformations, including the yield lines, remains distant to the boundaries of the partial structure model. Rather should no longitudinal fixa-

tion exist, as every such fixation rises doubt whether there may be any influence of it on the damage.

6.1. Comparison with a static collision model test

H. Ito et al. published in [6] about a model test with a rigid bulb model where both ends of the side wall model were fixed. It is reported that at a penetration depth of $\delta_0=63$mm (measure point 'P4') the supporting web frames started to collapse. P4 corresponded to a vertical length of the contact line of $2z_0=328$mm. The length of the model towards each end was 4.4-times of it. Necessary should be without end fixation about $0.65*\sigma_u/(\sigma_u-Y)\approx0.65*347/73=3.1$ –times. Thus may the test at least up to P4 have been without considerable influence of the fixation of the side model's ends.

To check that part of the analytic model that is represented in this paper, namely the resistance to deformation before the distribution of strain becomes most essential (particularly also with the determination of the crucial penetration depth when fracture is imminent), this test is nearly ideal. The rigid bulb model owns a parabolic stem contour with p=2 and (subscripts M and P for model and prototype) $c_M=0.00234$/mm. With a scale of the geometric lengths of $\lambda\approx10$, the latter corresponds to $c_P=0.25$/m. That is a typical quantity, see Table 1. The radius at the waterline's entrance was small: $R_M=70$mm or $R_P\approx700$mm. Shortcomings are that there were no longitudinal frames provided in the side model and that a 'collapse' of the web frames started already at P4, i.e. with $\delta_0=63$mm. The latter corresponds due to eq (29), see below, to $\varepsilon_0=0.189$. With no restraint to the boundary in vertical direction, i.e. $m^b=0$, is due to eq (24) $m_0=0.255$ and g=1. Then is $\varepsilon_e=0.233$, i.e. still less than $\varepsilon_{ue}(=0.264$, what to reach could have meant then that fracture would have been imminent).

In Table 2 are applied to the analytic models the material properties given in [6] for the side shell plating: $Y_e=274$N/mm², $\sigma_{ue}=347$N/mm² and $\varepsilon_{ue}=0.264$, i.e. $H_e=1.01$. The line with P5 is separated as, due to the mentioned reason, it does not suit for comparison. The results with an analytic model presented by H. Ito et al in [7] (there eq. (1), which simply gives $F_C=2.85$kN*δ_0/mm) and the deviations to the test results are added in the 4th and 5th column. (Incidentally—the results of the numerical model applied in [6] are up to P4 quite the same.) The deviations are not only of large quantity but the function is linear instead of parabolic.

Table 2. Comparison of the analytic collision reaction force with an experiment by H. Ito et al [6]

No.	δ_0/mm	F_C/kN test [6]	F_C/kN analyt. mod.[7]	deviations	F_S/kN	F_Y/kN	F_H/kN	F_C/kN new analyt. mod	deviations	ε_0
P1	18	22	51.3	+140%	6.6	16.9	0.8	24.3	+ 10%	0.054
P2	32.5	56	92.7	+ 66%	13.4	40.9	3.3	57.6	+ 3%	0.098
P3	50	103	142.5	+ 38%	22.6	77.6	9.6	109.8	+ 7%	0.150
P4	63	135	179.6	+ 33%	30.2	109.2	17.1	156.5	+ 16%	0.189
P5	78.5	163	223.9	+ 37%	40.0	151.2	29.4	220.6	+ 35%	0.236

In contrast, with the new analytic model are the deviations small and is the function of similar form—see the two last columns but one. The deviations with P1 to P3 are practically zero, if one regards that the largest difference of 6.8kN corresponds to only 1.3% of the maximal force and to 0.4mm in Fig. 8 of [6] (see also the scatter of material properties with 2.3mm thick plates for different parts of both models in [6]: Y=224N/mm² instead of 274N/mm², etc.). With P4 may the collapse of a web frame cause the deviation as with P5. F_H is calculated with ε_0 due to:

$$\varepsilon_0 \approx 0.675 \, \delta_0 \, \{t\sqrt{c} \, /[(t+t^*)x_1 k]\}^{2/3} H_e^{-1/3}. \quad (29)$$

This is an approximate relation ('semi-empirical' rather then purely analytic) as is eq (21). Both are derived from the already mentioned parameter study with t*/t=0.5, p=2 and $V_Y = V_H = 1.15$.

6.2. Possible effects with dynamic collision process

It should be emphasised that the quoted model test was statically conducted. The penetration depth at fracture ('crack') of the side shell was δ_{0M} =200mm ($\delta_{0P} \approx 2m$) or 93% of the width of the double hull. This meant with ships of 100kt DW each a critical collision speed of ≈2.4kn. One should be careful with a transfer of this rather favourable result to real ships. Namely, with mild steel does not only the yield stress increase but also the non-dimensional strain hardening modulus H *decrease,* and that dramatically. This can be derived for instance from the comparison of results with static and dynamic tensile test specimens in [5]. Due to eq (29) should the critical penetration depth at fracture of the side shell with a dynamic model test be much less as ε_0 varies with $H_e^{-1/3}$. Thus may ε_0 become large already with small δ_0. With a self-balancing membrane effect with tensile forces does the extension of the plastic membrane area depend essentially on the difference of the longitudinal forces in the central cross section and at the boundary. This difference decreases with ($H_e\varepsilon_0$).

7. CONCLUSION

The equations of solution which are derived in this paper give the response of a longitudinally stiffened panel plate to the penetration of a rigid bulb with any parabolic stem contour as function of the penetration depth with a good degree of accuracy. These equations show an only small influence of the strain hardening modulus and even a slightly minor contribution of the longitudinal frames to the force and the absorbed kinetic energy than an increase of the plate thickness would provide. Yet, to prevent any rush to conclusions it is most important to mention that the critical penetration depth at what fracture occurs to the panel is strongly increased by both, strain hardening and longitudinal frames. This means that the amount of energy absorbed before fracture increases dramatically with each of both. This second and most interesting result of the study could, due to lack of space, not be described in detail, but only be hinted at (see eqs 21, 24 and 29). Furthermore is pointed to the differences that will be met with the dynamic rather than static character of deformations with a ship collision. Generally does the analytic model possess a great potential to widen its applicability. Encouraged by the good agreement with the results of a static model test [6] will this be tackled.

REFERENCES

1. E. D. Egge and M. Böckenhauer, Marine Structures 4 (1991) 35.
2. L. Zhu and D. Faulkner, Marine Structures 9 (1996) 697.
3. A. Sano et al, paper no.8, SNAME/SNAJ- Conf., San Francisco, Aug. 1996.
4. P. Terndrup-Pedersen et al, Int J Impact Engng 13 (1993) 163.
5. O. Kitamura, paper no.9, SNAME/SNAJ- Conf., San Francisco, Aug. 1996.
6. H. Ito et al, J Soc Naval Arch Japan 156 (1984).
7. H. Ito et al, J Soc Naval Arch Japan 160 (1986).

Practical Design of Ships and Mobile Units
M.W.C. Oosterveld and S.G. Tan, editors.

A study on the improved tanker structure against collision and grounding damage

O. Kitamura[a], T. Kuroiwa[a], Y. Kawamoto[a] and E. Kaneko[b]

[a]Strength Laboratory, Nagasaki Research & Development Center, Mitsubishi Heavy Industries, Ltd. (MHI)
5-717-1, Fukahori-Machi, Nagasaki 851-0392, Japan

[b]The Association for Structural Improvement of the Shipbuilding Industry (ASIS)
2-5-1, Akasaka, Minato-Ku, Tokyo 107-0052, Japan

In this paper, the effectiveness and efficiency of the alternative structural designs of a VLCC are discussed from a view point of the increased energy absorption capacity (crashworthiness) against the collision or grounding damage.   The results of the comparative studies suggested a space for further improvement of the double hull design in its crashworthiness.   Guides to the side, bottom and bow structure of large ships are summarized for the prevention of oil spills.   This is a summary of the 7-year research project on "Prediction Methodology of Tanker Structural Failure and Consequential Oil Spill (1991-1995)" and "Development of New Hull Design with Improved Crashworthiness (1995-1997)".   The research project was sponsored by the Association for Structural Improvement of the Shipbuilding Industry (ASIS), Japan supported by the Safety and Technology Bureau of the Ministry of Transport.   With the spread of the ASIS methodology for the prediction of structural damage due to collision or grounding, i.e., the customized numerical simulation system, the crashworthiness and the safety of various kinds of ship designs, including innovative designs, can be evaluated systematically.

1. INTRODUCTION

Because of the massive oil spills due to tanker groundings and collisions around '90, preservation of the marine environment against crude oil tanker accidents became a matter of worldwide concern. In consequence, new tanker structure based on Double hull, Mid-deck and the Coulombi Egg concepts have been approved to comply with the renewed international regulation[1].   However among these options, only double hull design has been adopted to the newly built tankers of large size.

The Association for Structural Improvement of the Shipbuilding Industry (ASIS), Japan has completed a 7-year research project on the improved tanker safety early in '98.

In the first phase, a methodology to simulate the structural damage due to the collision or grounding was developed[1].   The methodology was realized mainly by the numerical simulation system based on the explicit finite element method (FEM).

The energy absorption capacity (crashworthiness) of the various tanker designs was evaluated in the second phase adopting the said methodology[2]. Based on the results of the numerical simulations associated with the large-scale experiments, a new type double hull design concept with improved safety against the collision and grounding damage was developed.

The ASIS research project has been supervised by the technical committee chaired by Prof. Ohtsubo from the University of Tokyo.   Mitsubishi Heavy Industries, Ltd. has exclusively been entrusted by ASIS to carry out these researches.

2. NUMERICAL SIMULATION SYSTEM

A numerical simulation system customized for the prediction of the damage to ship hull structure in case of the collision or grounding accident was developed and verified by numerous large-scale experiments as shown in section 3.

Subroutines to judge the rupture of steel plates and failure of welds were integrated to an explicit finite element core code (LS-DYNA). The criterion for the rupture of steel plates is based on the cumulative plastic strain considering the direction of principal strain and threshold failure strain, both of which are time-dependent[2]. The relationship between; (1) failure strain and gauge length and (2) true stress and true strain of the specific material should be defined a priori by the input data.

A subroutine to calculate the rigid body motion of ships was also developed and coupled with the calculation of the structural response. A schematic model of the coupling is shown in Fig. 2-1.

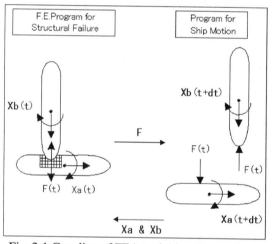

Fig. 2-1 Coupling of FEA to rigid motion analysis

3. LARGE-SCALE EXPERIMENTS

A numerical simulation system was developed and verified by the simulations following the large-scale experiments as listed below.

1) Joint Japanese-Dutch collision experiments in '91 employing 1,500 DWT inland waterway tankers[3]
2) Quasi-static grounding experiments in '91 with 1/3-scale bottom models of a VLCC
3) Quasi-static and dynamic collision experiments in '92, '93, '95 and '96 with 1/2-scale side models of a VLCC[2]
4) Joint Japanese-Dutch grounding experiments in '94 with 1/4-scale bottom models of a VLCC employing 1,000 DWT inland waterway tanker[3]
5) Joint Japanese-German-Dutch collision experiments in '97 with full-scale side models of a VLCC employing 1,500 DWT inland waterway tankers

Joint Japanese-German-Dutch collision experiments ('97:ASIS part) were carried out to confirm the reliability of the simulation system for the full-scale structures of a VLCC.

Fig. 3-1 shows the overview (numerical simulation model) of the collision experiments. Fig. 3-2 and 3-3 show the damage to the full-scale models of the conventional side shell structure (#1) and new type side shell structure (#2) of a VLCC respectively.

Both side shell structure models were fitted to the recessed side of the struck ship. A rigid bulbous bow model of 1 m in depth was fitted to the bow of the colliding ship. The colliding speed was approximately 9 knots (4.5 m/s).

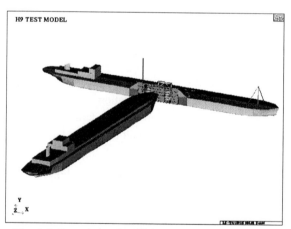

Fig. 3-1 Overview of collision experiments in '97

Fig. 3-2 Full-scale collision experiments (#1:'97)

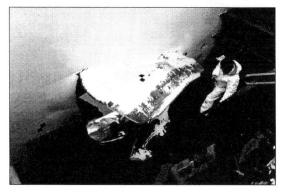

Fig. 3-3 Full-scale collision experiments (#2:'97)

4. NEW TYPE STRUCTURAL ELEMENT

As a result of a feasibility study to find out structural elements with improved crashworthiness, a Corrugated Plate and double plated panel (Frame Panel) were nominated for the options to the conventional structural elements. Fig. 4-1 shows a 1/2-scale model of the Frame Panel (6 mm in thickness and 150 mm in depth). These structural elements were tested for their crashworthiness by the dynamic collision experiments in '95. The results showed that the crashworthiness per unit weight of the Frame Panel is larger than that of Corrugated Panel by more than 50 %[2].

In '96, additional study on the steel material was carried out to enhance the performance of the Frame Panel. Two identical models made of ordinary mild steel and newly developed steel respectively were tested for their crashworthiness by dynamic

collision experiments. The results of experiments showed that the newly developed steel could increase the crashworthiness of the Frame Panel by more than 50 % as shown in Fig. 4-2 and 4-3.

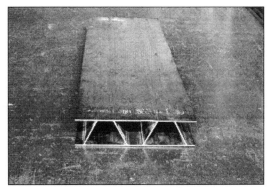

Fig. 4-1 1/2 model of Frame Panel

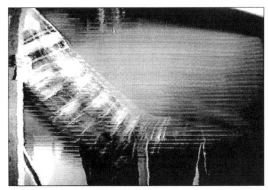

Fig. 4-2 Delayed fracture of new steel model

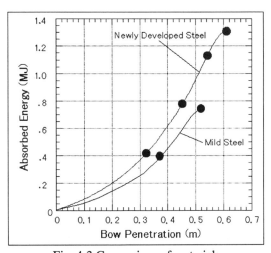

Fig. 4-3 Comparison of materials

5. COMPARATIVE STUDY

Comparative studies on structural arrangements in way of double side and double bottom were carried out using the simulation system developed in the first phase of the ASIS research project. The aim of the studies was to find out a better structural arrangement with increased crashworthiness.

5.1. Grounding

Assumed grounding accidents of Suezmax tankers with single bottom, standard double bottom and double bottom with unidirectional girder framing were simulated in '96. The double bottom design proved to be improved to a certain degree compared with single bottom design, while the unidirectional girder framing design proved to have reduced crashworthiness as shown in Table 5.1-1.

Table 5.1-1 Crashworthiness of bottom structure

Structure	Energy Absorption Capacity per Unit Length Ton-F*M/M	Net Steel Weight of Bottom Structure (per Unit B*L) Ton/M/M	Effective Energy Absorption Capacity per Unit Net Steel Weight Ton-F*M/M/Ton
Standard Single Bottom	1940	0.436	1(BASE)
Standard Double Bottom	3350	0.659	1.14
Unidirectional Girder Double Bottom	2350	0.720	0.73

5.2. Collision

Assumed collision accidents of a VLCC in laden condition with a Suezmax ship were simulated in '95 and '96. Seven kinds of double side design in total were calculated for their crashworthiness under a simplified collision condition[2].

Fig. 5.2-1 and 5.2-2 show the rough estimation of the crashworthiness of VLCCs when a Suezmax colliding ship is in ballast condition and laden condition respectively. As can be seen, the degrees of difference in crashworthiness are rather small.

The results of these comparative studies suggested; (1) the economical limit to the extensive strengthening of the side structure of a struck ship and (2) another approach such as the concept of a buffer bow structure to be considered.

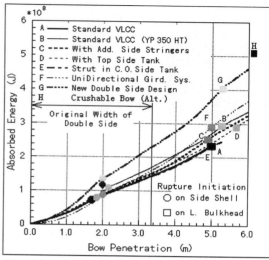

Fig. 5.2-1 Comparison of crashworthiness (Ballast)

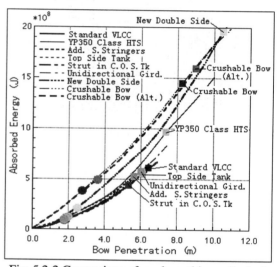

Fig. 5.2-2 Comparison of crashworthiness (Laden)

In order to estimate the crashworthiness of a VLCC accurately considering most of the factors, final numerical simulations of collision accidents were carried out as shown in Fig. 5.2-3. A Suezmax ship and VLCC navigating the crowded bay or straits, where the accidents are most likely to occur, at normal speed of 12 knots were assumed.

The comparison between the conventional and new type (Frame Panel) double sides was made. The standard bow (Fig. 5.2-4) was also compared with the prototype buffer bow (Fig. 5.2-5).

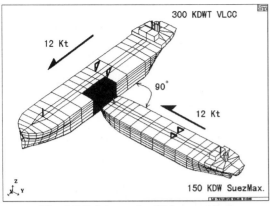

Fig. 5.2-3 Assumed collision accident

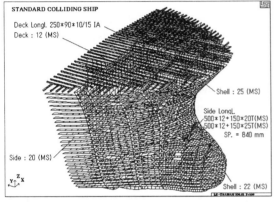

Fig. 5.2-4 Standard bow structure (skeleton)

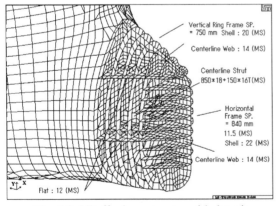

Fig. 5.2-5 Buffer bow structure (skeleton)

Damage to the standard and new type double side structures at a point of rupture initiation in the inner hull are shown in Fig. 5.2-6, 5.2-7 and Fig. 5.2-8 respectively. No rupture of inner hull or no cargo

oil spill can be expected if a Suezmax ship is in laden condition as shown in Fig. 5.2-9.

Considerable crushing of the bow is found if the double side is built with the new type structure. This is due to the enhanced strength of the double side and increased lateral bending moment caused by a struck ship in navigation. The point of rupture initiation at the side shell and/or inner hull of the new type structure may be earlier, since larger friction force is superposed on the in-plane tension force induced by the penetration of the bow.

Consequently, the overall crashworthiness of a VLCC with new type double side is judged to be more than doubled compared with a VLCC with standard one. It is also suggested that the crashworthiness of a standard double hull VLCC is about 3-4 times higher than that of a single hull VLCC, but further improvement is still required.

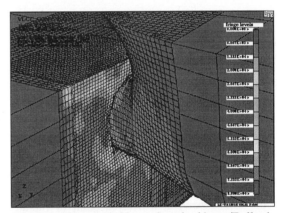

Fig. 5.2-6 Standard side vs. Standard bow (Ballast)

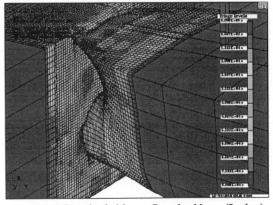

Fig. 5.2-7 Standard side vs. Standard bow (Laden)

178

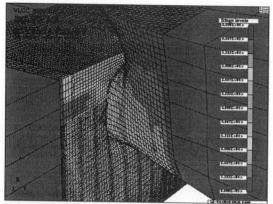

Fig. 5.2-8 New type side vs. Standard bow (Ballast)

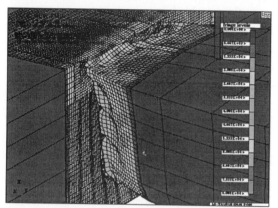

Fig. 5.2-9 New type side vs. Standard bow (Laden)

Fig. 5.2-10 and 5.2-11 show the accelerated crushing of the buffer bow (or increased penetration depth of the bow) at a delayed point of rupture initiation in inner hull, too.

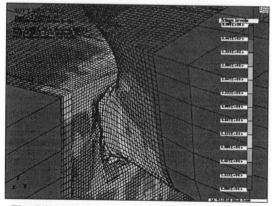

Fig. 5.2-10 Standard side vs. Buffer bow (Ballast)

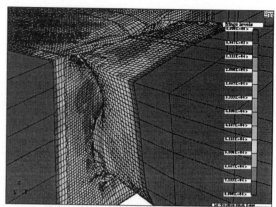

Fig. 5.2-11 Standard side vs. Buffer bow (Laden)

6. NEW TYPE TANKER DESIGN CONCEPT

Referring to the knowledge derived from the results of simulations and experiments up to '97, the development of a conceptual design of a new type VLCC with improved safety against collision and grounding was initiated. Placing priority to the realization, application of existing materials and fabrication techniques of those almost in practical use at present were precondition for this design.

The new type structure (Frame Panel) made of newly developed steel with increased energy absorption capacity is to be adopted to the side shell, inner hull, sloping top of hopper part and bottom shell. Top side tank is to be arranged to accelerate the crush of the bow structure and to reserve sufficient distance between the inner hull and raked stem of the colliding ship.

A triple bottom made of newly developed steel with double depth of the ordinary double bottom is to be fitted in way of foremost No.1 cargo tank. Primary transverses in the bottom part are to be arranged with half of ordinary spacing. These bottom designs can result in the reduced risk of oil spill and reduced extent of the damage.

A buffer box structure, bulbous bow with a flattened (deep) head and vertical ring frames (effective for the reduced buckling strength) are to be fitted for the protection of a struck ship. Profile and midship section are shown in Fig. 6-1.

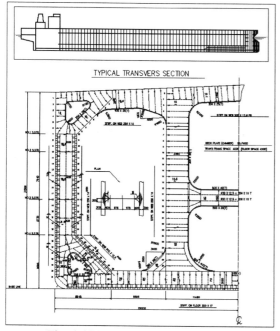

TYPICAL TRANSVERS SECTION

Fig. 6-1 New type VLCC (Concept)

7. IMPROVEMENT IN CRASHWORTHINESS

The maximum speed allowed for a colliding ship of various size, with which no oil spill should be expected from a struck VLCC, can roughly be derived from the results of the simulations. It is assumed that the energy absorbed by the damage should be proportional to the bow depth and the momentum and total energy among the colliding and struck ships should be conserved. The results are shown in Table 7-1 as one of the safety index.

Table 7-1 Allowable maximum collision speed

	Handy (Kt)		Suez Max. (Kt)		VLCC (Kt)	
	Ballast	Laden	Ballast	Laden	Ballast	Laden
Δ+A.W.M.	31130	55330	75000	173000	155100	379500
Standard Double Hull VLCC	6.2–7.1	6.7–7.7	4.8–5.7	4.9–5.8	4.3–4.7	4.5–4.9
New Type Double Hull VLCC	9.2–10.5 ↑	12 ↑	7.1–8.4 ↑	12 ↑	6.3–6.9 ↑	12?

Colliding Ship with Standard Bow

Struck Ship in Laden Condition (Δ+A.W.M. = 345000+91000 = 436000 ton)

	Handy (Kt)		Suez Max. (Kt)		VLCC (Kt)	
	Ballast	Laden	Ballast	Laden	Ballast	Laden
Δ+A.W.M.	31130	55330	75000	173000	155100	379500
Standard Double Hull VLCC	9.5	10.5	7.7	7.9	6.3	6.8

Colliding Ship with Crushable Bow

Struck Ship in Laden Condition (Δ+A.W.M. = 345000+91000 = 436000 ton)

8. CONCLUSION

It can be said that the crashworthiness of a standard double hull tanker is improved, however, there is still a space for further improvement. For a VLCC with a new type double side, the maximum allowable collision speed may be increased by approximately 50 %, whereas even 12 knots may be allowed for a Suezmax ship in laden condition.

The net steel weight is increased by about 1,000 tons if a new type double side is adopted to a VLCC. The total costs increase is estimated to be a few %. Moreover, breakthrough technologies such as an automatic butt welding procedure, leakage detecting system, etc. should be developed for the realization of the new type structure. However those may be, the degree and efficiency of the improvement (over 200 % by 1,000 tons) are to be worth considering. Since double hull regulation has been set into force, net steel weight of a VLCC has already been increased by about 10,000 tons with a resultant improved crashworthiness of 300 % and above.

The effect of a buffer bow design is considerable. The buffer bow structure may be more realistic, because this design requires almost no cost increase and no unproven fabrication technology.

REFERENCES

1. Ohtsubo, H., Astrup, O.C., Lehmann, E., Maestro, M.G., Paik, J.K., Spangenberg, S., Ximenes, M.C., Zhu, L., Yuhara, T., Kitamura, O. and Samuelides, M.S., Structural design against collision and grounding, Proceedings of the 13th International Ship and Offshore Structures Congress (ISSC97), Vol.2 (1997), pp. 83-116
2. Kitamura, O., Comparative Study on Collision Resistance of Side Structure, Marine Technology, Vol.34, No.4 (1997), pp. 293-308
3. Vredeveldt, A.W., and Wevers, L.J., Proceedings of ASIS Conference on "Prediction Methodology of Tanker Structural Failure & Consequential Oil Spill", (1992 and 1995)

Practical Design of Ships and Mobile Units
M.W.C. Oosterveld and S.G. Tan, editors.

Plastic buckling of rectangular plates subjected to combined loads

C. H. Shin, Y. B. Kim, J. Y. Lee and C. W. Yum

Research and Development Center
Korean Register of Shipping
23-7 Jang-dong, Yusung-ku, Taejon, Korea

In order to estimate the buckling strength of a rectangular plate, the analytical approach is used in this study. The plate is assumed to be simply supported on four edges and loaded by uniform stresses along the edges. If the plate is slender, the buckling is elastic. However, if the plate is sturdy, it buckles in the plastic range. Then, the instantaneous moduli in the constitutive equations depend on the external loading.

In this study, the elastic and plastic buckling equations are derived for rectangular plates under biaxial loading, and the corresponding interaction curves are presented. The influences of aspect ratios, load ratios and hardening factors on the buckling stresses are investigated for rectangular plates. From the plastic buckling analysis, the optimal combination of loads is given for the buckling strength.

1. INTRODUCTION

It is well known that ship's platings in most cases are subjected to combined in-plane loads. Therefore, biaxial loading conditions must be taken into account for the buckling strength assessment. In this study, using the analytical approach, general buckling equations are derived for simply supported rectangular plates under biaxial loading.

If the plate is slender, the buckling occurs in the elastic range and its buckling stress is less than the yield stress. For a sturdy plate, however, the yielding of plate material occurs before the plate buckles and the buckling load is smaller than the value given by an elastic buckling analysis [1]. Therefore, the plastic buckling load is an important factor which can not be overlooked in design [2]. If the plate buckles in the plastic range, the instantaneous stress-strain relationship depends on the external loading [3].

For elastic-plastic constitutive relations the deformation theory is employed and the material is assumed to be incompressible for simplification of buckling equations. The instantaneous moduli in the constitutive equations are obtained by modifying the Prandtl-Reuss stress-strain relation for a work-hardening material [4]. Then, analytical expressions for elastic and plastic buckling of biaxially loaded plates are obtained, and the corresponding interactive buckling curves are presented.

The plastic buckling load can be calculated using a plasticity reduction factor, i.e., it can be obtained by correcting the elastic buckling load. To this end, the formula for the plasticity reduction factor is developed.

In this study, the influences of aspect ratios, load ratios and material hardening factors on the buckling stresses are investigated for rectangular plates. From the plastic buckling analysis, the optimal load ratio, which results in the highest buckling load, can be found for a given plate.

2. ELASTIC BUCKLING OF A PLATE

A simply supported rectangular plate is loaded by uniform in-plane compressive stresses σ_x and σ_y, as shown in Fig. 1. From an equilibrium analysis of a plate, the following buckling equation can be obtained:

$$\frac{t^2}{12}\{D_{xx}dw,_{xxxx}+2(D_{xy}+2G_{xy})dw,_{xxyy}+$$
$$D_{yy}\,dw,_{yyyy}\}+\sigma_x\,dw,_{xx}+\sigma_y\,dw,_{yy}=0, \quad (1)$$

where dw is a small increment of the plate displacement due to buckling, t is the plate thickness and D_{xx}, D_{xy}, G_{xy}, D_{yy} are the instantaneous moduli for the case of plane stress:

$$d\sigma_x = D_{xx}d\varepsilon_x + D_{xy}d\varepsilon_y \quad (2a)$$
$$d\sigma_y = D_{xy}d\varepsilon_x + D_{yy}d\varepsilon_y \quad (2b)$$
$$d\tau_{xy} = G_{xy}\,d\gamma_{xy}\;. \quad (2c)$$

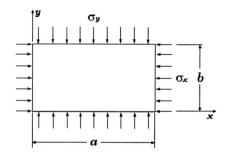

Figure 1. Rectangular plate subjected to biaxial loading.

It is well known that a solution of the form

$$dw = dA\sin\frac{m\pi x}{a}\sin\frac{n\pi y}{b}$$
$$m,n=1,2,3,\cdots \quad (3)$$

satisfies both the differential equation (1) and the simple-support boundary conditions. The parameters m and n indicate the number of half-waves in x and y direction in the buckled shape, respectively. By substituting Eq. (3) into Eq. (1), the following eigenvalue equation is obtained:

$$\left(\frac{m}{\beta}\right)^2\sigma_x+n^2\sigma_y=$$

$$\zeta\left\{\left(\frac{m}{\beta}\right)^4 D_{xx}+2\left(\frac{mn}{\beta}\right)^2\times\right.$$
$$(D_{xy}+2G_{xy})+n^4 D_{yy}\}, \quad (4)$$

where

$$\beta=\frac{a}{b}\,, \qquad \zeta=\frac{\pi^2 t^2}{12\,b^2}\,. \quad (5a,b)$$

If the plate is slender, the buckling is elastic and disappears when the load diminishes, since the buckling stress is below the yield stress. For a slender plate [5] defined by

$$\frac{b}{t}\sqrt{\frac{Y}{E}}>2.4, \quad (6)$$

the plate will mostly buckle in the elastic range with the following moduli:

$$D_{xx}=D_{yy}=\frac{E}{1-\nu^2} \quad (7a)$$
$$D_{xy}=\frac{\nu E}{1-\nu^2} \quad (7b)$$
$$G_{xy}=\frac{E}{2(1+\nu)}\,. \quad (7c)$$

where E is Young's modulus, ν is Poisson's ratio and Y is the uniaxial yield stress. Using the above moduli (7a, b, c), the eigenvalue equation (4) can be written as

$$\left(\frac{m}{\beta}\right)^2\sigma_x+n^2\sigma_y=\frac{E\zeta}{1-\nu^2}\left\{\frac{m^2}{\beta^2}+n^2\right\}^2. \quad (8)$$

Neglecting elastic compressibility, i.e., $\nu=1/2$, the elastic buckling equation is expressed as follows:

$$\left(\frac{m}{\beta}\right)^2\sigma_x+n^2\sigma_y=\frac{4}{3}E\zeta\left\{\left(\frac{m}{\beta}\right)^2+n^2\right\}^2. \quad (9)$$

In the above equation, it is seen that the buckling stresses σ_x and σ_y are interdependent and can be presented in terms of interaction curves:

$$\frac{S_x}{K_x}+\frac{S_y}{K_y}=1. \quad (10)$$

where

$$S_x = \frac{3}{4}\frac{\sigma_x}{E\zeta}, \qquad S_y = \frac{3}{4}\frac{\sigma_y}{E\zeta} \qquad \text{(11a, b)}$$

$$K_x = \frac{\beta^2}{m^2}\left(\frac{m^2}{\beta^2} + n^2\right)^2 \qquad \text{(11c)}$$

$$K_y = \frac{1}{n^2}\left(\frac{m^2}{\beta^2} + n^2\right)^2. \qquad \text{(11d)}$$

In applying the interaction formula (10), the number of half-waves m and n must be chosen to give the lowest buckling stress.

In Fig. 2, the interaction curves for elastic buckling of rectangular plates under biaxial loading are presented. This figure shows the stabilizing effect of tensile stress for three different plate geometries.

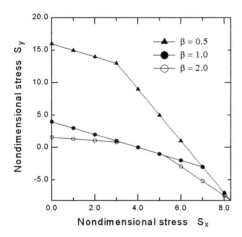

Figure 2. Elastic interactive buckling curves for biaxially loaded plates.

Using the load ratio ξ defined as

$$\xi = \frac{\sigma_y}{\sigma_x}, \qquad \text{(12)}$$

the buckling equation (9) can be expressed in the same form as for uniaxial loading:

$$\sigma_x = \frac{4}{3}K^* E\zeta. \qquad \text{(13)}$$

where K^* is a nondimensional elastic buckling coefficient:

$$K^* = \left(\frac{m^2}{\beta^2} + n^2\right)^2 \left(\frac{m^2}{\beta^2} + n^2\xi\right)^{-1}. \qquad \text{(14)}$$

Figure 3 shows the elastic buckling coefficient K^* with aspect ratio β. In this figure, negative values of ξ represent tensile loading in y direction. It is seen that the buckling stress decreases with increasing load ratio ξ.

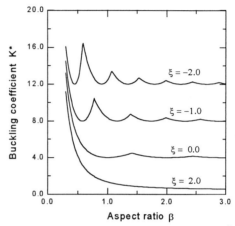

Figure 3. Elastic buckling coefficient K^* with aspect ratio β.

In Fig. 4, the elastic buckling coefficient K^* is presented with load ratio ξ for three different plate geometries. The buckling stress decreases with load ratio, regardless of plate geometries.

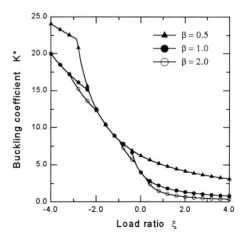

Figure 4. Elastic buckling coefficient K^* with load ratio ξ.

184

3. PLASTIC BUCKLING OF A PLATE

If the plate is sturdy, its material yielding occurs before it buckles. Then, the buckling load is smaller than the value given by an elastic stability analysis. For a sturdy plate [5] defined by

$$\frac{b}{t}\sqrt{\frac{Y}{E}} < 1.0, \tag{15}$$

the plate will mostly buckle in the plastic range. Then, the instantaneous moduli in Eqs. (2a, b, c) depend on the external loading. In this study, the deformation theory is employed for elastic-plastic constitutive relations. Thus, the instantaneous moduli can be written as follows:

$$D_{xx} = \frac{4(E_s + H_s) - 3E_s\frac{\sigma_x^2}{\sigma_e^2}}{D_0} \tag{16a}$$

$$D_{xy} = \frac{2(E_s + 2\nu_s H_s) - 3E_s\frac{\sigma_x\sigma_y}{\sigma_e^2}}{D_0} \tag{16b}$$

$$D_{yy} = \frac{4(E_s + H_s) - 3E_s\frac{\sigma_y^2}{\sigma_e^2}}{D_0} \tag{16c}$$

$$G_{xy} = G_s, \tag{16d}$$

where

$$D_0 = 2(1 - \nu_s)\left(3 + \frac{H_s}{G_s}\right) - (1 - 2\nu_s)\left(1 + \frac{3\sigma_x\sigma_y}{\sigma_e^2}\right) \tag{17a}$$

$$G_s = \frac{E_s}{2(1 + \nu_s)} \tag{17b}$$

$$\nu_s = \frac{1}{2} - \left(\frac{1}{2} - \nu\right)\frac{E_s}{E} \tag{17c}$$

$$\frac{1}{H_s} = \frac{1}{E_t} - \frac{1}{E_s}. \tag{17d}$$

In the above equations E_s is the secant modulus, E_t is the tangent modulus and σ_e is the equivalent stress:

$$\sigma_e^2 = \sigma_x^2 - \sigma_x\sigma_y + \sigma_y^2. \tag{18}$$

Neglecting elastic compressibility, $\nu = \nu_s = 1/2$, the plastic buckling equation is obtained from Eq. (4) as follows:

$$\left(\frac{m}{\beta}\right)^2\sigma_x + n^2\sigma_y = E_s\zeta\left\{\frac{4}{3}\left(\frac{m^2}{\beta^2} + n^2\right)^2 - \left(1 - \frac{E_t}{E_s}\right)\frac{1}{1 - \xi + \xi^2}\left(\frac{m^2}{\beta^2} + n^2\xi\right)^2\right\}. \tag{19}$$

When $E_s = E_t = E$, this buckling equation coincides with the elastic buckling equation (9).

In order to describe the material behavior, the following stress-strain law [4] is applied:

$$\frac{\sigma_e}{Y} = \left(\frac{E\varepsilon_e}{Y}\right)^\alpha \qquad \sigma_e \geq Y, \tag{20}$$

where α is a material constant. Then, the tangent and secant moduli can be calculated as follows:

$$E_t = \alpha E\left(\frac{\sigma_e}{Y}\right)^{\frac{\alpha-1}{\alpha}}, \quad E_s = E\left(\frac{\sigma_e}{Y}\right)^{\frac{\alpha-1}{\alpha}}, \tag{21a, b}$$

From Eqs. (19) and (21a, b), it can be seen that there is a nonlinear interaction between σ_x and σ_y. However, this interaction can be well described in terms of the nondimensional stresses:

$$\frac{R_x}{C_x} + \frac{R_y}{C_y} = 1, \tag{22}$$

where

$$R_x = \frac{Y}{E\zeta}\left(\frac{\sigma_x}{Y}\right)^{\frac{1}{\alpha}}, \quad R_y = \frac{Y}{E\zeta}\left(\frac{\sigma_y}{Y}\right)^{\frac{1}{\alpha}}, \tag{23a,b}$$

$$C_x = (1 - \xi + \xi^2)^{\frac{\alpha-1}{2\alpha}} \times \frac{\beta^2}{m^2}\left\{\frac{4}{3}\left(\frac{m^2}{\beta^2} + n^2\right)^2 - (1 - \alpha)\frac{\left(\frac{m^2}{\beta^2} + n^2\xi\right)^2}{1 - \xi + \xi^2}\right\}. \tag{23c}$$

$$C_y = \left(1 - \frac{1}{\xi} + \frac{1}{\xi^2}\right)^{\frac{a-1}{2a}} \times \frac{1}{n^2}\left\{\frac{4}{3}\left(\frac{m^2}{\beta^2} + n^2\right)^2 - \right.$$

$$\left. (1-a)\frac{\left(\frac{m^2}{\beta^2} + n^2\xi\right)^2}{1 - \xi + \xi^2}\right\}. \tag{23d}$$

As in the elastic interaction formula, the number of half-waves m and n must be chosen to give the lowest buckling stress.

Figure 5 shows the plastic interactive buckling curves for $a = 0.1$ and $\xi = 0.5$. For $R_y \cong 0$, the plastic buckling stresses for three different plate geometries have the same value.

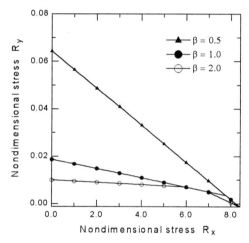

Figure 5. Plastic interactive buckling curves for biaxially loaded plates.

As before, the buckling equation in the same form as for uniaxial loading can be obtained from Eqs. (19) and (21a, b):

$$\sigma_x = C^* Y\left(\frac{E\zeta}{Y}\right)^a, \tag{24}$$

where C^* is a nondimensional plastic buckling coefficient:

$$C^* = (1 - \xi + \xi^2)^{\frac{a-1}{2}} \times$$

$$\left\{\frac{\frac{4}{3}\left(\frac{m^2}{\beta^2} + n^2\right)^2 - \frac{1-a}{1-\xi+\xi^2}\left(\frac{m^2}{\beta^2} + n^2\xi\right)^2}{\frac{m^2}{\beta^2} + n^2\xi}\right\}^a. \tag{25}$$

In Fig. 6, for various load ratios ξ, the plastic buckling coefficients C^* are presented with aspect ratio β. The buckling stress decreases with increasing transverse load.

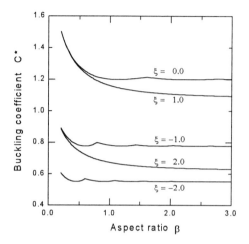

Figure 6. Plastic buckling coefficient C^* with aspect ratio β.

Figure 7 shows the variation of plastic buckling coefficient C^* with load ratio ξ for three different plate geometries. The curves have their own maxima in contrast to the elastic curves in Fig. 4. Thus, the optimal combination of loads can be found for a given plate geometry. In this figure, the highest buckling loads, which give optimal load combinations, occur with compressive stresses in transverse direction $(0.25 < \xi < 0.5)$. It is also seen that the value of load ratio for the highest buckling stress increases with decreasing aspect ratio.

In Fig. 8, the plastic buckling coefficients C^* with load ratio ξ for various hardening factors a are presented and the influence of hardening factors is investigated for a square plate. In this figure, the maxima of the buckling curves are clearly shown as in Fig. 7. The load ratios for the highest buckling stresses are in the range of $-0.5 < \xi < 0.5$. It is also seen that the value of optimal load ratio decreases with increasing hardening factor.

Thus, in this case, it can be predicted that for high hardening materials, the optimal load combinations are given by negative values of ξ, i.e., tensile loading in transverse direction.

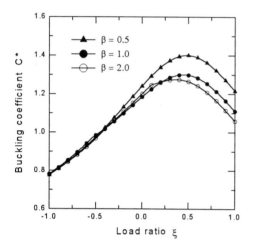

Figure 7. Plastic buckling coefficient C^* with load ratio ξ for various aspect ratios β.

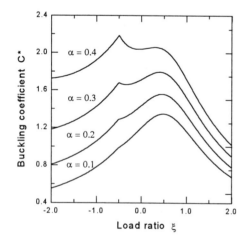

Figure 8. Plastic buckling coefficient C^* with load ratio ξ for various hardening factors α.

For the calculation of the plastic buckling load, it is useful to use a plasticity reduction factor, which is defined as the ratio of the plastic and elastic buckling loads. This plasticity reduction factor η can be calculated from Eqs. (13) and (24) as

$$\eta = \eta^* \left(\frac{E\zeta}{Y} \right)^{\alpha-1},\qquad(26)$$

where

$$\eta^* = \frac{3}{4}(1-\xi+\xi^2)^{\frac{\alpha-1}{2}} \left\{ \frac{\left(\frac{m^2}{\beta^2}+n^2 \right)^2}{\frac{m^2}{\beta^2}+n^2\xi} \right\}_{min}^{-1} \times$$

$$\left\{ \frac{\frac{4}{3}\left(\frac{m^2}{\beta^2}+n^2 \right)^2 - \frac{1-\alpha}{1-\xi+\xi^2}\left(\frac{m^2}{\beta^2}+n^2\xi \right)^2}{\frac{m^2}{\beta^2}+n^2\xi} \right\}_{min}^{\alpha}.\quad(27)$$

In the above equation the subscript min represents the smallest value for all integers of m and n. Figure 9 shows the nondimensional plasticity reduction factor η^* with aspect ratio β for $\alpha=0.1$. The factor η^* increases with increasing load ratio ξ.

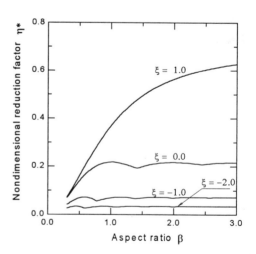

Figure 9. Nondimensional plasticity reduction factor η^* with aspect ratio β for $\alpha=0.1$.

4. CONCLUSIONS

In this study, elastic and plastic buckling equations for simply supported rectangular plates subjected to combined loads have been derived according to the deformation theory. The plate material was assumed to be incompressible for simplification.

For slender plates, the buckling is elastic and the buckling stress is lower than the yield stress. The interactive buckling curves were seen to be independent of the external loading. For sturdy plates, however, the buckling occurs in the plastic range and the moduli in the constitutive equations depend on external loads.

In this study, it was shown that the plastic interactive buckling curves were dependent on the load ratio. Using these curves, the buckling design formula for a biaxially loaded plate is expected to be accurately derived in the plastic range. It was also seen that for sturdy plates under biaxial loading, the optimal load combination could be found from the plastic buckling analysis.

Especially, the general expression for the plasticity reduction factor has been presented in this study. Then, using this factor, the plastic buckling load can be calculated with ease.

REFERENCES

1. D. O. Brush and B. O. Almroth, Buckling of Bars, Plates, and Shells, McGraw-Hill, New York, 1975.
2. C. H. Shin, Plastic Buckling Analysis of a Cylindrical Shell Considering Plastic Compressibility, Proc. 6th Int. Symp. PRADS (1995) 2.1203.
3. D. Durban, Plastic Buckling of Rectangular Plates under Biaxial Loading, in: I. Elishakoff et al. (Eds.), Buckling of Structures, Elsevier Sci. Pub., Amsterdam (1988) 183.
4. J. Chakrabarty, Theory of Plasticity, McGraw-Hill, New York, 1987.
5. O. F. Hughes, Ship Structural Design: A Rationally-Based, Computer-Aided, Optimization Approach, John Wiley & Sons, New York, 1983.

Practical Design of Ships and Mobile Units
M.W.C. Oosterveld and S.G. Tan, editors.

Investigations into the collapse behaviour of inland vessels

A. MEINKEN and H.-J. SCHLÜTER [a]

[a] Gerhard-Mercator-University-GH Duisburg; Institute For Shiptechnology (ISD); Department of Shipstrength; Bismarkstraße 69; 47048 Duisburg; Germany[*]

Modern inland vessels are open-top double-hull ships with an unusually large aspect ratio, a shallow draught and an extremely long cargo hold. These ships have a very low bending and torsional rigidity. Due to minor collisions, grounding in shallow water, corrosion and fatigue the state of the ship structure changes appreciably in course of time. These imperfections reduce the stability and strength of the structure. The safety against total collapse decreases. This paper will show the results of a systematic investigation of typical ship-constructions with the Finite-Element-Method (FEM).

1 Introduction

Newspapers, TV and the technical literature have reported that accidents and damages for ships are increasing. Not everytime human failure or bad weather was the cause. More often these damages lead back to the global structure failure of the ships. A spectacular example was the fracture of the „Carabella" in May 1996 on the river Rhine near Wesel.

It is to observe that damages as this happens for inland vessels after 15 - 20 working years. Principle the constructions are not too weak, but the structure features changes in course of time. The large need of repair confirms this statement.

The failures of a complex structure like a ship hull is not entails by one item. Usually more than one negative factors are responsible. For example these are an unfavourable cargo distribution, the reduction of the plate thickness and the large imperfections during service.

The following calculation results are in relationship to the DFG-research: „**The collapse behaviour of modern inland vessels with consideration to the structural imperfections**".

The first part of this paper will show different possibilities to calculate the structural stability and strength of an inland vessel. On account to the points mentioned above (imperfections, load distributions) the traditional calculation methods are not sufficient. Using the **Finite-Element-Method** (FEM) a new calculation concept will be introducing to achieve these elaborate analyses.

The application of the FEM for swimming structures needs a few special considerations to describe the boundary conditions. Further it is necessary to take notice of a few details that are in relationship to the calculation of structure stability. This induces that the expense for the calculations increases considerably. In this concept linear and non-linear investigations with the global model will be joining in an advantageous way.

* e-mail: meinken@nav.uni-duisburg.de
http://www.uni-duisburg.de/FB7/ISD/mitarbeiter/meinken

190

First, the simple example of a stiffened plate will be using to discuss the **influence of imperfections** on the structure behaviour.

Later the knowledge of these investigations will be utilizing for the analysis of typical inland vessels.

The FE- models of modern inland vessel will be introducing and their calculation results will be showing. These are deformations and stresses for various cargo distributions. Additionally the non-linear results for extreme loads will be presenting to simulate the collapse behaviour of the vessel. At least constructive modifications (e.g. high-tensile steel) will be discussing to show their effects.

2 Targets

Normally ships are construct for a relative long working season (20-30 years and more). In course of this long term the features of the structure are changing considerably.

To investigate the structure stability it is very difficult to take into account the imperfections (e.g. pre-deformations) with analytical calculation methods. This is a typical application for the FEM.

The heart of the following procedure is the registration and numerical consideration of typical damages. It is to investigate which imperfections are important to consider and which are negligible. A classification of the imperfections will support this.

At least new construction proposals will be developed. With this, the new ship should be lightweight, easy to build and with a maximum security against global breakdown.

2.1 Examples in practise

It is possible that the large deformations on particular components of the ship will cause a global fracture in the hull. Normally this is a constructive total loss. The damages in the ship structure are strong folds in inner bottom and trough side as well as fractures in the hatch coaming. Often this is inducing a

global bend up (Hogging) or set down (Sagging).

This relationship of local damages and global behaviour is calling interaction. In most cases a global failure or breakdown happens during loading and unloading in the harbour. Seldom on the journey e.g. due to unfavourable waves (see „Carabella"). The lecture will include a few examples of practice. These are available on the author's web site:

http://www.uni-duisburg.de /FB7 /ISD /mitarbeiter/meinken

The photos are showing different cases of slight and heavy damages. It is demonstrating that there can be a considerable hazard for ship and crew.

2.2 Classification of the imperfections

Imperfections are deviations from the ideal conditions. These are causing considerably bending effects. Non-linear behaviour in undercritical areas is occurring. The reason for imperfections are variations in the geometry (geometry imperfections) and material features (physical imperfections). Imperfections can refer to the new construction (in this case a new ship) or to the condition after a few working seasons.

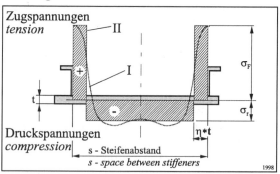

Fig. 2-1 Applied initial stress due to welding

In literature the imperfections are dividing into different items. For ships it is useful to relate them to their causes. The following text uses the designation *"imperfections due to manufacturing"* and *"imperfections during service"*.

The Fig. 2-1 and Fig. 2-2 represents the assumptions for the initial stress due to welding. A detailed description of these can get from /1/.

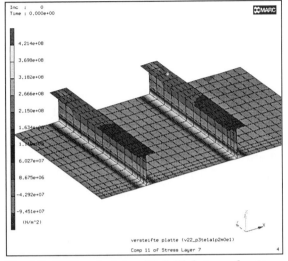

Fig. 2-2: Stiffened plate with initial stress due to welding

For shipbuilding the imperfections during service are notably important for structure stability and strength. On the one side the high life expectancy and on the other side the daily rough works are making it necessary that this is becoming a distinguished consideration.

In the following a few appearances of imperfections during service will be discussing.

Attrition and corrosion concern all plates and webs in the whole ship. A current average value for inland shipping is a material loss of 0,1 $^{mm}/_{year}$. Sometimes the values for the outer bottom, the inner bottom, the trough side and the bilge platting are higher.

Beside the mentioned items certainly the **ageing of the materials** is an other important point. It's quite possible that the values for young's modulus or the σ-ε diagram in the plastic area changes in course of time. Unfortunately the author does not know investigations about this. For this reason this point will be unconsidered in the following calculations.

The well-known **pre-deformations** in inland vessels are the corrugations of the inner- and outerbottom. Generally these are visible with the naked eye. They are calling *hungry horse* /2/ or *working folds*.

The structural deformations in seagoing ship panels are dividing in three categories /3/. This arrangement can be transferring to inland shipping.

First category: These is the global deformation of the whole ship hull. For example they are inducing from the cargo distribution that stresses the ship hull like a bending beam. The Fig. 2-3 explains that the deformations are only in the xz-plane and not in y-direction. They are parallel to neutral axis of the ship-structure.

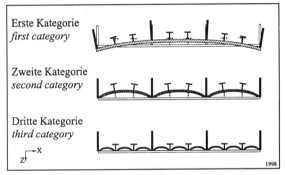

Fig. 2-3: Structure deformation at panels

Second category: It is including stresses and deformations at panels that are between two bulkheads. Loads are normal to the plate surface. The borders of a neighbouring panel, bulkhead or longitudinal girder are the boundary conditions. The deformations of the stiffener are adjusting to the plates. These are three-dimensional and not in a plane. This means that there are displacements in y-direction, too.

Third category: Here the deformations and stresses of a particular plate are describing. For example these are between two stiffeners of a panel. Once again the load is normal to the surface and the boundary conditions are creating from the stiffener. The deformations are three-dimensional and outside the planes of the stiffeners.

The stresses in the plates of the first and second category are membran-stresses and (nearly) constant through the plate thickness. The stresses in the plate of the third category are resulting from the local bending. Normally the deformations are greater than the plate thickness. Then the membran-stresses are not negligible.

The Fig. 2-3 should clear that the displacements and stresses for any point in the ship (e.g. on the inner bottom) consist of a lot of components. All have different reasons. There is only a less interaction among the three categories for most cases. In the calculations each category is particular establishing. The calculations based on the beam- and plate theories and the results were determining with superposition. Up to large displacements in the third category this simplification is not longer allowing. Then these have a crucial influence to the deformations of the first and second category.

The imperfections during service are mainly determining the deformations of the second and third category. The investigation of the relationship between these two and the first category is the target of the following calculations.

For the calculations it is necessary to define values for the pre-deformations. Only a few investigations /4/ are knowing, which research the imperfections during service. These (pre-) deformations are small local damages (third category) or extended to more than one panel. Depending of the components different causes are responsible (Table 2-1).

Table 2-1: Damaged components and the causes

	landing	striking the ground	slight collision	grab damages	local overloads	influence of cargo
working folds in outer bottom		X			X	
folds in bilge platting	X	X	X		X	
outer wall	X		X			
permanent fender	X		X	X		
gangboard				X	X	
hatch coaming				X	X	X
trough side				X	X	X
innerbottom				X	X	X
stern	X		X			
bow	X		X			

The shape of the geometric imperfections in the third category varies. At panels typical imperfections during service are sinusoidal corrugations between the stiffener respectively the floor rung. Other shapes are folds in outer-, innerbottom and trough side.

The investigations of these shapes are object of FEM-analysis in /1/. The sharp-edged folds weak the structure most as expected.

2.3   Assumptions for the analysis

The chapters ahead are allowing a few conclusions that are the foundation of the following calculations (local $^{and}/_{or}$ global) of inland vessels. First it is to comment that different models will be using.

The **model A** is the perfect structure of the inland vessel. To calculate the eigenvalue and the characteristic function (eigen-shape) this model will be using. This item is useful for checking the mesh.

The **model B** is including imperfections that are representing the new construction. Here are the deformations of rolling and initial stress due to welding.

The next models are considering the ageing. **Model C** is including the assumptions of model B and additionally a reduction of the plate thickness of 10%. The assumptions of the geometric deformations during service are in Table 2-2 and Fig. 2-4. **Model D** based on model C except that the reduction of the plate thickness is 20%.

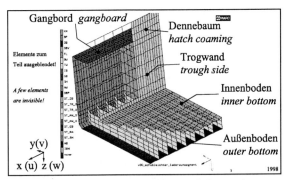

Fig. 2-4: Principal frame with pre-deformations

Table 2-2: Values for the deformations at the components

	plate thickness t	plate width b	w_0/t	model A $w_{0A} = 0{,}002b$	model B - D $w_{0B} = w_{0A} + w_0$
outer bottom	9,0 mm	0,6 m	1,0 *	1,2 mm	10,2 mm
bilge platting	11,0 mm	0,6 m	1,5 *	1,2 mm	17,7 mm
outer wall	9,0 mm	0,6 m	3,5 *	1,2 mm	32,7 mm
permanent fender	25,0 mm	0,6 m	2,0 *	1,2 mm	51,2 mm
gangboard	12,0 mm	0,6 m	0,8 *	1,2 mm	10,8 mm
hatch coaming	9,0 - 11,0 mm	0,6 m	3,5 **	1,2 mm	39,7 mm
trough side	9,0 mm	0,6 m	4,0 **	1,2 mm	37,2 mm
inner bottom	12,5 mm	0,6 m	4,0 **	1,2 mm	51,2 mm
stern	7,0 -9,0 mm		1,0 **	1,2 mm	8,2 mm
bow	7,0 -9,0 mm		1,0 **	1,2 mm	8,2 mm

* from /4/   ** own assumptions

3 The new calculation concept

To develop new constructions of future inland vessels the structure stability and strength is an important item. The aim is a durable safety against failure with a mass as small as possible. In the calculations a few items must be considering:

♦ Imperfections due to manufacturing (pre-deformations, initial stress and other failure due to manufacturing).

♦ Imperfections during service (damages, deformations, corrosion).

♦ Geometric and physical non-linearity.

♦ A realistic modelling of the loads due to the own weight, cargo distribution and water pressure.

♦ The changing of the buoyancy forces due the ship deformation.

♦ The interaction between local imperfections and global structure behaviour.

Due to this, the traditional calculation methods can not using the simplifications. Combined with the Finite-Element-Method (FEM) this paper will be introducing a new calculation concept. This concept is allowing these expensive analyses.

The application of the FEM in the traditional way, global and local FE-models are dividing the complete structure (e.g. a ship hull). Different jobs are calculating these models. The results of the global calculation are using as boundary conditions for the local run. This procedure will be calling *submodel-technique*. The consideration of interactions between local results and global structure behaviour is not possible.

Using the new calculation concept there will be only one global model and no local models. This one model owns the features of the global model and the possibilities of the local. The real buoyancy force at the deformed ship hull will be considering. There are further advantages. It is possible to consider the non-linear structure behaviour of the whole ship hull until the total collapse. This is including the imperfections during service. A reliability estimation is just possible with realtistic boundary conditions. To satisfy this demand an extended formulation of the boundary conditions will be necessary.

The application of the FEM for calculations of ship structure stability and strength is not new. Several authors use special techniques to reduce the necessary computer capacity. These are the

♦ Sub-model technique /5/ /6/.

♦ Substructure technique /6/.

♦ Superelements /7/,/8/,/9/,/10/,/11/,/12/,/13/

To explain these techniques in detail is not possible here. To get further information please use the author's e-mail address.

4 Analysis of a stiffened plate

The following calculation results will be using to discuss the influence of the imperfections. The results were estimate due to a non-linear FEM parameter investigation that was do with the FE-system **MARC / MENTAT**.

An investigation like this is really expensive, so it is useful to do this for a smaller model that represents the features of a ship typical construction.

4.1 Description of the structure

A longitudinal stiffened plate was model (Fig. 4-1). The plate was load with water pressure normal to the elements and a boundary displacement.

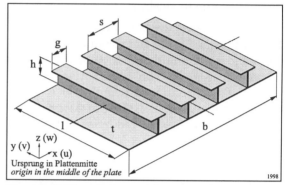

Fig. 4-1: Stiffened plate with geometrical data

Previous experimental and numerical investigations had done by *Kmiecik* /14/. The own calculation considers geometrical and physical non-linearity. A detailed description is in /1/.

4.2 Description of the FE-model

A striking feature of the model is the meshing of the stiffeners.

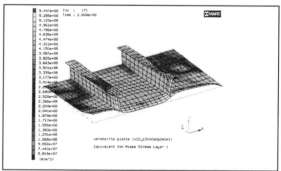

Fig. 4-2: FE-¼ Model

A shell element type is using to model plate, web and flange (Fig. 4-2). This is necessary to reproduce the complex failure of the stiffener. Similar statements are in /15/ and /13/.

4.3 Calculation results

A closer inspection of the initial stress due to welding shows that the relative high compression stresses in the midspan are resulting in a considerable reduction of the ultimate load. Especially for small displacements these are not negligible (Fig. 4-3).

The research of the imperfections during service shows that there are a few typical shapes of pre-deformations. The influences of sinusoidal corrugations and sharp-edged folds are especially to consider.

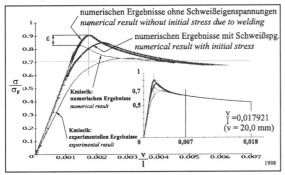

Fig. 4-3: Comparison of the resulting deformations with and without initial stress due to welding

For the depth of the pre-deformations the results are not surprising. An increasing depth is resulting in a reduction of the ultimate load. How badly the influence of the depth is depends from the shape of the imperfection.

Compared with other imperfections the corrosion has an extreme reduction of the ultimate load. The relatively thin plate thickness of an inland vessel and the relatively large material loss of $0,1 \frac{mm}{year}$ is an explanation for the high repair need of inland vessels. These calculations will be shown in the lecture.

5 Analysis of a pusher barge

Pusher barges have a large meaning for inland shipping. The following example is a type Europa IIa.

5.1 Description of the structure

The dimensions and data that are necessary to create the FE-model were get from technical drawings of a ship yard

Type:	Pusher barge EUROPA IIa
Length over all:	76,5 m
Width over all:	11,40 m
Moulded depth.:	4,00 m
Unloaded draught:	0,55 m
Load draught:	3,98 m
Load capacity:	2817 t

As usual for pusher barge the wing passage is build in a longitudinal framing system and the double bottom in transverse framing system (Fig. 5-1).

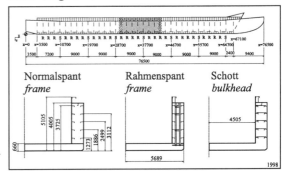

Fig. 5-1 Main dimensions of the FE-model

The space between the floor rung is 0,6m. There are no additional longitudinal girders in the double bottom. The web frames are on each 3rd constructional section (1,8 m). Each fifteen constructional section is a wing passage bulkhead (9 m). Fig. 5-1 is marking these with a R respectively with a S.

5.2 Description of the FE-model

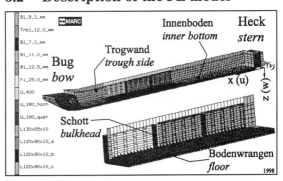

Fig. 5-2: Geometrical data

The plates were model by the **MARC** element type 75. Stiffeners in the wing passage made by the element type 78. The meshing will be showing in the Fig. 5-2. A detailed de-

196

scription of the structure and the model can
be get from the author /16/.

5.3 The load cases

The structure is load due to the water pressure, the own mass and the cargo. The various load distributions are organise in seven
load cases. **Load case 1** is only the own mass
and **load case 2** is a constant pressure on the
inner bottom. These are no critical loads for
the structure, but the are useful to verify the
model, new subroutines and the postprocessing. **Load case 3 and 4** are describing the
traditional Hogging and Sagging. It is to suppose that these are the most critical load case
for the structure. The loads in **load case 5
and 6** are various bulk materials. The loading
process for gravel and ore in one way is
simulate (GL-rule B). **Load case 7** investigates the loading with container. Especially
the possibility of torsion will be considering.
Load case 1 - 6 are pure bending stress. In
fact of the symmetry features it is possible to
use a half-model.

5.4 Calculation results

All load cases are calculate linear first and
only if necessary non-linear investigations
will be attaching. The extension of the model
and the difficulty of the non-linearity are
carefully to raise /17/.

5.4.1 Linear calculations

The following chapters will be including
only the results for load case 3. The deformations and stresses of load case 4 will be shown
in the lecture.

An area on the inner bottom in front and in
the back of the cargo hold is load with pressure if load case 3 is select. This is the traditional Hogging.

5.4.1.1 Deformation load case 3

Global plots of the ship hull are showing
the draught and deformations in natural
scale.

It is difficult to recognize the pure deformations. Due to this the y-displacement of the
hatch coaming (Fig. 5-3) and the z-displace-
ment of the outer bottom (Fig. 5-4) will be
showing in separately diagrams. The curves
are load-depended (1000 t, 2000 t, 2800 t).

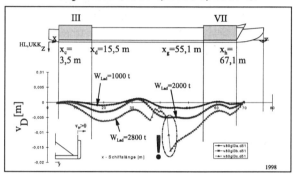

Fig. 5-3: Deformation of the hatch coaming (lc3)

Fig. 5-3 is showing that the hatch coaming
moves inside with an increasing load. This
results from the increasing water pressure.
The diagram is showing a discontinuity (x=
42 m) which conspicuous with the maximum
load. It is to suppose that the simplified assumptions of the linear analysis are not allowing. At later calculations this load case
will be non-linear investigate with a finer
mesh in the critical area.

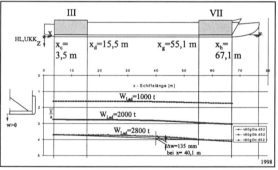

**Fig. 5-4: Immersion, trim and deformation of the
outer bottom (lc3)**

The maximum deflexion (135 mm) of the
outer bottom is reaching for 2800 t with a
draught of 4 m (Fig. 5-4).

5.4.1.2 Stresses load case 3

The Equivalent von Mises stresses in Fig. 5-
5 is relating to layer 1 of the elements. It is to

notice that the unit in the plot is $^N/_m{}^2$ and the deformations are 10 times scaled.

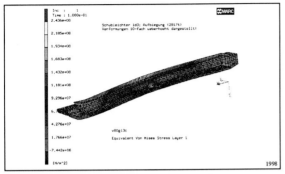

Fig. 5-5: Equivalent von Mises stress (lc3)

Load case 3 with maximum loading the Equivalent von Mises stress reaches approximately 244 $^N/_{mm}{}^2$. This value is over the yield point of normal low-tensile steel (R_{eh} = 235 $^N/_{mm}{}^2$).

5.4.2 Non-linear calculations

The chapters ahead showed that non-linearity in the ship structure is occurring. As mentioned before in the critical area a finer mesh is necessary. Problems with the capacity of the harddisc made it impossible to present the results in this manuscript. In the lecture this will be completing. There will be showing the influence of corrosion and pre-deformations.

6 Further way of acting

To reduce the costs for a FEM-analysis an effective usage of the soft- and hardware and especially the staff is necessary. The following chapters will describe the further way of acting to reach this aim.

6.1 Further development of the FEM

The tendency in the leading FEM software systems are the creation of universal multi-purpose software. The application of software that is only useful for one purpose is not longer sufficient enough. Today and especially in the future the FE-user will prefer user-kindly software. The costs for the development of such software are too high for a few

users. This is valid for the solver and the pre- and postprocessor. Every research should be awaring of this.

Some software systems offer user interfaces to implement own developments. This is a possibility to combine special user-problems with multi-purpose, user-kindly software.

The tendency of the element type is the using of high-order elements. To reduce the calculation time own element developments can be profitable. This means for structure stability that particular elements have the features of an imperfect plate. In the literature an element like this gets the designation "superelement". The next step is the implementation of an element like this to realize a coarse mesh in critical areas.

6.2 Expansion of the calculation concept

Modern FE-software offers a few special features. In future calculations these will be using to increase the effectiveness of the calculations.

6.2.1 Adaptiv meshing

Adaptive meshing is a local mesh refinement in critical areas during the calculation job. Probably this is useful in the high-loaded areas of the hatch-coaming.

6.2.2 Optimizing

An interesting software module is the optimizing. It is possible to influence particular dimensions (e.g. axis of inertia, plate thickness) to reduce stress peaks. Mostly the target is a low mass combined for a prescribed load. Modern software supports this, but a lot of development is necessary. Possible is a carefully combination of linear pre-investigation with optimizing and an adjacent non-linear check.

6.3 Modification of the constructions

In course of the mentioned DFG-research, construction proposals in favour of future

198

inland vessels are to develop. A few possibilities will be commenting in the next chapters.

6.3.1 High-tensile steel

The calculation results above are showing that for Sagging and Hogging the high-loaded component is the hatch coaming of the inland vessel. The stresses go beyond the yield point of normal steel (low-tensile stress). A reflection is it, to use high-tensile stress in this area.

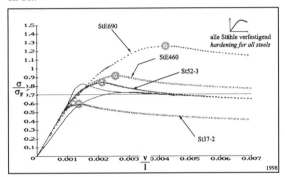

Fig. 6-1: Influence of yield point

Often it is mentioning that high-tensile stress has no advantages for structure stability. Using the introduced stiffened plate a non-linear FE-parameter investigation was do /1/. The comparison of the results (Fig. 6-1) shows that an increased ultimate load is possible.

6.3.2 Trapezoidal girder

Naturally the behaviour of a stiffened structure depends from the space between the stiffeners. It is possible that parallel flanged girders with slanting end posts increase the load limit. Similar investigations were describe in /15/.

6.3.3 Trough side

Normally the trough side of pusher barge is only between the hatch coaming and the inner bottom. A trough side that reaches down to outer bottom should increase the bending rigidity. A FE-analysis with and without these components can clear this.

7 Conclusion

The main aim of this study is to make it possible to use the FEM for the calculation of inland vessels. A special item is the consideration of imperfections. A calculation concept was introduce where a ship structure is divide in various global FE-models. These FE-models differ in the rank of the imperfections. The introduced models were investigate for different load distributions that are organize in various load cases.

To investigate which influences are important and which are negligible to the structure behaviour and the load limits a non-linear parameter FE-analysis was make. The FE-model was a ship typical stiffened plate. The investigated parameters were various imperfections during service and manufacturing and material features (yield point).

The knowledge of these results was use for the analysis of the global ship hull. The results of linear and non-linear analysis for typical inland vessels were show.

At least a few "special features" were describe to reduce the calculation time for future investigations. These are the implementation of a superelement and the using of adaptive meshing and optimizing software.

References

/1/ Andreas Meinken, H.-J. Schlüter,
Einfluß von Imperfektionen auf die Tragfähigkeit versteifter Platten - Nichtlineare Parameteruntersuchung mit der FEM an einer schiffbautypischen Konstruktion
will be appearing in „Schiffbauforschung" 1998,

/2/ Eike Lehmann
Analytische und halbanalytische Finite Elemente zur Konstruktionsberechnung schiffbaulicher Tragwerke
STG-Jahrbuch, Band 70 (1976) pp. 219-262

/3/ Robert Taggart and others
Ship Design and Construction
The Society of Naval Architects and Marine Engineers, (ISBN 0-9603048-0-0)

/4/ H. G. Payer
Neuere Erkenntnisse über die Festigkeit von Binnenschiffen
STG-Jahrbuch Band 73 (1979) S. 319-336

/5/ Svensen Tor E., Valsgard Sverre Thorkildsen Harald
Computational ship analysis of container vessels (DNV - Det Norske Veritas)
PRAD´S 95 17-22 September 1995, pp.2.1187-2.1202

/6/ Rüdiger Plum, Meinken, Schlüter,
Anwendung der Substrukturtechnik bei der Analyse von Schiffsstrukturen unter Nutzung des Programmsystems

MARC / MENTAT
Studienarbeit ISD (Nr.2), September 1997

/7/ Rüdiger Plum, Meinken, Schlüter,
Anwendung von finiten Superelementen zur Untersuchung des Kollapsverhaltens von Binnenschiffen
Diplomarbeit ISD, 1998

/8/ Ueda Yukio, Rashed Sherif M. H.,
The idealized structural unit method and its application to deep girder structures
Computers & Structures Vol. 18; No 2; pp. 277-293;

/9/ Ueda Y, Paik, J. K.,
The Idealized Structural Unit Method Including Global Non-linearities - Idealized Rectangular Plate and Stiffened Plate Elements
J. of the SNAJ; Vol. 159; June 1986 pp. 271

/10/ Ueda Yukio, Rashed Sherif M. H.,
Advances in the application of ISUM to marine structures
Advances in Marine Structures 2; 1991, pp.

/11/ Jeom K. Paik
Ultimate Longitudinal Strength-Based Safety and Reliability Assessment of Ship's Hull Girder - (1st report)
Society of Naval Architekts of Japan; Vol. 168; Nov. 1990, pp. 395-407,

/12/ Jeom K. Paik
Ultimate Longitudinal Strength-Based Safety and Reliability Assessment of Ship's Hull Girder- (2nd report Stiffened Hull Structure)
Society of Naval Architekts of Japan; Vol. 169; Jun. 1991; pp. 403-414

/13/ Committee III.1
Ultimate Strength
Proceedings of the 13th International Ship and Offshore Structure Congress (ISSC) 1997, Volume 1, pp. 235-283; Pergamon 1997, (ISBN 0-08-042829-0)

/14/ Marian Kmiecik, Helmut Pfau, Erno Wiebeck, Mieczyslaw Wizmur
Nichtlineare Berechnung ebener Flächentragwerke
Verlag für Bauwesen 1993; pp. 234-237;

/15/ Eike Lehmann, Leshan Zhang
Nichtlineares Verhalten von ausgesteiften Tragwerken mit schiffbaulichen, meeres- und anlagentechnischen Beispielen
Springer Verlag 1997 (ISBN 3-540-63444-4)

/16/ Andreas Meinken, 1997
Beschreibung des FE-Modells eines Schubleichters
Institut für Schiffstechnik Duisburg (ISD)
- institutsinterner Bericht Nr. 80

/17/ Klaus Jürgen Bathe
Finite-Elemente-Methoden
Springer - Verlag, 1986, ISBN 3-540-15602-X

Notation

Latin small letters

b	width of the plate
g	acceleration due to gravity
g	width of the flange
h	height of the web
l	length of the plate
p_w	water pressure
s	space between stiffeners
t	plate thickness
u, v, w	displacements
v_{max}	maximum boundary displacement

Latin capital letter

A	area.
F^{ges}	force
F_{krit}	critical load
L	length
L_{OA}	length over all
L_{WL}	waterline length
m	mass
R_{eh}	upper yield point
R_m	tensile strength
R_x, R_y, R_z	fixed rotation
T	draught
T_x, T_y, T_z	fixed translation
W_{Lad}	mass of the payed load

Greek letter

ν	poisson's ratio
ρ	density
ρ_w	density of the water
σ	stress
σ_r	compression stress (due to welding)
σ_F	yield point
ηt	width of the tension area

Authors

Prof. Dr.-Ing. habil. H.-J. Schlüter
Dipl.-Ing. A. Meinken
Gerhard-Mercator-University-GH Duisburg
Institute For Shiptechnology Duisburg (ISD)
Department of Shipstrength
Bismarkstraße 69 ; 47048 Duisburg; Germany

e-mail: meinken@nav.uni-duisburg.de

Internet   http://www.uni-duisburg.de/FB7/ISD/ mitarbeiter/meinken

Practical Design of Ships and Mobile Units
M.W.C. Oosterveld and S.G. Tan, editors.

The Role of Shipboard Structural Monitoring Systems in the Design and Safe Operation of Ships

F. H. Ashcroft [a] and D. J. Witmer [b]

[a] Vice President, Scientific Marine Services, Inc.
101 State Place, Suite N, Escondido, CA 92029, United States of America

[b] Engineer Superintendent/Naval Architect, BP Oil Shipping Co.
200 Public Square, Cleveland, OH 44114, United States of America

This paper describes the physical arrangement of the BP Oil Tanker Structural Monitoring System (BPSMS) including the suite of sensors used to measure ship response and hull girder bending stress. The bridge displays are illustrated and explained along with the use of the system for tactical guidance in heavy weather. Use of the data collected using this system to provide engineering data for the design of new vessels is discussed.

1   INTRODUCTION

BP Oil Shipping Company time charters a fleet of American flag tankers employed in the ocean transport of crude oil and petroleum products on the East, West, and Gulf coasts of the United States. The West coast tankers carry Alaska North Slope (ANS) crude oil from the Valdez Marine Terminal to various ports along the U. S. Pacific coast, Hawaii, and the Far East. These ships, ranging from 120,000 to 190,000 DWT, were constructed during the mid 1970's.   Beginning in 1991, ship response and structural monitoring instrumentation was installed aboard these ships as part of BP's world wide tanker structural management strategy.

These shipboard structural monitoring systems were designed to quantify in terms of *measured,* not *calculated,* values the effects of subjecting the ship to the typical loads and forces encountered both on passage at sea and in port during cargo operations. The goal was to develop a tool which would minimize the risks of incurring structural damage by providing the ship's officers with the means to actually measure the structural behavior of their ship.

2   SYSTEM DESCRIPTION

The BPSMS consists primarily of a suite of sensors and a computer to process, store, and display the data. A second computer is used to control automated communications and to acquire and

process a rolling ten day weather forecast. These subsystems work together to form the BP Structural Monitoring System which we refer to as the BPSMS.   A typical arrangement of the BPSMS hardware is shown in Figure 1.

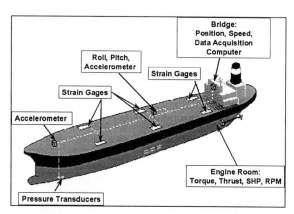

Figure 1: Typical Arrangement of BPSMS

Due to regulatory requirements in the United States, all sensors were either designed to be intrinsically safe or were housed in explosion proof enclosures.

Two pressure transducers are located in through hull penetrations at the bottom of the forepeak ballast tank adjacent to the centerline of the ship. They are used to determine the forward draft, relative bow motion, and the emergence of the forefoot.

Two inertial navigation quality accelerometers are mounted in explosion proof enclosures on the main deck. These are used to measure vertical acceleration at the bow and midship. A solid state dual axis gyro is located in the midship explosion proof box and is used to measure global hull girder roll and pitch. A view of the midship enclosure is given in Figure 2.

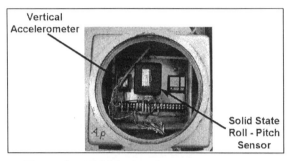

Figure 2: Midship Accelerometer and Gyro

Long base strain gages are used to determine the bending moments imposed on the ship. These sensors are located on the main deck above longitudinal bulkheads in order to ensure the measurement of global strain. For the vessels with a three tank transverse arrangement, the gages are typically located in pairs as shown in Figure 1. On ships with a centerline longitudinal bulkhead, a typical arrangement would be five or six gages on centerline at the transverse bulkheads. A typical long base gage (about 2 meters in length) is shown in Figure 3.

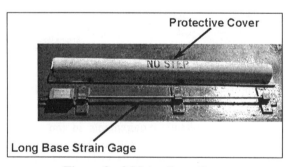

Figure 3: 2 Meter Long Base

During the design of these sensors care was taken to match the thermal expansion coefficient of the long base rod to that of the deck steel. During the first year of operation a special experiment was carried out which verified that the error induced in the strain gage measurements due to thermal effects was less than 1% of full scale (the gages are set to ±2000 microstrain).

In addition to the deck mounted long base gages, special studies have been conducted where weldable point type gages have been installed on some ships on deck and in both cargo and ballast tanks. The purpose of these gages was to collect data at critical structural details for use in engineering studies and the development of structural enhancements.

Ships which are equipped with the BPSMS normally also have engine room sensors which provide information on torque, thrust, horsepower, and shaft rpm. This information is transmitted from the engine room to the BPSMS computer on the bridge and integrated into the structural monitoring system for both display and storage.

These ships also have a GPS which will output a NMEA standard serial message which includes at least the ship's position, speed over the ground (SOG), and course over the ground (COG). This information is fed directly into a serial port on the BPSMS computer for display and storage.

During the initial design phase, the decision was made by BP personnel that the data acquisition computer and related hardware should be mounted in a dedicated console on the bridge of the ship. The standard BPSMS console is shown in Figure 4.

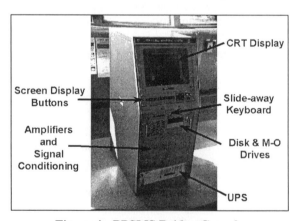

Figure 4: BPSMS Bridge Console

The console is a (US) standard 19 inch (48.26 cm) rack mount which provides space for the entire data acquisition and display portion of the BPSMS. The entire system including the sensors is powered through an un-interruptible power supply (UPS) located at the bottom of the rack. The UPS filters the power to the system and in the event of a main power failure the UPS provides sufficient energy to operate the entire system for approximately 30 minutes. Above the UPS is the power supply drawer where the power switches and fuses are located.

The next two drawers contain 16 channel signal conditioning boards for analog sensor input. The standard BPSMS configuration uses the first 16 channels of data acquisition. The second 16 channel board provides the capability to perform short term measurements such as in tank strain gaging without any modification to the BPSMS data acquisition system.

The next drawer up houses a PC compatible computer with 16 bit analog to digital converter, digital I/O, and network cards; a hard drive; a 3¼ inch floppy disk drive; and a 230 Mb magneto-optical drive. This computer controls the acquisition of analog and serial data as well as performing data analysis and providing information display functions. The hard drive is used to store the programs and to provide short term storage for the data as it is acquired. Every 4 hours, the data from the previous 4 hours is written from the hard drive to the magneto-optical disk for long term storage. The 3¼ inch floppy disk drive is used to provide program upgrades.

Above the computer is a standard computer keyboard in a pull out drawer and a custom keypad which allows easy access to the data display screens.

To allow the computer to simultaneously acquire, analyze, store, and display data while still remaining available to process watch officer requests for alternative data displays or input of data it was decided to construct the system around a real time multi tasking operating system. The computer was originally run under the TLX Executive, a real time multi tasking system which is compatible with MS-DOS. This operating system has been working successfully since 1991, however the BPSMS is presently under conversion to a more versatile WIN-32 application which will run under Microsoft Windows NT™, with change over scheduled to be completed by the latter part of 1998.

All the data acquisition and analysis software is written in the C language. Signals from the bow pressure transducers are sampled at 50 Hz, while all other data channels are sampled at 8.333 Hz. (500 samples per minute). Should one of the pressure transducers detect a forefoot emersion, the raw data from all the channels is stored in a special file called an "Event File", which includes data from 30 seconds prior to 90 seconds after the event.

The mean, minimum, maximum, standard deviation, root mean square, and average zero up-crossing period is processed for all channels in five minute intervals. Data from the strain gages is processed using a "Rainflow Method" developed by Wirshing and Shehata[1]. This method divides the measured strains into 50 microstrain bins and derives cycle count information over the range of 0 to 4000 microstrain which covers the range of compressive to tensile yield for the AH36 steel used in the decks of the instrumented tankers.

An additional feature of the BPSMS system software is the self checking routine. The system constantly compares the measured values from all sensors to allowable bound values for maximum, minimum and standard deviation. If a sensor signal goes out of bound, a "check diagnostics" warning is displayed on the console alerting the user to the run the diagnostics routine and check the sensors.

3   CALIBRATION

The calibration of any structural monitoring system is extremely important. During the design of the system analysis was performed to determine the measurement error associated with each of the sensors as well as the signal conditioning and data acquisition hardware. In addition, procedures were developed to calibrate the long base strain gages so that their output would match the still water bending moments predicted by the ship's on board loading calculator.

The BPSMS systems are inspected regularly and physically calibrated once per year by Scientific Marine Services personnel. If a sensor is damaged, it is replaced and calibrated insitu.

4   INFORMATION DISPLAYS

One of the most important uses of the BPSMS is to provide real time information to allow better structural management of the ship to the watch standing officers aboard ship. In order to provide this information, a series of data display screens was developed to inform the officers about the structural status of their ship. Data for all displays is continuously updated by the data acquisition computer, so no delay is experienced when switching from one display to another. All the displays have some features in common. These are:

- Information is updated every five minutes (except for the time series display).
- Every screen has a color version for day time and a red-for-night version.
- Every screen has a pop up "Help" window.
- The display title is in always in the top left corner.
- The top right corner contains a status display which includes: time, date, and system status.

Data displays are ordered from the most general to the most specific and may be accessed either from the keyboard or the custom keypad. The screens are shown and described in detail below.

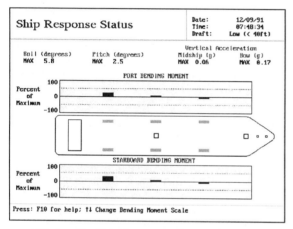

Figure 5: Ship Response Status Display

The Ship Response Status screen is the most general display. The upper part of the screen shows a digital display of the 5 minute maximum values of roll, pitch, and acceleration at midship and bow. The center portion of the screen shows a diagram of the ship with the location of each sensor outlined. Above and below the diagram are bar graphs which indicate the bending moment at each long base strain gage in real time. Bending moments are displayed as a percent total ABS allowable (at sea still water plus wave bending).

In the event that the system self checking routine finds an out of bound sensor, when the "check diagnostics" warning is displayed that sensor is also changed to a red color on this display.

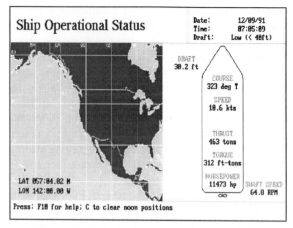

Figure 6: Ship Operational Status Display

This screen brings together information from several ship systems on one display. On the left is an area chart on which the noon (UTC) positions of the ship are displayed as a series of red crosses. In the lower left corner the instantaneous latitude and longitude from the GPS are shown.

On the right side is shown the forward draft (from one of the pressure sensors in the forepeak tank), the speed (SOG from the GPS), and thrust, torque, horsepower, and rpm (from the engine room instruments).

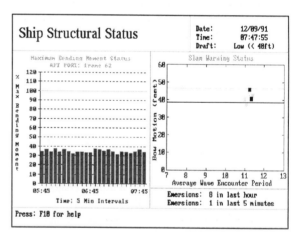

Figure 7: Ship Structural Status Display

The Ship Structural Status display provides a summary of the structural behavior of the ship and may be the most important display of the BPSMS. This screen is used to determine when action should be taken to ease the ship in heavy weather and how effective that action has been. On the left side is displayed a two hour trend graph of the maximum total bending moment. Every five minutes the system automatically determines which strain gage has experienced the highest bending moment and displays that value as a bar on the graph. If any value exceeds 60% of the total allowable bending moment, all the bars turn yellow. At 80%, all the bars turn red. At night, on the red screen, the bars become a much brighter red at 80%.

On the right side of the screen is a display showing the relative motion of the forefoot as determined from the pressure sensors located in through hull penetrations in the forepeak tank. The vertical scale shows motion in feet and the horizontal scale shows average wave period of encounter in seconds. The squares indicate the "most probable maximum" bow motion computed from the previous five minutes of data. A yellow line is provided for forefoot emergence warning and a red line for slam warning. The position of these lines is automatically adjusted for draft and in the laden condition becomes a submergence warning. The positions of the lines is determined theoretically based on the theory of emergence and slamming developed by Tick[2] and Ochi[3,4] with adjustment based on actual experience with these ships. The most recent data is plotted as a white square, becoming progressively darker shade of blue until disappearing after about 35 minutes. At the bottom, the actual number of emersions over the last five minutes and hour are digitally displayed.

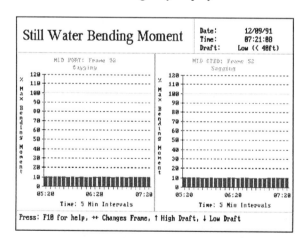

Figure 8: Still Water Bending Moment

This display contains two bar charts displaying the two hour trend graphs of still water bending moment for each of two long base strain gages. All the strain gages may be displayed in pairs by

scrolling with the left or right arrow keys on either the keyboard or custom keypad. The still water bending moment is determined from the five minute mean of the strain gages and is displayed as a percentage of the ABS allowable at sea still water bending moment at each instrumented frame. Whether the bending is hog or sag is indicated at the top of the display. If any bar goes above 80% of the ABS allowable at that frame, all the bars turn red (bright red at night).

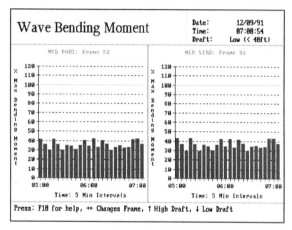

Figure 9: Wave Bending Moment

This screen shows the wave bending moment information displayed in a the same manner as the still water bending moment. The magnitude of the wave bending moment is calculated as 2.61 times the standard deviation over a five minute period. The 2.61 is a statistical estimating factor based on an average of 30 wave encounters in a five minutes. This display also has an 80% allowable warning.

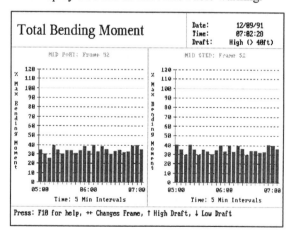

Figure 10: Total Bending Moment

This screen shows the total bending moment information displayed in a the same manner as the still water bending moment. The total bending moment is the sum of the still water and wave bending moments. This display is the same information which is displayed on the Ship Structural Status Display (highest values only), but here is available for all the strain gages. This display has a yellow warning at 60% of the allowable total bending stress and a red warning at 80%.

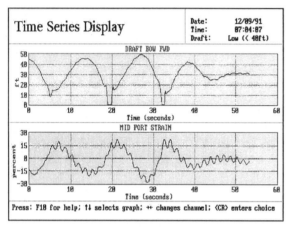

Figure 11: Time Series Display

This display allows the watch officers to view the output from any sensor in real time. This is very useful for studying the behavior of the ship in heavy weather or as a system diagnostic tool.

Statistics Display

Date: 12/09/91
Time: 07:21:40
Draft: Low (< 40ft)

CHANNEL NAME	UNITS	MAX	MIN	MEAN	STDEV	RMS	PERIOD
ROLL	deg	6.0	-4.1	0.7	2.0	2.1	11.25
PITCH	deg	3.4	-2.8	-0.0	1.0	1.0	11.76
MID HEAVE ACC	ft/s^2	2.60	-2.09	0.04	0.69	0.69	11.75
BOW HEAVE ACC	ft/s^2	6.75	-6.61	0.03	2.16	2.16	11.75
AFT PORT STRAIN	%	34.8	9.0	22.8	4.4	23.3	11.76
MID PORT STRAIN	%	17.3	-24.6	-4.7	7.4	8.7	11.29
FWD PORT STRAIN	%	2.8	-32.8	-16.8	5.7	17.7	11.29
AFT STBD STRAIN	%	35.0	7.9	23.0	4.3	23.4	11.76
MID STBD STRAIN	%	15.5	-27.6	-4.6	7.6	8.9	11.29
FWD STBD STRAIN	%	2.7	-33.2	-16.7	5.6	17.6	11.28
MID PORT S.G.	%	15.4	-24.1	-4.6	6.9	8.3	11.29
MID STBD S.G.	%	22.1	-26.2	-4.7	7.3	8.6	10.49
TANK STRAIN BLG	$\mu\epsilon$	-465	-470	-468	1	468	0.00
TANK STRAIN WL	$\mu\epsilon$	-466	-471	-469	1	469	0.00
DRAFT BOW AFT	ft	40.63	7.14	31.05	6.56	31.74	11.70
DRAFT BOW FWD	ft	52.70	-0.94	30.40	8.43	31.55	11.70

Press: F10 for help

Figure 12: Statistics Display

This display provides the results of the five minute statistical analysis for all the sensors. This display is useful for system troubleshooting. In the example screen shown, "Tank Strain BLG" and Tank Strain WL" are clearly not working.

DECK LOG ENTRY FORM

Date	12/09/91
Time	07:04:11
Voyage Number	9208
Doppler Speed (knots)	10.7
Gyro Heading (degrees)	323
Draft Forward (feet-inch)	35-00
Draft Aft (feet-inch)	35-00
Wave Height (feet)	4
Wave Period (seconds)	4
Wave Direction (STA Code)	2
Swell Height (feet)	4
Swell Period (seconds)	8
Swell Direction (STA Code)	2
Sea State (Beaufort Scale)	4
Air Temp (degrees F)	58
Wind Speed (knots)	12
Wind Direction (STA Code)	2
Shaft Speed (RPM)	64.7
Sea Temperature (degrees F)	56
COMMENTS:	

DIRECTIONS

<ESC> - Exit
<S> - Save Entry

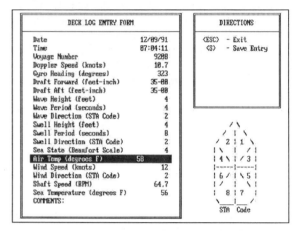

Figure 13: Deck Log Entry Form

This form is completed once per watch (or if there is a major change) using the keyboard built into the BPSMS console. It contains environmental and operational information which are used in shoreside engineering studies.

There are two other display plots available but rarely used by shipboard personnel. These are the Cross Plot Display, which allows the plotting of any two data channels against each other, and the Fatigue Display, which shows the results of the Rainflow Analysis for each strain gage. These displays are primarily intended for use by engineers when riding the ship during special experiments or trials.

5 ADDITIONAL GUIDANCE TOOLS

For each class of ship that is fitted with the BPSMS, a comprehensive theoretical seakeeping analysis is run using the U. S. Navy developed program SMP87 (or newer depending on system installation date). The results from this study are embedded in a function called WaveMax which is part of the standard BPSMS. WaveMax provides the vessel Master with a predictive tool with which he can check the probable result of a proposed change in speed, heading, or ballast condition on the motion and bending moment of the ship before ordering the change. The WaveMax screen is accessed from the Ship Structural Status Display by pressing the help key twice.

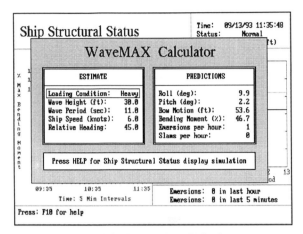

Figure 14: WaveMax Calculator

When first called the values on both sides of the table are based on the last 20 minutes of actual ship data. The value of any item on the "ESTIMATE" side can be altered using either the keyboard or custom keypad and the result will be immediately displayed on the "PREDICTIONS" side. If heavy (or heavier) weather is expected, the wave height and period may also be changed to show the effects of the worsening conditions on the ship and various courses of action "gamed" before they are necessary.

The other major tool uses the second computer of the BPSMS. This computer is programmed to automatically call the SMS offices via Inmarsat and retrieve a rolling 10 day weather forecast provided by WNI/Oceanroutes in Sunnyvale, California. This forecast is transmitted digitally and is processed by the onboard computer to provide clear color weather charts containing all the standard meteorological data which can then be printed out on a color printer for use on the bridge.

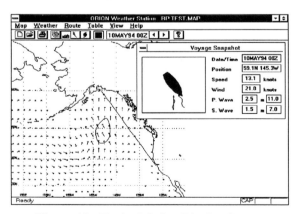

Figure 15: Typical Orion Display Screen

In addition, the Orion™ program allows the Master to enter his planned voyage on the chart so that he can see how the expected weather patterns relate to his projected track. The first step in avoiding structural damage due to heavy weather is to avoid heavy weather whenever possible. By combining the weather program with the structural monitoring system, the full BPSMS provides the shipboard personnel with the tools to avoid heavy weather where possible and minimize the effects on the ship when not.

6 USING THE BPSMS FOR DESIGN

The BPSMS has been in service on BP TAPS ships since 1991 and all of the data gathered has been archived for use in engineering studies.

One of the first uses of these data was to help develop a set of structural enhancements for a class of ships which had side longitudinal cracking problems[5].

More recently, data from these ships has been used to help develop TAPS trade specific sea spectra and provide valuable information on actual stresses absorbed by vessels trading this route. These data are being used to help design a new class of tankers which will be used in this trade.

7 BPSMS IMPROVEMENTS

The BPSMS is a system which is regularly updated and improved. Planned upgrades for 1998 include changing the software to a WIN 32 application to enhance the integration with other software and ship systems and installing a new version of the Orion weather program with enhanced voyage planning functions.

Future plans include greater integration of the BPSMS into the ship systems.

[1] Wirsching, P. H., and Shehata, A. M., "Fatigue Under Wide Band Random Stresses Using the Rainflow Method", Transactions of ASME, Journal of Engineering Materials and Technology, pp. 205 – 211 July 1977.

[2] Tick, L. J., "Certain Probabilities Associated With Bow Submergence and Ship Slamming in Irregular Seas", Journal of Ship Research, vol. 2, no. 1, 1958.

[3] Ochi, M. K., "Extreme Behavior of a Ship in Rough Seas – Slamming and Shipping of Green Water", SNAME Transactions, vol. 69, 1964.

[4] Ochi, M. K., and Motter, L. E., "Prediction of Slamming Characteristics and Hull Response for Ship Design", SNAME Transactions, vol. 78, 1973.

[5] Witmer, D. J., and Lewis, J. W., "Operational and Scientific Hull Structural Monitoring on TAPS Trade Tankers", SNAME Transactions, vol. 102, 1994.

Rough Weather Ship Performance
- A Quality to be Introduced into the Preliminary Design Process

by

J. Näreskog[a]. & O. Rutgersson[b]

KTH, Stockholm

Work on the development of the design process for cargo ships has been going on at KTH for a number of years. In the present paper results from two different studies within this work are presented. In the first study methodologies to implement the cargo owner's perspective into the design process is given together with the consequences of this approach on the ship speed on different routes. In the second study a method to include the ship performance in waves into the preliminary design process is presented.

[a]M.Sc., Naval Architecture at KTH in Stockholm.
[b]Professor of Naval Architecture at KTH in Stockholm.

1. INTRODUCTION

In the development of new vessels and new ship types the preliminary design process has become crucial. In this process contradictory requirements are balanced against the demand for effective cargo arrangements. Computer assisted, more or less automated, systems can be effective tools in this process, when a large number of alternatives has to be tested in a short time. At KTH, work has been going on for some time to develop such tools, which are to be used in the design work and in the important analyses of derived designs [1].

Ship performance in waves is important for the safe and effective operation of most ship types. However, it is usually not included in the optimisation process in the preliminary design due to difficulties to find useful criteria. In the present paper a method is proposed on how to include ship performance in rough weather in the preliminary design process and how to analyse the economic consequences of the performance of a chosen design.

2. CONCURRENT EGNINEERING IN THE DESIGN OF SHIPS

According to Suh [2] the conception of design considers the activities influencing the interaction between *what to create* and *how to create it*. The purpose of the design is stated in the functional domain as a number of functional requirements. The physical solution is generated in the physical domain through the establishment of a number of design parameters. The design process can be described as a mapping between the two domains at different hierarchical levels.

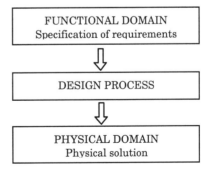

Regarding merchant vessels, the traditional description of the design task is to develop a vessel, which satisfies the requirements specified by the ship owner. The functional domain is restricted by the regulations and the specification of requirements, expressed in the form of service speed, cargo capacity, deck area and so on. The consequences of different design solutions are followed up and the vessel is developed and improved until the requirements are fulfilled. Yet, to reach as effective a solution as possible the functional domain should be expanded to include the operational consequences for all actors in the transport system due to changes of different design parameters. Otherwise, the operational differences between different design solutions are impossible to measure. According to Kusiak [3], an appropriate term of this more comprehensive perspective is concurrent engineering or integrated product development. Concurrent engineering is defined as a process that ensures consideration of how the product will be manufactured and used.

In concurrent engineering the design problem is approached by defining a multi-disciplinary design form observing aspects like functional requirements, production, quality assurance and the economic efficiency of the product. Generally, the term concurrent engineering is connected to production, but in the present work it is used to describe the consideration of the economy of the whole logistic system. Therefore, when applying concurrent engineering to the design of merchant ships, factors like economic efficiency are also included. In short, concurrent engineering recognises that the naval architect creates a vessel that is effective for both the ship owner and the cargo owner.

As stated by Suh [2], the design process is composed of the synthesis and the analysis, which map the physical domain to the functional. In the case of the preliminary design of merchant ships, it starts from the specification of requirements, obtained during the conceptual design, and it leads to a design

proposal, which satisfies all the requirements stated by the customer and by national and international authorities.

The synthesis comprises the determination of design parameters like: main particulars, weights, and building cost while the analysis aims to support selection of the preferable concept and constitute sensitivity analyses. Thus, guiding the naval architect in the further design work. During the synthesis the designer has at his disposal a large number of tools and integrated computer-based design systems, which have been developed through experience and research all over the world. The large variety of design tools to support the synthesis forms a sharp contrast to the few means available for the analysis.

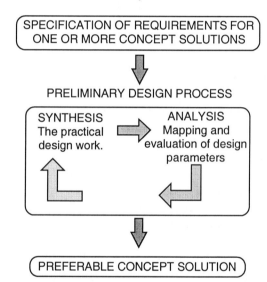

Figure 1. According to Suh [2], the preliminary design is composed of the synthesis and the analysis.

2.1. The synthesis

The synthesis resembles traditional design work. It is more or less to be regarded as a piece of craftsmanship ruled by the specification of requirements and by regulations stipulated by national and international authorities. The main objective of the synthesis is to obtain adequate design characteristics. It should be possible to estimate the cost of operation and utilisation of the vessel, thus provide information about the effectiveness of the ship, on basis of the information obtained during the synthesis. If adequate information is lacking, it is impossible to measure the quality of different design alternatives, implying that all attempts to reach an optimum solution would be fruitless.

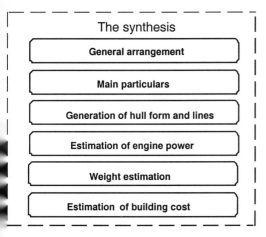

Figure 2. The design characteristics considered during the synthesis

2.2. The analysis

When applying concurrent engineering in the preliminary design the analysis constitutes an important part of the design work. The analysis is characterised by measuring the effectiveness of the result from the synthesis in economical terms. The goal is to facilitate the improvement of the vessel, by considering the cost of the operation and the utilisation of the ship. The different cost components indicate where efforts should be put to improve the vessel; thus they guide the naval architect in the further design work. The analysis considers aspects like the economy of the logistic system and the sensitivity of delay due to environmental circumstances in order to compare different design solutions. Another important aspect to consider during the analysis might be the

flexibility of the system, for instance the possibility to adjust the system with respect to fluctuations of cargo flow. The purpose of the analysis is to provide a qualitative assessment of different concepts and to guide the naval architect to the significant parts of the design process, by indicating where efforts should be put to increase the efficiency of the vessel.

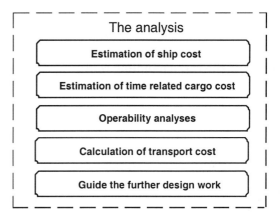

Figure 3. The significant parts of the analysis phase

One way to measure the effectiveness of a transport system, in economical terms, is to calculate the ship owner's cost to operate the vessel. Another more comprehensive approach is to investigate the total cost of the whole logistic system. In [1], the transport cost is considered to be a significant parameter. It is defined as the sum of the ship owner's cost of operating the ship, the ship cost, plus the time-related cargo cost during the trip. The different cost components included in the transport cost are described in the appendix.

3. A STUDY OF SHORT SEA CARGO FERRY SYSTEMS

For transports where the cargo is of considerable value, like RoRo ships on short sea traffic, the value time is important for the cost of the whole logistic system. In the present study the value time is defined as the time

between loading and unloading plus one third of the time between departures. Usually, a reduction in the value time implies an increase in the ship cost, although as long as the reduction in the time-related cargo cost is larger than the increase in the ship cost it is profitable to reduce the value time. There is a number of ways to reduce the value time, for instance by:
• increasing the speed of the carrying units
• increasing the frequency of the transport system
• enhancing the cargo handling routines.
• increasing the operability and the reliability of the transport system.
In [1], the significance of increasing the speed and transport frequency is discussed regarding short sea cargo ferry systems by analysing a number of routes and vessels to find the preferable level of speed and frequency. In [4], the importance of operability analyses, already during the preliminary design, is discussed by analysing the crossing time at different sea states and estimating the annual delay for a vessel at a certain route.

Näreskog et al. [1] discuss whether the consequences of high frequency cargo ferry operation, including high speed vessels, can solve the problem of service level and transport time consumption in a competitive way and with reasonable environmental impacts. The term service level is defined as the frequency of arrivals at each port in the logistic system. The result of the study was based on analyses regarding the gain in logistics, freight costs and environmental aspects for a large number of vessels operating on five routes in northern Europe. Figure 4 shows the ship costs and transport costs plotted as a function of the ship mean speed for two of the analysed routes. The cost levels are obtained through analyses of a number of different vessels operating at different speed levels. From this figure it can be concluded that the most economic speed in the investigated domain is about 25 knots. If the shipping cost alone is considered, the most

economic speed will be less than 20 knots.

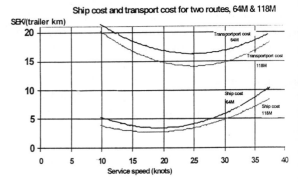

Figure 4. Principal cost distribution as a function of service speed [1]

The result of the study indicates that for trailer transportation and for distances up to 350 M the most effective service speed, when regarding transport cost, will be higher than when only regarding ship cost. Only when the bunker cost increases substantially, by about 50% from the level of today, the trend towards higher speeds will lose to an increased number of ships to keep the service level.

Between 350 - 550 M there are no significant differences between the preferable speed level with respect to ship cost and transport cost. One explanation might be that on such long routes the time related cargo cost is reduced due to the small number of runners. The time related cargo cost is considerably higher for runners than for drop trailers. Figure 5 shows the relation between ship cost and transport cost for a number of vessels operating at the five routes. The shaded areas represent the two cost approaches, the lower for ship costs and the upper for transport costs. The difference between ship cost and transport cost is the time-related cargo cost, depending on the value time. Each analysed route is represented by one line in each area. The marks along the lines are the individually analysed designs for each route.

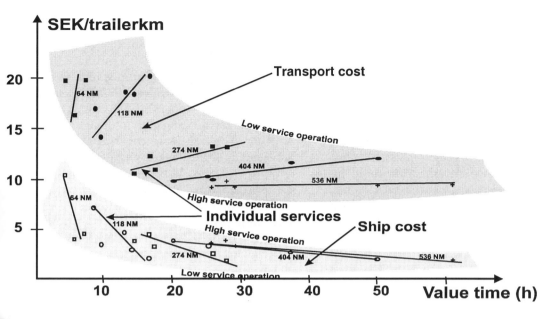

Figure 5. Principal cost distribution as a function of value time [1]

An increase in speed or frequency for a shipping system will reduce the value time. A reduction in the value time usually means an increase in the ship cost, but in many cases it results in a lower transport cost. Consequently, to understand the advantage of high-speed vessels and high service level in cargo traffic it is necessary to consider the transport cost. The gradients of the lines are less for increased length of service. For long service the lines are almost horizontal, implying that the advantage of high-speed will be less, due to the small numbers of runners. In short, when analysing RoRo-ships operating on routes below 400M, it is important to optimise the system with respect to the transport cost. For longer routes, the system with the lowest ship cost will also have the lowest transport cost.

4. ASSESSING ENVIRONMENTAL OPERABILITY IN TERMS OF TRANSPORT COST

Generally, it is difficult to improve the vessel with respect to the seakeeping performance since it is a rather diffuse property of the ship. Therefore, the hull is normally optimised with respect to the calm water performance, and whether the seakeeping performance is acceptable or not is analysed afterwards in the derived project. However, detailed changes of hull shape of cargo ships, such as small changes of the form coefficients, easing the curvature of the bilge or changing the deadrise angle at the bow, do not have any major influence on the seakeeping performance of a vessel. The designer seeking an improvement of a ship's operational performance in a seaway must think in terms of overall changes of the ship, rather than in local modifications [5]. It is therefore important to consider the seaway performance of a vessel already during the preliminary design. Attempts have been made to derive more clearly definable integrated numbers on the ability of ships to fulfil their missions in rough weather. As early as 1976 St. Denis suggested the *Seakeeping Performance Index* as being the fraction of time that a given ship in a given condition of loading can perform a specified mission. Especially for military vessels, with clearly defined missions and seakeeping performance criteria, this could be developed into *Seakeeping Operating Envelopes* and *Operating Indices* as demonstrated by Comstock and Keane [6]. St. Denis [7] introduced the term environmental operability, which is defined as and measured by the degree of attainment of calm air and still water mission performance. Näreskog and Rutgersson [4] discuss the subject of measuring the availability and reliability of a transport system. They suggest the usage of the transport costs of a given operation as the basis for comparison of performance in calm water and in rough seas. In this way, it is proposed that the seakeeping performance can be given the same priority in the preliminary design process as the calm water performance when comparing different hull configurations. The approach suggested in [4] aims to determine the environmental operability by calculating the annual delay for a vessel on a certain route due to rough weather. The outlined methodology is based on the seakeeping operating envelopes of the vessel and wave statistics at the route. It makes it possible to compare the seakeeping capacity of different design solutions, by measuring their annual delay and it is a convenient tool to apply at the preliminary design stage. To estimate the value of the seakeeping performance of a vessel, the transport cost per trailer kilometre, including environmental operability analysis, are compared to the transport cost assuming calm weather. The difference constitutes a monetary assessment of the seakeeping performance.

4.1. Estimation of ship response and seakeeping operating envelopes

The important elements in the process of estimating the seakeeping operating envelopes for a given operation are statistics giving the wave climate, predictions of the response in these waves and adequate criteria limiting the allowable motions in the seaway [8]. Information about wave climate can be obtained from BMT Global Wave Statistics [9].

The predictions of the ship response can be based on calculations of hydrodynamic coefficients and excitations using the linear strip theory [10]. Criteria on acceptable levels of the ship motions were discussed in the Nordic joint venture, NORDFORSK 1987-Conference [11]. Consideration was taken to comfort, hull safety, operation of equipment, cargo safety and efficiency. Some examples of the criteria are outlined in table 1.

The seakeeping operating envelope of a fast RoRo vessel, at sea state: $H_{1/3}$ = 3m, T_z = 7s is shown in Figure 6.

Table 1. NORDFORSK, 1987, comfort criteria concerning merchant ships

Criteria no i	Response	Merchant ships
1	Vertical acceleration at forward perpendicular (rms-value)	0.275 g (L<=100m) 0.05 g (L>330m)*
2	Vertical acceleration at bridge (rms-value)	0.15 g
3	Lateral acceleration at bridge (rms-value)	0.12 g
4	Roll (rms-value)	6.0°
5	Slamming criteria (probability)	0.03 (L<=100m) 0.01 (L>300m)*
	*The limiting criteria for lengths between 100 and 330m vary almost linearly between L=100m and 330m.	

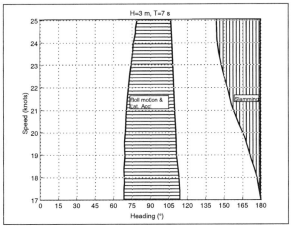

Combinations of speed and heading characterized by the risk of exceeding the criteria on roll motion and on lateral acceleration.

Combinations of speed and heading characterized by the risk of exceeding the criteria on slamming.

Figure 6. Seakeeping operating envelope for a fast RoRo vessel, at sea state: $H_{1/3}$ = 3m, T_z = 7s

4.2. Delay due to rough weather

The delay due to rough weather is influenced by two factors:
• Delay due to involuntary speed reduction, mainly caused by added resistance in waves.
• Delay due to voluntary speed reduction and course change, raised when the master decides to change course or speed to avoid hazardous response levels.

The relative significance of the two parts depends primarily on the sea state. The significance of the voluntary part increases with the severity of the sea state [12]. In [4], focus is put on the delay due to voluntary speed reduction and course change. The reason is that the delay due to involuntary speed reduction could be reduced by installing a certain power margin, while the delay due to the voluntary part is hard to influence after the preliminary design. Another reason is that generally, actions to avoid hazardous responses also reduce the involuntary speed reduction.

In [4], two different methods are suggested to get indications of enough power margin, for preliminary design purposes, to avoid involuntary speed reduction as much as possible. A method outlined by Strom-Tejsen et al [13] is used to estimate the added resistance in head sea, while the dependency of different headings is estimated by means of a method outlined by Towsin et al [14].

4.3. Delay due to voluntary speed reduction and course change

To avoid operating at hazardous combinations of heading and speed at a certain sea state, as shown by the seakeeping operating envelopes, the master should change course or speed, or sometimes both. In this way the crossing time will increase, thus causing delay. The basic strategy outlined in [4] is to study the shortest crossing time for each sea condition and wave direction using the optimum course change or speed reduction with the technique described below. The calculation of average crossing time at a certain speed and sea state is based on the seakeeping operating envelopes and the probability density function of different wave directions.

Transformation of the restricted headings into wave directions in global co-ordinates

The statistical information about wave directions is presented in global co-ordinates, while the restricted headings are calculated with respect to a local co-ordinate system. This implies that the restricted headings have to be transformed into global wave directions. Consideration has to be taken for the bearing from the point of departure to the destination point. The restricted headings may be transformed into global wave directions by means of equation 1.

$$\beta_{\text{Criteria i}} = 180° + C_c + H_{\text{Criteria i}} \qquad (1)$$

Where $\beta_{\text{criteria i}}$ is the restricted wave direction in global co-ordinates with respect to criteria i.

C_c is the bearing from the point of departure to the destination point at each straight route segment.

$H_{\text{criteria i}}$ is the restricted heading in local co-ordinates with respect to criteria i.

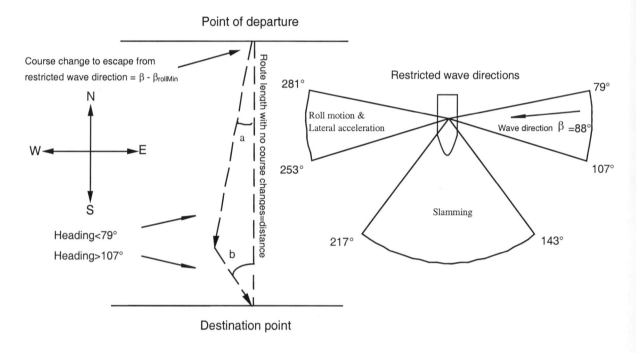

$\beta_{\text{rollMin}} = 79°$ $\beta_{\text{slamMin}} = 143°$ $\beta_{360°\text{-rollMax}} = 253°$

$\beta_{\text{rollMax}} = 107°$ $\beta_{\text{slamMax}} = 217°$ $\beta_{360°\text{-rollMin}} = 281°$, when $\beta=88°$ then $a=\beta-\beta_{\text{rollMin}}$, $b=\beta_{\text{rollMax}} -\beta$

Figure 7. The distance to be sailed as a function of wave direction. Here, the speed is 25 knots and the sea state is: $H_{1/3} = 3m$, $T_z = 7s$. The seakeeping operating envelope is shown in figure 6.

The distance to be sailed as a function of wave direction at a certain sea state

The seakeeping operating envelope provides $H_{\text{criteria i}}$ for each speed at a certain sea state. When the restricted wave directions are calculated the distance to be sailed for each wave direction is estimated according to Figure 7. The distance to be sailed for unrestricted wave directions is the straight distance between the point of departure and the destination point. If the wave direction, β, is within the restricted wave direction intervals for roll motion and lateral acceleration, then the distance to be sailed is:

$$\text{distance} \cdot \frac{\sin(a) + \sin(b)}{\sin(a + b)},$$

Where $a = \beta - \beta_{\text{rollMin}}$ and $b = \beta_{\text{rollMax}} - \beta$. and Distance = the straight distance between the point of departure and the destination point

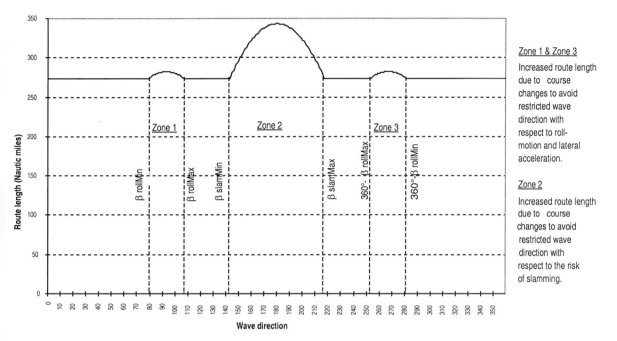

Figure 8. Distance to be sailed as a function of wave direction for a RoRo vessel, at speed 25 knots and at sea state: $H_{1/3} = 3m$, $T_z = 7s$. The straight distance between the ports is 274 M

The average crossing time at a certain sea state

The crossing time, i.e. the time between departure and arrival, with respect to course changes is calculated by dividing the distance to be sailed by the service speed. The seakeeping operating envelope gives the maximum allowable speed level assuming no course changes at wave direction β. Comparison of the crossing time with respect to course changes with the crossing time with respect to speed reduction provides the

preferable choice between speed reduction and course change. The average crossing time is calculated by integrating the distance to be sailed at wave direction β times the probability density function of β between 0° and 360°.

Equation 2 presents a general formula to estimate the average crossing time at a certain speed (V), distance and sea state. The delay due to different criteria, d_i, are calculated according to equation 3.

$$\text{Average crossing time (distance, V, sea state)} = \frac{\text{distance}}{V} \cdot \left(1 + d_1 + d_2 + d_3 + \dots + d_n - n\right) \qquad (2)$$

Where: V is the service speed, d_i, is delay due to criteria number i, and n is the number of restrictioning criteria. The delay, d_i, is calculated according to equation 3.

218

Due to course change   Due to speed reduction

$$d_i = \int_0^{\beta_{iMin}} \text{Prob}(\beta)d\beta + \int_{\beta_{iMin}}^{\beta_{iMax}} \text{MIN}\left[\frac{\sin(\beta_{iMax} - \beta) + \sin(\beta - \beta_{iMin})}{\sin(\beta_{iMax} - \beta_{iMin})}, \frac{V}{V_{max(\beta)}}\right] \cdot \text{Prob}(\beta)d\beta + \int_{\beta_{iMax}}^{360} \text{Prob}(\beta)d\beta \quad (3)$$

Where: i is the criteria number

$\beta_{i\,Min}$ is the minimum wave direction with respect to criteria i

$\beta_{i\,Max}$ is the maximum wave direction with respect to criteria i

Prob (β) = prob. density of wave direction β, $\int_{0°}^{360°} \text{prob}(\beta)d\beta = 1$

$V_{max(\beta)}$ is the maximum permissible speed at wave direction β

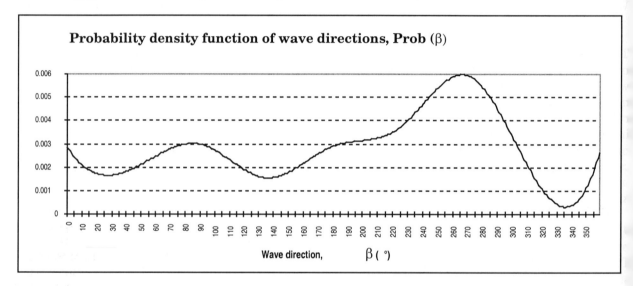

Figure 9. Probability density function of wave directions at the south part of the Baltic Sea

When the restricted wave directions with respect to different criteria overlap, for instance the restricted wave directions with respect to roll motion and lateral acceleration; then the merged restricted wave direction interval is defined as the wave directions between the minimum wave direction and the maximum wave direction of the overlapped region.

If there are abrupt changes in the seakeeping operating envelopes it might be preferable to both change speed and course. If so, the second integral of d_i has to be analysed for different speed reductions at restricted wave directions and the smallest d_i gives the preferable speed, according to equation 4. The annual reduced number of crossings is estimated according to equation 5.

$$d_i = \int_0^{\beta_{iMin}} \text{Prob}(\beta)d\beta + \int_{\beta_{iMin}}^{\beta_{iMax}} MIN\left[\frac{V}{V_{reduced}} \frac{\sin(\beta_{iMax} - \beta) + \sin(\beta - \beta_{iMin})}{\sin(\beta_{iMax} - \beta_{iMin})}, \frac{V}{V_{\max(\beta)}}\right] \cdot \text{Prob}(\beta)d\beta + \int_{\beta_{iMax}}^{360} \text{Prob}(\beta)d\beta \quad (4)$$

Where: V is the service speed and $V_{reduced}$ is the reduced speed at restricted wave directions. $\beta_{i\,boundaries}$ are the wave direction boundaries at speed $V_{reduced}$ in the seakeeping operating envelope.

$$\text{Annual reduction of crossings} = \text{Maximum possible} - \sum_{H_{1/3}Min}^{H_{1/3}Max} \sum_{T_zMin}^{T_zMax} \frac{\text{Annual duration of sea state } (H_{1/3}, T_z)}{\text{Average crossing time at sea state } (H_{1/3}, T_z)} \quad (5)$$

Where: The maximum possible is equal to the number of crossing assuming always calm weather. The annual duration of the sea states may be obtained from the wave statistics.

The duration of each sea state is estimated by means of the wave statistics. The annual delay will neither influence the absolute daily cost nor the capital cost of the ship. It will not reduce the absolute time cost of vehicles and interest cost of the cargo. However, the total number of trailer kilometres per year, the transport work, will be reduced implying an increase of the transport cost per trailer kilometre.

4.4. Procedure to assess environmental operability in terms of cost

The outlined delay estimation is based on wave statistics and seakeeping operating envelopes which, for each sea state, point out the hazardous combinations of heading and speed. It is suggested that the assessment of the seaway performance can be divided into three steps:

) Determination of the seakeeping operating envelops at each appropriate sea state and transformation of the restricted headings into global wave directions.

I) Calculation of the average crossing-time, for each sea state and speed of interest. Estimation of the number of crossings during the annual duration of each sea state and calculation of the annual delay.

III) Evaluation of the annual delay in terms of transport cost.

In [4] calculations are performed for two different RoRo vessels operating in the Baltic Sea. The result showed that the transport cost increased by about 2% when introducing environmental operability and delay due to rough weather. The transport cost would increase significantly if operating the vessels on other routes, characterised by more severe sea states.

5. CONCLUSIONS AND CONSIDERATIONS

During the synthesis, the naval architect has a number of tools to support the design work. The adequacy of the result depends on the accuracy of the applied tools. A significant part of the design work involves sensitivity analyses, which expose margins of error. When applying concurrent engineering, by continuously analysing the transport cost, the significance of the different transport cost components indicates where focus should be directed to improve the vessel. For instance, if the cost connected to the cargo owner is considerable, it is desirable to reduce the value time. In other circumstances, when the ship cost is significant compared to the time-related cargo cost and the travelling cost constitutes a major part of the ship cost, then the naval architect should consider reducing the speed of the vessel and the power prediction has to be made with care.

If the daily cost is a significant part of the transport cost, the designer should consider reducing the number of vessels in the system to decrease the crew number and simplify the maintenance of the vessels and the cargo handling routines. These kind of discussions focus the design work on its proper issues. There is no point in sharpening design tools whose purpose is of little importance to the efficiency of the system.

A vessel operating on a heavy loaded route needs strict operability requirements. Delay is to be considered as a large problem both for the ship operator and for the cargo owner. The cargo owner will suffer from the extra time-related cargo cost while the operator will suffer from reduced income, administrative costs due to cancellations and loss of confidence. On these routes the seakeeping performance of a vessel is of considerable significance. It might be more expensive to operate a vessel with a good seakeeping performance, considering only ship cost, but if the time related cargo cost is reduced more than the ship cost is increased then the transport cost is reduced and there might be a margin to rise the freight rates, still offering a transport service to a reduced cost for the cargo owner. The outlined procedure to estimate the reduced number of crossings per year due to rough weather and evaluate the environmental operability in terms of cost may easily be extended by connecting the fluctuation of cargo flow over the year to the wave statistics. For example the cargo flow in June might have a peak character while the sea state during this time of the year is rather mild. These kinds of extensions provide a more accurate prediction of operability and cost.

The time related cargo cost during the delay might be far higher than stated in the present paper. The cost due to loss in confidence or interruptions in the logistic system is hard to assess. But still, the essence of measuring the environmental operability of merchant ships is to focus on the economy of the whole logistic system.

6. REFERENCES

[1] Näreskog, J., Rydbergh, T. & Sjöbris, A. - "Future Short Sea Cargo Ferry Systems", ISSN 1103-470X, Skeppsteknik, KTH, 1997

[2] Suh, N. P. - "The Principles of Design", Oxford University Press, 1990

[3] Kusiak, A. - "Concurrent Engineering: Automation, Tools and techniques", John Wiley & Sons, 1993

[4] Näreskog, J. and Rutgersson, O. - "Operability in Rough Weather - an Important Preliminary Design Aspect", Int. Conference on High Speed Craft Motions & Manoeuvrability, RINA, London, 1998

[5] Lloyd, A.R.J.M - "Seakeeping: Ship Behaviour in Rough Weather", Ellis Horwood Series in Marine Technology, 1989

[6] Comstock, E. and Keane, R.G. - "Seakeeping by Design", Naval Engineering Journal, Vol. 92, No. 2., 1980

[7] St. Denis, M. - "On the Environmental Operability of Seagoing Systems", SNAME T&R Bulletin No. 1-32, 1976

[8] Andrew, R. N., Loader, P. R. & Penn, V. E. - "The assessment of ship seakeeping performance in likely encountered wind and wave conditions.", RINA, London, 1984

[9] N. Hogben, N. M. C. Dachuna, G. F. Olliver -"BMT Wave Statistics", British Maritime Technology Limited, 1986

[10] Salvesen, N., Tuck, E. O. & Faltinsen, O. M. - "Ship motions and sea loads", Trans.SNAME, 78, 250-90, 1970.

[11] NORDFORSK 1987, The Nordic Co-operative project. Seakeeping performance of Ships, Assessment of a ship performance in a seaway, Trondheim, Norway, 1987

[12] Lewis, E.V. - "Increasing the Sea Speed of Merchant Ships.", Trans. SNAME, 67, 1959.
[13] Strom-Tejsen, J.,Yen, H.Y. H. & Moran D. D. - "Added resistance in waves", Trans. SNAME, 81, 109-43, 1973
[14] Towsin, R. L. and Kwon, Y. J. - "Approximate Formulae for the Speed Loss Due to Added Resistance in Wind and Waves", RINA, 1982.

Appendix

Transport cost - a measure of effectiveness of a transport system

One way to measure the effectiveness, in economical terms, of a transport system is to investigate the total cost of the whole logistic system. It can be done by measuring the transport cost of the system. The transport cost is defined as the sum of the ship owner's cost of operating the ship plus the time-related cargo cost during the trip.

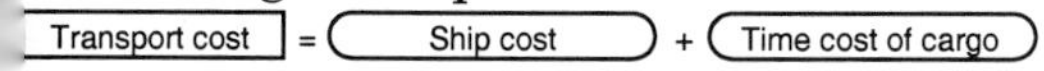

Ship cost

The ship cost covers the cost of operating the ship. It is composed of the travelling cost, the daily cost and the capital cost of the vessel.

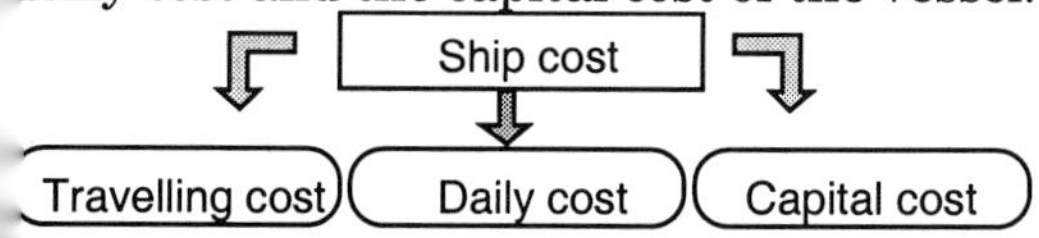

The travelling cost is the sum of the variable costs, like the fuel consumption, the harbour dues, the channel dues and the cost of loading and unloading. The daily cost is regarded to be independent of whether the ship is operating or not. The significant costs included in the daily cost are crew wages, maintenance cost, repairs, docking, classification cost, insurance and administration cost. The capital cost is estimated by a capital recovery factor, which enables an initial capital investment to be recovered as an annual capital charge, including both interest and depreciation of principal. In [1] approximate methods to estimate the three components of the ship cost are presented.

Time-related cargo cost

The time cost of load-carriers and the interest on cargo during the sea transport can be of considerable significance depending on the type and value of the cargo. For a conventional RoRo ship it might be of the same magnitude as the ship cost, and it is composed of two cost centres: the cost of the load-carrier, which includes the rent and the depreciation of the load-carrier, plus the wage to the driver, if appropriate.

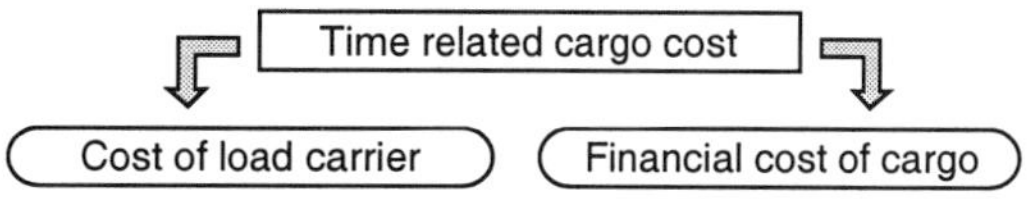

There is a large cost difference between runners[1] and drop trailers[2]. The financial cost of the cargo carried - i.e. the rent and the depreciation of the cargo. According to [1], the time-related cargo cost is depending on the value time, which is defined as the time between loading and unloading plus one third of the time between departures. Usually, it is preferable if there is a balance between the ship cost and the time-related cargo cost. If the time-related cargo cost is far more significant than the ship cost then there is certainly a margin to decrease the value time of the transport, by for instance, increasing the speed or the transport frequency

---

[1] Runners - Complete 16.5m Europe semi-trailer vehicles with both road truck and 13.6m trailer, loaded with 18 ton payload. Runners are driven onboard the ships by their drivers.

[2] Drop trailers - 13.6m Europe trailers, loaded with 18 ton payload, driven onboard with terminal tractors.

Steady Behavior of a Full Large Ship at Sea

Shigeru Naito[a] and Kenji Takagishi[b]

[a] Department of Naval Architecture and Ocean Engineering, Osaka University,
   2-1 Yamada-oka, Suita, Osaka 565-0871, Japan

[b] Ship and Marine Structure Lab., Tsu Research Center, NKK Corporation,
   1, Kumozukokan-cho, Tsu city, Mie 514-0301, Japan

Authors propose a practical method to estimate the time mean behavior of a full large ship at sea. The method includes the effects of wind, waves and main engine characteristics. At sea, there are disturbing external forces acting on a full large ship. These forces consist of a fluctuation part and a time mean part, that is steady part. The later is the added resistance, the lateral force and the turning moment. Also the hydrodynamic forces due to the hull, rudder and propeller have a time mean part. With those steady forces and the equilibrium condition of the engine and propeller torque, we can set up four equilibrium equations among hull, propeller, rudder and engine. Regarding the speed, drift angle, encounter rudder angle and the propeller revolution number as unknown variables, we can obtain those time mean values by solving the equilibrium equations with an iteration method.

1. INTRODUCTION

It has been reported that full large ships like tankers can not maintain their design speed even at the moderate sea. One of the reasons is that the speed loss caused by wind, waves and rudder motion is not estimated precisely. In that condition we need to know the speed drop, drift angle and encounter rudder angle etc.

Considering the forces acting on the ship as time mean values, equations of equilibrium for these forces can be formulated on the assumption that the motion of the ship in moderate sea is very small because the ship is quite large. Furthermore the equilibrium equation between the main engine and propeller torque can be formulated considering those torques as time mean values. Then the time mean behavior of a full large ship can be estimated by solving those equations using an iteration method.

The estimation method of forces acting on a ship by a propeller and a rudder is shown in reference[1]. The wind force can be estimated by Yamano and Saito [2]. The steady wave forces in short crested irregular waves are obtained by us and Mork [3]. The relation between main engine and propeller is known by reference[5]. By synthesizing the results from these methods, we can estimate the propulsive performance of ship at sea[6].

2. STEADY FORCES ACTING ON A SHIP

It is assumed that the motion of a full large ship is very small as the wave length in moderate seas is considered short. $O - XYZ$ is the space fixed coordinate system, and $G - xyz$ is the body fixed system as shown in Fig. 1 .

For a ship navigating with a forward speed V towards X, the velocity in the x and y direction are denoted u and v respectively. The wind (true wind direction γ) is an uniform flow and waves (true wave direction χ) are short crested irregular waves. Then three steady forces are acting on the ship, these are the added resistance(toward the x-axis), the side force (toward the y-axis) and the turning moment(round the z-axis). As these forces are steady, the equilibrium equations of a ship are described by the following equations (1).

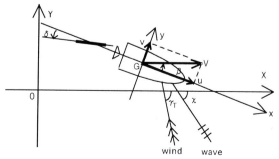

Figure 1. Co-ordinate system

$$\left.\begin{array}{l} F_{xP} + F_{xR} + F_{xH} + F_{xW} = 0 \\ F_{yP} + F_{yR} + F_{yH} + F_{yW} = 0 \\ N_{zP} + N_{zR} + N_{zH} + N_{zW} = 0 \end{array}\right\} \quad (1)$$

Here the suffix x, y and z indicate the forces of x, y components and the moment round z-axis. The suffix P, R and W indicate the propeller, rudder and disturbing external forces due to wind and waves, and the H indicates the ship hull.

2.1. Steady forces by propeller

A propeller causes a force toward the x-axis by the thrust F_{xP} with thrust deduction fraction of t. The side force F_{yP} and turning moment N_{zP} are zero. Thus,

$$\left.\begin{array}{l} F_{xP} = (1-t)\rho K_T n_p^2 D^4 \\ F_{yP} = 0 \quad ; \quad N_{zp} = 0 \end{array}\right\}, \quad (2)$$

where D, n_p and ρ are the propeller diameter, number of revolution per second and water density respectively. The thrust coefficient K_T is determined with the advance coefficient J as

$$K_T = a_T + b_T J + c_T J^2. \quad (3)$$

Where a_T, b_T and c_T are coefficients given by the propeller open chart. J is

$$J = \frac{(1-w)V}{n_p D}, \quad (4)$$

where $1 - w$ is a wake fraction.

2.2. Steady forces by rudder

For a ship navigating at sea with an encounter rudder angle δ, the forces, F_{xR}, F_{yR} and N_{zR} affecting the ship occur. Considering the interference between the rudder and hull, the three forces are presented as

$$\left.\begin{array}{l} F_{xR} = -(1 - t_R)F_N \sin(\delta) \\ F_{yR} = (1 + \alpha_H)F_N \cos(\delta) \\ N_{zR} = -l_R(1 + \alpha_H)F_N \cos(\delta) \end{array}\right\}. \quad (5)$$

Here t_R is the resistance coefficient of rudder, α_H the interference coefficient between rudder and hull. l_R is about 0.5L. The normal force acting on a rudder F_N is

$$F_N = \frac{1}{2}A_R f_\alpha U_R^2 \sin(\alpha_R). \quad (6)$$

Where A_R is the rudder area and f_α is given by Fujii[7] as

$$f_\alpha = \frac{6.13\Lambda}{2.25 + \Lambda}, \quad (7)$$

and U_R is the significant fluid velocity to the rudder and divided into its x and y components like

$$U_R^2 = u_R^2 + v_R^2 \quad . \quad (8)$$

By using the method of Yoshimura [8], u_R is

$$u_R = \frac{1 - w_R}{1 - s_R}V$$
$$\cdot \sqrt{1 - 2(1 - \eta\kappa)s_R + \{1 - \eta\kappa(2 - \kappa)\}s_R^2} \quad ,$$

$$s_R = 1 - \frac{u(1 - w)}{n_p P} \quad , \quad \eta = \frac{D}{h},$$

$$\kappa = \frac{0.6(1 - w)}{1 - w_R} \quad , \quad w_R = 0.25\exp(-0.4\beta^2).$$

Where p and h are the propeller pitch and the rudder height respectively. And v_R and β are given by

$$v_R = \Gamma_R v, \quad \beta = tan^{-1}(v/u). \quad (9)$$

The rectification coefficient Γ_R is about 0.4 for a full large ship. The effective angle, α_R, of inflow velocity to the rudder is given by:

$$\alpha_R = \delta + \sin^{-1}\frac{v_R}{U_R}. \quad (10)$$

2.3. Steady forces on a ship hull

The hydrodynamic forces acting on a ship in still water are given as

$$\left.\begin{array}{l} F_{xH} = \frac{1}{2}\rho L d V^2(X_0 - X_{\beta\beta}\beta^2) \\ F_{yH} = \frac{1}{2}\rho L d V^2(Y_\beta\beta - Y_{\beta\beta\beta}\beta^3) \\ N_{zH} = \frac{1}{2}\rho L^2 d V^2(N_\beta\beta - N_{\beta\beta\beta}\beta^3) \end{array}\right\}. \quad (11)$$

Where $X_0, X_{\beta\beta}, Y_\beta, Y_{\beta\beta\beta}, N_\beta, N_{\beta\beta\beta}$ are nondimensional hydrodynamic derivatives. The values are usually obtained by experiments. $X_{\beta\beta}, Y_{\beta\beta\beta}$ and $N_{\beta\beta\beta}$ can be omitted, as these values are small from a practical view point.

2.4. Steady external forces

Wind and wave as disturbing external forces are assumed that the wind is an uniform flow and waves are short crested irregular waves.

2.4.1. Steady wind forces

The steady wind forces are given by Yamano and Saito's experimental formula[2] as

$$
\left.
\begin{aligned}
F_{xw2} &= \frac{1}{2}\rho_a A_T U_{RW}^2 C_x(\gamma_R) \\[4pt]
F_{yw2} &= \frac{1}{2}\rho_a A_L U_{RW}^2 C_y(\gamma_R) \\[4pt]
N_{zw2} &= \frac{1}{2}\rho_a L A_L U_{RW}^2 C_z(\gamma_R)
\end{aligned}
\right\} \quad (12)
$$

Where ρ_a is the density of air. A_T and A_L are the projected area of the front and side view above the water line of the ship. $C_x(\gamma_R)$, $C_y(\gamma_R)$ and $C_z(\gamma_R)$ are wind resistance coefficients shown by

$$
\left.
\begin{aligned}
C_x(\gamma_R) &= \sum_{n=0}^{5} C_{xn}\cos(n\gamma_R) \\[4pt]
C_y(\gamma_R) &= \sum_{n=1}^{3} C_{yn}\sin(n\gamma_R) \\[4pt]
C_z(\gamma_R) &= \sum_{n=1}^{3} C_{zn}\sin(n\gamma_R)
\end{aligned}
\right\} \quad (13)
$$

Where C_{xn}, C_{yn} and C_{zn} are given as follows.[2]

$$C_{x0} = -0.0358 + 0.925\frac{A_L}{L^2} + 0.0521\frac{X_g}{L}$$

$$C_{x1} = 2.58 - 6.078\frac{A_L}{L^2} - 0.1735\frac{L}{B}$$

$$C_{x2} = -0.97 + 0.978\frac{X_g}{L} + 0.0556\frac{L}{B}$$

$$C_{x3} = -0.146 + 0.728\frac{A_L}{A_T} - 0.0283\frac{L}{B}$$

$$C_{x4} = 0.0851 + 0.0212\frac{A_L}{A_T} - 0.0254\frac{L}{B}$$

$$C_{x5} = 0.0318 + 0.287\frac{A_L}{L^2} - 0.0164\frac{L}{B}$$

$$C_{y1} = 0.509 + 4.904\frac{A_L}{L^2} + 0.022\frac{A_L}{A_T}$$

$$C_{y2} = 0.0208 + 0.230\frac{A_L}{L^2} - 0.075\frac{X_g}{L}$$

$$C_{y3} = -0.357 + 0.943\frac{A_L}{L^2} - 0.0381\frac{L}{B}$$

$$C_{z1} = 2.65 + 4.634\frac{A_L}{L^2} + 5.876\frac{X_g}{L}$$

$$C_{z2} = 0.105 + 5.306\frac{A_L}{L^2} + 0.0704\frac{A_L}{A_T}$$

$$C_{z3} = 0.616 - 1.474\frac{X_g}{L} + 0.0161\frac{L}{B}.$$

Where X_g is the length between $F.P.$ and the center of the gravity of A_L. U_{RW} and γ_R are the relative wind velocity and direction given by the true wind velocity U_{TW}, and the true wind direction γ_T. They are shown as

$$
\begin{aligned}
U_{RW}^2 &= \{U_{TW}\cos(\gamma_T - \beta) + u\}^2 \\
&\quad + \{U_{TW}\sin(\gamma_T - \beta) - v\}^2,
\end{aligned}
$$

$$\gamma_R = \frac{U_{TW}\sin(\gamma_T - \beta) - v}{U_{TW}\cos(\gamma_T - \beta) + u}.$$

2.4.2. Steady wave forces

A method of the practical estimation of the steady forces acting on a full large ships in short crested irregular waves was proposed as follows [3][4].

$$
\left.
\begin{aligned}
F_{xwa} &\simeq \frac{K(m)}{8}\rho g B H_{1/3}^2 \\
&\cdot \sum_{j=0}^{4} \epsilon_j\left(a_{1j} + 2\frac{V}{g}\frac{2\pi}{T_0}a_{2j}\right)\cos\{j(\chi - \beta)\}P_{jm} \\
F_{ywa} &\simeq \frac{K(m)}{8}\rho g B H_{1/3}^2 \\
&\cdot \sum_{j=0}^{4} \epsilon_j\left(b_{1j} + 2\frac{V}{g}\frac{2\pi}{T_0}b_{2j}\right)\sin\{j(\chi - \beta)\}P_{jm} \\
N_{zwa} &\simeq \frac{K(m)}{8}\rho g B L H_{1/3}^2 \\
&\cdot \sum_{j=0}^{4} \epsilon_j\left(c_{1j} + 2\frac{V}{g}\frac{2\pi}{T_0}c_{2j}\right)\sin\{j(\chi - \beta)\}P_{jm}
\end{aligned}
\right\} \quad (14)
$$

$$K(m) = \frac{1}{\sqrt{\pi}}\frac{\Gamma(1 + \frac{m}{2})}{\Gamma(\frac{1}{2} + \frac{m}{2})}, \epsilon_0 = 0.5, \epsilon_{j\neq0} = 1,$$

$$P_{jm} = \int_0^{\pi/2} \cos(j\theta)\cos^m\theta\, d\theta.$$

Where $H_{1/3}$ is the significant wave height, T_0 the mean wave period, m the parameter of wave directional distribution, χ the main wave direction and Γ the Gamma function. The Fourier coefficients a_{1j} and a_{2j} are determineded from the directional characteristics of the added resistance, b_{1j} and b_{2j} coeffecients from the side force, c_{1j} and c_{2j} coefficients from the turning moment in regular wave respectively [9][10][11].

For a more detailed explanation of this method the reader is reffered to [3] .

3. RELATIONSHIP BETWEEN THE PROPELLER AND MAIN ENGINE

The relationship of equibrium between the main engine torque (Q_e) and the propeller torque (Q_p) is now derived. From considering incremental changes in the engine torque and propeller torque the following equations are found as follows

$$\delta Q_p = P_{Qn0}\delta n_p + P_{Qu0}\delta u_p \tag{15}$$

$$\delta Q_e = E_{Qn0}\delta n_e + E_{Q\Lambda0}\delta\Lambda. \tag{16}$$

Where δu_p is the incremental change of the significant fluid velocity to the propeller toward x axis, and $\delta\Lambda$ is the fuel consumption per an hour and a revolution. And P_{Qn0}, E_{Qn0} are given by

$$\left.\begin{array}{l} P_{Qn0} = [\frac{\partial Q_p}{\partial n_p}]_0,\ P_{Qu0} = [\frac{\partial Q_p}{\partial u_p}]_0 \\[2mm] E_{Qn0} = [\frac{\partial Q_e}{\partial n_e}]_0,\ E_{Q\Lambda0} = [\frac{\partial Q_e}{\partial \Lambda}]_0 \end{array}\right\}. \tag{17}$$

Where the suffix 0 denotes the value in still water. The difference of each value at sea from in still water is small. When the gear ratio between the propeller axis and the engine one is r, the relations between Q_p and Q_e, n_p and n_e are shown as

$$rQ_e = Q_p, \quad n_e = rn_p. \tag{18}$$

Then equation(16) can be rewritten at the propeller axis as

$$\delta Q_p = r^2 E_{Qn0}\delta \dot{n}_p + r E_{Q\Lambda0}\delta\Lambda. \tag{19}$$

Thus, then combining equations(16) and (19), the equilibrium condition between Q_e and Q_p is shown as

$$(r^2 E_{Qn0} - P_{Qn0})\delta n_p + r E_{Q\Lambda0}\delta\Lambda - P_{Qu0}\delta u_p = 0. \tag{20}$$

In equation(20) P_{Qn0} and P_{Qu0} are derivative coefficients of the propeller torque with n_p and u_p shown as

$$P_{Qn0} = \rho D^4(2d_Q D n_{p0} + e_Q u_{p0}) \tag{21}$$

$$P_{Qu0} = \rho D^3(e_Q D n_{p0} + 2f_Q u_{p0}). \tag{22}$$

Where d_Q, e_Q and f_Q are coefficients of the torque coefficient K_Q given by

$$K_Q = d_Q + e_Q J + f_Q J^2. \tag{23}$$

Since n_p and u_p in still water are n_{p0} and u_{p0} respectively, δn_p and δu_p in the equation(20) can be replaced by

$$\begin{array}{l} \delta n_p = n_p - n_{p0} \\[1mm] \delta u_p = u_p - u_{p0} = (1-w)(u-u_0). \end{array} \tag{24}$$

The second term of the left-hand side of equation (20) is 0 when no command is given to inject fuel. $\delta\Lambda$ can be rewritten in term of the rate of opening the fuel injection valve, k,

$$\delta\Lambda = (\Lambda - \Lambda_0) = (k-1)\Lambda_0; (\Lambda = k\Lambda_0). \tag{25}$$

This is ordered by the captain. $E_{Q\Lambda0}$ is derivative coefficient of Q_e with the fuel consumption per an hour and a revolution, given with the power P_e and n_e as

$$Q_e = \frac{75}{2\pi}\frac{P_e}{n_e}. \tag{26}$$

The relation between P_e and $n_e\Lambda$ is a quasi-linear as

$$P_e \simeq a_\lambda(n_e\Lambda) + b_\lambda, \tag{27}$$

thus, equation(26) can be rewritten as

$$Q_e \simeq \frac{75}{2\pi}\frac{1}{n_e}\{a_\lambda(n_e\Lambda) + b_\lambda\}. \tag{28}$$

As the result, $E_{Q\Lambda0}$ is presented as

$$E_{Q\Lambda0} = \left[\frac{\partial Q_e}{\partial\Lambda}\right]_0 = \frac{75}{2\pi}a_\lambda. \tag{29}$$

With the equations(24)(25)(29), the equation(20) can be rewritten as

$$(r^2 E_{Qn0} - P_{Qn0})(n_p - n_{p0}) + (k-1)rE_{Q\Lambda0}\Lambda_0 \\ -(1-w)P_{Qu0}(u_p - u_{p0}) = R_{ep}(u, n_p) = 0. \tag{30}$$

Where E_{Qn0} is derivative coefficient of Q_e with n_e. These coefficients are determined with the characteristic of the main engines. Generally, the characteristics of the engine will depend on the type.

The following is the explanation of the torque constant, power constant and revolution constant. [5]

1. Torque constant mode : $E_{Qn0} = 0$ (31)

2. Power constant mode : $Q_e n_e = Q_{e0}\cdot n_{e0}$ (32)

Substituting $Q_e = Q_{e0} + \delta Q_e, n_e = n_{e0} + \delta n_e$ to equation(32), we get

$$\delta Q_e n_{e0} + Q_{e0}\delta n_e + \delta Q_e \delta n_e = 0. \tag{33}$$

In case of $\delta\Lambda = 0$, the equation(16) becomes $\delta Q_e = E_{Qn0}\delta n_e$. Substituting this equation into the equation(33), and neglecting the higher order term, we obtain

$$E_{Qn0} \simeq -Q_{e0}/n_{e0}. \tag{34}$$

3. Revolution constant mode : The first term of the left-hand side of the equation(30) becomes 0.

4. FOUR EQUILIBRIUM EQUATIONS

The equilibrium equations (1) and (30) are rewritten with variables as follows.

$$\left.\begin{array}{r}F_{xP}(u,v,n_p) + F_{xR}(u,v,\delta,n_p) \\ + F_{xH}(u,v) + F_{xW}(u,v) = 0 \\ F_{yR}(u,v,\delta,n_p) + F_{yH}(u,v) \\ + F_{yW}(u,v) = 0 \\ N_{zR}(u,v,\delta,n_p) + N_{zH}(u,v) \\ + N_{zW}(u,v) = 0 \\ R_{ep}(u,n_p) = 0 \end{array}\right\} \tag{35}$$

In these equations, u, v, δ and n_p are unknown variables. They are obtained with an iterated calculation method.

5. CALCULATION CONDITIONS

A case study is presented considering a VLCC Tanker. The details of the tanker and its operating condition are:

Ship :
$L_{pp} = 320(m)$, $\qquad B = 58(m)$, $\qquad d = 19.3(m)$
$A_T = 1130(m^2)$, $\quad A_L = 4000(m^2)$, $X_g = 188.2m$

Rudder : $h = 12.5(m)$, $c = 7.05(m)$

Propeller : $D = 9.3(m)$, $\quad p = 6.5(m)$
$a_T = 0.3085$, $b_T = -0.2571$, $c_T = -0.1688$
$d_Q = 0.3165$, $e_Q = -0.1100$, $f_Q = -0.2821$

Self-propulsion factor :
$1 - t = 0.797$, $1 - w = 0.643$, $\eta_R = 1.0$

Nondimensional hydrodynamic derivatives :
$X_0 = -0.02115$, $Y_\beta = 0.25536$, $N_\beta = 0.13204$

Main engine :
$a_\lambda = 7.857(\frac{Ps}{kg/h})$, $b_\lambda = 500(Ps)$, $r = 1$

Sea states :

B.N.	3	4	5	6
$H_{1/3}$ (m)	1.4	1.7	2.15	2.9
T_0 (s)	5.9	6.1	6.5	7.3
U_{TW} (m/s)	4.45	6.75	9.4	12.35

6. RESULTS AND DISCUSSION

Fig.2 shows calculated results of the speed drop ratio, drift angle, encounter rudder angle, revolution change ratio, torque change ratio and power change ratio under the torque, power and revolution constant mode respectively.

The abscissa in the figure is the true directions of wave and wind, which are the same direction. It is observed that the speed drop ratio increases in the order, revolution constant, power constant, torque constant.

Increasing k large implies that the ship speed increases. Thus, k and n_p for maintaining the set ship speed at sea (head wave and wind) are estimated under the torque constant mode. The results are shown in Fig.3. The abscissa is Beaufort Number (BN). It is observed that the set speed can not be maintained without increasing k by 10% in $B.N.3$ or 4 and over by 25% in $B.N.6$.

The relationships between speed drop ratio, drift angle, encounter rudder angle and sea state under the torque constant mode are shown in Fig.4. It is found from those results that each value is large in $B.N.6$ in comparison with that founded in $B.N.3 \sim 5$. Specifically, the speed drop ratio is over 10% in condition in which the wave and wind direction is from $0°$ to about $50°$ in $B.N.6$

The speed drop ratio is not always the largest in head wind and wave. It is observed largest with head wind and $30°$ wave direction.

7. CONCLUSION

A practical method to estimate the time mean behavior of a full large ship in moderate sea by unifying many research results on seakeeping of a ship has been presented and documented.

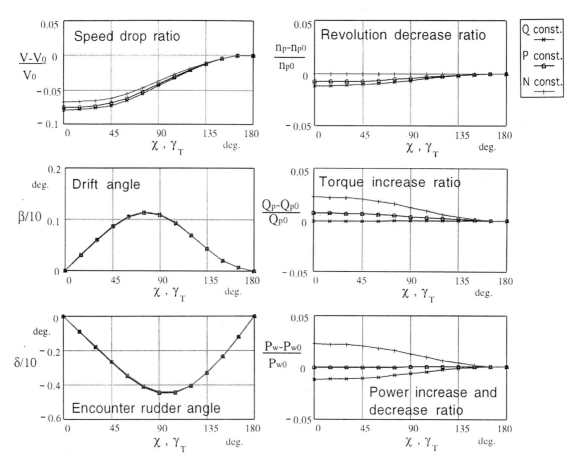

Figure 2. Calculated results of various quantities on performance under the conditions torque constant, power constant and revolution constant ($V_0 = 5.60 m/s$, $n_{p0} = 1.27 rps$, $B.N.5$)

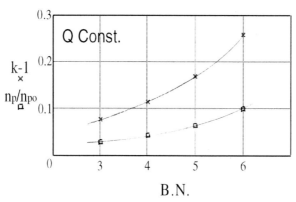

Figure 3. Calculated results of k and n_p for maintaining the set ship speed ($V_0 = 5.60 m/s$, Head wave and wind) at sea

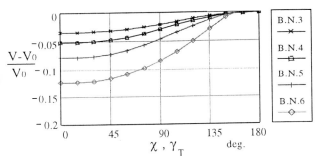

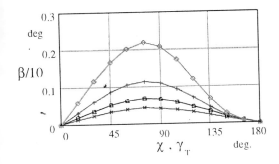

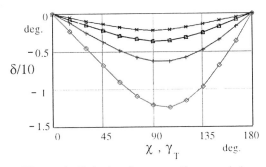

Figure 4. Relation between the speed drop ratio, the drift angle and the encounter rudder angle and the sea state

REFERENCES

1. Report of 221 Working Group : Research on the flow field around ships at manoeuvering motion, The Shipbuilding Research Association of Japan, 1995

2. Yamano, T. and Saito, Y., "An Estimation Method of Wind Forces Acting on Ships", J. of KSNAJ., Vol. 228 pp.91-100., 1971

3. Naito, S. and Takagishi, K. "Research on Added Resistance in Short Crested Irregular Waves and Bow Form of Water Line", J .of KSNAJ., Vol.225, 1996

4. Mork , B. ,"Ship Performance in Short Crested Low Seastates", In Siv.ing Thesis in Marine Hydrodynamics, NTH, Norway 1994

5. Nakamura , S. and Naito , S. "Nominal Speed Loss and Propulsive Performance of a Ship in Waves", J .of KSNAJ., Vol.166, 1977

6. Naito , S., Nakamura , S. and Hara , S. "On the Prediction of Speed Loss of a Ship in Waves", J. of SNAJ., Vol.145, 1979

7. Fujii,H. and Tuda,T. "Experimental Researches on Rudder Performance", J. of SNAJ., Vol.110., 1961

8. Yoshimura, Y. and Nomoto, K., "Modeling of Manoeuvering Behavior of Ships with a Propeller Idling Boosting and Reversing", J. of SNAJ., Vol.144, 1978

9. Faltinsen,O.M. , Minsaas,K.J. , Liapis,N. and Skjordal,S.O. : Prediction of Resistance and Propulsion of a Ship in a Seaway ,Proc. of 13th Symposium On Naval Hydrodynamics, pp.505-529, 1980

10. Naito, S. and Takagi, K. , "Practical Formula of Added Resistance in Short Crested Irregular Wave and Bow Form", Proc. of 16th PRADS, No.I-369p~381p, Sep. 1995

11. Report of 208 Working Group : Research on the elimination of the wave effect on the ship speed at sea trial, The Shipbuilding Research Association of Japan, 1993

Practical Design of Ships and Mobile Units
M.W.C. Oosterveld and S.G. Tan, editors.

Multiattribute Design Synthesis for Robust Ship Subdivision of Safe Ro-Ro Vessels

G. Trincas

Department of Naval Architecture, Ocean and Environmental Engineering, University of Trieste, Via A. Valerio 10, I-34127 Trieste, Italy

This paper gives a brief summary of the basic features of a fuzzy multiattribute method for the optimum concept design of ro-ro vessels. Technical and economic features are considered simultaneously to solve conflicting and uncertain issues of new compartmentation concepts and economic viability. Internal subdivision is associated to the technical subsystem to ensure compliance with damage survivability criteria. Evaluations, which have been carried out for alternative designs, are presented.

1. INTRODUCTION

The present paper deals with the multiattribute identification of the techno-economic performance of optimal ro-ro vessels where robust assessment of safe ship subdivision is associated. The related design method is mainly intended for application at concept ship design.

IMO compels to assess subdivision of ro-ro vessels by a probabilistic method requiring cumbersome computations that are to be repeated after enumerated changes in vessel compartmentation until the 'attained subdivision index' (A) equals the `required sub-division index' (R). But when the design process is envisaged as a spiral, even if A is greater than R the designer has no possibility to predict the overall impact of subdivision assessment on the other technical issues as well as on economics of his decisions. Work on optimisation of ship subdivision is increasing, but the subject must be further developed before it may be incorporated with the concept ship design process. Having this scope in mind and also because the optimum design problem cannot be solved as single-objective, a multiattribute decision-making (MADM) scheme was updated that showed to be the most powerful evaluation tool for the concept ship design [9].

Number of feasible solutions are generated in the design space, based on the revised mathematical design model which covers all main features of ro-ro vessels [7]. A damage stability model communicating with a relational data structure for ship compartmentation provides the ship subdivision. Feasible solutions are then mapped into attribute space where a nondominance structure is applied to select efficient, Pareto-optimal designs. Hence the most promising and preferred design is selected to arrive at a robust ro-ro vessel.

The first part of the paper describes the concept design strategy. The following sections discuss some of the pertinent aspects in the analysis and synthesis of the damage survivability modelling, and offer a short overview of the fuzzy MADM tool. The final section deals with an application of the method to a class of ro-ro vessels.

2. CONCEPT DESIGN STRATEGY

The overriding importance of giving due consideration to damage stability since concept design derives from the influence it has on the general vessel effectiveness. Sen and Gerigk [5] developed an expert system for safe ship subdivision more suitable to preliminary design. The basic difference between our approach and their work comes from the fact that decisions to be made at concept design must be fully comprehensive. By this we mean that this calls for a different design strategy when safety-related features, technical issues and economic aspects are considered simultaneously. Moreover, except for the bigger number of criteria, more difficulties arise when information is more imprecise then they are at subsequent stages.

Thus the concept design process forces the designer to decide in an uncertain environment where the techno-economic value of alternative ships should be described by a set of multiple

232

attributes that represent specification requirements to be achieved at desired levels. Most of them can be translated into soft constraints (*attributes*) to be treated by fuzzy-set theory which allows for normalised measure of levels attained of design x_i, modelled by membership grade functions $\mu_{Aj}(x_i)$ under attributes A_j. Fuzzy sets substantially enable the designer to control the convergence towards an acceptable compromise solution via weighted degrees of satisfaction. Hence safety-related ship subdivision can be easily put in close connection with other technical and economic issues depending on both hull form and subdivision. For instance, since there is no certain boundary capable to distinguish between capsize and non-capsizing when using A and R indices, fuzzy sets may help to avoid yes-or-no-type decision between feasible and infeasible design with respect to survivability and safety in damage condition.

The concept design process was structured according to this philosophy. It is broken down into the mathematical design (analysis) model consisting of both a technical and an economic subsystem, the nondominance analysis model, the fuzzy evaluation (synthesis) model, and the final sensitivity analysis to arrive at the most robust solution. In Figure 1, the structure of the concept design process is shown schematically. It is understood that the evaluation of feasible, hence nondominated designs, is to be done based on consideration of several modules of both technical (weight distribution, load capacity, service speed, seakeeping, intact and damage stability, vibration, manoeuvring, and so forth) and economic nature (acquisition, manning, and voyage cost).

When a significant number of feasible designs are randomly generated, the selection procedure is performed in the attribute space by firstly identifying the nondominated solutions through a fuzzy multi-attribute optimisation which requires designer's intra-attribute subjective preferences. Then the best possible ro-ro ship is selected, which is more distant from the *nadir* solution.

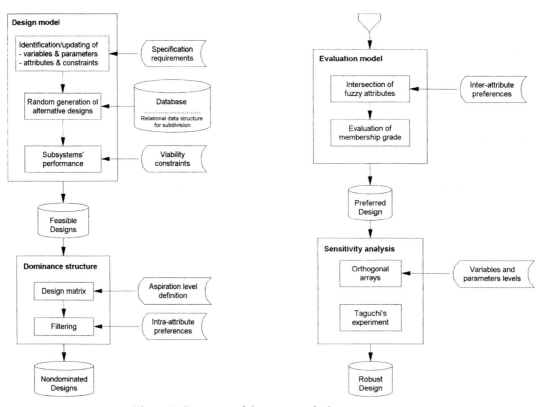

Figure 1. Structure of the concept design process

Finally, Taguchi's robustness concept [6] is applied to the best possible solution to identify the design which inherently tends to diminish the effect of uncertainty of economic scenario on global performance. The signal-to-noise ratio of some economic, uncertain attribute (fuel price, required freight rate, etc.) can be used as a measure of the robustness of nondominated ro-ro vessels generated around the best possible design. Different levels of the control factors (variables) and of the noise factors (economic parameters) are identified to set up the orthogonal array L_X which allows to find the optimal control factor levels towards the noise factors, that is, to identify the most robust design.

3. DAMAGE STABILITY MODELLING

As a result of the 1974 SOLAS Convention, subdivision arrangement of operated ro-ro vessels mostly shows relatively short and cross-connected compartments typically limited by dense transverse bulkheads and by longitudinal bulkheads inside $B/5$ line below the first deck above the deepest load line. The new probabilistic rules for these vessels reveal very low values of subdivision indices A. This fact confirms that the historical evolution of ro-ro vessel design has yielded hull form characteristics irrespective of damage stability safety concern. Catastrophes of the `European Gateway' (1982), the `Herald of Free Enterprise' (1987), the `Jan Heweling' (1993), the `Estonia' (1995), all of them without any subdivision on the vehicle deck, clearly illustrate that contemporary ro-ro vessels are vulnerable to flooding. The only way to increase markedly the A index usually has been either to increase freeboard or to apply removable transverse bulkheads in main deck which limits the main purpose of ro-ro arrangement. Such solutions clearly spoil operational economics of ro-ro ships since they add steel weight and increase operating costs, and often lead to reduction in payload and increase of turn-round time.

It follows that consideration of internal subdivision must be introduced since concept design. A damage stability model was therefore developed to help the designer in evaluating the influence of the subdivision arrangement in assessing damage survivability by probabilistic rules. Since some

compartments and their position could be subject to minor modifications afterwards, the constraints on the designer in placing bulkheads should be more vague than in the subsequent design phases. At the same time, alternative subdivision layouts, while complying enhanced survivability standards, should be combined at least with minimisation of loss of freight capacity as well as of loss of time during loading/unloading operations.

The geometric model built for different internal subdivisions allows for calculating volumes and centroids, as well as intact and damage stability. The description of the compartments is realised by mapping into a simplified relational data model based on the scheme proposed by Schumann-Hindenberg [4]. Alternative layouts were inserted into the design model to yield alternatives with enhanced survival capability. They involved some simplifications since the general arrangement of the ship cannot be documented in full detail: main hull under weather deck is considered only, whereas superstructure and floodable openings are not taken into account; permeabilities are maintained fixed. The vague elements of domains constituting the relational data structure are modelled by fuzzy sets and the relations by fuzzy relations.

The procedure proposed by Juncher Jensen et al. [2] to arrive at a boundary limit between A and R indices has been extended by considering that the total A is equally contributed by three draughts, e.g. for the ship loaded at the deepest subdivision load line and at two partial load lines taken at the lightship draught plus 20% and 60% of the difference between the lightship and the deepest load line draught, respectively. In this analysis, the Stockholm Regional Agreement was applied which accounts for the presence of a maximum 0.5 m height of water on the vehicle deck. The survival probability s contributing to the A index and that is a function of the damage stability parameter $Gm_f F_e/B$, was always calculated according to Pawlowsky [3] as

$$s = \sqrt[6]{GZ_d / GZ_m} \qquad (1)$$

where GZ_d is the maximum value of the residual GZ-curve and $GZ_m \cong 0.1$ m is a value of GZ_d yielding $s = 1$. The GZ-curve was calculated for the freely floating ship.

234

4. FUZZY MADM OPTIMISATION

A fuzzy MADM method has been implemented to select the optimum ship variables among a finite number of alternative designs under a finite number of technical and economic performance criteria. It is structured through three sub-fields: (i) identification of the nondominated designs; (ii) selection of the preferred solution by rating the overall performance of the feasible alternatives; and (iii) generation of the most robust design.

In order to arrive at nondominated solutions where the attribute-function values $f_i(x)$ had to satisfy crisp criteria, a weighted geometric mean of the global degree of satisfaction

$$\prod_{j=1}^{m}\left[\mu_j(x_i)\right]^{w_j} = \prod_{j=1}^{m}\left[\frac{f_j(x_i)-z_j^{\min}}{z_j^{\max}-z_j^{\min}}\right]^{w_j} \quad (2)$$

is maximised where $z_i^{\max}$ and $z_i^{\min}$ denote the maximum and minimum achievable value with respect to the j-th attribute, respectively. The ratio in equation (2) is the membership grade $\mu_j(x_i)$ of the j-th attribute under the i-th feasible design. The w_j, $j = 1, ... , m$, stand for normalised weights (*inter-attribute preferences*) assigned to the attribute functions and express designer's reluctance to approach the attribute values of the anti-ideal (*nadir*) design. Numerical computation of equation (2) is equivalent to the maximisation of a weighted geometric mean of the deviations from the *nadir* solution $\mu(0, 0, ...)$. Since the function (2) depends monotonically on the attribute-function values and its logarithm is concave, any maximum solution is Pareto-optimal.

Once the hypersurface of nondominated designs $X = \{x_1, ... , x_n\}$ is defined, then a *fuzzy set* $\tilde{A}$ in X for whichever vague attribute and constraint is a set of ordered pairs

$$\tilde{A}_j = \left\{\left[x_i, \mu_{\tilde{A}_j}(x_i)\right], x_i \subset X\right\} \quad (3)$$

The decision in a fuzzy environment can therefore be viewed as the intersection of fuzzy constraints and fuzzy attributes, as in the treatment of sets in classical set theory. Since the relationship between attributes and constraints in a fuzzy MADM environment is fully symmetric, there is no longer a

difference between the former and the latter. The decision is therefore defined as the intersection of all fuzzy criteria, that is

$$\tilde{D} = \tilde{A}_1^{w_1} \cap \tilde{A}_2^{w_2} \cap ..... \cap \tilde{A}_m^{w_m} \quad (4)$$

and the optimal alternative is defined as that achieving the highest degree of membership in $\tilde{D}$.

Finally, given the set of nondominated designs X and the membership grade $\mu_A(x_i)$ of all x_i representing the attributes, the MADM model by Yager [8] has been used as a basis for the development of the fuzzy multiattribute optimisation procedure consisting of four main stages:

1. Establish by pairwise comparison the relative importance of the attributes among themselves, i.e., .determine the weights w_j for each attribute by employing Saaty's eigenvector method.
2. Weight the degrees of attribute attainment $\mu_A(x_i)$ exponentially by the respective w_j and build the resulting intra-attribute fuzzy sets A_j.
3. Determine the intersection of all nondominated designs according to equation (4).
4. Select the solution x_i with the largest membership grade in $\tilde{D}$ as the preferred design.

So far, many characteristics have been held constant during the generation and selection process. But most of the economic parameters are probably subject to uncertain oscillation during the whole economic life of the vessel. Scope of the sensitivity analysis is to arrive at a design that is stable (*robust*) with respect to the economic environment changes. As a measure of design robustness, the metrics developed by Taguchi [6] has been modified so that the final solution is the one that gives the higher signal-to-noise ratio for the *RFR* as

$$SN_{RFR} = 10Log\left(\mu_{RFR}^2 / o_{RFR}^2\right) \quad (5)$$

where μ_{RFR} and σ_{RFR} denote the mean and the standard deviation of the *RFR*, respectively.

5. APPLICATION

The developed design tool was applied to the concept design of a class of ro-ro vessels intended for the Mediterranean shortsea shipping. Viable and feasible solutions were generated which, among the others, had to measure satisfaction level with respect to survivability safety. This was accomplished by introducing a new attribute for the A index into a previous list (Table 1), which also includes fuzzy constraints treated as attributes. Different membership grade functions (e.g., attracting, ascending, descending, averting) have been built [1]. The last attribute σ_{RFR} is the robustness attribute utilised in the process of further generation of nondominated solutions in `minicubes' built around the preferred design, by simultaneously evaluating RFR statistics and its measure of robustness.

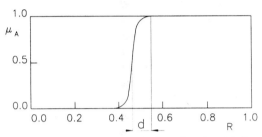

Figure 2 - Membership grade function for A index

An ascending, i.e. S-type, membership grade function was introduced to give flexibility required by vagueness of the A index at concept design stage (Fig. 2). It is defined as

$$A \leq R \Rightarrow \mu_A = 1/\left\{1 + [(R-A)/d]^n\right\} \atop A > R \Rightarrow \mu_A = 1 \right\} \quad (6)$$

where n is the slope of the function and d represents the negative deviation from R, i.e. the inflexion point where $\mu_{R-d} = 0.5$.

The design variables in the mathematical design model, i.e. the quantities that are optimised, are the following:

- length between perpendiculars (L_{PP});
- maximum beam amidships (B_X);
- design draught (T_X);
- longitudinal prismatic coefficient (C_P);
- maximum sectional area coefficient (C_X).

5.1 Generation of nondominated designs

For the time being, three subdivision layout concepts have been included in the design model, aimed at enhancing the damage survivability of ro-ro ships, namely, watertight bulkheads at $B/5$ with increased freeboard (S1), side casings (S2), centre casing (S3), all of them with partial transverse bulkheads on vehicle deck. Superstructures and ramps are identical for all the vessels generated by the design model.

A long-hauled Mediterranean route has been envisaged as case study. The optimum ro-ro vessel is searched for with 15-year economic life, 15 days off-hire, 1500 nautical miles endurance, 160 trailer and 200 pax minimum capacity, 23.5 minimum sustained speed at sea state 4. Powering is provided by twin-screw installation driven by 2 x 12500 kW medium speed diesel engine. As far as economic factors are concerned, capital cost with 60% credit at 5% interest rate over 8 years, with declining-balance depreciation over 15 years to reduce the amount of taxable income, have been assumed as the most influential parameters.

The design process for generating feasible designs, filtering nondominated designs and selecting the preferred solution flows through several sequential steps: (i) definition of min-max design subspace; (ii) generation of feasible designs via an adaptive Monte Carlo method; (iii) definition of intra-attribute fuzzy functions and interactive inter-attribute preference

Table 1 - Attributes in the ro-ro design model

Symbol	Attribute	Symbol	Attribute
V_T	Trial speed (kn)	N_{TR}	Number of trailers
T_{CYC}	Time to complete one cycle (h)	RFR	Required freight rate (MU/tr nm)
C_{ACQ}	Vessel acquisition cost (MU)	P_{TAI}	Turning ability index
V_{RED}	Speed loss in waves (kn)	N_{WET}	Number of deck wetness (1/h)
N_{SLM}	Number of slammings (1/h)	MSI	Motion sickness index (%)
A_{AV}	Average vertical acceleration (m/s^2)	μ_A	Attained subdivision index membership grade
μ_{VIB}	Vibration resonance membership grade	σ_{RFR}	Standard deviation of RFR

Table 2 - Comparison of variables and attributes between the preferred solution and other designs

Ship	L_{PP}	B_X	T_X	C_P	C_X	V_T	C_{ACQ}	V_{RED}	N_{SLM}	A_{AV}	N_{TR}	RFR	N_{WET}	MSI	μ_{VIB}	μ_A
N-S1	-----	-----	-----	-----	-----	23.7	92.0	1.3	16	1.5	166	2.51	15	10.4	0.94	0.50
V-S1	165.00	25.60	6.50	0.605	0.951	23.9	80.6	0.4	8	1.2	170	2.37	12	9.5	0.97	0.32
C-S2	180.65	26.60	6.16	0.544	0.922	24.5	86.9	0.5	11	0.9	177	2.03	10	8.8	0.96	0.89
P-S3	183.10	26.85	6.15	0.545	0.920	24.6	88.8	0.5	7	0.9	183	1.83	6	8.5	0.99	0.96

(iv) transformation of design attributes to fuzzy sets via membership grade functions; (v) structuring the nondominated design hypersurface by equation (3); (vi) selection of the 'best possible design'.

By 32,000 random tries, 232 feasible and 76 nondominated designs were generated. Although the value of the 'distance' in the membership grade function μ_A of equations (6) was kept very low ($d=0.05R$), most part of the preferred solutions resulted to be the ones of 'S2' subdivision-type. In spite of higher intact freeboard with related increment in acquisition and operating cost, all 'S1'-type vessels resulted with a membership grade value for the A index around and below 0.5.

In order to tune some features of the design model, to start the generation phase and to compare some characteristics of the non-dominated designs with the 'real world', a variant of a recently Italian-built ro-ro vessel now operated by 'Norse Irish Ferries' on the Belfast-Liverpool route has been considered. Its lines plan is given in Figure 3.

Table 2 illustrates the attribute values for the nadir solution (N-S1), the modified Italian ro-ro vessel (V-S1), the best solution of 'S3' type (C-S3), and the preferred design (P-S2). Attributes T_{CYC} and P_{TAI} have been discarded since their attained scores resulted quite similar to the aspiration levels also because exponent in the membership function formulation was forced to be high in both cases. All solutions present not optimal values for seakeeping attributes, showing that market demand for increasing compels designers to reconsider more globally hull form configurations of this class of vessels..

5.2 Sensitivity analysis

Analysis of economic-related attributes for non-dominated solutions showed that both RFR giving net present value equal to zero and C_{ACQ} show noticeable differences from design to design, mainly related to different subdivision arrangement. Therefore, a sensitivity analysis is required around the preferred design. Orthogonal arrays were used to generate 'minicube designs' suitable to sensitivity analysis of economic parameters. Four variables only were chosen as factors that influence the solution since variable C_X is almost constant over the nondominated hypersurface. Each variable has four levels according to Table 3 to be associated to the $\mathbf{L_9}$ orthogonal array with three economic parameter levels given in Table 4. The chosen uncertain parameters are the required interest rate (i), the ratio of borrowed-to-equity capital (B), and the loan repayment period (N). The extreme levels have been obtained by extrapolation from the present Italian financial market (level 2).

The orthogonal array $\mathbf{L_{16}}$ illustrated in Table 5 was then structured to determine the optimal combination of variable levels for simultaneous variation of parameter levels. All second-level nondominated designs are of 'S2' subdivision type, i.e. with side casings. The signal-to-noise ratio SN of

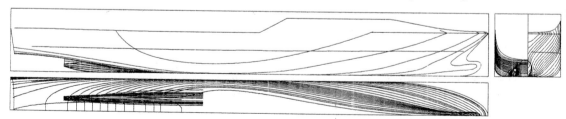

Figure 3 - Lines plan of a typical ro-ro vessel

Table 3 - Levels of variables

Variable	L_{PP}	B_X	T_X	C_P
Level 1	180.00	26.25	6.15	0.530
Level 2	181.50	26.50	6.30	0.540
Level 3	183.00	26.75	6.45	0.550
Level 4	184.50	27.00	6.60	0.560

Table 4 - Levels of parameters

Parameter	i	B	N
Level 1	0.08	0.30	5
Level 2	0.11	0.45	7
Level 3	0.13	0.65	8

RFR membership grade calculated according to equation (5) indicates the robustness of each new nondominated design. Trial 6 yielded the most robust design, that together with other robust designs (trials 6 and 11) show tendencies towards a reduction of L_{PP} and B_X with respect to the preferred design. Further analysis suggests the following main remarks:

- The values of robustness attribute at different trials indicate that large-sized ro-ro vessels are generally to be preferred, eventually depending on the fact that a high stowage factor was considered in the design experiment.
- The most robust vessel is less sensitive to economic uncertainties, thus ensuring higher operating revenues and net cash flows in spite of some additional design and acquisition cost.

6. CONCLUSIONS

In this paper it has been shown how the ship form and performance parameters of a ro-ro vessel can be optimised via a fuzzy multiattribute design process inclusive of assessment of internal sub-division. Achieved results indicate that solutions should be found in new subdivision layout and hull configurations.

The objective of the proposed approach is to modify the traditional design strategy so as to evaluate the techno-economic performance of the vessel while complying safety criteria imposed by damage survivability requirements since the real beginning of the design process.

REFERENCES

1. I. Grubisic, V. Zanic, and G. Trincas, Sensitivity of Multiattribute Design to Economy Environment: Shortsea Ro-Ro Vessels, IMDC'97, Newcastle, 1997, pp. 201-216.
2. R. Juncher Jensen, J. Bertram and P. Andersen, Collision Damage and Probabilistic Damage Stability Calculations, WEGEMT Workshop on Damage Stability of Ships, Lyngby, 1995.
3. M. Pawlowski and I.E. Winkle, Capsize Resistance through Flooding - A New Approach to Ro-Ro Safety, Ro-Ro'88, Gothenburg, 1988, pp. 250-261.
4. U. Schumann-Hindenberg, Interactive Design of Ship Compartmentation, ICCAS'85, Banda & Kuo ed., Trieste, 1985, pp. 343-352.
5. P. Sen and M.K. Gerigk, Some Aspects of a Knowledge-Based Expert System for Preliminary Ship Subdivision Design for Safety, PRADS'92, 1992, Vol. 2, pp. 1187-1197.
6. G. Taguchi, On Robust Technology Development: Bringing Quality Upstream, ASME Press, New York, 1993.
7. G. Trincas, V. Zanic and I. Grubisic, Comprehensive Concept Design of Fast Ro-Ro Ships by Multi-attribute Decision-Making, IMDC'94, Delft, 1994, pp. 321-333.
8. R. Yager, Fuzzy Decision Making Including Unequal Objectives, Fuzzy Sets and Systems, 1978, Vol. 1, pp. 87-95.
9. V. Zanic, I. Grubisic and G. Trincas, MADM System Based on Random Generation of Nondominated Solutions: An Application to Fishing Vessel Design, PRADS'92, 1992, Vol. 2, pp. 1443-1460.

Table 5 - Orthogonal array for selection of the most robust design

Trial	1	2	3	4	5	6	7	8	9	10	11	12	13	14	15	16
L_{PP}	1	1	1	1	2	2	2	2	3	3	3	3	4	4	4	4
B_X	1	2	3	4	1	2	3	4	1	2	3	4	1	2	3	4
T_{PX}	1	2	3	4	2	1	4	3	3	4	1	2	4	3	2	1
C_P	1	2	3	4	3	4	1	2	4	3	2	1	2	1	4	3
SN_{RFR}	24.2	28.0	26.4	28.2	28.8	28.6	23.7	26.3	26.2	27.0	28.6	26.8	27.2	27.1	27.7	27.8

Practical Design of Ships and Mobile Units
M.W.C. Oosterveld and S.G. Tan, editors.

On the Effect of Green Water on Deck on the Wave Bending Moment

Zhaohui Wang [a], Jørgen Juncher Jensen[a] and Jinzhu Xia[b]

[a]Dept. of Naval Architecture and Offshore Engineering,
Technical University of Denmark, DK-2800 Lyngby, Denmark

[b]Danish Maritime Institute,
Hjortekærsvej 99, DK-2800 Lyngby, Denmark

The aim of the present work is to investigate whether green water on deck in severe sea states have a notable effect on the maximum wave bending moments. The analysis is carried out for an S175 container ship for which results from model experiments are available. The static water head and a momentum term, using an effective relative motion calibrated with the model tests, model the green water load. The resulting loads are of the same magnitude as the slamming loads. The results show only a marginal influence of the green water load on the maximum wave bending moment, although the time signal may vary rather significantly. The reason is mainly the green water load being out of phase with the peak values of the bending moment.

1. INTRODUCTION

The extreme sagging wave bending moments in ships are usually determined by taking into account the non-linearities due to momentum slamming and hydrostatic restoring action. These non-linearities are very important to container ships with a large flare yielding extreme sagging moments twice as high as those obtained by a linear analysis, see Refs. [1], [2], [3].

However, the effect of green water on deck is seldom included in the calculations of the sectional loads, but if it is the associated vertical forces are typically based just on the static water head by which the relative motion exceeds the freeboard.

Recently, Buchner [4], [5] has shown by measurements that the actual pressure due to water on deck might be several times larger than the static water head. A much more accurate description of this load was obtained by including a term proportional with the change of the momentum of the water on deck.

In the present paper this approach is implemented in a non-linear time-domain strip theory [3]. The momentum of the green water on deck will be based on a modified relative motion, taking into account

the change of wave profile due to the flare and to the water flow over the deck. Because of the complexity of this flow, the modification is made empirically by introducing an instantaneous Smith correction and fitting the numerical calculations with the experimental results obtained by Watanabe et al. [1] for the S175 container ship. The outcome is an effective relative motion, non-linear in the nominal relative motion and depending on whether the deck is immersed or not.

The local loads due to green water on deck are certainly important to the design of protecting structures on the foredeck of for instance FPSOs and container ships. This aspect is treated in [4], [5].

The objective of the present investigation is to assess the influence of green water on deck on the global hull girder loads, more specifically the vertical wave bending moment. It may be expected that the phase lag between the maximum sagging wave bending moment and the green water on deck is small for wavelengths close to the ship length. However, as the green water load only acts for a short instant of time, even a small phase lag will make the maximum sagging bending moment insensitive to green water loads. This is investigated for the S175 container ship in regular and irregular waves.

2. GREEN WATER LOAD

The vertical load f_{gw} per unit length due to green water on deck in a longitudinal position x and at time t may be written as

$$f_{gw}(x,t) = -g m_{gw}(x,t) - \frac{D}{Dt}\left[m_{gw}(x,t)\frac{Dz_e}{Dt} \right] \quad (1)$$

directed positively upwards. Here m_{gw} denotes the instantaneous mass per unit length of green water, g is the acceleration of gravity, D/Dt the total derivative with respect to time t,

$$\frac{D}{Dt} = \frac{\partial}{\partial t} - U\frac{\partial}{\partial x} \quad (2)$$

with U being the forward speed of the ship. Finally, the effective relative motion $z_e(x,t)$ is taken to be a function of the nominal relative motion $z_n(x,t)$ based on the undisturbed wave elevation $\zeta(x, t)$:

$$z_n(x,t) = w(x,t) - \zeta(x,t) \quad (3)$$
$$z_e(x,t) = f(z_n(x,t)) \quad (4)$$

where $w(x,t)$ is the absolute vertical motion of the ship.

Eq. (1) is of the same form as suggested by Buchner [4], [5], except that the forward speed effect is included in the definition of velocities and accelerations. The present approach simply treats the green water load in the same way as the added mass of water for a submerged section. The change in wave profile due to the bow and the flow of water on the deck is accounted for by using z_e, rather than z_n. Before considering the functional relation between z_e and z_n, it is noted that the mass m_{gw} is taken to be proportional to the effective water height $h_e(x, t)$ on the deck:

$$m_{gw}(x,t) = \rho\, B_e(x)\, h_e(x,t) \quad (5)$$
$$h_e(x,t) = -z_e(x,t) - D_f(x) \quad (6)$$

where ρ is the density of the water, $D_f(x)$ the freeboard and $B_e(x)$ an effective breadth of the green water. In the present study B_e is taken to be half the sectional breadth $B_d(x)$ of the deck:

$$B_e(x) = 0.5\, B_d(x) \quad (7)$$

in accordance with the suggestion in [6].

In the time-domain calculation, Eq. (5) is applied to $h_e(x,t) \ge D_b(x)$, where $D_b(x)$ is defined as the height of the bulwark when the deck enters water, and $D_b(x) = 0$ when it leaves water. Otherwise m_{gw}, and thus f_{gw}, is taken to be zero.

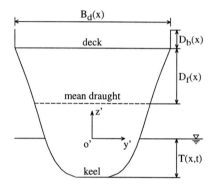

Figure 1. Definition of freeboard $D_f(x)$, deck breadth $B_d(x)$, bulwark height $D_b(x)$, instantaneous sectional coordinate system $o'y'z'$, instantaneous breadth $B(x,t)$ and instantaneous draught $T(x,t)$.

Mainly due to the positive component in the second term of Eq. (1), that is

$$-\frac{D}{Dt}(m_{gw})\frac{Dz_e}{Dt} = \rho\, B_e\left(\frac{Dz_e}{Dt}\right)^2 + \cdots \quad (8)$$

the force f_{gw} will be directed upward just at the moment when the water enters or leaves the deck. This unphysical behaviour is excluded by taking $f_{gw} = 0$ if $f_{gw} > 0$.

3. EFFECTIVE RELATIVE MOTION

Experiments, [1], [4], [5], have clearly shown that, although the wave-induced motions of the ship are determined quite accurately by linear theory, the relative motion exhibits strong non-linearities, especially in the bow region due to deformation of the incoming waves. Various modifications of the relative motion, including a dynamic swell-up coefficient and account of the static bow wave, have been discussed by Buchner [5] but found to be inadequate. In the present study an effective relative motion is derived in accordance with the experimental results presented in [1] for the S175 container ship. Both the original hull form and a model with increased flare were considered.

The present analysis is performed with the non-linear time-domain strip theory which has been validated by extensive model tests, [1], [7], for the S175. Very good agreement was obtained for the ship motion and acceleration for different speeds, wavelengths and wave steepnesses [3]. The experiments presented in [1] also contain measurements of the relative motion in the forepart. For this response the nominal relative motions deviate quite strongly from the measurements, see Fig. 2. Especially, it is seen that the nominal relative motion barely exceeds the freeboard, whereas significant water on deck was measured for a large frequency range. The relative motion amplitude is shown as a function of wave frequency ω for a ship sailing in head sea with a Froude number of 0.25. The position considered is FP and the wave amplitude a of the regular waves is $L/60$, where L is the length (175 m) of the ship. At this wave amplitude significant non-linearities are present in the responses. The relative motion amplitude z_a is defined as half the peak-to-peak value.

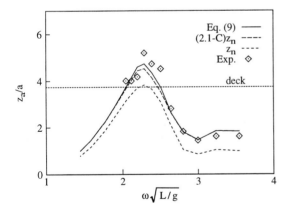

Figure 2. Relative motion amplitude z_a at FP for the S175 container ship sailing in regular head waves. Wave amplitude $a = L/60$. $Fn = 0.25$. Measured results from [1].

It is stated in [1] that the difference in relative motion comes from the deformation of the waves and that a larger bow flare reduces the relative motion, especially when the deck is immersed. Hence, the functional relation (4) must depend on the wave elevation and frequency together with the hull form.

No rational relation can be formulated at present due to the complicated flow pattern, but empirical curve-fitting, see Fig. 2, shows that the relation

$$z_e(x,t) = \begin{cases} (2.1-C)z_n(x,t); & h_e(x,t) \le D_b(x) \\ (3-2C)z_n(x,t); & h_e(x,t) > D_b(x) \end{cases} \quad (9)$$

yields a rather good agreement with the measurements over the whole frequency range. Here C is the instantaneous Smith correction

$$C(x,t) = 1 - \int_{-T(x,t)}^{0} \frac{2y'(x,z')}{B(x,t)} e^{kz'} d(kz') \quad (10)$$

where k is the wave number, $o'y'z'$ is the instantaneous sectional coordinate system, see Fig. 1, $B(x,t)$ the instantaneous breadth of the section, $T(x,t)$ the instantaneous draught. It should be noted that the Smith correction is introduced because it has been successfully used in the strip theories to account for the diffraction effect of the incident waves and it gives a plausible variation of z_e with the geometry of the submerged part of the section, wave elevation and frequency. For instance, an increasing bow flare will increase C and thus decrease z_e; when wave length is long or wave frequency is small, the effective relative motion is close to the nominal relative motion indicating a physically rational asymptotic behaviour of the dynamic wave deformation. In a stochastic sea an average value of C is applied with the individual wave amplitudes as weight factors.

A better fit has been obtained by including higher order terms (z_n^2) in Eq. (9), but this complication has not been judged to be necessary in the present study.

A similar analysis was performed for the S175 with the modified bow. However, the increased flare and overhang greatly reduced the measured relative motions at FP. Hence, the amount of green water on deck was limited and replaced by a significant spraylike water flow away from the hull. As the aim of the present study is to investigate the effect of green water loads on the wave bending moment, only the original hull form is considered.

Measurements in a stationary stochastic sea state were also performed in [1]. The average value $E[z_{pp}|\alpha]$ of the α-highest peak-to-peak values z_{pp} is shown in Fig. 3 as a function of the α-highest values included. Thus $\alpha = 1$ corresponds to the mean value and $\alpha = 1/3$ to the significant value. The sea state is characterised by an ISSC spectrum with a significant wave height $H_s = L/21 = 8.33$ m and a characteristic period $T_0 = 11.13$ sec. It is clear that the effective relative motion z_e, Eq. (9), fits much better with the measured results than the nominal relative motion z_n. In Fig. 4 the cumulative distribution function for the (negative) peaks z_p is shown. Freeboard exceedance corresponds to 0.06L. The measured results indicate a rather sharp change in probability distribution when the deck is submerged. This change is well predicted by the discontinuity in Eq. (9). The nominal relative motion z_n significantly underestimates the extreme relative motions, especially when the deck is submerged.

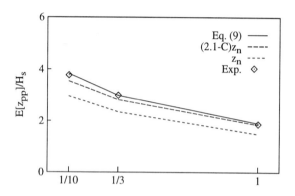

Figure 3. Average value of the α-highest peak-to-peak values z_{pp} in the relative motion at FP for the S175 container ship sailing in irregular head waves. $H_s = L/21$, $T_0 = 11.13$ sec. $Fn = 0.25$.

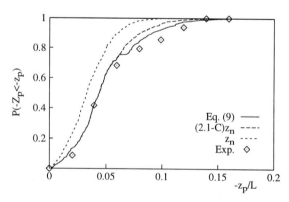

Figure 4. The cumulative distribution function for the negative peaks z_p in the relative motion at FP of the S175 container ship sailing in irregular head waves. $H_s = L/21$, $T_0 = 11.13$ sec. $Fn = 0.25$.

4. GLOBAL HULL GIRDER LOAD

According to [3], the nonlinear time-domain hydrodynamic force $F(x,t)$ can be expressed by

$$\begin{cases} F(x,t) = \dfrac{DI}{Dt} \\ \sum_{j=0}^{J} (B_j I - A_j \dfrac{D\bar{z}}{Dt})^{(j+1)} = 0 \end{cases} \quad (11)$$

where $()^{(j)} = \dfrac{\partial^j}{\partial t^j}$, $A_j(x,\bar{z})$ and $B_j(x,\bar{z})$ are the so-called frequency-independent hydrodynamic coefficients, $\bar{z}(x,t) = w(x,t) - \bar{\zeta}(x,t)$, where $\bar{\zeta}(x,t)$ is the wave elevation with Smith correction.

By integration of the higher order differential equation in Eq. (11) and by incorporation of the hydrostatic buoyancy force f_b and the green water force f_{gw}, the total non-linear external fluid force $Z(x,t)$ acting on a ship section can be expressed as

$$Z(x,t) = -\bar{m}\frac{D^2\bar{z}}{Dt^2} + U\frac{\partial \bar{m}}{\partial x}\frac{D\bar{z}}{Dt} - \frac{\partial \bar{m}}{\partial \bar{z}}(\frac{D\bar{z}}{Dt})^2 \\ - \frac{Dq_J}{Dt} + f_b + f_{gw} \quad (12)$$

where $\dfrac{Dq_J}{Dt}$ accounts for the 'memorial' hydrodynamic effect with q_J governed by the following set of differential equations

$$\frac{\partial q_j(x,t)}{\partial t} = q_{j-1}(x,t) - B_{j-1} q_J(x,t) - (\bar{m}B_{j-1} + A_{j-1})\frac{D\bar{z}}{Dt}$$
$$q_0 = 0, \qquad j = 1, 2, ..., J \quad (13)$$

In Eq. (12) $\bar{m}(x,\bar{z})$ is the added mass of the ship section when the oscillating frequency tends to infinity. The third term of $Z(x,t)$ in Eq. (12) is the momentum slamming force, hereafter denoted by f_{sl}.

The load due to green water on deck might influence the maximum wave-induced sagging bending moment if the two events are in phase. For two different wavelengths λ, the time histories for the wave elevation ζ, effective relative motion z_e, buoyancy force f_b, momentum slamming force f_{sl}, green water force f_{gw} and hydrostatic part of green water force f_s, all at FP, are shown in Fig. 5 together with the midship wave bending moment M_t. Positive values correspond to sagging moments. The incident waves are regular head-sea waves with the wave amplitude $a = L/60$; $Fn = 0.25$. The result for the bending moment assuming no green water is included for comparison. The hull is taken to be fairly rigid to suppress hydroelastic effects.

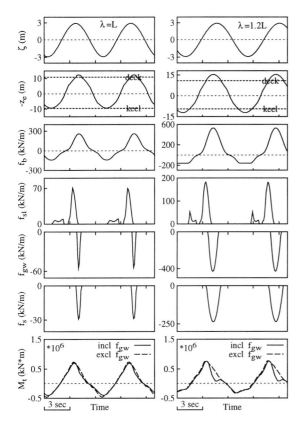

Figure 5. Time histories of wave elevation ζ, effective relative motion z_e, buoyancy force f_b, momentum slamming force f_{sl}, green water force f_{gw} and hydrostatic part of green water force f_s, all at FP, together with midship bending moment M_t for the S175 container ship in regular head waves. Wave amplitude $a = L/60$, $Fn = 0.25$, $\lambda = L$ (left), $\lambda = 1.2 L$ (right).

For $\lambda = 1.2L$ it is seen that the green water force f_{gw} is larger than the momentum slamming force f_{sl} at FP, but contrary to the momentum slamming force, f_{gw} appears slightly after the peak in the midship bending moment. Hence, the green water load only marginally influences the peaks of the midship bending moment. The magnitude of the green water force is seen to be about twice the hydrostatic value f_s given by the first term in Eq. (1). If the nominal

relative motion z_n has been used rather than z_e, no green water on deck appears, see Fig. 2, for this wave amplitude. In [4] the average value of the ratio between the local pressure in the centre line at FP and the static water height is found to be about 3.5 from measurements on a frigate at 20 knots. As the pressure must be expected to vary both in magnitude and phase over the breadth of the deck and with the highest value in the centre line the presently predicted value of f_{gw} is believed to be of the correct magnitude. The results for $\lambda = L$ show only marginally water on deck. Hence f_{gw} is small, but the phase nearly coincides with the peak of the midship bending moment. A small reduction in the peak sagging moment is therefore seen in Fig. 5.

Generally, the peak value of the green water force is given by the static and dynamic acceleration, whereas the velocity square term, see Eq. (8) reduces the duration of the force.

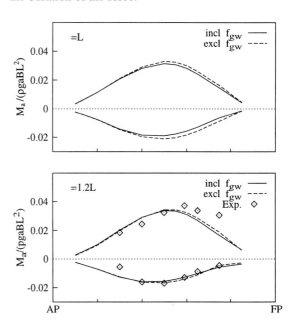

Figure 6. Sagging (positive) and hogging (negative) wave bending moments of the S175 container ship sailing in regular head waves, $a=L/60$, $Fn = 0.25$.

The peak bending moment distribution M_a along the length of the hull is shown in Fig. 6 for the same two regular wave conditions as in Fig. 5. Results obtained without the green water load are included for comparison. It is seen that effect of green water on the bending moment is located to the forward half of the ship and even here is rather small. The experimental values are taken from [1].

The significant value M_s of the wave bending moment in a stochastic seaway is shown in Fig. 7. The longitudinal variation is quite similar to that observed in regular waves, Fig. 6, and the effect of green water is negligible. The cumulative distribution function for the peak values M_p of the midship wave bending moment is presented in Fig.8. The results are obtained from a record with approximately 550 positive peaks and it is seen that the green water load only affects the extreme hogging moment. The time record with the largest sagging and hogging moments is shown in Fig. 9. It is seen that the green water load reduces slightly the sagging bending moment and increases significantly the subsequent hogging bending moment.

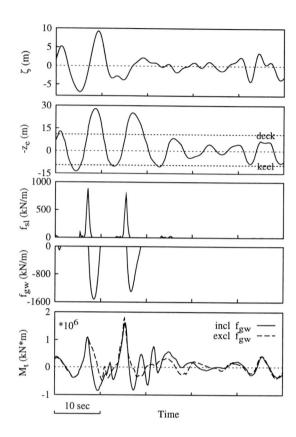

Figure 9. Time record corresponding to the largest sagging and hogging moments in Fig. 8.

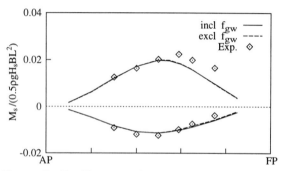

Figure 7. Significant sagging (positive) and hogging (negative) wave bending moments for the S175 container ship sailing in irregular head waves. $H_s = L/21$, $T_0 = 11.13$ sec. $Fn = 0.25$.

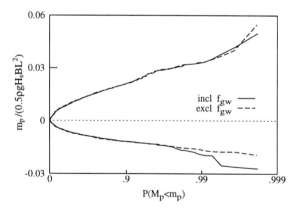

Figure 8. Cumulative distribution function of the sagging (positive) and the hogging (negative) midship bending moments M_p for the S175 container ship sailing in irregular head waves. $H_s = L/21$, $T_0 = 11.13$ sec. $Fn = 0.25$.

5. CONCLUSION

The effect of green water load on the global hull girder moment has been investigated and found negligible in the considered case of the S175 container ship. However, the analysis has provided some insight into the relative motion prediction and a formula, Eq. (1), for the green water load has been proposed, which is easy to be implemented in non-linear time-domain strip theories. The proposed effective relative motion formula, Eq. (9), seems reasonable. It may be worthwhile to verify and improve Eqs. (1) and (9) with other ship cases at different forward speeds.

The analysis has been carried out for a rigid hull. Due to the short duration of the green water load, hydro-elastic effects such as whipping may be more sensitive to green water on deck. This is being investigated.

REFERENCES

[1] I. Watanabe, M. Ueno and H. Sawada, Effects of Bow Flare Shape to the Wave Loads of a Container Ship, J. Society of Naval Architects of Japan, Vol. 166 (1989), pp. 259-266.

[2] A.E. Mansour and J. Juncher Jensen, Slightly Non-linear Extreme Loads and Load Combinations, Journal of Ship Research, Vol. 39, No. 2, June (1995), pp. 139-149.

[3] J. Xia, Z. Wang and J. Juncher Jensen, Non-linear Wave Loads and Ship Responses by a Time-Domain Strip Theory, Report of the Danish Centre for Applied Mathematics and Mechanics, No.569 (1998), submitted to Marine Structures.

[4] B. Buchner, On the effect of Green Water Impacts on Ship Safety (a pilot study), Proc. NAV'94, Vol. 1, Session IX, Rome, Italy (1994).

[5] B. Buchner, On the Impact of Green Water Loading on Ship and Offshore Unit Design. The Sixth International Symposium on Practical Design of Ships and Mobile Units, Seoul, 1.430 – 1.443, (1995)

[6] Peter Friis Hansen, Reliability Analysis of a Midship Section, Ph.D. thesis, Technical University of Denmark, Department of Naval Architecture and Offshore Engineering, Lyngby (1994), 200 pp.

[7] J. O'Dea, E. Powers and J. Zselecsky, Experimental determination of Non-linearities in Vertical Plane Ship Motions, Proceedings 19th Symposium on Naval Hydrodynamics, (1992).

Practical Design of Ships and Mobile Units
M.W.C. Oosterveld and S.G. Tan, editors.

247

Development of a formal safety assessment system for integration into the lifeboat design process.

P. Sen[a], R. Birmingham[a], C. Cain[a], R.M. Cripps[b].

[a] Department of Marine Technology, Newcastle University, UK.
[b] Royal National Lifeboat Institution (RNLI), UK.

1. INTRODUCTION

The Royal National Lifeboat Institution (RNLI) has provided a search and rescue service for the United Kingdom and the Republic of Ireland since 1824, and is recognised as being one of the most effective lifeboat services in the world. The RNLI, as with other search and rescue organisations, is primarily concerned with saving lives and as such expects its own volunteer crews to undertake hazardous tasks in all conditions of weather. Safety is therefore central to all aspects of the RNLI's operations. In the marine field safety in design has traditionally been intuitive, with safety analysis methods only being employed for verification purposes. This approach presents difficulties with complex and sophisticated designs: design for safety embodies the idea of systematically integrating well established formal safety assessment methods throughout the design process to identify and control high risk areas. This paper investigates the development of a design for safety procedure and describes its facilitation in the context of the RNLI's lifeboat design process through the creation of a formal safety assessment tool. The developed tool is able to model, manipulate and analyse safety information such that the consequences of design actions can be identified. The research is being undertaken by the Department of Marine Technology at Newcastle University, in collaboration with the RNLI.

2. DESIGN FOR SAFETY

Until the 1950s the incorporation of safety aspects in the design process was largely intuitive and based upon specific designers' experience [1]. Since then, many safety assessment techniques have been developed and are being used on a wide scale to control industrial hazards. In addition to using safety analysis methods for verification purposes the idea has evolved that a goal setting approach can be maintained to explicitly reduce or eliminate serious risks by continuous integration of safety aspects into the design process. This is the concept of design for safety [2], and advocates the use of formal safety assessment methods to assist decision making throughout the design process. As such, the emphasis on safety must shift from conforming to prescriptive regulations to a more analytical approach.

In the marine industry interest in improving the safety of engineering products through safety analysis from the initial design stages is growing, as accidents cannot be afforded, and rapid innovation is reducing the value of experience. The fact that in many cases deficiencies are only corrected after accidents have occurred has led to regulatory organisations increasingly advocating pro-active procedures which demonstrate levels of safety [3].

2.1. Formal safety assessment

Formal safety assessment is the mainstay of any design for safety procedure. It is devoted to the process of estimating the safety of a product and identifying appropriate measures to reduce system risks to an acceptable level. A risk is defined as the probability of a hazard occurring, combined with its associated effects or consequences. When design for safety is implemented throughout the design process various formal safety assessment tools can be applied individually or in combination to assess safety qualitatively or quantitatively, using either top down or bottom up approaches [4][5]. As the design process advances, and as the available information increases in detail, safety assessment can move from qualitative to quantitative analysis and can progress from being an initial assessment function to a

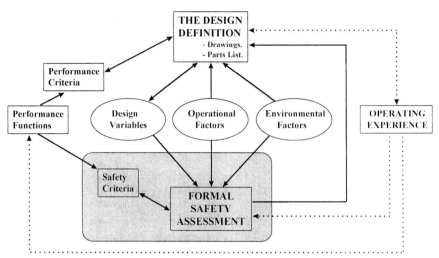

Figure 1. Conceptual representation of a design process incorporating formal safety assessment.

decision making function and finally to a verification function [2].

3. THE RNLI AND SAFETY

The RNLI operates approximately 200 vessels, this fleet being made up from various design classes, ranging from fast afloat lifeboats to inshore inflatable lifeboats. There is a regular updating of each class and as such the design activity is continuous. From a safety point of view the RNLI design process is of particular interest as the mission profile makes highly onerous and unique demands on these small vessels, requiring safety and reliability standards far in excess of any regulatory requirements. The RNLI's commitment to the combined maritime rescue services of the United Kingdom and the Republic of Ireland is to provide all-weather lifeboat cover for up to 50 miles offshore within $2\frac{1}{2}$ hours of launching [6].

3.1. Current practice

Safety has clearly always been given high priority in the RNLI lifeboat design process. However in the design of any product the designer must consider many different, and possibly conflicting, performance objectives in order to formulate the design requirements, from which possible solutions will be synthesised [7]. At present decisions taken during the design process concerning safety are treated as an aspect of performance and are based largely on the experience of the designers [8]. Whilst successful in the past there are increasing difficulties with this approach since the complexity of the vessels is continuously increasing: a modern lifeboat is a combination of many systems, subsystems and components, linked together in complex and sometimes unexpected ways. Some of these links may not be apparent to even the most experienced designers, raising the possibility that the total impact on safety of component failures may not be fully understood. This makes it difficult to systematically experiment with alternative design solutions or to explore the myriad interactions between systems, sub-systems, and aspects of safety throughout the vessel. The converse of potential design inadequacies is also possible as excess provision may be present in some areas compared with others, resulting in both fiscal and physical penalties. An appropriate analogy is that of a chain being at its optimum when the links are the same strength: weak links are unacceptable but overstrong links are unnecessary.

3.2. Conceptual integration of formal safety assessment into the lifeboat design process

The integration of formal safety assessment into the design process is shown conceptually in Fig.1. Different areas in the design process are indicated, as are the links between them, so modelling the flow of information. The shaded area demonstrates the scope of this work and highlights how the continuous integration of formal safety assessment occurs in parallel to the existing design procedures.

The formal safety assessment tool mirrors the conventional design process by storing and analysing a second continuously up-dated model of the vessel, simulating its safety characteristics. The methodology followed in developing the tool is shown in Fig.2. It was necessary to concurrently develop both a generic formal safety assessment framework and a lifeboat safety information model. By embedding the standard safety model in the safety assessment framework it is possible to create and manipulate alternative lifeboat safety information. The influence of each design decision on all aspects of the safety of the vessel can be rationally assessed and possible options for improvement can be explored when risks are judged to be unacceptable with respect to corresponding criteria.

4. CREATION OF A GENERIC FORMAL SAFETY ASSESSMENT FRAMEWORK

The generic formal safety assessment framework enables the storage, manipulation and analysis of appropriate qualitative and quantitative safety information. The modified Boolean representation method (MBRM) was chosen as the formal safety assessment method upon which to base the safety assessment system. MBRM is an inductive bottom up method for identifying system top events and associated causes [1]. Safety information such as the logic relationships between component, sub-system, and system failure events is held in a tabular format. Component or sub-system decision tables are combined into a single composite decision table containing all the possible system top events and the associated paths of basic failure events causing them [8]. Data held within the final MBRM table represents the system model: changes in design are implemented as changes in the individual MBRM tables, with the resultant effects being carried through to the final representation. The complexity of the inter-dependence between components and systems in a lifeboat requires that the storage, manipulation and analysis of MBRM tables be achieved with the aid of appropriate and practical software: this has been facilitated using the high level programming package 'LabVIEW' as the primary software development tool.

The safety information for each item (component, sub-system or system) is received from the designer and stored by the software as a cluster of data. The user interface by which the designer can input item safety data, or make changes to it during the evolution of a design can be seen in Fig.3. The MBRM table (No.1 in Fig.3) qualitatively describes the links (physical or otherwise) between item basic event failure occurrences, inputs from other items, and the item output event states.

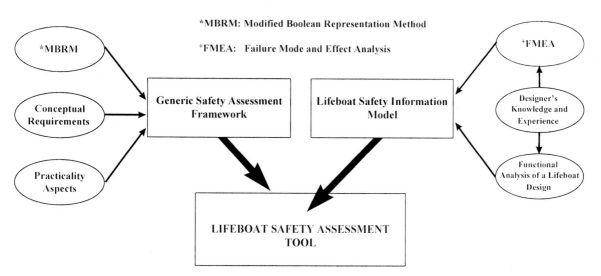

Figure 2. A methodology for creating a lifeboat formal safety assessment tool.

250

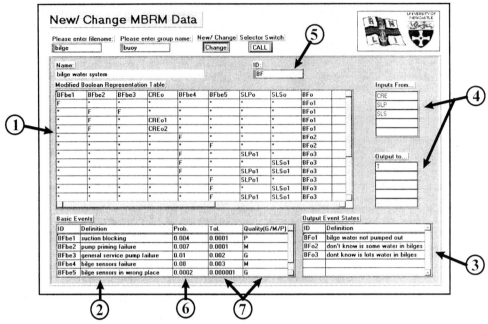

Figure 3. User interface allowing the designer to input or change item safety information.

The MBRM table contains abbreviations for basic events and output event states and these must be defined by the user (No.2 and No.3 in Fig.3). As such, inputs from other items appearing in an MBRM table are defined in those items' clusters as their output event states. The links between each item and other items in the vessel model are detailed (No.4 in Fig.3), and are cross-checked prior to analysing the whole model to ensure no dependencies between items are lost. The links are described in terms of IDs representing each item (No.5 in Fig.3). Quantitative failure data for an item basic event can be input initially in terms of a probability of occurrence (No.6 in Fig.3). However to account for the possibility of vague or scarce probabilistic data, a measure of confidence in the probability estimate can be associated with it. Each probability estimate is thought of in terms of a normal distribution spread about a mean. Values can be given to the lower and upper limits of that probability thus defining a band, and a measure of the quality of the estimate can also be specified (No.7 in Fig.3). The quality of the estimate can be specified as 'poor', 'moderate' or 'good', describing whether the given band spans one, two or three standard deviations from the mean.

Software has been written to manipulate a collection of item MBRM tables and so produce a complete model of safety in terms of a final simplified modified Boolean representation. This is automatically analysed and results are presented in the form shown in Fig.4. Qualitative analysis identifies chains of events leading to undesired top events, however the inclusion of probabilistic aspects significantly increases the usefulness of the assessment, as risk is the combined effect of likelihood and consequence. The probability estimates assigned to each basic event state making up the final MBRM table are processed in order to derive an estimate for the probability of a given system output event state (No.1 in Fig.4). The probability of a path 'R_1' leading to a general output event state can be found as shown in equation (1), where basic events in the path are represented as $B_1 \dots B_n$.

$$P(R_1) = P(B_1) x P(B_2) x \dots P(B_n) \qquad (1)$$

The total probability associated with the output event state of interest above may be calculated using equation (2), where $R_1 \dots R_N$ represent all the paths leading to that output state.

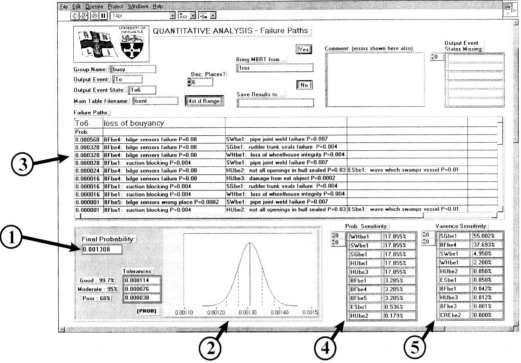

Figure 4. User interface showing the formal safety assessment results.

$$P(O) = \sum_{i=1}^{N} P(R_i) \qquad (2)$$

Basic event tolerance and quality estimates are also assigned to the final MBRM table, and using a Taylor series approximation [9], a value for the variance of the final probability (z) is calculated:

$$\sigma^2 \approx \sum_{i=1}^{n} \left(\frac{\delta z}{\delta x_i} \right)_\mu^2 \sigma_i^2 \qquad (3)$$

This is converted to tolerance and quality values which the software displays numerically and graphically (No.2 in Fig.4). The designer can decide how the distribution of the final probability estimate is interpreted: the important probability value of an output event state can either be the mean value, or a higher value such as that representing the ninety-ninth percentile.

By employing formal safety assessment methods within a design for safety procedure, the designer essentially aims to reduce the probability of a specified output event state. In practice this can be done by identifying the critical paths and basic events requiring attention on the basis of probabilistic estimates. Critical paths leading to a specified output event state are listed in order of probability of occurrence (No.3 in Fig.4), with the probabilities assigned to each event in a path being shown. This allows the probability of a path to be reduced by focusing on the events which contribute most. It is also possible to reduce the probability of the output event state by focusing on basic events which contribute to a large number of paths. A sensitivity analysis is performed to establish the influence of changes in final MBRM table basic events on the overall probability value [9]. Events are then ranked in terms of their percentage contribution to the sensitivity of the overall probability (No.4 in Fig.4):

$$\% \; \begin{array}{l} \textit{Contribution of} \\ \textit{Basic Event 'x'} \end{array} = \left(\frac{\left(\delta z / \delta x_i \right)_\mu}{\sum_{i=1}^{n} \left(\delta z / \delta x_i \right)_\mu} \right) x100 \qquad (4)$$

The designer also aims to improve the tolerance and quality of that probability estimate result: using equation (5) the software provides a list of the basic events ranked in terms of their percentage contribution to the output event state variance (No.5 in Fig.4).

$$\begin{array}{l} \% \ Contribution \ of \ Basic \\ \quad Event \ 'x' \ to \ Final \\ \Pr obability \ Varience \end{array} = \left(\frac{\sigma_i^2 \left(\delta z / \delta x_i \right)_\mu^2}{\sum_{i=1}^n \left(\delta z / \delta x_i \right)_\mu^2 \sigma_i^2} \right) x100 \quad (5)$$

5. DEVELOPMENT OF A LIFEBOAT SAFETY INFORMATION MODEL

The MBRM data storage, manipulation and analysis software is essentially generic and could be applied to any safety analysis situation. In order to create a dedicated lifeboat design tool the software framework must be populated with standard lifeboat safety information. The RNLI's all weather 'Trent' class lifeboat was used to provide this data. A rigorous functional analysis of the vessel was performed, resulting in explicit diagrammatic representations of the on-board physical systems, outside events that affect these systems, and operating procedures governing the use of the on-board systems [5]. System inter-dependencies were studied [8] and documented, including the influence of external environmental events and operational procedures. Failure mode and effect analysis (FMEA) [3][4] was used in liaison with RNLI designers to rigorously and systematically examine components, sub-systems and systems for all modes of failure, their causes and effects. In order to make the safety information model fully compatible with the generic safety assessment framework, the 'Trent' functional and FMEA analyses were directly converted to MBRM format.

6. A DEDICATED LIFEBOAT SAFETY ASSESSMENT DESIGN TOOL

The 'Trent' safety information model in MBRM format was embedded in the safety assessment framework using the format previously shown in Fig.3. The resultant tool can now be used to create and manipulate alternative lifeboat safety information, including that for a new design, whilst studying the influences of changes on different aspects of safety. The designer can thus compare various design options on the basis of formal safety criteria. This safety assessment tool can be used at any level throughout the development of a design. At the beginning of the design process, when data is scarce and only rough estimates are required, the software can provide a summary of which areas might be affected by a particular design change. For this to be possible, the safety model being analysed need only be comprised of MBRM tables. While a qualitative indication of areas of concern is useful, the addition of probability values to the assessment greatly increases its worth. Quantitative assessment based on probabilistic estimates becomes a more realistic proposition as the design progresses, with full analyses being possible for verification purposes when the design is complete. With this tool the designer can choose the level of safety assessment at which he wishes to work, with an option to increase complexity at any time.

7. FURTHER WORK

The future development of the design for safety procedure advocated in this paper centres firstly on the refinement of the formal safety assessment tool. The ongoing and future development of the software system is vital to its practical integration into the RNLI lifeboat design process, and is being achieved by continuous liaison with RNLI designers. Secondly, development of lifeboat safety information is paramount, with implications for both the RNLI and the marine industry in general. The acquisition of probabilistic data to extend the existing safety information model is a major task for the near future and will concentrate on a combination of RNLI logged failure data, RNLI designers' experience, and manufacturers' data. In creating any safety information model the acquisition of reliable data is problematic and in the long term there must be improvements in probabilistic data quality and availability: the emphasis falls on developing data collection and formulation procedures. The continued use of the tool for future designs will improve the situation, as a library of safety information can be developed (similar in concept to a CAD library), providing 'cut and paste' facilities. Developments providing an improvement in the ability to create realistic safety models will be the major contributor to the future implementation of design for safety procedures as protocol.

8. CONCLUSIONS

Traditionally safety assessment has not been a formal and integral part of the marine design process and has been incorporated in a largely intuitive manner: this is the case with the RNLI lifeboat design process. Difficulties are associated with this approach, as with the increasing complexity and sophistication of lifeboats, the design team may be unaware that a critical combination of individual system failures could present an unacceptable level of risk. As such, there is the need for a design for safety procedure allowing safety aspects to be incorporated from the initial design stages. This paper has described the development of a design for safety procedure and how it has been practically implemented in the context of lifeboat design.

A generic safety assessment framework based on the modified Boolean representation method (MBRM) was created to store, manipulate and analyse safety information. This framework, formalised into supporting software, was applied to the design of lifeboats by populating it with lifeboat safety data. This data was developed from the 'Trent' class lifeboat using functional analysis and failure mode and effect analysis (FMEA) procedures. By embedding the safety model in the generic safety assessment framework, a design tool is created which allows the designer to conveniently manipulate or input new data concerning the various systems on the vessel. As each change is made in a design, changes in aspects of safety can be readily monitored. Different design options can be compared and modifications performed when necessary.

It can be seen that design for safety is an expensive and time consuming procedure, as a second model of the entire vessel is required, which

is an additional design output not required in the conventional design process. In this a useful analogy can be drawn with the use of finite element procedures, which involve the creation and analysis of finite element models throughout the development of the design. Just as structural weaknesses can be identified and acted upon, the continuous assessment of safety and its resultant design implications will provide increased confidence in the final design. By following a design for safety ethic, and formally integrating a safety assessment process from the initial stages in the development of a design it is ensured that the high safety standards of the RNLI are maintained as complex technological designs planned for the future are executed.

REFERENCES

1. E.J. Henley and H. Kumamoto, Probabilistic Risk Assessment, IEEE Press, New York, 1992.
2. P. Sen, C.R. Labrie, J. Wang, T. Ruxton, J. Chan, A General Design for Safety Framework for Large Made-to-Order Engineering Products, Quality and its Applications, Newcastle, 1993.
3. IMO, International Code of Safety for High Speed Craft, London, 1995.
4. A. Villemeur, Reliability Availability Maintainability and Safety Assessment, Wiley and Sons, 1992.
5. R. Birmingham, P. Sen, R.Cripps, C. Cain, The Application of Formal Safety Evaluation into the Design of Lifeboats, SURV4: Surveillance, Pilot and Rescue Craft Conference, Gothenburg, 1997.
6. F.D. Hudson, I.A. Hicks, and R.M. Cripps, The Design and Development of Modern Lifeboats, IMechE Proceedings, Part A: Journal of Power and Energy, Vol.207, pp.3-22., 1993.
7. R. Birmingham, G. Cleland, R. Driver, D.Maffin Understanding Engineering Design Context, Theory and Practice, Prentice Hall, 1997.
8. P. Sen, R. Birmingham, R. Cripps, C. Cain, A Methodology for the Integration of Formal Safety Assessment into the Design of Lifeboats, IMDC, Newcastle, 1997.
9. J. Wolfram, Uncertainty in Engineering Economics and Ship Design, NECIES, 1979.

Practical Design of Ships and Mobile Units
M.W.C. Oosterveld and S.G. Tan, editors.

Reliability Based Quality and Cost Optimization of Unstiffened Plates in Ship Structures

Weicheng Cui[a], Alaa E. Mansour[b], Tarek Elsayed[b] and Paul H. Wirsching[c]

[a]China Ship Scientific Research Center, P.O.Box 116, Wuxi, Jiangsu, 214082, P.R.China

[b]Dept. of Naval Architecture and Offshore Engineering, University of California, Berkeley, CA94720, U.S.A.

[c]Dept. of Aerospace and Mechanical Engineering, P.O.Box 210119, The University of Arizona, Tucson, AZ 85721, U.S.A.

Reliability and total expected life cycle cost of a ship are a pair of conflict requirements, both of which depend on quality. In this paper, the idea to use the reliability based Economic Value Analysis (EVA) to determine the optimal quality level is proposed. An interval number representation for costs is recommended. Two methods to treat the interval cost estimates have been proposed. Both methods can complement each other. The proposed methods were illustrated by an example with regard to unstiffened deck plates in ship structures. From the example, it has been shown that the proposed methods would be very useful in handling the quality and cost optimization problems.

1. INTRODUCTION

Quality is defined as the ability to satisfy requirements. These requirements include those 3of serviceability, safety, compatibility and durability. Good quality in the design and construction of a ship can increase the safety and thus reduce the maintenance cost. However, too stringent quality requirements in the design and construction stages can drive up construction costs and this increase of initial construction cost may not be fully compensated by the operation income within the design life. The same argument can also apply to the other stages of ship life cycle. Therefore, for a particular structure, there always exists a tradeoff between quality and cost. The ultimate goal of this research is to take any possible measures to find the appropriate level or degree of quality that will minimize the total expected life cycle cost of the ship. The total expected life cycle cost includes the cost of design, cost of construction, cost of operation and cost of maintenance.

Decisions on quality requirements can be based on an Economic Value Analysis (EVA) or Utility Theory (UT) [1]. Such analyses must consider uncertainties associated with the design factors and cost estimates. To manage such uncertainty, these design factors can be treated as random variables in a formal reliability analysis and cost estimates can be treated with either the classical probability theory or the modern fuzzy set theory.

Some preliminary work has been done by Mansour and co-workers [1,2]. In [1], the scientific and technical merit and feasibility of the proposed concept has been determined and the basic framework of the methodologies and criteria was developed. Through several examples, it has been demonstrated that EVA can be a very effective tool in the decision making process. In [2], the method was applied to study the unfairness tolerances. The main purpose of this paper is to further develop the methodology. The particular problem on which the present paper will focus is how to treat interval numbers for cost estimates. In order to illustrate the methodology proposed in this paper, the unstiffened plate is chosen as the demonstration example. The reason for choosing unstiffened plates is due to its simplicity and significance in ship structures because unstiffened plates form more than 60% of the lightweight of ship structure.

256

2. BASICS OF ECONOMIC VALUE ANALYSIS (EVA)

The basic idea of EVA is that desirability of alternatives are expressed in monetary values. The final decision is based on that alternative which minimizes the total expected life cycle cost. A fundamental operational problem in executing EVA is that all costs must be included. Costs of inspection, repair and replacement, costs of quality requirements (e.g. tolerance requirements) as well as the costs of failure, including possible injury and death, must be quantified.

The general procedure and the required information for performing an EVA is defined by the following steps:

(1) Identify the structure to be considered.

(2) Identify the quality items to be considered for the structure. Here, this could be the discrete decision to grind a weld. Or it could be a tolerance level for plate thickness which would subsequently be translated into a random variable and included in the reliability analysis. In most of the existing design standards, tolerance is specified by its corresponding standard deviations.

(3) Identify the principal failure modes for the structural system to be considered. In general, there may be several failure modes needed to be considered for a complex structure like a ship or part of it (e.g. buckling, fatigue, fracture). The probability of failure for a structural system of n multiple failure modes can be computed based on the following formula:

$$P_{fs} = P(F_1 \cup F_2 \cup ... \cup F_n) \qquad (1)$$

(4) Write the limit state equations for each failure mode for the structural system. This is the equation that describe the failure condition. The main point of this step is that in limit state equations, the tolerances identified in step 2 must be explicitly considered.

(5) Collect all of the statistical data for each parameter in the limit state equations. This will consist of the distribution family (e.g. Weibull) and the distribution parameters (e.g. mean and standard deviation).

(6) Compute the probability of failure, P_{fs}, as a function of the quality measure.

(7) Define the cost of failure for the structural system, C_f.

(8) Compute the expected cost of failure E(Y) of the system during the service life as a function of the quality measure.

$$E(Y) = C_f P_{fs} \qquad (2)$$

(9) Define the initial costs of construction (C_0) as a function of the quality measure, e.g., thickness tolerance, material yield strength tolerance, unfairness tolerance, misalignment tolerance, and weld treatment.

(10) Perform the EVA, computing the optimal quality measure or tolerance that will minimize total expected life cycle costs, E(C). That is,

$$\text{Min. } E(C) = C_0 + C_f P_{fs} \qquad (3)$$

The above procedure can be transformed into the following mathematical problem:

Let $X_1, ..., X_k, X_{k+1}, ..., X_n$ be n random variables which affect the probability of failure of the structural system. The coefficients of variations of $X_1, X_2, ..., X_k$ are determined by their corresponding quality tolerances $T_1, T_2, ..., T_k$. All the other probability characteristics of random variables $X_1,...,X_k,...X_n$ are known. Then both C_0 and P_f will be functions of quality tolerances $T_1,...,T_k$. Now the optimization problem is as follows:

Min. $E(C) = C_0(T_1, T_2, ..., T_k) +$

$C_f P_{fs}(T_1, T_2, ..., T_k)$

Subject to

$$\begin{cases} T_1^L \leq T_1 \leq T_1^U \\ ...... \\ T_k^L \leq T_k \leq T_k^U \end{cases} \tag{4}$$

The probability of failure can be calculated with some existing packages such as CALREL and ISPUD and the optimization problem can be solved using some existing numerical packages such as NAG or even general PC software such as Excel and MATLAB. Compared with ordinary optimization problem, the only complication of the present problem is its interaction with the calculation of the probability of failure. That is, the two types of programs mentioned above should be combined together in order to derive the solution. In the present paper, the probability of failure is calculated either using the Mean Valued First Order Second Moment (MVFOSM) method or the lognormal format in order to avoid the combination problem. Since all the limit state equations used in this paper are very simple and close to linear, this will not have much effect on the results. However, if the nonlinearity in the limit state equation is high, then more accurate reliability methods should be used to calculate the probability of failure because the optimal solution is quite sensitive to the probability of failure.

3. TREATMENT OF INTERVAL COST ESTIMATES

In the previous section, the cost functions are assumed to be single numbers or functions. This requires that the cost must be known precisely. However, in practice, the accurate estimate of these costs is extremely difficult and very often only an interval can be provided. For example, the cost of failure, C_f, could be something between $5000 and $7000. In this case, an interval number such as $C_f = [5000, 7000]$ is used to represent the cost. In this

section, two methods to deal with the interval cost estimates are provided.

If we denote $C_0 = [C_0^L, C_0^U]$, $C_f = [C_f^L, C_f^U]$ and the original problem is translated to the following four problems:

Problem 1: Min. $E(C) = C_0^L + C_f^L P_f$
$\rightarrow X_{opt}^{(1)}, E(C)^{(1)}$

Problem 2: Min. $E(C) = C_0^L + C_f^U P_f$
$\rightarrow X_{opt}^{(2)}, E(C)^{(2)}$

Problem 3: Min. $E(C) = C_0^U + C_f^L P_f$
$\rightarrow X_{opt}^{(3)}, E(C)^{(3)}$

Problem 4: Min. $E(C) = C_0^U + C_f^U P_f$
$\rightarrow X_{opt}^{(4)}, E(C)^{(4)}$

$$\tag{5}$$

Because C_0 and C_f are both interval numbers, then the optimal solution could also be expected to be interval numbers. Using the idea of Vertex Method [3], the lower and upper bounds for X_{opt} and $E(C)$ can be determined from the following equation:

$$E(C)^L = Min(E(C)^{(i)})$$
$$E(C)^U = Max(E(C)^{(i)})$$
$$X_{opt}^L = Min(X_{opt}^{(i)})$$
$$X_{opt}^U = Max(X_{opt}^{(i)}) \tag{6}$$
$$i = 1, 2, 3, 4$$

The above solutions may be adequate for making decisions. However, if the bounds are found to be too wide for making the decision and a crisp solution is still preferred under this uncertain environment, then the idea of Fuzzy Quadratic Programming [3] can be used. In this method, a membership function is assigned to the initial cost C_0 and the cost of failure C_f, denoted as μ_{C_0} and

258

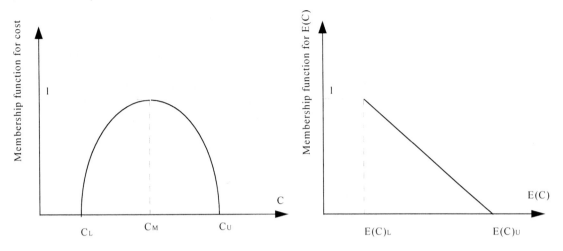

Fig. 1 Membership functions for Costs and objective functions

μ_{C_f} respectively. These membership functions could represent the decision maker's belief on the truth of the cost values. In this paper, a parabolic function is used to represent the membership function of these two costs, see fig.1. This assumes that the middle cost value has the highest truth. From solving the above four problems, we can assume that the objective (expected total cost) function E(C) will lie between $E(C)^L$ and $E(C)^U$. A membership function is also assigned to the objective function values, denoted as $\mu_{E(C)}$. This membership function can represent the decision maker's preference on the objective function values. In this paper, a linear membership function is used, see fig.1. This represents that the decision maker favors the smallest objective function value. The detail of the idea can be found in [3]. Then a crisp optimal solution can be found from the following optimization problem:

Max. $\quad \lambda = \mu_{C_0} + \mu_{C_f} + \mu_{E(C)}$

Subject to:

$$\mu_{C_0} = 1 - \frac{4(C_0 - C_0^M)^2}{(C_0^U - C_0^L)^2}, \quad \mu_{C_f} = 1 - \frac{4(C_f - C_f^M)^2}{(C_f^U - C_f^L)^2},$$

$$\mu_{E(C)} = \frac{E(C)^U - E(C)}{E(C)^U - E(C)^L}$$

$$E(C) = C_0(X) + C_f P_f(X)$$

$$C_0^L \leq C_0 \leq C_0^U, \qquad\qquad C_f^L \leq C_f \leq C_f^U$$

$$E(C)^L \leq E(C) \leq E(C)^U, \qquad X^L \leq X \leq X^U$$

$$\tag{7}$$

where C_0^M and C_f^M are the respective median values.

Example 1- Specifications on Misalignment Relative to the Fatigue Failure Mode

Fatigue strength of welded joints become more and more important in the design of ships. In this example, a butt-welded plate containing a misalignment of eccentricity, e, is considered. The misalignment produces a stress concentration which can have serious consequences with regard to fatigue. An empirical formula for the stress concentration, based on the experimental data, is given in :

$$K = 1 + 3\, e/t \qquad (8)$$

In this example, Miner's equivalent constant amplitude stress can be written as:

$$S_e = KS_n \qquad (9)$$

where S_{en} is the nominal Miner's equivalent stress in the plate.

The fatigue strength is represented as

$$N\, S^m = A \qquad (10)$$

where m and A are the fatigue strength exponent and coefficient, respectively. The reliability strength model used here follows that of Wirsching and Chen [1] in which m is assumed to be constant and A is assumed to be a lognormally-distributed random variable. The UK DEn C curve is used in this paper.

The probability of failure is calculated using the lognormal format [1], that is,

$$P_f = \Phi(-\beta), \qquad \beta = \frac{\ln(\tilde{N}/N_s)}{\sigma_{\ln N}} \qquad (11)$$

where $\tilde{N}$ is the median cycles to failure which is

$$\tilde{N} = \frac{\tilde{\Delta}\tilde{A}}{\tilde{S}_e^m} \qquad (12)$$

and

$$\sigma_{\ln N}^2 = \ln\left\{ (1 + C_\Delta^2)(1 + C_A^2)(1 + C_s^2)^{m^2} \right\} \qquad (13)$$

where C's are the coefficients of variation and $\tilde{\Delta}$ is the damage index at failure.

The values for all the variables used in this example are given in Table 1

Table 1 The values of all the variables for Example

Variable Name	Variable Type	Median	C.O.V.
m	Deterministic	3.0	0.0
A	Random	1.00E10	0.63
Δ	Random	1.0	0.3
S_{en} (ksi)	Random	4.0	0.20
N_s	Deterministic	10^7	0.0

The initial cost associated with misalignment is assumed to be

$$C_0 = C(1.0\text{-}1.2e) \qquad (14)$$

where C is a constant which lies between 4000 and 6000. The cost of failure is also assumed to be an interval number which is C_f=[30,000, 60,000]. The solutions for the four individual cases are given in Table 2.

Table 2 Lower and upper bound solutions for Example

C	C_f	e_{opt}	E(C)	β
4000	30,000	0.0185	3996	2.767
4000	60,000	0.0046	4051	3.033
6000	30,000	0.0279	5930	2.597
6000	60,000	0.0124	6030	2.880

Therefore, the optimal misalignment is between 0.0046 and 0.0279, and the expected total life cycle cost is between 3996 and 6030, the reliability index is between 2.597 and 3.033. If this information is regarded to be adequate for decision making, then there is no further calculation. However, if the solution is regarded to be too wide for making decisions, then the following optimization problem can be set out in order to seek for a crisp optimal solution.

Min. $\lambda = \mu_C + \mu_{C_f} + \mu_{E(C)}$

Subject to:

$$\mu_C = 1 - \frac{4(C-5000)^2}{2000^2}, \quad \mu_{C_f} = 1 - \frac{4(C_f-45000)^2}{30000^2}$$

$$\mu_{E(C)} = \frac{6030-E(C)}{2034}$$

$$E(C) = C(1.0-1.2e) + C_f P_f(e)$$

$$4000 \le C \le 6000, \quad 30{,}000 \le C_f \le 60{,}000$$

$$3996 \le E(C) \le 6030, \quad 0.0046 \le e \le 0.0279$$

$$(15)$$

The optimal solution for this problem is: C=4758, C_f=44882, E(C)= 4776, e=0.0136, β =2.858, and λ= 2.558.

5. SUMMARY AND CONCLUSIONS

Reliability and total expected life cycle cost of a ship are a pair of conflict requirements both of which depend on quality. If the quality is too high, then the reliability will be high, but the initial cost of construction will be high and this increase of initial construction cost may not be fully compensated by the operation income (or benefit) within the design life. This leads to an increase in the total expected life cycle cost. On the other hand, if the quality is too low, then initial cost of construction will be low, but the reliability will also be low. This means that the risk of ship failure will be high. If a ship does fail during its operation, it could cost a lot. Therefore, by relaxing the quality requirements to an extremely low level, it will also lead to an increase in the total expected life cycle cost. From these arguments, it can be seen that there exists an optimal choice for the quality requirements which could minimize the total expected life cycle cost of the ship.

Current design guidelines on quality requirements such as tolerance specifications are available from classification society rules or military standards. However, most of these guidelines are based on tradition and experience and have not been tested analytically or experimentally.

Ultimately, any specifications on the quality requirement should be based on minimizing the total expected life cycle cost of the ship. In this paper, the idea to use the reliability based EVA to determine the optimal quality level is proposed. The particular improvement over previous work is the treatment of interval cost estimates. Two methods to treat the interval cost estimates have been proposed. The first method is based on the idea of vertex method and it will also result in an interval number for the optimal solution. The second method is based on the idea of fuzzy quadratic programming and it will result in a crisp optimal solution. Both methods can complement each other.

The proposed methods were illustrated by an example of unstiffened deck plates in ship structures. This example is carefully chosen in order to cover the representative problem. From this example, it has been shown that the methods proposed in the paper would be very useful in handling the quality and cost optimization problems.

REFERENCES

1. A.E.Mansour,P.H.Wirsching and A.Thayamballi, Structural Fabrication and Structural Details: Phase I, U.S. Navy Contract #N00024-94-C-40621, 1994.
2. R.H.Vroman,Reliability-Based Optimization of Unfairness Tolerances in Welded Stiffened Panels, M.Sc thesis, Department of Naval Architecture and Offshore Engineering, University of California, Berkeley, CA94720, 1996.
3. W.C.Cui and D.I.Blockley, Decision Making with Fuzzy Quadratic Programming, Civil Engineering Systems, (1990), 140.

Practical Design of Ships and Mobile Units
M.W.C. Oosterveld and S.G. Tan, editors.

Hull girder safety and reliability of bulk carriers

D Béghin[a], G Parmentier[a], T Jastrzêbski[b], M Taczala[b], Z Sekulski[b]

[a]Bureau Veritas,

[b]Technical University of Szczecin

The paper deals with the hull girder safety and reliability of single side skin bulk carriers designed with topside and hopper tanks. The ship behaviour is examined for sea-going conditions as well as during loading and unloading operations.

The safety margin is calculated with respect to the ultimate bending capacity of the hull girder, using the computer code "RESULT". This code is a module of the VeriSTAR system developed by Bureau Veritas for computer-aided design and ship operation.

In a first chapter, the paper describes all assumptions of the method considered to calculate the ultimate bending moment. The second chapter presents the results of the deterministic and probabilistic calculations and the last chapter compares the hull girder reliability of bulk carriers to that of single hull oil tankers.

1. INTRODUCTION

At the end of the eighties and beginning of the nineties a series of casualties occurred on bulk carriers, leading to the loss of many lives. It concerns both ships at sea and in harbour during loading and unloading operations. Where the ship's loss is due to the ingress of water in one cargo hold, which occurs in most of the cases, hull girder collapse and loss of stability are the final sequential events of the accidental scenario. Overstressing of the hull girder may be aggravated by corrosion and mechanical deformations of structural elements caused by unloading devices.

Bulk carrier casualties and the well known "Energy Concentration" accident highlight the need for development of adequate design tools to assess the ultimate capacity of the hull girder, both for new and existing ships.

It is worth reminding that, in 1980, the "Energy Concentration" was broken in the port of Rotterdam due to errors in the discharging procedure. This accident initiated a series of research projects for identification of the actual ship structure resistance and development of computational methods for evaluation of the ultimate capacity and analysis of safety standards on longitudinal hull girder bending.

2. METHODOLOGY

2.1 Vertical bending

2.1.1 To date, the hull girder bending is addressed in the Classification Society Rules [1] considering only the yielding and buckling modes of failure with no reference to the ultimate bending capacity. For the extreme vertical wave bending moment corresponding to a probability of exceedance of 10^{-8}, it is ensured that the the normal stress at deck and bottom does not exceed the permissible value which is taken as 75 per cent of the yield stress.

More generally, in a rationally-based structural design the safety of the structure has to be examined for the various modes of failure

262

which may endanger the structure and, in particular, for the ultimate strength limit states.

Assessment of the hull girder ultimate strength has been discussed in many papers and, in particular, in the one addressing the case of the Energy Concentration [2]. In the present study the code "RESULT" developed jointly between Bureau Veritas and the Technical University of Szczecin has been used to calculate and compare the ultimate vertical bending moment of single side skin bulk carriers and single hull oil tankers.

The ultimate vertical bending moment is defined as the moment at which the flexural stiffness of the hull girder is equal to zero (refer to Figure 1), the flexural stiffness being the slope of the M – φ curve. In the elastic domain this slope is equal to EI and the M – φ relationship is given by :

$$M = E I \phi \qquad (1)$$

where :

ϕ = curvature of the hull girder,
E = Young's modulus,
I = inertia of the hull girder

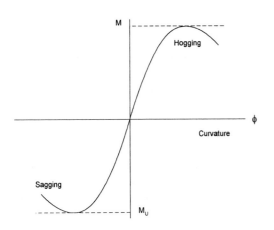

Figure 1. Moment - Curvature relationship

2.1.2 In references [3,4] the theoretical basis as well as a description of the code is given. However, for better understanding, it may be useful to remind that the method considered for development of the code "RESULT" is based on the simplified method of "components" which assumes that each cross section is made of an assembly of separate elements (plates and stiffened panels). Moreover, following assumptions are made :

- collapse occurs between two adjacent web frames,
- critical section remains plane after deformation,
- behaviour of the material is assumed perfectly elasto-plastic.

Assumption on the linear distribution of strain enables to evaluate the contribution ΔM_i of each component to the bending moment, provided that its load-end shortening curve (σ, ϵ) is known.

Load-end shortening curves of plates and stiffened panels are determined for the following modes of failures :

- plate induced failure mode,
- column buckling,
- tripping,
- local buckling.

assuming that the ultimate strength equations of plates and stiffeners may be generalized to any strain level, as suggested in [5].

2.2 Safety measures

2.2.1 To date, Classification Societies Rules are based on a deterministic approach, which means that safety factors are calculated empirically according to deterministic procedures and applied to conventional values of loads and resistance. Deterministic criteria may be expressed as follows :

$$f_R\ R \ge f_D\ D \qquad (2)$$

where f_D and f_R are the safety factors applied on loads and strength.

Though the efficiency of these deterministic safety factors is well proven, it is generally difficult to identify the type and nature of uncertainties which are taken into account and to be sure that the safety is optimum. To take into account the randomness of loads and strength, it is necessary to base the stuctural design on a probabilistic approach.

2.2.2 Probabilistic design necessitates to define, for each particular mode of failure, a limit state function $g(x) = g(x_1, x_2, ..., x_n)$ which characterizes the condition of the structure and defines two domains of safety :

$g(x) < 0$ in the unsafe domain,
$g(x) > 0$ in the safe domain,
$g(x) = 0$ on the limit surface.

Each variable x_i is generally a random variable having a known mean value and standard deviation.

Taking into account the various uncertainties on loads and resistance, especially when determining the ultimate capacity of the hull girder, the use of a probabilistic approach to assess the safety margin with respect to the hull girder collapse seems particularly appropriate.

2.2.3 In a probabilistic approach, the safety of the structure may be expressed in terms of a probability of failure or of a safety index β defined as follows :

- Hasofer-Lind safety index

This method enables to calculate the safety index β from a geometrical approach. The problem illustrated in Figure 2 is expressed into a standard space of normal, reduced and independent variables. Then, the safety index β is defined as the minimum distance from the origin to the failure surface.

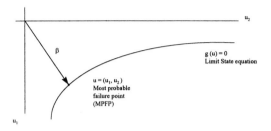

Figure 2. Geometrical definition of the safety index

- Cornell safety index

Assuming that the limit state function may be expressed as follows :

$$g(x) = C - D \qquad (3)$$

where C and D are random variables, Cornell proposed to characterize the structural reliability by its safety index given by :

$$\beta_C = \frac{\overline{g}}{\sigma_g} = \frac{\overline{C} - \overline{D}}{\sqrt{\sigma_C^2 + \sigma_D^2}} \qquad (4)$$

where :

$\overline{C}$ and $\overline{D}$ = mean values of the capacity and demand,
σ_C and σ_D = standard deviations of the capacity and demand,

Moreover, if the random variables are assumed normal and independent, the safety index and the probability of failure comply with the following relationship, as shown in Figure 3 :

$$P_f = 1 - \Phi(\beta) = \Phi(-\beta) \qquad (5)$$

where Φ is the standard normal cumulative distribution function.

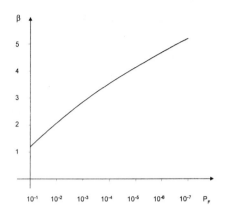

Figure 3. $P_f = \Phi(-\beta)$

2.2.4 For vertical bending of the hull girder the limit state function may be expressed by one of the following expressions :

- sea going conditions

$$g(X_i) = M_{ult} - (M_{sw} + M_{wv}) \qquad (6)$$

- harbour conditions

$$g(X_i) = M_{ult} - (M_{harb} + 0{,}2\,M_{wv}) \qquad (7)$$

where :

M_{ult} = ultimate vertical bending moment,
M_{sw} = still water bending moment for sea-going conditions,
M_{harb} = still water bending moment in harbour conditions,
M_{wv} = vertical wave bending moment in sagging or hogging condition.

3 CALCULATIONS

3.1 Ship model data

3.1.1 Investigation of safety factors and reliability indices has been performed for 10 single side skin bulk carriers with a standard structural configuration, i.e. with hopper and topsides tanks. For comparison, calculations have been carried out for 10 single hull oil tankers.

Main particulars of bulk carriers and oil tankers are given in Tables I and II.

3.1.2 Statistical modelling of random variables

Following assumptions are considered in this reliability analysis for statistical modelling of the ultimate bending moment M_{ult} and wave bending moment M_{wv} :

- normal and independent variables,

- $s = \dfrac{\overline{M}}{M_{nom}} = 1$,

- $COV = \dfrac{\sigma}{\overline{M}} = 0{,}1$, where σ is the standard deviation of the random variables. These values of the COVs have been selected according to the results of investigations published in the report of the International Ship Structure Congress [6].

The design still water bending moment M_{sw} is taken also as a random variable considering the following values for the coefficient of variation v :

- sea-going conditions $v = 0{,}1$

- harbour conditions :

bulk carriers	$v = 0{,}3$
oil tankers	$v = 0{,}1$

The coefficient of variation considered for harbour conditions of bulk carriers takes into account the uncertainties on the actual cargo loads in holds or tanks during the loading and unloading operations.

Based on these assumptions, the safety index is

expressed as follows :

- seagoing conditions :

$$\beta = \frac{M_{ult} - \left(M_{sw} + M_{wv}\right)}{\nu \sqrt{M_{ult}^2 + M_{sw}^2 + M_{wv}^2}} \qquad (8)$$

- harbour conditions :

bulk carriers

$$\beta = \frac{M_{ult} - \left(M_{harb} + 0,2\,M_{wv}\right)}{\sqrt{\nu^2 \left(M_{ult}^2 + 0,04\,M_{wv}^2\right) + \left(0,3\,M_{harb}\right)^2}} \qquad (9)$$

oil tankers

$$\beta = \frac{M_{ult} - \left(M_{harb} + 0,2\,M_{wv}\right)}{\nu \sqrt{M_{ult}^2 + M_{harb}^2 + 0,04\,M_{wv}^2}} \qquad (10)$$

When the random variables are assumed normally distributed, the Cornell and Hasofer-Lind safety indices are identical.

Moreover, taking into account the assumptions adopted for modelling of the random variables, formula (8) shows that the safety index for sea-going conditions is inversely proportional to the coefficient of variation :

- for an increase of the coefficient of variation of 50 percent, the safety index is reduced by 33 per cent.

3.2 Results of calculations

3.2.1 Deterministic analysis

Deterministic safety factors calculated according to formula (11) or (12) are given in Tables III to VI, considering :

- design values of the still water bending moments in hogging and sagging conditions,
- rule wave bending moments in hogging and sagging conditions,
- design harbour bending moments in hogging and sagging conditions

$$f = \frac{M_{ult}}{M_{SW} + M_{WV}} \qquad (11)$$

$$f = \frac{M_{ult}}{M_{harb} + 0,2\,M_{WV}} \qquad (12)$$

For each ship, the ultimate bending moment is calculated using the programme "RESULT" and based on the as-built scantlings.

3.2.2 Safety indices

Cornell safety indices, calculated according to formulae (8) to (10), are given in Tables III to VI :

- Bulk carriers
  Table III : seagoing conditions
  Table IV : harbour conditions

- Oil tankers

  Table V : seagoing conditions
  Table VI : harbour conditions

4. DISCUSSION OF THE RESULTS

4.1 This reliability analysis confirms the results of many other analyses and shows that structures belonging to the same class of ships have not necessarily the same level of safety (refer to Tables III to VI) though their scantlings are based on the same requirements. This may be due :

- either to existing rules which are not calibrated for a uniform reliability level, or
- to large sensitivity of the safety index to the modelling of stochastic variables.

Though the number of ships examined in this reliability analysis does not concern a large sample of ships, this can be used, however, as a basis for general conclusions.

4.2 Examination of Tables III to VI shows a relatively large dispersion of the safety coefficient f and safety index β within each class of ships. Table VII gives minimum, maximum, mean values and standard deviation of the safety coefficient and safety index for single side skin bulk carriers and single hull oil tankers. The dispersion may be explained as follows :

- while bottom or deck longitudinals comply with the same global and local section modulus requirements, their load-end shortening curves may differ significantly from one ship to another one, leading to a better or worst behaviour with respect to the ultimate strength of the hull girder.

For example, non-linear FEM calculations carried out for the bottom stiffeners of the Energy Concentration have shown that their governing buckling mode was a combination of tripping and local buckling. Improvement of the behaviour with respect to these two modes of buckling without modification of the cross sectional area of the bottom stiffeners should have increased the ultimate capacity of the hull girder in hogging conditions.

More generally, it may be concluded from these calculations that the section modulus requirements are not sufficient to reflect the actual safety margin with respect to the final collapse of the hull girder and that introduction in the Rules of ultimate strength criteria should

be a positive step forward to improve the ship safety.

In particular, such criteria will enable to calculate at any time of the ship's life the ultimate strength safety margin, taking into account the actual hull condition as obtained from thickness measurements carried out during the periodical surveys.

4.3 Though additional studies are needed prior to determining the permissible safety indices per type of ship, analysis of Tables III to VI gives some valuable information on the comparative ultimate vertical bending strength of bulk carriers and oil tankers while complying with the same longitudinal strength requirements :

- single side skin bulk carriers have generally a greater safety index than single hull oil tankers in both hogging and sagging conditions, due to the positive influence of the double bottom (hogging conditions) and to larger scantlings of the strength deck at side (sagging conditions) resulting from the presence of hatch openings. However, it is worth mentioning that this ultimate strength analysis does not take into account the effects of yielding at hatch corners.

- harbour safety indices of bulk carriers are generally less than sea-going indices, which may seem paradoxical. This is due to the assumption made on the coefficient of variation for the still water bending moment ($\nu = 0,3$), which is based on results of calculations [7] showing that a transfer of 10 per cent of the mass of a mid-cargo hold towards the foremost cargo hold may lead to an increase of the still water bending moment of 40 per cent.

5. CONCLUSIONS

In the paper the results of investigations of the safety margins with respect to the hull girder collapse of bulk carriers and single hull oil tankers are presented. This comparative analysis is based on the safety index concept, considering assumed characteristics for the random variables.

Within the same class of ships, there is a large dispersion of the safety indices which confirms that existing Rules are not calibrated for the same level of safety and emphasizes the need for reliability-based design.

For sea-going conditions, he mean and minimum safety indices of single side skin bulk carriers are generally greater than those calculated for single hull oil tankers.

Modelling of the random variables has a large influence on the safety index and, consequently, on the calculated probability of failure. It is therefore necessary to perform further systematic investigations with a view to defining an adequate modelling for the random variables.

REFERENCES

1. IACS - "Longitudinal Strength Standard, Unified Requirements of International Association of Classification Societies", UR S11, 1989.

2. S. B. Rutheford, J. B. Caldwell, "Ultimate Longitudinal Strength of Ships : A Case Study", SNAME Annual Meeting, San Francisco, 1990.

3. D. Béghin, Ph Baumans, T. Jastrzêbski, M. Taczala : "Some Considerations on Safety Margin of Ship Hull in Longitudinal Bending", ODRA'95, Szczecin 1995.

4. D. Béghin, T. Jastrzêbski, M. Taczala : "A Computer Code for Evaluation of the Ultimate Longitudinal Strength of Hull Girder", PRADS'95, Seoul 1995.

5. J. M. Gordo, C. Guedes Soares : "Approximate Load Shortening Curves for Stiffened Plates under Uniaxial Compression".

6. ISSC : Proceedings of the 13th Ship and Offshore Structures Congress, Trondheim 1997.

7. Bureau Veritas NI 402 : "Recommendations to avoid overstressing of Bulk Carriers Structures", March 1995.

Table I - Main particulars of Bulk Carriers

| Ship | L [m] | B [m] | D [m] | Design SWBM [kN.m] | | |
| | | | | Seagoing conditions | | Harbour |
				Hogging	Sagging	Hog/Sag
1	135	21,7	12,2	403200	403200	685380
2	152	24	13,10	547100	498030	846660
3	178.97	30,5	16,6	1013140	918460	1750820
4	180	24,4	15,1	821680	738430	1565810
5	185	25,3	15,4	907600	816300	1791060
6	210.49	32,2	18,3	1485560	1485560	2594170
7	211.36	32,2	17,6	1304920	1304920	2498860
8	230	32,25	19,95	1894550	1784260	3760030
9	230	41,6	20	2459590	2233230	4404460
10	256.57	43	23,9	3239700	2989700	5009700

Table II - Main particular of Oil Tankers

| Ship | L [m] | B [m] | D [m] | Design SWBM [kN.m] | | |
| | | | | Seagoing conditions | | Harbour |
				Hogging	Sagging	Hog/Sag
1	151,32	23,5	12,75	459730	459730	926530
2	162,92	28,4	13,70	635500	635500	1257840
3	310,89	56	29,4	6384350	5776330	10814760
4	323	53,6	26,4	6242780	6242780	11262210
5	323	53,6	26,4	6242780	6242780	11310460
6	324,95	53	28,3	6606975	6606975	12042020
7	327,3	51,82	27,35	6328460	6328460	11864840
8	327,3	51,82	27,35	6369920	6369920	11490970
9	400	63	37,13	11663640	10791000	19478750
10	470	80	43	19382050	18211690	35757000

Table III - Single Side Skin Bulk Carriers
Sea-going conditions

Ship	Hogging				Sagging			
	M_{design}	M_{ult}	f	β	M_{design}	M_{ult}	f	β
1	0,906 E06	1,681 E06	1,86	4,31	0,957 E06	1,199 E06	1,25	1,75
2	1,341 E06	2,478 E06	1,85	4,28	1,341 E06	1,798 E06	1,34	2,23
3	2,447 E06	4,572 E06	1,87	4,34	2,457 E06	3,954 E06	1,61	3,45
4	1,988 E06	3,519 E06	1,77	4,03	1,988 E06	2,911 E06	1,46	2,84
5	2,198 E06	3,748 E06	1,71	3,81	2,198 E06	3,610 E06	1,64	3,57
6	3,665 E06	6,112 E06	1,67	3,67	3,835 E06	5,076 E06	1,32	2,14
7	3,503 E06	5,696 E06	1,63	3,51	3,675 E06	5,238 E06	1,42	2,65
8	4,804 E06	9,400 E06	1,96	4,59	4,804 E06	8,772 E06	1,82	4,20
9	6,012 E06	1,080 E07	1,80	4,12	6,012 E06	1,014 E07	1,69	3,73
10	8,062 E06	1,405 E07	1,74	3,94	8,062 E06	1,068 E07	1,32	2,15

Table IV - Single Side Skin Bulk Carriers
Harbour conditions

Ship	Hogging				Sagging			
	M_{design}	M_{ult}	f	β	M_{design}	M_{ult}	f	β
1	0,786 E06	1,681 E06	2,14	3,37	0,796 E06	1,199 E06	1,51	1,69
2	1,005 E06	2,478 E06	2,46	4,15	1,015 E06	1,798 E06	1,77	2,51
3	2,037 E06	4,572 E06	2,24	3,64	2,058 E06	3,954 E06	1,92	2,88
4	1,799 E06	3,519 E06	1,96	2,93	1,816 E06	2,911 E06	1,60	1,98
5	2,049 E06	3,748 E06	1,83	2,59	2,067 E06	3,610 E06	1,75	2,38
6	3,030 E06	6,112 E06	2,02	3,11	3,064 E06	5,076 E06	1,66	2,16
7	2,938 E06	5,696 E06	1,94	2,93	2,973 E06	5,238 E06	1,76	2,47
8	4,342 E06	9,400 E06	2,16	3,44	4,364 E06	8,772 E06	2,01	3,08
9	5,115 E06	1,080 E07	2,11	3,33	5,160 E06	1,014 E07	1,96	2,99
10	5,974 E06	1,405 E07	2,35	3,92	6,024 E06	1,068 E07	1,77	2,52

Table V - Single Hull Oil Tankers
Sea-going conditions

Ship	Hogging				Sagging			
	M_{design}	M_{ult}	f	β	M_{design}	M_{ult}	f	β
1	1,192 E06	2,082 E06	1,75	3,95	1,254 E06	1,672 E06	1,33	2,19
2	1,670 E06	2,633 E06	1,58	3,32	1,765 E06	2,342 E06	1,33	2,15
3	1,557 E07	2,308 E07	1,48	2,92	1,557 E07	2,047 E07	1,31	2,09
4	1,584 E07	2,304 E07	1,46	2,80	1,643 E07	2,199 E07	1,34	2,22
5	1,584 E07	2,308 E07	1,46	2,81	1,643 E07	2,215 E07	1,35	2,27
6	1,610 E07	2,498 E07	1,55	3,22	1,674 E07	2,330 E07	1,39	2,50
7	1,574 E07	2,547 E07	1,62	3,49	1,637 E07	2,331 E07	1,42	2,65
8	1,578 E07	2,552 E07	1,62	3,49	1,641 E07	2,253 E07	1,37	2,40
9	2,905 E07	4,569 E07	1,57	3,31	2,905 E07	4,234 E07	1,46	2,81
10	4,903 E07	7,911 E07	1,61	3,47	4,903 E07	7,493 E07	1,53	3,12

Table VI - Single Hull Oil Tankers
Harbour conditions

Ship	Hogging				Sagging			
	M_{design}	M_{ult}	f	β	M_{design}	M_{ult}	f	β
1	1,073 E06	2,082 E06	1,94	4,42	1,085 E06	1,672 E06	1,54	3,06
2	1,465 E06	2,633 E06	1,80	3,99	1,483 E06	2,342 E06	1,58	3,22
3	1,265 E07	2,308 E07	1,82	4,08	1,277 E07	2,047 E07	1,60	3,31
4	1,318 E07	2,304 E07	1,75	3,83	1,330 E07	2,199 E07	1,65	3,50
5	1,323 E07	2,308 E07	1,74	3,82	1,335 E07	2,215 E07	1,66	3,53
6	1,394 E07	2,498 E07	1,79	3,97	1,407 E07	2,330 E07	1,66	3,51
7	1,375 E07	2,547 E07	1,85	4,16	1,387 E07	2,331 E07	1,68	3,60
8	1,337 E07	2,552 E07	1,91	4,33	1,350 E07	2,253 E07	1,67	3,56
9	2,296 E06	4,569 E07	1,99	4,57	2,313 E07	4,234 E07	1,83	4,11
10	4,169 E07	7,911 E07	1,90	4,30	4,192 E07	7,493 E07	1,79	3,96

Table VII -

	Sea-going conditions				Harbour conditions			
	Hogging		Sagging		Hogging		Sagging	
	Bulk	Tankers	Bulk	Tankers	Bulk	Tankers	Bulk	Tankers
f_{min}	1,63	1,46	1,25	1,31	1,83	1,74	1,51	1,54
f_{max}	1,96	1,75	1,82	1,53	2,46	1,99	2,01	1,83
f_{mean}	1,79	1,57	1,49	1,38	2,12	1,85	1,77	1,67
σ_f	0,10	0,09	0,19	0,07	0,19	0,08	0,16	0,09
β_{min}	3,51	2,80	1,75	2,09	2,59	3,82	1,69	3,06
β_{max}	4,59	3,95	3,73	3,12	4,15	4,57	3,08	4,11
β_{mean}	4,06	3,28	2,87	2,44	3,34	4,15	2,47	3,54
σ_β	0,332	0,358	0,823	0,33	0,476	0,252	0,443	0,316

Practical Design of Ships and Mobile Units
M.W.C. Oosterveld and S.G. Tan, editors.

Review of statistical models for ship reliability analysis

J. Parunov and I. Senjanović

University of Zagreb, Faculty of Mechanical Engineering and Naval Architecture,
Ivana Lučića 5, 10000 Zagreb, Croatia

The purpose of the article is to give an updated review of ship reliability methods and to examine their practical applicability. Uncertainties due to wave bending moment, still-water bending moment and resistance are analysed. The main problems involved in uncertainty assessment are mentioned, and a review of relevant literature is given. Finally, the safety indices calculated by various authors are compared.

1. INTRODUCTION

Reliability methods for ship structural design are used for two main purposes:
a) direct application to individual designs [1-3]
b) calibration of partial safety factors for a large number of ships [2-8].

The purpose of the direct application of reliability methods to certain ship structures is to establish rational safety margins between load an resistance in cases when accumulated past experience does not exist. Offshore production ships frequently designed in recent times can serve as an example of this application [2,3]. These are the offshore structures similar to ships. However, they are at fixed locations, always turned to head seas, thus the application of the Rules for the ocean-going ships to this type of structures is questionable.

The calibration of partial safety factors is predicted for standard ship types for which experience exists, in order to enable a more uniform safety level when a large number of ships is considered [4-9]. The rules based on the partial safety factors concept would provide the prescribed safety level for each failure mode. That would provide significant economic benefits without any loss of safety.

Ship reliability methods have been developing for already 25 years and they have not become standard engineering tool yet. The reason lies in the fact that it is not easy to establish rational and reliable statistical models for all significant variables: wave loads, still-water loads and resistance for different failure modes. Various assumptions for the statistical distribution of basic variables are possible leading to different calculated safety indices. However, reasonable comparison between various designs is possible only if similar statistical models are used.

In this article, a review of the types of statistical distributions and their parameters, which are used in reliability calculations of ship longitudinal strength, is presented with the aim to establish a more uniform frame for statistical models, which would enable more effective application of reliability methods.

2. WAVE BENDING MOMENT

For the evaluation of ship reliability, an appropriate statistical description of wave induced loads is of crucial importance, since greater part of the total extreme vertical bending load is due to wave bending moment (about 60%, [10]). Various probabilistic models were applied to describe wave load actions, leading to significant differences in calculated failure probabilities. It is important to observe that physical uncertainty, which is the consequence of the stochastic process of sea surface elevation, is only a minor part of total uncertainty in the prediction of long-term distribution of load effects. Recently, significant efforts have been done to quantify the approximated components of uncertainties which are due to model uncertainties of calculation tools, unreliable statistical data, unknown sailing roots, unpredictable human actions etc. [4,11]. In this section, the procedure for the calculation of long-term distribution of load effects on ship structures is presented and the main problems involved are cited from literature.

To calculate the long-term distribution of wave bending moments, it is assumed that the non-stationary sea state is composed of numerous stationary and ergodic short-term sea states which are defined with significant wave heights H_S and mean zero upcrossing periods T_Z. The sea surface elevation is considered to be a Gaussian, relatively narrow-banded process with zero mean, completely described with its energy spectrum $S_\eta(\omega, H_s, T_z)$. The spectrum of wave bending moments (or any other load effects) is usually calculated using the linear strip theory as in [4,12,13]:

$$S_R(\omega, \alpha, H_s, T_z, V, C) = H^2(\omega, \alpha, V, C) \cdot S_\eta(\omega, H_s, T_z)$$

where $H(\omega, \alpha, V, C)$ is the transfer function depending on wave circular frequency ω, relative heading angle α between wave propagation and ship velocity, ship velocity V and loading condition C. As a consequence of the linear transformation of the input spectrum, the wave bending moment during a short time interval (less than 30 minutes) is also considered as a Gaussian, relatively narrow banded process with zero mean. The response variance is calculated as the area under the spectrum curve:

$$R(H_s, T_z, \alpha, V, C) = \int_0^\infty S_R(\omega, H_s, T_z, \alpha, V, C) d\omega$$

If the assumption of narrow-bandness is adopted (which is often the case) then the amplitudes during the short time interval follow the Rayleigh distribution, which is completely described by the process variance R:

$$Q(x|R) = e^{\left(-\frac{x^2}{2R}\right)}$$

where $Q(x|R)$ denotes the probability of exceeding the specified level x in the short term period defined with response variance R.

To obtain the long-term distribution of wave bending moments, the response variance by itself should be considered as a random variable. The probability density function of the variance has generally a very complex form [4,11-13], because it depends on relative wave headings, manoeuvring in heavy weather, loading conditions and probability of encountering specific sea state. For general design purposes, however, a simplified discrete expression for the calculation of unconditional long-term peak distribution is available [7]:

$$Q(x) = \sum_{i=1}^{n_\alpha} \frac{\Delta\alpha}{2\pi} \left(\sum_{j,k}^{n_H, n_T} Q(x|H_{Sj}, T_{Zk}, \alpha_i) n_k(T_{Zk}) p_{jk}(H_{Sj}, T_{Zk}) \right)$$

This expression assumes a uniform distribution of relative wave heading angles α, by dividing polar plane in n_α equispaced sectors for which the response variance is calculated. The quantity $n_k(T_{Zk})$ is the ratio of the mean period in all sea states to the mean period of the sea state under consideration, accounting for the number of peaks during sea state duration. The term p_{jk} represents the probability of occurrence of a particular sea state following from the wave scatter diagram. Constant nominal ship speed V is often applied, although the Rules for the Classifications of Ships [14] propose a speed reduction for severe sea states. The described procedure can be applied for different loading conditions, implicitly including in that way the correlation which exists between the still-water and wave bending moments. Since the probability function Q(x) is not suitable for direct application in the reliability analysis, it is convenient to fit a theoretical distribution to it. The two parameter Weibull distribution with the exponent close to one proves to be suitable in most cases [4,6,15]. The methods for fitting the Weibull distribution to a series of points can be found in [15].

In early papers on ship reliability, the Weibull distribution was directly applied in the reliability analysis [1,6]. In [7] it is explained that such a distribution essentially represents a peak distribution, i.e. the probability that the bending moment amplitude exceeds the specified level x in any one of the cycles in a random time instant. But, more important in structural design is the extreme value distribution, showing the probability that the specified level is exceeded at least ones in the lifetime of the structure or some other time interval (one year, for example). If the assumption of mutually independent cycles is adopted, the extreme distribution can be obtained using the Poisson outcrossing or order statistics [7,10]. The obtained distribution converges rapidly to Gumbel (extreme type I) distribution [16]. The Gumbel distribution obtained in such a way represents the fundamental (stochastic) distribution of long-term wave bending moments and it has the coefficient of variation (COV) approximately between 0.06 and 0.1. It will be explained further that it represents only a small part of the total uncertainty of extreme wave bending moments.

The procedure described above requires some basic physical assumptions to be adopted, i.e.:
a) that the short-term process is narrow banded

b) that the wave cycles are statistically independent. The assumption a) means that the Rayleigh distribution is used instead of the Rice distribution as the peak probability density function. Such simplification has a negligible impact on the calculated failure probability [17]. The assumption b) is analysed in [11] where the approximate method of taking into account the statistical dependence between consecutive stress cycles is described. In most of the references, the assumption of statistical independence of cycles is adopted without further modifications, and is generally considered as conservative.

Once the assumptions mentioned above are adopted, there are no principal difficulties to calculate the transfer functions using the linear strip theory, and to obtain the long-term distribution of wave bending moments as well as its extreme value distribution. However, for the structural reliability calculation it is important to estimate the uncertainties of the method for the prediction of extreme values, i.e. to estimate the differences between the efforts to which the ship is exposed in reality and the calculated values, which represents an extremely difficult task. Here, a list of some of the most important sources of uncertainties is presented [4,8,10,11,18]:
1) shape of wave spectra
2) uncertainties of visual wave observations
3) shape of the transfer function, representing:
 a) model uncertainty, i.e.; differences between the actual and calculated transfer function for moderate wave heights
 b) non-linear effects, i.e. differences between the transfer functions for hogging and sagging for larger wave heights
 c) differences between the transfer functions calculated using various versions of programs for the implementation of the linear strip theory
4) uncertainty of operational philosophy
5) selection of wave scatter diagram.

The choice of wave scatter diagram probably introduces the greatest part of uncertainties, so it deserves special attention. Normally, the wave scatter diagram should be chosen from the wave data such as "Ocean Wave Statistics" [19], considering the ship operational area. Since the ship route is often not known a priori, it is usually assumed that the ship will sail in the North Atlantic, and the corresponding scatter diagram is constructed. However, it was already pointed out in [7] and [8]

that the calculated values of extreme wave bending moments, using directly the data from wave tables, significantly overestimate the measured values of wave bending moments. The Rule values for the vertical bending moment corresponding to a probability level of 10^{-8} are based on the accumulated experience, measurements and calculations of classification societies. The data from [15] and recently from [18], which were obtained using directly the wave tables from "Ocean Wave Statistics", show significant overestimation of the Rule values of vertical wave bending moments, and the reliability studies based on such a wave bending moment distributions risk to overestimate considerably the actual failure probabilities. It is shown in [7], where the comparison between the theoretical and the "measured" value of safety index β is done, that the latter one is much higher. According to the data from [18], the Rule value of the wave bending moment is overestimated for more than 30%, depending on the source of the wave scatter diagram. The COV of the most probable lifetime extreme wave bending moment, obtained using the wave tables from different sources, is approximately 10-15%, depending on the ship type. (It is to note that in the present work, the values of COV are estimated from the available data sets using elementary statistical formulae.)

In [13,18] the differences of the extreme values of wave bending moments due to different versions of the linear strip theory are also analysed. From the figures in [13,18] the COV of that type of uncertainty can be estimated to approximately 15%.

Other types of uncertainties can be considered once the wave scatter diagram and the calculation method are selected. When the uncertainties are assumed to be normally distributed random variables, the method described in [4] enables the calculation of long-term distribution and confidence interval. For the case of arbitrarily distributed variables, a sophisticated method for long-term prediction of peaks and extreme values is presented in [11]. In the mentioned two methods, various uncertainties are introduced together in the calculation, and the resulting variability is found to be relatively low.

Another question is the possible existence of correlation between sea states. The standard statistical procedure for long-term prediction implicitly assumes that short-term sea states are of equal duration and statistically independent, so in most of the papers that question is not considered. Only the method applied in [11] enables approximately the account for a correlation between two consecutive sea states. In [20], however, the method for calculating the encounter probability of specific sea state is presented, accounting for the correlation between sea states with the same significant wave heights occurring in different wave zones. One of the conclusions from the article is, that the existence of that type of correlation slightly reduces the encounter probabilities of extreme sea states.

It is interesting to compare various assumptions for the coefficient of variation (COV) of extreme wave bending moments which are used in some relevant reliability studies. In [7], 8.5% is calculated to be the lifetime stochastic uncertainty, while approximation variability is estimated to be 15%. Therefore, the total lifetime uncertainty is obtained as $COV = \sqrt{8.5^2 + 15^2} = 17.2\%$. As a statistical model, the Gaussian distribution is applied. In [10] log-normal distribution is used for the lifetime maximum value with the total COV of 20%. In a more recent paper [16], a multiple of three independent random variables is used as a statistical model for the action of wave bending moments on oil tankers. The first variable is distributed according to the Gumbel law representing the fundamental uncertainty, while the other two are normally distributed variables with equal COV of 15%, representing the uncertainty due to wave load calculations and non-linear effects. The total COV of that model can be estimated to be at least 23% for the yearly extreme value. In Classification Notes [2], based on full scale measurements on a floating production ship, the model uncertainty of the short term response is taken to be 10%. (It is to be noted that in this paragraph, only the values of COV are analysed, while the estimation of the bias of the model uncertainty requires separate considerations.)

The standard prediction approach described above is not the only possible way for assessing extreme wave bending moments. In [11], the extreme values during each short term sea state are determined using the Poisson outcrossing method. The extreme value distribution during vessel voyage is obtained considering outcrossings in each sea state as events of a new Poisson process. In Classification Notes [2], a similar approach is applied, but the extreme value distribution is obtained assuming that only a three-hour extreme storm is relevant to

calculate the vessel safety. Both mentioned methods involve the approximation of the wave scatter diagram with analytical distributions, as well as continuous approximation of transfer functions. The latter method described in [2] is compared with the classical approach and is found to give about 10% lower results for lifetime extreme vertical wave bending moment.

3. STILL-WATER BENDING MOMENT

Still-water bending moment is a static load effect whose magnitude depends on the loading condition and distribution of cargo. If the cargo distribution is known, the still-water bending moment can be calculated accurately. In the design stage, however, it is not possible to predict the distributions of cargo which would be realised during ship lifetime. Thus, the still-water bending moment should be considered as a random variable [21]. The statistical analysis of still-water data has shown, that the still-water bending moment is in most cases well below its design value, but in some cases the exceedings of the design value have been noted [21].

In the first papers on ship reliability, the still-water bending moment was considered as a deterministic quantity, although the methods to study the reliability under the Gaussian distribution of still-water bending moment were developed [1]. However, the parameters of the distribution were not determined by means of rational analysis.

The statistical analysis of still-water bending moments for various ship types was performed in [21], and practically all the reliability studies made later on refer to that paper. That study provides the mean value of still-water bending moments as the percentage of their maximum permissible value, as well as their COV. These information are used directly for the parameters of normal distributions in some reliability analyses [5,6,10].

In [16], the reliability analysis of oil tankers is performed. It is assumed, that the probability density function for still-water bending moment in one voyage is a Gaussian function. The parameters of the distribution are determined using the loading manual. The probability distribution of maximum values in n voyages is obtained using order statistics, assuming that outcomes are drawn from the same probability density function. Therefore, the Gaussian^n cumulative distribution is finally used in reliability calculations, where n indicates the number of voyages for one year in specified loading condition.

The statistical analysis of the still-water bending moments for an offshore production ship is presented in [22]. The Rayleigh distribution is found to fit the sagging still-water bending moments very well, while the exponential distribution is fitted to the hogging still-water bending moments.

The purpose of load combination studies is to take into account the probability that the maximum values of two processes occur simultaneously, which increases the calculated reliability [4,6,22,23]. To perform a load combination study, it is necessary to model loads not only as random variables, but also as time dependent stochastic processes [23,24]. Load combination studies for wave and still-water bending moments are presented in [23]. The results of that analysis are applied in [16] for the calculation of the failure probability. In load combination studies normal distribution is not directly used. Instead, the partially truncated normal distribution, which takes into account the load redistribution on board is used [23]. The direct application of Gaussian distribution would lead to considerable overestimation of lifetime extreme values for certain ship types [26,28]. In [22] a load combination study is performed for an offshore production ship, and a review as well as comparison of various methods for the calculation of load combinations is given. An advantage of short-term ship reliability, as defined for example in [17], is that load combination study is not required, and therefore, as a statistical model for the still water bending moment, the original Gaussian distribution can be employed without further modifications.

It should be noted that the calculation of sensitivity factors in [10,25] shows, that the safety index is more sensitive to the changes of strength and wave load than of still-water load parameters. The same conclusion is obtained in [2], where the reliability of a production ship is analysed. It simply means, that some assumptions and approximations in the statistical model for the still-water bending moment would have less influence on the results than the approximations on strength and wave loads.

4. SHIP LONGITUDINAL STRENGTH

The failure of the ship as a hull-girder can be initiated either due to yielding or due to buckling of the deck or bottom panels. After the failure of the first panel, there is some reserve strength before the ultimate failure occurs [26]. The stress redistribution after the first failure is called structural redundancy [2], and is a very important parameter for the ship safety [2,26]. The ultimate bending moment is defined as an extreme point in the moment-curvature diagram, where the slope of the curve is equal to zero [26] and there is no reserve of strength beyond this point. The ultimate strength can be calculated by the non-linear finite element method [10], but since that approach is very time consuming and expensive, approximate methods that are in good agreement with experimental results are developed [26,27]. Although the ultimate bending moment is a physically justified measure of ship safety, its application as the governing design criterion is not to be expected in the near future. The reason is given in [28], where it is explained that the ultimate ship failure is in fact a time dependant event, so time simulation methods should be applied to account correctly for the phase lags between global and local load effects. Such procedure has a large model uncertainty and is not convenient as a design tool at the moment.

4.1 First yield vertical bending moment

The variability of the bending moment which causes the first yield is primarily due to the variability of the plate thickness and due to the variability of the material yield stress. Corrosion effects, which can have a very important influence, are not considered in this paper.

The COV of the plate thickness is taken to be 4% in [1], but modern steel mills supply plates with a much smaller variability of about 1%, according to [25]. From the same source we can conclude, that the mean undersize of thickness is about 1%. Normal distribution can be used to represent that variability.

The mean yield strength is about two standard deviations higher than the characteristic material strength due to the acceptance procedure, consisting of rejecting the samples with the strength lower than the minimum specified value [25]. The COV of the material yield stress used by various authors is between 6 and 10% [1,5-7]. Classification Notes [2] suggest the COV for mild steel 8% and for higher tensile steel 6%. As the distribution type the log-normal distribution is used in [2,5], whereas in [25] the normal distribution is considered as more appropriate for steel from a single mill.

Table 1
Review of the calculated safety indices

Reference	Ship type	Failure mode	Strength model	Still-water bending	Wave bending	Reference time	Average safety index
[7]	warship	ultimate bending	normal	deterministic	normal	20 years	2.2
[6]	various ships	first yield	normal	deterministic	exponential	1 cycle	5.1
[5]	containership	first yield	log-norm.	normal	Weibull [n] *	1 year	2.9
[16]	oil tanker	ultimate bending	log-norm.	normal [n] *	Gumbel	1 year	2.9
[10]	oil tanker and bulk carrier	deck buckling	normal	normal	log-normal	20 years	2.0
[25]	oil tanker	deck and bottom buck.	normal	normal	Gumbel	1 year	2.5
[2]	production ship	deck buckling	log-norm.	normal	Poisson	1 year	3.7 (3.0)**
[33]	oil tanker	ultimate bending	log-norm.	normal	Gumbel	20 years	1.5

* n denotes the number of voyages per year   ** if design is based on the ship Rules

4.2 Statistical modelling of panel buckling

In a number of articles the buckling of the first panel at the deck or at the bottom is considered as a sufficiently serious event to represent the loss of ship's safety. There are numerous methods for calculating the ultimate buckling strength of the stiffened plates, out of which the methods according to Faulkner, to Carlsen, and to Hughes are oriented to marine structures [10,29,30,31,32]. The comparative analysis of the first two methods is given in [29], where the model uncertainty, using available experimental data, is determined. It is found that Carlsen's method has a mean bias of 0.895 and COV of 11%, while Faulkner's method has a mean value of 1.041 and COV of 12%. In Classification Notes [2], the mean value and standard deviation of Carlsen's method is given in the form of regression formulae, whose results do not substantially deviate from the values mentioned above. In [25], another method is compared with experiments, and it is found that a mean bias is 0.99, while COV is 9.63%. The important conclusion in that article is, that the normal distribution fits the model uncertainty very well. In [32] the mean bias of the method according to Hughes is calculated to be 0.91, while its COV is 15.6%. In [3] Carlsen's method is compared to the complex numerical calculations and the mean bias is found to be 0.91, while the COV is estimated to be 15%.

5. REVIEW OF SAFETY INDICES

From the very beginning of the application of reliability methods to ship structural design, a significant problem was the interpretation of the results. It has been recognised that obtained failure probabilities should not be considered as actual physical magnitudes which are expected to occur in practice, but as comparative measures between various designs. However, in order to make such an approach useful, a large number of designs should be calculated using similar statistical models. It is clear from the discussion in the preceding sections, that various types of distribution have been applied in ship reliability studies, resulting in a large scatter of calculated safety indices, which does not allow making a proper comparison. The problem is illustrated in Table 1, where average safety indices, as calculated in various articles, are presented. It is to be noted, that average values are calculated by the authors of this paper in order to enable a reasonable comparison, since a scatter of reliability within each article naturally exists. In the same table, the time periods for the calculation of reliability as well as the models applied for basic random variables are also presented.

The information about the statistical models presented in Table 1 is not by itself sufficient to explain the obtained safety indices. Assumed parameters used in probability distributions, particularly for the model uncertainties, could have a very important influence on the final result. The next question is whether the wave bending moment distribution is based on direct calculations or is, as given in some articles, obtained from the Rule values. Further, in some articles load combinations are included in the analysis, while in some others are not. Finally, due to its different nature, the ship type is also a very important parameter. All these considerations lead to the conclusion that one should be very careful, before drawing some general conclusions on the basis of the results given in Table 1. Only the methods referring to the same time interval and ship type, which consider the same failure mode and are based on the same or very similar assumptions, can be directly compared.

6. CONCLUSION

It can be concluded that there is a very large scatter in the long-term predictions of the wave bending moment. The recent investigations of the wave induced loads confirm very large variability of predicted extreme values and enable a more rational application of the reliability methods. The authors of this paper estimate, that, if the lifetime extreme wave induced bending moment is calculated using the standard calculation tools and long-term prediction methods, its total COV for the application in reliability studies is about 25%, depending on the ship type. The important question, which is still open, is why the Rule value of vertical wave bending moment is systematically overestimated for more than 30%, when direct calculation methods are applied, using directly the data from the wave tables. The possible answer is in the statistical methods for long-term predictions, which are based on some assumptions that are not completely justified, as discussed in the paper. Further research in that field is therefore necessary.

Concerning the still-water bending moment, two methods for the calculation of the annual failure probability are available. One of them uses the data from the statistical analysis, while the other one relies on the information from the loading manual. The latter method has been so far applied only to oil tankers. In both cases, order statistics for the calculation of extreme values should be applied. When calculating lifetime reliability, truncation effects should be taken into consideration, especially in load combination studies.

Two failure modes are analysed in this article: the first yield limit state and the first panel buckling. Both modes are to be considered as serious serviceability limit states. The ultimate hull girder limit state should also be part of the rational analysis to provide the information about structural redundancy. For both analysed failure modes there exists a rather uniform approach to define statistical models. It is interesting to observe that the data from several sources indicate a relatively similar COV for the model uncertainty of the calculation methods for panel buckling. Depending on the calculation method, its value is between 10% and 16%. Concerning the mean bias, the methods according to Carlsen and Hughes are conservative for about 10%, while Faulkner's method is slightely unconservative.

The deck buckling is considered as the most serious failure mode, especially when oil tankers are in question. The reliability index refers to the time period of one year, which has an advantage compared to the lifetime period in a more consistent calculation of wave and still-water induced loads, and enables the analysis of the influence of corrosion effects. Concerning the deck buckling failure mode, it can be roughly concluded that the yearly reliability index for oil tankers should be between 2.5 and 3, if recent statistical models, reviewed in this work, are applied.

ACKNOWLEDGMENT

The presented analysis was partially performed during the specialization of the first author in Bureau Veritas, Paris, thanks to the French Government Scholarship, which is gratefully acknowledged.

REFERENCES

1. Mansour, A.E., Probability Design Concepts in Ship Structural Safety and Reliability, Transactions SNAME, 1972, pp. 64-88
2. Det Norske Veritas, Classification Notes No. 30.6., Structural Reliability Analysis of Marine Structures, July 1992.
3. Wang, X., Jiao, G., Moan, T., Analysis of Oil Production Ships Considering Load Combination, Ultimate Strength and Structural Reliability, Transactions SNAME, Vol. 104, 1996.,pp. 3-30
4. Guedes Soares, C., Moan, T., Uncertainty Analysis and Code Calibration of the Primary Load Effects in Ship Structures, ICOSSAR, 1985, Vol. 3, pp.501-511.

5. Ostergaard, C., Partial Safety Factors for Vertical Bending Loads on Containerships, OMAE,1991, pp. 221-228.

6. Mansour, A.E. et al., Implementation of Reliability Methods to Marine Structures, Transactions SNAME, Vol. 92, 1984, pp.353-382.

7. Faulkner, D., Sadden, J.A., Toward Unified Approach to Ship Structural Safety, Transactions RINA, Spring Meeting, 1978, paper No. 3.

8. Faulkner, D., Semi-Probabilistic Approach to the Design of Marine Structures, Extreme Loads Response Symposium, SNAME, Arlington, VA, 1981, 213-230.

9. Madsen, H.O., Krenk, S., Lind, N.C. , Methods of Structural Safety, Prentice Hall, 1986.

10. Thayamballi, A. K., Chen, Y.-K., Chen, H.-H., Deterministic and Reliability Retrospective Strength Assessments of Ocean-going Vessels, Transactions SNAME, 1987, paper No.6.

11. Cramer, E.H, Friis Hansen, P., Stochastic Modeling of Long Term Wave Induced Responses of Ship Structures, Marine Structures 7(1994), 537-566.

12. Guedes Soares, C., Schellin, T. E., Long Term Distribution of Non-Linear Wave Induced Vertical Bending Moments on a Containership, Marine Structures 9(1996), 333-352.

13. Schellin, T. E., Guedes Soares, C., Uncertainty Assessment of Low Frequency Load Effects for Containerships, Marine Structures 9 (1996), 313-332.

14. Bureau Veritas, Rules and Regulations for the Classification of Ships, Part II, Hull Structure, 1996.

15. Guedes Soares, C., Moan, T., Model Uncertainty in the Long-Term Distribution of Wave Induced Bending Moments for Fatigue Design of Ship Structures, Marine Structures 4(1991), 295-315.

16. Casella, G., Dogliani M., Guedes Soares, C., Reliability Based Design of the Primary Structure of Oil Tankers, OMAE, 1996, Vol. II, pp. 217-224

17. Mansour, A.E., A Note on the Extreme Wave Load and the Associated Probability of Failure, Journal of Ship Research, Vol.30, No.2, June 1986, pp. 123-126.

18. Guedes Soares, C., On the Definition of Rule Requirements for Wave Induced Vertical Bending Moments, Marine Structures 9 (1996), 409-425.

19. Hogben, N., Lumb, F.E., Ocean Wave Statistics, London,1986.

20. Mansour, A.E., Priston, D.B., Return Periods and Encounter Probabilities, Applied Ocean Research 17(1995) 127-136.

21. Guedes Soares, C., Moan, T., Statistical Analysis of Stillwater Load Effects in Ship Structures, Transactions SNAME, Vol. 96, 1988, pp.129-156.

22. Wang, X., Moan, T., Stochastic and Deterministic Combinations of Still-Water and Wave Bending Moments in Ships, Marine Structures 9(1996)787-810

23. Guedes Soares, C., Combination of Primary Load Effects in Ship Structures, Probabilistic Engineering Mechanics 7(1992) 103-111

24. Guedes Soares, C., Stochastic Modeling of Maximum Still-Water Load Effects in Ship Structures, Journal of Ship Research, Vol. 34, No. 3, Sept. 1990, pp. 199-205

25. Hart, D.K., Rutheford, S.E., Wickham, A.H.S., Structural Reliability Analysis of Stiffened Panels, Transactions RINA, 1985, pp.293-310.

26. Rutheford, S.E., Caldwell, J.B., Ultimate Longitudinal Strength of Ships : A Case Study, Transactions SNAME, 1990, paper no. 14.

27. Hansen, A.M., Strength of Midship Sections, Marine Structures 9(1996), 471-494.

28. Nitta, A., On C. Guedes Soares Discussion of Paper by A. Nitta et al. 'Basis of IACS Unified Longitudinal Strength Standard', Marine Structures 7(1994), 567-572.

29. Guedes Soares, C., Soreide, T. H., Behaviour and Design of Stiffened Plates Under Predominantly Compressive Loads, International Shipbuilding Progress, Vol. 30, January 1983, No. 341., 13-27.

30. Carlsen, C.A., Simplified Collapse Analysis of Stiffened Plates, Norwegian Maritime Research, No. 4/1977., 20-36.

31. Hughes, O.F., Ship Structural Design, SNAME, New York, 1988.

32. Mansour, A. et al., Probability Based Ship Design : Implementation of Design Guidelines, SSC-392, 1996.

33. Mansour, A., Hovem, L., Probability-Based Ship Structural Safety Analysis, Journal of Ship Research, Vol. 38, no. 4, Dec. 1994,pp. 329-339.

Automatic hull form generation: a practical tool for design and research

Dr. R. W. Birmingham and T. A. G. Smith

Department of Marine Technology, University of Newcastle, Newcastle upon Tyne NE1 7RU, UK.

This paper describes how optimisation techniques can be used to automate the generation of preliminary hull forms. Presented are some of the theoretical and practical elements involved in the development of a demonstration system, and a discussion of further areas for development.

1. INTRODUCTION

In naval architecture the start of the iterative process of design is a preliminary definition of the external hull geometry. The influence of the hull geometry of the vessel is such that major changes made during subsequent iterations can cause severe disruptions to the convergence towards a final solution. It is therefore important that the initial hull form definition is well conceived.

The first stage of hull form synthesis involves the development of a symbolic model. This is a set of values for key parameters that correlate with various aspects of operational and economic performance, and largely based on empirical data. Some of these parameters are of a numeric nature, and may be dimensional (linear dimensions, areas, volumes and centroids) or non-dimensional (form coefficients such as the block coefficient, waterplane coefficient, and various other ratios of lengths, areas and volumes). Others are of a more abstract nature, referring to the presence and extent of more subtle form features (terms such as 'tumble home', 'flare' and 'hard bilge'). An important further requirement of any form is that it is fair, meaning the absence of any unwanted form features.

The task of the naval architect is to derive a geometrical definition (an iconic model) which corresponds to this parametric description. There are two alternative approaches that are usually considered. One is to apply some geometrical transformation to an existing form such that the resulting form satisfies some or all of the new parametric requirements. The other is to synthesise an original form, using any existing knowledge in an intuitive and informal way.

The advent of interactive graphical CAD has enhanced the original approach by allowing rapid re-evaluation of the design parameters when changes are made. However, the process is essentially still manual and time-consuming. Unfortunately CAD has not significantly aided the form transformation approach to design. Firstly there is the problem of transferring existing hull forms recorded as offsets to the B-Spline surface representation used by most modern ship CAD systems, and secondly B-Spline surfaces cannot be manipulated by the conventional transformation algorithms[1, 2].

The introduction of an element of automation into these approaches would provide benefits to both industry and academia. From a commercial viewpoint, the advantages are improved productivity and improved quality of designs, as a wider range of alternatives can be considered in a methodical manner with greater conformity to specified requirements. Additionally, the designer's understanding of the design limitations is enhanced, so bolstering expertise. From an academic perspective, there are various benefits. Methodical control of form parameters is central to a number of areas of research, a current example being the parametric study of vessels' hydrodynamic performance[3, 4]. The system also poses questions relating to the nature and philosophy of design, and suggests different ways of specifying a design and the routes taken to reach a particular definition.

2. DESIGN AS A SEARCH PROCESS

The synthesis of a hull's geometry is essentially a search process, consisting of three main elements. The first is a method of creating a controlled geometric form. The second is a means of evaluating this form in terms of specified requirements, and the third is a mechanism for using this information to move in a useful direction through the search space defined by the limitations on geometry. Iterations based on continual feedback progress until a satisfactory form is arrived at.

2.1. Progression

In the traditional process of manually drafting lines, the controls were the number and position of weighted ducks and the stiffness of the baton they held. Since manual evaluation of numerical form parameters is extremely time-consuming, aesthetic judgement played a major part in the guidance of the process. With interactive graphical CAD systems, form control is through the number and attributes (position and weightings) of B-Spline control points, and the orders of the surfaces they relate to. Rapid recalculation capabilities mean that numerical form parameters have a larger guiding influence than previously, although aesthetic judgement still plays a crucial role.

The impact of CAD on hull form design has been to significantly reduce the burden of modifying the form and entirely relieving the designer of evaluating it. However, the process of guiding the search using feedback information is still performed solely by the designer. Automating this process represents the next logical step of this progression, and is the key to automated hull form generation[5].

2.2. Original design and form transformations in a search context

A common approach to hull form generation is to perform a geometric transformation on an existing design[6, 7]. This approach has the advantage of being fast and straightforward. It is also reliable, as the characteristics of the parent vessel are known. However, it discourages innovation and tends to be restrictive, as the transformations used do not have the control needed to simultaneously satisfy all of the parametric requirements. This is because in order to work effectively, such methods must involve a move in the search space that is limited in both direction and magnitude.

A different problem presents itself with original design. Here it is time, not the method, that restricts the exploration of possible alternatives. Only a small region of the design space can feasibly be explored, regardless of the skill and expertise of the designer.

In optimisation terms, the difference between the two approaches is one of degree only, being the required proximity of the starting point in the search space to that corresponding to the final design. Form transformations are defined as a search starting from a specific point close to the final design. Therefore original design, starting from an arbitrary point, can be viewed as a generalised form transformation.

2.3. Surface offset fitting in a search context

The problem of fitting one or more B-Spline patches to a set of existing offsets has had numerous treatments, some analytical in nature, others being search-based[8]. One of the advantages of a search-based approach is that it lends itself to the linked problems of achieving both acceptable accuracy and fairness in the final form. Offset data will almost certainly have some errors present in it, even after it has been 'cleaned' using divided-difference type techniques[9]. Usually there will be more data points than the number of degrees of freedom necessary to generate the surface of sufficient accuracy and acceptable fairness. Thus attempting to interpolate all of the data points exactly with an analytical procedure leads to an excess of control points, which increases the likelihood of unfairness. This can be exacerbated during subsequent manual modification of the form. With a search-based approach a fair approximation of the data points (to a specified accuracy) can be sought with a reduced number of control points.

2.4. Automation of the search process

The success of any design with respect to the fulfilment of target parameters is determined by the success with which the search space is negotiated. There are two main factors that affect this. One is the way in which multiple requirements, often conflicting, are considered. How satisfactory a compromise solution is depends on how the trade-off between conflicting requirements is dealt with. The second factor is how the information from the evaluation of individual forms is used to negotiate the design space. In a manual design process both tasks are performed by the designer, skill being the chief factor in the success of a design. In the field of optimisation, the trade-off between multiple requirements is a problem of multi-criteria decision-making (MCDM)[10, 11]. The way in which feedback information is used to guide the search depends principally on the type of search strategy used.

There are many ways in which multiple requirements in a design decision-making context can be handled, and this is currently a rich area of research. The choice of scheme should reflect the nature of the problem in hand, with the main consideration being the way in which the desirability (or utility) of each objective relates to its own value and the values of the other objectives. The simplest scheme involves normalising each component objective and multiplying with a 'weight' factor, which represents the relative importance of the objective. These terms are then summed to form a single measure of merit. Similarly, extensive research into various search strategies has yielded many alternatives suited to different applications. The main factor to be considered in the choice of a search strategy is the nature of the search landscape. The problem in hand can be shown to be multi-modal, thus requiring a choice of search strategy that will not prematurely converge upon sub-optimal peaks.

3. IMPLEMENTATION

A demonstration system has been configured implementing the ideas discussed above. This system consists of two basic components. A hull modeller creates and evaluates forms, and a search strategy uses this information to move towards an acceptable form. The following discussion provides brief details of the methods chosen and how they have been developed in the creation of both elements.

3.1. Hull modeller

The hull modeller uses a number of B-Spline surfaces[8, 12, 13, 14] to represent the hull surface. The input consists of a set of control point attributes and the orders of the surfaces. Interrogation of this model yields the values of various form parameters, principally overall dimensions and non-dimensional coefficients. Additionally, values representing the fairness of the hull and the regularity of the control point net can be calculated.

3.2. Search strategy

The multi-modal nature of the search landscape requires the utilisation of an adaptive search strategy, if premature convergence is to be avoided. Various strategies that have emerged include Simulated Annealing, Tabu Search and Evolutionary Computing, a relation of which are Genetic Algorithms (GAs) which have been chosen in this case. The effectiveness of GAs in a variety of settings, including engineering design, has been extensively documented[11, 15, 16].

The mechanism of a simple GA is the evolution of a population of individuals, each member of the population being a set of numbers which are an encoding of the design variables (in this case the control points) representing a hull form, or a specific point in the design space. Each design's attributes can be evaluated and a measure of merit or 'fitness' established. In the simplest case, a single aggregate value for fitness is used. More sophisticated approaches consider the values of individual attributes. The starting point of the process is usually the creation of a randomised initial population. Each individual is then evaluated, and a selection process carried out on a probabilistic basis, whereby the fitter hulls have a greater chance of surviving. This selected population is then split into pairs of hulls on a random basis. Parts of the encoded variable groups are then swapped or 'crossed over' within each pair, according to a specified probability. The final stage involves making changes or 'mutations' to randomly chosen elements of the encoded variable groups. The population fitnesses are then re-evaluated, and the process repeated until some stopping criterion is reached, usually an indicator of convergence or an upper limit on the number of 'generations'.

Numerous additions to and variations of this basic scheme have been developed to improve convergence characteristics for various types of problem. In this research the main areas of customisation involve the crossover and mutation operations, as described below.

Crossover

Control points have a localised influence on the surface they define, that is a control point has the most influence on the part of the surface it is closest to. It is therefore reasonable to expect that within a whole B-Spline surface, there will be localised regions of varying merit, each region being principally influenced by a 'subnet' of control points. Intuitively, a crossover type operation could combine high-rating subnets in a single net, producing a superior surface.

In most GAs, the encoded design variables are arranged in a one-dimensional form as a 'string'. For a pair of strings, crossover is achieved by randomly defining one or more 'crossover points' which divide both strings into substring segments. Alternate segments are then swapped between the two strings.

The search variables in this case are the attributes of the vertices of the control net. The control net is topologically two-dimensional, thus must be 'unwound' in some way to form a one-dimensional string. However, this process is likely to adversely affect the preservation of high-rating subnets. This is because a subnet consists of points that are topologically neighbouring in two-dimensional sense, a property which is unable to be preserved in one dimension when the control net is unwound. The approach currently taken to this problem is to avoid unwinding the control net altogether. The notion of a crossover point (that divides a variable string) is extended to a line that divides up a variable matrix into two-dimensional sub-matrices, which are swapped in a similar fashion to the substrings in the one-dimensional case.

Mutation

In GA design, the type of mutation operator used depends principally on the type of variable encoding used. When real coding (simply representing the variables as real numbers) is used, a proven mutation method is gaussian mutation. A variable has a deviation added to it, with the deviation sizes having a gaussian distribution (with a mean of zero, and a controllable variance). This is the mutation operator being used in this application.

The deviation value distribution consists of two half distributions. These are approximately gaussian, one between zero and a specified negative limit, the other between zero and a specified positive limit. Both half-distributions have the same variance, specified independently of the negative and positive limits. The truncation of the half-distributions by the limits is achieved by replacing deviation values outside the limits with a value chosen randomly from between the limits. In this way, as the specified variance is increased, the deviation values tend towards a uniform distribution.

Individuals that are mutated with a higher variance are more likely to receive larger deviations, and so travel further through the search space, a process that could be viewed as exploration of new regions. Lower variances are more likely to produce more localised moves. If this occurs in a high-rating region of the search space, this could be viewed as exploitation of that region. The variance used is thus made a function of an individual's fitness. Weak individuals are best used to explore entirely new regions, whereas fit individuals are encouraged to exploit the reasons for their high fitness.

The system is undergoing continual improvement and modification, which is facilitated by the object-oriented approach to the implementation.

4. APPLICATION OF THE SYSTEM

To demonstrate the effectiveness of the system and to highlight areas that are still being developed, the application to two sample cases is considered. The first case is a generalised form transformation or original form synthesis. The second involves the fitting of a surface to a set of specified offsets.

The hull used for both cases is the canoe body of a yacht, defined by a single B-Spline surface. The corresponding control point net is relatively small, consisting of 5×5 points. The size of the search space is limited by specifying a positional accuracy on each point of 0.01 m, and placing global constraints on the control point positions. In order to provide an unbiased and comprehensive view of the process, these constraints are fairly relaxed. For all control points, the limiting envelope is a cuboid whose dimensions are slightly larger than the desired gross dimensions (L_{OA}, B_{MAX}, D) of the form. In a realistic situation, these constraints could be tightened, resulting in a smaller search space.

4.1. Generalised form transformation

The aim of this exercise is to recover the original form from a randomised population by specifying known form parameters. In a real design situation, the exact geometry of the form sought is not known in advance. Additionally, the parametrically specified form may not exist, meaning that there is conflict between requirements that must be resolved. However, since the purpose of this exercise is to examine the effectiveness of the search mechanism, it is useful to have a reference point to measure results against.

The search seeks to minimise an aggregated function based on the deviations of a form's parameters from the specified values. Six form parameters were specified, being a mixture of primary dimensions and secondary form coefficients, as shown in figure 2 below. An additional term represents the fairness of the surface. This measure is based on the familiar analogy with strain energy in a thin elastic plate[17]. The optimal (minimal) value of this term is zero, which leads to an element of conflict, as this value is only achievable if the surface is flat. It is therefore important that the weighting given to fairing is such that loss of shape due to over-fairing does not occur.

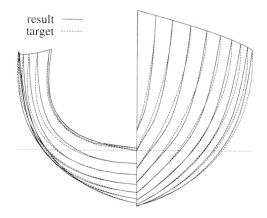

Figure 1. Comparison between form transformation result and target form

Parameter	Target value	Result of search
B_{WL} (m)	3.594	3.595
∇ (m³)	9.021	9.032
LCB (m)*	4.875	4.875
LCF (m)*	4.666	4.696
C_P	0.5472	0.5475
C_{WP}	0.6580	0.6567

* Forward of aft extremity

Figure 2. Parameter comparisons between form transformation result and target form

Figure 1 shows the target form overlaid with the result of a search performed with a population of 100 individuals for 200 generations. Equal weightings have been assigned to all parameter objectives. The performance of the best individual

in each generation, shown in figure 3, shows that the majority of improvement occurred in the first 50 generations.

There is considerable noise in the performance curve corresponding to the worst individual of each generation. Additionally, the standard deviation of the aggregate error values shown in figure 4 remains roughly constant for much of the duration. These two factors suggest that a reasonable balance between exploration and exploitation is being maintained, based on fitness as discussed earlier.

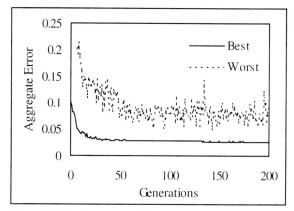

Figure 3. Performance of form transformation search

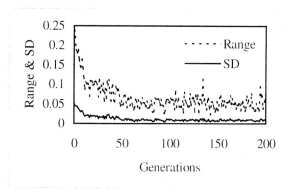

Figure 4. Range and standard deviation of aggregate error values during form transformation search

4.2. Offset surface fitting

The objective function in this case is simpler than that of the previous exercise, consisting of only two components. One is an indicator of surface fairness as before, the other being a sum of offset error terms, each term being a function of the deviation of an offset from the specified position. If the offset is within a specified tolerance of the specified position, the error term is zero. In this example, both components are given equal weighting.

Figure 5 shows the target form overlaid with the result of a search performed with a population of 100 individuals for 300 generations.

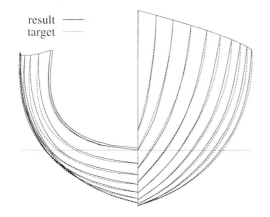
result ———
target ·········

Figure 5. Comparison between offset fit result and target form

The summed offset error in this case is zero, indicating that all of the offsets are within the envelope defined by the tolerances, in this case ±0.05 m for all offsets. The strain energy analogue value is in fact slightly lower than that of the target design, indicating that the search has located a fairer form than the original within the specified tolerance envelope. The search performance characteristics are not shown, but are of a similar nature to those from the previous exercise.

5. AREAS FOR IMPROVEMENT

The main improvement sought is a reduction in convergence time. Typically, a run involves upward of 30,000 hull evaluations. The running rate in real time is approximately 1 second per hull on a Sun UltraSPARC 2, and this is unlikely to be reduced substantially. Consequently, it is necessary to sample significantly fewer search space points to make any substantial impact. There are essentially two ways of achieving this.

5.1. Improving search efficiency

One approach is to increase the efficiency with which the search is guided, by improving the fitness function through the application of more sophisticated MCDM techniques[10, 11]. The principal shortcoming of the simple weighted sum approach currently adopted is that the merit of individual attributes of a form are not considered. This assumed independence of utility, whereby shortfalls in some attributes are compensated for by excesses in others, represents a loss of information. The problem is exacerbated by the fact that the designer is unlikely to be in a position to make meaningful weight assignments at the outset of a design.

5.2. Reducing size of search space

The second approach is to reduce the size of the search space. Discarding regions of the search space corresponding to unrealistic hulls has substantial benefits. In addition to reducing convergence time, reliability can be improved by removing potentially misleading information from the search. Obviously any such scheme should avoid discarding any potentially useful search space.

Search space size is defined in terms of the number of dimensions (the number of search variables) and the range in each dimension. Assuming the problem is well set up, the number of variables will be at a minimum. Therefore, reduction of the search space must be achieved by placing constraints on the variable ranges, in this case the control point positions. What is required is some criterion by which constraints are placed.

An implicit requirement of any form generated is that its net of control points is well ordered and regular. The interpretation of this varies between designers, and with the form requirements of different hulls. Most designers take a heuristic approach to control net 'tidiness', regarding factors such as the distribution, smoothness and alignment of control point rows and columns. Indeed, it is through this method that a designer working manually is able to keep good control over the developing design. However, in order to apply a similar principal to the problem in hand, a reconfiguration of control variables is required. Many elements of control net tidiness depend on the relative positions of control points. Currently, the only constraints that can be placed on control point positions are with respect to a common global frame of reference.

6. CONCLUSION

The process of hull form synthesis can be automated by replacing the designer's guidance of the developing design with optimisation techniques. The approach makes what are traditionally described as original and form transformation approaches part of a continuum. The same techniques can also be applied to the problem of fitting surfaces to sets of existing points.

This paper outlines how a genetic algorithm has been used to implement such a procedure. The main focus of ongoing research is the improvement of efficiency, principally by identifying rule-based procedures for reducing the search space.

ACKNOWLEDGEMENTS

The authors wish to thank Professor George Snaith for providing comments and insights throughout the project.

This research is funded by the Engineering and Physical Sciences Research Council (EPSRC) in the form of a postgraduate studentship.

REFERENCES

1. R. W. Birmingham & T. A. G. Smith (1995) "A practical approach to form transformation for software generated designs", *proc. CADAP '95* (Southampton, UK).
2. R. W. Birmingham & T. A. G. Smith (1997) "Interpreting design intent to facilitate automatic hull form generation", *proc. IMDC 97* (Newcastle, UK).
3. G. E. Hearn et al (1992) "Practical seakeeping for design: An optimised approach", *proc. PRADS 92*, UK.
4. G. E. Hearn et al (1992) "Seakeeping for design: The demands of multihulls in comparison to monohulls", *proc. ICCAS 94*, Bremen.
5. R. W. Birmingham & T. A. G. Smith (1997) "Automating the interpretation of design intent", *proc. ICED 97* (Tampere, Finland).
6. H. Lackenby (1950) "On the systematic geometrical variation of ship forms", *trans. INA*, **92** 289-316.
7. H. Schneekluth (1987) *Ship design for efficiency and economy*, Butterworths, London.
8. L. Piegl & W. Tiller (1995) *The NURBS Book*, Springer-Verlag, Berlin.
9. S. A. Berger et al (1966) "Mathematical ship lofting" *J. of Ship Research*, **10**(4) 203-222
10. P. Sen & J-. B. Yang (1995) "Multiple criteria decision making in design selection and synthesis", *J. of Engineering Design*, **6**(3) 207-230.
11. D. S. Todd (1997) *Multiple criteria genetic algorithms in engineering design and operation*, Ph.D. thesis, University of Newcastle Engineering Design Centre.
12. G. Farin (1988) *Curves and surfaces for computer aided geometric design*, Academic Press, San Diego.
13. R. C. Beach (1991) *An introduction to the curves and surfaces of computer-aided design*, Van Nostrand Reinhold, New York.
14. H. Nowacki et al (1995) *Computational geometry for ships*, World Scientific, Singapore.
15. D. E. Goldberg (1989) *Genetic algorithms in search, optimization and machine learning*, Addison-Wesley, Reading MA.
16. M. Srinivas & L. M. Patnaik (1994) "Genetic algorithms: A survey" *IEEE Computer*, **27**(6) 17-26.
17. H. Nowacki & D. Reese (1983) "Design and fairing of ship surfaces", *Surfaces in computer aided geometric design*, 121-134, North-Holland, Amsterdam.

Practical Design of Ships and Mobile Units
M.W.C. Oosterveld and S.G. Tan, editors.

Hull Form Modelling using NURBS Curves and Surfaces

M. Ventura[a] and C. Guedes Soares[a]

[a]Unit of Marine Technology and Engineering, Technical University of Lisbon
Instituto Superior Técnico, Av. Rovisco Pais, 1096 Lisboa, Portugal

The NURBS based hull modelling system CadSHIP is described, together with the methodology for the design of a ship hull surface on which the system is based. The approach adopted starts from the knowledge of some few main lines and basic geometric parameters and then applies different methods of surface generation and analysis. The main methods for surface generation and analysis used in the system are presented. A review is provided of NURBS main characteristics, highlighting those which make them an efficient replacement to the wireframe and surface models used so far in ship hull design.

1. INTRODUCTION

The mathematical description of the ship hull geometry needs formulations capable of representing free-forms, planar regions and conic shapes.

In the last 30 years, many types of parametric curves and surfaces have been used to model hull forms. The earlier ones, such as cubic splines, Coons patches, Gordon's patches and splines in tension, interpolated all the defining points. In general, their main drawbacks are their global behaviour which implies that any local changes affect the complete shape. Also a problem is dealing with some quantities, such as cross-derivatives and others, whose influence in the shape is not obvious for the designer. The use of Bézier curves and surfaces introduced the concept of control polygons and meshes that provide a more intuitive geometric control of the shape.

B-splines curves and surfaces have been widely used for the representation of ship hull geometry. The properties of B-splines, namely the local control, stronger convex hull and the possibility of introducing discontinuities by increasing multiplicity in control points have proved to be more suitable for the task than the Bézier formulation. B-splines can contain Béziers as a particular case and also are not able of representing exactly conic shapes. Fog [1] represented the entire hull by a single fourth order non-uniform tensor product B-spline surface. Beyer [2] used B-spline curves on the design tool DCM, developed for the interactive modelling of hulls, which allowed to select the type of continuity between curve segments (C^0, C^1 or C^2) The system HULLSURF [3] represents the hull form by bi-cubic B-spline patches defined over boundaries approximated by B-spline curves, both using uniform knot vectors.

Jensen [4] developed an automatic procedure for generating a single B-spline surface to represent a ship hull surface. First the longitudinal and any knuckle lines are interpolated by cubic splines. Then, for each section, a user defined number of control points is obtained by least-square approximation. The grid composed by the section control points is then used to generate a tensor product B-spline surface. Standersky [5] combines the interactive capabilities of the B-spline tensor product surface with the variational approach to the shape generation of ship hulls.

More recently, Bardis and Vafiadou presented a model [6] that tries to combine the B-spline formulation with the local control of the patch boundaries obtained by concepts borrowed from the Beta-spline formulation. First, longitudinal boundary lines are approximated by B-spline curves interpolating selected points. Then, transverse sections and the longitudinal parametric first derivatives are approximated by B-spline curves fitted to section offset points and longitudinal tangent values, respectively. Finally B-spline surface patches are generated between each pair of consecutive curves, using first derivative values from the tangent values on the boundaries multiplied by bias functions β_1, similar to those used in Beta-spline formulations. Beta-splines are a generalisation of B-splines which added two new variables, the shape parameters β_1 and β_2, called *bias* and *tension*, respectively, allowing the capability of controlling the degree of continuity at the joints between curve segments without interfering with the order or the number of control vertices. However, this extra control is obtained at the cost of replacing the requirement of second degree *parametric continuity*, C^2, between curve segments used in B-splines, by the requirements of the so called *geometric continuity*, G^2, of the unit tangent and curvature vectors.

The brief review presented here shows that most of the formulations in the recent past are based on B-splines, which have the limitation of not being able to deal in an exact way with conic and quadric shapes, which are essential in the hull geometry. This limitation is solved by NURBS, (Non-Uniform Rational B-Splines) which are a generalisation of B-Splines, in the rational form, that use an extra degree of freedom, the control point weight. As a superset of B-splines, they retain all their properties, with a stronger convex hull and the additional capability of representing conic and quadric shapes exactly, which previously was only possible using the analytical expressions. This set of properties makes NURBS particularly attractive as a mathematical formulation for hull form modelling systems and was the motivation to develop the present system based on that formulation.

2. CADSHIP SYSTEM

The CadSHIP system is a hull form modeller, based on an unified formulation, entirely based on NURBS curves and surfaces. The idea behind this option was to reduce the number of algorithms to develop and implement, to simplify the design of the database and to reduce the data storage requirements, obtaining a system easier to implement and maintain. NURBS, have been established as an industrial standard in CAD systems, and are included in standard graphics libraries like PHIGS PLUS and OpenGL and in standards for data exchange like IGES and STEP. For a brief description of NURBS, refer to the Appendix.

Although very convenient to represent the hull surface, the shape control of a 3D surface is not always a trivial task and so the system uses both curves and surfaces, in order to combine the easy of use from shaping the hull from a set of characteristic curves (midship section, longitudinal profiles, tangency lines), closer to the traditional approach, with the advantages obtained from the final shape representation by surface patches.

The CadSHIP system is composed of three main components, a graphical editor, a database and a STEP processor. The graphical editor provides the user interface and allows the interactive creation, edition and interrogation of NURBS curves and surfaces of any degree and with several possible types of parameterisations. From the resulting surfaces, the typical sections, waterlines and buttock contours can be obtained. The system is intended to be flexible enough to allow the user to decide the sequence of operations as result of the available data and the required target.

2.1 Modelling Methodology

Hull surface modelling can have two distinct targets: to create a new form from scratch or to represent an existing one. Although in the last case more data is available, the excess of information makes the task more restrictive and therefore more difficult to accomplish. In the typical case, of an existing offset table or draft body plan, a possible sequence could be as follows:

- To digitise a set of transverse sections, and the stern and bow contours, converting them into polylines
- To filter each polyline, to a given tolerance, to eliminate redundant points, reducing the volume of data
- Approximate each polyline by a NURBS curve
- Analyse the curves curvature and edit them interactively, if necessary
- Divide the hull into a set of regions with common geometrical characteristics, avoiding the existence of internal knuckle lines, and split the curves at the boundaries
- Generate surface patches for each region, using the available methods, described in section 4.
- Interrogate the surface patches using the methods described in section 5 and edit them interactively, if required
- In the patches where the continuity across boundaries is not acceptable, boundary extended polylines, defined by 3D points and the corresponding normal vectors are computed and made compatible. The patches are deleted and new ones are generated by lofting, using the boundary conditions supplied by the extended polylines.

2.2 Database

The system database was specified to store all the data considered necessary to describe geometrically the shape of a ship hull. It contains both generic information and the description of all the geometric entities used in the model. The generic information includes the ship designation and reference, main dimensions, the co-ordinate reference system and the type of units used and the frame spacings. The geometric entities considered, are the following:

- Polylines, defined by ordered lists of 3D points.
- Extended polylines, defined by ordered lists of 3D points and normal vectors
- NURBS curves
- NURBS surfaces
- Embedded polylines, defined in a surface 2D parametric space (u,v)
- Embedded curves, defined in a surface 2D parametric space (u,v)

2.3 STEP Processor

Due to the requirements of data exchange and re-use, during the ship design process, a STEP translator was integrated into the system, capable of importing and exporting B-spline and NURBS surface patches together with other non-geometrical data, such as the ship main characteristics, dimensions and frame spacings, in accordance to the AP 216. The STandard data Exchange Protocol (STEP) is a standard (ISO 10303) for product data exchange in neutral format.

3. SURFACE GENERATION

In the CadSHIP system, surfaces can be created directly (spheres, cones, cylinders) or generated from existing curves (Fig. 1).

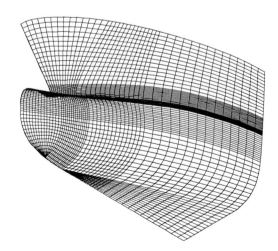

Figure 1. Bow modelled with NURBS surface patches

Curves can be created directly, defined by parameters, such as conic primitives (circular or elliptical arcs) or by least-square interpolation or approximation of given polylines. Due to the peculiarities of a ship oriented system, a tool is also provided to generate directly curves representing ship transverse sections, from a set of parameters.

The surface generation methods implemented are the extrusion, ruling, lofting, sweeping, revolving and blending, which are briefly described in the following.

3.1 Extrusion

Given one curve of order k defined by the control points B_{1i} and the knot vector x_j and the vector $\vec{d}$, a surface can be defined by the grid obtained joining the control points of the curve and the control points obtained by

$$B_{2i} = B_{1i} + \vec{d}$$

The surface will have order k and the knot vector x_j in v direction and order 2 and the knot vector $\{0,0,1,1\}$ in the u direction.

3.2 Ruling

A ruled surface $S(u,v)$ is generated by sliding a straight-line segment between two curves. Mathematically it corresponds to a linear interpolation between the two given ruling curves and it can be defined by

$$S(u,v) = b_1(v)\gamma_1(u) + b_2(v)\gamma_2(u)$$

in which the blending functions are given by

$$\gamma_1 = (1-u)$$
$$\gamma_2 = u$$

3.3 Lofting

Lofting [7], is the process of generating a surface that interpolates a set of cross section curves oriented in the same parametric direction.

The lofting process can be divided into two main steps. First, all the cross section curves are made compatible and next, isoparametric points of the curves are interpolated in the opposite parametric direction. To be made compatible, all curves will have the degree raised to a common value. Then, a knot vector common to all the curves is computed and knots are inserted as required in each curve, obtaining the new corresponding control points. The quality of the interpolation of the isoparametric curve control points is fundamental for the quality of the surface obtained. In CadSHIP this is made by an algorithm taking into consideration boundary conditions in the extreme points.

3.4 Sweeping

Sweep surfaces are obtained by sweeping a planar *profile* curve along a *trajectory* space curve. This is done by locating copies of the profile curve (cross-sections) along the trajectory curve, which requires the correct definition of local reference frames. One convenient and well known frame is the one due to Frenet, defined by the unit tangent T, the principal normal N and the binormal, B. The unit tangent vector is obtained from the first parametric derivative of the curve

$$T = \frac{D_1}{\|D_1\|}$$

The principal normal N is defined in the direction of the curvature, K, that can be obtained from the first and second curve derivatives, D_1 and D_2

$$K = \frac{D_1 \times D_2 \times D_1}{\|D_1\|^4}, \qquad N = \frac{K}{\|K\|}$$

and the binormal is obtained from the cross-product $B = T \times N$.

Although easy to compute analytically, the Frenet frame dependence on the curvature brings some inconveniences: it is undefined at points where the curvature is infinite, that is, in straight line segments, and changes suddenly in direction at inflection points. This last problem is of particular importance when dealing with the location of cross-sections in a surface sweeping process. The system use the Frenet frame only to locate the first cross-section.

The following frames are obtained by computing T_1 at each new location P_1. The new N_1 and B_1 vectors can be obtained by a rotation of the previous frame (T_0 - N_0 - B_0). The rotation axis A is obtained by [8]:

$$A = T_0 \times T_1$$

and the rotation, δ is

$$\delta = \cos^{-1}\left(\frac{T_0 \cdot T_1}{|T_0||T_1|}\right) \qquad (1)$$

If the increment of the position vector is small enough, the orientation of the sections will be the same (Fig. 2) and the twist effect will be reduced.

To locate the cross-sections along the trajectory curve, the problem reduces now to a rotation about

an arbitrary axis A defined by the direction cosines (c_x, c_y, c_z) and the trajectory curve point (x_0, y_0, z_0).

In the CadSHIP system, sweeping was implemented with an optional scale factor that affects continuously the dimensions of each cross-sections. The system also provides a second type of sweeping using two trajectory curves and one profile curve.

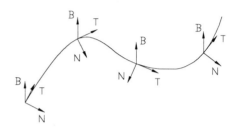

Figure 2. Reference system used for sweeping

3.5 Revolving

The creation of a surface of revolution by revolving a planar section curve around an arbitrary axis can be seen as a particular case of sweeping in which the trajectory curve is a circular arc and so the positioning of the cross-section is always centripetal.

3.6 Blending

Surface patches interpolating four boundary curves are known as blending surfaces. This type of surface are also known as Coons surfaces. To represent them as a tensor-product NURBS surface, a two step process was adopted. First, the two pairs of opposite curves are made compatible as described for lofting, and then the surface can be obtained as a Boolean-sum [9] of three surfaces – one interpolating each pair of opposite curves (S_1 and S_2) and the third one (S3) interpolating the only corner points

$$S(u,v) = S_1(u,v) + S_2(u,v) - S_3(u,v) \qquad (2)$$

The grid control points can then be obtained by

$$C_{i,j} = C_{1_{i,j}} + C_{2_{i,j}} - C_{3_{i,j}}$$

and the knot vectors and orders in each of the parametric directions are the ones obtained for the compatible curves.

3.7 Basic Algorithms

Most operations applied on NURBS entities such as the surface generation methods described before, are based on a set of algorithms that change some entity attributes without changing, or at least producing minimal alterations to the shape. The basic algorithms, that can be implemented in a similar way for both curves and surfaces, are the following:

- **Degree Raising** - The purpose of this algorithm [10] is to increase the degree of the B-spline without changing the shape. It is used mainly to make curves and surfaces compatible.
- **Knot Insertion** - These algorithms allow the introduction of new values in the knot vector without changing the resulting shape. Two general knot insertion algorithms are adopted: the Bohem algorithm inserts one single knot at a time and the other, the Oslo algorithm inserts a single control point [11].
- **Knot Removal** - Is the inverse operation of knot insertion and it removes knots with a minimum change in the resulting shape. The knot removal is an exact operation only when the knots removed were redundant to start with. So, and unlike knot insertion, knot removal is an approximation process, which means that the shape is actually changed to some extent, which should be controlled by a tolerance value. Efficient algorithms for knot removal for curves and surfaces were presented by Tiller [12].

4. CURVE AND SURFACE INTERROGATION

The fairness of the shape is a very important issue of the ship hull design. The mathematical representation allows the use of analysis tools that provide the designer with some measures for evaluating shape quality during the design process. The CadSHIP system provides a set of shape analysis tools that include curvature display for curves and curvature (normal, Gausian and mean) and isophotes display for surfaces. Surface curvatures can be represented as color maps or as iso-contour lines.

4.1 Curve Curvature

The curvature κ of a space curve is positive by definition and is defined by [13]:

$$\kappa(t) = \frac{|x'(t) \times x''(t)|}{|x'(t)|^3} \tag{3}$$

In the particular case of a planar curve, and in order to be able to detect inflection points, a signed curvature may be defined by [13]:

$$\kappa(t) = \frac{\ddot{x}(t)\dot{y}(t) - \ddot{y}(t)\dot{x}(t)}{\left[(\dot{x}(t))^2 + (\dot{y}(t))^2 \right]^{3/2}} \tag{4}$$

To make the interpretation of the analysis results easier, it is normal practice to use graphical representations either by plotting the values of curvature or using the so called "porcupine" representation (Fig. 3).

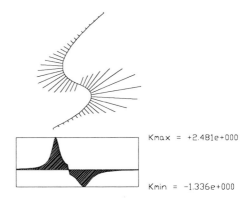

Kmax = +2.481e+000

Kmin = -1.336e+000

Figure 3. Curvature porcupine and plot display

In the curvature plot, the curvature values can be plotted against the parameter value, but as these values depend on the type of parameterisation, a better practice is to plot them against the curve length. On the porcupine representation, the *curvature vectors*, with modulus proportional to the curvature values, are normal to the curve at each point. By convention, they are oriented to the side opposite to the centre of curvature and are plotted directly over the corresponding points on the curve.

4.2 Surface Curvatures

For surfaces several curvatures can be defined. A surface $r(u,v)$ can be determined from two intrinsic quantities called the first and second fundamental forms. The *first fundamental form*, gives the infinitesimal arc length ds between two points (u,v) and $(u+du, v+dv)$, measured in the tangent plane of the surface at (u,v) and is defined by

$$
\begin{aligned}
ds^2 &= r_u \cdot r_u \, du^2 + 2 r_u \cdot r_v \, du dv + r_v \cdot r_v \, dv^2 \\
&= E du^2 + 2F du dv + G dv^2
\end{aligned} \tag{5}
$$

294

where

$$E = r_u.r_u \quad F = r_u.r_v \quad G = r_v.r_v \qquad (6)$$

The *second fundamental form*, gives twice the component of the displacement dh between (u,v) and $(u+du, v+dv)$ perpendicular to the tangent plane at (u,v):

$$dh = Ldu^2 + 2Mdu.dv + Ndv^2 \qquad (7)$$

where

$$L = n.r_u \quad M = n.r_{uv} \quad N = n.r_{vv} \qquad (8)$$

and n is the surface unit normal at (u,v)

$$n = \frac{r_u \times r_v}{|r_u \times r_v|} \quad when \ |r_u \times r_v| \neq 0 \qquad (9)$$

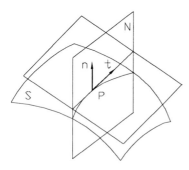

Figure 4. Normal curvature

The *normal curvature* of a surface S in a point P and in a given direction t, is the curvature of the normal section curve, i.e., the curvature of the intersection of the surface with a plane N in that direction containing the normal to the surface at the point (Fig. 4) and is defined [14] by

$$\kappa = -\frac{L(du)^2 + 2Mdudv + N(dv)^2}{E(du)^2 + 2Fdudv + G(dv)^2} \qquad (10)$$

The sign convention used in equation (10) gives a positive curvature when the centre of curvature and the surface normal lie on opposite sides of the surface.

The *mean curvature* H and the *Gaussian curvature* K are defined at each point (u,v) by

$$H = \frac{2FM - (EN + GL)}{2(EG - F^2)}$$

$$K = \frac{LN - M^2}{EG - F^2} \qquad (11)$$

The *principal curvatures* κ_{min} and κ_{max} represent the minimum and maximum values of the curvature at that point

$$\kappa_{min} = H - \sqrt{H^2 - K}$$
$$\kappa_{max} = H + \sqrt{H^2 - K} \qquad (12)$$

When at a certain point $H^2 - K = 0$, then $\kappa_{min} = \kappa_{max}$ and so κ at this point is constant in all directions. Such points are called *umbilic points*. If at a umbilic point $\kappa=0$, then at that point the surface is approximately plane. If $\kappa \neq 0$, then the surface is approximately spherical.

The Gaussian (K), the mean (H) and also the absolute (κ_{abs}) curvatures can be expressed in terms of the principal curvatures

$$K = \kappa_{min}\kappa_{max}$$

$$H = \frac{1}{2}(\kappa_{min} + \kappa_{max}) \qquad (13)$$

$$\kappa_{abs} = |\kappa_{min}| + |\kappa_{max}|$$

If the Gaussian curvature $K = 0$, that means that one of the principal curvatures κ_{min} or κ_{min} is zero. In this case, the surface is *developable*, i.e., can be unrolled into a plane along the principal direction, without stretching or distortion. If $K > 0$ the principal curvatures have the same sign, whether positive, (the surface is convex at the point) or negative (the surface is concave). If $K < 0$, the principal curvatures have opposite signs in the principal directions, which corresponds to a surface with a saddle shape.

4.3 Isophotes

Isophotes are lines of constant light intensity on a surface, created by a parallel light source with a given direction, L. For a surface, an isophote is a line along which the quantity

$$n \cdot L = \cos\alpha \qquad (14)$$

is constant, n being the normal unit vector and α is the angle of incidence, so that $0 \leq \alpha \leq 90$. If the surface is C^r continuous, then the isophotes are C^{r-1} continuous curves.

Isophotes can be used to check the surface continuity across the boundaries of the patches. In the particular case of $n \cdot L = 0$, the corresponding isophotes are called *silhouettes*.

In (Fig. 5) a light direction of $[1.0, 1.0, 1.0]$ generates on surface (A) the isophotes (B) and the silhouette (C).

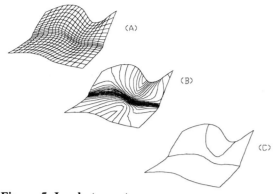

Figure 5. Isophote contours

5. CONCLUDING REMARKS

NURBS have the properties that make them a solid formulation for hull surface modelling applications and the methods described here allow this to be accomplished. Future developments should improve the basic surface modelling methods and some specific tools for ship modelling. In the first case, better methods for increasing control over the surface continuity across patches and improved surface intersection. In the second case, new tools to speed up the creation of certain parametrically definable hull regions such as decks (with sheer and camber), bows, bulbs, and so on.

REFERENCES

1. Fog, Nils G. , "Creative Definition and Fairing of Ship Hulls using a B-Spline Surface", *Computer Aided Design*, Vol.16, No.4, July 1984.

2. Beyer, Klaus-Peter "Direct Curve and Surface Manipulation for Hull Form Design", *Proceedings of Ship Meeting/STAR Symposium*, SNAME, 1988, pp. 247-256.

3. "A Description of the BMT Hull Design Software", British Maritime Technology, 1985

4. Jensen, J. J. and Baatrup, J. "Transformation of Ship Body Plans into a B-Spline Surface", The Technical University of Denmark, *Report No.373*, August 1988.

5. Standersky, Nelson Bianco "The Generation of Ship Hulls with Given Design Parameters Using Tensor Product Surfaces", *Proceedings of Theory and Practice of Geometric Modelling*, University of Tubingen, 1988.

6. Bardis, L. and Vafiadou, M. "Ship Hull Representation with B-spline Surface Patches", *Computer Aided Design*, Vol.24, No.4, 1992, pp. 217-222.

7. Woodward, C. D. "Skinning Techniques for Interactive B-spline Surface Interpolation", *Computer Aided Design*, Vol.20, No.8, 1988, pp. 441-451.

8. Bloomenthal, J. "Modeling the Mighty Maple", Proceedings of SIGGRAPH 1985, *Computer Graphics*, 19, pp. 305-311.

9. Lin, Fenqiang and Hewit, W.T. "Expressing Coons-Gordon Surfaces as NURBs", *Computer Aided Design*, Vol.26, February 1994, pp. 145-155.

10. Cohen, E., Lyche, T. and Schumaker, L. "Algorithms for Degree-Raising of Splines", *ACM Transactions on Graphics*, Vol.4, No.3, July 1985, pp. 171-181.

11. Bohem W. "Inserting New Knots into B-spline Curves", *Computer Aided Design*, Vol.12, No.4, July 1980, pp. 199-201.

12. Tiller, W. "Knot Removal Algorithms for NURBS Curves and Surfaces", *Computer Aided Design*, Vol.24, No.8, 1992, pp. 445-453.

13. Farin, G. *Curves and Surfaces for Computer Aided Geometric Design - A Practical Guide*, Academic Press, 1988.

14. Beck, J. M., Farouki, R. T. and Hinds, J. K. "Surface Analysis Methods", *IEEE Computer Graphics and Applications*, December 1986, pp. 18-36.

15. Rogers, D. F. and Adams, J. A. *Mathematical Elements for Computer Graphics*, MacGraw-Hill, 1990.

16. Lee, E.T.Y. "Choosing Nodes in Parametric Curve Interpolation", *Computer Aided Design*, Vol.21, July/August 1989, pp. 363-370.

17. Piegl, L. "A Menagerie of Rational B-spline Circles", *IEEE Computer Graphics & Applications*, September 1989, pp. 48-56.

APPENDIX – Non-Uniform Rational B-Splines

A rational B-spline curve of order k is defined by the expression [17]:

$$P(t) = \frac{\sum_{i=1}^{n+1} C_i . w_i . N_{i,k}(t)}{\sum_{i=1}^{n+1} w_i . N_{i,k}(t)} \qquad (15)$$

where C_i are the vertices of the control polygon, w_i are the weights and $N_{i,k}(t)$ are the B-spline basis functions of order k defined by the Cox-de Boor recursive expressions:

$$N_{i,1}(t) = 1 \quad if \quad t_i \le t < t_{i+1}$$
$$= 0 \quad otherwise$$

$$N_{i,k}(t) = \frac{t - t_i}{t_{i+k} - t_i} N_{i,k-1}(t) + \qquad (16)$$

$$\frac{t_{i+k} - t}{t_{i+k} - t_{i+1}} N_{i+1,k-1}(t)$$

defined over the knot vector:

$$X = \left\{0,...,0, t_1, t_{s,},..., t_{n+k-1}, t_{n+k},..., t_{n+k}\right\} \qquad (17)$$

A rational B-spline surface is given by the Cartesian product defined by:

$$S(u,v) = \frac{\sum\limits_{i=1}^{n+1} \sum\limits_{j=1}^{m+1} C_{i,j}.w_i.N_{i,k}(u) M_{j,l}(v)}{\sum\limits_{i=1}^{n+1} \sum\limits_{j=1}^{m+1} w_{i,j}.N_{i,k}(u) M_{j,l}(v)} \qquad (18)$$

where $C_{i,j}$ are the vertices of the control grid, $w_{i,j}$ are the weights and $N_{i,k}(u)$ and $M_{j,l}(v)$ are the B-spline basis functions of order k and l, defined as in (16).

The knot vector must be non-decreasing, that is, $t_i \le t_{i+1}$. The knots can be repeated r times, (*multiplicity r*), or be *simple* (multiplicity one). The multiplicity must not be greater than the order of the spline. The knot vectors can be classified as *uniform*, if all the knots have an equal spacing, *periodic* if that spacing is the unity, or as *open non-uniform*, if the knots have multiplicity equal to the order of the spline at the extremities and the internal knots are not necessarily equally spaced, or have multiplicity greater than one. The *centripetal* parameterisation obtained by (19) has proven [16] to give good results, for *p=0.5*.

$$t_1 = 0$$

$$t_i = \frac{\sum\limits_{j=2}^{i} |C_j - C_{j-1}|^p}{\sum\limits_{j=2}^{j=n} |C_j - C_{j-1}|^p} \quad i = 2,3,...,n \qquad (19)$$

For *p=1.0* the normalised accumulated *chord length* between control points is obtained, which is still the parameterisation used in most systems.

If all the control points have *w=1* then the shape of a NURBS coincides with the shape of an integral (non-rational) B-spline. Generally, to ensure the non-negativity of the basis functions, it is assumed that $w_i \ge 0 \qquad i = 1,2,.....,n-1$

When w_{i-1} and w_{i+1} are fixed, for decreasing values of w_i in the interval [0,1] the NURBS is pushed away from the control point while increasing the weight to values $w_i \ge 1$ has the opposite effect. For $w_i = \infty$ the control point will be interpolated. Finally, for $w_i = 0$, the respective control point is ignored.

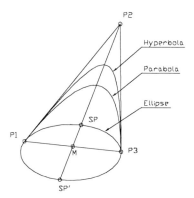

Figure 6. Conics representation by NURBS

NURBS have the capability to represent exactly conical shapes. Considering a curve of degree two with 3 control points as shown in Fig. 6, if it represents a conic, then, although the weights w_1, w_2 and w_3 can have different values, the ratio

$$k_c = \frac{w_1.w_3}{4.w_2^2} \qquad (20)$$

called the *conic shape invariance*, remains constant for each type of conic, as follows:

$$4k_c < 1.0 \Rightarrow ellipse$$
$$4k_c = 1.0 \Rightarrow parabola$$
$$4k_c > 1.0 \Rightarrow hyperbola$$

For circular arcs, not only the condition $k_c>1$ must be fulfilled, but also the triangle $\Delta[P_1P_2P_3]$ must be isosceles. Full circles can be obtained by patching together circular arcs [3].

© 1998 Elsevier Science B.V. All rights reserved.
Practical Design of Ships and Mobile Units
M.W.C. Oosterveld and S.G. Tan, editors.

A New Transformation Method for The Designed Waterline

ZHANG Jun   SHENG Hongcui   CHENG Mingdao

China Ship Scientific Research Center

The paper presents a new method for transforming the designed waterline. It combines the proportional variation along longitudinal direction with the generalized LACKENBY's method. The design demands of designed waterline is satisfied by an iteration process. The present method is not limited to variation of one or two parameters of the designed waterline. It need not look for directly the variation function for transforming transverse section lines ,while the derived new ship form keeps fair.

1. Introduction

With the development of ship form research ,the investigation on the synthetic performances such as ship resistance ,sea-keeping, etc. becomes more and more concerned. Therefore not only the effect of area curve of transverse section on ship performance ,but also that of the designed waterline is studied. So the needs arise naturally to vary the designed waterline independently from the view of ship lines design.

As a basis ship form transformation method, the LACKENBY's method is usually applied in ship lines design to vary block coefficient Cb and longitudinal position of center of buoyancy Lcb, etc. However, in LACKENBY's method ,while shifting the transverse sections along longitudinal direction, the designed waterline is also changed .In other words the LACKENBY's method can not be used to vary the designed waterline independently.

The method which tries to vary directly the each transverse section line in someone form of mathematical function is only applicable in a very small range. The mathematical method based on the draft functions may be used ,but it should first represent the parent ship form at high accuracy, so it is too complicated and not convenient .

To overcome the disadvantage of conventional basis ship form transformation method, the paper develops a new method to vary the designed waterline independently. The variation is defined as :

(1)It must keep the area curve of transverse section unchanged;

(2)Provided the required designed waterline is practicable ,the method is suitable;

(3)The demands for varying the contours of stern and stem are also considered.

2. Transformation method

2.1.Transformation of the designed waterline

Fore- and afterbody are analogous. The paper therefore can consider them separately .The separation is made at the midship section .Suppose the parent ship form and new designed waterline are given, then follows these steps:

(1)Proportional transformation along longitudinal direction :

$$x_1 = x_0$$
$$y_1 = y_0 \cdot y_w(x_0) / y_{w0}(x_0)$$
$$z_1 = z_0 \tag{1}$$

where (x_0, y_0, z_0) is the parent ship form , $y_{w0}(x_0)$ the designed waterline of parent ship form, $y_w(x)$ the new required designed waterline . The new ship form (x_1, y_1, z_1) derived by this step is temporarily called ship 1, the designed waterline of which must be the same as the required. But the area curve of transverse section has been changed from the parent ship form to ship form 1. i.e.

$$A_1(x) = A_0(x) \cdot y_w(x) / y_{w0}(x) \tag{2}$$

Where $A_1(x)$ is the area curve of transverse section of ship 1, $A_0(x)$ the area curve of parent ship .

(2)The generalized LACKENBY's variation

In order to draw the area curve of ship 1 back to that of parent ship, the generalized LACKENBY's method is applied:

$$\begin{cases} y_2 = y_1 \\ z_2 = z_1 \\ x_2 = x_1 + f(x_1) \end{cases} \tag{3}$$

Where (x_2, y_2, z_2) represents a new ship form 2, $f(x)$ is just the longitudinal shift function, it suits:

$$A_0(x_2) = A_1(x_1) \tag{4}$$

In LACKENBY's method, longitudinal shift of transverse section is executed in term of the second order polynomial. But in the paper, it is impossible to shift transverse section in term of a certain form of function given in advance. For this reason , the method is called the generalized Lackenby's method.

In this step ,the designed waterline is also changed while shifting the transverse sections of ship form 1 along longitudinal direction. Therefore the design requires is still not satisfied.

By these two variations ,ship form 2 is derived from the parent ship .The area curve of transverse section keeps unchanged. Although the designed waterline is not the same as the required, but it has been different from that of parent ship .

In order to make the designed waterline to satisfy the design demands and in the meanwhile to keep the area curve of transverse section unchanged , the iteration process is applied. It takes the ship form 2 as the parent ship form, just repeats the above-mentioned steps, until to suit the iteration precision .

The iteration control error may be written as:

$$sum = \left[\frac{1}{n} \sum_{i=1}^{n} (y_w^*(i) - y_w(i))^2 \right]^{\frac{1}{2}} \quad (5)$$

where y_w is offset of the required designed waterline , y_w^* the designed waterline of new ship attained in each iteration step.

By iteration process, the new ship form is derived from the parent ship form. Its designed waterline is approached to the required one, and the area curve of transverse section keeps unchanged.

2.2. Variation of contours of stem and stern

On the other hand, the contours of stem and stern must be transformed while longitudinal shifting transverse sections in the generalized LACKENBY's variation. Besides it, the designer may sometimes expect to vary contour of stem or stern .The paper also presents a method to circumvent the question.

For the convenient of express, the new ship form attained in the foregoing paragraph is temporarily called intermediate ship. The derived ship after varying contours of stem and stern is called designed ship.

This process includes following two steps:
(1)Vertical variation of transverse section:

It transforms the intermediate ship form in vertical direction as follows:

$$z_1 = k \cdot z_0$$
$$y_1 = y_0 \quad (6)$$

where $k = (t - hl)/(t - hl_0)$, t is design draft, hl_0 and hl are keel height of intermediate ship and designed ship, respectively.

(2)Transforming each transverse section in breadth:

$$z_2 = z_1$$
$$y_2 = y_1 + g(z_1) \quad (7)$$

Where $g(z)$ should satisfy the following conditions:

$$\int_{hl0} g(z)dz = (1-k) \cdot s_0$$
$$g(z_1) = yl - yl_0, \quad z_1 = hl$$
$$g(z_1) = 0, \quad z_1 = t \quad (8)$$

Where s_0 is the area of transverse section.

By these two variations ,the designed waterline and area curve of transverse section of intermediate ship are kept unchanged .The design requires of varying contours of stem or stern is satisfied.

3. Application

In order to show the variation ability of present method, the paper gives an example which varies a V-shape ship form to U-shape in great difference. The designed waterline of the parent ship form is represent by mathematical method as:

$$y(x) = \begin{cases} y_p, 0 \le x < x_p \\ \sum_{i=1}^{4} a_i . x^{g(i-1)}, x_p \le x < x_k \\ \sum_{i=5}^{8} a_i . x^{(i-5)}, x_k \le x < x_j \\ \sum_{i=9}^{12} a_i . x^{(i-9)}, x_j \le x \le x_n \end{cases} \quad (9)$$

Where x_p is length of straight part; x_n longitudinal position of end; y_p and y_n are half breadth of shoulder and end, respectively. xk and xj are longitudinal position of divisions ;g is a internal parameter to moderate the form of transition .

The above unknown coefficients $a_i (i = 1, \dots, 12)$ are decided by form parameters of waterline such as area coefficient of waterline Cwl, longitudinal position of geometric center of waterline Xw and half angle of entrance Ie (or run Ir) ,etc. The new waterline can be generated fairly freely by modifying these form parameters .Referring to table 1,the parameters are modified in large amplitude purposely.

In table 1, $Xp_f, Xp_a, Xw_f, Xw_a, Xp$ and Xw are all normalized based on ship length L. The lower script "f" labels fore-body, "a"

afterbody. Cw_f, Cw_a and Cw are waterline area of fore-body 、 after-body and whole body ,respectively.

From the comparison between section lines(fig.1) of the parent ship and designed ship, the form of designed waterline has a big change .Owing to keeping the transverse section area curve unchanged ,the above variation is also a kind of UV-shape change of transverse sections.

It is notable in the example the contours of stem and stern of the designed ship form is the same as the parent ship .It just shows the ability of the method for varying the contours of stem and stern. Because the contours of stem and stern have been changed from the parent ship form to the intermediate ship during ship form transformation.

4.Conclusion

The paper develops a new method for varying the designed waterline, which combines the longitudinal proportional variation with the generalized LACKENBY's method. The design demands of designed waterline is satisfied by iteration process.

Table 1 The comparison between the designed waterlines of parent ship and designed ship

	Xp_f	Cw_f	Xw_f	Ie (°)	Xp_a	Cw_a	Xw_a	Ir (°)	Xp	Cw	Xw
parent ship	0.01	0.3188	0.1801	5.94	0.065	0.3513	0.1866	19.97	0.075	0.6701	-0.012
designed ship	0.01	0.2914	0.1715	4.89	0.030	0.3289	0.1794	16.63	0.040	0.6203	-0.015

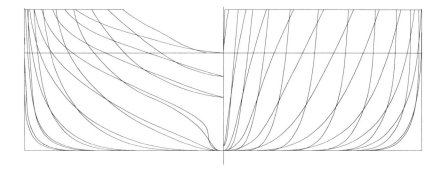

Fig.1 The comparison of ship lines between parent ship(V-shape) and designed ship(U-shape)

The design demands of contours of stem and stern are satisfied by the vertical variation and variation in breadth of transverse section. The area curve of transverse section keeps unchanged.

The proportional variation along longitudinal direction and the generalized LACKENBY's method can all be used to vary the area curve of transverse section and designed waterline .The conjunction of these two variations can make the design waterline change in some extent while keeps the area curve of transverse section unchanged. By iteration process, this kind of change can be accumulated gradually, and finally satisfy the design demands. It is the reason that the present method can be applied to transform the design waterline.

The application indicates :

(1)The method is not limited to variation of one or two parameters of the designed waterline. Provided the required design waterline is practicable, the design demand can be satisfied.

(2)It need not look for directly any form of mathematical function to vary each transverse section ,the derived ship lines is also kept fair;

(3)Not only fore-body, but also after-body, the method can be applied to. Its variation ability is rather good;

(4)The design demands of varying contours of stem and stern can also be satisfied.

Above all, the paper develops a new method to vary the designed waterline .It can satisfy the design demands of designed waterline ,while keeping the area curve of transverse section unchanged. Aided by computer, the present method is almost convenient as the LACKENBY's method. It further richens the contents of basis ship form transformation method .

References

1 H.Lackenby, "ON THE SYSTEMATIC GEOMETRICAL VARIATION OF SHIP FORMS",INA 1950

2 G.Kuiper, "Preliminary Design of Ship Lines by Mathematical Methods", JOURNAL OF SHIP RESEARCH ,March,1970,vol 14.No.1.

Practical Design of Ships and Mobile Units
M.W.C. Oosterveld and S.G. Tan, editors.

Multiple Criteria Design Optimisation of RO-RO Passenger Ferries with Consideration of recently proposed Probabilistic Stability Standards

K. W. Hutchinson[a], P. Sen[b], I. L. Buxton[b] and W. Hills[c]

[a]Armstrong Technology.
Swan Hunter House, Station Road, Wallsend, Tyne & Wear, NE28 6HQ, United Kingdom*

[b]Department of Marine Technology, University of Newcastle upon Tyne.
The University, Armstrong Building, Newcastle upon Tyne, NE1 7RU, United Kingdom*

[c]EPSRC Engineering Design Centre, University of Newcastle upon Tyne.
The University, Armstrong Building, Newcastle upon Tyne, NE1 7RU, United Kingdom

This paper briefly discusses research currently being undertaken to create design solutions for new-build Roll-On / Roll-Off Passenger Ferries with improved survivability. This aspect is being assessed using the probabilistic standard recently proposed by the Joint North-West European Project [1]. Consequently this complex preliminary design problem cannot call on previous experience or legacy information. Therefore the optimisation methodology applied is that of a Genetic Algorithm (GA) which inherently facilitates extensive searches of unknown and potentially multi-modal design spaces. The system utilised is a Multiple Criteria GA which can simultaneously consider a range of potentially conflicting performance criteria. Using the database of optimal designs, solutions can be compared and selected and knowledge in the form of performance boundaries and design lanes derived.

1. INTRODUCTION

In the marine industry, as with others engaged with large made-to-order (MTO) products, design knowledge is almost exclusively built up over time through experience obtained from designing, building and operating similar products with comparable functionality and performance.

There is a potential "knowledge gap" facing naval architects involved in the design of passenger vessels when, as it is widely accepted, the current deterministic regulations are superseded by a harmonised probabilistic standard applicable to both passenger and cargo vessels. The probabilistic approach for the assessment of the standard of survivability provides a more sound basis for the

evaluation and comparison of different design configurations than the traditional deterministic one. Therefore, ship designers, builders and operators, who will have to apply such probabilistic regulations, will need to rapidly gain experience in their application and also an appreciation of their influence on their current design knowledge and practices.

This paper describes a methodology which addresses this deficiency in the area of survivability and subdivision through a critical examination of the current and potential future regulations and their application during the preliminary design phase. The specific class of ship being considered is that of Roll-On / Roll-Off (RO-RO) Passenger Ferries.

The design and optimisation of subdivision arrangements for RO-RO Passenger Ferries is

* This research work is being conducted at the EPSRC Engineering Design Centre (EDC)[c] for Marine and Other Made-To-Order Products, based at the University of Newcastle upon Tyne, under the auspices of Industrial Research Project R6 *"Decision Support in the Evaluation and Selection of Safer RO-RO Ship Designs"*. This project was initially a joint collaboration between Argonautics Maritime Technologies, a cluster of Marine Consultancies, and the Engineering Design Centre and supported by the European Regional Development Fund (ERDF) through North Tyneside Council. The current phase of work is being exclusively funded by Armstrong Technology and the Engineering Design Centre.

304

discussed with particular reference to the concept of local (partial) indices of subdivision together with a recently proposed probabilistic framework. This method incorporates dynamic effects due to wave action, utilising the concept of a boundary stability curve to account for flooding of the RO-RO decks, and other considerations which have tragically resulted in the rapid capsize of a number of vessels.

As the design and optimisation problem is a complex techno-economic one, depending on many potentially conflicting factors, not just survivability, a Multiple Criteria Decision Making (MCDM) approach [2] is adopted using an evolutionary Multiple Objective Decision Making methodology for design synthesis and suitable Multiple Attribute Decision MakingMultiple Attribute Decision Making (MADM) methodologies for selection.

2. PROBLEM: DESIGN OF SAFER RO-RO's

Recent events have illustrated the vulnerability of this ship type to relatively minor and sometimes exceptional incidents of damage and as a result there is considerable pressure to improve their stability and survivability. Therefore, the aim of this research is to investigate new-build solutions and develop optimal or near optimal but robust arrangements for passenger RO-RO ferries. The primary goal is to enhance their standard of survivability after damage using the methods outlined below and assess the effect of these proposed regulations on design practice.

2.1 Current and Future Stability Regulations
There are two approaches to the assessment of survivability - deterministic and probabilistic - and the salient features of both are outlined below as a prelude to examining their influence on design.

2.1.1 Deterministic Approach
Traditionally passenger ships, including those incorporating the concept of RO-RO cargo handling, have been designed on the basis of the deterministic approach of assessing damage stability utilising floodable length curves, factor of subdivision etc. However, for some considerable time it has been appreciated that this approach is fundamentally flawed as many of the factors affecting survivability are random in nature and influence the safety of ships of different types and sizes differently.

As stated in the introduction, damaged stability is currently almost exclusively assessed using the probabilistic approach, neglecting the current Safety of Life at Sea (SOLAS) regulations [3] as even the North West European Agreement [4], which incorporates 50cm of water on the bulkhead deck, could still prove to be inadequate in certain damage scenarios and sea conditions.

2.1.2 Probabilistic Approach
The probabilistic approach to the assessment of a vessel's "safety" after damage was first proposed over forty years ago [5] and for the last quarter of this century the A.265 probabilistic regulations [6] have existed as an "equivalent" to SOLAS for passenger ships. However these have not been widely applied to the design of new vessels because of their complexity and associated computational implications and also the fact that for passenger RO-RO vessels, at least, the present SOLAS 90 deterministic stability standard, not to mention SOLAS 60, probably represent a lower and therefore less onerous safety standard.

In the probabilistic approach survivability is represented by the attained index A which reflects the average degree of subdivision for the whole ship and is therefore the probability of collision survival.

$$A = \sum p_i s_i \text{ for } i \in J$$

where:
p_i - probability of compartment(s) being flooded
s_i - (conditional) probability of survival of flooding
J - all feasible flooding cases, singly or as groups

Both IMO probabilistic regulations [3][6] specify a minimum level for the A index, namely the required subdivision index R, and therefore it follows that $A \geq R$ for all acceptable designs.

The International Maritime Organisation (IMO) recognise that the deterministic approach is flawed and have stated that in the future the assessment of damage stability should be undertaken exclusively using the probabilistic approach, as demonstrated by the introduction, six years ago, of probabilistic rules to assess subdivision and damage stability for cargo ships [3]. Much work is currently being undertaken by and on behalf of the IMO into developing a standardised probabilistic procedure to replace the present deterministic SOLAS one for passenger ships. Such future regulations will harmonise the existing passenger [6] and dry cargo ship [3] ones, and will also embrace other ship types e.g. tankers. It is likely that all ships will have to comply with a common assessment approach but for ship types which are susceptible to certain damages a modification to the damage survival factor, s, will apply. This would be the case with freight and passenger RO-RO's which are susceptible to the effect of entrained water on the large open cargo decks, as discussed below. Alternatively, different ship types could have different R values, recognising the differential in their survivability needs.

2.2 Dynamic Effects and Cases of Minor Damage

Over forty RO-RO vessels have capsized in recent times [7], most due to the dramatic loss of stability from flooding of the normally large and undivided RO-RO cargo / vehicle decks and so highlighted the vulnerability of some existing designs to minor collisions, operational errors or mechanical failure.

2.2.1 Accumulated Water on RO-RO Decks

The speed of response and the relatively small amount of flood water required to induce large angles of heel and capsize together with the general increase in vessel size and capacity (and therefore the potential number of persons involved) has prompted substantial effort into researching the causes and the development of regulations providing a reliable survival criterion.

In addition to static flooding due to damage the dynamic action of waves causes an elevation of the flood water h, as in Figure 1, so inducing heel and trim until inflow and outflow are balanced. Using this, a generalised damage stability criterion (boundary stability curve) [8] was developed [7]:

$$h/Hs = f(f/Hs)$$

where:

Hs - significant wave height for mean critical sea state

Such a curve will obviously be fuzzy rather than distinct due to the random nature of the critical sea state. It can be seen that at this capsize boundary the factors considered dominant were h the depth of water accumulated on the deck (the critical elevation h_{crit}) and f the residual freeboard, as illustrated in Figure 1

Obviously for practical applications any damage stability criterion must be simple and universal. The

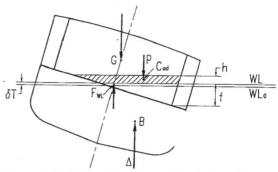

Figure 1. Stability with Water on Deck, from [8]

boundary stability curve is easy to apply using the static equivalent method [8] developed to calculate h_{crit} and it has also been shown [9] not too sensitive to vessel size, subdivision, loading or damage scenario. At that time f was found not to influence the capsize resistance, therefore the following was derived for a universal boundary stability curve [9].

$$h_{crit} = 0.085 Hs_{crit}$$

2.2.2 Minor Damage Factor and Local Indices

The A index, which can be thought of as the global index of subdivision, only reflects the overall safety of a vessel with respect to flooding. It is therefore conceivable that two designs may have the same A index but quite different capabilities for withstanding damages at various positions. It has been found that ships designed to comply with the recently introduced probabilistic cargo ship regulations [3] can sink given certain one compartment damage cases because such minor damages are not accounted for, as they are for passenger ships [6] - but in a purely deterministic way. Therefore a tentative probabilistic approach has been recently proposed [10] together with a minimum value for the minor hull damage survival factor s.

The concept of local (partial) indices of subdivision A_l [11] is also an attempt to address this problem in a probabilistic manner by measuring the distribution of survivability along the length of a ship:

$$A_l = \frac{\sum p_i s_i}{\sum p_i} \text{ for } i \in K$$

where:

K - all feasible flooding cases in a specified zone

This local index of subdivision A_l represents the survivability contribution of all possible combinations involving a given zone; in other words local vulnerability. To ensure that no zone is unacceptably vulnerable a design should ideally produce a uniform or low variation of A_l indices over the length of the ship. This therefore precludes designs from incorporating compartments with excessive lengths and breadths, in a similar way to floodable length curves. Therefore A_l indices are a useful tool in the configuration of subdivision arrangements when applying probabilistic damage stability regulations.

2.3 The Joint North-West European R&D Project

Recently draft probabilistic stability regulations for passenger vessels, known as the NG-Standard (Nordic Group), were proposed at IMO [10]. In this the survival capability, s factor, was directly based on the characteristics of the damaged (residual) righting lever (GZ) curve rather than an approximation based on freeboard and metacentric height (GM) which is the case with the current regulations [6].

This was built on by the Joint North-West European (NWE) Research and Development Project on the Safety of Passenger RO-RO Ships [1]. This framework incorporated, for the first time, dynamic effects due to wave action utilising the concept of the boundary stability curve [8]. The project developed a modified formulation of the damage survival factor s, as did others [9] but in two parts:

$$s_i = sa \cdot sw$$

Firstly, a factor sa based on the static GZ curve, representing the probability to survive pure loss of stability, heeling moments, cargo shift, angle of heel (θ) and progressive flooding [12]. But significantly, also a factor sw, given in Figure 2, which represents the probability to survive water on deck (Figure 1).

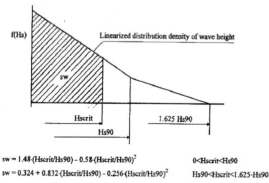

$sw = 1.48 \cdot (\text{Hscrit/Hs90}) - 0.58 \cdot (\text{Hscrit/Hs90})^2 \qquad 0 < \text{Hscrit} < \text{Hs90}$
$sw = 0.324 + 0.832 \cdot (\text{Hscrit/Hs90}) - 0.256 \cdot (\text{Hscrit/Hs90})^2 \qquad \text{Hs90} < \text{Hscrit} < 1.625 \cdot \text{Hs90}$

Figure 2. Calculation of sw factor, figure 7 from [12]

Note that this is only applicable to vessels with large open spaces such as RO-RO decks. Also the formulation of this sw factor is derived from the previous boundary survivability curve, based on theoretical and experimental results for two ships [9]. However recent theoretical work [13], allowing a survival time of one hour for evacuation, proposed:

$$h_{crit} = 0.088 Hs_{crit}^{0.97 + 0.43f}$$

Therefore the distribution in Figure 2 will be subject to modifications as research continues.

The NWE project also proposed new distribution functions for damage extent [14], and hence modified formulations for the damage location a, length (vulnerability) p and penetration (transverse reduction) r factors and the inclusion of a vertical extent of damage (reduction) factor v for horizontal subdivision (not currently [6] incorporated), so that:

$$p_i = f(a \cdot p \cdot r \cdot v)$$

This framework is applied for the assessment of a design's standard of survivability after damage, the attained subdivision index A. Importantly for RO-RO ships, the concept of a capsize index C was introduced for the first time, the total probability of capsize is:

$$C = \sum a_i p_i (1 - s'_i) \text{ for } i \in J$$

where:
s' - probability of survival allowing for θ up to $30°$

Therefore the probability of capsize is $(1 - s'_i)$. This situation is characterised by the development of a large θ and therefore the criterion for capsize can be connected to a maximum θ, tentatively set to $30°$ [12].

Obviously the A index is to be maximised and the C index minimised, therefore increasing the vessels standard of survivability and ensuring that if the vessel sinks it is in a controlled manner represented by the controlled sinking index $S = 1 - A - C$.

As the NWE Framework does not specify a minimum level for the A index that presented in the NG-Standard [10] is used, which is an improvement over the current R index [6] used previously [15].

3. OPTIMISATION MODEL

There are four levels to the developed model (Figure 3 overleaf) as described in Table 1. The first two (0 and 1) are manual operations but the second two (2 and 3) are automated and therefore coded as a computer program under an optimisation shell. In Figure 3, the steps in levels 3 and 4 relate to Figure 4, and the modules in the evaluation model of level 3 to Figure 5. As this design and optimisation problem is a complex techno-economic one depending on many potentially conflicting factors a Multiple Objective Decision Making (MODM) approach is adopted. Design synthesis on levels 2 and 3 is achieved in an evolutionary manner through the utilisation of a Multiple Criteria Genetic Algorithm (MCGA) which creates optimal or near optimal solutions considering a range of design variables, constraints, performance criteria and objectives. Coupling this with a Preliminary Design model, to formulate and evaluate designs, a population of optimal new-build RO-RO passenger ferry design solutions, which lie on the multi-dimensional technical and economic performance surface, is generated.

3.1 Multiple Criteria Genetic Algorithm

A Genetic Algorithm (GA) is a search technique based on Darwinian principles of evolution, such as survival of the fittest, mating and mutation [16]. It is a directed probabilistic search methodology which,

from an initial random population, progressively builds on combining the strengths of pairs of designs over a number of generations to provide a range of optimum or near optimal solutions and hence develops and defines the Pareto surface of efficient solutions. As it requires only a "black box" evaluation mechanism and no measure of gradient, it can tackle problems that traditional optimisation methods, such as hill climbing techniques, cannot deal with. It is therefore inherently capable of avoiding local maxima and minima and consequently facilitates extensive searches of unknown and potentially multi-modal design spaces. Also GAs are generally very efficient in applications with a large number of possible solutions but where each solution takes a relatively modest effort to evolve and evaluate.

The GA being applied to this parametric problem, is a multiple criteria one which can simultaneously consider a range of potentially conflicting performance criteria [17]. This is obviously necessary for the design of a complex and novel marine vehicle to regulations for which no previous experience can be called upon. The MCGA uses a standard GA structure (Figure 4) but with several modifications.

The major difference is that instead of dealing with one criterion as a fitness value, as with a single

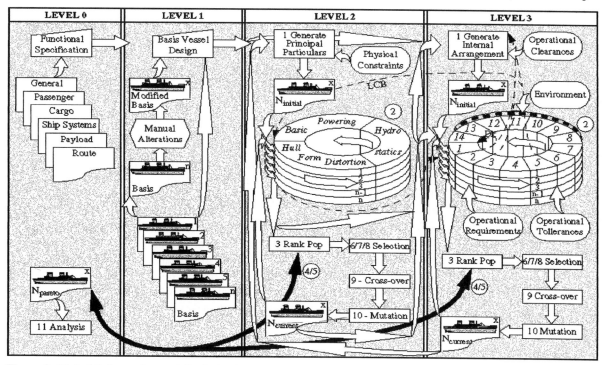

Figure 3. Global Information and Optimisation Model

criteria GA, the MCGA uses the degree of dominance (step 3) to rank each individual after conversion to a fitness value (step 6). Dominance is defined by a better value of at least one criterion without worse values of any. The collection of non-dominated solutions define the Pareto surface. Fitness sharing (step 7) is also undertaken to prevent genetic drift and therefore concentration on one area of the Pareto surface by penalising members of the population that gather in close proximity. Crossover (step 9) is restricted to members within the appropriate "niche" as crossover between diverse individuals can destroy the characteristics of both.

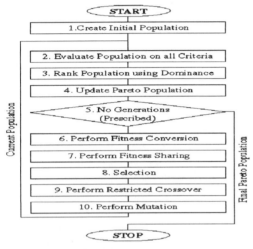

Figure 4. MCGA Search Algorithm (levels 2 and 3)

3.2 Evaluation Model for RO-RO Ferries

The developed evaluation model, shown in figure 4 overleaf, creates balanced working solutions given a number of independent preliminary design parameters as inputs (Table 1) together with their assumed extreme limits.

The model utilises an adaptive design approach and incorporates modules to initialise and assess the basis ship design data, decode the designs string (gene) produced by the MCGA and assign the principal input particulars, undertake the design creation and evaluation calculations (Figure 4 overleaf). Each design is also assessed for compliance with a number of physical constraints on form, size and configuration, such as those given in Table 2, and if non-compliant (i.e. not balanced) penalised. The values for the performance criteria evaluated together with the input variables are then returned to the MCGA. This exercise is carried out for each member

of the population which is then assessed and ranked with respect to the chosen criteria.

If a design is feasible its chance of survival in the population (gene pool) depends on its fitness in terms of its overall performance. The foregoing process is repeated over a prescribed number of generations to generate the final Pareto population for which both performance and input data are available. This population is then analysed and statistics regarding its evolution and make-up calculated. Post processing can then be applied to this population of optimal or efficient design solutions to select designs and also understand the Pareto (trade-off) surface.

Table 1
Optimisation Hierarchy and Variables / Constraints

Level	Description	Action
0	Functional Specification	Platform, Payload, Standards etc., Operational Environment and Restrictions etc.
1	Basis Vessel Design	Select and alter parent vessel manually
2	Principal Particulars and Proportions (Hull Form)	
2.1	Principal Particulars and Ratios are within rigid limits / good practice	If not Penalise or Re-limit
-		Distort Hull Form envelope to suit
-		Evaluate Hull Design
-		Evaluate for Arrangement(s) (3)
-		Repeat for other Particulars (2)
3	Internal Arrangement	Assign Bhd / Dk pos'ns
3.1	Freeboard Assessment	Deficient Penalise or raise Bhd Dk
3.2	Passenger Area, Lane Length / Deadweight, Range	If not within tolerances Penalise
3.3a	Can the design be Trimmed	Trim not within tolerances Penalise or Distort Hull Form (*LCB*)
3.3b	Metacentric Height (GM_T)	Too high / fluctuates Penalise
-		Continue Evaluation of Design
-		Repeat for other Arrangements (3)
-		Repeat for other Particulars (2)

3.3 Computational Implications

Compared to other search and optimisation methods GA's are quite computationally expensive as by their nature they evaluate a large number of solutions most of which are rejected for one reason or another. However for a large design space the number of solutions that have to be created and evaluated to achieve reasonable quality solutions is relatively small. Depending on the size of the Pareto population the computational expense is largely dependent on the evaluation model.

In the case of this problem the most complex and computationally demanding module is that concerned with the damage stability calculations (10 in figure 5).

This is due to the large number of damage cases. Also the deterministic calculations which provide the damaged waterline and associated residual *GZ* curves, required as input for the probabilistic assessment, are computationally demanding and therefore time consuming.

To ease the computational burden the existing passenger ship regulations [6] assume fixed trim. However this assumption does not give a true reflection of the behaviour of a vessel and is fundamentally flawed as there can be a significant deviation between damaged *GZ* curves derived under fixed (artificial) and free (natural) trim conditions particularly at the ends of a vessel. It is not explicitly quoted in the NWE Framework [12] but, because of the above, the calculations assume free trim.

Inevitably this requirement will greatly increase computation time. Also calculations to obtain damaged waterlines are iterative and traditionally consist of integrating the areas of transverse sections longitudinally to obtain the underwater volumes. Obviously this is time consuming and therefore the Krylov-Dargnies Constant Displacement method is applied [18]. This utilises the properties of equi-volume waterplanes and with this technique the derivation of damaged waterlines and stability

characteristics under free trim is largely simplified [19]. As the large number of damage stability cases are independent, parallel computing can be used to further reduce run times. Due to the availability within the Engineering Design Centre (EDC) of high performance multi-processor computers parallel processing (on one machine rather than that of Parallel Virtual Machines (PVM) linked across a network, which is subject to network constraints/traffic and overnight runs) is employed.

3.4 Robust Design

Robust design techniques can also be combined into the optimisation procedure to incorporate operational aspects which involve significant uncertainty [20]. Robust design [21] is a procedure by which designs can be developed such that the influence of uncontrollable (uncertain) parameters is minimised. For the problem in question such parameters include the physical environment, route characteristics and economic factors. Obviously a more robust design is a more flexible one which is less route dependent and therefore more able to operate on alternative routes and will be less sensitive to changes in regulations and the economic climate. These will also reduce its depreciation in value.

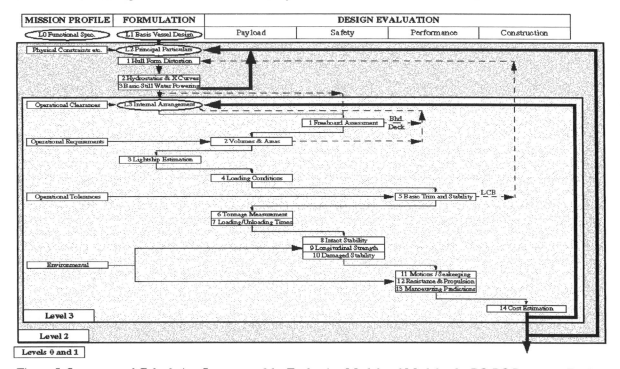

Figure 5. Structure and Calculation Sequence of the Evaluation Model and Modules for RO-RO Passenger Ferries

Table 2
Possible Optimisation Criteria / Constraints

	Criteria	Aim
Design	*Hydrostatics*	*Min./Max. appropriate*
	Stability (KN Curves)	*Maximise*
	Still Water Resistance	*Minimise*
	Lightmass	Minimise
Payload	Numbers of each type of vehicle	Min. deviation from req. Less req. penalise more than excess
	Number of Passengers	Min. deviation from req. Less req. penalise more than excess
	Number of Crew	Min. deviation from req. Less req. penalise more than excess
	Public Space	Max. area per person
	Service Space	Max. area per person
	Fuel Range	Min. deviation from req. Less req. penalise more than excess
	Fresh Water Endurance	Min. deviation from req. Less req. penalise more than excess
	Max. Cargo Deadweight	Max. inc. flexibility, cargo mix
	Tonnage (ICTM 1969)	Min. therefore tolls etc.
	Loading / Unloading	Min. therefore turn around time
Safety	Freeboard (ICLL 1966)	Max. excess of Rule If deficient raise bulkhead deck
	Intact Stability (A.749(18))	Max. excess of Rule If any deficiency then penalise
	Longitudinal Strength	Minimise *BM* and *SF*
	Damage Stability	Max. therefore survivability
	A index (NWE)	Max. excess *R*. Deficient penalise
	C index (NWE)	Minimise
	Local Indices, A_l index	Min variation across trans. zones
	Minor Damage *s* factor	Max excess Rule Deficient penalise
	Water Ballast	Min. therefore max. deadweight
Performance	Op. for Load Conditions	Maximise
	Adequate Trim attainable	Min deviation / head trim penalise
	Metacentric Height (GM_T)	Min deviation (target) / adverse roll
	Deviation in GM_T over conditions	Minimise
	Motions	Min. so max up-time, comfort etc
	Powering	Min. therefore operating costs
	Manoeuvring	Max. therefore up-time etc.
	Turning Diameters	Minimise
	Velocities & Accelerations	Maximise
	Capital (First) Cost	Min. both materials and labour

4. POST PROCESSING

The MCGA optimisation shell in conjunction with the appropriate mathematical generation and evaluation models (levels 2 and 3) facilitate the development, in an evolutionary way, of configurations which provide improved overall vessel safety. This can be applied over an exhaustive design range and for a representative sample of displacement mono hull RO-RO Passenger Ferry types (levels 0 and 1) including inter-island, coastal, short and long haul designs. The outcome of this process is an extensive database of feasible, balanced and optimal designs for the specified combinations of performance criteria, containing both the principal particulars and performance predictions. Using this database of efficient solutions post processing can be undertaken to select designs and distil out design knowledge and identify the interrelationships and attainable performance envelope.

4.1 Reduction of Population for Decision Making

As the Pareto population may be very large it may be prudent to discard some solutions prior to any analysis by limiting criteria, by boundaries or by proximity to target values, therefore restricting the database to a small number of designs. Inevitably the overall fitness of the designs will be altered by the preference judgements made and therefore this process can be undertaken a number of times with different boundaries to explore different subsets of the population.

4.2 Selection of Designs

Evaluations of different design options for any large MTO engineering product often need to take into account many performance attributes so that the economic and technical aspects can be comprehensively assessed. Design selection is therefore achieved using Multiple Attribute Decision Making (MADM) methodologies [22] which facilitate the handling of potentially conflicting design and performance attributes simultaneously reflecting a designer's and / or owner's preferences in a comprehensive and rigorous manner. A whole range of techniques exist [2][22], two appropriate ones [23] are briefly discussed below.

1. Multiple Attribute Utility Theory (MAUT). The utility analysis approach adopts a simple variation of the weighting technique to obtain the overall utility value for each option. The options are then ranked on the basis of the magnitude of their utility values. The Additive Utility (UTA) function method provides a rational way to elicit preference information from the designer (decision maker). This is achieved from the subjective ranking of a subset of designs based on their evaluation criteria values. From this and using Linear Programming and Sensitivity Analysis utility functions (variable weights) can be derived and applied.

2. Evidential Reasoning based Hierarchical Analysis. Some attributes are inherently subjective and it is therefore more natural to evaluate them using judgements with uncertainty. This method which is based on Dempster-Shafer (D-S) theory of evidence combination can deal with problems consisting of both quantitative and qualitative attributes in a robust and rational mannor.

4.3 Knowledge Extraction

The population of developed designs are easily compared with present day vessels of comparable capacity and functionality designed and constructed according to existing rules and regulations. A designer can assess the overall impact on vessel performance not just in terms of safety but a range of performance criteria. Further more, quantitive and subjective assessments can be made as to their impact on vessel design, construction and operation.

After further post processing, a designer can identify and explore the feasible design space thereby gaining an understanding of the multi-dimensional Pareto surface and hence potential performance limits applicable to these potential new survivability requirements. This involves a more scientific and rigorous critical investigation of the relationships between design variables and relevant performance criteria, and requires the application of various multi-variate statistical techniques, such as Cluster Analysis, to undertake analysis of the principal variable and performance components. Techniques such as Neural Networks can also be applied to 'learn' the surface and therefore identify the interrelationships between the variables and criteria. Using these methods design rules and lanes pertaining to the design of RO-RO vessels to new probabilistic regulations with acceptable overall performance characteristics can be formulated.

5. CONCLUDING REMARKS

As the pace of technological advances increases there will be an increasing need for marine and other designers to more frequently embrace potentially novel concepts which are beyond their current knowledge base and sphere of competence.

In particular naval architects who are tasked with the creation of solutions for new or unconventional operational requirements need to rapidly gain experience regarding both feasible and optimum designs. They will also need to quickly acquire an appreciation of performance boundaries compared to and contrasted with their current design knowledge and practices, which may only be applicable to existing vessels or regulations.

It appears inevitable that, early in the next millennium, IMO will introduce harmonised probabilistic damage stability regulations applicable to all ship types to supplant the existing SOLAS ones. Regarding cargo ships this is potentially not too much of a problem as the recently introduced regulations, which did not supersede any existing ones, are probabilistic in nature. In contrast however, passenger ship subdivision still tends to be designed to comply with the existing deterministic rather than the equivalent probabilistic regulations. This paper has presented an approach to addressing what, without doubt, will be an area in which most passenger and RO-RO ship designers will not be able to draw on previous experience or legacy data available either within their organisation or in the public domain. It also provides a rapid and rational method for establishing efficient subdivision arrangements using suitable design criterion.

The methodology in question is of wide applicability. Hence, this evolutionary multiple criteria approach to the design of large MTO products has also been applied to marine vehicles at the less computationally demanding Concept Design stage prior to geometric definition. These studies have concerned Fast Mono Hull RO-RO Ferries and also Surface Warships, namely a Future Escort (FE) for the Royal Navy based on the novel Trimaran Concept utilising Integrated Full Electric Propulsion (IFEP).

ACKNOWLEDGEMENTS

The work reported here is being supported by Armstrong Technology and the Engineering Design Centre. This assistance is gratefully acknowledged as is that previously given by Argonautics Maritime Technologies and its member companies and also North Tyneside Council and Government Office North East with respect to the ERDF award.

REFERENCES

1. "The Safety of Passenger RO-RO Vessels - Presenting the Results of the North-West European Research and Development Project" Int. Seminar, RINA, London, 7 June, 1996.

2. Sen, P. and Yang, Y-B.: Multi-Criteria Decision Support in Engineering Design. Springer, 1998.

3. "Consolidated Text of the International Convention for Safety of Life at Sea, 1974, and its Protocol of 1978" IMO, 1992.

4. Allan, T.: "The new survivability requirements for RoRo passenger vessels from the 1995 SOLAS conference and the 1996 Stockholm regional conferences" RO-RO 96, Lubeck, 21 to 23 May, 1996.

5. Wendel, K.: "Subdivision of ships" Proceedings 1968 Diamond Jubilee International Conference - 75th Anniversary, SNAME, New York, 1968.

6. "Regulations on Subdivision and Stability of Passenger Ships as an equivalent to Part B of Chapter II of the International Convention for the Safety of Life at Sea, 1960" IMCO, 1974.

7. "Ad Hoc Panel on RoRo Safety: Final Report" Technical and Research Symposium Paper, Trans. SNAME, Vol. 104, 1996.

8. Pawlowski M "Outline of the Probabilistic Concept of Ship Subdivision" Workshop on the Practical Application of Probabilistic Subdivision Regulations in Ship Design, RINA, Newcastle upon Tyne, 9 November, 1995.

9. Vassalos, D., Pawlowski, M., and Turan, O.: "Criteria for Survival in Damage Condition" The Safety of Passenger RO-RO Vessels, RINA, 7 June, 1996.

10. SLF 38/5/2 (29-10-1993) and SLF 38/INF.5 (22-11-1993) "Draft Probabilistic Damage Stability Regulations for Passenger Ships" Agenda Item 5, SLF 38, IMO, London, 16 December, 1993.

11. Pawlowski, M. and Sen, P.: "Probabilistic Concept of Ship Subdivision: Theory and Application" Department of Marine Technology, University of Newcastle upon Tyne, 1992.

12. Rusaas, S., Jost, A.E.E. and Francois, C.: "A New Damage Stability Framework based upon Probabilistic Methods" The Safety of Passenger RO-RO Vessels, RINA, London, 7 June, 1996.

13. Vassalos, D., Jasionowski, A., Dodworth, K., Allan, T., Matthewson, B. and Paloyannidis, P.: "Time Based Survival Criteria for RO-RO Vessels" Spring Meetings, RINA, London, 1998.

14. Terndrup Pedersen, P., Friis Hansen, P. and Nielsen, L.P.: "Collision Risk and Damage after Collision" The Safety of Passenger RO-RO Vessels, RINA, London, 7 June, 1996.

15. Svensen, T.E. and Rusaas, S.: "A New Stability Standard for Passenger / RO-RO Vessels" Watertight Integrity and Ship Survivability, RINA, London, 21 and 22 November, 1996.

16. Goldberg, D.: Genetic Algorithms in Search Optimisation in Machine Learning. Addison-Wesley Publishers, 1989.

17. Todd, D.S. and Sen, P.: "The Multiple Criteria Genetic Algorithm" ACTCC, Rutherford Appleton Laboratory, Didcot, 12 March, 1997.

18. Semyonov-Tyan-Shansk, V.: Statics and Dynamics of the Ship: Theory of Buoyancy, Stability and Launching. Peace Publishers, 1963.

19. Pawlowski, M. "Advanced Stability Calculations for a Freely Floating Rig" PRADS '92, University of Newcastle upon Tyne, 17 to 22 May, 1992.

20. Sen, P., Rao, Z. and Wright, P.N.H.: "Multicriteria Robust Optimisation of Engineering Design Systems under Uncertainty" ICED '97, Tampere, 19 to 21 August, 1997.

21. Fowlkes, W.Y. and Creveling, C.M.: Engineering Methods for Robust Product Design. Addison-Wesley Publishing Co., 1995.

22. Sen, P.: "Marine Design: The Multiple Criteria Approach" Trans. RINA, Vol. 134, 1992.

23. Hutchinson, K.W., Byrne, D., Hewitt, A.D. and Sen, P.: "Retrofitting RO-RO Ferries - Selecting Safer Systems" Cruise + Ferry 97, London, 13 to 15 May, 1997.

Is Tonnage Measurement Still Necessary ?

Dipl. Ing. Roman Albert;
Albert Consulting; Jaburgstr 37; 28757 Bremen

Abstract:

Tonnage measurement rules/regulations are maybe one of the oldest aspects in the history of shipping and as a consequence in shipbuilding itself. In 15th century documents were written, which still exist today, giving royals and / or landowners right to levy one part of cargo as tax/duty for using port or rivers for the transport of cargo. This practice has influenced the development in shipbuilding, due the fact, that owner have ordered vessels with certain cargo capacity keeping in mind one part of cargo or corresponding value must be for duty. With the unification of rules, at end of the last century, gross and net tonnage values were born, which was also the end of levy of duty based on cargo volume. This practice today acts as a brake in the technical development of several vessel types. For this reason we must put the question "Is Tonnage Measurement Still Necessary"?.

Introduction:

Today international bodies, such as IMO, national authorities and classification societies are very active in improving safety standards of vessels and in solving environmental threat in such a way as to remove human error. Shipowners and the shipbuilding industry supported these steps. A common consequence of such actions is an increase in the vessels size due to additional space for ballast water or voids. Increases in a ship's volume owing to safety or environmental reasons will not normally be considered under today's tonnage measurement rules. A consequence of this is, that shipowners will be penalised due through a higher gross tonnage value, when they improved safety and environment protection on the vessels.

In many countries we have contradictory situation where two governmental departments, sometimes in same governmental ministry, where one is working to improving safety and environment protection and the other to maximise gross tonnage owing to the vessels consequent increase in size. Although shipowners are ready to accept higher investment costs for new tonnage they will try to avoid the continuous higher levies due to the consequent increase in gross tonnage.

It goes without saying that the current tonnage measurement regulations are the result of continuos improvements to reflect IMO and other safety and environmental rules and regulations. For this reason tonnage measurement regulations have been the subject of radical change especially in the last few years, and today a part of the ballast water capacity or open structure maybe deducted from the gross tonnage value. However tonnage measurement regulations or rules are still not flexible enough to accommodate future developments to improve vessel, safety security of crew or environmental considerations.

Tonnage measurement today only considers "total volume of all enclosed spaces of the ship in cubic meters". Certain flexibility under the old regulations has been deleted in the new. In the past it was also possible to reduce tonnage figures by cutting openings in the main deck, so called tonnage opening or tonnage well on smaller vessels which in today's view results in absolutely unsafe vessels and it is perfectly correct that such measures to so reduce tonnage measurement are no longer permitted. But on the other hand in the past we could deduct the full capacity of ballast tanks and void spaces above the freeboard deck and ballast water volume in the double bottom, if the double bottom had a height in accordance with the tonnage measurement rules which was to the advantage of the vessels and past rules.

As already stated tonnage measurement today is based on the total enclosed volume of the vessel, this premise leads to very strange calculation, as per example. Hatch cover volume will be incorporated in the gross tonnage, if the hatch covers are open to the holds. If they are closed on the underside as well as on top, their volume will not be included in the gross tonnage calculation. Such practices in the calculation of the gross tonnage value do not of course meet with the agreement of shipowners who naturally try to reduce the figure to a minimum in order to lower costs and to obtain other benefits such as a reduction in the number of crew, and deletion of certain equipment, etc.. This conflict of interest which arose in past has resulted today in vessel designs which are on the limits of safety and hazardous to the environment.

For this reason we must today put the question are tonnage measurement regulations necessary or not? If in today's form tonnage measurement regulation act as brake for development of new vessels type which have more safety on the one hand and on other hand less risk to the environment we must have tonnage measurement in this form stopped. Obviously duties, taxes, levies, must be paid – port fees, - supervision cost of classification society or authorities, - transition for canals and locks, - national taxes, - crewing cost (number of crew), - etc., but could this not be based on other tonnage regulations which are more flexible to support such development or must another form of vessel's parameter be looked for or should we again take the original idea and use cargo volume as a basis for calculation of duties and other levies, which give shipowners and not least the designer more freedom to develop vessels without gambling with safety and environment.

To clarify this problem further we should consider some specific vessel types in which today we may see the influence of the current tonnage regulation on safety or environment or both. Due to the fact to that the value of gross tonnage has more influence on smaller vessel than larger in the following examples we will concentrated on the former.

Container vessel

Typical negative influences of tonnage measurement first appeared during the development of container vessel. Cargo is no longer stowed directly in the vessel itself in purposed designed spaces. Cargo is packed in the container somewhere in the countryside and the container stowed in the vessel later, a position specially developed. This place "container cell" or "container stack" is below deck and in comparison with a general cargo vessel cargo hold there may only be 30% to 40% of the full container capacity utilised. This is not because the cargo officer has been negligent, but because it was not physically possible to put more container in the hold. Tonnage measurement has not considered this new development and the tonnage of a container vessel is the same as that for a general cargo vessel of the same size. Also the relationship between gross and net tonnage is the same as that for a general cargo vessel.

In particular in Germany where many rules and regulations regarding design and equipment, but most importantly crewing regulations are based on gross tonnage. It is important for the shipowner of smaller vessels to keep the value of gross tonnage as low as possible. This resulted in the development of a fleet of feeder vessels below 1000 BRT, below 1500 BRT or below 6000 BRT, which have twice the number of containers (ie. paying cargo) on deck than below

deck together with an absolute minimum of freeboard and possibly also tonnage opening or tonnage well in the main deck. It is perhaps only due the fact that German stability regulations are more stringent than those of SOLAS that a minimum of stability related accidents have occurred. Those vessels were developed to a minimum safety standard. Due to the very high numbers of container on deck, the lashing of containers was reduced to a minimum which to be carried out by the crew. In practical it was not possible to lash all deck containers between Hull and Rotterdam owing to lack of time, in particulars if the containers were stacked six high.

The solution to this problem is probably the open top container vessel which in comparison with a vessel with hatch covers suffers large disadvantages in gross tonnage.

To improve understanding here is an example of Container Feeder Vessels with same nominal carrying capacity of 1110 TEU containers and with same deadweight of 13350 t. Farther to improve comparison the main dimensions of the vessel length overall of 147,87 m; length between perpendicular is 139,0 m and breadth of 23,1 m are the same.

Vessel Type	Full Scantlings	With Freeboard	Open Top
Depth of Main Deck (H) m	11,6	11,6	18,2
Freeboard Deck (Hf) m	11,6	9,1	9,1
H immerse in Water; Angel)°	14,1	17,2	45,0
BRZ	10300	10300	13000
BRT (Oslo)	7250	5999	*)

Note: *) as irrelevant not calculated

All three vessels fulfil the damage stability requirements according to SOLAS and accordingly they could have minimum MG of 15 cm (IMO) if other intact stability conditions of SOLAS are fulfilled. But open top container could sail with MG of 15 cm only, because of very large reserve buoyancy above freeboard deck. This is also evident with comparison of immersion angle of main deck. However due to the very high gross tonnage figure shipowners have little interest in this type of vessel in spite of its many advantages as follow:

- In loading condition vessel could carry more containers due to the lower MG
- Due to the lower MG the movement of a vessel in a seaway is more gentle; resulting in less cargo damage
- More safety for crew or waterside workers due to the deletion of lashing, because it is not necessary, the containers are in cell guides.

Of course these vessels will be more expensive, about 5% and shipowners are ready to invest for the technical advantage, but they are not willing to pay a continuous penalty for the larger gross tonnage.

From the shipowners point of view gross tonnage is the same both for a full scantling vessel or a vessel with freeboard. Although today shipowners give preference to a full scantling vessel as they can obtain higher deadweight and with higher deadweight although they can carry more containers.

The small stimulus to build safer vessels to the shipowners give by the Oslo tonnage regulation disappears with the new London tonnage measurement regulation as maybe seen from the above table.

Ferry boat / Passenger vessel

All studies subsequent to tragical accidents involving ferries or passenger ships have shown that in each case had they had some additional buoyancy below the freeboard deck, they would have remained afloat after the incident. With today´s knowledge we could built such safe ship which, would have enough additional buoyancy that in the event of a severe collision, or water entering the (via the bowdoor) car deck, or grounding the ship could still survive.

This would result in an increase in gross tonnage of about 25% for a vessel with the same pay load capacity. This as the spaces below the freeboard deck are increased the spaces above the freeboard deck and will be corresponding larger. Of course investment costs will be also higher but not in proportion as these additional spaces are void spaces (that mean steel without large outfit).

The importance of additional safety measures on passenger vessels of though not keeping gross tonnage as low as possible, maybe seen from the following example.

In 1982 MS "Europa" was built with the following dimensions Loa/Lpp x B x H till 9th deck = 199,92/170,50 x 30 x 25,98; Passenger number 650; BRT 34500 which are very high figure in comparison to vessel size and number of passenger. In those days, the vessel was one of the largest passenger ships if we compare gross tonnage.

The reason for this is as follows:

The shipowner's and shipyard's idea was to build a very safe vessel. For this reason it was decided that it should have a double bottom height of 2,0 m, but according to the tonnage measurement regulations deductible double bottom height was 1,85 m, any height above this could not be set against a claim to reduce gross tonnage volume, in a addition you could not claim also double bottom part below 1,85 m. The consequence was higher gross tonnage. The shipowner and shipyard had very heated discussions with the authorities to reduce correspond gross tonnage but without success.

Later during service, MS "EUROPA" was grounded near Greenland and without help came free and returned to Bremenhaven to dry dock. After inspection it was found that the double bottom was damaged in parts up to a height of 90%. After preliminary repairs the vessel returned to service for about three months, until preparations for final repair were completed.

If the vessel had had a double bottom height in accordance to tonnage measurement rules the tank top would have been penetrated and two watertight compartments flooded. This would have meant that the vessel could not have come free without assistance and would have been laid up till full repair work was completed in Bremenhaven for about 3 months.

This practical example shows the importance of having the correct depth and size of spaces which protect the vessel's vital areas.

Today's tonnage measurement rules have no alternative to deduct volume of spaces in a vessel, even though human lives maybe saved We must put the question: is this correct?

Tanker

In 1980 according to the MARPOL rules/regulations and amendments concerning prevention of pollution by tankers a new era started for shipowners and shipyards. Due to the fact that these rules were not to increase safety for crew on a vessel. These rules were for the protection of the environment.

In connection with tonnage measurement we must concentrate on the rules concerning segregated water ballast. At first the size of vessel which must have segregated water ballast was reduce from 40000 t deadweight to a lower value. Irrespective of the size of vessel, the volume of segregated water ballast was about 30% to 40% of the cargo tank volume and the corresponding tanker size was increased accordingly and of course gross tonnage.

It is odd that the tonnage measurement regulations did not encourage through the deduction of segregated water ballast tank volumes commencing with the introduction of MARPOL rules for environmental protection. Not until

318

1993 did an IMO paper appear where gross tonnage could be reduced by the volume of segregated water ballast, but even then only partly.

It is interesting that the MARPOL rules coming into force was accepted by shipowners as all tanker shipowners were involved in the same problem.

Bulkcarrier

During the last PRADS meeting in Seoul the advantage of double skin for bulkcarrier was discussed and it was expected that many such type vessels would be built. But very few have been built according to my knowledge.

Despite the statistical evidence of classification societies regarding the total losses of bulkcarriers older than 15 years arising from the collapse of side framing or bulkheads, shipowners are not prepared to build double skin bulkcarriers.

One of the reasons why bulkcarriers with box shape holds have not been built is increase in gross tonnage. In particular vessels below 25000 dwt are influenced by increased tonnage measurement arising from the double skin and this is main reason why shipowners avoid building double skin vessels. Only shipowners with the intention of carrying cargoes such as timber products and paper are interested at the moment in building double skin vessels.

Perhaps through changed tonnage regulations classification societies or insurance companies could influence shipowners to build more double skin

bulkcarriers to improved structural safety standards.

Conclusion

With these few examples covering several types of vessel we have tried to show the negative influence of tonnage measurement upon the development of various improved types of vessel, having enhanced safety for crew, increased structural safety and reduce threat of environmental pollution. For this reason tonnage measurement rules must be change and change in such a way as to be in harmony with current technical development and to permit further development in future.

Perhaps one solution would to be deduct from the total volume spaces as void spaces, cofferdams and ballast tanks.

If such a solution is not acceptable we must find another vessel dimension or dimensions which maybe used as a basis for assessment. In this connection we must questions is it necessary to have gross tonnage as a parameter for payment of duties/royalties/taxes/levies for:

- usage of harbour? It is not necessary because cost of harbour will be paid by length of quay and depth of water in front of berth. For this reason it would be logical to pay by **L x T**

- supervision and approval of classification society or authorities? No! due to the fact that each design office works on man-hour cost basis for design work and inspection a cost estimation based man-hour would be more justifiasble.

- transit through locks? Payment based on **LxBxT** would be better, answer due to the fact that lock cost would be based of the dimension locks.

- usage of canal? Canal cost will be taking in account breadth and depth of canal for this reason it would logical to use **BxT**.

- payment of insurance fee? It would be logical for insurance of vessel to be based on prices of the vessel on delivery and for cargo insurance deadweight best reflecting payload.

- payment of tax? Deadweight as earning value for shipowner would be best answer.

- pilot fee? Fee based on **LxBxT** would be logical answer.

- cost for tugs? If this cost was not incorporated in pilot fee, fee based on **LxBxH** to be the correct way.

- number of crew? Some countries base the number of crew in accordance with gross tonnage which is difficult to understand. From the above table of container vessel according to the rules the vessel with full scantling must have 4 crew members more than the vessel with freeboard curing to the change of gross tonnage. For this reason it is preferable to take **L** as a basis for determining.

- limits to regulation? Some countries required if the gross tonnage is above a certain limit than a swimming pool or other equipment for crew must be provided. Such a requirement should be based on the number of crew and not on gross tonnage.

As maybe seen from the above the current tonnage measurement could be replaced a more logical basis for payment of duties found.

One thing is certain in discussions with shipowner today they are ready to accept an improved vessel with greater safety and increased environmental protection, they will also accept higher investment and higher gross tonnage. However no shipowner will do this alone, owing to competition in shipping which is always present. All shipowners try to keep costs as low as possible therefore without IMO, governmental, insurance or classification society pressure no shipowner can be expected to opt for an improved vessel in respect of safety of life or the environment when paying a continuous levy for this step. Perhaps the deletion or modification of tonnage measurements is one way out of this dilemma.

References:

1. E. Vossnack n.a., C.T.Buys: Fatal Influence of Gross Tonnage; TU Delft
2. Tonnage Measurement Rules 1899
3. Tonnage Measurement Rules; Oslo
4. Tonnage Measurement Rules; London

PRODUCTION

Product Modelling for Design and Approval in Shipbuilding

U. Rabien and U. Langbecker[*]

Ship Data Processing Dept., Germanischer Lloyd,
Vorsetzen 32, D-20459 Hamburg , Germany

Co-operation between shipyard or design office and classification society is facilitated by exchange and sharing of geometry and functional models instead of paper drawings. Computer supported approval techniques - including rule scantlings and direct calculations - are taking advantage of the 3D data models generated by various CAD systems. The calculation results can be returned to the design system to support design decisions.

STEP technology is shown to be a solid basis for the definition ship product models as well as for their implementation. Most relevant in the context of class approval are the data models for *Ship Moulded Forms* and *Steel Structures* developed within ISO as well as for *Hull Cross Section Approval Data*, developed by members of the European Marine STEP Association.

1. INTRODUCTION

Today ship designs are submitted almost exclusively as paper drawings for approval by the classification society. This is due to the capabilities of the design systems, the traditional way of ship design, and also to approval and redlining procedures, including archiving and legal aspects. In shipbuilding, modelling usually was associated with CAD or CAE. However, due to years of experience with the large variety of application software systems, product modelling has become a discipline of its own. Since in today's engineering practice 3D models are used for design and related work, the increasing exchange of such information will also affect approval techniques. Often a design is not submitted in a single step, rather than changes and updates are added to the original documents on request. This requires continued communication and co-operation between design office and classification society leading to an approved design.

Data modelling techniques provide a means not only for the product definition but also for all kinds of information related to approval procedures, results and documents.

2. CURRENT SITUATION

In ship design and construction normally a model is developed within CAD or CAE systems. However, these systems cover only certain phases of the production process. Exchange and conversion of information between them is difficult and time consuming for two major reasons. At first, the scope of the transferred data is closely related to the capabilities and the modelling philosophy of the originating system and therefore variable. The second point is, that for accessing, extracting, editing and converting of data each user has to develop his own tools without any chance of reuse. Such interfaces are not easy to maintain due to changing software and hardware environments.

2.1. Electronic documents and drawings

There are some international as well as de-facto standards for exchange of drawings and documents. While, for written documents, orders, invoices, piece part descriptions, etc. the EDIFACT standard is widely used, the de-facto standard DXF directly supports exchange of electronic drawings, which allows processing of the received data by appropriate applications.

---

[*] {rbllgb}@hamburg.germanlloyd.de

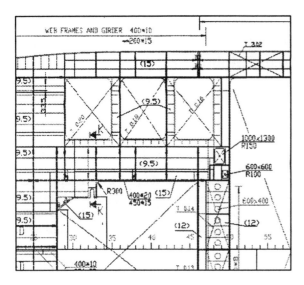

Figure 1. Interpretation of electronic drawings

However, there is not always a common under-standing of object representations and meaning of some attributes; e.g., colours and line styles may be interpreted ambiguously (Figure 1). To solve these problems, a German maritime inter-est group has been set up. It works on a proposal for commonly agreed semantics of symbols and styles in DXF.

2.2. Geometry models

For ship design approval by direct calcula-tion, Finite Element Models are generated from drawings, from 3D geometry or from both. Most commercial CAD systems are based on geomet-ric kernels as Parasolid or ACIS. Therefore di-rect data exchange between different systems would be possible if both are using the same kernel. This also applies, if different installa-tions of the same system are involved in the exchange. Things become more difficult when data shall be transferred between different sys-tems, e.g., between NAPA and TRIBON, al-though this is relevant at bigger shipyards or if subcontractors are involved. For 3D geometry data, there are a few standards, such as VDAFS and IGES already in use, but problems may occur due to missing capabilities, loose toler-ances or mismatching conformance classes. Fur-ther, these models are not suited to transfer results of strength calculation, vibration and noise propagation etc.

2.3. Bilateral Interfaces

Bilateral interfaces are the traditional way of exchanging information between different appli-cation programs. An example is the very popu-lar link between NAPA hull fairing and TRI-BON HULL. Interfaces become more difficult as soon as the number of partners and systems is growing. This is critical for a classification soci-ety since they have to communicate with many customers using a wide range of systems.

2.4. Data sharing concepts

Another way of communicating between in-dependent application systems is the concept of distributed object models, which has become popular as CORBA (Common Object Request Broker) and DCOM (Distributed Component Object Model). The underlying software archi-tecture provides client-server communication. It uses object classes and an Interface Definition Language (IDL) to make data structures, meth-ods, and formats understandable to applica-tions. The interfaces defined in IDL can be un-derstood as a contract between communicating software modules [1]. The technology that sup-ports both data sharing and data exchange can be applied across the Internet as well as a local environment. The need for common semantic objects shows a clear link to product models.

3. SHIPBUILDING PRODUCT MODELS

Product data models cover a variety of prop-erties that go far beyond the scope of pure ge-ometry. In particular, they may contain addi-tional information on

- topological relations (associative and adja-cent elements, holes, compartments),
- functional descriptions (inner bottom, side shell, knuckle lines, flat of bottom),
- calculation results, stresses, design loads,
- material (weights, properties),
- manufacturing aspects (welding, seams), and
- administrative information (version man-agement, approval information).

In practical exchange, the receiver has to know the structure of the model to make any use of the data. This structure must be de-scribed in terms of data elements and their relations. A common approach to formally represent product models is provided by data description language EXPRESS [2], as part of the International *Standard for Exchange and Sharing of Product Data* ISO 10303 - better known as STEP. Based on this technology, various data models are currently available, while others are still under development. Some of them are part of the standard, some are not officially standardised. The usage of product model in daily work of a classification society requires

- obtaining a product model data from the shipyard or design office already during early design,
- transferring data from various CAD systems to in-house formats and vice versa,
- setting up software tools for viewing, updating, and completing of data as well as for consistency checks, and
- accumulating information emerging from different phases of the ship's life cycle, especially from design, production and operation.

Since product models permit permanent and consistent storage of data independent of a particular application format, they may serve as neutral data base for analysis, class approval and other applications.

3.1. STEP Data Models

The STEP standard consists, among other important parts, of a suite of domain specific models called *Application Protocols* (APs). In the shipbuilding domain, there are 5 APs currently under development:

- Ship Arrangements (AP 215),
- Ship Moulded Forms (AP 216),
- Ship Piping (AP 217),
- Ship Structures (AP 218), and
- Ship Mechanical Systems (AP 226).

3.1.1. AP 216 - Hull Form Description

The *Ship Moulded Forms* model [3] was developed with special focus on preliminary design. The model describes the moulded hull form, major inner structures (decks, bulkheads), and appendages of a ship. It covers a set of major design parameter, general characteristics and also hydrostatic properties. It is applicable to commercial ships and navy ships as well as to standard monohull and multi-hull ships. Each moulded form element may be individually specified, but any number of elements may be collected into a group - called region. Any element has positioning information and geometric shape representations in form of offset tables, wireframes or surfaces. If naval architectural terms (waterline, flat of side, etc.) have been additionally assigned to an element, those may be associated to geometry by references.

AP 216 was already used successfully for several purposes. For instance, moulded geometry was visualised to examine a design and to perform consistency checks, see Figure 2. STEP based viewers as well as converters to the virtual reality format VRML were employed for that purpose. An implementation of AP 216 is also used to transfer hull forms into a CFD panel generation module [4] and for data conversion into, e.g., VDAFS and in-house formats. The interface to POSEIDON scantling package will be discussed later.

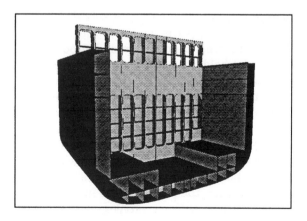

Figure 2: Hull and inner structures extracted from a CAD system using AP 216

Hydrostatic properties which also belong to the scope of this AP, are particularly interesting to classification societies, to exchange stability calculations, to transfer **hydro-mechanic results,** or to receive measurement data from towing tanks.

3.1.2. AP 218 - Steel Structures

As shown above, the exchange of idealised - and sometimes simplified - moulded forms is useful during the early design stage. However, in the classification process the level of detail required is extensive. One needs, for instance, the description of stiffeners, plating arrangement and plate thickness as input for doing the scantlings.

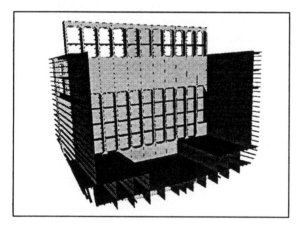

Figure 3: Inner structures and profiles can be represented by AP218

The *Ship Structures* model [5] was developed to support mainly the activities and applications concerning main design and production. The model holds not only definitions for general characteristics, but also designations, rules, regulations and standards applicable to a ship. External objects, documents, standards, rules, and regulations may be referenced from any data element. All parts have their own shape representation, which is given either by a wireframe or a surfaces model. Further, the AP 218 scope includes

- hull plating, stiffener profiles, definition of features and holes
- product structure in terms of systems and assemblies

- cargo assignments and deck load for the purpose of design and design approval
- design loads, shear forces bending moments for determination of longitudinal strength
- transverse cross sections for approval of strength
- welding, material properties, and weights
- configuration management including versioning and approval data

3.2. Industrial Business Cases

Due to the complexity of the APs, first partial implementations are currently underway, and they are not yet complete. These models are not exactly the STEP APs, they are rather subsets selected according to industry needs to validate methodology by implementations.

3.2.1. European Maritime STEP Association

To assist the standardisation process of STEP as well as to support its industrial deployment, the *European Maritime STEP Association* (EMSA, [6]) was formed in 1994. Today it has some 20 members, including major European shipyards, classification societies, research institutes and model basins. Areas of common interest are being identified as EMSA business cases, describing scenarios in which the usage of data models is considered most productive or urgently needed. Due to priorities EMSA is currently looking into four business cases, but more have already been drafted and will be launched as soon as resources become available.

3.2.2. Hull Cross Section Approval Data

During early design shipyards and design offices submit typical cross sections of a ship to a classification society for approval. After evaluation the design is either approved, conditionally approved or rejected and the approval results are returned. This may include notes, comments, red-markings and calculations. Analysis packages for, e.g., rule scantlings and direct calculations are used as key applications during approval, complementary to the traditional redlining of paper drawings. They need detailed descriptions of the cross sections in question as input. Today these cross sections are usually collections of two dimensional descriptions of

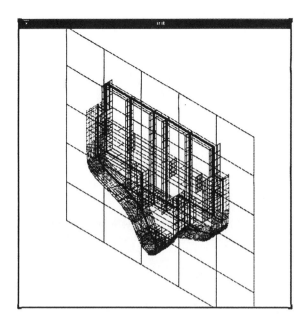

Figure 4. Hull cross section generated from AP 216 and AP 218 data

structural parts, see Figure 4. It was decided by EMSA to start with this implementation as a subset of AP 218, specifically customised to the needs of class approval. After complete implementation 3D models will be processed.

The advantage over pure drawings becomes obvious since they are readily available for a number of software applications

The hull cross section data model [7, 8] is based on unified company requirements in EMSA. Current development and implementation are undertaken jointly by EMSA and the Seasprite project. The model covers all information necessary for or produced by design assessment according to longitudinal strength as there are:

- general characteristics, functional aspects and material properties of structural elements
- groups of plates, stiffeners or other longitudinal elements contributing to the ship structure at the considered hull cross section
- transverse structural elements
- compartments necessary to define tank loads and deck loads
- design loads
- topological relations between elements

- frame positions in the ship coordinate system
- hull class notations
- symmetry information for hull cross sections and functional elements
- shape representation of structural elements (surface representation or 3D curves)

The model may be used to describe and transfer any number of individual cross sections. Major parts were already implemented and successfully tested [9].

4. SUPPORT OF DESIGN APPROVAL

Design approval increasingly makes use of software tools both for Rule Scantling and for Rational Structural Design by direct calculations. The POSEIDON system, eveloped by Germanischer Lloyd, is supporting these tasks.

4.1. Evaluation of structural design using POSEIDON

POSEIDON was originally developed for initial design and analysis of ship structures. Only a minimal set of information is necessary to start modelling and sizing, e.g., of a midship section. The package integrates tools for different tasks under an easy to use graphical user interface [10]. Its latest version supports also rational design techniques and safety assessment of the ship structure over the whole lifetime. POSEIDON performs not only checks for compliance with classification requirements, but also supports alternative design decisions. Due to its specific view on the data model for compliance with classification requirements POSEIDON often needs information than cannot be provided by CAD systems, e.g. topology and material. Figure 5 shows topological relations in a cross section model, which was derived from a 3D model generated by the NAPA system (see Figure 2) and transferred to POSEIDON via a direct interface.

328

POSEIDON introduces the notion of *functional elements* to represent aggregations of components which fulfil a particular function. Changing the geometry of one functional element causes an automatic update of all depended ones.

Creation of a new hull cross section is assisted by wizards providing predefined skeletons for various ship types. Due to the topological definitions the structural skeletons can easily be adjusted to the current needs by changing a few parameters, see Figure 7 and [11].

Then, from all cross sections a model is created, which can be used also to generate finite element meshes automatically for rational structural design by direct calculation, see Figure 6. Modules for assessment of buckling, fatigue, and structural response using the partial safety factors concept are incorporated into the system.

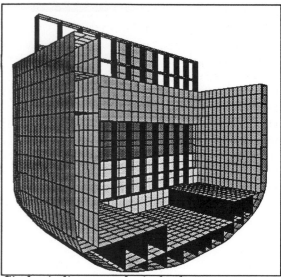

Circles indicate topological relations
Figure 6: FEA model generated by POSEIDON

4.2. Interfaces to Design Systems

One possible approach of interfacing is the direct interchange of information between CAD system and POSEIDON. Data from different

CAD systems related to the various stages of design at the shipyard are used as input for POSEIDON. Interfaces to CAD systems NAPA, TRIBON, and E4 are under development or already available. It is intended to return re-

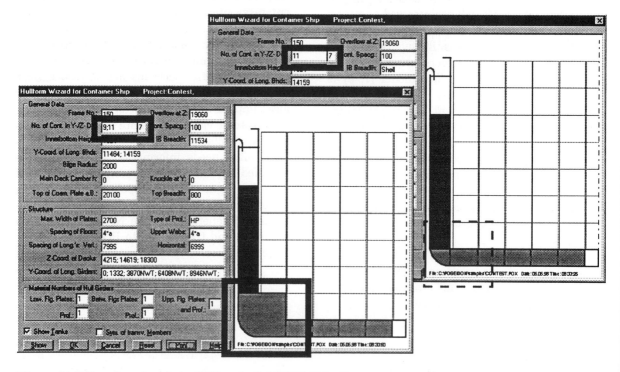

Figure 7: Wizard assisted modelling in POSEIDON. Changing the number of containers leads to modified layout of the bottom tank.

sults of evaluations and calculations in POSEI-DON to the respective CAD system.

In parallel to this practice, CAD vendors and classification societies work on a neutral data model described in section 3.2.2. Thus future communication will be based on a standardised Hull Cross Section Model rather than on bilateral interfaces. A schematic illustration is given in Figure 8, showing STEP product model data as well as specialised data structures in the data storage. POSEIDON will be able to work on both types of information.

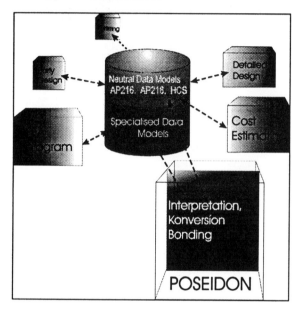

Figure 8: Schematic view on the implementation of a product model environment

5. IMPLEMENTATION

The validation of data models can be achieved by their pilot implementation. This is one of the primary goals of EMSA in order to increase industrial acceptance and to prepare for usage in a production environment. Sample applications of a company-internal implementation have been presented in the previous chapters. However, developments have not yet reached a final status.

5.1. Toolkit Support
Assuming, somebody has got a data model which is formally described in EXPRESS, then he may want to use this model to write an interface to a particular system. There are a number of commercial toolkits available to generate a computer-internal representation of this model (data repository) and possibly source code (Application Programming Interface) to access and manipulate data instances of that model. This so-called *Standard Data Access Interface* (SDAI) is already available for C and C++, and further language bindings are under development for Java and CORBA/IDL.

The tools provide functionality to manipulate, i.e. to create, edit, extract, change, correct and visualise data elements a any product model.

5.2. Security and Legal Aspects
The transfer and processing of electronic information leads to security questions and related legal considerations. Encryption, secure communication channels, and authentication techniques including electronic signature and Trust Centers for key certification have to be introduced into daily practice. The relevant authorities and bodies have to recognise the new forms of approval, and finally, easy tools and mechanisms must be made available to support such mechanisms and transactions. Only some of these problems are already solved in the shipbuilding community. Property rights concerning data, location of master data, and conditions for accessing data, e.g., archived at a classification society, must be clarified.

6. CONCLUSION

In the process of approval of a ship design the digital product representation is a key element to increase productivity, to shorten response times and to improve quality of information exchange and related services. To overcome the shortcomings of currently used exchange formats, the international standard STEP has been developed for various industrial domains. By now, the shipbuilding APs have reached a stage of development, which permits practical implementations. First pilot implementations of AP

216 Ship Moulded Forms and AP 218 Ship Structures were successfully tested among EMSA members.

STEP-based product data models offer new potentials for engineering information exchange and cooperation in the maritime sector with respect to cost effectiveness, time savings, and quality. By providing not only the product definition but all kinds of information related to approval procedures, calculation results and documents, they contribute to the introduction of computerised 3D approval methods. The POSEIDON software package supports this tendency, combing a number of advanced application modules.

REFERENCES

1. P. E. Chung et al., *DCOM and CORBA Side by Side, Step by Step, and Layer by Layer*, www.bell-labs.com/~emerald/dcom_corba/ Paper.html

2. ISO 10303-11 Industrial automation systems and integration - Product Data Representation and Exchange - Part 11: Description method: EXPRESS language reference manual, TC184/SC4/,1996.

3. ISO 10303-216 Industrial automation systems and integration - Product Data Representation and Exchange - Part 216: Application Protocol: *Ship Moulded Forms*, TC184/SC4/WG3/N702 (T23),1998.

4. U. Rabien and U. Langbecker, Practical Use of STEP Data Model in Ship Design and Analysis, 9th International Conference on Computer Applications in Shipbuilding, Yokohama, 1997.

5. ISO 10303-218 Industrial automation systems and integration - Product Data Representation and Exchange - Part 218: Application Protocol: *Ship Structures*, TC184/SC4/WG3/N590 (T23),1997

6. European Marine STEP Association, www.oss.dk/EMSA/index.html

7. R. Bronsart, *Requirements on a Model Representing Ship Hull Cross Sections*, Technical Report ISO TC184/SC4/WG3/N380, 1995

8. J. Haenisch, *Exchange of hull definition data between shipbuilder and classification society - Protocol specification*, Seasprite Document D332, Det Norske Veritas, Hovig, Mar. 1998

9. *EMSA News* Vol. 5, No. 1, Jan. 1998, see also http://www.oss.dk/EMSA/index.html

10. *POSEIDON Reference Manual*, Germanischer Lloyd, Hamburg, 1997

11. *POSEIDON User´s Guide*, Germanischer Lloyd, Hamburg, 1997

Design for Production

George Bruce[a,] Bill Hills[b], and Richard Storch[c]

[a]**Department of Marine Technology, University of Newcastle,**
Newcastle upon Tyne, NE1 7RU

[b]**Engineering Design Centre, University of Newcastle,**
Newcastle upon Tyne, NE1 7RU

[c]**Mech. Engineering Building, FU-20, University of Washington**
Seattle, WA 98195, USA

Making a ship production friendly has often been regarded by designers as an additional, possibly unnecessary, activity.   In order to make an effective contribution to cost reduction, production requirements need to be taken into account from the earliest stage of design. A Build Strategy is the key, and the use of CAD to create a product model has provided an opportunity to improve design/production integration. The increased use of sub-contractors complicates the process, but also offers an opportunity for additional cost savings.

INTRODUCTION

The pace of technological change in the shipbuilding industry is accelerating. Although it remains a labour-intensive business for the most part, and therefore tends to migrate to areas of low labour cost, these do not remain low cost for long. Builders in high labour cost areas use technology to redress their cost disadvantage, both in terms of the production hardware and also the organisation of work.

New technology is in large part readily adopted once it is proven, and is thus available to successive builders as their cost structures change, typically as labour costs increase. The result is a continuous pressure on the leading, high cost area builders to develop more advanced technology in a shorter timescale.

The rapid development of computer technology has had a major impact on the industry, in such areas as Computer-aided Engineering, but also in less overt ways in terms of database technology and communication. The capability of shipyard managements to manage the high variety process which is ship construction has been enhanced dramatically. The technology of production has developed less in terms of the hardware, although some automation and new processes are in use. The "soft" side of production has developed more, primarily in the ability of computers to assist management in realising the potential of principles such as hierarchical planning and integration of design and production.

The increasing cost of production, against a background of low prices which are driven by long-term-over-capacity in the ship production industry, necessitates all means of cost reduction. The appropriate way to consider design is to take into account not only the function of the artifact, but also:

- the cost of production
- the cost of maintenance
- in some cases, the cost of alteration   during its working life
- potentially, the cost of disposal

The initial reaction to the need to take into account all these additional factors is that it will add greatly to the initial expense of design, will add to the time and most importantly that it will degrade the capability of the finished artefact. This belief is firmly held by many designers, who confidently assert that design for production results in inferior products.

It is true that taking production (and maintenance) into account will increase the workload on design. On the other hand, as the need to reduce costs demands shorter building times and fewer man-hours, the increase is unavoidable. Design for production does have a dramatic effect on the production hours and performance, which makes the investment in additional design time very worthwhile.

It is therefore of increasing importance that the methods of ship production and construction which the shipyard is planning to use are fully documented. The well-organised shipyard has a Production (Shipbuilding) Strategy, which largely defines all the procedures and methods in use in the shipyard. It also includes any plans for change, such as development of new facilities or adoption of new (for example) computer systems. The Strategy provide an umbrella for all the shipyard operations, down to detailed procedures. These will include work-station design and capabilities, welding procedures, standards for interim products, quality assurance and so on.

1. BUILD AND REPAIR STRATEGIES

A build, or repair strategy, should be the application of the company's shipbuilding, or ship repairing, strategy to a specific ship. Ideally it should be a formally written down document which has the approval of, and cannot be changed without the authority of, the executives of the yard.

A build/repair strategy provides the framework for the effective development and co-ordination of the design and ensures that it is produced in accordance with the current, or projected methods to be used by the production departments. Emphasis should be placed upon the outfitting and machinery elements of the ship as these are the areas in which most problems occur and in which the major savings

can be achieved. Efforts should be made to reduce work content by use of standards and the application of production engineering techniques. A build/repair strategy will identify any problem areas to which special attention will have to be paid in order to avoid bottlenecks in the production of the ship.

The build/repair strategy should:

- Show in outline form how and where the ship will be built/repaired

- Indicate any special requirements in terms of facilities, manpower, skills, etc.

- Highlight any problem areas and specify how they will be overcome

- Show adopted block breakdown, sequence of erection on the berth and initial process engineering for the blocks

- Identify zones, machinery arrangements, outfit units and main service routes

- Define purchasing requirements and drawing schedule required to support the needs of production.

2. INTEGRATION OF BUSINESS FUNCTIONS

Considerable benefits can be gained by ensuring that businesses are fully integrated.. The primary objective of such an approach being the delivery of a superior quality ship to the customer at minimum cost to the shipbuilder.

Whilst there is little need for the concurrent design of the ship with its related production system, improved integration between product engineering and production engineering should be one of, if not the major objective of any strategy for improving competitiveness. Through such integration the costs associated with unnecessary design and production activities, and the associated rework that arises as a result of poor communication and a lack of shared knowledge between product and production engineering, can be reduced. In addition, the cost reductions associated with new production technology such as robotic welding, can be fully realised through the creation of new ship designs that are designed to enable maximum advantage to be taken of such production technology. A number of tools and techniques for achieving improved integration include;

- New organisational and team structures that promote improved communication between all disciplines and functions
- Improved integration of distributed multi-agent design activities using IT applications, such as Internet and Intranet
  New formal and informal partnerships between suppliers, customers and classification societies
- Improved methods of formal feedback from production engineering to product engineering
- A CAD/CAM based product modelling tool that, in addition to having a 3D geometry modelling capability, can interact with a set of object-orientated user libraries of *standards,* including;

    - Equipment modules and components that are both out-sourced to suppliers and produced in-house

    - Documentation of the production system and its facilities

    - Guidelines for product engineering, documenting preferred production methods and working procedures. Computer-based *methods engineering* is becoming increasingly popular, for example there is growing interest in the use of simulation programs to evaluate assembly and erection sequences, and in particular are used to check for welding access.

3. SOME RELATED METHODS & TECHNOLOGIES

Design and Production Engineering

Shipbuilding is a Made-to-Order (MTO) industry with short lead times. This fundamental fact significantly influences the design and production processes of building a ship. In order to generate enormous volumes of information for each ship in a relatively short delivery period, it is necessary that the design of hull-structures, machinery, equipment and electrical systems have to be carried out concurrently and consistent/detailed information for the production process has to be prepared through the collaborative involvement of teams of designers and production personnel. It is of paramount importance to integrate the whole information from initial design through to production and also to support concurrent work. Of concern is the avoidance of re-input or re-definition of data since this is tedious, inefficient and thus expensive, and can lead to errors and inconsistencies.

CIM systems for shipbuilding support the increase of productivity during the production stage by linking the design system with the production support system. This includes the incorporation of production details and information into the design process together with the associated work instructions.

Many advanced CIM systems used in shipbuilding incorporate advanced production s upport systems. Such systems lead to improvements in the quality of production planning/scheduling, consequently, enabling improved production flow. The systems also enable the introduction of automated facilities/robots by electrical data transfer of the design information.

These systems support:

- Automatic generation of design information
- Concurrent engineering
- Functions to generate the necessary information for robots and automated equipment
- Advanced production management system
  - Covering all production stages
  - Optimization function for scheduling etc.
- Work instructions for production personnel

The Application of Concurrent Engineering

The classical definition of concurrent engineering , coined by the Institute for Defence Analysis (IDA), is the structured and systematic approach to the integrated, concurrent design of products and their related production systems. The approach is intended to cause engineering companies to consider, from the very outset, all of the inter-related issues associated with all phases of the product life-cycle from concept through to disposal. With this definition in mind there exists within shipbuilding an anomaly that needs to be addressed before going on to discuss how the practice of concurrent engineering can impact upon the design-production cycle of a ship.

Whilst ships may vary in their configuration and size, on the whole they are made up of similar components. In this way, a shipbuilding production system is designed for the construction of a range of ship types, configurations and sizes, rather than one specific ship. As a result, a new ship is designed to be compatible with an existing production system, and whilst the shipbuilding production system should be developed and refined through a well managed process of continuous improvement, there is almost no requirement for the concurrent design of a ship to be concurrent with its related production system.

Whilst the classical definition of concurrent engineering is not directly applicable, shipbuilders can apply a number of techniques incorporated within the philosophy in order to achieve benefits. Most specifically, these benefits relate to a reduction in;

- the elapsed time between the signing of a contract-to-build, and delivery.
- the total cost of design and production of the ship.

However whilst the benefits of adopting such techniques can be significant, shipbuilders should always analyze the costs, as well as the benefits associated with adopting such techniques. In particular, management should always remember that concurrent engineering is a philosophy and as such involves strong cultural change as well as a change to working practices. The adoption of the tools and techniques of a new technology without a culture aligned to maximum integration and improved communication will greatly reduce the benefits and may even increase costs as a result of dynamic conservatism, manifested the "Ah! But…" syndrome. In this way it is essential that any new initiative receives the full and noticeable backing of all levels within the management hierarchy.

Integrated Design and Manufacturing using CAE

Typical applications contain special functions for Product Information Model viewing and the development of drawings and 3D modelling. The Product Information Model should be used by all applications to store model objects which have been created.

In many organisations the concept design stage and detailed definition stage are viewed as separate and distinct processes. The tools traditionally used at these stages are also very different and there is limited application of the concept of information sharing. It is therefore imperative that new design tools can be integrated into the product information model, which is company-wide in scope.

Introduction of Product Planning Earlier into the Design Process.

This enables companies to adapt their processes to reflect changing technologies, roles and market demands by supporting on-line production planning. This can be used to focus the design teams on high priority tasks and can be used by management to achieve good work flow targets without having to undertake a potentially expensive business process re-engineering exercises.

Work Reporting

Reporting is an essential activity for management that facilitates the collection and distribution of information regarding the availability of personnel and the status of production in a particular product.

The reporting tool should be flexible and provide information in any format which is required. Further, it should be able to retrieve information from many sources, such as design, purchasing, production etc.

Materials Management

Materials management should be integrated with all the other applications and extract information from the Product Information Model.

The materials management function is responsible for the trading of materials within an organisation and act as a repository for information and allow the identification of products which have materials allocated to them. The information contained at this level is required for Integrated Logistics Support (ILS).

The Introduction of the Product Information Model Concept

The product information model provides facilities for information storage, process management and configuration management. However a detailed process flow information model of working practices is required before such a model can be constructed and implemented.

An information flow analysis or business process re-engineering (BPR) exercise will need to be performed. This is required to elicit a detailed description of the information, its properties and owners within the organisation. A computerised model can then be built using a Computer Aided Systems Engineering (CASE) tool. The model built using this tool then becomes the dynamic model which can be updated as the business process changes. Data models generated by the CASE tools are then used to construct the company database using established database technologies such as Relational Database Management Systems (RDBMS) e.g. Sybase, Oracle etc., or more recently Object-oriented Database Management Systems (ODBMS) e.g. Object Store, Versant etc.

An information flow analysis or business process re-engineering (BPR) exercise will need to be performed. This is required to elicit a detailed description of the information, its properties and owners within the organisation. A computerised model can then be built using a

CASE tool. The model built using this tool then becomes the dynamic model which can be updated as the business process changes. Data models are generated by the CASE tools are then used to construct the company database using established database technologies such as Relational Database Management Systems (RDBMS) e.g. Sybase, Oracle etc., or more recently Object-oriented Database Management Systems (ODBMS) e.g. Object Store, Versant etc.

Configuration Management

Proper implementation of configuration management can lead to shorter lead times and faster delivery of made-to-order products. Configuration management also provides the infrastructure that enables engineers from different disciplines to work concurrently on the same design.

4. Influence of the Business Environment: Distributed Multi-Agent Activities

The increasing trend towards collaboration between companies together with the growth in new network technologies is significantly altering the way design is conducted. Strategic alliances and partnerships often require the design activity to be carried out simultaneously by a number of widely distributed design agents. This is particularly the case with large and complex products, such as ships.

An effective product development process, supported by scientifically validated design theories and tools, is consequently becoming an increasingly useful asset, reducing lead-times and costs as well as improving quality. Engineers and managers in shipbuilding need to be able to integrate both new theory and new practice into their product development process. They require a means by which to match their activities with available design methods.

This requirement demands that networks move from the ad hoc proprietary based solutions of the past to become standardized open architectures which support component based middleware technology. Such systems will facilitate the requirement for greater communication between domain design experts working in various areas and the need for global product knowledge and data exchange.

Product Model Management

Given the size and complexity of marine artefacts it is unlikely that all the information pertaining to the product model will be stored in a single database. However, it is difficult to give details as to how a proposed federal structure will keep up with the configuration and constraint management of such complex products. **A strategy should be adopted which will incorporate a compromising variant in which the product model is logically centrally located but the real-time data storage is still controlled locally at each design agent. This allows separate design agents to participate in the total product model whilst retaining control of their own data.**

The product model for a ship is unique to each vessel and contains the information that describes the product in a standard form. For a ship the product model contains information relating to the following :

- Specification
- Concept design
- Contract design
- Detailed definition for for production
- Production
- Logistic support
- Overhaul planning
- Overhaul
- Decommissioning

It is important that the utility gained from the use of a product model at all levels of design and production is recognised. The problems associated with technical data transfer are proportional to the size and complexity of the product, the time associated with the design/build/operation of the product and the differences between computer systems in organisations. One constraint with this process is the tremendous cost associated with transferring, maintaining and validating manual databases. The costs associated with this process are generally hidden in the day-to-day business practice.

The aim of the product model is to simplify the administration problems while at the same time opening up the data resource within the shipyards. Working from a single data source will mean that data is timely and current. This one fact will help to alleviate the inconsistencies in released information between design and production.

5.   SUBCONTRACTING: THE MAKE OR BUY CHOICES

The decision on whether to make or buy particular components is part of the strategic decision-making process for the shipyard. The main driver for the decision is generally the cost of the component in question. Whether or not a particular component should be regarded as "strategic" is also taken into account. A strategic component is one which is of sufficient importance to the organisation that its production must be undertaken in house. For most shipyards, the steel hull would be regarded as the ultimate strategic component, in that it must be made in the shipyard because most other activities depend on the hull, and it effectively "defines" the ship.

Although the key driver is production/acquisition cost, it is also important to take into account what are termed "transaction costs". Where a component is outsourced, the transaction costs include:

- the cost of setting up and administering a contract
- the cost of external inspections and monitoring of progress
- the costs of transport, insurance and any others which would not be incurred for in-house production
- the risk of non-delivery

The last of these may be difficult to quantify, but includes the possibility that a particular supplier may not be available when needed, so forcing a shipyard to go to another, possibly higher cost supplier. The transaction costs are a function of the loss of control associated with external supply.

In the context of design for production, the main interest in the effects of external production, as a consequence of the make or but decisions, is in producibility.

However, the contractual relationship with a supplier, especially if the external supply decision is motivated by cost reduction alone, makes the integration of design difficult. If there are particular producibility requirements, these must be specified contractually, and monitored. If there is interaction, as the design develops, between the externally supplied and internally produced components, then this can be a basis of claims for extra payments and time delays. These are extreme examples of transaction costs.

It is important to develop a good supplier relationship, in which there is co-operation and a degree of trust. The supplier will then be willing to collaborate on design, with the expectation of a long-term agreement

In some ways, the need to be very specific about a component design could be beneficial. If

a component is made within a shipyard, then late changes to improve producibility may be taken but the cost of these may be hidden. On the other hand, if a supplier is asked to make changes, then an explicit cost will be incurred. (Even if the relationship is good, it is still contractual). So outsourcing may force a shipyard to improve its *product definition and planning*.

6. APPLICATION OF DESIGN FOR PRODUCTION PRINCIPLES

The design development process is illustrated by the examples from a double-hulled tanker. In figure 1, the preliminary design is developed around the proposed block breakdown, to ensure compatibility with and good utilisation of facilities.

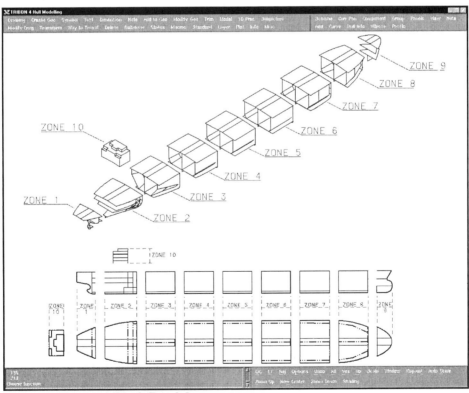

Figure 1 Proposed Block Breakdown

338

As the structural design is developed into more detail, the preferred interim products are created, figure 2, again to make best use of available facilities. Outfitting is also integrated into the design, using pre-designated locations.

If a sub-contractor is to be used, for assemblies or piece parts (figure 3), the detail must be complete and also compatible with the production facilities.

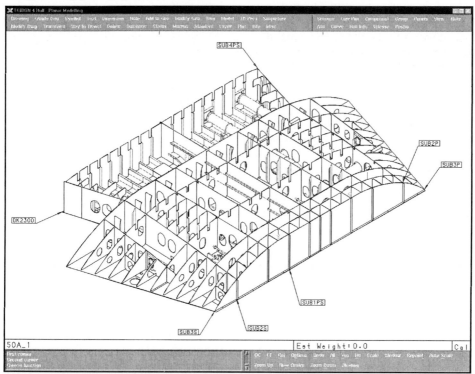

Figure 2 Example of an Interim Product

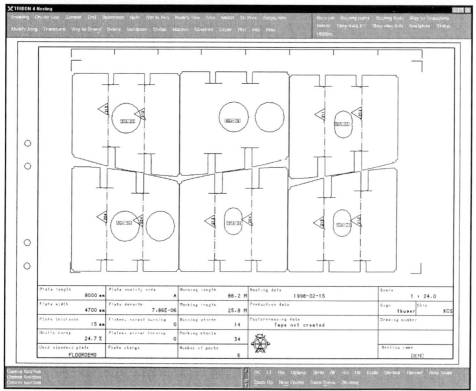

Figure 3 Nested Piece Parts

Figure 4 illustrates the product hierarchy which is the basis for the development of the detailed design of the vessel.

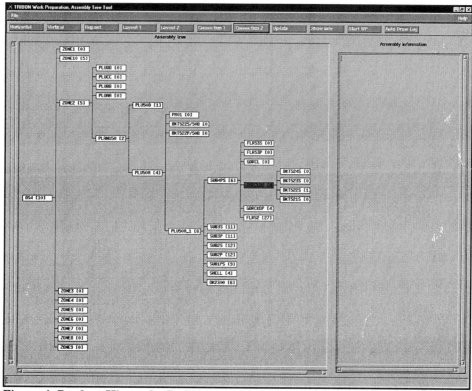

Figure 4 Product Hierarchy Definition

7. CONCLUSIONS

The capabilities of modern computer systems allow large and complex product models to be built. The capabilities contain inherent dangers, particularly that if not carefully co-ordinated there will be a massive increase in the variety of information produced. In that case, the potential for improvement in contract performance may be swallowed in simply managing the large increase in information to be handled.

The ability of the designer to explore more options and create alternatives must paradoxically be constrained in order to manage the practical realities of constructing a ship to the design. This particularly applies where a distributed manufacturing base is used. At the same time, the need to constrain variety cannot be allowed to lose opportunities for product improvement.

The need for a formal production strategy, and linked to that a contract build strategy, is increasing rather than diminishing. The ship can then be regarded as a specific ensemble of standard and semi-standard elements.

Practical Design of Ships and Mobile Units
M.W.C. Oosterveld and S.G. Tan, editors.

Ship Hull Surface Fairing System

T.K. Yoon[a], D.J. Kim[b], Y.W. Chung[a], S.Y. Oh[a], H.K. Leem[a] and N.J. Park[a]

[a]Hanjin Heavy Industries co, Ltd.
29-5 Pongnae-Dong, Youngdo-Gu, Pusan, 606-065, Korea.

[b]Dept. of Naval Architecture and Marine System engineering,
Pukyong National University,
599-1 Daeyeon3-Dong Nam-Gu, Pusan, 608-737, Korea.

This paper describes a CAD system for generating the faired hull surface rapidly. It consists of two major parts, surface modeling and surface fairing. Surface model is constructed by geometric continuous(GC^1) composite surface modelling technique. For surface modelling of non-rectangular patch the new interpolation method is proposed. In order to get the good global fairness various fairing methods are examined. Modified direct curvature manipulation in curve fairing, energy minimization of normal vector line curve-net and solving the inverse problem by constrained reflection line using optimization technique gives quite good results. Interactive graphic system with user friendly GUI is implemented using the Phigs+ library in UNIX environment.

1. INTRODUCTION

The rapid generation of faired hull surface is very important. Especially, early fairing and modeling of ship hull in preliminary design stage has more effective advantages in many sides, for example, performance analysis, production design and process management etc. Hull form fairing process, however, still has been a very iterative and time consuming job.

For automatic fairing, hull form must be modelled in computer. In the field of surface modelling of hull form, geometric complexity of hull form gives many difficulties in adopting surface modelling technique which can describe the irregular topological characteristics precisely.

Many methods have been developed to generate surfaces from a set of data points or one group of parameter curves[1-3]. To smooth existing surface a lot of methods have been applied, for example, mesh fairing method[4], reflection line method[5], FANGA curves method[6], minimizing a sum of the squares of principal curvatures[7].

In this paper, initial hull form definition scheme which can exchange the data with the commercial design package is firstly discussed, and description on modified GC^1 surface modelling technique is followed. The modified interpolation method for non-rectangular patch is proposed and shown the good fairness for hull form. The automatic fairing strategy combined proposed fairing scheme for each step to determine the final hull form for productional use is discussed. Configuration of system is described and results of application to actual ship is shown at the end part of this paper.

2. HULL FORM DEFINITION

Hull form is defined by plane sectional curves (station/waterline/buttock) and boundary curves (center profile/side and bottom tangential etc.) to discriminate planar surface with sculptured surface. In this system, all curves are defined using non-uniform B-spline curve due to it's well-known usefulness and converted into the Bezier curves for surface generation.

Matrix form of non-uniform B-spline is defined as follows[8]

$$r'(u) = UN_c^i V^i ; \quad 0 \le u \le 1 ; \quad i = 0, 1, \ldots, n-1$$

where,

$$U = [1 \quad u \quad u^2 \quad u^3],$$

$$V^i = [V_i \quad V_{i+1} \quad V_{i+2} \quad V_{i+3}]^T,$$

$$N_c^i = \begin{bmatrix} \dfrac{(\nabla_i)^2}{\nabla_{i-1}^2 \nabla_{i-2}^3} & (1 - n_{11} - n_{13}) & \dfrac{(\nabla_{i-1})^2}{\nabla_{i-1}^3 \nabla_{i-1}^2} & 0 \\ -3n_{11} & (3n_{11} - n_{23}) & \dfrac{3\nabla_i \nabla_{i-1}}{\nabla_{i-1}^3 \nabla_{i-1}^2} & 0 \\ 3n_{11} & -(3n_{11} + n_{33}) & \dfrac{3(\nabla_i)^2}{\nabla_{i-1}^3 \nabla_{i-1}^2} & 0 \\ -n_{11} & (n_{11} - n_{43} - n_{44}) & n_{43} & \dfrac{(\nabla_i)^2}{\nabla_i^3 \nabla_i^2} \end{bmatrix}$$

$$n_{43} = -\{\tfrac{1}{3} n_{33} + n_{44} + (\nabla_i)^2 / (\nabla_i^2 \nabla_{i-1}^3)\},$$

$$\nabla_i^k = \nabla_i + \nabla_{i+1} + \ldots + \nabla_{i+k-1}.$$

V_i ; control vertices

∇_i ; knot span

The system has a function to read data file formated in commercial package. On the other hand, since the initial hull form data read from commercial package is usually very rough, the system has additional functions to modify an existing curve or create a new curve. By those functions, the curve network needed in generating hull surface is produced.

To keep the positional consistency between two dimensional section curves during fairing process, all these curves are connected at common node points(mesh points).[see fig. 1] So, if one curve is modified in curve network, all of the connected curves are also modified.

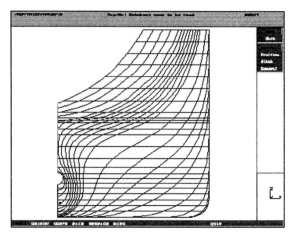

Figure 1. Initial Hull Form

3. SURFACE GENERATION SCHEME

Surface model used in CAGD field is often classified by regular and irregular topology of the boundary curves surrounded surface. It is known that the regular surface defined in u, v domain is easy to handle and has a good quality since curvature continuity (C^2) is guaranteed. Unfortunately, hull form has an irregular topological characteristic, because it is represented by an union of plane sectional curves. Besides, the boundary between planar and non-planar surface is also defined as a plane curve. It is an unavoidable obstacle in geometric representation of hull form to satisfy those conditions.

In this paper, in order to express correctly the above special features of hull form surface, we adopt the GC^1 composite surface interpolation method based on Gregory's surface patch. The advantages of this method is to reflect the geometric condition of the boundary curve. In case of non-rectangular surface patch, the smoothness of interpolation surface is not good as compared with rectangular patch, because of errors occurred in assuming the internal dividing curves(one triangular surface patch is divided into three rectangular surface patch). Hence, a new interpolation method for non-rectangular surface patch is proposed in this paper.

3.1 Gregory's Surface Patch Interpolation Methods

Gregory's surface patch interpolation method by Chiyokura and Kimura[9] using non-standard rational bi-cubic Bezier surface patch is expressed as following formula.

$$\mathbf{r}(u,v) = \sum_{i=0}^{3} \sum_{j=0}^{3} V_{i,j}(u,v) \; B_i^3(u) \; B_j^3(v) \quad u,v \in [0,1]$$

where, $B_i^3(u)$; Bernstein polynomial,
$V_{i,j}(u,v)$; control vertices
(function of u,v).

The difference of Gregory's patch from general Bezier surface patch is that the control vertices are function of surface parameter u, v. It is composed of interior control vertices Pi,j, Qi,j and boundary control vertices Vi,j.[see fig. 2]

$$V_{11}(u,v) = \{ vQ_{11} + uP_{11} \} / (u+v),$$

$$V_{12}(u,v) = \{ (1-v)Q_{12} + uP_{12} \} / \{ u+(1-v) \},$$

$$V_{21}(u,v) = \{ vQ_{21} + (1-u)P_{21} \} / \{ (1-u)+v \},$$

$$V_{22}(u,v) = \{ (1-v)Q_{22} + (1-u)P_{22} \} / \{ (1-u)+(1-v) \}.$$

$$V_{ij}(u,v) = V_{ij} = P_{ij} = Q_{ij} \quad i=0,3 \quad j=0,3.$$

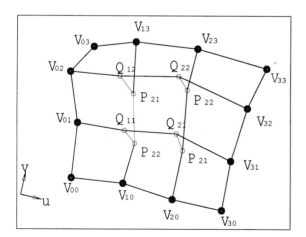

Figure 2. Control vertices of Gregory Patch

3.2 Non-rectangular surface patch interpolation

Generally, the triangular patch is divided into three rectangular patches for interpolation. In this case determination of the magnitude of cross boundary tangent vector at dividing point P_i is very important for surface fairness.[see fig.3] So, it is determined by minimization of surface strain energy through the sequential quadratic programming(SQP)[10][11] in this paper.

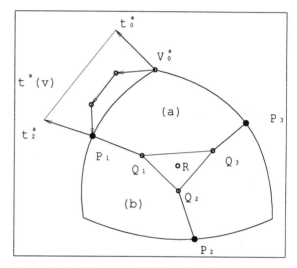

Figure 3. Subdivision of Triangular Patch

For a pentagonal patch it is same as a triangular patch to be divided into five rectangular patches. In case of ship hull curve-net, some patches of pentagonal patch may have large/small aspect ratio. In this case the internal dividing curve has improper shape which is distorted in three dimensional space and the co-planar condition is not satisfied at the internal dividing point[12]. The modified interpolation procedure are as follows[13].

-. Connecting the one corner point(V_3) and the dividing point(V_0) of one boundary curve in view of obtaining the adequate aspect ratio of two divided rectangular patches.[see fig. 4]
-. Assuming the dividing curve is the cubic

344

Bezier curve.

-. Determining the cross boundary tangent vector(C).
-. Determining the intersection point(Q) between the vector(C) and the plane made by three points(A, B, V₃)
-. Finding two control points(V₁, V₂) in view of minimizing surface strain energy of two divided rectangular patches. V_1, V_2 are positioned on the line V_0Q, V_3Q.

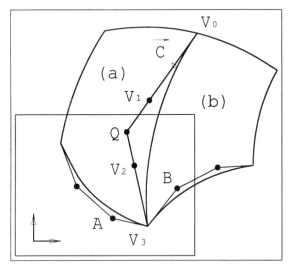

Figure 4. Subdivision of Pentagonal Patch

4. AUTOMATIC FAIRING METHODOLOGY

A smooth curve-net can give a visually smooth composite surface interpolating to the curve-net data. From this concept mesh curve fairing method has been developed by many researchers. But applying mesh curve fairing method to curve-net directly can not give a good result comparing with the result applying to normal vector line or constrained reflection line, because the irregularity on the derivative curve should be magnified. So, mesh curve fairing method by the random search optimization technique is applied to normal vector line for global fairing. To adjust the local fairness of curve-net, constrained reflection line(CRL) is introduced. After modification of desired CRL interactively, the new

curve-net points are found by SQP.

Hull form fairing is accomplished by the sequential manner as follows.

step 1
boundary curve fairing by DCM(Direct Curvature Manipulation) technique[14].

step 2
global fairing of normal vector line curve-net by random search optimization technique.

step 3
local fairing of constrained reflection line by sequential quadratic programming with boundary constraints.

4.1. Modified Direct Curvature Manipulation

Recently, Lu[14] proposed the inverse fairing method which can get a faired curve by direct manipulation of the curvature distribution. Since curvature distribution is the best fairness criterion, good curvature distribution guarantees the fairness of curve.

The object function and constraints in Lu's paper are as follows.

Object function : $Minimize \sum_{j=1}^{m}(Q_j-P_j)^2$

where, Q_j : the new control vertices
P_j : the original control vertices

constraints : $k'_i \le k_i + tol$

$k'_i \ge k_i - tol \quad (i=0,1,...,n)$

where, k'_i ; the calculated curvature value at the target points
k_i ; the desired curvature value at the target points.

To minimize the differences of two curvature curves, the object function should be the area bounded two curves. Evaluating the area, however, is not desirable due to the increasing calculating time. So, Lu used the sum of the distances between two control vertices at the target points instead of the area. But this approximation can not be improper in some cases. In case of same distance shifting of the control vertices, the area bounded two curves can not be same as shown in fig. 5. Solid line means original vertex position and curve, and dashed line is the result of moving two intermediate vertex positions to +x-direction together, and the dotted dash line is the result of moving lower vertex position by same distance along the starting tangent direction and moving upper vertex position by same distance along the reverse of end tangent direction. The sum of moving distances is the same in two cases. But, the resulting curves are shown considerable deviation between them.

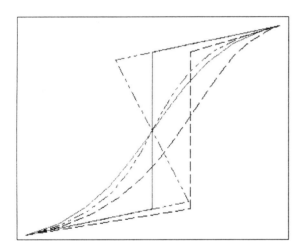

Figure 5. Counter Example of Lu's Approximation

As a result, the curve can not be desired one. So the modified object function and constraints are proposed in this paper.

Object function : $Minimize \sum_{j=1}^{m}(Q_j-P_j)^2$

where, Q_j : the minimum distance point from P_j

P_j : the target point of original curve.

constraints : $k'_i \leq k_i + tol$

$$k'_i \geq k_i - tol \quad (i=0,1,...,n)$$

where, k'_i ; the calculated curvature value at the target points

k_i ; the desired curvature value at the target points.

The modified object function is the sum of the minimum distances between the target point of original curve and the minimum distance point from the target point. To preserve the direction of end vector, which is very important characteristics in lines fairing, an additional constraint is proposed as follows.

$$1 - tol \leq \left(\frac{Q_2-Q_1}{|Q_2-Q_1|}\right) \cdot \left(\frac{P_2-P_1}{|P_2-P_1|}\right) \leq 1 + tol$$

where,
P_1,P_2 ; end point and the nearest point on the new curve
Q_1,Q_2 ; end point and the nearest point on the old curve

Optimization technique(SQP) is used to find the given curvature distribution of the curve.

4.2. Desired curvature distribution

The most important thing of making the desired curvature distribution is not to make big change of curvature plot to get the solution by optimization technique. In such a case, it is very difficult to find search direction because of cross intersection of the control vertices due to large offset of original curve. Moreover, it is quite tedious and not easy to make the desired curvature distribution considering above mentioned problem. Therefore initial desired curvature distribution curve is made by least square method.[see fig 6., fig. 7] The newly fitted curve has a fewer control vertices. So it is much smoother than original one. The result after DCM is shown in fig. 8

346

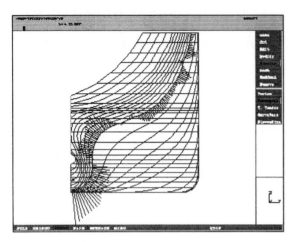

Figure 6. Curvature Plot of Original Curve

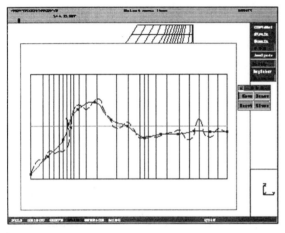

Figure 7. Modification of Curvature Plot

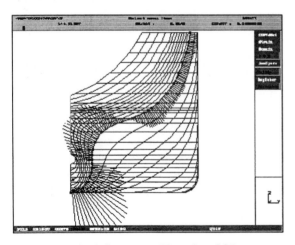

Figure 8. Faired Curvature Plot after SQP

4.3. Normal vector line

Normal vector line is a curve which has the following data points at the node position of curve net.[see fig. 9]

$$Q = R + \alpha N$$

where,

R ; the position vector at the node

N ; the surface normal vector at the node of curve net

α ; scaling factor

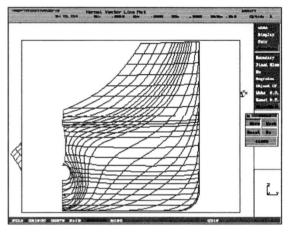

Figure 9. Normal Vector Line Curve-net

In order to smooth the curve-net, the sum of the strain energy of normal vector line should be minimized. The design variable of this optimization problem is node position of the original curve-net. The random search technique is used to find the optimum solution.

Object function ; $Minimize \sum_{i=1}^{n} \int_{i} (P'')^2 dt$

Constraint ; $x^i_{lower} \leq x_i \leq x^i_{upper}$

It is known that random search algorithm is more effective than SQP algorithm in case of using the finite difference method for derivatives and large numbers of design variables.

Random Search Algorithm :

```
begin
  S(X): = initial solution S(X₀)
  while (stopping criterion is not satisfied ) do
    X_L = MAX (X_min , X− X_D/2)
    X_U = MIN (X_max , X+ X_D/2)
    begin
      X′ = X+(X_U− X_L) random(0,1)
      S(X′) ; random neighboring solution
      △ = S(X′) − S(X)
      if △ <0 then S(X) : =S(X′)
    end
  end
end

where,
X ; best solution
X_min, X_max ; global range of design variable
X_L, X_U ; local range of design variable
X_D ; random range
```

4.4. Constrained reflection line

Constrained reflection line is a curve which has the following data points at the node position of curve net.

$$C_R = \alpha(N \cdot S)$$

where,

N : unit normal vector at the node

S : given vector

α : scaling factor

In order to smooth the local curve-net, the differences between the desired CRL and the calculated reflection line should be minimized as same as DCM. The design variable of this optimization problem is node position of the original curve-net. SQP is used to find the optimum solution.

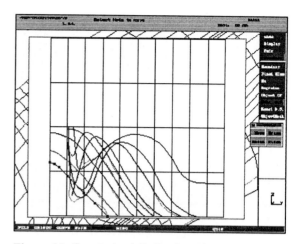

Figure 10. Constrained Reflection Lines

Object function ; $Minimize \sum_{i=1}^{n} \int (Q_i - P_i)^2$

Constraints ; $C_R^i \le C_R^i + tol$

$$C_R^i \ge C_R^i - tol \quad (i=0,1,...,n)$$

where,

Q_i : the new design variable

P_i : the original design variable

C_R^i : the calculating reflection value

C_R^i : the desired reflection value

The result of final fairing after applying proposed automatic fairing strategy is shown in fig. 11 and fig.12.

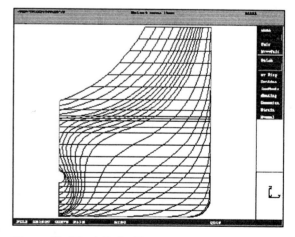

Figure 11. Final Body Plan

348

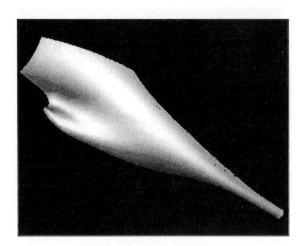

Figure 12. Rendering Image of Final Surface

5. SYSTEM CONFIGURATION

All source programs of this system are coded by Fortran, and as a graphic kernel Phigs+ library in X-window system is used. Currently, the source code is not portable to any other system, but conversion to another graphic system is going to proceed.

Main functions of this system are summarized as follows.

. Interface with commercial design package (SIKOB/SHIPFLOW).
. Standard data exchange format support.
. Modify curve and surface.
. Curve and surface fairing.
. Calculate mass property of curve and surface
. Surface quality analysis using rendering technique.
. Various output
(Lines drawing/offset table etc.)
. User friendly GUI (multi-view, one-step windowing, rubber-band, show/no-show etc.)

6. CONCLUSIONS

Interactive graphic system with user friendly GUI, which can be used at initial design stage, is implemented, and proposed surface modelling technique and fairing scheme can generate a faired hull surface satisfying designer's demands easily. Final lines from this system shows good fairness and can be used for production stage.

Since proposed automatic fairing strategy is based on optimization technique, considerable calculating time is needed. If we can get a faired normal vector line distribution, step 2 can be omitted, so the working time can be reduced. This will be our further research topic.

REFERENCES

[1] Rogers, D.F. and Fog, N.G., "Constrained B-spline Curve and Surface Fitting", ICCAS88, 1988.

[2] Nowacki, H., Liu, D. and Lu, X., "Mesh Fairing GC1 Surface Generation Method", Proceedings of Theory and Practice of Geometric Modeling Symposium, 1988.

[3] Jensen, J.J. and Baatrup, J., "Transformation of Body Planes to a B-spline Surface", ICCAS88, 1988.

[4] Rong, H., Chen, G. and Zhang W., "Nonuniform B-spline Mesh Fairing Method", ICCAS91, 1991.

[5] Kaufmann, E. and Klass, R., "Smoothing Surfaces Using Reflection Lines for Families of Spline", CAD, Vol. 20, No. 6, 1988.

[6] G. Liden, S. K. E. Westberg, "Fairing of Surface with optimization techniques using FANGA Curves as the quality criterion", CAD Vol. 25 No. 7, 411-420, 1993.

[7] Lott, N. J. and Pullin, D. I., "Method for Fairing B-splines Surfaces", CAD, Vol. 20, No. 10, 1988.

[8] Choi. B.K., "Surface modeling for CAD/CAM", Elsevier, 1991.

[9] Chiyokura, H. and Kimura, G., "A New

Surface Interpolation Method for Irregular Curve Models", Computer Graphics Forum3, 1984.

[10] Vanderplaats, G.N., "Numerical optimization techniques for engineering design with applications", McGraw- Hill, 1984.

[11] Rao, S.S., Engineering Optimization theory and practice, 3rd. edition, Wiley Interscience, 1996.

[12] Yim, J.H., A Unified Surface Modeling Technique using the Bezier Curve Model (de Casteljau Algorithm) and Visualization, MS. Thesis, Dept. of Naval Architecture and Ocean Engineering, Seoul National University, Seoul, Korea, 1997.

[13] Yoon, T.K., Development of CAD System for Ship Shell Modeling, MS. Thesis, Dept. of Naval Architecture and Marine System Engineering, Pukyong National University, Pusan, Korea, 1998.

[14] Yun Lu, Direct Manipulation of Curve and Surface Properties using a Piecewise Polynomial Basis Function, Ph. D. Dissertation, Dept. of Naval Architecture and Marine Engineering, The University of Michigan, Ann Arbor, Michigan, 1995.

Practical Design of Ships and Mobile Units
M.W.C. Oosterveld and S.G. Tan, editors.

An Evolutionary Approach to the Scheduling of Ship Design and Production Processes

J.A. Scott[a] D.S Todd[a] and P. Sen[b]

[a]EPSRC Engineering Design Centre, University of Newcastle upon Tyne.
Armstrong Building, The University, Newcastle-upon-Tyne, NE1 7RU, United Kingdom

[b]Department of Marine Technology, University of Newcastle upon Tyne.
Armstrong Building, The University, Newcastle-upon-Tyne, NE1 7RU, United Kingdom

This paper summarises ongoing research which is focused on the development of optimisation-based techniques for the scheduling of design and production processes. The Multiple-Criteria Genetic Algorithm (MCGA) is introduced, and its application to scheduling problems is described. To illustrate the functionality of the MCGA Scheduler, two examples are documented. The first example relates to a production job-shop, whilst the second relates to the scheduling of design activities.

1. INTRODUCTION

Scheduling can be defined as matching processes to be completed with (limited) resources in order to meet some objective. This objective can be to minimise elapsed time, maximise customer satisfaction, maximise utilisation of resources, and minimise costs.

Historically, scheduling techniques have focused on the creation of work plans which schedule processes with the sole objective of minimising elapsed time. In such cases only the precedence relationships between processes are considered whilst resource availability is effectively assumed to be infinite. However, whilst production planners and project managers are required to cut production and pre-production lead-times, it is likely that they will have to achieve this objective with a limited amount of resource. This is because, as a consequence of cost cutting initiatives by many shipbuilders, the availability of resources in many yards has been reduced over recent years. As a result, in most cases the assumption of unlimited resources cannot be justified. In reality only a finite amount of resources will be available whilst the short term cost of acquiring additional resources may be prohibitive. As demonstrated by Just and Murphy [1], the difference between scheduling that is resource-

constrained, and that which is not, can be significant in terms of unanticipated project overtime, schedule delays, and the resulting cost overruns.

As a result there is a need for scheduling techniques which are sufficiently flexible to be capable of modelling both production and pre-production processes at a level of detail where the resulting schedules represent an acceptable reality, upon which sound and effective management decisions can be made.

Whilst there are a significant number of proprietary software packages that can be used to plan and schedule processes, they are invariably based on heuristic scheduling rules. Heuristics can be used quickly and efficiently to develop acceptable schedules, however they tend to be inferior to schedules derived using optimisation-based techniques.

In the past, the time taken by optimisation-based techniques has limited their application to relatively small problems. However, as computer hardware becomes increasingly more powerful, the range of application of such techniques is becoming wider. One such technique, under increased investigation, and subject to growing application, is genetic algorithms.

The remainder of the paper is structured as follows. Section 2 describes the mechanism of the standard genetic algorithm and introduces the *multiple-criteria genetic algorithm (MCGA)*. Section 3 describes the *MCGA Scheduler* and concludes by reporting on two applications of the *MCGA Scheduler*. The first application focuses on a job-shop scheduling problem, whilst the second focuses on design project scheduling.

2. THE MULTIPLE CRITERIA GENETIC ALGORITHM (MCGA)

The genetic algorithm is an evolutionary search technique based on the axioms of natural selection and genetics. The technique, pioneered by Holland [2] and developed by Goldberg [3], combines survival of the fittest with some of the intuition of human search. The genetic algorithm begins with a single string. Using the terminology of genetics, a string is analogous to a *chromosome*. Each position in the string is analogous to a *gene* and is used to represent one of a range of *parameters* that have been chosen to define the problem. The parameter associated with each string position can take on a value (*allele*) such that any one string of parameter values, i.e. any one *chromosome of alleles*, can be mapped to a single distinct solution

Using the single input string, *a population* of new strings is developed by randomly resequencing the original string. This *initial population* should be spread over enough of the search space to represent as wide a variety of solutions as possible. The diversity so introduced prevents premature convergence to a local optimum. This population of randomly generated strings represents the *first generation*. The size of this population, in terms of the number of generated strings, is defined by the user, and once set, the *population size* remains constant throughout the search.

The first generation of strings then goes on to receive *genetic operation*, with the resulting development of a *second generation* population of equal size to the first. As a result of genetic operation, this second generation should contain fitter strings, ie strings that more closely match the search objective. In turn, the second generation is operated on to develop a *third generation* that should contain even fitter strings. This process continues developing *successive generations* of fitter strings, converging towards the global optimum

until some stopping condition is met. At this stage a population of the fittest strings is returned.

How closely a string matches the search objective, or more specifically the *fitness* of a string, is derived by first decoding the string into the combination of parameter values it represents. These parameter values are then plugged into a pre-defined *search objective function*. The resulting value represents the fitness of the associated string. The search objective function is so called since it represents a mathematical expression of the *search objective*. Using the objective function, any one solution can be evaluated and ranked according to how well it satisfies the search objective.

A genetic algorithm that yields good results in many practical problems is composed of three basic genetic operators *Selection*, *Crossover*, and *Mutation*. Figure 1 illustrates the basic genetic algorithm.

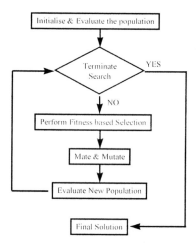

Figure 1. The basic genetic algorithm.

By representing search criteria, to be optimised as search objectives, in terms of simple 'black-box' functions, the basic algorithm can be easily modified to search the solution space using a range of varied and sometimes conflicting criteria.

Recently, scheduling techniques based on multiple criteria have begun to emerge [4]. However, examples of such work are still relatively scarce. The procedure used for the applications detailed in Sections 3.1 and 3.2 is a modified version of the standard genetic algorithm and is called the *Multiple Criteria Genetic Algorithm (MCGA)*.

Instead of using a single search criterion, the MCGA generates solutions on the basis of a range of different and sometimes conflicting criteria. The search criteria are maintained separately at all times, allowing the simultaneous consideration of separate criteria during the search. When individual solutions have been evaluated and assessed against each of the criteria, the individual fitnesses are converted to a single fitness based on *dominance*. A solution is dominated by another only if the latter is superior to the former with respect to at least one criterion and equal with respect to all other criteria. In this way, the MCGA finds a set of solutions that define the trade-off surface between search criteria. These solutions form a so-called *Pareto set*, from which the solution that most closely matches the decision-maker's preferences can be selected and, either analysed further, or implemented. For a complete description of how the MCGA deals with multiple criteria the reader is referred to Todd & Sen [5].

3. SCHEDULING USING THE MCGA

As a new development, the MCGA search technique has been combined with a so-called *Schedule Builder*. This combined tool - *The MCGA Scheduler*, produces work schedules that are optimal over a range of different and sometimes conflicting criteria, and can take into account constraints such as resource unavailability, set-up times and constraints on individual activity start and finish times.

In this case, the Schedule Builder replaces the previously described objective function. Using strings, each of which represent a distinctive schedule, the Schedule Builder first decodes the string to create the corresponding schedule. The schedule is then evaluated in terms of the pre-defined criteria which may include lead-time, utilisation, and tardiness. The string and its associated fitness is then returned to the MCGA. The *MCGA Scheduler* formulates a scheduling problem as follows:

(1) The Precedence Matrix defines the processes and their precedence relationships

(2) The Resource-Efficiency Matrix lists all resources and details their relative efficiency against a list of operation-types. For the job-shop problem,

the resources relate to individual machines (e.g. Machine 01, Machine 02, etc), whilst the operation-types relate to job descriptors (e.g. milling, turning, drilling etc). For a design organisation, the resources are the individual design engineers, whilst the operation-types are activity descriptors (e.g. stress analysis, engineering drawing, etc). The matrix elements represent, for each of the pre-defined operation-types, the relative efficiencies of each resource,. The resource-efficiency matrix can be updated over time such that resources can be added and deleted, and their efficiencies updated, as a result of maintenance (job shop), training and experience (design project organisation)

(3) The Process-Operation Matrix lists all processes, detailing for each process, the operation-type that describes the process. For added flexibility, each process can be broken down into a sequence of stages to allow representation of variable resource allocation over the duration of a process.

(4) Constraints include periods of time during which processors are not available, as well as constrained process start and finish times.

Using this input data, and based upon a given string, the *Schedule Builder* builds the corresponding schedule by trying, at each decision point in time, to schedule all unscheduled processes that satisfy precedence relationships, resource availability, down-time considerations and any pre-defined process start/finish constraints. In this way, the Schedule Builder builds a complete schedule by successively adding unscheduled processes to a partial schedule.

The variation in the derived schedules results from the variable choice of resource allocation as encapsulated within each distinctive string. The specific allocation of resource to a process represents an operational mode.

The Schedule Builder has been designed in such a way that it will tackle a broad range of scheduling problems. Two examples are documented in the following sub-sections. The first example relates to a flexible manufacturing job-shop problem where the processes relate to manufacturing jobs.

Because the *Schedule Builder* can handle precedence relationships between processes, the MCGA Scheduler can also be used to tackle resource-constrained project scheduling problems. In such cases, a workload of project activities is only a modified version of a job-shop workload where, in the basic formulation, the processes are design activities linked by precedence relationships.

3.1 Flexible manufacturing scheduling

The aim of a scheduler is to determine the job sequences on each machine in order to minimise the total makespan of all jobs. Whilst the *MCGA Scheduler* has been used on much larger scale problems [6], a simple 4 job, 3 machine problem formulation is used to demonstrate how the MCGA Scheduler can be applied. The variables and constraints of the example are detailed in Table 1. The table can be read as follows: Job 0 has three stages, Stage 0 of Job 0 uses machine M1 to complete the stage in 5 time units. In this case it is assumed that precedence relationships do not exist between jobs, however, sequential precedence relationships do exist between successive stages of the same job.

Table 1
Formulating a job-shop scheduling problem

	Stage				
	0	1	2	3	4
Job 0	M1 .5	M0, 3	M1, 4	-	-
Job 1	M2, 3	M1, 6	M0, 5	M1, 3	M2, 4
Job 2	M0, 4	M2, 3	M1, 6	-	-
Job 3	M1, 4	M2, 5	M1, 3	M0, 5	-

Based on this problem formulation, an optimal schedule can be derived. Figure 2 illustrates the resulting schedule which is optimised to minimise the make-span of all jobs.

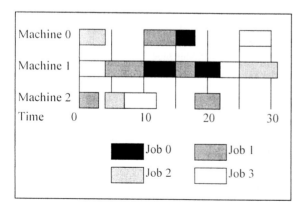

Figure 2. An MCGA-derived optimal schedule

In order to highlight the latest developments to the MCGA Scheduler, consider the job-shop production system as a sub-set of the whole ship-construction system. With this systems approach in mind, the relationships between the job-shop and its related up-stream and down-stream production sub-systems must be considered. Job-shop start-times may be influenced by upstream sub-systems, whilst finish-times may be constrained by strict deadlines imposed by downstream requirements. There are also cases when certain manufacturing jobs must wait for the completion of other jobs. Whilst most classical job-shop schedulers, reported in the literature, consider the precedence relationships between job stages, few consider the precedence relationships which may exist between the jobs themselves, and even fewer consider constraints both internal and external to the job-shop.

Furthermore, the application of flexible manufacturing practices, precipitated by the development of more flexible machinery, means that jobs are no longer tied to a single machine. Instead, jobs can often be processed on one or more of a choice of machines. Such machines are capable of performing several manufacturing operations with varying time-based operational efficiencies. This case is commonly referred to as *multiple-mode resource allocation*. Because such machines require tool change and reset between different operational modes, set-up times become an integral part of any derived schedule.

In addition, in order for schedules to be more realistic, schedulers must be capable of incorporating down-time periods, during which machines are taken off-line for routine maintenance.

Based on these factors, which all affect the development of an effective production schedule, the requirements of a job-shop scheduler become more complex. Makespan is no longer the only objective, and it may be necessary to minimise penalty costs for late deliveries, reduce the amount of jobs in progress because of limited space requirements, maximise machine utilisation or reduce queues. The *MCGA Scheduler* is able to handle this complexity and develop near optimal solutions, allowing the production planner to experiment with different machine configurations and consider options relating to multiple-mode resource allocation. For example, if a machine was identified as a bottleneck, a reconfiguration could be implemented by adding a second machine or by upgrading the efficiency of the existing machine through repair and maintenance. The benefits of this change can be assessed by comparing the new trade-off surface with the surface derived previously, comparing the gains derived against the cost of such an implementation. Figure 3, based on the previous problem formulation, shows that significant gains, in terms of reduced makespan, can be made if Machine 2 were to be modified to allow it to complete some of the jobs previously allocated to Machine 1.

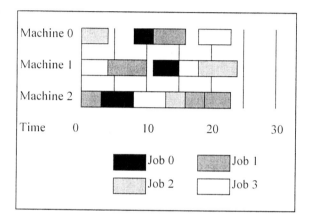

Figure 3. The revised MCGA-derived optimal schedule

3.2 Resource-constrained project scheduling

The following example illustrates an application of the previously described *MCGA Scheduler* to a design project scheduling problem. Table 2 illustrates a partial *resource-efficiency matrix* which maps the skills of five engineers (resources) to five operation-types. Figure 4 illustrates a combined *precedence matrix* and *process-operation matrix*, which details the design activities, their durations, their precedence relationships, and the operation-type associated with each activity.

Critical path (CP) analysis, which assumes the infinite availability of engineers, yields a critical path length (project lead time) equal to 230 time units. Using the five engineers detailed in the partial skills matrix (*See* Table 2), the *MCGA Scheduler* returns the schedule summarised in Table 3. In this

case, the schedule has been optimised using two criteria: (1) project lead time, and (2) the average utilisation of engineers. Naturally, the respective search objectives are (1) to minimise project lead time, and (2) to maximise the average utilisation of engineers. The resulting schedule yields a project lead time equal to 307 time units, with an average utilisation of the five engineers equal to 60%.

Obviously the increased lead time (33% longer than the CPM derived schedule) is due to the constraints of resource availability. However, whilst the lead time for the resource-constrained schedule is longer, it is much more realistic. In this respect, scheduling in this manner can help the shipbuilder avoid the unanticipated project overtime, schedule delays, and the cost overruns which sometimes result from schedules which are derived without adequate consideration of resource limits.

Table 2.
A partial resource-efficiency matrix

	Operation Type				
	NA0	ED0	MD0	WD0	DG0
D.Styles	100%	0%	100%	0%	25%
P.Davis	50%	100%	50%	25%	0%
D.Wood	50%	75%	100%	50%	50%
C.Ewan	25%	75%	0%	100%	25%
A.Brown	0%	25%	25%	0%	100%

It is possible that an experienced project manager would be able to create a schedule similar to the one derived by the *MCGA Scheduler*. However, such an informal approach has three major disadvantages:

(1) Without a formalised approach to project scheduling, the experience of project managers is not analogous to the experience of the company. As a result, this experienced is lost when project managers leave the company.

(2) As the size and complexity of the project increases, there is an ever increasing chance that project managers will not be able to derive optimal schedules using their own 'in-built' heuristic scheduling capabilities. Even for relatively small projects, the unavailability of certain engineers, and constrained due dates on the output of certain

356

activities can result in a scheduling problem where the project manager is simply faced with too many variables. In such cases, the absence of a formal scheduling procedure can result in the creation of schedules that are far from optimal.

(3) Of course this problem is further exacerbated in a multi-project environment, where different project managers are faced with the problem of resourcing activities using scarce resources across a range of different projects.

4. CONCLUSIONS

In today's business climate of fierce international competition, there is a need for shipbuilders to reduce costs and deliver high quality ships faster. Production planners and design project managers are increasingly looking for robust mathematical modelling techniques that can be used to develop flexible and efficient work schedules. Such schedules need to be optimised on the combined objectives of minimising lead-time, minimising overall production and project costs and maximising the use of scarce resources such as design engineers.

This paper has introduced the notion of resource-constrained scheduling applied to the scheduling of production and pre-production activities, and has summarised a multiple-criteria genetic algorithm-based approach to resource constrained scheduling.

The methodology described here is under continuous development and is supported by a growing suite of programs. Current research is focused on improving the efficiency of the procedures and improving the functionality of the MCGA Scheduler to incorporate the assignment of multiple resources to activities. In addition, real-life industrial case studies continue with a view to validating the methodology further, whilst increasing its practical usefulness.

Most recently the MCGA Scheduling technique has formed the basis of a research and development contract undertaken by the University on behalf of a major UK shipbuilder.

Table 3.
An optimised resource-constrained project schedule

Activity	Start	Finish	Engineer
PD001	0	10	David Styles
PD002	10	22	Deborah Mulroy
PD003	22	47	Craig Ewan
PD004	47	63	Deborah Mulroy
PD005	47	77	David Styles
PD006	110	112	David Styles
PD007	63	101	Deborah Mulroy
PD008	10	20	Philip Davis
PD009	76	98	Philip Davis
PD010	101	131	Deborah Mulroy
PD011	112	119	David Styles
PD012	131	205	Deborah Mulroy
PD013	20	40	Philip Davis
PD014	47	61	Craig Ewan
PD015	22	32	Deborah Mulroy
PD016	134	148	Philip Davis
PD017	40	47	Alice Schofield
PD018	40	76	Philip Davis
PD019	77	87	David Styles
PD020	87	110	David Styles
PD021	161	193	Alice Schofield
PD022	148	200	David Styles
PD023	148	202	Philip Davis
PD024	205	245	Alice Schofield
PD025	87	127	Craig Ewan
PD026	129	161	Alice Schofield
PD027	116	134	Philip Davis
PD028	127	147	Craig Ewan
PD029	205	223	Deborah Mulroy
PD030	147	175	Craig Ewan
PD031	202	220	Philip Davis
PD032	223	231	David Styles
PD033	61	129	Alice Schofield
PD034	98	116	Philip Davis
PD035	193	203	Craig Ewan
PD036	245	286	David Styles
PD037	286	301	Philip Davis
PD038	301	307	Deborah Mulroy

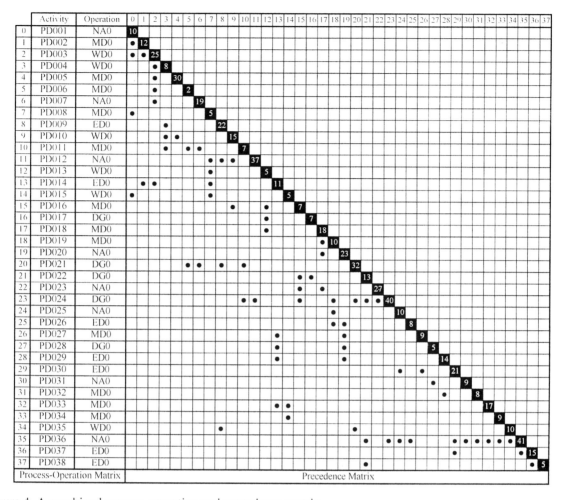

Figure 4. A combined process-operation and precedence matrix.

REFERENCES

1. Just, M.R. and Murphy, J.P, The effect of resource constraints on project schedules, Transactions of the American Association of Cost Engineers, pp. E1.1-E1.6, 1994.

2. Holland, J.H., Outline for a logical theory of adaptive systems, Journal of the Association for Computing Machinery (ACM), Vol 3, pp 297-314, 1962.

3. Goldberg, D.E., Genetic algorithms in search, optimisation and machine learning, Addison-Wesley Publishing, 1989.

4. Slowinski, R., Multiobjective scheduling under multiple category resource constraints, in Slowinski, R. & Werglarz, J. (Eds.), Advances in Project Scheduling, Elsevier, 1989.

5. Todd, D.S. and Sen, P., A multiple criteria genetic algorithm for containership loading, Proceedings of the 7th International Conference on Genetic Algorithms, East Lansing, MI, Morgan Kaufmann Publishers, 1997.

6. Todd, D.S. and Sen, P., Multiple criteria scheduling using genetic algorithms in a shipyard environment, Proceedings of ICCAS 97, the International Conference on Computer Applications in Shipbuilding, Yokohama, Japan, October,1997.

A study on the production–oriented structural design information system of panel blocks

Joo-Sung Lee[a] and Gu-Gun Byun[b]

[a]School of Naval Architecture & Ocean Engineering, Univ. of Ulsan,
Ulsan P.O.Box 18, Ulsan, Rep. of Korea, 680-749

[b]Production System Division, Hyundai Heavy Industries Co.
1, Cheonha-Dong, Dong-Ku, Ulsan, Rep. of Korea, 682-792

This paper is concerned with development of the production–oriented structural design information system to predict the inaccuracy level of panel blocks and to consider the result at the structural design stage. Emphasis is placed on that the inaccuracy during production should likely be considered at the structural design stage to reduce the undesirable adjusting work and therefore to enhance the productivity. The primary goal of the present study is to consider the productivity and the efficient design at the same time for a high quality product of panel block. Usefulness of the developed information are illustrated through some application examples.

1. INTRODUCTION

In shipbuilding industry accuracy control takes place in the important part[1-4]. In order to produce a high quality products, the accuracy control should be probably managed from the subassembly stage of panel block. In addition the concept of accuracy control is to be also incorporated with the structural design so that the designer can produce a better design result accounting for the geometric inaccuracy. For this purpose, it is necessary to develop the production–oriented structural design information system (POSDIS here after), which is aimed at producing a better design result accounting for the information (or knowledge) about the inaccuracy level of panel block assembly.

In this paper emphasis is placed on that accuracy control is to be performed at the subassembly stage to keep higher geometric accuracy level at the following assembly stage and that it is important to keep the weld deformation be uniform along the joint line as possible to reduce the adjusting work at the next assembly stage. This paper does not go into detail of the developed POSDIS and the general concept of the present system is illustrated. Since fillet weld takes more portion than butt weld in panel block

assembly, presented is how deformation due to fillet weld is affected by parameters such as weld leg length, initial deflection due to butt weld and span length (longitudinal space). Some application examples of the present system to subassembly of panel block are followed. Throughout the analysis the geometric inaccuracy arising from plate cutting, transportation, stock and so on is neglected.

This paper ends with discussion on the application results of the present system and extension of the present study aimed at enhancing productivity.

2. PRESENT SYSTEM

2.1. General

The POSDIS presented in this paper is aimed at reducing the undesirable man power during production process at the structural design stage with referring to the information about inaccuracy level of panel block assembly and not aimed at obtaining the optimal structural design result. For this it is necessary to provide the information about how the deformation during block assembly would be kept both as low level and uniform as possible at the same time.

With such system designer can grasp the resultant deformed shape or inaccuracy level along the joint line and then, one can try to change design (including rearrangement of structural members) to reduce the inaccuracy level. The present system consists of following program modules.

(1) Data input module
  – geometric data of panel block
  – joint line data
  – data for structural design
(2) Structural design module
(3) Predicting residual deformation
  module

Data of joint line include the joint type (fillet or butt weld) and welding condition for each weld line. Structural design module performs the structural design according to the Classification Society Rule when design conditions and parameters are changed. Predicting residual deformation module is to simulate the welding deformation which consistits of two steps : butt and fillet weld simulation. Simplified formulae of predicting various welding deformation described in the next section are used. Structural analysis is performed by the finite element method and equivalent load and/or displacement are applied to obtain the deformed shape. Data for structural analysis are automatically generated with information about panel block and structural design. Since butt weld is followed by fillet weld at the subassembly stage for panel block, the deformation during butt weld process is treated as the initial displacement in simulating fillet weld process.

2.2. Formulae of predicting weld induced deformation

Weld induced deformations are usually classified as follow [5] :

1) transverse shrinkage perpendicular to the weld line
2) longitudinal shrinkage parallel to the weld line
3) angular distortion around the weld line

The formulae for the above deformation are

well summarized in many references [5,6]. As it has been mentioned, fillet weld takes a more portion than butt weld in panel block assembly, especially at the subassembly stage since many stiffeners are to be welded on panel. Concerning with the formulae of predicting the angular distortion due to fillet weld, most formulae do not account for the change of angular distortion along the weld line. Referring to the past experimental and numerical studies as well as the measured results for the real panel blocks in ship yard, it does significantly vary along the weld line. In this paper a modified formula is proposed to account for the change of angular distortion along the weld line, which has been derived based on the results of thermo elasto-plastic analysis[7]. Models for numerical analysis are listed in Table 1. In reference [7] following formula was proposed as the average angular distortion for FCAW.

$$\phi_{fo} = 1.2427 \; p^{1.8943} \; e^{-0.165p} \; (\times 10^{-3} \text{rad.}) \quad (1)$$

where p is heat input parameter defined as

$$p = Q / \; t^{1.5} \qquad (2)$$

with t and Q are plate thickness and heat input per unit length, respectively. To account for the change of the angular distortion along the weld line it is expressed as:

$$\phi_f = \; \phi_{fo} \; F(p, \; x/ \; w_l) \qquad (3)$$

where ϕ_{fo} is given as Eq.(1) and $F(\bullet)$ is the correction factor reflecting for the change along the weld line and w_l is weld length. The correction factor is assumed to be a linear function.

$$F(p, \; \frac{x}{w_l}) = \; C_0 + C_1 \; \frac{x}{w_l} \qquad (4)$$

x is the distance from starting point of weld. Coefficients C_0 and C_1 are derived through the regression analysis based on the numerical analysis results for the models shown in Table 1 as follow [8]:

$$C_0 = 0.94 - 0.026p + 0.00094\ p^2 \qquad (5)$$
$$C_1 = 0.13 + 0.050p - 0.00170\ p^2$$

Comparison of the present formula with the numerical analysis results is illustrated in Figure 1. It can be seen that the proposed formula well agrees with the numerical analysis results.

For other deformation type appropriate formulae are used [5-7].

Table 1 Models for numerical analysis

Model	t (mm)	heat input, Q (cal/mm)	heat input parameter, p
N1	7.0	213.6	11.5
N2		356.5	19.2
N3	8.0	279.6	12.4
N4		367.1	16.2
N5	13.0	321.2	6.9
N6	15.0	298.3	5.1
N7		354.0	6.1
N8		298.3	3.9
N9	18.0	321.2	4.2
N10		354.0	4.6

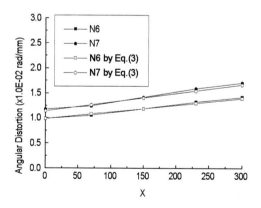

Figure 1. Comparison example

2.3. Welding deformation simulation

Weld induced deformations have been measured for several panel blocks in shipyard and the present system has been applied to show the validity the elementary parts consisting the present system. Table 2 shows the models measured in shipyard and the simulation results. Model A and Model B series denote the case that 28 pole and 10 pole automatic welding machine were used, respectively. The deflection at mid-span between stiffeners were measured. Figure 2 shows the modelling for the present simulation. Angular distortion and shrinkage are considered by applying the equivalent forces. Simulation results are summarized in the same table. Simulation results in general well agree with the measured results such that mean of the ratio between measured and simulated results is 1.076 as far as the measured models in Table 2 are concerned. It implies that the present system can reasonably predict weld induced deformation.

Table 2 Models measured in yard

model	t (mm)	s (mm)	Q (cal/mm)	measured result	simulated result
A1	12.5	840	190.9	1.30	1.01
A2	14.0	840	190.9	3.50	3.56
A3	15.0	700	190.9	2.00	1.94
A4	15.0	810	187.5	4.10	3.97
A5	17.0	830	183.0	2.90	2.66
A6	17.0	830	192.1	2.70	2.52
A7	17.0	830	188.2	1.80	1.72
A8	17.0	885	173.4	2.80	2.55
A9	17.0	830	224.3	1.45	1.44
A10	17.5	840	204.4	2.40	2.26
A11	18.5	840	166.9	1.70	1.49
A12	19.0	875	187.5	3.60	3.09
A13	19.0	840	167.8	1.85	1.59
A14	16.0	830	131.2	2.95	2.69
B2	18.0	850	319.5	3.10	3.12
B3	18.5	840	347.0	1.80	1.77

362

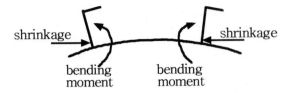

Figure 2. Analysis model for fillet weld

From the designer's point of view quantitative information how design parameters affect the welding deformation is useful at the decision making stage. Referring to the past researches the important parameters affecting the weld deformation during fillet weld process are not only weld leg length and span length (longitudinal space) but also the initial deflection occurring during butt weld process. Several parametric studies have been carried out to quantitatively show the affecting level. For the present parametric study plate thickness is 12.5, 14.0, 17.0, 19.0mm and weld length is 4000mm. Let δ, ϕ and s be the initial deflection due to butt weld, weld leg length and span length. Following three cases are considered.

Case 1 : δ = 0.0, 0.5, 1.0, 2.0mm
with ϕ = 6mm and s = 840mm
Case 2 : ϕ = 6.0, 6.5, 7.0, 8.0mm
with δ = 0.5mm and s = 840mm
Case 3 : s = 700, 750, 800, 850, 900mm
with δ = 1.5mm and ϕ = 6.0mm

The same heat input of Q = 190cal/mm is assumed. Results of parametric studies are presented in Figure 3. It can be seen that deflection is generally proportional to the above three parameters regardless of plate thickness and that the initial deflection due to butt weld is the most affective parameter followed by weld leg length and span length in turn. This implies that accuracy control should be made at the previous assembly stage from production side and selection of span length is relatively open to the designer such that it gives the optimal structural design just satisfying the requirements specified in Classification Society Rules.

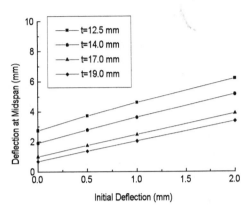

(a) effect of initial deflection

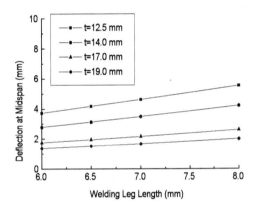

(b) effect of weld leg length

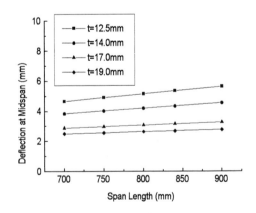

(c) effect of span length

Figure 3. Result of parametric studies

3. APPLICATION TO PANEL BLOCK ASSEMBLY

In this section presented is the application of the present system to subassembly stage with varying longitudinal space. Figure 4 shows a panel block model with five stiffeners. Following three panel block models are considered as for illustrating the present POSDIS.

M1 : s = 800mm, no. of stiffeners = 5
M2 : s = 750mm, no. of stiffeners = 6
M3 : s = 850mm, no. of stiffeners = 5

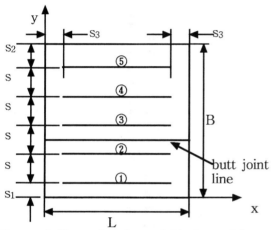

Figure 4. Example of panel block model

For all models L and plate thickness are 2250 and 14.0mm, and s_1, s_2, s_3 are 200, 600, 150mm. B depends on the number of stiffeners, and so B = 4000, 4550, 4200mm for model M1, M2 and M3, respectively. The same welding conditions, say the same heat input of Q = 190.0cal/mm is assumed and it is also assumed that there is a but joint line at mid of no.2 and 3 stiffeners.

The deformed shape magnified fifty times is shown in Figure 5 for M1, in which dotted line denotes the original panel. Deflection along joint lines, x = 0 and L are shown in Figure 6. As it can be easily expected, deflection significantly varies along side perpendicular to the weld line although its magnitude is not great while it is nearly uniform along the free edge joint lines (y =

0 and B).

There are a couple of ways to produce less and uniform deflection along the joint lines perpendicular to the weld line. One possible way may be attaching the temporary stiffener perpendicular to the weld line. This is not, however, expected to reduce deflection at mid of weld line, say along x = L/2. For model M1 the initial deflection between two stiffeners due to butt weld was 0.9, −0.4, 0.8, 0.6mm and from Figure 5 it is noticeable that the deflection between no.2 and 3 stiffeners having negative initial deflection is much lower than others. Giving the negative initial deflection is therefore another possible way to achieve both the deflection as uniformly distributed along the joint line and low geometric inaccuracy as possible. In the case that the all initial deflection at mid-span between stiffeners are −0.5mm for model M1, the deformed shape is shown in Figure 7. Comparing Figures 5 and 7, it can be seen the apparent reduction of deflection along the joint line except the free edge side, for which another way can be applied to reduce the deflection.

This kind of simulation can be rapidly carried out at the design stage with the present system. At this time it can be said that providing the quantitative information about how to reduce the inaccuracy level must be important at the design stage and very useful in enhancing the productivity.

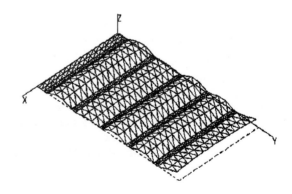

Figure 5. Deformed shape for panel block model M1

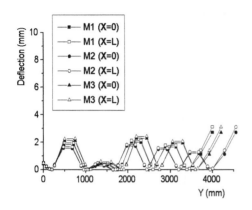

Figure 6. Deflection along joint lines

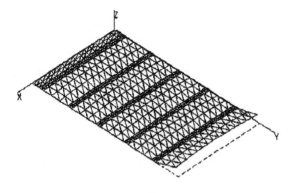

Figure 7. Deformed shape for panel block
model M1 with negative initial
deflection between stiffeners

4. CONCLUSIONS

In this paper proposed is the general concept of the production-oriented structural design information system toward reducing the undesirable additional work at panel lock assembly stage. A prototype system has been developed and example application of the system is illustrated. As far as the results illustrated in this paper are concerned, it can be said that the present POSDIS can provide the useful information about the geometric inaccuracy to designer and hence one can produce a better design result with accounting for the productivity. The system can be extended to application the panel block before erection stage. The application example will be presented at the judicial proceedings in the near future.

REFERENCES

1. M. Ichiji et al., Computer Applications to Accuracy Control in Hull Construction, Proc. ICCAS'85, Trieste (1985).
2. K. Aoyama, T. Nomoto and S. Takechi, Basic Studies on Accuracy Management System for Shipbuilding, Proc. ICCAS'97, (1997) 323.
3. T. Okumoto ,Study on Accuracy Control of Hull Structure, J. Ship Production, SNAME (1994)
4. M. Yuzaki et al., An Approach to a New Ship Production System Based on Advanced Accuracy Control, J. Ship Production, SNAME (1993)
5. K. Masubuchi, Analysis of Welded Structures - Chap.7 Distortion in Weldments, Pergamon Press (1980)
6. K. Satoh and T. Terasaki, "Effects of Welding Conditions on Welding Deformations in Welded Structural Materials, J. Japanese Welding Society, Vol.45, No.4 (1976) 187
7. S.I. Kim, S.I., J.S. Lee et al ,A Study of the Accuracy Control of Block Assembly in Shipbuilding - simulation of residual deformation due to fillet welding, Proc. ICCAS'97 (1997) 367
8. J.S. Lee, Development of the Computer Software System for Evaluation of Deformation and Shrinkage due to Welding, Final Report of UOU-HHI Corporative Research (1998)

Practical Design of Ships and Mobile Units
M.W.C. Oosterveld and S.G. Tan, editors.

THE ASSESSMENT OF SHIP HULL WEIGHT UNCERTAINTY

K. Žiha[a], I. Mavrić[a] and S. Maksimović[b]

[a] Faculty of Mechanical Engineering and Naval Architecture, University of Zagreb,
   Ivana Lučića 5, 10000 Zagreb, Croatia

[b] Jadroplov, International Maritime Transport, Branimirova obala 18, 21000 Split, Croatia

The paper considers the assessment of uncertainties of the ship hull steel weight with respect to uncertainties of built-in steel products. First, a general procedure for worst-case approach based on tolerances is presented. A review of tolerances on rolled plates and sections provided for shipbuilding, according to the Classification Societies, international standards as well as the tolerances defined by the steelworks, are given herein. Using tolerances of plates and sections, the deviations of weights of the hull, structural parts and substructures from the nominal weights are assessed by application of the worst-case approach. Additionally, the influence of the geometrical tolerance in production, like cutting, welding and other workmanship, on the weights of the ship hull has been checked. Five typical stiffened panel types, most frequently used in the ship construction are subject to weight deviation analysis. Next, a typical bulk carrier midship section has been investigated from weight deviation point of view. Finally, for a bulk-carrier built in Croatian shipyards, a review of overall ship hull weight distribution, according to plate thickness and dimensions of sections as well as the maximal possible deviations of the nominal weights due to tolerances, are summarised.

1. INTRODUCTION

Complex technical problems are always faced with objective and/or subjective uncertainties of their components as well as with inaccuracies of the modelling, calculation, numerical, operational and production procedures. The consequences of the uncertainties are that the technical products differs from their planned, designed or simply desired features, sometimes denoted as the nominal characteristics. A widely used method for prediction of the uncertainties is based on statistical data on the component level and usage of statistical inference to predict the uncertainties of the complex system. Statistical methods requires a large amount of data, the collection of which can be time consuming and costly. The application of statistical methods requires significant experience in application and interpretation of results.

There are some complex technical problems with essential lack of statistical data. For those reasons the paper will not deal with statistics but with tolerance. In many technical problems the use of tolerances is much more suitable and simpler, due to the fact that tolerances of components are either known, or given, or can be assumed by using common engineering reasoning, and, last but not least, the tolerances of components can in general be easily controlled in the design and production process. The characteristics of components are in general represented by their nominal values. The tolerance represents the bounds of acceptable uncertainties and usually represents the deviations from nominal values. The amount of tolerance can also be expressed as fractions of the considered component characteristic. Tolerance can in some problems be expressed in terms of standard deviations, e.g. tolerance equals threefold the standard deviation.

A reasonable assessment of component tolerance can contribute to predict the deviation of complex system characteristics or to define an acceptable tolerance level using a minimal and maximal tolerance procedure, e.g. for linearised non-linear functions, the worst-case approach [1] or exact non-linear procedure [2].

The idea underlined in the paper is to investigate the effects of tolerances of steel products on the deviation of the ship's hull weight. The procedure can in general case lead to a non-linear analysis.

Presented approach is illustrated by numerical examples.

2. TOLERANCE OF NONLINEAR FUNCTION

Let us denote the component characteristics as x and the given tolerance denoted as t_x. The deviations from nominal values in a positive and/or a negative sense can be denoted by t_x^{upp} and t_x^{low}.

Consider a function $Y = f(x_1, x_2, \ldots, x_n)$.

If there are some uncertainties in the function value, they can be taken into account by additional (possibly subjective) tolerances t_f^{low} and t_f^{upp}.

Consider first a non-linear function Y where each of the variables can be separated into "n" single derivable terms, e.g. $Y = f(x_1, x_2) = x_1^5 - x_2^2$.

The tolerance limits of such a non-linear function of "n" variables, given with their upper and lower tolerance, can be approximated by the first order Taylor's series expansion in a given linearisation point X*, [1], as presented below:

$$Y \approx f(x_1^*, x_2^*, \ldots, x_n^*) + \sum_{i=1}^{n} \left(\frac{\partial f}{\partial x_i} \right)_{X^*} t_{x_i} + t_f \quad (1)$$

The problem can be solved exactly, with slightly more efforts [2], using following relations:

$$Y = f(x_1^* + t_{x_1}, x_2^* + t_{x_2}, \ldots, x_n^* + t_{x_n}) + t_f \quad (2)$$

where the tolerances t_x for Y^{upp} are $t_f = t_f^{upp}$ and:

$$t_{x_i} = t_{x_i}^{upp} \text{ if } \frac{\partial f}{\partial x_i} > 0 \text{ or } t_{x_i} = t_{x_i}^{low} \text{ if } \frac{\partial f}{\partial x_i} < 0 \quad (3)$$

and for Y^{low} are $t_f = t_f^{low}$ and

$$t_{x_i} = t_{x_i}^{low} \text{ if } \frac{\partial f}{\partial x_i} > 0 \text{ or } t_{x_i} = t_{x_i}^{upp} \text{ if } \frac{\partial f}{\partial x_i} < 0 \quad (4)$$

If $\left. \frac{\partial f}{\partial x_i} \right|_{X^*} = 0$ and $t_{x_i}^{upp} \neq t_{x_i}^{low}$, discontinuity in function tolerance limits in point X* occur.

The partial derivatives in upper terms can be regarded as rates of changes i.e. sensitivity factors with respect to the considered function.

More general case is a non-linear function Y where "n" variables can be separated in more than "n" single derivable terms of a single variable, e.g.

$$Y = f(x_1, x_2) = (x_1 - x_2^2)(x_1 - 2x_2^2).$$

The problem can be solved by introducing an additional notation for same variables contributing to different terms, e.g.

$$Y = f(_1x_1, _2x_1, _1x_2, _2x_2) = (_1x_1 - _1x_2^2)(_2x_1 - 2 \cdot _2x_2^2)$$

The prefix in the above notation denotes the sequence number of considered terms. Such a function can be considered as a function of more then "n" variables, maximum $m_1 + m_2 + \ldots m_n$, as presented below:

$$Y = f(_1x_1, _2x_1, \ldots, _{m_1}x_1, \ldots, _1x_2, _1x_2, \ldots, _{m_2}x_2, \ldots, _1x_n, _2x_n, \ldots, _{m_n}x_n)$$

where $_jx_i = x_i$ for $j = 1, 2, \ldots, m_i$.

3. TOLERANCES ON STEEL PARTS

The underdimensions of ship hull parts jeopardise the structural integrity of the ship construction and therefore the reduction on ship scantlings is a serious problem in shipbuilding and a subject of rules of Classification Societies.

In contrary to the underdimensions, the overdimensions of scantlings in general improves the structural capabilities, or strength and no special considerations of Classification Societies are provided. On the other hand overdimensioning increases the weight of built in steel products being in this sense a serious problem for shipbuilders and shipowners which should be held under control.

3.1. Tolerance of plates

In the case of plates steelworks should manage the constant underthickness tolerance of about -0.3 mm, related to the full range of thickness. The classification societies has defined earlier the underthickness of plate tolerance somewhat differently, usually relating the underthickness tolerance to the thickness. The underthickness tolerance provided recently by Classification societies should be in accordance to EN10029 standards using four classes [3] as shown in Table 1.

In case of overdimensions the relevant data are those which are declared by standards and by the steelworks, see also in Table 1.

Table 1
The tolerances of plate thickness (mm) according to EN10029 (also DIN1543)

Nominal thickness	Permissible nominal thickness deviation			
	Class A	Class B	Class C	Class D
≥ 6 < 8	-0.4/+1.1	-0.3/+1.2	0/+1.5	-0.75/+0.75
≥ 8 <15	-0.5/+1.2	-0.3/+1.4	0/+1.7	-0.85/+0.85
≥15 <25	-0.6/+1.3	-0.3/+1.6	0/+1.9	-0.95/+0.95
≥25 <40	-0.8/+1.4	-0.3/+1.9	0/+2.2	-1.10/+1.10
≥40 <70	-1.0/+1.8	-0.3/+2.5	0/+2.8	-1.40/+1.40

3.2. Tolerance of sections

In case of sections, none of present international (and as suppose also national) standards accept constant underthickness tolerances. The underthickness tolerances of sections as set forth by standards are related to their width and depth but not to their thickness, e.g. DIN [4]. Sections of the same thickness and various dimensions have different underthickness tolerances, e.g. [5].

Table 2a

Cross-section tolerances of bulb / Jumbo bulb (mm) according to INEXA-PROFIL

Width	Width tolerance	Thickness tolerance
60 – 80	+ 2.0 / - 1.0	+ 0.8 / - 0.2
100 – 120	+ 1.5 / - 1.5	+ 0.7 / - 0.3
140 – 180	+ 2.0 / - 2.0	+ 1.0 / - 0.3
200 – 300*	+ 2.0 / - 2.0	+ 1.0 / - 0.3
320 – 430*	+ 2.0 / - 2.0	+ 1.2 / - 0.3

Note: Closer tolerance than DIN standards.

Table 2b

Cross-section tolerances of T-sections (mm) according to INEXA-PROFIL

Flange width	Toler- ance	Web flange th	Toler- ance
100–300	+2.5/-2.5	12 – 20	+0.5 /- 0.3
(300)–600	+3.0/-3.0	(20) – 40	+1.0 /- 0.3

Table 2a and 2b, shows cross-section tolerances of bulb flats, Jumbo bulb flats and T-sections as defined by INEXA-PROFIL. Closer tolerances can be delivered by special agreement.

4. PANELS WEIGHT DEVIATION

Orthogonally stiffened plates constitute about 50% of steel hull structural weight and dominate the total cost and production time. Consequently, their structural integrity and total weight and cost must be carefully analysed [6]. Shipyard practice has established several widely used types of stiffened panels. Five typical configurations of orthogonally stiffened plates were identified by surveying shipyards [7]. The paper will apply the EN standard tolerance class B to five panel types, denoted as A, B, C, D and E, presented in Fig. 1, see also Table 3 for principal scantlings and results.

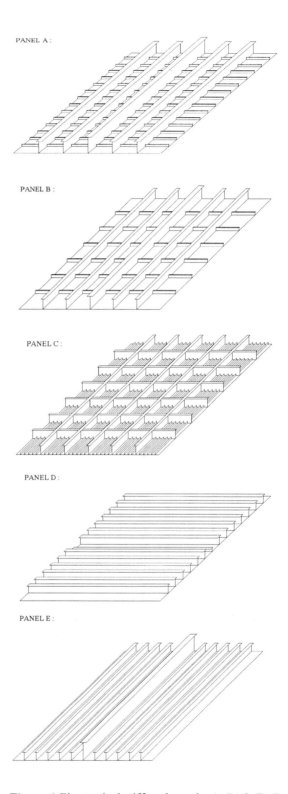

PANEL A :

PANEL B :

PANEL C :

PANEL D :

PANEL E :

Figure 1 Five typical stiffened panels: A, B, C, D, E

Table 3
The five orthogonally stiffened panels principal scantlings (mm) and weights (kg)

Panel	A			B			C			D			E		
deviation	min	nom	max	min	nom	max	min	nom	max	min	nom	max	min	nom	max
ITEMS	SCANTLINGS (mm)														
Length of panel	10699	10700	10704	10699	10700	10704	10699	10700	10704	10699	10700	10704	10699	10700	10704
Width of panel	9499	9500	9504	9499	9500	9504	9499	9500	9504	9499	9500	9504	9499	9500	9504
Thickness of plating	10.7	11.0	12.4	20.7	21.0	22.6	7.7	8.0	9.4	10.7	11.0	12.4	10.7	11.0	12.4
No. of main longitud. girders	1	1	1	1	1	1	5	5	5	0	0	0	1	1	1
Web height	1197	1200	1203	897.	900	903	597	600	603				1197	1200	1203
Web thickness	11.7	12.0	12.5	9.7	10.0	10.5	5.7	6.0	6.5				11.7	12.0	12.5
Flange width (keel)	397.	400	403	397	400.	403	197	200	203				397	400	403
Flange thickness (keel)	29.7	30.0	31.0	29.7	30.0	31.0	7.7	8.0	8.5				29.7	30.0	31.0
No. of longitudinal girders	4	4	4	4	4	4	flat30	flat30	flat30	0	0	0	10	10	10
Web height	867	870	873	697	700.	703	111	114	117				547	550	553
Web thickness	11.7	12.0	12.5	9.7	10.0	10.5	11.7	12.0	12.5				9.7	10.0	10.5
Flange width	357.	360	363	247	250	253							247	250	253
Flange thickness	29.7	30.0	31.0	29.7	30.0	31.0							19.7	20.0	21.0
No. of main trans. girders	bulb14	bulb14	bulb14	bulb6	bulb6	bulb6	6	6	6	1	1	1	0	0	0
Web height	148	150	152	198	200	203	597	600	603	247	250	253			
Web thickness	10.7	11.0	12.0	7.7	8.0		7.7	8.0	8.5	7.7	8.0	8.5			
Flange width (keel)						31.0	147	150	152	147	150	152			
Flange thickness (keel)						21.0	7.7	8.0	8.5	14.7	15.0	15.5			
Number of transverse girders	0	0	0	0	0	0	0	0	0	13	13	13	0	0	0
Web height										597	600	603			
Web thickness										9.7	10.0	10.5			
Flange width										247	250	253			
Flange thickness										19.7	20	21			
ITEMS	WEIGHTS (kg)														
Plating weight	8536	8777	9902	16514	16757	18048	6143	6383	7506	8536	8777	9902	8536	8777	9902
Longitudinal stiffening weight	9136	9353	9763	6461	6636	6968	5397	5632	6058				10717	11036	11656
Transverse stiffening weight	2096	2193	2341	920	966	1045	2566	2684	2872	10644	10981	11652			
Total stiffening weight	11232	11546	12104	7381	7602	8013	7963	8316	8930	10644	10981	11652	10717	11036	11656
Total welds weight	40	50	60	24	30	36	88	110	132	28	35	42	24	30	36
Total panel weight	19808	20373	22068	23919	24389	26097	14194	14819	16568	19208	19793	21596	19277	19843	21594
	-2.7%		+8.3%	-1.9%		+7.0%	-4.0%		+12%	-2.9%		+9.1%	-2.8%		+8.8%
ITEMS	WEIGHTS DUE TO SPECIFIC CHANGES OF SCANTLINGS (kg)														
Changes panel length/width	20320	20323	20355	24375	24359	24375	14695	14699	14707	19750	19758	19770	19810	19814	19825
Changes plates/stiff. thickness	19846	20323	21896	23951	24359	25996	14125	14699	16295	19263	19758	21440	19333	19814	21456
Change web height/flng width	20228	20323	20418	24292	24359	24426	14633	14699	14765	19677	19758	19839	19728	19814	19900

5. MIDSHIP SECTION WEIGHT DEVIATION

For a bulk-carrier of about 48000 dwt built in a Croatian shipyard, with following main particulars: Loa=192m, Lpp=183m, B=32m, H=16.7m, T=10.7, Cb=0.836, a weight deviation analysis using EN standard tolerance class B, has been performed in order to find out the minimal and maximal cross-sectional properties of ship-s longitudinal structural elements, i.e. weight per unit of ship length. The distribution of plate thickness in the midship section are given in Table 4a. The distribution of bulb flat dimension in midship section are given in Table 4b.

The calculation of the section characteristics is performed by the computer program BV-Mars [8] and the results are presented in Table 4c.

Table 4a
The distribution of plate thickness (mm) of a bulk-carrier midship section

Thickness			
min	nom	max	length (m)
23.7	24.0	25.6	23.182
19.2	19.5	21.1	23.400
18.7	19.0	20.6	1.980
18.2	18.5	20.1	3.272
17.7	18.0	19.6	16.823
17.2	17.5	19.1	32.194
16.7	17.0	18.6	26.020
12.7	13.0	14.4	24.000
11.7	12.0	13.4	14.089

Table 4b
The distribution of bulb flat dimension (mm)
of a bulk-carriermidship section

Hight (mm)			Thickness (mm)			No
min	nom	max	min	nom	max	
348	350	352	23.7	24	25.2	20
338	340	342	11.7	12	13.2	24
318	320	322	11.7	12	13.2	20
308	310	312	10.7	11	12.2	6
298	300	302	10.7	11	12.0	18
278	280	282	11.7	12	13.0	6

Table 4c
Characteristics of a bulk-carrier midship section

Characteristics	Min	Nom	Max
Cross area of longitud. (m^2)	0.5837 (-2.26%)	0.5972	0.6376 (+6.76%)
Cross area of plating (m^2)	2.8562 (-1.52%)	2.9004	3.1567 (+8.84%)
Cross area total (m^2)	3.4400 (-1.64%)	3.4976	3.7944 (+8.48%)
Neutral axes above BL (m)	11.437 (-1.66%)	11.631	12.566 (+8.04%)
Mom. of iner. about NL (m^4)	163.32 (-1.98%)	166.63	179.30 (+7.60%)
Mom. of iner. about CL (m^4)	440.33 (-1.54%)	447.24	485.12 (+8.747%)
Sec. mod. at deck (m^3)	17.520 (-1.61%)	17.805	19.111 (+7.33%)
Sec. mod. at bottom (m^3)	22.134 (-1.80%)	22.540	24.501 (+8.70%)
Single mom. above NL (m^3)	11.437 (-1.66%)	11.631	12.566 (+8.01%)
Single mom. half sec. (m^3)	17.740 (-1.56%)	18.024	19.546 +8.44%

6. SHIP HULL WEIGHT DEVIATION

For a 38199/42600 dwt bulk carrier built in a Croatioan shipyard, with following principal characteristics: Loa=187.60m, Lpp=179.00m, B=15.40m, H=15.40m, d=10.10, Class LR., a weight deviation analysis based on EN standard tolerance class B has been performed.

The plate weights by their thickness and deviations are indicated in Table 5a, and presented on Fig. 2a. The rate of change is the total weight of the plate divided by the appropriate plate thickness.

Table 5a
Distribution of overall built-in plate weights (kg)
for considered ship

Thk. mm	Toler. (mm)	Rate of changes (kg/mm)	WEIGHT (kg)		
			min	nom	max
5	-0.3/+1.2	170	798	849	1053
6	-0.3/+1.2	12742	72628	76450	91740
6.5	-0.3/+1.2	5805	35989	37730	44696
7	-0.3/+1.2	15582	104402	109077	127776
7.5	-0.3/+1.2	5119	36856	38392	44535
...	...	...	...	...	...
36	-0.3/+1.9	59	2096	2114	2226
40	-0.3/+2.5	3239	128693	129565	137663
50	-0.3/+2.5	522	25929	26086	27390
60	-0.3/+2.5	961	57384	57672	60075
95	-0.3/+2.5	55	5204	5220	5379
TOTAL		6204808	6341536		7033548
		-2.15%			+10.9%

The weight of sections by their scantlings as well as the weight deviations are indicated in Table 5b and presented on Fig. 2b. The rate of change is the total weight of sections divided by the appropriate cross-area.

Table 5b
Distribution of overall built-in section weights (kg)
for considered ship from INEXA-PROFIL

Scantl. (mm)	Rates of changes kg/cm²	WEIGHT (kg)		
		min	nom	max
BP 120x6	1019.4	9235	9480	10458
BP 160x7	2798.5	39962	40858	44440
BP 160x8	110.1	1747	1783	1923
BP 180x8	621.0	11513	11737	12631
BP 180x9	390.4	7940	8081	8643
.....	....		.....	..
BP 300x13	155.9	8093	80234	8795
BP 320x12	907.2	48007	49169	53813
BP 340x14	752.6	48271	49294	53388
BP 370x13	323.7	22047	22527	24443
BP 430x15	162.6	15019	15299	16417
TOTAL		729102	743861	802852
		-1.98%		+7.93%

The total nominal steel weight equals 7085397 kg.
The minimal steel weight is 6933910 kg (-2.14%).
The maximal steel weight is 7836400 kg (+10.6%).

370

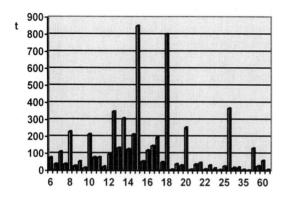

Figure 2a.Distribution of weights of plates (tons)

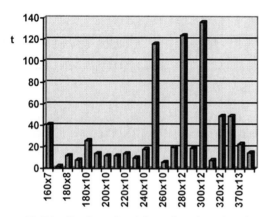

Figure 2b Distribution of weights of sections (tons)

7. CONCLUSION

The presented report on deviations of the weight of ship's steel hull is a part of a lasting investigations on effects of tolerances to different aspects of ship structural capabilities, like yielding strength or buckling strength under different loads.

The deviation of weights of stiffened panels due to tolerances of plates and stiffeners has been investigated on five typical panels and the amount of deviation of total panel weight was -4% to +12%. The deviations due to tolerances in thickness of plating and sections are found dominantly greater (about twenty times) than due to other geometrical tolerances such as length and width of the panel (under 0.1%) or height of web or width of flange (under 0.5%). The inaccuracies in weights of welded joints are assessed according to tolerances of Classification Societies relatively high to ±20% but

their ovarall effect on the total weight as well as on the weight deviation are small.

Considering the deviation in weight of longitudinal structural members per unit of ship's length by investigating a midship section of a bulk carrier, a range of about-2% to about +9% has been found. The other cross sectional properties has been changed too, e.g. the deck module has changed from about -2% to about +8.

The total weight of steel of a ship built in Croatian shipyard consituted of about 89.5% plating and 10.5% sections, has changed from about -2% to about +11%.

The aim of these investigations were in finding out the maximal deviations of weight of ship steel hull with respect to the tolerances of steel plates and sections as defined by Classification Societies or given by steelworks in order to predict more rationally the acceptable limits on the assessment of ship hull steel weight. The presented procedure providing rates of changes can render, on one hand, the tolerances which can assure that in no way the weight of steel will not exceed the defined upper limit. On the other hand, the application of the procedure to specific ships can suggest an upper limit of steel weight which can be applied for rational assessment of the contractual penalty range of the displacement.

REFERENCES

1. C. M. Creveling, Tolerance Design, Addison-Wesley, 1996.
2. K. Ziha, Post-optimal uncertainty analysis, Preceeding of the ITI'97 Conference, eds. D. Kalpic, V. Hljuz, Pula, Croatia, 1997.
3. BSI - Standards, (BSEN 10029), London, 1991
4. DIN Standards, DIN, 1985.
5. The INEXA-PROFIL Product Range, Lulea, Sweden, 1997.
6. N. Hatzidakis and M. M. Bernitsas, Comparative Design of Orthogonally Stiffened Plates for Production and Structural Integrity-Part 1: Size Optimisation, J.of Ship Production, Vol. 10No.3, Aug 1944, pp. 146-155.
7. I. Winkle and D. Bird, Towards More Effective Structural Design Through Synthesis and Optimisation of Relative Fabrication Cost, Trans. RINA, Nov., 1985.
8. Buerau Veritas, Mars User Manual, Paris, 1983.

SHIP HYDROMECHANICS

Practical Design of Ships and Mobile Units
M.W.C. Oosterveld and S.G. Tan, editors.

The CALYPSO Project: Computational Fluid Dynamics in the Ship Design Process

J. Tuxen[a] , M. Hoekstra[b] , H. Nowacki[c] , L. Larsson[d], F. van Walree[b], M. Terkelsen[e]

[a] Odense Steel Shipyard, Denmark
[b] MARIN, The Netherlands
[c] Technical University Berlin, Germany
[d] Flowtech Inc., Gothenburg, Sweden
[e] Maersk Data, Denmark

1. INTRODUCTION

In many aspects, design of ship hull forms is the most important process in ship design. The hull form serves a number of important life cycle aspects such as cargo capacity, speed and propulsion power, manoeuvring characteristics etc.. The hull form also strongly influences the geometry and topology of the inner hull structure and, as an indirect effect, the block division and production methods for the ship. In a life-cycle perspective, hull form design probably commits more costs than any other process in the creation of a ship.

The hull form design process of today is mainly based on experience and experiments, not on exact methods. It is also a lengthy, distributed and mostly sequential process, characterized by a number of non-integrated tools. The need for Concurrent Engineering in addition to fast, efficient and reliable tools is obvious, but so are the obstacles.

Computational Fluid Dynamics (CFD) provides an attractive answer to some of these problems. CFD codes have now been developed for almost four decades and are widely accepted as an important addition to model tests. Although CFD is by no means a proper replacement for model experiments, there is no longer any serious doubt that the use of CFD is a positive contribution to the hull form optimisation process. On the other hand, there are major obstacles to maximising the applicability of CFD. The methods are far from being mature, particularly due to the free surface problem, which distinguishes the ship case from most other applications of CFD. Due in part to this problem, the codes are mainly used by experts. In short, the accuracy and reliability of the predictions are still limited and the results are obtained in a process separated from the core ship design process.

But even if these problems can be overcome, CFD is only part of the answer. The work involved in using CFD codes is too high to allow an interactive process. The geometry modellers of today are non-objective oriented and rather time consuming. In addition, the transfer of information between geometry modellers, CFD codes, visualisation engines and design software is cumbersome.

The objective of the CALYPSO project is to provide a cost-effective solution to these problems simultaneously. CALYPSO is a 3-year industrial research project under the auspices of the CEC Brite-Euram programme. The approach of CALYPSO is to validate state-of-the-art CFD codes, while increasing their reliability and applicability, all with a particular view to enable ship designers to make optimal use of the codes in their own process. This is supplemented by the creation of a structured and documented design methodology, leading to a more systematic design process with less trial-and-error. The developments include a Designer's Workbench with plug-and-play mechanisms to support a flexible design environment, into which alternative codes can be integrated without major adaptations. To underline this openness, transfer of hull form data is largely based on ISO 10303 STEP AP216 (Ship Moulded Forms). These developments will be detailed in the subsequent sections.

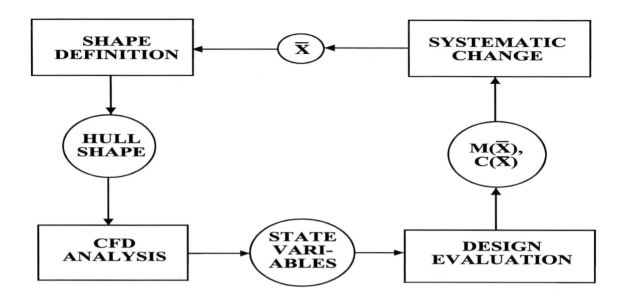

Fig. 1: Hull Shape Optimization Process

2. DESIGN METHODOLOGY (WP1)

2.1 Objectives

In CALYPSO the tools for fluid dynamic analysis are embedded in a computational environment, the CALYPSO Designers Workbench, which supports the hydrodynamic design process of hull shapes. The design process is viewed as a rational decision-making process organised in a sequence of steps, which are to be realised as modular functions in CALYPSO.

Figure 1 presents an overview of this design process, consisting of the steps:

- Shape definition
- CFD analysis
- Design evaluation and review
- Systematic change of hull shape

These stages of the process are to be supported by the following design methodology:
The *shape definition* can in principle be performed by any available front-end ship geometry modelling system capable of representing the hull shape uniquely, preferably at the level of a surface model. The hull shape data are transferred to the CFD system via a neutral, STEP based interface where at the receiving end a discrete data set is extracted to suit the CFD system's needs, e.g., for mesh generation. The CFD results, mainly the state variables of the flow, are systematically evaluated and, if desired, supplemented by forming merit criteria and feasibility constraints. This is done at the *design evaluation* stage. A *review function* extends these capabilities to include systematic comparisons with previous designs and other reference shapes.

Given this information, the designer can now form an opinion in which direction to improve the shape. This decision may be guided in CALYPSO by a design reference data base and some knowledge based recommendations.

Much priority is placed in CALYPSO on the ability for *systematic changes* of flow relevant form parameters of hull shape. Thus an alternate shape representation is made available in CALYPSO based almost exclusively on characteristic hull form parameters from which a desired hull form change can be constructed without trial and error and without violating any other independent hull form constraints. This is done in the spirit of parametric design of hull shapes based on a form parameter approach [1]. Thus a few systematic changes are intended in principle to be sufficient for exploring the most flow relevant form characteristics.

The basic premise for this design methodology in CALYPSO is that the system will make enough quantitative information available to let the designer appreciate the cause and effect relationships between hull form change and flow phenomena. The system must provide enough guidance and clues to enable the designer with some experience to concentrate on the promising changes.

2.2 Developments

The TRIBON Initial Design (TID) system, provided by CALYPSO partner Kockums Computer Systems (KCS), was chosen as the starting platform for the hull shape geometric modelling component for the project. In order to adapt TID to the requirements of hydrodynamic design of the underwater hull shape in CALYPSO, specifications were written which further functions will be needed. These functions will include several operations for systematic global and local shape control, e.g., stretchings, volume relocations and centroid shifts. Although such operations enable the designer to take a macroscopic view of the desired shape changes, the system always also maintains its internal data structure based on a spline mesh and a corresponding surface. The flexible control of minor local features (via offsets) is not sacrificed either. According to these requirements a significant extension of the TID modeller is under development by KCS and in CALYPSO.

For the resulting surface geometry of the hull a set of interface processors to the CFD systems was specified and in large measure developed in CALYPSO by KCS. The approach for conversion of ship form data is based largely on the ISO 10303

Standard STEP, AP 216 (Ship Moulded Forms). The CFD systems chosen in CALYPSO (WP2) receive the hull geometry file and extract whatever discrete data they need for deriving the CFD discretized geometry model.

The results of any CFD analysis run can be displayed in a CALYPSO design session by presenting tabular data, diagrams and flow visualisations for a given active design and potentially a desired reference design. A review function will provide options for systematic comparisons, including reference to a design data base and physical interpretation of flows by means of simple reference shapes.

The functions for hull shape modification have been specified and will be controlled in the same form parameter oriented mode as used in the initial hull form definition.

The results of the design session in CALYPSO are saved in the database and administered by version, variant, and ship count. Thus gradually the users and user communities will build up an archive which can serve as a long-term memory of the organisation.

The workflow in the CALYPSO Design Cycle is concisely summarised in Figure 2. The majority of these functions is still under further development.

2.3 Status and Outlook

CALYPSO has completed its system specification and many developments are now taking place concurrently. Among the chief results in design methodology CALYPSO is aiming at a flexible, systematic hull shape control and a deliberate working style guided by cause and effect evidence on the relation between shape and flow. The methods and software tools for this approach have been specified and are now being developed.

The new capabilities which CALYPSO will provide in its Workbench environment go well beyond the pure analytical capabilities of CFD codes. They will offer the designer many new options and insights to support effective and rational design work.

376

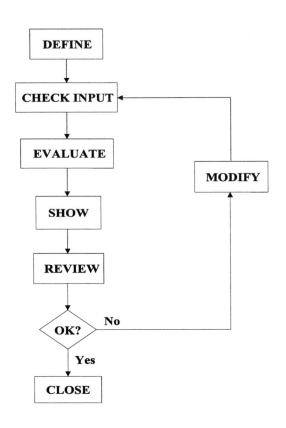

Fig. 2: Workflow in Design Cycle

3. CFD SYSTEM ADAPTATION AND INTEGRATION (WP2)

3.1 Objectives
Although CFD is becoming more and more popular as a tool in the early ship design process, its usefulness is still a bit hampered by some problems related to the available software. The objective of workpackage 2 is to improve the CFD software in several respects to enable more reliable calculations to be carried out for a wider class of ships than previously possible. Important links to the geometry modeller are also included.

The codes used in the project are SHIPFLOW from FLOWTECH International and RAPID and PARNASSOS from MARIN. SHIPFLOW is based on a zonal approach, where three flow codes are integrated: a non-linear potential flow panel method, a boundary layer method and a Reynolds-Averaged Navier-Stokes (RANS) method. RAPID is a non-linear panel method and PARNASSOS is of the RANS type. In principle, all resistance and flow properties of a ship may be computed using these codes, but to increase the robustness, efficiency and applicability several improvements are being made within CALYPSO.

3.2 Developments
Since the basic idea of the Designer's Workbench is a plug-and-play approach, the CFD software used must be able to read the geometry of the hull surface from many different CAD packages. As explained in Section 2.2 this is accomplished through the use of the international STEP standard for transferring geometry data. In principle, any CAD system which can generate a STEP file may be used for providing the CFD input through a special Application Program Interface (API) developed by KCS.

A weak point of all CFD methods is their inability to predict resistance components accurately. The information about flow properties and generated waves provides a good guide to the designer on how to optimise his design, but most ship designers also request reasonably accurate resistance predictions. This is difficult, since most methods suffer from discretization errors, which are sometimes of the same order as the resistance component in question. Obtaining wave resistance from a pressure integration over the hull surface means adding a large number of positive and negative contributions, whose sum may be as small as one of the contributions. One way to improve the prediction is to compute the wave resistance from the generated waves, and in this workpackage a method for calculating the wave resistance based on several transverse cuts through the wave system behind the hull has been developed [2].

The viscous resistance prediction often suffers from a distortion of the RANS grids at the fore and aft ends. Single block grids offer limited capabilities to follow the hull contour accurately with unskewed cells. Since the larger part of the viscous pressure resistance comes from the ends this is a serious disadvantage. Within CALYPSO a new RANS method, based on overlapping structured grids, is being developed. The method, called CHAPMAN, will become part of the SHIPFLOW system. By covering the stern end with overlapping grid patches a very accurate representation is possible.

Several research tasks aim at extending the range of applicability of the CFD codes and to improve their accuracy. Thus, an investigation into the possibilities of improving the potential flow results by a displacement thickness correction is under way. Also, some problems related to the potential flow prediction behind hulls with very flat sterns are being resolved by a new technique for handling the stern flow. Twin skeg hulls have always been tough to compute in a RANS method, due to the difficult grid generation. The new CHAPMAN solver, with its flexible grid generator, will remedy this problem. Also an extensive investigation is carried out of different turbulence models, suitable for predicting stern flows with strong bilge vortices. The turbulence model investigation is part of a PhD project at Chalmers University of Technology [3]. Other developments of interest within the workpackage include an investigation into the possibilities of representing appendages by body forces in a RANS solver, improved models for representing the propeller as an actuator disk and a feasibility study on full-scale RANS predictions.

3.3 Status and outlook
By the time of the PRADS Conference most of the developments described above will be finished. STEP Interfaces to CAD packages will be available, the potential flow codes improvements related to the wave resistance prediction and the flat stern approach will have been implemented and tested and the new CHAPMAN code will have been in use for a few months.

The developments in workpackage 2 will enhance the present capabilities of ship hydrodynamics CFD codes, and the integration into the Designer's Workbench will considerably increase their usefulness as a tool in the ship design process.

4. VALIDATION OF CFD CODES (WP3)

The main objective of the validation work is to provide the user of CFD codes with information from which he or she can judge the applicability and accuracy of CFD codes without the assistance of CFD experts. An additional objective is the establishment of an archive system containing experimental and CFD data, which can be used for validation, as a starting point for new designs and for judging the quality of current and future designs. Furthermore, a study on the feasibility of the generation and use of empirical correction functions to be applied to CFD code results was to be performed. The correction functions are intended to enhance the accuracy and applicability of CFD results and to define validity bounds of the codes.

Expected benefits of the validation work are the building of confidence for the user in applying CFD codes and in judging their results: validation indicates the validity and error bounds. By letting the users do part of the validation work themselves, knowledge of how to apply the CFD codes is gained. Validation further indicates possible improvements of CFD codes and thus gives directions for future CFD developments. Finally, an infrastructure and a strategy for validation of future versions of the CFD codes are established.

Validation is a process that can not be completed, if ever, within the scope of the CALYPSO project. This is due to the complexity of the CFD computational models, the large number of parameters and validation items involved, the limited amount of reference data available and the required time for carrying out the computations. To limit the scope of the validation work the following validation items have been selected:

Potential flow:
- numerical validation: investigate the sensitivity of CFD results to panelling particulars,
- wave patterns: longitudinal cuts at various section including wave behind transom sterns,
- wave profile along the hull,
- sinkage and trim,
- wave resistance.

Viscous flow:
- numerical validation: investigate the sensitivity of CFD results to grid particulars,
- viscous resistance,
- streamlines along the hull,
- wake fields,
- propeller-hull interaction.

These items have been selected in order to focus on the use of the CFD tools for hull form design work at shipyards. In this respect it is of importance to investigate how well CFD results can be used to predict flow field effects due to local hull form modifications.

A series of dedicated model tests has been carried out at MARIN to generate experimental data on the effects of hull form modifications on flow field particulars. In view of the relatively large amount of data available for tanker hull forms, the model tests were focused on container ship and ferry/cruise liner hull forms. Hull form modifications were obtained from adjustments of the geometry of the hull form (bulbous bow, stern sections), but also by changing the draft and/or trim of the models. The experiments comprised the following measurements for 12 hull forms:
- longitudinal wave cuts next to and behind the model,
- wave profiles along the hull,
- resistance and propulsion,
- wake surveys,
- paint smear tests.

In a feasibility study on the empirical correction functions for CFD results, a comparison between the wave making resistance obtained from model tests and the RAPID potential flow code has been made. For 24 modern hull forms, ranging from low Froude number (0.18) tankers to relatively high Froude number (0.38) ferries and frigates, the "experimental" wave making resistance was determined by means of a three-dimensional form factor method, based on MARIN statistics. The wave making resistance coefficient C_{wm} follows from:

$$C_{wm} = C_{tm} - C_F (1 + k) \qquad (3.1)$$

where C_{tm} is the total model resistance coefficient, C_F is the frictional resistance coefficient according to the ITTC-57 formulation and k is the form factor.

RAPID results were readily available as this code has been used in past commercial projects for hull form optimization prior to model testing, for all cases considered here. The RAPID wave resistance

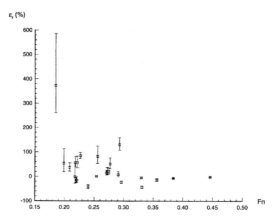

Fig. 3   Relative difference in measured and calculated wave resistance coefficients

results are based on pressure integration over the actual wetted surface with a correction for the hydrostatic pressure.

For cases with a low wave making resistance, at low Froude numbers, a small variation in the form factor may have a large effect on the magnitude of the wave making resistance. In order to get some insight in the uncertainty of the experimental wave making resistance values, these values were also determined for (1+k) values plus or minus 2.5%. This 2.5% is an estimate of the standard deviation for (1+k) observed from cases present in the MARIN data.

Figure 3 shows the percentual relative differences in the wave making resistance coefficient plotted versus the Froude number, together with the uncertainty due to the variation in the form factor. The relative error ε_r is defined as

$$\varepsilon_{rt} = \frac{100 \ (C_{wr} - C_{wm})}{C_{wm}}$$

where C_{wr} is the wave making resistance coefficient according to RAPID and C_{wm} is the experimentally obtained wave making resistance coefficient.

It is seen that the discrepancies are indeed large at low Froude numbers, where the wave resistance is small (implying that in absolute sense the deviations may not be dramatic). With increasing Froude number, and thereby decreasing fullness of the hull forms, the differences reduce in magnitude, but are still hard to fit in a correction function.

No obvious single reason for the large discrepancies can be given. Possible reasons are:
- the ambiguity in the definition of wave resistance, derived from total resistance measurements;
- the well-known scatter of experimental data for total resistance at low speeds;
- the heterogeneity and smallness of the sample of hull forms;
- the inaccuracy of numerical pressure integration for full hull forms;
- the neglect of viscosity effects in the computations, leading in general for full hull forms to an overestimation of the stern wave system and thereby of the wave making resistance.

The latter reasons are valid for all potential flow panel methods. In relation to the inaccuracy of pressure integration, it is worth mentioning that in CALYPSO a successful alternative based on a transverse wave cut analysis has been developed.

5. DESIGNER'S WORKBENCH

5.1 Objectives
As can be seen from previous sections of this paper, proper inclusion of CFD into the hull design process imposes that a significant number of different but interdependent computer based tools has to be managed. From this, the need for a dedicated workflow system supporting the hull designers has emerged. Workpackage 4 of the CALYPSO project was set up in order to meet this need by development of a prototype of such a workflow management system.

From analyses of today's hull form design practise at the yards as well as of the requirements imposed by design methodologies (see section 2), a specification and design of the Designer's Workbench have been made, which addresses the basic architecture as well as detailed functional requirements. At this moment, the types of systems, which are relevant for the hull design process, include advanced graphical modelling and review tools, mesh generation tools, CFD codes and different types of reference and validation tools. It is obvious, though, that existing tools will be improved and in some cases replaced in the near future, and that future developments in different fields will bring even more types of tools into play.

Due to the fact, that no single computational platform is recognised as suitable for all kinds of tools, the systems used by the designer are likely to be distributed in a heterogeneous environment. This may encompass computer classes ranging from dedicated number crunchers at Computing Centres to the workstation at the designer's own desk. In summary, a modular (i.e. Plug-and-play like) system is required.

The goal is to develop a simple graphical user interface (GUI) to control the total computational environment as well as to achieve a consistent and easy accessible history of the total set of data produced during a whole design project. Hereafter, the prototype of the system is referenced as the Designer's Workbench.

5.2 Developments
So far, all developments in workpackage 4 have been based on an object-oriented methodology, and a "sufficient but minimal approach" has been taken to the resulting products in terms of the numbers of documents etc. All formal documents are based on the widely accepted UML notation, and the specification is built up in a dynamic CASE model. In short, the following has been achieved:
- An early analysis has resulted in a so-called Rich Picture and a set of Role Definitions. By these simple and informal means the hull design process has been captured in a way which satisfies the traditional approach to hull design as well as the advanced and formalised approaches given by workpackage 2 of the CALYPSO project.
- At the architectural level, a general system structure has been developed, which is capable of dealing with true distribution of functionality as well as data. The architecture covers abstraction levels from the very general approach down to a specific configuration that

encompasses all categories of tools within the scope of CALYPSO.

- A simple Design Project Framework has been established in order to support the hull designer when managing a large set of data in an experimental design process. This approach is made operational as a specification of detailed functionality of the Designer's Workbench, which is produced by the end-users themselves (naval architects), in terms of a coherent set of Use Case descriptions. This specification technique has proved to be easy to work with and at the same time to be very powerful too.

- The specification made is met by a technical design, which provides the runtime structure of data in terms of object classes, as well as design of a set of so-called agents for automating complex tasks for the users. The class structure will be in near correspondence to the actual code for implementing the prototype itself, as well as to the formalised structures derived from the STEP (AP216) hull geometry format description. The class design provides the static structures and relations between objects as well as the dynamic behaviour of the structure at runtime — in fact, a close correspondence exists between the functional Use Case descriptions and the dynamic class model.

- A prototype is under development. This will be implemented incrementally throughout the project life time, and the specific features for each increment are given by the end-user's prioritisation of Use Cases descriptions. At the current stage the overall layout and functionality of GUI features are agreed on, and simple integration enables a fundamental set of tools, which is sufficient for proving the feasibility of the overall approach and to gain experience with specifics of the derived design methodology.

5.3 Status and outlook

Future developments shall improve the prototype, especially focussing on the following issues:
- The (Plug-and-play) integration mechanisms: As mentioned above, a very flexible system is to be developed. In CALYPSO we have adopted the term Plug-and-Play for describing this, although we realise that the process of "plugging in a tool" may still be more complex than just to install the tool itself. We aim at an

approach, where any new tool may be integrated nicely, just by providing a standard "wrapper object", that is an object which provides proper communication and access features between the Designer's Workbench and the specific tool.

- Inherent in the above, general communication mechanisms have to be implemented between distributed systems, which in nature are implemented as stand-alone solutions. It shall be possible to define relationships between different types of systems and their I/O flow as well as satisfying I/O requirements for specific systems by general means. A significant part of this is accomplished by implementing a so-called Object Bus, on which data are processed under control of the Designer's Workbench. It has been decided, that this object bus will be implemented in compliance with the CORBA 2.0 standard given by the Object Management Group.

- At least, but not last, general parameter mechanisms are to be developed. Our analyses have shown, that one of the most important obstacles in employing CFD is the very complex parameter-sets required by the CFD codes. As a result, one of the main targets of the development is to automate parameter creation and transfer, and to some extent to validate the parameters in order to enable comparable results from one case to another. This will be accomplished by implementing Agents or Wizards, that shall guide the user through a couple of standard cases, which are expected to cover most cases relevant for "normal" hull design practice.

REFERENCES

[1] H.Nowacki, M.I.G.Bloor, B.Oleksiewicz, Computational Geometry for Ships, World Scientific, Singapore, 1995
[2] H.C.Raven, H.J.Prins, Wave pattern analysis applied to nonlinear ship wave calculations, 13th International Workshop on Water Waves and Floating Bodies, Delft 1998
[3] S.U.Svennberg, B.Regnstrom, L.Larsson, A test of turbulence models for vortices, 3rd Osaka Colloquium on Advanced CFD Applications to Ship Flow and Hull Form Design, Osaka 1998

Practical Design of Ships and Mobile Units
M.W.C. Oosterveld and S.G. Tan, editors.

Computing free surface ship flows with a volume-of-fluid method

C. Schumann

Hamburg Ship Model Basin (HSVA),
Bramfelder Str. 164, 22305 Hamburg, Germany

A finite volume method for free surface flows is described. A VOF-method is used to determine the shape of the free surface. Computed test cases are carried out for the flow around a Series 60 hull and around a tanker with a breaking bow wave.

1. INTRODUCTION

We would like to compute the flow around a steady advancing ship to obtain the wave resistance. Potential flow codes, e.g. *Jensen* [1] or *Raven* [2], are often capable of computing this flow, but if the flow cannot be approximated by a potential (because of breaking waves or spray) these codes fail. In this study, well established methods for solving the Navier-Stokes equations and the Volume-of-Fluid method (VOF) from *Hirt & Nichols* [3] to calculate the motion of the free surface are used. VOF-like methods are able to handle arbitrary free surface shapes. This study is focused on this phenomenon, ignoring the viscosity to avoid additional numerical problems arising from turbulence modelling and resolution of the boundary layer at rigid walls. Most codes which calculate ship flows with free surface use a different kind of technique, in which the grid fits the free surface at each time step. This allows an easy implementation of the boundary conditions but leads to problems in cases of breaking waves and complicated ship forms (e.g. pertruding bulbous bow, barge stern) because the method requires a smooth free surface and smooth changes in the ship geometry. It is expected to avoid these problems by using the described VOF method.

2. MATHEMATICAL MODEL

The transient, incompressible and inviscid flow is computed by solving the so-called Euler and continuity equations. The liquid phase is considered only and the surface tension is neglected.

The unknown variables are the velocities u, v, w and the pressure p. It is defined a cartesian coordinate system with x in longitudinal, y in sideway and z in vertical direction.

The motion of the free surface is described by the convection transport equation of the VOF function F'. This function is defined as 1 in the liquid region and 0 otherwise. The free surface is defined at the border between 1 and 0. At the free surface the dynamic boundary condition has to be fullfilled (constant pressure, $p = 0$).

The equations are solved in a finite domain and boundary conditions are imposed at the ship hull (no flux through it), at the inflow (undisturbed flow), at the plane of symmetry, at the outer lateral boundary, and at the outflow (hydrostatic pressure).

The transient formulation is an aid to solve the flow around a steady advancing ship using a time marching procedure.

3. SOLUTION ALGORITM

A finite volume method is used to discretise the governing equations by subdividing the domain into small cells. The time is split in time steps Δt. The velocities, pressure and F' are defined in each cell centre of the numerical grid. To solve the resulting surface integrals over every cell, the values at the cell faces are interpolated as follows:

- Linear interpolation is used for the pressure.

- The volume fluxes are calculated by a scheme similar to the idea of *Rhie & Chow* [4] (linear

382

interpolation plus a term depending on the pressure derivatives).

- The convected velocities of the momentum fluxes are approximated by the linear upwind differencing scheme.

- The value F' is computed with the upwind scheme plus a correction which is derived from the geometry of the free surface. This is described below.

The linear interpolation is approximated by using only the values of the two adjacent cells of a cell face. In the case of straight grid lines the interpolation is original linear.

The SIMPLE-algorithm is used to solve the coupled system of equations of velocities and pressure. A detailed description can be found in *Cura Hochbaum* [2] or *Ferziger & Perić* [3]. The domain consists of empty cells, full cells and cells which are partially filled and contain a part of the free surface. In partially filled cells, the velocity is extrapolated from adjacent full cells and the pressure is set according to the dynamic boundary condition, instead of enforcing the continuity equation. The equations are solved for each time step as follows:

1. Determining new values of the VOF function F' explicitly.

2. The following items describe the iterative solution of the momentum and continuity equation.

    (a) Set the pressure and extrapolate velocity from full cells into free surface cells.

    (b) Sole the linearised momentum equations with given pressure and fluxes.

    (c) Calculate the new volume fluxes at the cell faces.

    (d) Correct the velocities, fluxes and pressure by the pressure correction scheme and continue with item (a) until the required accuracy is achieved. This procedure results in an implicit approach for the velocities and the pressure. The continuity and momentum equations are iterated in every time step

until their initial error (sum of the absolute errors of all cells) is reduced by the factor ϵ (accuracy).

The first item will be described in more detail, because it differs from usual solving strategies.

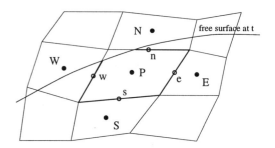

Figure 1: Grid with partially filled cells and the free surface

Determination of the VOF function F': In order to solve the transport equation for the VOF function it is integrated over each cell taking into account the discontinuity at the free surface. This leads to a different definition of the VOF function F'. Now the function represents the volume of liquid in the current cell. To find a similar definition as before F' is divided by the volume of the cell and this is called F. F can take values <u>between</u> 0 and 1 (or equal). In a cell with a value in the range of 0 and 1 (free surface cell), a free surface shape can be constructed. For simplification it is considered a two dimensional grid with free surface position at time t, figure 1. Using a first order explicit approach to solve the VOF function F in the time domain one obtains an equation for F_P in each cell:

$$F_P^1 = F_P^0 + \Delta t \cdot \left(\sum_{f=e,w,n,s} m_f^0 \cdot F_f^0 \right)/V_P . \qquad (1)$$

The subscripts 0 and 1 denotes the time t and $t + \Delta t$ respectively. The volume flux at a cell face is called m and the subscript f marks the current cell face; V_P is the volume of the cell. The summation has to be done over all cell faces (in 3D six faces). According to this equation the VOF function F_P^1 can become lower than 0 or greater than 1. In such cases F_P^1 has to be corrected. For example, if $F_P^1 > 1$, the 'wrong' amount $F_P^1 - 1$ is distributed into the neighbouring cells. These could be the cells W and N. The cells S and E are filled up to 1 and cannot absorb the correction. Between the cells W and N the distribution of the correction is weighted by the fluxes. The correction for $F_P^1 < 0$ works similarly. The VOF function at the cell faces F_f is calculated as follows:

- The free surface is assumed to be linear (plane in three dimensions) and its slope is approximated using the values of the VOF function F of the surrounding cells (see *Hirt & Nichols* [3]).

- With the value F and the slope in the current cell, the position of the free surface can be calculated, such that the liquid volume beneath the free surface represents the actual fluid volume in the cell. The position is solved iterativly. The required value F_f is approximated by the length of the wetted part of the cell face (bold lines in fig. 1) divided by the cell face length (area in three dimensions).

- At each cell face two different approximations of F_f can be done depending on which of the two adjacent cells is considered. The value of the upstream cell is taken.

The first two items represent the most effort in programming, but they are not very time consuming because the procedure is only necessary for the free surface cells.

4. RESULTS OF COMPUTATIONS

4.1. Series 60 ship

Calculations are performed for the Series 60 ship (*Toda & Stern & Longo* [7]) with length l_{PP} =121.92m, block coefficient c_B=0.6 and a Froude number F_n =0.316 with fixed trim and sinkage condition. Three different numerical grids are used as shown in table 1. Systematic grid refinement was not possible because of the requirement of cpu time, but nevertheless some characteristics can be recognised.

The resolution of the grid shown in the third column of table 1 has to be read in longitudinal, sideway and framewise direction. The computational domain was shaped like a quarter of a cylinder with length $3 \times l_{PP}$ and radius l_{PP}. Fig.2 left

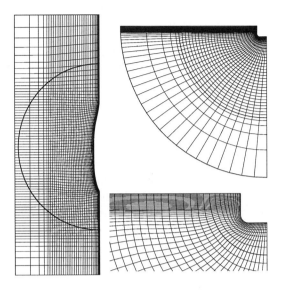

Figure 2: Top view and frame view of medium numerical grid, Series 60

shows the top view of the medium grid illustrating the quite regular resolution lateral and behind of the hull. The figure right shows views of the grid around the main frame.

The figure also shows the circle which indicates the border of a 'numerical beach'. Outside of the circle the vertical velocity component w is dampend by adding a force b. Inside of the circle the momentum equations are unchanged. The beach should dampen waves which are trapped interior the computational domain. The force b in a cell is

$$b = -c_b \, \rho \, V \, \frac{w}{\Delta t} \, \left(\frac{r - r_b}{r_m - r_b}\right)^2 \; \text{if } r > r_b, \quad (2)$$

$$b = 0 \text{ if } r \le r_b, \quad r = \sqrt{(x_0 - x)^2 + (y_0 - y)^2}\,.$$

x and y are the coordinates of the cell centre, x_0 and y_0 are the coordinates of the circle centre, $r_b \approx l_{PP}$ is the radius of the beach, ρ is the fluid density, V is the cell volume, $r_m \approx 1.5 \cdot l_{PP}$ is a constant which controls the increase of damping and c_b is a damping constant. For this calculations c_b =0.5 is used.

The calculations start at rest and are continued until 150s. The sinusoidal acceleration of ship speed from 0 to 10.91m/s is made during 40s. The time steps Δt are adjusted every step to exceed not a *Courant*-number of 0.7. From 0 to 130s

Table 1: Grid dimensions

	grid	resolution	resolution hull	time steps	passed way/l_{PP}	cpu time
	coarse	62×30×32	30×32	3420	10.8	9h 12min
Series 60	medium	74×38×46	35×46	4309	10.8	26h 26min
	fine	107×50×48	61×48	7110	10.8	62h 46min
tanker		93×54×59	44×59	3718	2.9	37h 09min

the calculations are made with the limitation of one cycle (a) to (d) per time step, i. e. with a quite coarse solution of the momentum and mass conservation equations in every step. From 130s to 140s the equations were solved with accuracy $\epsilon = 0.001$ and from 140s to 150s with $\epsilon = 0.0001$. This needs about 4 to 6 cycle per time step or 5 to 8 respectivly. The initial phase of the computation leads a solution which is already near to the steady state, but it may contain large errors, because of the coarse solution. To eliminate these errors the accuracy is increased at the end of the simulation. The last 20s of the simulation time needs about two third of the required cpu time. The computations are carried out on a HP J282 Workstation.

Top of fig.3 shows the resistance coefficient calculated by pressure integration over the hull versus time for the fine grid. It can be seen high and low frequency oscillations of the curve. The first one due to sudden changes of the boundary conditions in cells near to the free surface: Nearly full cells fill up and the continuity equation has then to be fulfilled in the cell leading to a small change in pressure near the free surface. The low frequency oscillation is caused by long waves in the domain, which travel in the direction of the ship and are reflected at the inflow boundary and then again at the outflow boundary. The amplitude of the long waves are damped by the numerical beach and vanish with time. The high frequency oscillation remains still at the end of the computation. For better comparison, it is valid to smooth the resistance curve averaging every 10 time steps. Fig.3 bottom shows the resulting curve and the results of the coarse and medium grid. The finer the grid, the smaller the oscillations.

The mean resistance of the computation on

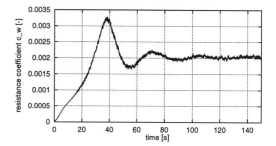

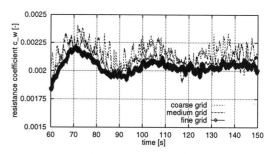

Figure 3: Resistance coefficient C_W for Series 60; top: fine grid and original data, bottom: smoothed results of different grids

the fine grid can be read to $C_W = 2.05 \cdot 10^{-3}$, but the solution is still far from a grid independent solution. The error caused by 'numerical diffusion' is expected to be the largest one and can roughly be estimated by a calculation without free surface on the same grid, because the resistance should be zero in theory. For the considered grid it was found $C_{\text{dif}} = 0.18 \cdot 10^{-3}$. Thus the computed C_W should contain at least an error of similar magnitude. The resistance computed on a much finer grid can be expected to be about $C_W - C_{\text{dif}} = 1.87 \cdot 10^{-3}$. Measurements [7] give a residual resistance of $C_R = 2.5 \cdot 10^{-3}$ using ITTC57 method. *Kajitani* [8] gives smaller C_R in the range from $1.8 \cdot 10^{-3}$ to $2.0 \cdot 10^{-3}$.

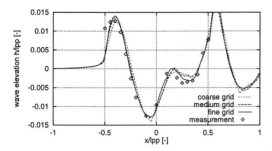

Figure 4: Wave elevation at the hull, Series 60, calculations and experiment [7]

Figure 4 compares the computed and measured wave profile at the hull. The profiles are averaged over the last 10s, because they still oscillate somewhat at the end of the simulation. The wave profiles are insensitive with respect to the grid resolution and the results of the different grids are close to the measurement. Figure 5 shows wave pattern of computations and measurement. The result of the fine grid is very close

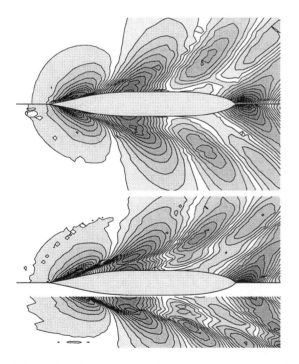

Figure 5: Wave pattern of Series 60, increment of isolines 0.15m; coarse, medium and fine grid and measurement [7], from top to bottom

to the measurement in the region from the bow to behind the main frame. The three grids show convergence to the measurement, finding finer and finer structures.

4.2. Tanker

Eckert & Sharma [9] investigated the flow around a tanker model with and without a bulbous bow. A computation is performed for the ship without bulbous bow, the ballast draft and a Froude number of $F_n=0.2047$. The ship length is $l_{PP}=232.54$m, the draft 6.391m and the block coefficient $C_B=0.758$. The experiments were made using fixed trim condition. The computation simulates a time of 100s and the acceleration to ship speed 9.77m/s is from 0s to 40s. From 0s to 90s only one iteration cycle per time step is used, but from 90s to the end it is used the accuracy $\epsilon = 0.0001$.

Figure 6 shows the computed C_W versus time. At the end, the resistance can be approximated by $C_W=2.0\cdot10^{-3}$. A computation without free surface gives a 'false diffusive' resistance of $C_{\text{dif}}=0.49\cdot10^{-3}$ and the resistance on a much finer grid should be near $C_W - C_{\text{dif}} = 1.51\cdot10^{-3}$. From the experiment the wave resistance is calculated $C_W = C_T - C_V = 1.20 \cdot 10^{-3}$, were C_T means the total resistance coefficient and C_V the resistance coefficient measured in an additional experiment with a double body model beneath the water surface.

Figure 7 shows the computed wave elevation at the hull. It is averaged over the last 10s. From a photo of the bow, some elevations could be read, which are also given in the figure. The bottom of the figure shows the wave elevation at every time step from 90s to 100s. The wave at the bow oscil-

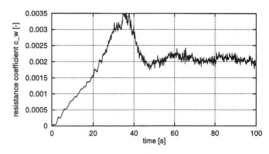

Figure 6: Resistance coefficient C_W of the tanker

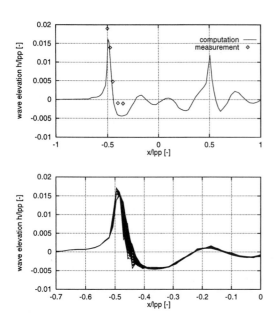

Figure 7: Wave elevation at the hull, tanker; top: averaged results and experiment [9]; bottom: history of wave elevation from 90s to 100s

Figure 8: Wave pattern of the tanker, increment of isolines 0.20m

lates. In the experiment a breaking bow wave was clearly observed and the computed phenomenon seems to be similar.

The wave pattern are given in figure 8. Very high waves can be seen at the bow and at the stern, but the free wave amplitudes are small.

5. CONCLUSION

With the described finite volume method the wave pattern for slender fast ships are computed accurately with acceptable effort. The wave resistance is too inaccurate yet, because of the used relativ small number of cells. The method also seems to be able to predict breaking bow waves, but some more research is needed in this subject.

The 'numerical diffusion' caused by the coarse discretisation of convective terms in the momentum equations leads to the inaccurate computation of the resistance.

REFERENCES

1. Jensen G., Bertram V., Söding H.: Ship wave-resistance calculation; 5th Int. Conf. Num. Ship Hydrodyn., Hiroshima 1989.

2. Raven H.C.: Nonlinear Ship Wave Calculations Using RAPID Method; 6th Int. Conf. Num. Ship Hydrodyn., Iowa City 1993.

3. Hirt C.W., Nichols B.D.: Volume of Fluid (VOF) Method for the Dynamics of Free Boundaries; Journal of Computational Physics 39 (1981).

4. Rhie C.M., Chow W.L.: Numerical study of the turbulent flow past an airfoil with trailing edge separation; AIAA J. 21/11 1983.

5. Cura Hochbaum A.: A Finite-Volume Method for Turbulent Ship Flows; Ship Technology Research 41/3 1994.

6. Ferziger J.H., Perić M.: Computational Methods for Fluid Dynamics; Springer-Verlag Berlin Heidelberg 1996.

7. Toda Y., Stern F., Longo J.: Mean-flow Measurement in the Boundary Layer and Wake and Wave Field of a Series 60 c_B=0.6 Ship Model for Froude Numbers .16 and .316; Iowa Institute of Hydraulic Research Report No. 352 August 1991.

8. Kajitani H.: A Wandering in Some Ship Resistance Components and Flow; Ship Technology Research 34 1987.

9. Eckert E., Sharma S.D.: Bulbous bows for slow, bluff ships; Part II (in German), Jahrbuch der Schiffbautechnischen Gesellschaft 1970, 64. Band, Springer Verlag 1971.

Practical Design of Ships and Mobile Units
M.W.C. Oosterveld and S.G. Tan, editors.

Development of computational system for flow around a ship and its validation with experiments

Wu-Joan Kim, Suak-Ho Van, Do-Hyun Kim, Geun-Tae Yim

Ship Performance Department
Korea Research Institute of Ships & Ocean Engineering
171 Jang-Dong, Yusung-Gu, Taejon 305-343, KOREA

A computational system "WAVIS" for the hull form evaluation is developed. The system includes a pre-processor, flow solvers, and a post-processor. The pre-processor is composed of hull form presentation, surface mesh and field grid generation. The flow solvers are for potential and viscous flow calculation. The post-processor has graphic utility for result analysis. All the programs are integrated in a GUI-launcher package. To validate the developed CFD programs, the measurements of wave patterns and local velocity fields are carried out in the towing tank for two modern commercial hull forms. The calculated results are in good agreement with the experiment, illustrating the accuracy of the numerical methods employed for WAVIS.

1. INTRODUCTION

Computational Fluid Dynamics(CFD) technology has spread throughout the entire field of fluid engineering. It was a decade ago that a few ship hydrodynamicists had applied CFD to calculate flow around a ship, however, such an application was limited only to basic research. Recently some ship-building yards are trying to utilize computational tools for performance prediction in the basic design stage. But it is not easy to apply such commercial packages to a modern practical hull form in spite of their versatility, since those are developed for general flow calculation. The successful application of CFD tool into the design of hull form depends upon user-friendliness as well as accuracy. The needs for a reliable numerical tool, which is concentrating upon the computation of flow around a ship hull and serving for basic hull form design, are emerging.

To cope with the aforementioned request, a computational system(WAVIS: WAve and VIScous flow analysis system for hull form development) for flow around a modern commercial ship is developed with a very user-friendly preprocessor, a robust turbulent flow solver as well as potential panel code, and a graphic post-processor. The most time-consuming task to use a flow solver is known to be preparing surface and field meshes based on a given hull geometry, even though modern commercial hull forms have a typical regularity. A program for the presentation of hull surface using station offsets and centerline contour has been developed. This novel method employs non-uniform parametric splines with predetermined waterline end-shapes. The field meshes can be also obtained from a solution of the Poisson equation with surface meshes and prescribed boundary topologies. A potential panel method of Rankine source is developed to solve wave-resistance problem with the linear and nonlinear free surface condition. A robust Reynolds-averaged Navier-Stokes solver is employed to provide all the turbulent flow information such as surface pressure and velocity field at the propeller plane, etc. All of the above programs with a post-processor having graphic utilities can be accommodated with the menu-driven capability in a 128MB workstation.

To validate the developed computational tool, it is essential to confirm the calculated results against the reliable experimental data. A few well documented experimental data are available for flow around a ship, however, the chosen hull forms are not practical and the data is usually partial. Therefore, in order to provide the reliable and complete flow information, flow measurement has been carried out for a 3600 TEU container ship and a 300K VLCC in the towing tank to document local velocity field as well as generated wave pattern. This experimental data can be used for the validation of any CFD tools developed for the computation of flow around a practical commercial ship.

In the sequel, the details of developed computational system for flow around a ship are presented and the calculated results are compared with the experimental data for the container ship and the VLCC.

2. COMPUTATIONAL SYSTEM: WAVIS

2.1 GUI-Launcher

For the integration of the developed CFD programs, a Windows/UNIX-based launcher program is utilized. The main objective of the launcher is to provide user-friendly environment for computation and analysis. Seven Fortran programs are integrated for hull form display, surface mesh and field grid generation, potential and viscous flow calculation, and result analysis. Input data can be put via the input dialog window. Each program module has different input window. Radio buttons are operated to set parameter or to execute the program for each module. Tool bars are located at the top of the window.

After the offset table is prepared in proper format, the 'Offset View' module can be facilitated to see the body plane and side profiles to confirm the data set. The 'Surface Mesh' module will provide the surface mesh generated as panels for the potential flow solution, or boundary grids for the viscous flow calculation. The 'Potential Flow' module provides the potential flow solution with linear or nonlinear free surface condition for wave-making resistance prediction. Turbulent flow solution can be obtained by using the 'Viscous Flow' module. In the end, two postprocessors will facilitate result analysis and provide graphic display. The skeleton of WAVIS is shown in Fig. 1.

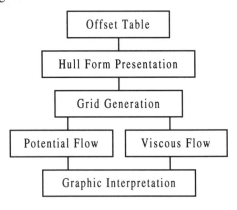

Fig. 1 Contents of WAVIS

2.2 Pre-processor
2.2.1 Surface mesh generation

It is cumbersome to describe a modern hull form by using only the offset table, since the complexity of bow and stern shapes always require more informa-tion. To make it easy and quick to present hull form and generate surface meshes with acceptable accuracy, the user-friendliness is emphasized in the newly developed program. In the present study focus is laid only on commercial ships such as tanker, bulk carrier, and container ship with bulbous bow and stern end bulb, although the same procedures can be applied to other hull forms. The procedures for surface mesh generation are described in the followings.

First of all, three-dimensional space curves are presented by using non-uniform parametric spline with the slope controlled at any given points. The parameter chosen for interpolation is the accumulated chord length. The Ferguson basis, utilizing position vectors and tangent vectors(slopes) at the ends, are chosen since it is useful to handle the discontinuities of slope or curvature existing in hull surface profiles[1]. If the curvature or slope at the point is not specified from the offset table, the condition of continuous slope and curvature is applied, which is equal to the cubic spline interpolation.

The most important input data is an offset table provided by the designer. The offset table usually contains the offset points of 25~30 stations and the centerline profiles of bow and stern. The developed program converts automatically the offset table into the proper form containing slope-controlling integers at every special offset points of discontinuous slope or curvature. After the file conversion is completed, non-uniform parametric spline is utilized to generate sufficiently many interpolated points for the body plan and the side profile.

As it is well known, more information than in the offset table is required to construct complicated bow and stern shapes. Thus, it is necessary to determine additional station offsets near the bow and the stern. In the present method, at first, waterlines are defined to specify additional station offsets. But the shapes of waterline ending are usually not clarified in the original station offset. However, it is possible to define the end shapes of waterline near the bow and the stern by using the predetermined parameters, which choose waterline endings from natural spline, normal spline, ellipse, parabola, hyperbola, and their combination. After the waterlines are obtained, 50 each of additional stations near the stern and the bow are generated using the waterline intersection points at the constant longitudinal location. These 100 additional station offsets carry the information on complex stern and bow shapes.

To make the surface mesh flexible, the longitudinal and transverse distribution can be arbitrary adjusted with the girth length ratio. Once the transverse distribution is given, grid nodes at each station can be found, utilizing the length ratio of fine station curves. The nodes at the same transverse length ratio, are connected to complete the longitudinal node lines. The longitudinal grid length ratio of node lines will finally provide the hull surface meshes. It is noteworthy that all the above procedures can be processed within 10 minutes on PC when the original offset table is given. This efficient hull surface mesh generator will improve the usability of available computational tools for the hull form evaluation. Fig. 2 shows the generated hull surface mesh for the 3600TEU container ship. For the details of the present procedures, see [2],[3].

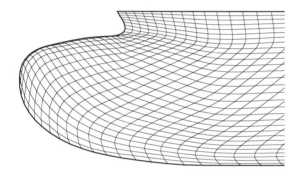

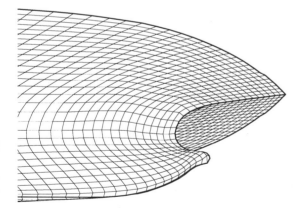

Fig.1 Surface mesh for the 3600TEU container ship

2.2.2 Field grid generation

As described in the previous section, the surface mesh contouring bow and stern profiles constitutes one boundary surface of three-dimensional field grid system for viscous flow calculation. To implement the turbulence model and near-wall specifications for high-Reynolds number flow, it is necessary to determine the distance and intersecting angles of the first grid points off the wall. With the aid of O-O topology of field grid system and the surface mesh contouring bow and stern profiles chosen in the present study, the above requirement can be easily satisfied.

Among six boundary surfaces, the hull surface mesh is already given and the outer surface mesh can be defined algebraically. The two-dimensional grid systems on four other planes of symmetry are obtained by solving 2-D Poisson equations. After constituting six boundary surfaces, it is possible to use the algebraic method like trans-finite interpolation(TFI) to fill out the inside grid system. It is rather simple, but does not guarantee the normality of grid on the hull surface. In the present study 3-D Poisson equation is solved to meet the requirement of grid orthogonality and controllability. Sorenson's method, which is also known as GRAPE[4], is extended into three-dimension to define the grid-controlling function of Poisson equations. Since the initial guess for the Poisson iterative solver for grid generation is usually very poor, the main focus for grid generator should be laid upon the robustness of Poisson solver. In the present study the Poisson equation is discretized by using the weighting function scheme[5] and resulting linear equations are solved by the Modified Strongly Implicit Procedure(MSIP)[6]. The combination of the weighting function scheme and the MSIP linear solver provides the robust and efficient solution of Poisson equation for three-dimensional grid generation. The generated field grid system for the 300K VLCC is shown in Fig. 3.

2.3 Flow solvers
2.3.1 Potential flow solver

The wave resistance and generated wave pattern are predicted using the potential panel method. It is possible to perform viscous flow calculation with free surface, however, it is too expensive to use in daily basis. Thus, in WAVIS, the wave resistance problem is treated by the following potential flow solver.

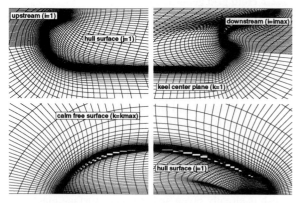

Fig. 2 Generated field grids for the 300K VLCC

The Rankine source method(RSM) is thus adopted to perform the potential flow calculation of practical hull forms and the raised panel approach[7] is used for handling the nonlinear free surface condition. To enforce the radiation condition the present RSM employs 4-point upwind-difference operator. Furthermore, upstream collocation points are shifted to smooth out the source strengths and to prevent upstream waves at high speeds. Pressure integration and longitudinal cut method are implemented to calculate the wave resistance. However, the longitudinal cut method is more preferable since it avoids the unreasonable wave resistance for tankers at low speeds. For the application to practical hull forms with the large bow flare, flat stern overhang and transom stern at low speeds, special care is taken to provide reasonable solution. The details of the present panel method can be found in [8].

2.3.2 Viscous flow solver

For the computation of turbulent flow, the Reynolds-Averaged Navier-Stokes(RANS) equations are numerically solved by the explicit finite-difference method. For the turbulence closure, the Baldwin-Lomax model is exploited. The numerical method employs the second order upwind scheme for convection terms, and the central-difference scheme for diffusion terms. The discretized momentum equations are integrated in time by using the four stage Runge-Kutta method. To obtain the pressure field, the pressure-Poisson equation is solved. The non-staggered grid system is chosen and a specific scheme for the discretization of the pressure-Poisson equation is introduced to suppress the pressure oscil-

lations. To improve the robustness, local time stepping with residual smoothing is applied. More details of the numerical methods for the solution of RANS equations can be found in [9,10].

2.4 Post-processor

After the solution of potential and viscous flow calculation is achieved, it is necessary to investigate the results. The WAVIS includes the simple graphics tools for the display of calculated results. For potential flow, the wave elevation along the hull surface is obtained. Wave profiles along a longitudinal cut are also available. Wave resistance is calculated by using surface pressure integral, the momentum approach, and the longitudinal cut method.

For the viscous flow solution, friction and pressure drag can be calculated. Flows in thick boundary layer and wakes around the stern region should be examined since they will determine the propulsive efficiency of the hull and decide the inflow at the propeller plane. Since the grid system does not coincide with the constant x-planes, interpolation is carried out. In the present study, for given x-planes, 2-D TFI is applied to supply grids on which the values are interpolated. Tri-linear interpolation is used to provide the interpolated plane values from 3-D velocity field of O-O topology.

For the graphic display of the post-processed results, simple mesh plot and X-Y plot are included in WAVIS. It is also possible to use the advanced graphic tools like Tecplot, since the output ASCII files are saved in the same format.

3. VALIDATION WITH EXPERIMENTS

To validate the developed computational tool, the documentation of local mean velocity field as well as generated wave pattern is carried out in the KRISO towing tank. A 3600TEU container ship and a 300K VLCC with bow and stern bulbs, recently designed by KRISO, are selected for the test. The developed CFD programs are applied for the above practical hull forms and the calculated results are compared with experimental data. The Reynolds numbers in the test are 1.4×10^7 and 4.6×10^6, while the Froude numbers are 0.260 and 0.142 for the container ship and the VLCC, respectively.

3.1 Experiments

To document the full details of flow characteristics around practical hull forms, the wave patterns and local flow velocity components are measured in the towing tank. The servo-needle type wave-height gauges are used to measure wave patterns of full spectrum on the free surface. First, the wave profiles along longitudinal cuts at 36~39 lateral locations are measured. Next, wave-height gauges are connected to three-dimensional traversing mechanism to survey the waves around the bow and stern region.

A rake of five-hole Pitot tubes is utilized to measure the local velocity fields at six stations near stern including the propeller plane. Traversing mechanism is used to locate the rake in 150~250 positions, resulting in 700~1200 points of measurement, since the rake is composed of five 5-hole Pitot tubes. The measured velocity fields clearly show the effect of viscosity on the stern flow, including the formation of bilge vortex.

The experimental data contained in this paper can provide the valuable information on the flow characteristics of two types of modern practical hull forms: one of small C_B with high speed and the other of large C_B with low speed. The present experimental data can also provide good test cases for the validation of the CFD code for both inviscid and viscous flow calculations[11].

rection. The origin of the coordinates is located at the midship and calm free surface. All the quantities are non-dimensionalized by the speed and the length between perpendiculars(L_{pp}) of the model ship.

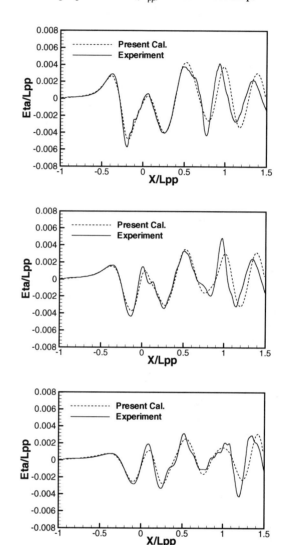

Fig. 5 Comparison of wave profile along longitudinal cuts for 3600TEU container ship (Fn=0.260, from the top: Y/Lpp=0.1024, 0.1509, 0.2167)

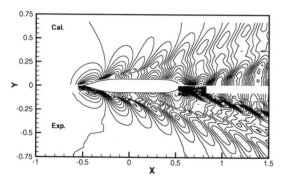

Fig. 4 Comparison of wave pattern for the 3600TEU container ship (Fn=0.260)

3.2 Comparison of results

The calculated wave patterns and local mean velocity components are compared with the experiments. The Cartesian coordinates (X,Y,Z) are used for displaying the data, where X denotes the downstream direction, Y starboard, and Z the upward di-

3.2.1 Wave patterns

The developed potential panel method is applied to calculate wave resistance and generated wave patterns for two practical hull forms, namely the KRISO 3600TEU container ship and the 300K VLCC. The calculation is performed after modification of transom stern, since the effect of transom does not appear clearly for the present case. First, the potential flow around the container ship at Fn=0.260 with the non-linear free surface condition is performed. 80x16 panels are distributed on the hull surface, while 118x16 panels are used on the free surface, corresponding to 20 panels per a fundamental wave length. The calculated wave pattern is compared with the measured data in Fig. 4. The prediction is in close agreement with the experimental data. Fig. 5 shows the comparisons along the longitudinal cuts at Y/Lpp= 0.1024, 0.1509, 0.2167, respectively. A fairly good agreement is illustrated in these figures except peaks near crest and trough.

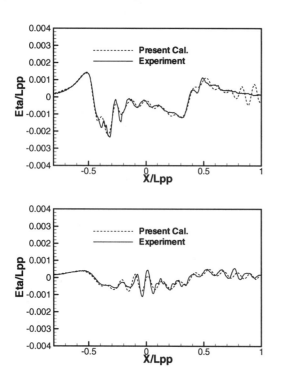

Fig. 6 Comaprison of wave profile along longitudinal cuts for the 300K VLCC (Fn=0.142, from the top: Y/Lpp=0.1169, 0.2086)

In the next, the wave pattern around the VLCC at Fn=0.142 is calculated. 80x16 panels are used for the hull surface, but 213x16 free surface panels are distributed, since the Froude number is very low. As a result, 15 panels per a fundamental wave length are used. The calculated results are compared with the measured data in Fig. 6 for Y/Lpp=0.1169, 2086. The bow wave systems are well predicted, although a little bit of discrepancies near parallel middle body are seen. It is believed that the agreement shown in the present paper is one of the best among the published potential flow solution for practical hull forms.

3.2.2 Turbulent flow

The grid generator and turbulent flow solver described in the preceding sections are applied for two practical hull forms. 101x41x50 grids are used with the first grid interval of 1.0×10^{-5}. The axial velocity distributions at stern stations are compared with experiments, as shown in figures 7 and 8. It should be noted that the measurements were carried out in towing tank, thus, flow field was certainly affected by wave generation on the free surface. However, the present calculation ignores the effect of waves, since the inclusion of free surface condition is rather cumbersome for the viscous flow calculation around a practical hull form with flat stern overhang. Instead, Neumann condition is applied on the calm free surface. Calculated axial velocity contours of both ships agree well with experiments in spite of the complicate stern shape. The container ship (3600TEU) of the present study does not have strong distortion of wake contours, since the bilge vortex formation is very weak. On the other hand, wake contours of VLCC (300K) at propeller plane shows the so-called hook-like shape, which implies that bilge vortex formation is much stronger than that of the container ship.

4. CONCLUDING REMARKS

A computational system(WAVIS) for the hull form evaluation is developed to provide an efficient numerical tool in the basic hull form design. The system includes a pre-processor, flow solvers, and a post-processor. Fortran programs for hull form presentation, surface mesh and field grid generation, and potential and viscous flow calculation are integrated in one unit with GUI launcher. To validate the developed CFD programs, the measurement of wave patterns and local velocity fields are carried out in the towing tank for two practical hull forms. The calculated results are in good agreement with experiment, proving the accuracy of the numerical methods. It is believed that the developed programs can provide a computational tool for the hull form evaluation in the basic design stage of commercial ships.

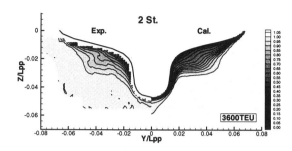

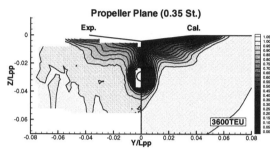

Fig. 7 Comparison of axial velocity contours for the 3600TEU container ship (Re=1.4x10^7, from the top: X/Lpp=0.4000, 0.4825)

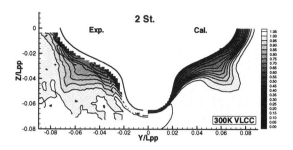

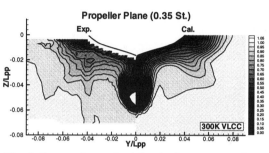

Fig. 8 Comparison of axial velocity contours for the 300K VLCC(Re=4.6x10^6, from the top: X/Lpp=0.4000, 0.4825)

REFERENCES

[1] Nowaki, H., Bloor, M.I.G., Oleksiewicz, B., Computational geometry for ships, World Scientific Publishing Co., 1995

[2] Kim, W.J. et al., "A computational study on turbulent flow around a practical hull form with efficient grid generator," Proceedings of The 3rd Osaka Colloquium on advanced CFD application to ship flow and hull form design. 1998. 5.

[3] Van, S.H., Kim, W.J., Kim, D.H., Yim, G.T.,"A practical method of generating hull surface mesh for commercial ships," Proceedings of China-Korea Marine Hydrodynamics Meeting, 1997. 8.

[4] Sorenson, R.L., "A computer program to generate two-dimensional grids about airfoils and other shapes by the use of Poisson equation," NASA TM 81198, 1980.

[5] Hsu, K, Lee, S.L., "A numerical technique for two-dimensional grid generation with grid control at all of boundaries," Journal of Computational Physics, Vol. 96, pp. 451-469, 1991.

[6] Schneider, G.E., Zedan, M., "A modified strongly implicit procedure for numerical solution of field problems," Numerical Heat Transfer, Vol. 4, pp. 1-19, 1981.

[7] Raven, H.C., "A Practical Nonlinear Method for Calculating Ship Wavemaking and Wave Resistance," 19th Symposium on Naval Hydrodynamics, Seoul, Korea, 1992.

[8] Kim, D.H., Kim, W.J., Van, S.H., Kim, H., "Calculation of potential flow around modern commerical ships using nonlinear free surface condition," Proceedings of 3rd International Conference on Hydrodynamics, Seoul, 1998.

[9] Van, S.H., Kim, H.T.,"Calculation of turbulent flow around HSVA and Mystery tankers," Proceedings of CFD Workshop Tokyo 1994. 3.

[10] Van, S.H., Kim H.T., "A computational study on turbulent flow characteristics around full form tankers," Journal of Hydrospace Technology, Vol. 2, No. 2, 1996.

[11] Van, S.H. et al., "Experimental investigation of the flow characteristics around practical hull forms," Proceedings of the third Osaka colloquium on advanced CFD applications to ship flow and hull form design, 1998.

Practical Design of Ships and Mobile Units
M.W.C. Oosterveld and S.G. Tan, editors.

A new hull form for a Venice urban transport waterbus: Design, experimental and computational optimisation.[1]

H.C.Raven[a], M.van Hees[a], S.Miranda[b], C.Pensa[b]

[a] Maritime Research Institute Netherlands (MARIN), P.O.Box 28, 6700AA Wageningen, Netherlands

[b] Università degli Studi di Napoli "Federico II", Dipartimento di Ingegneria Navale (D.I.N.), Via Claudio 21, 80125 Napoli, Italy

A combination of naval architects experience, modern CFD tools and model testing has been used in the hydrodynamic design of a new waterbus for use in the city of Venice, required to have minimum wave generation at low speed in shallow water, and minimum resistance at higher speed. The free-surface panel code RAPID has been extended for wave making in shallow channels with sloping sidewalls, and has been found to adequately predict the effect of hull form changes and the phenomena occurring in shallow water.

1. INTRODUCTION

With the expanding use of fast passenger ferries and an increasing awareness of environmental issues, the wave wash of ships is a topic of growing interest worldwide. Ship waves may disturb or damage moored or sailing ships, waterway banks, or marine life. A particular example is found in Venice, Italy. The passenger transport system in the city and lagoon of Venice is mostly based on public services, managed by Azienda del Consorzio Trasporti Veneziano (A.C.T.V.). The A.C.T.V. fleet consists of 54 waterbusses or "vaporetti" and 59 motor crafts, plus a few larger units. Almost all of the vessels are steel built, equipped with fixed axis propellers and manoeuvring rudders fitted at the stern, and powered by diesel engines. Since the transport service takes place along the main canals of the city, the vessels operate in restricted waters, sometimes very shallow. The size of the public water traffic flows (5-10 minutes between subsequent passes) is not very large, but together with the freight traffic and private transport it causes a very high risk of erosion to monumental buildings, as a result of wave impact on the channel walls, pressure fields and flow along the walls, and propeller slipstreams.

The EC sponsored Brite-Euram project "LIUTO" (Low Impact Urban Transport water Omnibus) has two major innovative aspects in this regard:

- the design and development of a novel urban waterbus to be used on the Venice waterways in the next decade.
- the extension, application and validation of the latest tools for predicting and minimising wave wash.

The new vessel will be the prototype of the Venice urban transport fleet for the 2000's, and shall substantially improve the passenger comfort and reduce the overall impact of transport through Venice. It will be built in GRP, will have a hybrid diesel-electric propulsion and steerable twin-propeller units.

While other aspects of the project are discussed in other papers in this symposium [1,2], the present paper addresses the hull form development of the LIUTO M/B and the wave generation and wave impact issues which posed additional requirements to the design. A basic hull form design has been made by the D.I.N. and has been refined with model tests (Section 2). The next stage was a further optimisation study by MARIN, based on wave pattern calculations using the CFD-code RAPID for several variations of the DIN design (Section 3). The same code has then been extended to include the conditions critical for erosion (Section 4). Further tank tests have then been carried out at MARIN, first in deep water, then in shallow water and in a typical channel cross section, both to study the performance of the vessel and to validate the predictions by the

---

[1] This research project was sponsored by the EC under contract BRPR-CT96-0210. This support is gratefully acknowledged.

extended RAPID code. Altogether the design problem, the tools used and the experimental validation have several features justifying their presentation.

2. THE BASIC HULL FORM DESIGN

The basic hull form design has been made by the D.I.N. based on the following data supplied by A.C.T.V.:

1. M a i n   d i m e n s i o n s :
   $L_{OA} \leq 25.00$ m, $B_{MAX} \leq 5.00$ m
2. P a y l o a d : 250±5 passengers ($\cong 17.5$ t)
3. Speed requirements and water depth (h):
   - Urban speed $V_1 = 5.94$ kn; h = 4.5 m
   - Max full load speed $V_2 = 10.0$ kn; h = 10.0 m
   - Max half load speed $V_3 = 10.8$ kn; h = 10.0 m
4. Maximum power: 147 kW (power margin 5% at 10.8 kn);
5. Transverse metacentric height ≥ 0.30 m;
6. Free board ≥ 0.20 m, half of the passengers staying on one side with a crowd factor of 4 / m^2;
7. Minimum hydrodynamic pollution, i.e. low wave generation and weak propeller slipstream.
8. Manoeuvrability as high as possible
9. Minimum shallow water effects on wave pattern characteristics and resistance.

Regarding points 7 and 9, the CMO (Wave Motion Committee of Venice Municipality) imposes a maximum value of the residual resistance. This somewhat unusual requirement is meant to limit the wave wash intensity.

Item 3 asks for a compromise between best performance in two, very different conditions: low speed in the confined waterways inside the city, and high speed in the deeper water of the Venetian lagoon. At low speed the focus is on wave wash and on the overspeed of the flow next to the hull. For high speed the required propulsive power should be minimised, since most of the fuel is consumed during navigation on the lagoon.

In a first step an analysis of the A.C.T.V. ships currently in service and D.I.N. database was carried out. Afterward the new hull was designed starting from a probable value of the ship displacement and with the following guidelines:

- To keep as high as possible the ratio L_{WL}/B_{WL} in order to reduce the height of the waves generated and the wave resistance;
- To keep as high as possible the ratio L_{WL}/L_{OA}

- To keep as high as possible the difference between L_{CB} and L_{CF} in order to oppose the tendency to trim at high Froude numbers;
- To minimise the slope of buttocks astern to improve the ship's ability at high relative speeds;
- To minimise the inertia moment of ahead half waterplane area in order to reduce the height of the second transverse wave;
- To reach a compromise between the best values of C_P for maximum speed and for lower speed in a canal;
- To optimise the ratio between the values of afterbody and forebody prismatic coefficients;
- To minimise the lateral area coefficient in order to improve the manoeuvrability and still to obtain a ship length larger than that of the ships currently in service;
- To adopt a sectional area curve that minimises shallow water effects.

On the basis of these considerations, three hulls (LIUTO 1.3, LIUTO 1.4, LIUTO 1.5) were designed. Two of these were derived from the third one characterised by the lowest value of C_P. Particularly the transverse sections were moved by Lackenby method in order to obtain the given ratio between the values of astern and ahead prismatic coefficients.

	LIUTO 1.3	LIUTO 1.4	LIUTO 1.5
C_P	0.532	0.556	0.580
C_{PF}	0.518	0.560	0.602
C_{PA}	0.560	0.560	0.560

Three models with scale ratio 12 were made and resistance tests in deep water were carried out in the ship speed range 5.5 ÷ 12 kn, in the towing tank of D.I.N. This has a length of 145 m, the width is 9 m and the depth 4.5 m. For the shallow water experiments a false bottom has been constructed on a length of 60 m.

In these tests, the LIUTO 1.3 hull showed the best performance. Based on this result and a closer examination of the stability requirements, two new hulls (LIUTO 1.3.1 and LIUTO 2.1) were designed. The LIUTO 1.3.1 hull was derived from LIUTO 1.3 by affine transformation. Both the breadths and the length were slightly expanded so the value of transverse metacentric radius was increased and the length was adapted to the maximum value allowed by A.C.T.V. LIUTO 2.1 was designed by increasing both the waterplane area and the maximum draught at ballast (i.e. higher metacentric height); and minimising the consequent changes of the sectional area curve.

The two models, with scale ratio 5.6, were tested in deep water and in shallow water at two load conditions. Additionally several tests were carried out to evaluate the sensitivity to trim and load variations. During the tests, wave elevations were measured using three capacitive probes located in the tank at 20%, 40%, 60% of model L_{WL} from the hull centreplane.

The analysis of the experimental results extrapolated to the ship was made by comparing:
- The performance of the new hulls;
- The performance of the chosen hull (LIUTO 2.1) with the available data of the boats in service, i.e.: E-1 hull in shallow water and Serie 80 hull in deep water (Fig.2)
- The residual resistance curve of the chosen hull with the maximum curve imposed by the CMO;
- The ship performances at different displacements and trim angles;

After these comparisons LIUTO 2.1 was selected on the basis of its best stability characteristics and powering performance very close to LIUTO 1.3. The results obtained, as showed by the figures, were good, the values of the effective power of the new hulls are lower by 20÷35% than the values of the hulls currently in service.

The main characteristics of LIUTO 2.1 hull are as follows: L_{OA} = 25.045 m;
B = 4.708 m; L_{WL} = 23.395 m
B_{WL} = 4.564 m; C_B = 0.379;
C_P = 0.529; C_{WP} = 0.669.
The general form of the vessel is shown in Fig.6.

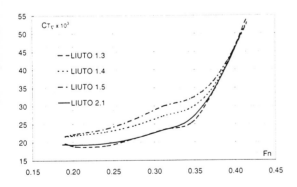

Fig. 1 Experimental resistance curves for initial design variations.

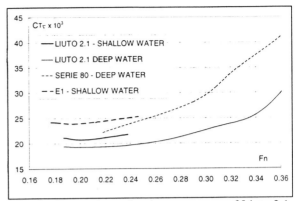

Fig. 2. Comparison of resistance curves of Liuto 2.1 and existing vessels.

3. COMPUTATIONAL OPTIMISATION
3.1 The RAPID method

The main features of the design thus having being fixed, the second stage was started: a refinement of the hull form using CFD tools. This study was performed by MARIN through a succession of RAPID calculations and hull form modifications.

The well-known code RAPID [3,4] is a panel method that computes the inviscid flow and wave pattern generated by a ship at steady forward speed. Source panels are distributed over the wetted part of the hull and over a plane at a specified distance above the wave surface. An iterative procedure, with repeated adjustments of the wave surface, the boundary conditions, the panellings and the attitude of the hull, leads to the final solution satisfying the complete nonlinear boundary conditions. The resulting streamline pattern and pressure distribution on the hull, the wave pattern, and the wave resistance, dynamic trim and sinkage provide much insight in the flow and its relation with the hull form, and indicate how further improvements can be achieved. RAPID thus has become the main workhorse for hydrodynamic hull form optimisation at MARIN, hundreds of calculations being carried out per year.

4.2 Optimisation study

Calculations were made both for the low-speed shallow water operation and for the high-speed deep-water conditions. These are partially conflicting, and a decision had to made on the primary focus of the study. The first data obtained on LIUTO 2.1 showed very moderate wave generation at low speed: According to the Naples experiments, at 6 knots waves of about 0.06 m (peak-to-trough) are generated at 14 m distance from the path of the vessel. These values are very moderate in comparison with the fleet currently in operation. Figure 3 compares longitudinal cuts through the computed low speed wave patterns for LIUTO 2.1 and two existing designs, at 4.65 m off the vessel centreplane. The wave height divided by ship length is here plotted against a non-dimensional length coordinate X/L, which is zero at the mid ship section (1/2 Lpp) and +0.5 at the stern. The bow wave system, at the left side of the figures, is responsible for the largest wave heights. According to these initial calculations, the E-1 prototype (C0312) generates waves of about 0.15 m, the ACTV Series 80/A M/b (C0313) a little more, compared to 0.11 m for LIUTO 2.1.

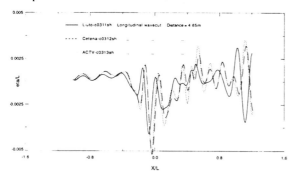

Fig. 3. Low speed wave cuts of C0311, C0312 and C0313, water depth 4.5 m.

In the actual situation in Venice, significant wave heights of 0.05-0.06 m are measured in the centre of the Canal Grande close to Rialto Bridge. Values up to 0.1m and higher are frequently observed due to passing boats. The waves generated by LIUTO 2.1 at 5.94 knots are well within the highest one third of the spectrum and will hardly increase the energy contained in the wave spectrum. Therefore it was decided to focus the further optimisation primarily on the high-speed condition.

Even so, througout the following process the low-speed wavemaking has been monitored. As an example, Figure 4 shows the calculated wave profile along the hull, for C0311 (DIN-LIUTO 2.1) at 5.94 knots. The most prominent wave system predicted is the stern wave; but from experience and the initial comparisons with the DIN tests it was obvious that in reality this will be strongly reduced by viscous effects. At this speed the flow detaches ahead of the transom edge, a phenomenon quite sensitive to viscous effects and hard to model in a calculation anyway. Therefore the low-speed stern wave

predictions were disregarded in the optimisation. On the other hand, the distinct second wave crest of the bow wave system is to be taken seriously and asks for reducing the waterline entrance angle or fitting a bulbous bow.

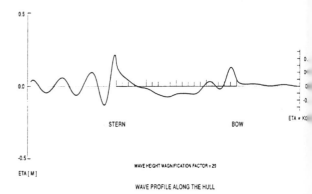

Fig. 4. Low speed wave profile along C0311

Turning to the 10 knots calculations now, these in the first place indicated the transom to be too high above the still water surface. Fig.5 (top) clearly shows that the wave surface is drawn upwards by the rising afterbody before being released at the transom. In the second variant C0311A, the transom was therefore lowered and the buttock slope aft was reduced. Fig.5 (bottom) shows that this produces a much smoother stern wave system, and it has therefore been basically adopted in all following variations.

With the new afterbody, the calculations indicated that the additional lift caused by the more concave buttocks resulted in a small increase of the forward dynamic trim and thus of the bow wave height at high speed. In the next versions the buttock curvature in the afterbody was therefore carefully adjusted to find the right balance between suppressing the stern wave system and limiting the bow-down trim.

At the same time, for C0311A the waterline entrance was made sharper by introducing slightly hollow waterlines and lowering the bow contour, thus lengthening the submerged hull. Together with the modified stern, the calculated wave resistance coefficient at 10 kn decreased by 15%. Since the resistance predicted by RAPID corresponded well with the DIN measurements, this figure may be considered realistic.

On the basis of statistical data and experience, a

small bulbous bow was expected to be advantageous at high speed without noticeable negative effects at low speed. An initial bulbous bow design, with sharp waterlines and little rounding, appeared to yield some improvement at high speed but to be unfavourable for lower speeds. The next version was given a slightly larger and more conventional bulbous bow, which was lifted towards the still water surface. At high speed it performed similarly to the previous version, at low speed it was marginally better than without bulb.

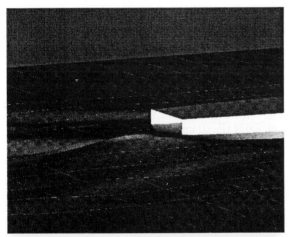

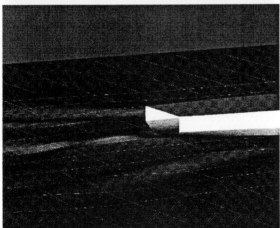

Fig. 5. Computed wave pattern around the stern, for LIUTO 2.1 (top) and C0311F (bottom).

Fig.6 (above). Hull form of C0311F

Statistics suggested that more could be gained at high speed. Therefore, in C0311D the bulbous bow was slightly enlarged by further rounding of the bulb waterline entrance. Evidently all these adjustments involved a careful tuning of the bulbous bow and forebody design, for which visualisations of the detailed flow field were indispensable.

This final form meant a considerable improvement at high speed, whereas the low speed wave making was comparable with C0311A. From a propulsive performance and wave making point of view, the bulbous bow version C0311D therefore is preferred for the design condition at both high and low speed. It was then to the owner ACTV to decide whether this would justify the larger construction costs, vulnerability and sensitivity to operational draft and trim variations. The decision was then taken not to pursue the bulbous bow. Prime reason were the expected loading variations, and the unfeasibility to study the performance for the entire spectrum of operational conditions that emerged from the design work.

The challenge was now to further improve the latest non-bulbous form (C0311A) in one final attempt. The fore body was given a somewhat deeper forefoot and the beam at the water line was marginally reduced. The fore body sections were given more S shape, moving the section centroid a little downward. The combined effect of the S curvature and beam reduction resulted in a finer waterline entrance. Also a last adjustment of the afterbody form was made. The resulting final form and the hull panelling used are illustrated in Fig.6.

This hull form C0311F appeared to indeed outperform C0311A, in particular at high speed. In Fig. 7 the 10 knots wave profile of C0311F is compared with that of the LIUTO 2.1 design and of the bulbous bow design C0311D. The latter obviously has the lowest bow wave and a shallower wave trough next to the hull. The significantly lower stern wave of C0311F reflects the last successful

400

adjustment to the afterbody and transom.

The versions A to F discussed above represent a sequence of attempts to improve on the DIN design. The combination of advanced CFD tools and expert analysis of the results again permitted a directed refinement of an already quite good design. While this design C0311D was eventually abandoned, also the non-bulbous final version (C0311F), a derivative of versions A and D, showed a significant improvement at high speed and comparable results at low speed and was chosen to be model tested.

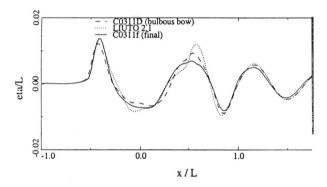

Fig.7 Calculated hull wave profiles for Liuto 2.1, C0311D and C0311F. 10 kn, deep water.

One aspect that played a role in the analysis of the calculated results was that an accurate prediction for the low speed was not easy, as initial validations with the DIN measurements demonstrated. The slender hull, low speed and relatively small hull form modifications required considering minor changes in minor waves and optimising in the margins of the code. At higher speed the assessment was easier and more accurate, and more representative of a common design problem for a merchant vessel. Although the calculations for C0311F indicate improvements at both low and high-speed conditions this is deemed more reliable at 10 knots than at 6 knots.

3.3 Experimental validation
For the hull form thus found, resistance and propulsion tests and wave height measurements were done with a 1:3.4 model in the Deep Water Towing Tank at MARIN; and with a 1:5 model in the Shallow Water Basin, in a simulated waterdepth of 4.5 m. These provided an interesting validation of the ability to predict the effect of hull form changes on the wave making.

While a precise comparison with the tests for LIUTO 2.1 required corrections for the different displacements, model scales and appendages, it was found that the new design had a 10% lower total model resistance at 5.94 kn, and 7.3% at 10 kn. If the latter figure is entirely attributed to a reduced wave resistance, that reduction would amount to a further 29% compared to LIUTO 2.1. The reduction predicted by RAPID was 26 % !

Fig. 8 illustrates the typical level of agreement of measured and computed longitudinal wave cuts for the higher speeds. The deviations are limited to some local differences in amplitude of the shortest wave components; and a minimal overestimation of the amplitude of the stern wave system due to the neglect of viscous effects. Fig. 9 shows that also the difference between the wave patterns generated by LIUTO 2.1 and C0311F is very well predicted, save for some local deviations in the first wave trough

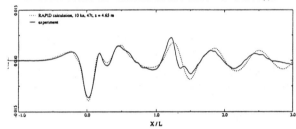

Fig. 8. Calculated and measured longitudinal wave cut. C0311F, 10 kn, deep water, z = 4.65 m.

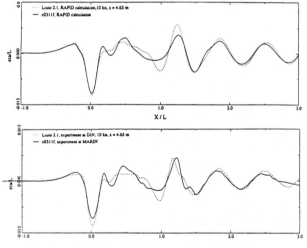

Fig. 9. Comparison of longitudinal wave cuts for Liuto 2.1 and C0311F; as calculated (top) and as measured (bottom). 10 kn, deep water, z = 4.65 m.

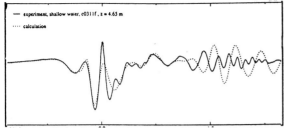

Fig. 10. Calculated and measured longitudinal wave cut for C0311F. 5.94 kn,waterdepth 4.5m, z=4.65 m.

and the first stern wave crest. These results reconfirm that a nonlinear inviscid flow model may be most useful also for afterbody design, provided there is a smooth flow off the transom.

Fig. 10, for 5.94 knots in 4.5 m of water, gives a less convincing agreement. The stern wave system is again overestimated, as discussed above. Unexpected, however, was the underestimation of the second bow wave crest. Inspection of the experimental data indicates that this has been a tiny wave on the verge of breaking, quite sensitive to experimental conditions and hard to fully resolve in the calculations. The precise cause for the inability of the method to correctly indicate the presence or absence of this wave peak and its relation with hull form details is still to be found. However, it should be noted that the peak-to-trough wave heights found amount to only some 0.1 m at full scale.

Otherwise, comparison of the measured wave heights at 5.94 kn in deep and shallow water indicates just a quite small effect of the water depth, as expected for a depth Froude number of 0.46.

4. CHANNEL AND SLOPING BOTTOM EFFECTS

4.1 Extensions of the CFD code

The final stage was an evaluation of the wavemaking and performance in confined waterways. The wider scope of the EC project asked for the development and validation of a prediction tool of more general application, requiring some extensions of the RAPID code.

While shallow water effects on the wave making and flow conventionally are included by mirroring the source distribution in a bottom, this only works well for uniform water depth. However, in many instances of severe wave wash effects, wave propagation from deep into shallow water and the resulting amplification play an important role. In the case of the Canal Grande, the depth typically decreases from 4.5 m in the centre to 1.0 m near the banks. At 5.94 kn, the longest (transverse) waves just become critical near the banks (depth Froude number = 0.98). The possibly resulting wave amplification therefore is to be taken into account by including the sloping bottom and bank in the calculation.

To do this, the channel wall and bottom are covered with additional source panels, and a boundary condition of zero normal velocity is imposed. The flow field then follows from the solution of all boundary conditions simultaneously: the zero normal velocity conditions on ship hull, channel wall and bottom, and the free surface condition on the wave surface. Numerical experiments were carried out to determine the required panel densities and length of the channel to be represented.

A second complication was that the most critical situation for erosion is one with the vessel out of the centre of the channel. It was decided to choose a distance of 15 m from the bank as representative. The other bank then is 30 m away, and the resulting asymmetry also needs to be taken into account, since supposing a symmetrical situation would exaggerate the blockage. Therefore, the calculations had to be done for asymmetric flows, such that just the nearby channel wall and slope would be included.

The asymmetry of the flow, the low speed (Fn=0.196), the sharply diverging waves, the required resolution of the wave steepening near the banks and of the reflection at the channel walls all added up to a calculation with 21000 panels, requiring some 15-20 min CPU per iteration, on a single processor of a CRAY C916 computer; more than 20 times usual CPU times for this method. Obviously, even larger problems will ask for a less 'brute-force' approach.

4.2 Experimental validation

Experiments have been carried out in the MARIN shallow water basin in which the channel effects were simulated. Besides permitting to estimate the required propulsive performance in such conditions, these tests provided unique information and validation material regarding the pronounced channel effects on the wave propagation. A sloping bank was constructed along one side over a part of the length of the tank. Wave height probes were mounted at 4 lateral positions, both in the rectangular and in the trapezoidal part of

402

the tank. The model was towed at a distance from the bank corresponding to 15 m full scale.

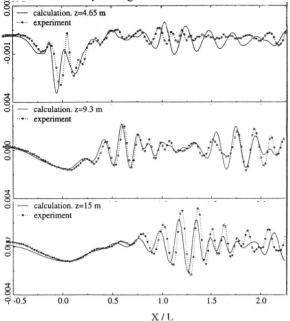

Fig. 11. Calculated and measured longitudinal wave cuts for C0311F in shallow channel with sloping bank. 5.94 kn. z = 4.65m, 9.30 m and 15 m(at the channel wall).

Fig. 12 (below). Calculated wave pattern for C0311F in shallow channel with sloping bank. Speed 8 kn.

Fig. 11 compares calculated and measured longitudinal wave cuts for this situation at 5.94 kn. At 4.65 m off the ship centreline, the agreement is reasonable but displays the underestimation of the second crest and the overestimation of the stern wave system discussed before. At z=9.3 m the agreement is actually quite good, with almost equal maximum wave height and primary disturbance next to the hull. Closer to the channel wall the prediction has a somewhat lower wave height but basically the right shape. The wave amplification in the shallower area can only just be observed in the experiments.

Much stronger effects were observed at 8 knots. While propagating into the shallower area near the bank the wave pattern strongly deforms, wave crests become essentially transverse, and the wave amplitude increases drastically, as the computed wave pattern in Fig. 12 illustrates. The measured waves along the channel wall are radically different from that in a rectangular channel section, with a wave amplitude more than twice as large. This illustrates the dominant effect of wave propagation into shallow water on erosion. Fig.13 shows that this phenomenon is quite well predicted; the reflection, however, is somewhat incomplete, reducing the wave amplitude further aft; and some slight phase differences occur. It goes without saying that in this condition the erosion effects are drastically stronger

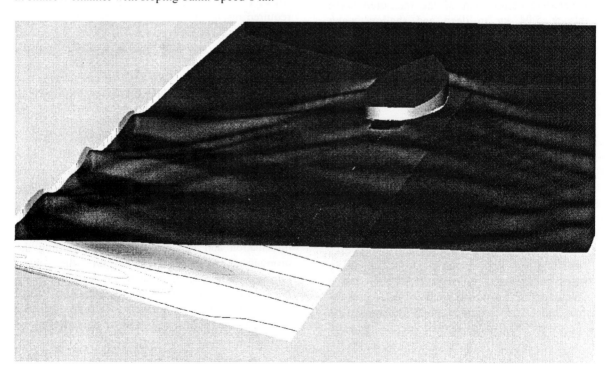

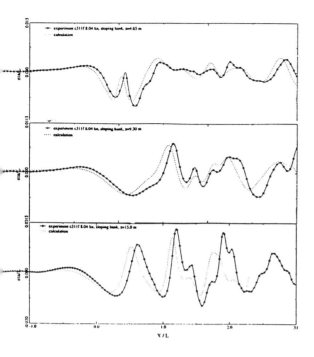

Fig. 13. Calculated and measured longitudinal wave cuts for C0311F in shallow channel with sloping bank. Speed 8 kn. z = 4.65m, 9.30 m and 15 m(at the channel wall).

than at 5.94 kn, and the speed limit imposed appears to be a sensible choice for this channel profile.

At 10 knots in the rectangular channel, pronounced shallow water effects were found, and the prediction was quite good. With sloping bank a giant breaking wave was found at this speed, making validation a useless affair.

4.3 Comparison of erosional effects

To assess the achieved reduction of erosional effects with the new design compared to existing vessels, it is of interest to compare the wave amplitudes and flow disturbances caused. Full scale data still lacking, at present we can only base ourselves on the calculations; with the qualitative, and in many cases also quantitative agreement found between calculations and experiments, such comparisons seem justified.

It was found that according to the RAPID calculations for 5.94 kn in the channel with sloping sidewalls, the wave heights generated at the channel wall are some 15-20% lower for the new design, and the flow velocity along the walls and bottom as well.

5. CONCLUSIONS

The combination of a classical approach for the initial design, CFD-supported optimisation, and model testing has resulted in a new Venice waterbus design with quite substantial power savings compared to existing vessels, and a significant improvement in wave making both at low and at high speed.

The project has, however, a much wider scope, as a demonstration of the application of the best available techniques to a partly unusual design problem. Moreover, the further extension of the RAPID code and the validation material collected have contributed to making it a versatile and accurate tool for predicting wave wash effects and their dependence on hull form and channel cross section, of potential use in many other situations of current interest.

REFERENCES

1. R. Schultze, S. Kaul, A. Brighenti, "LIUTO - Development and optimisation of the propulsion system; study, design and tests," 7th PRADS Symposium, The Hague. Netherlands, 1998.
2. F. Balsamo, A. Paciolla, F. Quaranta, "A system for the experimental determination of M/B's operating in Venice," 7th PRADS Symposium, The Hague. Netherlands, 1998.
3. H.C. Raven, H.H.Valkhof, "Application of nonlinear ship wave calculations in design," PRADS'95 symposium, Seoul, Korea, 1995.
4. H.C.Raven, "A solution method for the nonlinear ship wave resistance problem", Doctor's Thesis, Delft Univ. Techn., 1996.

Practical Design of Ships and Mobile Units
M.W.C. Oosterveld and S.G. Tan, editors.

A system for the experimental determination of the hydrodynamic impact of M/Bs operating in Venice.

F. Balsamo[a] , A. Paciolla[a] , F. Quaranta[a]

[a]Dipartimento Ingegneria Navale , University Federico II of Naples
Via Claudio 21   80135 NAPOLI ITALY

The Liuto Project has the main aim of designing the hull of the water bus of the future for the city service in Venice. The project itself must be sided by a series of parallel studies with the aim of investigating into the real impact of the lagoon environment of both the old and the innovative hulls.
The W.P.4.2 has to set up a methodology capable of measuring the main parameters involved in the environmental impact of the operations of the M/B working on the lagoon.
A first campagn of data logging was carried out some months ago and this paper reports the characteristics of the system used for measurements, the tests carried out, the problems arisen during tests to be solved to obtain a reliable measuring system for the final testing.

1. INTRODUCTION: THE PROBLEM OF THE DECAY OF THE LAGOON ENVIRONMENT

The transformation processes of the morphology of the Venetian lagoon have achieved such a level as to require a continuous observation and a recognition of the necessary operations to contrast the decay. The simple rising of the land level, in the last one hundred years, has been evaluated in about 24 cm, which substantially modifies the morphology of the areas [ref. 1].

The environmental damage is a consequence both of long term causes (subsidence, immersion of the shoal areas, modification of the streams etc.) and of temporary or artificial situations generally due to human presence, including the destruction of the submerged walls surrounding Venice.

In particular, the effects of the radiated waves due to the passing of motor boats in the canals are well known together with those associated to the operations of vessels having transversal dimensions similar to the ones of the canal they pass through.

Since the necessity to limit the damages of the bank erosion has been always felt, the phenomenon of the radiated waves in the canals of the Venetian lagoon has been studied for a long time. In the 1907, the foundation of the Magistero alle acque (Magistrate of the Waters) and of the *Ufficio Idrografico* (Hydrographic Office) [ref. 1,2], with the aim of inquiring into the causes of the changes of the lagoon and of following its evolution, showed the interest and experience of the Venetian people as regards the lagoon conservation. More recently, with the help of new technologies for the monitoring of phenomena connected to the evolution of the lagoon, the *Commissione Moto Ondoso* (Wave Motion Committee) was created ad hoc, with the task of studying the genesis, consequences and possible solutions of such undesired and serious phenomenon for the devastating effect it can generate. The presentation of the data acquired [ref. 1, 3, 4, 5] clearly showed that the main factor is the speed.

2. THE ENVIRONMENTAL IMPACT DUE TO THE OPERATIONS OF THE M/B IN THE VENETIAN LAGOON

The lagoon navigation causes two kinds of damages to the submerged structures.

1. The waves radiated by hulls and propellers interfere with the walls generating the wash effect; because of the permeability of the bricks, with the passing of the wave top, the structure gets filled with water then drained when the bottom passes [ref. 2]. This creates a

406

pulsating action that draws out incoherent materials included in the structure. With time, the less stable elements are first separated from the solid parts and then drawn down by the water with the consequence of weakening and emptying the structures, thus letting the water pass through, with all the imaginable consequences.

2. The perturbations of speed and pressure, connected with the fluid streams generated by the propellers, create a consumption of the material of the submerged walls that, in time, may become relevant. In particular, the component of the velocity in the direction normal to the wall has an undermining effect because of the crash; the tangential component draws out the less coherent material by friction, once it has been weakened by the first effect.

The two phenomena have a different nature and create different effects on the surrounding environment. The wave motion by itself does not create translation of material since the trajectories of particles are orbital and only the perturbation has its own celerity. The effects of radiated waves are sensible even at remarkable distances from where the vessel passes and therefore can influence also far away areas. Therefore, such phenomenon is basically connected to the vessel navigation.

The perturbations created by streams are destined to dissipation due to the friction and can also last in time due to the principle of the conservation of the angular momentum (applied, in this case, to the vortexes created by the fluid stream). The latter case regards the manoeuvring of the vessels while the propeller remains in proximity of the submerged walls for a relevant time emitting the flow in their direction.

3. THE ACTION OF THE LIUTO PROJECT

The main goal of the LIUTO project is to realise an innovative water bus having a reduced environmental impact. The vessel in project should be qualified by defining its hydrodynamic characteristics and by the impact of its operations on the surrounding environment. The task of the WP 4.2 is to determine a valid criterion for the evaluation of one of the effects of the service of the

M/B hull and propulsion system to be kept under control: pressure and wave perturbations whose negative influence damages the integrity of submerged walls surrounding the lagoon city.

In order to compare the behaviour of the present motor boats and LIUTO, it is necessary to determine some test procedures which would allow a real comparison between the influences of the two hulls — and related propulsion systems — on the environment they operate in; such procedures, practically, should analyse the behaviour of such elements showing the aspects of the hydrodynamic impact on submerged walls and foundations.

4. THE METHODOLOGY OF THE TESTS CARRIED OUT IN JULY, 1997

The first campaign of data logged carried out in Venice (July, 1997) was destined to a first screening of the possibility of registering data about the environmental impact of the operations of water busses; in particular, these tests had to screen the operations of the old hull working in the lagoon. The main parameters involved in these phenomena were determined in:
- velocity of the water;
- pressure of the water;
- mean value of the water (due to the tide) in the period of the tests;
- position of the ship;
- speed of the ship.

In order to log these parameters, commercial sensors were used except in the case of the water velocity; indeed, for the foreseen field of velocity of the water (coping with the budget available for this research), no commercial sensors can meet the required aim. Thus, special sensors - based on a plate moved by a sphere sensitive to the flow of the water - were set up capable of giving a signal proportional to the velocity of the water in the zone corresponding to the centre of the sphere.

Of course the tests were set with the aim of simulating the real operating conditions of a water bus in service.

Thus, tests included:
- *tests in canal* i.e. measurement of perturbations due to the passage of water busses and other kinds of vessel in the

canal; this test allows the determination of the standard perturbation in the canals due to the normal traffic;

- *tests in parallel route* i.e. measurement of the perturbation due to the passage of a water bus in protected waters (a canal without traffic, where the only perturbation present is due to the tested M/B); in this case the M/B is controlled from the point of view of its position and speed;

- *tests during manoeuvring* i.e. evaluation of the perturbations due to the operations of the M/B during manoeuvres (mooring and unmooring)

5. THE RESULTS OF THE CAMPAIGN OF TESTS CARRIED OUT IN VENICE (JULY 1997)

Before the tests an evaluation of the possible values of the parameters involved in the phenomena under study was carried out; more details are available in [ref. 6]. The velocity and wave height sensors were set on a support consisting in a pole driven in the bottom of the canal; the sensors of velocity of water measured the longitudinal and the transversal component of the velocity of the water at a depth of about 0.1 m; the capacitive probes were set on the same pole at a distance of about 0.6 m (internal) and 1.1 m (external) from the axis of the support.

As regards the results of the tests, some considerations must be stated beforehand: indeed, since the campaign of July was a sort of initial setting of a new (and enhanced) instrumentation, some malfunctioning lowered the value of the final results. In particular, an absolute sensor for the evaluation of the pressure broke just before the start.

The sensors worked well but, due to problems of initialisation and zeroing, the final figures are not always reliable. In the following some example of the results are given; for further details, please, see [ref. 7].

Fig. 1 to 4 show an example of the data logging related to a passage in parallel route; fig. 1

gives the overall arrangement of the working post; fig. 2 shows the position and speed of the ship during the test, fig. 2 contains the tracks of the velocity of the water (longitudinal and transversal components) measured by the self-made currentometers, fig. 4 gives the values of the wave height measured by a couple of capacitive probes. Of course, times on fig. 2 to 4 are synchronised.

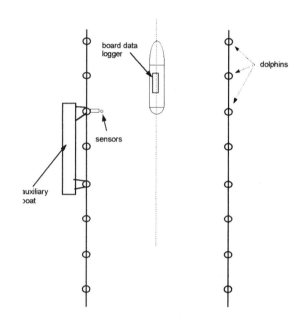

Figure 1- general layout of tests in parallel route

6. . WHAT THE TESTS SUGGESTED FOR THE IMPROVEMENT OF THE DATA LOGGING SYSTEM

Apart from the pressure sensors - that, unfortunately, showed some inconveniences and required a maintenance operation -, the main problem occurring in each part of the tests was the difficult zeroing of the sensors due to the impossibility of obtaining a condition of zero velocity of water and no change in the water level. Moreover, although the moving of water prevented from a correct zeroing, a procedure of zero for sensors was used both for velocity and wave height sensors. For velocity sensors, a tube was introduced

408

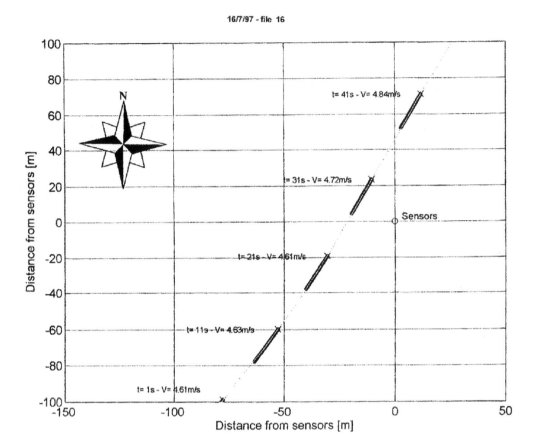

Figure 2-position and velocity of the M/B (test in parallel)

sensors was used both for velocity and wave height sensors. For velocity sensors, a tube was introduced in the cylinder keeping the rod and the sphere so avoiding the effect of the water velocity on the sphere and maintaining the sensor in position of zero.

For the capacitive probes a true calibration was made by keeping the probe in the water and reading directly the positions of zero and span (meanly, due to the waves). The zero position was read in a zone where the water was completely calm and while keeping the probe in the same position (and height) it had on the support. Spans were revealed by immersing and de-immersing the probe at its maximum position and reading the value given by the station of conditioning of the signal coming from the probe. These values (due to the linearity of the system) allow the determination of the scale of conversion from the voltage out

from the station and the corresponding value of the height of the water.

In both cases, during calibrations an acquisition was performed in order to record the values of the voltage corresponding to each condition of zero for every sensor; nevertheless, the files containing the read values show the low efficiency of this system of revelation of the zero. The system used could not prevent the presence of fluctuation in velocity and height of water so a correct determination of the zeros and spans involved in each measurement was impossible.

For future campaigns, an improved laboratory method to determine the procedure of zero for the instruments should be set up which must be repeated in the zone where the measure has to be done.

In particular, for the velocity sensors, a system capable of blocking the group rod - sphere must be

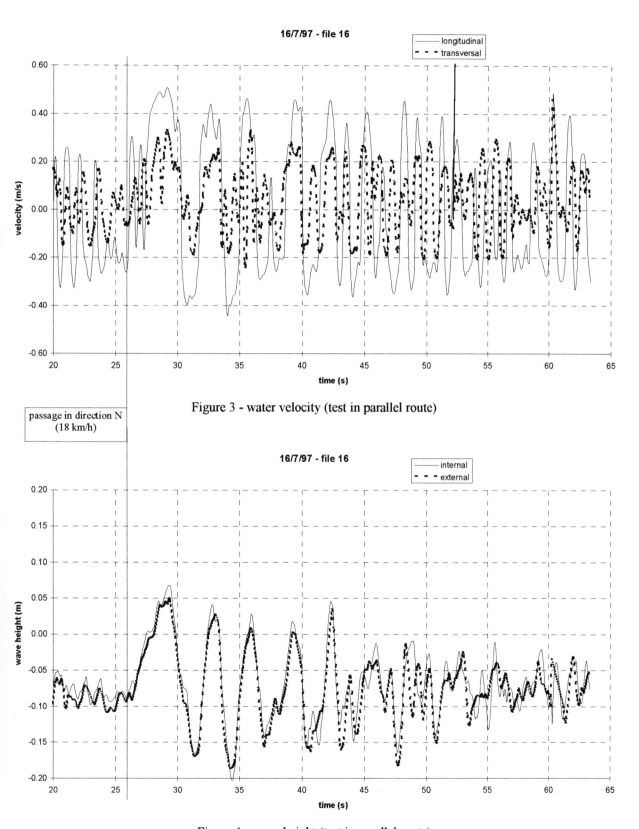

16/7/97 - file 16

longitudinal
transversal

Figure 3 - water velocity (test in parallel route)

passage in direction N
(18 km/h)

16/7/97 - file 16

internal
external

Figure 4 - wave height (test in parallel route)

410

set in order to allow the phase of zero before the tests. As regards the capacitive probes, the preliminary set must be obtained by creating a zone of calm water near the support and measuring the maximum and minimum immersion of the probes (span + and -) and the position corresponding to the level of undisturbed water (0).

Moreover, other serious problems came from the choice of the sites of tests and from the construction of the structures for the support of the sensors. Practically in each case the agreed conditions were not respected and the structures were not suitable for the use they were destined to; in the occasion of the final tests a preliminary inspection of the sites and a follow up of the setting of the chosen sites will be necessary.

Finally, due to the continuous variation of tide, a better revelation of the mean level of the water is required; this will be done by using an absolute pressure transducer capable of reading the level of the water with high precision. Such a transducer is already available in our Department but it could not be used because of the abovementioned malfunctioning

7. CONCLUSIONS

Finally, it is possible to say that the campaign of data logging carried out in July in the frame of LIUTO WP 4.2 gave important indications about the development of the very complex instrumentation necessary to perform the final comparative tests of environmental behaviour of both the present hull and the innovative one. Although a final word about the possibility of using the sensor nowadays built and tested is still to be said, the scene where tests will be made is substantially clearer so as the necessary

improvement of the whole measurement system.

Of course, the future of the WP 4.2 of the LIUTO project will show the opportunity of organising another step of testing on the described system of measurement ; in any case , the information gathered in the occasion of this pre-testing activity will simplify the organisation of the final testing.

REFERENCES

1. A. Rusconi *Rilevamento di moto ondoso in alcuni importanti canali lagunari* Ufficio Idrografico del Magistrato alle Acque, Venice.
2. L. D'Alpaos *Idraulica e conseguenza dei moti ondosi nei canali* Università degli Studi di Padova.
3. Gianfranco Liberatore: *Prove sperimentali di onde generate da natanti, elaborazioni preliminari*, Comune di Venezia, Centro stampa LL.PP., December 1988.
4. F. Costa, G.P. Nadali: *Prove sperimentali su natanti per la rilevazione del moto ondoso*, Commissione per lo studio del moto ondoso, Comune di Venezia.
5. *Relazione finale della Commissione per lo studio del moto ondoso*, Comune di Venezia, May 1994.
6. F. Balsamo, A. Paciolla, F. Quaranta: *Preliminary considerations about the full tests to determine the environmental impact of the present M/Bs and the LIUTO hull.-*.Report n° DINLIUTO-WP42-1-V1
7. F. Balsamo, A. Paciolla, F. Quaranta: *Final report on the preliminary full scale tests carried out in Venice in July 1997* - Report n. DINLIUTO-WP42-5-V1

An Inverse Geometry Design Problem in Optimizing the Hull Surfaces

[a]Shean-Kwang Chou, [b]Cheng-Hung Huang , [b]Cheng-Chia Chiang and [c]Po-Chuan Huang

[a]United Ship Design and Development Center
5th FL. Tai-Tze Bldg., 20 Sec. 3, Pa-Teh Road, Taiwan, R.O.C.

[b]Department of Naval Architecture and Marine Engineering, National Cheng Kung University
1 Ta-Hsueh Road, Tainan, Taiwan, R.O.C.

[c]Department of Industrial Technology, Ministry of Economic Affairs
6th FL. 15, Fu-Chou Street, Taipei, Taiwan, R.O.C

The techniques of Inverse Design Problem for optimizing the hull surfaces from a specified pressure distribution is presented. This desired pressure distribution can be obtained by suitable modification of the existing pressure distribution of the parent ship. The surface geometry of the ship is generated using B-spline surface method which enables the shape of the hull to be completely specified using only a small number of parameters (i.e. the control points). The technique of parameter estimation for inverse problem is thus chosen. Results show that the accuracy of the final desired ship form depends on the number of polygon used in B-spline surface fitting, only when enough number of polygons are used, good final geometry that was calculated based on given pressure distribution can be obtained.

1. INTRODUCTION

The hydrodynamic performance of a vessel is strongly dependent on the shape of the vessel's hull. It is thus important that the form of the hull be carefully designed to achieve as optimal a performance as the constraints (i.e. the pressure distribution for the present study) will permit. This should be done at an early stage in the total vessel design schedule since any subsequent changes to the hull form may incur large costs resulting from other associated design modifications, for instance, the need to redesign bulkhead.

Traditionally, naval architects have based new hull designs on hulls already in service and known to perform well (i.e. parent ship), with any changes to the design being investigated using expensive model towing tank tests. In recent years advances in computational fluid dynamics have made possible the analysis of new, possible novel, hull forms at a fraction of the cost of model tests, with good estimates of the hydrodynamic forces acting on the vessel being obtained by Van Oortmerssen [1].

The use of computational techniques, however, requires a numerical description of the hull shape. Various methods of defining the complex free-form shape of hulls for use in design optimization methods can be found in the literature [2-5]. However two user friendly methods of surface representation commonly used in the field of computer-aided design are Bezier and B-spline surface patches [6]. Due to their simplicity, B-spline surface technique is used in this study.

Typically, the surface to be represented is broken into a mesh of mainly rectangular curvilinear regions, for example, the areas formed by the section lines and waterlines of a vessel. A surface patch is then defined over each region, its shape being determined by a set of control points. These points form a polyhedron which the surface approximates.

412

The shape parameters in this formulation are thus the coordinates of each control point, i.e. these limited number of control points become the parameters in controlling the hull surface geometry. In the present study the task is to redesign the hull form of the ship based on previous shape and new desired pressure distribution.

With this in mind, the technique of inverse design problem should be used to design the new hull form in accordance with the desired pressure distribution on the hull This desired pressure distribution can be obtain by modifying the existing pressure distribution of the parent ship whenever one found that there exists a unfavorable pressure pattern on the hull surface.

The direct problem involves the determination of the hull surface pressure distribution when the hull form is given. On the other hand the inverse design problem is concerned with the determination of the modified hull form from the given desire pressure distribution.

The present work addresses the development of an efficient method for parameter estimation, i.e. the Levenberg-Marquardt algorithm [7], in estimating the new hull form that satisfies the desired pressure distribution. This method has proved to be a powerful algorithm in inverse calculations [8-10], especially in parameters estimation.

The method of hull surface generation and B-spline surface fitting is described in Section 2. In Section 3 the method used to calculate the hull surface pressure distribution , i.e. the direct problem, is explained. The inverse design problem involving the definition of cost function and Levenberg-Marquardt algorithm is addressed in Section 4. Finally a computational procedure is summarized in Section 5.

2. HULL GENERATION AND B-SPLINE SURFACE FITTING

2.1 Hull generation
Consider a Cartesian product parametric

B-spline surface [6] given by

$$Q(u,w) = \sum_{i=1}^{n+1}\sum_{j=1}^{m+1} B_{i,j} N_{i,k}(u) M_{j,l}(w);$$

$$2 \le k \le n+1; 2 \le l \le m+1$$

where

$$N_{i,1} = \begin{cases} 1 & if \quad x_i \le u \le x_{i+1} \\ 0 & otherwise \end{cases}$$

$$N_{i,k}(u) = \frac{(u-x_i)N_{i,k-1}(u)}{x_{i+k-1}-x_i} + \frac{(x_{i+1}-u)N_{i+1,k-1}(u)}{x_{i+k}-x_{i+1}}$$

(2a)

and

$$M_{j,1} = \begin{cases} 1 & if \quad y_j \le w \le y_{j+1} \\ 0 & otherwise \end{cases}$$

$$M_{j,l}(w) = \frac{(w-y_j)M_{j,l-1}(w)}{y_{j+l-1}-y_j} + \frac{(y_{j+1}-w)M_{j+1,l-1}(w)}{y_{j+1}-y_{j+1}}$$

(2b)

Where the x_i , y_j are the elements of a uniform knot vector, k and l are the order of the B-spline surface in the u and w directions and n and m are one less than the number of polygon net points in the u and w directions, respectively. Here Q(u,w) are the surface data points. The N and M basis functions can be determined from the knot vector and the parameter values u and w. The $B_{i,j}$ are the require polygon net points (control points). If $B_{i,j}$ are given, the surface data points Q(u,w) can be calculated from equation (1).

2.2 B-spline surface fitting
When the surface is described by external data it is convenient to obtain an initial non-flat B-spline surface approximating the hull for subsequent real time interactive modification. This requires determining the defining polygonal net from an existing network of three dimensional surface data points.

For each known surface data point equation (1) provides a linear equation in the unknown $B_{i,j}$'s and similarly for all the surface data points. In matrix notation this can be written as

$$[Q] = [C][B] \tag{3a}$$

More explicitly, for the case when k = 4, l = 3 and using 16 control points, the following matrix system can be obtained.

$$
\begin{bmatrix} Q(u_1, w_1) \\ \cdot \\ \cdot \\ \cdot \\ \cdot \\ Q(u_{16}, w_{16}) \end{bmatrix} =
$$

$$
\begin{bmatrix} N_{14}(u_1)M_{13}(w_1) & N_{14}M_{23} & . & N_{44}M_{43} \\ \cdot & & & \cdot \\ \cdot & & & \cdot \\ \cdot & & & \cdot \\ \cdot & & & \cdot \\ N_{14}(u_{16})M_{13}(w_{16}) & \cdot & \cdot & N_{44}M_{43} \end{bmatrix}
\begin{bmatrix} B_{11} \\ B_{12} \\ \cdot \\ \cdot \\ \cdot \\ B_{44} \end{bmatrix}
\tag{3b}
$$

Since for any arbitrary r × s topologically rectangular surface point data, [C] is not normally square, a solution can only be obtained in some mean sense. In particular

$$[B] = \left[[C]^T[C] \right]^{-1} [C]^T[Q] \tag{4}$$

The v and w parametric values for each surface data are obtained using a chord length approximation [6].

3. THE DIRECT PROBLEM

The wave resistance of a surface ship in calm water is calculated by a high-order panel method using Rankine-source [11]. The Cartesian coordinate system o-xyz is defined so that the x-y plane laid on the still water surface with x-axis in the direction of constant advance speed U and z-axis upwards. The fluid is assumed to be inviscid, incompressible and irrotational. The velocity field of the fluid can be defined as $\nabla\phi(x, y, z)$ and the velocity potential ϕ must satisfy the

following condition

$$\nabla^2\phi = 0 \tag{5}$$

in the fluid domain and the solid body boundary condition

$$\phi_n = 0 \tag{6}$$

on the hull surface and the kinematic and the dynamic boundary condition on the free surface $z = \varsigma(x, y)$ are specified as

$$\phi_x \varsigma_x + \phi_y \varsigma_y - \phi_z = 0 \tag{7}$$

$$g\varsigma + \frac{1}{2}(\nabla\phi \cdot \nabla\phi - U^2) = 0 \tag{8}$$

The radiation condition that there be no waves far upstream of the ship and the waves always travel downstream in the far field.

The free surface boundary conditions (3) and (4) are linearized in term of double-model velocity potential, Φ, suggested by Dawson [12]. The resulting free surface boundary condition becomes

$$\left(\Phi_\ell^2 \phi_\ell\right)_\ell + g\phi_z = 2\Phi_\ell^2 \Phi_{\ell\ell} \tag{9}$$

where the subscript ℓ denotes partial derivatives alone a streamline of Φ.

Once the flow potential is determined, the dynamic pressure at each panel can be found using Bernoulli's equation

$$p_i = p_\infty + \frac{1}{2}\rho(U^2 - \nabla\phi \cdot \nabla\phi) \tag{10}$$

where ρ is the fluid density and p_∞ the pressure at infinity. A dimensionless pressure coefficient C_{pi} can be defined as

$$C_{p_i} = \frac{p_i - p_\infty}{\frac{1}{2}\rho U^2} = 1 - \left(\frac{|\nabla\phi_i|}{U}\right)^2 \tag{11}$$

and will be used as the desired reference in designing the new hull form.

4. THE INVERSE DESIGN PROBLEM

For the inverse problem, the hull form of the bow is regarded as being unknown and controlled by a set of control points, in addition, the desired distribution of dimensionless pressure coefficients Cp_i on the bow surface are considered available.

Let the desired pressure coefficients on the bow surface be denoted by $Cp(x,y_i,z) \equiv Cp_i$, $i = 1$ to I, where I represents the number of panel for the redesign portion (bow) of hull. Then the inverse problem can be stated as follows: by utilizing the above mentioned desired pressure coefficients C_{pi}, design the new hull shape for the bow.

The solution of the present inverse design problem is to be obtained in such a way that the following functional is minimized:

$$J\left[\hat{\Omega}(\hat{B})\right] = \sum_{i=1}^{I}\left[\hat{C}_{p_i}(\hat{B}_j) - C_{p_i}\right]^2 = U^T U ; \quad (12)$$

$$j = 1 \text{ to } J$$

here, $\hat{C}_{p_i}$ are the estimated or computed pressure coefficients on the bow locations (x, y_i, z). These quantities are determined from the solution of the direct problem given previously by using an estimated hull form $\hat{\Omega}(\hat{B})$, J represents the number of control points, i.e. $J = (n+1) \times (m+1)$. Here the hat " ∧ " denotes the estimated quantities.

4.1 The Levenberg-Marquardt method for minimization

If the redesigned hull shape is discretized into I panels and J control points are used, equation (20) is minimized with respect to the estimated parameters B_j to obtain

$$\frac{\partial J[\hat{\Omega}(\hat{B})]}{\partial B_j} = \sum_{i=1}^{I}\left[\frac{\partial \hat{C}_{p_i}}{\partial B_j}\right]\left[\hat{C}_{p_i} - C_{p_i}\right] = 0 ;$$

$$j=1 \text{ to } J \quad (13)$$

where I should be equal to or greater than J, otherwise an underdetermined system of equations will be obtained and it is impossible to calculate the inverse solutions under this situation. Equation (13) is linearized by expanding $\hat{C}_{p_i}(B_j)$ in Taylor series and retaining the first order terms. Then a damping parameter m^n is added to the resulting expression to improve convergence, leading to the Levenberg-Marquardt method (Marquardt 1963) given by

$$\left(F + \mu^n I\right)\Delta B = D \quad (14a)$$

where

$$F = Y^T Y \quad (14b)$$

$$D = Y^T U \quad (14c)$$

$$\Delta B = B^{n+1} - B^n \quad (14d)$$

here the superscript n and T represent the iteration index and transport matrix, respectively, I is the identity matrix and Y denotes the Jacobian matrix defined as

$$\Psi \equiv \frac{\partial C_p}{\partial B^T} \quad (15a)$$

the Jacobian matrix defined by equation (15a) is determined by perturbing the unknown parameters B_j one at a time and computing the resulting change in pressure coefficients from the solution of the direct problem, equation (5).

Equation (15a) is now written in a form suitable for iterative calculation as

$$\mathbf{B}^{n+1} = \mathbf{B}^n + \left(\mathbf{\Psi}^T\mathbf{\Psi} + \mu^n\mathbf{I}\right)^{-1}\mathbf{\Psi}^T\left(\hat{\mathbf{C}}_\mathbf{p} - \mathbf{C}_\mathbf{p}\right) \quad \text{(15b)}$$

When $\mu^n = 0$, the Newton's method is obtained, as $\mu^n \to \infty$, the steepest-descent method is obtained. For fast convergence the steepest-descent method is applied first, then the value of m^n is decreased, finally the Newton's method is used to obtain the inverse solution. The algorithm of choosing this damping value m^n is described in detailed in (Marquardt 1963), so it is not repeated here.

5. COMPUTATIONAL PROCEDURE

The iterative computational procedure for the solution of this inverse problem using Levenberg-Marquardt method can be summarized as follows:

Choose the initial guess for control points **B** (obtained by using original hull form and B-spline surface fitting) at iteration n to start the computation.

Step 1. Solve the direct problem given by equation (5) to obtain computed pressure coefficient **Cp**.

Step 2. Construct the Jacobian matrix in accordance with equation (15a).

Step 3. Update **B** from equation (15b).

Step 4. Check the stopping criterion, if not satisfied to Step 1 and iterate.

6. RESULTS AND DISCUSSIONS

The accuracy of the B-spline surface fitting plays a very important role in this inverse design problem. If the hull surface can not be reproduced by B-spline surface fitting method, the inverse solutions can never be obtained accurately. The first task is thus to show the validity for the present B-spline surface fitting from the surface data of a known hull form.

The original bow surface is given in Figure 1 while Figure 2 represents the contour plots of y for the open, uniform B-

spline surface fitting by using 10(u-direction) × 20(w-direction) surface data points and 8×8 control points.. From this Figure we learn that the fitted surface approaches the given bow surface as the number of control points is increased to 8×8. For the cases studied here, 8×8 control points seems to result in a very good fitting. Therefore the number of the unknown control points B $_j$ is chosen as 8×8 for the rest of this paper.

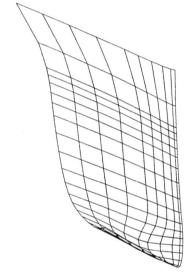

Figure 1. The original bow surface that to be fitted by B-spline surface fitting method.

Solid Lines: Original surface
Dash Lines : Fitting surface

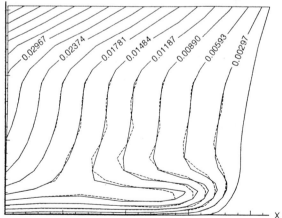

Figure 2. Contour plots of y for the original and fitted bow surface (8×8 control points)

In order to show the validity of our inverse design algorithm in estimating the optimal hull form from the desire pressure distribution, two different parent hull forms, i.e. a series-60 ship and a container ship, are considered here. The procedures of the present numerical experiments will be stated as follows:

(1). Distorting the parent hull form to obtain an objective hull from.

(2). Calculate the pressure distribution for the objective hull form and retain the pressure distribution around bow (or any blocks) as the desire pressure distribution **Cp**.

(3). Perform the B-spline surface fitting for parent hull form to obtain the initial guess values for B_j.

(4). Using the initial guess of B_j to calculate estimated bow pressure $\hat{C}_P$.

(5). Apply the Levenberg-Marquardt method to perform this inverse design problem until the specified stopping criterion is satisfied.

Let us now begin our numerical analysis.

EXAMPLE 1 ----- Series 60 ship:

In the first example, the parent ship is series-60 (C_B=0.6) shown in Figure 3. The panel system for the whole ship is 51 (x-direction) × 17 (z-direction) below the water line, therefore total of 51×17 pressure coefficients can be used as the design reference to estimate the optimal hull form.

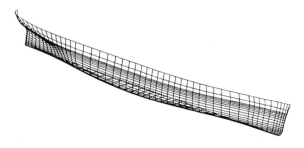

Figure 3. The series-60 ship used as parent ship in numerical example 1.

The objective hull form is obtained by extending the bow shape (the first 10 columns of panels) of the parent ship outwards. The comparison of the bow shape between the initial (parent) and objective (desired) ship is shown in Figure 4. The pressure distribution for the bow can thus be calculated and used as the design criterion. Figure 5 shows the contour plot of pressure coefficient of bow under the water line for initial and objective ship when Froude number equals to 0.28.

BODY PLANE

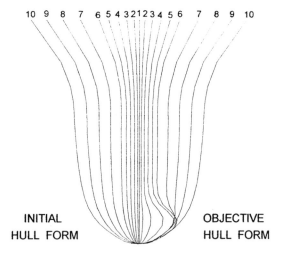

Figure 4. The comparison of the initial (parent) and objective (desired) bow shape.

Solid Lines: Pressure of Objective hull form
Dash Lines : Pressure of Initial hull form

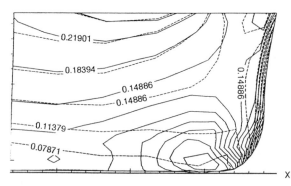

Figure 5. Contour plots of Cp on the initial (parent) and objective (desired) bow surface.

The inverse calculations are then performed by following the Levenberg-Marquardt method. The initial guesses of B_j are obtained by using B-spline surface fitting for the parent ship. With only 4 iterations a very accurate solution can be obtained. The bow form for the exact and estimated ships and contour plots of pressure coefficient are shown in Figures 6 and 7, respectively. From those Figures we conclude that the Levenberg-Marquardt method has been applied successfully in estimating the optimal hull form in this numerical example.

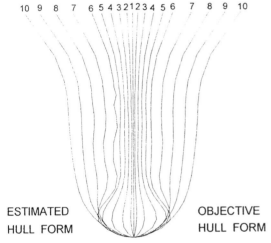

ESTIMATED HULL FORM   OBJECTIVE HULL FORM

Figure 6. The comparison of the estimated (optimal) and objective (desired) bow shape.

Solid Lines: Pressure of Objective hull form
Dash Lines : Pressure of Estimated hull form

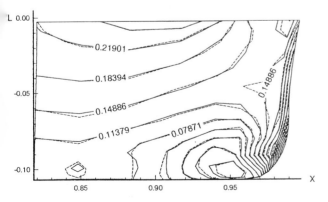

Figure 7. Contour plots of Cp on the estimated and objective (desired) bow surface.

EXAMPLE 2 ----- Container Ship:

In the second example, the parent ship is a container ship shown in Figure 8. The whole ship is divided into 4 patches and the 3rd patch is our target. By observing the calculated pressure distribution on that patch (F_N=0.21), a modification of low pressure zone near the bilge is performed and this modified pressure distribution is given as the design criterion. The panel system for that patch is 22 (x-direction) × 13 (z-direction), therefore a total of 22 × 13 pressure coefficients are used as the design reference to estimate the optimal hull form that is controlled by 8 × 8 control points. The comparison of pressure distribution of the 3rd block between the initial (parent) and objective (desired) ship is shown in Figure 9.

Figure 8. The container ship used as parent ship in numerical example 2.

Solid Lines: Pressure of Objective hull form
Dash Lines : Pressure of Initial hull form

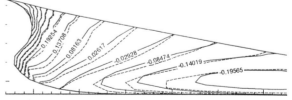

Figure 9 Contour plots of Cp on the initial and objective hull shape of the container ship.

The inverse calculations are performed again by using the Levenberg-Marquardt method. The initial guesses of B_j are obtained by using B-spline surface fitting for the parent ship. After only 4 iterations, a very accurate solution can be obtained and is

418

shown in Figure 10 for the ship form at the 3rd block for initial and estimated ships. A modified hull form is then generated by smoothing the estimated shape and the experimental results of C_R vs F_N for the parent and modified hull form are also shown in Figure 11. It is obvious that the modified hull form performs better in calm water resistance than the parent ship. From those Figures we conclude again that the Levenberg-Marquardt method has been applied successfully in estimating the optimal hull form in this numerical example.

-●- Estimated hull form
------ Parent hull form

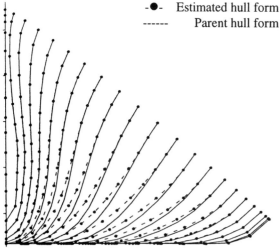

Figure 10. The comparison of the initial and estimated hull shape.

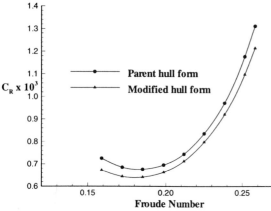

Figure 11. The experimental C_R values vs F_N for the parent and modified container ships.

CONCLUSIONS

An inverse design problem in estimating the optimal hull form from the knowledge of desired pressure distribution by the techniques of B-spline surface fitting and Levenberg-Marquardt method has been developed and applied successfully.

Results show that the present algorithm needs only a few iterations to obtained the optimal hull form if enough number of control points are given. One should note that even though more control points describe the unknown surface more accurate, on the other hand, it takes more computer time to obtain the inverse solutions.

The advantages of using the technique of inverse design problem in preliminary design stage lies in the fact that the hull form which satisfies a desired pressure distribution can be found automatically and efficiently.

ACKNOWLEDGMENT

This work was supported in part through the United Ship Design and Development Center, R. O. C., Contract number, S864023 and the National Science Council, R. O. C., Grant number, NSC-87-2611-E-006-026.

REFERENCES
(1). G. Van Oortmerssen (ed.), *CFD and CAD in Ship Design*. Elsevier, Amsterdam, (1990).
(2). W. C. Lin, , W. C. Webster, and J. V. Wehausen, Ship of Minimum Total Resistance, Proceedings. *International Seminar on Theoretical Wave-Resistance*, University of Michigan, Ann Arbor, 907-953, (1963).
(3). D. C. Wyatt, and P. A. Chang, Development and Assessment of a Total Resistance Optimized Bow for the AE-36. *Proceedings of the International Symposium on CFD and CAD in Ship Design*, Wageningen, The Netherlands,

(1990).

(4). L. Larsson, K. J. Kim, B. Esping, and D. Holm, Hydrodynamic Optimization using Ship flow. In Practical Design of Ships and Mobile Units. J. B. Caldwell and G. Ward, Eds., Elsevier Applied Science, London, 1.1-1.16, (1992).

(5). T. W. Lowe, M. I. G. Bloor, and M. J. Wilson, The Automatic Functional Design of Hull Surface Geometry. *Journal of Ship Research*, **38**, 319-328,(1994).

(6). D.F. Rogers and J. A. Adams, *Mathematical Elements for Computer Graphics*. 2nd Edition, McGraw-Hill,(1990).

(7). D. M. Marquardt, An Algorithm for Least-Squares Estimation of Nonlinear Parameters. *J. Soc. Indust. Appl. Math.*, **11**, 431-441,(1963).

(8). C.H. Huang, and M. C. Huang, Inverse Problem in Determining the Normal and Tangential Drag Coefficients of Marine Cables. *Journal of Ship Research*, **38**, 296-301,(1994).

(9). C. H. Huang, and M. N. Ozisik, A Direct Integration Approach for Simultaneously Estimating Temperature Dependent Thermal Conductivity and Heat capacity. *Numerical Heat Transfer*. Part A, **20**, 95-110,(1991).

(10). C. H. Huang, and D. M. Wang, Statistical Consideration for the Estimation of Spatially Varying Sound Velocity and Water Density in Acoustic Inversion. *Inverse Problems in Engineering*, **4**, 129-151,(1996).

(11). C. Y. Lu and S. K. Chou, A Computational Method for Calculating Ship Wave Resistance, Proceedings of CFD Workshop TOKYO 1994, Japan, (1994).

(12). C. W. Dawson, A Practical Computer Method for Solving Ship Wave Problems. *Proceedings, 2nd International Conference on Numerical Ship Hydrodynamics*, Berkeley, CA, U. S. A., 30-38,(1977).

Practical Design of Ships and Mobile Units
M.W.C. Oosterveld and S.G. Tan, editors.

Optimum Hull Form Design using Numerical Wave Pattern Analysis

Akihito HIRAYAMA[a], Tatsuya EGUCHI[a], Koyu KIMURA[a], Akihiko FUJII[b]
and Moriyasu OHTA[b]

[a]AKISHIMA LABORATORIES (MITSUI ZOSEN) INC.
1-1-50 Tsutsujigaoka, Akishima, Tokyo, Japan

[b]MITSUI ENGINEERING & SHIPBUILDING CO., LTD.
5-6-4 Tsukiji, Chuou-ku, Tokyo, Japan

A new method for minimizing the wave-making resistance on ship hull forms has been developed, and is now being used in our design work. This method uses computed wave patterns instead of experimental ones to improve ship hull forms, and the computed wave patterns are obtained by the modified Rankine Source Method. In this point, this method differs from conventional hull form improvement methods using experimental results. The advantages of using this method enable one to shorten time and labor for hull form improvement. This paper presents the optimization method using the modified Rankine Source Method.

Some examples of application using the present method are shown. Three types of hull forms, namely, Series60(Cb=0.6), an over-panamax type container ship and a high-speed car ferry, are chosen as the objects of optimization. The optimization results show that the significant reductions of wave-making resistance are achieved

1. INTRODUCTION

The design of ships with low resistance satisfying certain design restrictions has been one of the most important subjects in ship design. This aim can not be achieved without reducing wave-making resistance. Especially, for ships such as container ships and car ferries which are operated at speeds more than Froude number = 0.2, it is more important to reduce wave-making resistance because of the fact that the wave resistance of these faster ships occupies a bigger part of the total resistance as compared with other slower ships.

Various methods to reduce the wave-making resistance of ship forms have been proposed. Baba introduced a method using experimental wave pattern analysis results in 1972 [1]. Matsui introduced a method of another type using wave pattern analysis in 1980 [2]. These methods and similar methods are readily available, and have been used in the design of ships. However, due to the necessity of experimental results to obtain wave amplitude functions, these methods require large amounts of time and labor for hull form improvement. This is the biggest obstacle in

practical uses.

On the other hand, an estimation method of wave-making characteristics of ships is also important to predict the reduction of wave-making resistance. Remarkable developments have been made regarding computation methods of wave-making characteristics during the last 20years. Especially, the Rankine Source Method (RSM), which was introduced by Dawson in 1977 [3], is particularly effective. After the publication, many improvements have been made to the RSM. Various modified RSMs have been also proposed, and many researchers are now trying the improvements of RSM. However, there are not many practical methods that can be applied to practical hull forms and give accurate estimation of wave-making characteristics.

If the accuracy of the computation method for wave-making characteristics is enhanced to the level of the tank test, it is possible to realize an effective optimization method which does not need significant amounts of time and labor, because a computation method can replaces a tank test. The present optimization method has been developed based on this idea. Investigations have been made concerning the improvement of RSM and the combination of RSM with a ship optimization method. As a result, a measure of success in making a practical optimization method has been achieved.

2. OPTIMIZATION METHOD

The principle of optimization technique is based on the interference of free waves, in the same way as the improvement methods using tank test results. It is considered that the modification to the original hull is led by the addition of a volume distribution. Assuming that the original hull form and the added volume distribution float on the water surface and move at same speeds, these volume distributions generate free waves and the interference between these free waves is induced. The wave-making resistance of the ship that consists of these volume distributions depends on the interference, namely, the suitable volume distribution reduces the wave-making resistance, and inversely the inadequate added distribution increases the resistance. In the present optimization method, the added volume distribution is defined in order to reduce the height of the free wave. The modified hull form with low wave-making resistance is shaped by adding the volume distribution to the original hull form.

2.1. Numerical wave pattern analysis

The present optimization method, as mentioned above, uses a computed wave pattern instead of experimental one. This technique to obtain wave amplitude function from computed wave pattern was named "numerical wave pattern analysis" [4]. The availability of hull optimization methods using numerical wave pattern analysis, which is like the present method, depends on the accuracy of the computation method to estimate wave-making characteristics. In the present optimization method, one kind of Panel Shift Rankine Source Method (PSRSM) is chosen as the method to predict wave-making characteristics. It is widely known that a PSRSM has a high accuracy for estimation of wave pattern around ship. However, it also has the unstableness of numerical calculation, in fact, there are some cases that a conventional PSRSM is not able to solve wave pattern around ships which have large fineness coefficient or small ratio of length to breadth. In order to obtain a practical method of computation for wave-making characteristics, PSRSM was modified by one of the authors, and the problem of unstableness was solved by a contrivance of numerical calculation technique [5] [6]. In the present optimization method, using this modified PSRSM, accurate wave patterns and wave-making resistance can be obtained without fail.

Fn = 0.289

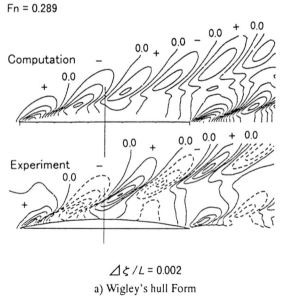

$\Delta \zeta / L = 0.002$

a) Wigley's hull Form

Fn = 0.316

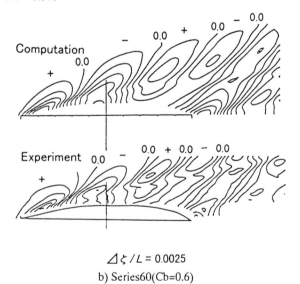

$\Delta \zeta / L = 0.0025$

b) Series60(Cb=0.6)

Figure 1. Comparison of wave patterns between computation and experiment

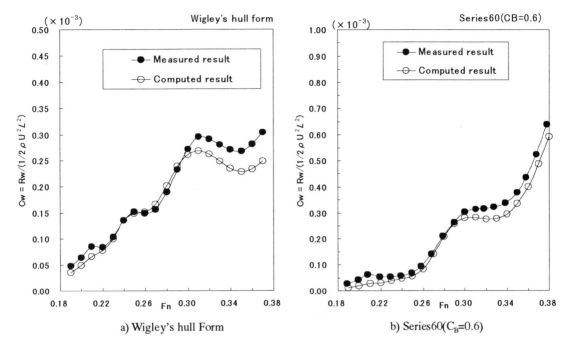

a) Wigley's hull Form

b) Series60(C_B=0.6)

Figure 2. Comparison of wave-making resistance coefficients

The results of the modified PSRSM are compared with experimental ones. Figure 1. shows the comparisons of wave patterns of Wigley's hull form and Series60(C_B=0.6). Figure 2. shows wave making resistance coefficient about these ships. These Figures reveal that the computed results of the modified PSRSM show a good agreement with experimental ones.

The modified PSRSM plays an important role in the present hull optimization method. Through the optimization process, the modified PSRSM provides wave patterns to calculate wave amplitude functions, and wave-making resistance to confirm the reduction due to the hull optimization.

2.2. Procedure

The concrete optimization process of the present method is as follows:

1) The constraint conditions are determined, considering the design requirements such as speed, displacement and water plane area.

2) The wave amplitude functions of the original hull form are obtained by numerical wave pattern analysis employing the modified PSRSM.

3) The added volume distribution that reduces

the wave resistance of the original hull is obtained by the optimization process employing the technique of calculus of variations.

4) Adding the volume to the original hull form creates the new hull form.

5) The wave-making resistance of the new ship form is evaluated by the modified PSRSM.

6) The above process is iterated until the new hull form satisfies the requirement of resistance or a reduction of wave resistance is saturated.

7) Finally the optimal hull form is obtained.

In the above explanations, the optimization process is iterated many times. From the viewpoint of mathematical optimization problem, the optimum solution is obtained to execute the above process only once. However, the solution is only a result of a variational problem and the obtained hull form slightly differs from actual optimum one, because a difference between free waves of the corrected linearised theory and actual ones are remained. The optimization process is iterated in order to reduce the influence of the difference between the theory and the actuality. As the number of iteration of the process increases, the change of optimization becomes smaller and the free waves of the corrected theory become closer to the actual ones.

2.3. Optimization technique

The detail of the optimization process to find the added volume distribution is shown.

Figure 3 shows the coordinate system. The origin is located at midship in still water. The x-axis is in direction of the uniform flow U, y-axis is directed to starboard side from the centerline, and z-axis directs vertically upwards.

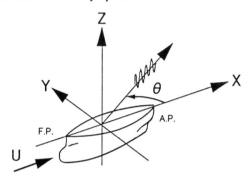

Figure 3.   Coordinate system

The range of an optimization is determined as $x_1 \leqq x \leqq x_2$, $z_1 \leqq z \leqq z_2$.

The form of added volume distribution $y(x,z)$ is defined using function $f_i(x), g_j(z)$, as follows:

$$y(x,z) = \sum_i \sum_j a_{ij} \cdot f_i(x) g_j(z) \quad (i,j = 1,2,3,4,\ldots,)$$

where, a_{ij} are parameters of the added volume distribution. The function $f_i(x)$ determines the longitudinal distribution of volume, and the function $g_j(z)$ determines the vertical distribution of volume. These functions are defined with consideration of shape and constraint conditions, mostly the functions based on trigonometrical series are taken.

The wave amplitude functions S, C of the added volume distribution which are described using Michell-Havelock thin ship theory is as follows:

$$S = S_c \times \frac{2k_0}{\pi} \sec^4 \theta$$
$$\times \int_{x_1}^{x_2} \int_{z_1}^{z_2} y(x,z) \sin \left(k_0 \cdot x \cdot \sec \theta \right)$$
$$\times \exp \left(k_0 \cdot z \cdot \sec^2 \theta \right) dx dz$$

$$C = -C_c \times \frac{2k_0}{\pi} \sec^4 \theta$$
$$\times \int_{x_1}^{x_2} \int_{z_1}^{z_2} y(x,z) \cos \left(k_0 \cdot x \cdot \sec \theta \right)$$
$$\times \exp \left(k_0 \cdot z \cdot \sec^2 \theta \right) dx dz$$

$$k_0 = g/U^2$$

S_c, C_c are correction factors, and the difference of wave amplitude function between the linearised theory and the actual wave are corrected using these correction factors. The factors are determined comparing linearised theory with stored experimental results of similar ship forms.

If the wave amplitude functions of original ship S_0, C_0 are obtained, the difference of wave-making resistance between original and new hull form can be given by

$$\Delta C_{wp} = \frac{4\pi}{L^2} \int_0^{\pi/2} \left(S_0 \cdot S + C_0 \cdot C \right) \cos^3 \theta \, d\theta$$
$$+ \frac{2\pi}{L^2} \int_0^{\pi/2} \left(S^2 + C^2 \right) \cos^3 \theta \, d\theta$$

where, L is the ship length. The original amplitude functions S_0, C_0 are obtained by numerical wave pattern analysis in original hull forms.

Considering requirements in ship design, the restraint are determined, and constraint functions D_k are defined. As examples, some typical cases of function D_k are as follows:

1) The hull form is unchangeable at the boundary :
$$y(x,z) = 0 \ at \ x = x_1, x_2, \ z = z_1, z_2$$

$$D_1 = \sum_i \sum_j a_{ij} \cdot f_i(x_1)$$

$$D_2 = \sum_i \sum_j a_{ij} \cdot f_i(x_2)$$

$$D_3 = \sum_i \sum_j a_{ij} \cdot g_j(z_1)$$

$$D_4 = \sum_i \sum_j a_{ij} \cdot g_j(z_2)$$

2) The hull form is unchangeable at the midship
$$y(x,z) = 0 \quad at \ x = 0$$

$$D_5 = \sum_i \sum_j a_{ij} \cdot f_i(0)$$

3) The variation of volume by optimization

$$\nabla_{new} - \nabla_{original} = \nabla_{add}$$

$$D_6 = \sum_i \sum_j a_{ij} \cdot \int_{x_1}^{x_2} \int_{z_1}^{z_2} f_i(x) g_j(z)\, dx dz$$
$$-\frac{1}{2}\nabla_{add}$$

Optimization functional T is defined with Lagrange multipliers λ_k as follows:

$$T = \Delta C_{wp} + \lambda_1 D_1 + \lambda_2 D_2 + \cdots + \lambda_k D_k + \cdots$$

The derivative of functional with optimization parameters is as follows:

$$\frac{\partial T}{\partial a_{ij}} = 0 \;,\quad \frac{\partial T}{\partial \lambda_k} = 0 \qquad (i,j,k = 1,2,3,4,....,)$$

The parameters of the added volume distribution are obtained by solving above simultaneous equations, and an optimized hull form, which has minimum wave-making resistance under given conditions, is obtained.

3. APPLICATIONS

Some optimization examples of the present method are shown. Three types of hull forms, namely, Series60(C_B=0.6), an over-panamax type container ship and a high-speed car ferry, are chosen as application examples.

3.1 Series60

Series60(C_B=0.6) are optimized at Froude number = 0.25. There are 2 steps in the optimization.

During the 1st stage, the optimum bulb is designed. Considering Series60 is a ship with a bulb of zero in thickness, an added volume is attached to its bow by the present method. The length of the bulb is determined 3%Lpp. At 2nd stage, the whole hull form is optimized keeping the displacement volume.

In this case, the improvement of the longitudinal volume distribution is chosen for the aim of optimization, and the following simple functions are taken as functions of added volume distribution.

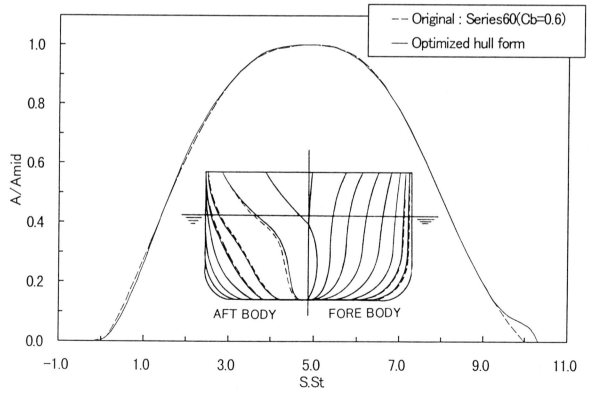

Figure 4. Change of hull form and sectional area curve by optimization

426

$$f_i(x) = \begin{cases} \sin\left(\dfrac{2i-1}{2}\pi \cdot \dfrac{x_1-x}{x_1}\right) & (\text{for fore body and bulb}) \\ \\ \cos\left(\dfrac{2i-1}{2}\pi \cdot \dfrac{x}{x_2}\right) & (\text{for aft body}) \end{cases}$$

$$g_j(z) = 1$$

F.E. : $x=x_1$, A.P. : $x=x_2$

These function have two features as follows:

(i) $f_i(x)=0$ at $x=x_1, x_2$ (ii) $\dfrac{d}{dx}f_i(x)=0$ at $x=0$

thus the optimization can be carried out, even if a few constraint conditions are set. Figure 4. shows the comparisons of sectional area curves and hull forms.

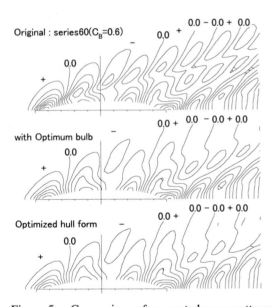

Figure 5.　Comparison of computed wave patterns

The comparisons of computed wave pattern are shown in Figure 5. between original hull form and optimized one, and the comparisons of the amplitude spectrum and the wave pattern resistance coefficient by numerical wave analysis are shown in Figure 6. and Figure 7.. Due to adding optimum bulb to the bow, the wave resistance is reduced by about 10%, and due to the optimization of whole hull form, the wave resistance is further reduced by about 35%. The wave resistance of final hull form is

about 55% of the original hull form's one.

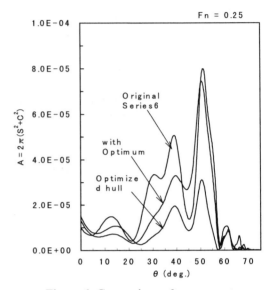

Figure 6. Comparison of wave spectrum

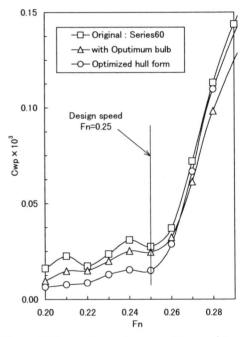

Figure 7. Comparison of wave pattern resistance coefficients

The reduction to wave resistance is large in spite of small change in hull form. This result shows the possibility that a large reduction of wave resistance can be obtained even if the change of hull form is small.

3.2 Container ship

An over-panamax type container ship is optimized at Froude number = 0.24, keeping the displacement. The optimization was carried out for the hull form, which has a hump near design speed. The comparisons of computed wave spectrum between the original hull form and optimized one are shown in Figure 8. And the comparison of wave resistance coefficients is shown in Figure 9. , and the tank test results are also plotted. Due to the optimization, the wave making resistance is reduced by about 15%.

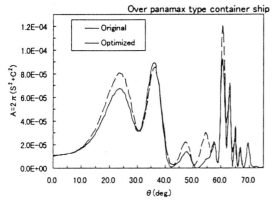

Figure 8. Comparison of computed wave spectrum

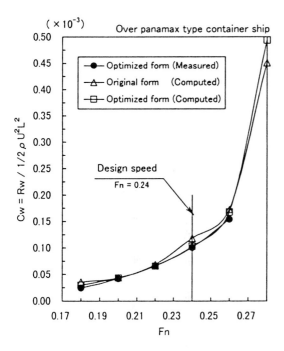

Figure 9. Comparison of wave making resistance coefficients

3.3 Car ferry

A high-speed car ferry is optimized at Froude number = 0.36. To obtain a practical hull form, the function of added volume distribution and the constraint conditions are determined with care. The towing test is carried out to validate optimization gains of the present method. Wave patterns around original ship and optimized one are measured in the towing tank, and wave pattern analysis are done. Figure 10 shows the comparison of wave pattern between computation and experiment. The comparison of the wave spectrum between original hull form and optimized one is shown in Figure 11. The comparisons of wave-making resistance coefficient between original hull form and optimized one are shown in Figure 12. Due to the optimization, the wave resistance is reduced by about 15%.

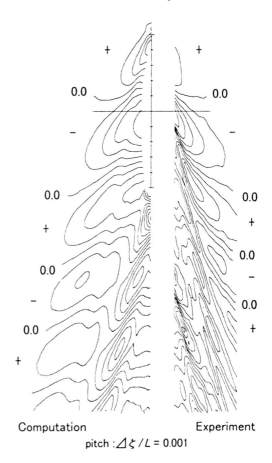

Computation   Experiment

pitch : $\Delta \zeta / L$ = 0.001

Figure 10. Comparison of wave pattern between computation and experiment

428

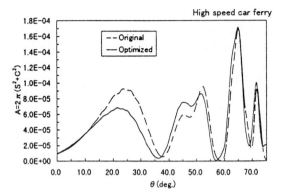

Figure 11. Comparison of measured wave spectrum

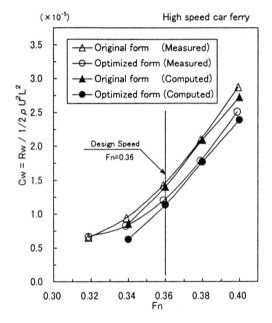

Figure 12. Comparison of wave making resistance
coefficients

4. CONCLUSION

The conclusions are as follows:

(1) A new optimization method of ship hull forms
to minimize wave-making resistance using the
modified Rankine Source Method is introduced.

(2) It is shown that the modified PSRSM is
available as the method to predict wave-making
characteristics of ships in this optimization
method. The computed results of the modified
PSRSM show a good agreement with the
experimental results.

(3) The application examples of the present

optimization method for series60, an over-
panamax type container ship and a high-speed
car ferry are shown. The examples show that
significant reductions in wave-making
resistance have been achieved without
considerable changes in hull form.

(4) The application example and tank test results
show that the present optimization method is
adequate to accomplish the purpose of hull form
improvement.

ACKNOWLEDGEMENT

The authors wish to express their gratitude to Mr.
S. Soejima, Dr. N. Ishii and Mr. M. Fukushima for
their kind aid and encouragement in writing this
paper.

REFERENCES

1. Eiichi Baba, An Application of Wave Pattern
Analysis to Ship Form Improvement, Journal
of society of naval architects of Japan No.132
(1972)

2. Masahiro Matsui, Tatsuo Tsuda, Katsuhiro
Ohkubo and Seiichi Asano, A Method for
Optimization of Ship Hull Forms based on
Wave-pattern Analysis Data, Journal of society
of naval architects of Japan No.147 (1980)

3. C. W. Dawson, A Practical Computer Method
for Solving Ship Wave Problems. 2nd
International Conference on Numerical Ship
Hydrodynamics, Berkeley, (1977)

4. Akihito Hirayama and Tatsuya Eguchi, An
Application of Wave Pattern Analysis
employing Rankine Source Method for Ship
Form Optimization, Transactions of the West-
Japan society of naval architects No.92 (1996)

5. Tatsuya Eguchi, Numerical Analysis of Rankine
Source Collocation Method for the Steady Wave
Making Resistance Problem, Journal of society
of naval architects of Japan No.177 (1995).

6. Tatsuya Eguchi, A study on Power Estimation
for High-speed Vessels using Panel Shift
Rankine Source Method, Transactions of the
West-Japan society of naval architects No.95
(1998)

Practical Design of Ships and Mobile Units
M.W.C. Oosterveld and S.G. Tan, editors.

Tankers: Conventional and Twin-Gondola Hull Forms

Eduardo Minguito[1], Henk H. Valkhof[2] and Eric v.d. Maarel[2]

[1] Astilleros Españoles
    Ochandiano 14-16, 28023 EL PLANTIO, Madrid, Spain

[2] Maritime Research Institute Netherlands
    P.O. Box 28, 6700 AA Wageningen, The Netherlands

ABSTRACT

For a series of Shuttle Tankers constructed or still under construction at the PuertoReal Shipyards and Sestao Shipyard of the ASTILLEROS Group, two concepts are being studied, i.e., a conventional hull form concept and a twin-gondola hull form concept. The study is not only related to operational and propulsive performance aspects, but also concerns safety and environmental aspects and the possible contribution in the design process of computational fluid dynamics (CFD) tools, such as potential and viscous flow codes developed at MARIN.

1 INTRODUCTION

In September 1992, ASTILLEROS ESPAÑOLES signed a contract with KNUTSEN OAS Shipping for the building of two 126,500 TDW Single Screw Shuttle Tankers. The ships, named *Hanne Knutsen* and *Jorunn Knutsen*, were deliveried by Puerto Real shipyard (PR66/67) in respectively march and may, 1995. During 1992 and 1993, an extensive series of model tests were performed [1,2] to achieve the best propulsion performance. In September 1997, ASTILLEROS ESPAÑOLES Sestao shipyard delivered a new 125,000 TDW Twin Skeg Shuttle Tanker for KNUTSEN OAS Shipping, named *Elisabeth Knutsen* (SS290) (another five shuttle tankers are under construction at PuertoReal and Sestao Shipyard for NAVION, with a slightly different general arrangement). The hull lines of this ship were also extensively optimised at the MARIN facilities throughout the second half of 1995 and the first quarter of 1996 [3].

Since, the main particulars and the operational demands are equal for the single- and twin screw ships, it was felt to be interesting to study the advantages and/or disadvantages of the operational, environmental and safety aspects of the two different ship concepts. The operational aspects described in this paper, are the more interesting considering the present higher demands with regard to safety and related environmental issues. Hence, the advantages and disadvantages of both concepts from that point of view will be presented and discussed. This study will be presented in section 2 of this paper. After that the differences in performance between the two hull form concepts are described, as determined by means of an extensive series of model tests. Special attention is paid to the flow in the propeller plane and the results of the hull pressure measurements and the cavitation observations. Moreover, a comparison of the propulsive performance of both ship concepts will be presented and discussed in section 3.

Next, the predictive value of CFD tools in the present context is considered in more detail. The last decade, it has become custom practice at MARIN to conduct a shape-optimization cycle based on computational fluid dynamics, prior to any towing test experiment. The potential flow codes DAWSON (linearized treatment of free surface boundary) and RAPID (non-linear treatment) have proven to be excellent tools in this shape-optimization cycle. However, any possible flow separation and related increase in drag for a ship under certain conditions, cannot be assessed with potential flow calculations. The prediction of such a separation and possible removal by adaptation of the shape, prior to any towing test, can be a major step in ship hull form design. In section 4 some aspects concerning viscous-flow computation with PARNASSOS in domain decomposition mode for a twin-gondola hull form are discussed. PARNASSOS is proving to be of major importance during the design of single-screw, full-block ships and is capable to accurately predict the onset of flow separation, for both model-scale and full-scale Reynolds numbers [6].

Since, the LCB position of twin-gondola ships is shifted more aftwards when compared to a single screw ship, the availability of viscous-flow codes for these type of ships even become more important. Finally, and described in section 5 the conclusions of this study are drawn followed by some recommendations. Here it is shown that the twin-gondola concept has the advantages of a twin-screw ship from a safety and environmental point of view, while from a propulsive and hence, fuel consumption point of view the ship is comparable with or even better than the single-screw concept. The use of modern CFD tools, such as DAWSON, RAPID and PARNASSOS already have and certainly will further reduce the number of model tests.

2 COMPARISON OF OPERATIONAL, SAFETY AND ENVIRONMENTAL ASPECTS

Both hull concepts are double hull and double bottom segregated-ballast crude oil tankers, designed and built for unrestricted world-wide service, prepared to carry crude oil and fitted with the following equipment for offshore operations: mooring equipment forward, bow loading system, dynamic positioning system and emergency towing. At present the *Hanne Knutsen* and the *Jorunn Knutsen* are being operated at Heydrum Field, in the Norwegian Sector of the North Sea, near the Arctic Circle, while the *Elisabeth Knutsen* is being operated at fields located a little more southward. The cargo capacity of both projects is about the same: 140,000 m^3, distributed respectively in 18 (PR66/67) and 12 (SS290) cargo tanks. Apart from the Bow Loading System (BLS) above mentioned, all ships are also fitted with a Submerged Turret Loading (STL) system.

From an operational point of view the main differences between both projects are on the main propulsion system, the engine rooms arrangement and the dynamic positioning system. *Hanne Knutsen* and *Jorunn Knutsen* were the first two Shuttle Tankers built with a diesel-electrical main propulsion system, consisting of four Sulzer diesel engines 9ZAL40S of MCR 6,598 kW each. These engines are coupled to four ABB electrical generators with a total energy power capacity of 25.1 MW at 60 Hz and 6.6 kV. The ships are propelled by one 8.00 m diameter fixed pitch propeller, directly coupled to a ABB synchronous electrical engine of MCR 19 MW at 98 RPM.

The *Elisabeth Knutsen* is driven by two MDE MAN B&W 7S50MC main engines of MCR 10,010 kW at 127 RPM each, directly coupled to two variable pitch propellers of 5.75 m diameter. The full electrical power required aboard is produced by two 8,000 kW ABB shaft alternators powered by the main engines through a pneumatic coupling, which produced energy at 60 Hz and 6.6 kV. An important part of the engine room equipment is located inside the two big skegs. The skegs have been designed to improve as much as possible the ship's performance from a resistance (no flow separation phenomena were detected in the model test results) and propulsion point of view (a larger hull efficiency has been achieved as compared to the single-screw concept). It is important to realise that the available space of the two engine rooms has been reduced as much as possible to achieve the afore mentioned good propulsive performance. This however implied that it has been quite difficult to find room enough to fit any further equipment. Hence, the engine room arrangement nearly completely changed compared to the usual diesel engine room arrangement.

The third main difference between the design concepts is related to the dynamic positioning system: the lateral thrust capacity of *Hanne-* and *Jorunn Knutsen* (three bow and two stern thruster units of 1,700 kW each) is larger than the one fitted on *Elisabeth Knutsen* (two bow thruster units of 2,200 kW each one and two stern units of 500 kW each one), because of the Heydrum Field environmental sea and wind conditions, which can be rather extreme. On the contrary, *Elisabeth Knutsen*

dynamic positioning system is more modern, since the ship has been designed to be redundant on all energy sources (two main propellers, two flap rudders, two main engines, two shaft generators, etc).

A wide description of *Hanne Knutsen* and *Jorunn Knutsen* can be found in [1], whereas *Elisabeth Knutsen* is extensively described in [2] and [3]. The general arrangements of the concepts are given in Fig. 1 and 2.

3 COMPARISON OF PROPULSIVE PERFORMANCE

Propulsion

In order to compare the propulsive characteristics of both design concepts a few tables are presented to show the differences between said concepts starting with the comparison between the main particulars of the single screw- and twin-gondola ships:

		Single-screw	Twin-Gondola	
Length on waterline	LWL	260.50	260.50	m
Breadth moulded	B	42.50	42.50	m
Draught on FP	TF	15.00	15.00	m
Draught on AP	TA	15.00	15.00	m
Displacement volume moulded	∇	142312	142288	m^3
LCB position aft of FP	FB	122.04	124.58	m
Block coefficient on Lwl	CB	0.857	0.857	

Note: Except for an aftward shift of the LCB position no differences between the main particulars can be observed.

The following table shows the main particulars of the designed propellers of both concepts:

Diameter	D	8000	5750	mm
Pitch ratio at 0.7 R	$P_{0.7}/D$	0.701	0.760	
Boss-diameter ratio	d/D	0.165	0.289	
Expanded blade area ratio	AE/AO	0.670	0.527	
Number of blades	Z	4	4	
Number of propellers		1	2	
Type		FPP	CPP	

432

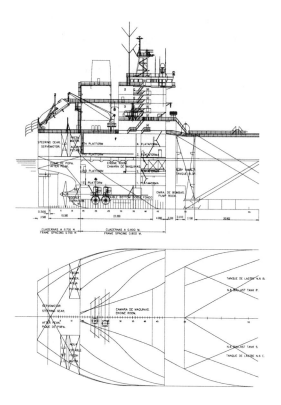

Figure 1.   General arrangement of single screw ship

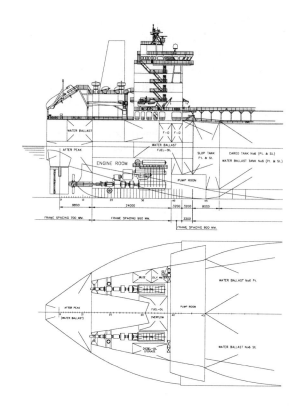

Figure 2.   General arrangement of twin screw ship

Though, the differences between the ships main particulars and their operational demands are marginal it always remains difficult to compare all aspects. For example the effects of the difference in propeller design condition, the difference between a CP- and an FP propeller, etc., though marginal, have not been taken into account. Still an attempt is made to compare the concepts, starting with the differences in required power at constant speed. The presented results have been obtained by model tests carried out in MARIN's Depressurized Towing Tank and have been extrapolated to ideal ship trial conditions, implying unrestricted deep water of 15.0 degrees Celsius, a clean hull and propeller blades and no effects of wind and waves.

| Speed in knots | Draught 15.00 m | | | |
| | Single-screw ship | | Twin-gondola ship | |
	Shaft power P_s in kW	Per cent	Shaft power P_s in kW	Per cent
15	15185	100	15825	104.2
16	19870	100	19460	97.9
17	26595	100	24765	93.1

It is obvious that the aftward shift of the LCB position for the Twin-gondola concept plays in important role. The wave making resistance becoming more and more dominant at the higher speeds is significantly reduced for the twin-gondola hull form at those higher speeds.

On the other hand at the lower speeds where the wetted surface and the related frictional resistance is more important the Single Screw concept shows the best results.

Wake

To study the wake survey, wake measurements were carried out in the propeller plane. Below a comparison of the circumferential distribution of the axial velocity components of both hull concepts is presented.

The results of the single-screw (symmetrical hull form) show an axial wake peak of about 45 per cent in the 12 o'clock position (180 degrees) with very good circumferential gradients. The results of the twin-screw (a-symmetrical hull form) show a local rather steep wake peak with a top-valley value of about 50 per cent. The circumferential gradients still are considered good. For both concepts no cavitation problems are expected assuming well-designed wake adapted propellers.

Hull pressure

Since for both concepts hull pressure measurements have been carried out, a comparison of the highest-pressure amplitudes and the hull excitation forces is made. The ship models were equipped with 20 pressure transducers in the hull region above the propeller. The recorded pressure signals are expressed in harmonic components.

Single screw concept   Twin-gondola concept

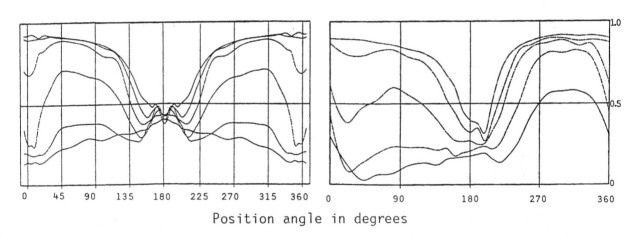

Position angle in degrees

Figure 3. Comparison of axial velocity components

Propeller-induced pressure fluctuations in kPa:

Harmonic components							
1st		2nd		3rd		4th	
Single	Twin	Single	Twin	Single	Twin	Single	Twin
3.48	0.48	1.26	0.13	0.20	0.07	0.10	0.11

Propeller-induced hydrodynamic forces on the hull in KN:

Harmonic components							
1st		2nd		3rd		4th	
Single	Twin	Single	Twin	Single	Twin	Single	Twin
97.8	35.9	66.2	11.6	8.7	5.9	2.9	5.5

Furthermore, the pressure field was integrated over the afterbody to hull excitation forces. The resulting forces and moments were expressed in harmonic components too.

The measurements were performed at more or less equal conditions. Still, it has to be noted that the propeller loading of the Twin-gondola is much lower compared to the Single-screw concept; i.e., $K_T/J^2 = 0.990$ for the single-screw and 1.209 for the twin screw. Both concepts show low to very low hull pressures and related hull excitation forces. Hence, no vibration problems are to be expected for both ship concepts, provided that resonance in the ship's structure does not occur.

Cavitation

By studying the cavitation patterns of both concepts it became clear that the differences found were only marginal. The Twin-gondola concept shows a somewhat reduced extend of the sheet cavitation compared to the Single-

Figure 4. Comparison of cavitation pattern at equal blade position

screw concept, despite the ballast draught of 5.90/8.50 m for the twin gondola and of 8.50/9.60 m for the single-screw concept, which is somewhat in favour for the single-screw concept. The test conditions were such that a service condition of the ship was assumed, taking into account 15 – 20 per cent of sea margin, while 100 per cent MCR of the engine was taken. Moreover, an increase of the wake fraction of about 0.04 was taken due to fouling. Given the results no erosion problems were expected on full scale.

Manoeuvring

The manoeuvring properties of both projects were studied by means of model tests [4, 5]. *Hanne Knutsen* and *Jorunn Knutsen* are fitted with one flap type rudder of 60 m^2 area size (1.56% lateral area), whereas *Elisabeth Knutsen* is fitted with two flap type rudders of 42 m^2 area size each (2.11% total lateral area). By comparing both sets of experimental results, the following main conclusions can be drawn:

- The average port/starboard advance to Lpp ratio for $\delta=35^\circ$, are about the same: 2.3 (PR66/67) and 2.2 (SS290).
- The average port/starboard tactical diameter to Lpp ratio for $\delta=35^\circ$, are also about the same: 1.9 (PR66/67) and 1.85 (SS290). However, the difference between port and starboard is obviously larger for the single screw ships (~ 0.4) than for the twin skeg one (less than 0.1).
- The *Elisabeth Knutsen* 10/10 and 20/20 overshoot angles are slightly smaller than those of the *Hanne Knutsen* and *Jorunn Knutsen* ones: about 1 to 2 degrees.
- Both ships are sligthly dynamically course unstable, but less than other ships with the same size. So, the hysteresis loop width of *Hanne Knutsen* and *Jorunn Knutsen* is about 4 degrees, whereas the *Elisabeth Knutsen* only shows a value of about 1 to 2 degrees.

From the above, it is easy to conclude than the manoeuvring properties of the ships are almost the same. Given the two rudders of the two shaft lines arrangement, the course checking ability of the Twin gondola Shuttle show somewhat better results. The two shaft concept also implies that in case of an emergency the Twin-gondola ship still is capable to manoeuvre and hence can return home safely.

4 NOTE ON VISCOUS-FLOW CALCULATION

For the analysis of the steady, viscous flow along full-block, single-screw hull forms, MARIN has available a CFD tool called PARNASSOS. This tool has proven to reliably predict onset of flow separation and can be used for model-scale and full-scale Reynolds number calculations [6]. Viscous-flow computations with PARNASSOS require the domain of fluid to be divided into a large but finite number of hexahedra, arranged in an ordered fashion. This set of hexahedra is called the grid. For each node in this structured grid, unknown flow quantities are defined (velocity and pressure). A large number of equations is set up through the discretization of the Navier-Stokes equations. From this large set of equations the unknown flow quantities in each point are computed. Grid properties such as the sizes of the hexahedra and their skewness greatly influence the stability of the computation itself and of the accuracy of the computed result. The construction of a grid that has satisfactory properties for the PARNASSOS computation and which gives proper accurate results, becomes increasingly difficult as the geometry gets more difficult.

Recently MARIN has been working on extending PARNASSOS into a tool that allows for more complex geometries than single-screw, full-block hull forms and will relax on the grid generation phase of PARNASSOS computations. This extension is implemented by means of a domain-decomposition method. In such a method the fluid domain is split into a number of separate subdomains.

A grid generation tool is separately applied to each subdomain. This allows a grid for each domain to be independent of the grid in any other subdomain. As we then have a series of subdomains we also have a series of sub-problems.

This series of sub-problems is a coupled system. It is solved by treating the problem for each subdomain as a separate flow problem. The coupling between the sub-problems (one for each subdomain) is restored through an iteration process within the calculation.

This iteration process requires flow data be transferred between the sub-problems across the sub-problem interfaces. Since the grid on these interface have been generated separately, the grid points (that hold the fluid-flow data) on the boundary of a subdomain do not necessarily coincide with grid points on the boundary of an adjacent subdomain. Therefore, the transfer of data between the sub-problems on each subdomain requires an interpolation step.

An interesting item is the data itself that is transferred. The basic implementation of the domain decomposition method only allows the unknown data itself to be transferred between the sub-problems. Alternatively, derivatives of the data itself may be transferred and a possible better alternative is to transfer a combination of unknowns and derivatives of unknowns between the sub-problems.

The first alternative has been successfully implemented and can be used with sub-problems, which do have points that coincide on a common boundary. The alternatives, which also transfer derivatives of the data are subject to investigation.

We have started work to apply the PARNASSOS code in domain decomposition mode on the twin gondola hull form, which is the subject of this paper. This uses the transfer of the fluid flow quantities (velocity and pressure) only, and grids that have grid points that do not coincide on the common boundaries.

So far, the computations have been somewhat disappointing. The coupling iteration of the calculation fails to converge for the complex case. The reason of this behavior has so far remained unclear. This method with the new type of boundaries (non-coinciding grid points) may be considered a generalization of the method with coinciding grid points on the common boundaries, which has worked on a test case. Therefore we have good hope that our method will eventually work on this generalization. Until then we unfortunately must postpone the application of PARNASSOS to the twin-gondola problem to a later data.

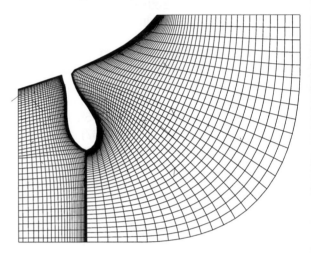

Figure 5. Configuration of domain decomposition grids in a streamwise station, for twin gondola hull

5 CONCLUSIONS

In general the conclusion seems to be justified that the twin-gondola concept shows the best overall hydrodynamic performance. Since, also from an environmental and safety point of view this concept is to be advised, only economical aspects, such as the building costs and maintenance costs, are remaining aspects playing an important and probably dominant role during the decision taking. Still it has to be noted that the pollution of the environment cannot be expressed in terms of money, while finally our responsibility to the future generation even is of greater importance.

REFERENCES

[1] Valkhof H. H., Minguito E. *Shuttle Tanker Design problems solved by CFD-Code Dawson.* SNAME TRANSACTIONS. 1994.

[2] Infomarine ship report. *Elisabeth Knutsen 265 m Loa, 125,000 DWT shuttle tanker, built by Astilleros Españoles Sestao Yard for Knutsen OAS.* INFOMARINE. November 1997.

[3] Petrolero Shuttle "Elisabeth Knutsen" para la companía Knutsen OAS. Ingenieria Naval. October 1997.

[4] Berg, Tor Einar. *Manoeuvring tests. 140,000 m³ Shuttle Tanker.* MARINTEK report nº 5149009.00.01. May 1993.

[5] Rem A., Visser C., Valkhof H.H. *Manoeuvring tests for a 1340,000 TDW Twin Skeg Shuttle Tanker.* MARIN report nº 13226-2-BT/GT. May 1996.

[6] Hoekstra, M. and Eça, L. *PARNASSOS: An efficient method for ship stern flow calculations,* to be published in the proceedings of the '3rd Osaka Colloquium on advanced CFD applications to ship flow and hull form design', Osaka, 1998.

Practical Design of Ships and Mobile Units
M.W.C. Oosterveld and S.G. Tan, editors.

439

Experimental and computational study on resistance and propulsion characteristics for Ro-Ro passenger ship of twin propellers

Suak-Ho Van[a], Do-Hyun Kim[a], Bong-Ryong Son[b], Jung-Kwan Lee[b], Dong-Yul Cha[b], Jae-Kyoung Huh[b]

[a] Ship Performance Department, Korea Research Institute of Ships & Ocean Engineering
171 Jang-Dong, Yusung-Gu, Taejon, 305-343, Korea

[b] Hanjin Heavy Industries, Co., Ltd.
5-29 Bongrae-Dong, Youngdo-Gu, 606-065, Pusan, Korea

The powering performance and flow characteristics of Ro-Ro passenger ship of twin propellers are investigated. Three hull forms(two of skeg-type and one of open-type stern) are designed and the resistance, self-propulsion, propeller open-water, wake measurement and flow visualization test are performed. It is observed that there exists a rotational velocity component of the inflow in the propeller plane and it differs with the stern shape and affects significantly on the propulsive efficiency. The relations between the rotational component and the resistance and propulsion characteristics are explored based on the experimental and computational results for the inward and outward rotation of propeller.

1. INTRODUCTION

After the diplomatic relationship between Korea and China was established in 1993, the trade and tourism exchanges have been increased greatly between them. More than 98% of trade goods are transported by sea and the passenger ferry services have taken charge of an important role in the transportation system for both countries.

Hanjin Heavy Industries Co., Ltd.(HHIC) has examined the major sea routes between Korea and China and associated ports for the development of a Ro-Ro passenger ship suitable for those routes. The draft is confined by the water depth of the ports and this type of ships usually require a large volume and deck space above the waterline. In order to meet the above requirements, the after-body shape with broad transom is chosen. The major requirements are given in Table 1 and HHIC developed several hull forms to meet the major requirements and selected three hull forms(2 of twin skeg-type after-body and 1 of open-type after- body) to study their resistance and propulsion characteristics. The model tests are performed at the Towing Tank of Korea Research Institute of Ships and Ocean Engineering(KRISO). The flow characteristics and powering performances including the

generation of rotational inflow, wake distribution in the propeller plane, effect of propeller rotation direction on the propulsion factors, convergence of limiting streamlines are discussed based on the results of the resistance, propeller open water, self-propulsion, wake measurement and paint tests.

Table 1 Major Requirements

Item	Range(value)
LOA	130~140(m)
Draft	less than 6.0(m)
Cruising speed	22.0(knots)
Cruising range	500(N miles)
Number of containers	160(TEU)
Number of cars	50
Number of passengers	400

2. CONCEPT OF HULL FORMS

2.1. Principal particulars

The principal particulars of the hull forms are given in Table 2 and outboard profile of the 125m Ro-Ro passenger ship is shown in Fig. 1. The

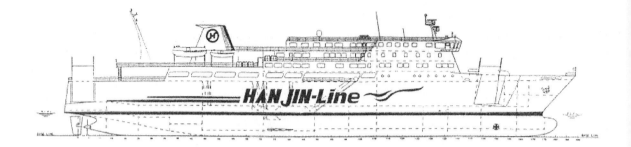

Fig. 1 Outboard Profile of Ro-Ro Passenger Ship

containers are to be loaded by trailers and forklifts and cars roll-on, roll-off by themselves.

Table 2 Principal particulars of three hull forms

Hull forms	I (S1)	II (S2)	III (S3)
LBP(m)		125.0	
Beam(m)		21.0	
Draft(m)		5.650	5.685
$\nabla(m^3)$		8360.0	
C_B		0.563	0.560
C_P		0.592	0.588
C_M		0.952	
LCB(%)	-4.696	-4.686	-4.468

2.2. Characteristics of hull forms

Fig. 2 shows the body plans of the three hull forms at the typical stations, S1 and S2 are twin skeg-type hull forms and S3 is an open-type after body hull form that has a center skeg at the centerline of the hull. The principal particulars of S3 are slightly different from S1 and S2. These ships have a very small half entrance angle(13°) and no parallel middle body, however, the position of longitudinal center of buoyancy is unusually located behind the mid-ship to satisfy the cargo loading requirements.

(a) Fore-body : The three ships have the same "V" shaped fore-body in order to achieve good seaworthiness. Bulbous bow is designed based on the Hagen and Kracht bulb design methodology. The designed bulb has its center of volume near the top of bulb(∇-shape) to maximize the ship performance at the design speed of 22.0knots in full load condition.

(b) After-body of Hull Form I(S1): The skegs are perpendicular to the base line and the centerline of the skegs and propeller shafts are parallel to that of the hull. The waterline shape of each skeg is symmetric to the centerline of the skeg below the contact line between the hull and the skeg. The bottom of the skeg is rounded in order to reduce the drag induced by the cross flow. The distance between the skegs is

Table 3 Test items and conditions

Test Item	Measuring Range	Remarks
Resistance	15.0~25.0(knots)	Vertical displacements measured
Self-Propulsion	15.0~25.0(knots)	Propeller rotation : inward and outward
Propeller Open Water	0~0.9 (advance ratio, J)	RPS fixed at 15
Wake Measurement	Design Speed 22.0(knots)	5 radial positions × 32 circumferential positions
Paint	Design Speed 22.0(knots)	Oil paint + polywax + petroleum

2.6 times the propeller diameter to avoid congestion effect and overlapping of propellers. The buttock profile at the centerline begins to rise at the station 6 and its buttock angle is approximately 10° in the propeller plane.

(c) After-body of Hull Form II(S2): The displacement, hull form parameters and distance between skegs are kept same as those of S1. The volume and sections of skegs are similar, however the vertical plane of the skeg is inclined 10° outward (distance between the bottom of skeg is wider than the top) with keeping the same position for the center of shafts.

(d) After-body of Hull Form III(S3): S3 hull form is developed to compare the resistance and propulsion characteristics of an open-type hull form with the twin skeg hull forms, S1 and S2. The principal particulars and the position of LCB are a little different from those of S1 and S2, however, kept as same with them as possible. Frame sections and buttocks of after body are flat pram-type. S3 has multi-flow section to achieve the sufficient propeller clearance at the propeller plane. The center skeg is box-type and each propeller shaft is supported by a pair of V-struts, respectively.

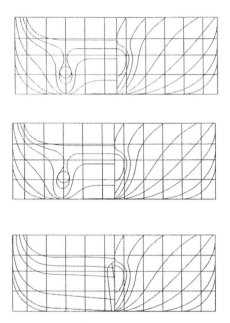

Fig. 2 Body Plan of Hull Forms(from top: S1, S2, S3)

3. NUMERICAL CALCULATION

Based on the panel method, preliminary calculation is performed to estimate the flow characteristics for S1 and S2 hull forms. The free surface effect around the skeg part is assumed to be negligible and the double body solutions are provided. The pressure contours around the stern part are compared in Fig. 3. For both hull forms, the pressure distribution in the outside of the skeg bottom is bigger than in the inside. The difference of pressure between outside and inside is larger for S1 than S2. This implies that the cross flow will be more significant and the inward rotational components in propeller plane are stronger for S1.This conjecture is confirmed from the experiments performed afterwards.

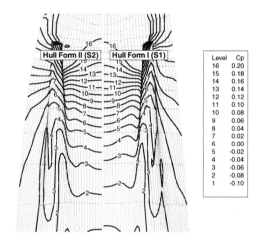

Fig. 3 Comparison of Cp Distribution

4. EXPERIMENTS

Three ship-models of 1/16 scale are manufactured of wood and a pair of stock propellers is chosen. The studs of 1.2mm diameter with 2.0mm height, 10.0mm spacing are attached at the station 19 and middle of bulb to generate the turbulent flow. All the experiments are carried out at the towing tank of KRISO. The towing tank has the dimensions of 200m in length, 16m in width and 7m in water depth and has an electronically controlled towing carriage with the maximum speed of 6.0m/s. The test items and conditions are summarized in Table 3 and more details on the experiments can be found in Van et al.[1].

442

Table 4 Comparison of propulsion characteristics for 3 hull forms

Hull Form	S1		S3		S3	
Rotation	In	Out	In	Out	In	Out
EHP(%)	100		97.6		105.7	
W_N	0.129		0.130		0.076	
W_E	0.213	0.101	0.160	0.094	0.120	0.093
T	0.165	0.139	0.157	0.149	0.188	0.170
η_H	1.060	0.958	1.004	0.979	0.923	0.915
η_r	0.975	0.972	1.004	0.992	1.017	1.008
η_O	0.635	0.652	0.647	0.653	0.645	0.652
η_D	0.657	0.607	0.652	0.608	0.605	0.601
Rpm	201.4	215.0	206.7	215.4	217.8	220.3
DHP(%)	100.0	108.1	98.3	105.3	114.6	115.4

5. RESULTS AND DISCUSSION

5.1. Resistance tests

The residuary resistance coefficients that can be calculated based on the Froude's assumption, are compared for three hull forms in Fig. 4. For the S3, the principal dimension parameters, such as LCB, could not be optimized for the open-type stern shape to keep the parameters as same as possible with those of S2 and S1 for comparison. As a result, the resistance of S3 is the largest among three hull forms and there might be a room for improvement for this hull form.

S2 hull form undergoes the least residuary resistance. It can be considered that the inclined skeg of S2 is more consonant with the pressure distribution around the stern area and the flow is expected to follow the after body smoothly. The longitudinal vortex originated from the bottom part of skeg is weaker for S2 than S1 because the pressure difference between the inner and outer region is less for S2 than S1. This flow characteristics around the skeg are also discussed by Jonk[2] and coincident with the other test results, i.e., flow visualization and wake measurements which will be explained later. The formation of the longitudinal vortex is followed by the expenditure of energy and will cause the increase of the resistance. Those features elucidate the increase of the resistance for S1 hull form whose cross flow is more significant than S2.

During the resistance tests, the vertical displacements at FP and AP are measured and the trim and sinkage in model scale are plotted in Fig. 5. Sinkage is increased as the ship speed increases, especially at AP. In lower speed range, bow trim is observed and changed to stern trim as the speed increases. The trends are almost same for the three hull forms.

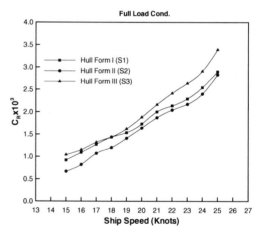

Fig. 4 Residuary Resistance Coefficients (C_R)

5.2. Self-propulsion tests

The self-propulsion tests are performed with the propeller rotating inward and outward for each hull form. As it can be shown clearly from the wake measurement, the tangential component of inflow in the propeller plane is not symmetric about the centerline of skeg, especially for S1 and S2 and the propulsion factors are expected to be different due to the direction of propeller rotation. The results are summarized in Table 4.

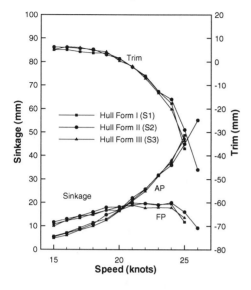

Fig. 5 Comparison of Trim and Sinakge

For S3, the asymmetry in the wake distribution is not significant because the stern of this hull form is open-type and there is slight difference for propulsion factors for different rotational direction of propeller. However, the difference of propulsion factors due to the change of the rotational direction of propeller for S1 and S2 is notable because of the strong outward rotational inflow. If the propeller rotates inward, although the axial velocity into the propeller blade remains the same but the relative rotational velocity into the propeller blade is increased because of the rotational component in inflow. Consequently angle of attack into the propeller blade is increased and the propeller RPM decreased. On the contrary, angle of attack is decreased if the propellers rotate outward and RPM increased. This will produce the larger value of effective wake for inward rotation of propeller than outward rotation. For example, at the design speed(22.0knots), the effective wake with inward rotation is 0.210 which is larger than that with outward rotation(0.101).

Actually, the axial velocity into the propeller blade changes not so much and the increase of the effective wake for inward rotation is mainly due to the counter rotational component of inflow. This value of effective wake is even larger than the nominal wake (0.129) obtained from the wake measurement for the same speed. This obviously contradicts to the definition of effective and nominal wake. From this point of view, to reflect the physical phenomena in the analysis of powering performance, there should be an alternative method which can take into account of the rotational component correctly, such as proposed by Van[3, 4]. However, the full-scale prediction in this paper is made by the "1978 ITTC method for single propeller ships" without any modification.

The propulsion efficiency for S1 and S2 is much better for the inward rotation of propeller than outward as it can be seen in Table 4. When the propeller rotates inwardly, the inward rotational component is generated by the propeller which is opposite to that of inflow. Therefore, those rotational components (generated by hull and propeller) can be canceled each other and the propulsion efficiency increases as a result. In the analysis procedure of self-propulsion tests, this physical phenomena can be represented as an increase of hull efficiency($\eta_H = 1-t/1-w$) with the larger value of effective wake.

5.3. Wake Measurements

The velocity distribution in the propeller plane is measured using the 5-hole Pitot tubes for each hull form. 5 Pitot tubes are tied up into a rake which can rotate in the propeller plane. Measured points are 0.38, 0.5, 0.7, 0.9 and 1.1 of propeller radius in the radial direction and 0° to 360° with 5° or 10° increment in the circumferential direction.

The distribution of axial velocity and transverse velocity vectors viewed from behind for starboard side are shown in Fig. 6. The axial velocity of S3 is almost 1.0 except the angular positions between 150° and 210° where the flow due to the V strut is in effect. The lowest peak is found for the outer radius(0.9 and $1.1r_p$) which is located inside of the boundary layer of the hull surface. The transverse components are symmetric along the vertical plane because the after body of this hull is open-type.

The axial velocity distribution of S1 and S2 are similar to each other and symmetric except that the lowest velocity for S2 is observed at the angular position of 170° instead of 180°. This is natural because the skeg of S2 is inclined outward by 10°.

444

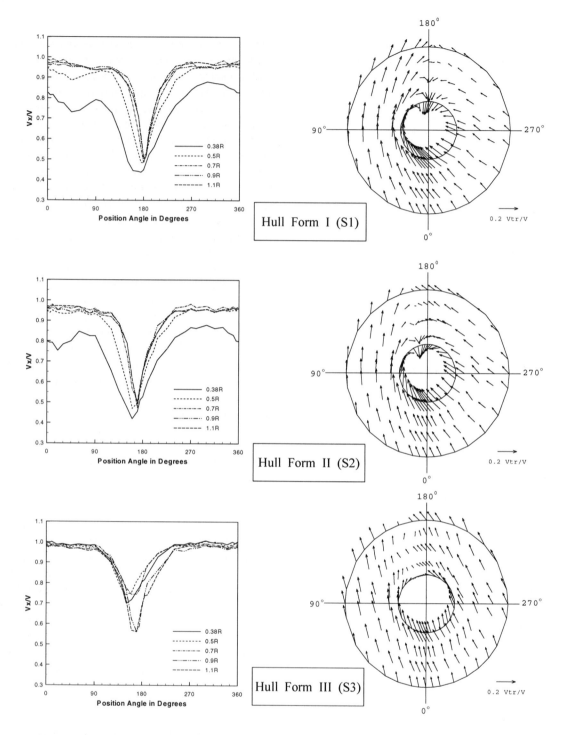

Fig. 6 Axial Velocity Distribution & Transverse Vectors
(Viewed from behind for Starboard side)

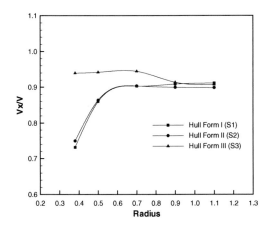

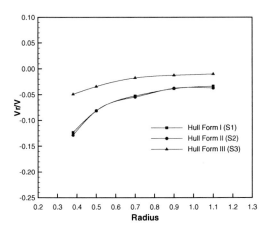

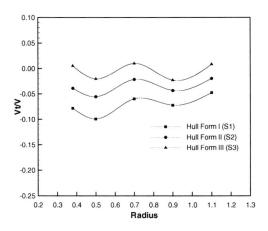

Fig. 7 Radial Distribution of Circumferential Mean of Axial, Radial, and Tangential Velocity Components

However, the transverse vectors are notably asymmetric for both hull forms. In the outer region, the flows are directed to inside especially underneath the skeg, but in the inner region, the flows are directed upward and as a result, strong outward rotational components exist in the propeller plane. The existence of tangential components for S1 and S2 is accompanied by the pressure difference between inner and outer region of skeg and can be thought as an indication of the formation of longitudinal vortex. Again, it can be supposed that the resistance of S1 will be greater than that of S2 as is confirmed from the resistance tests because the tangential components of S1 are more significant than those of S2.

The radial distribution of circumferential mean for the axial, radial and tangential velocity components is compared in Fig. 7. It can be easily seen that the tangential velocity component rotating outwardly is most significant for S1 and almost negligible for S3. The difference of propulsion characteristics (especially RPM and effective wake) between inward and outward rotation of propeller will be largest for S1 and almost same for S3 as discussed in Section 5.2. This configuration of wake distribution for skeg-type twin hull is also explained by Jonk[2]. The nominal wake obtained from the integration of the axial velocity component in the propeller plane is included in Table 4.

5.4. Flow visualization

The flow visualization test with oil paint on the hull surface to investigate the near wall (limiting) streamlines for three hull forms is performed. For bow and middle part, there is no difference for three hull forms because the fore parts of them are identical. However the flow lines around the skeg are quite different for S1 and S2 as compared in Fig. 8. The flows are moving from outside to inside of skeg especially underneath skeg and it can be understood that pressure distribution on the outside is greater than that on the inside. This convergence of flow from outside to inside implies the formation of longitudinal vortex rotating clockwise at starboard side and counter-clockwise at port side, respectively, viewed from backward. In Fig. 8, the cross flow around the bottom of skeg is more remarkable for S1 and it coincides with the results of wake measurements.

446

6. CONCLUSIONS

For three hull forms of 125m Ro-Ro passenger ship with twin propellers, the powering performances and flow characteristics are analyzed. For skeg-type stern hull forms, the pressure difference between inner and outer region of skeg generates the outward rotational components of inflow in the propeller plane. It is considered that the resistance characteristics are closely related with the generation of rotational component and S2 hull form, whose skeg is inclined by 10° outward, shows the least resistance within the extent of present work. The propulsion factors are also changed greatly with the rotational direction of propeller and the better propulsion efficiency can be obtained if the propellers are rotating inward which is opposite to that of inflow. The experiments show consistent results and flow features in line with the generation of longitudinal vortex could be clearly explained. The present work, although not enough for complete understanding of complicated stern flow, will be a helpful guidance for the design of the skeg-type hull form with twin propellers. More experiments for the hull forms with the various configurations of skegs are recommended for future study.

REFERENCES

[1] Van, Suak-Ho, Kim, Moon-Chan and Lee, Jin-Tae, *Some Remarks on the Powering Performance Prediction Method for a Ship Equipped with Preswirl Stator-Propeller System*, Proceedings of 20th ITTC, vol. 2, 1993.

[2] Jonk, A., *The Use of Non-Viscous Flow Calculation in Hull Form Optimization*, Proceedings of Hull Form Design Workshop, vol. 1, 1985.

[3] Van, Suak-Ho et al., *A Powering Performance Prediction Method Considering the Hydrodynamic Characteristics of a Preswirl Stator Propeller System*, Proceedings of 1st ICHD, Wuxi, China, 1994.

[4] Van, Suak-Ho et al., *Hull Form Development for Hanjin Ro-Ro Passenger Ship*, KRISO Report, IS121-1564ED, 1996.

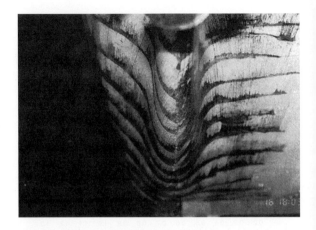

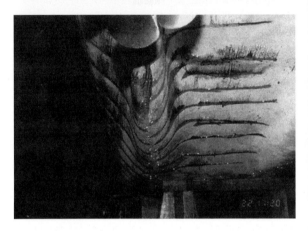

Fig. 8 Comparison of Limiting Streamlines near Stern Area (Portside, from top : S1, S2)

Practical Design of Ships and Mobile Units
M.W.C. Oosterveld and S.G. Tan, editors.

Geosim experimental results of high-speed catamaran: co-operative investigation on resistance model tests methodology and on ship-model correlation

Cassella P.[a], Coppola C.[b], Lalli F.[b], Pensa C.[a], Scamardella A.[a], Zotti I.[c]

[a] Dipartimento di Ingegneria Navale (DIN) University "Federico II" of Naples
Via Claudio, 21 - 80125 Naples (ITALY)

[b] Istituto Nazionale per Studi ed Esperienze di Architettura Navale (INSEAN)
Via di Vallerano 139 - 00128 Rome (ITALY)

[c] Dipartimento di Ingegneria Navale del Mare e per l'Ambiente (DINMA)
University of Trieste - Via Valerio 10 - 34127 Trieste (ITALY)

The paper summarises the results of a high speed catamaran model tests program jointly performed at INSEAN (Rome Towing Tank) and at Naples University facilities.

Systematic calm water tests were carried out on three geosims (model scale λ = 7, 13 and 26) of a hard chine symmetric catamaran operating in the Naples Gulf, whose dimensions at designed load condition are: length L = 35.867 m, breadth B = 11.334 m (demihull B = 3.30 m); draught 1.58 m, displacement Δ = 137 t, separation-length ratio s/L = 0.224.

The greatest model of 5.12 meters was tested in the Rome Towing Tank (453 m x 13.5 m x 6.5 m), the other two models were tested in the Naples Towing Tank (140 m x 9 m x 4.20 m).

The experimental research carried out on the catamaran and on its demihull was undertaken with the goal to obtain indications on the effect of towing line position, on the resistance components scale effects, on the resistance interference phenomena, on the most suitable ship-model correlation and on the methodology for the evaluation of additional resistance due to the appendices.

The results obtained, analysed and discussed, can be considered a contribution to the solution of the above mentioned problems.

1. INTRODUCTION

At present, in many geographic areas, there is an increasing demand for higher speeds in marine transportation, especially in passenger/vehicle ferries. Therefore, various new types of fast hull forms have been developed.

However, comparing the different operative crafts, it appears that the catamaran is now the leading commercial high speed vehicle.

In fact, due to a large deck area, to high transverse stability, to shallow draught and to hydrodynamic resistance characteristics, the catamaran offers a better transport efficiency than the other high speed crafts.

Although many experimental and theoretical investigations have been carried out till now, fast catamarans are still presenting some problematic questions due to inadequate information on the splitting of the total resistance into different components, on the scale effects of these components, on the interference phenomena between the demihulls, on the model tests procedure and on ship-model correlation.

This paper deals with an experimental project on three geosim models of a high-speed hard chine catamaran, which intends to provide a better understanding of the above mentioned problems.

Since a significant presence of spray was not observed during all the resistance model tests, we assumed that also for the high speed crafts, as well as for the traditional ships, the hydrodynamic resistance can be split only into the viscous and the wavemaking resistance.

2. EXPERIMENTS

The experiments were carried out on three geosim models of both the catamaran and its demihull.

The tested Froude number range was $F_n = V/\sqrt{g \cdot L} = 0.50 \div 1.10$ in order to determine the total hydrodynamic resistance, the wave pattern resistance, the running trim and sinkage, and to investigate the interference phenomena between the demihulls.

Model scales were carefully chosen in order to avoid the laminar flow on the smallest model and the blockage effects on the largest model.

Tank blockage and shallow water effects, estimated by practical formulas, were found to be negligible.

The wave pattern analysis was performed with the above mentioned three models by using the Sharma-Newman longitudinal cut procedure and the wave resistance was calculated by means of the expression

$$R_{WP} = \int_0^\infty \left(C'^2 + S'^2\right) \frac{du}{w^2\left(2w^2 - 1\right)} \quad (1)$$

where C', S' are the modified Fourier transforms of the longitudinal cut, w and u are the circular wave numbers in the longitudinal direction x and in transverse direction y.

Capacity probes were used for the wave profiles measurement. The wave probes showed a constant calibration factor before and after each experiment.

The wave interference phenomenon was also investigated by placing a probe between the two demihulls of catamaran.

Resistance model tests of the catamaran with four different towing line positions were carried out and compared: shaft line direction with running trim correction, horizontal direction on base line, on main bridge and on superstructures.

The resistance results obtained with shaft line direction were analysed by both ITTC'57 and ITTC'78 methodologies.

For this latter purpose resistance tests were carried out also at low speed by running the models with transom emerged and with turbulence stimulators suitable to the model scale in order to calculate the form factor.

3. RESULTS AND ANALYSIS

3.1. Model Towing Line Position

For two different draughts, corresponding to designed load condition of 137 t and to overload condition of 165 t of the catamaran in full scale, the bare hull resistance results obtained with the above mentioned four different towing line positions were compared.

The analysis of these results shows that the differences of the trim and consequently of the total resistance among the different towing directions are not negligible in the operative speed of the catamaran.

On the other hand these differences are negligible in the operative Froude range $F_n < 0.30$ of the traditional ships, but in the range of the higher Froude numbers, due mainly to the different running trim, they are of $3 \div 4$ per cent among the shaft line direction and other positions.

3.2. ITTC '57 Correlation Methodology - Global Interference

The total hydrodynamic resistance obtained from the towing tank model tests (shaft line towing position) was reduced to coefficient by means of the usual relation

$$C_T = \frac{R_T}{0.5\rho S V^2} \quad (2)$$

where R_T is the total model resistance, ρ the fresh water density, S the wetted surface at rest, V the model speed. Using the classical Froude methodology of resistance component subdivision, and using specifically the ITTC'57 ship-model correlation we obtain the residual resistance coefficient:

$$C_R = C_T - C_{Fo} \quad (3)$$

being C_{Fo} the frictional resistance coefficient according to ITTC'57 frictional line.

As it is well known, we have two types of interference between the demihulls of the catamaran:
- A viscous interference, due to the variation of the wetted surface and to the modification of the pressure and velocity fields between the demihulls;
- A wave interference, due to the interaction between the transverse and the divergent wave systems of the

two demihulls. Therefore, if C_R' and C_R are the residuary resistance coefficients for the catamaran and for its demihull respectively, the difference ΔC_R could be considered as the global resistance interference coefficient due to both the viscous flow and the waves between the demihulls.

The ratio $\alpha = C_R'/C_R$ could be assumed as the global interference factor.

The factor α decreases rapidly in the range of $F_n < 0.90$ and afterward its value is practically constant and nearly equal to unity.

Fig. 1 highlights that the scale effect on α is not negligible and the trend is opposite in two different ranges of Reynolds number R_n. At lower values of R_n, α decreases as the model size is increasing, whereas the opposite occurs at higher values of R_n.

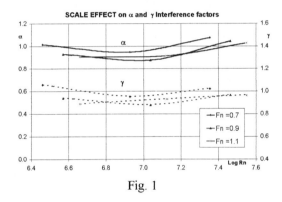

Fig. 1

3.3. ITTC'78 Correlation Methodology - Viscous and Wave Interference

As it is well known, in the 1978 ITTC Prediction Method, the model resistance is extrapolated to the full scale ship by using the form factor $(1+k)$.

For high speed crafts the presence of transom stern and of spray should preclude the satisfactory application of the form factor for ship-model correlation.

However, since a significant spray formation was not observed during the model resistance tests, in order to have an indication of the interference phenomenon for the examined catamaran, the form factor for both the catamaran and the demihull was determined by running the model bow down (transom emerged), being this condition that one observed in Froude number range of model resistance tests. So, the model resistance data were analysed by means of the following relations, which give the total resistance coefficients for the demihull

C_{Tm} and for the catamaran C_{Tm}' respectively as sum of the viscous component $C_v(R_n)$, function of Reynolds number only, and the wave component $C_w(F_n)$, function of Froude number only:

$$C_{Tm} = C_w(F_n) + C_v(R_n) = C_w + (1+k)C_{Fo} \qquad (4)$$

$$C_{Tm}' = C_w' + C_v' = C_w' + (1+k')C_{Fo} =$$

$$= \gamma\, C_w + \tau\,(1+k)C_{Fo} \qquad (5)$$

where $(1+k)$ and $(1+k')$ are the form factors.

By adopting a formulation similar to that suggested by Insel and Molland (1992) the ratios

$$\tau = \frac{1+k'}{1+k} \quad ; \quad \gamma = \frac{C_w'}{C_w} \qquad (6)$$

can be considered as viscous and wave interference resistance.

The form factor for both the demihull and the catamaran was determined by means of the following two different procedures:
- The Prohaska method, based on a linearization of the measured resistance values in the Froude number range $0.12 \div 0.20$, by considering the exponent for F_n equal to four;
- The Hughes method, which assumes $C_w = 0$ at very low Froude numbers and consequently the form factor value has been evaluated by enveloping versus Reynolds number R_n the curve of the total resistance coefficient obtained by the model tests.

The form factor determination for all the three geosims of both the catamaran and the demihull was obtained by transom emerged model tests with trip wire diameter of 1.00 mm, as turbulence stimulator.

In order to investigate the effect of the trim and of the turbulence stimulator on the value of the form factor, some model tests were carried out also for level trim and for bow up conditions with trip wire diameter of 1.80 mm and 0.50 mm.

All the results so obtained of the form factor, based on ITTC friction line, are given by the table I.

In the operative Froude number range the form factors of the catamaran with bow down and trip wire of 1.00 mm (model scales 7 and 13) or with trip wire 0.5 mm (model scale 26) are nearly equal to the mean value of those ones obtained by iso-Froude

Table 1

Form factors with different trims and different diameters of trip wires as turbulence stimulators: prohaska method $(1 + k)_p$ and hughes method $(1 + k)_h$

Φ (mm)	TRIM	τ°	CATAMARAN Model scale						DEMIHULL Model scale					
			7		13		26		7		13		26	
			$(1+k)_P$	$(1+k)_H$	$(1+k)_P$	$(1+k)_H$	$(1+k)_P$	$(1+k)_H$	$(1+k)_P$	$(1+k)_H$	$(1+k)_P$	$(1+k)_H$	$(1+k)_P$	$(1+k)_H$
1.0	level	0	1.576	1.589			1.264	1.300					1.446	1.500
	bow down	-3.6	1.088	1.075	1.115	1.120	1.211	1.200	1.051	1.032	1.135	1.130	1.424	1.400
	bow up	+3.6					1.846						1.926	2.000
1.8	level	0			1.752				1.636	1.600	1.492	1.600		
	bow down	-3.6			1.263	1.300			1.028	1.100	1.209	1.230		
	bow up	+3.6			2.547	2.600			2.295		2.511	2.550		
0.5	level	0					1.212							
	bow down	-3.6					1.072							
	bow up	+3.6					1.879							

determination being the values of the form factor 1+k' so determined equal to 1.042; 1.11 and 1.18 respectively for F_n = 0.70; 0.90 and 1.10.

As far as the viscous interference factor is concerned, the value of τ for the considered catamaran is nearly equal to the unity. Therefore it can be deduced that the viscous interference resistance is very low at each F_n value.

However the ITTC'78 ship-model correlation of the catamaran with the form factor so determined seems to give good results, as shown in fig. 2, where the curve of total resistance coefficient $C_T(7)$ obtained directly by the tests carried out by means of the scale model λ = 7 has been compared with the curves of $C_T(13 \rightarrow 7)$ and $C_T(26 \rightarrow 7)$ determined by adopting the ITTC'78 correlation applied to the data of the two smaller models.

The differences among these curves are very negligible in the operative speed range of the catamaran.

The significant differences for F_n < 0.80 between the experimental C_T of the model scale λ = 7 and that one of the model scale λ = 26 are due probably to presence of the laminar flow on the entrance body of the smallest model, as the resistance model tests were carried out without turbulence stimulator, and we have, for the smallest model, Reynolds numbers lower than $3.5 \cdot 10^6$ in the range of F_n < 0.90.

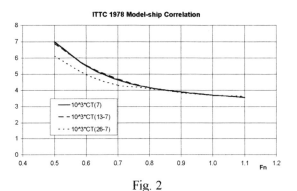

Fig. 2

3.4. The Wave Pattern Analysis

The wave pattern experimental analysis was carried out by using four capacity probes placed in different transverse positions to measure the wave profile. For all the three models one probe was placed near the models at a distance y = 0.262 L (catamaran) and y = 0.162 L (demihull) from the model centerplane, to calculate the wave pattern resistance by using the Sharma-Newman longitudinal cut method.

Fig. 3 shows the wave pattern resistance coefficient C_{wp} for each of three geosim models. From the figure the wave breaking resistance coefficient can be also estimated by the difference $C_w - C_{wp}$ between the wave resistance coefficient $C_w = C_{Tm} - (1 + k) C_{Fo}$ obtained by means of routine model tests with form factor determination (transom emerged) and the wave pattern resistance coefficient C_{wp} obtained by longitudinal cut method.

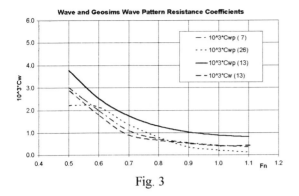

Fig. 3

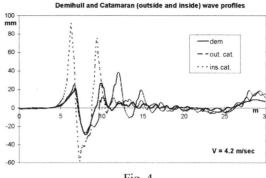

Fig. 4

In order to assess the accuracy of the experimental wave pattern resistance evaluation, an investigation on the longitudinal cut method was carried out with scale model of the catamaran $\lambda = 13$ by considering the wave profiles of four probes placed in different transverse positions in the range $0.262 < y/L < 0.875$.

The results showed a good agreement among the data obtained by the different probes.

From the values of C_w and of C_{wp} given versus F_n in Fig. 3 for the examined catamaran we can deduce the following considerations:

- The global wave resistance coefficient and the wave pattern resistance coefficient show the same trend versus Froude number, and the wave breaking coefficient $C_{WB} = C_W - C_{WP}$ is nearly constant versus F_n ;

- The energy loss due to the wave breaking is a high percentage of the global wave resistance and it is increasing with the Froude number;

- The scale effect on C_{wp} among the smallest model and the other two models shows opposite trend in two different ranges of Froude number: at values of $F_n < 0.75$ the C_{wp} of the smallest model is higher than the C_{wp} of the other two models, the opposite occurs at values of $F_n > 0.90$.

This different trend of the scale effect can be due to measure mistake of very small wave profiles for the smallest model in the range of $F_n < 0.75$.

Interference phenomena were also investigated by placing a probe between the two demihulls of the scale model catamaran $\lambda = 13$ to record the wave profile at the speeds corresponding to $F_n = 0.50$; 0.80 and 1.10.

Fig. 4 shows the comparison among the wave profile of inner region of the catamaran and those measured outside both the catamaran and the demihulls at the same transverse positions and at $F_n = 1.10$.

In particular the wave profile inside the demihulls highlights that the interference of the waves generated by the two demihulls in the inner region leads to a wave high stepness and consequently to a wave breaking.

3.5. Appendages Resistance

Two streamlined flow oriented appendages (one stern flap and one bow gyrofin) have been fitted to each demihull of the operative catamaran.

It is well known that scale effects are present on the drag of appendages fitted to ship-model.

However, for the traditional ships, the scale effect on the drag of the flow oriented appendages can be well accounted by the so-called beta method, using ITTC'78 ship-model correlation as in the following

$$C_{Ts}^* = C_{Ts} + \beta (C_{Tm}^* - C_{Tm}) \qquad (7)$$

where:

- C_{Ts}^* and C_{Ts} are respectively the total resistance coefficient of the ship with appendages and without appendages;

- C_{Tm}^* and C_{Tm} are respectively the total resistance coefficient of the model with appendages and without appendages;

- $\beta = C_{Fs}/C_{Fm}$ the reduction factor, being C_{Fs} and C_{Fm} respectively the frictional resistance coefficient of the ship and of the model.

In order to check the validity of this method in the case of high speed catamarans, resistance tests have been carried out on the three geosim models with appendages.

The beta method seems to give good results, as shown by the negligible differences in the operative Froude range among the total resistance with appendages R_{Tm}^* obtained by the experimental tests of model scale $\lambda = 7$ and those one determined for the same model by means of beta method applied to experimental data of the two other models.

4. CONCLUSIONS

A set of experiments on three geosim models of a high speed catamaran has been carried out and some results are given in this paper:
- The comparison of different towing positions shows that the difference among the resistance results is not negligible in the operative speed range of high speed catamaran. Therefore, the necessity arises to adopt the shaft line direction in resistance model tests;
- The significant presence of spray due to turbulence stimulators does not allow to use the stimulators in resistance model tests of high speed catamarans;
- The conventional and the wave pattern resistance tests carried out on the geosims give us some indications on the global, viscous and wave interference phenomena. These tests can be also used to study scale effects on the components of the hydrodynamic resistance. Particularly, the wave pattern analysis highlights that the scale effect on the wave pattern resistance of the different models obtained by means of longitudinal cut method could be significant;
- The ITTC'78 ship-model correlation applied to routine resistance model tests seems to give good results, if the form factor determination is carried out with transom emerged condition and with turbulence stimulator suitable to model dimensions. Good results we also have for the prediction of the additional resistance due to appendages by means of the beta method;
- The comparison among the wave profiles measured by the probes in the inner region of the demihulls and those measured outside them, in demihull tests, and outside of the catamaran at the same transverse positions and at the same speeds, highlights the wave breaking phenomena due to the interference of the wave generated by both demihulls.

The authors, however, believe that the interference phenomena, the ship-model correlation and the scale effects on the resistance components of high speed catamaran are not yet completely clarified.

Therefore, further experimental and theoretical studies are necessary in the next future in order to examine the validity of the interpretation given in the present paper.

ACKNOWLEDGEMENTS

This work was supported by the Italian Ministry of Transport and Navigation in the frame of INSEAN research plan 1991-1993.

The authors are very grateful to Ca.Re.Mar. SpA for providing them with hull form of catamaran ACHERNAR.

REFERENCES

1. Eggers K.W.H., Sharma S.D., Ward W., "An Assessment of Some Experimental Methods for Determining the Wave Making Characteristics of a Ship Form" Trans. SNAME, Vol. 74 pp. 112-157, 1967.
2. Myazawa M., "A Study on the Flow Around a Catamaran" Journal of the Society of the Naval Architects of Japan, vol. 145, pp.46-53, 1979.
3. Insel M., Molland A.F. "An Investigation into the Resistance Components of High Speed Displacements Catamarans", Trans. RINA, vol. 134, pp. 1-20, 1992.
4. Cordier S., Dumez F.X., "Scale Effects on the Resistance Components of a High Speed Semi-Displacement Craft", Proceedings of the FAST'93 Conference, Yokohama, vol. 1, pp. 409-419, 1993.
5. Cassella P., Miranda S., Pensa C., Russo Krauss G., "Comparison between the Catamaran and the Monohull Resistance Characteristics", Proc. ISSH, S. Petersburg, pp, 167-173, 1995.
6. Molland A.F., Wellicome J.F., Couser P.R., "Resistance Experiments on a Systematic Series of High Speed Displacement Catamaran Forms: Variation of Length Displacement Ratio and Breadth-Draught Ratio" Meeting of the Royal Institution of Naval Architects, 1995.
7. Cassella P., Pensa C., Russo Krauss G., "Sur les Characteristiques de Resistance des Catamaran a Grande Vitesse" ATMA Paris, 1996.
8. Bruzzone D., Cassella P., Miranda S., Pensa C., Zotti I., "Wave Pattern and Wave Resistance: Validation of a Numerical Method Using Geosim Experimental Results", Proc. of the 2° Int. Conf. on Hydrodynamics, Hong Kong, pp. 127-132, 1996.

Practical Design of Ships and Mobile Units
M.W.C. Oosterveld and S.G. Tan, editors.

Influence of the submergence and the spacing of the demihulls on the behaviour of multi-hulls marine vehicles: a numerical application

Daniele Peri, Marco Roccaldo, Stefano I. Franchi

I.N.S.E.A.N.- Italian Ship Model Basin
Via di Vallerano 139 – 00128 Roma, Italy

The numerical prediction of the total resistance of a SWATH vehicle and a fast ferry catamaran in steady motion has been attempted by using a numerical approach. An iterative solver, able to automatically adjust the grid of a model free to sink and trim [1], has been used to perform a deep analysis of the combined effects of varying hull submergence and space between demihulls. The validation of the code has been carried out by comparing present results with the experimental data for a modular SWATH vehicle tested at the HSVA [2] and a fast ferry catamaran tested at the INSEAN.

1. INTRODUCTION

In the design of multi-hulls vessels two parameters are to be considered with a great attention: the displacement and the spacing of the hulls. The displacement is equivalent to the payload, while the spacing is important in determining the resistance at a certain speed.

The mutual influence of the hulls has different effects on the wave and total resistance, and this behaviour changes with the spacing and the displacement. Moreover, the effect of an increasing displacement on the total resistance may be balanced by a particular choice of the spacing, so that parameters must not considered in an independent way.

In this work the evaluation of the resistance of a multi-hull vessel has been carried out by using a numerical code based on the potential flow model, with linear free surface conditions. Viscous effects have been considered in a simplified way. Finally, the numerical solver is able to automatically simulate the displacements of the hull in the free-model condition [1].

The comparison with the experiments shows that this tool seem able to catch most of the information needed in the early stage of the hydrodynamic part of the design, and allow hull form data to be easily used to computed forces and displacements.

2. THE SWATH MODEL

The SWATH model analyzed in this paper was tested at the HSVA tank [2], and consists in a modular structure, capable to assume about 150 different configurations. In the numerical simulation the configuration that has been considered has the following principal parameters:

Prismatic coefficient of the hull C_{PH}	0.77
Relative diameter of the hull D_H/L	0.06
Relative width of the struts t_S/D_H	0.50
Relative spacing of the demihull B_S/L	0.32
Relative submergence of the hull d/D_H	0.10

The experimental setup is also reported in [2]. The model is 2.5 m long. It is important to note that the model was free only to sink, and not to assume a trim angle. Indeed, these vehicles, because of the small waterplane area, cannot naturally oppose the trim moment, and an active control is essential.

3. THE CATAMARAN MODEL

This model was tested at the INSEAN. Its main dimensions (ship scale) are reported in the following table:

L	125 m
B	9.64 m
T	3.6 m
L_{wl}/B_{wl}	12.95
Velocity	40-60 knots
Displacement	$4000 m^3$

Since the Froude number Fn ranges from 0.6 to 0.9, the hull is a *Deep-V*-type, with deadrise about 20°, low block coefficient and high L/B ratio and C_p between 0.6 and 0.7. The hull is a symmetric one. The main characteristics linked to this choice are that it gives lower resistance, but a greater interference in the space between the hulls with respect to an asymmetric or a semi-symmetric hulls. A perspectical view of the model is reported in fig. 1.

Figure 1: *The catamaran model: a perspective view.*

3.1. The experimental tests

The model used for the experimental test has a scale factor $\lambda = 25$. After the check of the equivalence of the hulls from the geometrical and hydrodinamical point of view, 10 steady towing series of tests have been performed. Each test corresponds to a pair of submergence and spacing measured between the centerplane of the hulls.

Test $1 \div 5$ have been performed with spacing of 1.5 meters and with the following values for the static submergence (model scale): 0.203; 0.14152; 0.121; 0.10452; 0.09952 meters respectively. Test $6 \div 10$ have been performed with spacing of 2.0 meters and the same static submergence series.

4. THE NUMERICAL MODEL

A potential linear code, based on the Dawson linearization, was developed at the INSEAN [3],[4] to predict the inviscid flow with free surface around the hull. Starting from this CFD tool, an iterative method, that take into account the effective sinkage and trim of the model, was developed [1]. However, during the computation, to reproduce the experimental conditions of the SWATH test case, one of the degree of freedom can be fixed. The convergence on the displacements is mostly reached within 4 or 5 iterations, according to the desired precision.

In view of using this numerical solver to automatically optimize the shape of the hull, a simple and fast boundary layer model was introduced in order to compute the total resistance of the ship. The frictional resistance is evaluated taking into account the local value of the Reynolds number on the single panel. The use of a linearized potential flow solver results in an underprediction of the bow wave but this has been considered unessential in the present study. Another well known effect of the linearization [5] is that the wave resistance is accurate only for slender ships, but fortunately both the SWATH model and the catamaran belong to this class.

5. COMPUTATIONAL RESULTS

5.1. The SWATH model

As a first step, the single-hull configuration was considered and tested at the main static submergence. The results are reported in fig. 2. 1000 quadrilateral panels have been placed on the hull, while up to 3200 on the free surface. The free surface is 1.5 L_{pp} wide, 2.75 L_{pp} long (0.25 L_{pp} in the fore region and 1.5 L_{pp} astern).

It is possible to observe a good agreement for the lowest velocities, up to a speed of $4\,m/s$; as the speed increases, the linearized model is no more suitable, and the wave resistance is underpredicted. Sprays, non-linear interactions and many other phenomena comes to play, and growing discrepancies are encountered between numerical results and measured data. During the towing tank tests the wave resistance was not measured. Hence, here the wave resistance is compared with the residual resistance. The computed frictional resistance is compared with the resistance

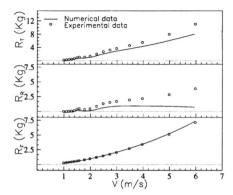

Figure 2: *Total, wave and frictional resistance of a single hull of the SWATH model ($\frac{d}{D_H} = 1.0$).*

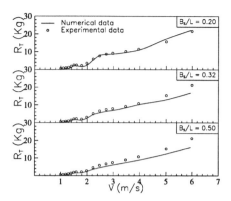

Figure 4: *Total resistance of the SWATH model for three different spacing of the hulls.*

obtained from the ITTC formula used in the experiments to obtain the residual resistance.

In the following, starting from the main statical submergence, variations of the spacing and the submergence are considered. The comparison between the experimental and the numerical results may be regarded as satisfactory. Differences are quite small in the low *Fn* range, while gradually increase in the high *Fn* range.

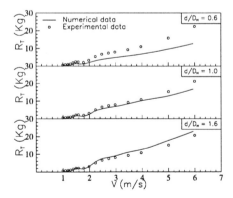

Figure 3: *Total resistance of the SWATH model for three different static submergence.*

The interference factor, defined as:

$$IF = \frac{R_t - 2\,R_t^m}{2R_t^m} \qquad (1)$$

where R_t is the total resistance of the SWATH model and R_t^m is the total resistance of a single hull alone, has been computed for these configurations, and reported in fig 5.

The interference factor decrease obviously to zero for high *Fn*, but in the field of velocity $1 \div 2.5\ m/s$ sometimes it becomes less then zero. In these cases it is evident that the optimum spacing, relative to that velocity and displacement, was reached. Nevertheless, as will become evident in the next case, the IF factor is largely influenced by the numerical and experimental uncertainties.

In figure 6 the numerical comparison of the interference coefficients for several conditions is reported. It may be observed that, for this choice of the geometrical parameters, the spacing is more important than the submergence. This behaviour reflect the particular form of this hull: the shape of the struts is constant, so there is no variation of the waterline area when the hull sinks, and the wave resistance undergoes only small variations. The submerged hull have some influence only for the smaller value of the depth.

5.2. The catamaran model

The complete set of the towing tank series can be found in [6], together with the complete results.

456

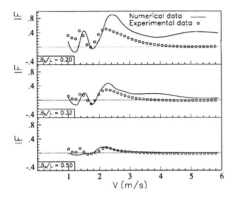

Figure 5: *Interference coefficient computed for the SWATH model in the main static sinkage condition for three different spacing conditions.*

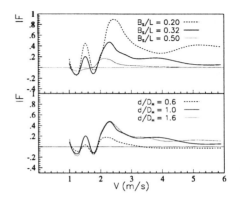

Figure 6: *Interference coefficient computed for the SWATH model for three different spacing and submergence conditions: numerical results. $\frac{d}{D_H} = 1.0$ when the spacing is constant. $\frac{B_S}{L} = 0.32$ when the submergence is constant.*

1000 quadrilateral panels have been placed on the hull, while up to 5000 on the free surface. The free surface is 1.5 L_{pp} wide, 2.75 L_{pp} long (0.25 L_{pp} in the fore region and 1.5 L_{pp} astern). In order to validate grid dependency effects on numerical results, a convergence test for the numerical solution have been performed on a single experimental configuration. Some results are reported in figure 7.

Three different numerical grids have been used in the convergence tests with increasing refinement level. The total resistance is quite independent from the grid, showing a value almost identical since from the coarsest grid. For the dynamic trim and sinkage (fore and aft part) convergence is obtained on the medium grid.

The estimates of grid uncertain have been reported in the following table that shows the grid convergence metric ε of the numerical code [7].

v (m/s)	$\varepsilon_{12} = \frac{R_2 - R_1}{R_1}$	$\varepsilon_{23} = \frac{R_3 - R_2}{R_2}$
1.75	0.0012	0.0017
2.10	-0.0047	0.0638
2.45	0.0378	-0.0103
2.80	0.2602	-0.0278
3.50	-0.0486	-0.0049
4.20	-0.0266	-0.0060
4.90	-0.0100	-0.0077
5.60	-0.0043	-0.0086
6.30	-0.0044	-0.0054

R_1, R_2 and R_3 are the computed total resistance with the coarse, medium and fine grid respectively. The mean value $\bar{\varepsilon}_{12}$ is 8.97 %, while $\bar{\varepsilon}_{23}$ 2.40 % and this decreasing trend indicates grid convergence.

The convergence order of the method, defined as:

$$Ord = -\frac{\log \frac{|R_3 - R_2|}{|R_1 - R_2|}}{\log \sqrt{2}} \qquad (2)$$

is reported in the following table:

v (m/s)	Ord
1.75	-1.00
2.10	-7.54
2.45	3.65
2.80	5.79
3.50	6.75
4.20	4.38
4.90	0.79
5.60	-2.01
6.30	-0.56

The bigger the convergence order, the faster the method converges to an unique value for the considered quantity. Negative values of the convergence order indicate that the numerical solution has not reached the asymptotic region of convergence. For

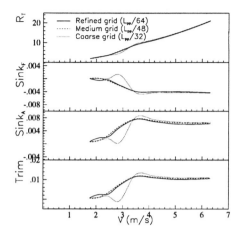

Figure 7: *Convergence test for the catamaran model. From head to bottom: total resistance, sinkage fore, sinkage aft, trim angle. Distance between the hulls = 1.5 m. Static submergence = 0.09952 m*

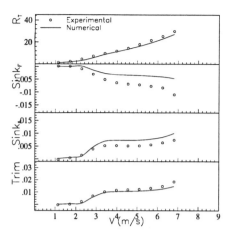

Figure 8: *From head to bottom: total resistance, sinkage (fore and aft) and trim for the catamaran model. Distance between the hulls = 1.5 m; submergence = 0.10452 m.*

low Fn values this may be caused by an insufficient grid refinement, unable to catch the smallest wavelength. For high Fn values this may be due to an insufficient extension of the grid, unable to catch the longest wavelength.

Numerical simulations has been performed for all the tested configurations. In figure 8 an example of the computed quantities, compared with the experimental data, is reported. Even though differences occur in the aft and fore sinkages, the trim is quite well predicted. This behaviour is observed in all the tested cases.

In figure 9-10 some numerical-experimental comparison for the total resistance are reported. The agreement with the experimental data is quite good. Differences between the numerical and experimental data are comparable for both the demihull spacing.

In figure 11, the relative difference δ between numerical and experimental values, defined as:

$$\delta = 100 \cdot \frac{R_t^n - R_t^e}{R_t^e} \qquad (3)$$

has been reported. R_t^n is the numerical total resistance, and R_t^e is the experimental total resistance. The relative difference is about 10%. If δ_i is the per-

centual difference computed at the speed v_i, we define the mean difference as

$$\bar{\delta} = \sqrt{\frac{1}{N}\sum_{i=1}^{N}\delta_i^2} \qquad (4)$$

For some tested cases, the mean difference is reported in the following table. The error decreases for growing depth of the model.

Test n.	$\bar{\delta}$
Test 2	4.620
Test 3	8.862
Test 4	16.112

It must be observed that no information about the accuracy of the experimental tests is available, both for the SWATH and for the catamaran model. Owing to this, it is hard to get, from the evaluation of the relative difference δ, any information about how close the numerical simulation is to the experimental values. Anyway, the numerical model is capable to put in the correct order the various configurations, as shown in figure 12, indicating that the numerical method is useful for design purposes.

458

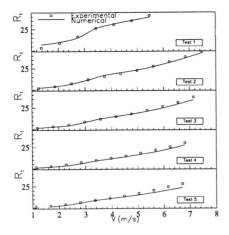

Figure 9: *Total resistance for the catamaran model. Distance between the hulls = 1.5 m*

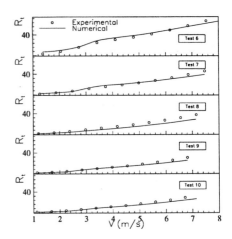

Figure 10: *Total resistance for the catamaran model. Distance between the hulls = 2.0 m*

A last consideration on the interference factor can be done. The mean grid convergence metric ε of the numerical code was founded to be about 2%. This value is of the same order of the interference factor IF. On this base, it is clear that the convergence of the numerical method must be improved in order to get information useful for the IF, i.e., the risk of making errors bigger than the quantity we need to evaluate is real. This problem can be avoided improving both the number of elements of the numerical grid employed in the computation or the order of the numerical scheme.

On the other side, some difficulties also occur in the experimental measurements, and the knowledge of the precision index of the measure and the entity of the error introduced by the data elaboration are crucial: as an example, if we need to use the interference factor based on the wave resistance [8], the method used for its deduction is another source of errors.

6. CONCLUSIONS

A numerical analysis of the influence of some geometrical parameters on the resistance of fast catamarans has been performed, and it was shown how this numerical method meets the tasks of the design process. Validation technique have been applied to

the numerical results in order to understand the convergence and to assess the accuracy of the method, Although, this kind of analysis is crucial for the assesment of the usefulness of CFD techniques in the design process, its application is still far from being standard. Improvements of the numerical code can be reached, in the frame of the considered model: a simplified wake, in order to include the vortex shedding influence, can be applied in the case of little spacing values.

7. ACKNOWLEDGEMENTS

The research was suppported by Italian Ministry of Merchant Marine in the framework of INSEAN research plan 1991-1993.

The authors are particularly grateful to Mr. P. Schenzle of the HSVA tank for his kindness in giving essential informations about the model and the experimental results.

References

[1] CAMPANA E.F., PERI D., 1997, Simulazione numerica delle prove di rimorchio in acqua calma. Parte II: prove con assetto libero, *IN-*

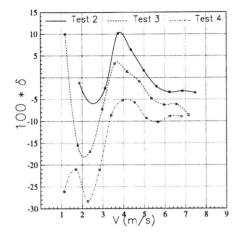

Figure 11: *Percentual difference on the total resistance for the catamaran model. Distance between the hulls = 1.5 m*

SEAN report 1993-35.

[2] PETER SCHENZLE, 1995, "The HSVA Systematic SWATH Model Series" *Fast '95 Lubek Travemunde, Germany C.F.L. Kruppa*

[3] CAMPANA E.F., PERI D., 1997, Simulazione numerica delle prove di rimorchio in acqua calma. Parte I: prove con assetto bloccato, *IN-SEAN report 1993-30.*

[4] BASSANINI P., BULGARELLI U., CAMPANA E.F., LALLI F., 1994, "The Wave Resistance Problem in a Boundary Integral Formulation", *Surveys on Mathematics for Industry*, n.4, 1994

[5] RAVEN H., VALKHOF H.H., 1996, "Application of nonlinear ship wave calculations in ship design" *Prads '96*, Seul.

[6] RANOCCHIA D., 1997, Prove sistematiche di rimorchio su un catamarano per alte velocitá, *IN-SEAN report 1993-37.*

[7] COLEMAN H. W., STERN F., 1997, Uncertainties and CFD code validation, *J. Fluids Eng.*, *Vol. 119*

[8] TURNER H., TAPLIN A., 1968, The resistance of large powered catamarans, *Tr. SNAME, Vol. 76*

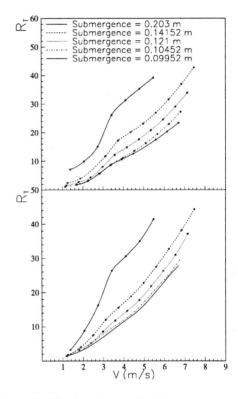

Figure 12: *Total resistance for the catamaran model. Distance between the hulls = 1.5 m. Upper part: numerical results. Lower part: experimental data.*

Practical Design of Ships and Mobile Units
M.W.C. Oosterveld and S.G. Tan, editors.

Experimental investigation on the drag characteristics of a high speed catamaran

R. Natarajan[a] and Malle Madhu[b]

[a] Associate Professor, Ocean Engineering Centre, Indian Institute of Technology Madras, Chennai-600 036, India

[b] Research Scholar, Ocean Engineering Centre, Indian Institute of Technology Madras, Chennai-600 036, India

In this paper, experimental investigation on the drag characteristics of a high speed catamaran has been presented. The model tests were carried out in a 82m x 3.2m x 2.5m towing tank at Ocean Engineering Centre, Indian Institute of Technology Madras, Chennai-600 036. The drag and trim angle of the model were measured using a load cell and potentiometers respectively. The experiments were conducted for three displacement conditions with different gaps. Suitable spray deflectors and wedge were fitted on to the model and their influence on the drag has also been studied.

1. INTRODUCTION

In the recent years, there is a tremendous demand worldwide for fast marine vehicles. Due to this, different types of high speed crafts such as planing crafts, SWATH vessels, catamarans, hydrofoil crafts etc., are being developed to meet the specific functional requirements. Usually, the hull form of these vessels are evolved from the data of the existing vessels during the preliminary design process. But these hull forms are further optimised critically to obtain desired high speed with minimum power, since the drag characteristics of the vessel are governed by various hull parameters. The drag of any vessel is predicted theoretically or experimentally. The experimental method is preferred over the analytical method, since the drag of the vessel can be estimated accurately through extensive model tests. Hence, in the present study, an experimental method has been illustrated to investigate the drag characteristics of a typical catamaran. The influence of fitting wedge and spray deflectors on the drag of the vessel is also studied.

2. DESCRIPTION OF THE MODEL

The catamaran model was fabricated in PVC material with the following demihull dimensions:

Length of the demihull	: 0.96 m
Breadth	: 0.24 m
Depth	: 0.21 m
Dead rise angle	: 15^0

3 mm white transparent PVC sheets are used for making the demihull model. The hull form of the demihull is of prismatic and planing type. The catamaran was finished by suitably connecting the demihulls with aluminium angles. Two hooks were provided at the bow and stern of the model. The details of the catamaran model is shown in Figure. 1.

3. EXPERIMENTAL SET UP

3.1. Towing tank

The particulars of the towing tank at Ocean Engineering Centre, Indian Institute of Technology, Chennai are as follows:

462

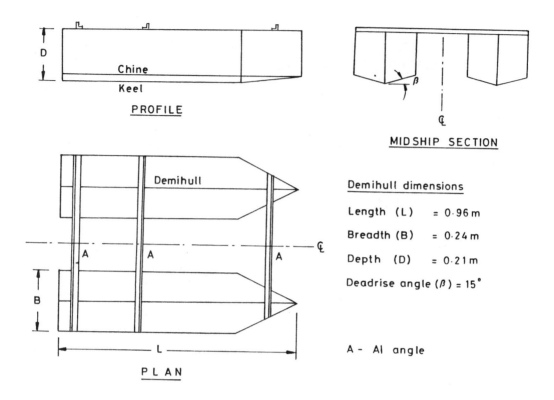

Chine
Keel
PROFILE

β
₵
MIDSHIP SECTION

Demihull

Demihull dimensions

Length (L)   = 0.96 m

Breadth (B)   = 0.24 m

Depth (D)   = 0.21 m

Deadrise angle (β) = 15°

A A A   ₵

B

L

PLAN

A - AI angle

Figure 1. Description of the catamaran model

Length	: 82.0 m
Breadth	: 3.2 m
Depth of water	: 2.5 m

The towing tank is fitted with a towing carriage of speed range of 0.05 - 5.5 m/sec. Suitable clamping arrangements are provided in the carriage to fix the model .The model is restrained to have sway, roll and yaw motions. The drag of the model is measured, using a load cell. The model is connected to the load cell by means of a steel tape through a pulley. A digital type electronic counter is used to measure the speed of the carriage.

3.2. Instrumentation

A proving ring strain gauge type load cell was made to measure the drag of the model. It was then suitably calibrated to measure the drag upto 180N. One end of the load cell was fixed on a steel channel section. The channel section was clamped properly on the

towing carriage. The other end of the load cell was connected to the model using a pulley and steel tape arrangement. Two potentiometers were fitted in the carriage to measure the bow and stern displacements of the model. The load cell and potentiometers were connected to the data acquisition system of the computer through carrier frequency amplifier and D.C. power unit respectively as shown in Figure 2.

4. MODEL TESTS

First, the experiments were carried out on the demihull for three displacement conditions to estimate the drag of the demi hull. Then the two demihulls were joined together to form the catamaran. The catamaran model was fixed on the towing carriage and the drag of the model was measured for three displacement conditions

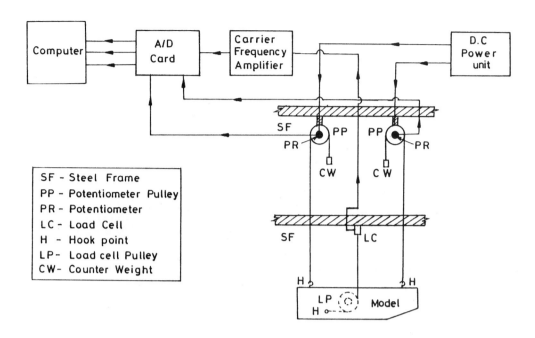

Figure 2. Instrumentation of the model

i.e., 106.94 N, 128.52 N and 148.14 N. For each displacement, the drag was obtained by varying the towing carriage speed from 0.9 to 3.82 m/s. The gap between the demihulls was changed and the experiments were repeated for three gaps i.e., 0.212m, 0.312m and 0.512m. The model was then fitted with wedge and spray deflectors and the drag was measured similar to the naked hull conditions. Wooden blocks and steel batten weights were used as ballasts to simulate the three displacement conditions.

5. RESULTS AND DISCUSSIONS

Using the measured values of the displacements at the bow and stern from the experiments, the trim angle has been computed as:

$\tan \tau$ = (Stern displacement + Bow displacement)/(Distance between hook points)   (1)

where τ= Trim angle

From the trim angle, the pressure drag (D_P) is obtained as:

$D_p = W \tan \tau$ (2)

Then the interference factor (IF_{wp}) will be calculated from:

$IF_{wp} = (D_{wp} - 2 D_{wpd})/ 2 D_{wpd}$ (3)

where IF_{wp} = Interference factor due to wave pattern, D_{wp} = Drag of the catamaran due to wave pattern and D_{wpd} = Drag of the demi hull due to wave pattern.

464

The various parameters governing the drag characteristics of the demihull and the catamaran for a typical displacement condition at a particular gap with and without spray deflectors and wedge are shown in Tables 1 to 4.

The flow pattern around the catamaran for a typical test condition is also shown in Figure.3.

5.1. Naked hull

The drag of the catamaran for different displacement conditions vary abruptly with the speed and as well as the spacing between the demihulls. The drag increase with the the displacement relative to the hull spacings. For the same speed, the drag is highly influenced by the large gap between the demihulls. It is observed that the drag is

Figure 3.   Flow pattern around the catamaran

The total drag (D) of the catamaran is normalised with the displacement (W) of the vessel and its variation with the Froude number (Fn) at three displacement conditions for three gaps with and without spray deflectors and wedge are illustrated in Figure 4.

Also, the variation of interference factor with Froude number for different test conditions is shown in Figure 5.

From the analysis of experimental results, the following inferences are obtained:

reduced by 16% to 50% for the large gap with the increase of displacement for the first hump speed.

The variation of interference factor with the speed of the catamaran also indicates that the interference effect is very much pronounced by the large gap high displacement vessels. For the large gap between the demihulls, the interference factor varies from -0.325 to +0.625.for the high and low displacement vessels with the same first hump speed speed respectively

Table 1
Drag characteristics of the demihull for the naked hull condition
(Displacement=64.26 N)

Fn	Total drag (N)	Trim angle (Deg)	Frictional drag (N)	Pressure drag (N)
0.62	7.25	4.77	1.89	5.36
0.78	8.74	5.44	2.61	6.13
0.94	9.97	5.94	3.26	6.71
1.10	10.19	5.46	4.03	6.16
1.24	11.02	5.50	4.81	6.21

Table 2
Drag characteristics of the demihull fitted with wedge and spray deflectors
(Displacement=64.26 N)

Fn	Total drag (N)	Trim angle (Deg)	Frictional drag (N)	Pressure drag (N)
0.62	6.26	4.12	1.63	4.63
0.78	4.29	2.68	1.28	3.01
0.94	2.43	1.46	0.79	1.64
1.10	4.43	2.39	1.75	2.68
1.24	1.22	0.62	0.53	0.69

Table 3
Drag characteristics of the catamaran for the naked hull condition (Displacement=128.52 N, Gap=0.312 m)

Fn	Total drag (N)	Trim angle (Deg)	Frictional drag (N)	Pressure drag (N)	Interference factor
0.62	20.38	7.55	3.78	17.05	+0.548
0.78	22.75	7.77	5.22	17.53	+0.403
0.94	16.08	4.25	6.52	9.56	-0.290
1.10	22.88	6.58	8.06	14.82	+0.203
1.24	21.79	5.41	9.62	12.17	-0.019

Table 4
Drag characteristics of the catamaran fitted with wedge and spray deflectors
(Displacement=128.52 N, Gap=0.312 m)

Fn	Total drag (N)	Trim angle (Deg)	Frictional drag (N)	Pressure drag (N)	Interference factor
0.62	17.05	6.12	3.26	13.79	+0.490
0.78	10.91	3.72	2.56	8.35	+0.387
0.94	3.88	1.03	1.58	2.30	-0.300
1.10	7.50	1.78	3.50	4.00	-0.253
1.24	2.29	0.55	1.06	1.23	-0.110

466

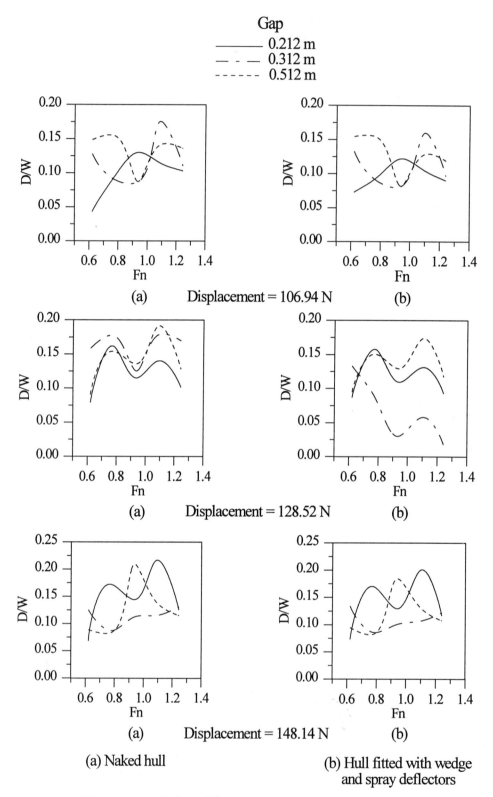

Figure 4. Variation of drag with speed of the cataraman

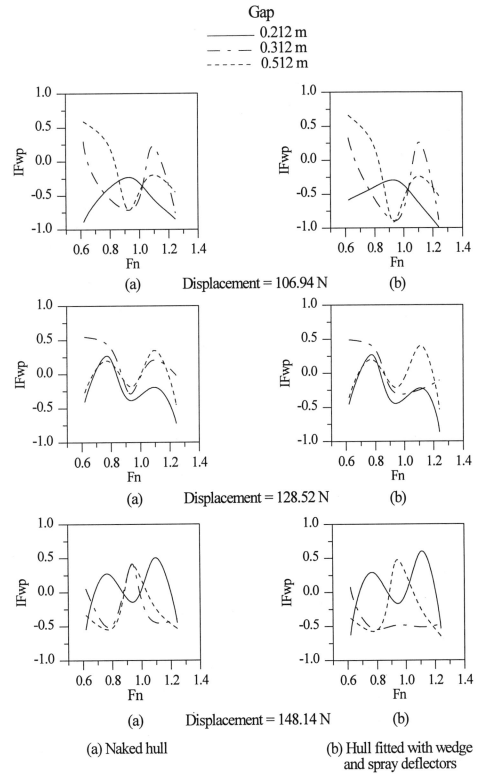

Figure 5. Variation of interference factor with speed of the cataraman

5.2. Hull fitted with wedge and spray deflectors

When the hull is fitted with wedge and spray deflectors, the drag is reduced by 3% and 6% for the low and high displacement vessels with large gap respectively for the same first hump speed, as compared to the naked hull condition.

In the case of interference factor, a relative increase has been observed as -0.375 to +0.675 for high and low displacement vessels respectively for the similar naked hull condition.

In view of above facts and on considering the drag characteristics, it is recommended to provide large gap between the demi hulls for high displacement catamarans with and without the wedge and spray deflectors.

6. CONCLUSIONS

The present experimental study has been carried out with an indent to obtain insight knowledge about the interference effect between the demihulls on the drag characteristics of the catamaran with and without wedge and spray deflectors. Moreover, physical modelling obviates the complex simulation problems that are involved in the numerical solution techniques. Hence, as compared to the analytical method, the experimental method will be a better tool to predict the power accurately for the new generation high speed catamarans for fast marine transportation.

REFERENCES

1. E.P. Clement and D. Blount, Resistance Tests of a Systematic Series of Planing Hull forms, SNAME, Vol. 71 (1963).
2. E. D. Fry and T. Graul, Design and Application of Modern High Speed Catamarans. Journal Marine Technology, Vol. 9 (1972).
3. A. Incecik, B.F. Morrison and A.J. Rodgers, Experimental Investigation of Resistance and Seakeeping Characteristics of a Catamaran Design, FAST '91, Vol. I (1991).
4. M. Insel and A. F. Molland, An Investigation into the Resistance Components of High Speed Displacement Catamarans, TRINA (1992).
5. J. A.. Keuning, and J. Gerritsma, Resistance Tests of a Series of Planing Hull Forms with 25 degrees Deadrise Angle. International Journal of Ship Building Progress, Vol. 29 (1982).
6. H. Turner and A. Taplin, The Resistance of Large Powered Catamarans. SNAME, Vol. 76 (1968).
7. J.D. Van Manen and P. Van Oossanen, Principles of Naval Architecture, Vol. II, SNAME (1988).
8. Robert Latorre, Study of Prismatic Planing Model Spray And Resistance Components, Journal of Ship Research, Vol. 27 (1983).

Practical Design of Ships and Mobile Units
M.W.C. Oosterveld and S.G. Tan, editors.

A Study for Improvement in Resistance Characteristics of a Semi-Planing Ship

Yong-Jea Park [a], Seung-Hee Lee, Young-Gill Lee, Sung-Wan Hong [b]

[a]Korea Research Institute of Ships and Ocean Engineering,
 Taejon 305-343, Korea

[b]Department of Naval Architecture & Ocean Engineering, Inha University,
 Inchon 402-751, Korea

In the present paper, flow fields around a semi-planing hull with a transom stern have been studied experimentally and numerically. Flow fields around the hull are measured at the towing tank of Korea Institute of Ships and Ocean Engineering(KRISO). And the flow field are simulated by an numerical method utilizing a MAC type Finite Difference Method and compared to the measurements at the design speed of 21knots. The effect of the hull attitude is considered in the numerical computation to show good agreements with experimental ones. Based on the results so obtained, original hull form has been modified to have an improved resistance characteristics. Emphases are given to the reductions in wave and spray resistance of the ship. Through the model tests and numerical simulations, it will be shown that considerable improvement is achieved in the present study.

1. INTRODUCTION

Planing type high speed ships are one of the well developed high speed vessels which is designed to exploit hydrodynamic lifts generated on the bottom planing surface when running. A semi-planing type vessel exploits an identical concept but has a conventional bow shape to yield less changes in hull attitudes than the full planing ships.

In the present paper, numerical and experimental results on the flow fields around a semi-planing ship with a transom stern are reported and analyzed. The results, then, are utilized for modification of hull form to reduce the resistance of the ship, especially, wave and spray resistances.

Flow fields around the hull are measured at the towing tank of KRISO. Wave elevations and velocity fields as well as the total resistance are measured at cross-sections located downstream of the stern. Wave pattern analyses and paint tests are performed and surface pressure distributions are also measured to isolate resistance components and to investigate the flow characteristics in more details[1].

The flow fields are simulated by an numerical method utilizing a MAC type finite difference method. The results are used for preparation of experiment and for analyzing measured data at the

design speed of 21knots. The influence of the changes in hull attitude to the flow characteristics is considered in the numerical computations.

Based on the results so obtained, original hull form has been modified to have an improved resistance characteristics. Emphases are given to the reductions in wave and spray resistance of the ship. Through the model tests and numerical simulations, it is shown that considerable improvement can be achieved.

2. MODEL TESTS

2.1. Model

A 23 meter long planing type hull is selected in the present study. The model has been studied already in an international cooperative experimental program organized by the High-Speed Marine Vehicle Committee of ITTC. And the results have been reported at the 20th ITTC held in 1993.

The model has been constructed in accordance with the lines from Setouchi Craft Co. in Japan[2]. The scale ratio is 1/11.6 and the model is made of fiber reinforced plastic to keep it's weight light enough to sustain a resistance dynamometer and trim gauges. The lines of model is shown in Fig. 1. The accuracy of the model has been carefully checked with an procedure established by KRISO and the maximum deviations are found to be within ±0.2% of the given offsets.

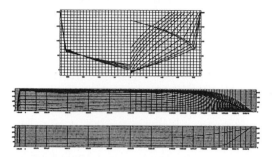

Figure 1. Lines of the model

2.2. Resistance

The model has been towed parallel to the thrust line of the propeller shaft inclined eight degree positively with respect to the still water level. The resistance was measured by a resistance dynamometer connected to the bar lying along the thrust line starting from the towing point located at L.C.B[3].

At the full load condition, resistances were measured at thirteen different speeds in the range of Fn = 0.307 (9knots) to Fn = 1.023 (30knots). Relative displacements of the model ship with respect to the still water level are also measured at F.P. and A.P.

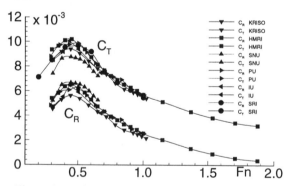

Figure 2. Resistance components of the semi-planing ship (full load)

The results for resistance tests are shown in Fig.2 where results of the other institutions including SRI (Ship Research Institute of Japan), HMRI(Hyundai Maritime Research Institute), SNU(Seoul Natiinal University), PU(Pusan National University) and IU(Inha University) are also plotted to show fairly good agreements[4].

2.3. Wave Pattern Resistance

Wave elevations are measured at two longitudinal planes located at y/L=0.325 and 0.5 at the full load condition and wave pattern resistance are computed. The resulting wave pattern resistance coefficients are shown in Fig. 3 where SRI's result is also included for comparison. A hump is present in the curve near Fn=0.5 similar to the measured resistance curve.

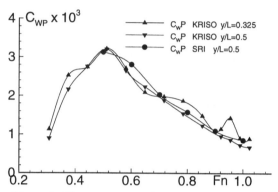

Figure 3. Wave pattern resistance of the semi-planing ship (full load)

2.4. Limiting Streamlines

At the speeds of 21.0, 25.0 and 28.0 knots, limiting streamlines on the hull surface are traced with an oily paint pasted on the surface. The results are shown in Fig. 4. At the bow of the ship, the streamlines begin steeply climb up the surface until spray forms and then, slide down with a considerable slope until the 7.0 station. Behind the station where chine line is present, the limiting streamlines are almost parallel to the towing direction.

Figure 4. Results of the paint tests - limiting Streamlines

2.5. Hull Surface Pressure Distributions

An extra model is made of FRP from the same mould. It has thicker wall comparing to that of the one used for the resistance measurements so as to safely punch holes on the surface. A total of 442 stainless tubes are carefully inserted in the holes perpendicular to the hull surface to measure pressure[5].

At the various trim conditions measured in the resistance tests, the unsteady hull surface pressure distributions are measured with the fifteen pressure transducers manufactured by Validyne Engineering Corporation.

The results are shown in Fig. 5 where the magnitude of the pressure are very low except the region near the stagnation point and which become negative around the bottom of the 9.0 station and the region behind the 2.0 station.

The region of thin color near the midship is getting broader as speed increases. The Cp values in this region is in the range of 0.0 to 0.05 and cause the hull to rise. Hence is can be concluded that the ship will rise further if the speed increases. The fact has been confirmed in the resistance test. And since the regions is located at the rear part of the ship it is clear that the lift comes from the region behind the midship.

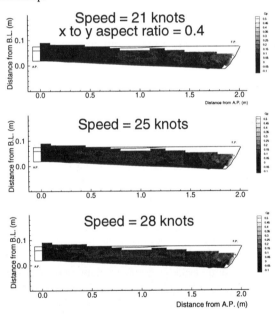

Figure 5. Hull surface pressure distributions

2.6. Wave Elevations

To measure wave elevations, a three-axis traverse unit is designed and manufactured. The unit is driven by a servo motor and moves at a rate of 1 cm per 500 pulses. It also have the two modes, slow and fast, and a jog shuttle to allow arbitrary input. A servo type wave-height gauge manufactured by West Japan Fluid Engineering Laboratory Co, LTD

are attached to the traversing mechanism. At the speed of 21.0 knots, free surface elevations behind the stern is measured as in Fig. 6 where a photograph taken at the same time is also shown.

It is found that the most dominant wave system is generated near the stagnation point located at about 7.0 station and the resultant shape of the wave looks like a triangle near the transom. Among the waves, the wave generated near the transom experience the largest variation in magnitudes.

In general, when a planing hull moves, a wake region develops behind the transom stern where streamlines close in to the center-plane. The low pressure in the region yield negative wave elevations right behind the stern which drops as low as to fully dry up the transom stern at the speed of 21.0 knots. Far behind the transom, wave height regains a positive value.

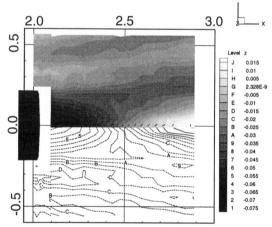

Figure 6. Wave elevation measured behind the transom stern

2.7. Velocity Components

On the three-axis traverse unit, a rake assembly consists of 5 Pitot tubes is attached to measure the velocity component on three cross-sections behind the stern. The sections are chosen to be located at A.P., 2.0 cm behind and 40.0 cm behind the stern at the speed of 21.0 knots. It is found that Pitot tubes bend slightly if towed in the water and the amount varies with the submerged depths. The measured data have been corrected to account for effect of the bendings and the results are shown in Fig. 7.

The measured velocity components on the cross-section located at A.P. show that the flow is directed downward following the bottom surface and a small vortex exists at the side of the hull near the free surface. Axial velocity contours show that an accelerated flow region exists where flow velocity is higher than the towing speed. At the cross-section 2.0㎝ behind the transom, the flow goes upward since the effect of the bottom plane is insignificant and since streamlines are closing in at the region.

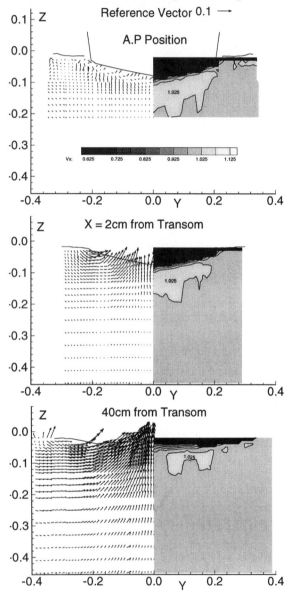

Figure 7. Velocity fields on the cross-sections behind the transom stern

Axial velocity components indicate that the region of accelerated flow exists as before and wake is confined in a very narrow region. At the cross-section 40.0㎝ behind the stern, the flow goes upward even faster, causes free surface rising and there still remains the region of accelerated flow.

2.8. Numerical Computation

Most of the available theories for prediction of wave resistance are not able to predict nonlinear ship waves and resulting resistances. In the present study, an effective finite difference method is used for the simulation of the flow field and nonlinear free surface wave around the model.

Boundary-fitted coordinate systems are widely employed in the numerical simulations of the flow fields around a body with complex geometry such as a ship and their effectiveness is well known. In the present computations, however, a rectangular grid system is chosen since it is not easy to generate an efficient and reasonable curvilinear grid system encompassing a complex hull geometry and non-linear free surface.

The MAC(Marker And Cell) method is one of the efficient procedure to solve the nonlinear free surface flow around a body with complex geometry in rectangular grid system[6,7] and adopted in the present computation. The velocity-pressure coupling has been achieved through a staggered mesh system with variable cell sizes to improve numerical accuracy.

2.8.1. Computational Method

A ship is considered to be a fixed floating body placed in an uniform stream where free surface are present. The fluid has been assumed incompressible and the viscosity neglected. Then, the Euler and continuity equations are solved numerically by a finite difference method based on the MAC method in the Cartesian coordinate system.

The computational domain is appropriately discretized into rectangular cells with variable sizes. Free surface is represented by a succession of segments and the hull surface is approximated by layers of rectangular cells. The governing equations are represented in finite difference forms by first order forward differencing in time and second order centered in space except for the convection term in which a higher order hybrid scheme is used.

The discretized unsteady momentum equations along with the pressure Poisson equation are solved by the SOR(Successive Over Relaxation) with appropriate boundary conditions. On the body boundary cell, the velocity and pressure are iterated simultaneously to satisfy a zero divergence condition in each cell[8].

2.8.2. Results of Computation

The computation is performed for the ship speed of 21.0 knots. Details of the computational conditions are listed in the Table 1.

Table 1. Conditions for numerical computation

Axis	x	y	z
Number of cells	150	45	30
Computational Domain	5.139m	1.020m	0.383m
Minimum cell	0.017m	0.016m	0.008m
Froude number		0.7097	
Non-dimensional Time		17.85	

The computed wave height contours are shown in Fig. 8, which resemble the wave height measured at 2cm behind the transom as shown in Fig. 6: the minimum wave height occurs at right behind the transom is about -7 cm at the centerline, the shapes afterward are parabolic. However, there are little differences also: the location where wave height regains positive values are 80cm and 70cm behind of the transom, respectively. Moreover, in the calculation the wave is developed near the 7 station where the chine meets the still water level, presumably because grid spacings used in the calculation are too coarse to represent the chine line exact enough .

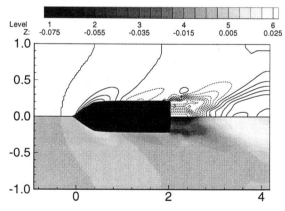

Figure 8.   Computed wave height contour (21.0 knots)

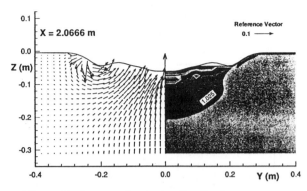

Figure 9. Computed velocity distributions (21.0 knots)

The computed velocity components are shown in Fig. 9. Comparing to the measured ones in Fig. 7, y and z components of the velocity are in good coincidence qualitatively. The x component of the acceleration seems reasonably predicted but there exist some discrepancies in their distribution. Pressure is integrated to give the pressure resistance coefficient but differs considerably from the $Cr = 3.9 \times 10^{-3}$ found in the resistance tests.

3. MODIFIED HULL FORM

3.1. Lines

The hull form of the model has been modified based on the above results with special emphases on the reduction of the spray, frictional and wave making resistance.

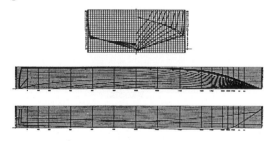

Figure 10. Lines of the modified hull form

The stem of the new model becomes thinner to reduce excessive spray observed at the bow region. The volume taken out near the bow is added to the region near the 4 station to maintain the same displacement base on the observation of surface pressure distribution measured. The width of the midship is unchanged and the chine line was moved down to rise the hull more than before.

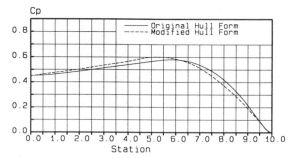

Figure 11. Comparison of Cp curves

The lines of the modified hull form are shown in Fig. 10 and the Cp curve is compared to that of the original one in the Fig. 11 where positions of the LCB are 7.39%(14.78cm) and 4.74%(9.49cm) of the model length downstream of the midship, respectively.

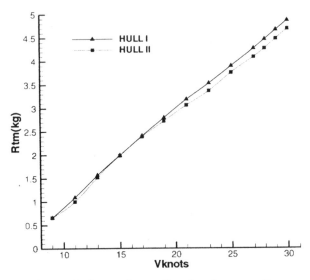

Figure 12. Comparisons of total resistances of two hulls (full load)

3.2. Comparative model tests.

To compare the resistance performance of the two hull forms, resistances of the models are measured in a same day with identical GM's determined by inclining tests, Fig. 12 show comparisons in which the resistance of the modified hull is lower through out the whole velocity ranges except for 9

knots. Resistance coefficients curves shown in Fig. 13 include the effect of changes in wetted surface area.

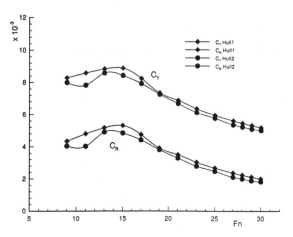

Figure 13. Comparisons of resistance coefficients (full load)

Figure 14. Photographs of the free surface elevation (21knots) top : original hull form, bottom : modified hull form

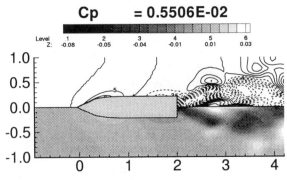

Figure 15. Computed wave height contour (21.0 knots)

The numerical simulation on the flow field around the modified hull form is also performed at the identical conditions with those for the original hull as stated in chapter 2. The results are shown in Fig. 15 in which the wave height restores positive value at 40cm behind the stern which is much closer than that of original ship. The stern wave system is weaker in general but pressure resistance coefficient calculated is 5.506 x 10^{-3}, almost same with the original ones.

4. CONCLUSION

Through the experiments and numerical calculations on the flow field and resistance characteristics of a semi-planing hull form, following conclusions are drawn.

1. The peak of hull surface pressure is located at the very narrow region near the bow, but it diminish sharply behind and so the pressure is almost negligible in the rest. But the small region in the middle in which pressure is positive makes the ship rise. The region becomes wider if towing speed is increased. However the peak pressure does not change too much and the increase of the velocity make a little difference in hull attitude

2. The resistance components found by the wave pattern analyses are similar to those measured in qualitative sense, but the deviation in the magnitudes are significant .

3. Near the A.P, a region in which flow speed is faster than the speed of advance exists near the bottom. The accelerated flow is separated into the flow upward and diverging at the stern and that closing in to the center plane to cause triangular wave observed in the down stream of the stern.

4. Significant differences exist between the values of the residual resistance coefficient C_R measured and pressure resistance C_P obtained from the numerical computations.

5. At the region right behind the transom, the measured velocity components indicate that the flow there moves upward and inward as confirmed in numerical computations. This explain usual occurrences of triangular waves behind the transom stern of high speed ships.

6. Further development in numerical methods are necessary since, for example, with the rectangular grid system employed in the present computations,

influences of the chine lines and the spray at the stem of the high speed planing ships cannot be fully included.

7. Modified hull form has been constructed and tested to yield 3.86% reduction in the total resistance at the design speed of 21 knots by modifying the bow thinner.

5. ACKNOWLEDGMENTS

The experiment has been performed in KRISO and sponsored by the Ministry of Science and Technology. The authors would like to thank to all staff members of the towing tank.

The numerical works have been done at Inha University and funded partly by the KOSEF under grant of 97-02-00-01-01-3 and the Regional Research Center for Transportation System of Yellow Sea, Inha University.

REFERENCES

1. Yong-Jea Park, Seung-Hee Lee, Young-Gill Lee and Sung-Wan Hong, A Study on the Flow Field around a Semi-Planing Hull Form, 1997 HSMV Proceedings, 1977.
2. H. Tanaka, et. al., Cooperative Resistance Tests with Geosim Models of High-Speed Semi-Displacement Craft, Journal of the Society of Naval Architects of Japan, vol. 169,1991
3. S.H. Kim et. al., New Test Method for the Resistance Prediction of a Planing Boat (in Korean), KIMM Report No. UNC131D-271 D, Dec. 1982
4. Seung-Il Yang, et. al., Report on the Cooperative Resistance Test (23m Class Planing Hull), Korea Towing Tank Conference High-Speed Marine Vehicle Committee, Dec. 1995.
5. Susumu Hirano, Suguru Uchida and Yoji Himeno, Pressure Measurement on the Bottom of Prismatic Planing Hulls, J. Kansai Soc. N.A., No213, 1990.
6. Welch. E., Harlow F.H., Shannon J.P & Daly B.J., The MAC method Los Alamos Scientific Lab. Report. 1966
7. Nishimura S., Miyata H., Finite Difference Simulation of Ship Waves by the TUMMAC-IV Method and its application to Hull-Form Design, Journal of Society of Naval Architects of Japan, vol. 157,1985
8. Chan, R, K. et. al., SUMMAC-A Numerical Model for Water Waves, Dept. of Civil Eng., Stanford Univ., Technical Report No. 135,1970.

Practical Design of Ships and Mobile Units
M.W.C. Oosterveld and S.G. Tan, editors.

On Optimal Dimensions of Fast Vessel for Shallow Water

Milan Hofman

Department of Naval Architecture, Faculty of Mechanical Engineering, University of Belgrade, 27 marta 80, 11000 Belgrade, Yugoslavia

When designing a vessel for shallow waterways, its parameters (length and speed) should be selected in a way to avoid the critical region. That is, however, not always possible. In such cases, when the vessel has to travel by its critical speed or pass the critical region, it would be of great practical importance to find some design measures (some optimal form and dimensions) that would reduce the expected negative effects. The present investigation aims to find such beneficial measures by evaluation and analysis of the critical wave-making resistance for vessels which form and dimensions are systematically changed in number of numerical experiments.

The results indicate (unfortunately) that there is no efficient way to reduce the expected unfavorable effects, but to avoid the critical region itself. The common measures known from deep-water hydrodynamics could be applied, but their effect in shallow water is largely reduced. That is the consequence of opposing impact of vessel lengthening on deep-water resistance (longer the better) and the shallow water effects (longer the worse). Although in a way disappointing, the analysis did demonstrate number of interesting and important issues that any engineer, starting to design the shallow water object should be well aware of.

1. INTRODUCTION

The effects of shallow water on ship resistance, although known for over a century and analyzed in number of papers and textbooks, are commonly considered to be of marginal importance in ship design. That is natural, as most of the ships, most of time travel in deep waterways. However, a number of vessels are intended for service in shallow water. One of the examples are vessels for inland transportation. The other (less common) are new fast coastal ferries, for which, because of their speed and length, many of the important sea routes should be considered shallow (see Radojcic 1998). Not only that the influence of water depth on hydrodynamics of such fast vessels should not be ignored, but it can crucially and unfavorably change their resistance compared to the deep-water value.

A typical comparison of shallow to deep-water wave-making resistance (R_h and R_∞) is presented in Fig. 1[a]. The ratio of these values $r = R_h / R_\infty$ (so-called shallow-water resistance ratio) is presented in Fig. 1[b]. The curves are given as functions of Froude number $F_L = v / \sqrt{gL}$, where v is speed, L is length of vessel, and g is gravitational acceleration. The crucial feature of these diagrams is the critical shallow water resistance peak, connected to the increase and disappearance of ship's transverse wave system. It is well known that position and height of that peak depend mostly on water depth h (more precisely, ratio L/h), and that it can be highly increased in extremely shallow water.

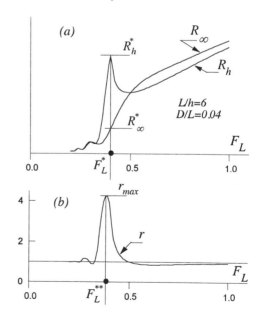

Fig.1. Typical shallow water influence on resistance.
a) Wave-making resistance in deep and shallow water R_∞, R_h.
b) Shallow water resistance ratio $r = R_h / R_\infty$.

It is obvious that an engineer, when planning and designing a vessel for shallow water, should try to avoid the unfavorable (critical) region. Properly, the vessel should be designed for sub-critical region, where the influence of water depth is negligible (slow vessel), or for super-critical region, where the influence of water depth is small but favorable (fast vessel). Unfortunately, that is not always possible. Often, the vessel has to pass the critical region (for instance to achieve the super-critical speeds), or it even has to travel continuously by its critical speed. In such cases, it would be of great practical importance to find a way to reduce the expected negative effects. Can that be done? Are there some designs measures, for instance some changes of vessel form or dimensions that are optimal for the critical region? What are these measures? The present investigation tries to answer these questions in aim to help the designer in adjusting the vessel properly to its shallow waterways.

The critical wave-making resistance can be found by number of different methods. It can be obtained by classical semi-empirical methods, summarized e.g. in Basin and el. 1976, it can be obtained by regression analysis (Xu-tao, 1990), or it can be found experimentally. Also, the critical wave-making resistance can be found theoretically by means of so called Srettensky integral (Srettensky 1937). That last method is applied in the present analysis. Although it gives only a thin ship approximation (such as Michell integral for deep water), it was shown (Millward 1984, Hofman and Radojcic 1997) that it has a reasonable accuracy for the practical applications, at least as far as the shallow-water resistance ratio r is concerned. On the other hand, the method enables a simple change of vessel form and dimensions, so offers a unique opportunity to perform a great number of numerical experiments on systematically varied vessel hull.

In all the calculations that follow, the conditions in shallow waterways are supposed to be the worst possible - the vessel is supposed to travel exactly by its critical speed. Under this condition, the critical wave-making resistance R_h^*, the corresponding deep-water resistance R_∞^* and the maximum of shallow-water resistance ratio r_{max} (see Fig. 1) are found and their changes are followed and analyzed for two different groups of problems.

First, it is assumed that the vessel has unchanged displacement ($\Delta=const$), and under this general condition, the ratios of main dimension (length to breadth L/B, breadth to draught B/D, draught to length D/L) are systematically changed. The situation corresponds to some very early stages in vessel design, when the engineer assessed the preliminary weight, and tries to find the optimal vessel dimensions for its given speed.

In the second group of problems, the displacement is supposed to change, but in two different ways. In the first example the hull remains geometrically similar, and in the other the displacement is changed on behalf of only one of the main dimensions (length, breadth or draught), while the other two are kept constant. The problems of variable displacement aim to simulate a possible situation in vessel design when the engineer finds the assessed weight wrong, and looks for the favorable way of changing the hull dimensions so it could carry the new, corrected weight.

The results of the performed "what would happen if..." type of analysis are, of course, left for the end of the paper. Still, it should be noted that they turned to be somewhat disappointing. The calculations showed generally that, if the vessel travels by its critical speed, there is little the designer can do to reduce the negative shallow-water effects, So, the results emphasize again how important is to avoid the critical region. However, the analysis gives also some unexpected and interesting results. It indicated that, in number of cases, the variation of form and dimensions impacts the shallow water resistance very differently from to the corresponding deep-water value. Such a response, surprising to someone used to classical resistance calculations, should be known to every engineer starting to deal with the fast shallow-water objects.

2. BASIC THEORY

A linear approximation of wave-making resistance for a thin vessel advancing with speed v in inviscid water of depth h is given by the integral

$$R_h = \frac{2\rho g F_h}{\pi h} \int_{\gamma_0}^{\infty} \frac{\left(I_h^2 + J_h^2\right)\gamma}{\cosh^2 \gamma \sqrt{F_h^2 \gamma^2 - \gamma \tanh\gamma}} \, d\gamma \quad .$$

This is Srettensky integral (Srettensky, 1937). Its lower border γ_0 is defined by

$$\gamma_0 = 0, \quad \text{for } F_h > 1, \quad \frac{\tanh\gamma_0}{\gamma_0} = F_h^2, \quad \text{for } F_h < 1,$$

where $F_h = v/\sqrt{gh}$ is Froude number based on water depth, and I_h, J_h are given by the integrals

$$I_h = \iint_S \frac{\partial y}{\partial x} \cos\left(\frac{x}{h}\sqrt{\frac{\gamma}{F_h^2}}\tanh\gamma\right) \cosh\left[\gamma\left(1+\frac{z}{h}\right)\right] dx\,dz$$

$$J_h = \iint_S \frac{\partial y}{\partial x} \sin\left(\frac{x}{h}\sqrt{\frac{\gamma}{F_h^2}}\tanh\gamma\right) \cosh\left[\gamma\left(1+\frac{z}{h}\right)\right] dx\,dz .$$

In these expressions ρ is water density, x and z are longitudinal and vertical coordinate respectively, $y(x,z)$ is a function defining the vessel hull and S is its lateral surface. In the deep-water case $(h \to \infty)$ the above expressions reduce to well known Michell integral

$$R_\infty = \frac{4\rho g^2}{\pi v^2} \int_1^\infty \frac{\left(I_\infty^2 + J_\infty^2\right)\gamma^2}{\sqrt{\gamma^2 - 1}} d\gamma$$

with

$$I_\infty = \iint_S \frac{\partial y}{\partial x} \cos\left(\frac{gx}{v^2}\gamma\right) \exp\left(\frac{gz}{v^2}\gamma^2\right) dx\,dz ,$$

$$J_\infty = \iint_S \frac{\partial y}{\partial x} \sin\left(\frac{gx}{v^2}\gamma\right) \exp\left(\frac{gz}{v^2}\gamma^2\right) dx\,dz .$$

The derivation of Srettensky integral or its detailed analysis will not be presented here. It is given e.g. in Basin and el. 1976. Presently, we just concentrate on the systematic numerical calculation of the above expressions, in aim to obtain the relationships between the vessel form and dimensions and its shallow water resistance. All the results presented are calculated by the self-made computer program *Sretenski*.

3. CALCULATIONS, RESULTS, ANALYSIS

The analysis that follows is restricted to the mathematical, so - called Wigley ships of the form

$$y(x,z) = \pm\tfrac{1}{2}B\cdot\left[1-(z/D)^2\right]\cdot\left[1-(2x/L)^2\right] .$$

Such symmetric forms, with parabolic sharp-edged waterlines and parabolic cross sections, were selected in aim to simplify the numerical treatment of the problem. Still, it is believed that such simplification does not effect the main conclusion that will be reached. This especially implies to the results for resistance ratio r. It is known that the ratio r evaluated by the thin ship approximation (Srettensky and Mitchell integral) is much more reliable then the corresponding values of R_h or R_∞. The ratio r calculated for Wigley form can even be successfully used for practical prediction of shallow water resistance for other hull forms of the same dimensions (see Millward 1984, Hofman and Radojcic 1997). That is, presumably, due to the physical fact that the increase of wave making in critical region is mostly governed by water depth (the ratio L/h) and not by the details of the hull shape.

In each of the examples to be analyzed, the same arbitrary parent hull is selected (notified by index *0*). It has $L_o/B_o=6$, $D_o/L_o=0.04$, $D_o/B_o=0.24$ and it travels by the critical speed in a shallow waterway of depth h, satisfying $L_o/h=6$. Note that the resistance curves of that basic problem are already presented in Fig.1, as a typical example. All the other results, for the hulls of different dimension ratios, different displacements or different water depths (depending on the type of problem analyzed) are calculated for the same (parent) speed and divided by the corresponding parent value. Another words, all the presented non-dimensional results are normalized to give the change of resistance experienced by the "children" hulls, in aim to maintain the same, critical speed.

Concerning the critical speed, some further explanation seems necessary. There are (at least) three different speeds in shallow water that are sometimes called "critical". First, it is the maximum speed of transverse waves in water of depth h, $v = \sqrt{gh}$. Expressed as Froude numbers, it is the critical Froude number based on water depth $F_h=1$ and the critical Froude number based on vessel length $F_L = \sqrt{h/L}$. Second, it is the speed corresponding to the peak value of wave-making resistance curve, or the Froude number F_L^*, as presented in Fig.1[a]. Third, the critical speed is sometimes called the speed corresponding to the maximum of shallow water resistance ratio, or to Froude number F_L^{**}, as presented in Fig 1[b]. All the three speeds have the close (but different) values, it is always valid $F_L^{**} < F_L^* < \sqrt{h/L}$, and, as the water depth decreases, they tend to the same limit $F_L^{**}, F_L^* \to \sqrt{h/L}$. In the present investigation, naturally, wave-making resistance (R_h^* and R_∞^*) are calculated for F_L^*, and shallow water resistance ratio (r_{max}) is calculated for F_L^{**}.

Before continuing, it seems necessary to note that the curves of deep-water resistance presented in the following figures should (mainly) serve to stress the shallow water effects. However, some of these curves may not seem familiar (at least some of them surprised the author), so the reader is advised to look them carefully, before continuing to the shallow water analysis.

3.1 Constant displacement vessel

In the examples to be analyzed, it is supposed that the vessel has a constant displacement ($\Delta=const$) and that it travels by constant speed ($v=const$), which is exactly the shallow water critical speed. Under these assumptions, the ratios of main dimensions are changed: first the ratio L/B with $D=const$, then the ratio D/B with $L=const$, and finally the ratio D/L with $B=const$. The calculated critical wave-making resistance in shallow water R_h^*, the corresponding deep-water value R_∞^*, and the maximum of shallow water resistance ratio r_{max} (all normalized to show the increase/decrease compared to the chosen parent form) are presented in Fig. 2. The analysis, as explained, aims to simulate some early stage in vessel design, when the designer has assessed the preliminary weight, and tries to find the favorable vessel dimensions for the critical shallow water region. Note that (for the simplicity) in all the figures to follow, the earlier introduced notation $R_h^*, R_\infty^*, r_{max}$ is kept for the corresponding normalized values.

First, we have to analyze the deep-water results. Although obtained by the thin ship theory (the Mitchell integral), so containing all the doubts of such an approximation, they are (at least qualitatively) natural and expectable: the increase of length always reduces the resistance, and oppositely,

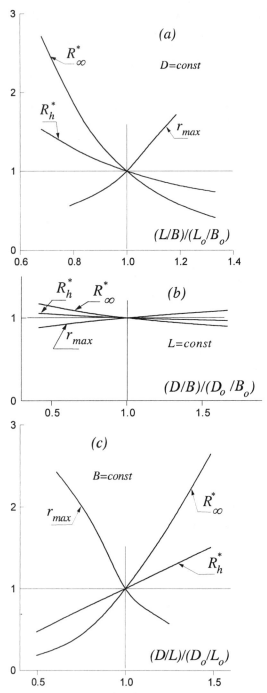

Fig.2. Variation of $R_h^*, R_\infty^*, r_{max}$ for vessels of constant displacement.
a) Influence of L/B with $D=const$.
b) Influence of D/B with $L=const$.
c) Influence of D/L with $B=const$

the increase of breadth always increases it. Analyzing the obtained shallow water results, one can notice that the shallow water resistance always follows the corresponding deep-water value: it increases if the deep-water resistance is increased and decreases if the deep-water resistance is decreased. However, the shallow water curves are always less steep then the corresponding deep-water curves. Another words - the shallow water resistance is less sensitive to the changes of main dimensions. The shallow-water resistance ratio has (of course) the opposite trend then the resistance curves. It indicates that every favorable change of dimensions (the change that would reduce the resistance), would be followed by the increase of unfavorable influences of water depth.

From the practical point of view, the obtained results demonstrate how little the designer can do to improve the behavior of the vessel, if it travels in the critical shallow-water region. Every measure, effective in deep water, becomes less effective in the shallow water. That especially implies to the most common measure: the increase of vessel length. Because of the increased ratio L/h, such a measure most directly stresses the negative influence of water depth. A short example shows the following. An increase of ratio L/B for 10% would in deep-water decrease the wave-making resistance by approximately 25% (Fig 2^a). In shallow water, however, the decrease of resistance would be by some 10% only.

Although disappointing, the above results seem to indicate at least one (somewhat unexpected) practical suggestion. As it is shown in Fig. 2^b, the critical resistance is quite insensitive to the changes of ratio D/B (notice, in that case L/h remains constant). The extreme increase of D/B for more then 300% reduces the shallow-water critical resistance only slightly, not more then 10%. As the small draught is often crucial for the fast inland vessels, the result shows that it can be achieved by making the vessel flat (wide and shallow), practically with no expense in wave-making resistance.

3.2 Variable displacement vessel

If the engineer estimated the preliminary weight wrong, the displacement has to be changed. In the following analysis it is supposed that the displacement is increased, as such a measure seems to be more applicable for the engineering practice.

However, it should be no problem for the reader to follow what happens if, for some reason, the displacement has to be decreased.

An increase of the displacement can be achieved in number of different ways. One of the possibilities is to keep the vessel geometrically similar to its parent hull. The consequences of such hull enlargement on the resistance are presented in Fig. 3. Deep-water calculations (obtained by thin-ship approximation) show that the wave-making resistance R_∞^* is influenced only slightly and that it decreases in the most part of the diagram. Such a decrease is somewhat surprising, so it was checked by some other engineering methods. Although not given here, this additional checking indicated that the influence on deep-water resistance is indeed small, but unfavorable, so that (in reality) R_∞^* slowly increases with the increase of the displacement.

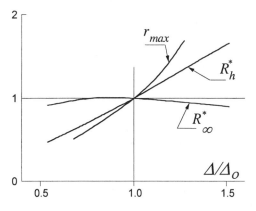

Fig.3. Variation of $R_h^*, R_\infty^*, r_{max}$ with displacement for the geometrically similar vessels.

The evaluated resistance in shallow water R_h^*, that shows practically a linear increase with the displacement, can be doubted on the same ground. It seems that the line, actually, should be somewhat steeper. But, in spite of some doubts concerning R_h^*, R_∞^*, the calculated shallow-water resistance ratio r_{max} (as explained earlier) is believed to be reliable. And its steep increase shows a strong and unfavorable influence of the analyzed, geometrically similar hull enlargement. Is there some other, more favorable way to increase the vessel displacement?

The other possibilities are presented in Fig. 4. There, the displacement is changed on behalf of only one of the dimensions, while the other two are kept constant. In Fig. 4^a draught is changed. That is the most common case of heavier, but otherwise unchanged hull, which has to sink deeper into the water. As often experienced in practice, such an increase of displacement effects unfavorable the vessel resistance. It is presented here by a steep rise of the curve R_∞^*. The shallow water curve R_h^* is even somewhat steeper. The small difference between these curves indicates that the increased draught increases the unfavorable effects of shallow water only slightly, in accordance with the flat curve r_{max}. It should be noted that such a small change of r_{max} is a consequence of constant ratio L/h in the presented example.

In the second example, presented in Fig. 4^b, the displacement is increased on behalf of breadth. It is a measure no designer would suggest, at least from the powering point of view. The results, of course, show a steep rise of the resistance curves. However, it should be noted that it is the only example where the deep and shallow water curves coincide, and the value r_{max} remains constant. Naturally, it is a consequence of constant values L/h and D/L.

In the last two examples (Figs 4^a and 4^b), the hull enlargement did not influence significantly the shallow-water resistance ratio r_{max}. Although that may seem promising, it is of little practical importance, as such an enlargement (in both cases analyzed) is already found unfavorable for the deep-water resistance.

The last example to be analyzed is the enlargement of displacement on behalf of length (Fig. 4^c). The thin-ship approximation shows a steep decrease of R_∞^* curve in the most part of the diagram. Such a result (unexpected to the author) was checked by some other "engineering approach" methods. They showed that the resistance is indeed decreased, but much less steeply. Roughly, the slope seems to be only half of the one presented in the figure. Note that this is the second time the Michell integral is found to overestimates the beneficial influence of the hull lengthening. Such its behavior may be an interesting subject for some future investigation.

The shallow water resistance R_h^* is found to decrease much less steeply then the corresponding

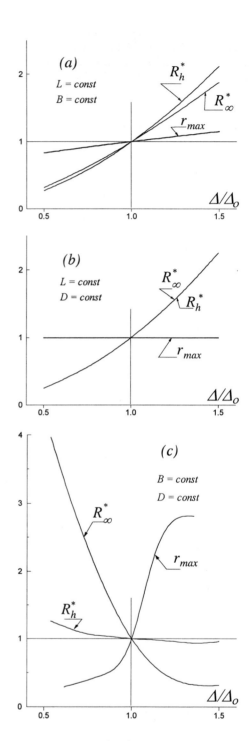

Fig.4. Variation of R_h^*, R_∞^*, r_{max} with displacement.
a)   Displacement changes with draught.
b)   Displacement changes with breadth.
c)   Displacement changes with length.

R_∞^* curve. Comparing to the extreme decrease of deep-water resistance for more then *1000%*, it decreases for some *30%* only, in whole domain analyzed. Because of the doubts concerning the thin-ship approximation, it seems that, in reality, it even increases. The results for resistance ratio r_{max} (that are again believed to be reliable) show an extreme increase of the unfavorable shallow water influences that would follow the lengthening of the hull.

The analysis of increased ship weight (Figs. 3 and 4) showed that, in all the cases where the influence on deep-water resistance was found favorable, the effects of shallow water (presented by r_{max}) are pronouncedly negative, canceling all the benefits. The changes of r_{max} were found small (or did not appear) only if the effects on deep-water resistance are already unfavorable, so are of no practical interest. So, it turned out again that there exist no beneficial practical measure (some change of hull shape) able to reduce effectively the unfavorable influence of the critical speed.

4. CONCLUSION

The present investigation aimed to find some effective way to reduce the unfavorable influence on resistance if the vessel travels by its critical speed. The theoretical approach was used, and, in number of numerical experiments, vessel dimensions were systematically varied. The impact of these changes on the resistance was followed and analyzed. All the examples were chosen to simulate some early stages in vessel design, to follow designer's dilemmas and help him to adjust the vessel properly to its shallow waterway.

The results obtained show, generally, that the same principles known from deep-water hydrodynamics should be followed (increase of length, decrease of breadth), but that the effects of these common measures would be largely reduced by the shallow water influences. In some of the cases, the shallow water influences even cancel all the benefits. So (figuratively), the full price has to be paid for the vessel to travel by its critical speed.

Physically, it is the consequence of opposite influences of vessel length on deep and shallow water resistance. The increase of L is the most efficient (and common) way to reduce the deep-water resistance. It is well known - longer the better. But (because of the increase of L/h) it is also the

most efficient way to increase the negative shallow water effects. Therefore, for the shallow water influences it is - longer the worse.

Although failed to find some efficient way to reduce the negative shallow water influences, the investigation did indicate some interesting and important practical issues. It emphasized, again, the significance of shallow water critical speed. That speed should be detected in the most early stages of vessel design and the proper vessel parameters should be chosen in aim to avoid the whole critical region. One of the ways to accomplish that is the use of shallow water resistance charts, as proposed by Hofman and Kozarski 1998. On the other hand, the investigation demonstrated how very differently the changes in vessel form and dimensions (especially the length) impact the resistance in deep and shallow water. Any engineer dealing with shallow water objects should be very familiar with these opposing influences. If not, some unpleasant surprises are likely to happen.

REFERENCES

1. Basin A.M, Velednicky I.O. and Lahovicky A.G., *Hydrodynamics of Ships in Shallow Water* (in Russian), Sudostroenie, Leningrad, 1976.
2. Hofman M. and Radojcic D., *Resistance and Propulsion of Fast Ships in Shallow Water* (in Serbian), Faculty of Mechanical Engineering, Belgrade 1997.
3. Hofman M. and Kozarski V., Shallow water resistance charts for preliminary vessel design, Submitted for the publication.
4. Millward A., The effect of hull cross section on the theoretical wave resistance of a fast ship in shallow water, International Shipbuilding Progress, Vol.31, No.354, 1984, pp. 28-33.
5. Radojcic D., Power prediction procedure for fast sea-going monohulls operating in shallow water, 19th Duisburg Colloquia on Ships for Supercritical Speeds, 1998.
6. Srettensky L.N., Theoretical investigation of wave-making resistance (in Russian), Central Aero-Hydrodynamics Institute Report 319, 1937, pp. 3-55.
7. Xu-tao Q., Study on the optimal coordination between ship and restricted waterway, International Shipbuilding Progress, Vol.37, No.411, 1990. pp. 273-288.

Practical Design of Ships and Mobile Units
M.W.C. Oosterveld and S.G. Tan, editors.

A simple surface panel method to solve unsteady wing problems

K. Nakatake[a] , J. Ando[a] and S. Maita[b]

[a]Department of Naval Architecture and Marine Systems Engineering, Kyushu University, Hakozaki 6-10-1, Higashi-ku, Fukuoka 812-8581, Japan

[b]West Japan Fluid Engineering Laboratory Co. Ltd., Kozima 339-30, Kosaza-cho, Kitamatsuura-gun, Nagasaki-ken 857-0401, Japan

This paper proposes a simple surface panel method to solve unsteady wing problems mainly for 2-D wing. After describing the method, we show some numerical results for Wagner problem, sinusoidal gust problem, heaving oscillation and pitching oscillation problems, and confirm the usefulness of this method by comparing the numerical results with analytical results and experiments.

1. INTRODUCTION

The flow field around an unsteady wing can be analysed by panel methods, such as Morino's method [1], Hoshino's method [2], Koyama's method [3] and so on.

At the last PRADS'95 we presented a simple surface panel method (SQCM)[4] to solve steady wing problems. Since then SQCM has been succesfully applied to steady state problems such as hydrofoils, marine propellers, etc. SQCM uses source distributions (Hess and Smith type[5]) on the wing surface and discrete vortex distributions arranged on the camber surface according to Lan's quasi-continuous vortex lattice method (QCM)[6].

In this paper, we extend SQCM to the unsteady problems for 2-D and 3-D wings by combining the source panels and unsteady QCM which was succesfully developed. We consider the unsteady Kutta condition that the pressures must coincide with each other on the upper and lower surfaces at the trailing edge. In the steady problem, SQCM satisfies the Kutta condition by setting zero normal velocity on the camber surface at the trailing edge. But it is not usually equal to zero in the unsteady problem but a finite normal velocity exists. In the unsteady 2-D wing problems, we obtain this normal velocity by iteration for the equal pressure condition at the trailing edge.

Moreover, we extend this Kutta condition to

unsteady 3-D wing and show that SQCM gives reasonable results for unsteady 2-D and 3-D wing problems (Wagner problem, sinusoidal gust, heaving oscillation and pitching oscillation problems). In the 3-D heaving oscillation, we compare about the lift characteristics between calculation and experiment and confirm a good agreement.

2. FORMULATION

We describe the formulations using SQCM for an unsteady 2-D wing, i.e., the induced velocity, the Kutta condition, boundary conditions, numerical calculations and equations of unsteady pressure and lift.

2.1. Induced velocity

Let us consider an oscillating 2-D wing in the uniform flow U (see Figure 1). We express the wing by the vortex distributions on the camber surface and the source distributions on the wing surface and the shed vortex on the ξ- axis. Denoting the coordinate along the x- axis, the chord length, the strength of vortex on the camber and the strength of the shed vortex by ξ, c, $\gamma(\xi,t)$ and $\gamma_w(\xi,t)$, respectively, the velocity V_γ induced by these vortices at an arbitrary point P is expressed by

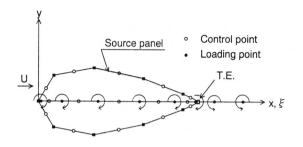

Figure 1. Coordinate system and panel arrangement.

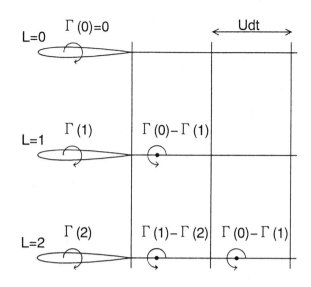

$$V_\gamma = \int_0^c \gamma(\xi,t)\boldsymbol{v}(P,\xi)d\xi$$
$$+ \int_c^{c+Ut} \gamma_w(\xi,t)\boldsymbol{v}_w(P,\xi)d\xi, \tag{1}$$

where $\boldsymbol{v}(P,\xi)$ and $\boldsymbol{v}_w(P,\xi)$ are induced velocities due to the bound vortex and the shed vortex of unit strength, respectively.

Since the circulation of whole system remains constant by Kelvin's theorem, the following relation exists.

$$\int_0^c \gamma(\xi,t)d\xi + \int_c^{c+Ut} \gamma_w(\xi,t)d\xi$$
$$= Const. \qquad (=0) \tag{2}$$

Next, discretizing Eqs. (1), (2) and combining them, we get the velocity $\boldsymbol{V}_{\gamma i}$ induced at the i-th control point by these vortices as follows.

$$V_{\gamma i} = \sum_{\nu=1}^{N_\gamma} \gamma(\xi_\nu, t_L)(\boldsymbol{v}_{i\nu} - \boldsymbol{v}_{wi1})\Delta\xi_\nu$$
$$+ \sum_{\ell=1}^{L-1} (\boldsymbol{v}_{wi\ell} - \boldsymbol{v}_{wi\ell+1})\Gamma(t_{L-\ell}) \tag{3}$$

where

Figure 2. Model of Circulation.

N_γ : number of camber discretization,

L : number of shed vortices (time steps)

$\boldsymbol{v}_{i\nu}, \boldsymbol{v}_{wi\ell}$: induced velocities due to bound and shed vortices of unit strength

$$\Gamma(t_\ell) \cong \sum_{\nu=1}^{N_\gamma} \gamma(\xi_\nu, t_\ell)\Delta\xi_\nu$$

: circulation at time t_ℓ

$$\Delta\xi_\nu = \frac{\pi c}{2N_\gamma}\sin\frac{2\nu-1}{2N_\gamma}\pi$$
$$\xi_\nu = \frac{c}{2}\left(1 - \cos\frac{2\nu-1}{2N_\gamma}\pi\right)$$

: loading point on camber $(\nu = 1, 2, \cdots, N_\gamma)$

Figure 2 shows the model of the shed circulations and then shed vortices are assumed to shift on the x - axis by $U dt$ retaining each strength during one time step. As to the first position of the shed vortex X_w, we adopt $X_w = c + 0.25c/N_\gamma$ according to Hoshino [7] and Murakami [8]. The circulation of vortex $\Gamma_w(t_L)$ shed at time t_L becomes

$$\Gamma_w(t_L) = \Gamma(t_{L-1}) - \Gamma(t_L) \qquad (4)$$

In calculating the induced velocities, we use the two kinds of expressions, that is, Eq.(3) is used for the control points on the camber surface and the following equation is used for the control points on the wing surface.

$$V_{\gamma i} = \sum_{\nu=1}^{N_\gamma} \gamma(\xi_\nu, t_L) \left(\int_{\xi_{a\nu}}^{\xi_{b\nu}} v_{i\nu} d\xi - v'_{wi1} \Delta \xi_\nu \right)$$
$$+ (v'_{wi1} - v_{wi2}) \Gamma(t_L) \qquad (5)$$
$$+ \sum_{\ell=2}^{L-1} (v_{wi\ell} - v_{wi\ell+1}) \Gamma(t_{L-\ell})$$

where

$$v'_{wi1} = \frac{1}{U dt} \int_c^{c+U dt} v_{wi1} d\xi \qquad (6)$$

As the wing thickness becomes small, the evaluating error of $V_{\gamma i}$ becomes large due to discretized vortex. Therefore we treat it by the vortex distribution on the camber and the wake until the first shed vortex. On the other hand, the velocity induced by the source distribution V_{si} is expressed by

$$V_{si} = \sum_{j=1}^{N_\sigma} \sigma(s_j, t_L) v_{sij} \qquad (7)$$

where N_σ is the number of source panels on the wing surface, $\sigma(s_i, t_L)$ is source strength at time t_L and v_{sij} is the induced velocity due to a line source of unit strength.

Summing up the inflow velocity V_{Ii}, $V_{\gamma i}$ and V_{si}, we obtain the total velocity V_i at the control point as

$$V_i = V_{\gamma i} + V_{si} + V_{Ii} \qquad (8)$$

2.2. Kutta condition

As the Kutta condition we adopt the so-called "pressure condition"[9] that the pressure distribution is continuous at the trailing edge (T.E). In the steady SQCM, we set at T.E the zero normal velocity to the camber surface ($V_N = 0$). But, in

the unsteady SQCM, we consider that the finite normal velocity exists so as to eliminate the pressure difference between the upper and the lower surfaces at T.E. Denoting the pressure difference, the disturbed velocity potential and the magnitude of the velocity by $\Delta p, \phi$ and V, respectively, we can express Δp from the unsteady Bernoulli's equation

$$\frac{\Delta p}{\rho} = \frac{\partial \Gamma(t_L)}{\partial t} + \frac{1}{2}(V_+^2 - V_-^2) \qquad (9)$$

where $\Gamma(t_L) = \phi_+ - \phi_-$ and the suffixes $+, -$ mean the values at the nearest control points to T.E on the upper and lower surfaces. Therefore, the Kutta condition in the unsteady problem becomes

$$\Delta p = 0 \qquad (10)$$

2.3. Boundary conditions and numerical calculations

The boundary conditions of the steady SQCM are zero normal velocity across the wing and camber surfaces. In the case of unsteady SQCM, we adopt simultaneously

$$V_i \cdot n_i = 0 \qquad \text{on wing and camber (except T.E) surfaces}$$
$$V_i \cdot n_i = V_N \quad \text{at T.E.} \qquad (11)$$

where n_i expresses the unit normal vector to the wing or camber surface. In Eq.(11), the unknowns are $\gamma(\xi_\nu, t), \sigma(s_j, t)$ and the normal velocity V_N which should be determined so as to satisfy Eq.(10). If V_N is given, we can solve the linear simultaneous equations for $N_\gamma + N_\sigma$ singularities. The condition ($V_N = 0$) of the steady SQCM is included also as a special case of the unsteady SQCM.

Next we explain how to determine V_N by iteration of n times.

- For $n = 1$, we assume $V_N^{(1)} = 0$ and obtain $\gamma(\xi, t_L)$ and $\sigma(s_j, t_L)$ and calculate $\Gamma(t_L)$ and $\Delta p^{(1)}(\neq 0)$

- For $n = n$, we correct $V_N^{(n)}$ as

488

$$V_N^{(n+1)} = \Delta p^{(n)} \beta/(\rho U) + V_N^{(n)} \qquad (12)$$

and then calculate $\Delta p^{(n+1)}$ and repeat the above procedures until $\Delta p^{(n+1)}$ converges.

2.4. Unsteady pressure and lift

The unsteady pressure on the wing is obtained as follows. Describing the unsteady Bernoulli's equation in the frame of the coordinate systems fixed to the wing, the pressure equation becomes

$$p - p_0 = -\frac{1}{2}\rho(q_r^2 - v_F^2) - \rho\frac{\partial\phi}{\partial t} \qquad (13)$$

where

p_0 : base pressure

ρ : density of water

v_F : relative speed to the coordinate systems fixed to the space at the point in the coordinate systems fixed to the wing.

q_r : speed in the coordinate systems fixed to the wing

ϕ : disturbed velocity potential in the coordinate systems fixed to the wing

In Eq.(13), $\partial\phi/\partial t$ is obtained as follows. Firstly we divide ϕ into the source component ϕ_σ and the vortex component ϕ_γ as

$$\phi = \phi_\sigma + \phi_\gamma \qquad (14)$$

and obtain each component analytically at each time step. Next we calculate the time derivative by the two-point upwind difference scheme. Then the lift L acting on the wing in the vertical direction to U is calculated by the pressure integration around the wing, and the pressure and the lift coefficients (C_p and C_L) are expressed by

$$C_p = (p - p_0)/\frac{1}{2}\rho U^2, \qquad C_L = L/\frac{1}{2}\rho c U^2 \qquad (15)$$

3. NUMERICAL EXAMPLES (2-D WING)

After many numerical tests, we adopt $N_\sigma = 60$, $N_\gamma = 31$ and $dt = 0.1c/U$ and perform

lift calculations for Wagner problem, the heaving oscillation problem, the sinusoidal gust problem and pitching oscillation problem for wings. For comparison, we also show the analytical results for thin wings using Wagner function [10], Theodorsen function [11] and Sears function [12].

3.1. Wagner problem

As the first example, we calculate about the wings advancing with constant speed U abruptly from rest state with angle of attack of 5 degrees. Figure 3 shows for NACA0001 wing the source distribution σ and vortex distributions γ, γ_w at the tenth time step and Figure 4 expresses the corresponding pressure distributions with and without consideration of $\partial\phi/\partial t$ term. We understand that the Kutta condition is satisfied at T.E if we consider $\partial\phi/\partial t$ term. We show in Figure 5 C_L values versus time step number with and without consideration of $\partial\phi/\partial t$ term which is important at the initial stage. A good agreement is obtained with the analytical result. Figure 6 shows the time step solution of C_L of NACA wings with different thickness. C_L converges to the values of the steady condition with time and C_L of thinner wing approaches to the thin wing results. These results confirm the usefulness of our method.

3.2. Heaving oscillation problem

Let us consider a wing advancing with a constant speed U and doing heaving oscillation of the vertical displacement $h = h_0 e^{i\omega t}$, where $h_0 (= 0.1c)$ is the amplitude of heaving and ω the circular frequency. Defining the reduced frequency $k = \omega c/2U$, smaller k means slow heaving oscillation. Figure 7 shows the time history of C_L of wings with different thickness. The result of NACA0001 agrees well with thin wing result. The phase of C_L lags behind the wing heaving. Figure 8 expresses the relation between the normal velocity V_N and strength of shed vortex. These values are in phase. Figure 9 shows the relations between the maximum lift amplitude $\overline{C}_L$ and k comparing with experimental results [13] and thin wing results. In the range of $k(> 0.4)$, $\overline{C}_L$ becomes small with increase of thickness. This tendency is seen from calculation and experiment.

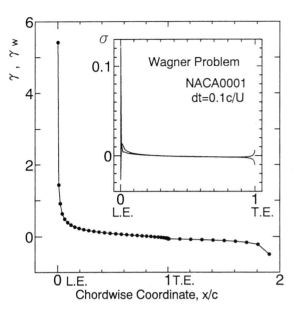

Figure 3. Vortex and source distributions.

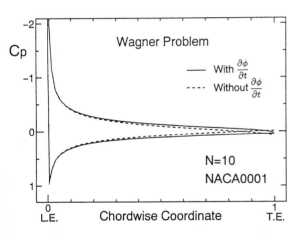

Figure 4. Comparison of pressure distributions.

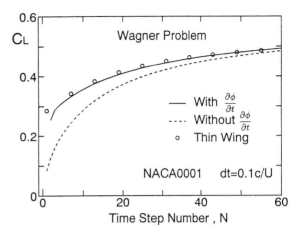

Figure 5. Time step solution on lift coefficients.

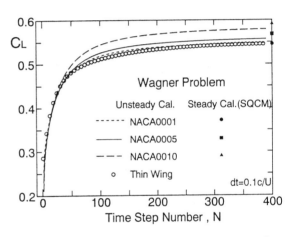

Figure 6. Time step solution on lift coefficients of wings with different thickness.

On the contrary, in the range of $k(< 0.2)$, thicker wing produces larger $\overline{C}_L$. This result coincides with the tendency that thicker wing produces larger $\overline{C}_L$ in the steady problem. Figure 10 shows the pressure distributions of the cases of maximum lift and zero lift. The corresponding vortex and source distributions are shown in Figure 11.

They are obtained as smooth curves.

3.3. Sinusoidal gust problem

We calculate the lift of wings with differerent thickness advancing in the sinusoidal gust whose vertical speed $v(xt)$ is $v_0 e^{i\nu(t-xU)}$, where $\nu = 2\pi U/\ell$, ℓ is wave length of the gust and v_0 is gust amplitude. In this case, the reduced frequency k is given by $\pi c\ell$. Figure 12 shows the time history of C_L comparing with thin wing result. A good agreement is obtained between NACA0001 and

490

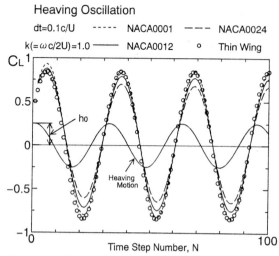

Figure 7. Time history of lift coefficients.

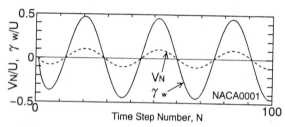

Figure 8. Normal velocity V_N and shed vortex γ_w (Heaving oscillation).

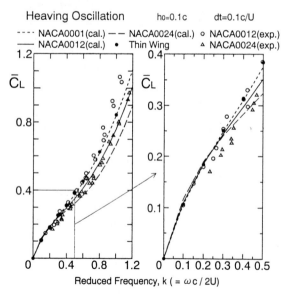

Figure 9. Unsteady lift coefficients in heaving oscillation.

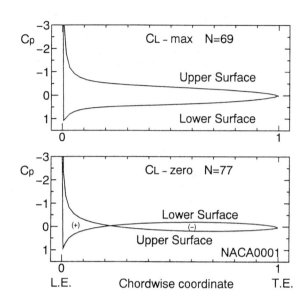

Figure 10. Comparison of pressure distributions (Heaving oscillation).

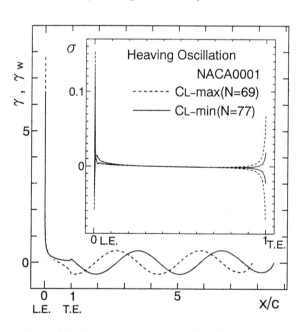

Figure 11. Vortex and source distributions.

491

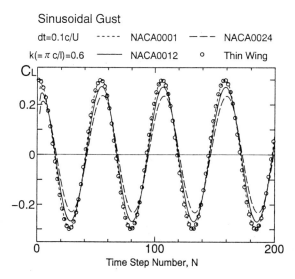

Figure 12. Time history of lift coefficients.

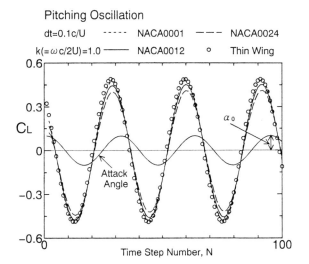

Figure 13. Time history of lift coefficients.

thin wing.

3.4. Pitching oscillation problem

Let us consider a wing pitching about mid-chord with pitching angle $\alpha(t) = \alpha_0 e^{i\omega t}$, where α_0 is amplitude. In this case the reduced frequency k is defined by $\omega c/2U$. Figure 13 shows

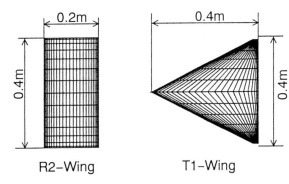

Figure 14. Model wings.

the time history of C_L comparing with thin wing result. A good agreement between NACA0001 and thin wing confirms the usefulness of our method.

4. 3-D PROBLEM

The extension of 2-D SQCM to 3-D SQCM is not difficult and the basic procedure is quite similar[14]. Since we need to consider the span-wise variation of source and vortex distributions, formulation becomes more complicated and the computing time increases considerably. In the 3-D problem, the drag D acts on the wing in addition to the lift L. These forces are indicated by the lift and drag coefficients as

$$C_L = L/\frac{1}{2}\rho AU^2, \qquad C_D = D/\frac{1}{2}\rho AU^2 \qquad (16)$$

where A is the area of the plan form. The definition of the reduced frequency is same as 2-D case and the advancing speed U is assumed to be 1.0m/s.

4.1. Numerical examples (3-D wings)

We adopt 3-D wings (R2, T1)[15] with two kinds of plan forms shown in Figure 14 and perform calculations for the similar cases to 2-D wing. The wing surface is divided into 14 strips in the spanwise direction and into 50 segments around the wing surface, and the camber surface is divided into 26 segments. At first, we show the time history of C_L of R2-wing in Figure 15 as

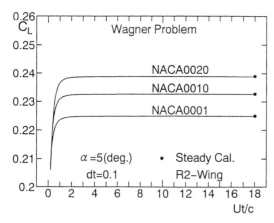

Figure 15. Time history of lift coefficients of wings with different thickness.

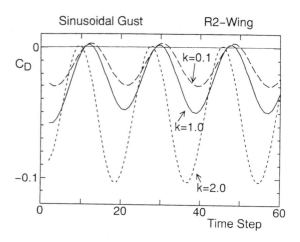

Figure 17. Time history of drag coefficients.

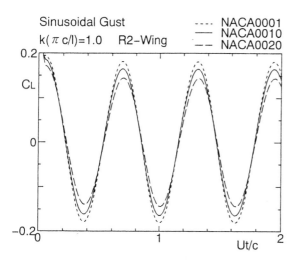

Figure 16. Time history of lift coefficients of wings with different thickness.

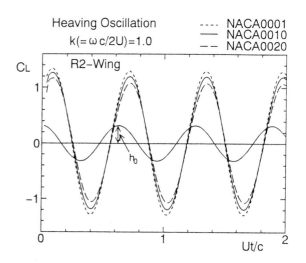

Figure 18. Time history of lift coefficients of wings with different thickness.

an example of Wagner problem. C_L converges to the values of the steady case with time. As the sinusoidal gust problem, let us consider R2-wing advancing in the sinusoidal gust whose vertical speed $w(x,t) = 0.1Ue^{i\nu(t-x/U)}$ and the reduced frequency k equals c/ℓ, where ℓ is wave length of the gust. Figure 16 shows the time history of C_L of R2-wing with different thickness at $k = 1.0$. In this case, the amplitude of C_L increases with

decrease of thickness. Figure 17 shows the time history of C_D of the same wing at different k values. With increase of k (decrease of ℓ), minus C_D (thrust) acts on the wing due to the leading edge suction force.

As the heaving oscillation problem, we treat R2-wing advancing with vertical displacement $h = h_0 e^{i\omega t}$ where $h_0 = 0.04m$. Figure 18 shows the time history of C_L of R2-wings with different

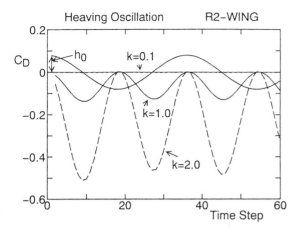

Figure 19. Time history of drag coefficients.

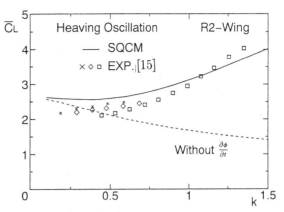

Figure 20. Amplitude of unsteady lift coefficients.

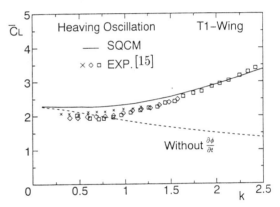

Figure 21. Amplitude of unsteady lift coefficients.

thickness and Figure 19 does that of C_D at different k values. The k value changes the drags drastically. In Figures 20 and 21, we compare the amplitude $\overline{C_L}$ of two wings between SQCM and experiment[15]. Both results agree well if we take into consideration the time derivative term.

Lastly we consider C_L and C_D of R2-wing advancing and pitching around the leading edge with the angle of attack $\alpha = \alpha_0 e^{i\omega t}$ where $\alpha_0 = 5$deg. Figure 22 shows the time history of C_L of R2-wing with different thickness and Figure 23 does that of C_D at three k values. We notice the thickness of wing does not affect C_L so much but k value changes C_D drastically.

5. CONCLUSION AND ACKNOWLEDGMENT

We proposed a simple surface panel method (SQCM) to solve unsteady 2-D and 3-D wing problems. Through comparison of calculated results with analytical or experimental ones, we confirmed the usefulness of SQCM.

We wish to express our deep thanks to Dr. K. Kataoka, Mr. A. Yoshitake and Mrs. Y. Yamasaki for their help in completing this manuscript.

A part of this research was supported by the aids from Grant-in-Aid for Scientific Research(A), the Ministry of Education, Sciense and Culture.

REFERENCES

1. L. Morino, L.T. Chen and E.O. Sciu, Steady and Oscillatory Subsonic and Supersonic Aerodynamics around Complex Configurations, AIAA Journal, Vol. 13 (1975)
2. T. Hoshino, Hydrodynamic Analysis of Propellers in Unsteady Flow Using a Surface Panel Method, Journal of the Society of Naval Architects of Japan, Vol. 165 (1989)
3. K. Koyama, Application of Panel Method to Unsteady Hydrodynamic Analysis of Marine

494

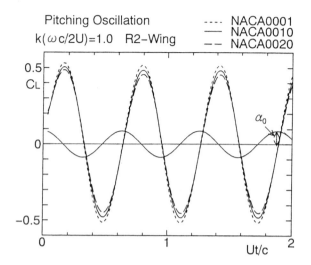

Figure 22. Time history of lift coefficients of wings with different thickness.

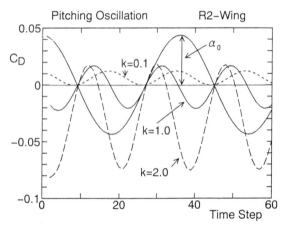

Figure 23. Time history of drag coefficients.

propellers, 19th Symposium on Naval Hydrodynamics, in Korea (1992)

4. K. Nakatake, J. Ando, K. Kataoka, and A. Yoshitake, Practical Calculation Method for Thick Wings, 16th International Symposium on Practical Design of Ships and Mobile Units, (1995)

5. J.L. Hess and A.M.O. Smith, Calculation of Nonlifting Potential Flow about Arbitrary Three-Dimensional Bodies, Journal of Ship Research, Vol. 8 No.2, (1964)

6. C.E. Lan, A Quasi-Vortex-Lattice Method in Thin Wing Theory, Journal of Aircraft, Vol. 11 No.9, (1974)

7. T. Hoshino, Application of Quasi-Continuous Method to Unsteady Propeller Lifting-Surface Problems, Journal of the Society of Naval Architects of Japan, Vol. 158 (1985)

8. M. Murakami, M. Kuroi, K. Ando and K. Nakatake, Practical Quasi-Continuous Method to Estimate Unsteady Characteristics of Propeller, Transactions of the West-Japan Society of Naval Architects, Vol. 84 (1992)

9. B.C. Basu, and G.J. Hancock, The Unsteady Motion of a Two-Dimentional Aerofoil in Incompressible Inviscid Flow, Journal of Fluid Mechanics, Vol. 87, Part 1 (1978)

10. H. Wagner, Dynamisher Auftrieb von Tragflügeln, Zeitschrift für Angewandte Mathmatik und Mechanik, Band. 5, Heft 1 (1925)

11. T. Thedorsen, General Theory of Aerodynamic Instability and Mechanism of Flutter, NACA Technical Reports, No.496 (1935)

12. W.R. Sears, Some Aspects of Non-Stationary Airfoil Theory and Its Practical Application, Journal of Aeronautical Sciences, Vol. 8 (1941)

13. Y. Kyozuka, M. Kurahashi and W. Koterayama, Hydrodynamic Forces Acting on a Oscillating Wing in Uniform Flow, Transactions of the West-Japan Society of Naval Architects, Vol. 76 (1988)

14. S. Maita, J. Ando and K. Nakatake, A Method to Solve Unsteady Three-Dimensional Wing Problems by a Simple Surface Panel Method (SQCM), Transactions of the West-Japan Society of Naval Architects, Vol. 94 (1997)

15. Y. Kyozuka, T. Hori and W. Koterayama, Unsteady Hydrodynamic Forces on a Lifting Body, Journal of the Society of Naval Architects of Japan, Vol. 167 (1990)

1998 Elsevier Science B.V.
Practical Design of Ships and Mobile Units
M.W.C. Oosterveld and S.G. Tan, editors.

TIME-DOMAIN ANALYSIS OF LARGE-AMPLITUDE RESPONSES OF SHIPS IN WAVES

N. Fonseca and C. Guedes Soares

Unit of Marine Technology and Engineering, Technical University of Lisbon,
Instituto Superior Técnico, Av. Rovisco Pais, 1096 Lisboa, Portugal

ABSTRACT

The motions and loads induced on ships by waves are studied in the time domain. The method takes into account the most important sources of the nonlinear behaviour associated with the large amplitude motions.

The paper describes an experimental programme carried out on a seakeeping towing tank, with a model of the S175 containership. The objective was to obtain a reliable data base of systematic measurements of motions and sectional loads in a wide range of wave lengths and wave heights, extending into the nonlinear range.

Comparisons between numerical and experimental results are presented and analysed. The importance of the nonlinear terms is assessed by comparing these responses with the corresponding linear ones.

1. INTRODUCTION

The expected maximum structural loads induced by waves during the lifetime of a ship, are essential parameters for the design of the hull structure. For most ships these maximum loads will occur when the ship is operating in heavy seastates, and usually the vertical bending moment at midship results on the critical tensions applied to the ship structure.

It is recognised nowadays that above certain seastate level the ship responses are nonlinear, and specially the wave induced structural loads may present a strong nonlinear behaviour.

In fact, some full-scale measurements, which show this nonlinear behaviour, can be found in the literature. In his discussion of Jensen and Pedersen [1] paper, Hachmann presents results of the peaks of full scale measurements of vertical bending moment at midship on a containership. The results show that the ratio of the sagging moment / hogging moment is greater than 1 and the difference increases with the seastate.

The asymmetry of the peaks of sagging and hogging moments was already known for some years from the results of experiments with ship models. Dalzell [2] presented measurements of the wave induced loads in models of the mariner ship, a tanker and a destroyer. The results showed an important nonlinear behaviour, and that sagging/hogging ratio tended to be larger for the models with smaller block coefficients.

More recently, Watanabe et al [3] presented results of tests carried out with two models with different bow shapes in regular and irregular waves. The objective was to assess the influence of the bow flare on the deck wetness and asymmetry of the vertical wave bending moment.

Numerical methods for the direct calculation of the linear wave induced motions and loads on ships are in use since the strip theory developments (Salvesen et al., [4]). More recently the three-dimensional panel methods improved the linear solution, specially for lower frequencies and bluff ships (Chang, [5]).

The former methods provide reliable tools for predicting the seakeeping of conventional ships in the linear range of conditions, however in the moderate to severe seastate conditions the nonlinear effects become important and linear results are no longer valid.

Much research work has been carried out in the recent years in order to obtain solutions where some nonlinear effects are introduced. Lin et al. [6] applied a transient free surface Green function panel method to calculate large amplitude motions and wave loads, where the non-linear body condition was used. Kring et al [7] extended the time domain Rankine source method to non-linear predictions of ship motions and wave loads.

The desingularized method is under development to tackle the fully non-linear three dimensional free surface problem. Results by this method were presented by Beck et al. [8].

While the former methods are certainly the most advanced ones, the numerical solutions are very complex and the computational effort to obtain simulations is huge.

In the present paper a practical time domain strip method is used to predict the vertical motions and sectional induced loads of ships in large amplitude regular and irregular waves. The solution is obtained in the time domain using convolution to account for the memory effects related to the free surface oscillations. The radiation forces and diffraction exciting forces are kept linear. The buoyancy forces (hydrostatic and Froude-Krylov) are evaluated over the instantaneous wetted surface of the hull to account for the large amplitude motions and hull flare. The linear and non-linear terms are used in the equations of motion to compute the motions and structural loads. The formulation is presented in detail in Fonseca and Guedes Soares [9]. In the present paper the predictions of the method are compared with experimental results and show a generally good agreement.

2. FORMULATION

A coordinate system fixed with respect to the mean position of the ship is defined, $X=(x,y,z)$, with z in the vertical upward direction and passing through the centre of gravity of the ship, x along the longitudinal

direction of the ship and pointing to the bow, and y perpendicular to the later and in the port direction. The origin is in the plane of the undisturbed free surface.

Considering a ship advancing in waves and oscillating as an unrestrained rigid body, the oscillatory motions will consist of three translations and three rotations. Although the theory is valid for arbitrary headings relative to waves, the present work is restricted to head waves, thus the oscillatory motions to be studied are the heave displacement and the pitch rotation.

For the solution of the hydrodynamic problem the flow is assumed inviscid, the hull is slender the forward speed is small and the amplitudes of the incident waves and unsteady motions are small enough.

2.1 Motions

Equating the hydrodynamic external forces to the mass internal forces one obtains the equations of motion. These equations, which have nonlinear terms, are solved in the time domain by a numerical procedure. For heave and pitch the equations are:

$$\left(M + A_{33}^{\infty}\right)\ddot{\xi}_3(t) + \int_{-\infty}^{t}\left[K_{33}^{m}(t-\tau)\dot{\xi}_3(\tau)\right]d\tau +$$

$$C_{33}^{m}\xi_3(t) + A_{35}^{\infty}\ddot{\xi}_5(t) + \int_{-\infty}^{t}\left[K_{35}^{m}(t-\tau)\dot{\xi}_5(\tau)\right]d\tau \quad (1)$$

$$+ C_{35}^{m}\xi_5(t) + F_3^{H}(t) - Mg = F_3^{E}(t)$$

$$\left(I_{55} + A_{55}^{\infty}\right)\ddot{\xi}_5(t) + \int_{-\infty}^{t}\left[K_{55}^{m}(t-\tau)\dot{\xi}_5(\tau)\right]d\tau +$$

$$C_{55}^{m}\xi_5(t) + A_{53}^{\infty}\ddot{\xi}_3(t) + \int_{-\infty}^{t}\left[K_{53}^{m}(t-\tau)\dot{\xi}_3(\tau)\right]d\tau \quad (2)$$

$$+ C_{53}^{m}\xi_3(t) + F_5^{H}(t) = F_5^{E}(t)$$

where ξ_3 and ξ_5 represent respectively the heave and pitch motions and the dots over the symbols represent differentiation with respect to time. M is the ship mass, g is acceleration of gravity and I_{55} represent the ship inertia about the y-axis.

The hydrostatic force and moment, F_3^{H} and F_5^{H}, are calculated at each time step by integration of the hydrostatic pressure over the wetted hull under the undisturbed wave profile.

The exciting forces due to the incident waves, F_3^E and F_5^E, are decomposed into a diffraction part and the Froude-Krilov part.

The diffraction part, which is related to the scattering of the incident wave field due to the presence of the moving ship, is kept linear. It results from the solution of the hydrodynamic problem of the ship advancing with constant speed through the incident waves and restrained at her mean position.

Since this is a linear problem and the excitation is known a priori, it can be solved in the frequency domain and the resulting transfer functions be used to generate a time history of the diffraction heave force and pitch moment. The transfer functions are calculated by a strip method.

In the case of regular waves the representation in the time domain of the diffraction forces is:

$$F_K^D(t) = \zeta^a d_k^a \cos(\omega_e t - \theta_k) , \quad k = 3,5 \quad (3)$$

where the indices 3, 5 stand for heave force and pitch moment respectively, d_k^a is the amplitude of the diffraction force or moment due a unit amplitude wave, ω_e is the encounter frequency between the ship and the waves and θ_k is the phase angle representing the delay of the force or moment.

The Froude-Krilov part is related to the incident wave potential, and results from the integration at each time step of the associated pressure over the wetted surface of the hull under the undisturbed wave profile. For regular waves travelling on the negative x-direction, the Froude-Krilov pressure is:

$$p^I(t,x) = -\rho g \zeta^a e^{kz} \cos(\omega_e t + kx) \quad (4)$$

where ρ represents the density of the fluid, g is the gravity acceleration and k is the wave number. It is assumed that the wave potential above the still water line is given by the expression representing the linear wave potential of gravity waves under the still water line.

The Froude-Krylov as well as the hydrostatic pressures are integrated over the whole wetted cross section contour, which includes the deck when the wave is above the deck. Thus the effects of the water on deck are partially accounted, although in a simplistic way that represent only hydrostatic and incident wave components. The inertial effects, the turbulence and the speed effects are not accounted.

The radiation forces are represented in the time domain by infinite frequency added masses, A_{kj}^∞, radiation restoring coefficients, C_{kj}^m, and convolution integrals of memory functions, $K_{kj}^m(t)$. The radiation restoring forces, associated with the restoring coefficients, represent a correction to the hydrodynamic steady forces acting on the ship due to the steady flow.

The convolution integrals represent the effects of the whole past history of the motion accounting for the memory effects due to the radiated waves.

The memory functions and the radiation restoring coefficients are obtained by relating the radiation forces in the time domain and in the frequency domain by means of Fourier analysis:

$$K_{kj}^m(t) = \frac{2}{\pi} \int_0^\infty \{B_{kj}(\omega)\cos\omega t\} d\omega \quad (5)$$

$$C_{kj}^m = \omega^2[A_{kj}^\infty - A_{kj}(\omega)] - \omega \int_0^\infty K_{kj}^m(\tau)\sin(\omega\tau)d\tau \quad (6)$$

in these equations $A_{kj}(\omega)$ and $B_{kj}(\omega)$ represent frequency dependent ship added masses and damping coefficients, which are calculated using a strip method.

Details of the derivation of A_{kj}^∞, C_{kj}^m and $K_{kj}^m(t)$ can be found in Fonseca and Guedes Soares (1998).

2.2 Dynamic Loads

The dynamic loads at a cross section are the difference between the inertia forces (or moments) and the sum of the hydrodynamic forces (or moments) acting on the part of the hull forward of that section. In the case of vertical loads, the vertical shear force and vertical bending moment are given respectively by:

$$V_3(t) = I_3(t) - R_3(t) - D_3(t) - K_3(t) - H_3(t) \quad (7)$$

$$M_5(t) = I_5(t) - R_5(t) - D_5(t) - K_5(t) - H_5(t) \quad (8)$$

where I_k represent the vertical inertia force (or moment) associated with the ship mass forward of the cross section under study. As assumed for the calculation of the ship motions, the radiation (R_3, R_5) and diffraction (D_3, D_5) hydrodynamic contributions for the loads are linear, and the Froude-Krylov (K_3, K_5) and hydrostatic (H_3, H_5) contributions are non-linear since they are calculated over the "exact" hull wetted surface at each time step.

The convention for the loads is such that the sagging moment is negative and the hogging moment is positive. The exciting moment over the portion of the hull forward of the cross under study section is given by the diffraction and Froude-Krylov contributions, where the first is linear and the second is nonlinear.

The hydrostatic contribution from each cross section is given by the difference between the static equilibrium hydrostatic force and the actual hydrostatic force calculated on the "exact" wetted surface.

The radiation vertical shear force and bending moment may be decomposed into contributions from the forced heave and pitch motions:

$$R_3(t) = R_{33}(t) + R_{35}(t) \qquad (9)$$

$$R_5(t) = R_{53}(t) + R_{55}(t) \qquad (10)$$

these contributions for bending moment need special care since, again, these forces have a time dependency of the history of the fluid motion. As used for the calculation of the global radiation forces, the radiation contribution for the vertical bending moment is given by infinite added mass terms, convolution integrals of memory functions and radiation restoring coefficients:

$$R_{33}(t) = -\int_{L'} a_{33}^{\infty} \ddot{\xi}_3(t) dx - \int_{-\infty}^{t} \left\{ K_{33}^{l}(t-\tau)\dot{\xi}_3(\tau) \right\} d\tau - C_{33}^{l} \xi_3(t) \qquad (11)$$

$$R_{35}(t) = -\int_{L'} a_{33}^{\infty} x \ddot{\xi}_5(t) dx - \int_{-\infty}^{t} \left\{ K_{35}^{l}(t-\tau)\dot{\xi}_5(\tau) \right\} d\tau - C_{35}^{l} \xi_5(t) \qquad (12)$$

$$R_{53}(t) = -\int_{L'} \left\{ (x-x^*) a_{33}^{\infty} \ddot{\xi}_3(t) \right\} dx - \int_{-\infty}^{t} \left\{ K_{53}^{l}(t-\tau)\dot{\xi}_3(\tau) \right\} d\tau - C_{53}^{l} \xi_3(t) \qquad (13)$$

$$R_{55}(t) = -\int_{L'} \left\{ (x-x^*) x a_{33}^{\infty} \ddot{\xi}_5(t) \right\} dx - \int_{-\infty}^{t} \left\{ K_{55}^{l}(t-\tau)\dot{\xi}_5(\tau) \right\} d\tau - C_{55}^{l} \xi_5(t) \qquad (14)$$

where a_{33}^{∞} represent the sectional infinite frequency added mass for heave oscillation, and integration of these coefficients are over the length of the ship forward of section in question 'x^*'. K_{kj}^{l} and C_{kj}^{l} are the memory functions and radiation restoring coefficients associated with the loads, and the superscript l is used to distinguish these coefficients from those used in the equations of motion (1) and (2).

3. EXPERIMENTAL PROGRAMME

An experimental programme was carried out at the El Pardo Model Basin (CEHIPAR) in Madrid. The objective was to obtain a reliable data base of systematic measurements of motions and sectional loads in a wide range of wave lengths and wave heights, extending into the nonlinear range. The authors were particularly interested in the asymmetry of the responses, specially the loads, and the influence of the wave steepness on the nonlinear behaviour. This way the tests were carried out in regular head waves, for two Froude numbers (Fn=0.15 and Fn=0.25), and for several wave lengths along the range of interest. For each wave length four wave heights were tested.

The tank has a length of 145m, width of 30m and depth of 5m. The wave generator is composed of 60 flaps controlled individually and hydraulic driven. It is capable of generating regular and irregular long-crested waves, as well as short-crested irregular waves.

A three-part segmented model of the S175 containership was wooden manufactured. The cuts were placed at midship and a quarter from the bow. A rigid bar fixed to the model along its length connected the segments. Strain gauges were mounted on the bar at midship and a quarter from bow, prepared to measure the vertical shear force and bending moment at these positions. The vertical motions at three positions along the ship length were measured by potentiometers.

Accelerometers were installed at stations 20 (bow), 15, 10, 5 and 0. The relative motions were measured by capacitive wave gauges fixed to the sides of the model at stations 20, 18, 10 and 2. In addition, one more wave gauge was installed on the bow deck to identify and measure the height of water on deck. Finally, a wave gauge measured the wave elevation close to the bow of the model.

4. DISCUSSION OF NUMERICAL AND EXPERIMENTAL RESULTS

In this section, part of the experimental results are presented and compared with corresponding numerical results computed using the formulation described. All results are for head regular waves and for a Froude number of 0.25.

All results representing the experiments are obtained by subtracting from the measured responses, the steady responses with the model running with the same Froude number in still water.

Figure 1 represents the time history of the measurements together with the numerical simulations of: wave elevation at the ship centre of gravity, heave motion, pitch motion, relative motion at the forward perpendicular and vertical bending moment at midship (VBM). The measurements are represented by the dashed lines and the numerical results by the solid lines. The VBM is non-dimensionalized by $\rho g L_{pp}^2 B \zeta_a$ and the convention is negative moment for sagging condition. The regular

wave has an amplitude of 1.96m, a ratio wave length / ship length given by $L_w / L_{pp} = 1.6$ and a ratio wave height / wave length of approximately $H_w / L_{pp} = 1/80$.

The graphs show that the delay of the calculated responses with respect to the wave elevation is smaller than that of the measured responses.

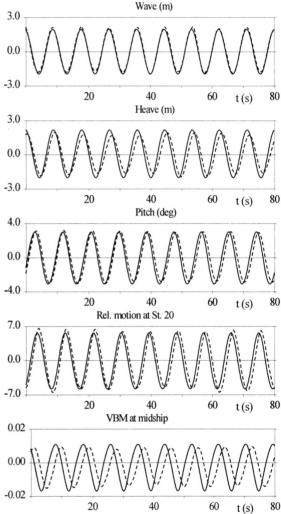

Fig. 1 Measurements and numerical simulations. Head waves with $\zeta_a = 1.96m$ and $L_W / L_{pp} = 1.6$

The results of heave and pitch for all wave frequencies are summarised in the form of transfer functions and presented in figures 2 and 3. Three sets of experimental results are presented corresponding to the wave steepnesses of $H_w / L_w = 1/80$, $1/60$ and $1/40$. The calculated results are for the wave steepness of 1/80, and the dashed lines represent linear results while the solid lines represent nonlinear results.

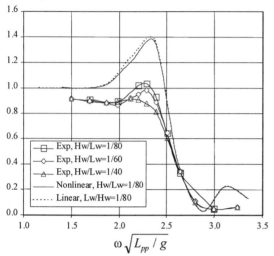

Fig. 2 Heave amplitude $\left(x_3^a / \zeta^a \right)$

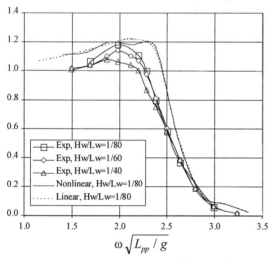

Fig. 3 Pitch amplitude $\left(x_5^a / k\zeta^a \right)$

The experimental results show that the nondimensional amplitudes of the heave and pitch motions near the resonance are dependent of the wave steepness. In fact these amplitudes tend to decrease with the wave steepness. For higher and lower frequencies the heave and pitch amplitudes are independent of the wave steepness.

Experimental results with the S175 model obtained by several organisations were reported by the 15th and 16th ITTC Seakeeping Committee, with a slightly higher Froude number. In fact the present motion calculations agree better with the experiments reported by the ITTC than with the ones presented here. A more indept analysis of the experiments is necessary to clarify the reason of the differences.

The curves of heave numerical results follow the tendency of the experiments, however close to the ressonance frequency the predictions largely overestimate the responses. It is believed that there are two main reasons for this; the damping due to radiation of waves alone is not sufficient to represent the real damping on the model, and the interaction between the steady and unsteady flow is represented in a simplistic way on the numerical model. As for the pitch, the agreement between predictions and experiments is good, although the theory tends to overpredict the responses. For the vertical motions the linear and nonlinear predictions are very simmilar.

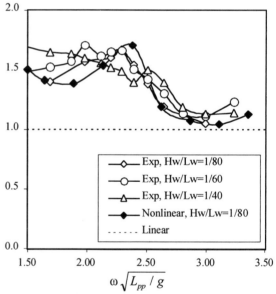

Fig. 4 Ratio between sagging and hogging amplitudes for vertical bending moment at midship

One of the objectives of the experimental programme was to assess the importance of the assymetry of the positive and negative peaks of the vertical bending moment. Fig. 4 represents the ratio between sagging and hogging amplitudes (Sag/Hog).

The ratio between sagging and hogging amplitudes is presented in figure 4. The experimental results are plotted together with the nonlinear and linear results. One may observe that for long and medium length waves the sagging amplitudes are 60 to 70 % higher than the hogging amplitudes. For shorter waves the ratio Sag / Hog tends to one. This relation between the amplitudes is well predicted by the nonlinear method.

5. CONCLUSIONS

The quality of the non-linear predictions of a time domain strip method was assessed by comparing the results with results obtained experimentally.

A practical engineering approach was followed to solve the problem of large amplitude motions and induced loads in head waves. The radiation and diffraction forces are kept linear, and it is assumed that the predominant nonlinear contribution for the responses is due to the buoyancy forces.

The experimental results show that the nondimensional amplitudes of the heave and pitch motions tend to reduce with the wave amplitude near the resonance frequency. The numerical predictions tend to overestimate the motions. As for the loads the experiments showed that the ratio between the sagging and hogging peaks reaches values of 1.6 – 1.7. This asymmetry is well predicted by the numerical model.

6. ACKNOWLEDGEMENTS

The experimental work was developed in the project "Nonlinear Responses of Ships in Waves" which has been funded by the Commission of the European Communities, through the Spanwave Project under contract ERBFMGECT950074. The authors are indebted to the personal of the Seakeeping Laboratory of CEHIPAR for their valuable assistance during the course of the tests.

The first author is grateful to the Fundação para a Ciência e a Tecnologia for having provided a scholarship which has supported him during the development of the theoretical work.

7. REFERENCES

1. Jensen, J.J. and Pedersen, P.T. (1979),"Wave-Induced Bending Moments in Ships - a Quadratic Theory", *Trans. RINA*, Vol. 121, pp. 151-165.
2. Dalzell, J.F. (1964), "An investigation of midship bending moments experienced in extreme regular waves by a Mariner type ship and three variants", Report SSC-155, *Ship Structure Committee*.
3. Watanabe, I., Ueno, M. and Sawada, H. (1898), "Effects of Bow Flare Shape to the Wave Loads of a container ship", *Journal of the Society of Naval Arch. of Japan*, Vol. 166, pp. 259-266.
4. Salvesen, N., Tuck, E. O., and Faltinsen, O. (1970), "Ship motions and sea loads", *Trans. Soc. Nav. Arch. Mar. Eng.*, Vol. 78, pp 250-287.
5. Chang, M. S. (1977), "Computations of three-dimensional ship motions with forward speed", *Proc. 2nd Int. Numer. Ship Hydrodyn.*, pp. 124-135. University of California, Berkeley.
6. Lin, W.M., Zhang, S., and Yue, D.K.P. (1996), "Linear and nonlinear analysis of motions and loads of a ship with forward speed in large-amplitude waves", Proc., *11th Int. Workshop on Water Waves and Floating Bodies*, Hamburg, Germany.
7. Kring, D.C., Huang, Y.F., Sclavounos, P.D., Vada, T., and Braathen A. (1996), "Nonlinear ship motions and wave-induced loads by a Rankine panel method", *Proceedings 21st Symp.. on Naval Hydrodynamics*, Trondheim.
8. Beck, R., Cao, Y., Scorpio, S., Schultz, W., (1994), "Nonlinear Ship Motions Computations Using the Desingularized Method", *Proc, 20th Symp. on Nav. Hyd.*, Santa Barbara, California.
9. Fonseca, N. and Guedes Soares, C. (1998), "Time-Domain Analysis of Large-Amplitude Vertical Motions and Wave Loads", *Journal of Ship Research,*Vol.42, No. 2, pp. 100-113.

Wave-Induced Motions and Loads for a Tanker. Calculations and Model Tests

J. Lundgren[a], M.C. Cheung[b] and B.L. Hutchison[c]

[a] SSPA Maritime Consulting AB, P.O. Box 24001, SE-400 22 Göteborg, Sweden
[b] MCA Engineers, Inc, 2960 Airway Ave, Suite A-103, Costa Mesa, CA 92626, USA
[c] GLOSTEN Associates, Inc., 600 Mutual Life Building, 605 First Avenue, Seattle, WA 98104-2224, USA

ABSTRACT

Results of theoretical calculations and model tests in regular waves and irregular seas for a tanker are presented. Calculations and measurements of motions, accelerations, vertical and horizontal bending moments, vertical shear force, hydrodynamic water pressure along the side hull, bottom slamming, water on deck and added resistance were carried out.

1. INTRODUCTION

ARCO Marine, Inc. (a wholly owned subsidiary of the Atlantic Richfield Company) engaged SSPA to assist in the methodical development and optimization of the hull form for their new Millennium Class twin-screw, twin-rudder, double hull tanker now under construction at Avondale Shipyard and intended for service between the Trans-Alaska Pipeline terminal located at Valdez, Alaska and US West Coast ports in Washington and California. The development program included computational fluid dynamic (CFD) analysis using SHIPFLOW, resistance and self-propulsion tests, paint streak flow visualization and wake surveys, all at full load and ballast drafts, at a scale of 1:34.118 in the deep water towing tank at SSPA. Following tests of the initial hull form minor modifications were introduced in the stern to shift the LCB, improve the wake in the propeller plane, and reduce flat surfaces at the stern that may have been vulnerable to slamming. This modified hull was then fully re-tested, verifying that the objectives were achieved. Following the design of wake adapted propellers this 1:34.118 model was installed in the

large cavitation tunnel at SSPA and the design propellers were tested for cavitation performance, including propeller induced pressure on the hull.

A second model was then constructed of the modified hull at a scale of 1:48.33 for testing in the Maritime Dynamics Laboratory (MDL) at SSPA. Testing programs in the MDL included extensive free model maneuvering tests and the seakeeping tests which are the subject of this paper. Key objectives of the model test program were to measure and observe performance to determine operability in seas characterized by a 5.5 m significant wave height, verify survivability behavior in head seas characterized by a 13.0 m significant wave height, and to collect data that would be useful in verifying analytical seakeeping analysis tools to be used for operability analysis (SCORES) and for the ABS dynamic loads analysis (SPLASH).

Critical to the ABS dynamic loads analysis (DLA), and particularly to the spectral fatigue analysis component of the DLA process, are the unsteady hydrodynamic pressures acting on the hull. Experience with tankers in the Trans-Alaska Pipeline trade (TAPS) has been that an abnormal number of structural failures are located near the waterlines on the weather side - near the load waterline on the starboard side and near the ballast waterline on the port side. This is because the weather predominates from the south-west along this essentially north-south trade route. The tanker travels from north to south loaded and from south to north in ballast. Thus, the deep loadline on the starboard side is

504

exposed when loaded and the port side is exposed in ballast. The failures are clearly the consequence of combinations of primary, secondary and tertiary loading. The seakeeping test program included measurement of pressure along the side at nine locations which were used to correlate and verify the predictions from SPLASH. Primary loads were also measured during the tests and these were correlated with both SCORES and SPLASH.

2. SHIP PARTICULARS

A model in scale 1:48.33 of an 125 000 dwt twin skeg tanker was manufactured. The main particulars of the loaded ship are shown in Table 1.

Length, between pp	L	m	258.16
Beam	Bmax	m	46.2
Draught	Tf	m	16.0
Draught	Ta	m	16.0
Block coefficient	CB	-	0.80
Displacement	∇	m³	152660
Roll gyration	Kxx/B	-	0.30
Pitch gyration	Kyy/L	-	0.24
Vert. centre of gravity	KG	m	13.7
Metacentric height, corr	GM	m	4.2

Table 1. Main particulars

3. TEST ARRANGEMENT AND INSTRUMENTATION

The tests were carried out in the Maritime Dynamics Laboratory (MDL) at SSPA. The MDL has a basin with the dimensions 88 m x 39 m with variable water depth up to 3 m. At the tests the water depth was 2.6 m. Wave generators for regular and irregular long-crested waves are installed on two perpendicular sides of the basin. A multi-motion carriage, used for data logging and model control, spans the whole basin.

The tests were carried out with a free-running (free-sailing) model. In a free-running test the model is accelerated by the carriage and at the proper speed the model is disconnected from the carriage and continues self-propelled and controlled to keep the

course with an auto-pilot. The model is now free to move in all degrees of freedom.

The following parameters were measured:

■ surge, sway, heave, roll, pitch, and yaw
■ vertical and horizontal bending moment at L/2
■ accelerations in x-, y-, and z-directions
■ vertical shear force at L/2
■ relative motions
■ slamming pressures
■ water pressures along the side hull

4. TESTS IN REGULAR WAVES

Test were carried out in regular waves with wave periods from 7 seconds to 19 seconds and in six wave directions: 180° (head sea), 150°,120°, 90°, 60° and 30°. The wave height was about 5 meters at all wave periods. For one wave period, 15 seconds in head sea, tests were also carried out for two additional wave heights, 2.5 m and 7.5 m, to study the linearity.

The results of the tests are compared with results from theoretical calculations with the SCORES program for the vertical bending moment amidships, see Figs. 1a-c. The agreement between measurements and calculations is very good.

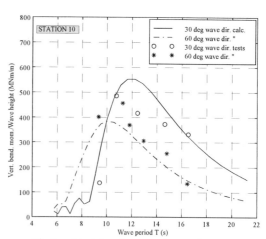

Fig 1a: Vert. bend. mom. in regular waves
Full load; 15 knots

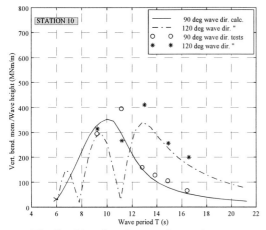

Fig 1b: Vert. bend. mom. in regular waves
Full load; 15 knots

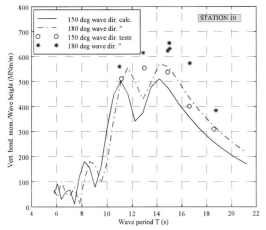

Fig 1c: Vert. bend. mom. in regular waves
Full load; 15 knots

5. TESTS IN IRREGULAR WAVES

Tests were carried out in irregular waves composed of two wave spectra generated simultaneously and perpendicular to each other. Both spectra were ITTC (International Towing Tank Conference) standard spectra with the following characteristics:

Significant wave height H1/3 (m)	Zero-crossing wave period Tz (sec)	Spectrum peak period Tp (sec)
4.4	9.94	14.0
3.3	9.94	14.0

The significant wave height for the combined irregular sea was 5.5 m.

The ship speed was about 15 knots.

The following combinations of irregular waves were used:

Main wave dir Degrees	H1/3 m	Perpendic. wave degrees	H1/3 m
30	4.4	120	3.3
60	”	330	”
90	”	0	”
120	”	30	”
150	”	240	”
180	”	270	”

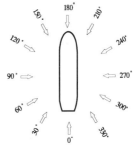

The results of the measurements in combined seas are shown in Figs. 2a-f. In these figures are also shown results of calculations in long-crested irregular waves, with significant wave height H1/3 5.5 m, based on calculations in regular waves.

Furthermore, these figures also show results of theoretical calculations in combined seas by adding the squared calculated significant responses in the two wave directions:

$$Heave_{1/3}(180° + 270°) = \sqrt{Heave_{1/3}^2(180°) + Heave_{1/3}^2(90°)}$$

The results show that this method can be used to estimate motions and loads in combined seas.

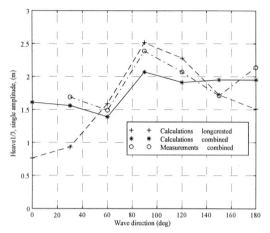

Fig 2a: Heave from combined seas

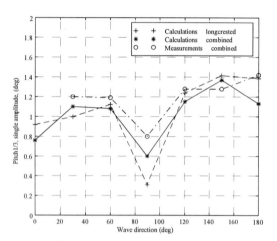

Fig 2b: Pitch angle from combined seas

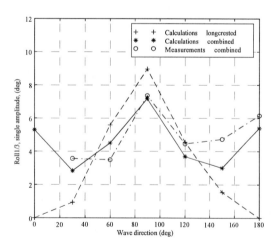

Fig 2c: Roll angle from combined seas

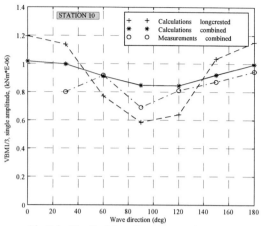

Fig 2d: Vertical bending moment
from combined seas

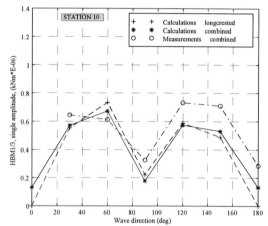

Fig 2e: Horizontal bending moment
from combined seas

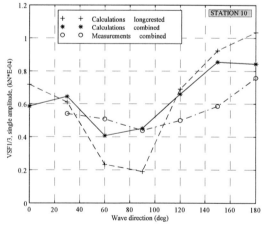

Fig 2f: Vertical shear force
from combined seas

6. TESTS IN SURVIVAL HEAD SEAS

Tests were carried out in survival head seas, JONSWAP spectrum, γ=3.3, significant wave height 13.0 m, Tp= 14.0 sec at a forward speed of 5 knots. The results are shown in the table below.

Significant values, single amplitude

		Measured	Calculated
Heave	M	2.61	2.88
Pitch	Sec	3.71	3.66
VBM	kNm*E+06	3.02	3.18
VSF	kN*E+-04	2.05	2.40

7. HYDRODYNAMIC ANALYSIS

SPLASH (Ref. 1 and 2) is a hydrodynamic numerical computational code that uses potential flow theory to describe the steady and non-steady linear, inviscid flow about three-dimensional solid bodies (panels). The code has been used extensively for a number of America's Cup yacht designs for various syndicates since 1987, and has been repeatedly verified by model testing. In 1994, the SPLASH code was applied for the first time to a 165,000 deadweight ton (dwt) single-hull tanker operating in the TAPS trade. Fatigue estimates on the side shell structure were improved by using the pressure profiles generated by SPLASH. The SPLASH-augmented structural analysis explained why damage on the starboard side was located higher than on the port side, corresponding to differences in the southbound and northbound draft under predominately south-westerly waves. Since then, several more tanker classes have been analyzed using SPLASH. Unfortunately, none of the tanker analyses could be confirmed by model testing until now.

From classical hydrodynamics, the fluid flow around a solid boundary is assumed to be incompressible, inviscid and irrotational, and therefore can be characterised by the velocity potential. Figure 3

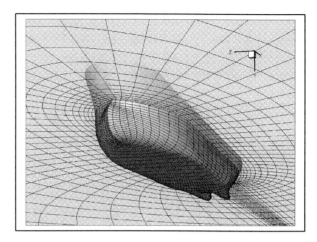

Fig 3. SPLASH model at loaded draft

shows the fluid model of the ARCO 125,000 dwt tanker with the element grid painted on the free surface and the submerged hull. The bulb bow and the twin skeg are clearly visible. From the continuity condition, the potential must satisfy the Laplace equation:

$$\nabla^2 \Phi = 0$$

where Φ = Nondimensional Velocity Potential
$= V / V_B$
V_B = Steady Forward Speed.

The equilibrium of the fluid system is governed by the Bernoulli's equation. The steady viscous drag is estimated by a post-processor using a viscous "stripping" approach, which is a semi-empirical handbook-type of viscous drag estimate. By iterative process, the strength and distribution of the point sources and doublets (representing the velocity potential that satisfies all boundary conditions) can be computed. The mean waterline is determined from each iteration by adjusting the sink and trim. Upon updating the sink and trim, a new fluid model mesh is generated. The final acceptance of the solution is such that the equilibrium of heave and pitch must achieve pre-determined convergence tolerances that include varying hydrostatic forces and the updated hydrodynamic forces.

The pressure coefficient C_p at various points along the body can be calculated by the Bernoulli equation:

$$C_p = \frac{p \: - \: p\infty}{\frac{1}{2}\,\rho V_B{}^2} = 1 - \nabla \Phi \bullet \nabla \Phi - (2/V_B)\Phi_t + (2g/V_B{}^2)\,y$$

where: Φ_t = Partial derivative of Φ with respect to time.

y = Vertical co-ordinate, positive down from free surface.

In general, the motion responses calculated by SPLASH agree well with the model test results as reported in Section 4 of this paper. What is rare in model testing is the measurement of Relative Motion (RM, freeboard) and hydrodynamic pressure (P) along the side shell. SSPA installed six RM gauges from stern to bow. The one gauge of most interest in this report is RM4 located near amidships on the port side. SSPA also installed 14 pressure sensors, including several at the bottom of the stern and bottom of the bow to measure possible slam pressure. Nine pressure sensors were arranged to form a 3 x 3 matrix around the mid-ship area. The highest sensor was located slightly below the loaded draft. Three sensors of particular interest to this report are P13, P10 and P7, arranged vertically on the port side, closest to amidships and RM4. In the loaded configuration, the vessel floats even-keeled in still water at a 16-meter draft. P13 is 12 meters above the base line, P10 is 7.75 meters ABL and P7 is 3.5 meters ABL.

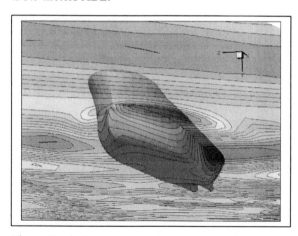

Fig 4. Typical pressure contours / oblique waves

Figure 4 illustrates a typical pressure contour at the free surface and the hull body computed by SPLASH in an oblique wave condition.

Figures 5 and 6 summarise the Pressure RAO profiles along a station near amidships, calculated by SPLASH for an 11 second bow wave (180°) and beam wave (90°). The pressures measured by sensors P13, P10 and P7 are also shown for comparison.

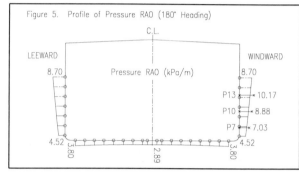

Fig 5. Profile of Pressure RAO (180° Heading)

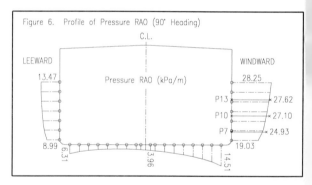

Fig 6. Profile of Pressure RAO (90° Heading)

The general trend agrees reasonably well, but the sensor results are slightly higher than the SPLASH predictions. The pressure profiles for oblique and beam seas are asymmetrical from side to side. This is because the velocity potentials on both the windward and leeward sides are the sum of the incident, radiation and diffraction wave potentials. The magnitude and phase of the diffraction potential is such that it increases the wave amplitude and pressures on the windward side, while the diffraction potential acts to reduce and cancel the incident wave potential on the leeward side, thereby reducing wave amplitudes and pressures on that side.

The typical procedure for analysing hull structural strength, buckling and fatigue is to map the hydrodynamic pressure profile directly onto the finite element structural model. This approach is far more rigorous than the quasi-static approach and produces dynamic equilibrium for the hull body as a whole, but is incorrect when evaluating the side shell structure at the waterline. Unless the mean stress is zero, it leads to errors for two reasons. The first reason is the "Total Stress". The dynamic approach should be performed with minimum and maximum total external pressure separately. For each case, the hydrostatic pressure is included. This procedure would get the proper maximum stress, mean stress, and stress-range. Although a number of procedures have been developed in the mechanical and naval engineering communities for dealing with the effect of mean stress on fatigue prediction, there is no uniformly accepted procedure within the marine community. Most approaches provide for cyclic stress reduction if the mean stress is compression.

The second reason is "compensation at the splash zone." Since suction pressures do not occur at the side shell when the wave drops below the mean draft, directly applying the dynamic profile to the side shell underestimates the pressure cycle above the mean draft line and overestimates it below. Depending on vessel characteristics, the wave period, and wave direction, the height fluctuation of the wetted surface at the side shell could be several times the wave amplitude. This makes the splash zone compensation very critical.

Figure 7 illustrates the pressure profile assuming a 3-meter amplitude / 11-second bow wave. The vertical axis represents the side shell of the hull with the origin "O" at the base line. Point "A" is at the mean draft line.

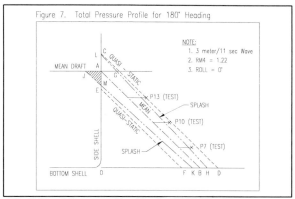

Fig.7 Total Pressure Profile for 180° Heading

Line AB represents the mean hydrostatic pressure. Lines CD and EF represent the quasi-static pressure envelopes without an exponential decay adjustment, assuming the free surface varies 3 meters up and down from "A". Similarly, lines GH and JK represent the hydrodynamic pressure envelopes calculated by SPLASH, superimposed to the mean pressure. The pressure measured in the test at P13, P10 and P7 are bound between the quasi-static and SPLASH envelopes. In this instance, the quasi-static loading is acceptable and qualitatively represents the nature of the splash zone. The SPLASH pressure profile, however, needs to be extrapolated up the side shell from point "G" to point "L" and truncated from point "J" down to point "M". Effectively, the true pressure amplitude at the mean draft is only half of the calculated SPLASH amplitude. The dynamic pressure oscillates from a mean position at the midpoint between "A" and "G", not around point "A" as the SPLASH analysis specifies. The double amplitude of the dynamic pressure is equal to the value of AG and not JG. Because of the omission of the triangle "JAM", the mean pressure and the double amplitude of dynamic pressure below the mean waterline must be modified until below point "M".

510

Similar pressure profiles for beam waves are shown in Figure 8.

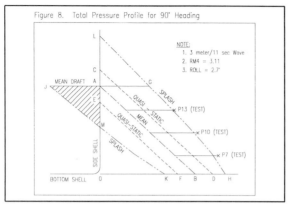

Fig 8. Total pressure Profile for 90° Heading

Using the same nomenclature as defined in Figure 7, the quasi-static envelopes (CD and EF) are unchanged from the previous case. The SPLASH envelopes (GH and JK) are drastically different from those calculated for the bow wave (Figure 7). The test-measured values are very close to those calculated by SPLASH. The quasi-static representation in this condition is clearly inaccurate. The correction above and below the draft line for SPLASH becomes crucial because of the magnitude of the pressure and the broadness of the region affected. The RM4 value measured at this location is 3.11. That is to say that the wetted surface moves up and down the side shell more than 3 times the wave amplitude as represented by points "L" and "M".

Figures 9 and 10 show the pressure RAOs at P13 as functions of wave period for head and beam seas. The maximum Pressure RAO in each of these headings could be as much as eight times their corresponding minimum values. This can be due to motion resonance occurring with maximum wave reflection versus the other extreme with motion cancellation. Generally, the correlation of the results between SPLASH and the tests is extremely good.

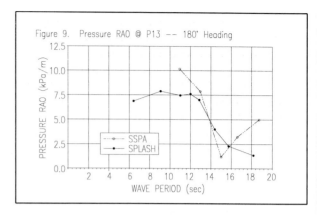

Fig 9. Pressure RAO @P13 - 180° Heading

The large fluctuation of the pressure profiles confirms the importance of using a 3-dimensional fluid dynamic code, such as SPLASH, to capture the true magnitude of side shell pressure loading and therefore calculates adequate side shell scantlings.

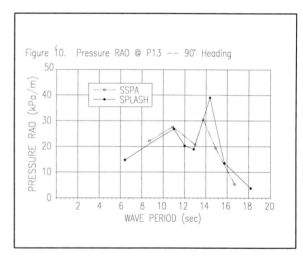

Fig 10. Pressure RAO @P13 - 90° Heading

8. CONCLUSIONS

A modern twin-screw, twin-rudder, double hull tanker hull has been developed and optimised using CFD and physical model tests. Seakeeping model tests have been correlated with the strip theory program SCORES and the 3-D panel code SPLASH. Good agreement in motions and primary loads was found with both programs. Correlations were also carried out for pressures on the hull, comparing SPLASH with pressure measurements from the model tests. The agreement was quite good, verifying the suitability of SPLASH for dynamic loads analysis, including the secondary and tertiary loading processes which are critical to the spectral fatigue analysis of tanker side structure. The physical model tests also verified the linearity with wave height of primary loads for this tanker hull.

9. REFERENCES

1. "SPLASH: FREE SURFACE PANEL CODE COMPUTER PROGRAM", Developed by South Bay Simulations, Inc., Babylon, NY 11702

2. "SPLASH Nonlinear and Unsteady Free-Surface Analysis Code for Grand Prix Yacht Design", Bruce Rosen and Joseph Laiosa, 13th Chesapeake Sailing Yacht Symposium.

ACKNOWLEDGEMENT

The authors would like to recognise ARCO Marine Inc. for granting approval to publish the findings. Particularly, our heartfelt thanks goes out to AMI's management, Messrs. John Sullivan, Bob Levine and Jim Read, for their genuine support and encouragement. Suggestions, comments and critiques from the staff of SSPA, Glosten, MCA and Mr. Bruce Rosen are deeply appreciated.

Practical Design of Ships and Mobile Units
M.W.C. Oosterveld and S.G. Tan, editors.

Practical Time Domain Simulator of Wave Loads on a Ship in Multi-Directional Waves

Hisaaki Maeda and Chang Kyu Rheem

Institute of Industrial Science, University of Tokyo
7-22-1 Roppongi, Minatoku, Tokyo 106, Japan

This paper describes the reliable and practical numerical simlator for ship motions and wave loads in general multi-directional ocean waves. The simulator calculates and displays the 6 degree freedom ship motions, wave loads such as vertical and horizontal shear force and bending moment or torsional moment, and pressure distribution on an arbitrary section. The analytical method is based on the consistent Strip Method. The time domain hydrodynamic forces are expressed by the convolution integral of the impulse response function and the corresponding responses.

The simulator of the multi-directional waves is based on the superposition of component sinusoidal waves coming from various direction which correspond to a given directional wave spectrum. For the sake of avoiding unrealistic steep waves, an wave breaking process is introduced in the incoming multi-directional wave simulator the criteria of which is the vertical acceleration of water particle less than 0.5g. This simulator can be applicable to the preliminary design for bulk carriers, oil skimmers etc., and to the assistance of a ship operator in severe sea conditions.

1. INTRODUCTION

The analytical method of advanced ship motion in time domain was proposed by Cummins and Oglivie in 1960s. However, practical applications of it have been scarce, because it's computation algorism is very complicated. Saito & Azuma, and Fukazawa reported some time domain computation results of ship motions and wave loads in uni-directional regular waves based on the Strip Method. However, these applications had some limitations. Saito and Azuma treated only linear ship motion, and Fukazawa took only into consideration non-linear effect of hydrodynamic force due to the change of submersed section area.

In this paper, the time domain simulator of 6 degree freedom ship motions, wave loads such as vertical and horizontal shear force and bending moment or torsional moment, and pressure distribution on an arbitrary section are described. The analytical method is based on the consistent Strip Method developed in Japan by several researchers. The time domain hydrodynamic forces (radiation and wave exciting forces) are expressed by the convolution integral of the impulse response function and the corresponding responses. The impulse response functions are derived by the Fourier transformation of the frequency domain response functions.

The simulator of the multi-directional waves is based on the superposition of component sinusoidal waves coming from various directions which correspond to a given directional wave spectrum. For the sake of avoiding extraordinary steep waves, an wave breaking process is introduced in the incoming multi-dirctional wave simulator the criteria of which is the vertical acceleration of water particle less than 0.5 g.

This simulator can be applicable to the preliminary design for various kinds of ships such as bulk carriers, oil skimmers etc, and to the assisstance of a ship operator in sevear sea conditions by the computer graphics and the related video system.

2. FORMULATIONS

The co-ordinate system is the right hand, body fixed Cartisian system. The impulse response functions are derived from the Fourier transformation of the corresponding frequency domain response functions.

The time domain fundamental equation of motion of a ship in waves is expressed as follows;

$$\sum_{j=1}^{6}\left[\begin{array}{l}A_{ij}\ddot{\varsigma}_j(t)+B_{ij}\dot{\varsigma}_j(t)+C_{ij}\varsigma_j(t)\\+Ak_{ij}(t)*\ddot{\varsigma}_j(t)+Bk_{ij}(t)*\dot{\varsigma}_j(t)+Ck_{ij}(t)*\varsigma_j(t)\end{array}\right]$$

$$= h_i(t)*\varsigma_W(t)$$

$$(i=1,2,...,6) \quad (1)$$

where $Ak_{ij}(t)*\ddot{\varsigma}_j(t)$ etc. indicates the convolution integral of the impulse response function and the acceleration of the motion of a ship, respectively, while $\varsigma_j(t)$ denotes the motion amplitude of a ship and $\dot{\varsigma}_j(t)$, $\ddot{\varsigma}_j(t)$ are its corresponding velocity and acceleration, respectively. $h_i(t)$ is the impulse response function of wave exciting force, while $\varsigma_W(t)$ is incident wave amplitude.

Once the motion amplitude is solved, the wave loads on the ship are derived from the forces acted on each strip of the ship, considering the hydrodynamic forces on each strip $f_{xsi}(t)$ as follows;

Shear force

$$f_{si} = \int_{AP}^{x_i} f_{xsi}(t)dx \quad (2)$$

Bending moment

$$m_b(t) = \int_{AP}^{x_i}(x-x_i)f_{xsi}(t)dx \quad (3)$$

Torsional moment

$$m_t(t) =$$

$$\sum_j\left[\begin{array}{l}A_{x4j}\ddot{\varsigma}_j(t)+B_{x4j}\dot{\varsigma}_j(t)+C_{x4j}\varsigma_j(t)\\+Ak_{x4j}(t)*\ddot{\varsigma}_j(t)+Bk_{x4j}(t)*\dot{\varsigma}_j(t)\\+Ck_{x4j}(t)*\varsigma_j(t)\end{array}\right]$$

$$-h_{x4}(t)*\varsigma_W(t)$$

$$(j=2,4,6) \quad (4)$$

The time domain pressure is also derived from the time domain motion of a ship and incident wave amplitude considering the corresponding time domain velocity potential as follows:

$$P(t) =$$

$$\sum_j\left[\begin{array}{l}A_{Pj}\ddot{\varsigma}_j(t)+B_{Pj}\dot{\varsigma}_j(t)+C_{Pj}\varsigma_j(t)\\+Ak_{Pj}(t)*\ddot{\varsigma}_j(t)+Bk_{Pj}(t)*\dot{\varsigma}_j(t)\\+Ck_{Pj}(t)*\varsigma_j(t)\end{array}\right]$$

$$+h_{PW}(t)*\varsigma_W(t)$$

$$(j=2,3,4,5,6) \quad (5)$$

3. TIME DOMAIN SIMULATOR

The authors developed the time domain simulator of responses of a ship in multi-directional waves in which both radiation and diffraction hydrodynamical forces are calculated strictly in time domain and which is applicable to multi-directional ocean waves with any wave spectrum. This method is called as the Convolution Integral Method(C.I. Method).

While, there is another well known time domain simulation method, so called Constant Coefficient Method(C.C. Method), in which constant coefficients of radiation forces are used. The equation of motion for C.C. Method is described as follows;

$$P(t) = \sum_{j=2}^{6}\left[\begin{array}{l}A_{Pj}(\omega_P)\ddot{\varsigma}_j(t)+B_{Pj}(\omega_P)\dot{\varsigma}_j(t)\\+C_{Pj}(\omega_P)\varsigma_j(t)\end{array}\right]$$

$$+h_{PW}(t)*\varsigma_W(t)$$

$$(6)$$

Though this equation is strictly held only in regular waves, this equation is widely used especially in maneuvring field because of simple algorithm.

3.1 Time domain computer graphics simulator of wave loads

The authors also developed the time domain computer graphics simulator of wave loads in multi-directional waves. This kind of simulator must be very useful at a preliminary design stage of ships and must be a good assistant for a ship master in rough seas.

This simulator takes account of the breaking wave criteria in the multi-directional wave simulator. Multi-directional waves with any wave spectrum are available, that is to say, any real irregular ocean waves can be realized on the simulator. Ship motions and multi-directional waves are shown in a CRT, while shear forces, bending moments etc are shown in the same CRT independently. The pressure distribution at any section of a ship can also be drawn on the same screen.

3.2 Validation of time domain simulator

In order to guarantee the validation of the time domain simulator developed here, the comparison with frequency domain results due to the C.C. Method is done in regular waves.

The frequency domain calculated results of pitch, horizontal bending moment at midship, and pressure distribution at midship are shown in Fig. 1 to Fig. 3 respectively. In these figures, Reff(*) indicates the frequency domain numerical results based on the original Strip Method developed by the SR-108 research committee. The generic ship model is the SR-108 container ship.

It is confirmed that the C.I. Method is fairly well and practically enough, even though this method has complicated algorithm comparing with the C. C. Method.

3.3 Validation due to experiment

The experiment of wave loads on a ship in bi-directional irregular waves with ISSC spectrum was carried out in the seakeeping basin of the University of Tokyo. The ship model is a four block back-born type one. Time series of ship motions, wave loads and pressure distribution were measured.

The experimental results of heave, vertical bending moment at midship and pressure at midship portside are shown in Fig. 4 to Fig. 6, respectively, with the corresponding numerical results due to the C. I. Method. Since the comparison of both results is fairly good, then it is confirmed that this simulator is useful in multi-directional irregular waves at least in the linear range.

4. NUMERICAL EXPERIMENT

In order to investigate the advantage of C.I. Method and the effect of directionality of ocean

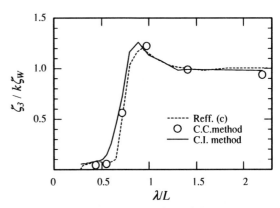
Fig. 1 Response function of pitch (Fn=0.15, wave incident angle=135°)

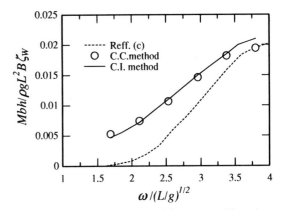
Fig. 2 Response function of horizontal bending moment at midship (Fn=0.2, wave incident angle=120°)

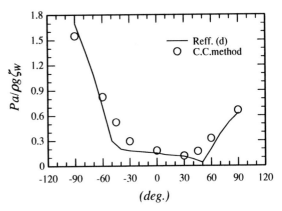
Fig. 3 Pressure distribution at midship (Fn=0.2, wave incident angle=150°)

waves, the following numerical experiments are intended. The example of 5-directional waves is shown in Fig. 7. The numerical experiment of C.C.

516

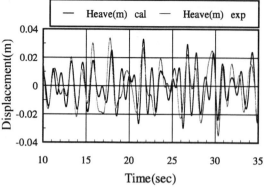

Fig. 4 Time history of heave in 2-directional irregular waves (Fn=0.15)

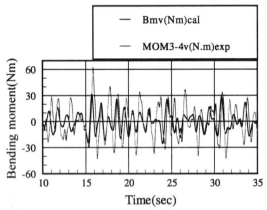

Fig. 5 Time history of vertical bending moment at midship in 2-directional irregular waves (Fn=0.15)

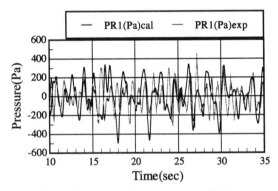

Fig. 6 Time history of pressure at midship portside waterline in 2-directional irregular waves (Fn=0.15)

Method and C.I. Method is carried out in this 5-directional waves with the same single peak ISSC spectrum.

Figure 8 shows the spectrum of heave re-

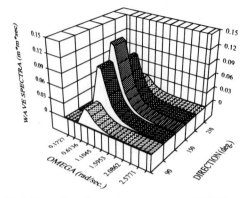

Fig. 7 5-directional wave spectrum of ISSC

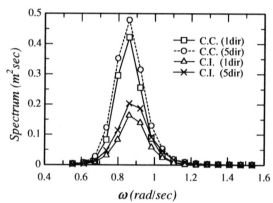

Fig. 8 Heave spectrum in ISSC irregular waves (Fn=0.2, wave incident angle=150°)

sponse in 1- and 5-directional waves calculated by both C.C. Method and C.I. Method. In this case the wave directionality does not have big effect on the heave response, while C.C. Method gives larger value which is unreliable. Since the gradient of heave response function with regard to frequency is very steep in the dominant frequency range of incident waves, it is difficult to predict reliable estimation based on C.C. Method with the constant radiation hydrodynamic coefficients.

Figure 9 shows the spectrum of pressure at port side water line of the midship. In this case the wave directionality has large effect on the pressure response.

According to the example of the measured ocean wave data, at least one third of the wave spectra have multi-peaks in the frequency range, and the others are simple mono-peak spectra. Then twin peak spectrum is rather common ocean wave spectrum as reality. In these twin peak spec-

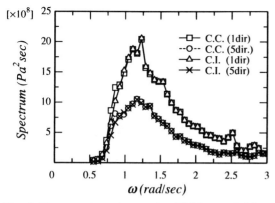

Fig. 9 Pressure spectrum at midship portside waterline in ISSC irregular waves (Fn=0.2, wave incident angle=150°)

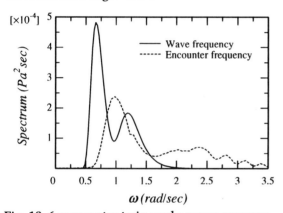

Fig. 10 6-parameter twin peaks wave specrum

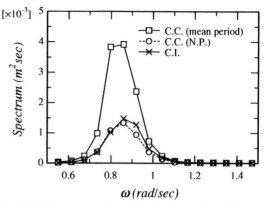

Fig. 11 Heave spectrum in 6-parameter twin peaks irregular waves (Fn=0.2, wave incident angle=150°)

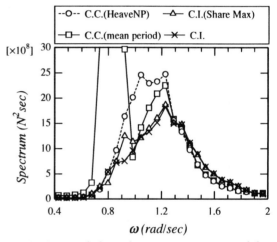

Fig. 12 Vertical shear force spectrum at midship in 6-parameter twin peaks irregular waves (Fn=0.2, incident wave angle=150°)

trum waves, the authors wonder whether C.C. Method is held or not. 6-parameter twin peaks spectrum and the corresponding simple single peak spectrum which has the same energy as that of the twin peak one are shown in Fig. 10. Figure 11 and 12 are the heave and vertical shear force spectra calculated by the C.C. Method and C.I. Method, respectively. It is easily known that the C.C. Method does not give reliable results if the mean or peak frequency is adopted for the constant coefficients.

5. NONLINEAR EFFECTS

In this section, nonlinear effects of the restoring force on roll motion and the hydrodynamic force on heave motion inregular waves are discussed.

Figure 13 shows the GZ-curve of the SR108 container ship. Figure 14 shows the time history

of calculated roll motion of the SR108. The solid line denotes the roll motion taking into consideration the change of the restoring force of each sections due to the motion. The non-linear restoring force has an effect on amplitude of roll motion as well as the period of it. The authors expect this result will be applied to explain the capsizing of ship in following oblique waves.

The authors calculated the second order hydrodynamic force and made the quadratic impulse response function of SR108 midship section as shown in Fig. 15. Figure 16 shows the heave motion of SR108 in regular heading waves using the quadratic impulse response function of each sections, the nonlinear phenomena of heave motion is well shown.

518

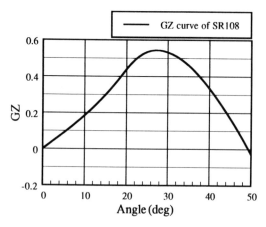

Fig. 13 GZ curve of SR108

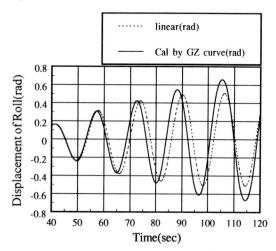

Fig. 14 Time history of roll with nonlinear
(wave incident angle=90°, height=20m, angular
velocity=0.4rad/sec)

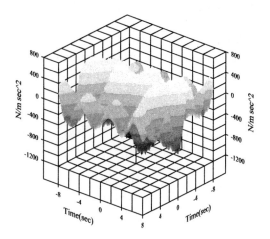

Fig. 15 Quadratic impulse response function of
SR108 midship section

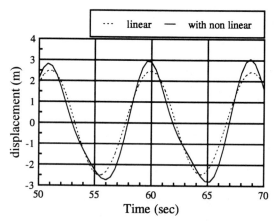

Fig. 16 Time history of heave with nonlinear
(wave incident angle=180°, height=4m, angular
velocity=0.7rad/sec)

6. CONCLUSIONS

The authors developed the reliable and practical numerical simulator for ship motions and wave loads taking into consideration the non-linear effects of ship motion which is applicable even in general multi-directional and multi-peak ocean waves.

The C.I. Method (convolution integral method for hydrodynamic forces) which is adopted in this simulator is the most reliable even in general ocean waves, even though the C.C. Method (constant coefficient method for hydrodynamic forces) saves much computational time and is applicable

to some single peak wave spectrum and some responses.

This simulator can be applicable to the preliminary design for various kinds of ships such as bulk carriers, oil skimmers etc, and to the assisstance of a ship operator in sevear sea conditions by the computer graphics and the related video system.

This research is limited only to the linear and partial nonlinear problem at this moment, however, the extension to the fully nonlinear problem will be intended in near future, because the time domain simulator for ship responses is the most powerful for the reliable estimation of extreme values.

ACKNOWLEDGEMENTS

The authors appreciate very much the assistance of Mr. J. Okuyama and Mr. S. Takeda for computation and experiments.

REFERENCES

1. Bruzzone, D., Pittaluga, A. and Bonvino C. P.: Feasibility of a Second-Order Strip-Theory for the Longitudinal Strength of Ships, 6th Intern. Symp. on Practical Design of Ships and Mobile Units, 1995

2. Fukazawa, T.: Time domain numerical calculation of nonlinear ship motions and wave loads in oblique waves, Journal of Society of Naval Architects of Japan, 167, 1990

3. Kobayashi, K.: Transverse motions and wave loads, 1st Symposium of Ship Dynamics, Society of Naval Architects of Japan, 1984

4. Maeda, H.: Wave exciting forces on arbitrary ship section, Journal of Society of Naval Architects of Japan, 126, 1969

5. Mizoguchi, J., Hirayama, T., Kobayashi, K., Ikeda, Y. and Watanabe, I.: Improved Strip Method, 1st Symposium of Ship Dynamics, Society of Naval Architects of Japan, 1984

6. Saito, M. and Azuma, H.: Time domain analysis of ship motions in waves (part2), Journal of Society of Naval Architects of Japan, 174, 1993

7. SR-108 Committee: Performances of high speed cargo vessels in waves, Report of Japan Ship Research Association, 110, 1970

8. Takagi, M. and Saito, M.: Non-periodic wave problems in frequency domain(part 1)-Memory effect function on 2D body, Journal of Society of Naval Architects of Kansai, 182, 1981

9. Takeda, Y.: Ocean wave data measured by ocean research vessel Umitakamaru, 6th Intern. Symp. on Practical Design of Ships and Mobile Units, 1995

10. Watanabe, I., Toki, N. and Ito, A.: Strip Method, 11th Symposium of Ship Dynamics, Society of Naval Architects of Japan, 1994

Practical Design of Ships and Mobile Units
M.W.C. Oosterveld and S.G. Tan, editors.

Added Resistance of a Ship Moving in Small Sea States

Sverre Steen[a] and Odd M. Faltinsen[b]

[a] Department of Marine Vehicles, The Norwegian Marine Technology Research Institute (MARINTEK)
P.O. Box 4125 Valentinlyst, N-7002 Trondheim, Norway

[b] Institute of Marine Hydrodynamics, The Norwegian University of Science and Technology (NTNU)
N-7034 Trondheim, Norway

The added resistance of slender ships in small sea states has been studied numerically and experimentally. Added resistance in small sea states is important for the overall fuel economy of the ship, since most ships operate most of the time in sea states that must be considered small in this context. The added resistance for ships with slender bows has been found to be of significance for significant wave heights as small as 1% of the ship length. The current method for computing added resistance in short waves, uses only the wave reflection effect. It has been found to give correct results for a ship with a blunt bow and vertical sides in the waterline, but has been found to seriously underestimate the added resistance of slender ships. On the other hand, measuring added resistance in small sea states in the towing tank is very difficult, mainly due to stability problems of the small waves in question. Thus, an improved numerical scheme for computing added resistance in small sea states is needed.

1 INTRODUCTION

Real ships travelling in real seas seldom encounter calm water. For most ships, the most common sea states will be small sea states where the wavelengths are only a fraction of the ship length and the wave heights only a small percentage of the ship length. Such small sea states give very small ship motions and hull loads, and modest added resistance. Since the small sea states usually stand for a large portion of the total voyage, knowledge of the nature and magnitude of the added resistance is important even if it is not very large. Maybe ship resistance should be optimised in small sea states rather than in absolutely calm water?

The objective of the study reported here was to establish the typical level of added resistance in small waves, to verify the reliability of current numerical methods and to see if optimisation of hull forms with respect to added resistance in short waves was practically possible. Towed model tests with five different merchant ship models were carried out in head sea waves. Added resistance was also computed using second order theory.

2 MODEL TESTS

The model tests were carried out in the large towing tank at MARINTEK (L/B/D=260/10/5 m). The towing tank is fitted with a computer controlled double flap wave maker. The models were towed in the arrangement normally used for calm water tests, where a spring and damper system allows for a limited surge motion. The models were free to heave and pitch. The wave height was measured 10 m from the wave maker, 80 m from the wave maker (towards the end of typical run) and by one probe following the model at FP about two meters off centre. The time histories of the towing force, heave, pitch, horizontal acceleration, vertical acceleration fore and aft and wave elevations were recorded on the computer. All models were run at zero trim. Model C was run at an additional displacement, where it had vertical sides in the waterline at the bow. Model A and B are equal, except that Model B has had a bulb added. Table 1 shows model particulars. Figure 1 shows the bow profiles and waterlines.

The waves were selected as steep as possible, provided that they were stable at a sufficient distance from the wave maker. 11 different regular waves and one irregular wave spectrum were applied. Wave data are given in Table 2.

The models were run in calm water immediately before the wave tests, and the results from that run subtracted from the resistance measured during the wave runs to find the added resistance. The actual wave amplitude was found by analysing the wave measured by the probe following the model, since it was found

522

that the wave train was slightly changing down the tank.

Table 1 Main particulars and velocities of models

Model		A	B	C	D	E
Type		Reefer	Reefer	Chem. tanker	Cont-ainer	LPG
Scale		27	27	28.18	37.92	33.04
L_{WL}	[m]	5.891	5.965	6.316	5.978	6.660
Lpp	[m]	5.926	6.000	6.210	6.145	6.507
B	[m]	0.889	0.889	1.143	0.849	1.090
$T_{LPP/2}$	[m]	0.299	0.299	0.376	0.290	0.340
∇	[m³]	0.884	0.881	2.082	0.866	1.881
C_p	[-]	0.577	0.568	0.783	0.589	0.782
C_B	[-]	0.561	0.553	0.78	0.572	0.779
C_M	[-]	0.973	0.973	0.997	0.971	0.997
LCB	[m]	-0.264	-0.281	0.167	-0.140	0.072
S	[m²]	6.219	6.362	10.09	6.318	9.904
Fn	[-]	0.2 0.25 0.3	0.2 0.25 0.3	0.175	0 0.25	0 0.15 0.175 0.2

Table 2 Waves applied in model tests

Regular waves			λ/Lpp [-]				
T [s]	H [m]	λ [m]	A	B	C	D	E
0.8	0.03	1	0.17	0.17	0.16	-	-
0.9	0.03	1.26	0.21	0.21	0.2	0.21	0.19
1	0.04	1.56	-	-	-	0.25	0.24
1	0.05	1.56	0.26	0.26	0.25	-	-
1.1	0.03	1.89	-	-	-	0.31	0.29
1.1	0.05	1.89	0.32	0.32	0.3	0.31	0.29
1.1	0.07	1.89	-	-	-	0.31	0.29
1.2	0.05	2.25	0.38	0.38	0.36	0.37	0.35
1.5	0.05	3.51	0.59	0.59	0.57	0.57	0.54
1.5	0.1	3.51	0.59	0.59	0.57	0.57	0.54
1.5	0.15	3.51	0.59	0.59	0.57	0.57	0.54
JONSWAP T_P=1.0 s, H_S=0.04 m,λ=1.0			yes	yes	Yes	no	no

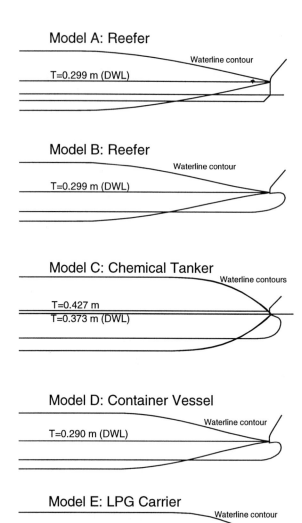

Figure 1 Bow profiles and waterlines of models

3 NUMERICAL CALCULATIONS

Added resistance has been calculated by the computer program MASHIMO (MArintek SHIp MOtion program). For small wave lengths the ship motions can be neglected, and MASHIMO computes the added resistance, $\overline{F_n}$, exclusively as a wave reflection effect, according to the following asymptotic equation (Faltinsen et. al., 1980):

$$\overline{F_l} = \int_{L_l} \overline{F_n} \sin(\theta)\, dl$$

$$\overline{F_n} = \frac{1}{2}\rho g \zeta^2 \left([\frac{1}{2}\frac{k_1}{k_0} - \frac{1}{2}\cos^2(\theta+\alpha)] + \frac{1}{2}\frac{k_2}{k_0}\sin(\theta+\alpha) \right)$$

$$k_1 = \frac{[\omega_e - V k_0 \cos(\theta+\alpha)]^2}{g}$$

$$\text{(1)}$$

$$k_2 = \sqrt{k_1^2 - k_0^2 \cos^2(\theta+\alpha)}$$

$\overline{F_n}$	force per unit length normal to the hull
ζ	wave amplitude
θ	angle between the tangent of the waterline and the fore-and-aft axis (x-axis)
α	wave propagation direction with respect to the x-axis
L_1	the part of the water line that experience the incoming waves
ω_e	circular frequency of encounter
V	horizontal steady velocity parallel to the ship side
k_0	wave number

Hence, the added resistance is expected to be proportional to the wave amplitude squared. Further, Eq. (1) is based on moderate Froude numbers and blunt ship bows. It assumes vertical sides in the water line. For larger wave lengths, where ship motions have influence, the formulas given by Gerritsma and Beukelman (1972) is used together with the strip theory by Salvesen, Tuck and Faltinsen (1970).

4   RESULTS

Added resistance is needed mainly in order to correct speed trials and in routing studies. In addition, it might be possible to utilise knowledge of added resistance in resistance optimisation studies. Common for all these applications is that one do not consider one particular sea state, rather a large number of sea states. Thus, what is needed is primarily transfer functions for added resistance. Such transfer functions are obtained best in experiments by testing in regular waves. Figure 2 shows dimensionless added resistance as function of ship length divided by wavelength for models A and B. This is a reefer with a slender bow. Thus, the added resistance computed with the asymptotic formula in equation (1) is very small, since the waterline entrance

angle θ is small. The measured added resistance is not very small. For model A, which has no bulb, the added resistance is found to rise notably when the wavelength decreases. The same tendency is not that clear when the bulb is added. It should be noted that for the two shortest wavelengths the added resistance is a very small percentage of the calm water resistance. Furthermore, the wave quality, especially for the shortest wave is not quite satisfactory. The dimensionless added resistance as presented in Figure 2 and 3 is very sensitive to errors in the estimated wave height, since it is made dimensionless by wave amplitude squared. Estimation of the wave amplitude is found to be the major source of error. However, any reasonable interpretation of the wave height will give

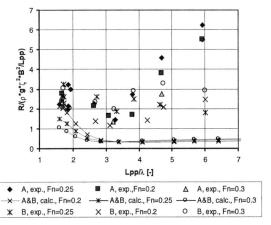

Figure 2 Non-dimensional added resistance for model A and B at DWL, from model tests and calculations.

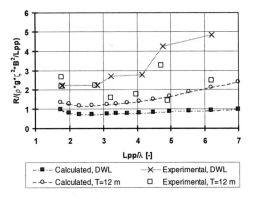

Figure 3 Non-dimensional added resistance for model C at Fn=0.175, from model test and calculations.

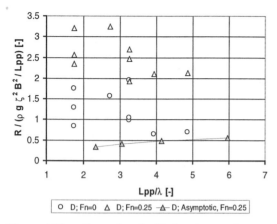

Figure 4 Non-dimensional added resistance for model D, from model tests and calculations

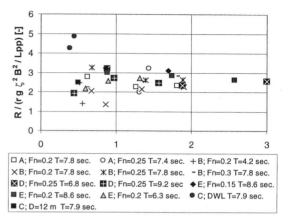

Figure 7 Experimental dimensionless added resistance as function of full scale wave amplitude

the same main conclusion: that the asymptotic theory seriously under-predicts the measured added resistance for this slender bow shape.

In Figure 3, the dimensionless added resistance from model tests and computations for model C is presented. Model C is a chemical tanker that has a blunt bow and moderate bulb. At the full-scale draught of 12 m, which is larger than the design draught of 10.5 m, the ship has perfectly vertical sides in the waterline in the bow region. Thus, it fulfils all the conditions of Eq. (1). It is seen from Figure 3 that the agreement between calculations and experiments is quite good for the 12 m draught. The agreement for the design waterline is less good, but still much better than for the more slender A and B models. From this, it seems that the theory of Faltinsen et. al. (1980) is quite sensitive to the formal conditions being fulfilled, and that for a

slender bow there are other phenomena than wave reflection that are most important for added resistance.

Figure 4 shows dimensionless added resistance for model D. It shows the same trends as the A and B models, where the asymptotic theory of Eq. (1) under-predicts the experimental results. This is consistent with the A and B model results, since the bow is very slender also for model D.

Figure 5 shows dimensionless added resistance for model E. The experimental and theoretical results are at the same level for Lpp/λ=3, but the trend in the experimental results is steeper towards shorter wave lengths than the theoretical results. Thus, the results are similar to those of model C on DWL.

Figure 6 shows the experimental dimensionless added resistance as a function of wave amplitude. Since the level of the dimensionless added resistance is

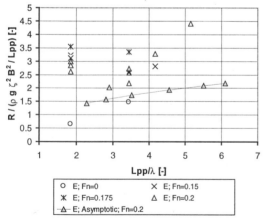

Figure 5 Non-dimensional added resistance for model D, from model tests and calculations

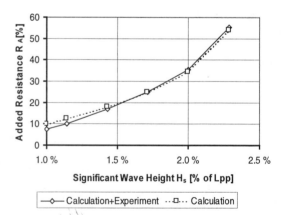

Figure 6 Added resistance as fraction of full scale resistance for Model C, as function of significant wave height

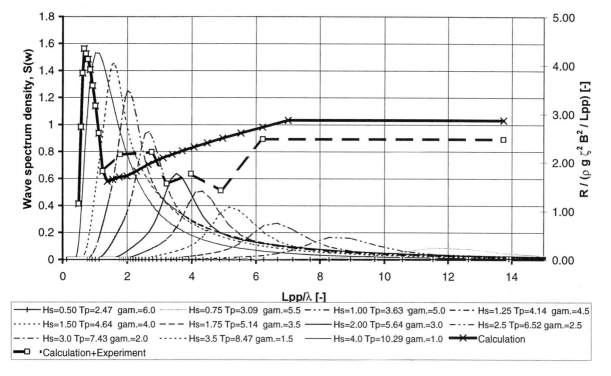

Figure 8 Combination of transfer functions with wave spectra to calculate added resistance as function of significant wave height (sea state)

nearly independent of wave amplitude, it is confirmed that the added resistance for the short waves in question is proportional to the wave amplitude squared. It is assumed that the larger spread for the lowest wave amplitudes shows the problems related to model testing in such small waves.

Since model C best fulfils the assumptions in Eq. (1), the results will be studied in some more detail. In Figure 7 the added resistance of model C is presented as function of significant wave height. It is seen from Figure 7 that the added resistance is quite significant already at a significant wave height of 1% of the ship length. The added resistance is calculated by combining the transfer function with wave spectra of different significant wave height and wave period, as shown in Figure 8. Other wave periods are of course possible for a given significant wave height. The chosen wave period is a frequently occuring period. Two transfer functions have been used; one purely calculated and one where the transfer function derived from the model tests has been used for wave lengths $\lambda < 0.5 * Lpp$. It is found that the difference in computed added resistance is small. For $Lpp/\lambda > 7$ there is no

experimental results, and the transfer function is assumed to have a constant value for shorter wavelengths. The reason why the added resistance has not been computed for even smaller wave heights is that the main part of the response is calculated from the part of the transfer function where we have no experimental results.

5 CONCLUSIONS

The practical lower limit for wavelengths in experiments is found to be about $Lpp/\lambda = 7$. It found that a wave spectrum with such a peak wavelength might still give considerable added resistance. In addition, extreme care had to be taken both during experiments and analysis to get useful results at the shortest wavelengths. Even then, the spread in the experimental results was too large to be satisfactory. Thus, a numerical method would be the more practical way to estimate added resistance in small sea states. The asymptotic formula by Faltinsen et. al. (1980) has been found to represent correctly the wave amplitude dependence of added resistance. The formula has also been found to correctly predict the overall level of added resistance when all the formal conditions of the formula is satisfied, that means for a blunt bow with vertical sides in the waterline. However, the formula has been found to be very sensitive to deviations from the formal conditions, and to under-predict the added resistance if the ship sides are not vertical (which they rarely are), or if the bow is not blunt. Especially for a slender bow with a small waterline entrance angle, the asymptotic theory will seriously under-predict the added resistance in small sea states. Thus, it is concluded that a new numerical method is needed to predict added resistance in small sea states, especially for slender ship forms.

REFERENCES

Faltinsen, O.M., Minsaas, K.J., Liapis, N., Skjørdal, S.O. (1980) 'Prediction of Resistance and Propulsion of a Ship in a Seaway' Proc. Thirteenth Symp. On Naval Hydrodynamics pp. 503-530. Tokyo, Japan.

Gerritsma, J., Beukelman, W. (1972) 'Analysis of the Resistance Increase in Waves of a Fast Cargo Ship' Intern Shipbuilding Progress, Vol. 19, 217, 285-293.

Salvesen, N., Tuck, E. O., Faltinsen., O. M. (1970) 'Ship Motions and Sea Loads' Transactions SNAME, Vol. 78, pp. 345-356.

Stansberg, C. T. (1993) 'Propagation-dependent Spatial Variations Observed in Wavetrains Generated in a Long Wave Tank' MARINTEK report 490030.01 (MT49 A93-0176).

NOMENCLATURE

B	Ship (model) beam
C_B	Block coefficient
C_M	Midship coefficient, C_M=Midship area/(B·T)
C_p	Prismatic coefficient
DWL	Design waterline
F_n	Froude number $F_n = V/\sqrt{g \cdot Lpp}$
$\overline{F}_n$	Force per unit length normal to the hull
g	Acceleration of gravity
H	Wave height
H_s	Significant wave height
k_0	wave number
L_1	the part of the water line that experience the incoming waves
LCB	Longitudinal centre of gravity (relative to Lpp/2)
Lpp	Length between perpendiculars
L_{WL}	Length of the waterline
R	Added resistance
S	Wetted surface
T	Wave period of a regular wave, or draught of ship or model
$T_{Lpp/2}$	Draught amidships
V	Horizontal steady velocity parallel to the ship side
T_p	Peak period of a wave spectrum
α	Wave propagation direction with respect to the x-axis
ζ	Wave amplitude
θ	Angle between the tangent of the waterline and the fore-and-aft axis (x-axis)
ω_e	Circular frequency of encounter
∇	Volume displacement
λ	Wavelength

Practical Design of Ships and Mobile Units
M.W.C. Oosterveld and S.G. Tan, editors.

BEAK-BOW to Reduce the Wave Added Resistance at Sea

Koichiro Matsumoto[a], Shigeru Naito[b], Ken Takagi[b], Kazuyoshi Hirota[a] and Kenji Takagishi[a]

[a] Ship and Marine Structure Lab., Tsu Research Center, NKK Corporation
   1, Kumozukokan-Cho, Tsu-City, Mie-Pref., Japan
[b] Dept. of Naval Architecture and Ocean Engineering, Osaka University
   2 − 1, Yamadaoka, Suita-City, Osaka-Pref., Japan

This paper presents a new concept of bow shape on full form ships such as tankers and bulk carriers. The purpose of this concept is to reduce the added wave resistance for energy saving at sea. The bow shape above the design load waterline is peaked as the beak of birds, which is therefore called the BEAK-BOW. The reduction of the added wave resistance in regular head waves was confirmed by experiments on the model of 150,000DWT bulk carrier. The reduction ratio of wave added resistance was between 20 to 30%. This corresponds to about 10% horsepower reduction or 0.5knots reduction of the speed loss at Beaufort 6 sea condition.
For designing such bow shapes, a new practical procedure to calculate the added wave resistance is also presented, which can consider the hull shape above still waterline.

1. INTRODUCTION

In these days, to reduce the energy consumption is indispensable in the wade range of the industry and human activities. In the field of marine transportation, it is also requested to reduce the fuel oil consumption of ships. For this purpose considerable effort has been spent on improving the hull shape of ships and/or developing the energy saving devices fitted on the ship.[1] By these efforts, in the last two decades the horsepower necessary to the ship has been considerably reduced.

Such horse power reduction on ships has so far been focused on that in still water. The actual ships are, however, not always sailing in still water. The resistance increase due to waves is not small and leads the increment of total fuel oil consumption and/or the speed loss of ships. To reduce the fuel oil consumption increase in waves is also important, but it has not so far been tried so much.

The purpose of the present study is to develop the ship shape to reduce the resistance increase due to waves, or added wave resistance. The added wave resistance on full form ships such as tankers or bulk carriers is mainly generated by the diffraction of the incident waves at the blunt bow. An idea to decrease this diffraction of waves at bow is to reduce the bluntness of the bow shape especially above the load waterline. From this viewpoint the BEAK-BOW was developed and its effect to reduce the added wave resistance was investigated by model tests as described in the following sections.

2. CONCEPT OF BEAK-BOW

As mentioned above the added wave resistance acting on a full form ship such as tankers or bulk carriers is mainly due to the diffraction and wave breaking at its blunt bow. Such wave diffraction or breaking generates the reaction force backward on the ship's bow, especially when the incident wave surface elevates at the bow. A simple idea to reduce the added wave resistance can, therefore, be to sharpen the bow shape. The blunt bow above the load waterline was peaked and sharpen on the model of an oil tanker, and the resistance tests in waves on this model were performed in the towing tank of Osaka University. The wave added

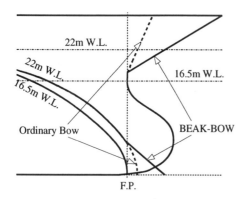

Fig.1 Comparison of Bow Shape

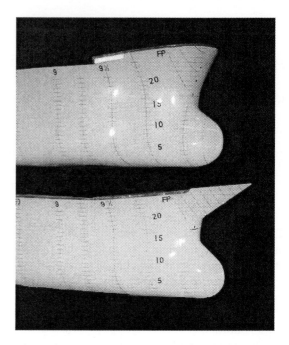

Photo.1 Comparison of Bow Shape

resistance in regular head waves on this sharpen bow was reduced by the ratio of 20 to 30% in comparison with that on the ordinary blunt bow.[2]

This sharpen bow is named BEAK-BOW, because of its shape's looking like the beak of a bird. Fig.1 shows a comparison between the ordinary bow and a BEAK -BOW shape applied on a 150,000DWT bulk carrier. In this BEAK-BOW figure, the resistance can be smaller with sharpener and more peaked bow waterline shape. The sharpener and more peaked bow extends the length of the ship. But in the practical viewpoint, the ship length is limited by for instance the port regulations. In the case of the present 150,000DWT bulk carrier, the BEAK-BOW length was designed so that the ship length over all is satisfied with Dunkirk port maximum length, i.e. 289m. Under this restriction, the BEAK-BOW profile shape is decided as shown in Fig.1. And the waterline shape at each draft level above the load waterline was sharpened as much as possible, which means as straight as possible. In the case of Fig.1, the BEAK-BOW is assumed to attach to an existing ship around its fore perpendicular (FP) position. Photo 1 shows the comparison between the ordinary bow and the BEAK-BOW on the model of the bulk carrier given in Fig.1.

3. EXPERIMENTAL INVESTIGATION

Two scale models of a 150,000DWT Type Bulk Carrier were used for an experiment in waves. One is the ordinary bow model and the other is the BEAK-BOW model. The hull form below the

design load waterline of each model was identical. These models are 4.0m long (Lpp) and 1/69.75 scale. The added wave resistance tests were performed at Tsu Ship Model Basin of NKK Tsu Research Center, Japan. The model was connected to a measuring device fitted to a towing carriage allowing freedom for surge, heave and pitch motions. The model was towed at the ship speed of 13.0knot in ship scale in regular head waves. The wave height was 3.0m in ship scale and the range of wave length was $\lambda/Lpp = 0.4 \sim 1.6$. (where, λ = wave length, Lpp = ship length) The measurement items were ship motions, resistance and pressure distributions at the bow.

The added wave resistance was obtained by deducting its resistance in still water from the total resistance in waves. The difference of the resistance in still water between the ordinary bow model and the BEAK-BOW model was negligible. And the static swell up at the point of FP in still water for both bow models was observed to be the same.

Fig.2, Fig.3 and Fig.4 show the added wave resistance and ship motions obtained experimentally and theoretically. Circle marks in these fig-

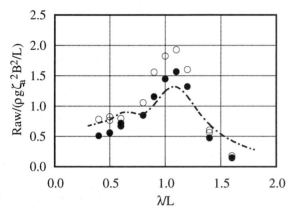

Fig.2 Added Resistance in Regular Wave
(Full Load, 13knot, Head Sea)

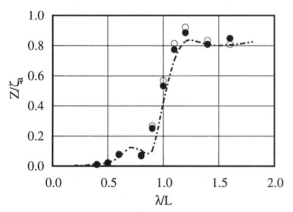

Fig.3 Heaving Motion
(Full Load, 13knot, Head Sea)

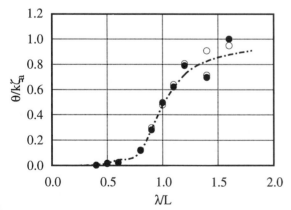

Fig.4 Piching Motion
(Full Load, 13knot, Head Sea)

ures show the experimental results and dashed line shows the theoretically calculation result. The calculation result of the added wave resistance in regular waves is obtained from the strip theory in the range of ship motion dominant [3] and from Fujii & Takahashi's Method [4] in the range of shorter wavelength. Ship motions are calculated by the strip theory.

Fig.2 shows the comparison of added wave resistance in regular head waves between the BEAK-BOW($\bullet$) and the ordinary bow($\circ$). The vertical axis shows the non-dimensional value of added wave resistance (Raw) and horizontal axis shows the ratio of wavelength and ship length (λ/L). BEAK-BOW gives the smaller added wave resistance than that of ordinary bow by the ratio of 20~30%. The theoretical calculation result in Fig.2 corresponds to that for the ordinary bow ($\circ$). It gives smaller value than the experiments in the range of longer wavelength where the ship motion's influence is dominant. This will be discussed in the next section. Experimentally obtained heave and pitch motions given in Fig.3 and 4 do not have much difference between the two bow models and the theoretical calculations give good agreement to the experimental values.

In order to estimate the horsepower reduction due to the BEAK-BOW in the actual sea condition, the short-term prediction of the horsepower increase in long-crested irregular waves was performed by using the experimental results of the added wave resistance in the regular head waves. The results of horsepower curves in Beaufort 5 and 6 sea conditions are shown in Fig.5 and Fig.6 respectively together with the horsepower curves in still water. Significant wave heights ($H_{1/3}$) and mean wave periods (T_0) of Beaufort 5 and 6 conditions are as follows:

$Beaufort$	5	6
$H_{1/3}(m)$	2.4	6.0
$T_0(sec.)$	3.7	7.5

For the shot-term predictions, the following ISSC 1960 type wave spectrum is used.

$$S(\omega) = \frac{1}{2\pi} 0.11 H_{1/3}^2 T_0 \left(\frac{T_0}{2\pi} \omega \right)^{-5}$$
$$\times exp \left\{ -0.44 \left(\frac{T_0}{2\pi} \omega \right) \right\} \qquad (1)$$

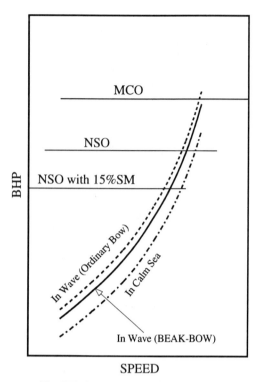

Fig.5 Estimate Power Curves in Wave
150BC/BF5 (Head Sea)

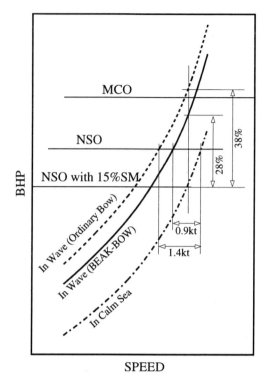

Fig.6 Estimate Power Curves in Waves
150BC/BF6 (Head Sea)

In Beaufort 6 sea condition given in Fig.6, the horsepower increase on the ordinary bow at the main engine output of NSO with 15% Sea Margin is about 38%. The horsepower increase on the BEAK-BOW in the same condition is about 28%. The horsepower reduction by the BEAK-BOW in the Beaufort 6 condition is, therefore, estimated to be 10%. From the viewpoint of speed loss in the same sea condition, the speed loss at NSO for the ordinary bow is about 1.4knots. The speed loss for the BEAK-BOW is about 0.9knots. Therefore, the reduction of speed loss by adopting the BEAK-BOW is about 0.5knots in Beaufort 6 sea condition.

The above estimation is only for the case of head waves. The effect of BEAK-BOW to reduce wave added resistance in oblique waves is to be further investigated.

4. THEORETICAL INVESTIGATION

Added wave resistance acting on a full form ship is normally obtained by the sum of the resistance due to ship motions and that due to wave diffraction at the blunt bow. The resistance due to ship motions can be calculated by Maruo's method and that due to wave diffraction at the bow can be calculated by Fujii & Takahashi's method[4] or Faltinsen's method.[5]

Maruo's method is based on the slender ship theory by using source distributions due to the ship hull shape below the load waterline. Fujii & Takahashi's method is based on calculating the reaction force in consideration of the load waterline shape at the blunt bow in accordance with the wall side assumption. Therefore, both of these methods can not consider the effect of bow shape above the load waterline as given the present BEAK-BOW.

Recently, some recommendations using a nonlinear theory are made to consider the effect of hull shape above the load waterline on the added wave resistance due to diffraction on the bow.[6, 7] But they are still complex and need much computing time so that they are not so far suitable for the practical design.

In order to estimate the added wave resistance on the BEAK-BOW practically, the following modifications are applied to the conventional method (Fujii & Takahashi's method) for calculating the added wave resistance due to reflection at the bow.

1) Instead of design load waterline shape in still water, the shape of waterline at the water surface in consideration of the dynamic swell up at the bow in still water was considered. When a ship is sailing in still water, the water surface is elevated at the bow, which is called the dynamic swell up. Incident wave motion is occurred around this swell upped water level. Therefore, to use the bow's waterline shape at this swell upped water level seems to be more reasonable in order to calculate the wave added resistance than to use that at still water surface level.

2) Instead of incident wave amplitude, the relative wave amplitude at the bow was considered. In Fujii & Takahashi's method, the wave added resistance due to the diffraction of the incident wave at the bow is calculated by the following equation.

$$R_{aw} = \alpha_1(1 + \alpha_2)\frac{1}{2}\rho g \zeta_a^2 B \overline{\sin^2 \beta} \qquad (2)$$

where,

α_1 : reflection ratio of vertical barrier
α_2 : experimental value
ρ : density of water
ζ_a : incident wave amplitude
β : encounter angle

When the ship motions do not occur in the range of short wave length, the wave added resistance due to the wave diffraction at bow is generated by the incident wave amplitude as given in the equation (2). But, when the ship motions in longer waves, wave diffraction can be caused by the relative motion amplitude at the bow. The present proposal, therefore, is to use the relative

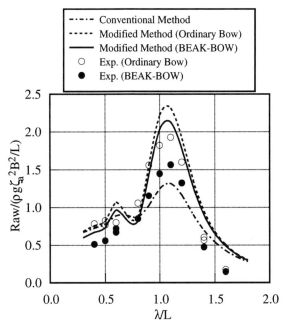

Fig.7 Added Resistance in Regular Wave
(Full Load, 13knot, Head Sea)

wave amplitude (ζ_r) instead of the incident wave amplitude (ζ_a) in the equation (2).

Fig.7 shows the comparison between the theoretical result of added wave resistance in regular head waves given by the conventional method and that by the modified method given above. Fig.7 gives the theoretical results by the above modification for both of the ordinary bow and the BEAK-BOW together with their experimental results.

As shown in Fig.7 the present method modified by the above item 1) can describe the difference of added wave resistance between the ordinary bow and the BEAK-BOW. The difference of calculated wave added resistance between the ordinary bow and the BEAK-BOW is small in comparison with that obtained by experiments. In order to describe the experimentally obtained difference of wave added resistance between the ordinary bow and the BEAK-BOW, further investigation on improving the calculation method is necessary.

In the range of longer wavelength around $\lambda/Lpp = 1$, the conventional method gives smaller added

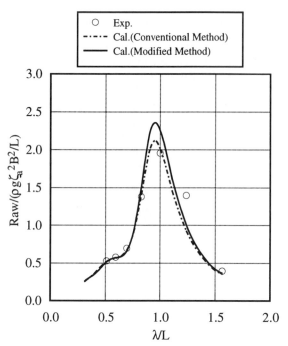

Fig.8 Added Resistance in Regulae Wave
(Container Ship, Full Load, Fn=0.2, Head Sea)

wave resistance than the experimental values. In this wave length range about $\lambda/Lpp = 1$, the influence of ship's heave and pitch motions to the added wave resistance is large.

The present method considering the above modification 2) can give more accurate results to the experiments (o) than the conventional method. It indicates that the added wave resistance due to the wave diffraction at the blunt bow is large also when the ship motions are large. Therefore, the wave diffraction at the blunt bow does not seem to be only due to the incident wave amplitude but due to the relative motion amplitude between the incident wave and the ship motion, and the idea given in the above modification 2) can be valid.

The influence of this relative motion amplitude at the bow to the resistance can be large in the case of the blunt bow as given in Fig.7, because the blunt bow generates large wave diffraction regardless of the magnitude of ship motions.

In the case of the sharp bow shape such as that for container ships, the wave diffraction at the bow itself is rather small. In such case, the consideration of the relative motion amplitude at

the bow becomes small as shown in Fig.8, and both of the conventional and the present modified method can predict the experiments accurately.

5. CONCLUSIONS

Conclusions of the present study are as follows:

1. A new bow shape, BEAK-BOW, to reduce wave added resistance is developed.

2. Model tests on a 150,000DWT bulk carrier in regular head waves show the reduction ratio of added wave resistance by 20 to 30%.

3. Based on the model tests results, the horse power reduction in Beaufort 6 head sea condition is estimated to be 10%, which corresponds to be 0.5 knots reduction of speed loss.

4. For designing BEAK-BOW's, a practical procedure of calculating added wave resistance is proposed. This procedure can consider the bow shape above the load waterline and diffraction of incident wave at bow with ship motions.

Further investigations are necessary for the following items.

- Improvement of the theoretical procedure of added wave resistance for the optimum design of the BEAK-BOW.

- To perform the experimental investigation to obtain the effect of BEAK-BOW in oblique, beam and follow sea conditions.

- To confirm the effectiveness of BEAK-BOW on a ship at the actual sea.

- Based on the above investigations, a design procedure of a ship hull shape to be developed, which will be used for an optimum ship hull design from the viewpoint of its long life sailing at sea.

REFERENCES

1. Watanabe,T., et al. : Recent Development on Energy-Saving Technology for Actual Ships, Proc. 4th International Marine Systems Design Conference, Vol.1, Kobe, 1991

2. Naito,S., et al. : An Experimental Study on the Above-Water Bow Shape with a Small Added Resistance in Waves, J. Kansai Soc. N. A., No.226, 1996 (in Japanese)

3. Maruo,H. : The Theory of the Wave Resistance of a Ship in a Regular Seaway, Bulletin of the Faculty of Engineering, Yokohama University, Vol.6, 1957

4. Fujii,H. and Takahashi,T. : Experimental Study on the Resistance Increase of a Ship in Regular Oblique Waves, Proc. 14th ITTC, Vol.4, Ottawa, 1975

5. Faltinsen,O.M., et al. : Prediction of Resistance and Propulsion of a Ship in a Seaway. Proc. 13th Symp. on Naval Hydrodynamics, Tokyo, 1980

6. Song,W. and Maruo,H. : Bow Impact and Deck wetness : Simulations Based on Nonlinear Slender Body Theory, Proc. 3rd ISOPE , 1993, Vol.3, pp.34-38

7. Lin,W.M., Newman,J.N. and Yue,D.K.P. : Nonlinear Forced Motions of Floating Bodies., Proc.15th ONR, 1984, pp.33-49

1998 Elsevier Science B.V.
Practical Design of Ships and Mobile Units
M.W.C. Oosterveld and S.G. Tan, editors.

A Prediction Method for the Shipping Water Height and its Load on Deck

Yoshitaka Ogawa, Harukuni Taguchi and Shigesuke Ishida

Ship Dynamics Division, Ship Research Institute, Ministry of Transport, Japan
6-38-1, Shinkawa, Mitaka, Tokyo, 181-0004, Japan

A practical prediction method for shipping water height, its load and pressure on the bow deck is proposed. The inputs of this method are relative water height at bow, pitching angle and bow vertical velocity, those can be calculated by strip theory and so on. The theory of "flood waves" was applied for evaluating the shipping water height distribution on deck. This theory gives a better estimation than the conventional model of "dam breaking" because it can include the effect of ship forward speed. On the shipping water load and pressure, not only the static component but also the time derivative of the momentum of the shipping water was considered. It was clarified that the momentum component is important to estimate the peak of shipping water load and pressure. H aving compared with measured data, it was confirmed that the combination of these two methods is practical enough for predicting shipping water load and pressure.

1. INTRODUCTION

It is well known that deck wetness sometimes causes serious damage to the bow deck itself and structures on the deck. Many studies have been carried out on this phenomenon. It was found that the occurrence of deck wetness could be determined by the comparison of relative water height at bow with bow top height. However, only a few of the studies [1-2] gave us quantitative information about water volume on deck, impulsive load and the relation between them and ship motion. A practical prediction method, which can be made use of by designers and rule-making people, is needed.

The authors carried out an experiment using a Japanese domestic tanker model in regular head and bow seas in order to measure the shipping water height distribution, the load and pressure due to deck wetness. The limiting condition, e.g. model speed, wave steepness, of the occurrence of deck wetness was clarified. The relations among the key quantities relating to deck wetness mentioned above are discussed. Using those relations a prediction method for shipping water height, load and pressure is proposed.

2. MODEL SHIP AND MEASURING INSTRUMENTS

The experiment was carried out for a Japanese

domestic tanker in regular head and bow seas. The principal particulars and the body plan are shown in Table 1 and Fig.1 respectively. The setup of measuring instruments around the bow deck is shown in Fig.2. A load cell was attached under the bow deck that is separated from the main body of the model for measuring shipping water load directly. An accelerometer was attached beside the load cell to exclude the effect of inertia of the bow deck plate. Five mean water height probes, combinations of wave probes at a cross section, were attached on the bow deck from F.P. to S.S. 9. Shipping water volume was estimated by integrating the mean water height at each cross section. Four pressure gauges were instrumented along the centerline. A wave probe was also attached at stem. A video camera was attached to observe the behavior of shipping water.

3. SOME RESULTS OF EXPERIMENT

At first, relative water height amplitude and freeboard height, both at stem, were compared in Fig.3. The calculated RAO by strip theory (NSM) is also shown. Bow top height f' is corrected by static swell up measured in the still water test and shown as f'-h_s in Fig.3. In the experiment severe deck wetness was observed at λ/L=1.0, 1.25 and 1.5 (λ :wave length, L:ship length). It is found that deck wetness occurred when relative water height is larger than the corrected bow top height.

536

Table 1 Principal Particulars

	Ship	Model
Lpp	72.0m	4.00m
B	11.5m	0.638m
D	5.25m	0.292m
d	4.74m	0.263m
Disp.	2720ton	466.4kg
GM	1.37m	0.076m
Cb	0.68	0.68
Tr	7.8sec	1.83sec

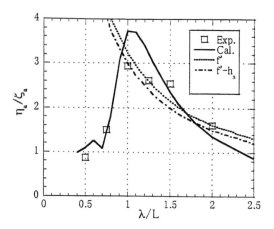

Figure 3 RAO of relative water height at stem in head sea (χ =180°)

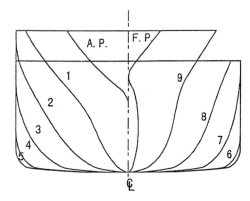

Figure 1 Body Plan

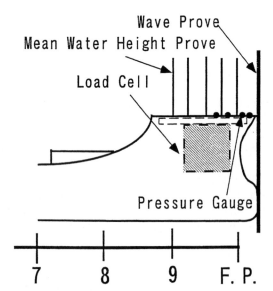

Figure 2 Setup of measuring instruments around the bow deck

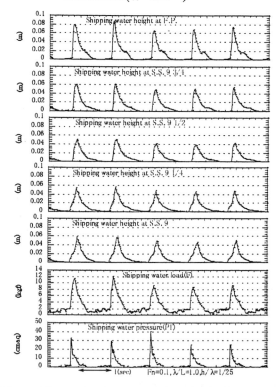

Figure 4 Example of time histories of shipping water height, load and pressure in head sea (χ =180°)

An example of time histories of shipping water height, load and pressure is shown in Fig.4. It is found that inflow of shipping water is very rapid and its load and pressure contain impulsive components. Experimental results in various wave steepness and ship forward speed are shown in Fig.5 and Fig.6 respectively. The relation between shipping water volume and relative water height at stem is also shown in Fig.7. It is found that wave height, ship forward speed and relative water height are much related to deck wetness.

4. PREDICTION FOR SHIPPING WATER HEIGHT DISTRIBUTION AND ITS LOAD

In this section a prediction method for shipping water height, load and pressure is proposed. This method is separated to two parts. One is the prediction of shipping water load and pressure from shipping water height, the other is the prediction of shipping water height from relative water height at bow. The validation of this method was conducted using the experimental data.

4.1. Prediction for shipping water load

Concerning the shipping water pressure, it was assumed that the pressure of the shipping water on deck was mainly due to the static component. Recently, Buchner proposed a shipping water pressure P can be evaluated by equation (1) [3].

$$P = \frac{d(\rho h \cdot W)}{dt} + \rho g h \cos\theta$$
$$= \rho\left(\frac{\partial h}{\partial t}\right)W + \rho\left(g\cos\theta + \frac{\partial W}{\partial t}\right)h \qquad (1)$$

where, ρ is a density of water, h is the shipping water height on deck, W is the vertical velocity of the deck, g is the gravity acceleration, θ is a pitching angle. Equation (1) is constructed by two components. The first term of equation (1) presents the rate of change of the momentum of the shipping water, whether the second term presents the static pressure corrected by the vertical acceleration of the deck. Buchner showed the first term is important for predicting the peak shipping water pressure based on the comparison of measured pressure and calculated. Following his approach, we investigated the relation of measured shipping water volume with measured shipping water load. Fig. 8 shows the comparison of the time history of measured shipping water load and calculated one by shipping water volume m using equation (1). Each components of equation (1) is also illustrated in Fig. 8. The calculated values are in good agreement with measured one. Comparison of shipping water load at various wave steepness and ship forward speed was also made and it is confirmed the rate of change of the shipping water is important to predict the peak of shipping water load and pressure.

4.2. Prediction for shipping water height distribution

It is difficult to describe the behavior of shipping water flow strictly and dam breaking model [4] was often used for describing it [5]. But, from observation by the video camera attached on the model ship, inflow of shipping water is quite different from the one assumed by dam breaking model. So, we consider that dynamic effects by ship forward speed should be considered for describing shipping water phenomenon accurately.

The same result was concluded from the observation by Mizoguchi [6] and he calculated shipping water distribution using Glimm's method by assuming shipping water behaves like shallow water. In this paper, the theory of "flooded waves" [7] is applied to describe shipping water.

The equation of continuity and motion in the two-dimensional water channel are expressed:

$$\frac{\partial h}{\partial t} + \frac{\partial q}{\partial x} = 0 \qquad (2)$$

$$\frac{\partial q}{\partial t} + \frac{\partial vq}{\partial x} + gh\frac{\partial h}{\partial x} = gh(i_0 - i_f) \qquad (3)$$

where t is time, x is the coordinate axis taken along the channel and positive downstream, h is the depth of the channel, q is the volume of the flow, v is the mean velocity of flow, i_0 is the slope of the channel bed, i_f is the friction of the slope.

Supposing that the disturbance is given on x=0 in the steady flow where the depth of the channel is h_0 and velocity of the steady flow is v_0, we examine the propagation of waves by this disturbance.

538

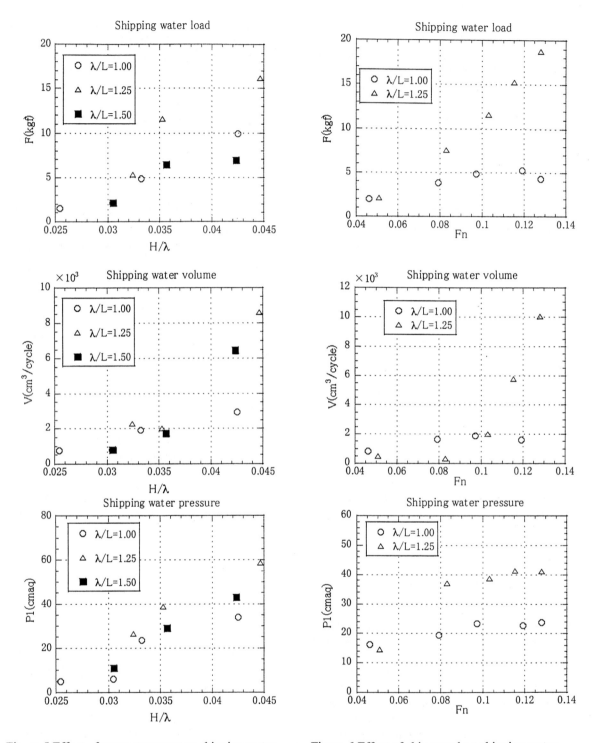

Figure 5 Effect of wave steepness on shipping water
(χ =180° , Fn=0.1)

Figure 6 Effect of ship speed on shipping water
(χ =180° , H/λ =1/30)

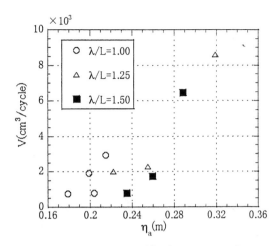

Figure 7 Relation between shipping water and relative water height at stem ($\chi = 180°$)

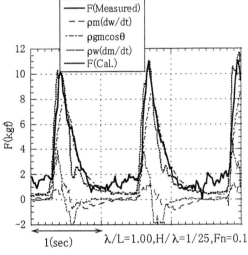

Figure 8 Comparison between measured and estimated shipping water load ($\chi = 180°$)

In order to obtain approximate solution, equation (2) and (3) is substituted by two parameters, $\alpha = h_0 / Li_0$, $\beta = v_0 / \sqrt{gh_0}$ where $L \equiv Tv_0$ and T denotes time scale. By assuming $\alpha \approx 1$ and $\beta^2 \approx 0.1$, equation (2) and (3) become

$$\frac{\partial h}{\partial t} + \frac{5}{3} v \frac{\partial h}{\partial x} = \frac{vh}{2\left(i_0 - \frac{\partial h}{\partial x}\right)} \frac{\partial^2 h}{\partial x^2} \qquad (4)$$

with initial condition $t = 0; h = h_0$ and boundary condition in the upstream.

The first approximate solution of equation (4) is $\phi_1(x,t)$

$$= \frac{x}{2\sqrt{\pi D}} \int_0^t \frac{F(\tau)}{(t-\tau)^{3/2}} \exp\left\{\frac{-(x-\omega_0(t-\tau))^2}{4D(t-\tau)}\right\} d\tau \qquad (5)$$

where $D = v_0 h_0 / 2i_0, \omega_0 = 5v_0 / 3$. This equation is used for analysis of flood waves in a river.

We define a coordinate system shown in Fig.9. By substituting $F(t)$ as $F(t) = f(t) - h_0$, where $f(t)$ is the relative wave height at the stem and h_0 is the bow top height at stem, and v_0 as the ship forward speed, equation (5) can be applied for the shipping water problem.

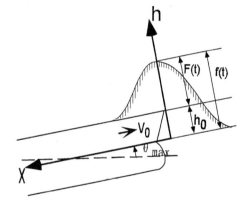

Figure 9 Model of flood waves

Taking the effect of the breadth on the deck into consideration, equation (5) can be expressed as

$$\phi_1(x,t) = \frac{B_0}{B(x)}$$

$$\cdot \frac{x}{2\sqrt{\pi D}} \int_0^t \frac{F(\tau)}{(t-\tau)^{3/2}} \exp\left\{\frac{-(x-\omega_0(t-\tau))^2}{4D(t-\tau)}\right\} d\tau \qquad (6)$$

where B(x) is a breadth at x, the longitudinal distance from stem, and B_0 is an inflow breadth of shipping water.

540

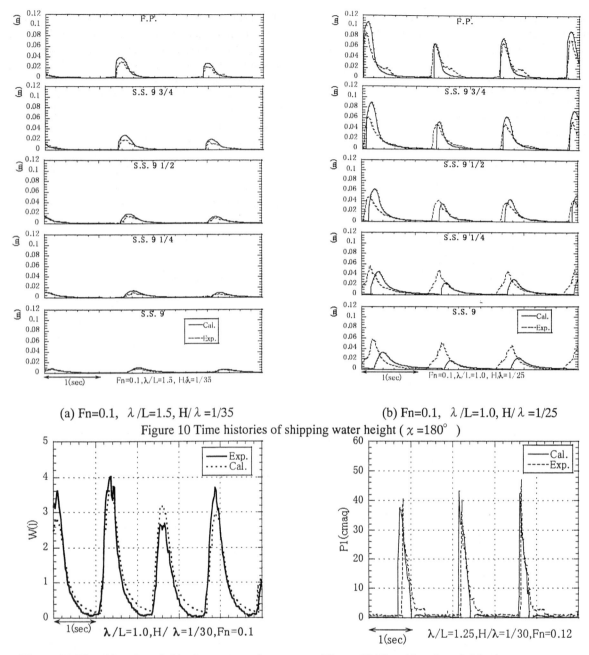

(a) Fn=0.1, λ/L=1.5, H/λ=1/35   (b) Fn=0.1, λ/L=1.0, H/λ=1/25

Figure 10 Time histories of shipping water height (χ =180°)

Figure 11 Time histories of shipping water volume (χ =180°)

Figure 12 Time histories of shipping water pressure (χ =180° , 120mm fore from F.P.)

Tasaki [8] reported that an inflow breadth of shipping water is in proportion to exceeded height of relative water height $\delta = f(t)_{max} - f'$ and determined that an inflow breadth of shipping water B_0 equals 0.8 δ based on the volume of shipping water per a

encounter period. From our experimental result, B_0 is determined as B_0=1.1 δ . It is considered this discrepancy is probably due to the difference of the bow shape.

For examining the validity of this model, the

input data of relative water height is given by measured one. In the calculation, we assume that the slope of the channel bed i_0 is same as $\sin\theta_{max}$ where θ_{max} is maximum pitching angle and the inflow of shipping water begins just after the relative water height level is maximum, so the start time of calculation, t=0, is set on the relative water height level is maximum.

The comparison between measured and calculated of time histories of shipping water is shown in Fig.10. The agreement is good in every section. From these results, it is found that this method is useful for predicting shipping water. But,

as shown in Fig.10 (b) when the wave height becomes large, the agreement between calculated and measured shipping water height is not so good at the backward of the end of Bulwark, S.S.9 1/2. Predicted peak value of shipping water height decreases along the deck, because the effect of the change of the deck breadth is included in the calculation. However, measured data of shipping water height does not decrease as predicted value. From the observation with the video camera, it is concluded that these differences are due to the inflow from beside of the deck and our predicting method is based on that inflow of shipping water occurs only from the stem.

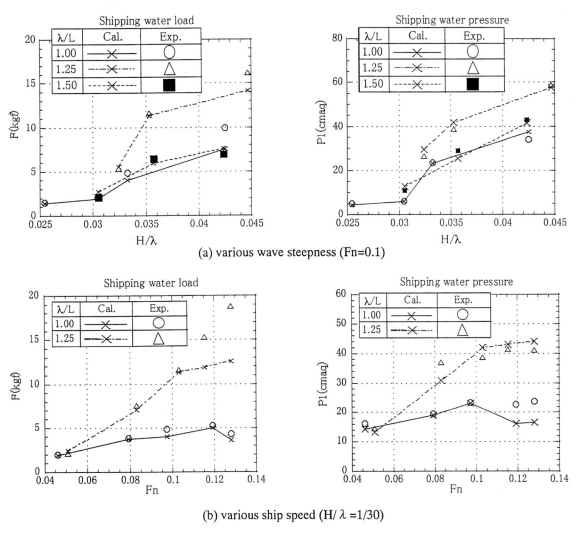

(a) various wave steepness (Fn=0.1)

(b) various ship speed (H/λ =1/30)

Figure 13 Predicted shipping water load and pressure in head sea (χ =180°)

5. DISCUSSION

The comparison of time histories of shipping water volume and pressure is shown in Fig.11 and Fig.12 respectively. Predicted data is well agreed with measured one. From the result of shipping water volume, it is found that the rate of change of shipping water volume, that is important for predicting shipping water load, is also well agreed. It is confirmed that the combination of these two models mentioned in the previous chapter is practical enough for predicting shipping water load and pressure.

The comparison of shipping water load and pressure in head sea at various wave steepness and ship forward speed is shown in Fig.13. Predicted data is well agreed with measured one. The agreement between predicted and measured is not so good when wave height and ship forward speed become large. The difference is due to the inflow from beside of the deck mentioned in the previous section.

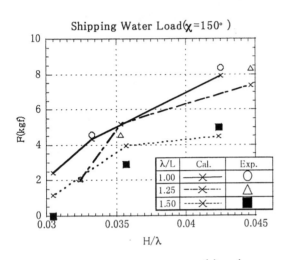

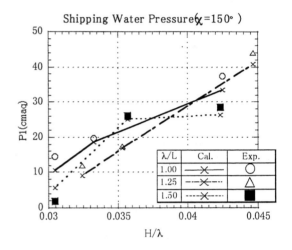

(a) various wave steepness (Fn=0.1)

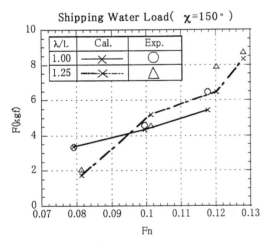

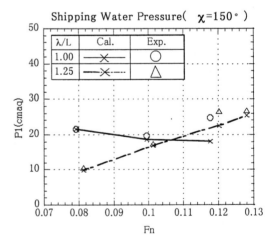

(b) various ship speed (H/λ =1/30)

Figure 14 Predicted shipping water load and pressure in bow sea (χ =150°)

The comparison of shipping water load and pressure in bow sea at various wave steepness and ship speed is shown in Fig. 14. Predicted data is well agreed with experimental one. In head sea, the inflow from beside of the deck is not negligible when wave height and ship speed becomes large. But, from the observation results by the video camera, it is found that little shipping water strands on the deck. It is confirmed inflow from stem is dominant and inflow from beside of the deck is negligible for predicting shipping water. The same result is concluded by Kitagawa [9]. So, it is concluded inflow from stem is dominant for deck wetness in bow sea.

From these results, it is confirmed the combination of these two methods is practically useful for predicting shipping water height, its load and pressure.

6. CONCLUSION

A model test was conducted for a Japanese domestic tanker in regular head and bow seas in order to measure the shipping water height distribution, the load and pressure due to deck wetness. Making use of this data a prediction method for these quantities is proposed and validated.

The conclusions are as follows.

(1) The static component of shipping water is dominant for the load and pressure on deck. But the momentum component is indispensable to estimate the peak value, which is very important for design and rule. That component can be correctly calculated by equation (1).

(2) The theory of "flood waves" gives a good estimation of the time history and distribution of shipping water height on deck. This theory is better than the conventional model of "dam breaking" because it can include the effect of ship forward speed.

(3) The combination of these two models can predict the quantities relating to deck wetness. The necessary data of these methods is relative water height at bow, pitching angle and bow vertical velocity. This method can be a practical tool for estimating the quantities as long as strip theory gives a good estimation of the required input data.

REFERENCES

1. M. Kawakami and K. Tanaka, Stochastic Prediction of Impact Pressure due to Shipping Green Sea on Fore Deck of Ship (in Japanese), Transaction of The West Japan Society of Naval Architects, vol.50 (1976)
2. S. Takezawa, K. Kobayashi, I. Hagino, K. Sawada, On deck wetness and impulsive water pressure acting on the deck in the head seas (in Japanese), Journal of The Society of Naval Architects of Japan, vol. 141 (1977)
3. Buchner, B., On the impact on green water loading on ship and offshore unit design, PRADS'95 (1995)
4. Stoker, J., Water waves, Pure and Applied Mathematics–Volume Ⅳ, INTERSCIENCE PUBLISHERS INC., (1957)
5. K. Goda, T. Miyamoto, A Study of Shipping Water Pressure on Deck by Two Dimensional Ship Model Tests (in Japanese), Journal of The Society of Naval Architects of Japan, vol. 140 (1976)
6. S. Mizoguchi, Analysis of Shipping Water with the Experiments and the Numerical Calculations (in Japanese), Journal of The Society of Naval Architects of Japan, vol. 163 (1988)
7. Shoichiro Hayami, On the Propagation of Flood Waves, Bulletin of Disaster Prevention Research Institute, Kyoto University, No. 1 (1951)
8. R. Tasaki, On Shipping Water (in Japanese), Monthly Report of Transportation Technical Research Institute, vol. 11, No.8 (1961)
9. H. Kitagawa, A. Kakugawa, Experimental Study on Shipping Water of the Ship in Waves (Part 1; On Shipping Water onto Deck of the Ship in Oblique Regular Waves) (in Japanese), Journal of The Society of Naval Architects of Japan, vol. 143 (1977)

1998 Elsevier Science B.V.
Practical Design of Ships and Mobile Units
M.W.C. Oosterveld and S.G. Tan, editors.

A STUDY ON MOTION ANALYSIS OF HIGH SPEED DISPLACEMENT HULL FORMS

Predrag Bojovic[*] and Prasanta K. Sahoo[**]

Australian Maritime Engineering CRC Ltd.

PO Box 986, Launceston, Tasmania 7250, Australia

ABSTRACT

Motion analysis can be rationally included in the ship design process. This paper attempts to outline a round bilge hull form systematic series approach to determine the seakeeping performance due to changes in hull form geometry. This paper presents different mathematical models, suitable for regression analysis, so as to predict the motion characteristics of round bilge hull forms. It has been shown that combining nonlinear estimation techniques with different loss functions gives more accurate predictions than the 'classical' multiple regression analysis approach.

The importance of the mathematical model's selection has been discussed and a data presentation technique suitable for motion analysis prediction has been suggested. The resulting prediction procedure illustrates that an effective design tool has been developed for use by designers in the preliminary ship design process.

1. Introduction

1.1 Background

Over a ten year period, starting in 1979, a major research project on combatant vessels design was conducted at Maritime Research Institute Netherlands (MARIN). This program was initiated by the growing belief that a significant improvement in the performance of transom stern, round-bilge monohulls could be obtained, especially with regard to their seakeeping characteristics. The project was jointly sponsored by the Royal Netherlands Navy, the United States Navy, the Royal Australian Navy and MARIN.

Extensive testing in calm water and waves were carried out on a systematic series of high speed displacement hull forms (HSDHF), as described in [1] [2] [3] and [4]. The test data of forty models were analysed and included in a powerful computer system. Except for the parent hull, the results of the tests and the analysis were not published.

1.2 AMECRC Systematic Series

The AME CRC systematic series is based on the HSDHF systematic series. The research work on this project started in 1992, as described in [5]. The parent model

---

[*] Research Fellow, AMECRC Ltd.
[**] Lecturer, Australian Maritime College, PO Box 986, Launceston, TAS 7250, Australia, on part secondment to AMECRC Ltd.

is very similar to that of the HSDHF series and has the following parameters: L/B = 8.0, B/T = 4.0 and C_B = 0.396. The series transformation procedure is based on the variation of L/B, B/T and C_B parameters and are as follows:

L/B from 4.0 to 8.0
B/T from 2.5 to 4.0
C_B from 0.396 to 0.5

This 'parameter space' or series 'cube' is shown in Figure 1.1. The parameters of each of fourteen models can be identified from this figure. All models have the same length of 1.6 m and the influence of change of the series' parameters on the hull shape is illustrated in Figure 1.2, where all models' body plans are presented in same scale.

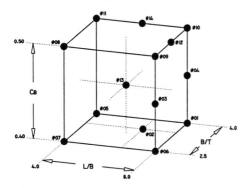

Figure 1.1 : AMECRC Systematic Series

1.3   Regular Wave Tests

The standard set of regular head wave tests, performed alongside the calm water resistance tests, for each model is as follows:

- standard design displacement, level trim
- pitch radius of gyration: 0.25 L (L = 1.6 m, for all models)
- forward model speeds: 1.13, 2.26 and 3.39 m/s (Fn = 0.285, 0.57 and 0.856)
- model wave frequency range: 0.6 to 1.4Hz
- nominal model wave height: 30 mm

The regular head wave testing is based on an important assumption that the linear motion responses, meaning that the motion amplitude is proportional to the wave amplitude. Additional information regarding

that assumption can be found in Ref. [9] and [10].

The overview of the testing performed is illustrated in Table 1.1. It illustrates that some models were not tested at the highest speed and that only about 50% of the tests included the measurement of the drag in waves.

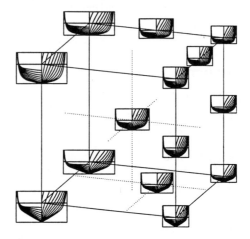

Figure 1.2: AMECRC Systematic Series Body Plans

2   Test Data Presentation

2.1   Data Formats

Data reduction was performed according to the following formulae:

$$\text{Non-dimensional Pitch } (P') = \frac{x_{50}}{k \cdot \zeta_0} \quad (1.1)$$

$$\text{Non-dimensional Heave } (H') = \frac{x_{30}}{\zeta_0} \quad (1.2)$$

Non-dimensional Vertical Acceleration

$$(A') = \frac{\ddot{x}_{30} \cdot L}{g \cdot \zeta_0} \quad (1.3)$$

Non-dimensional Added Resistance

$$(R') = \frac{R_{RW} - R_{CW}}{\zeta_0^2 \cdot \rho \cdot g \cdot \dfrac{B^2}{L}} \quad (1.4)$$

The vertical acceleration transfer function used in reference [11] has the following form

$$A' = \frac{\ddot{x}_{30}}{\zeta_0 \cdot \omega_e^2} \quad (1.5)$$

547

Model Speed (m/s)	#01	#02	#03	#04	#05	#06	#07	#08	#09	#10	#11	#12	#13	#14
1.13	◎	O	◎	O	◎	◎	◎	◎	◎	O	O	◎	◎	O
2.26	◎	O	O	O	◎	◎	◎	◎	◎	O	O	◎	O	O
3.39	◎	O	O	O	◎				◎	O		◎	O	O

O - Drag not recorded
◎ - Drag recorded

Table 1.1 : Testing overview

While Equation 1.3 is considered to be the standard way of presenting the seakeeping test data, Ref. [12], Equation 1.5 provided direct link between heave transfer function and vertical acceleration function at LCG. But in order to keep consistency the vertical acceleration, data was presented according to Equation 1.3.

2.2 Data Analysis

In order to enable regression analysis of the test data, it was necessary to transform the data into the tabulated form. Therefore, a tedious manual process of approximating the test data with best-fitting transom-function curves was performed. Subsequently, the necessary tabulated values were read-out from these curves. Using so-called 4D data presentation, developed earlier in the work. The trends of parameter variation were visualized and any outstanding test values were carefully checked so as to ensure that test data (sometimes containing a significant scatter) was not misinterpreted during the initial tabulation. If found that the tabulated point in question corresponds well with the test data, no attempt was made to change its value, even when it didn't correspond to other trends noticed within the series' space.

This paragraph briefly describes 4D charts and how they should be read. The "cube-like" structure in Figure 2.1 represents the series' parameter space. The model #07, with the lowest combination of L/B, B/T and C_B parameters (4, 2.5 and 0.396 respectively) is marked in the figure. Knowing that all trend lines going away from this model present an increase of appropriate parameter, all series' models could be easily identified on this chart.

The "x-values" of the models from Figure 2.1 were consistently used on all 4D charts developed in this project. Just "y-values" were replaced by the corresponding value of the dependent variable. Until the user becomes familiar with this way of presentation, Figure 1.1 should be used to identify the relationship between the series' models. Table 2.1 presents the order (from left to right) in which models are presented on all 4D charts for easy understanding.

For each model, charts of pitch, heave, vertical accelerations at LCG and added resistance (where available) were prepared as functions of model wave frequency, together with the tabulated curves. The charts for vertical acceleration have an extra data series added to them derived from the heave transfer function according to Equation 2.3, based on Equation 1.5. For a periodic heave function, Equation 2.1, at LCG is:

$$x_3 = x_{30} \cdot \sin(\omega_e \cdot t + \varepsilon_3) \quad (2.1)$$

and the second derivative (vertical acceleration) is given by:

$$\ddot{x}_3 = \omega_e^2 \cdot x_3 \cdot \sin(\omega_e \cdot t + \varepsilon_e)$$
$$= \ddot{x}_{30} \cdot \sin(\omega_e \cdot t + \varepsilon_3 + \pi) \quad (2.2)$$

Therefore non-dimensional vertical acceleration could be shown to be:

$$A' = \frac{\ddot{x}_{30} \cdot L}{g \cdot \zeta_0} = \frac{\omega_e^2 \cdot x_{30} \cdot L}{g \cdot \zeta_0}$$
$$= \frac{x_{30}}{\zeta_0} \cdot \frac{\omega_e^2 \cdot L}{g} = H' \cdot \frac{\omega_e^2 \cdot L}{g} \quad (2.3)$$

A very good correlation between the "measured" accelerations (actually calculated from the forward and aft post movement, as described in [8]) and the ones obtained using Equation 2.3 can be seen. The approach outlined in Equation 2.3 enabled successful "retrieving" of the missing data and subsequently more complete data analysis.

548

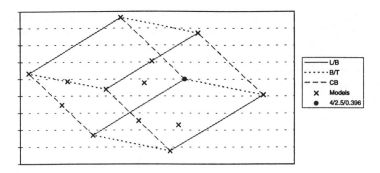

Figure 2.1: 4D chart explanation

#09	#03	#12	#06	#10	#08	#04	#13	#14	#01	#02	#07	#11	#05

Table 2.1: Models' sequence on 4D charts (from left to right)

3. Regression Analysis

3.1 Speed/Frequency Dependent/ Independent Regression Model

Ship resistance regression models may be broadly categorized into two groups: speed-independent and speed-dependent models. In speed-independent regression models, ship's speed is not included as an independent variable, and separate regression equations must be generated at a series of discrete speeds covering the range of interest. Fung [13] discusses that the major shortfall of speed-independent regression models is that the predicted resistance curves do not always vary properly with speed, despite the high statistical correlation which may be achieved at any individual speed. This is

because the resistance computed at one speed is not directly linked to that at another speed since the speed variable is not explicitly included in the regression. In speed-dependent regression models ship speed is explicitly included as an independent variable, providing direct control over the nature of variation of resistance with speed. Detailed discussion of these two types of models can be found in [14]. According to [15], the speed-independent model provides a superior analysis as it allows the different contributions of the various hull-form parameters, at different speed, to come into play.

A discussion similar to the above could be undertaken in regard to frequency-

dependent and frequency independent regression model for the analysis of seakeeping data. The authors view is that frequency-independent approach provides an additional benefit, as consistency between independent predictions provides ultimate proof of the reliability of the developed prediction model.

3.2 Overview of the General Regression Methods

The Multiple Regression Analysis (MRA) and Nonlinear Estimation (NLE) modules of StatSoft's data analysis package, Statistica for Windows v.5.0, are the most suitable for the type of analysis intended here. The MRA module provides a more robust statistical analysis, while NLE model enables a more flexible, fine-tunning analysis, but without the control over the statistical significance of the parameters used in the equation. Both modules were used in [7], by first determining the form of the most influential parameters by MRA and then further exploring that form using the NLE's flexible control over the error function. In this report, only Multiple Regression module was used, as test data scatter didn't justify usage of a fine-tuning tool like NLE.

4. Regression Analysis of Systematic Series Seakeeping Data

4.1 Regression Analysis Mathematical Models

As mentioned earlier separate analysis was performed for each frequency in the range of 0.6 to 1.4 Hz, with increment of 0.1 Hz. The data was tabulated in model wave frequency domain, as all models at all speeds were tested over this range of frequency. Otherwise, it would be necessary to use different frequency ranges for different speeds, in accordance to Equation 4.1, valid for head seas:

$$\omega_e = \omega + \frac{\omega^2 \cdot V}{g}$$

$$= 2 \cdot \pi \cdot f + \frac{(2 \cdot \pi \cdot f)^2 \cdot V}{g} \qquad (4.1)$$

The dependant variables were used in the format outlined in Equations 1.1 to 1.4 . Table 1.1 illustrates that all fourteen models were tested only at two lower speeds. Unfortunately, the limited data sample at the highest speed was found to be too small for the development of the reliable prediction method. The analysis of added resistance found that the reasonable predictions could be obtained for the lowest speed only.

4.2 Selection of Independent Variables

Table 4.1 presents the parameters which have a constant value for all series' models. The variables not included in the regression model should either have a constant value throughout the data or should have an insignificant influence on the dependent variable. Selection of the independent variables should include all possible influential parameters. On basis of the experience from calm water data analysis, independent variables were generated using the following form:

$$\left(\frac{L}{B}\right)^{m_1} \cdot \left(\frac{B}{T}\right)^{m_2} \cdot (C_B)^{m_3} \qquad (4.2)$$

LCB	5.4% aft from midships
C_P	0.626
C_{WL}	0.796
A_T/A_X	0.296
B_T/B_X	0.964
C_M	$C_P \times C_B$

Table 4.1: Hull parameters common for all systematic series models

The limitation of Statistica for Windows package is 100 variables per analysis. It influenced the derivation of the independent

variables. The first 27 independent variables were obtained as all possible combinations of m_i exponents from the following table:

m_1	0	-2/3	-4/3
m_2	0	-2/3	-2/3
m_3	0	+1/3	+2/3

Table 4.2: Exponents used for the first set of independent variables

The exponents were selected in such a way that some of the possible combinations present $L/\nabla^{1/3}$ and $(L/\nabla^{1/3})^2$ values, according to Equation 4.3. The $L/\nabla^{1/3}$ ratio is known to be the single most important parameter for calm water performance of this type of vessels:

$$\frac{L}{\nabla^{1/3}} = \sqrt[3]{\frac{\left(\frac{L}{B}\right)^2 \cdot \frac{B}{T}}{C_B}}$$

$$= \left(\frac{L}{B}\right)^{2/3} \cdot \left(\frac{B}{T}\right)^{1/3} \cdot C_B^{-1/3} \qquad (4.3)$$

m_1	-	-	0	-1	-2
m_2	-	-	0	-1	-2
m_3	-2	-1	0	1	+2

Table 4.3: Exponents used for the second set of independent variables

The additional 45 variables were generated using the m_i exponents from Table 4.3. These exponents provide variables which correspond to linear and quadratic basic series parameters, and their couplings. From both sets of variables the variables corresponding to all zero exponents were excluded, as they represent the intercept which is already present in the mathematical model used. So, the final set of variables had 70 independent variables. The role of the multiple regression analysis was to identify the most influential parameters among them and to quantify their influences (determine the appropriate coefficients).

Regression analysis of vertical acceleration data was not performed. The relationship between heave and vertical acceleration at LCG, Equation 2.3, was used to predict vertical acceleration transfer functions, from the predicted heave transfer function. This is believed to be at least as accurate as the regression analysis of vertical accelerations would be.

Table 4.4 presents the R^2 values obtained in the analysis. Some pretty low values of R^2 values could be seen there. Some test results contain opposite trends for the same change of basic parameter (i.e. change of L/B between two models, for example, suggests an increase in the motions, while the same change between other two models may suggest the opposite). Also, some models experienced stronger pitch-heave coupling (two pronounced peaks on their response curves).

5. Software Implementation

The spreadsheet implementation of the systematic series prediction software was used for calm water performance prediction software described in Ref [6] and [7]. Following the same practice, the seakeeping prediction software was also developed as MS Excel 5.0 spreadsheet. Spreadsheet development enables its use together with previously developed spreadsheets. Adding worksheets to calm water prediction software could be done simply by copying all of them (using the "Move or Copy Sheet..." command from Edit menu) and by linking cells with L/B, B/T and C_B values For convenience sake, all worksheet names in this spreadsheet start with "SK".
Figure 5.1 shows the main worksheet of the spreadsheet. At the bottom of the screen, a worksheet structure can be seen:

- **SK Prediction** - contains brief instructions, data input and prediction data

- **SK Pitch** - plot of predicted non-dimensional pitch vs encounter wave frequency for Fn=0.285 and Fn=0.57

- **SK Heave** - plot of predicted non-dimensional heave vs encounter wave frequency for Fn=0.285 and Fn=0.57

- **SK Vert Accel** - plot of predicted non-dimensional vertical acceleration vs encounter wave frequency for Fn=0.285 and Fn=0.57

- **SK Added Res** - plot of predicted non-dimensional added resistance vs encounter wave frequency for Fn=0.285

- **SK Data** - contains charts' data

- **SK Functions** - contains listings of user-defined functions

User-defined functions in MS Excel spreadsheet enable programming within the spreadsheet. With just a minor syntax of changes, they could be transported to any the popular programming languages.

Wetted surface area estimation for any model within the series' parameter space can be evaluated as per Ref [7].

6. Concluding Remarks

This report provides a comprehensive presentation of the test data together with the developed predictions. The tabulated test data was analysed using the multiple regression analysis method previously used. It is strongly believed that the quality of the analysis was of the higher order than the quality of the test data analysed. The test results contained some scatter, which may be explained by the fact that the testing was done over the period of five years, by numerous people, most of them without previous towing tank testing experience. However, it is believed that the prediction method developed provides reasonable predictions for the early design. It is to be noted that this regression model has not been tested on any random hull form data falling within the AMECRC systematic series.

7. Acknowledgments

The efforts of numerous people within AMECRC are much appreciated. The support of Australian naval industry, namely the industry participants of ADI, ASC, NES and TDS, throughout the task life is gratefully acknowledged. The availability of Dr. Patrick Couser and Damien Holloway for discussions, their time and expertise are very much appreciated. Special thanks go to Gregor Macfarlane and his team for self initiated compiling of the seakeeping test data and its presentation in Ref [8].

Frequency (Hz)	Heave (Fn=0.285)	Heave (Fn=0.57)	Pitch (Fn=0.285)	Pitch (Fn=0.57)	Vertical Accel. (Fn=0.285)	Vertical Accel. (Fn=0.57)	Added Resistance (Fn=0.285)
0.6	0.896	0.967	0.836	0.394	0.896	0.967	0.945
0.7	0.947	0.285	0.971	0.820	0.947	0.285	0.984
0.8	0.931	0.955	0.976	0.974	0.931	0.955	0.636
0.9	0.481	0.898	0.944	0.972	0.481	0.898	0.849
1	0.656	0.979	0.827	0.985	0.656	0.979	0.894
1.1	0.694	0.308	0.951	0.961	0.694	0.308	0.988
1.2	0.936	0.954	0.968	0.399	0.936	0.954	0.910
1.3	0.942	0.939	0.895	0.870	0.942	0.939	0.000
1.4	0.940	0.866	0.939	0.941	0.940	0.866	0.000

Table 4.4: Regression analysis R^2 values

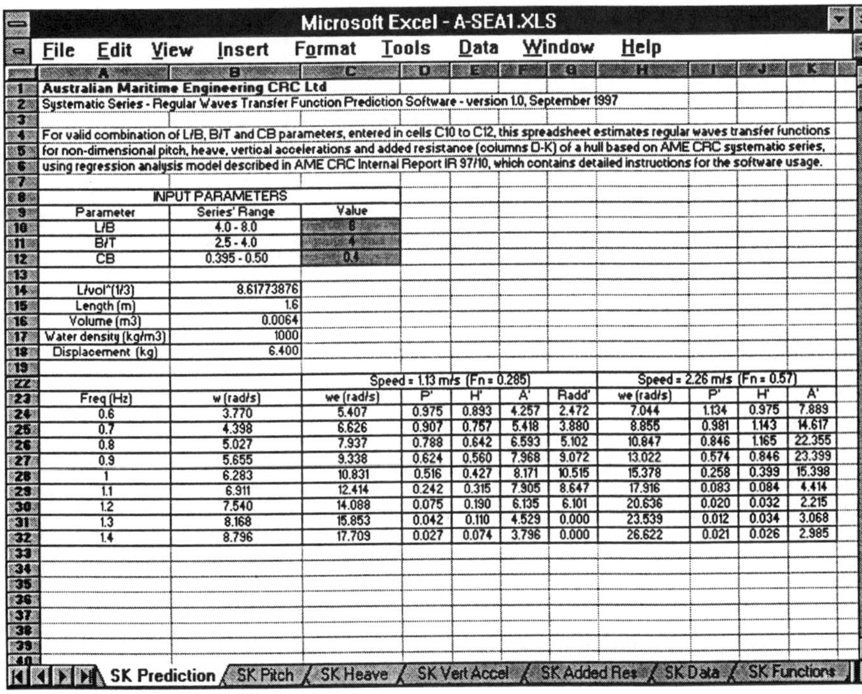

Figure 5.1: Seakeeping Performance Prediction Software

552

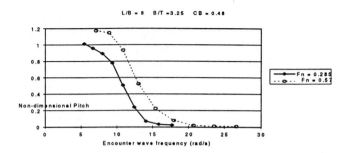

Figure 5.2: Prediction example:'SK Pitch' worksheet

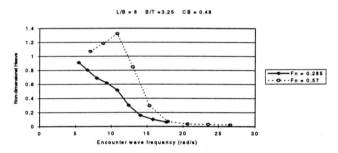

Figure 5.3: Prediction example: 'SK Heave' worksheet

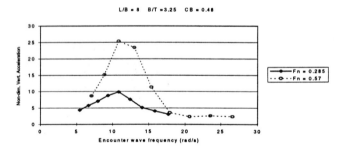

Figure 5.4: Prediction example: 'SK Vert Accel' worksheet

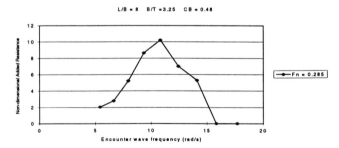

Figure 5.5: Prediction example: 'SK Added Res' worksheet

8.   References

1. Blok, J.J., Beukelman, W., "TheHigh Speed Displacement Ship Systematic Series Hull Forms - Seakeeping Characteristics", Trans. SNAME, Vol. 92, pp 125-150, (1984).

2. Van Oossanen, P., Pieffers, J.B.M., "NSMB-Systematic Series Of High Speed Displacement Forms", Workshop on Developments in Hull Form Design, Wageningen, 16pp, (1985).

3. MARIN, "MARIN High Speed

Displacement Hull Form Designs", MARIN Report 30, December, (1987).

4. Robson, B.L., "Systematic Series Of High Speed Displacement Hull Forms For Naval Combatants", *Trans. RINA*, Vol. 130, pp 241-252, (1988).

5. IR 92/1 Rikard-Bell, M.,"Report of Research - October/November 1992", AME CRC Internal Report 92/1,(1992).

6. Bojovic, P., "Regression Analysis of AME CRC Systematic Series - Calm Water Testing Results", AME CRC IR 95/24, (1995)

7. Bojovic, P., "Re-Analysis of AME CRC Systematic Series - Calm Water Testing Results", AMECRC IR 96/10,(1996).

8. Macfarlane, G., Shaw, M., Lees, M., "Presentation of the AME CRC Systematic Series Regular Wave Testing Results (Project 2.1.2 Interim Report)", AME CRC IR 95/25,(1995).

9. Macfarlane, G.J., Renilson, M.R. "A

Note on the Effect of Non Linearity on the Prediction of the Vertical Motions of a Small High Speed Craft", Proceedings RINA International Conference on Seakeeping and Weather, also AMECRC CR 95/1,(1995)

10. Boyd, J.C.M., "Analysis of Non-Linear Vessel Motions Using Modified Linear Strip Theory", Master's Thesis, Curtin University of Technology,(1995).

11. Wellicome, J.F., Temarel, P., Molland, A.F., Couser, P.R., "Experimental Measurements of the Seakeeping Characteristics of Fast Displacement Catamarans in Long-crested Head-seas", Ship Science Report 89, University of Southampton, Department of Ship Science,(1995).

12. Lloyd, A., "Seakeeping - Ship Behaviour in Rough Weather", Series in Marine Technology, Ellis Hprwood Limited,(1989).

13. Fung, S.C., "Resistance And Powering Prediction For Transom Stern Hull Forms During Early Stage Ship Design", Trans SNAME, Vol. 99, pp29-73,(1991).

14. Fung, S. C., Leibman, L., "Statistically-Based Speed-Dependent Powering Predictions For High-Speed Transom Stern Hull Forms", NAVSEA 051-05H3-TN-0100,(1993).

15. MacPherson, M., "Reliable Performance Prediction : Techniques Using A Personal Computer", Marine Technology, Vol.30, No.4, pp243-257, October, (1993).

Hydrodynamic Development for a Frigate for the 21 Century

G.K. Kapsenberg[1] and R. Brouwer[2]

[1] Maritime Research Institute Netherlands
P.O. Box 28, 6700 AA Wageningen, The Netherlands

[2] Royal Netherlands Navy, Department of Maritime Technology
P.O. Box 20702, 2500 ES The Hague, The Netherlands

Abstract

The paper discusses the hydrodynamic development for a new type of frigate which is a first step in the replacement of the current Kortenaer-class frigates of the Royal Netherlands Navy. The development was based upon a systematic series of hull forms and a numerical evaluation of the hydrodynamic characteristics. A major task appeared to be the quantification of the seakeeping behaviour.
The paper presents the results of the study and identifies the hull form parameters with a large impact on the hydrodynamic characteristics. A new type of hull form evolved from this study which is characterised by the centre of floatation being located on an extreme aft position (COFEA hull form). The numerical predictions were verified with model tests.

1. INTRODUCTION

Considering the total number (16) and the different types (4) of frigates within the Royal Netherlands Navy fleet today, there is a continuous process of study and development into new concepts and designs for frigate type ships. Several studies were carried out to look into the applicability of different concepts like the Swath, SES and more recently the trimaran. The study described here is directed to the original concept of a monohull, since the monohull is still considered to be a cost effective solution for a frigate capability. The idea was to look into a wide range of hull form parameters and thus finding novel shapes for which a considerable decrease of propulsion power and an increase of the seakeeping performance could be obtained. In other words: to find the Holy Grail of naval architecture. The goal was set rather ambitiously at reducing the required propulsion power with a factor two and increasing the seakeeping capabilities with a factor two also. This goal was set to force the

research to tread outside the existing paths and come up with an innovative hull form design.

2. SET-UP OF THE SYSTEMATIC STUDY

2.1 Comparison of hull forms

In order to differentiate between good and bad hull forms from a hydrodynamic point of view, the powering performance and the seakeeping capabilities have to be quantified. Although the intention was to combine resistance requirements and seakeeping capabilities in a single figure expressing the total hydrodynamic quality of a design, this did not appear to be useful in this part of the study, therefore two separate figures are being used. Where the comparison of power requirements for different designs is very straightforward (see 4.2), comparing seakeeping capabilities is a more complex task. Within NATO (North Atlantic Treaty Organization) a common procedure for the assessment of seakeeping in ship design by means of a so called operability figure was developed [1]. The method was used in this study and is shortly discussed hereafter.

2.2 Seakeeping assessment procedure

Operability can be defined as the proportion of time the ship is able to successfully accomplish its missions for all given combinations of sea areas, speeds, headings and sea-states. A mission or operation will be considered successful, if all systems required for that mission are operational. The statistical nature of ship motions (irregular seas) and naval ship operations (changing missions, speeds etc.) requires a probabilistic approach [2] for the assessment of seakeeping performance. Keeping this in mind the operability figure for seakeeping as used in this study is defined as follows:

$$\text{Operability} = \sum_{j,k,l,m} p_j * p_k * p_l * p_m * \Gamma_{(jklm)}$$

Where p_j, p_k, p_l, p_m are the probabilities of respectively a naval mission, ship speed, heading relative to the waves and a sea state. The factor "Γ" describes the capability of the design to successfully perform the given mission "j", with the ship motions resulting from sailing with speed "k" and heading "l" in sea state "m". The summation for all combinations of missions, speed, headings and sea states gives the seakeeping operability figure for the design.

In this study the following naval missions were considered:

- Transit and Patrol
  (TaP, $p_{j=1}$ = 32%)
- Replenishment at Sea
  (RAS, $p_{j=2}$ = 8%)
- Anti-Submarine-Warfare
  (ASW, $p_{j=3}$ = 20%)
- Anti-Air-Warfare
  (AAW, $p_{j=4}$ = 20%)
- Anti-Surface ship-Warfare
  (ASuW, $p_{j=5}$ = 20%)

The probability for each of these missions was based on the expected time of the ship being involved in these missions as a fraction of total sailing time. A so called mission speed profile was used to determine the relevant ship speeds and their probability for each of the missions. The probability of the heading relative to the waves was considered to be uniformly distributed over head to following seas, except for RAS operations, where head to beam seas only were considered. The probability distribution of sea states was derived from the available wave statistics for the North-Atlantic Ocean during wintertime. No other areas or seasons were considered.

The calculation of the ship performance factor "Γ" for any combination of mission, speed, heading and sea state is described below.
First all "systems" required to accomplish a given mission were determined. The following subdivision was used:

- Ship hull, propulsion and auxiliary systems
- Helicopter
- Personnel at several locations
- Weapons & sensors

Secondly ship motion criteria for these systems were derived from literature [1]. Thus, so called "mission criteria sets" could be defined for each mission. The types of criteria used were varying from general motion criteria (heave, roll, pitch, accelerations, green water, slamming etc.) to human performance criteria such as Motion Sickness Incidence (MSI) based on the vertical accelerations and Motion Induced Interruptions (MII) based on roll, vertical and lateral accelerations. Finally these "mission criteria sets" were compared with the ship motions calculated for all given combinations of speed, heading and sea state. If all criteria were satisfied, "Γ" was defined to be "1", else "Γ" was assumed to be zero for that condition. So no gradual performance degradation as a function of ship motions was considered in this study.

2.3 Computational tools

The powering performance of each hull form was predicted using the DESP program [8]. This program is based on statistical analysis of a very large number of models tested at MARIN. The hull forms range from tankers to frigates, tugs and ferries; so they cover a large range of hull form parameters and coefficients. The program was tuned for the parent hull form, the so called Point Design in this study, for which model test results were available. For the speed range of interest, 18 to 30 knots (i.e. cruise to maximum speed), the maximum difference between predictions and model tests was 6%. The ship motions were calculated with the MARIN strip theory program SHIPMO. This program has an option to calculate added mass and damping using a 2-D diffraction method. This method was used since some of the hull forms in this project have SWATH-like sections over a large part of the hull; the generally used conformal mapping method is not suitable for such a hull form. The operability of the vessel for the specified missions, speeds and headings in the wave climate of the North-Atlantic was determined using the WASCO program [4]. This program calculates motions, velocities, accelerations MSI and MII at any location on the ship using the RAO's of the 6 motions at the centre of gravity as calculated by SHIPMO.

3. DEVELOPMENT OF THE SYSTEMATIC SERIES

3.1 The Point Design

As a parent hull form for this systematic series, the so-called Point Design (PD) of the MO2015 Navy study into the design of frigates for the next century was chosen. A body plan of this hull form is given in Figure 1, the main hull form parameters are presented in Table 1. This hull form could be described as a modern type frigate design. This design serves as a benchmark for all its derivatives. The PD was equipped with bilge keels with a length of 30.0 m and a height of 0.90 m. Also 6.3 m² active anti-roll fins were added to provide roll damping. These appendages were used to determine the speed dependent roll damping.

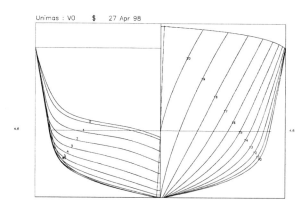

Figure 1.   Body plan of Point Design

Table 1.   Main dimensions and main hull form parameters of the Point Design

Length perpendiculars	120.00 m
Beam (max)	17.80 m
Draft	4.60 m
Displacement	4250 ton
LcB (ref Station 10)	-2.28 m
LcF (ref Station 10)	-7.73 m
GM	1.73 m
C_{WP}	0.7867
C_B	0.4747
C_P	0.6272

3.2 Stability requirements

The static stability criteria as used in this study are based on the comparison of a wind heeling moment and the ship's righting moment in calm water as described by Sarchin and Goldberg [5] and are generally used for navy ships throughout the world. More recently guidelines were developed for an assessment of dynamic stability when sailing in large stern quartering extreme seas [6]. The result of this study is a requirement for the range of positive stability and a minimum value for the total dynamic stability (i.e. range and area of GZ-curve). Since these two dynamic requirements could be met by including a part of the superstructure as watertight compartment, they were not being used further for this systematic series. The remaining stability requirements were used to calculate the maximum allowable KG, which was used with a small margin as an input for the seakeeping calculations.

3.3 Derived hull forms

The systematic series were built without too much prejudice on good and bad hull forms. All sorts of parameters were used; in general hull forms having lower and higher values of that parameter than the PD were designed. It was however accepted that, especially to influence the seakeeping behaviour, rather large variations of each parameter were to be considered. Changing the value of a parameter usually leads to changes in the displacement. The basic idea behind not keeping the displacement constant was the fact that frigate type ships tend to grow dramatically during the design process. So therefore the objective of this study was not in the first place to design a hull form with the same displacement as the Point Design, but to provide the ship designer with guidelines, indicating the best way to realize a required change of ship size. In this way weight calculations were avoided for each derivative, the change of displacement as a direct consequence of the parameter variation was accepted.

Using this approach, hull forms were designed with:

Increased length,	+ 20% and + 40%
Variation in beam,	± 20%
Variation in C_B,	± 20%
Variation in C_P,	± 10%
Variation in LcB,	± 100%
Variation in LcF,	± 100%

Note: The LcB and LcF variation is relative to Station 10.

The main dimensions and hull form coefficients of this series is given in Table 4 at the end of the paper. The increased length hull forms cannot be regarded as a step towards the Enlarged Ship Concept (ESC) as introduced by Keuning and Pinkster [7]. The displacement increases in this case linear with the length; the idea behind the ESC is to increase the length without essentially increasing the displacement.

4. RESULTS OF THE STUDY

4.1 Performance of the Point Design

The powering performance of the PD was determined for speeds of 18 and 30 kn in calm water, for a speed of 28 kn in Ss 3 head seas and for 25 kn in Ss 5. The last two quantities were determined to check whether the available propulsion power, which is normally determined by the 30 kn requirement, is sufficient to overcome the added resistance in wind and waves. The effect of voluntary speed loss in waves, due to slamming and green water on deck was incorporated in the calculation of the operability. The result of the powering calculations are:

Condition	Speed V_s [kn]	Required power P_D [kW]
Cruise speed	18	4210
Maximum speed	30	34511
Sea state 3	28	25705
Sea state 5	25	17926

Figure 2 gives the operability for the different missions. Using the defined mission profiles, a mean operability of 52.9 % was determined for the Point Design. The calculated operability for the vessel is relatively low, because the selected wave scatter-diagram was rather extreme and moreover no gradual degradation of the performance was accepted.

Figure 3 shows more detailed results for the TaP mission; it shows the operability as a function of heading and speed. For TaP the operability is lowest for head to bow quartering seas at high speeds.

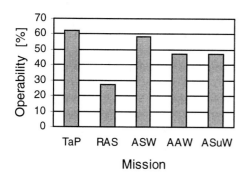

Figure 2. Operability of all five missions.

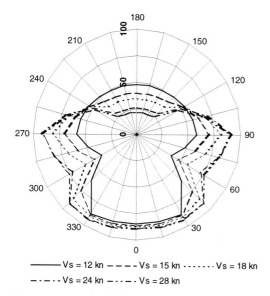

——— Vs = 12 kn – – – – Vs = 15 kn ······· Vs = 18 kn
– · – · · Vs = 24 kn – · · – · – Vs = 28 kn

Figure 3. Operability plot for the Transit and Patrol (TaP) mission (180° = head seas).

The Motion Sickness Incidence (MSI) criterion on the flight deck (10% of personnel) is responsible for the low operability in these conditions. The roll amplitude (2.5° RMS) is the limiting factor for the range of wave directions from beam to stern quartering seas, while the pitch amplitude (1.5° RMS) is limiting in following seas. The latter is hardly a matter of concern since the operability is more than 80% for these headings.

The RAS mission is carried out in only a limited range of conditions: the speed is in between 12 and 18 knots and the wave direction in between head and beam seas. It is the mission with the highest downtime. The vertical replenishment operations by means of a helicopter allow only very low pitch motions (0.75° RMS) and vertical displacements at the flight deck (0.7 m RMS); the pitch amplitude is limiting for the range of headings from head to bow quartering seas. This means that a relocation of the helicopter deck on the ship will not have an effect on the operability. The effect of the low operability for the RAS mission on the overall operability is not very large; since its small probability of occurrence of only 8% of total sailing time.

Most downtime for the ASW mission is caused by the MSI criterion in the Command and Control Centre (10% of personnel). The operability reduces to only 18% for the highest speed in head seas. A change of course to beam seas considerably increases the operability. For very low speeds, below 10 kn., the roll angle (2.5° RMS) is the most important criterion, since the lack of speed reduces the roll damping to very low values.

For The AAW and ASuW mission the vertical velocity criterion (0.5 m/s RMS) at the position of the gun on the fore deck reduces the operability from 16% at 15 kn to 11% at 28 kn in head seas. Again a significant higher operability (45%) is obtained when changing course to beam seas.

In conclusion, it appears that the vertical plane motions are the dominant factor for the low operability of this frigate hull form, so the optimisation should be focused on this parameter which makes the head seas condition the most important one.

4.2 Powering results for the systematic series

It appeared that the required power for a speed of 30 kn in calm water was highest for all cases. Therefore the results of the added resistance calculation in waves is not taken into account in this discussion.

The ranking of the different designs with regard to the powering requirement is based on a fuel consumption basis, with an adopted simple speed profile described by sailing at 18 kn for 80% of time and sailing at 30 kn for 20% of time. Table 2 below uses this profile as a way to add the results of the powering calculation for the two speeds. The results are normalised using the results for the PD, they are based on the fuel consumption per ton displacement.

The results show that an increase in length gives a very large reduction of the fuel consumption. It must be noted that this increase in length is an increase of $L_{PP} = 120 \rightarrow 168$ m.

Table 2. Ranking of hull forms based on fuel consumption. Score compares fuel consumption / ton displacement.

Variation	Score
L ++	0.688
L+	0.752
Cp -	0.907
T +	0.930
LcF +	0.956
B -	0.969
LcB -	0.975
Point Design	1.000
Cm +	1.016
Cm -	1.017
LcB +	1.038
T -	1.071
LcF -	1.084
B +	1.091
Cp +	1.299

A second point is, that a reduction of the fuel consumption of 30% will never be attained. The displacement of the 168 m vessel will be higher than that of the PD because of the increased construction weight; this makes the absolute fuel consumption higher.

An interesting result is, that a fuel reduction of 9% can be achieved by reducing the prismatic coefficient (and hence the displacement) while keeping the main dimensions the same. The loss in displacement could be compensated by an increase of the draft. The prismatic coefficient appears to be a very strong parameter with respect to the fuel consumption. If all hull form changes are considered 'equal', it has the largest effect, even larger than the change in length. This is illustrated in Figure 4 below.

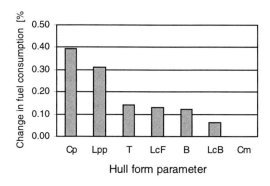

Figure 4. Effect of hull form parameters on the fuel consumption.

4.3 Seakeeping results for the systematic series

The calculated operability as described in Chapter 2 is used for the comparison of the seakeeping qualities of the different designs in these systematic series. If the results are normalised, based on the PD, a ranking similar to that for the powering results can be made.

The results, Table 3 and Figure 5, show a very large effect of the separation between centre of buoyancy and centre of floatation. Contrary to 'common seakeeping knowledge', this separation must not be made as large as possible, but as small as possible (LcB aft and LcF forward). In fact, the hull form at the top of the operability ranking has the centre of floatation right on top of the centre of buoyancy.

Table 3. Ranking of hull forms based on operability for all 5 missions. Score normalized on operability of PD.

Variation	Score
LcF -	1.20
LcB +	1.11
L ++	1.09
T -	1.08
Cp -	1.06
B -	1.05
Cm +	1.05
Cm -	1.04
L+	1.04
B +	1.00
Point Design	1.00
Cp +	1.00
T +	0.98
LcB -	0.98
LcF +	0.78

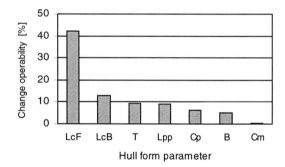

Figure 5. Effect of hull form parameters on the operability.

One important publication which promoted a large separation between centre of floatation and centre of buoyancy is that of Blok and Beukelman [8]. In re-analyzing their results, it appeared that their optimum hull form not only has a large LcB-LcF separation, but maybe more importantly a very large water plane area, so a very low vertical prismatic coefficient.

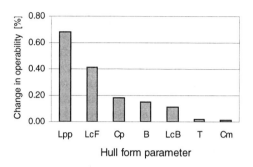

Figure 6. Effect of hull form parameters on the operability in head seas only.

The result of this study is very much in line with the last sentence in Lloyd's book on seakeeping behaviour [2] 'A large forward water plane area coefficient is always beneficial'.

Considering the analysis of the seakeeping performance of the PD, and the conclusion that the vertical plane motions in head seas are of paramount importance, a second seakeeping ranking was made based on head seas only. The result, shown in Figure 6, demonstrates again the sensitivity of the LcF parameter; it also shows an increased importance of the length. The heave and pitch RAO's as shown in Figures 7 and 8 clearly show the superiority of the LcF– hull form. The large water plane area creates a large amount of damping, which reduces the RAO's to a large extent.

The hull form of the LcF– design is shown in Figure 9. Looking at this body plan, the human factor began to take the overhand in relation to the strict numerical approach used to this point. Besides the fact that the powering performance of this hull form was not very good, the authors just did not like the hull form resulting from this study. So bearing in mind the high sensitivity for the LcB-LcF separation, a large step into the (wrong) LcF+ direction was taken and a new hull form developed. This new hull form is denoted LcF++ or COFEA (Centre Of Floatation Extreme Aft); a small scale body plan is presented in Figure 10.

562

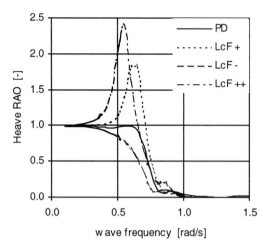

Figure 7. Heave RAO in head seas, Vs = 24 kn.

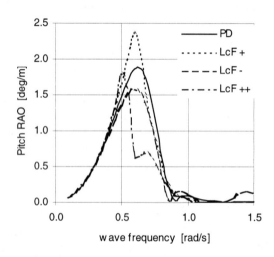

Figure 8.   Pitch RAO in head seas, Vs = 24 kn.

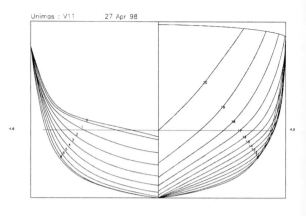

Figure 9.   Body plan of LcF- hull form.

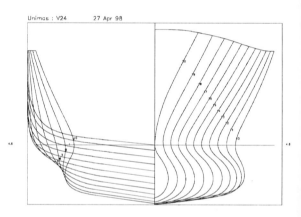

Figure 10. Body plan of LcF++ (COFEA) hull
form.

This new hull form has some interesting features. The heave RAO shows a very high resonance peak, but the pitch RAO shows a sharp drop for frequencies higher than $\omega = 0.55$ rad/s. The phase difference between the heave and the pitch motion is totally changed compared to a normal hull form; this results in the vertical acceleration being at minimum at Station 11 instead of Station 8.

The distribution of the vertical acceleration is also much flatter than for a normal vessel, see Figure 11. Taking into account the large sensitivity of human performance for vertical accelerations (MSI), this increases the useful length onboard the vessel to a large extent. According to the predictions with DESP, the required power of this vessel is higher. It must be noted, however, that the reliability of a power prediction method, which is based on regression analysis for such an unusual hull form, is not very high.

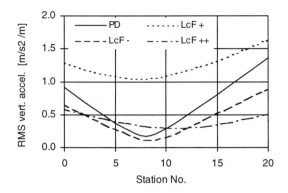

Figure 11. Distribution of the RMS vertical acceleration over the length of the vessel, speed 24 kn, sea state with $T_2 = 7.5$ s.

Therefore some additional calculations with the CFD code RAPID were done and they showed that the expected increase in frictional resistance due to the increased wetted surface could probably be compensated for by the lower wave making resistance. However, model tests were required to verify these expectations.

5. EXPERIMENTAL VERIFICATION

5.1 Resistance tests

To verify the calculations, model tests to measure the resistance and seakeeping characteristics of the COFEA hull form were carried out. The model was too small to do a reliable propulsion test. A surprising result of the resistance test was the low resistance. Although the wetted surface of this hull form is 13% higher than for the PD (for the same displacement), the total resistance at 18 kn appeared to be 4% higher and the total resistance at 30 kn 10% lower. This low resistance can totally be attributed to the low wave making resistance, partly a feature of the extremely narrow waterlines in the fore ship.

The model tests also showed that there was still room for improvement, considering some local deficiencies in the flow around the hull near the transition zone from fore to aft body.

5.2 Tests in head seas

Considering the extreme non-linearity of the hull form, verification of the 2D-linear seakeeping calculations was required. The seakeeping tests also showed some surprises. At first it appeared that the heave RAO was much lower than predicted, Figure 12. However, when the wave amplitudes were reduced to very low values the RAO became higher and showed a good correlation with the predictions as illustrated in Figure 13. Surprisingly, the non-linearities, which are so apparent in the heave motion, have no effect on the pitch motion; the calculated results are confirmed by the model tests as is shown in Figure 14. Based on the results it was concluded that the non-linear effect on the heave motion – due to the rapid change in hull shape just above and below the waterline – reduced the sharp resonance peaks to acceptable levels. This will increase the operability of the COFEA vessel in relation to the linear predictions.

Photograph showing the COFEA hull form in Sea state 6, speed 15 kn, head seas.

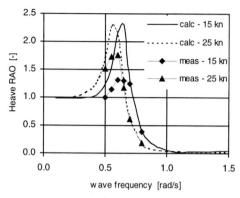

Figure 12. Predicted and measured heave RAO in head seas, speeds 15 and 25 kn. Wave height 2.0 m.

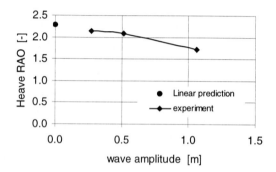

Figure 13. Tests at different wave amplitudes, head seas, speed 25 kn, wave frequency 0.55 rad/s.

5.3 Tests in quartering seas

A major problem was encountered when testing the model in stern quartering seas. Very large roll amplitudes were recorded for some wave frequencies. The vessel appeared to lose a lot of its stability when the stern, with its very wide and shallow sections, was lifted out of the water. It appeared that the minimum stability based on the well-established stability standards was insufficient for this hull form. This resulted in extreme roll angles for moderate sea conditions. The wave condition for which the largest roll angles were recorded had a projected wavelength ($\lambda/\cos \mu$) in the order of the ship length.

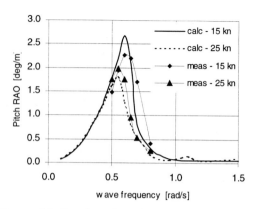

Figure 14. Predicted and measured pitch RAO in head seas, speeds 15 and 25 kn. Wave height 2.0 m.

The phenomenon appeared to be independent of the wave encounter frequency, which was varied by changing the speed of the vessel. Some additional tests were done with an increased GM value and this appeared to give lower roll angles.

6. NON-LINEAR CALCULATIONS

The stability problem in stern quartering seas is considered to be a major problem for the proposed concept, so additional seakeeping calculations were carried out. The program used is FREDYN, which calculates ship motions in the time domain and covers the frequency range from manoeuvring up to ship motions at the wave encounter frequency. This last component is non-linear in the sense that the pressures are integrated up to the actual water surface – hull interface. With FREDYN it is possible to predict extreme roll motions up to capsizing and time domain related phenomena such as broaching, surf-riding and roll resonance. This program has been used to develop the referenced guidelines for dynamic stability of frigates [6]. A time domain run with FREDYN is shown in Figure 15. The condition is a regular stern quartering wave (ω = 0.9 rad/s, ζ_A = 1.0 m) at a speed of 15 kn, this was the worst condition during the tests. The calculations reproduce the problems encountered when sailing in stern quartering seas. The very non-linear behaviour of the phenomenon can be noted; the roll angle builds up over several wave encounters until it reaches a very large value; after this it starts all over again. The effect of increasing the metacentric height is also shown in the figure; the non-linear behaviour disappears for GM = 1.30 m. The roll angle is still quite large; however, a detailed investigation into a range of wave frequencies and directions is necessary before this design can be approved.

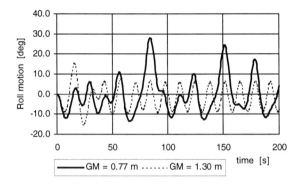

Figure 15. Time domain simulation with non-linear program FREDYN in regular stern quartering seas.

7. FINAL REMARK

Although the driving factor for the development of this hull form was more oriented towards improvement of the seakeeping behaviour than reduction of the calm water resistance, it is interesting to note that the COFEA hull form shows some similarities with the hull forms developed by Breslin and Eng [9] more than 30 years ago. Breslin and Eng tried to develop a hull form with low resistance characteristics. The gain they found by model testing was about 4% at 30 kn, but they found much improved seakeeping characteristics in head seas.

8. CONCLUSIONS

Based on the work presented in this paper, the following conclusions were drawn:

1. A significant step forward could be made in the hydrodynamic performance of a monohull frigate using existing computer programs (resistance prediction based on regression analysis and seakeeping prediction using linear strip theory).
2. An essential aspect of the success of this study is the human intervention in the numerical optimization process.
3. An experimental verification of the new hull form showed the limitations of the prediction tools used. The new hull form was well outside the range used for the resistance regression analysis; for seakeeping the vessel showed a strong non-linear motion behaviour.
4. Although the performance in head seas is most critical, it appeared to be very dangerous to limit the optimization to this condition only.
5. The performance of the vessel as designed in stern quartering seas is unacceptable; the non-linear roll motion lead to very large roll angles. The experiments showed that the existing stability requirements are insufficient for this new hull form; a larger stability is necessary.
6. Non-linear hydrodynamic tools are essential in the design of this vessel. It appears that a new study needs to be done to determine the minimum allowable stability.

REFERENCES

1. STANAG 4154 (Edition 3), *NATO Standardisation Agreement, Common procedures for seakeeping in the ship design process*, 1998, unclassified.
2. A.R.J.M. Lloyd, *Seakeeping - ship behaviour in rough weather*, Ellis Horwood Limited, 1989.

3. J. Holtrop, *A statistical resistance prediction method with a speed dependent form factor,* SMSSH '88, Varna 1988.
4. R.P. Dallinga, *Selection on seakeeping,* MARIN Jubilee workshop, 1992.
5. T.H. Sarchin and L.L. Goldberg, *Stability and Buoyancy criteria for US Naval Surface Ships,* Trans. SNAME, 1962.
6. J.O de Kat, R. Brouwer, K. McTaggart, and W.L. Thomas, *Intact ship survivability in extreme waves: new criteria from a research and navies perspective,* STAB '94 Conference, Melbourne, FL, Nov. 1994.
7. J.A. Keuning and J. Pinkster, *Optimisation of the seakeeping behaviour of a fast monohull,* FAST '95 Conference, L•beck-Travemunde, 1995.
8. J.J. Blok and W. Beukelman, *The High-Speed Displacement Ship Systematic Series Hull Forms - Seakeeping characteristics,* SNAME Annual Meeting, November 1984.
9. J.P. Breslin and K. Eng, *Resistance and seakeeping performance of new high speed destroyer designs,* Davidson Laboratory Report 1082, June 1965.

Table 4. Main dimensions and hull form coefficients of the designs in the systematic study

ID	L_{PP} [m]	B_{WL} [m]	T [m]	Δ [ton]	GM [m]	C_P [-]	C_M [-]	L_{cB} [m]	$(L_{cB}-L_{cF})$ / L_{pp}
PD	120.00	15.80	4.60	4204	1.725	0.621	0.757	-2.846	0.0422
L+	144.00	15.80	4.60	5044	0.765	0.621	0.757	-3.415	0.0422
L++	168.00	15.80	4.60	5885	0.765	0.621	0.757	-3.984	0.0422
B-	120.00	12.64	4.60	3363	0.956	0.621	0.757	-2.846	0.0422
B+	120.00	18.96	4.60	5044	0.637	0.621	0.757	-2.846	0.0422
T-	120.00	15.80	3.68	3363	0.899	0.621	0.757	-2.846	0.0422
T+	120.00	15.80	5.52	5044	0.669	0.621	0.757	-2.846	0.0422
C_P-	120.00	15.80	4.60	3388	0.949	0.498	0.758	-2.832	0.0421
C_P+	120.00	15.80	4.60	5125	0.627	0.741	0.754	-2.898	0.0412
C_M-	120.00	15.80	4.60	3785	0.849	0.622	0.680	-2.808	0.0418
C_M+	120.00	15.80	4.60	4732	0.679	0.626	0.830	-2.871	0.0413
LcB-	120.00	15.80	4.60	4208	0.764	0.630	0.753	-0.008	0.0421
LcB+	120.00	15.80	4.60	4210	0.763	0.610	0.763	-5.733	0.0418
LcF-	120.00	15.80	4.60	4199	0.765	0.614	0.750	-2.830	0.0008
LcF+	120.00	15.80	4.60	4198	0.766	0.600	0.770	-2.906	0.0830
COFEA	120.00	20.00	4.60	4242	0.765	0.633	0.925	-0.336	0.1343

Note: LcB is defined with respect to Station 10 (midships), a forward location is positive.

Practical Design of Ships and Mobile Units
M.W.C. Oosterveld and S.G. Tan, editors.

Theoretical Validation of the Hydrodynamics of High Speed Mono- and Multi-Hull Vessels Travelling in a Seaway

P.A. Bailey[a], D.A. Hudson[a], W.G. Price[a] and P. Temarel[a]

[a]Department of Ship Science, University of Southampton,
Highfield, Southampton, SO17 1BJ, U.K.

Mono-hull and catamaran heave, pitch, sway and yaw hydrodynamic data and a selection of waveloads are presented illustrating dependence on forward speed and frequency of oscillation. These results are determined from two distinct Green's function solutions with their respective separately developed numerical schemes of study to solve seakeeping problems mathematically modelled by translating, pulsating sources. The two sets of evaluated data show remarkable qualitative and quantitative agreement and therefore provide a measure of validation and confidence that these reported findings are solutions of the posed seakeeping problem subject to the constraints of the developed mathematical models.

1. INTRODUCTION

This paper presents a theoretical validation of the solution to the problem associated with a surface piercing mono- or multi-hull form travelling with forward speed and oscillating with prescribed frequency in calm water or in sinusoidal waves. The mathematical formulation of this problem is well documented assuming the fluid ideal (Newman (1978)). Under this assumption the dynamic characteristics of the hull and fluid are defined in terms of a velocity potential function and its solution sought subject to free surface and bottom boundary conditions and appropriate radiation conditions.

One of the first mathematical descriptions of the velocity potential function was given by Wehausen and Laitone (1960). This is of a complicated nature involving single and double integrals of functions dependent on operating conditions (i.e. forward speed and frequency of encounter between hull and waves) and these numerical integrations prove difficult to evaluate. By modifying the mathematical formulation simplifications to the integrals are possible and numerical solutions to this problem appeared in the literature (Chang (1977)) though they were computer intensive and time consuming to achieve. To reduce the latter overheads, continued mathematical developments provided means to reduce the necesssity to undertake the troublesome double integrals (Inglis and

Price (1981), Guevel and Bougis (1982), Wu and Eatock-Taylor (1987)).

Bessho (1977) presented an alternative velocity potential solution involving a single integral formulation. However, more emphasis was placed on the method of solution of certain terms which cause additional difficulties in the development of a suitable numerical scheme of study (Iwashita and Ohkusu (1989)), but the overall saving of computer effort could be significant. This was shown in a preliminary study by Du et al. (1998) into the behaviour of an ellipsoid in waves. This body shape is again chosen in the present study in mono- and multi-hull forms. It is devoid of local curvature effects, a transom stern, etc. as may occur in a ship or catamaran but it allows focus on the mathematical treatise of the problem rather than specific hull characteristics.

This paper therefore presents heave, pitch, sway and yaw hydrodynamic coefficient data and a selection of wave load data for an ellipsoid mono-hull and catamaran configuration travelling at various Froude numbers and oscillating over a wide range of frequencies. These results, determined from two fundamentally different formulations of the velocity potential functions given separately by Inglis and Price (1981) and Du and Wu (1998), are compared and discussed. The developed numerical schemes of study are also fundamentally different. This investigation is an extension of the preliminary study undertaken by Du et al.

(1998), where great care was taken to validate intermediate steps (i.e. convergence of solution, field-source point characteristics, etc.) in the calculation process. The interested reader is advised to consult this publication for additional information since a description of the intermediate validation process is omitted here.

This study shows the remarkable agreement in the numerically derived data from the two separate mathematical formulations and numerical schemes of study over wide ranges of forward speed and frequencies. This therefore provides a measure of validation of the solution to the posed problem. Having confidence in the calculation of hydrodynamic coefficients and wave loads, using either method, provides a firm foundation for the subsequent calculation of vessel response to regular waves of arbitrary heading and frequency. This in turn allows the use of statistical techniques to obtain operational envelopes for the chosen vessel in a variety of operating conditions (i.e. vessel speed, wave height, modal wave period, etc.). The applicability of either method of solution to both monohulls and catamarans, as demonstrated here, provides a means to compare quite different vessel types using the same, validated, theoretical method.

2. GENERAL FORMULATION AND SOLUTION TO THE PROBLEM

Let us define a right-handed equilibrium axis system $Oxyz$ such that the origin O is in the plane of the calm water surface vertically above the centre of gravity of the vessel. The z-axis is positive upwards and the x and y axes lie in the plane of the undisturbed free-surface. The axis-system translates with constant velocity $\overline{U}$ in the positive x-direction.

The Green's function at field point (x, y, z) is defined as the velocity potential

$$G(x, y, z; x_0, y_0, z_0)e^{i\omega t}$$

of a translating, pulsating source at (x_0, y_0, z_0). This velocity potential ϕ satisfies the Laplace equation throughout the fluid domain, the linearised

free-surface condition:

$$\left(i\omega\phi - \overline{U}\frac{\partial\phi}{\partial x}\right)^2 + g\frac{\partial\phi}{\partial z} = 0 \qquad \text{on } z = 0 \qquad (1)$$

and the usual radiation condition ensuring outgoing waves at infinity.

The form of the speed dependent velocity potential satisfying these conditions for water of infinite depth is given by Wehausen and Laitone (1960) as:

$$G(x, y, z; x_0, y_0, z_0) = \frac{1}{R_1} - \frac{1}{R_2} + \frac{2g}{\pi}\left\{\int_0^\gamma \int_0^\infty\right.$$
$$\left. + \int_\gamma^{\frac{\pi}{2}}\int_{L_1} + \int_{\frac{\pi}{2}}^\pi \int_{L_2}\right\} f(\theta, k)\, dk\, d\theta \qquad (2)$$

where,

$$
\begin{aligned}
R_1 &= \sqrt{(x - x_0)^2 + (y - y_0)^2 + (z - z_0)^2}, \\
R_2 &= \sqrt{(x - x_0)^2 + (y - y_0)^2 + (z + z_0)^2}, \\
f(\theta, k) &= h(\theta, k)/\left[gk - (\omega + \overline{U}k\cos\theta)^2\right], \\
h(\theta, k) &= k\exp\left[k\{z + z_0 - i(x - x_0)\cos\theta\}\right] \\
&\quad \cos\{k(y - y_0)\sin\theta\}, \\
\gamma &= \begin{cases} 0 & \text{if } \beta = \frac{\omega\overline{U}}{g} < \frac{1}{4} \\ \cos^{-1}(\frac{1}{4\beta}) & \text{if } \beta = \frac{\omega\overline{U}}{g} \geq \frac{1}{4} \end{cases}
\end{aligned}
$$

As it stands, the Green's function in equation (2) is troublesome to evaluate numerically, as there are two singular points in the second integral (i.e. $k = k_1$ and $k = k_2$) and two in the third (i.e. $k = k_3$ and $k = k_4$). These singularities are given by:

$$\left.\begin{array}{c}\sqrt{(gk_1)} \\ \sqrt{(gk_3)}\end{array}\right\} = \omega\left[\frac{1 - \sqrt{(1 - 4\beta\cos\theta)}}{2\beta\cos\theta}\right],$$

$$\left.\begin{array}{c}\sqrt{(gk_2)} \\ -\sqrt{(gk_4)}\end{array}\right\} = \omega\left[\frac{1 + \sqrt{(1 - 4\beta\cos\theta)}}{2\beta\cos\theta}\right].$$

The integration path for the contours L_1 and L_2 extends over k, ranging from $0 \to \infty$ and at the singular points, the integration path is defined as $k_1 + i\delta$, $k_2 - i\delta$, $k_3 + i\delta$, $k_4 + i\delta$ as $\delta \to 0$.

3. METHOD A SOLUTION

To reduce numerical difficulties, the Green's function in equation 2 can be rewritten as a Cauchy principal value integral (see, Chang (1977), Inglis and Price (1981)). After the necessary manipulations, the Green's function solution takes the form

$$G(x, y, z; x_0, y_0, z_0) = \frac{1}{R_1} - \frac{1}{R_2}$$
$$+ \frac{2g}{\pi} \int_0^{\gamma} \int_0^{\infty} \frac{h(\theta, k) \, dk \, d\theta}{gk - (\omega + \overline{U}k \cos\theta)^2}$$
$$+ \int_{\gamma}^{\frac{\pi}{2}} \{Q_1(w_+) + Q_1(w_-) - Q_2(w_+) - Q_2(w_-)\} \, d\theta$$
$$+ \int_{\frac{\pi}{2}}^{\pi} \{Q_3(w_+) + Q_3(w_-) - Q_4(w_+) - Q_4(w_-)\} \, d\theta$$
$$- 2i \int_{\gamma}^{\frac{\pi}{2}} \frac{\{h(\theta, k_1) + h(\theta, k_2)\} \, d\theta}{\sqrt{(1 - 4\beta \cos\theta)}}$$
$$- 2i \int_{\frac{\pi}{2}}^{\pi} \frac{\{h(\theta, k_3) + h(\theta, k_4)\} \, d\theta}{\sqrt{(1 - 4\beta \cos\theta)}}$$
$$= \frac{1}{R_1} - \frac{1}{R_2} + T_1 - iT_2. \tag{3}$$

where,

$$Q_j(w_\pm) = \frac{k_j}{\pi\sqrt{(1 - 4\beta \cos\theta)}} \{i\pi \exp(q_{j\pm})\mathrm{sgn}(w_\pm)$$
$$- q_{j\pm}^{-1} + \exp(q_{j\pm})\mathrm{E}(q_{j\pm})\}$$
$$w_\pm = (x - x_0)\cos\theta \pm (y - y_0)\sin\theta,$$
$$q_j = -k_j(|z + z_0| - iw) \text{ for } j = 1, 2, 3, 4.$$

Here, $q_{j\pm}$ implies $w_\mp$ respectively and $\mathrm{E}()$ denotes the exponential integral.

For $\beta < 1/4$, this formulation requires less computational effort to evaluate but for $\beta > 1/4$ the double integration over the semi-infinite k-range remains.

4. METHOD B SOLUTION

Bessho (1977) expressed the Green's function as

$$G(x, y, z; x_0, y_0, z_0) = \frac{1}{4\pi}\left(\frac{1}{R_1} - \frac{1}{R_2}\right)$$
$$- \frac{i}{2\pi}K_0 T(X, Y, Z) \tag{4}$$

where,

$$T(X, Y, Z) =$$
$$\int_{\alpha-\pi}^{\frac{\pi}{2}+\psi-i\epsilon} \frac{[k_2 \exp(k_2 w) - \mathrm{sgn}(\cos\theta)k_1 \exp(k_1 w)] \, d\theta}{\sqrt{1 + 4\beta \cos\theta}}$$
$$k_1 = \frac{1}{2\cos^2\theta}\left(1 + 2\beta\cos\theta + \sqrt{1 + 4\beta\cos\theta}\right),$$
$$k_2 = \frac{1}{2\cos^2\theta}\left(1 + 2\beta\cos\theta - \sqrt{1 + 4\beta\cos\theta}\right),$$
$$\psi = \cos^{-1}\frac{X}{\sqrt{X^2 + Y^2}},$$
$$\epsilon = \sinh^{-1}\frac{|Z|}{\sqrt{X^2 + Y^2}},$$
$$\alpha = \begin{cases} \cos^{-1}(\frac{1}{4\beta}) & \text{if } \beta \geq \frac{1}{4} \\ -i\cosh^{-1}(\frac{1}{4\beta}) & \text{if } \beta < \frac{1}{4}, \end{cases}$$
$$w = Z + i(X\cos\theta + Y\sin\theta),$$
$$\beta = \frac{\overline{U}\omega}{g}, \quad K_0 = \frac{g}{\overline{U}^2},$$
$$X = K_0(x - x_0),$$
$$Y = K_0|y - y_0|,$$
$$Z = K_0(z + z_0).$$

This is a genuine single integral formulation in which the integrand only contains expressions involving elementary functions.

As can be seen the $1/R_1$ and $1/R_2$ terms, defining the Rankine source and its image, are common to the three solutions but in solutions A and B their contributions are proportionally different.

Equation (4) is a more mathematically involved function than the corresponding form in solution A due to the dependence of the limits of integration on the position of the source and field points. Iwashita and Ohkusu (1989), Iwashita and Ohkusu (1992), Du (1996), Du and Wu (1998), and Du et al. (1998) describe detailed approaches for the evaluation of this solution.

5. MONO- AND MULTI-HULL CONFIGURATIONS

In this validation exercise, a simple ellipsoid was chosen in preference to an actual hull form to minimise the influence of local curvature, bilges,

transom stern, etc. Mono- and multi- hull configurations as illustrated in figure 1(a,b) were examined. Each surface-piercing ellipsoid has a length
to beam ratio of 8:1 and a panel discretisation of
the mean wetted surface area by 336 elements for
the monohull and by 400 elements for the whole
multihull. In the catamaran configuration the
spacing S between the two longitudinal axes is
40% of the hull length L (i.e. $S = 0.4L$).

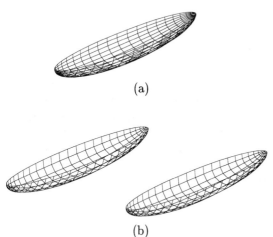

(a)

(b)

Figure 1: (a) Mono- and (b) Multi-hull ellipsoid
surface idealisations.

6. NUMERICAL SOLUTIONS AND COMPARISONS

In the non-dimensional data presented in this section it is noted that the length L of the mono-hull
and multi-hull forms are the same and the displacement ∇ of the catamaran is twice that of the
mono-hull. Thus when comparing corresponding
predictions for mono- and multi-hull forms relative differences in magnitudes based on the monohull displacement provides an indicative measure
of the interaction between the hulls with forward
speed and frequency of oscillation.

Figures 2-6 show selections of non-dimensional
data evaluated from the separate mathematical
models. Based on the same idealisation of the hull
forms and executed on the same computer, the
agreement between individual sets of predicted
values show excellent qualitative and quantitative
similarity to one another over all chosen forward
speeds and frequency range.

6.1. Heave-pitch hydrodynamic coefficients

Figure 2 illustrates the heave, pitch and their
cross-coupling hydrodynamic coefficients for the
ellipsoid and figure 3 shows the corresponding
data for the twin ellipsoid configuration.

The influences of forward speed and frequency are
observed in both sets of data with more variations
occurring in the catamaran data sets due to hull
interactions. This is most noticeable at very low
speeds but reduces with increasing speed when
the catamaran behaves more akin to the monohull (e.g. compare A_{33} at Fn=0.5 and at Fn=0.2,
say). At Fn=0.0025 an irregular frequency occurs
at $\omega_e' \approx 5.5$, both for the mono- and multi-hull,
but this appears to disappear from the hydrodynamic data at higher speeds. Also prominent at
the lowest speed for the catamaran are the influences of hull interactions, which can be attributed
to standing waves between the hulls at frequencies of the order of $\omega_e' = 4.8$ and to a vertical
oscillation of the fluid trapped between the hulls
at frequencies around $\omega_e' = 2.4$.

Predictions using a simpler, pulsating, Green's
function model closely resemble the zero speed
predictions illustrated here, irrespective of forward speed.

6.2. Sway-yaw hydrodynamic coefficients

Figure 4 illustrates the sway, yaw and their cross-
coupling hydrodynamic coefficients for the monohull ellipsoid form and figure 5 shows the corresponding data for the catamaran ellipsoid configuration.

In contrast to the symmetric coefficients, the sway-
yaw coefficients show a more pronounced dependence on forward speed over a wider frequency
range.

At the lowest speed considered, an irregular frequency is observed in the mono-hull results at
$\omega_e' \approx 7.0$ but this again appears to disappear
with increasing speed. This irregular frequency
also appears in the multi-hull results with additional standing wave effects caused by the twin
hull configuration. At Fn=0.2, significant variations with frequency are observed in coefficients

A_{26}, A_{62} which become less pronounced at higher speeds. At higher Froude numbers the multi-hull data again reflect those of the mono-hull implying a diminishing influence of hull configuration.

6.3. Wave loadings

Figure 6(a,b) shows the heave and pitch excitation wave loadings on the mono- and multi-hull forms travelling in sinusoidal head waves of unit amplitude. These selected solutions determined from both Green's function approaches show remarkable numerical agreement over the chosen speed and frequency ranges. Similar comparisons are achieved at other headings and for all wave exciting loads.

7. CONCLUSIONS

The numerical findings display excellent agreement between the predicted values derived from the separate and distinct Green's function solutions of the oscillating, forward speed problem fundamental to seakeeping analyses. This has been achieved by developing, in isolation, two numerical schemes of study. Intermediate step calculations were performed and compared previously by Du et al. (1998) and overall, the exercise has proved fruitful providing a measure of theoretical validation and confidence in the solution to the posed seakeeping problem. Confidence in the solution is enhanced because in the intermediate study the dependence of solution on hull idealisation proved of little significance though this remains an element of uncertainty in any numerical calculation in contrast to a pure analytical solution.

From the results presented, either solution with its numerical scheme is suitable for the analysis of a mono- or multi-hull travelling in waves. This comparison was undertaken with the prediction of motions for a high speed vessel in mind and the development of operational envelopes to enhance safety of passage. The excellent agreement over the wide range of chosen forward speeds and frequencies is particularly encouraging and such linear based analyses provide a yardstick by which to judge the contributions of higher order linear and non-linear theories.

REFERENCES

Bessho, M. (1977). On the fundamental singularity in the theory of ship motions in a seaway. *Memoirs of the Defence Academy of Japan*, 17(3):95–105.

Chang, M. S. (1977). Computations of three-dimensional ship-motions with forward speed. In *2nd Intl. Conference on Numerical Ship Hydrodynamics*, pages 124–135, Berkeley.

Du, S. X. (1996). *A Complete Frequency Domain Analysis Method of Linear Three-Dimensional Hydroelastic Responses of Floating Structures Travelling in Waves*. PhD thesis, China Ship Scientific Research Centre, Wuxi, China.

Du, S. X., Hudson, D. A., Price, W. G., and Temarel, P. (1998). Comparison of numerical evaluation techniques for the hydrodynamic analysis of a ship travelling in waves. *Submitted for publication*.

Du, S. X. and Wu, Y. S. (1998). A fast evaluation method of the Bessho form translating and pulsating Green's function. *China Shipbuilding*, 2.

Guevel, P. and Bougis, J. (1982). Ship motions with forward speed in infinite depth. *International Shipbuilding Progress*, 29:103–117.

Inglis, R. B. and Price, W. G. (1981). Calculation of the velocity potential of a translating, pulsating source. *Trans. RINA*, 123:163–175.

Iwashita, H. and Ohkusu, M. (1989). Hydrodynamic forces on a ship moving with forward speed in waves. *Journal of the Society of Naval Architects of Japan*, 166:187–206.

Iwashita, H. and Ohkusu, M. (1992). The Green function method for ship motions at forward speed. *Ship Technology Research*, 39(2):3–21.

Newman, J. N. (1978). The theory of ship motions. *Advances in Applied Mechanics*, 18:221–283.

Wehausen, J. Y. and Laitone, E. V. (1960). Surface waves. In *Encyclopædia of Physics*, volume 9, pages 446–778. Springer Verlag.

Wu, G. X. and Eatock-Taylor, R. (1987). A Green's function form for ship motions at forward speed. *International Shipbuilding Progress*, 34(389):189–196.

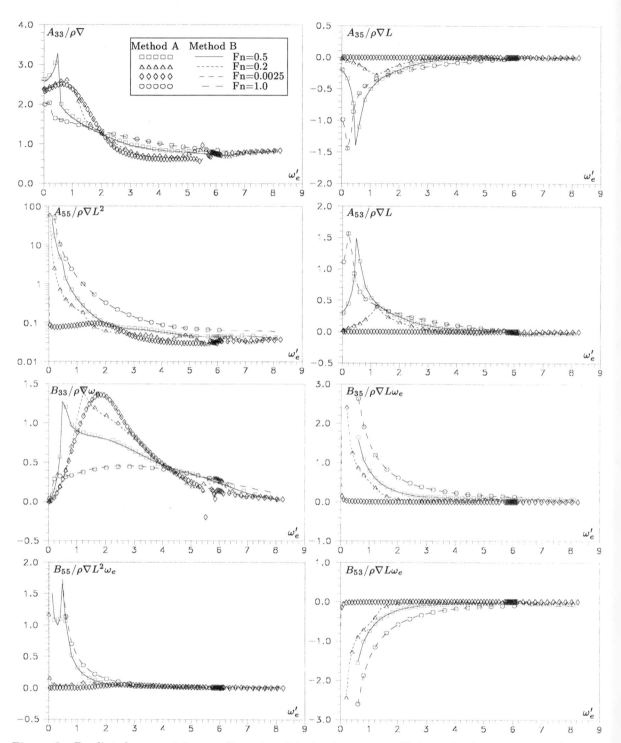

Figure 2: Predicted symmetric non-dimensional hydrodynamic coefficients (3=heave, 5=pitch) for a mono-hull ellipsoid travelling with forward speed, Fn=0.0025, 0.2, 0.5, 1.0, and oscillating with non-dimensional frequency $\omega_e' = \omega_e\sqrt{L/g}$ in calm water.

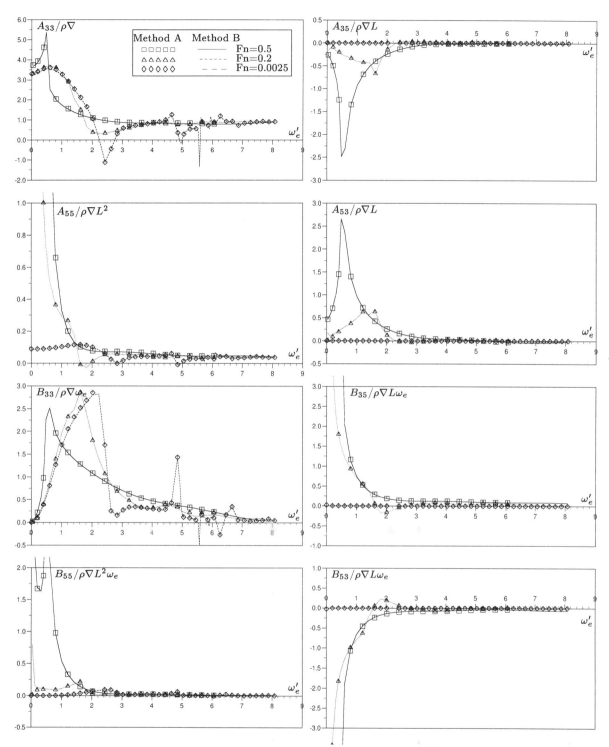

Figure 3: Predicted symmetric non-dimensional hydrodynamic coefficients (3=heave, 5=pitch) for a multi-hull ellipsoid ($S/L = 0.4$) travelling with forward speed, Fn=0.0025, 0.2, 0.5, and oscillating with non-dimensional frequency $\omega_e' = \omega_e\sqrt{L/g}$ in calm water.

574

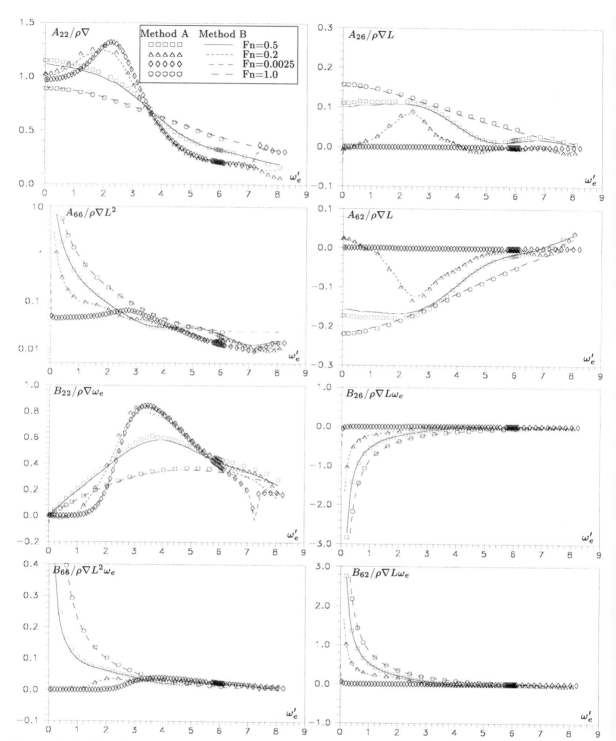

Figure 4: Predicted antisymmetric non-dimensional hydrodynamic coefficients (2=sway, 6=yaw) for a mono-hull ellipsoid travelling with forward speed, Fn=0.0025, 0.2, 0.5, 1.0, and oscillating with non-dimensional frequency $\omega'_e = \omega_e \sqrt{L/g}$ in calm water.

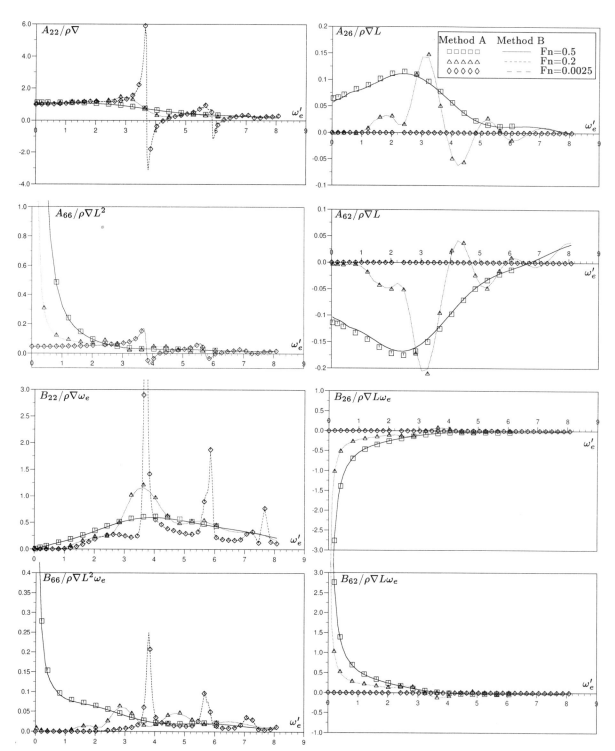

Figure 5: Predicted antisymmetric non-dimensional hydrodynamic coefficients (2=sway, 6=yaw) for a multi-hull ellipsoid ($S/L = 0.4$) travelling with forward speed, Fn=0.0025, 0.2 0.5, and oscillating with non-dimensional frequency $\omega'_e = \omega_e \sqrt{L/g}$ in calm water.

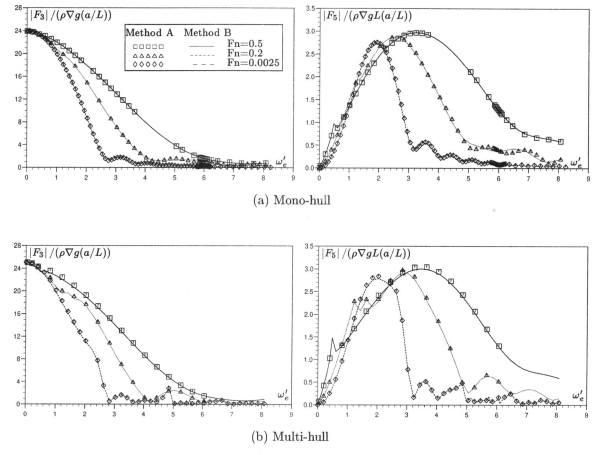

Figure 6: Predicted non-dimensional wave exciting loads in heave $|F_3|$ and pitch $|F_5|$ for a mono- and multi-hull ($S/L = 0.4$) ellipsoid travelling with forward speed, Fn=0.0025, 0.2, 0.5 in sinusoidal waves of amplitude ($a/L = 0.025$) and non-dimensional frequency $\omega'_e = \omega_e\sqrt{L/g}$.

Practical Design of Ships and Mobile Units
M.W.C. Oosterveld and S.G. Tan, editors.

Issues in the Assessment of Design Slamming Pressure on High Speed Monohull Vessels

Jianbo Hua

Div. of Naval Architecture, Dept. of Vehicle Engineering,Royal Institute of Technology
SE-100 44 Stockholm, Sweden

ABSTRACT

The slamming pressure is decisive for the structure scantling of high speed vessels up to medium size. The linear wave model and long-crested heading sea condition are usually used for direct assessment of the design slamming pressure using computer codes for seakeeping calculation. The design experience has shown that the assessment result is uncertain and hardly useful without feed back experiences of similar vessels in service.

A Monte-Carlo procedure based on a linear strip theory is used for probabilistic assessment of slamming pressure on a medium-sized V-shaped high speed vessel of about 70 m in length over all. Through demonstration of some calculation examples, the validity of the most used slamming pressure evaluation method by Ochi, originally for conventional ships is discussed, and its theoretical shortage is pointed out while applied for high speed vessels. The effects of bow waves, nonlinear waves, wave breaking and combined seas on slamming pressure are then investigated. The result shows considerably additional effects to the slamming pressure level in heading waves.

1. INTRODUCTION

One of the purposes of ship technology research is to be able to evaluate design loads for new ship concepts based on the physics' first principle. Numerous seakeeping theories have been successfully utilised for analysis of ship motions in waves and the wave loads and implemented into practical design processes. However, the progress in the field has not yet reached to a such level that a complete direct method is available in the sense that a new ship construction could be optimally designed without referring to the previously accumulated service experiences.

Light construction materials such as composite material and aluminium are more and more utilised for high speed ships of unconventional construction. A question arises how far the semi-empire rules from service experiences of conventional ships are applicable to the new ship constructions. Our experience from a project work of a medium-sized offshore patrol vessel build of composite sandwich, see [1-3], shows that there are risks that the hull strength can be insufficiently dimensioned according to the design rules of ship classification associations.

The application of seakeeping theories is expected to an important tool in the design process. However, a rational methodology for determining design slamming pressure by applying these theoretical tools is still lacking. For example, the linear wave model and long-crested heading sea condition are usually considered for the direct assessment of design slamming pressure, while the slamming problems of high speed vessel are more complex.

The linear wave model is a simplification of the real ocean waves and can give good prediction of wave-induced ship motions. But that does not mean accurate prediction of slamming pressure magnitude, since the slamming impact is sensitive to the local wave form and wave-induced particle velocity. As well, it is not self clear that the heading sea condition gives the severest slamming impact.

In this paper, a Monte-Carlo procedure based on a linear strip theory is used for probabilistic assessment of slamming pressure on a medium-sized V-shaped high speed vessel of about 70 m in length over all. Through the demonstration of some calculation examples, the validity of the most used slamming pressure evaluation method by Ochi, is discussed, and its theoretical shortage is pointed out while applied for high speed vessels. By the author's opinion, the complexity of ocean wave phenomenon is not sufficiently considered in today's design process. Therefore, the effects of bow waves, nonlinear waves, wave breaking and combined seas on slamming pressure are investigated. The result is compared with the one in heading sea condition described with the linear wave model.

The main particulars of the vessel is:

L_{oa}	(m):	70
B	(m):	10.4
T	(m):	2.35
Displ	(m^3):	593

The line drawing is shown in Fig.1.1. The V-shaped hull form has a deadrise angle of about 18^0 at the midship section so that the vessel is able to run in near planning condition. Due to the high speed in combination with large vessel size, high acceleration level and severe slamming impact pressure are expected in moderate wave conditions. Therefore it is important in the design stage to determine the expected maximal values during a vessel's service life in order to achieve an safety construction.

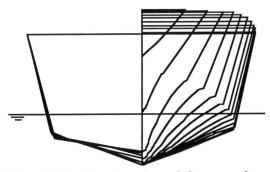

Fig.1.1 The line drawing of the vessel.

2. CALCULATION METHOD

2.1 Calculation of ship motions in waves

The motions of a high speed vessels in waves and the wave loads are nonlinear, and the nonlinearity increase with increased speed and wave severity. Numerous nonlinear mathematical models have been developed based on hydrodynamic momentum theory and nonlinear strip approach. The problem analysis is usually carried out by means of computer simulation, see [4-5]. In general, the simulated results show good agreement with the model measured even in severe wave conditions. Fig.2.1 is comparisons between the simulated and the model-measured accelerations at bow and midship respectively [6] for the actual example vessel.

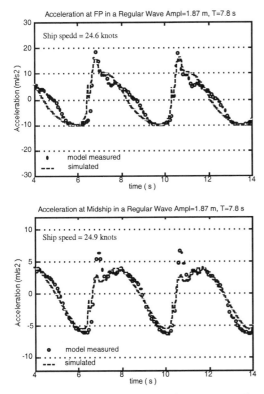

Fig.2.1 The simulated acceleration at bow and at midship in comparison with the model measured. The wave amplitude is 1.87 m and wave period 7.8 s.

However, time-domain simulation of a nonlinear model demands large computation capability, particularly when making probabilistic analysis which usually require a simulation corresponding to at least four to six hour real time. On the other side it is not always necessary to apply this method for some kind problem analysis. The fact is that a high speed vessel can not run in waves as fast as in calm water, because of voluntary and involuntary speed reduction with consideration to the comfort ability, safety and increased resistance. Consequently, the moderate sea should be used as design condition, which is realistic and optimal with aspect to the building cost. For lower speeds and moderate wave conditions, the linear strip theory can give reasonable accuracy for motion prediction. Fig.2.2

shows the comparison of the relative velocity at bow of the example vessel in two regular waves from [6], calculated by means of the linear strip theory and the nonlinear theory respectively. The nonlinear effect is apparent in the case of 2 m wave amplitude in the time histories of the relative velocity. Still, the maximal relative velocities at bow are very near to each other.

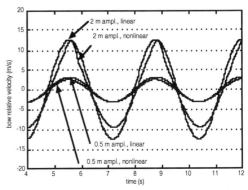

Fig.2.2 Comparison of the nonlinear theory with the linear by the time histories of the relative velocities at bow of the vessel in two heading waves of wave period 7s, wave amplitude 0.5 and 2 m respectively. The vessel speed is 25 knots.

Our investigation, [1] shows that the calculation result of the linear strip theory is in general in good agreement with the simulated one of the nonlinear strip theory in the motion time variation, fasten discrepancies in magnitude. That means that the linear result has the ability to qualitatively reflects the effects of different ship load conditions and wave conditions on the local slamming loads, even the result is overestimated. This is an important feature because slamming is a combined effect of the simultaneous wave motion and ship motions.

The linear strip theory is a frequency-domain analysis, therefore has the advantage of fast numerical evaluation. The result can then be easily transferred into time-domain so that Monte-Carlo simulation procedure can be applied for probabilistic analysis of slamming impact on ship structures, see [7] for detail description. The extraordinary linear strip theory [8] is therefore used here for the slamming pressure calculation of the actual vessel.

2.2 Slamming pressure calculation

For the water entry of a wedge shaped cylinder with a deadrise angle α, see Fig.2.3, the maximum pressure is

$$P_{max} = \frac{1}{2}\rho \cdot C_{P_{max}}(\alpha) \cdot V^2 \qquad (2.1)$$

where the slamming factor $C_{P_{max}}(\alpha)$ is

$$C_{P_{max}}(\alpha) = \frac{0.25 \cdot \pi^2}{\tan^2 \alpha_e}$$

according to Wagner's theory, which is in good agreement with the numerical result by Zhao and Faltisens [9].

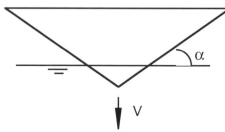

Fig.2.3   Water entry of wedge shaped cylinder

In general, the stresses in a detail structure due to slamming impact are not so sensitive to the maximum slamming pressure at the impact front because of its narrow band characteristics. The stress analysis by nonlinear FEM [3], has shown that the mean slamming pressure can be used as a measure for the slamming effect. The derivation of the mean slamming pressure is as followed.

The added mass of a wedge [10] is;

$$a_{33} = \frac{\rho \cdot \pi \cdot z^2}{2} \cdot f^2(\alpha) \qquad (2.3)$$

where $f(\alpha) = \frac{\pi}{2 \cdot \alpha} - 1$

The slamming force on the wedge is;

$$F = \frac{\partial}{\partial t}(a_{33} \cdot V) = \frac{\partial a_{33}}{\partial z} \cdot V^2$$
$$= \rho \cdot \pi \cdot z \cdot f^2(\alpha) \cdot V^2 \qquad (2.4)$$

according to the hydrodynamic momentum theory.

The mean pressure becomes;

$$P_{mean} = \frac{1}{2} \cdot \pi \cdot \rho \cdot f^2(\alpha) \cdot \tan \alpha \cdot V^2$$
$$= \frac{1}{2}\rho \cdot C_{P_{mean}}(\alpha) \cdot V^2 \qquad (2.5)$$

where $C_{P_{mean}} = \pi \cdot f^2(\alpha) \cdot \tan \alpha$

The derivation of the above formula takes the pilling-up effect into consideration.

For slamming pressure in heading waves, the calculation method by Stavovy and Chuang [11] is applied. In bow waves, the method is extended by introduction of an effective deadrise angle and an effective relative velocity in order to make equivalence to the deadrise angle and the entering velocity respectively in (2.5). The effective deadrise angle is defined as the angle between the hull surface and the momentary wave surface, in which the effect of the wave-induced roll, pitch and yaw motion on hull surface slope is taken into account for every moment. The effective relative velocity vector is calculated with consideration to both the ship forward speed, time derivatives of wave-induced sway, heave, roll, pitch and yaw motion and the water particle velocity due to the wave motion.

The short-term irregular wave motions and the induced ship motions are calculated by the Monte-Carlo simulation procedure. Thereby, the relative velocity normal to the body surface can be obtained, which is the product between the relative velocity vector and the normal vector of the body surface. The relative velocity tangential to the body surface gives a planing pressure effect and insignificant in comparison with the slamming pressure. The slamming pressure is calculated according to (2.5).

3. RESULT AND DISCUSSION

Since the probabilistic distribution of the slamming pressure is far from the Normal or Rayleigh function, the time-domain simulation has to been made sufficiently long in order to get an insight into the probabilistic nature of the problem. For every short term sea, a four-hour real time simulation is carried out, which is fair for probabilistic analysis. The four-hour time simulation is divided into 24 time series. The maximal slamming pressure in each time series is picked up and the result is then presented in terms of **minimum**, **mean**, **maximum** and **standard deviation** of the selected 24 values and used for comparison of different assumptions and effects.

In the following five subchapters, slamming calculation results are presented and discussed with aspects to 1) effect of relative velocity evaluation, 2) slamming in bow waves, 3) nonlinear wave effect, 4) wave breaking effect and 5) combined sea effect.

3.1 Effect of relative velocity calculation

As (2.1) and (2.5) show, both the peak and mean slamming pressure are proportional to the relative velocity in square. Actually, there exists two approaches generally for calculation of the relative velocity.

The vertical motion of a ship relatively to a irregular wave surface at a location somewhere along the ship can be expressed as followed

$$\eta_R = \sum_i a_w^i \cdot \sin(\omega_e^i \cdot t + \beta_w^i) - \sum_i a_m^i \cdot \sin(\omega_e^i \cdot t + \beta_m^i) \tag{3.1}$$

where a_w^i is the wave component amplitude and a_m^i ship motion amplitude which is the combined effect of the heave and pitch motion in heading waves. The time derivative of (3.1) is then

$$\dot{\eta}_R = \sum_i a_w^i \cdot \omega_e^i \cdot \cos(\omega_e^i \cdot t + \beta_w^i) - \sum_i a_m^i \cdot \omega_e^i \cdot \cos(\omega_e \cdot t + \beta_m^i) \tag{3.2}$$

Equation (3.2) is physically the motion velocity relatively to the wave surface, and can be measured under a model test. In fact, the measured bottom slamming pressure is mostly analysed using the relative velocity defined in (3.2) as the main parameter, thereby the slamming factor can be determined, see [12] and [13]. This definition is also used for the calculation of sectional slamming force by Chiu and Fujino [5] in their mathematical model for semi-planing boat in heading waves.

In fact, the wave-induced water particle's velocity can not be affected by the ship forward speed. Instead of (3.2), the vertical relative velocity should be

582

$$\dot{\eta}_R = \sum_i a_w^i \cdot \omega^i \cdot \cos(\omega_e^i \cdot t + \beta_w^i)$$
$$- \sum_i a_m^i \cdot \omega_e^i \cdot \cos(\omega_e \cdot t + \beta_m^i) \quad (3.3)$$

(3.3) is the definition of velocity relatively to the water particle, which can be derived according to the velocity potential theory. (3.3) is used by Stavovy and Chang [11] for slamming pressure calculation, and by Meyerhoff and Schlachter [4] and Martin [10] for motions of high speed ship in heading waves. The difference between (3.2) and (3.3) is in the first term on the right side of the equations where the encounter frequency ω_e^i in (3.2) is changed to wave frequency ω^i in (3.3).

Large discrepancy can be resulted in the numerical results calculated by (3.2) and (3.3) respectively. Fig.3.1 shows the response operator of the vertical relative velocity near FP as function of mean wave period in heading irregular waves. The ship forward speed is 30 knots, i.e. Fn=0.63. The result by (3.2) shows that the vertical relative velocity increases with decreasing mean wave period, while the one by (3.3) has it maximum at the mean wave period of about 6 s and decreases again with decreasing mean wave period. Because the slamming pressure is proportional to the relative velocity in square, a difference in the magnitude with a factor of 1.41 means a factor of 2 in the difference of slamming pressure. For a same wave height, the slamming pressure calculated by (3.2) will be more than 2 greater than the one by (3.3) for the mean wave period equal or less than 6 s.

Tab.3.1 shows the difference between the two mean slamming pressures at a position 43.67 m from the stern perpendicular in a long-crested irregular wave. (3.2) and (3.3) are used for the calculation of relative velocity. A P-M spectrum has been used with significant wave height 4 m and mean

period 6 s for the wave condition. The ship speed is 30 knots and the course angle 165⁰. The difference is great with a factor of about 2. Obviously, this discrepancy is not acceptable in practice.

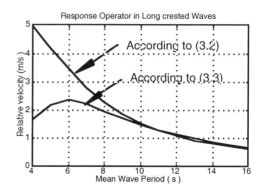

Fig.3.1 Comparison of relative velocities near FP, calculated according to equation (3.2) and equation (3.3) respectively. The ship speed is 30 knots in heading waves.

Tab.3.1 Slamming pressures with the two methods

Method	Min. (MPa)	Mean (MPa)	Max. (MPa)	Std (MPa)
(3.2)	0.523	0.618	0.831	0.086
(3.3)	0.249	0.302	0.356	0.037

It should be pointed out that the bottom slamming pressure estimation method was derived by Ochi for ships with Froude number less than 0.2. He found through the model measurements that the form coefficient is insensitive to the forward speed with Fn<0.2. Within this speed range, the discrepancy between the two vertical relative velocities calculated according to (3.2) and (3.3) respectively will be considerable lower than the one shown in Fig.3.1, since the effect of ship speed on the encounter frequence is lower.

However, when semi-planning high speed monohull vessels are considered, the flare slamming is the design load because of their V-formed forebody. At a fore section, the vertical relative velocity reaches its maximum as the wave surface is passing the calm water line. At the moment, high slamming pressure is applied on a large area of the hull surface, causing high stress in the hull structure, particularly as large panel structure is considered which is quite commonly with composite materials.

3.2 Slamming in bow waves

Slamming pressure in heading sea is usually considered in structure design. If only vertical relative velocity is regarded, the heading sea may be the severest case if equation (3.2) is used for the definition of the vertical relative velocity. If equation (3.3) is used, the magnitude of the vertical relative velocity will change slightly as the relative course angle changes from the heading wave to a bow wave of 120^0, comparing Fig.3.1 with Fig.3.2 and Fig.3.3. In fact, the slamming pressure is not only dependent upon the vertical relative velocity, but also the deadrise angle between hull surface and the momentary wave surface, which has a strong effect on the slamming factors for both the peak and mean value, see (2.2) and (2.3). For a ship in bow waves, the momentary wave slope may be up to about 23^0 according to the linear wave model. The effective deadrise angle gets a reduction of some degrees depending on the relative course angle and the phase relationship of the ship motions to the wave surface.

Apparently, several stochastic variables are involved in the slamming problem in bow waves, such as time histories of wave surface, the momentary ship position in waves besides of the vertical relative velocity.

Tab.3.2 shows the result from the time-domain simulation of slamming pressure on the vessel with a forward speed of 30 knots in two long crested waves with relative course angle of 150^0 and 180^0

respectively. The wave height is 4 m and mean wave period 6 s. The statistical measures shows that the slamming pressure in the bow wave is in general at least 25% higher than in the heading wave.

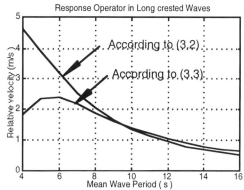

Fig.3.2 Comparison of relative velocities near FP, calculated according to (3.2) and (3.3) respectively. The ship speed is 30 knots and the heading angle is 150 degrees.

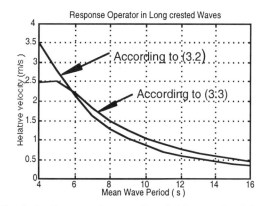

Fig.3.3 Comparison of relative velocities near FP, calculated according to (3.2) and (3.3) respectively. The ship speed is 30 knots and the heading angle is 120 degrees.

Tab.3.2 Comparison of slamming pressure in two different wave directions

Wave direction	Min (MPa)	Mean (MPa)	Max. (MPa)	Std (MPa)
150^0	0.273	0.361	0.488	0.055
180^0	0.218	0.278	0.389	0.038

584

3.3 Nonlinear wave effect

In the previous section, it is shown that the wave slope has a perceptible effect on the slamming pressure magnitude in the linear wave model. However, as the wave height increases, the nonlinear wave effect on the wave surface feature increases. The wave slope becomes therefore greater.

The second order Stock's wave is:

$$\eta(x,t) = \sum_{i=1}^{N} a_i \cdot \cos \psi_i +$$

$$+\frac{1}{4}\sum_{i=1}^{N}\sum_{j=1}^{N} a_i \cdot a_j \cdot \left(k_i + k_j\right) \cdot \cos\left(\psi_i + \psi_j\right)$$

$$-\frac{1}{4}\sum_{i=1}^{N}\sum_{j=1}^{N} a_i \cdot a_j \cdot \left|k_i - k_j\right| \cdot \cos\left(\psi_i - \psi_j\right)$$

(3.4)

where

$$\psi_i = k_i \cdot x - \omega_i \cdot t + \theta_i$$

from Juncher and Pedersen [14]:

When studying the second order effect on the slamming pressure, we assume that the vessel motions is slightly affected by this effect, because the sum frequencies in the second order part of the wave are far away from the frequencies governing the vessel motions. So the linear transfer function of the ship motions are used in the time-domain simulation. The simulated result is shown in Tab.3.3 in comparison with the one in the linear waves. Significant wave height of 4 m and zero-cross mean period of 6 s are used as the wave condition. The ship speed is 30 knots in a course angle of 150° relatively to the wave propagation direct. The result in statistic term shows that the slamming pressure in the second order Stoke's wave is in general over 15% greater than in the linear wave. The increased wave slope and water particle velocity due to the nonlinear effect seems to be the two main causal factors. The effective deadrise angle between the hull surface and wave

surface decreases while the effective relative velocity increases.

Tab.3.3 Slamming pressure due to different wave models

Wave type	Min. (MPa)	Mean (MPa)	Max. (MPa)	Std (MPa)
lin. wave	0.273	0.361	0.488	0.055
2nd wave	0.327	0.425	0.587	0.038

3.4 Wave breaking effect

Wave breaking is a common phenomenon during the wave growing process in stormy weathers due to the nonlinear free surface condition. The wave crest becomes steeper and the wave trough flatter as the wave height grows. However, the wave crest can not be too large. Theoretically, the wave height of a regular wave can not be bigger than 14% of its wave length. When the critical value is exceeded, the wave will be unstable and wave breaking will take place, because the horizontal velocity of water particles near the wave crest becomes greater than the wave propagating speed g/ω. The horizontal velocity of water particles near a wave crest can be more than 10 m/s, far greater than the wave-induced water orbit velocity. The wave crest becomes strongly deformed under wave breaking.

To determine the probability of breaking wave occurrence, a crest-acceleration threshold method can be used. Wave breaking occurs as the water particles at the wave crest has a down acceleration in excess of a threshold value α*g. A number studies have been conducted and the α value has been determined between 0.388 and 0.5 depending on chosen condition for wave breaking, see [15-17].

In Tab.3.4 is shown significant crest-accelerations calculated for various extreme wave conditions with mean zero-cross period from 5 to 10 seconds, according to the following formula:

$$a_{1/3} = 2 \cdot \sqrt{\int_0^\infty \omega^4 \cdot S(\omega) \cdot d\omega} \qquad (3.6)$$

P-M spectrum is used. Hence, we can calculate the probability of crest-acceleration exceeding the threshold value α g for wave breaking at a certain location. For an example, for an irregular wave with H1/3 of 6 m and T2 of 6 s, the significant acceleration will be about 3 m/s^2. Supposing $\alpha = 0.5$, the probability for wave breaking will then be 0.0048, according to

$$P(a \geq a^*) = e^{-2 \cdot \left(\frac{a^*}{a_{1/3}} \right)^2}$$

(3.6)

It is to say that the wave breaking will take place in average about three times per hour at a same location in the actual area. if $\alpha = 0.388$, the probability for wave breaking will then be 0.04, i.e. 24 times per hour. For a ship in bow waves with a forward speed, the time span for the ship to experience wave breaking will be shorter because of more waves encountered.

Tab.3.4 Significant crest-accelerations for various extreme wave conditions

T2 (s)	5	6	7	8	9	10
H$_{1/3}$ (m)	5	6	7	8	9	10
a$_{1/3}$ (m/s^2)	4.07	3.74	3.44	3.17	2.95	2.75

How the wave breaking effect on the slamming pressure is dependent upon both the severity of the wave breaking and the ship motions at the breaking moment. In the previous section the nonlinear wave effect is studied, and the increased wave slope due to the nonlinear effect is an important factor affecting the slamming pressure. As a matter of fact, wave breaking is the extreme case of the nonlinear effect.

Wave breaking occurs only as the significant wave height is sufficiently

large relatively to the mean period. For instance, 5-6 m significant wave height should be for a mean wave period of 6 s if the occurrence probability of wave breaking will be considerable. Certainly, it is not possible to run the actual vessel in a such severe wave condition with a forward speed of 30 knots. However, the vessel should be functional in low speeds. Therefore, the wave breaking induced slamming pressure is an item of interesting for investigation.

The condition that the wave breaking will cause structural damages is that sufficiently large area of the hull surface is simultaneously exposed to a high slamming pressure. That means that wave breaking should take place moment before the wave front has reached the hull and at the same time the wave-induced vessel motions are so that the sufficiently large part of the forebody should be submerged under the momentary wave surface.

Fig.3.4 shows the simultaneous vertical wave acceleration, vertical relative motion and velocity at a position with a quarter ship length from the bow during a 15-second time interval in an irregular wave with $H_{1/3}$ =5 m, T_2 =5s. The ship speed is 10 knots and the relative course angle 120^0. The vertical wave acceleration means here the wave-induced vertical acceleration of water particles at the mean water level. The vertical wave acceleration are calculated according to the linear wave model. The relative motion is referred to the draught line of the ship in calm water. A zero relative motion means that the wave surface at the moment coincides with the draught line at the actual position and a relative motion of 2 m means that the actual body part is submerged by the wave surface from 2 m above the draught line.

Fig.3.5 shows the same kind of results as Fig.3.4 with the same input data except the mean zero-cross wave period are changed to

7 s and significant wave height 7 m. It is shown in Fig.3.4 and Fig.3.5 that large down wave acceleration takes place simultaneously as a relatively large part of the body surface is submerged. In all the two cases the down wave acceleration exceed 5 m/s2. According to the crest-height acceleration threshold criteria, wave breaking will take place.

Obviously, the large submergence of the actual body part can occur at the same time as a breaking wave is propagating towards it. An impact load on the hull surface can be expected. However, it is not easy to evaluate the slamming load magnitude because the wave front can have various shapes dependent upon the wave breaking severity, which in turn can result in different slamming impact mode. The hydrodynamic aspect can be very complex in a slamming process and needs further investigations.

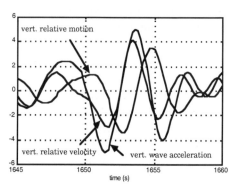

Fig.3.5   Time histories of vertical wave acceleration, vertical relative motion and velocity at a position with a quarter ship length from the bow. Speed is 10 knots in a course angle of 120^0 relatively to the wave direction. $H_{1/3}$ is 7 m and T_2 7 s.

3.5 Combined sea effect

It is not seldom that a wind wave is accompanied by a swell from an early wind wave. The importance of this kind wave is discussed for the design of conventional ships and ocean structures by Ochi [18]. Here we make an investigation into the effect of combined wave on the slamming pressure.

As known, the magnitude of slamming pressure is sensitive to the effective deadrise angle besides of the velocity of a hull body relatively to the wave surface. Suppose a vessel running in a combined sea is encountered by the wind wave from bow while the swell from beam. The wind wave has a such energy spectrum that the vessel is subjected to large vertical motions and large relative velocity against wave surface, in our case, the zero-cross mean wave periods is about 6-8 seconds dependent upon the wave spectrum bandwidth. At the same time the mean period of the swell is near the natural roll period of the vessel, and considerable roll motion is induced due to the resonance effect. Obviously, the roll motion can affect the variation of the

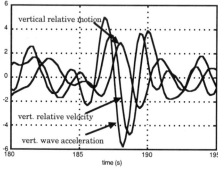

Fig.3.4   Time histories of vertical wave acceleration, verical relative motion and velocity at a position with a quarter ship length from the bow. Speed is 10 knots in a course angle of 120o relatively to the wave direction. $H_{1/3}$ is 5 m and T_2 5 s.

deadrise angle between the hull surface and wave surface. The slamming pressure will be magnified, if the deadrise angle at the instance is reduced due to the roll motion.

Witt the purpose to point out the significant effect of this kind wave condition, a sea condition is chosen which gives an illustrative result. The wind wave is represented by P-M wave spectrum with mean zero-cross period of 6 s and significant wave height of 4 m. The swell is represented by a mean JONSWAP wave spectrum with significant wave height of 2 m, modal period of 8 s and γ equal to 5. The relative course angle to the wind wave is 150° and to the swell 90°. Comparing with the result in Tab.3.2 for the slamming pressure in the wind wave, the maximal one becomes almost twice due to the swell effect, while the total significant wave height of the combined sea is increased with about 10%. If the second order wave effect is taken into account, the slamming pressure will be 10% further more.

Tab.3.5 Slamming pressures in combined sea

Wave type	Min. (MPa)	Mean (MPa)	Max. (MPa)	Std (MPa)
lin. wave	0.351	0.538	0.917	0.142
2nd wave	0.423	0.597	1.020	0.150

4. CONCLUSION

(1)   The slamming pressure magnitude is very sensitive to the definition of vertical relative velocity. The vertical relative velocity of the forebody to the momentary wave surface gives considerably higher slamming pressure with a factor of about 2 than the one relative to the momentary water particle velocity for the actual vessel with Fn=0.63 in the design wave condition.

(2)   Running in heading waves is usually considered as design condition. However, our calculation has shown that the actual vessel in bow waves results in higher slamming pressure than in heading waves due to the wave surface slope effect .

(3)   The nonlinear wave effect results in increased wave surface slope so that the effective deadrise angle between the hull surface and wave surface decreases. The simulated result shows that the slamming pressure in second order Stoke's wave is in general over 15% greater than in linear waves.

(4)   Large submergence of the forebody part can occur at the same time as a breaking wave is propagating towards it. A severe impact load on the hull surface can be expected. Further investigations should be made in order to quantify the slamming pressure.

(5)   The swell in a combined sea may have a considerably additional effect on the slamming pressure. This effect is due to the increased roll motion by the swell. The simulated result shows that the slamming pressure level can in some cases increase with more than 50% by adding the swell effect. The total significant wave height increases only with 12% while the significant wave height is 4 m for the wind wave.

To determine the probably maximal slamming pressure due to combined seas, to which a vessel will be subjected under her service life, it requires statistics over the combined seas in the service areas including wave direction correlation. Such kind wave statistics has not yet appeared.

588

REFERENCE

1. J. Hua, Result of Computer Simulation of YS2000 in Irregular Waves, in Swedish, Div. of Naval Architecture, Dept. of Vehicle Engineering Royal Institute of Technology, Sweden May 1996.

2. J. Hua, Slamming Pressure for Construction Design of YS2000" in Swedish, Div. of Naval Architecture, KTH, 1996Restricted distribution

3. T. Milchert, J. Hua and K. Mäkinen, Design Slamming Loads for a Semi-Planning Naval Craft Build of Composite Sandwich, restricted distribution

4. W.K. Meyerhoff and G. Schlachter, An Approach for the Determination of Hull-Girder Loads in a Seaway Including Hydrodynamic Impacts, Ocean Engng. Vol.7, pp. 306-326, 1977. 7

5. F. Chiu and M. Fujino, Nonlinear prediction of vertical motions and wave loads of high-speed vessels in head sea, International Shipbuilding Progress, Vol.36, No.406, Sep. 1989

6. J. Hua, Computer Simulation of Motions High Speed Vessels in Heading Waves and the Wave Loads,TRITA-FKT Report 9645 ISSN 1103-470X, ISRN KTH/FKT/SKP/FR--96/45--SE Div. of Naval Architecture, Royal Institute of Technology, Sweden 1996.

7. J. Hua, Wave Load Mechanism to Avoid Bow Visor Damage, The Naval Architect, January 1996.

8. N. Salvesen, E.O. Tuck and O. Faltinsen, Ship Motion and Sea Load, Trans SNAME, Vol.78, 1970

9. Zhao, R. and Faltinsen, O., Water entry of two-dimensional bodies, Journal of Fluid Mechanics, Vol. 246, 1993

10. M. Martin, Theoretical Prediction of Motions of High-Speed Planning Boats in Waves, Journal of Ship Research, Vol.22, No.3 Sept.1978, pp.140-169

11. A. Stavovy and S.-L. Chuang, Analytical determination of slamming pressures for high-speed vehicles in waves, Journal of Ship Research, Vol. 20, No. 4, 1976

12. M. Ochi and L. Motter, Prediction of Slamming Characteristics and Hull Responses for Ship Design, SNAME Transection Vol.81 1973

13. C. Colwell, I. Datta and R. Rogers, Head Seas Slamming Test on a Fast Surface Ship Hull Form Series, International Conference on Seakeeping and Weather, Feb. 1995 London

14. J.J. Juncher and P.P. Pedersen, Wave-Induced Bending Moments in Ships - a Quadratic Theory, TransRINA, Vol.121, pp.151-165,1979

15. R.L. Snyder and R.M. Kennedy, On the formation of Whitecaps by a Threshold Mechanism. Part I. Basic Formalism, J. Phys. Oceanogr. Vol13 pp 1482-1492. 1983

16. R.L. Snyder and R.M. Kennedy, On the formation of Whitecaps by a Threshold Mechanism. Part II. Field Experiment and Comparison with Theory, J. Phys. Oceanogr. Vol13 pp 1506-1518. 1983

17. M.A. Strokosz, On the Probability of Wave Breaking in Deep Water, J. Phys. Oceanogr. Vol16 no.2 pp 382-385. 1986

18. M. Ochi, Wave Statistics for the Design of Ships and Ocean Structures, SNAME Transactions, Vol. 86, 1978 , pp.47-76

1998 Elsevier Science B.V.
Practical Design of Ships and Mobile Units
M.W.C. Oosterveld and S.G. Tan, editors.

A coupled approach for the evaluation of slamming loads on ships

A. Magee[a] * and E. Fontaine[b] †

[a]Bassin d'Essais des Carènes, Chaussée du Vexin, 27100 Val-de-Reuil, FRANCE.

[b]Ocean Engineering Laboratory, UCSB, Santa Barbara, CA 93106-1080, USA.

The behavior of a ship navigating in waves and subject to slamming is studied within the framework of potential flow theory, by mean of a time-domain simulation. The global ship motion is computed by a 6 degrees-of-freedom large amplitude motion seakeeping code. When a slamming event is detected, the local flow and the force due to the impact are computed separately. A coupled approach is developed where the feedback between the ship motion and the slamming force is taken into account, allowing consistent estimates of both the motion and the slamming loads to be obtained simultaneously. The methods and the coupling procedure are first described. The coupled approach is then validated against experiments in the case of free-fall drop tests. Subsequently, simulations are carried out for a ship in irregular seas. The results show a reduction in local mean and maximum slamming pressures using the coupled method as compared with the uncoupled method.

1. INTRODUCTION

Along with green water on deck, slamming generally sets the operational limit for Naval and commercial vessels operating in near head seas conditions. In heavy seas and at higher forward speed, the probability of occurrence of slamming increases and a prudent ship master must voluntarily reduce speed since severe impacts may cause damage to the ship's structure. A practical tool for Naval Architects would be the ability to quickly and accurately assess the effects of slamming on different proposed designs, and to aid in rationally designing the ship's structure to resist expected loads.

Slamming forces can become very large compared with others forces acting on the ship. However, because their duration is short, their effects on the global ship motions are usually neglected. Such a decoupled approach was taken in [1]. In that study, time-domain simulations were used to compute the relative velocities between the hull and free surface. Slamming pressures were then computed, neglecting the body retardation due to the slamming forces, giving useful first approximations. The present approach improves the predictions by including the slamming forces in the

computation of the ship motions. Due to the slamming forces, the body slows down as it impacts the water. The relative acceleration and velocity and hence the slamming loads are reduced as compared with the uncoupled approach. The coupled approach represents a step forward in fast, realistic and practical calculations for design evaluation.

2. DESCRIPTION OF THE COUPLED METHOD

In the present model, the fluid is assumed to be perfect and incompressible, the flow to be potential, and the body to be rigid. Structural deformations and entrapped air that could occur during slamming are neglected. It is felt that before attempting the hydroelastic problem, it is important to consider the case of the rigid body first, especially for typical ships which are relatively more rigid than light fast monohulls or catamarans. The large amplitude motion seakeeping code RATANA, used to compute the global ship motion, and the code IMPACT, used to model the local flow when slamming occurs are first described. It is then shown how a coupling procedure provides simultaneously an estimate for both the motions and the impact loads.

*E-mail: magee@becvdr.dga.fr
†E-mail: emmanuel@vortex.ucsb.edu

RATANA [2] is a large-amplitude, 6 d.o.f. sea-keeping simulation code. The Froude-Krylov and hydrostatic forces are calculated on the instantaneous immerged hull surface at each time step. The Euler equations are solved in a coordinate system fixed to the ship's center of gravity. Non-linear external forces may be taken into account directly in the simulation without the need to linearize either the forces or the responses.

RATANA uses added masses, damping and diffraction force coefficients as inputs. In the present calculations these were obtained using a linear three-dimensional frequency domain code, DIODORE [3]. RATANA transforms the frequency domain coefficients into an equivalent time-domain model as explained in [4]. As shown in that study, in the case of small-amplitude regular waves, the motion and relative motion predictions obtained using the time-domain calculations are equivalent to those of the frequency-domain calculations. However, in the case of large-amplitude irregular waves, when slamming events occur, the temporal model of RATANA allows the most important non-linearities to be taken into account. In addition, the effects of the instantaneous slamming force on the motion can be properly accounted for only within the time-domain model.

The code IMPACT [1] is used to compute accurately and rapidly the force acting on the bottom of the ship as it re-enters the free-surface. The model assumes the body is slender and flat, and the vertical impact velocity is sufficiently high, so that the effect of gravity is negligible. Under these assumptions, the method of matched asymptotic expansions justifies the development of a strip theory which can be viewed as the limiting case of the $2D + t$ or $2D + 1/2$ theory [5],[6] as the draft goes to zero. Since gravity effects are neglected, there is no interaction between the strips. The flow around each cross section is similar to that around a flat plate of equivalent length and a jet solution is introduced at the edges of the plate to remove the singularity. A proper treatment of the jet is essential since, during the impact, half of the energy transfered from the body to the fluid is kinetic energy within the spray [7].

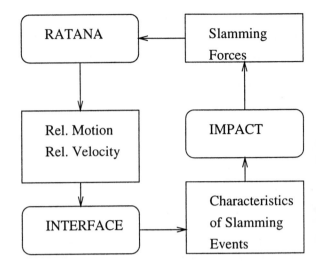

Figure 1. Schematic diagram of the coupled method.

The previous two-dimensional asymptotic results [8] have been extented so that shapes of arbitrary cross section and time-varying velocity can be accounted for. Relative velocity is used to derive the impact pressure distribution and the cross section slamming loads. Since most of the results are analytic, the CPU times are insignificant and the resulting force can be taken into account directly in RATANA, during the global simulations.

The coupling procedure is schematically described in figure 1. The solution proceeds by iteration, with the first estimate equivalent to the decoupled approach. To begin with, RATANA is used to compute the ship motion without taking into account the slamming forces. The relative motions and velocities between the hull and free surface are calculated at each timestep. An interface program identifies individual slamming events and extracts the relevant time histories at a number of ship sections. These are used as input to the IMPACT module which calculates the slamming force on each section. The sectional forces are then summed to obtain the total slamming force and moment, which are included in the next iteration of the equations of motion calculation using RATANA. Converged and consistent results for both motions and slamming forces are obtained very rapidly, typically after only two iterations. Slamming events appear to be localised in space and time. However, a large number of cross sections is required to properly account for the spatial and temporal evolution of the pressure distribution, in computing transient slamming forces. Thus, the advantage of the rapidity of IMPACT is essential.

3. VALIDATION

The coupled method is first validated against experiments for the case of two-dimensional geometries in free-fall drop tests [9]. Due to the relative simplicity of the problem, a fourth order Runge-Kutta method is used to resolve the vertical body motions due to combined forces of gravity, hydrostatics and slamming. The first iteration is chosen to be a constant drop velocity. In such tests, the slamming force can be large with respect to the mass of the body, leading to large accelerations. Force and motion are strongly coupled and five iterations are needed to get converged predictions for both quantities. A 30 degree wedge and a bow-flare body are tested.

In the case of the wedge, the slamming force largely dominates the others so that retardation is observed almost immediately (see fig.2). The body retardation is well reproduced when compared to experiments. The difference between predicted and measured velocities is less than 2%, which is excellent. In figure 3, the converged force per unit length on the wedge is compared to the measured values. The force is slightly over estimated, but quite close to the experimentally obtained values. The relative maximum difference is less than 10% which again is exellent considering the amount of approximations that have been made.

In the case of the flare body, the vertical velocity, force per unit length, and pressure coefficients are presented in figures 4 to 6. The flare body has a portion of nearly vertical sides and thus it experiences a lower slamming force than the wedge. Even though the body penetrates the water surface, the acceleration remains al-

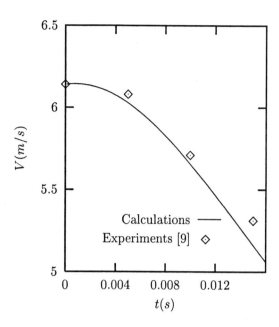

Figure 2. Time history of the vertical velocity of a 30° wedge in free-fall. Calculation is performed using the coupled method.

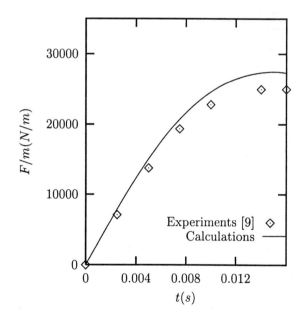

Figure 3. Time history of the vertical force per unit length on a 30° wedge in free-fall.

592

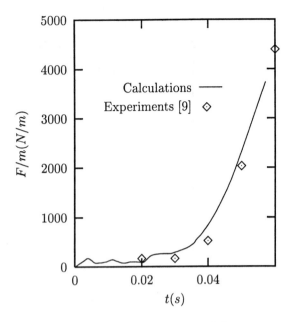

Figure 4. Time history of the vertical force per unit length on a 2-D flare body.

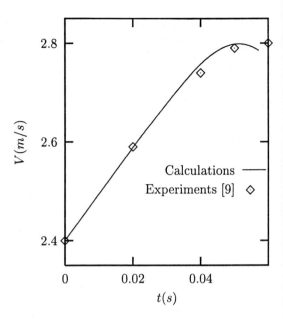

Figure 5. Time history of the vertical velocity of a 2-D flare body.

most constant (equal to gravity) during the initial stage. Retardation is observed when the water level reaches the flared part of the body. Again, excellent agreement is obtained between calculated and measured values . At the end of the simulation, the flare body is fully submerged and flow separation occurs. This physical phenomenon, which is not yet included in the model used to compute the slamming load, will lead to a decreasing force at later times.

4. RESULTS OF THE COUPLED METHOD FOR A SHIP

The probability of occurence of slamming events increases with rough weather and higher forward speed. For the numerical simulation, the case of a frigate is considered which is navigating in a head sea with 25 knots forward speed in a seastate 5 to 6. The corresponding input wave spectrum is of the Pierson-Moskovitz type with $H_s = 4m$ and $Tpic = 10s$. For this condition of navigation, the encounter spectrum peak frequency $\omega_e \simeq 1.15 rad/s$ corresponds to the maximum of the vertical relative velocity transfer func-

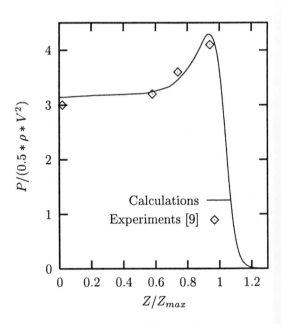

Figure 6. Distibution of pressure coefficient C_p on the flare body just before body submergence (t=0.0575 s for the computation and t=0.06 s for the experiment).

tion for points in the bow region of this ship. (See fig. 7). As a result, the relative motions between the hull and the water surface can be large and violent. For example, the RMS value of the vertical acceleration at the bridge is about $0.2g$, which is considered to be a limiting value [10].

Different criteria can be used to detect slamming events. According to [11], the condition for the occurrence of a slamming event is that the keel enters the water with an impact velocity $V_i \geq 3.3m/s$. In the present simulation, about 100 slamming events per hour occur using this criteria, i.e. about 15 slams out of 100 waves. This result exceeds five times the usual criteria [10] for 'voluntary speed reduction' and thus confirms that this is not a tolerable condition of navigation for a long period of time. The criterion [11] is derived from a statistical approach which ignores the phasing between the relative motion and relative velocity. A simplified method has been proposed by [12] to take into account the phase lag between the instant of the maximum velocity and the instant when the section touches the surface. However, the resulting criterion has been shown to under-predict the impact velocities, when compared with temporal simulations [1].

Figure 8 shows the response spectrum of relative vertical bow velocity obtained using linear theory and by a Fourier transform of the time signals calculated using RATANA, with and without the coupled slamming forces. The time-domain simulations were carried out for one hour. With respect to the linear calculations, the large-amplitude method predicts lower relative velocities because the ship is slowed as it enters large waves due to the instantaneous hydrostatic and incident wave forces. For example, using the large-amplitude model, but not yet including the slamming forces (RATANA uncoupled), the significant values of the relative velocities are reduced by about 15% for $V_{1/3}$ and by about 19% for $V_{1/10}$ with respect to the linear values. Coupling the slamming forces with the equations of motion further reduces the predictions of these quantities (and thus of the slamming forces), be-

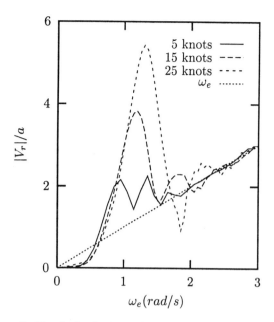

Figure 7. Vertical relative velocity transfer function versus encounter frequency for three forward speeds. This results applies at a point on the bow at station STA. 20, i.e. at the intersection of the forward perpendicular with the calm waterline.

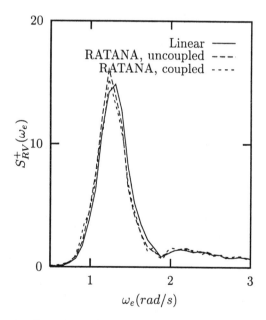

Figure 8. Relative velocity spectrum versus encounter frequency for 25 knots forward speed.

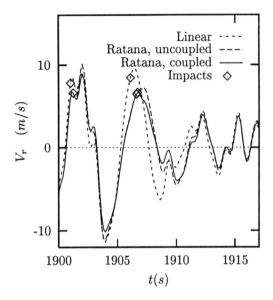

Figure 9. Time history of the vertical relative velocity at STA. 17.5. Note the occurence of two consecutive slamming events indicated by symbols on each curves.

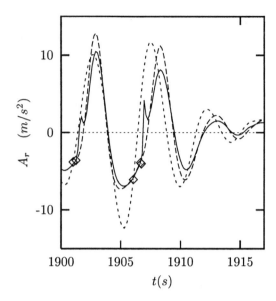

Figure 10. Time history of the vertical relative acceleration at STA. 17.5. The key is the same as in fig. 9. Note the appearence of peaks and the reduction of level whith the coupled approach.

cause the body also slows down as it impacts the water. The higher the impact velocity, the larger the slamming force and the stronger the influence of the coupling effect will be. However, very long simulations must be carried out to obtain precise estimates of the maximum relative velocities. This is an important point since design values may be chosen based on a single representative event, i.e. of the most rarely occurring events.

Figures 9 and 10 show a portion of time histories of the vertical relative velocity and acceleration at station 17.5. Three calculations are shown: the linear theory, the large-amplitude results with and without the inclusion of the coupled slamming forces. The slamming criteria is reached at $t \simeq 1901s$ and $t \simeq 1906s$ so that the ship slams on two successive large waves. After each impact, the slamming loads tend to limit the relative vertical acceleration and this leads to the appearence of peaks on fig. 10. The subsequent oscillations are reduced due to the memory effect which is properly accounted for in the time-domain model. This effect is more visible after

the second impact and this also holds for the vertical relative velocity. As expected for that heavy ship, inertia forces dominate and the coupling of the slamming force leads only to a slight reduction of the vertical relative velocity after each slamming event. However, the coupling should be more significant for light, shallow draft vessels with small deadrise angles at high Froude numbers.

Figure 11 shows snapshots of the pressure contours calculated using the uncoupled and the coupled approach. These snapshots occur at different instants corresponding to when the maximum pressure is reached. This is due to the fact that the trajectories are slightly different for the different approaches. The use of the coupled approach leads to a reduction in the surface over which a given high pressure is exerted, as well as, the mean and maximum slamming pressures. In the present case, the very localized maximum pressure is decreased by about 20%.

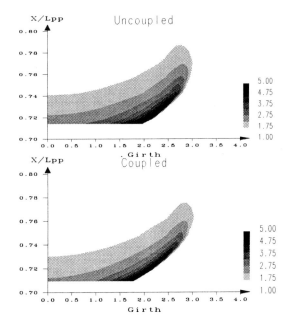

Figure 11. Pressure contours (in bars) over the forward hull surface at the instants of the maxima.

5. CONCLUSIONS

A coupled approach has been developed where the feedback between the ship motion and the slamming force is taken into account. The asymptotic model used to compute the slamming loads and the coupling procedure have been validated succesfully against experiments in the case of two-dimensional free-fall drop test. Comparisons show that the coupled approach allows consistent estimates of both the motion and the slamming load to be obtained simultaneously. Taking into account the slamming load leads to a body retardation and a decrease in the predicted mean and maximum slamming pressure, as well as the area over which the high pressure applies. By systematically using the coupled approach as soon as the forward part of the bow is out of the water, the severity of each slamming event can be determined. This approach can therefore be used to check and derive precise slamming criteria for a given ship, thus allowing different designs to be compared.

REFERENCES

1. E. Fontaine, L. Boudet, J.-F. Leguen and A.R. Magee, 1997. Impact Hydrodynamique d'un Corps Allongée et Plat : Application au tossage. *Proc. of the 6th Journées de l'Hydrodynamique,* Nantes, France.

2. A.R. Magee, 1997. Applications using a sea-keeping simulation code. *Proc. of the 12th International Workshop on Water Waves and Floating Bodies,* Carry-le-Rouet, France.

3. T. Coudray and J.-F. Le Guen, 1992. Validation of a 3-D seakeeping software. *Proceedings of CADMO'92,* Madrid.

4. A.R. Magee, J.-F. LeGuen and X. Dupouy, 1997. A time-domain simulation method for stabiliser fin evaluation and design. *STAB'97,* Varna, Bulgaria.

5. H. Maruo and W. Song, 1994. "Nonlinear Analysis of Bow Wave Breaking and Deck Wetness of a High-Speed Ship by the Parabolic Approximation", *Twentieth Symposium on Naval Hydrodynamics,* USA.

6. E. Fontaine, O.M. Faltinsen and R. Cointe, 1998. Some new insight in the generation of bow waves, submitted to *J. Fluid Mech.*

7. E. Fontaine, B. Molin and R. Cointe, 1998. On energy arguments applied to hydrodynamic impact force, subm. to *J. Fluid Mech.*

8. R. Cointe and J.-L. Armand, 1987. Hydrodynamic Impact Analysis of a Cylinder, *J. Offshore Mechanics and Artic Engineering,* Vol 9, pp. 237,243.

9. R. Zhao and O.M. Faltinsen, 1996. Water entry of arbitrary two-dimensional sections with and without flow separation. *Proc. of the 21st Symp. on Naval Hydrodynamics,* Norway.

10. NORDFORSK, 1987. The Nordic Cooperative project. Seakeeping performance of ships, *Assesement of a ship performance in a seaway,* Tronheim.

11. M.K. Ochi and L.E Motter, 1973. Prediction of slamming characteristics and hull responses for ship design. *SNAME Transactions,* Vol. 81, pp 144-190.

12. Lloyd, A.R.J.M. 1989. *Seakeeping, Ship Behaviour in Rough Weather,* Ellis Horwood Limited.

Practical Design of Ships and Mobile Units
M.W.C. Oosterveld and S.G. Tan, editors.

The Effect of Forward Speed on the Hydroelastic Behaviors of Ship Structures

S.-X Du and Y.-S. Wu

China Ship Scientific Research Center
P. O. Box 116, Wuxi, Jiangsu 214082, P. R. China

The present paper concentrates on the influence of forward speed on the structural responses.   The contributions from the velocity field of non-uniform steady flow to the interface boundary condition, and the generalized hydrodynamic forces are included in the three-dimensional hydroelasticity analysis.   A fast numerical approach is also developed to allow for the translating-pulsating source Green's function to be used for the calculation of the radiation and diffracted potentials for a deformable body. Numerical results of an ellipsoid moving beneath wave surface, and a flexible surface ship with semi-ellipsoid shape traveling in waves are illustrated. The corresponding mathematical models which influence the prediction results of the forward speed effect are discussed.

1. INTRODUCTION

Similar to most of the seakeeping theories, the existing 2-D and 3-D hydroelasticity theories of ships[1-3] assume that the advancing of the ship does not generate non-uniform steady flow around and behind it.   This assumption is acceptable for a slowly moving slender, or thin body. If the ship is neither slender, nor thin, and its forward speed is not small, the steady state disturbance can not be omitted.   According to the three-dimensional potential theory of a traveling flexible body[2], the influence of the steady flow field on the structural responses appears in the interface boundary condition on the wetted surface, the kernels of the integral formulae of the hydrodynamic coefficients and the wave exciting forces.   The following questions therefore need to be clarified:

--- To what extent does the steady flow field give effect on the structural dynamic responses of a ship?

--- How is the effect quantitatively dependent on the geometric ratio of slenderness and the forward speed?

To take all of these into account a revised approach of the three-dimensional hydroelastic analysis is presented in this paper. This approach follows the full form of the linear three-dimensional hydroelasticity theory of ships[2,3], and employs the translating-pulsating source Green function, so as to allow for the steady-state disturbances to be considered in a rational manner.   The terms containing the non-uniform steady flow retain in all the corresponding equations and formulae. Although the free-surface condition with forward speed terms is linearized, no further assumptions is made to restrict the geometric form of the floating body.

To calculate the three-dimensional non-uniform steady flow potential and its high-order derivatives a Galerkin technique with double integrals is introduced, together with a desingularized source distribution approach.   Also introduced is a fast algorithm for calculating the Bessho-form translating-pulsating source Green's function[4], where the steepest descent path method[5], and a self compatible integration method are used

The effect of forward speed on the hydroelastic behavior of ship structures is investigated and illustrated by calculations of an ellipsoid moving beneath wave surface, and a surface ship of the semi-ellipsoid form traveling in waves. Numerical comparisons of the hydrodynamic coefficients, wave exciting forces, rigid body motions , and flexible body distortions are made for different forward speeds, different slenderness parameters, and different heading angles.

598

2. THE LINEAR FREGUENCY-DOMAIN HYDROELASTICITY THEORY OF SHIPS

The ship structure is assumed linear with small motions and distortions about its equilibrium position. The displacements of the structure may be expressed as

$$\vec{u} = \{u, v, w\} = \sum_{r=1}^{m} \vec{u}_r p_r(t) \quad (2.1)$$

where p_r (r=1,2,...,m) represent the m principal coordinates associated with the principal modes $\vec{u}_r = \{u_r, v_r, w_r\}$ (r=1, 2,... m) of the structure. The first six principal modes are the rigid body modes defined as surge, sway, heave, roll, pitch and yaw respectively.

The fluid is assumed inviscid, incompressible, and the flow is irrotational. If the body travels with a constant forward speed U in x direction, the velocity potential may be decomposed in the form[6]

$$\Phi(x,y,z,t) = U\overline{\Phi}(x,y,z) + \phi(x,y,z,t) \quad (2.2)$$

with the unsteady component expressed as

$$\phi(x,y,z,t) = \phi_0(x,y,z,t) + \phi_D(x,y,z,t)$$
$$+ \sum_{r=1}^{m} \varphi_r(x,y,z)p_r(t) \quad (2.3)$$

$\overline{\Phi}$ denotes the velocity potential for the steady motion of the body in calm water. The velocity of the steady flow relative to the moving equilibrium frame[2] of reference is

$$\vec{W} = U\nabla(\overline{\Phi} - x) \quad (2.4)$$

ϕ_0 and ϕ_D denote the incident and diffracted wave potentials respectively. φ_r (r=1,2,...,m) represent the radiation wave potentials corresponding to all the dry modes of the structure.

The interface boundary condition on the mean wetted surface of the structure may be represented in the form[7]

$$\frac{\partial}{\partial n}\phi = \vec{n}\cdot[\dot{\vec{u}}+\vec{\theta}\times\vec{W}-(\vec{u}\cdot\nabla)\vec{W}+\vec{W}\cdot\varepsilon] \quad P\in\overline{S} \quad (2.5)$$

where ε denotes the strain tensile of the structure, and $\vec{\theta} = \frac{1}{2}\nabla\times\vec{u}$. The boundary condition for ϕ_D is

$$\frac{\partial}{\partial n}\phi_D = -\frac{\partial}{\partial n}\phi_0 \quad \text{on } \overline{S} \quad (2.6)$$

By employing (2.5), (2.6) and the three-dimensional translating-pulsating source Green's function $G(P,q;-U)$, the solution of the radiation and diffracted wave potentials may be simplified to a boundary value problem defined on the mean wetted surface $\overline{S}$ and water-line $\overline{C}$ as follows

$$\phi(P) = \iint_{\overline{S}}\sigma(q)G(P,q;-U)dS_q$$
$$+ \frac{1}{g}U^2\oint_{\overline{C}}n_1(q)\sigma(q)G(P,q;-U)dl \quad (2.7)$$

The generalized equation of motion may be presented in the form

$$(\mathbf{a}+\mathbf{A})\ddot{\mathbf{p}}+(\mathbf{b}+\mathbf{B})\dot{\mathbf{p}}+(\mathbf{c}+\mathbf{C})\mathbf{p} = \Xi \quad (2.8)$$

where $\mathbf{a}$, $\mathbf{b}$, $\mathbf{c}$ are the matrices of generalized modal inertial, modal damping and modal stiffness; $\mathbf{A}$, $\mathbf{B}$ and $\mathbf{C}$ are matrices of the hydro-dynamic added mass, damping and restoring coefficients, with their components of the following forms

$$A_{rk}(\omega_e) = \frac{1}{\omega_e^2}R_e \left[\rho\iint_{\overline{S}}\vec{n}\cdot\vec{u}_r(i\omega_e+\vec{W}\cdot\nabla)\right]$$
$$B_{rk}(\omega_e) = \frac{i}{\omega_e}I_m \left[\quad \varphi_k(\omega_e)dS \quad \right]$$

$$C_{rk} = \rho\iint_{\overline{S}}\vec{n}\cdot\vec{u}_r[gw_k+\frac{1}{2}(\vec{u}_k\cdot\nabla)W^2]dS \quad (2.9)$$

The matrix $\mathbf{p}$ is the vector $\{p_1(t), p_2(t), ..., p_m(t)\}$; the components of the generalized wave exciting force matrix Ξ_r are defined as

$$\Xi_r(t) = \rho\iint_{\overline{S}}\vec{n}\cdot\vec{u}(\frac{\partial}{\partial t}+\vec{W}\cdot\nabla)[\phi_0(t)+\phi_D(t)]dS \quad (2.10)$$

Evidently these hydrodynamic coefficients and wave exciting forces may be readily calculated by employing the pre-obtained distribution of the velocity field of the steady flow, and the radiation and diffracted wave potentials.

In (2.5), (2.9) and (2.10) the steady flow velocity $\vec{W}$ and its derivatives are required. According to Dawson[8], $\overline{\Phi}$ may be represented as

$$\overline{\Phi} = \overline{\phi}_0 + Fn^2\overline{\phi}_W \quad (2.11)$$

where the subscript "o" denotes the double-body solution, while the subscript "w" denotes the wave generated component. $Fn = U/\sqrt{gL}$ is the Froude number. The governing equations of $\bar{\phi}_o$ and $\bar{\phi}_W$ may be found in [8].

In solving $\bar{\phi}_o$ and $\bar{\phi}_W$ by Hess-Smith constant panel method, the Rankine source distributed over the wetted and imaginary body surface (for $\bar{\phi}_o$ and $\bar{\phi}_W$), and on the free surface (for $\bar{\phi}_W$ only) are employed. To overcome the difficulties caused by the singularities of the Rankine source Green's functions and its up to the second order derivatives, a Galerkin form of the integral equation for solving the source strengths $\sigma_q(x_1, y_1, z_1)$ together with the virtual source distribution approach which the sources are distributed over an imaginary (virtual) surface in a vicinity inside the body's wetted surface is proposed[9]. By employing this approach, the efficiency and accuracy of calculating the three-dimensional non-uniform steady flow field is well increased. Fig. 1 illustrates the comparison of the numerical predictions and the analytical results of the spatial derivatives of the fluid velocity components for a floating hemisphere advancing in calm water. It is shown that the virtual source distribution approach presents fair results.

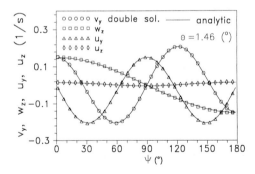

Figure 1. Comparison of the second order derivatives of the steady flow potential at the wetted surface of a hemisphere of radius 1.0m (U=1.0m/s).

A scheme using the steepest descent integration method proposed by [5] and a self-compatible integration method is developed[9,10] to ensure an efficient and accurate calculation of the Bessho form translating-pulsating source Green's function in frequency domain[4]. The details may be found in [9,10]. This scheme was numerically verified for a wide range of forward speed (U=0.5~15.0 m/s) in the major wave frequency bandwidth from 1.0 to 20.0 rad/s. The computational efficiency was demonstrated by the comparison of the computing times spent by using the present method and another fast calculation method developed by [11]. On a Sun Ultra-Sparc work station of a 200 MHz processor without special code optimization at compiling, the present method took 0.013s ($\tau < 0.25$) and 0.005~0.025 ($\tau > 0.25$) of the average CPU time for the calculation of one pair source and field points of Green function and its first derivatives at each forward speed and frequency. The method of [11] required about 0.1s for each pair of real and imaginary parts of the Green function and 0.17~0.2s for the first derivatives on an ALLIANT FX40 (4 processors). The numerical accuracy was confirmed by comparison with the results obtained by the existing methods. For example the numerical differences between the present method and [12] are less than 0.5×10^{-4}.

3. NUMERICAL PREDICTIONS OF THE FORWARD SPEED EFFECT

The influence of forward speed is examined by three factors in the mathematical models. The first is the treatment of the steady flow, namely considering the non-uniform steady flow by employing the calculated $\vec{W}$ or neglecting the non-uniformity by assuming $\vec{W} = -U\vec{i}$ in (2.5), (2.9) and (2.10). These are denoted as the $\vec{W}$-flow solution or the U-flow solution respectively. The second is the choice of Green's functions, namely using the translating-pulsating source Green's function $G(P, q; -U)$, or simply the pulsating source Green's function $G(P, q)$ in (2.7). The third is the contribution of the line integral in (2.7).

The effect of these factors on the hydrodynamic and hydroelastic behaviors of ships is investigated and illustrated by calculations of responses of rigid ellipsoid bodies with different slenderness (length to beam) ratios L/B moving under wave surface, and a flexible surface ship with semi-ellipsoid shape traveling in waves.

Fig. 2 illustrates the comparison of the hydrodynamic coefficients of the moving ellipsoids

600

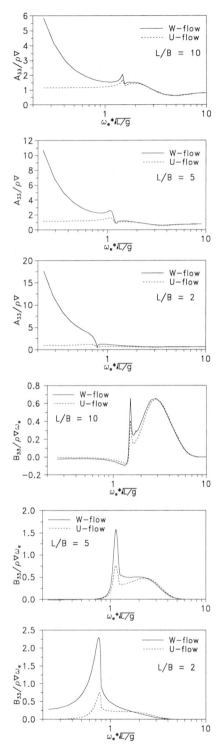

Figure 2. Comparison of the hydrodynamic coefficients of the moving ellipsoids with the slenderness ratios.

with the slenderness ratios L/B=2, 5 and 10. It is shown that in the low encounter frequency region of $\omega_e \sqrt{L/g} < 1.5$ the non-uniform steady flow greatly increases the added mass. The similar effect of the steady flow also exists on the added damping. These effects get stronger when the body is blunter. It should be noted that for a slender body, L/B=10, for example, the differences between the added damping predictions by the $\bar{W}$-flow solution and the U-flow solution are not significant. However it appears that the U-flow solution still underestimates the added mass even if the body is slender.

The numerical simulation of the forward speed effect on the hydroelastic behavior of a semi-ellipsoid surface ship of length 200m, beam 40m and draft 20m advancing in waves with the forward speed of 18kn was performed. The structure of the ship together with its seven transverse bulkheads is modeled as a combination of beam and membrane elements for the finite element analysis of the dry modes. Fig. 3 shows the 7th to 10th mode shapes of the 'dry' structure with the first six not included being the rigid body modes. The mean wetted surface of the ship is discretized as 348 panels for hydrodynamic analysis. To investigate the effects of the above mentioned three factors on the numerical results, six cases defined in Table 1 are calculated.

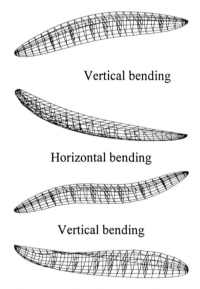

Vertical bending

Horizontal bending

Vertical bending

Horizontal bending and twisting

Figure 3. Flexible mode shapes of 'dry' structure.

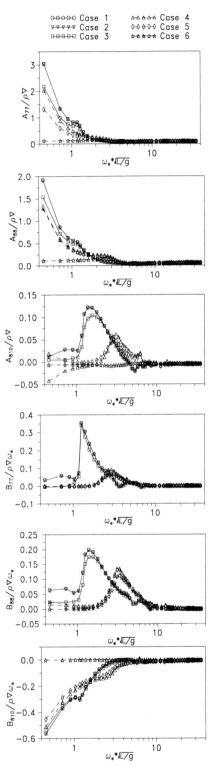

Figure 4. Comparison of the hydrodynamic coefficients corresponding to six cases.

Table 1
The Calculate Conditions

No.	U (kn)	$\vec{W}$-flow	line integral	Green's function
1	18	included	included	$G(P,q;-U)$
2	18	included	excluded	$G(P,q;-U)$
3	18	excluded	included	$G(P,q;-U)$
4	18	included	included	$G(P,q)$
5	18	excluded	included	$G(P,q)$
6	0	----	----	$G(P,q)$

Fig. 4 exhibits the comparisons of the hydrodynamic coefficients corresponding to the six cases. The following aspects may be noticed: (1) Almost no differences are shown between the results for Case 1 and Case 2. This means that the line integral in (2.7) gives little contribution to the hydrodynamic coefficients provided that the body's wetted surface is vertical wall-sided along its water line. (2) The added mass coefficients for the cases with forward speed (Cases 1-5) are quite different from those for zero speed case (Case 6). (3) The variations of the added damping coefficients B_{ij} ($i=j$) with respect to the encounter wave frequency obtained with the pulsating source Green's function (Cases 4,5), no matter whether $\vec{W}$-flow solution or the U-flow solution is considered, are of the same tendency as that of the zero speed case. (4) It appears that the forward speed effect on the added mass and added damping coefficients exhibited by employing the translating-pulsating source Green's function $G(P,q;-U)$ is much more pronounced than by the $\vec{W}$-flow solutions with the pulsating source Green's function $G(P,q)$. (5) The tendencies of the variations of the diagonal added damping and the off-diagonal added mass coefficients with respect to frequencies are totally different for the two cases mentioned in (4) over a wide frequency region. At the peak frequencies their results given by $\vec{W}$-flow solution are about 10% higher than those given by the U-flow solution. (6) The variations of the diagonal added mass coefficients and the off-diagonal added damping coefficients predicted by Cases 1-4 are of the similar trends. However the predictions by the $G(P,q;-U)$ and $\vec{W}$-flow solution are about 30% higher than those by the $G(P,q)$ and $\vec{W}$-flow solution, and are about 25% higher than

602

those by the $G(P,q;-U)$ and U-flow solution for the diagonal added mass coefficients at low frequency region. It may then be concluded that the forward speed effect reflected by the translating-pulsating source Green's functions is important in the predictions of the hydroelastic responses of a ship traveling in waves.

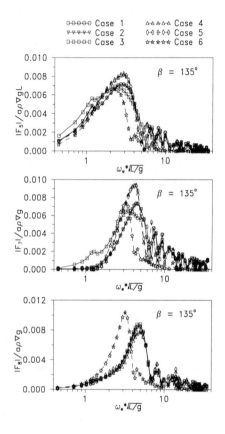

Figure 5. Wave exciting forces.

Fig.5 provides the comparisons of the generalized wave exciting forces corresponding to the six cases for the ship traveling in quartering seas. Similar conclusions as those of (1) and (2) drawn from Fig. 4 may be applied to Fig. 5 for the wave exciting forces. Except the differences of the magnitudes and the slight frequency shift of the peaks, the tendencies of the predicted results for Cases 1~5 are quite similar. For pitch and vertical bending modes the magnitudes of the forces obtained by employing the Green's function $G(P,q)$ (Cases 5,6) are over- estimated for about 10% to

30% in comparison with those of (Cases 1-3) based on the Green's function $G(P,q;-U)$. The magnitudes of the horizontal bending moment $|F_8|$ predicted by employing $G(P,q)$ however are under-estimated for about 14%.

Fig. 6 represents the responses of the ship traveling in quartering seas with forward speed of 18 kn. Evidently the prediction of Cases 1-5 have the similar tendency and are different from that of Case 6. The results given by using either $G(P,q)$ or the U-flow solution, generally provide over-estimated responses at the resonant frequencies.

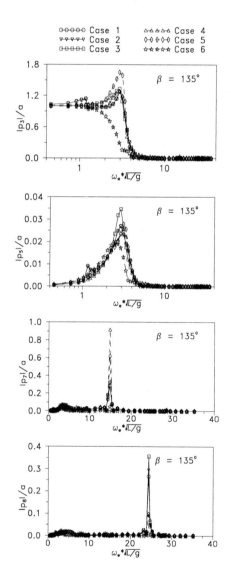

Figure 6. Responses of the principal coordinates.

5. CONCLUDING REMARKS

The influence of the steady flow on the motions and structural loadings of a ship traveling in waves has attracted attentions of naval architects for years. An efficient method of numerically solving the linear three-dimensional hydroelasticity theory which rigorously retains all the forward speed terms is described in this paper. The efficiency and accuracy of the numerical approaches developed in the present method has been confirmed by comparison with the existing results.

The numerical example of a ellipsoid moving beneath wave surface has shown that the influence of the non-uniform steady flow on the unsteady potential flow (thus on the variations of the hydrodynamic coefficients) increases when its slenderness ratio gets smaller. For a body of the slenderness ratio even equal to 10, evident discrepancies in low frequency region still exist between the curves of the diagonal added mass coefficients predicted by including or excluding the effect of the non-uniform steady flow.

Based on the present method a flexible surface ship with semi-ellipsoid shape is also investigated to show how the numerical results are influenced by the treatment of the steady flow, the choice of the Green's functions, and the contribution of the line integral. It seems that the use of translating-pulsating source Green's functions is the most important to reveal the forward speed effect on the motions and structural responses of the ship. The velocity field $\bar{W}$ of the non-uniform steady flow provides notable influence on the hydrodynamic coefficients and the wave exciting forces in comparison with the U-flow assumption. The contribution from the line integral can be omitted.

To thoroughly understand the forward speed effect detailed numerical simulations and model tests with careful flow and structure measurements are needed.

REFERENCES

1. R.E.D. Bishop and W.G. Price, Hydroelasticity of ships, Cambridge Press, 1979.
2. W. G. Price and Y.S. Wu, Hydroelasticity of marine structures, Theoretical and Applied Mechanics, ed. F.I. Niordson and N. Olhoff, Elsevier Science Publishers B.V., The Netherlands, 1985.
3. R.E.D. Bishop, W.G. Price and Y.S. Wu, A general linear hydroelasticity theory of floating structures moving in a seaway, Phil. Trans. of Royal Society, London, A316 (1986).
4. M. Bessho, On the fundamental singularity in the theory of ship motions in a seaway, Memoirs of the Defense Academy of Japan, Vol.17, No.3, 95-105 (1977).
5. H. Iwashita, and M. Ohkusu, The Green function method for ship motions at forward speed, Ship Technology Research, Vol.39, No.2, 3-21 (1992).
6. J.N. Newman, The theory of ship motions, Advances in Applied Mechanics, Vol.18, 221-28 (1978).
7. J. Z. Xia and Y. S. Wu, A general interface boundary condition for fluid-structure interaction problems of ships, Research of Ship Behavior, No.2, 73-79 (1993). (in Chinese)
8. C.W. Dawson, A practical computer method for solving ship-wave problem, Symp. of the Second Int. Conf. on Numerical Ship Hydrodynamics (1977).
9. S. X. Du, A complete frequency domain analysis method of three-dimensional hydro-elastic responses of floating structures traveling in waves, Ph.D. Thesis, China Ship Scientific Research Center, China (1996).
10. S. X. Du and Y. S. Wu, A fast evaluation method for the Bessho form translating-pulsating source Green's function, Shipbuilding of China, 141, No.2 (1998).
11. M. Ba and M. Guilbaud, A fast method of evaluating for the translating and pulsating Green's function, Ship Technology Research, Vol.42 (1995).
12. R.B. Inglis and W.G. Price, Calculation of the velocity potential of a translating, pulsating source, Trans. Royal Inst. of Naval Arch., Vol.123 (1981).

Practical Design of Ships and Mobile Units
M.W.C. Oosterveld and S.G. Tan, editors.

The influence of fixed foils on seakeeping qualities of fast catamaran.

W. Wełnicki

Ship Design and Research Centre, Ship Hydromechanics Division ul. Szczecińska 65, 80-392 Gdańsk, Poland
Some aspects of insufficient seakeeping qualities of fast catamarans are commonly known. The stabilizing abilities of hydrofoils are known too. The intention of this investigation was to obtain quantitative data about the inluence of such foils on seakeeping qualities of fast catamaran. Two types of foils were investigated: single foil spread between the hulls and double T - foils under the bows. Three locations of single foil over the ship length and three angles of attack were tested. The comparison of resistance, accelerations, heave and pitch in irregular head waves between the ship fitted out with and without foils was conducted. Quantitative effects of foils in different configurations and in different sea and operational conditions were estimated.

1. INTRODUCTION

Fast catamarans able to reach 30÷50 knots dominated passenger-ferry shipping on short and middle distances during last fifteen years, Lastly, there are also trends to apply them in cargo trade. Very slender hull forms cause that, the seakeeping qualities of those ships are not too good - specially to account for high accalerations in head and oblique waves. Damping abilities of hydrofoils located under or between the hulls of catamaran are already approved but, there is a lack of quantitative information useful for desingers.

There are two main types of stabilizing foils adapted for catamarans at present:
- foils with automatically steered angles of attack (ride control system), relatively complicated and expensive,
- fixed foils, technologically simple and cheap, but not so effective.

The seconds are the object of our investigation carried out within the framework of a research project financed by Polish Committee of Scientific Researches.

Taking into account technical and financial possibilities, two types of fixed foils were tested: a single foil extented between the hulls and two smaller T - foils foils located under the bows of both hulls. Two main parameters were investigated: the location of single foil on the length of ship and its angle of incidence (for single foil only).

The seakeeping qualities are expressed by fo llowing reactions of ship on wave excitations:
- vertical accelerations at the bow and at LCB
- heave and pitch
- added resistance in waves.

The stabilizing effect of foil is expressed here by the ratio of given reaction of the ship with foil to the similar reaction of the ship without foil. Up to date, there is no universal index of seakeeping qualities being know then, all these characteristics are compared separately.

2. GEOMETRICAL CHARACTERISTICS OF TESTED SHIP AND FOILS.

The investigations of influence of stabilized foil on seakeeping qualities of catamaran were conducted with the model F of fast catamaran's hull shown in fig 1. and in table 1.

Table 1.

No	Main particulars	Value
1.	Length of DWL L_w m	38,59
2.	Breadth of a hull B_H m	2,385
3.	Draught T m	1,197
4.	Block coeff. C_B -	0,581
5.	Displacement ∇ m^3	128,2
6.	$L_w/(\nabla/2)^{1/3}$ -	9,64
7.	Dist. of sym. planes B_S m	9,25
8.	LCB from sect.0 x_v % L_w	40,1
9.	Radius of long inertia k_{yy} -	0,288

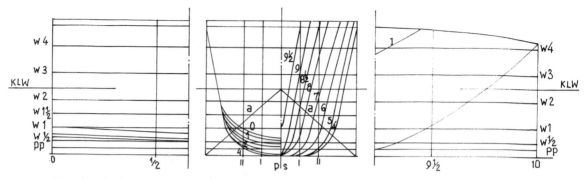

Fig. 1. Shape of tested hull.

Single foil spread between the hulls was installed as it is shown in fig. 2. It was fixed on two vertical struts to simplify the changing of its location and angles of attack. In the same figure there are shown tested positions denoted as F_1, F_2, and F_3. Twofold bow foils of type ⊥ (model F4) are shown in fig. 2. too. Their span was designed in order not to cross the breadth of ship's hull; then their area was smaller than this of single foil. All foils and struts were made of the same type of asymmetrical circular profile of relative thickness equal 0,06. From some technical reasons the dimensions of foil on small model were different than these of large model. The main particulars of foils are given in table 2. in full scale.

Table 2.

Dimensions of foils

	Span l [m]	Chord c [m]	Area S [m]	Asp. r. λ [-]
Single foil small model	6,50	0,75	4,87	8,67
Single foil large model	5,32	0,76	4,04	7,00
Double foil ⊥ small model	1,675	0,75	1,25	2,24

Parameters of foils which were variable during tests are given in table 3.

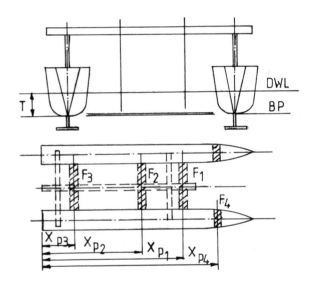

Fig. 2. Location of foils on the model

Table 3.
Variable parameters of foils.

	Location of foils x_p / L_w			Angles of attack α_i [°]		
Single foil small model 1:25	F_1 0,648	F_2 0,412	F_3 0,071	F_1 + 2,5	F_2 + 2,5	F_3 + 2,5
Single foil large model 1:9,5	F_{50} 0,648	-	-	F_{51} + 3,1	F_{52} + 1,3	F_{53} - 0,8
Double foil ⊥ small model	F_4 0,872	-	-	F_4 + 2,8	-	-

3. DOMAIN OF INVESTIGATION AND FORMS OF RESULTS PRESENTATION.

The model tests were conducted in two stages and in two model tanks. The first part of investigation was carried out in small tank (30 x 3 x 1,5 m) of Technical University of Gdansk. The tests were conducted in calm water and in regular head waves aiming to find optimal location of single foil in the length of ship (F_1, F_2, F_3) and additionally, to estimate comparable effect of twofold foils ⊥ localized under ship bows. These tests were conducted in scale 1:25. The second part of research, aiming to find the influence of angle of attack on stabilizing effect of foil, was conducted in large tank of Ship Design and Research Centre in Gdansk (240 x 12 x 6 m) with the model 1:9,5. The reason was that, the needed exactness of measurements should be larger in this case because of smaller differences in measured values. There was also a possibility to check the behaviour of model in irregular waves which could be generated in this tank directly for given spectrum. The investigation in small tank, conducted with the hull without foils and with all configurations of foils, comprised following tests:
- in calm water: measurements of resistance R_T, trim θ and emergence z in the range of speed 12÷40 knots (0,3 < Fn<1,05);
- in 6 regular head waves: measurements of resistance R_{TW}, pitch θ , heave z and vertical accelerations at FP and at CG; the particulars of waves were changed in the range:

$0,7<\omega_s<1,7$ [sec/rad]
$3,26>\lambda_s/L_w>0,55$ [-]
$9,0>T_s>3,7$ [s]
$2\xi_A\approx1,0$ [m]

All these tests were repeated for five ship speeds (0,385<Fn<0,9).

The investigatios in large tank were conducted with the hull without foil and with single foil situated in bow position like F_1 for three angles of attack in calm water and in the same waves as given above but, for two ship speeds only (Fn=0,53 and 0,9). Additionally, two tests in irregular head waves were made. The wave spectrum was modelled according to ISSC standard for T_1 = 4,9 s and $H_{1/3}$ = 1,135 m for ship speeds 20 and 34 knots. Direct results of tests [1,2,3] were obtained in the form of RAO's = f(ω), where RAO =

$\dfrac{\alpha_A}{g}$ / $\xi_A(\omega)$, $\theta_A/\xi_A(\omega)$, $z_A/\xi_A(\omega)$) respectively,

for Fn = const and then transformed to the characteristics in irregular waves using the common method based on superposition principle and ISSC wave spectrum. These characteristics were originally presented in the form of relative variances of particular responses to wave excitations, where the relative variance equals:

$$\overline{m}_{oi}(T_1) = \frac{m_{oi}(H_{1/3}, T_1)}{H_{1/3}^2} \qquad (1)$$

and $m_{oi} = \int\limits_{o}^{\infty} |RAO(\omega)|^2 S(\omega)d\omega \qquad (2)$

Added resistance in waves is expressed be nondimensional coefficient:

$$r^*_{AW}(T_1) = \frac{\overline{R}_{AW}(H_{1/3}, T_1)}{2\rho g L_{ws} H^2_{1/3}} \qquad (3)$$

For the aim of comparison between ship characteristics with an without foils we decided to use increments of amplitudes X_{Ai} of particular responses (acceleration, pitch etc) of the ship with

608

foil in relation to these without foil (X_{ao}):

$$\Delta X_{Ai} = \frac{X_{Ai} - X_{Ao}}{X_{Ao}} 100 \,[\%], \qquad (4)$$

where the values of ΔX_{Ai} were calculated as follows:

$$\Delta X_{Ai} = \sqrt{\frac{m_{0i}}{m_0}} - 1 \qquad (5)$$

The increments of added resistance are presented as:

$$\Delta r^*_{AW} = \frac{r^*_{AWi} - r^*_{AW0}}{r^*_{AW0}} x100\,[\%] \qquad (6)$$

A comparison of particular characteristics of seakeeping qualities is also difficult because their maxima occur at different speeds and at different frequencies. Therefore, it was decided to compare relative increments of particular responses

ΔX_{Ai} for four combinations of operational conditions:
- maximal values in whole of tested area
- maximal values at the speed of 34 knots (Fn=0,9, different frequencies)
- the values at Fn = 0,9 and T_1 = 4,4 s
- the values at 20 knots (Fn = 0,51) and T_1 = 4,4 s
Period T_1 = 4,4 s was chosen as typical for Baltic waves at the wind of 6°B ($H_{1/3} \approx 1,2$ m).

4. THE RESULTS OF INVESTIGATION

After the model tests were completed it appeared that,not all parameters of motion were consiste nt in both tanks.

Repeatability of results was quite good for resistance without foils, but motions in waves were smaller in smaller tank. It was scale effect partially but, as we think, it was probably the effect of different kind of towing method in both cases. In small tank the point of towing force was placed above the model deck and in large tank - on the hight of shaft line. From this reason the results in both tanks must be compared separately between themselves only.

4.1. Resistance in calm water

The comparison of resistance in calm water is interesting, because it shows the influence of foil location and its angle of attack on resistance directly. The increments of total resistance for all foil combinations are given in table 4.
As it is to see, positions F_1 and F_2 are almost equivalent from this point of view and position F_3 is clearly the worst. At these parameters of foil its lift force was too small to give a decrease of hull resistance. Fully free model F_{50} in large tank gave really reliable results and showed that, resistance increments in the case of foil at the bow (F_{51}) are small and even negative at high speed.
From this point of view the foil should be located at the bow part of ship.

4.2. Influence of foil location on seakeeping characteristics

The influence of foil location on particular seakeeping characteristics in irregular seas shows fig. 3.
Additionally, the effect of double T - foils is presented in the same figure. All foil positions F_1, F_2, F_3 and F_4 were tested at the same angle of attack.

Table 4.
Relative increments of total resistance in calm water.

V [kn]	20	25	30	34	40
R_{To} (F_0=F_{50}) [kN]	68,7	83,8	102,1	121,0	162,4
$\Delta R_T =[(R_{Ti}-R_{To})/R_{To}]x100[\%]$					
F_1	18,0	10,0	17,5	19,0	10,8
F_2	16,4	11,0	17,5	18,5	11,0
F_3	30,3	24,7	27,3	28,1	19,1
F_{51}	7,6	7,2	5,6	3,6	- 2,2
F_{52}	7,4	6,7	5,1	2,9	- 5,7
F_{53}	9,1	9,3	8,9	8,4	4,6
F_4	16,4	9,5	12,6	15,7	11,4

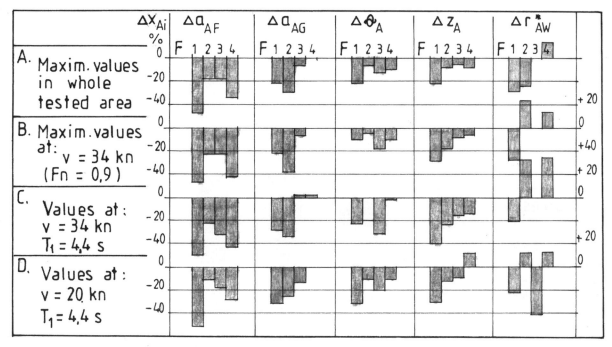

Fig. 3. Relative increments of amplitudes of ship's responses at different foil location.

The following observations can be made:
- highest damping of accelerations at FP is obtained at location F_1; decrease of amplitudes $\Delta\alpha_{AF}$ is about 50 %.
- for accelerations in CG, the location F_2 is the best one - $\Delta\alpha_{AG} \approx 25 \div 40$ %; however, the damping at position F_1 is not much worse - at smaller speed is even better;
- from the pitch point of view, the worst position is F_2; in some cases, foil at the bow is the best one (A, D), in others - foil at the stern (B, C);
- on account of heave, position F_1 is the best in all cases - $\Delta z_A = 20 \div 40$ %;
- decreasing of added resistance in waves can be obtained by position F_1 only : negative $\Delta r^*_{AW} \approx 20 \div 30$ %; position F_3 influences added resistance minimally.
It is evident that, at other angles of attack effects can be different; it is explained in next chapter.

4.3. Influence of foil angle of attack on seakeeping characteristics.

The effect of angle of attack of single foil was investigated for three angles given in table 3. The range of angle variation was limited, because higher values ($\alpha_i > 3°$) caused too large trim in calm water and smaller angles ($\alpha_i < -1,0°$), made the trim being negative.
The results of tests transformed to irregular sea and presented in the form of relative increments of amplitudes of particular ship responses in relation to ship without foil, are shown in fig. 4.
Generally speaking, it can be stated, that the influence of angle of attack on seakeeping characteristics is not significant.
In particular:
- accelerations at the bow are damped practically equally at all angles in the range from -1° to +3°
- damping of accelerations at C.G, of heave and pitch as well, grows when α_i diminishes;

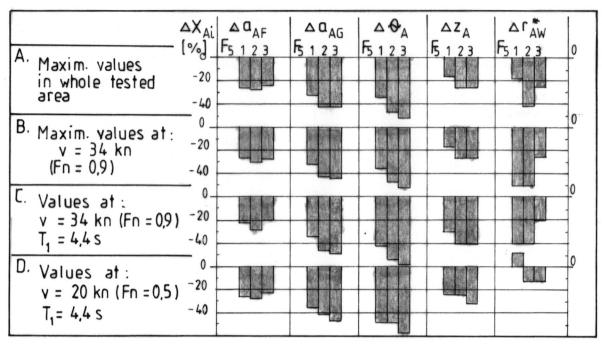

Fig. 4. Relative increments of amplitudes of ship responses at different foil angles of attack.

- lowest increments of added resistance in waves occur at small angles of attack;
- observed influence of angles of attack on ship motions does not depend on ship speed.

5. OTHER REMARKS.

The above presented investigation gave an opportunity to make some other observations. It is worth to say about two of them. Firstly, it is the influence of longitudinal moment of mass inertia on ship's seakeeping characteristics. Model F was tested in waves with two radii of inertia: $\rho = 0,25$ and $\rho = 0,296$. Generally one can say, that version with ρ_2 shows better seakeeping qualities than ρ_1. In case of resistance and pitching its predominance is the higher the lower is sea state and the higher is ship speed. As regards accelerations and heaving, version ρ_2 is „better" in the speed range of $0,5 < Fn < 0,8$ and $3,5\ s < T_1 < 5,5\ s$. Broadly speaking, the higher ship speed and wave length, the smaller is effect of moment of

inertia on seakeeping characteristics.

Secondly, the results of tests made in regular waves and transformed to irregular ones by means of RAO's were compared with these obtained in direct trials in irregular waves. The results of this comparison are given in table 5 in the form of relative differences between amplitudes of motion characteristics received in both cases (in percentages).

Table 5.

	Δz_A	$\Delta \theta_A$	$\Delta a_A(FP)$	$\Delta a_A (CG)$
V = 20 kn	+ 4,7	- 7,3	- 8,0	- 9,7
V = 34 kn	- 4,4	- 17,7	-11,0	-13,3

Sign (-) means, that values obtained directly in irregular waves are greater than these taken from regular waves. These differences could be explained by approximations in transformation method and inaccuracies in simulation of wave spectrum in model tank. But, it is interesting, that all these differences are of the same sign.

6. CONCLUSIONS.

Summarizing results of this investigation, it can be concluded as follows:
- a fixed foil spread between the hulls of catamaran, omitting its known defects, can be an effective tool to damp vertical accelerations and ship motions in waves;
- the most efficient location of such foil upon ship length is in its bow part, between LCB and 0,75 L_w;
- the angle of attack of foil has rather small influence on its damping effectiveness;
$\alpha_i = 0$ seems to be the best one:
- the double $\perp$ - foils, even at twice smaller area than single foil, are of almost the same efficiency in damping vertical accelerations at the bow; however, their damping influence on other seakeeping characteristics is much smaller.

REFERENCES

1. E. Brzoska et al: The influence of stabilizing foil location on hydrodynamic characteristics of fast catamaran. TUG, F.O.E.Sb. Rep. no: 6/97 (in polish)
2. E. Brzoska et al: Hydrodynamic characteristics of fast catamaran with double bow foils.T.U.G, FOESb, Rep no: 25/97 (in polish)
3. P. Grzybowski: Fast catamaran with stabilizing foil. S.D.& R.C.Rep. no: RH-97/T-121 (in polish).

Seakeeping Design of Fast Monohull Ferries

L.Grossi, S.Brizzolara
FINCANTIERI Divisione Costruzioni Militari - Via Cipro 11 - 16121 Genova - ITALY
G.Caprino, L.Sebastiani
CETENA S.p.A. - Via Savona 2-16121 Genova - ITALY

ABSTRACT:

The Fast Ferry market is presently requesting relatively large vessels capable to run at speeds up to or above 40 knots. This constitutes a challenging demand to the ship designer, who is asked to meet an outstanding calm-water performance ensuring at the same time a satisfactory comfort to the passengers even in worst weather.

The present paper addresses the seakeeping aspects of fast monohull ferries design, illustrating the methodology developed in Fincantieri to assess/improve passenger comfort already at the very preliminary design stage.

In particular the influence of major design parameters on the seaworthiness of fast ferries has been investigated by means of the systematic use of computational tools. The results of this theoretical study have been successfully confirmed by specific seakeeping model-tests comparing the original configuration of an existing Fast Ferry with the devised optimal configuration.

INTRODUCTION

Fast Ferry is without doubt a most dynamic sector of shipbuilding industry. In 1996 Fast Ferries were employed on almost 40 different ferry services around the world and during the 1997 the Fast Ferries production had a remarkable growth, both in terms of vessels delivered and outstanding orders. Due to the technological innovation they experienced in these last few years, monohull Fast Ferries resisted the competition from catamarans and now possess a share of about 30% of the Fast Ferry market.

At the beginning of the nineties the Military Division of Fincantieri Shipyards designed and built *Destriero* that, by winning the *Blue Ribbon*, proved the possibility to achieve and sustain a very high cruise speed even in adverse weather condition with a monohull ship. Following this success a new generation of monohulls high speed car/passenger ferries has been designed and built by Fincantieri and other shipbuilders all over the world, which make the cargo capacity of monohulls to increase with respect to multihulls. To date the largest existing Fast Ferry, built at Fincantieri yard for the Italian ship operator *Tirrenia*, is a monohull 146m long able to carry 1800 passengers and 460 cars at a service speed of 40 knots.

The level of comfort enjoyed by the passengers on board of Fast Ferries is a major concern for the ship owner, both to reduce the down-time of the ship and to have an edge in the competition with other operators.

Fincantieri has devoted special care to introduce comfort considerations already at the early stage of the design process by means of a global integrated approach which combines the use of computations, model-tests and sea-trials.

To assess the margin for further seaworthiness improvement a parametric investigation has been performed on the seakeeping behaviour of a typical Fincantieri's Fast Ferry, namely MDV 1200. The study was based on seakeeping calculations and the final verification was carried out in seakeeping basin at DERA.

HUMAN FACTORS AND SEAKEEPING OPERABILITY OF SHIPS

In the past for seakeeping assessment designers were accustomed to utilise criteria based on wave-induced motions, accelerations and undesired effects (like deck-wetness, slamming and propeller emergences). Today acceptable limits on motions and other aspects of behaviour in rough seas are related to their effect on people, ship systems (engines, propulsion) and on the ship itself.

SEAKEEPING CRITERIA BASED ON HUMAN FACTORS

The wave-induced ship motions may affect human effectiveness either through a direct mechanical action on the human body (i.e. a lateral tipping moment causing a people up-standing on the deck to lose balance) or indirectly acting at a physiological level (i.e. a disturbance of the vestibular system causing the so-called sea-sickness).

The endurance to wave-induced ship motions, the so-called *sea-sickness* or more exactly *motion-sickness*, is the major concern in the seakeeping design of Fast Ferries so that attention will be hereafter focused on this factor.

The sea-sickness seems to be caused by a number of external factors among which :
- the level of vertical accelerations on the ship ;
- their frequency contents;
- the exposure time to the accelerations.

The MSI (Motion Sickness Incidence) statistically measures the percentage of people who will be likely sea-sick due to a given exposure time to the ship motions in rough sea and is based on systematic laboratory tests conducted on volunteers in ship motions simulators.

The MSI depends on both the level and the frequency of the vertical accelerations as illustrated in Fig. 1 below :

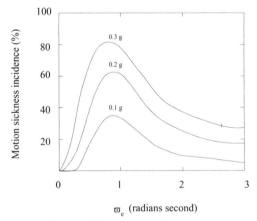

Fig. 1. MSI vs wave frequency

In the interval between 0.01 Hz and 10 Hz, which accounts for the whole frequency spectrum of the typical ship motions, the following international regulations apply for an evaluation of the MSI :

- ISO Standard 2631 fatigue-decreased proficiency boundaries (1 Hz a 80 Hz);
- ISO Standard 2631/3 severe discomfort boundaries (0.1 Hz a 0.63 Hz);
- O'Hanlon & Mc Cauley MSI method.

Based on such recommendations it is possible to derive a global proficiency boundary, for each pair of MSI level and exposure time to the ship motions.

In the Fig. 2 it is reported the boundary curve for a MSI level of 10 % over 2 hrs of exposure time.

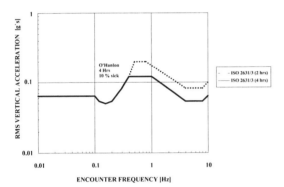

Fig. 2. Limiting curve for comfort

In order to compare the actual performance of the ship with the boundary curve it is first necessary to sample the encounter-frequency spectrum of the vertical accelerations on the ship (either measured or calculated) in third-octave band and to plot the resulting RMS values (in g) on a bi-logarithmic paper. By selecting the desired exposure time, in relation to the expected duration of a typical point-to-point voyage of the vessel, and by adjusting the MSI value so that the acceleration spectrum lays below the limiting curve it is possible to determine the actual level of MSI on-board.

It must be noted that the MSI refers to a specific position on the ship. To produce an overall assessment of the MSI it is therefore necessary to evaluate the MSI at representative points within the passenger area and to take the weighted average with respect to the number of seats.

METHODOLOGIES FOR COMFORT ASSESSMENT

In the case of Fast Ferries it is of particular importance to assess the effect of ship motions on the comfort on board and its influence on the overall operability of the ship in terms either of reduced service speed or reduced service time.

To this regard Fincantieri worked in close co-operation with CETENA in order to set-up a software package specifically addressed to the assessment of the operability of sea-going ships.

The result of such co-operation is the *operability suite* SOAP (Seakeeping Operability Analysis Program) conceived to act as a post-processor of a generic seakeeping code in frequency domain and allowing to perform the following tasks:

1. *short-term* statistics of all the relevant Seakeeping Performance Criteria (SPC);
2. *long-term* statistics of motions, accelerations and global dynamic loads;
3. *operability analysis* based on the Seakeeping Operability Envelope (SOE) or Percentage Time of Operability (PTO) concepts.

The *Operability Suite* accounts for the following SPCs:

1. Effective power due to added resistance in wind/waves ;
2. Amplitude of rigid-body motions ;
3. Occurrence of undesired effects (slamming, submergence, emergence) ;
4. Motions Sickness Incidence (MSI) ;

5. Subjective Motion Magnitude (SMM) ;
6. Longitudinal/Lateral Force Estimator ;
7. Motion Induced Interruptions (MII) ;
8. Effective Gravity Angle (EGA) ;
9. Lateral sliding occurrence .

More details on the theoretical background and the characteristic features of the suite can be found in [1].

IMPACT OF DESIGN PARAMETERS ON SHIP MOTIONS

In the following some general findings on the influence of design parameters on the seakeeping behaviour of mono-hull ships are provided, which were used as guidelines for the set-up of the study.

The rationale for understanding the effects of design parameters on seakeeping is to consider the reciprocal interaction of the following three factors:

- natural period;
- hydrodynamic damping;
- wave exciting forces.

The design parameters considered in the present investigation pertain to three main groups :

- parameters associated with the main ship proportions which must be defined at the very preliminary design stage such as length, beam, draft, displacement;
- parameters which are expected to play a primary role but which affect main ship proportions only indirectly such as water-plane geometry and weight distribution;
- parameters which effect is deemed to be secondary and can be finalised at the optimisation design stage such as bow-flare angle, bulbous bow and transom stern forms.

VERTICAL PLANE MOTIONS

Attention will be restricted to heave and pitch motions as surge does not contribute to the vertical accelerations which are the governing factor for motion sickness on board.

The effect of design parameters on the vertical motions is often investigated in head sea condition, assuming that such heading would produce the worst vertical motions. This assumption is no longer valid in the case of fast vessels which, due to the high frequencies of encounter, are sensitive to headings coming from the bow-quartering sector.

Hull size is undoubtly a major factor in determining the ship behaviour in waves. It can be defined in terms of displacement Δ or length L. From a general standpoint it can be stated that increasing the size of the ship will lead to improved seakeeping performance.

On the other hand, within the range of variation reasonable for a given design, displacement is not a major parameter; as a matter of fact the displacement was kept constant in the present parametric investigation in order to consistently compare the various configurations.

As for ship length, although increasing ship length usually results in reduced ship motions and accelerations, the effect of such variation should be considered not in an absolute way but in relation to the wave lengths λ likely to be encountered by the ship. Long waves with respect to ship length are in general an unfavourable condition.

When considering the combined effect of synchronism with the wave period, hydrodynamic damping and wave excitation, it must be noted that the latter two terms are often conflicting, in the sense that a large damping may imply a large exciting force.

For instance pitch damping is greater the greater is the fore-aft asymmetry of the hull geometry but such asymmetry generally means a larger wave pitch moment. The same consideration hold for V bow-forms which produce a good amount of pitch damping but also a greater wave exciting force as the immersed area is closer to the still-water level. Near synchronism however damping effects usually prevail on excitation.

As ship response in the vertical-plane is more sensitive to long waves, it comes out that to reduce vertical motions is generally convenient to keep the natural period as low as it is possible. The effect of ship speed, represented in non-dimensional terms by the Froude number, will tend to decrease the wave encounter period and thus to shift synchronism towards longer waves resulting in higher motions: this means that fast vessels will be more sensitive to synchronism.

In order to relate synchronism to the design parameters recourse can be made to the following non-dimensional expressions for heave and pitch natural periods :

$$T'_{33} = \frac{2\pi}{\sqrt{\dfrac{C_{VP}}{\sqrt{(L/T)}}}} , \quad T'_{55} = 2\pi \frac{k'_{55}}{\sqrt{GM'_L}}$$

having indicated with GM'_L the non-dimensional longitudinal metacentric height:

$$GM'_L = \left(\frac{L}{T}\right)\left(\frac{1}{C_{VP}}\right)\left[k'^{2}_L + \left(\frac{LCB - LCF}{L}\right)^2\right]$$

with k'_L the non-dimensional pitch gyradius of the water-plane (referred to LCF) and with k'_{55} the non-dimensional

pitch gyration radius of the ship (referred to the LCG).

The condition for heave/pitch synchronism with the waves is thence :

$$k'_{55} = \sqrt{k'^{2}_L + S^2_H} \approx k'_L$$

Unfortunately k'_L usually lays in the typical range of k'_{55} , that is between 0.2 and 0.3. The following items should be further considered:

- little direct control can be exerted by the designer on the pitch gyradius;
- a variation larger than 2 % in the pitch gyradius is not to be expected unless extreme measures are taken;
- the precise actual value of the pitch gyradius is difficult to ascertain.

The effect of the horizontal separation s_H on the water-plane inertia is not very significant, as it is of an order of magnitude smaller than the intrinsic water-plane gyradius k'_L . On the other hand s_H plays an important role in heave/pitch coupling through the heave-into-pitch / pitch-into-heave hydrostatic restoring term:

$$C_{35} = C_{53} \propto A_{WP} (LCF\text{-}LCB)$$

It can be shown that the effect of this cross-coupling is to split the heave/pitch resonance frequency ω_n apart of a factor $\omega_n \left(S_H / k'_L\right)^2$. As heave resonance is *wide-banded* whereas pitch resonance is *narrow-banded*, the effect of the cross-coupling term seems to increase the heave resonance peak (detrimental effect) whilst shifting the pitch resonance pitch.

The effects of a variation of the design parameters on the vertical-plane motions are summarised in Table 1, where the sign

+ means an increasing of the parameter would increase the performance; the symbol o means that the parameter is not very significant

Design Parameter Symbol	Design Parameter Denomination	Varia tion	Note
L/B	length-to-beam ratio	+	overall beneficial effect on heave/pitch motions
B/T	beam-to-draft ratio	+	beneficial effect on damping and detrimental effect on excitation and keel emergence related to V bow-forms
L/T	length-to-draft ratio	+	beneficial effect on heave/pitch synchronism, may be detrimental on wave excitation and keel emergence related to draft
k_L/L	non-dimensional water-plane gyration radius	+	beneficial effect on pitch synchronism and on heave/pitch coupling
s_H	horizontal separation	+	beneficial effect on pitch synchronism
C_{WP}	water-plane area coefficient	+	beneficial effect on pitch synchronism related to higher local B/T values, particularly effective in the fore-body
C_B	block coefficient	o	not very significant, a lower value may be beneficial
C_{vp}	vertical prismatic coefficient	-	beneficial effect on pitch synchronism
$L/\nabla^{1/3}$	slenderness	o	not very significant, a higher value may be beneficial
C_p	prismatic coefficient	o	not very significant, a higher value may be beneficial
k_{55}/L	non-dimensional gyration radius	-	independent on ship geometry, little control by the designer, beneficial effect on pitch synchronism and on heave/pitch coupling
LCG	longitudinal center of gravity	o	not very significant, a shifting forward may be beneficial for pitching and detrimental for heaving

Table 1. Effect of design parameters on heave/pitch motions

It must be noted that the parameters s_H, k_L and C_{WP} are closely related to the geometrical characteristic of the water-plane.

This means that the shape of the water-plane area is a major factor in the vertical plane motions, especially in the case of fast ships where the pitch synchronism effects are magnified by the high operational speeds.

Increasing C_{WP} can be obtained by acting locally on the aft-body or fore-body ship forms: in the fore-body this generally means a favourable change of the sections shape from U-type to V-type, in the aft-body this can be obtained (when applicable) by designing a large transom stern. It must be however noted that most fast vessels are already characterised by a transom-width ratio B_T/B only slightly smaller than unity, so that little margin exists for improvement of C_{WP} in the aft-body ; the relevant parameter seems therefore the forward water-plane area coefficient C_{WF} . Filling the waterlines at the ends has also the beneficial effect of increasing the gyradius of the water-plane k_L and shifting the LCF forward. In the case of fast ships, it is possible to take advantage of this effect, considering that they are usually characterised by a LCF forward of LCB.

These major design parameters apart, attention should be paid also to other factors not amenable of reduction into a simple design parameter such as above-water forms, shape of fore-body sections, bulbous bow and transom stern.

LATERAL PLANE MOTIONS

Attention will be focused on roll as it is the only later-plane motion which may contribute to vertical accelerations and therefore to motion sickness incidence.

The effect of design parameters on rolling can be generally assessed based on the same approach used for the vertical plane motions, provided that the role of ship length is replaced with that of ship beam. On the other hand a feature peculiar to rolling is the small amount of hydrodynamic roll damping so that roll motion is ruled by resonance to a much greater extent than heave/pitch motions.

To avoid synchronism roll natural period should be in principle kept as long as it is possible. It must be however considered that this implies a low transverse metacentric height and thus a reduction of the transverse stability of the ship: stability requirements pose therefore a limit to what can be achieved.

The effect of ship speed will tend to decrease the wave encounter period thus shifting synchronism towards longer waves ; besides increasing ship speed will increase roll damping resulting in an even more significant reduction of ship rolling. This means that fast vessels will be less sensitive to roll with respect to conventional ships.

The relevant formulas for roll synchronism are :

$$T'_{44} = 2\pi \frac{k'_{44}}{\sqrt{GM'_T}}$$

$$GM'_T = \left[(B/T) \cdot (1/C_{VP}) \cdot k'^2_T\right] - \left(\frac{KG - KB}{B}\right)^2$$

having denoted with k'_T the non-dimensional roll gyradius of the water-plane and with k'_{44} the non-dimensional roll gyration radius of the ship, with respect to ship beam.

The effects of a variation of the design parameters on roll motion are summarised in Table 2, where the same conventions if Table 1 are applied.

It can be seen that in general parametric variations that are beneficial for heave/pitch are detrimental for roll and vice versa. When addressing the task of optimising the seaworthiness of a ship in oblique waves a compromise must be therefore achieved between the opposite requirements of vertical-plane motions and lateral-plane motions.

Design Parameter Symbol	Design Parameter Denomination	Variation	Note
B/T	beam-to-draft ratio	-	beneficial effect on synchronism and detrimental effect on damping related to more U-type forms
k_T	water-plane gyration radius	-	beneficial effect on synchronism, little control in design
s_V	vertical separation	+	beneficial effect on synchronism, detrimental for stability
C_{WP}	waterplane area coefficient	-	beneficial effect on synchronism and detrimental effect on damping related to more U-type forms
C_B	block coefficient	+	beneficial effect on damping in relation to fuller sections
C_{VP}	vertical prismatic coefficient	+	beneficial effect on synchronism
CM	midship section area coefficient	+	beneficial effect on damping in relation to fuller sections
C_p	prismatic coefficient	o	not very significant
k_{44}	gyration radius	+	independent on ship geometry, little control by the designer, beneficial effect synchronism
KG		+	beneficial effect on synchronism, detrimental for stability

Table 2. Effect of design parameters on roll motions

PARAMETRIC INVESTIGATION

In order to produce practical guidelines for the selection of the major design parameters of fast monohull ferries it was felt necessary to supplement the above outlined general recommendations with a specific parametric investigation on a typical representative of the new generation of monohull Fast Ferries.

Such investigation was carried out by means of the seakeeping operability package described in the previous section.

PARENT HULL

The hull of Fincantieri's Fast Ferry MDV 1200 was chosen as the Parent Hull for the parametric investigation on the seakeeping characteristics.

MDV 1200 represents the new generation of car ferry monohulls,

620

according to the increasing market interest in high speed applied to passenger\car fast ferries service.

The vessel has a service speed of 36 knots and a transport capacity of 450 passengers with luggage and 120 cars, arranged on three different decks.

The adoption of the Deep Vee concept led to good hydrodynamic characteristics both in calm-water and rough-sea.

SELECTION AND VARIATION OF THE DESIGN PARAMETERS

The following design parameters were selected for the parametric study :

- ship main proportion parameters L, B and T ;
- form of the water-plane;
- LCB ;
- bow-flare .

To compare the various configurations it was decided to keep the displacement constant at 1200 t through the whole study with the only exception of VD1 which displacement was 1470 t.

In order to quickly assess the influence of the ship main proportions on the seakeeping behaviour, the ship dimensions were varied in *affinity* relation that is the co-ordinates x, y, z of the offset of the Parent Hull were uniformly stretched by a proper scale factor along the three directions:

$$x' = \alpha \cdot x$$
$$y' = \beta \cdot y$$
$$z' = \gamma \cdot z$$

This means that global form parameters such as C_B , C_p , C_{vp} and C_M are not changed by these variations. The

requirement of constant displacement implies that $\alpha \cdot \beta \cdot \gamma = 1$.

Local form variations were obtained by drawing specific design alternatives, based on the examination of the outcome of the *affinity* variations. In the LCB variations it was assumed that the ship was even-keel. The KG was kept constant for all the parametric variations.

A total of 23 design alternatives was considered, according to Table 3.

Version	Type of variation
P.H.	MDV 1200
V1	Affinity variation of P.H.- 10% longer and 5% thinner
V2	Affinity variation of P.H. - 10% longer and wider
V3	Affinity variation of P.H. - 20% longer and wider
V4	Affinity variation of P.H. - 20% longer and 10% wider
V5	Affinity variation of P.H. - 20% longer and 10% thinner
V6	Affinity variation of P.H. - 10% shorter and 5% wider
V7	Affinity variation of P.H. - 10% thinner at constant length
V8	P.H. with a straight transom WL
V9	V8 with a larger bow-flare
V10	V9 with further bow-flare
V11	Affinity variation - V10+V14
V12	V10 with LCB fore-shifted
V13	V12 with a smaller bow-flare
V14	Affinity variation of V10 - 10% longer and thinner
V15	Affinity variation of V12 - 8% longer and correspondingly thinner
V16	Affinity variation of V13 - 8% longer and correspondingly thinner
VA	Revision A of P.H. design
VB	Revision B of P.H. design
VC	Revision C of P.H. design
VD	VC with a straight transom WL
V-Light	MDV Light Alloy
VD1	Revised design of P.H. - Δ = 1470 t
VD2	Revised design of P.H. - Δ = 1200 t

Table 3. Summary of parametric variations

It must be noted that :

- Revision A of P.H. derives from a modification of the longitudinal distribution of the sectional areas, shifting the LCB forward ;
- Revision B of P.H. is quite a drastic change of the original design, with a keel-line inclined fore, LCB shifted aft and rounded hull lines ;

- Revision C of P.H. is a further variation of Revision A ;
- V-Light is the light alloy existing version of the MDV1200 ;
- Revisions VD1, VD2 of P.H. derived from the same hull geometry at two different drafts, corresponding to 1470 t and 1200 t respectively. LCB and LCF have been shifted fore, the bow-flare has been enlarged.

The main characteristic of the configurations are reported in Table 4.

Selection of the seakeeping indicator

For a consistent comparison of the various design alternatives, a seakeeping indicator was to be defined able to fulfil the following requirements :

- to be closely related with the global level of comfort aboard ;

- to account for the changes in wave period and ship speed ;

- to account for the effect of oblique waves.

The peak value of the RMS of the vertical acceleration in third-octave band was selected as the proper indicator ; it is in fact such peak value which determines the actual MSI level aboard.

As the vertical accelerations depend on the position, representative points were considered within the passenger deck and a weighted average of the results was taken based on the number of seats. The points were located half-way off the center-line of the deck to account for the effect of roll on the vertical accelerations in oblique waves.

Two ship speeds were considered, namely 30 and 35 knots, and several heading between head and following sea.

Version	/ Parameter	L	B	T	Bwl	AWP	LCB	LCF	GMt	L/Bwl	Bwl/T	CWP	SH	LCB/L	LCF/L	CB	Cvp
	Units	[m]	[m]	[m]	[m]	[m^2]	[m]	[m]	[m]	[-]	[-]	[-]	[%]	[%]	[%]	[-]	[-]
P.H.		82,0	16,0	2,7	13,7	829,0	30,9	32,2	5,2	6,0	5,1	0,7	1,5	37,7	37,7	0,39	0,52
V1		90,2	15,2	2,6	13,0	869,0	34,0	35,4	4,6	6,9	5,1	0,7	1,5	37,7	37,7	0,39	0,52
V2		90,2	17,6	2,2	15,0	877,0	34,0	35,4	8,8	6,0	6,7	0,6	1,5	37,7	37,7	0,39	0,60
V3		98,4	19,2	1,9	16,4	1168,0	37,0	38,6	13,9	6,0	8,7	0,7	1,6	37,6	37,6	0,39	0,53
V4		98,4	17,6	2,0	15,0	1071,0	37,0	38,6	9,9	6,6	7,4	0,7	1,6	37,6	37,6	0,39	0,54
V5		98,4	14,6	2,5	12,4	886,0	37,0	38,6	4,2	7,9	5,0	0,7	1,6	37,6	37,6	0,39	0,54
V6		73,8	16,8	2,8	14,4	784,0	27,8	28,9	5,6	5,1	5,1	0,7	1,5	37,7	37,7	0,39	0,53
V7		82,0	14,4	3,0	12,4	746,0	30,9	32,2	3,0	6,6	4,2	0,7	1,5	37,7	37,7	0,39	0,53
V8		82,0	16,0	2,7	13,8	840,0	30,4	31,8	5,4	5,9	5,2	0,7	1,7	37,1	37,1	0,39	0,52
V9		82,0	16,0	2,7	13,8	859,0	30,2	32,3	5,9	5,9	5,1	0,8	2,6	36,8	36,8	0,38	0,50
V10		82,0	16,0	2,7	13,8	878,0	30,2	32,9	6,2	5,9	5,1	0,8	3,3	36,8	36,8	0,38	0,49
V11		98,4	17,6	2,0	15,2	1110,0	37,5	40,1	10,9	6,5	7,5	0,7	2,6	38,1	38,1	0,38	0,52
V12		82,0	16,0	2,7	13,8	877,0	31,8	32,9	6,0	5,9	5,1	0,8	1,3	38,8	38,8	0,38	0,49
V13		82,0	16,0	2,7	13,8	877,0	31,9	32,9	6,0	5,9	5,1	0,8	1,2	38,8	38,8	0,38	0,49
V14		90,2	14,4	2,7	12,4	869,0	33,2	36,2	4,3	7,3	4,6	0,8	3,3	36,8	36,8	0,39	0,50
V15		88,0	14,9	2,7	12,8	877,0	34,1	35,2	4,8	6,9	4,7	0,8	1,3	38,8	38,8	0,38	0,49
V16		88,0	15,4	2,6	13,2	906,0	34,1	35,2	5,5	6,7	5,0	0,8	1,3	38,8	38,8	0,38	0,49
VA		82,0	16,0	2,7	13,6	795,0	31,2	31,6	4,0	6,0	5,0	0,7	0,5	38,0	38,0	0,39	0,55
VB		82,0	16,0	2,7	13,4	722,0	31,0	29,5	2,8	6,1	5,0	0,7	-1,8	37,8	37,8	0,39	0,60
VC		82,0	16,0	2,7	12,7	818,0	32,7	33,7	3,8	6,5	4,7	0,8	1,2	39,9	39,9	0,42	0,53
VD		82,0	16,0	2,7	12,7	818,0	32,7	33,5	3,8	6,5	4,7	0,8	1,0	39,9	39,9	0,42	0,53
V-Ligth		88,0	17,1	2,6	14,3	892,0	32,8	33,7	5,2	6,2	5,4	0,7	1,0	37,3	37,3	0,37	0,52
VD1		89,2	16,2	2,8	13,7	918,0	34,2	35,5	4,9	6,5	4,9	0,8	1,5	38,3	38,3	0,42	0,56
VD2		89,2	16,2	2,6	13,4	885,0	34,0	35,0	5,1	6,7	5,3	0,7	1,1	38,1	38,1	0,38	0,52

Table 4. Main characteristics

A reference significant wave height H_S of 2 m (SS4) was adopted and three different zero-crossing periods T_Z of 4.5, 5.5, 6.5 s were considered. The waves were supposed to be long-crested and the sea state was modelled by a JONSWAP spectrum with an enhancement factor of 3.3, typical for fetch-limited seas.

The basis for the development of the comfort indicator consisted of polar plots versus wave heading. A comfort indicator was associated to each diagram as the area limited by the polar plot. By taking the arithmetic average of this quantity with respect to the wave period it is possible to eliminate the dependency on the wave period. By further taking the arithmetic average with respect to ship speed an overall comfort indicator is obtained.

Selection of the optimal hull

The comparison of all the design alternatives is shown in Fig. 1 in terms of a histogram of the percentage difference of the comfort indicator with respect to the value for the Parent Hull ; a positive difference means an improvement with respect to the Parent Hull.

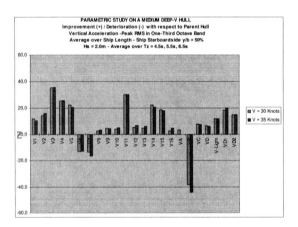

Fig. 1. Comparison of all the parametric variations

A first selection of the various configurations was thence performed, as shown in Fig. 2, where the extreme parametric variations reported in Fig. 1 have been eliminated ; namely V2, V3, V4, V5, V11 which are characterised by an unrealistically high value of the metacentric height.

It can be seen that the worst configurations are, in pejorative order: V6 , the only shorter version ; V7, with the lowest value of B/T ; VB, with LCB shifted aft. On the other hand the best configurations are, in meliorative order: V1, 10% longer ; V-Light, which is the Light Alloy version about 10% longer at equal displacement ; VD2, which is the final design revision at 1470 t displacement ; V15, 8% longer with straight transom WL, LCB shifted forward and very large bow-flare ; VD1, which is the final design revision at 1200 t displacement ; V14, which is similar to V15 but with a smaller bow-flare.

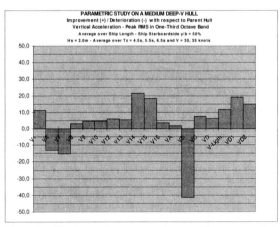

Fig. 2. Reduced comparison of parametric variations

By ruling out VD1, which has a larger displacement, the best configurations were therefore, in meliorative order VD2, V15 and V14. On the other hand V14 and V15 were characterised by an exaggerated flare in the fore-body which, even if beneficial from the point of view of wave-induced vertical accelerations, was considered to be potentially very dangerous as regards bow-flare slamming with consequent induced loads and vertical accelerations. This left VD2 as the optimal balancing between practical design requirements and seaworthiness considerations.

FINAL VERIFICATION IN SEAKEEPING BASIN

The performance of the original hull and of the *optimal* configuration as indicated by the theoretical parametric study, i.e. version VD2, has been compared at DERA seakeeping basin in terms of vertical acceleration along ship length. The following conditions were tested :

- ship speed : 25, 35 knots
- sea state 4 : $H_S = 2$ m, $T_0 = 7.1$ s
- sea state 5 : $H_S = 3$ m, $T_0 = 8.3$ s

The results in head sea are presented in Figs. 3 and 4 . The comparison between the two configurations is reported in Fig. 5 in terms of decreasing percentage with respect to the Parent Hull.

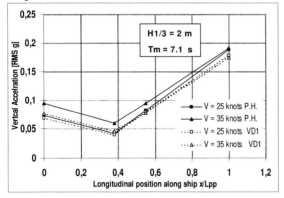

Fig. 3. Vertical accelerations in Sea State 4

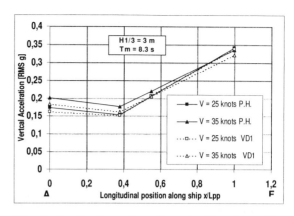

Fig. 4. Vertical accelerations in Sea State 5

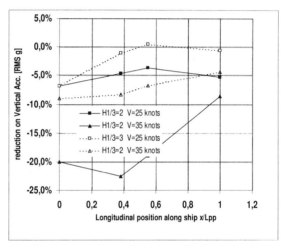

Fig. 5. Comparison between Parent Hull and Final Design

CONCLUSIONS

In order to assess the margin for further improvement in the seaworthiness of monohull Fast Ferries, a parametric study has been performed on the hull of MDV 1200, which is a typical representative of the latest generation of Fast Ferries designed by Fincantieri. The various configurations were numerically scored based on a seakeeping indicator related to motion sickness incidence.

Apart the expected benefit in increasing the length up to about 10%, some specific local variation of the hull forms was found to be quite effective ; namely the increasing of the water-plane area, by straightening the water-line at the transom stern and by increasing the local B/T ratio in the fore-body, and the shifting of LCB fore.

It must be considered that MDV 1200 hull is typically characterised by a strongly asymmetric longitudinal distribution of the sectional areas with much of the displaced volume concentrated aft and a very fine long fore-body. These features imply a larger wave-exciting pitch moment and a smaller pitch damping, both resulting in relatively large pitch motions.

Reducing the aft-fore unbalance of the immersed volume, thus shifting the LCB fore, resulted in fact in a lower pitching.

The same fore-aft asymmetry is present in the water-plane area, so that increasing the water-plane area can be mostly achieved by increasing the forward water-plane area coefficient thus shifting the LCF forward. Such measure, which affects the heave/pitch synchronism resulting in lower vertical motions, is particularly effective when obtained through increasing the local B/T ratio. Obvioulsy there is some practical restraint to the latter action, as an excessive flare may lead to dangerous slamming loads in the fore-body and high induced vertical accelerations.

As the original design is characterised by a B_T/B ratio less than unity, some gain is also obtained by drawing a straight water-line up to the transom stern ; the margin for improvement in this direction is however much limited.

These guidelines were utilised to re-draw the hull of MDV 1200. The effectiveness of the measures taken has been confirmed by specific model-tests carried out at DERA seakeeping basin on the original and re-designed versions of MDV 1200 hull. Preliminary results in head sea showed that the *optimal* hull provides an improvement of up to 20 % in terms of vertical accelerations.

LIST OF REFERENCES

[1]   E.Pino, V.Rossi, L.Sebastiani
        'Operability Suite
        CETENA's Technical Report n. 5864, March 1996.

Practical Design of Ships and Mobile Units
M.W.C. Oosterveld and S.G. Tan, editors.

Prediction of excessive rolling of cruise vessels in head and following waves

H.R. Luth and R.P. Dallinga

Maritime Research Institute Netherlands (MARIN), the Netherlands

This paper discusses the results of a study into the prediction of large roll angles of a cruise ship in head and following seas by a non-linear, six degrees of freedom time domain code. The results are compared with results of dedicated model tests.

1. INTRODUCTION

Modern large cruise vessels have high natural roll periods, typically in the order of 25 seconds. Since this is way outside the range of prevailing wave periods, no resonance induced large roll amplitudes are expected for this type of ships. Nevertheless, large roll motions were observed during recent seakeeping experiments.

To further investigate these motions, a dedicated model test program was carried out in MARIN's Seakeeping Basin, which included tests with a typical cruise vessel in head and following seas at zero and low forward speed. Large roll motions, up to 40 degrees single amplitude were observed, and considerable insight in the background of this behaviour was gained.

Since a non-linear numerical tool predicting the ship motions in the time domain in six degrees of freedom was available, the question was raised to which extent this tool is able to predict the occurrence and magnitude of the observed large roll motions. Therefore, an extensive numerical test program was carried out. The results of this study are given and discussed in the present paper.

In section 2, the background of the large roll motions in head and following seas is described in more detail. The seakeeping experiments and their results are described in section 3. Section 4 contains a description of the numerical tool and the numerical test program. In section 5, the results of model experiments and calculations are being compared and section 6 provides the conclusions drawn from the study carried out.

2. BACKGROUND

In general it is assumed that the transverse metacentric height (GM) is a quantity of constant value. However, when sailing in waves, the relative wave height along a ship introduces variations in the water line geometry. Especially for modern ship types with wide and shallow aft contours, a wave trough located at the stern reduces the local water plane width significantly. Due to the strong dependence of the metacentric height on the width of a ship, this could result in large variations in the metacentric height.

As discussed in [1], resonance roll with large roll amplitudes can occur when the wave encounter period is close to half the natural period of roll of the ship, especially if this unfavourable tuning is associated with relatively large stability variations. This phenomenon is called "parametric roll" and can occur unexpectedly in head or following seas.

Dunwoody [2,3] provided a link between the stability variations and reduction of the roll damping. According to his theory, parametric roll occurs if the effective damping, defined as the roll damping minus the damping reduction due to stability variations, is negative. This explains the emphasis on the prediction of roll damping by the numerical tool in the present paper.

In a previous paper on parametric roll of the same authors [1], some results of calculations were presented. Those results were the first available results from the present study. The present paper is based on all results of the numerical study.

3. EXPERIMENTS

A series of model tests was performed with a 1:50 scale model of a 240 m cruise ship. The body plan is shown in Figure 1, the main particulars are:

Lpp = 240.00 m
B   = 32.20 m
T   =   7.75 m
D   = 37,956 Tf
KG = 16.40 m

The hull form and weight distribution are typical for modern large cruise ships. During the tests, the model was self propelled and steered by means of an autopilot. It was equipped with bilge keels and one pair of active stabilisers. Measurements comprised the motion response and the forces acting on the individual rudders, fins and bilge keels.

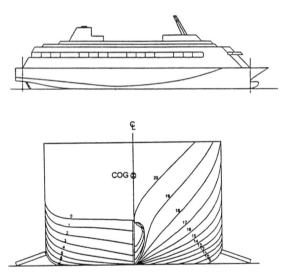

Fig. 1 General arrangement and body plan

All tests were performed in realistic irregular head and following waves. A JONSWAP type wave spectrum characterised the irregular wave conditions during the tests. For a more detailed description of the tests and general conclusions about the influence of the various parameters investigated, reference is made to [1].

In addition to the tests in waves, calm water roll decay tests were carried out to determine roll damping values of the ship at various speeds, with different bilge keel heights and with active and passive fins. The results of a selection of these tests are presented in Figure 2.

The results of the tests in waves considered at this moment are only the maximum, minimum and standard deviation of the roll amplitude of a run. Results are given as a function of wave height, see for example Figure 4. From the results was concluded that parametric roll is a threshold phenomenon, with negligible roll motions up to a certain wave height and gradually increasing roll angle at increasing wave height above the threshold level. This threshold level is also given for a certain set of tests as a function of wave height and period, see for instance Figure 5.

4. CALCULATIONS

4.1. Method of calculation

For the time domain simulations use was made of the non-linear, 6 degrees of freedom time domain code FREDYN, which was developed for the Cooperative Research Navies Dynamic Stability working group to simulate the motion behaviour of a steered ship in moderate to extreme waves and wind. The mathematical formulation includes physical aspects of wave making and of viscous origin in a six degrees of freedom model. It includes rudder, bilge keel and fin effects. The code includes the following force components:

- wave excitation forces: determined from integration of the wave-induced pressure in the time domain up to the instantaneous free surface, thus taking into account variations in stability due to waves;
- diffraction forces: calculated using a linear strip theory approach;
- added mass and damping: calculated using a linear strip theory approach and converted to retardation functions in the time domain;
- linear and non-linear drag (manoeuvring) forces in the transverse plane: estimated using empirical relationships;

- resistance and propeller forces: from full scale results, model tests or numerical predictions;
- rudder forces and forces of (active) fins: coefficients from model test data;
- roll damping moment: from the Ikeda model, reported by Himeno [4].

Non-linearities which are covered are: the effect of large angles (roll, yaw) on the Froude-Krylov part of the excitation forces; rigid body dynamics with large angles; drag forces associated with hull motion; wave orbital velocities and wind and the integration of the wave-induced pressure up to the free surface. See [5] and [6] for more background and applications of FREDYN.

4.2. Roll damping calculations

As was explained in section 2, the roll damping is considered to be an important parameter for the inception of parametric roll. Therefore, the first series of simulations were calm water roll decay runs. During these simulations the ship was free sailing, steered by an autopilot with an initial roll angle of 30 degrees. After the start of a simulation, the ship was free to roll and performed a roll decay motion.

The presentation of the roll damping results is based on the concept of the relative roll decrement, see Figure 2. On the vertical axis, the mean roll amplitude of two consecutive roll amplitudes is plotted. On the horizontal axis, the relative decrement is plotted, being the difference of two consecutive roll amplitudes divided by the mean roll amplitude. For each consecutive pair of roll amplitudes, the relative decrement was determined. A straight line was fitted through the results. Only this line is given in the figure. This line can be characterised by its slope and its intersection with the vertical axis. The intersection is a measure for the linear damping. The slope is a measure for the non-linear roll damping.

The standard roll damping routine in FREDYN is based on the work of Ikeda, reported by Himeno [4]. Figure 2 presents the results of the model tests and the simulations with the standard Ikeda routine. As can be seen from this figure, at zero speed the linear part of

the roll damping is under-predicted, while the non-linear part is over-predicted. At 5 knots, both parts are over-predicted. This trend is seen more often when the Ikeda formulation is applied for relative fine hull forms, such as cruise vessels or pleasure yachts.

To obtain a better correlation with the experiments, the roll damping routine was adjusted until it produced results similar to the model experiments, see also Figure 2. The results will be referred to as the calibrated results. The calibrated damping is expected to give better results than the Ikeda option.

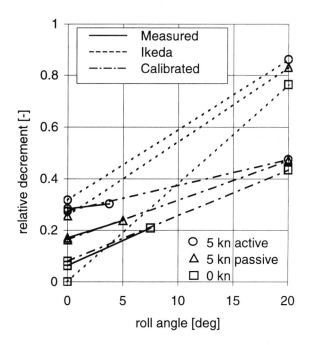

Fig. 2 Roll decay results

At roll angles above 5 degrees, the Ikeda method predicts higher roll damping values for zero speed than the model test results. Therefore, it is expected that the Ikeda option predicts smaller roll angles above 5 degrees roll amplitudes than the calibrated roll damping values

4.3. Irregular wave runs

Simulations were performed for a whole range of wave height and wave period combinations in irregular head and following waves at both zero and 5 knots forward speed with active and passive fins.

Because of the small effect of the rudder at the low speeds of simulation, it is difficult to keep the ship on its desired course. Therefore, a numerical spring in the ship fixed yaw direction was attached to the ship. The stiffness of the spring was chosen such that the introduced natural yaw period was out of the range of wave encounter periods.

For numerical reasons it is desirable to have a small disturbance in the roll motion during the simulations. Therefore, simulations started with a roll angle of 3 degrees. To maintain a small disturbance throughout the run, the desired course in the autopilot was set on 1 degrees off head seas or following seas. Each simulation had a duration of 30 minutes, from which the first 100 seconds were discarded to minimise start-up effects.

From the results of the simulations, the minimum, maximum and standard deviation of the roll angle of each run were stored. The largest roll angle (being the greater of the maximum and minimum roll angle) was plotted against the wave height, as can be seen in Figure 4.

5. COMPARISON AND DISCUSSION

Although only some results are presented as examples in Figure 4 of this paper, the overall comparison is based on a larger range of conditions, including zero speed and 5 knots forward speed, active and passive fins and two methods for the roll damping calculation. A comparison between measured and calculated standard deviation of roll angles showed the same trends and agreements as a comparison between measured and calculated maximum roll angles.

5.1. Roll angle as function of wave height

Contrary to what was expected, the Ikeda roll damping option in general provides a better agreement between measurements and calculations over the whole range of wave heights. The simulations with the calibrated damping over-estimate roll angles above 10 degrees significantly. Obviously, linear

extensions of the roll damping below 10 degrees is not suitable to make a prediction for the roll damping at high roll angles. This can be caused because the fins and bilge keels emerge from the water at about 25 degrees roll angle. The roll damping at large roll angles should be investigated further.

The results show that the best overall correlation is obtained for the 12.1 seconds wave period. Since this period is at zero speed about equal to half the natural roll period of the ship and the wave length is about equal to the length of the ship, the maximum response can be expected at this wave period. This is confirmed by the measurements and the calculations: the lowest critical wave height and the largest roll motions were indeed found at this period.

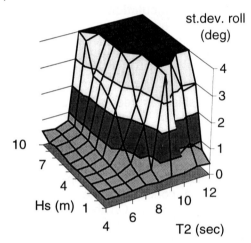

Fig. 3 Contour plot of results
(0 kn, following waves)

The results for the lower wave period show that the roll motions tend to be overpredicted. As can be seen in Figure 3, the roll angles at the lower wave periods strongly depend on the wave period. Therefore, a small difference in wave peak period or a small difference in shape of the wave spectrum can result in a large difference in roll angles. As such, this left part of the diagram is more difficult to predict than the right part.

The results of the longer waves, with a peak period of 14.1 seconds, show a fair agreement with the measured results.

5.2. Critical wave height

The critical wave height defines the boundary in the scatter diagram below which parametric roll does not occur.

From the results of the tests, contour plots were generated for the maximum roll angles and the standard deviation of the roll motion. An example of these plots is given in Figure 3. In order to obtain the critical wave height from these plots, a criterion should be defined for the maximum roll angle and the standard deviation.

For the roll amplitude, a criterion of 3 degrees was chosen. This is equal to the initial roll angle of the simulations. Since the first 100 seconds of the simulation were discarded from the results, this value can not be exceeded when no parametric roll occurs. For the standard deviation of roll, a criterion of 1 degree was chosen. This value correlates well with visual observations of the model experiments.

An influence of the roll damping option on the critical wave height was found, see for instance the left three diagrams of Figure 4. The Ikeda option provides a better prediction at the higher

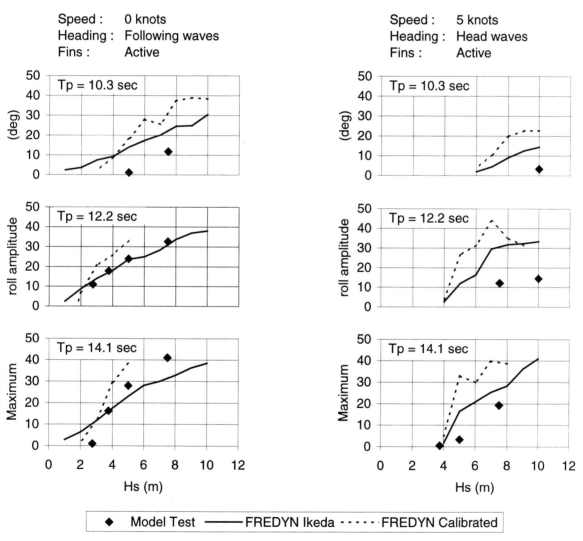

Fig. 4 Results of model tests and calculations
(max. roll ampl. vs. significant wave height)

630

waves, but the calibrated results seem to give a better prediction of the results below 10 degrees roll angle. Since the Ikeda formulation predicts for most ships a damping close to zero at small roll angles, using the Ikeda option will in general result in a conservative estimation of the critical wave height.

In Figure 5, the measured and calculated (calibrated) critical wave heights are compared. Although the shape of the curves look similar, a distinguishable difference between downtimes is found. The model tests show downtimes of 10% for zero speed and 5% for the 5 kn case, while the calculations show about twice those values.

6. CONCLUSIONS

The following conclusions can be drawn from the results of the study presented in this paper:
- FREDYN is capable of predicting the phenomenon parametric roll;
- in general, the results of the simulations show a fair agreement with the measured results;
- for some cases substantial differences were found between measured and calculated results;

- measured damping coefficients provide better predictions of the critical wave heights;
- roll damping values at low angles, say below 10 degrees, might not be suitable to predict damping values at large roll angles, say up to 40 degrees.

Based on these conclusions, FREDYN is seen as a tool that can be used to give a first impression of the occurrence and extent of parametric roll. It can not replace model tests before a more substantial validation is carried out and until more knowledge about parametric roll itself is obtained. It can be used to extrapolate the results of a limited set of model tests to a wider range of wave periods and wave heights.

REFERENCES

1. Dallinga, R.P, Blok, J.J. and Luth, H.R.: "Excessive rolling of cruise ships in head and following waves". RINA International Conference on Ship Motions & Manoeuvrability. London, Feb. 1998.
2. Dunwoody, A.B: "Roll of a Ship in Astern Seas – Metacentric height spectra" Journal

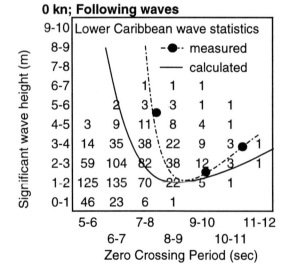
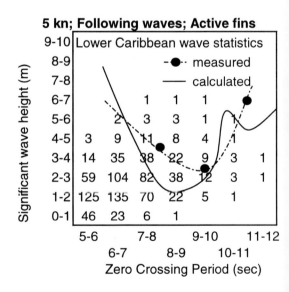

Fig. 5 Measured and calculated critical wave height

of Ship Research". Vol. 33, No. 3, September 1989, pp 221-228.

3. Dunwoody, A.B: "Roll of a Ship in Astern Seas – Response to GM Fluctuations". Journal of Ship Research, Vol. 33, No. 4, December 1989, pp 284-290.

4. Himeno, Y: "Prediction of ship roll damping – State of the art". Dept. of Naval Arichtecture and Marine Engineering, Report No. 239, September 1981.

5. De Kat, J.O, Brouwer, R., McTaggert, K.A. and Thomas, W.L.: "Intact Ship Survivability in Extreme Waves: New Criteria from a Research and Navy Perspective". STAB '94, May 1994.

6. De Kat, J.O: "Irregular Waves and Their Influence on Extreme Ship Motions. 20th Naval Hydrodynamics Symposium, Santa Barbara, August 1994.

The Prediction of ship's manoeuvring performance in initial design stage

Ho-Young Lee[a] and Sang-Sung Shin[a]

[a]Hyundai Maritime Research Institute, R & D Division, HHI
1 Cheonha-Dong, Dong-Ku, 682-792, Ulsan, Korea

Most of fuel the economic hull forms, nowadays, have stern bulbs at their afterbodies. The adoption of stern bulb, even though good for the vessel's propulsion performance, has been known to deteriorate the course keeping stability of a vessel. The change of the concept of stern hull form requires the modification of the empirical formulae of various hull-oriented hydrodynamic derivatives used in the manoeuvring prediction programs.

In this study, PMM(Planar Motion Mechanism) tests and rudder open water tests were carried out for 19 models of low-speed blunt ship with stern bulb and horn type rudder. The MMG model has been used as a basic mathematical model for manoeuvring equation. Then the regression analyses were performed using selected principal parameters.

The results of present study have been compared with those of Kijima's formulae and PMM model tests. From those, It is found that the present prediction give improvement of prediction for ship's directional stability.

1. INTRODUCTION

International Maritime Organization adopted the interim standard A.751(18)[1] for ship manoeuvrability in November 1993. In order to cope with the manoeuvring standard well, it is required for a ship designer to have tools to accurately predict the ship manoeuvrability at the preliminary design stage of a vessel. In general, low-speed blunt-ships with stern bulb are known to have bad manoeuvring characteristics because of the full hull form with large block coefficient and small length to beam ratio.

The captive model test is seen to give correct solution about ship's manoeuvring performance but designer can not entrust to towing tank test at initial design stage because of the limitation of time and cost.

Resultly, the numerical simulation with the principal parameters derived from PMM database is taken a measure for ship's manouevrability prediction at initial design stage.

Kijima et al.[2] proposed the approximate formulae for hydrodynamic forces on a ship with closed stern. These formulae are obtained semi-empirically by the results of model test and of numerical calculation by lifting surface theory. Whereas the Kijima 's formulae are limited for a ship with closed stern, those could not be adopted manoeuvring prediction for a ship with stern bulb.

In previous study, Lee et al.[3] derived regression equations for ships with stern bulb tested at Hyundai Maritime Research Institute by nondimensionalized principal parameters which Kijima used to predict hydrodynamic derivatives for ships with closed stern and simulated ship's manoeuvring performance by making use of formulae derived by regression analysis.

At the present paper, we studied to improve Kijima's model used to predict ship's manoeuvrability at initial design stage. The mathematical model is adopted as Kijima's model and the regression analyses are carried out for hydrodynamic derivatives and hull-propeller-rudder interaction coefficients. Finally, we simulate ship's manoeuvrabillity to validate the present MMG model and compare those with results of PMM test and Kijima's method.

2. MATHEMATICAL MODEL

By reference to the ship-fixed coordinate system shown in Fig.1, O-XYZ is space fixed coordinate and G-xyz is the body fixed coordinate.

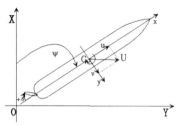

Fig.1 Coordinate systems

Referring the body fixed coordinate system, G-xyz, the basic equations of the ship's manoeuvring motion can be written in the following form.

$$m(\dot{u} - vr) = X = X_H + X_P + X_R$$
$$m(\dot{v} + ur) = Y = Y_{HP} + Y_R \qquad (1)$$
$$I_{zz}\dot{r} = N = N_{HP} + N_R$$

where the terms with subscript H represent hull forces and the terms with subscripts P and R denote propeller forces and rudder forces respectively.

The mathematical model of hull forces in the paper was developed on the basis of the MMG model. Details of the model tests, their analyses and derivation of the mathematical model can be found in Lee et al.[4]. The forces acting on ship hull can be written as follows.

$$X_H + X_R = -m_x\dot{u} + (m_y + X_{\beta r})\beta r + X_{uu}\cos^2\beta$$
$$Y_{HP} = -m_y\dot{v} - m_x ur + Y_{HO}(\beta, r) \qquad (2)$$
$$N_{HP} = -J_{zz}\dot{r} + N_{HO}(\beta, r)$$

The hull forces and moment are as follows.

$$Y_{HO} = \frac{1}{2}\rho LdU^2[Y'_\beta\beta + Y'_r r' + Y'_{\beta\beta}\beta\,|\beta| + Y'_{rr}r'\,|r'| + (Y'_{\beta\beta r}\beta + Y'_{\beta rr}r')\beta r'] \qquad (3)$$

$$N_{HO} = \frac{1}{2}\rho L^2 dU^2[N'_\beta\beta + N'_r r' + N'_{\beta\beta}\beta\,|\beta| + N'_{rr}r'\,|r'| + (N'_{\beta\beta r}\beta + N'_{\beta rr}r')\beta r']$$

According to regression analysis in this paper, hydrodynamic derivatives are obtained as follows.

$$X'_{\beta r} + m' + m'_y = 0.998 - 14.6991A'_pd/B - 1.0925k - 4.5865(A'_pd/B)^2 - 21.9443k^2 + 119.2277kA'_pd/B$$

$$Y'_{\beta rr} = 1.8779 - 30.8615k + 153.6857k^2 \qquad (4)$$

$$Y'_\beta = -9.5114 + 30.278C_b - 36.8419k - 22.1929C_b^2 + 20.3124k^2 + 40.3232C_bk$$

$$Y'_r - (m' + m'_x) = 0.2443 - 0.1962C_b - 1.854B/L$$

$$Y'_{\beta\beta} = -31.3506 + 3.622C_b + 318.2181B/L + 29.1844C_b^2 - 290.0526(B/L)^2 - 262.1299C_bB/L$$

$$Y'_{rr} = 0.5578 - 8.3636A'_pd/B - 10.281(2lcb/L) + 11.8301(A'_pd/B)^2 + 30.8604(2lcb/L)^2 + 120.0584(2lcb/L)(A'_pd/B)$$

$$N'_\beta = 0.0024 + 1.0272k + 0.2218kA'_p(1 - C_b)$$

$$Y'_{\beta\beta r} = 7.9058 - 28.8732A'_p - 69.5545B/L - 4.431A'^2_p + 152.0147(B/L)^2 + 169.9388(A'_pB/L)$$

$$N'_r = -0.0416 - 0.0006A'_pd/B + 0.29B/L - 2.29(A'_pd/B)^2 - 2.01(B/L)^2 + 2.22(A'_pd/B)(B/L)$$

$$N'_{\beta\beta} = -0.2149 - 0.1991(A'_pd/B) + 4.6127k + 13.38(A'_pd/B)^2 - 22.53k^2 - 9.59k(A'_pd/B)$$

$$N'_{\beta rr} = -1.5678 - 5.41A'_p + 23.16B/L + 2.38A'^2_p - 78.96(B/L)^2 + 25.33A'_pB/L$$

$$N'_{\beta\beta r} = -1.5678 - 5.41A'_p + 23.16B/L + 2.38A'^2_p - 78.96(B/L)^2 + 25.33(A'_pB/L)$$

where $k = 2d/L$. The variable of $Ap'(=100\ A_p/Ld)$ represents side area of stern bulb and the definition of that is shown in Fig.2

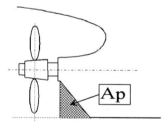

Fig.2 Configuration of A_p

The propeller thrust is as follows[2].

$$X'_P = (1-t)n^2 D_P^4 K_T(J)/\frac{1}{2}\rho L U^2 d \qquad (5)$$

The rudder force and moment including the hydrodynamic force and moment induced on ship hull by the rudder action can be written in the following forms[2].

$$X_R' = -(1-t_R)F_N'\sin\delta$$
$$Y_R' = (1 + a_H) F_N' \cos\delta \qquad (6)$$
$$N_R' = (x_R' + a_H x_H') F_N' \cos\delta$$
where
$$a_H = -11.4036 + 40.94C_b - 81.11k$$
$$\quad -31.69C_b^2 + 90.76k^2 + 79.47C_b k$$
$$x'_H = -6.054 - 0.101A'_p + 58.18B/L +$$
$$\quad 3.4A'^2_p - 148.44(B/L)^2 - 8.73(A'_p B/L)$$

The rudder normal force is expressed as follows.

$$F_N' = (U_R/U)^2 (A_R/Ld)f_\wedge \sin\alpha_R$$
$$f_\wedge = (5.5426\wedge)/(\wedge + 2.4280)$$
$$U_R = \sqrt{u_R^2 + v_R^2} \qquad (7)$$
A_R ; moveable rudder area
$\wedge$; rudder aspect ratio

where the rudder coefficient($f_\wedge$) can be estimated for horn type rudder. The effective inflow angle and velocity at a rudder can be described as follows.

$$\alpha_R = \delta + \delta_0 - \gamma(\beta - l_R' r')$$
$$\delta_0 = -(\pi s_0/90)$$
$$s_0 = 1 - u(1 - w_p)/nP \qquad (8)$$
$$u_R = \varepsilon u_P \sqrt{\eta_H \left\{ 1 + x \left(\sqrt{1 + \frac{8K_T}{\pi J^2}} - 1 \right) \right\}^2 + (1 - \eta_H)}$$
$$x = 0.6/\varepsilon \text{ where } \eta_H = D_P/H$$

The ε and γ that represent the rudder normal force in Eqn.(8) can be described below.

$$\varepsilon = -2.3281 + 8.697C_b - 3.78k$$
$$\quad + 1.19C_b^2 + 292.k^2 - 82.51(kC_b) \qquad (9)$$
$$\gamma_1 = 6.8736 - 16.77C_b + 3.5687k + 4.68C_b^2$$
$$\quad - 253.14k^2 + 74.83kC_b \text{ , where } \beta \leq 0$$
$$\gamma_2 = 23.708 - 83.84C_b + 173.72k + 71.64C_b^2$$
$$\quad + 157.01k^2 - 261.11kC_b \text{ , where } \beta \geq 0$$

3. REGRESSION ANALYSIS AND SENSITIVITY STUDY

3.1 Regression analysis

The selection of regressive parameters must to be appropriated for the accurate prediction of hydrodynamic derivatives and hull-propeller-rudder interaction coefficients.

The coefficients of determination($R2$) and regression errors checked by regression analyses are shown in Fig. 3 ~Fig. 4. For the purpose of reducing the error of regression analyses at the start, the regression analyses were carried out to selecting regressive parameters beyond five numbers. Owing to lack of experimental datas and mathematical uniqness due to the selection of many variables, the accuracy of manoeuvring analysis is decreased rather than regression analysis by Kijima's nondimensioal parameters[3]. Therefore, the parameters of principal dimensions and A_p' that express side area ratio of stern bulb are introduced for regression analysis. The 1st and 2nd polynomial regression analyses are carrried out to selecting two nondimensional parameters that the accuracy of correlation is the highest, correlation coefficient is the largest and the regression error is the smallest. Fig.3 show the results of regression analysis for linear hydrodynamic derivatives in comparison with experimental datas. In these figures, the plotted symbols such as a circle show the measured results and the lines show the estimated results. The estimated values for linear hydrodynamic derivatives agree well with the measured results of PMM tests.

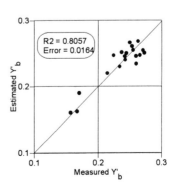

636

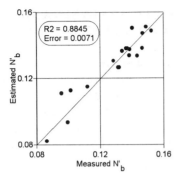

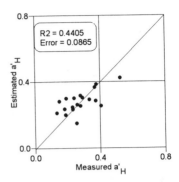

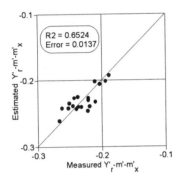

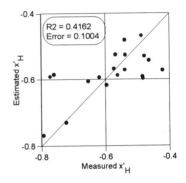

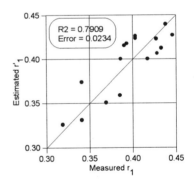

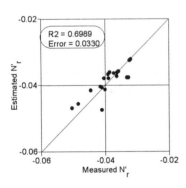

Fig.3 Comparison of measured and estimated
coefficient(Y'_β, $Y'_r - m' - m'_x$, N'_β, N'_r)

Fig.4 show the results of regression anal
ysis for $a_H, x'_H, \gamma_1, \gamma_2, \varepsilon$ and $f_\wedge$. In these
figures, it can be seen that the errors of
flow-straightening coefficients are smaller
than those of other variables.

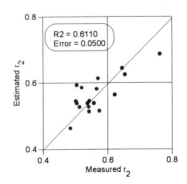

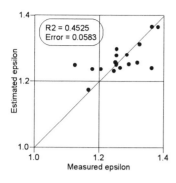

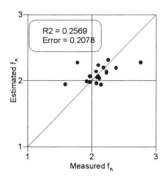

Fig.4 Comparison of measured and
estimated coefficient(a_H, x'_H, γ_1, γ_2 and ε)

3.2 Sensitivity study on simulation parameters

For the purpose of improving the predicti
on accuracy, the most technique is to conce
ntrate the improvement on such parameters
that more strongly affect the predicted valu
es. The sensitivity study has been perform
ed with the three ships as used in the pre
sent study. The full scale ships used for the
sensitivity study are a chemical carrier (Ship
A), oil tankers(Ship B and D) shown in
Table 3.

Table 3. Dimension ratios of ships

	Ship A	Ship B	Ship C	Ship D
L/B	5.6	5.57	6.52	5.52
B/T	2.88	2.82	2.90	2.56
L/T	16.15	15.70	18.92	14.09
C_B	0.7855	0.8166	0.8152	0.8149
A_R/Ld(%)	1.84	1.92	1.47	1.75

The items to be evaluated are those
specified in the 1st and 2nd overshoot angle
of 10°/10° zig-zag manoeuvre. The relative
sensitivity here represents the ratio of change
in estimated results when each parameter is
individually increased 10%[5].

Fig.5 show the results of sensitivity study
on the 1st and 2nd overshoot angle of
10°/10° zig-zig manoeuvre. It is found that
ε and N'_β are paticularly dominant in the
prediction of the overshoot angle of 10°/10°
zig-zig manoeuvre.

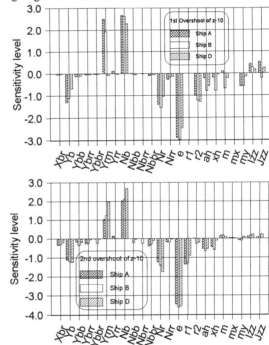

Fig.5 Relative sensitivity of parameters on
1st and 2nd overshoot angle of 10°/10°
zig-zig manoeuvre

4. MANOEUVRING SIMULATION

The manoeuvring simulation has been
carried out with MMG mathematical
modelling program developed in this paper
for the case of three ships. The full scale
ships used for the manoeuvring simulation
study are a chemical carrier(Ship A), a chip
carrier(Ship C) and a oil tanker(Ship D)
shown Table 3.

The results of numerical simulation on
manoeuvring motion by using regressed

638

hydrodynamic forces are shown in the following figures, which the plotted symbols such as circle show the measured results, solid line show the present predicted results and dotted line show the Kijima's predicted results. Fig.6 show the trajectories of turning motion with rudder angle of 35° to starboard and port. They are nondimensionalized by LBP. It is clearly shown that turning performance of present prediction definitely agree well with the results of Kijima's and PMM model tests. Additionally, the case of ship A is better agreement with other results. Fig.7 show comparative plots of the time histories of heading angle for the 10°/10° zig-zag manoeuvres, respectively. It is found that the present prediction of 1st and 2nd overshoot angle is better correct than any other methods in case of ships with stern bulb. Fig.8 show comparative plots of the time histories of heading angle for the 20°/20° zig-zag manoeuvres.

The trends are similar to those of the 1st and 2nd overshoot angles 10°/10° zig-zag manoeuvres.

In case of unstable ship, the improvement is definitely proved by overshoot angle appeared in the present prediction.

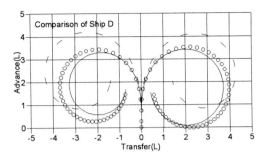

Fig.6 Comparison of turning trajectories by present prediction, Kijima's results and PMM results

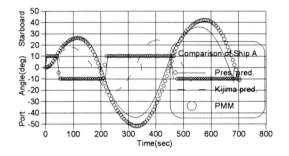

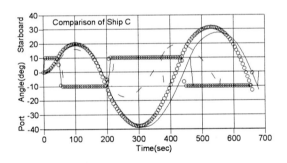

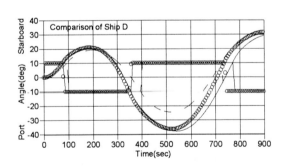

Fig.7 Comparison of 10°/10° zig-zag manoeuvre by present prediction, Kijima's results and PMM results

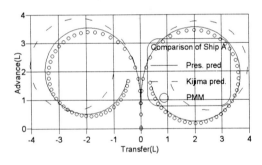

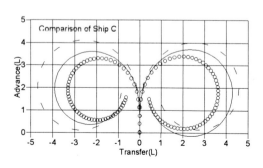

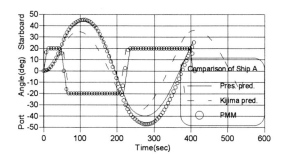

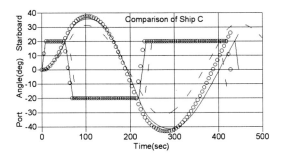

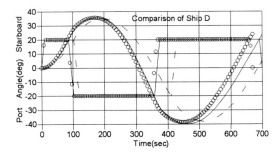

Fig.8 Comparison of 20°/20° zig-zag manoeuvre by present prediction, Kijima's results and PMM results

5. CONCLUSION

To predict ship's manoeuvrability at early design stage, the prediction method by semi emperical technique based on model tests is developed and the capability of present technique is validated by Kijima's method and PMM test.

The present study is applied to Kijima's

mathematical model, and is proposed the approximate formulae to estimate the hydrodynamic derivatives acting on a ship and hull-propeller-rudder interaction coefficients. The approximate formulae proposed in this paper are obtained by the data on PMM model tests at HMRI. From the simulation results, the prediction method gives satisfied results for ships with stern bulb of bad course keeping stability and we partially consider hull form parameters such as side area of stern bulb in database system.

Additionally, it must be to accumulate the data for more correct prediction of ship's manoeuvrability, and the formulae for prediction of high speed vessels equal to container must be also obtained by accumulation of model test datas.

REFERENCES

[1] IMO A 18/Res. 751, 22 November, 1993
[2] Kijima K., et,al.,"On a Prediction Method of Ship Manoeuvring Characteristics", MARSIM 93, 1993.
[3] Lee, H.Y. et al., "Improvement of Prediction Technique of the ship's Manoeuvrability at Initial Design Stage", Journal of SNAK, Vol.35, No.1, 1998.
[4] Lee, H.Y. et al.,"The Prediction of Manoeuvrability using PMM Model Tests-The Comparative Study of Mathematical Models-", Journal of SNAK, Vol.34, No.2, 1997.
[5] Ishiguro, T., et al., "A Study on the Accuracy of the Prediction Technique of Ship's Manoeuvrability at Early Design Stage", MARSIM 96, 1996.

Practical Design of Ships and Mobile Units
M.W.C. Oosterveld and S.G. Tan, editors.

An Experimental Study on the Effects of Loading Condition on the Maneuverability of Aframax-type Tanker

In-Young Gong, Sun-Young Kim, Yeon-Gyu Kim, Jin-Whan Kim

Korea Research Institute of Ships & Ocean Engineering
171 Jang-Dong, Yusung-Gu, Taejon 305-343, KOREA

Soo-Cheol Shin, Sa-Weon Kang, Yoon-Soo Kim

Samsung Heavy Industries Co. Ltd. Koje Shipyard

530, Jangpyung-Ri, Shinhyun-Up, Koje-Si Kyungnam, 656-800, Korea

The maneuverability of a ship is becoming more important with the application of IMO maneuvering standards, which is to be evaluated during her sea trials at scantling draft(summer load line draft) condition[1]. In case that the sea trials cannot be carried out at this draft, it is required to predict the maneuverability of a ship at scantling draft condition as well as at the draft of trial condition, and to extrapolate the sea trial results to the scantling draft condition. Therefore, it is often required to predict the maneuvering performance of the ship at several drafts during the initial design stage.

HPMM (Horizontal Planar Motion Mechanism) tests have been carried out with the Aframax type tanker at four draft conditions and computer simulation of standard maneuvers have been made with hydrodynamic coefficients obtained from HPMM test results. The effects of loading condition on the hydrodynamic coefficients and maneuverability is discussed in this paper. Based on the results of HPMM test and sensitivity test, a practical method is proposed to predict the maneuvering performance at the draft other than the draft where reliable hydrodynamic coefficients are available. The proposed method is validated by the comparison of the prediction with the results of model tests.

1. INTRODUCTION

With the introduction of IMO maneuvering standards, ship designers have to consider maneuvering performance as a critical criteria which should be fulfilled at the initial design stage. IMO proposed standard maneuvers such as turning test, zig-zag test and stopping test to check overall maneuvering performance of the ship[1]. Shipyards can demonstrate compliance with IMO standards by model tests and/or computer predictions using mathematical models at the design stage or full-scale trials.

IMO requires that the standard for the ship maneuverability is applied at the full load, even keel condition. In practice, however, many sea trials will be made at a different load condition from full load due to the difficulty of loading and high costs. In that case, the maneuverability of a ship at the full load condition should be evaluated by extrapolating the sea trial results using a mathematical model. Computer predictions using a mathematical model based on hydrodynamic database is the most popular and preferable method for shipyards at the design stage in its simpleness and low cost. But most of the prediction programs based on empirical formula and hydrodynamic database, at present, does not give reliable predictions for the recent full tankers with a poor maneuverability due to their insufficient hydrodynamic database. More accurate hydrodynamic coefficients through captive model tests would be still necessary for reliable predictions. Especially, the maneuverability of the ship and hydrodynamic data in ballast condition are further necessary to be able to extrapolate the trial results at the ballast condition to the full load condition.

The main purpose of this study is to give some guide of predicting maneuvering performance of the ship at a load condition with the information of the maneuvering performance. For this purpose, we carried out HPMM(Horizontal Planar Motion Mechanism) tests with the model ship of the Aframax type tanker at four draft conditions. Using the hydrodynamic coefficients obtained from the HPMM test, computer simulations of standard maneuvers have been made and the maneuvering performances at the different loading conditions have been compared.

Based on the results of HPMM test and sensitivity test, a practical method is proposed to predict the maneuvering performance at the draft other than the draft where reliable hydrodynamic coefficients are available. The proposed method is validated by the comparison of the prediction with the results of model tests. In the proposed method, the hydrodynamic coefficients at a trial draft are estimated from the experimental hydrodynamic coefficients at the other draft. It is shown that the prediction by the proposed method agrees well with the maneuvering performance simulated by using full experimental data at the ballast and the scantling draft. The proposed method in this paper can be used usefully when the maneuvering prediction is necessary at the draft other than the draft at which HPMM tests have carried out.

2. MODEL TESTS

HPMM tests have been carried out in the towing tank of KRISO whose size is 200m in length, 16m in width and 7m in depth. Details of HPMM system and HPMM test procedure are described in the reference[1].

HPMM tests were conducted with 1/38.7 scaled model of the Aframax type tanker at the following four loading conditions.

a. Ballast Even : 7.5 m even draft
b. Ballast : 7.5m draft with 3.0m stern trim
c. Design : 12.19m even draft
d. Scantling : 14.6m even draft (full load)

The principal geometrical characteristics of the ship are summarized in Table 1 and schematic drawing of four loading conditions is given in Fig.1.

Table 1 Principal Geometrical Characteristics of the Aframax Tanker

Lpp(m)	238.0		B(m)		44.0
Draft Condition	BT	BE	D		S
D (m)	7.5	7.5	12.19		14.6

Trim(m)	3.0	0.0	0.0	0.0
C_B	0.776	0.786	0.819	0.834
LCB/Lpp (%)	4.12	4.47	3.5	2.83
Rudder Area(m^2)	64.5	58.5	76.5	76.5

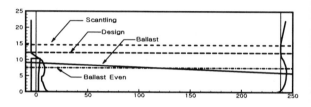

Fig.1 Schematical Drawing of the Loading Conditions

Table 2 shows the full test program. At Ballast Even condition, however, only Brief-HPMM test(static drift and pure yaw test) have been made just to predict the stability lever of the ship.

Table 2 Test Programs

Group	Test Item
HPMM Test	Static Drift Speed and Rudder Drift and Rudder Pure Sway, Pure Yaw Combined Yaw
Resistance & Propulsion Test	Resistance Propulsion Propeller Open Water
Rudder Open Water Test	Rudder Open Water

During HPMM tests, the model was free in heave and pitch but was fixed in roll. The model propeller revolutions were kept constant at the ship propulsion point except for the propeller overload and underload HPMM test.

3. MATHEMATICAL MODEL

The mathematical model of the ship maneuvering motion in this paper is based on MMG model[3]. In the coordinates system fixed at the midship of the ship shown in Fig.2, the equations of maneuvering motion is written as follows.

$$m'(\dot{u}' - v'r' - x_G r'^2) = X'_H + X'_P + X'_R$$
$$m'(\dot{v}' + u'r' + x_G \dot{r}') = Y'_H + Y'_P + Y'_R$$
$$I'_{ZZ}\dot{r}' + m'x_G(\dot{v}' + u'r') = N'_H + N'_P + N'_R \quad (1)$$
$$2\pi I'_{PP}\dot{n}' = Q'_E + Q'_P$$

where the terms with subscripts H, P and R represent the hull, propeller, and rudder forces and moments, respectively. Q_E and Q_P represent the engine and propeller torques. x_G represents the x-coordinate of the center of gravity of the ship. And the dots on u, v, r, and n represent the time derivatives of each variable. In eq. (1), primed symbols are used to designate the nondimensional form of each term[3].

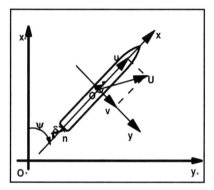

Fig. 2 Coordinate System and Sign Convention

Hull Forces and Moment :
Hydrodynamic forces and moment acting on the hull are described as follows:

$$X'_H = X'_{\dot{u}}\dot{u}' + X'_{vv}v'^2 + X'_{vr}v'r' + X'_{rr}r'^2 + X'(u)$$
$$Y'_H = Y'_{\dot{v}}\dot{v}' + Y'_{\dot{r}}\dot{r}' + Y'_v v' + Y'_{v|v|}v'|v'| + Y'_r r'$$
$$\quad + Y'_{r|r|}r'|r'| + Y'_{vrr}v'r'^2 + Y'_{vvr}v'^2 r' \quad (2)$$
$$N'_H = N'_{\dot{v}}\dot{v}' + N'_{\dot{r}}\dot{r}' + N'_v v' + N'_{v|v|}v'|v'| + N'_r r'$$
$$\quad + N'_{r|r|}r'|r'| + N'_{vrr}v'r'^2 + N'_{vvr}v'^2 r'$$

where X(u) represents the resistance of a ship and is obtained from the resistance test. Unless indicated otherwise, the hydrodynamic coefficients used in these equations are those for the ship propulsion point.

Propeller Forces and Engine Torque :
Hydrodynamic forces and moments due to the propeller can be written as follows:

$$Y'_P = 0$$
$$N'_P = 0 \quad (3)$$
$$Q'_P = -2\pi J'_{PP}\dot{n} - \rho n^2 D_P^5 K_Q(J_P)/(1/2\rho L^3 U^2)$$

where $J_P = \{u(1 - w_P)\}/(n\ D_P)$. L means L_{PP} of the ship and U means the resultant velocity of the ship. n is propeller rps and D_P represents the diameter of the propeller.

Rudder Forces and Moment :
Hydrodynamic forces and moment due to rudder are modeled as follows:

$$X'_R = -(1 - t_R)F'_N \sin\delta$$
$$Y'_R = (1 + a_H)F'_N \cos\delta \quad (4)$$
$$N'_R = (x'_R + a_H x'_H)F'_N \cos\delta$$
$$F'_N = (U_R/U^2)(A_R/L^2)f_\alpha \sin\alpha_R$$

where

$$U_R = \sqrt{u_R^2 + v_R^2}$$
$$\alpha_R = \delta - \delta_R$$
$$\delta_R = \delta_0 + \gamma_R(v' + l'_R r')(U/u_R) \quad (5)$$
$$u_R = \varepsilon u_P \sqrt{\eta\{1 + \kappa(\sqrt{1 + \frac{8K_T(J_P)}{\pi J_P^2}} - 1)\}^2 + (1 - \eta)}$$
$$v_R = -u_R \tan\delta_R$$

η : propeller diameter / rudder height

4. EFFECTS OF LOADING CONDITION ON THE DIRECTIONAL STABILITY AND THE HYDRODYNAMIC COEFFICIENTS

4.1 Directional Stability

Fig.3 shows the variation of the directional stability depending on the loading conditions. The damping lever, l'_r, does not change so much with the loading conditions but the static stability lever, l'_v, increases significantly as the draft decreases. The stern trim has a similar but a little larger effects on the static stability lever. The dynamic lever, l'_d, consequently, increases as the draft decreases and the stern trim increases. With this results, we can say that decreasing draft and stern trim makes the ship stable and the ship is the most unstable at the full load condition. This is somewhat well known and expected results and the same conclusion can be seen in the previous study[4].

644

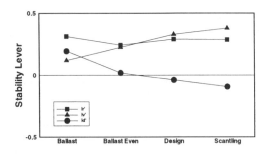

Fig.3 Variation of Directional Stability with Loading Conditions

4.2 Hydrodynamic Coefficients and Interaction Coefficient

Fig.4 shows the effects of the loading conditions on some hydrodynamic force coefficients and propeller-hull-rudder interaction coefficients which have a significant influence on the maneuvering performance of the ship. To show the relative change of the coefficients with loading conditions, all the coefficients are divided by the corresponding coefficient at the design draft. So, the values in Fig.4 represents the ratio of the coefficient at the corresponding draft to that at the design draft. It should be noted that the hydrodynamic coefficients in Fig.4 are nondimensionalized by using draft in order to represent the effects of loading more clearly. That is, force coefficients and moment coefficients are nondimensionalized by LT and L^2T respectively. On the whole, the change of coefficients is larger at Ballast than at Scantling. Especially, N_v is remarkably reduced in magnitude at Ballast. As draft increases, Y_r decreases but N_r increases. Y_v does not change so much with the loading conditions compared with the other control derivatives. The added mass, $Y_{\dot{v}}$ and $N_{\dot{r}}$, is much reduced at Ballast but there is no noticeable change at Scantling. For the interaction coefficients, a_H and γ increases as draft increases but ε and f_α change little with loading conditions.

5. EFFECTS OF LOADING CONDITIONS ON THE MANEUVERING PERFORMANCE

With the hydrodynamic coefficients obtained from the HPMM tests, computer simulation of the standard maneuvers have been made at three loading condi-

tions; Ballast, Design and Scantling. In the simulation, approach speed is 15.4knots at Ballast and 15.0knots at Design and 13.5knots at Scantling.

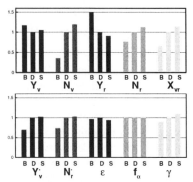

Fig.4 Relative Change of Hydrodynamic Coefficients With Loading Conditions

Fig.5 shows the turning trajectories with loading conditions. At Ballast, the tactical diameter, advance and transfer has the biggest values. The tactical diameter is almost same at Design and Scantling but the advance and transfer is bigger at Scantling.

Fig.6 and Fig.7 show the time histories of heading changes for 10°/10° zig-zag maneuvers and the variation of overshoot angles with loading conditions respectively. It can be seen that the characteristics of 10°/10° zig-zag maneuver is changed significantly with the loading conditions. At Ballast, the ship shows good course stability. As the draft increase, the ship becomes more unstable and the overshoot angles increase greatly. These figures show that the maneuvering performance at the ballast condition is quite different from that at the full load condition.

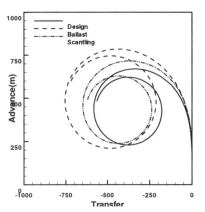

Fig.5 Trajectories of 35° Port Turn

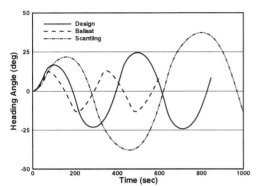

Fig.6 Heading Change of 10°/10° Zig-Zag Maneuvers

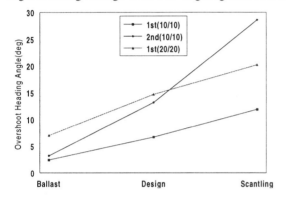

Fig.7 Variation of Overshoot Angles with Loading Conditions

Fig.8 shows spiral maneuvers. It is seen from this figure that the ship is highly stable at Ballast but becomes unstable as the draft increases.

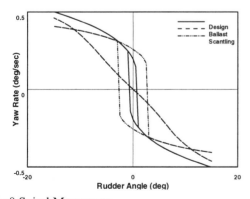

Fig.8 Spiral Maneuvers

6. ESTIMATION OF LOADING CONDITION EFFECTS ON THE MANEUVERING PERFORMANCE

As shown in section 5, the maneuvering performance of the ship at different loading conditions can be predicted well if all the hydrodynamic coefficents at each loading condition are obtained from HPMM tests. But it is very expensive and time consuming work to carry out HPMM test for several loading conditions so that HPMM test has been carried out generally for one load condition. So it would be very desirable to be able to make a reliable estimation of the maneuvering performance at the draft other than the draft at which the HPMM tests have been made.

Two methods can be considered to predict the maneuvering performance at trial draft with the informations of maneuvering performance at test draft. The first method is to extrapolate the maneuvering performance at the trial draft directly from that at the experimental loading condition by using mathematical model based on the empirical formula. The maneuvering performance can be overshoot angles in zig-zag test, advance and tactical diameters in turning test. It is assumed here that the mathematical model based on the empirical formula can predict the maneuvering performance with different loading conditions at least qualitatively although it does not give a good quantitative prediction. In the second method, firstly the hydrodynamic coefficients are calculated both at the trial draft and at the test draft by using some empirical formula. If the ratio of the hydrodynamic coefficients at the trial draft to ones at the test draft is calculated by empirical formula, the hydrodynamic coefficients at the trial draft can be calculated just by multiplying the ratio by the coefficients at Design obtained from the HPMM test. In the next, the computer simulation at the trial draft is made by using coefficients, thus obtained. Then, the maneuvering performance at the trial draft is estimated from the simulation results.

The first method was applied to estimate the maneuvering performance at Ballast and at Scantling. MMG model [5,6,7,8] was used as a mathematical model of the simulation. The estimated maneuvering performances were compared with the ones which were obtained from the simulation with the HPMM test results. Generally the agreement was good and estimation seemed to be acceptable. But the estimated overshoot angles were too low for $10^\circ/10^\circ$ zig-zag maneuver at Scantling. This is a very disappointing result because the second overshoot angle at full load condition is the most critical criteria among IMO maneuvering standards. It seems that the first method cannot be applicable until the empirical formula in mathematical model improves further.

Before applying a second method, the sensitivity test has been performed to find out the effects of each hydrodynamic coefficient on the maneuvering performance. The simulation has been made at Design by using experimental data, and its maneuvering performance is set as a reference REF. Subsequently simulations have been made by changing only one coefficient +20% and -20% while keeping the other coefficients as experimental data. If we call the difference of the maneuvering performance between two simulations as ΔP, ΔP is the measure of the coefficient's influence on the maneuvering performance. Then the sensitivity of the coefficient on the maneuvering performance is defined here as the ratio of ΔP to REF. Fig.9 and Fig.10 show the sensitivity of hydrodynamic coefficients on the turning characteristics and overshoot angles, respectively. From these figures, similar conclusions to Kose's sensitivity study[9] can be drawn. That is, N_v, N_r, ε, γ and Y_v is most influential. f_α has a large sensitivity on the overshoot angles but it has a small sensitivity on turning circle characteristics.

In view of the sensitivity results, a good estimation of maneuvering performance at Ballast and at Scantling can be expected if we could estimate influential coefficients well. Hydrodynamic coefficients at Ballast and Scantling are estimated based on the hydrodynamic coefficients at Design obtained from the HPMM tests. Principally, most of hydrodynamic coefficients are kept same with the ones at Design, except some hydrodynamic coefficients "influential coefficients" which influences the maneuvering performance with loading conditions. This influential coefficients are modified by using empirical formula. Influential coefficients can be chosen by referring Fig.4, Fig.9 and Fig.10. Although f_α and ε shows a large sensitivity on the maneuvering performance, they have not been modified because they change little with loading conditions. γ may influence the maneuvering performance but it is not modified because there is no appropriate formula. Kijima[11] proposed a formula for gamma but his formula gives opposite trend with our test results.

Inoue's empirical formula[7] and Motora chart[10] are used to extrapolate the value of experimental coefficients at Design. In Fig.11, finally, chosen influential coefficients are shown and their presented results are compared with experimental values. The value of coefficients represented here are relative values to the value of coefficients at Design as same with Fig.4. They show good agreements except Y_r.

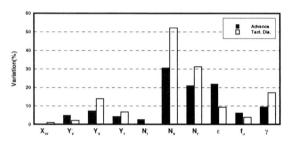

Fig.9 Relative Sensitivity of Hydrodynamic Coefficients on 2nd overshoot angle of $10^\circ/10^\circ$ Zig-Zag test

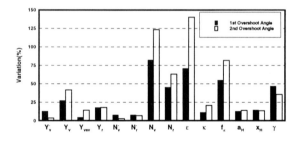

Fig.10 Relative Sensitivity of Hydrodynamic Coefficients on Tactical Diameter of 35° turning test

With modified influential coefficients, the computer simulations have been made to predict the maneuvering performance at Ballast and at Scantling. Figs.13-18 show the results of simulation compared with the results of simulation based on HPMM test data. In these figures, HPMM represent the simulation with test data and Prediction A with modified influential coefficients mentioned above. The predictions for Scantling are very satisfactory. For Ballast, the prediction shows some difference in the initial motion and the motion at large rudder angle. To check the effects of γ, the measured γ at each draft has been used instead of the value at Design. The simulated results are shown in Figs.13-18 with the name of Prediction B. The difference between Prediction A and Prediction B is very little but Prediction B gives a much better prediction of zig-zag and spiral characteristics for Scantling. Reliable empirical formula for γ seems to be necessary for improving predictions.

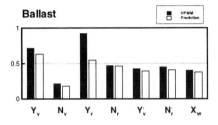

Fig.11 Predicted Hydrodynamic Coefficients at Ballast

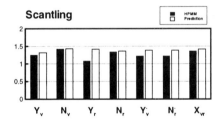

Fig.12 Predicted Hydrodynamic Coefficients at Scantling

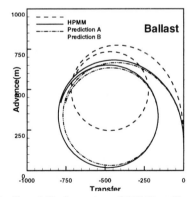

Fig.13 Predicted Trajectories of 35° Port Turn at Ballast

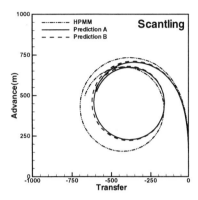

Fig.14 Predicted Trajectories of 35° Port Turn at Scaltling

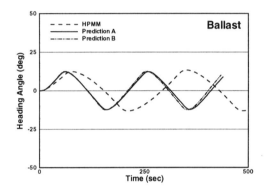

Fig.15 Predicted 10°/10° Zig-Zag Maneuvers at Ballast

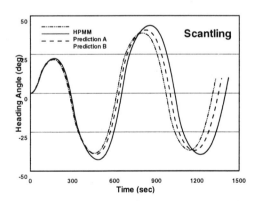

Fig.16 Predicted 10°/10° Zig-Zag Maneuvers at Scantling

648

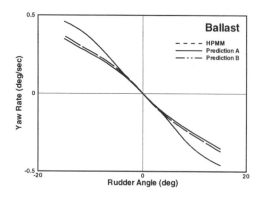

Fig.17 Predicted Spiral Maneuvers at Ballast

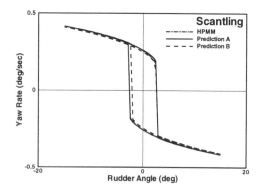

Fig.18 Predicted Spiral Maneuvers at Scantling

7. CONCLUSION

Based on the HPMM tests with Aframax type tanker at four different drafts and computer simulations of standard maneuvers, the effects of loading condition have been investigated in this paper. The main conclusions drawn in this paper are summarized as follows.

a. The directional stability of the ship increases as the draft decreases.

b. At ballast condition with stern trim, N_v reduces significantly.

c. If reliable maneuvering coefficients are available at design draft, the maneuvering performance at the other draft can be practically estimated to a good degree by the proposed method in this paper.

ACKNOWLEDGEMENT

This work has been supported by Samsung Heavy Industries Co., Ltd.. The authors wish to acknowledge the support of this organization.

REFERENCES

[1] "Explanatory Notes to the Interim Standards for Ship Manoeuvrability", MSC Circ.644 JUN. 1994.

[2] Kim,S.Y., "Development of Maneuverability Prediction Technique", KIMM Report No. UCE.337-1082.D, March 1988.

[3] "MMG report I,II,III,IV,V", Journal of Society of Naval Architects of Japan, No.575(1977), No.577(1977), No.578(1977), No.579(1977), No.616(1980).

[4] Kose, K. et al. , "Studies on the Effects of Loading Condition on the Maneuverability of Ships", Trans. of the West-Japan Society of Naval Architects, No. 82, 1992.

[5] Kang,C.G., Kim,Y.G., "A Prediction Method of Maneuverability Including the Effects of Side Profile at the Initial Design Stage", 1995, PRADS.

[6] Hirano,M., "On Calculation Method of Ship Maneuvering Motion at Initial Design Phase:, J. of the Society of Naval Architects of Japan, Vol.147, 1980.

[7] Inoue,S., Hirano,M. and Kijima, K., "Hydrodynamic Derivatives on the Ship Maneuvering", Int. Shipbuilding Progress, Vol.28, No.321, 1981.

[8] Inoue,S., Hirano,M., Kijima, K. and Takashina,J., "A Practical Calculation Method of Ship Maneuvering Motion", Int. Shipbuilding Progress, Vol.28, No.325, 1981.

[9] Kose, K. and Misiag. W.A., "A systematic Procedure for Predicting Maneuvering Performance", MARSIM '93, 1993.

[10] Motora, S., "On the measurement of Added Mass and Added Moment of Inertia of Ships in Steering Motion", DTMB Report No. 1461, 1960.

[11] Kijima, K. et al., "On a Prediction Method of Ship Maneuvering Characteristics", MARSIM 93, 1993.

Prediction of crabbing in the early design stage

Ir. F.H.H.A. Quadvlieg[a] and Ir. S.L. Toxopeus[a]

[a]Maritime Research Institute Netherlands, Wageningen
P.O. Box 28, 6700 AA Wageningen, The Netherlands

Crabbing is the ability of ships to move sideways without having a forward speed. Crabbing can be induced by the use of a combination of main propellers, rudders and lateral thrusters. The crabbing ability is a useful operating mode and a ship owner can have considerable profit from this ability because of a reduction of the time and costs in the harbour. Consequently, the crabbing ability is a design criterion, and needs to be assessed in early design stage. Ship yards need to know the power and amount of lateral thrusters that are to be installed in the ship. Some used criteria for crabbing are given. When the crabbing operating mode is chosen, a complex amount of circulation and fluid flows occur around the ship, which is effected considerably by the environment (quay and water depth). Using an example ship, it is shown that simple calculations are not sufficiently adequate to estimate the crabbing capability, so that advanced calculations or model tests are needed.

1 INTRODUCTION

Twin screw ships have due to their ability to use both main propellers in a different mode of operation, a special feature on board. By reversing one propeller and giving normal thrust with the second propeller, a circulating flow around the aft ship is produced. By steering with the rudder in the slipstream and by using eventual lateral thruster units, this can give the ship the ability to move sideways without having a forward speed. This ability has been called crabbing, traversing or the "push-pull" mode.

This ability is a very powerful one. It means that in the aft ship, a very powerful transverse force can be generated, able to manoeuvre the ship in complete lateral motion or to withstand currents, waves and wind.

This application is therefore the most interesting with ships who benefit from these large forces such as ferries and cruise ships. It saves the use of harbour tugs up to certain environmental conditions. Other types of ships which may benefit are the workboats. Supply vessels, cable laying vessels are for example vessels where this system can be used in the DP system, in addition to stern thrusters and bow thrusters.

Hydrodynamically seen, this push-pull mode operation of the propellers gives a complex flow around the aft ship. A relative simple calculation technique is not sufficient to tackle the above problem. A large propeller-rudder-hull interaction is to be expected, yielding large transverse forces, which may be directed opposite of the desired forces. This system is again influenced by the working of stern thrusters and even by bow thrusters. The circulation of water around the aft ship is also influenced a lot by the presence of the quay if a berthing or unberthing procedure is carried out. The presence of the sea bottom is also a very important factor.

In this paper, crabbing as criterion in early design is emphasised. Some possible criteria are given, a calculation technique is introduced and shortcomings are sketched. It is shown that a simple calculation technique will not always work.

2 CRABBING CRITERIA

Already in the design process, the crabbing ability should be investigated. Several examples are inventoried here.

In [1], the example is given that a transverse speed of 0.25 m/s should be maintained against 22 knots of wind (Beaufort 6). For this transverse speed, all applicable means can be applied. The distance to the quay and the shallowness of the water for which the crabbing ability should be investigated are not specified.

Another criterion can be: the ability to withstand a transverse current of 3 knots.

For cruise ships, the design criterion is often the ability to berth and unberth in a wind speed corresponding to Beaufort 7 of any direction without help from tugs. This means that the vessel has to move sideways parallel to or from the quay against a 35 knots beam, wind, assuming this to be the worst direction, without any forward speed.

This illustrates that in these cases, the importance of the influence of the quay and the shallow water are indicated.

The US Coast Guard has some specific requirements in which they indicate that the ability to safely manoeuvre has to be proven once the lateral area above the waterline is larger then 3 times the lateral area below the waterline (LxT) [2]. This is the case with most modern cruise ships and ferries.

A fourth criterion which is emphasised more and more recently is the impact of the crabbing procedures on the environment. During berthing and unberthing procedures, the circulation in the harbour basins can be so large that damages can occur to the quay or other ships in the harbour. Therefore, optimisation can be required with respect to a minimum of environmental impact [3].

The speed which can be used during harbour manoeuvring is very important. If the ship has a good crabbing ability, very fast berthing procedures can be conducted. In [4], the example is given of a ferry whose berthing procedure has been speeded up so that the service speed could be lowered, resulting in a lower fuel consumption.

It is shown in [5] and [6] that if the berthing/unberthing procedure is optimised by a set of crabbing model tests and computer simulations where the crew is trained on this system, a considerable profit can be obtained.

3 BASICS OF CRABBING

As put forward in [1], the following actuators can be used during the crabbing process for ordinary twin screw ships:
- Main propellers
- Rudders
- Stern thrusters
- Bow thrusters

When the main propellers work in different loadings, a strong propeller rudder hull interaction is induced,

yielding a pressure field around the aft ship.

In addition to this, the neighbourhood of the quay is of importance. It is also observed that the water depth is of large influence on the flow of the water.

In the attached figure, it is illustrated which water flows occur around the ship.

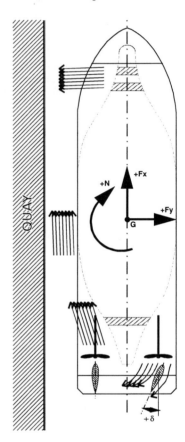

The major transverse force in the aft ship will come from the lift of the rudder in the jet stream of the balancing propeller. Also, it is clear that it is possible that a strong flow due to the backing propeller can exist between the quay and the ship, resulting in a negative pressure, which causes a suction force towards the quay.

4 STANDARD MODEL EXPERIMENTS TO INVESTIGATE CRABBING

Crabbing performance is investigated at MARIN by means of captive model tests: the ship is attached to a towing carriage by means of force transducers. This enables the measurement of forces developed by the ship propulsive gear on the ship. For instance, all combinations of main propellers, stern thrusters and bow thrusters can be combined with any particular rudder angle. The quay and the water depth are of major importance towards the crabbing behaviour. If the neighbourhood of the quay is desired, this is simulated by the solid side wall of the basin. The water depth is adjusted to the desired scaled value. It is of importance that the basin is wide and long enough. During the crabbing, a strong interference is present with any "quay" in the neighbourhood. This can be observed in harbour basins and locks in real life.

5 CALCULATION PROCEDURE WHILE NEGLECTING INTERACTIONS

Using a simple calculation procedure, it may be possible to calculate the forces working on the ship in the zero speed case. The following calculation scheme is focussed on assessing the forces due to working propellers, rudders and thrusters. This simple calculation procedure can be created in a spreadsheet. It contains a summation of the forces of the lateral thrusters, the main propellers and the lift effect of the rudder due to the propeller slipstream.

5.1 Thruster forces
The forces generated by the lateral thruster can be calculated as a function of the available power. According to Brix [7], a specific lateral force of the lateral thrusters of 100 N/kW may be assumed.

5.2 Propeller forces
The thrusts of the propellers can be calculated based on the 4 quadrant - curves of propellers [8], given propeller design data. The propeller thrust is given by:

$$T_P = C_T(\beta_{PR})\tfrac{1}{2}\rho\tfrac{\pi}{4}D_P{}^2\left(u_P{}^2 + (0.7\pi n D_P)^2\right) \qquad (1)$$

in which:
   T_P is the propeller thrust
   D_P is the propeller diameter
The required power is given by:

$$P_P = c_Q(\beta_{PR})\rho\frac{\pi^2}{4}D_P{}^3\left(u_P{}^2 + (0.7\pi n D_P)^2\right)n \qquad (2)$$

in which n is the number of propeller revolutions per second and D_P is the propeller diameter. The angle

β_{PR} in the above equation is the indicative angle of incidence of the flow to the propeller blade:

$$\beta_{PR} = \arctan\left(\frac{u_P}{0.7\pi n D_P}\right) \qquad (3)$$

in which u_P is the longitudinal component of the undisturbed flow velocity through the propeller. In the crabbing case, this speed is zero, therefore β_{PR} is zero for the balancing propeller and π for the backing propeller. Values of c_T and c_Q can be obtained for forward and astern bollard pull conditions. Using these coefficients together with the maximum available backing power P_p, the maximum astern RPM can be calculated.

Having established the RPM, a thrust is found. This thrust will be compensated with a forward thrust of the balancing propeller and the longitudinal force of the rudder. $T_{p,balancing}$ will be calculated as:

$$T_{p,balancing} = F_{x,rud} - T_{p,backing} \qquad (4)$$

Given the required forward thrust, the RPM and required power of the forward propeller can be calculated. The forward thrust will be used to calculate the rudder forces.

5.3 Rudder forces
Because the rudder behind the backing propeller is not located in the propeller slipstream and therefore the inflow velocity will be small, the forces on this rudder are supposed to be negligible.
Using the thrust of the balancing propeller, the inflow velocity at the location of the rudder behind the balancing propeller can be calculated. In case of a zero forward speed, the induced speed at the rudder u_R is found to be:

$$u_R = c_{UR}\sqrt{\frac{8T_{P,balancing}}{\pi\rho D_P{}^2}} \qquad (5)$$

A value of 0.7 D_P/H_R is suggested for c_{UR} according Inoue et al [9].
Using the above calculated inflow velocity at the rudder, a rudder lift force can be calculated using the lift formulas according [10]:

Rudder	Stern thr	Bow thr	Reversing propeller (port)		Balancing propeller (starboard)	Rudder		Total	
Angle	F_y	F_y	Power	Thrust	Thrust	u_R	Lift	F_y	M_z
[deg]	[kN]	[kN]	[kW]	[kN]	[kW]	[m/s]	[kN]	[kN]	[kNm]
0	85	300	5300	-279	279	2.92	0	385	11076
5	85	300	5300	-279	280	2.93	21	406	9517
10	85	300	5300	-279	286	2.96	42	426	7916
15	85	300	5300	-279	295	3.01	65	447	6270
20	85	300	5300	-279	307	3.07	89	469	4580
25	85	300	5300	-279	323	3.15	115	490	2873
30	85	300	5300	-279	341	3.23	144	510	1200
35	85	300	5300	-279	361	3.33	175	529	-347
40	85	300	5300	-279	381	3.42	207	544	-1634

Table 1: Sample calculation of crabbing forces

$$L = \tfrac{1}{2}\rho C_{L\delta} A_R u_R^2 \sin\delta \qquad (6)$$

In this formula, the following nomenclature is adopted:

A_R is the rudder area
λ_R the rudder aspect ratio
δ is the rudder angle

$C_{L\delta}$ is the lift coefficient which can be calculated for conventional rudders [9] as:

$$C_{L\delta} = \frac{6.13\lambda_R}{2.25 + \lambda_R} \qquad (7)$$

Together with the lift induced drag on the rudder, the longitudinal and transverse forces of the rudder on the hull can be calculated. The longitudinal force of the rudder on the hull is used again in equation (4), so that this last series of equations is an iterative process.

For an imaginary cruise ship [11], this can result in the calculation results as shown in Table 1. In this example, the ship is equipped with a stern thruster and three bow thrusters and the main propellers are used to balance the ship for zero forward speed. In the table, it is seen that for a sideways unberthing manoeuvre, a transverse force F_y of 520 kN can be generated when the starboard rudder is set to 35°.

6 FORCES TO BE OPPOSED

The crabbing capability can be calculated using techniques sketched in the previous paragraph. The forces which are to be opposed, follow directly from the crabbing criteria as put forward in paragraph 2. In general, a wind resistance and a speed through the water are important.

The wind resistance can be estimated using the techniques put forward by Brix et al. [7]. A relatively simple calculation technique can give the force and moment generated by the wind. The wind force is calculated by:

$$F_{wind} = \tfrac{1}{2}\rho_{air} v_{wind}^2 A_{wind} c_{D,wind} \qquad (8)$$

while for the wind moment a similar formula is used. The forces and moments generated by a speed through the water can be calculated using the cross flow drag theory [12].

For the example ship in paragraph 5, having a criterion of 22 knots and a pure sideways motion of 0.25 m/s, a side force F_y of 270 kN is required.

7 COMPARISON WITH MEASUREMENTS

In this paragraph, the calculation method presented in paragraph 5 is used to calculate the lateral force F_y and yawing moment M_z for a cruise vessel, using the main propellers and rudders, one bow thruster and without stern thruster. This is illustrated in Table 2 where the calculation results are compared with corresponding model tests.

Rudder	Total Calculated		Total Measured	
angle	F_y	M_z	F_y	M_z
[deg]	[kN]	[kNm]	[kN]	[kNm]
0	100	-1545	58	20090
15	217	-10596	153	13832
25	297	-16991	225	10036
35	370	-23055	280	4500
45	417	-27097	330	-1000

Table 2: Comparison with measurements.

It can be seen that although the trend and forces in the lateral force seem to conform, the moment is completely out of range. This has the following consequences. Based on the simple calculations, a lateral force of approximately 100 kN can be resisted when the yawing moment should be approximately zero. By installing additional bow thruster capacity, the moment M_z is increased and by doubling the bow thrust capacity, a pure transverse force of about 280 kN can be generated. However, the model tests show that already a force of some 320 kN is obtained in the original configuration.

Hence, to be able to fulfil the criterion in paragraph 6, based on the simple calculation an additional bow thruster would be purchased, while based on the model tests this will not be necessary. The reason for the discrepancies is that certain interaction forces are omitted, which have influence, not only on the attained lateral force, but also on the moment and the longitudinal force. This has its effects on the thrust of the balancing propeller and hence the side forces. An erroneous prediction on certain aspects can therefore be more critical than expected.

8   INTERACTIONS

Although in literature reference is given to the different interaction processes, see [6] and [13], it is not quantified how large the impact of neglecting the interactions actually is. The interactions are described quantitative below.

8.1 Bow thruster forces
In addition to the thrust generated by the bow thruster, a bow thruster-hull interaction takes place. For crabbing in harbours, two environmental aspects are considered to be important. The water depth causes an increase of the thruster effective force with decreasing water depth. The distance between the outlet of the thruster and the quay is a reason for retardation effects and increased thruster hull interaction.

The assumption of a specific lateral force value of for example 100 N/kW [7] is questionable. The value is dependent on the relative loading of the lateral thruster and on the interaction between any other lateral thrusters. When for example three bow thrusters are present, the generated lateral force will be less than three times the lateral force of only one bow thruster.

8.2 Rudder forces
When a strong circulation is initialised around the aft ship, a change of the flow around the rudder is found. It is observed that in most cases stall of the rudder does not occur. Due to the highly instationary flow, there is no dramatic decrease in lift force at higher rudder angle.

A second aspect is that due to the changed flow towards the rudder(s) the neutral rudder angle will change, depending on the circulation around the ship. The form of the circulation itself is dependent on the underwater shape of the aft ship, the size of the skeg and the water depth. Even the shape of the neighbouring quays is of influence.

As mentioned before, the rudder which is located behind the backing propeller has almost no influence on the generated forces.

8.3 Main propeller interaction forces
As put forward in [14] and [15], by reversing one propeller, and balancing the other, several physical phenomena are introduced. The "shoulder moment" is a direct moment generated by the different thrusts. The "lateral stream effect" is caused by the different pressure fields on port and starboard at the hull [7] due to the fact that the streamline of the reversed propeller "hits" the hull of the ship. When the ship is equipped with fixed pitch propellers, the "paddle wheel effect" is introduced, caused by reversing the rotation of the backing propeller.

All these main propeller effects are affected by: the choice of propeller; fixed pitch or controllable pitch; outward turning or inward turning propellers [13]. Furthermore, we see that even more interactions are taking place. The "blockage" of the flow has an impact on the propeller coefficients c_Q and c_T, so that

654

we can state that the rudder angle, the water depth and the distance to the quay have influence on the propeller characteristics.

8.4 Quay interactions

As already indicated in the previous paragraph, a strong quay interaction exists due to suction between ship hull and the quay.

Due to the working of the backing propeller in the neighbourhood of the quay, a strong current is generated between the ship's hull and the quay. This results in a large suction force, causing the ship to be pulled towards the quay, instead of a correct unberthing procedure. This is then the reason for the fact that the unberthing condition is often the most severe condition for a crabbing ship. Based on model tests it was found that the most critical situation is the situation with the ship located at half its breadth from the quay.

9   CONCLUSIONS

A summation of some common practice crabbing criteria is given. Several criteria are related to the specific target operational area of the ship, some are general criteria.

A recapitulation is given of a calculation technique to assess the crabbing capability, using elementary thruster, propeller and rudder forces.

Although in literature, reference is given to the different interaction processes, it is not quantified how large the impact of the interactions is. In this paper, the significance of the interactions is shown, based on an example ship for which calculations are compared with measurements. Apparently, neglecting the interactions leads to erroneous conclusions. Therefore, more detailed calculation procedures or model tests have to be carried out to predict the crabbing ability properly and reliably.

As the influence of quay and water depth are important for the interactions and hence for the crabbing ability, the distance to the quay and the water depth should be mentioned in the crabbing criterion.

REFERENCES

1. Hooren, C.M. van and Huisman, J.M.; Crabbing performance of ferries. 6th Lips symposium, Drunen, 1985
2. Barr, R. et. al.; Technical basis for manoeuvring performance standards. Hydronautics Inc. Report CG-M-8-81, December 1981
3. Raven H.C. et. al.; A new hull form for a Venice urban transport waterbus: design, experimental and computation optimization. PRADS '98, The Hague.
4. Kristensen, H.O., Chislett, M.S. and Schilder, A.; Low speed manoeuvring of ferries - problem identification, solutions and service experience. Cruise and ferry 1993
5. Fabietti, V., Payne, S.M., Elzinga, T. and Rem, A.; Ship handling simulators for optimisation of manoeuvring strategies of cruise ships in ports. Cruise and Ferry 1989
6. Nienhuis, U and de Joode, W.; Handling of Roroships and ferries in ports. RoRo '90, Trieste 1990.
7. Brix, J.; Manoeuvring Technical Manual. Seehafen Verlag GmbH. Hamburg, 1993.
8. Kuiper, G.; The Wageningen B-screw series. MARIN publication 92-001.
9. Inoue, S., Hirano, M, Kijima, K, and Takashina, J.; A practical calculation method of ship manoeuvring motion. International Shipbuilding Progress, Vol. 28, No. 235, Sept. 1981 pp. 207-222
10. Hooft, J.P. and Nienhuis, U.; The prediction of the ship's manoeuvrability in early design stage. SNAME Transactions 1994, Vol. 102, new York, 1995
11. Dallinga, R.P., Blok, J.J. and Luth, H.R., Excessive rolling of cruise ships in head and following waves. RINA Conference on ship motions and manoeuvrability, February 1998
12. Hooft, J.P.; The cross flow drag on a manoeuvring ship. Ocean Engineering, Vol. 21, No. 3, 1994
13. Jonk, A. and Rem, A.; Hydrodynamic optimization in ship design with regard to manoeuvrability. PRADS '89, Varna, Bulgary, 1989
14. Oltmann, P. and Sharma, S.D.; Simulation of Combined Engine and Rudder Manoeuvres Using an Improved Model of Hull-Propeller-Rudder Interactions. Proc. 15th ONR Symp. on Naval Hydrodynamics, 1984, pp. 83-108
15. Janke-Zhao, Y. Manoeuvring motion simulation of twin screw ships. Schiffstechnik Bd. 41, 1994 / Ship technology research, Vol 41, 1994

Practical Design of Ships and Mobile Units
M.W.C. Oosterveld and S.G. Tan, editors.

655

Improvement in resistance performance of a barge by air lubrication

Jinho Jang , Hyochul Kim [a] and Seung-Hee Lee [b]

[a]Department of Naval Architecture and Ocean Engineering, Seoul National University, Seoul, 151-742, Korea

[b]Department of Naval Architecture and Ocean Engineering, Inha University, Inchon, 402-751, Korea

Improvement in resistance performance of a barge by air lubrication has been sought numerically and experimentally. Various shapes of the air supplying nozzles are tested to find the optimum configurations. It is found that air cavity length is sensitive to the slope of the nozzle and height of the slope also plays an important role. Increase in height and slope of the nozzle, however, increases the form resistance and may cause increase in the total resistance. In the present study, careful selection of the air supplying device and conditions have achieved about 10% reduction in the total resistance of the barge with a single air nozzle. Further reductions in resistance are possible with the present method. However, more refinement in experimental and numerical method is necessary before used for resistance reduction of a practical ship.

1. INTRODUCTION

Researches to reduce frictional components of ship resistance have been widely performed in these days and reported to have achieved a certain extent of success. Among those, the approaches in which two phase flow, a mixture of micro air bubbles and the water, is injected into the boundary layer on the hull surface to effectively lessen fluid viscosity received special scientific attention for their effectiveness in reducing frictional drag. It has been reported that up to 80% of the frictional resistance of a flat plate can be reduced by the method.[1,2]

However, if the identical method is to be applied to the drag reduction of a general three dimensional body such as a ship whose curvature is not negligible, characteristics of the flow field around it should be understood before effectively supplying air into the boundary layer. Since air injection into the boundary layer of a conventional ship may reduce the frictional drag but, in some cases, excessive increase in form drag cause increase in to-

tal resistance.[3,4]

Use of a stepped hull and formation of an air cavity behind the step can resolve the problem.[5,6,7,8,9] The air cavity formed right behind the step can smoothen the flow around and so reduce the form drag. The local friction of the region covered by the cavity will be reduced as well. Kim et.al[5] reported 20% reduction in the total resistance of a planing hull has been achieved accordingly. Similar progress has been achieved by many researchers worldwide and in some cases it has been already applied for practical purposes.

The aim of the present study is to improve resistance performance of a practical ship with exploitation of the recent achievements in the relevant researches. A barge self-propulsive with outboard engines in inland waters has been selected for the purpose and air is supplied under the flat bottom of her through the attached nozzles to reduce the total resistance. Experimental results indicate that improvement in resistance characteristic of the barge can be attained with proper choice of parame-

Figure 1. Body plan of the selected model

ters related to nozzle shape, location and *etc*.

2. MODEL

The model ship adequate for the present experimental investigation has to satisfy the following conditions.

1. To yield large reduction in frictional resistance and to efficiently supply air, hull form should have a flat bottom with large breadth and small draught.

2. To produce larger air cavity, the ship should have small changes in hull attitudes and relatively high design speed.

3. To sustain air cavity steadily, the ship should have small motion responses and operate in relatively calm water.

The hull form shown in Fig.1 has been selected as the one satisfying the above three conditions. As shown in the Fig.1, the ship has a large flat bottom area, a small deadrise and a small draught comparing to its breadth. The hull form changes behind the station 2 to ensure propulsion performance when operating with an outboard engine at the full loading condition.

The principal dimensions of the ship are listed in Table 1.

3. AIR SUPPLYING NOZZLE

3.1 Location of nozzle

To maximize the reduction in resistance of a ship, an air supplying device should be designed to generate the largest air cavity possible and to minimize the resulting increase in form resistance. For the purposes nozzles for

Table 1. Principal Characteristics of the Ship

	Ship	Model
LOA(m)	55.0	2.2
LBP(m)	53.0	2.12
Breadth$_{MLD}$(m)	10.0	0.4
Depth$_{MLD}$(m)	2.0	0.08
Disp. Vol. (m^3)	396.735	0.025393
WSA(m^2)	506.875	0.811
Draft(m)	1.15	0.046
LCB(m):f$^+$	-1.645	-0.0658
C_B	0.6561	0.6561
Scale Ratio, λ		25

Figure 2. Limiting streamlines on the bottom of bare hull(V_M=1.543 m/s)

air supply should be attached normal to the flow direction and flow characteristics around the selected ship should be understood.

It is necessary to limit the heights of the nozzles so as to confine them inside the boundary layer to reduce increase in form resistance. Then, useful information for determining the locations of air supplying nozzles can be found by inspecting limiting streamlines traced in paint tests.

Fig.2 shows limiting streamlines for the model speed of 1.543 m/s which corresponds to the design speed of 15 *knots*. In the figure, limiting streamlines become almost parallel to the bottom at the region downstream of the station $8\frac{1}{2}$. Considering this fact and that parallel middle body begins from the station 6 as shown in Fig.1, the location of the air supplying nozzle is chosen to be the station 6.

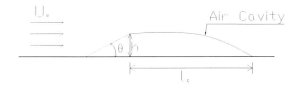

Figure 3. Flow around a nozzle

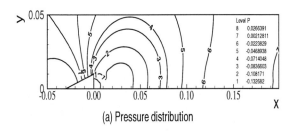

(a) Pressure distribution

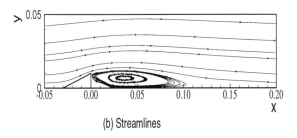

(b) Streamlines

Figure 4. Pressure contour and streamlines without bubble model($\theta = 16.6^o$, h =10 mm)

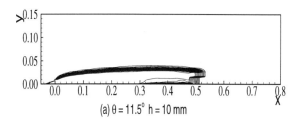

(a) $\theta = 11.5^o$ h = 10 mm

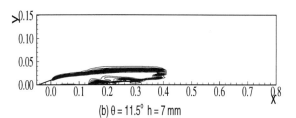

(b) $\theta = 11.5^o$ h = 7 mm

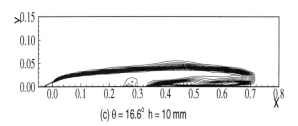

(c) $\theta = 16.6^o$ h = 10 mm

Figure 5. Void fraction contour

3.2 Shape of Nozzle

3.2.1 Numerical simulation

Numerical computation is performed to determine the optimum configuration of the nozzle. Air nozzle is approximated as a three dimensional wedge of unit thickness attached on the flat plate as shown in Fig.3.

Thee Navier-Stokes and continuity equations are solved with a finite volume method and pressure field is found by a MAC method.

Air cavity is assumed to be represented as a mixture of water and air bubbles. Then, Rayleigh's equation is used to account for the growth and collapse of the air bubbles as in computing cavitating hydrofoils[10].

Fig.4 shows the pressure distribution and streamlines around the wedge when air is not supplied. Fig.5 show the changes in air cavity length for different slopes and heights of air nozzle. It is found that the length of the cavity is most sensitive to the slope of the nozzle but if slope is identical it also depends on height. The results are in good coincidence with the experimental ones qualitatively.

3.2.2 Design of nozzle

In accordance with basic understandings for the numerical results, nozzles of various heights and slopes are manufactured and attached at the station 6.

Without supplying air, increase in total resistance caused by the presence of the nozzles are measured. And then, changes in cavity

Table 2. Performance of Various Nozzle
($V_M = 1.543m/s$, $Q_{Air} = 100l/min$)

No.	h(mm) $\theta(^o)$	$\frac{R_{TM,A}}{R_{TM,NA}}$	$\frac{R_{TM,A}}{R_{TM,B}}$	l_C (mm)	Strip
I	17 20	0.85	1.27	150	no
II	10 11.5	0.95	1.06	110	no
III	10 16.6	0.92	1.08	130	no
		0.83	0.98	340	yes
IV	7 11.5	0.94	1.01	90	no
V	10 30	0.92	1.14	130	no

NA : w/ nozzle & No Air, *A* : w/ Nozzle & Air
B : Bare hull

length and reduction in total resistance due to air injection are measured for various speeds, rates of air supply and nozzle shapes. Table 2 shows the results of measurement at the model speed of 1.543 m/s in which the higher nozzle generate the longer air cavity but experiencing higher increase in total resistance. The slopes of the nozzles have similar effect and the reduction in frictional resistance due to the air cavity cannot compensate increase in the total resistance in most cases. It necessitates employment of an air nozzle which has the least slope and height practicable.

Fig.6(a) shows the limiting streamlines near the hull surface when nozzle is attached. It shows formation of a strong vortex right behind the nozzle. This explains why increase of air flow above a certain limit does not increase the length of cavity further. Then, the excessive air will leak through the vortex core at the both end of the nozzle. Three dimensional effects are also apparent near the both ends of the nozzle which eventually shrink the reattachment length.These facts suggest necessity of end strips at the sides to prevent air from escaping and so increase cavity length.

Flow measurements in a suddenly expanding rectangular duct[11] also indicate that suppression of three dimensional phenomena at the ends of the step increases reattachment length and weakens secondary vortex downstream.

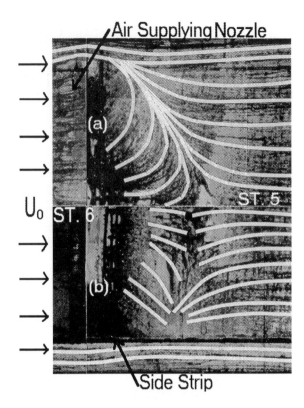

Figure 6. Limiting streamlines around air nozzle: (a) without strip (b) with strip

stream. Fig.6(b) shows limiting streamlines for the case where reattachment lengths are significantly increased and secondary vortex weakened.

As shown in Table 2, nozzle III , the cavity length and the reduction in resistance are considerably increased if end strips are employed at the both ends.

4. RESISTANCE REDUCTION WITH AIR

Based on the numerical and experimental results discussed above, a nozzle of $10mm$ height and 3° slope angle is selected for the final configuration and attached at the station 6 to occupy 80% of the breadth of the model as shown in Fig.8.

Fig.8(a) shows resistance reduction and air cavity length for various speeds when air is supplied at the fixed rate of 100 l/min. The faster the model speed become, the longer air

Figure 7. Final configuration of air supplying nozzle

Table 3. Measured Pressure & Skin Friction (V_M=1.543 m/s)

x	C_p		C_f	
(mm)	No Air	Air	No Air	Air
-107	-0.049	-0.058	2.969E-3	2.781e-3
13*	-0.115	-0.016	-	-
161*	0.016	-0.033	1.690E-3	2.508E-4
321*	0.033	-0.066	2.655E-3	2.508E-4
482	-0.082	0.082	3.466E-3	3.590E-3

x : Distance from ST 6(*mm*)

* : location inside air cavity(il_c= 385 *mm*)

cavity generated. Fig.8(b) shows changes in total resistance for a range of air pressure at the model speed of 1.543 it m/s which indicate that cavity length is insensitive to the air pressure. Chanson[12] found similar results and it is possible to minimize the power necessary for air supply by selecting low air pressure. Furthermore, it is found that at the rate of 50*l/min*, relatively large reduction in the resistance occur as shown in Fig.8(c) where air pressure maintained at 0.04 *MPa* and air flux changes. From the above discussions, conditions for air supplying has been selected as 0.04 *MPa* of air pressure at 50 *l/min* of air supply rate.

Results for the measurements with the final configurations and conditions as stated above are shown in Fig.9. Considerable reductions in total resistance are obtained as expected. The shapes of air cavity for different speeds are presented in Fig.10.

It is also found that the size of air cavity is directly related to the amount of reduction

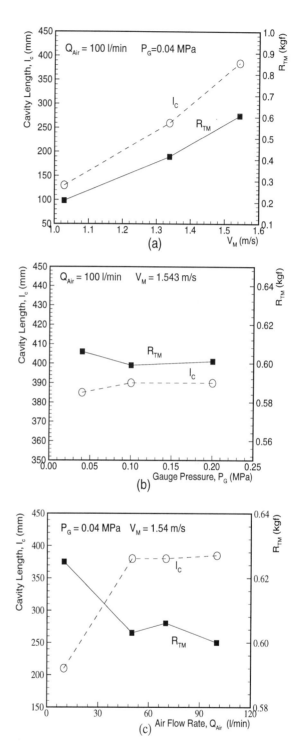

Figure 8. Variations of air cavity length & R_{TM}

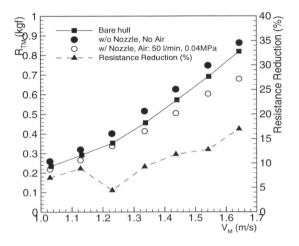

Figure 9. Resistance reduction due to air cavity

in resistance. The reductions occur mainly at the regions covered by the air cavity and presumed that main contribution comes from reductions in viscous resistance since supplied air does not affect wave making resistance significantly.

To isolate the main cause of resistance reduction, pressure distribution is measured with Preston tubes at the holes aligned along the centerline and the result is shown in Table 3 and Fig.9. The results show that pressure changes occur in the region covered by the air cavity but the regions are located at the parallel middle body and so they cause no significant influences on the pressure resistance. Table 3 also presents the surface friction measured with Preston tubes which rapidly decrease in the region covered by the cavity.

5. CONCLUSIONS

Experiments and numerical simulations are performed to select the optimum configuration of air supplying device and conditions for air supply. The selected device is attach to the bottom of the model and reductions in total resistance are measured for a range of model speed. The following facts are found as a result:

1. About 10% of the total resistance is re-

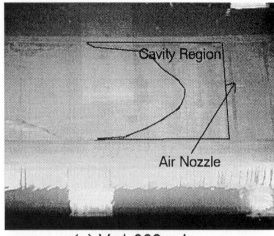

(a) V=1.029 m/s

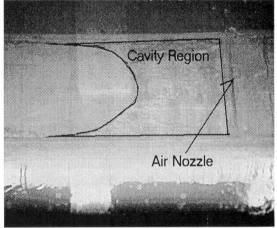

(b) V=1.338 m/s

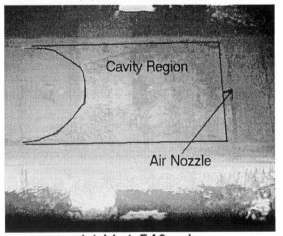

(c) V=1.543 m/s

Figure 10. Shapes of air cavity

duced with formation of air cavity with carefully selected shape of the nozzle and conditions such as pressure and flow rate of supplying air.

2. The length of an air cavity is sensitive to the slope of the nozzle. If the slopes are identical then higher nozzle generates longer air cavity. However, the increase in slope and/or height of the nozzle generally accompanies increase of the form resistance. And reductions in total resistance are not always achievable.

3. The faster the ship moves, the longer air cavity is generated. Cavity length is proportional to the speed squared.

4. The side strips attached at the both ends of the nozzle enhance air cavity formation by preventing air from escaping and by suppressing cross flow components.

5. The numerical model adopted to represent the air cavity yield reasonable results at least qualitatively but more improvement is necessary before being used for design purposes.

6. Improvement in air supplying mechanism will be necessary to reduce the resistance further. Influence of the hull attitudes upon the formation of air cavity and effect of air bubbles upon the performance of propulsor should be investigated.

ACKNOWLEDGMENT

The present research has been supported partly by KOSEF under grant of 97-02-00-01-01-3 and by internal research fund of Inha University in 1997.

REFERENCES

1. N.K.Madavan, S.Deutsch, and C.L.Merkle, Measurement of Local Skin Friction in a Microbubble-Modified Turbulent Boundary Layer, J. Fluid Mechanics, vol.156, 1985

2. M.M.Guin, and H.Kato, Reduction of Skin Friction by Microbubbles and Its Relation with Near-Wall Bubble Concentration in a Channel, J. Marine Science and Technology, 1996

3. G.Yim, and H.Kim, On the Variation of Resistance Components due to Air Bubble Blowing on Bulb Surface of a Ship, J. SNA Korea, Vol.33, No.1, 1996

4. Y. Doi, K. Mori and T. Hotta, Frictional Drag Reduction by Microbubbles, J. SNA Japan, Vol.170, 1991

5. G.Kim, H.Kim, and S.Lee, Effects of Air Injections on the Resistance Reduction of a Semi-Planing Hull, J. Hydrospace Technology, Vol.2, 1996

6. B.Han and H.Kim, A Study on the Reduction of the Frictional Resistance of a Bottom Plate by Air Supply, MS Thesis, Seoul National University, 1998

7. S.Go, B.Han, H.Rhyu, and H.Kim, Formation of Air Cavity on the Bottom of Stepped Semi-Planing Boat and its Effects on Resistance, Proc. CKMHM'97, Shanhai, China, 1997

8. A.A.Butuzov, A Review of the State-of-the-Art in Air Lubricated Bottom Ship Development, Report No. KSRI/STEPI-03-F, Krylov SRI, March, 1997

9. A. N. Ivanov, V. G. Kalyuzhny and A. N. Pavlenko, Problem of Hydrodynamic Resistance Reduction by Artificial Gas Cavities on the Vessel's Hull, Krylov SRI, 1994

10. A. Kubota, H. Kato and H. Yamaguchi, A New Modelling of Cavitating Flows; a Numerical Study of Unsteady Cavitation on Hydrofoil Section, J. Fluid Mechanics, 1992

11. G.papadopoulos and M. Ötügen, Separating and Re-attaching Flow Structure in a Suddenly Expanding Rectangular Duct, J. Fluids Engineering, Vol.117, March, 1995

12. H.Chanson, Air Bubble Entrainment in Free Surface Turbulent Shear Flows, Academic Press, 1996

Hydrodynamic Design of Integrated Propulsor/Stern Concepts by Reynolds-Averaged Navier-Stokes Techniques

Rich Korpus[a], Bryan Hubbard[a], Paul Jones[a], Chel Stromgren[b], and James Bennett[c]

[a] Science Applications International Corporation, Ship Technology Division
134 Holiday Court, Suite 318, Annapolis, Maryland 21401 USA
[b] Newport News Shipbuilding
Commercial Ship Eng., E56, Bldg. 600/1, Newport News, VA USA
[c] Bath Iron Works Corporation
46 Church Road, Brunswick, Maine 04011 USA

A new tool has emerged to assist ship designers with the difficult task of propulsor/stern integration. Viscous flow computational methods, particularly Reynolds-Averaged Navier-Stokes (RANS) techniques, have left the realm of academics and entered practical service. When used in conjunction with traditional towing tank tests, the new capability has the potential to greatly improve the design process. This paper presents a number of practical examples to demonstrate this potential. The first demonstrates the effect of stern shape modifications on propeller inflow for a traditional single screw product carrier. The second utilizes RANS to investigate the efficiency of a podded propulsion system with tractor propeller for a high speed "wave piercer" hull form.

1. INTRODUCTION

Proper integration of hull and propulsor is essential for both the operating efficiency and vibratory behavior of a given design. Naval architects have traditionally utilized empirical techniques for this integration, and often use model testing to find the "effective wake" or "thrust deduction" needed by the propeller designer. Unfortunately, model tests require scaling assumptions, and only account for the presence of a propeller in very general ways.

Computational alternatives to model tests can circumvent these limitations by including the propeller's presence in simulations performed at full scale Reynolds numbers. Since propulsors are embedded in the hull boundary layer, however, any selected computational approach must rely at least on fully three-dimensional, Reynolds-Averaged Navier-Stokes (RANS) techniques.

While RANS techniques have traditionally remained limited to university or government research applications, recent improvements to their efficiency and reliability now allow them to meet the demands of ship design cycle time frames. This is not to say that RANS will replace the towing tank, but rather that the ship design process can benefit by the intelligent application of both. One advantage of RANS is that less lead time is required than model tests, and a greater number of design alternatives can be investigated. RANS can also help guide the design process because both local flow and global measures of merit result from the solution.

Although RANS does reliably trend different design alternatives, it lacks sufficient fidelity for quantifying resistance of a final design. Tow tank testing is still required for this crucial step. An application where RANS can assist the tanks is for scaling of test results to full scale.

This paper is intended to serve two functions: 1) introduce the RANS system employed for hydrodynamic assessment at SAIC; and 2) demonstrate RANS on two practical applications. Development of the SAIC RANS system was initiated by more than ten years of Department of Defense support, and is heavily utilized for a number of submarine technology programs. It is unique, however, in that its later development has been driven mostly by industry, and has therefore become a practical design analysis tool. The applications presented are actual jobs performed for Newport News Shipbuilding and Bath Iron Works, and have been selected to demonstrate the utility of RANS for ship design.

2. APPROACH

The time-dependent viscous flow solutions presented in this study were obtained by solving the incompressible RANS equations in conjunction with a $k\varepsilon$ turbulence model. When non-dimensionalized by a characteristic length L, velocity V_0, and density ρ, the Cartesian form of these governing equations can be written

$$\nabla \bullet \mathbf{V} = 0 \tag{1}$$

$$\frac{\partial \mathbf{V}}{\partial t} + \mathbf{V} \bullet \nabla \mathbf{V} + \nabla p - \frac{1}{\mathrm{Re}} \nabla^2 \mathbf{V} - \nabla \bullet \tau = \mathbf{F} \tag{2}$$

$$\frac{\partial k}{\partial t} + \mathbf{V} \bullet \nabla k - (\frac{1}{\mathrm{Re}} + v_t)\nabla^2 k - P + \varepsilon = 0 \tag{3}$$

$$\frac{\partial \varepsilon}{\partial t} + \mathbf{V} \bullet \nabla \varepsilon - (\frac{1}{\mathrm{Re}} + \frac{v_t}{1.3})\nabla^2 \varepsilon - \frac{\varepsilon}{k}(c_{\varepsilon 1} P_{sol} + c_{\varepsilon 3} P_{irr})$$

$$+ c_{\varepsilon 2} \frac{\varepsilon^2}{k} = 0 \tag{4}$$

where the Reynolds stress components τ_{ij} are defined by the Boussinesq approximation.

$$\tau_{ij} \equiv -\frac{2}{3}k\delta_{ij} + v_t S_{ij}, \qquad S_{ij} = \frac{\partial u(i)}{\partial x^j} + \frac{\partial u(j)}{\partial x^i} \tag{5}$$

and $[x^1, x^2, x^3]^T$ represents the Cartesian position vector. $\mathbf{V}$ represents the Cartesian velocity's vector $[u(1), u(2), u(3)]^T$, p the pressure, k the turbulent kinetic energy, and ε the turbulent dissipation rate. $\mathbf{F}$ is an arbitrary body force used to represent propulsor effects. The quantity v_t is defined as the linear eddy viscosity $.09\ k^2/\varepsilon$, and Re the Reynolds number LV_0/v. The rate of production of k is represented by P, and production in the ε equation has been split into its solenoidal and irrotational components following Hanjalic and Launder (1980):

$$P = P_{sol} + P_{irr} \tag{6}$$

$$P_{sol} = 4\left[S_{12}^2 + S_{13}^2 + S_{23}^2 \right] \tag{7}$$

$$P_{irr} = 2\left[S_{11}^2 + S_{22}^2 + S_{33}^2 \right] \tag{8}$$

The modeling coefficients $C_{\varepsilon 1}$, $C_{\varepsilon 2}$, $C_{\varepsilon 3}$ are taken as constants set equal to $(1.44, 1.92, 2.4)$, respectively.

The usual near-wall stiffness problem associated with Eq.(4) has been circumvented herein by using the two-layer approach of Chen and Patel (1988,1989). The approach utilizes the $k\varepsilon$ model outlined above for most of the flow field, but a one-equation kl model in the viscous sublayer and buffer zone. Switching between ε and l dissipation models is performed automatically when the wall Reynolds number $\mathrm{Re}_{wall} = \mathrm{Re}\sqrt{k}\delta$ (δ being the nodimensional distance to the closest wall) becomes less than 300 (Chen and Korpus, 1993). Details of the l dissipation model can be found in Chen and Patel (1989) and will not be repeated here.

Computations about complex geometries requires that Equations (1) through (4) be first transformed into body fitted coordinates. This is accomplished by defining a curvilinear system (ξ^1, ξ^2, ξ^3) such that any physical boundary in the domain coincides with one or more surfaces of constant ξ^i. The system need not necessarily be orthogonal, and can be defined separately for each block of the grid. All independent variables in Equations (1) through (4) are then transformed to (ξ^1, ξ^2, ξ^3) space by chain rule while leaving the dependant variables Cartesian.

With the equations in their curvilinear form, the discretization is accomplished by linearizing each equation, over a computational element, then solving analytically by separation of variables. Evaluation of the analytic solution at the interior node of a computational element provides a stencil for the center point in terms of its nearest neighbors. Time derivatives are handled by the Euler implicit method, and unknowns from the previous time step are lumped into the source term. The resulting implicit system equations is solved by the alternating direction implicit (ADI) method in each cross-flow plane, and then swept repetitively in the streamwise direction. Detailed expressions for the coefficients of the finite-analytic stencil can be found in Chen et al. (1990).

Pressure coupling is supplied using a modified SIMPLER/PISO algorithm (Chen and Patel, 1989) that uses the strong conservation from of Eq. (1)

$$\frac{\partial \sqrt{g} U^1}{\partial \xi^1} + \frac{\partial \sqrt{g} U^2}{\partial \xi^2} + \frac{\partial \sqrt{g} U^3}{\partial \xi^3} = 0 \tag{9}$$

where U^i is the contravariant velocity $U(1)\partial \xi^1 / \partial x^i +$

$U(2)\partial\xi^2 /\partial x^i + U(3)\partial\xi^3 /\partial x^i$ and g is the determinant of the covariant fundamental metric tensor. The technique defines pseudo-velocities from the discretized form of Eq.(2) as:

$$U^i = \hat{U}^i + E^{ii} \frac{\partial p}{\partial \xi^i} \qquad (10)$$

where $\hat{U}^i$ and E^{ii} necessarily involve the finite-analytic coefficients, and will not be repeated here (see Chen and Korpus 1993).

A Poisson equation for pressure is then derived by substituting equation (10) into (9), and discretizing using central differences. By centering the differences in equation (10) on staggered grid locations, the resulting Poisson equation involves unknowns only at the grid nodes. The technique is thus unique in that it does not require staggered velocity and pressure variables.

To facilitate calculations around complex or moving geometries, the discrete forms from Eqs. (1) through (4) are embedded in a Chimera, multi-block environment. The solver works on one block at a time, and the only grid connectivity requirement is that the union of blocks spans the entire computational domain. Individual blocks are allowed to overlap arbitrarily, and inter-block communication is handled by conservative triquadratic interpolation. The overall approach has been extensively validated for both steady and unsteady three-dimensional applications (Korpus, 1995; Chen and Korpus, 1993; Weems and Korpus, 1994).

The advantages of Chimera grids are well demonstrated. Figure 1, which shows the tanker stern to be used as the first demonstration case. While the complete mesh is three-dimensional, the figure shows only these planes coincident with the hull surface, the water surface, and the hull centerplane. Note that each piece of the stern can be resolved to whatever level is locally required, and that blocks of extra resolution can be easily pieced in when needed (e.g. in way of the propeller)

It is this use of Chimera grids that has greatly increased the practicality of RANS. The fully 3D hull grid shown in Figure 1 can now be obtained in just a few days, as compared to conventional grids that could require many man-months.

Propelled simulations are performed by adding a non-zero body force vector **F** in Equation (2) to represent propeller acceleration and swirl. Both propeller design studies and propeller analysis studies are possible. In the former a body force designed to represent the load distribution chosen by a designer is input into the RANS code, and the solution allowed to converge. The analysis case (i.e. where a propeller geometry is known, but not its load), is more complex. In that case, a vortex lattice propeller code is used to compute the body forces for an assumed inflow. The RANS code is then run using these forces for a small number of iterations. Since the body forces affect the inflow, the vortex lattice code must be re-run and the process iterated to convergence.

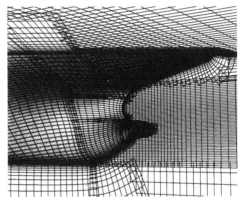

Figure 1: Chimera Grid Around Stern of Newport News Tanker

The entire described system, including Chimera grid scheme and propeller force method, has been integrated into an automated package to minimize user input. Coding is ANSII standard FORTRAN and C, and is vectorized for efficiency on large supercomputers. Simulations up to about 2.5 million grid points can be run on workstations while larger jobs typically run on CRAY supercomputers. Run times are in the order of 70 μsec/iteration/cell on a 2 processor Silicone Graphics Inc. (SGI), OCTANE, and 14 μsec/iteration/cell on a CRAY C90.

3. APPLICATIONS

Application of the described method will be demonstrated using calculations performed to support actual ship design processes. The first example describes a comparison of two preliminary hull form candidates for Newport News Shipbuilding's "Double Eagle" product tanker hull form. The second describes an investigation of a podded propulsor option for Bath Iron Works' high speed "Wave-Piercer" hull form. Both examples focus on propeller/hull interaction simulations, and were chosen to demonstrate the utility of RANS for supporting practical design decisions.

3.1. Newport News Shipbuilding Hull Study

At one point during preliminary design of its "Double Eagle" product carrier, Newport News Shipbuilding (NNS) entertained a customer request to investigate low construction cost options to its existing stern design. One option called for maximizing the use of flat and rolled plate (as opposed to formed) in the stern. Calculation of life-cycle costs for such an option requires that the fuel efficiency of the "developable" stern be compared to the baseline design. Newport News decided to use RANS for quantifying these differences.

The two candidate stern forms are shown in Figure 2. The hull labeled NNS Model #333 represents the baseline case for comparison, whereas Model #334 represents the developable option.

Unpropelled calculations were made in each case to study the effect on nominal wake, and to support a rough propeller design for quantifying efficiency. Each hull was gridded using a similar mesh topology and about 450,000 points. Propelled

calculations were also made for model #333 to compare with experiments. A grid of 1,300,000 points was used in that case because both sides of the boat are included, and because a high resolution patch (about 300,000 points) was added in way of the propeller (Figure 1). Grid generation using the Chimera scheme described above required about three weeks turn-around time.

All runs were started from uniform flow, and required approximately 2000 iterations to converge. Run times on an SGI Octane workstation are approximately 38 CPU hours for the unpropelled cases and 100 CPU hours for the propelled one. A model scale Reynolds number of 9.1 million was used to facilitate comparison to experiment, but full scale calculations can be performed at the cost of additional iterations. The propeller was loaded radially similar to B-screw series distribution, and chordwise using a sine distribution. The propeller's J, Kt, and Kq are 0.733, 0.1811, and 0.0237 respectively.

Sample post-processed results are given in Figures 3 through 6. Figure 3 shows near-surface streamlines similar to what would be seen in an experimental paint smear, and demonstrates that Model #334 exhibits greater separation than Model #333. Separation on the baseline hull is limited to very near the deadwood and under the bossing, but on the developable stern extends further forward and up.

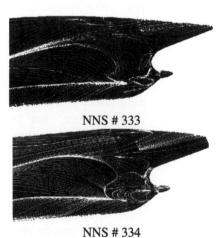

NNS # 333

NNS # 334

Figure 3: Streamlines for Newport News Model # 333 and # 334

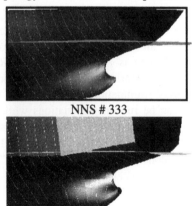

NNS # 333

NNS # 334

Figure 2: Comparison of Newport News Models # 333 and #334

The strong "streamline attraction" lines shown in each figure indicate the attachment point of the bilge vortices.

Figure 4 shows computed versus measured wake fractions in the plane of the propeller. Note that the measurements do not extend to the hull surface, but that the comparison is fairly good in the outer regions. The only significant difference seen is directly under the bossing where measurements show steeper gradients than calculations.

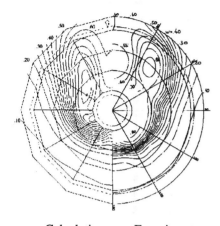

Calculation   Experiment

Figure 4: Comparison of Computed and Experimental Propeller Plane Nominal Wakes.

A comparison of computed to measured forces shows that drag prediction is off by about 10%, but this is fairly typical for streamlined bodies with little drag. While RANS can reliably trend design alternatives, it can not yet compete with the towing tank for accurate final drag and powering numbers.

Figure 5 also shows wake fractions, but for a comparison between Model #333 and #334. It is immediately obvious that the differences are greater than those observed in Figure 4, and that Model #334 has a significantly greater wake. This result is also borne out by a comparison of computed drags which show Model #334's Cd to be 7% greater than Model #333's. The resulting increase in fuel bills over the life of the ship were computed to more than offset construction savings, and the customer was convinced not to pursue a developable stern option.

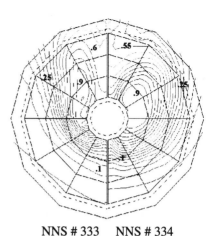

NNS # 333   NNS # 334

Figure 5: Comparison of Models # 333 and #334 Propeller Plane Nominal Wakes.

Figure 6 shows the streamwise velocities of a propelled Model # 333, at the propeller plane. When compared to the left sides of figures 4 and 5, the contours provide a graphic example of how propeller inflow cannot be derived from nominal wakes using simple extrapolation techniques. True effective wakes can be derived from RANS, however, by subtracting propeller potential flow velocities from the data shown in Figure 6. The resulting effect on propeller design will be substantial.

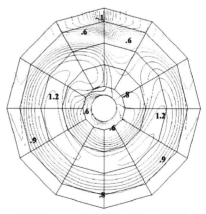

Figure 6: Streamwise Velocities of Model #333

668

3.2. Bath Iron Works Podded Propulsor Analysis

The "Wave-Piercer" hull forms now appearing hold great promise for both commercial and naval applications. Bath Iron Works Shipbuilding (BIW) plans to tap this potential, and has begun a program aimed at coupling Wave-Piercer hulls with advanced non-conventional propulsion systems. One option found worthy of detailed study is a twin installation of podded propulsors utilizing tractor propellers. The lack of historical experience to judge such new designs, however, has caused BIW to call on RANS for ranking the merits of such options. Particular concerns for the proposed pod combination include verifying the system's efficiency, and demonstrating that the prop wash will not stall the strut.

RANS simulations were therefore initiated for a commercially available motor pod geometry including strut, fillet, and lower stabilizer fin. The pods were aligned on an idealized hull surface, and yawed about their rotation axis to a two degrees bow out attitude. The inboard turning tractor propellers are represented by three-dimensional body forces, but only the starboard pod is included because of symmetry. Figure 7 shows the resulting geometry.

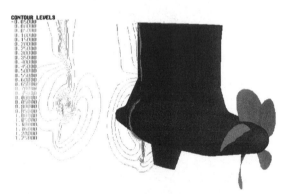

Figure 7: Contours of Axial Velocity for Podded Propulser with Tractor Propeller.

The computational grid for this case required approximately 800,000 points, and about two weeks grid preparation time. A model scale Reynolds number of 0.9 million (based on pod length) was used to facilitate comparison to experiment. Convergence was achieved in about 60 CPU hours on an SGI Octane workstation, and additional

solutions (attitudes or speeds) with the same geometry can be generated in just a few days.

Sample results are shown in Figures 7 through 9. Figure 7 depicts contours of axial velocity in two planes perpendicular to the shaft axis. The first cut is taken just downstream of the stabilizer fin, and shows the fin tip vortex rolling up to starboard (i.e. the fin is lifting away from the ship centerplane).

Computed side force coefficients on the strut and fin (based on total wetted surface) are 0.0012 and 0.0069 respectively, with both being to starboard. This is indicative that the current pod attitude is not an optimal one. The main advantage of a tractor configuration is that the strut and fin can recapture rotational energy and convert it to thrust, thus improving total propulsive efficiency. This requires that both strut and fin lift in the direction of propeller rotation, and to approximately equal magnitudes. The current results indicate that the strut operates opposite to the propeller. Decreasing the pod yaw would change the strut lift back towards the centerline, and simultaneously decrease the fin's net lift to starboard. Fin redesign may be needed to achieve a side force balance between strut and pod.

Figure 8 shows off-body streamlines beginning upstream of the pod, and clearly demonstrates the propeller induced swirl.

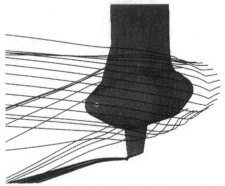

Figure 8:Off-Body and Stabilizer-Tip Streamlines for Podded Propulser with Tractor Propeller.

The very strong fin tip vortex indicated confirms the side-force overloading of the fin mentioned above, and closer examination of the detailed surface flow reveals the fin to be partially stalled. Computed drag indicates the fin contributes 14% to the total even though it comprises 7% of the wetted surface.

Although the design is not necessarily optimal with the current propeller, Figure 9 indicates that the system is recapturing some of the rotational energy in the propeller race.

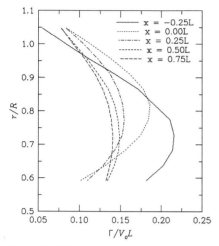

Figure 9: Circulation Behind Tractor Propeller

The figure shows distributions of total wake circulation (swirl) versus radius at five locations downstream of the propeller. The location labeled x=-0.25 is very near the propeller trailing edge, and indicates the total swirl induced in the flow. The curve labeled x=0 is at the pod yaw axis, and indicates that maximum circulation has already been reduced 20%. The remaining contours are all downstream of the strut, and indicate that approximately 40% of the propeller rotational energy has been recaptured.

4. CONCLUSIONS

Recent progress in RANS development has finally matured these complex methods into practical ship design assessment tools. Chimera gridding schemes combined with robust algorithms and turbulence models now enables complete propeller/hull flow simulations within practical design cycle turn-around times. When combined with the latest computers, RANS analyses for complete ship cases can be run on desktop workstations thereby bringing total analysis costs in line with model tank tests.

RANS applications to the test cases described herein have demonstrated the potential for these tools to greatly improve the overall design process. The efficiency and vibration performance of single screw propellers, for example, could be increased through the a-priori knowledge of effective wake demonstrated in the Newport News example. Strut efficiency for podded propulsors (or any tractor combination) could likewise be improved using the knowledge of race swirl depicted by the second demonstration case.

As demonstrated, RANS may be effectively used in the design process to assess complex stern/hull flow. The RANS system provides the designer a fast, inexpensive assessment tool, which compliments towing tank tests and improves the overall design process. Furthering the interaction between the ship design community and RANS developers will help utilize this assessment tool to the best advantage of the ship industry.

REFERENCES

Chen, H.C., and Korpus, R.A., 1993, "A Multi-Block Finite-Analytic Reynolds-Averaged Navier-Stokes method for 3-D Incompressible Flows." ASME *J. of Fluids Engineering.*

Chen, H.C., and Patel, V.C., 1988, "Near-Wall Turbulence Models for Complex Flows Including Separation," *AIAA Journal*, Vol. 26, No. 4, pp. 641-648.

Chen, H.c., and Patel, V.C., 1989, "The Flow Around Wing-Body Junctions," *Proceedings, 4th Symposium on Num. And Phys. Aspects of Aerodynamic Flows,* Long Beach, CA.

Chen, H.C., Patel, V.C., and Ju, S., 1990, "Solutions of Reynolds-Averaged Navier-Stokes Equations for Three-Dimensional Incompressible Flows," *J. of Computational Physics,* Vol. 88, No.2, pp. 305-336.

Hanjalic,K., and Launder, B.E., 1980, "Sensitizing the Dissipation Equation to Irrotational Strains," ASME *J. of Fluids Engineering,* Vol. 102.

Korpus,R., 1995, "Six Years of Progress Under the ARPA SUBTECH Program," SAIC Report No. 95/1143.

Korpus,R., and Falzarano, J.M., 1997, " Prediction of Viscous Ship Roll Damping by Unsteady Navier-Stokes Techniques," *J. of Offshore Mech. and Arctic Eng.*, vol. 119, no 2., pp. 108-113.

Weems, K., and Korpus, R., et al., 1994, "Near-Field Flow Predictions for Ship Design," *Proceedings, 20th Symposium Naval Hydrodynamics,* Santa Barbara, CA.

1998 Elsevier Science B.V.
Practical Design of Ships and Mobile Units
M.W.C. Oosterveld and S.G. Tan, editors.

Marine propeller hydroelasticity by means of the finite/boundary element method - a preliminary approach

Bogdan Ganea[a]

[a] Ship Hydrodynamics Department, ICEPRONAV-SA (Research and Design Institute for Shipbuilding)
Str. Portului, nr. 19A, 6200 Galați, ROMÂNIA

Marine propeller blade displacement under its own hydrodynamic load depends strongly on the blade skew. For the highly skewed propellers the blade displacement may affect the hydrodynamic characteristics (thrust, torque, rate). Moreover, the very dangerous hydroelastic unstable phenomena, like the flutter and the divergence, may occur, especially during the crash-stop or reverse operation. Using the finite element method (FEM) for the structural calculation and the boundary element method (BEM) for the hydrodynamic one, a mathematical model taking into account the blade flexibility is here presented. Quasi-static calculations are performed for the Boswell highly skewed propeller research series. Dynamic hydroelastic aspects are also briefly presented.

1. MATHEMATICAL BACKGROUND

Using FEM, the mathematical model taking into account the blade flexibility is:

$$[M_s]\{\ddot{d}\} + [C_s]\{\dot{d}\} + [K_s]\{d\} = \{F_H(t, \{d\})\} \quad (1)$$

$[M_S]$, $[C_S]$, $[K_S]$: structural inertial, dumping and stiffness matrices;
$\{F_H(t,\{d\})\}$: hydrodynamic force depending on time and blade elastic displacement.

Adopting the incompressible and inviscid fluid hypothesis, the mathematical model of the potential flow is the continuity equation with its boundary condition:

$$\Delta\varphi = 0\big|_{D'} \quad , \quad (2)$$

$$\vec{n}'(\vec{d})\vec{V}(\vec{d}) = \vec{n}'(\vec{d})\dot{\vec{d}}\big|_{S'} \quad . \quad (3)$$

φ : perturbation potential;
D' : fluid domain influenced by elastic displacement
S' : fluid domain surface influenced by elastic disp.;
n' : S' outward normal;
V : total velocity on S';
d : fluid domain surface elastic displacement.

BEM solves the equivalent 2nd kind Fredholm integral equation:

$$2\pi\varphi(\vec{y}) - \int_{S'} \varphi(\vec{x}) \frac{(\vec{x}-\vec{y})\vec{n}(\vec{x})}{|\vec{x}-\vec{y}|^3} dS' - \int_{S'_w} \delta\varphi_w(\vec{x}) \frac{(\vec{x}-\vec{y})\vec{n}(\vec{x})}{|\vec{x}-\vec{y}|^3} dS'_w = \quad (4)$$

$$= \int_{S'} \frac{\frac{\partial\varphi}{\partial\vec{n}}\big|_{\vec{x}}}{|\vec{x}-\vec{y}|} dS' \quad , \quad \vec{y}\in S' \quad ,$$

S'_w : elastically deformed propeller wake surface;
$\delta\varphi'_w$: wake surface associated potential difference.

The equations (2) , (3) are general but (4) is particularised to the lifting body case, because it contains the specific wake surface term.

It is expressed on the elastic deformed propeller surface. To simplify the calculation, it must express it on the initial (non-deformed) propeller surface but still taking into account its elastic displacement. Therefore the small elastic displacement hypothesis will be adopted.

The initial propeller surface S normal is:

$$\vec{n} = \vec{\xi}_1 \times \vec{\xi}_2 \quad , \quad (5)$$

ξ_1 , ξ_2 : distinct tangent directions on S.

Due to the elastic displacement, ξ_1, ξ_2 become:

$$\begin{aligned}\vec{\xi}'_1 &= \vec{\xi}_1 + \delta\vec{\xi}_1 = \vec{\xi}_1 + (\vec{\xi}_1\nabla)\vec{d} \\ \vec{\xi}'_2 &= \vec{\xi}_2 + \delta\vec{\xi}_2 = \vec{\xi}_2 + (\vec{\xi}_2\nabla)\vec{d}\end{aligned} \quad (6)$$

Similar to (5), using (6) and neglecting the d quadratic terms, the elastically deformed propeller surface normal is:

$$\vec{n}'(\vec{d}) = \vec{n} + \vec{\xi}_1 \times (\vec{\xi}_2\nabla)\vec{d} - \vec{\xi}_2 \times (\vec{\xi}_1\nabla)\vec{d}, \quad (7)$$

synthetically:

$$\vec{n}'(\vec{d}) = \vec{n} + \Xi(\vec{d}) \quad . \quad (8)$$

An equivalent form of (7) is:

$$\vec{n}' = \vec{n} + (\text{rot}(\vec{d}))\times\vec{n} + \vec{\xi}_1 \times (T_\varepsilon\vec{\xi}_2) - \vec{\xi}_2 \times (T_\varepsilon\vec{\xi}_1) \quad (9)$$

T_ε : elastic deformation tensor.

If the hydrodynamic calculation is performed adopting the lifting surface method and the structural one adopting a thick shell finite element, (9) may be simplified:

$$\vec{n}' = \vec{n} + (\text{rot}(\vec{d}))\times\vec{n} \quad , \quad (10)$$

because, on the blade mean surface, the elastic displacement is neglected. By difference, if the hydrodynamic calculation is performed by BEM, the boundary condition (3) is expressed on the actual blade surface.

The elastic displacement affects the velocity on the propeller surface also:

$$\vec{V}(\vec{d}) = \vec{V}_i + \vec{V}_a'(\vec{d}) \quad , \quad (11)$$

$$\vec{V}_i = \nabla\varphi \quad , \quad (12)$$

V_i : induced velocity by the perturbation potential;
V'_a: attacking velocity,

$$\vec{V}'_a(\vec{d}) = \vec{V}_a + (\vec{d}\nabla)\vec{V}_a \quad , \quad (13)$$

$$\vec{V}_a = \vec{V}_\infty + \vec{\omega}\times\vec{r} \quad , \quad (14)$$

V_∞ : upstream velocity;
ω : propeller angular rotation velocity.

Replacing (8) and (11)-(13) in (3), it results:

$$\begin{aligned}(\vec{n} + \Xi(\vec{d}))(\nabla\varphi(\vec{d}) + \vec{V}_a + (\vec{d}\nabla)\vec{V}_a) = \\ = (\vec{n} + \Xi(\vec{d}))\dot{\vec{d}}\end{aligned} \quad (15)$$

Neglecting once again the d quadratic terms and suitably arranging the terms in equation (15), it results:

$$(\vec{n} + \Xi(\vec{d}))\nabla\varphi(\vec{d}) = \vec{n}[\dot{\vec{d}} - \vec{V}_a - (\vec{d}\nabla)\vec{V}_a] - \Xi(\vec{d})\vec{V}_a \quad (16)$$

$$\frac{\partial\varphi(\vec{d})}{\partial\vec{n}'} = \vec{n}[\dot{\vec{d}} - \vec{V}_a - (\vec{d}\nabla)\vec{V}_a] - \Xi(\vec{d})\vec{V}_a \quad . \quad (17)$$

The equations (8) and (17) will be applied in the integral equation (4). It results:

$$\begin{aligned}&2\pi\varphi(\vec{y}) - \int_S \varphi(\vec{x})\frac{(\vec{x}-\vec{y})[\vec{n}(\vec{x}) + \Xi(\vec{d}(\vec{x}))]}{|\vec{x}-\vec{y}|^3}dS - \\ &- \int_{S_w}\delta\varphi_w(\vec{x})\frac{(\vec{x}-\vec{y})[\vec{n}(\vec{x}) + \Xi(\vec{d}(\vec{x}))]}{|\vec{x}-\vec{y}|^3}dS_w = \\ &= -\int_S \frac{\vec{n}(\vec{x})\vec{V}_a(\vec{x})}{|\vec{x}-\vec{y}|}dS + \qquad , \quad \vec{y}\in S. \\ &+ \int_S \frac{\vec{n}(\vec{x})[\dot{\vec{d}}(\vec{x}) - (\vec{d}(\vec{x})\nabla)\vec{V}_a(\vec{x})] - \Xi(\vec{d}(\vec{x}))\vec{V}_a(\vec{x})}{|\vec{x}-\vec{y}|}dS\end{aligned} \quad (18)$$

This equation models the potential flow past the propeller surface, taking into account its elastic displacement also. As an advantage, it is expressed on the initial propeller surface. Its left side contains high order $\varphi\cdot d$ terms Assuming that φ, as the perturbation potential, is small, this terms may be neglected. A simpler equation results:

$$\begin{aligned}&2\pi\varphi(\vec{y}) - \int_S \varphi(\vec{x})\frac{(\vec{x}-\vec{y})\vec{n}(\vec{x})}{|\vec{x}-\vec{y}|^3}dS - \\ &- \int_{S_w}\delta\varphi_w(\vec{x})\frac{(\vec{x}-\vec{y})\vec{n}(\vec{x})}{|\vec{x}-\vec{y}|^3}dS_w = \\ &= -\int_S \frac{\vec{n}(\vec{x})\vec{V}_a(\vec{x})}{|\vec{x}-\vec{y}|}dS + \qquad , \quad \vec{y}\in S. \\ &+ \int_S \frac{\vec{n}(\vec{x})[\dot{\vec{d}}(\vec{x}) - (\vec{d}(\vec{x})\nabla)\vec{V}_a(\vec{x})] - \Xi(\vec{d}(\vec{x}))\vec{V}_a(\vec{x})}{|\vec{x}-\vec{y}|}dS\end{aligned} \quad (19)$$

In (18) and (19) it was assumed that there is no important difference between S'_w and S_w.

The upstream velocity is often non-uniform and the problem is an unsteady one. The upstream velocity may be decomposed in a constant part and a variable one, e.g. by means of the Fourier series. So, the attacking velocity may be also decomposed:

$$\vec{V}_a = \vec{V}_{a0} + \vec{V}_{aV} \qquad , \qquad (20)$$

and consequently the right member of the integral equation (19) too. The high terms $d \cdot V_{aV}$ will be neglected assuming the small perturbation hypothesis. So, the equation (19) becomes:

$$2\pi\varphi(\vec{y}) - \int_S \varphi(\vec{x}) \frac{(\vec{x}-\vec{y})\vec{n}(\vec{x})}{|\vec{x}-\vec{y}|^3} dS -$$

$$- \int_{S_W} \delta\varphi_W(\vec{x}) \frac{(\vec{x}-\vec{y})\vec{n}(\vec{x})}{|\vec{x}-\vec{y}|^3} dS_W = \qquad (21)$$

$$= -\int_S \frac{\vec{n}(\vec{x})\vec{V}_{a0}(\vec{x})}{|\vec{x}-\vec{y}|} dS - \int_S \frac{\vec{n}(\vec{x})\vec{V}_{aV}(\vec{x})}{|\vec{x}-\vec{y}|} dS + \quad , \quad \vec{y}\in S.$$

$$+ \int_S \frac{\vec{n}(\vec{x})[\dot{\vec{d}}(\vec{x}) - (\vec{d}(\vec{x})\nabla)\vec{V}_{a0}(\vec{x})] - \Xi(\vec{d}(\vec{x}))\vec{V}_{a0}(\vec{x})}{|\vec{x}-\vec{y}|} dS$$

Each of the right side terms of (21) integral equation corresponds to a problem which me be individually solved:
- stiff propeller working in uniform flow;
- stiff propeller working in non-uniform flow;
- flexible propeller working in uniform flow.
Evidently, the integral equation (21) was obtained in the frame of a linear theory.

The first two problems are classical and were approached in [5]. The third problem is a hydroelastic indeed one. Solving the first two problems, it results the time dependent but not elastic displacement dependent part of the hydrodynamic force of (1). The last one is obtained solving the third problem. Now (1) becomes:

$$\left([M_H] + [M_s]\right)\{\ddot{d}\} + \left([C_s] + [C_H]\right)\{\dot{d}\} + \qquad (22)$$
$$+ \left([K_s] + [K_H]\right)\{d\} = \{F_H(t)\}$$

$[M_H]$, $[C_H]$, $[K_H]$: hydrodynamic inertial, dumping and stiffness matrices;

The dynamic hydroelastic aspect means the resolution of the equation (22) and its solution stability study. Since it is a very complex task, for the instance the calculations are performed in a quasi-static manner only.

So, the dynamic terms in equation (22) will be neglected, also the hydrodynamic stiffness matrix. The hydrodynamic force is computed as in [5] and its corresponding elastic displacement as in [4] . Having the propeller deformed shape, new hydrodynamic calculation is performed. Some iterations are performed until a stop criterion will be fulfilled.

2. QUASI-STATIC RESULTS

The Boswell skewed propeller research series, [2], it was used. It includes 4 propellers, named NSRDC4381...NSRDC4384, having the same main data but different skew angle. All propellers are designed in the same point and so, having the same hydrodynamic characteristics but different skew angle, this series is an ideal benchmark for the studies aiming the skew angle influence. Its main data are:

Section thickness distribution: NACA66, NSRDC
                              modified nose and tail
Section meanline : NACA a=0.8
Diameter   D   =308.4mm (1 ft)
Blade number   Z   =5
Aspect Ratio A_E/A_0 =0.725
Pitch Ratio(0.7R)$P/D_{0.7R}\cong1.2$, slightly varying
                              according to θ_S
Rake θ_R = 0°
Skew θ_S = 0° (0%) NSRDC4381
                          36° (50%) NSRDC4382
                          72° (100%) NSRDC4383
                          108°(150%) NSRDC4384
Prop. rot. speed n   =7.8 s^{-1}
Design Point: J=0.889 ; K_T=0.213 ; 10K_Q=0.447 .

The assumed propeller material is nickel-aluminium bronze having: ρ=7600kg/m^3, E=120600 N/mm^2, ν=0.33, σ_Y=245N/mm^2, σ_B=590N/mm^2.

The extreme propellers, NSRDC4381 has a straight symmetric blade and NSRDC4384 is a highly skewed propeller. The remained two are moderately skewed propellers.

Al the propellers produce, at least theoretically, the same hydrodynamic force but, due to the skew angle, the correspondent elastic displacement are different. The most important consequence is the pitch angle modification, figure 1. For NSRDC4381 the pitch angle increases, and for NSRDC4382... NSRDC4384 propellers the pitch angle decreases.

Table 1. Boswell propeller series hydrodynamic characteristics modification consequent to the blade elastic deflection

Propeller	Δn [%]	ΔQ [%]	ΔP [%]	ΔJ [%]	ΔK_T [%]	ΔK_Q [%]	$\Delta \eta$ [%]
4381	-0.93	-0.51	-1.44	0.90	1.83	1.38	1.29
4382	0.03	-0.88	-0.85	-0.06	-0.05	-0.96	0.83
4383	1.24	-1.83	-0.62	-1.24	-2.44	-4.22	0.60
4384	8.36	-2.34	5.83	-7.76	-14.84	-16.83	-5.56

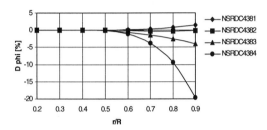

Figure 1. Boswell propeller series pitch angle modification under the design point hydrodynamic load.

The elastic propeller is actually an other propeller, slightly different from the undeformed one, and its hydrodynamic characteristic differs too. So, the actual running point differs from the stiff propeller design point.

The elastic deflection occurs until a new equilibrium geometry is gained as a complex interaction between the engine characteristics (engine load diagram), ship resistance, propulsive coefficients and the propeller characteristics, elastic and hydrodynamic. Briefly, decreasing the pitch, the propeller rate increases (increasing the pitch, the rate decreases) until the equilibrium condition is fulfilled.

In those bellow, the conventional thrust equality condition was adopted, meaning that the propeller rate increases (decreases) until the deflected propeller thrust becomes equal to the undeflected one. A similar torque equality condition may be used, more appropriate to $J \cong 0$ propeller working regime. Using the thrust equality condition, the hydrodynamic characteristics modification was calculated, table 1.

Analysing the figure 1 and table 1, some conclusions may be expressed:
-the skew angle has an important effect on the blade elastic deflection;
-For $\theta_S = 0°$ and a symmetric blade the elastic deflection determines the pitch angle to increase and

the hydrodynamic load increases too. So, the hydroelastic divergence may occur;
-for the highly skewed propellers the elastic deflection substantially increases, leading to the pitch decreasing Consequently, the propeller rate increases. Moreover, the propeller efficiency decreases and the blade stress increases dramatically. These are sufficient reasons for a carefully design when such kind of propeller is adopted;
-a moderate (50%) skew angle seams to prevent significant elastic deflection occurrence and so the hydrodynamic characteristic remains nearly unchanged. The blade stress are even lower than that for $\theta_S = 0°$.

Those above were deduced assuming the normal forward operation condition. Inversely, in the astern operation condition, generally speaking, the conclusions may be also inversely expressed. So, a highly skewed propeller deflects increasing the pitch angle. Consequently, the hydroelastic divergence may occurs for such a propeller but not for a symmetric blade, no skew, no rake one.

The elastic deflection depends on not only the propeller main data but its whole geometry. Therefore a proper study must be done for each propeller susceptible to important elastic deformation.

However, the quasi-static manner exposed above is not suitable for the study of the hydroelastic phenomena divergence and flutter, characteristic to the lifting body, which the marine propeller belongs.

3. DYNAMIC HYDROELASTICITY

We here present only the basic elements of a dynamic hydroelastic study. Its development will be the subject of an other paper.

The purpose is to solve the general equation (22) and to evaluate the stability of its solution.

The matrices $[M_H]$, $[C_H]$, $[K_H]$ of (22) are computed solving the third problem of the linearization presented in §1:

$$2\pi\varphi(\vec{y}) - \int_S \varphi(\vec{x})\frac{(\vec{x}-\vec{y})\vec{n}(\vec{x})}{|\vec{x}-\vec{y}|^3}dS -$$
$$- \int_{S_W} \delta\varphi_W(\vec{x})\frac{(\vec{x}-\vec{y})\vec{n}(\vec{x})}{|\vec{x}-\vec{y}|^3}dS_W = \quad , \quad \vec{y}\in S \quad . \tag{23}$$
$$= \int_S \frac{\vec{n}(\vec{x})[\dot{\vec{d}}(\vec{x})-(\vec{d}(\vec{x})\nabla)\vec{V}_{a0}(\vec{x})]-\Xi(\vec{d}(\vec{x}))\,\vec{V}_{a0}(\vec{x})}{|\vec{x}-\vec{y}|}dS$$

After the BEM specific discretization, the integral equation (23) becomes a linear equation system:

$$[A]\{\varphi\} = [b]\{\dot{d}\} + [c]\{d\} \tag{24}$$

Its solution may be expressed synthetically:

$$\{\varphi\} = [A^{-1}][b]\{\dot{d}\} + [A^{-1}][c]\{d\} \tag{25}$$

The hydrodynamic force is computed integrating the pressure due by the unsteady Bernoulli law:

$$p(t) = \frac{1}{2}\rho\left[V_{a0}^2 - |\vec{V}_{a0}+\nabla\varphi|^2\right] - \rho\frac{\partial\varphi}{\partial t} \quad . \tag{26}$$

$$\left\{\frac{\partial\varphi}{\partial t}\right\} = [A^{-1}][b]\{\ddot{d}\} + [A^{-1}][c]\{\dot{d}\} \quad . \tag{27}$$

The induced velocity $\nabla\varphi$ is computed from (23):

$$\nabla|_{\vec{y}}(\varphi(\vec{y})) = \frac{1}{2\pi}\left[\int_S \varphi(\vec{x})\nabla|_{\vec{y}}\left(\frac{(\vec{x}-\vec{y})\vec{n}(\vec{x})}{|\vec{x}-\vec{y}|^3}\right)dS + \right.$$
$$+ \int_{S_W} \delta\varphi_W(\vec{x})\nabla|_{\vec{y}}\left(\frac{(\vec{x}-\vec{y})\vec{n}(\vec{x})}{|\vec{x}-\vec{y}|^3}\right)dS_W + \tag{28}$$
$$+ \int_S \left(\vec{n}(\vec{x})[\dot{\vec{d}}(\vec{x})-(\vec{d}(\vec{x})\nabla)\vec{V}_{a0}(\vec{x})]-\Xi(\vec{d}(\vec{x}))\,\vec{V}_{a0}(\vec{x})\right)$$
$$\left. \nabla|_{\vec{y}}\left(\frac{1}{|\vec{x}-\vec{y}|}\right)dS\right] \quad , \quad \vec{y}\in S \quad ,$$

synthetically, after the BEM discretization:

$$\{\nabla\varphi\} = [\vec{E}]\{\varphi\} + [\nabla b]\{\dot{d}\} + [\nabla c]\{d\} \quad , \tag{29}$$

and, finally, by (25):

$$\{\nabla\varphi\} = \left([\vec{E}][A^{-1}][b]+[\nabla b]\right)\{\dot{d}\} + \left([\vec{E}][A^{-1}][c]+[\nabla c]\right)\{d\} \quad . \tag{30}$$

So, neglecting the high order d terms since the small perturbation hypothesis was initially adopted, it results:

$$\left\{|\vec{V}_{a0}+\nabla\varphi|^2\right\} = 2\left[\left([\vec{E}][A^{-1}][b]+[\nabla b]\right)\vec{V}_{a0}\right]\{\dot{d}\} + $$
$$+ 2\left[\left([\vec{E}][A^{-1}][c]+[\nabla c]\right)\vec{V}_{a0}\right]\{d\} + \tag{31}$$
$$+ \{V_{a0}^2\} \quad .$$

Using (27) and (31), the unsteady pressure (26) may be expressed in terms of d:

$$\{p(t)\} = -\rho[A^{-1}][b]\{\ddot{d}\} - $$
$$- \rho\left[\left([\vec{E}][A^{-1}][b]+[\nabla b]\right)\vec{V}_{a0}\right]\{\dot{d}\} - \tag{32}$$
$$- \rho\left[\left([\vec{E}][A^{-1}][c]+[\nabla c]\right)\vec{V}_{a0}\right]\{d\} \quad .$$

By (32) integration on the propeller surface it results the hydrodynamic inertial, dumping and stiffness matrices of (22):

$$\int_S \{p(t)\}dS = -[M_H]\{\ddot{d}\} - [C_H]\{\dot{d}\} - [K_H]\{d\} \tag{33}$$

The differential equation (22) resolution interests the propeller vibration and its stability the hydroelastic aspect. Both requires the (22) characteristic equation determination:

$$\left|[M]\lambda^2 + [C]\lambda + [K]\right| = 0 \quad , \tag{34}$$

equivalent:

$$A_n\lambda^n + A_{n-1}\lambda^{n-1} + \ldots + A_1\lambda + A_0 = 0 \quad . \tag{35}$$

To avoid the hydroelastic divergence and flutter means the equation (22) to have a stable solution, equivalent al the (35) eigenvalues have a negative real part. The Routh-Hurwitz-Frazer criterion expresses an equivalent condition:

$$B_1 = A_{n-1} > 0$$

$$B_2 = \begin{vmatrix} A_{n-1} & A_n \\ A_{n-3} & A_{n-2} \end{vmatrix} > 0 \tag{36}$$

$$B_3 = \begin{vmatrix} A_{n-1} & A_n & 0 \\ A_{n-3} & A_{n-2} & A_{n-1} \\ A_{n-5} & A_{n-4} & A_{n-3} \end{vmatrix} > 0$$

676

$$B_4 = \begin{vmatrix} A_{n-1} & A_n & 0 & 0 \\ A_{n-3} & A_{n-2} & A_{n-1} & A_n \\ A_{n-5} & A_{n-4} & A_{n-3} & A_{n-2} \\ A_{n-7} & A_{n-6} & A_{n-5} & A_{n-4} \end{vmatrix} > 0$$

$$\vdots$$

$$A_n > 0$$

$$A_0 > 0 \qquad . \qquad (36)$$

The condition of stability expressed by (36) is a strict one. The flutter limit condition is:

$$B_{n-1} \geq 0 \qquad , \qquad (37)$$

and the divergence limit condition is:

$$A_0 \geq 0 \qquad . \qquad (38)$$

The (36) criterion does not require the characteristic equation calculation but, involving the determinants calculation, it is not easy to handle it.

Those above are only theoretic statements. Obviously, further numerical validation as those exposed in §2 are necessary.

4. CONCLUSIONS

A marine propeller hydroelastic mathematical model using FEM/BEM is here presented.

Quasi-static calculations are performed for the Boswell highly skewed propeller research series.

The highly skewed propellers showed important influence of the elastic deflection on the propeller hydrodynamic characteristics.

As a matter of future, the dynamic mathematical model must be solved to approach the propeller vibration, also the flutter or divergence occurrence.

REFERENCES

1. P. Atkinson and E.,J. Glover, Propeller Hydroelastic Effects, Propellers '88 Symposium, paper no. 21, Virginia Beach, Virginia, September 20-21, 1988

2. R.J. Boswell, Design, Cavitation Performance, and Open-Water Performance of a Series of Research Skewed Propellers, Naval Ship Research and Development Center, Washington D.C., 2034, Report 3339, March 1971

3. R.A. Cumming and W.B. Morgan and J.R. Boswell, Highly Skewed Propellers, Transactions SNAME, Vol. 80, 1972, pp. 98-135

4. B. Ganea, Program pentru calculul rezistenței palei elicelor navale prin metoda elementului finit, Buletinul tehnic RNR , Nr. 3 , 1991, pp. 3-8

5. B. Ganea and D. Ghioca and D. Leroux, Steady and Unsteady Marine Propeller Hydrodynamic Calculation by Means of the Direct Boundary Element Method, 22nd ITTC Propulsion Committee Propeller RANS/Panel Method Workshop, Grenoble, 5/6 April 1998

6. T. Koronowicz and T. Tuszkowska, Strength Aspects of the Design of Highly Skewed Propellers, Polish Maritime Research, No.1(7) , March 1996 , Vol. 3 , pp. 6-13

7. J.-F. Kuo and W. Vorus, Propeller Blade Dynamic Stresses, Spring Meeting STAR Symposium, paper no. 4, May 21-24, 1985, Norfolk Va.

8. H.-J. Lin and J.-J. Lin, Nonlinear hydroelastic behavior of propeller using a finite-element method and lifting surface theory, Journal of Marine Science and Technology, Vol. 1, 1996, pp. 114-124

9. I.-S. Nho and C.-S. Lee and M.-C. Kim, A Finite Element Dynamic Analysis of Marine Propeller Blades, PRADS'89 Symposium Proceedings, Vol. 3, pp. 114-1 - 114-10 , Varna, 1989

10. A. Petre, Teoria aeroelasticității - fenomene dinamice periodice, Editura Academiei R.S.R., București , 1973

11. Z.Q. Suo and R.X. Guo, Hydroelasticity of Marine Propeller Blades, ISMS'91 Symposium Proceedings, pp. 151-156, Shanghai, Sept. 13-14, 1991

12. ** , ITTC'90 Proceedings, Vol. 1, Madrid, 16-22 Sept. 1990

1998 Elsevier Science B.V.
Practical Design of Ships and Mobile Units
M.W.C. Oosterveld and S.G. Tan, editors.

U.S. Navy Sealift Hydrodynamic Investigations

Siu C. Fung[a], Gabor Karafiath[a] and Donald McCallum[b]

[a] Naval Surface Warfare Center, Carderock Division,
   9500 MacArthur Boulevard, West Bethesda, MD 20817-5700, U.S.A.

[b] Consultant, 720 Hillsboro Dr., Silver Spring, MD 20902, U.S.A.

1. ABSTRACT

The U.S. Mid-Term Sealift Ship was designed as part of an Advanced Technology Development Program focusing on advances in efficiency, producibility, and commercial utility. The ship is designed to perform as a commercially operated container vessel, with the proviso that during a national emergency it can be transformed into a Ro-Ro for military sealift in a very short period of time. The Naval Sea Systems Command (NAVSEA) and Naval Surface Warfare Center, Carderock Division (NSWC-CD) have worked as a team to design and test a series of low resistance hull forms, and bow and stern variants beginning in 1994. The baseline hull form of this study was derived from an ammunition ship (AE 36) design which has an outstanding resistance characteristic but was never built [1]. Other hydrodynamic innovations, such as producible skeg, extended-skeg, podded propulsion, pre-swirl and reaction fins, and split-rudders were explored. In addition to the conventional propulsors, contra-rotation and large-diameter/low tip clearance propellers were also designed and tested. The results from these R&D efforts were very encouraging; significant savings in power were found when compared to an already excellent baseline design. This paper discusses the design rationale and presents some significant test data. The design is presented to the U.S. shipbuilding industry as a prototype to be used in the 21st century.

2. BACKGROUND

The 1992 Operational Requirements Document (ORD) for the Mid-Term Sealift Ship [2], specifies the deck area, draft (35 ft), and speed (20 knots) which became the governing parameters for the principal design characteristics of the ship. Prior to the selection of the ship's principal dimensions, the design team based on [3] conducted a commercial operational scenario investigation. The results indicated that the most potential or profitable commercial trade route for the Mid-Term Sealift Ship would be the one between North and South America along the eastern seaboard. This finding ultimately restricted the maximum overall length of the ship to be 200 meters as determined by the common berthing length for container ships in South America.

3. FIRST BASELINE HULL FORM DESIGN

Several technologies specifically called out for in this series of investigation were the use of diesel-electric drive and advanced propulsors, e.g., low tip clearance propellers and contra-rotating propellers. The main propulsion diesels for the Mid-Term Sealift ship were located far forward (station 2) and high up on B deck, see Figure 2.1-3, [4]. Such innovative arrangements help eliminate the need for a conventional main machinery room, except to house the electric propulsion motor. The motor itself was placed far aft on E deck, between stations 16 and 17.5. The location and size of the electric motor dictated the need of a sizable centerline skeg for the hull shape. The parent hull of the baseline design for this study was the AE 36 design, which also served as the progenitor for the other recently designed U.S. Navy ships, e.g., Strategic Sealift [1] and LPD 17.

The design requirements cause the Mid-Term Sealift ship to have a fuller form when compared to its parent. A McNaull transformation [5] was used to increase the parent hull prismatic coefficient from 0.62 to 0.67 and shift the longitudinal center of buoyancy from midships to 1.2% aft. Two new features, a tunnel stern and a bulbous skeg, were introduced to the new hull. Their respective functions were to provide enough clearance for a large diameter propeller (single shaft)

678

and room for the electric motor (16-ft diameter). In addition, the new hull also received a parallel middle body, which is 8.8% of the design waterline length. The baseline design also received an elliptical bulbous bow, which was proportionately derived from the AE 36 bulb. Bulb parameters were adjusted based on [6] and [7]. The baseline bulb was mainly designed to have a good all-round performance. The hull would be powered by a single centerline propeller coupled to a diesel-electric drive train and with a conventional spade rudder positioned on the centerline in the propeller wash. The body plan of this 1st Baseline design is shown in Figure 1 and the bare hull resistance characteristics are shown in Figure 2. The derived Taylor worm curve factor (WCF) values are less than 0.9 for most of the operating speed range, which clearly indicates a very favorable residuary resistance characteristic, despite the existence of the tunnel and centerline skeg.

3.1. Goals

The main objective of the program was to develop a design with low required power through the use of advanced hydrodynamic features, which promote efficiency, producibility, and commercial utility. The program as a whole was a design / test / design cycle; the insights gained from the earlier phases were used to aid the improvements for the subsequent designs. Both empirical and analytical (numerical computation flow codes (CFD)) methods were used to assist the designs. However, intuition and experience played the most important roles in this three-year multi phase effort.

4. PHASE I - OBJECTIVES

The first phase of this program was to develop and test alternative bow bulb and afterbody variants to the 1st Baseline hull form. During the investigations, the forebody was kept unchanged for the after body series and vice versa for the bulb series, so that qualitative assessment for each of the configurations could be identified. The forebody of each model was identical from the bow to station 11 except for a removable bow bulb insert.

4.1. Phase I -Afterbody Series

For the after body series, a removable skeg insert was cut into the 1st Baseline hull with the Bulbous Skeg (BS) for the investigation of different skeg designs. The following alternative skeg inserts were

designed and evaluated; Asymmetric Skeg (AS), Pre-Swirl Skeg (PS), and Producible Skeg (PR). In addition, two new afterbodies, the Extended Skeg (ES), and Twin-Podded Design (PD) were designed and tested. Self-propulsion tests using stock propellers were used to predict the performance of each of the designs. The selected configurations, which advanced to the next phase, were PR, ES, and PD. They are discussed below and in [8]. The body plans of the selected designs are shown in Figure 3.

Producible Skeg (PR): This design is a modification of the Bulbous Skeg. The main objective was to design an easily fabricated skeg. All double curvature plating and fairings were replaced with either flat plate or constant radius single curvature plating.

Extended Skeg (ES): The Extended Skeg is a lengthened version of the Bulbous Skeg. The trailing edge of the skeg was stretched by three-quarter of a station (from stations 18.75 to 19.5) so that finer skeg endings could be achieved. The objectives of this design were to reduce the likelihood of separation at the skeg, while still maintaining the required longitudinal center of buoyancy and to provide a larger propeller aperture. The edges of the tunnel were also bounded by two tunnel knuckles (ventral fins) which extended from station 16 to the transom, see Figure 3. The aft location of the propeller aperture left inadequate space for conventional rudder installation, so a pair of flanking rudders were designed and the tunnel knuckles became convenient places for landing the rudders. The function of the tunnel knuckle is twofold: (1) to provide an end plate effect for the propeller (which is enhanced by the rudders) and (2) to direct more flow from the side of the hull into the upper portion of the propeller plane. The alignment of the tunnel knuckle was guided by potential flow calculations. The use of flanking rudders was expected to make the Extended Skeg well suited for contra-rotating propellers.

Twin-Pod Design (PD): The afterbody shape for the Twin-Pod design differs substantially from the previous designs. It was completely redesigned from station 11 aft. The tunnel was filled, the deadrise angle in the afterbody was increased substantially and with more "Vee" shaped afterbody sections. The integral centerline bulbous skeg was removed and replaced by an appended small docking skeg. The pods

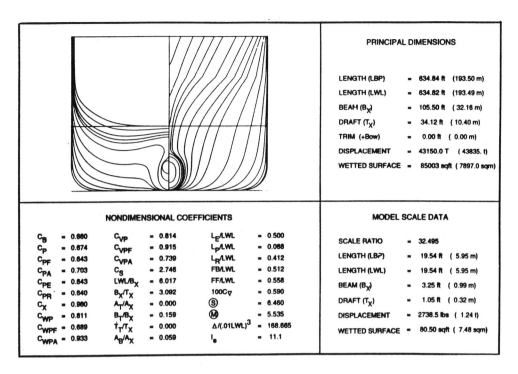

Figure 1 Hydrostatic characteristics of the First Baseline Mid-Term Sealift
Hull at the design displacement

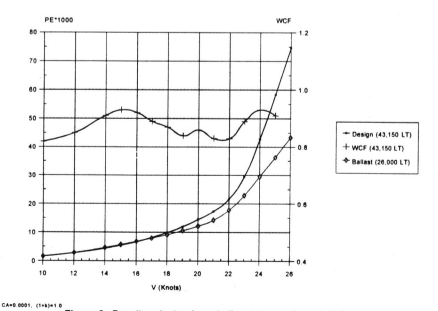

Figure 2 Baseline design bare hull resistance characteristics

accommodate an electric motor and its drive train and they also swivel for steering. The pods help eliminate the conventional machinery spaces for motor/reduction gear/shaft, steering gears and rudders.

Each pod was designed to accommodate an electric motor with the output of 15,000 HP at the propeller. The resulting maximum pod diameter was 8 feet and the shape was derived from a Series 58 body of revolution with its after sections truncated to mate with the propeller hubs. A strut supported each pod with its maximum strut thickness of 50 inches or 22% of its chord length, see Figure 4. This thickness should be ample enough to serve as an access trunk from the hull to the pod for inspection. The section shape of the strut followed the contour of the Series 58.

4.2. Phase I - Forebody Series

In addition to the Baseline Bulb (BB), four different bow bulbs, namely Nabla Bulb 1 (NB1), Nabla Bulb 2 (NB2), Nabla Bulb 3 (NB3), Producible Bulb (PR), plus a Cutaway Bow (CB) were designed and tested. The design rationale for NB1 and NB2 are given in the following paragraphs due to their better performances. The body plans for these two designs are given in Figure 3. More design rationales for the other bulbs (CB, NB3 and PR) can be found in [8].

Nabla Bulb 1 (NB1): This bulb is a moderate size (A_{BT}/A_x=6.1%) near-surface bulb. The transverse section shape of the bulb is rather slender with its upper surface reaching the 30-ft waterline. The majority of the volume is distributed at the upper part of the bulb and the waterlines in the lower portion are kept fine. The bow/hull intersections sweep downward. The vertical distribution of the bulb volume and the downward sweeping bulb/hull intersection are meant to provide maximum wave cancellation. Like the Baseline Bulb and the rest of the bulb designs in this study, the maximum forward protrusion of the bulb is 17 ft forward of the forward perpendicular. This maximum protrusion was dictated by the allowable maximum LOA for the ship.

Nabla Bulb 2 (NB2): This bulb is designed mainly for the full load condition and design speed of 20 knots. Ballast condition was also considered. The nose of the bulb was relatively high when compared to the upper surface of the bulb (24 ft above the baseline), so that the lower bulb waterlines could be made fine as possible. The main objective of this design was to have maximum wave cancellation in the full load

condition. For producibility reasons, constant bulb cross section shape (A_{BT}/A_x=8.7%) was given to sections at FP and aft. Neither fairing nor flow sweep down was designed for the bulb/hull intersection.

4.3. Phase I - Energy Enhancement Devices

The energy enhancement devices investigated in this phase were limited to rudders and fins. The investigations were focused on their effectiveness with respect to generating contra-guide flow, pre-swirl flow rotation to propellers, and in recovering losses from the propeller slipstream.

Pre-Swirl and Reaction Fins: Three different sizes of Pre-Swirl Fins were installed on the Extended Skeg. Fins were either installed on one side or both sides of the skeg. The installation and cross section shape of the fins are shown in Figure 5. For simplification, the fins were designed to have one side flat and the other side to have a simple camber. The fin was installed on a rotatable mounting plate so that the incidence angle of the fin could be adjusted. The function of the fin is to generate pre-swirl flow into the propeller as well as generating a thrust (negative drag) to the hull. The same sets of fins were used as the Reaction Fins on the rudder of the Bulbous Skeg configuration, see Figure 5. The fin(s) can generate thrust and recapture some of the rotation loss from the propeller.

Split Rudder: The baseline Mid-Term Sealift ship was a single screw design. A constant section shape, NACA 0018 was used for both the centerline rudder on the baseline hull and for the flanking rudders on the Extended Skeg, see Figure 6.

For the single centerline spade rudder, the rudder is spilt at 1.4 ft (ship scale) above the shaft line. Either the upper or the lower portion of the rudder could be rotated independent of the other. The idea was to have the rotated portion of the rudder recover some of the propeller swirl.

Since the flanking rudders of the Extended Stern were placed to each side of the propeller they were out of the slipstream. Therefore the total rudder area was made 10% larger than that of the centerline spade rudder in order to compensate for the expected side force loss due to missing the propeller slipstream. Each of the flanking rudders has three segments and each of them can be twisted at different angles along the rudderstock. The objective was to align the rudder sections for minimum power. The arrangement of the

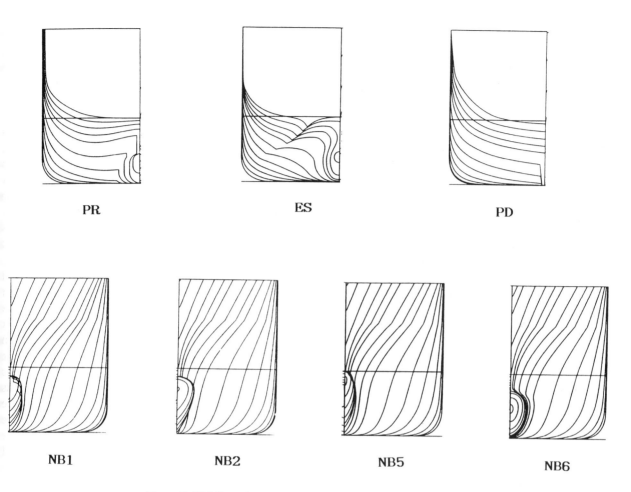

PR   ES   PD

NB1   NB2   NB5   NB6

Figure 3 Mid-Term Sealift stern and bulbous bow series body plans

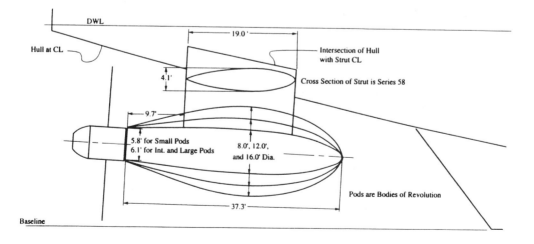

DWL

Hull at CL

19.0'

Intersection of Hull
with Strut CL

4.1'

Cross Section of Strut is Series 58

9.7'

5.8' for Small Pods
6.1' for Int. and Large Pods

8.0', 12.0',
and 16.0' Dia.

Pods are Bodies of Revolution

37.3'

Baseline

Figure 4 Profile of the podded designs

682

flanking rudders with the Extended Stern is shown in Figure 6.

4.4. Phase I Model Tests

Bare hull resistance, appended resistance, wake survey, and self-propulsion tests with stock propellers were conducted for the 1st Baseline hull and Afterbody Series during this phase of study. Since the effects from bulbs on propulsion coefficients are relatively small, only bare hull resistance tests were performed for the Bulbous Bow Series.

Stern Series: At the design displacement of 43,150 tons, among all the stern variants, the Producible Skeg (PR) showed the lowest resistance at the design speed, probably because of its good wake characteristic and fine ending along the trailing edge. The bare hull resistance of the Twin-Pod (PD) design was extremely good and its appended resistance was also quite remarkable, being less than 11% of the bare hull resistance at the design speed. Its low appended resistance was mainly caused by the small 8-ft diameter pod size, which was designed for a geared electric drive system. Different pod sizes designed for a direct drive system were investigated in the second phase. An 26.25-ft (8.0-m) right hand rotating stock propeller (Model 5037) was used in the powering tests for the single screw afterbodies. The same propeller and its matched left hand unit was used for the podded tests. The comparative self-propulsion performance for all the configurations are shown in Figure 7. At the 20-knot design speed, the best performers were the Twin-Pod, Producible Skeg and Extended Skeg. At speeds above 20 to 22 knots, both the Extended Stern (ES) and the Twin-Pod Design show significant savings (17 to 24 percent at 24 knots) when compared to the Bulbous Skeg. The better high-speed performance of the Twin-Pod Design could result from its lighter propeller loading when compared to its single screw counterparts. The longer and finer centerline skeg inherent by the Extended Skeg provides better inflow to the propeller and lower resistance at high speed when compared to the other single screw designs.

Bow Series: Bare hull model tests for all the bow designs were conducted at the design and ballast displacement of 43,150 tons and 26,000 tons, respectively. The predicted bare hull performance comparisons with the Baseline Bulb are shown in

Figure 8. At the design displacement, Nabla 2 Bulb has the best performance at 20 knots. Nabla 1 Bulb also performs very well even though its size is smaller than the Nabla 2.

At the ballast condition, the best performer at the design speed is the Baseline Bulb. It is interesting to note that in the low speed range in ballast, the Nabla type bulbs performed better than the Baseline design. Apparently the fine entrances provided by the Nabla type bulbs were able to cut through the bow waves in this low speed regime. The reverse trend in the higher speed range is mainly caused by the bow wave elevations that confronted the blunter, upper part of the bulb.

Energy Enhancement Devices: All self-propulsion tests for the energy enhancement devices were conducted a the design speed only. In general their savings are relatively small and in some cases are within the repeatability of the model tests. The savings from the split rudders are on the order of 1% when compared to the untwisted rudders.

The maximum power reductions due to the Pre-Swirl Fins and Reaction fins relative to no fin cases were 2% and 3%, respectively. Both of them occurred with fins installed on the port side only. The span of the Pre-Swirl Fins also played a major role in the delivered power reduction. In general the power decreased as span increased. Unlike the Pre-Swirl Fins, the Reaction Fins seemed insensitive to fin size. Different arrays of fins were found to have measurable effect on the total performance. Only fins installed on the port side of the skeg or rudder produced any powering reduction. The optimum fin angles (reference to the baseline, positive as trailing edge points downward) for the port side fins were 3 degree and -5 degree for the Pre-Swirl and reaction Fins respectively.

4.5 Phase I – Wake

Wake surveys were conducted on all the stern configurations of interest. The following table shows the Vx/V at the 0.82 r/R , zero degree circumferential position.

Configuration	Vx /V
Bulbous Skeg (BS)	0.42
Extended Skeg (ES)	0.45
Producible Skeg (PR)	0.60
Twin- Pods (PD)	0.68

683

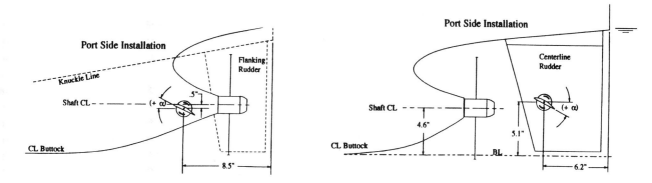

Figure 5 Pre-Swirl and Reaction fin geometry

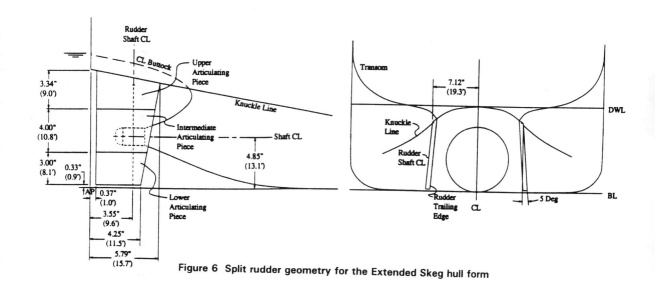

Figure 6 Split rudder geometry for the Extended Skeg hull form

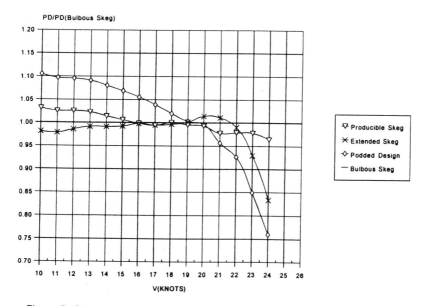

Figure 7 Stern series stock propeller delivered power (PD) comparisons

Examination of other wake characteristics such as the local wake gradient factor and the circumferential variation of inflow angles showed that the Twin Pod (PD) wake characteristics, followed by the Producible Skeg wake characteristics were the best.

4.6. Phase I - Recommendations

The test results from the phase 1 tests clearly indicated that the best afterbody design with respect to the 20-knot powering and wake characteristics was the Producible Skeg (PR). The unusually good high-speed characteristics of the Extended Skeg and Twin-Pod designs were also very attractive, and thus they were continued as candidates for the next phase of studies. The all around good performance of the Baseline Bulb was not surpassed by the other bulb designs. This led to the decision to retain the Baseline Bulb for the next phase of investigation.

The marginal savings from fins led to the speculation that a better fin section design might yield additional savings. However, with design propellers neither the old fins nor the new fins could be made to work because the flow condition in way of the fins changed. Thus there will be no further discussions in this paper regarding the fins.

5. Phase II - Objectives

The main objective of the Phase II study was to determine the powering characteristics with different type of design propellers for the selected designs which advanced from the previous phase of studies. In addition the powering effect due to different pod sizes was also investigated.

5.1. Phase II - Hull Form and Pods

The excellent performance of the hull form with Producible Skeg (PR) and Baseline Bulb (BB) evolved and became the new baseline (or 2nd Baseline) for the Phase II investigation. The initial Twin-Pod design had focused on hydrodynamic efficiency of an 8-ft diameter pod with epicyclic-geared electric drive. In order to reduce pod cost and complexity, two additional pods were designed, one with a 12 ft diameter to accommodate two direct drive electric motors mounted in tandem, and the other with 16 ft diameter to accommodate a single direct drive electric motor.

5.2. Phase II - Propellers

A total of five propeller blade rows were designed for the three selected hullforms. They were all fixed pitch propellers and custom-designed to their wake patterns for the corresponding hull forms. The propeller particulars for the design propellers are given in Table 1. The hull/tip clearances for the single screw configurations are 8-10% and 19% for the Twin-Pod design.

5.3. Phase II - Model Tests

New Baseline Hullform PRBB (Model 5501PR):

The following propellers were designed for this configuration: (1) a conventional, low tip clearance (8%), 28.02 ft (8.54m) fixed pitch propeller (Model 5228) and (2) a contra-rotating propeller set (Model 5229 and 5230), with forward and aft propeller diameters of 28.02 ft (8.54m) and 23.59 ft (7.19m) respectively. The forward to aft contra-rotating propeller design rpm ratio is nf/na=0.925.

The powering performance of the fixed pitch propeller turned out to be only marginally better than that of the 26.25 ft (8.0m) stock propeller which was used in the Phase I model tests, despite its increased diameter (6%) and increased disk area.

The contra rotating propellers were tested at three settings of forward to aft RPM ratio, nf/na=0.85, 0.925 and 1.00. At the design speed of 20 knots with the full load displacement of 43,150 LT, the lowest shaft horsepower was recorded not at the design rpm ratio but at nf/na value of 1.0. The performances of the contra-rotating propellers are relatively insensitive to the nf/na ratios between 0.85 and 0.95, however up to 3% powering differences were found between nf/na of 1.0 and 0.95. The performance of the contra-rotating propellers (nf/na=1.0) offers an average of 6% decreased power over the entire speed range when compared to conventional propeller (Model 5228), Figure 9. This improvement is in general several percent short of the anticipated improvement.

Extended Stern ESBB (Model 5502ES:

A 30.45-ft (9.28m) fixed pitch propeller (Model 5242) was designed and tested on the Extended Skeg. In addition, the Extended Skeg was also tested with the model contrarotating propellers designed for the producible skeg stern. With either propeller, the Extended Skeg benefits are greatest at the high speed. At 24 knots, a 15% reduction in power was achieved

685

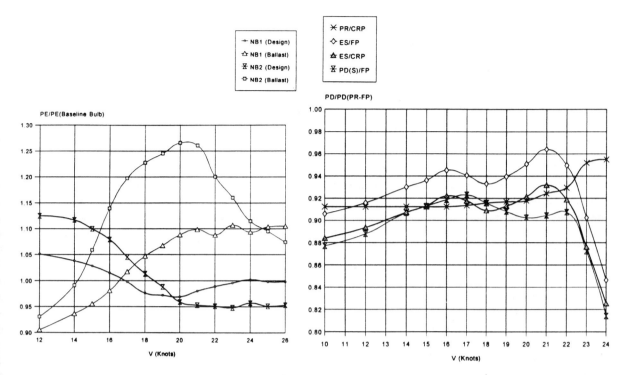

Figure 8 Bare hull effective horsepower (PE) comparisons for the Phase I bulbous bows

Figure 9 Phase II delivered horsepower (PD) comparisons with different design propellers

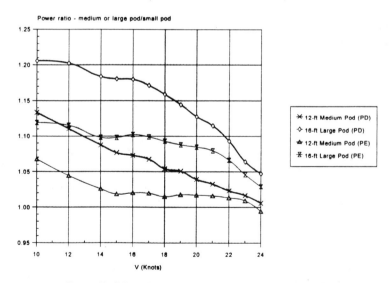

Figure 10 Effect of pod size on powering performance

686

by the conventional fixed pitch propeller and an 18 % reduction was achieved by the contrarotating propellers. The powering performance between the fixed pitch propeller and CRP at nf/na=1.0 are within 3% through out the speed range, see Figure 9. It is believed, that with custom designed CRP propellers even better powering performance would have been achieved with the Extended Skeg.

Twin Pod PDBB (Model 5506):

A pair of outward turning, 28.74-ft (8.76m) diameter fixed pitch propellers (Models 5240 and 5241) was designed for the podded hullform with the 8 ft diameter pods. At the design displacement, this configuration achieved close to 10% powering reduction at the 20-knot design speed, relative to the 2nd Baseline hull with a 28.02-ft (8.54m) propeller. The power reductions approached 13% to 18% as speeds increased to 23 and 24 knots, respectively. The superior hydrodynamic performance is attributed to (1) the excellent afterbody design, which is low in resistance and well suitable to large diameter propellers, (2) low appendage drag and favorable wake mainly due to the slenderness of the pusher type pods, and (3) extremely efficient propellers.

There is a direct trade off between hydrodynamic efficiency and the commensurate mechanical complexity of the pods, with the simpler mechanical arrangements resulting in increased pod diameter. At the design speed of 20 knots, the 12-ft and 16-ft pods require 4% and 13% more horsepower than that of the original 8-ft pods. The penalty in hydrodynamic efficiency tends to decrease as speed increases, see Figure 10. Despite the significant increase in required delivered power, the medium pod (12-ft) still performs better than the new Baseline for the entire speed range. Machinery studies undertaken after the completion of the entire model testing hint at the possibility of a direct drive conventional electric motor propulsion system that might be housed within the confines of the 8 ft diameter pod.

5.4. Phase II Recommendations

The Extended Skeg, Small and Medium Pods show reduced power relative to the 2nd Baseline hull, particularly at speeds above 20 knots. These findings are significant for ships that require operations at the higher speed regime. The performance of the designed conventional and contra-rotating propellers for the 2nd Baseline hull were slightly short of the designers'

expectation. The general findings from this phase of study led to a conclusion that small power reductions may be achievable by fine tuning the propulsors.

6. Phase III - Objectives

In this final phase, the objective was to give a recommendation on the best hull/propulsor combination that could be pursued by the shipbuilders using current technology. Despite the superior performances of the ES and PD designs, it was felt that more research and development might be needed. The rudders of the ES were not placed behind the slipstream of the propeller, and therefore the maneuvering performance could be an issue that would require additional model evaluation. There was some uncertainty with regard to the size of pod that would be required, and the pods were not an off the shelf item from U.S. shipbuilders. The PR skeg in essence presented a well balanced conventional design in all aspects, therefore, it was chosen as the after body configuration for the final phase of study. It was recognized that the propeller performance for the PR skeg was somewhat less than expected and it was hoped that a new propeller design would help reduce the 5 % powering disadvantage of the PR relative to the ES at design speed.

At this time, the ballast condition operations were reviewed and a new ballast displacement and trim were established as 37,169 LT with 5-ft trim by the stern. These changes allowed for the use of a taller bow bulb and therefore additional bow bulbs were designed.

6.1. Phase III - Bulbous Bow Series

Two new bulbs were designed, (Nabla Bulb 5 and Nabla Bulb 6) and tested along with two of the previous designs, (Baseline Bulb and Nabla Bulb 2). During this series of tests, the Producible Stern (PR) and a new design 4-bladed, right hand turning 27.99-ft (8.53m) design propeller (Model 5298) were used to quantify the performance of the final bulb/hull designs.

Nabla Bulb 5 (NB5): This bulb is a more slender and taller version of the NB1 bulb design. The bulb transverse section area ratio (A_{BT}/A_x) was reduced from 6.1% to 5.1% and the top of the bulb was stretched up to the design waterline. The "gooseneck" features of the NB1 bulb were still retained but the bulb hull intersection sweep down angle was significantly reduced. The objective of the NB5 design was to maximize the high-speed performance at full load condition without sacrificing the ballast condition performance, see Figure 3.

<u>Nabla Bulb 6 (NB6)</u>: The Baseline Bulb (BB) performed well at various loading conditions and the good high-speed performance of the nabla bulbs inspired the design team to combine the nabla bulb and the baseline bulb characteristics in order to obtain maximum performance in the design condition and not to compromise the ballast performance, particularly in the low speed range. The design guideline was to have nabla shape cross section and to have the top of the bulb just barely submerged beneath the free surface at the ballast condition. The bulb transverse section area ratio was increased from 5.8% for the Baseline Bulb to 8.7% for NB6, see Figure 11.

5.2. Phase III -Design Propeller

The performance of propeller Model 5228 that was previously designed for the Producible Skeg did not live up to expectations. Therefore, a decision was made to design a second fixed pitch propeller (Model 5298), which is described in Table 1.

5.3. Phase III - Model Tests

The bare hull effective power comparisons for the different bulbs are shown in Figure 12. The performance of the NB6 clearly surpasses that of the other designs, with up to 3 to 5% improvements relative to the NB2 and Baseline Bulb in the design condition.

Both the Baseline and NB6 Bulbs were tested with the first and final designed propellers. The self-propulsion tests for the final configuration, which consists of NB6 Bulb, Producible Skeg (PR), Propeller Model 5298 at the design displacement is shown in Figure 13. The delivered power comparisons for the different bulbs and propellers are given in Figure 14. The performances of the NB6 Bulb and the final design propeller were found to be very good. At the design displacement, the NB6 Bulb reduces the delivered power by 5.1% relative to the ship with Baseline Bulb (at 20 knots). The final design propeller (Model 5298) reduces ship delivered power by approximately 1.4% when compared to the previous designed propeller (Model 5228) throughout the entire speed range. The combined effect of the NB6 Bulb and design propeller (Model 5298) is to reduce the design displacement 20 knot delivered power by 6.5%

relative to the second baseline design (with Baseline Bulb, Producible Skeg, and Propeller 5228).

7. CONCLUSIONS

Numerous hull forms, propulsors, and energy enhancement devices were designed and tested during this 3-year study. The final recommended configuration is the hull form with Producible Skeg and Nabla 6 Bulb, propelled by the four-bladed fixed pitch propeller design represented by Model 5298. The performance of this recommended configuration could no doubt be furthered enhanced by using contra-rotating propellers. The podded propulsion concept and the extended stern concept offered additional performance benefits, especially at speeds above the 20-knot design speed.

Most of the energy enhancement devices that were investigated in this study (including those that were not mentioned in this paper, e.g., Costa Bulb, stern bulb, off centerline propeller, etc.), were not able to achieve the designers' expectations. It is believed that these devices were not effective because of the excellent stern designs and careful propeller designs that led to minimal wasted flow energy around the hull.

The authors wish to give their special thanks to Messieurs William Beaver, Kenneth Forgach, John Slager, and Andrew Kondracki who provided assistance with the hydrodynamic design efforts and model testing.

REFERENCES

[1] FUNG, S.C., D. McCallum, G. Karafiath, and J. Lee, "U.S. Navy Hull Form Investigations For Strategic Sealift," PRADS 95 Vol.1, 1995

[2] NAVSEA Operational Requirements Document (ORD) for Strategic Sealift Commercial Roll-On/Roll-Off Ship (COM-20) dtd: 10 June 1992.

[3] CHERRIX C.B. and M.P. Lasky, "A Commercial Cargo Ship for Sealift in the Year 2000," SNAME 1992 Annual Meetings.

[4] FILLING, J.C., Jay Howell, C.F. Snyder III, J. Vasilakos, and J. Watts, "Mid-Term Sealift Ship Technology Development Program," SNAME 1996 Annual Meeting.

[5] McNaull, R., "Generating New Ship Lines from a Parent Hull using Section Area Curve Variation,"MARAD..

[6] KRACHT, A., "Design of Bulbous Bows," SNAME Transactions Vol. 86, 1978.

[7] HAGEN, G, and S. Fung, "A Guide for Integration Bow Bulb Selection and Design into the U.S. Navy's Surface Ship Hull Form Development Process," NAVSEA TN No.

835-55W3-T N0003, April 1983. (Official Use Only).

[8] ALMAN, P, D. McCallum, and S. Fung, "Mid-Term Sealift: Hydrodynamic System Improvements for the Bow and Stern," ASNE 1998.

Table 1 Design Propeller Description

Prop. No.	Diameter	Hull Clearance	Hull Configuration for Design	Type, Rotation & Blade No.	P/D @ 0.7R	E.A.R	Max ETAO
5228	8.54m	8%	PRBB	F.P. R.H. 5	0.976	0.421	0.749
5229	8.54m	8%	PRBB	Fwd CRP Nf/Na = 0.9245 5	1.203	0.285	0.756
5230	7.19m	20%	PRBB	Aft CRP Nf/Na =0.9245 4	1.443	0.349	0.801
5240	8.76m	19%	PDBB	F.P. R.H. 4	1.196	0.268	0.831
5241	8.76m	19%	PDBB	F.P. L.H. 4	1.196	0.268	0.825
5242	9.28m	10%	ESBB	F.P. R.H. 4	1.012	0.333	0.776
5298	8.53 m	8%	PRNB6	F.P. L.H. 4	0.913	0.379	0.754

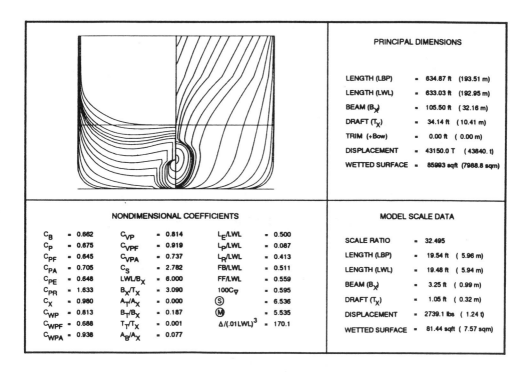

PRINCIPAL DIMENSIONS	
LENGTH (LBP)	= 634.87 ft (193.51 m)
LENGTH (LWL)	= 633.03 ft (192.95 m)
BEAM (B_X)	= 105.50 ft (32.16 m)
DRAFT (T_X)	= 34.14 ft (10.41 m)
TRIM (+Bow)	= 0.00 ft (0.00 m)
DISPLACEMENT	= 43150.0 T (43840. t)
WETTED SURFACE	= 85993 sqft (7988.8 sqm)

NONDIMENSIONAL COEFFICIENTS

C_B	= 0.662	C_{VP}	= 0.814	L_E/LWL	= 0.500
C_P	= 0.675	C_{VPF}	= 0.919	L_P/LWL	= 0.087
C_{PF}	= 0.645	C_{VPA}	= 0.737	L_R/LWL	= 0.413
C_{PA}	= 0.705	C_S	= 2.782	FB/LWL	= 0.511
C_{PE}	= 0.648	LWL/B_X	= 6.000	FF/LWL	= 0.559
C_{PR}	= 1.633	B_X/T_X	= 3.090	$100C_\nabla$	= 0.595
C_X	= 0.980	A_T/A_X	= 0.000	ⓢ	= 6.536
C_{WP}	= 0.813	B_T/B_X	= 0.187	Ⓜ	= 5.535
C_{WPF}	= 0.688	T_T/T_X	= 0.001	$\Delta/(.01LWL)^3$	= 170.1
C_{WPA}	= 0.938	A_B/A_X	= 0.077		

MODEL SCALE DATA	
SCALE RATIO	= 32.495
LENGTH (LBP)	= 19.54 ft (5.96 m)
LENGTH (LWL)	= 19.48 ft (5.94 m)
BEAM (B_X)	= 3.25 ft (0.99 m)
DRAFT (T_X)	= 1.05 ft (0.32 m)
DISPLACEMENT	= 2739.1 lbs (1.24 t)
WETTED SURFACE	= 81.44 sqft (7.57 sqm)

Figure 11 Hydrostatic characteristics of the recommended Mid-Term Sealift Hull with NB6 Bulb and Producible Skeg at the design displacement

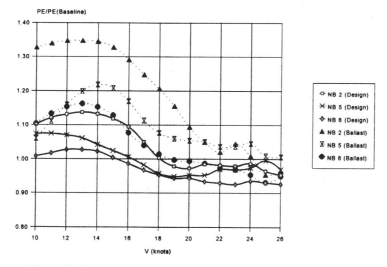

Figure 12 Bare hull effective horsepower comparisons at design and ballast displacement for the Phase III bulbous bows

690

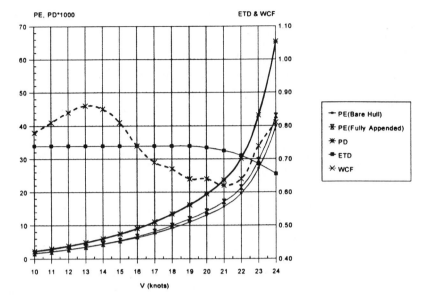

Figure 13 Powering performance of the recommended Mid-Term Sealift ship

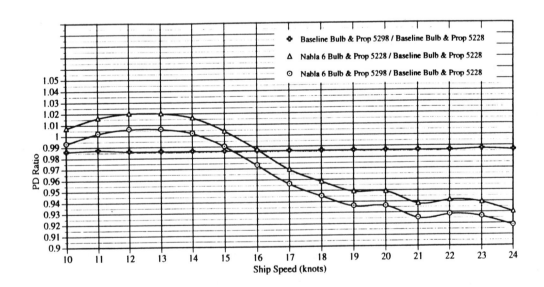

Figure 14 Delivered power comparison: Mid-Term Sealift with different bulbous bows and
propellers at the design displacement

Practical Design of Ships and Mobile Units
M.W.C. Oosterveld and S.G. Tan, editors.

The Influence of the Stern Frame Shape for a High Speed Container Ship on the Powering Performance

Kuk-Jin Kang[*], Ki-Sup Kim[*], Young-Jea Park[*], Chun-Ju Lee[*],
In-Haeng Song[*] and Il-Sung Moon[*]

[*]Ship Performance Department, Korea Research Institute of Ships and Ocean Engineering
Yusung P.O. Box 101, Daejeon, 305-343, KOREA

Present paper describes the influence of the stern frame shape for a high speed 3,600TEU class container ship on the powering performance. One fore-body and three after-bodies of barge shape, conventional U shape and medium shape were systematically designed. The resistance, propulsion, wake and cavitation characteristics are compared by using the model tests and calculations. The ship with a barge shape shows the best resistance performance, and the ship with a conventional U shape shows the best propulsive efficiency among them. As for delivered horse power, the medium shape is slightly inferior to the barge shape which is about 3.4% superior to the conventional U shape. The medium shape is recommendable as compromised shape in the view of the powering performance and the available space for engine room. The fluctuating pressure levels for the three hull forms are considered acceptable.

1. INTRODUCTION

As a consequence of increasing demand of rapid cargo transportation, the design speed around 24 knots is often required for large container ships. After-body shape is very important in the view of hull form resistance, propulsive efficiency and cavitation characteristics.

Up to now, many efforts have been paid to the development of after-body shape. However, ship designers are still eager to improve the powering performance of ships.

The object of the present research is aimed at finding out some relations between the stern frame shapes and the powering performance characteristics for a high speed container ship, and to share informations about the hull forms and model test results.

A 3,600TEU class container ship is selected as a target ship. Three different after-bodies of barge shape, conventional U shape and medium shape for one fore-body are systematically varied.

Resistance, propulsion, wake characteristics are compared through the systematical model tests. And the cavitation characteristics and the fluctuating pressures induced by a propeller are also compared.

2. HULL FORM DESIGN

The main particulars of the selected target ship from KRISO data bank are shown in table 1.

Table 1 Principal particulars

$L \times B$	230.0 m $\times$ 32.2m
Design Draft	10.8 m
Block Coeff.	0.65
Design Speed	24 knots
Engine	36,550 BHP $\times$ 96.0 rpm
Propeller	Dia.=7.9 m, Z=5, P/D=0.945, EAR=0.814

2.1 Fore-body shape

Fore-body shape(F1) was modified from the parent hull form to have a small wave resistance as possible using a Rankine source method. A bulbous bow of high nose shape was adopted to reduce the wave resistance, especially the wave breaking resistance.

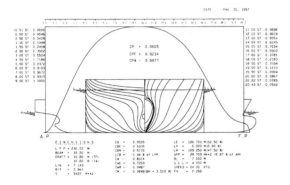

Figure 1. Lines and hydrostatics of the hull form(F1+A2)

Figure 1 shows the lines, cp-curve and hydrostatics of the designed fore-body shape combined with the after-body of conventional U shape.

Figure 2 shows the calculated wave profile compared with experimental results for the hull form(F1+A2).

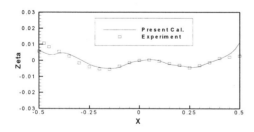

Figure 2. Wave profile

2.2 After-body design

After-body shape should be designed to consider the hull form resistance, propulsive efficiency and cavitation

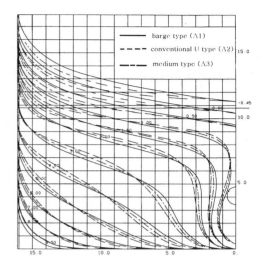

Figure 3. Comparison of stern shapes

characteristics carefully. The position of longitudinal center of buoyancy(lcb), cp-curve shape, frame shape, designed load water line(DLWL) and stern profile are included in the important design concepts. Among them, stern frame shape may be one of the most important design parameter.

In order to investigate the influence of the stern frame shape on the resistance, propulsive efficiency and cavitation characteristics, three after-bodies of barge shape(A1), conventional U shape (A2) and medium shape(A3) were systematically varied, and shown in Figure 3.

In the meantime, the cp-curve is kept constant and the tip clearance between hull and propeller is assured to be 24% to reduce the hull vibration induced by the propeller. And transom is 0.7m apart from the design draft in height to minimize the drag force due to the stern vortex.

3. MODEL TESTS AND EVALUATION

The resistance, self-propulsion and wake survey tests for three hull forms(F1+A1, F1+A2, F1+A3) with a propeller (KP393) were performed at the design draft. The length of ship models is 7.3m, and the scale ratio is 1/31.6. Based on the Froude's principle and the ITTC 1957 model-ship correlation line, the full scale values were predicted from the model test results. The scale effect correction were carried out based on the 1978 ITTC Performance Prediction Method[1].

The cavitation on the propeller blade was observed and the propeller-induced fluctuating pressure was measured for the medium shape(A3) in the cavitation tunnel. Numerical predictions of the fluctuating pressures were executed to evaluate the pressure levels for three hull forms qualitatively.

3.1 Resistance characteristics

Figure 4 shows the tendency of resistance characteristics according to the after-body variation. The barge shape(A1) shows the best resistance performance. The resistance curve of the medium shape(A3) follows the middle line below the 24 knots and coincides with the barge shape(A1) over the speed.

3.2 Self-propulsion characteristics

Figure 5 shows the comparison of propulsive coefficients analyzed from the test results. The conventional U shape(A2) gives slightly better propulsive efficiency than the others. There is no difference in the relative rotative efficiency η_R and the propeller open water efficiency η_o. But the wake fraction w_s and the thrust deduction factor t are varying on the after-body shapes. Powering performances of the three hull forms are summarized in table 2.

Figure 6 shows the curves of the delivered horse power(DHP) and the revolution of propeller(RPM) for the three hull forms. The required delivered power of the barge shape(A1) is 3.4%(1,217PS) less

Table 2 Comparison of Test Results

Hull	$Cr \times 10^3$	EHP(PS)	DHP(PS)	N(rpm)
A1	0.750	25,130	35,064	99.79
A2	0.832	26,291	36,281	99.81
A3	0.760	25,317	35,351	99.25

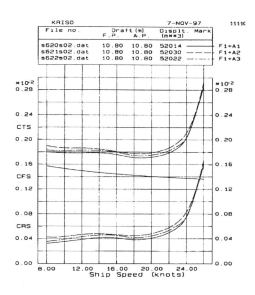

Figure 4. Resistance coefficients

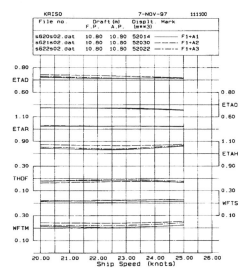

Figure 5. Propulsive Coefficients

694

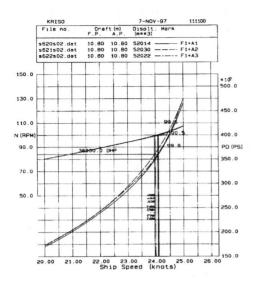

Figure 6. Curves of DHP and RPM

than that of the conventional U shape(A2) at the design speed 24 knots. And the powering performance of the medium shape(A3) is slightly inferior to the barge shape(A1).

3.3 Wake survey results

Figures 7, 8 and 9 show the iso-axial velocity contours at the propeller plane for the three after-bodies. Nominal wake fraction values are 0.223 for the barge shape, 0.24 for the medium shape and 0.261 for the conventional U shape, which show the fact that the nominal wake is highly affected by the stern bulb shape in front of the propeller.

3.4 Cavitation observation and propeller-induced fluctuation

In order to evaluate the cavitation characteristics of the design propeller for the medium shape(A3), a cavitation on the propeller blade was observed and the propeller-induced fluctuating pressure was measured in a cavitation tunnel. The measuring section of the cavitation tunnel has a sectional area of 0.6m × 0.6m with

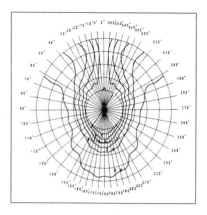

Figure 7. Iso-Axial Velocity Contours (Barge shape : A1)

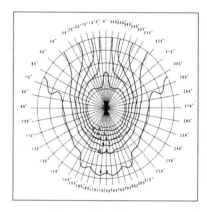

Figure 8. Iso-Axial Velocity Contours (Conventional U shape : A2)

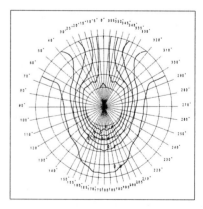

Figure 9. Iso-Axial Velocity Contours (Medium shape)

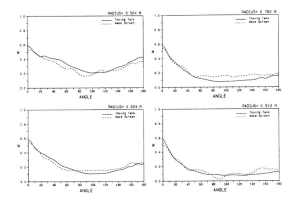

Figure 10. Simulated wake distribution in the cavitation tunnel

maximum water velocity of 12m/sec. The model wake simulated by using wire meshes was based on the result of the wake survey test carried out in the towing tank, which is compared with the measured one in Figure10.

Cavitation observation and measurement of fluctuating pressure for the medium shape(A3) were carried out at the full load condition(NCR, 20% power margin). The propulsive conditions of experiment and numerical prediction are summarized in Table 3, which are based on the analysis results of self-propulsion test.

Table 3 Propulsive conditions

Item	A1	A2	A3
Advance coeff. : Ja	0.710	0.700	0.701
Propeller rpm	104.33	103.37	103.52
Thrust coeff. : Kt	0.1805	0.1858	0.1851
Cavitation no. : σ	1.424	1.450	1.446
Ship speed : knots	24.06	23.92	24.03

Measurement of the fluctuating pressure acting on a finite flat plate(95cm × 34cm) with the tip clearance of 0.24D(60.13mm) was carried out at the same conditions as the cavitation observation. Using the

Table 4. Measured fluctuating pressures induced by the propeller

	Cent.	FWD	STD	Aft.	Port
1st BF	12.30	8.05	9.83	8.14	7.04
2nd BF	9.48	6.19	7.92	7.47	6.32
3rd BF	4.25	2.90	4.71	3.82	2.90
4th BF	1.95	1.38	2.51	1.53	0.97

BF : Blade frequency

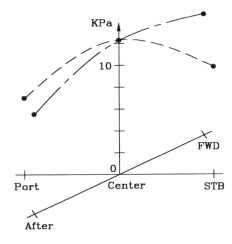

Figure 12. Measured fluctuating pressures induced by the propeller

method[2] recommended by ITTC, fluctuating pressures for full scale were predicted from the model scale experiment[3].

Cavitation patterns on the propeller for the medium shape(A3) were sketched in Figure 11 with a schematic conventions. The cavitation was observed on the suction side at blade angles between -20° and 60°. Behavior of the cavitation for the ship wake is considered good.

The measured fluctuating pressure are shown in Table 4 and Figure 12. Numerical predictions of the fluctuating pressures induced by the propeller were executed to compare the pressure levels for

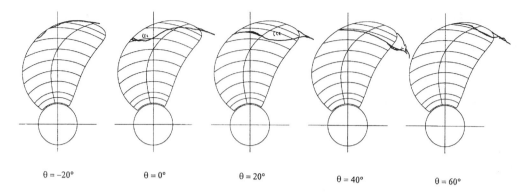

$\theta = -20°$ $\theta = 0°$ $\theta = 20°$ $\theta = 40°$ $\theta = 60°$

Figure 11. Cavitation pattern of the design propeller for medium shape(A3)

the three hull forms qualitatively.

A code XForSHIP based upon a lifting surface theory was used for these numerical predictions[4]. The fluctuating pressures calculated for the 3-D wake distribution measured at the towing tank are 10.4 KPa, 10.3 KPa and 10.2 KPa for the hull forms A1, A2 and A3 respectively. The flat plate similar with the experimental system was placed above the propeller in this calculation.

The solid boundary factor(SBF), which is the ratio of a fluctuating pressure acting on a body surface to a pressure at the free space, should be considered to predict the fluctuating pressure level for the hull surface with a curvature. The experimental situation corresponds to $(SBF)_{finite\ plate}$ of 1.90 because it has a finite flat plate[5].

M.C. Kim[5] had estimated the value of $(SBF)_{hull}$ from the calculations of fluctuating pressure on a hull surface of a container ship and at the free space. And then he extracted some correlation factor from those SBFs as follows;

$$(SBF)_{hp} = (SBF)_{hull} / (SBF)_{finite\ plate}$$

The estimated value of $(SBF)_{hp}$ is between 0.65 and 0.70 for this kind of ship. This value should be considered to predict the fluctuating pressure level for a ship from the results of this model experiment. When the SBF is considered, the characteristics of fluctuating pressure levels for these ships are assumed to be almost identical and acceptable.

4. CONCLUSIONS

The powering performance characteristics of the three after-bodies for a 3,600TEU class container ship were surveyed.

(1) The barge shape(A1) showed the best resistance performance and better powering performance by 3.4% than the conventional U shape(A2) at the design speed 24 knots.
(2) The medium shape is considered as an recommendable after-body shape since it shows almost same powering performance and it is more favorable than the barge shape in the view of available engine room space.
(3) The fluctuating pressure levels for the three hull forms are considered acceptable.

All the information about the stern frame shapes and analysed results of model tests will be very helpful to the designer of high speed container ships.

5. ACKNOWLEDGMENT

The present research was sponsored by the Ministry of Science and Technology. The authors would like to thank to all staff members of the towing tank and cavitation tunnel in KRISO.

REFERENCES

1. Eun-Chan Kim and Seung-Il Yang, "The Development of the analysis Program for the Resistance and Propulsion Test Results", Bulletin of KIMM Vol.17, October 1987.
2. Report of the performance committee, Proceedings of the 15th ITTC, the Hague,1978.
3. Chang-Sup Lee and Ki-Sup Kim, "Study on prediction method of fluctuating pressure on hull surface induced by a propeller", Report No. UCN 166G-347.D, KIMM, Dec. 1983.
4. Chang-Sup Lee, "Comparative studies on propeller-induced pressure fluctuation", Report No. PTL 10-87, Ship Rearch Station, KIMM, Sep. 1987.
5. Moon-Chan Kim, Ki-Sup Kim and In-Haeng Song, "A study of a Correlation between Experiments and Calculation of Pressure Fluctuation on Hull Surface", Vol.33, No. 2, The Society of Naval Architects of Korea, 1996.

Practical Design of Ships and Mobile Units
M.W.C. Oosterveld and S.G. Tan, editors.

Some Aspects in Designing Shaft Brackets for High-Speed Vessels.

A. Jonk[1] and J. P. Hackett[2]

[1] Maritime Research Institute Netherlands
     P.O. Box 28, 6700 AA Wageningen, The Netherlands

[2] Ingalls Shipbuilding, Inc., Pascagoula, Mississippi, USA

ABSTRACT

Many modern high speed vessels, such as containerships, ferries, and RO/ROs have an open shaft arrangement. In cases where the speed of these vessels is larger than about 26 knots, there is a risk of bracket cavitation. At model basins, test-techniques have been developed to determine the correct alignment of the brackets with the flow. In spite of these techniques, bracket cavitation still occurs on full scale. Therefore, improvement of the existing test techniques was necessary. At MARIN, methods have been developed to improve the accuracy of the alignment tests.   In addition investigations have been carried out to study parameters, which influence bracket alignment. This paper reviews the test methods used; describes the implications of model instrumentation, model appendage construction, and installation procedures to the measurements; and shows results of some measurements. It is concluded that bracket location is an important parameter; therefore, a better understanding of the flow field in the region of the bracket is required.

1. INTRODUCTION

Twin screw vessels have been built for many decades. Years ago the shaft supports were enclosed in bossings. However, bossings produce relatively pronounced wake distributions and therefore are less desirable for high-powered high speed vessels.

Consequently, many modern high speed containerships, RO/ROs, ferries, patrol vessels, and naval support and combatant ships have an open shaft arrangement. Figure 1 shows a typical open shaft arrangement.

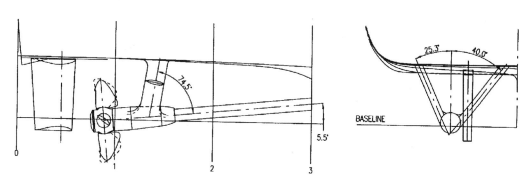

Figure 1.   Open shaft arrangement with "V" strut.

Large brackets, referred to as struts, support the shafting and propeller. Located just forward of the propeller, these are typically "V" struts for larger vessels. The strut arms for high speed vessels are normally streamlined foil sections that have a good tolerance to angle of attack variations before cavitating. Strut foil sections not aligned with the flow will cavitate. The higher the ship's speed, the greater the danger of strut cavitation. Cavitation is undesirable because it leads to strut erosion, noise, reduced performance of the propeller, and an increase in drag. Additionally, under certain conditions, propeller erosion will occur.

When ships suffer from strut cavitation, the only possible fix, to avoid future cavitation, is to wrap the existing strut arm casting with new plating in such a way as to change the angle of incidence with the oncoming flow. It has been proposed in some cases to cover the strut arms with polyurethane, which has a very good resistance to cavitation erosion. It appears, however, that the adhesion of the polyurethane is not resistant to the forces occurring during the collapse of the cavitation cavities.

The angle between the brackets of the "V" strut is chosen to adequately support the shaft, and integrate into the hull structure. It sometimes has been stated that the angle between the strut arms should not coincide with the angle between the propeller blades to avoid transverse vibration of the shaft. This statement, however, has to be considered in relation to the clearance between the strut and the propeller, the propeller load, and the skew of the propeller. Propellers with skew back can theoretically tolerate a smaller clearance than propellers with forward skew at the tip.

The clearance between the struts and the propeller is important with regard to interaction effects. In cases where the clearance is small, the induced velocity of the propeller in combination with the wake distribution can cause difficulties with strut alignment. When less advanced testing techniques are used for strut alignment there is a greater risk of strut cavitation. However, while performing model measurements for strut alignment on an Ingalls Shipbuilding Inc. design, for the inboard strut arm, a large rate of change in the strut alignment angle along the length of the strut arm was observed. Due to fabrication issues this strut location was deemed unacceptable. A new location was selected, and the test was repeated, producing an alignment with a moderate rate of change.

Normally variations of the inflow angle are relatively small over the length of the strut. Strut alignments have traditionally not been adjusted for scale effects associated with differences between model and full scale boundary layers. To improve the reliability of the Ingalls model experiments at Maritime Research Institute of the Netherlands (MARIN), all aspects of the strut alignment tests were reconsidered. After determination of the proper strut alignment, observations in MARIN's Depressurized Towing Tank with strut sections, designed with extremely small cavitation buckets highlighted the need to re-evaluate MARIN's procedures for installation of pitot tubes for strut alignment, manufacturing of the struts, and installation of struts on the hull model

2. PROPELLER-STRUT INTERACTION

The character of the nominal wake distribution of an open shaft arrangement is primarily induced by the short bossing, the distance between the short bossing and the propeller, the diameter of the shaft and strut barrel, and the clearance between the propeller and the struts.

In the axial wake peak, the influence of the struts arms is discernible at the inner radii. The strut "shadow" is the result of omitting the effect of the induced velocity of the propeller on the incident flow, and hence demonstrates the interaction between the propeller and the strut. Figure 2 shows the axial wake distribution.

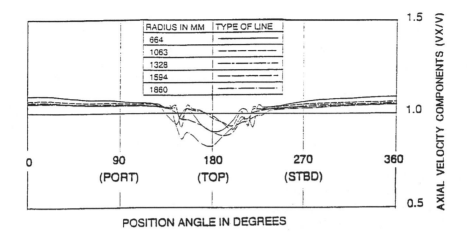

Figure 2.   Velocity components in the propeller plane of an open shaft arrangement

The smaller the clearance between the propeller and the strut, the more pronounced the wake peak would be. Increasing the clearance by moving the strut arms forward and lengthening the cylindrical part of the barrel, however, will increase the Von Karman vortices on the wake.

This effect can be reduced by extending the aft strut arm fillet to a small fin on the barrel, producing a "Kutta condition". Figure 3 shows the difference in axial wake distribution on two models. The upper set of curves has the strut arm located farther forward on the barrell.

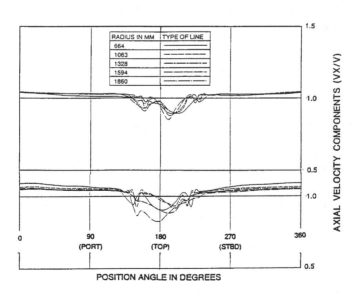

Figure 3.   Effect of propeller-strut clearance on axial wake peak

702

"V" strut arms may be connected to the barrel either normal to the barrel or tangential to the barrel. It is recommended to arrange "V" strut arms tangentially to the barrel in order to reduce the wake peak. However, this leads to asymmetrical sections at the intersection between the barrel and the strut. This is the most difficult area and the addition of a fillet from the strut arm to the barrel should be considered. The sensitivity of this area with regard to cavitation is demonstrated in Figure 4. This Figure shows strut cavitation on the inner side of the outboard strut of a twin screw container vessel while rolling slightly in a seaway. Similar problems occur on small patrol vessels, which typically use "I" struts.

Figure 4.   Strut cavitation due to swell on a container vessel.

On some vessels the shafting will require additional support, so another set of struts is installed between the stern tube bossing and main propeller strut. These intermediate struts can be either "V" struts or "I" struts. The flow into the intermediate struts with a normal intersection with the barrel, especially "I" struts, is more influenced by the hull form and the bossing in the vicinity of the hull. At the barrel the flow will be more parallel to the shaft centre line. As a consequence a relatively strong variation over the length of the strut arm can occur.

To reduce the risk of cavitation of the strut arms, sections are chosen which have a wide

cavitation bucket and hence, can handle a large angle of incidence variation. Apart from the well-known NACA family, many combatant vessels use EPH sections. Recent calculations have been carried out to arrive at new section shapes, which show even a wider cavitation bucket at maximum operating speed. Figure 5 shows the relation between the minimum pressure coefficient as function of the angle of incidence for several typical sections.

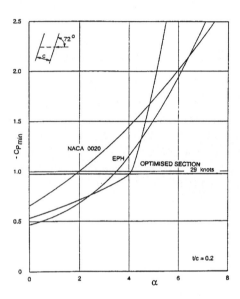

Figure 5.   Cavitation bucket of different strut sections.

3. TESTING TECHNIQUES

A traditional testing technique was to install the scale strut arms parallel to the vessel's center plane, attaching small plates perpendicular to the strut arm in a transverse direction and with a small angle to the direction of the inflow. Wet paint was then applied to the plate, and the model was placed in the tank and towed at a constant speed to determine alignment. The resulting paint streaks indicated the flow direction. As the displacement effect of the strut causes a flow disturbance in front of the strut, relatively large plates had to be used. An improvement was obtained by replacing the scaled strut arms with two thin rods. With this method it is generally possible to determine the flow direction at five locations on a 7 m model: three over the strut arm length, one on the barrel, and one on the hull. It is to be noted that in this case all the flow orientation results must be brought into a single strut reference plane. Generally an average strut twist angle from the hull to the barrel was determined by this method. Figure 6 shows the results of such a test.

Figure 6.   Results of strut paint smear test

For high-powered vessels this method was not adequate to determine the neutral strut angle as a function of strut arm length. Therefore, swivel vanes were introduced to replace the model strut. Typically 10 swivels could be placed along the span of the strut arms. After the vanes were set by the flow to a certain angle, they were locked while the model was still moving. This produced an average twist angle over the span length covered by each swivel segment. Problems with this method are the balance of the vanes, friction in the rotation point, and measurement of the angle. A somewhat better method was introduced by using a small cylindrical force transducer, which was moved along the leading edge of the strut section. It was then possible to take readings at each centimeter along the strut arm. The direction of the resultant force in a plane perpendicular to the leading edge of the strut arm coincides with the flow direction.

All of these methods fall short of providing an accurate and economical determination of the neutral flow angle.

To arrive at a higher accuracy a new type of three-hole pitot tube was developed. The new pitot tube was built around a stainless steel pipe of 20 mm diameter consisting of four rows of three holes. The three holes were perpendicular to the axis of the pipe and used to determine the flow angle in the plane of the holes. The pipe was able to extend through the hull and trace the leading edge of the strut arm from hull to barrel. This instrument allowed four measurements per run and provided a two-dimensional picture of the flow.

Later a smaller (10 mm diameter) pitot tube with four rows of three holes was developed. This instrument appeared to be a very useful instrument for a fine grid of measurements along the strut arm.

These three-hole pitot tubes revealed some unusual alignment angles; therefore, it was felt necessary to obtain more information about the oncoming flow field. Hence, a prototype of a five-hole spherical head pitot tube was introduced to obtain a three-dimensional picture of the flow field. The head of this pitot tube is 10 mm in diameter. After a number of measurements for Ingalls with the five-hole pitot tube and mutual discussions, it was felt necessary to procure a smaller pitot tube with a 5 mm diameter head. The smaller pitot tube will reduce the mutual interaction phenomena between the model propeller and the pitot tube.

4. MODEL OUTFITTING

The most difficult and laborious part of the test procedure is the positioning of the pitot tube on the ship model, followed by positioning of the struts on the model. The traditional two-

704

dimensional appendage drawing shows dimensional data such as the inclination angle of the strut in section view and a profile view designating a strut sweep angle with reference to a vertical line. When positioning the pitot tube using both angles, the tube rotates over an angle γ. This angle, however, is sequence dependent; that is, if the section inclination angle is used first followed by the profile view sweep angle, the γ angle arrived at may not be the same as if the profile view sweep angle is used first followed by the section view inclination angle. To standardize current practice at MARIN, the angle γ is zero, if a vector which is perpendicular to the plane containing all strut section nose-tail lines, is also in a plane perpendicular to the longitudinal axis of the ship model. The model shop has an alignment drilling jig. For a three-hole pitot tube model installation, the model shop is provided with a reference line in a plane perpendicular to the tube's axis and which is at a right angle to the nose-tail line of the strut. The holder can then be oriented in the proper plane perpendicular to the ship's longitudinal center plane by setting up a laser sheet

For a five-hole pitot tube this procedure cannot be used, as orientation has to be done around the pitot tube head and not the vertical rod of the pitot tube. However, by using AutoCAD R13, a three-dimensional computer model of the strut including the local hull section is made, and the proper location of the vertical rod of the pitot tube is determined. To ease the adjustment procedure, the rod with the spherical head is set parallel to the longitudinal axis of the ship model. In this way the pitot head is moved along the leading edge of the strut and the results are given in a plane perpendicular to the leading edge and a plane through the leading edge. Figure 7 shows the computer model of the strut with the five-hole pitot tube. In this procedure it was very helpful to have the correct information in the form of the traditional three-view appendage arrangement drawing appropriately dimensioned in hard copy, as well as in electronic media as submitted by Ingalls.

Figure 7. Computer model of pitot tube and strut

A laser alignment tool is used to assist in positioning the pitot tube on the model. To assure that the pitot tube hole through the model is aligned properly, a drill support jig was developed. The boring angles are provided from the AutoCad three-dimensional model. An oversized hole is bored through the model hull, then a plastic sleeve that acts as a bearing and watertight fitting for the pitot tube is installed. The pitot tube moves in and out of the model through this sleeve. Figure 8 shows the setup.

Figure 8. Instrumentation jig

5. TEST RESULTS

During measurements for Ingalls small combatant vessel with a three hole pitot tube, an unusual 8 degree twist angle variation over about 0.5 m length near the mid-span of the inboard strut was observed. Such a phenomenon has not been observed on frigates and commercial vessels. Usually the strut twist angle variations are relatively small over the strut length and therefore, strut alignments have traditionally not been adjusted for scaling effects. Scaling effects can exist with regard to the differences in boundary layer between model and full scale.

In an attempt to explain these results, measurements were taken without propellers, but with the propeller shaft rotating. In comparing the results between the rotating propeller and the no propeller case, a relative large was observed for the inboard strut. The inflow angles for the outboard strut show just a shift. Figure 9 shows the results for the inboard and outboard strut.

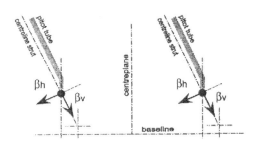

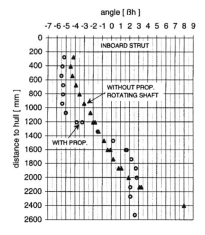

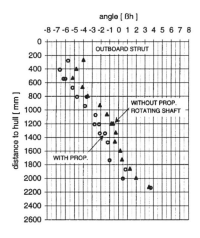

Figure 9.   Distribution of inflow angles.

The authors are of opinion that the observed phenomenon can not totally be explained by scaling effects of boundary layer, as the difference in boundary layer thickness would generally result in a more or less vertical displacement along a "uniformly" twisted strut. Moreover, to reduce the large twist angle

variation, the inboard strut was relocated further inboard and the strut alignment test repeated. The results show that the large twist angle variation has disappeared. The authors are of opinion the induced velocity of the propeller in conjunction with the axial wake distribution is responsible for the observed phenomenon. The effect of the propeller on the outboard strut is as expected.

Conducting five hole pitot tube measurements on an Ingalls patrol vessel with open shafts and outward over the top turning propellers and shafts, an inflow angle variation was found on the outboard strut. In the vicinity of the barrel this variation was about 4 degrees over approximately 40 – 50 cm strut length. The inboard strut, however, showed a variation over the same length of about 1 degree. Figure 10 shows the results.

was quite small. The difference between inward and outward rotating propellers for the outboard strut was about 5 degrees.

However, as can be expected, the inboard strut shows the reverse: the inflow angle variation was larger with inward rotating propellers than with outward rotating propellers. These variations were in line with those found at the outboard struts.

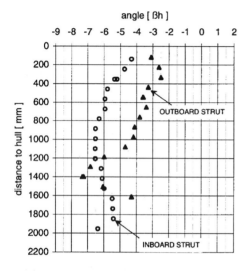

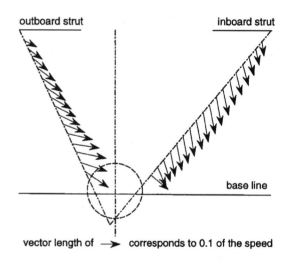

Figure 10.   Distribution of inflow angles.

New tests were conducted with the outboard strut tying into the hull at a more inboard location. These flow measurements showed a variation of about 5 degrees in almost the same area. This lead to the hypothesis that the direction of propeller and shaft rotation, may have an influence on the results. Therefore, strut alignment measurements were conducted with inward turning propellers. The outboard strut was located again at the first position. The resulting variation in inflow angles for the outboard strut

Measurements with a covered shaft (not rotating) and outward rotating propellers showed inflow angles for the outboard strut close to those found with inward rotating propellers. As the inflow angles are affected by the shaft rotation, when scaled by Froude's law, the alignment of the struts is influenced by scale effects. To better estimate the influence of the rotating shaft, measurements were made with the propeller and shaft rotating at a reduced RPM. This RPM is based on attaining equal rotational momentum on model and full scale. To accomplish this the propeller pitch was increased so as to maintain the same thrust as before but at the lower RPM.

The differences in inflow angles for the outboard strut due to the effect of shaft rotation are shown in Figure 11.

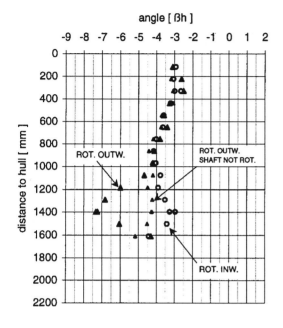

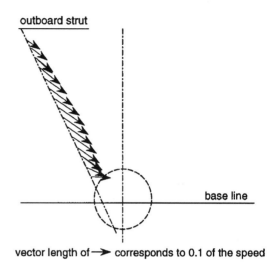

Figure 11.   Inflow angle variation due to direction of shaft rotation.

CONCLUSIONS

-   For high-speed vessels more advanced testing techniques than paint tests are necessary to determine the inflow angle variation in way of the struts. Five-hole pitot tubes are preferred because of their accuracy and ability to provide better insight into the flow field which are influencing the measurements.
-   The intersection of the barrel and tangentially fitted "V" strut tends to be susceptible to cavitation. This is partly influenced by the ability to align the strut with the flow, which in turn is dependent on the accuracy of the model flow alignment measurements.Selection of cavitation tolerant section shapes and installation of barrel fins and fillets can help reduce cavitation.
-   The interaction of the propeller induced velocity and the wake distribution can give rise to large variation of the inflow angle.
-   Shaft rotation has a strong influence upon the variation of inflow angles in the vicinity of the barrel.
-   At MARIN the outfitting of the ship models with the measuring equipment for strut alignment and the positioning of the struts has been substantially improved in close co-operation with Ingalls Shipbuilding Inc.

Practical Design of Ships and Mobile Units
M.W.C. Oosterveld and S.G. Tan, editors.

A Powering Method for Super High-Speed Planing Ships

Tadao Yamano[a] , Takeshi Ueda[b] , Isao Funeno[b] , Tetsuro Ikebuchi[b] and Yoshiho Ikeda[c]

[a]Mechanical Engineering Laboratory, Practical Life Studies,
Hyogo University of Teacher Education
942-1 Shimokume, Yashiro-cho, Kato-gun, Hyogo 673-1494, Japan

[b]Fluid Dynamics Research Department, Akashi Technical Institute,
Kawasaki Heavy Industries, Ltd.
1-1 Kawasaki-cho, Akashi, Hyogo 673-8666, Japan

[c]Department of Marine System Engineering, Osaka Prefecture University
1-1 Gakuen-cho, Sakai, Osaka 599-8531, Japan

Hull attitude of planing ships at high speed much affects its hull resistance. On the other hand, water jet pump has some issues to be cleared such as hull and water jet pump interaction. These make the powering for the ships difficult, and the powering method for such ships has not yet established. This paper proposes a powering method based on model test on hull and water jet pump, and its effectiveness is evaluated by applying it to full scale ships.

1. INTRODUCTION

Planing ships with water jet pump (abbreviated WJP hereafter) whose Froude number is over 1.0 are increasing in number. Hull attitude of the ships at high speed is much different from that at 0-ship speed, and the hull attitude much affects its hull resistance. On the other hand, the WJP still has some issues to be cleared such as hull and WJP interaction[1]. Therefore it can be said that the powering method for such ships including model tests to be conducted for the powering has not yet established, though some proposals[2],[3],[4],[5] have been made.

This paper proposes a powering method based on model test on hull and WJP, and its effectiveness is evaluated by applying it to full scale ships.

2. MODEL TEST

Test on hull without WJP and test on WJP itself were conducted independently.

2.1. Test on hull

Model ships without WJP were tested in the towing tank at Osaka Prefecture University.

The applied model test procedure has been explained by one of the authors[6][7]. A model ship was arrested to a towing carriage through a three-component load cell as shown in Fig.2.1. Resistance, lift and trimming moment were measured by the three-component load cell with varying test condition. The test conditions (ship speed, draft and trim angle) were systematically changed to make a database to solve the equilibrium equation described in chapter 3.2.2. Principal dimensions of the tested model ships are shown in Table 2.1.

In order to grasp the effect of the water flow through the duct on the above mentioned hydrodynamic forces and moment, the forces and moment were measured on each model ship, one without a water duct and another with it. Fig. 2.2 shows the water duct for WJP on a hull.

The amount of the water flow through the water duct was varied by changing inner diameter of nozzle situated at the aft end of duct. Fig. 2.3 shows the water duct with nozzle.

An examples of the measurement results on hull without a water duct is shown in Fig. 2.4, and that on hull with a water duct in Fig.2.5.

Model	Water duct	Nozzle	A_j/A_d
Model-A (a)	without	-	-
Model-A (b)	with	with	0.29
Model-A (c)	with	with	0.57
Model-A (d)	with	without	1

A_d : sectional area of water duct
A_j : sectional area of nozzle exit

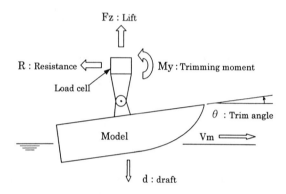

Fig.2.1 Schematic view of model test

Table 2.1 Principal dimensions of tested model ships

Model	Length over all (m)	Breadth (m)	Depth (m)
A	0.72	0.27	0.12
B	0.73	0.28	0.10

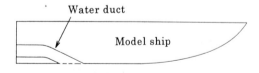

Fig. 2.2 An example of a water duct

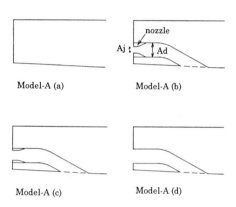

Fig. 2.3 Water duct with nozzle (model-A)

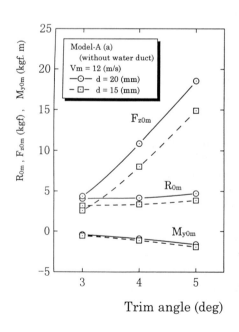

Fig. 2.4 An example of measured resistance, lift and trimming moment (model-A)

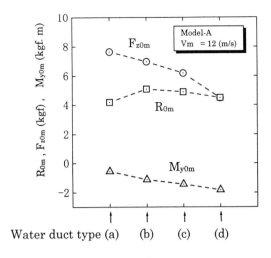

Water duct type (a)   (b)   (c)   (d)

Fig.2.5 An example of measured resistance,
lift, and trimming moment with varying
water duct and nozzle(model-A)

2.2. Test on WJP

WJP without duct was tested at the facility
of Kawasaki Heavy Industries Ltd. An example
of obtained characteristic curves of WJP is
shown in Fig.2.6. Scale ratio of tested WJP is
1/1. Non-dimensional coefficients shown in
Fig.2.6 are defined as follows:

$$K_g = q/(n \cdot D^3) \tag{1}$$
$$K_h = g \cdot H/(n^2 \cdot D^2) \tag{2}$$
$$K_m = M/(\rho \cdot n^2 \cdot D^5) \tag{3}$$

where
K_g : flow rate coefficient
K_h : head coefficient
K_m : torque coefficient
q : flow rate (m³/sec)
D : impeller diameter (m)
H : head (m)
M : torque(kgf·m)
g : gravity acceleration (m/sec²)
n : impeller revolution rate (rps)
ρ : water density (kgf·m⁻¹·sec²)

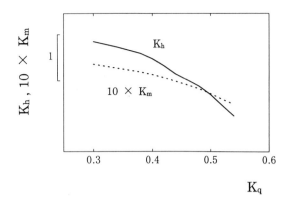

Fig. 2.6 An example of measured WJP
characteristic curves (model-A)

3. A POWERING METHOD

3.1. Hull and WJP interaction

Fig 2.5 shows that water duct and nozzle
diameter affect the values of measured hydrodynamic forces and moment. In other words,
these results contain forces acting on hull
surface and on water duct itself; moreover, they
also contain the hull and WJP interaction (the
indirect effect of water flow through a water
duct on hull forces and moments).

3.1.1. Separation of Hull and WJP interaction from measured forces and moment

In order to clarify the hull and WJP interaction, forces and moment due to the water duct
itself are subtracted from the measured forces
and moment on hull with a water duct. The
procedure is as follows:
· Resistance : equation(4)
Frictional resistance due to the hull surface
and the water duct inside is subtracted from
the measured resistance.
· Lift : equation(5)
Water duct center line and keel line are
approximately parallel, so that vertical momentum change through the water duct (direct lift
change) is considered negligibly small.

712

Therefore measured lift is considered to include only the indirect effect of the water duct.

· Moment : equation(6)

Water flow through a curved water duct generates trimming moment on hull. The trimming moment caused by the curve of water duct is subtracted from the measured moment.

$$R_{1m} = R_{0m} - R_{fm} - R_{fdm} \quad (4)$$
$$Fz_{1m} = Fz_{0m} \quad (5)$$
$$My_{1m} = My_{0m} - My_{cdm} \quad (6)$$

where
R : resistance (kgf)
Fz : lift (kgf)
My : trimming moment (+ : bow up) (kgf·m)
suffix $_0$: measured
suffix $_m$: model scale
R_f : frictional resistance of hull (kgf)
R_{fd} : frictional resistance of water duct (kgf)
My_{cd} : trimming moment due to
curve of water duct (kgf·m) .

Fig. 3.1 shows the relation between non-dimensionalized hydrodynamic forces and moment and IVR (= V_i/V : inlet velocity ratio, where V_i= speed of water flow through duct at impeller, V = ship speed) for model-A and Model-B.

Resistance coefficient r_1, lift coefficient fz_1 and trimming moment coefficient my_1, which are based on the data corrected by the above mentioned procedure, are defined as follows:

$$r_1 = R_{1m}/(\rho_m \cdot A_{dm} \cdot V_m^2) \quad (7)$$
$$fz_1 = Fz_{1m}/(\rho_m \cdot A_{dm} \cdot V_m^2) \quad (8)$$
$$my_1 = My_{1m}/(\rho_m \cdot A_{dm} \cdot L_{oam} \cdot V_m^2) \quad (9)$$

where
A_d : sectional area of water duct (m²)
L_{oa} : hull length (m)
V : ship speed (m/s)

$$\Delta r_1 = r_1' - r_1'' \quad (10)$$
$$\Delta fz_1 = fz_1' - fz_1'' \quad (11)$$
$$\Delta my_1 = my_1' - my_1'' \quad (12)$$

suffix ' : one of water duct types (a) ~ (d)
suffix " : water duct type (a)

Water duct types (a) ~ (d) are illustrated in Fig.2.4.

Δr_1, Δfz_1, and Δmy_1 represent the hull and WJP interaction.

IVR at model test shown in Fig.3.1 was determined as follows:

Main flow speed against a hull at duct exit position was almost same to ship speed at every duct. Taking this result ($V_{jm} \fallingdotseq V_m$) into consideration, IVR at model test is approximately calculated as follows:

$$IVR_m = V_{im}/V_m \fallingdotseq V_{im}/V_{jm} = A_{jm}/A_{dm} \quad (13)$$

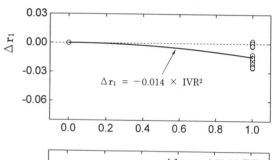
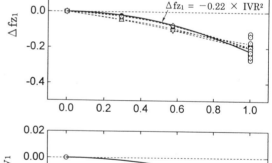
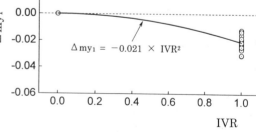

Fig. 3.1 Relation between hydrodynamic forces and moment, and IVR (model test result,model-A and B)

where

A_j: sectional area of nozzle exit

V_j: speed of water flow through duct at nozzle exit

$$A_{dm} \cdot V_{im} = V_{jm} \cdot A_{jm}$$

3.1.2. Formulation of Hull and WJP interaction

For practical use, the authors try to formulate Δr_1, Δfz_1, and Δmy_1 in a simple form by theoretical consideration and by using test results. The hull and WJP interaction is caused by flow through the water duct, therefore, it is considered that the force is proportional to square of water flow speed through the duct. For example, above mentioned tendency is seen in Δfz_1 shown in Fig.3.1. Therefore, these non-dimensional coefficients can be approximately expressed as follows as functions of IVR:

$$\Delta r_1 = C_1 \cdot IVR^2 \qquad (14)$$
$$\Delta fz_1 = C_2 \cdot IVR^2 \qquad (15)$$
$$\Delta my_1 = C_3 \cdot IVR^2 \qquad (16)$$

where

C_1, C_2, C_3: coefficients to be determined by test data as shown in Fig.3.1

In Δr_1 and Δmy_1 of water duct types (b) and (c) (IVR_m =0.29 and 0.57), still remains effect of nozzle loss, and this effect seems not to be negligible. Therefore, coefficients C_1, C_3 were determined mainly based on data at IVR_m =0.0 and 1.0.

As seen in Fig.3.1, the scattering of data remains at IVR_m=1.0 due to difference of hull form and the test conditions (ship speed, draft and trim angle). Therefore, Mean value was used for the determination of the values C_1, C_2, C_3.

3.2. Full scale powering
3.2.1. Scaling up of measured hydrodynamic forces and moment

Using the data obtained above, hydrodynamic forces and moment for a full scale ship can be calculated by the following formulae.

Resistance is scaled up based on two dimensional extrapolation method which is usually used for resistance prediction for high-speed ships, and is added to our proposing formulae including the hull and WJP interaction.

$$R_s = \rho_s \cdot \nabla_s^{2/3} \cdot V_s^2 \cdot [\{r_{1(IVR=0)} + \Delta r_1\} \cdot (A_{ds}/\nabla_s^{2/3}) + r_{fs} + \Delta C_f \cdot \{S_s/(2 \cdot \nabla_s^{2/3})\}] \quad (17)$$

$$Fz_s = \rho_s \cdot \nabla_s^{2/3} \cdot V_s^2 \cdot [f_{z1(IVR=0)} + \Delta fz_1] \cdot (A_{ds}/\nabla_s^{2/3}) \quad (18)$$

$$My_s = \rho_s \cdot \nabla_s^{2/3} \cdot L_{oas} \cdot V_s^2 \cdot [\{m_{y1(IVR=0)} + \Delta my_1\} \cdot (A_{ds}/\nabla_s^{2/3}) + m_{ycd} + (h_s/L_{oas}) \cdot (r_{0m} - r_s)] \quad (19)$$

where

∇: displacement volume (m³)

S: wetted surface area (m²)

h: vertical distance from center of gyration to mid-line of water line and keel line (m)

ΔC_f: friction correction

suffix $_s$: full scale

$r_{1(IVR=0)}$, $fz_{1(IVR=0)}$, $my_{1(IVR=0)}$: data of the hull without a water duct

R_s contains resistance on hull and that due to hull and WJP interaction, but it does not contain resistance due to water duct itself. The reason for this is that resistance due to water duct itself is included in WJP thrust T_s.

3.2.2. Equilibrium equation

At a ship speed V_s, by solving the following equilibrium equations for horizontal force, vertical force, and trimming moment (20) ~ (22)[6)7)], we can obtain the resistance, lift, trimming moment, draft and trim angle at the equilibrium condition.

$$R_S = T_s \qquad (20)$$
$$Fz_s + \rho_s \cdot \nabla_s = W_s \qquad (21)$$
$$My_s + T_s \cdot h_s - \rho_s \cdot \nabla_s \cdot l_s = 0 \qquad (22)$$

where

W: weight of ship (kgf)

T: thrust (kgf)

l: longitudinal distance from center of gyration to center of buoyancy (m)

3.2.3. Powering procedure

Now we have a full scale ship speed-resistance curve, the hull and WJP interaction coefficients and WJP characteristics curves. From these data, we can get ship speed ~ power and impeller revolution curves. The calculation procedure is as follows:

1) WJP operating point

The following equation can be derived from equations (1) and (2).

$$K_h/K_q^2 = g \cdot H_S \cdot D_S^4 / q_S^2 \qquad (23)$$

At a ship speed V_S, K_h/K_q^2 value can be calculated from required thrust $T_S (=R_S)$ using the following relations:

$$H_S = (V_{jS}^2 - \varepsilon \cdot V_S^2)/2g \qquad (24)$$
$$q_S = A_{jS} \cdot V_{jS} \qquad (25)$$

V_{jS} can be calculated from eqation (26).

$$\begin{aligned} T_S &= \rho_S \cdot q_S \cdot (V_{jS} - V_S) \\ &= \rho_S \cdot A_{jS} \cdot V_{jS} \cdot (V_{jS} - V_S) \end{aligned} \qquad (26)$$

where
A_j : sectional area of nozzle exit
ε : obtained by inlet duct wind tunnel test

K_q value where WJP operates can be obtained from above obtained K_h/K_q^2 value and $K_q \sim K_h/K_q^2$ curve as shown in Fig.3.2. Then we can obtain K_m value from the obtained K_q value and $K_q \sim K_m$ curve.

2) Calculation of BHP and N

BHP(PS) and N(rpm) can be calculated from above obtained K_q and K_m values, using equations (27) ~ (30).

$$n = q/(Kq \cdot D^3) \qquad (27)$$
$$M = K_m \cdot \rho \cdot n^2 \cdot D^5 \qquad (28)$$
$$BHP = 2\pi \cdot n \cdot M/75/\eta_T \qquad (29)$$
$$N = 60 \cdot n \qquad (30)$$

where
η_T (transmission coefficient) = 1.0

3.3. Application of the developed powering method to full scale ships

Fig.3.3 shows a full scale speed-resistance curve at the equilibrium condition for model-A.

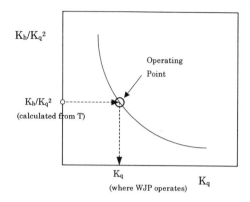

K_h/K_q^2

Operating Point

K_h/K_q^2
(calculated from T)

K_q

K_q
(where WJP operates)

Fig.3.2 Illustration of $Kq \sim Kh/Kq^2$ curve and WJP operating point

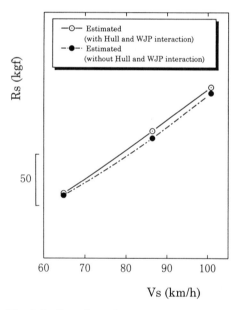

Rs (kgf)

—○— Estimated
(with Hull and WJP interaction)
—●— Estimated
(without Hull and WJP interaction)

50

60   70   80   90   100

Vs (km/h)

Fig.3.3 Speed-resistance curve at equilibrium condition (full scale, model-A).

Fig.3.4 (a) and (b) show the comparison of the estimated power curves according to the above proposed powering method with the measured ones on full scale ship for model-A and B respectively. Estimated result without taking the hull and WJP interaction into consideration ($C_1 = C_2 = C_3 = 0$) is also shown on

715

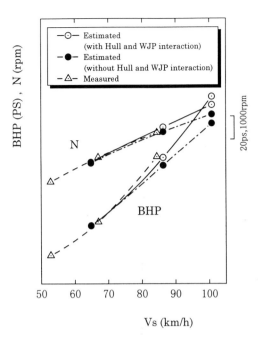

Fig. 3.4(a) Comparison of measured power
curves with estimated ones
(full scale , model-A)

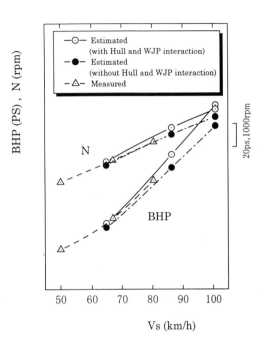

Fig. 3.4(b) Comparison of measured power
curves with estimated ones
(full scale , model-B)

each of Fig.3.4 (a) and (b). For the estimation,
ΔC_f=0 is adopted because scale ratio of model
ship is as large as 1/4. Fig.3.4(a) and(b) show
that the estimated result agrees fairly well with
the measured one, and also that the amount of
hull and WJP interaction is rather small at the
subject ship *IVR* range 0.4~0.5 and at rather
low ship speed range, and even the estimation
without taking the hull and WJP interaction
into consideration has practically enough
accuracy at the *IVR* and ship speed range.

4. CONCLUSIONS

Through this study, the following conclusions
have been obtained:
(1) A powering method for super-high speed
planing ships with WJP based on model test
on hull and that on WJP was proposed.
(2) The effectiveness of the proposed powering

method was evaluated by applying it to full
scale ships, and its effectiveness was confirmed.
(3) It was found that the amount of hull and
WJP interaction is rather small at the *IVR*
range 0.4 ~ 0.5 and at rather low ship speed
range, and even the estimation without taking
the interaction into consideration has
practically enough accuracy at the *IVR* and
ship speed range.

ACKNOWLEDGEMENTS

The authors would like to thank Prof.
H.Miyata of the Department of Environmental
and Ocean Engineering, University of Tokyo
for his discussion and encouragement.
The authors also would like to thank Dr.
R.Ogiwara, director of Akashi Technical
Institute, Kawasaki Heavy Industries for his
understanding the importance of this study and

716

encouragement.

Thanks are also extended to the staff of Fluid Dynamics Research Department, Akashi Technical Institute, Kawasaki Heavy Industries and the staff of Osaka Prefecture University who cooperated with the authors in this study.

REFERENCES

1) Funeno, I., et al., "Three-dimensional Flow Analysis around the Inlet of a Water-jet Propulsor" Kawasaki Technical Review No.135, Oct., 1997, P. 32

2) Hoshino, T. et al., "Determination of Propulsive Performance of Waterjet in Model and Full Scales" ITTC 1996, Supplement to the Report of the Waterjets Group

3) van Terwisga, T., "On the Description of Waterjet Powering Performance" ITTC 1996, Supplement to the Report of the Waterjets Group

4) Aren, P. et al., "Review of Test Methods Adopted for Water Jets at the KaMeWa Marine Laboratory" ITTC 1996, Supplement to the Report of the Waterjets Group

5) Waterjets Group "Final Report and Recomendations to the 21st ITTC"

6) Ikeda, Y., et al. "Development of an Experimental Method to assess the Performance of a High Speed Craft (1st Report)" J. Kansai Soc. N. A. , Japan, No.223, March 1995, P. 43

7) Ikeda, Y., et al. "Simulation of Running Attitude and Resistance of a High-Speed Craft Using a Database of Hydrodynamic Forces Obtained by Fully Captive Model Experiments" Proceedings of the Second International Conference on Fast Sea Transportation, Dec.1993. P583

Practical Design of Ships and Mobile Units
M.W.C. Oosterveld and S.G. Tan, editors.

LINEAR-Jet: A propulsion system for fast ships

M. Bohm[a] and D. Jürgens[b]

[a] Propellerhydrodynamics , Schiffbau-Versuchsanstalt Potsdam GmbH
Marquardter Chaussee 100, 14469 Potsdam, Germany

[b] Manoeuvring and propulsion department, JAFO Technologie, Member of the Blohm+Voss Group
Am Elbtunnel 6, P.O. Box 11 13 07, 20413 Hamburg, Germany

The JAFO-Technologie Hamburg and the Potsdam Model Basin (SVA) developed and investigated a propulsor for fast shallow-draught ships, unconventional ships especially SES, submarines and marine ships. The propulsor is called LINEAR-Jet. It works like a water jet. A rotor and a stator are ducted by a nozzle. The upstream and downstream pipes are cut off and the propulsor stands in the wake like a ducted propeller.

The propulsion system LINEAR -Jet has a couple of advantages especially for fast and flat going ships. The velocity in the area of the rotor is less than the velocity at the trailing edge of the nozzle. The inception of cavitation moves to higher advance coefficients. The cavitation behaviour was investigated in oblique flow in a range up to 5 degrees. The cavitation behaviour was nearly independent from the oblique flow. Propulsion tests have shown a reduction of the required power of 2.1% for a patrol boat at a ship speed of $V_S = 21.6$ kn. The thrust loading coefficient of a LINEAR-Jet can be up to ten times higher than the thrust loading coefficient of a ducted propeller, without cavitation problems. For that reason it is possible to reduce the diameter, compared with a free running propeller or reduce the propeller induced noise by reduction of the number of revolutions.

The stream of the propulsor is nearly spin free. If the LINEAR-Jet is powered by a Z-drive, the shaft is covered by a blade of the stator. For shaft-powered LINEAR-Jets working with a rudder, the danger of rudder cavitation is low because the stream is nearly spin free.

For these reasons the LINEAR-Jet is an attractive alternative for fast ships; especially with a design speed between $V_S = 20$ and 30 kn and for ships with a high thrust loading coefficient.

1. INTRODUCTION

Especially for the development of fast and/or unconventional ships suitable propulsion systems are necessary. The LINEAR-Jet is an alternative to the well known propulsion systems. The alternative propulsors will be discussed for a comparison with the LINEAR-Jet.

Depending on the ship design, propeller configurations for fast ships are difficult. Usually the propeller is powered by an inclined shaft. The shafts are supported by a boss and struts. The designers have to find a compromise between the optimum diameter of the propeller and the angle of the inclined shaft. The inclined upstream increases the risk of cavitation of the propeller. For shallow-draught ships there are relative small and high loaded propellers. For higher speeds it is impossible to get no cavitation. Because of cloud and bubble cavitation the propeller could be destroyed by erosion. Pressure pulses increase and there could be thrust deduction.

Water jets are often used for high speed ships. Usually the water enters the ship through an inlet in the bottom. Pipes lead the water to a pump. The pump increases the pressure level and the accelerated stream leaves the ships through a nozzle. The total efficiency of the system is influenced by the pump and the inlet. A good configuration delivers a high efficiency. For an unfavourable shape of the inlet flow separation and cavitation is possible. There is a risk of air suction and thrust reduction for surface effect ships.

The blades of surface piercing propellers are partly or full ventilated. To optimise the propeller in different conditions it is possible to make the pitch controllable. The thrust is completely generated by the pressure side, so that the thrust coefficient is low for a high efficiency.

2. CONSTRUCTION OF THE LINEAR-JET

The propulsor is a synthesis of a water jet and a ducted propeller. It is developed for fast ships in a speed range of $V_S = 20$ to 40 kn. Because the streamlines pass the propulsor linear, the propulsor is called LINEAR-Jet. The LINEAR-Jet was be divided into three parts: a rotor, a stator and a nozzle (Fig. 1).

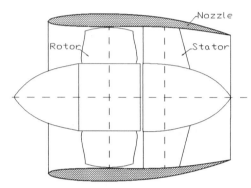

Figure 1. LINEAR-Jet

The construction is based on a water jet. The upstream and downstream pipes are cut off and the propulsor is positioned in the wake.

This construction has a lot of advantages in comparison with alternative propulsors:

- There isn't any loss of friction and jet deflection in the pipes.
- The nozzle reduces the stream from the leading edge to the rotor. So the risk of cavitation is reduced and moved to higher ship speeds. It reduces the pressure equitation from the pressure side to the suction side and reduces the noise because of the stationary pressure field. In addition to this the nozzle saves the rotor of damage.
- The combination of the rotor and the stator delivers a nearly spin free stream. This increases the efficiency and decreases the noise. A relative high thrust is available due to the induced velocities of the stator to the rotor In combination with the reduced velocity in the rotor plane it is possible to reduce the diameter and/or reduce the number of rotations.
- The optimal LINEAR-Jet has a relative big hub. This reduces the risk of hub cavitation.

A configuration with a LINEAR-Jet powered by a Z-drive gives some further advantages:

- No additional resistance by shafts, struts and rudders.
- No cavitation on rudder and struts. Higher efficiency in fact of axial inflow. The big ratio of the hub diameter to the rotor diameter gives enough space to integrate a gear box. The shaft of the gear box is covered by an enlarged stator blade.

Several arrangements of the LINEAR-Jet at the ship are possible. Powered by a shaft, by a Z-drive or electric.

3. CALCULATION METHOD

JAFO Technology developed a design tool for the LINEAR-Jet [1]. It is a re-calculation method. Giving the design point and a starting point, the propulsors data will calculate iterative by variation of single parameters.

The interaction of the rotor and the stator is calculated after Gibson and Lewis [2].

To calculate the flow through the LINEAR-Jet, a powerful vortex panel method was developed. The singularities are located on the surfaces of the blades. The strength of the singularities are calculated with a numerical program. So the normally used correction factors are not needed [3]. The rotor is simulated by a simple model. It is sufficient to replace the rotor with a semi-infinite vortex cylinder [4-5].

Fig. 2 shows the scheme of the program. In the following chapters, the program is described more in detail.

3.1 Calculation of the flow round the nozzle

The calculation of the flow round the nozzle is based in majority on Gibson and Lewis [4].

Following assumptions are made:

- The flow is symmetrical to the axis.
- The effect of the rotor-stator combination is simulated by a vortex cylinder.
- The surface of the nozzle is simulated by vortex rings with a constant circulation around the circumference.
- For the nozzle, the pressure equitation at the tip of the rotor blades is neglected.
- The influence of the finite number of blades is corrected after Tachmindji and Milam [6], the boundaries are corrected after Kopeeckij [7].

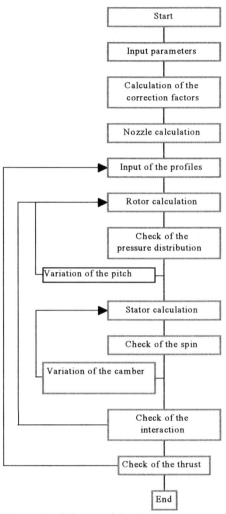

Figure 2. Scheme of the design program for
LINEAR-Jets

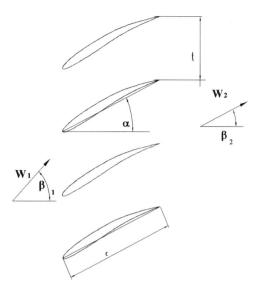

Figure 3. Geometrical and kinematical parameters
of the grid flow

The strengths of the circulation are calculated
with the requirement on the surface of the nozzle
and the Kutta condition at the trailing edge. For the
strengths of the circulation we get the following
Fredholm formula.

$$-\frac{\gamma}{2} + \oint_{S_D} tu_i^* \gamma_D(s)ds + tu_P^* \gamma_P + tu_a = 0 \qquad (3)$$

The solution of this equation is similar to the
solution of the grid flow in chapter 3.3.

The pressure coefficient of the nozzle is:

$$c_p = 1 - \left(\frac{\gamma_D}{U_\infty}\right)^2 \qquad (4)$$

By integration with the x-component of the normal vector n_x, we get the thrust of the nozzle:

$$T_D = \int_{S_D} c_p n_x \, {}^\rho\!/_2 \, u^2 ds \qquad (5)$$

The influence of the nozzle is considered by a
global correction.

The hub is modelled by ring vortexes. For a given
velocity in the plane of the rotor and the stator we
make demands for a minimum resistance of the
nozzle.

A vortex panel method was developed for the
calculation of the pressure distribution on the nozzle
and the induced velocity in the planes of the rotor
and the stator. N control points are located on the
surface of the nozzle, in the middle of two contour
points (Figure 3). A vortex ring around the nozzle is
located on each control point.

$$x_i = 1 - \sin^n(\pi\frac{i-1}{N}) \qquad y_i = f(x_i) \qquad (1)$$

The requirement in the control points is:

$$u_{ti} = 0 \qquad (2)$$

3.2 Influence of the rotor on the nozzle

The modelling of the rotor as a semi-infinite vortex cylinder leads to considerable simplifications.

The induced velocities of the vortex cylinder are calculated by the formula of <u>Biot-Savart</u> [4]. For the ring vortexes the integration of the Biot-Savart formula is necessary .

3.3 Calculation of the grid flow

The profile grids are calculated with the potentional theory. The presented design program calculates with the finite thickness of the blades. The following parameters are relevant for the program (Fig3.):
- Shape of the profile.
- Ratio of the gap t of the profiles and the chord length c.
- Geometrical angle of attack of the profiles α.
- Hydraulic angle of attack β_1.

The strength of the circulation of a vortex sheet with the length ds is:

$$\gamma ds = (u_t - u_{ti})ds \qquad (6)$$

The vortexes are located on the surfaces of the blades. There are no sources. If the velocity inside the blade u_{ti} is zero, the strength of the circulation on the surface of the blade γ is identical to the tangential velocity u_t on the blade.

To determine the strength of the circulation we calculate the velocity for each point on the surface. With $u_{ti} = 0$ we get the Fredholm formula for the distribution of the surface.

$$-\frac{\gamma}{2} + \oint u^* t\gamma ds + u_a t = 0 \qquad (7)$$

The factor for the velocity u^*, calculated with the Biot-Savart formula, depends on the geometry. T is the tangential vector of the profile contour and u_a the upstream velocity.

We calculate the unknown strength of the circulation with a numerical program. The induced velocities from the single panels are integrated by the trapezoidal formula. The induced velocities

from the single vortexes are known and described analytic.

Based on the flow around a single profile, the profile grid will be calculated. The components of the factors of the velocities are:

$$u_x^* = \frac{1}{2\pi} \frac{(y_a - y_w)}{(x_w - x_a)^2 + (y_w - y_a)^2} \qquad (8a)$$

$$u_y^* = \frac{1}{2\pi} \frac{(x_a - x_w)}{(x_w - x_a)^2 + (y_w - y_a)^2} \qquad (8b)$$

For a row of vortexes the induced velocity on a control point could be calculated by an analytical formula. It is calculated in the complex plane (Fig. 4). For a row of vortexes with a gap t between the single vortexes, located on the imagine axis, the conjugated complex velocity is:

$$\bar{u}^* = -\frac{i\Gamma}{2t} \frac{\cos\left(\dfrac{\pi D}{t}\right)}{\sin\left(\dfrac{\pi D}{t}\right)} = -\frac{i\Gamma}{2t} \cot\left(\frac{\pi D}{t}\right) \qquad (9)$$

z = x+iy describes the location of the <u>single point</u>. The Fredholm formula (7) will be transformed to a linear equation system:

$$\sum_{j=1}^{N} u_{ij}^* \gamma = -u_a t_i \qquad (i = 1,N) \qquad (10)$$

The distribution of the control points and the contour points is similar to the distribution of the nozzle (Fig. 5) . To fulfil the Kutta requirement we set the pressure on the pressure side equal to the pressure on the suction side at the trailing edge.

$$\gamma(N) = -\gamma(1) \qquad (11)$$

Insert formula (11) in (10); the equation system is solvable and we get the unknown strength of circulation.

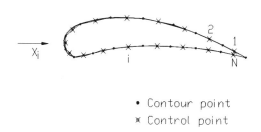

Figure 4. Distribution of contour and control points

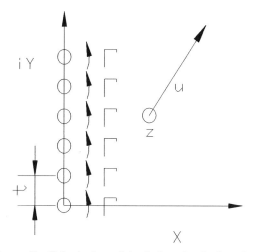

Figure 5. Calculation of the induced velocity of a
vortex row on a point

3.5 Design concept

The design conditions implicit a high volumetric
discharge rate and a low discharge head. To decrease the risk of cavitation, the nozzle is designed
as a diffusor from the leading edge to the rotor.

The calculation method considers all interactions
of the single components. A powerful singular method gives a good simulation stream characteristic,
especially the pressure distribution. Factors correct
the influence of the finite number of blades and the
boundaries.

The nozzle is replaced in the calculation by vortexes with constant circulation around the circle.
The rotor is simulated by a semi-infinite vortex
cylinder.

This design tool was validated in different investigations [8-10]. A comparison of the latest investiga-

tions with the predicted behaviour is given in Fig. 6.
The induced velocities of the rotor, the stator and
the nozzle were confirmed. The calculated thrust of
the nozzle was corrected with former experimental
data. So it is possible to design every rotor-stator
combination for an investigated nozzle with an
acceptable accuracy. The cavitation behaviour will
be predicted by a comparison of the local pressure
with the vapour pressure.

4. DESIGN OF A LINEAR-JET FOR A PATROL BOAT

The KBN-Konstruktionsbüro Nord in Bremen developed a patrol boat for German country Saxon-
Anhalt. After consultation of KBN a LINEAR-Jet
was designed for this small and fast ship. The
design speed is $V_S = 21.6$ kn. A model of this ship
was investigated for the shipyard Genthin in the
SVA Potsdam. The original design with two
propellers at inclined shafts was compared with the
new design with two LINEAR-Jets powered by
Z-drives. For this design the shafts, the struts an the
rudder were taken away. The tunnels of the boat
were partly refilled to adapt the upstream and the
downstream for the new propulsion system. The
nozzles were partly integrated. Based on the geometrical and hydrodynamic mean data in Table 1
and 2 we get the mean design parameters in
Table 3. The mean parameters of the rotor, the stator and the nozzle are summarised in table 4 and 5.

Table 1
Mean data of the patrol boat

			Propeller	LINEAR-Jet
Length	L_{PP}	[m] :	13.20	
Breadth	B	[m] :	4.00	
Draught	T	[m] :	0.86	
Displacement	∇	[m³] :	14.42	14.67
Wetted surface	S	[m²] :	46.47	45.19

Table 2
Hydrodynamic data

Thrust	T [kN] :	11.111
Ship velocity	V_s [kn] :	21.6
Wake number	w [-] :	0.049
Diameter of the rotor	D [m] :	0.55
Number of rotation	n [1/s] :	13.9
Draught of the shaft	h_0 [m] :	0.57

722

Table 3
Design data

Advance coefficient	J^*	[-] :	1.383
Cavitation number	σ^*	[-] :	3.66
Total thrust coefficient K_{TT}^*		[-] :	0.628

Table 4
Mean data of the LINEAR-Jet

			Rotor	Stator
Diameter	D	[m] :	0.55	0.55
Pitch ratio	P/D	[-] :	2.086	25.95
Numbers of blades	Z	[-] :	5	7
Diameter ratio	d_H/D	[-] :	0.48	0.47

Table 5
Mean data of the nozzle

Inner Diameter	D	[m] :	0.551
Length ratio	L/D	[-] :	1.2
Inflow diameter	Di	[m] :	0.589
Outflow diameter	Da	[-] :	0.482

Two LINEAR-Jets were manufactured in a scale $\lambda = 2.2$ in SVA Potsdam. For the propulsion tests with the stock propellers, Gawn propellers 3.80 with following mean data were used.

Table 6
Mean data of the model propeller

Diameter	D	[m]	0.130
Pitch ratio	P/D	[-]	1.4
Numbers of blades	Z	[-]	3
Area ratio	A_E/A_0	[-]	0.8

5. MODEL TESTS

Model tests were carried out to validate the prediction of the open water characteristic and to check the influence of the oblique flow on the open water characteristic and the cavitation behaviour. Additionally the influence of the strut for a Z-drive was investigated. Finally propulsion tests were carried out to compare a LINEAR-Jet configuration with a propeller configuration for a patrol boat. The cavitation tests were carried out in the cavitation tunnel of the SVA Potsdam. The torque and the thrust of the rotor were measured by a dynamometer located upstream. The propulsion tests were made in the towing tank of the SVA.

5.1 Open water characteristic

The LINEAR-Jet has a high efficiency for a width area of advance coefficients. The maximum efficiency of the model is about $\eta = 70\%$ (Fig. 6). Due to the induction of the velocities of the single components of the LINEAR-Jet, the inflow to the rotor is shockless for a width area of advance coefficients. Depending on the design, the combination of the nozzle and the stator delivers small resistance or small thrust in the design point. With a good adjustment of the components the stream of the LINEAR-Jet is nearly spin free (Fig. 7). Only behind the hub in the interior of the jet are higher tangential velocities. There is no flow separation at the nozzle.

For the characteristic of the engine, the torque characteristic has a good, flat shape. For these reasons, the LINEAR-Jet is suitable for fast ships with medium or fast rotating mean engines.

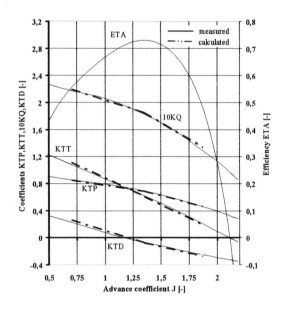

Figure 6. Calculated and measured open water characteristic

The rotor is normally powered by a gear box. So it is possible to drive backwards. Investigations show adequate backward behaviour of this system. The reason for that is the moderate contraction of the nozzle and the nearly axial orientation of the stator blades.

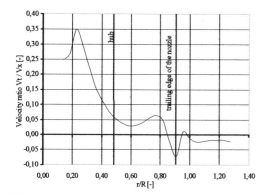

Figure 7. Tangential velocity behind the stator

5.2 Influence of the oblique flow

The influence of the oblique flow on the open water characteristic was investigated up to an angle of 5 degrees. The thrust and the torque of the rotor were nearly independent of the inflow angle in the investigated range. Only the thrust of the nozzle decreases, which was measured together with the stator. It decreases ca. $K_{TD} = -0.014$ per degree (Fig. 8). The cavitation behaviour is nearly independent of the angle. This is the effect of the nozzle, which guarantees an axial inflow for the rotor.

So it is possible to install the LINEAR-Jet at an incline shaft, without getting cavitation problems at least up to the investigated angle of 5 degrees.

5.3 Influence of the strut of the Z-drive

To simulate the influence of the partial blocking of the nozzle by the shaft of a Z-drive, a blade of the stator covered by a profile. The dimension of the profile is calculated by the minimum shaft diameter for full scale.

The influence of the covered shaft of the Z-drive on the open water characteristic is low. The torque coefficient increases from $K_Q = 0.179$ to $K_Q = 0.182$. The thrust coefficient of the propulsor decreases from $K_{TT} = 0.574$ to $K_{TT} = 0.568$ (Fig. 9).

5.4 Propulsion tests

For a comparison, propulsion tests were carried out with two CPPs type Gawn 3.80 (Table 6). The propellers were powered by an inclined shaft of 8 degrees. For the tests with the LINEAR-Jets, the lines were adapted. The tunnels were partly refilled to get an optimal inflow to the LINEAR-Jets.

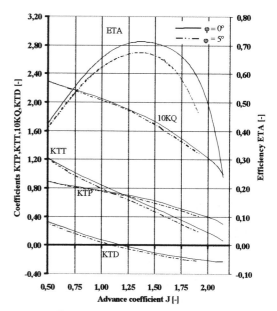

Figure 8. Influence of the oblique flow on the open water characteristic of the LINEAR-Jet

The Jets were partly integrated in the tunnels. For backward investigations, the LINEAR-Jets were rotated by 180 degrees.

The tests were carried out for the same draught. For the little difference in the displacement, the results of the tests were referenced to the Froude number related to the displacement.

$$Fn = \frac{v}{\sqrt{g\nabla^{1/3}}} \qquad (12)$$

The patrol boat with the LINEAR-Jets needs more power for low ship speeds. From $V_s = 14$ kn and higher, the configuration needs less power. For the maximum design speed $V_s = 21.6$ kn the new propulsion system needs 2.1 % less power as the conventional system with the propellers (Fig. 10).

There was not made any prognosis for the full scale ship. The difference of the needed power will increase for the LINEAR-Jet configuration. Due to a correction of the Reynolds number, the efficiency will increase for both systems. For the LINEAR-Jet system more, because rotor, stator and nozzle have to be corrected and for the propeller system only the blades.

724

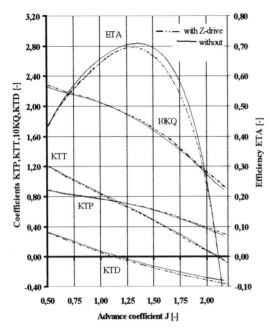

Figure 9. Influence of the Z-drive strut on the open
water characteristic

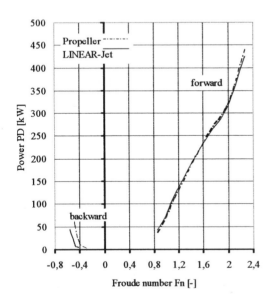

Figure 10. Propulsion tests

6. CONCLUSIONS

The propulsion system LINEAR-Jet was developed in different R&D projects as an alternative propulsion system for fast ships.

The LINEAR-Jet system has following mean advantages:

- High efficiency for a wide region of advance coefficients.
- Flat shape of the torque coefficient.
- Minimum loss because of low tangential velocity components.
- High safety against cavitation at the hub for a great hub diameter ratio.
- Rotor diameter can be reduced.
- Number of rotation can be lower.
- Lower noise level because of the nozzle, the nearly spin free stream and the low number of rotation.
- No additional resistance by inclined shafts, struts and rudders.
- Protection of the rotor by the nozzle.
- No cavitation on struts and rudders.
- Safety against cavitation by axial inflow to the rotor.
- The LINEAR-Jet can be used as a rudder propulsor.

Symbols

c	m	chord length
c_p	-	pressure coefficient
dn	m	diameter of the hub
D	m	diameter
Fn	-	Froude number
g	m²/s	gravity constant
J	-	advance coefficient
K_Q	-	coefficient of the moment
K_{TD}	-	coefficient of the nozzle and the stator thrust
K_{TP}	-	coefficient of the rotor thrust
K_{TT}	-	coefficient of the total thrust
n	m	vector normal to the surface
nx	m	x component of n
S_D	m²	Surface of the nozzle
t	m	tangential vector
t	m	gap between two profiles
T	N	trust
T_D	N	trust of the nozzle
V_s	kn	Ship speed
u	m/s	velocity
u_a	m/s	velocity vector of the inflow
u_{ti}	m/s	velocity inside the blade
u_∞	m/s	velocity vector of the inflow
u*	-	factor for the velocity
v	-	velocity
x_i	-	normalised co-ordinate
y_i	-	normalised co-ordinate
ρ	kg/m³	density of the water
γ	m/s	strength of circulation
α	deg	geometrical angle of attack
β	deg	hydraulic angle of attack
η	–	efficiency
σ	–	cavitation number
Γ	m²/s	circulation
∇	m³	displacement

Indicees

a	-	position of an investigated point
w	-	position of a vortex

REFERENCES

1 Thieme, C. ; Jürgens, D. ; Delius, K.
Antriebssystem für flachgehende Hochgeschwin-
digkeitsfahrzeuge, JAFO Technologie 1994.

2 Gibson, B.S., Lewis, R.I.
Ducted Propeller Analysis by Surface Vorticity
and Actuator Disc Theory, Symposium on
ducted propellers, RINA, 1973, S.1-10.

3 Morgan, W.B., Silovic, V., Denny, St. B.
Propeller Lifting Surface Corrections, HYA
Paper 11, Lyngby, Denmark

4 Gibson, B.S., Lewis, R.I.
Ducted Propeller Analysis by Surface Vorticy
and Actuator Disc Theory, Symposium of ducted
propellers, RINA, 1973, S. 1-10

5 Philipp, O.
Untersuchungen der Wechselwirkungen
zwischen einem Propeller und einer axialsym-
metrischen Düse beim Entwurf von Düsen-
propellern, Schiffbauforschung 21 (1982)

6 Tachmindji, A.J. , Milam, A. B.
The Calculation of the Circulation Distribution
for Propellers with Finite Hub having Three,
Four, Five and Six Blades, ISP 4 (1957), No. 37

7 Kopeeckij, V. V.
Gidrodinamika vinta w trube krugovov secenija
Sudopromgiz, Leningrad 1956

8 Heinke,H.-J., List, S.
Hydrodynamische Untersuchungen mit einem
LINEAR-Jet, SVA Report 1980, Febr. 1993.

9 Bohm, M. , Heinke, H.-J.
Hydrodynamische Untersuchungen mit Modell-
varianten eines LINEAR-Jets, SVA Report
2053, March 1994.

10 Bohm, M. , Jürgens, D.
Entwicklung des Antriebssystems LINEAR-Jet
für Yachten, SVA Report 2272, Dec. 1996.

Practical Design of Ships and Mobile Units
M.W.C. Oosterveld and S.G. Tan, editors.

A Dynamic Model for the Performance Prediction of a Waterjet Propulsion System

Giovanni Benvenuto[a], Ugo Campora[b], Massimo Figari[a], Valerio Ruggiero[a]

[a]Dipartimento di Ingegneria Navale e Tecnologie Marine, Università di Genova
Via Montallegro, 1 - I - 16145 Genova - Italy

[b]Istituto di Macchine e Sistemi Energetici, Università di Genova
Via Montallegro, 1 - I - 16145 Genova – Italy

In designing the propulsion plant of a high-speed vessel it is important to investigate, in addition to the steady behaviour, the dynamic performance of the system over a wide range of operating conditions.

This paper presents a mathematical model for the dynamic simulation of a waterjet propulsion system. The model, developed in a Matlab-Simulink software environment, is structured in modular form, in order to describe the various elements of the system as individuals blocks (hull, prime mover, gear, waterjet, etc.), taking into account their interactions. In this way it is possible to characterise the dynamic behaviour of both the single components and of the whole propulsion plant.

The present work represents an extension to the high-speed vessels of a simulation model already developed and published by some of the authors for the performance prediction of marine propulsion plants. In its earlier version the model was provided with a block representing a controllable pitch propeller. In the present version this block has been substituted by a waterjet block, whose inclusion has required, however, some further modifications of the simulation scheme.

A first validation of the model has been obtained by comparing the results given by the simulation with the experimental data surveyed on board of a built vessel, showing a good agreement between them.

1. INTRODUCTION

As it is well known, the waterjet propulsion can offer many advantages over conventional open propeller arrangements particularly at speeds near or in excess of about 35 knots, that is in the range where the waterjet efficiency may reach its optimum value. At present there is a growing number of waterjet installations in different types of fast vessel such as sports boats, naval ships, passenger ferries, surface effect ships. Certainly the great development of fast passenger and car ferries has contributed to the success of waterjet propulsion.

Because of the increasing number of installations and to their growing size and power, it becomes most important a better knowledge of these systems, which can be obtained by investigating their working by means of experimental tests and with the aid of simulation techniques.

In this context the paper describes a dynamic simulation model, developed in a dedicated software environment, for the performance prediction of a waterjet propulsion system.

The different elements of the propulsion plant, such as hull, prime mover, waterjet, etc., have been modelled using blocks, whose behaviour is described by tables, mathematical operators and algebraic or differential equations.

The model may be used to optimise the choice of the propulsion system components at steady-state design conditions, but especially to analyse the system response at off-design and transient conditions.

In particular the developed computer simulation code may be considered as an useful tool both to facilitate the correct matching of the prime mover (diesel or gas turbine) to waterjet and ship in the proper range of operating conditions and to foresee the propulsion plant behaviour in presence of critical situations or when changing the normal working conditions. The simulation may be used, for instance, to evaluate fully all manoeuvring situations which can be envisaged in service (stop–full-ahead, half-ahead–full-ahead, crash–stop, etc.), or to predict the influence on the ship propulsion of adverse environmental conditions (heavy sea, wind), leading to an increase of the hull resistance.

The model, which can be applied to different plant configurations, requires, as input data, a set of information regarding the hull type and geometry, the prime mover performance and operating range, the waterjet arrangement and pump characteristics. In turn the model gives, as output values, the time histories of all the considered variables (such as, for instance, ship speed, waterjet shaft r.p.m. and torque, waterjet thrust, pump mass flow rate, engine power and fuel consumption, etc.) which define the dynamic behaviour of the system.

2. PROPULSION SYSTEM SIMULATION

The application of simulation techniques to marine propulsion systems represents a research field developed at the University of Genova since some years, which has already allowed to obtain interesting results, as shown in [1], [2]. Within this research activity a mathematical model has been developed able to describe the various elements of the propulsion chain (hull, propulsors, diesel engines, etc.) and their interaction. Recently the code was upgraded by the implementation of a

multiple zone combustion model into the engine block, allowing to obtain additional information on the exhaust gas emissions.

The software used for the propulsion plant simulation is SIMULINK, a powerful toolbox of MATLAB. Within this software environment it is possible to simulate the dynamic behaviour of a system using several types of pre-defined or custom blocks connected together in a graphical scheme.

The differential equations describing the process can be solved by numerical integration in the time domain by means of integration blocks with proper initial conditions.

Among the various algorithms available in SIMULINK [3] for the solution of differential equations, the Euler and Gear methods have been adopted, giving good performance with the present model, characterised by many non-linear blocks (look-up tables, rate and level limiters, delay blocks) and by elements with time constants very different from each other, like those related to ship mass and engine inertia ("stiff system"). Since the MATLAB is an interpreter language, in order to save computational time, some parts of the code are written in C++ and compiled and linked to MATLAB as Mex files. All of the physical variables of interest are stored in memory (Work-Space) for monitoring.

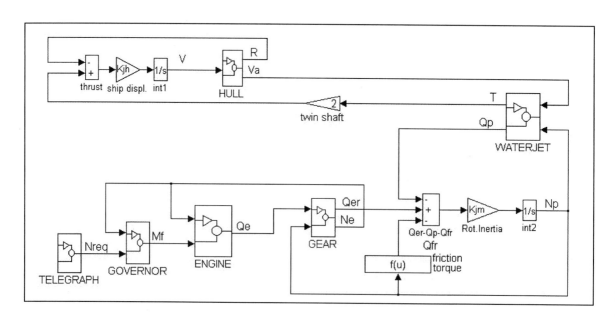

Figure 1. Waterjet propulsion system simulator

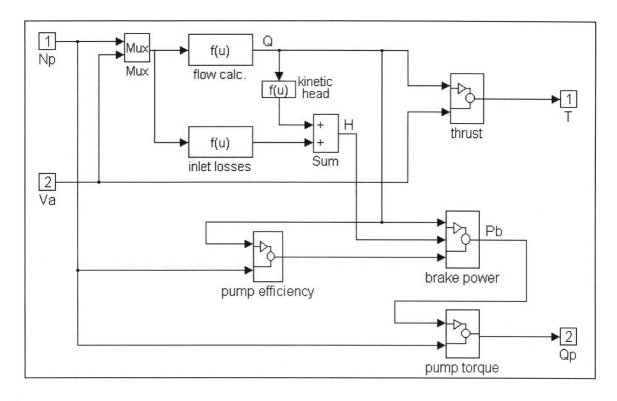

Figure 2. Waterjet block

Some post-processing programs were developed for plotting the simulation results or printing all the monitored variables as functions of time, or in correspondence of a particular simulation step.

In its original version the model was applied to simulate the propulsion system of a naval vessel, consisting of a twin shaft arrangement with controllable pitch propellers, driven, trough reduction gears, by turbo charged medium-speed diesel engines.

The main elements of the propulsion plant, modelled as separated blocks, were: the diesel engine with its turbocharging system, the engine governor, the hull, the controllable pitch propellers and the telegraph. The interested reader is referred to [1] and [2] for a detailed description of the different blocks. Here it will be pointed out only that the main differential equations appearing in the whole scheme are those relative to the ship acceleration and to the engine shaft dynamics.

In the modified version of the simulation model presented in this paper, the controllable pitch propeller block has been removed and replaced by a new block, representing the waterjet behaviour, as shown in fig. 1. However, the general structure of the simulation scheme has been maintained so that, for instance, the main differential equations are the same as before and the waterjet block, even if very different from the propeller block, may be considered, at a first sight, as a black box with the same input and output variables.

3. THE WATERJET BLOCK

The SIMULINK graphical representation of this block is shown in fig. 2, which is an expansion of the same block appearing in compressed form in fig. 1. The input variables entering the waterjet block are the effective velocity of approach Va, as defined in [4], and the pump rotational speed Np. The output variables are the thrust T of the waterjet and the torque Qp required by the pump.

According to [4], the effective velocity of approach (fig. 3), evaluated at a station in front of

730

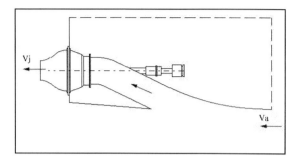

Figure 3. Velocities definition

the intake ramp tangency point and supposed uniform, is given by Va = (1-w)V, where (1-w) is the effective Taylor wake fraction and V is the ship velocity.

Neglecting the difference between Va and Vm (momentum velocity), the thrust T may be calculated, as shown in [4], [5], from the change in momentum:

$$T = \rho Q \, (Vj-Va) \qquad (1)$$

where Q = Vj Aj is the pump flow rate, ρ is the water mass density and Vj, Aj are the jet velocity and the jet area respectively. According to [5], a thrust deduction fraction t = 1-R/T may be defined in analogy with the screw propulsion coefficients, R being the ship resistance at ship speed V.

The total head rise H of the pump is equal to the head needed to produce the jet and to overcome the ducting losses minus the recovered head of the incoming flow:

$$H = Vj^2/(2g\eta_N) - (1-k) \, Va^2/2g + \Delta h \qquad (2)$$

with η_N nozzle efficiency, k loss coefficient of inlet ducting and Δh elevation of jet.

3.1. Pump flow rate calculation

In order to apply the thrust equation (1) it is necessary to calculate the pump flow rate Q. For a given pump the operational point follows from the intersection of the head capacity curve of the pump with the head curve required by the system.

If the nozzle area is known, it is a simple matter to evaluate the jet area Aj; then, taking into account that the jet velocity may be written as Vj = Q/Aj , it is possible to express the total head rise required by the ducting system, given by eq. 2, as function of Q:

$$H = Q^2/(2g\eta_N Aj^2) - (1-k) \, Va^2/2g + \Delta h \qquad (3)$$

The head curve expressed by the above equation, once Va is given and k is evaluated, has to be compared with the characteristic head capacity curve of the pump corresponding to the given rotational speed Np. This curve has been obtained, in the present model, by using the following analytical expression, based on a quadratic regression, able to fit the pump experimental data taking into account the similarity laws:

$$H=(Np/Np_0)^2(a(Np_0/Np)^2Q^2+b(Np_0/Np)Q+c) \qquad (4)$$

with Np_0 reference speed of rotation and a, b, c constants to be determined.

By comparing eq. 2 and eq. 3, a quadratic equation is obtained, whose closed form solution gives the required value of the pump flow rate Q.

3.2. Waterjet thrust and pump torque calculation

Once determined the pump flow rate, the waterjet thrust may be easily calculated by using eq. 1. The pump torque may be obtained from the power demand at the prime mover:

$$P_B = \rho g \, QH/(\eta_P \, \eta_r \, \eta_m) \qquad (5)$$

with η_P pump efficiency, η_r relative rotative pump efficiency and η_m mechanical efficiency ([5]). If the pump rotational speed Np is expressed in r.p.m. than the pump torque is given by:

$$Qp = 60 \, P_B/(2\pi Np) \qquad (6)$$

This torque is compared, in the simulation scheme, with the engine torque, taking into account the presence of the reduction gear.

4. STEADY STATE RESULTS

The developed simulation model has been tested in order to verify its validity in predicting the propulsion plant performance both in steady state conditions and during transient situations. Although based on some simplifying assumptions the model gives substantially good results, as will be proved by means of a comparison with available experimental data.

One of the advantages of the considered methodology is that its implementation requires a limited amount of information on the different components of the propulsion system. Among the necessary input data, the following items have to be

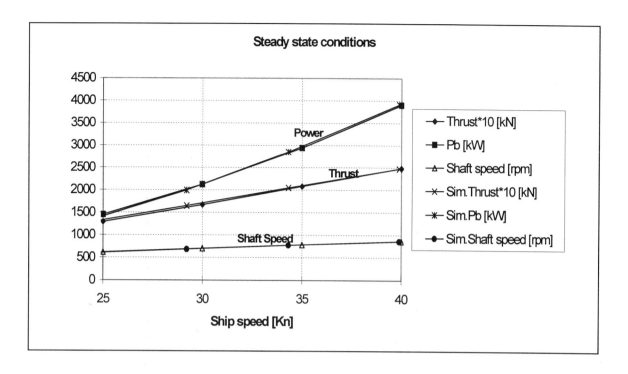

Figure 4. Comparison of simulated and experimental results

mentioned:
- ship displacement and hull resistance curve;
- waterjet configuration and main dimensions; nozzle area;
- head capacity curves of the pump;
- engine and turbocharging system data;
- reduction gear data.

However the main merit of the model is that it is able, in authors' opinion, to capture the physics of the phenomena under investigation and to give useful indications on the influence of the various parameters involved.

Fig. 4 shows a comparison of the steady state results given by the simulation with those provided by an important waterjet manufacturer with reference to a twin waterjet propulsion system. Four conditions have been analysed, corresponding to different values of the brake power. As it may be observed, the agreement is good for all of the considered parameters (thrust, shaft speed, power) and over a wide range of operating conditions. The same conclusion may be drawn by examining the behaviour of other variables, not reported in the

paper, for which the simulated values are characterised by percent errors less than about 2%.

5. TRANSIENT RESULTS

The model has been applied to study also the transient behaviour of the propulsion system during typical ship manoeuvres. In particular the presented results refer to a stop–full-ahead manoeuvre of the same ship considered in the steady state calculations.

Fig. 5 shows a typical sequence of events which can occur under the action of the speed demand law (engine r.p.m. law) imposed by the control system when the bridge control lever (telegraph) is moved from the stop to full-ahead position. Clearly, if a different speed demand law is imposed, the system transient response will change, but the steady state values reached by the different variables will be the same.

The time histories represented in fig. 5 regard only the main variables of the propulsion system (brake power, thrust, shaft speed, ship speed), but actually much more parameters, concerning the

732

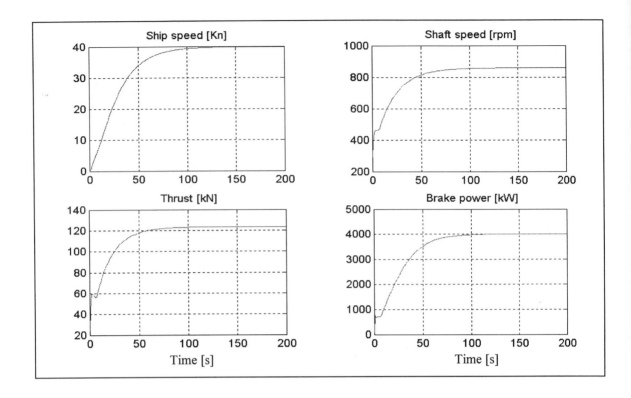

Figure 5. Propulsion system transient results

different components of the propulsion plant, are calculated and monitored by the simulation model.

As an example, some results describing the diesel engine dynamic response are reported in fig. 6, in dimensionless form, as a fraction of the different variables referred to nominal conditions.

The transient engine load influences not only the air/fuel ratio and the turbocharger speed but also parameters inherent to the thermodynamic cycle inside the cylinders, such as the maximum cycle pressure and temperature. Also the exhaust gas emissions, that the present version of the code, with the inclusion of a multiple zone combustion model, is able to calculate, are influenced by the transient engine load.

The presented results are only an example of the amount of information that can be obtained by the described simulation model, which may easily applied to study different situations and, because of its modular arrangement, may be extended also, with minor implementations, to other propulsion plant configurations.

6. CONCLUSIONS

This paper has illustrated a newly developed simulation model for the behaviour prediction of a waterjet propulsion system.

Though based on some simplifying assumptions, the model is able to capture the physics of the phenomena under investigation, as demonstrated by comparing the simulated results with available experimental data. In particular the considered test case shows that the accuracy of the results is good over a wide range of operating conditions.

The modular arrangement of the model and the adopted techniques of solution contribute to the good stability of the code and to its ability to handle quite severe transient conditions, with reasonable computer time.

The developed methodology gives a remarkable set of information on the behaviour of the propulsion plant and of its components, but above all it allows an improved physical understanding of the system, giving useful indications on the influence of the

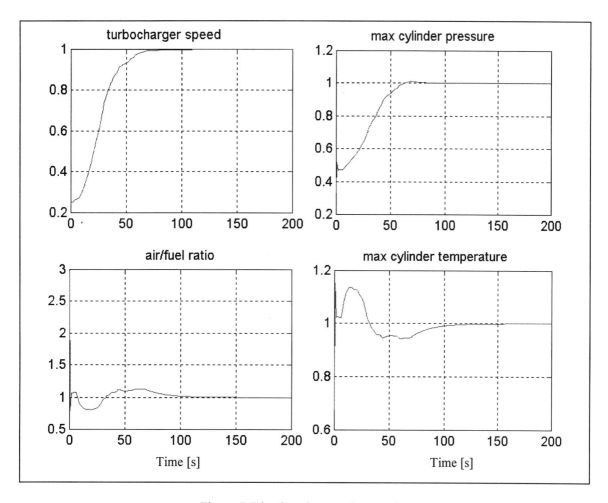

Figure 6. Diesel engine transient results

various parameters involved.

It is believed that the present model could be an useful tool both to favour the correct matching of the prime mover to waterjet and ship and to optimise the control system of the propulsion plant.

REFERENCES

1. Benvenuto, G., Carrera, G., Rizzuto, E., Dynamic Simulation of Marine Propulsion Plants, NAV 94, International Conference on Ship and Marine Research, Rome, October 2-7, 1994.

2. Benvenuto, G., Campora, U., Carrera, G., Rizzuto, E., A Dynamic Diesel Engine Model for Ship Manoeuvrability Prediction Studies, MARIND '96 – First International. Conference. on Marine Industry, Varna, Bulgaria, June 2-7, 1996.

3. SIMULINK User's Guide - The MathWorks, Inc.,1993.

4. Allison, J. L., "Marine Waterjet Propulsion", Transactions of SNAME, Vol. 101, 1993, pp. 275-335.

5. ITTC (1987) Report of the High Speed Marine Vehicle Committee, International Towing Tank Conference 1987, pp. 304-313.

Practical Design of Ships and Mobile Units
M.W.C. Oosterveld and S.G. Tan, editors.

HYDRODYNAMICS IN PRE-CONTRACT SHIP DESIGN

Janusz T. Stasiak

Technical University of Gdańsk, Faculty of Ocean Engineering & Ship Technology
Hydrodynamic Division
80-952 Gdańsk, G.Narutowicza Str. 11/12, POLAND

ABSTRACT

Bringing out the role of hydromechanics in the ship designing process some complex procedure regarding both seakeeping and resistance tools for developing the ship hull form is introduced. In particular, an essence of the algorithm of the OPTIMALSHIP computer program intended to the container vessel pre-contract design stage is presented. This program generates a ranking of the number of ship hull variants for specified voyage route and for required value of the vessel transportation capability. Thus, the program is a typical CAD tool which will never replace the ship designers but it should support them in designing better ships more efficiently.

The work reported here has been performed by the team of Hydrodynamic Division TUGdańsk within the research project sponsored by State Committee for Scientific Research - Ministry of Polish Government.

1. INTRODUCTION

There is no doubt that the hydromechanic qualities are the most important properties of the seagoing ship. They decide of a ship essence and determine its operational success : profitability and safety in particular. For this reason the role of hydromechanics in ship design should be a fundamental one. Unfortunately, this need has not been fulfiled yet.

For decades within the naval architecture a conventional piecemeal approach based on the Cartesian - mechanistic paradigm has still obliged. Under that approach a problem of interest is divided into elementary parts and next each detail is separately solved. In particular the seakeeping evaluation is, at most, considered as a secondary problem of ship designing following the choice of ship form considering the minimum calm water resistance and powering. Most often, this was reduced to conclusive evaluation only. In result of that an optimal ship can be found only by chance. However, the "bad ships" are, in general, made the performances of which are usually reduced as a result of even not large worsening of the environmental conditions.

There are at least three reasons for which the rough sea has to be included into the designing process of seagoing ships:

Firstly, the rough sea is the natural environment of these ships. As the statistics show at the open seas a wind and waving are the dominant phenomena whereas a smooth water is rather a rare event. Secondly, it is evident that both the wind and waves considerably affect almost all the aspects of ship behaviour. They reduce the ship performance and ship ability to carry out a specified duty. In particular, the rough sea creates a threat to ship safety. Thirdly, the sea waving not only produces the quite new ship features such as the seakeeping, but also turns all the ship proprieties into a system the features of which we should neither identify nor the more approximate with the calm water ones. By way of example, whereas the calm water ship speed is homogeneous problem determined only by resistance-propulsion ship characteristics, the rough sea speed is also the result a number of heterogeneous influences such as: ship stability, cargo securing, freeboard, body strength, etc. Thus, all the hydrodynamic aspects of seagoing ships should be investigated jointly - within the hull shape designing the piecemeal approach must be replaced by overall-system one.

Both experience and theoretical analysis have shown that calm water ship proprieties are more sensitive to the local hull geometry changes than seakeeping ones. Thus, if ship hull is optimized first for seakeeping, it is likely to have enough flexibility for additional modifications in local geometry to reduce calm-water resistance e.g., but not vice versa. These denote, that improvement of the calm-water ship resistance e.g. should be preceded by ship hull optimization in respect of the seakeeping..

The ship designing is, as a rule at least, a decision making process of hierarchic type within of which we can distinguish the three main phases following one after the other:
- a fleet optimization, where type of the ship and its transportation capability (i.e. both carrying capacity and mean voyage speed) are determined;
- a pre-contract (concept) design the aim of which is, among other things, a selection of dimensions, proportions and coefficients of the ship hull for a given ship transportation capability;
- a post-contract (detailed) design where the details of hull shape are improved and also the problems of ship structure and equipment are solved.

It is evident, that as the design progresses and decisions are made, the freedom to make changes as one proceeds is reduced. It is also estimated ([1] e.g), that the concept design which is generally less than 5% of total design costs commits 40% to 60% of later spending and that cost of design changes increases by an order of magnitude at each major stage of design.

In order to make the pre-contract design container ships more rational and more efficient, the computer program called OPTIMALSHIP has been done. This program is especially intended for quick studying a wide variety of hull forms to assist the shipping company in precise statement of the contract requirements. It is here considered that the principal parameters of ship hull should be, a part of these requirements and hence they should be determined under the ship owner surveillance. The fact of the matter is that, the shipyard is solely interested in fulfilment of the contract requirements but not in improvement of the ship's form.

2. AN ESSENCE OF THE PROGRAM

The OPTIMALSHIP program belongs to the CAD area. Its task is aid of a decision-maker in proper selection of the global parameters of hull form of the container ship to be investigated. The inverse analysis, necessary to this end, is realized by means of number of the forward analyses of design variants automatically created in accordance with principle of the factorial design. Strictly speaking, the program generates the ranking of 23 variants of ship hull for specified voyage route and for required value of the ship transportation capability.

Each hull variant is characterized by the values of 7 hull parameters $X\{x_i\}$ definied as follows:
$$x_1 = \nabla \quad \text{(ship displacement)}$$
$$x_2 = T/L \quad \text{(draught to length ratio)}$$
$$x_3 = B/T \quad \text{(breadth to length ratio)}$$
$$x_4 = C_B \quad \text{(block coefficient)}$$
$$x_5 = C_W \quad \text{(waterplane coefficient)}$$
$$x_6 = LCB \quad \text{(longitudinal centre of buoyancy)}$$
$$x_7 = H/L \quad \text{(depth to length ratio)}$$

The variants are ordered according with the numerical values of the variant measure of merit (MOM) defined as:

$$C(X) = \frac{P_N(X) \cdot N_{TEU}^*}{DWT(X) \cdot N_{TEU}(X)} \quad (1)$$

where:
$P_N(X)$ - shaft horse-power (SHP) of variant,
$N_{TEU}(X)$- container capacity (number of TEU) of variant,
$DWT(X)$- variant's deadweight,
N_{TEU}^* - required container capacity.

To above end the following standarized index is also adopted:

$$W(X) = \frac{C(X) - C_{min}}{C_{max} - C_{min}} \;;\; W(X) \in <0,1> \quad (2)$$

where:
C_{min}, C_{max} are the MOM values of the best and the worst variant respectively.

3. THE PROGRAM WORKING

Two following questions are worked out within the program OPTIMALSHIP:
- distribution of the design variants $X\{x_i\}$ inside the determined space of hull form variability;
- calculation of the $C(X)$ values for each specified variant.

Both the procedures transform the information taken from input data and base data into the orderly, considering to $C(X)$ values, set of the identified variants $X\{x_i\}$).

The input data consist of :

- the required ship transportation capability i.e.: both TEU capacity N^*_{TEU} and expected voyage speed V^*;
- the provided two-way voyage route.

The data base consist of :
- the list of historic container ships where each ship is described by relation: $N_{TEU} \leftrightarrow X\{x_i\}$. This list contains dozens of ships and has still been updated;
- the weather-geometric profiles of following seven two-way routes : 1) Bishop Rock - New York, 2) Bishop Rock - Rio de Janerio, 3) Bishop Rock - Yokohama, 4) Bishop Rock - Melbourne, 5) New York - San Francisco, 6) Sydney - New York, 7) Yokohama - San Francisco.

All the profiles are identified on basis of information taken from [2] and [3] . Each route approximated by the line segments is described by the following parameters:
- the overall length of two-way route: S_{ABA},
- the line segment to overall route ratio :

$$p_i = S_i / S_{ABA} \text{ where: } S_{ABA} = \sum_i S_i$$

- the conditional distribution : $p_{i,k} = P[W_k / S_i]$ which determines the probability of appearance of the concrete seawave situation W_k within the S_i route segment.

Each seawave situation W_k ($H_{1/3}$, T_1, β) is charakterized jointly by : significant wave height - $H_{1/3}$, mean wave period - T_1 and the relative ship course - β.

The number of different seawave situations is on the whole determined as a product of: 8-wave heights, 13-wave periods and 7-ship-wave courses.

Because each route is divided into i = 1,2, ... n segments and within each segment k = 1,2, ... m seawave situations are distinguished, one should notice that each two-way route consists of n×m fragments the number of which is 10^4 order.

3.1. Procedure for determination of the design variants distribution

The design variants have been each time distributed within 7-dimensional decision space in accordance with principle of the factorial design ([4]). Concretely, the five-level composition factor design has been here adopted. This design consists of:
- $2^{7-4} = 8$ points of the plan's nucleus which are placed in 8 vertexs of 14-sided, space form;
- $2 \times 7 = 14$ stellar points which are symetrically situated on the 7 main axis of the decision space;

- 1 central point placed in the centre of the design space.

In order to ensure generality of the procedure of distributions a standarized variables t_i have been applied here. The design matrix $T\{t_i\}$ recorded by these variables is presented in Table 1.

During the concrete realization the vector $T\{t_i\}$ is transformed into the vector $X\{x_i\}$ in accordance with the dependency:

$$x_i = 0.03 \mid x_i^{(o)} \mid t_i + x_i^{(o)} \text{ for } i = 1,2,...7 \quad (3)$$

where $x_i^{(0)}$ for i = 1,2, ... 7 are the natural hull parameters of the central (initial) design variant. This variant is identified by means of linear interpolation of the historic ships which are holded in the data base. A discriminant of the interpolation is the required container capacity N_{TEU}^*.

As a result of the expression (6) as well as the values from Table 1 the following domain of the decision space has been determined:

$$D^7 : x_i \in \langle 0.95 x_i^{(0)} ; 1.05 x_i^{(0)} \rangle \quad \text{for} \quad i = 1,2,3,......7$$

3.2. Procedures for calculation of the measure of merit

The calculation of C(X) in form determined by expression (1) resolves itself into independent determination of the shaft horse-power $P_N(X)$, dead-weight DWT(X) and container capacity $N_{TEU}(X)$ for all of the design variants determined by the design matrix. The relationship DWT(X) and $N_{TEU}(X)$ have been identified on basis of the existing container ships delivered after 1988. The forms of the relationship are as follows:

$$DWT(X) = 1.025 x_1 - 0.10 \frac{x_1 \cdot x_7}{x_2 \cdot x_4} - 1025$$

$$N_{TEU}(X) = (0.25)^3 \frac{x_1}{x_2 x_4} \cdot$$
$$\cdot [x_2 \cdot x_4 + (x_7 - x_2 + 0.05) x_5] + 400$$

$$(4)$$

In order to calculate the variant propulsion power $P_N(X)$ a voyage simulation procedure is applied. An operation of the design variant of ship is modelled on a selected trading route and various

TABLE 1. STANDARIZED DESIGN MATRIX.

Experi-met.	t_i							REMARKS
	t_1	t_2	t_3	t_4	t_5	t_6	t_7	
1	-1	-1	-1	-1	-1	-1	-1	NUCLEUS OF PLAN
2	+1	+1	+1	+1	-1	-1	-1	
3	+1	+1	-1	-1	+1	+1	-1	
4	+1	-1	+1	-1	-1	+1	+1	
5	+1	-1	-1	+1	+1	-1	+1	
6	-1	+1	+1	-1	+1	-1	+1	
7	-1	+1	-1	+1	-1	+1	+1	
8	-1	-1	+1	+1	+1	+1	-1	
9	-1.668	0	0	0	0	0	0	STELLAR POINTS
10	1.668	0	0	0	0	0	0	
11	0	-1.668	0	0	0	0	0	
12	0	1.668	0	0	0	0	0	
13	0	0	-1.668	0	0	0	0	
14	0	0	1.668	0	0	0	0	
15	0	0	0	-1.668	0	0	0	
16	0	0	0	1.668	0	0	0	
17	0	0	0	0	-1.668	0	0	
18	0	0	0	0	1.668	0	0	
19	0	0	0	0	0	-1.668	0	
20	0	0	0	0	0	1.668	0	
21	0	0	0	0	0	0	-1.668	
22	0	0	0	0	0	0	1.668	
23	0	0	0	0	0	0	0	CENTRAL POINT

factors affecting the ship's behaviour are considered in combination rather than on individual basis only. The environment is characterized by the height, period and direction of wave and also by the speed and direction of wind. The seakeeping proprieties as well as the voluntary and involuntary speed losses are taken into account. In order to evaluate the voluntary ship speed the following seaworthiness criteria have been assumed here:
- the constraint of ship bow acceleration:

$$a(X,V) = [a_V^2 + a_H^2 + a_L^2]^{\frac{1}{2}} \le a^* = 0.6g \quad (5)$$

where: a_V, a_H, a_L are respectively the significant amplitudes of vertical, horizontal and longitudinal ship bow accelerations whereas g denotes the gravity acceleration;
- the constraint of slamming probability:

$$p(X,V) \le p^* = 0.02 \quad (6)$$

To evaluate of various hydrodynamic performances the following models have been adopted:
To calculate calm-water resistance the Holtrop (1984) method has been applied. The wave added resistance has been calculated in agreement with Gerritsma and Beukelman method [5] generalized to oblique sea conditions as in [6]. The wind resistance has been calculated with regard to the dependences taken from Brix ([7]). The absolute wind speed V_W is correlated with the significant wave height $H_{1/3}$ in accordance with the modified here Scot ([8]) dependence:

$$V_W = 6.384(H_{\frac{1}{3}} - 1.0)^{\frac{2}{3}} \; for \; H_{\frac{1}{3}} \ge 1m \quad (7)$$

The calm-water ship propulsive efficiency is for each design variant the same: $\eta = 0.7$. However, the propulsive efficiency in W_k seawave situation is calculated as:

$$\eta(W_k) = 0.7(1 - k_H \cdot k_T) \quad (8)$$

where: k_H and k_T are the reduction coefficients the values of which have been dependent respectively on $H_{1/3}/L$ and T_{1E}/T_θ (T_{1E} - encounter characteristic wave period, T_θ - free period of pitch).
All seakeeping characteristics have been computed on the basis of strip theory and superposition methods. Within this a spectral description of seawaves has been modeled by the two-parameters function in form

recommended both by ISSC and ITTC.

In order to use the strip theory and superposition procedures the hull lengthwise distributions of beam - b(x), draught - T(x) and cross sectional area - A(x) must be at least attainable. The point is that both the variance of each ship-wave response and the mean added resistance of ship in wave depend, among the other, on these distributions. Thus it was necessary to apply the suitable procedures for identyfication of the hull characteristics: b(x), T(x) and A(x) on the basis of the decision variables X{x_i}.

The variant propulsion power $P_N(X)$ is defined as:

$$P_N(X) = \max_{i,k}[P_{N_{i,k}}(X,V_{i,k})] \quad (9)$$

where: $P_{Nik}(X,V_{i,k})$ is the variant power necessary to obtain the ship speed $V_{i,k}(X)$ suitable for k-situation of i-voyage route segment. This power is calculated as:

$$P_{Ni,k}(X,V_{i,k}) = \frac{R_{TWi,k}(X,V_{i,k}) \cdot V_{i,k}(X)}{\eta(W_k)} \quad (10)$$

where:
- $R_{TWi,k} = R_{CW} + R_{AW} + R_W$ is the variant total resistance consist of: calm-water resistance R_{CW}, added resistance in seawave R_{AW} and wind resistance R_W;
- $\eta(W_k)$ is propulsive efficiency defined by equation (8).
The $V_{i,k}(X)$ value has, however, to satisfy the following two conditions:
- the ship safety condition what denotes that:

$$V_{i,k}(X) = \min_l\{V_{i,k,l}(X)\} \quad (11)$$

where: $V_{i,k,l}(X)$ for l=1 is the maximum possible speed for which the acceleration criterion (5) is satisfied where as for l=2 is the one for which the slamming criterion (6) is satisfied;
- the strategic condition of maintenance of the required two-way voyage time t_{ABA}^*:

$$t_{ABA}^* = \frac{S_{ABA}}{V_{ABA}^*} = t_{ABA}(X) =$$

$$= S_{ABA} \cdot \sum_i \frac{p_i}{\sum_k p_{i,k} \cdot V_{i,k}(X)} \quad (12)$$

where: p_i and $p_{i,k}$ are the probabilistic characteristics of the voyage route profile, S_{ABA} is the overall length of the specified two-way voyage route, V_{ABA}^* is the required value of mean voyage speed determined within the input data.

The speed values are in fact calculated by the successive approximation method until the following inequality will be satisfied:

$$|t_{ABA}(X) - t_{ABA}^*| \leq 5 \cdot 10^{-3} \, t_{ABA}^* \qquad (13)$$

If $t_{ABA}(X)$ is greater than t_{ABA}^* only the calm water speed ($V_{i,k}$ for k=0) is incremented.

If $t_{ABA}(X)$ is less than t_{ABA}^* the local speed $V_{i,k}(X)$ for which the uppermost powers are required are in turn reduced.

One should notice, that the design variant power $P_N(X)$ applied to calculate the measure of merit C(X) is moreover calculated as the arithmetic mean:

$$P_N(X) = \frac{1}{2}[P_N^{(1)}(X) + P_N^{(2)}(X)] \qquad (14)$$

where: $P_N^{(1)}(X)$ and $P_N^{(2)}(X)$ are the calculated by expression (9) variant powers for $GM_1 = 0.03B$ and for $GM_2 = 0.1B$ respectively.

This denotes that both the loading condition (GM_1) and ballast condition (GM_2) have been taken into consideration with the same probability of appearance.

TABLE 2. INPUT / OUTPUT DATA

INPUT DATA

Voyage route : Yokohama - San Francisco - Yokohama ;   Required transport capability:
Circular voyage distance - 9437 NM   - required container capacity - N^*_{TEU} = 3000
    - required mean ship speed - V^{**} = 24.00 kn

OUTPUT DATA - Ranking of the design variants

No	∇	T/L	B/L	Cb	Cw	LCB	H/L	C(X)	W(X)
	x_1	x_2	x_3	x_4	x_5	x_6	x_7		
1	49808.8	0.044	0.127	0.567	0.771	-2.739	0.080	1.6785	0.0000
2	49808.8	0.042	0.135	0.567	0.726	-2.739	0.084	1.7041	0.0637
3	48358.1	0.043	0.131	0.555	0.748	-2.824	0.082	1.7116	0.0823
4	50777.9	0.043	0.131	0.584	0.748	-2.824	0.082	1.7322	0.1334
5	46907.3	0.042	0.127	0.567	0.726	-2.909	0.080	1.7613	0.2058
6	48358.1	0.043	0.124	0.584	0.748	-2.824	0.082	1.7622	0.2080
7	49808.8	0.042	0.127	0.602	0.771	-2.909	0.084	1.7877	0.2714
8	48358.1	0.041	0.131	0.584	0.748	-2.824	0.082	1.7950	0.2895
9	48358.1	0.043	0.131	0.584	0.748	-2.824	0.086	1.8482	0.4214
10	48358.1	0.043	0.131	0.584	0.748	-2.683	0.082	1.8549	0.4382
11	48358.1	0.043	0.131	0.584	0.748	-2.824	0.082	1.8554	0.4395
12	48358.1	0.043	0.131	0.584	0.748	-2.824	0.082	1.8555	0.4396
13	48358.1	0.043	0.131	0.584	0.748	-2.965	0.082	1.8560	0.4410
14	48358.1	0.043	0.131	0.584	0.711	-2.824	0.082	1.8586	0.4473
15	48358.1	0.043	0.131	0.584	0.748	-2.824	0.078	1.8655	0.4646
16	48358.1	0.045	0.131	0.584	0.748	-2.824	0.082	1.9223	0.6055
17	46907.3	0.044	0.135	0.567	0.771	-2.909	0.084	1.9337	0.6339
18	48358.1	0.043	0.138	0.584	0.748	-2.824	0.082	1.9501	0.6745
19	49808.8	0.044	0.135	0.602	0.726	-2.909	0.080	1.9948	0.7857
20	45938.2	0.043	0.131	0.584	0.748	-2.824	0.082	1.9961	0.7888
21	46907.3	0.044	0.127	0.602	0.726	-2.739	0.084	2.0194	0.8466
22	48358.1	0.043	0.131	0.613	0.748	-2.824	0.082	2.0489	0.9200
23	46907.3	0.042	0.135	0.602	0.771	-2.739	0.080	2.0811	1.0000

C(X) - variant's measure of merit   W(X) = 0   denotes the best variants
W(X) - standarized measure of merit   W(X) = 1   denotes the worst variants

3.3. Numerical example

The printouts of the input / output data placed in Table 2 is an example of the program results form rather than results itself. To recive those results 2.5-hours of program calculation was necessary using a PC DOS based compatibile computer. This time is closely related to the number of distinguish fragments (see chapter 3) of the specified voyage route. For the voyage route taken by way of the example this number is equal 8400. For the other routes these numbers are as follows: 1) Bishop Rock - New York: 7000, 2) Bishop Rock - Rio de Janerio: 9800, 3) Bishop Rock - Yokohama: 22400, 4)Bishop Rock - Melbourne: 16800, 5) New York - San Francisco: 9800, 6) Sydney - New York: 12600.

4. FINAL COMMENTS

The OPTIMALSHIP computer program characterized here is to determine the initial feasible ship main dimensions and hull form global parameters for a design which satisfies values of container capacity (TEU-number) and ship speed together with specific operational constraints. This program as a typical CAD tool performs a role of an Intelligent Knowledge Base System which provides design advice for use at the pre-contract stage of design. In particular, the program is intended to state precisely the contract requirements which should be determined under the shipping company surveillance.

5. REFERENCES

1. Johnson B.:"On the Integration of CFD and CAD in Ship Design", Proceedings of the International Symposium on CFD and CAD in Ship Design Wageningen, September 1990.

2. "Ocean Passages for the World" - No 136 publi shed by the Hydrographer of the Navy - third edition 1973.

3. Hogben N., Dacunha M.C. and Olliver G.: "Global Wave Statistics" - Unwin Brothers - London 1986.

4. Cochran W., Cox G.: "Experimental Design" - John Wiley and Sons, New York Champan and Hall, London 1957.

5. Gerritsma J. and Beukelman W.: "Analysis of the Resistance Increase in Waves of a Fast Cargo Ships", Inter. Shipbuilding Progress, Vol. 31, 1984.

6. Woytkunskij J.:"Handbook of Basic Ship Theory", Shipbuilding, Leningrad, 1985.

7. Brix J. and group of experts:"Manoeuvring Technical Manual", Schafen Verlag, Hamburg, 1993.

8. Motte R.:"Weather Routeing of Ships", Maritime Press, London, 1972.

Practical Design of Ships and Mobile Units
M.W.C. Oosterveld and S.G. Tan, editors.

Sea Trial Experience of the First Passenger Cruiser with Podded Propulsors

R. Kurimo [a]

[a]Kvaerner Masa-Yards Inc., Helsinki New Shipyard
P.O.Box 132, FIN-00151 Helsinki, Finland

Propulsive performance and manoeuvring characteristics of the well proven cruise ship type with a novel propulsion arrangement were tested during the sea trial of M/S Elation. The extensive test program included also cavitation observations and pressure pulse measurements. The excellent results of these tests are described in the paper with some comparisons to the model test predictions. The superior performance of the Elation is also compared to the performance of the sister ships with conventional diesel-electric propulsion system.

1. INTRODUCTION

The Carnival Cruise Lines' Fantasy-class will consist of eight 70,000 GT cruise vessels, when the last ship in the series will be completed by Kvaerner Masa-Yards in late 1998. This series, with the first delivery in the beginning of 1990, is the result of the most extensive cruise ship building program in the history of passenger ships.

Innovations in the field of ship propulsion made attractive new concepts available, and in October 1995 Carnival Corporation selected Azipod™ propulsion for the seventh and the eighth ship of the series. In this propulsion system the conventional electric propulsion motors and the twin propeller-rudder-arrangement were replaced by two steerable thrusters, in which the electric propulsion motor

was located inside the 360° azimuthing pod and a fixed pitch propeller directly in front of the pod. The Azipod device was jointly developed by Kvaerner Masa-Yards Inc. and ABB Industry Oy in Finland. The hydrodynamic development of this podded propulsor, its tailoring for the Fantasy-class and related model tests are described in the reference [1].

The seventh Fantasy-class ship M/S Elation is the first passenger cruiser equipped with podded propulsors. The Elation's 14 MW thrusters are also the first large podded units optimised for other than icebreaking conditions. Carnival Corporations pioneering selection of Azipod units was a clear signal to the market of propulsion systems, and today also other manufacturers are marketing their own versions of podded drive units.

Kvaerner Masa-Yards' Helsinki New Shipyard

Figure 1. M/S Elation on the sea trial

Figure 2. The podded propulsors of the Elation

delivered the Elation to Carnival Corporation in February this year. The sea trial of the ship took place in the beginning of December 1997 on the Gulf of Finland at the sea area close to Helsinki. Some results obtained during this sea trial are presented in this paper. The scope of the paper is limited to general hydrodynamic performance of the ship with this novel propulsion arrangement. Speed trials, cavitation observations and pressure pulse measurements of the main propellers and manoeuvring tests are handled. Emphasis has been put to the manoeuvring tests, because most of all the manoeuvring performance of the ship exceeded the predictions and expectations.

2. SPEED TRIALS

Recording of the speed was based on the determination of the ship's position by means of a differential GPS receiver. At full propulsion power of 2 x 14,000 kW two double runs were performed. The elapsed time in each of the four runs was 15 minutes, so that sufficient accuracy in the measurement of the distance could be achieved.

Unfortunately the environmental conditions were not ideal for speed trials. Wind speed varied around 10 m/s and wave height was circa 1.2 m. Relative wind speed and direction detected by the ship's wind anemometer were both recorded during each speed run and the recordings were checked by comparing them to the registrations of an automatic weather station close to the trial area. Short term "wind history" recordings of the weather station and the actual wind fetch information were used together with visual observations in the determination of wave height and period.

After environmental corrections it was found out that the final speed of the Elation at full power of 2 x 14 MW was circa 0.55 knots higher than the mean value of the corresponding speeds of the six previous Fantasy-class ships with conventional diesel-electric propulsion arrangement. The Elation could be compared also to the three "fastest" vessels in the series. In this comparison the speed of the Elation was more than 0.45 knots higher than the mean speed of those three vessels. The recorded speed increase of 0.45 to 0.55 knots of the Elation corresponds a reduction of circa 7.5 to 9 per cent in the need of propulsion power to achieve the original

speed of the conventional sisterships. In a case of typical cruise ship operation profile this reduction in the power need means a fuel saving of circa 40 tons during one week.

The excellent powering performance of the selected podded propulsors with pulling propellers was thus finally confirmed. The ship's speed even slightly exceeded the prediction of the shipyard presented in the reference [1]. Nevertheless, in a case of propulsion systems with pulling propellers, the speed predictions of the different model basins and the related extrapolation methods of propulsion model test results should be taken with some care in order to avoid too optimistic expectations.

3. CAVITATION OBSERVATIONS

For the observation of propeller cavitation seven windows above the port propeller were installed. The windows can be seen in Figure 2. The registration of the cavitation was performed by Krylov Shipbuilding Research Institute (KSRI), which was responsible for the design of the propeller as a subcontractor to the propeller manufacturer, Zvezdochka Propeller Plant in Severodvinsk, Russia.

Registration was made using a video camera and special lighting equipment with a powerful flash lamp. The flash control device allowed to adjust the phase of the flash in relation to the angular position of the propeller shaft, and thus it was possible to register the cavitation pattern in different angular positions of the propeller blades. The registrations were also monitored by means of a portable TV set.

Visibility from the windows varied remarkably during the whole sea trial. Sometimes the water was quite turbid, sometimes the content of air bubbles spoilt registrations. Fortunately sufficient water quality was occasionally encountered, too. As a supplement to video registrations, patient visual observations of the research team of KSRI in the cramped and not so comfortable double bottom had an important role in the cavitation study.

At full speed only tip vortex cavitation was detected on the propeller blade. All other cavitation types were absent. The volume of the tip vortex was approximately the same as was observed in the cavitation model tests at MARINTEK in Trondheim. The angular range of the tip vortex cavitation existence was circa from -40°...-50° to +70°...+80°, and thus it was clearly wider than in the cavitation tunnel. This observation result was in good agreement with the theory of the scale effect in the tip vortex cavitation inception. According to this theory the tip vortex cavitation starts in full scale conditions at higher cavitation number than in the model scale at the same instant advance ratio.

One of the most important observation results at full speed was that short-term steering angles of the propulsion units up close to ±10 degrees didn't cause any changes in the cavitation pattern of the propeller. This angle range covers well steering angles used by the autopilot in course keeping and the manoeuvring angles used in the normal operation when the course of the ship shall be changed at full speed on the open sea.

Cavitation observations were performed also at lower speeds during the loading tests of the propulsion system. At 40% loading only a very thin tip vortex at the suction side was recorded in the range of blade angular position from -10° to +30°...+40°. In stabilised conditions the propeller was thus free of any cavitation almost up to this power loading which gave to the ship a speed of 18 knots.

4. PRESSURE PULSE MEASUREMENTS

The pressure pulses above the starboard propeller were measured using five pressure transducers flush mounted to the hull. Three of the transducers were installed in the propeller plane with one transducer just above the propeller centre. The pressure impulses at different frequencies were obtained after Fast Furier Transformation analysis.

The prediction of the maximum pressure impulses at blade frequency was 1.32 kPa, based on the pressure pulse measurements in the cavitation tunnel [1]. The measured pressure pulses during the sea trial were notably lower than this prediction. One reason could be the longer existence of the tip vortex cavity in angular passage of the blade (and thus a smoother variation of its volume).

The relative decrease of the pressure pulses above the propeller centre as a function of the loading of the propeller is presented in Figure 3.

The transversal distribution of the recorded blade frequency pressure pulses in the propeller plane at full speed is shown in Figure 4. In the same figure

PRESSURE PULSE AMPLITUDE (%)

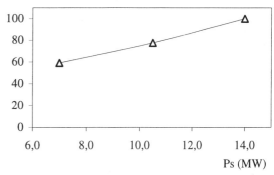

Figure 3. Recorded pressure pulses on the sea trial

PRESSURE PULSE AMPLITUDE (%)

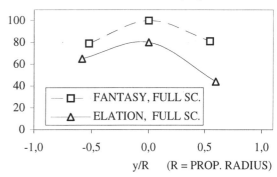

Figure 4. Transversal distribution of the pressure pulses above the propeller centre.

there is also presented a relative comparison to the corresponding pressure pulses obtained by propeller manufacturer on the sea trial of M/S Fantasy, the first ship of the series. In the comparison it should be noted, that the new propulsion arrangement of the Elation made possible to locate the propellers slightly more aft than in the Fantasy, and this resulted a minor increase in the hull clearance in the Elation. On the other hand, it shall also be noted, that the excitation level recorded on the Fantasy, due to the successful propeller design with extremely high skew angle, was already very low.

An essential factor in the reduction of the Elation's excitation level is the minimal cavitation pattern of the Elation's propeller. This cavitation pattern is the result of the good propeller design and specially of the excellent wake field, in which the pulling propeller operates.

746

5. MANOEUVRING TESTS

One of the most interesting subjects of the sea trial was the manoeuvring performance of the ship. The test program included a wide range of different manoeuvres, and in addition to normal turning circle, zigzag and crash-stop tests also some special manoeuvres like turning tests with one active propulsion unit and combined turning and stopping were performed. Crabbing performance of the ship was tested as well.

The manoeuvring tests of the Elation suffered from the similar non-ideal weather conditions as the speed trials. The all turning circle and zigzag tests were however started systematically the ship heading directly to the wind.

The recording system was the same as used in the speed trials. The speed and the track of the ship were measured by means of a DGPS (Differential Global Positioning System) and recorded continuously to PC with relevant data on the propulsion units and the ship's heading. In the following some results of this wide manoeuvring test program are highlighted.

5.1. Turning ability

The main results of the basic turning circle tests are presented in Table 1, together with the model test predictions for the Elation and the corresponding test results obtained on the sea trial of the "sister" ship Fantasy. The presented nondimensional parameters are the mean values of the turning circles performed to starboard and to port. The improvement in the turning ability as a result of the adoption of the

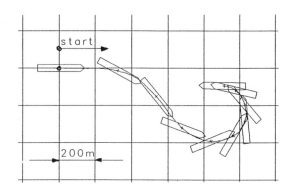

Figure 5. Williamson turn at full speed.

podded propulsors is very clear, for example the reduction in tactical diameter is circa 38%. This improvement was as clearly underpredicted in the computer simulations after model tests with a Planar Motion Mechanism (PMM).

The speed loss in the turning manoeuvres was remarkable due to the large drift angle of the ship. In the full speed turning circles (with pod angles of 35°) the speed dropped down close to 6 knots when the stabilised rate of turn was circa one degree per second.

The excellent turning ability of the Elation can be realised from the track plot of the full speed "Williamson turn", which is presented in Figure 5 with ship figures in the same scale. In this manoeuvre the helm was first ordered 35° to starboard and after heading change of 25° it was

Table 1
Turning circle test results

	M/S Fantasy on sea trial		Model test prediction for M/S Elation (PMM+simul.)	M/S Elation on sea trial	
Initial speed	10.2 knots	Full speed	19.5 knots	11.2 knots	Full speed
Rudder/pod angle	40°	40°	35°	35°	35°
Advance	2,73	3,11	2,99	2,01	2,35
Transfer	1,67	1,55	1,00	0,73	0,71
Tactical Diameter	3,14	3,05	3,14	1,94	1,89
Steady Diameter	3,28	3,15	3,07	1,81	1,77
Speed loss	31%	51%	42%	60%	> 70%

Parameters nondimensionalised with the perpendicular length.

Table 2.
Overshoot angles in zigzag tests (degrees)

Ship's speed	Model test predict. for Elation (PMM-tests + simul.)		Elation on sea trial		
	19.5 knots		19 knots	22 knots	
Test type	10/10	20/20	10/10	10/10	20/20
1st oveshoot angle	5.8	12.1	7	5	11
2nd overshoot angle	9.7	14.7	9	8	15

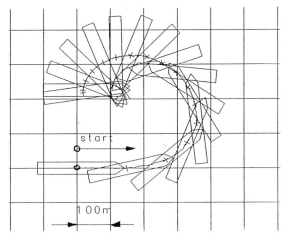

Figure 6. Turning with the pods 60° to port.
(Starboard pod active in propulsion)

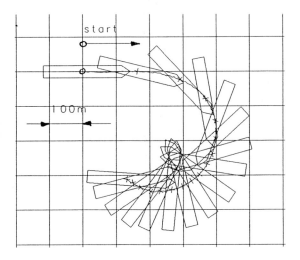

Figure 7. Turning with the pods 60° to starboard
(Starboard pod active in propulsion)

ordered 35° to port. The turning of the ship was stopped using contra helm of 20° after heading change of 160°, because the target of this manoeuvre was to reach the contra course on the original track line.

Effectiveness of the selected podded propulsors for turning was also demonstrated in special manoeuvres performed at speed of circa 10 knots. In these manoeuvres the propeller of the starboard side propulsor only was actively running for propulsion (with a constant output of 7 MW) while the port side propeller was windmilling. However, the steering mechanisms of both thruster units were active in the tests. The track plots of these turning manoeuvres with the pod angles of 60° are presented in Figure 6 and 7. In spite of the unsymmetrical loading of the propellers, the turning ability to both directions is very good. The traditionally defined turning circle parameters are so small, that a more relevant description of the resulting turn of the ship should rather be based on the dimensions of the "sweep area" of the ship's path.

5.2. Course stability and controllability

Although the Elation had extremely good turning ability, she performed also well in steering without any problems in course stability. Course-keeping by the helmsman and the autopilot succeeded both well and only small pod angles were needed in steering.

The recorded overshoot angles in the zigzag tests are presented in Table 2. In the same table there are also shown the corresponding model test predictions for the Elation. It can be seen, that zigzag manoeuvres were predicted very well, the overshoot angles during the trial differing hardly more than one degree from the predicted ones. One factor in the

748

good course-checking ability of the Elation was the remarkable lateral area of the actual podded propulsor type. The large active control surface contributed to the effective stopping of the fast rate of turn.

5.3. Special manoeuvring tests

When the ship was running full astern with reversed rotation of the propellers (without turning the pods), a speed of over 16 knots was achieved. At this speed the ship was steerable with the azimuthing propulsion units. The course changing and the stopping of a slow rate of turn succeeded as well.

A track plot of a special emergency stop manoeuvre is presented in Figure 8. In this manoeuvre stopping was combined with a tight turn. Helm order 35° to starboard was executed with reversing the rotation of the propellers. Traditionally defined nondimensional maximum advance (headreach) and maximum transfer were 2.78 and 1.65 respectively. Still the distances x and y, (which could be called "sweep reach" and "lateral sweep" respectively, see Figure 8) have maybe more importance from the point of navigation. In the presented manoeuvre the "sweep reach" was 2.40 and the "lateral sweep" 2.24 (both values nondimensionalised with the perpendicular length of the ship).

The podded propulsors thus introduce a manoeuvring and steering possibility to the situations, where quick stopping of the ship is required.

5.4. Crabbing performance

In the "sidling" tests the recorded maximum transverse speed of 2.4 knots was higher than the speed reached with one previous Fantasy-class ship when she used all her six 1.5 MW tunnel thrusters. In that ship the net thrust of the stern tunnel thrusters was reduced due to the flow interaction with the propeller shaft and the bossing and for the thrust balance it was thus necessary (if the main propellers were not in operation) to reduce slightly the power used in the bow thrusters.

In the Elation, which was built without stern tunnel thrusters, the balance without rate of turn was reached using a full power of 1.5 MW in all three bow tunnel thrusters and a power of 4 MW in one propulsion unit. In this condition the noise levels

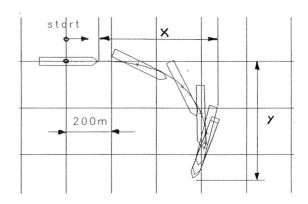

Figure 8. Combined turning and stopping from full speed

in the aft ship were remarkably lower in the Elation than in the previous Fantasy-class ships.

In the turning test at zero speed a turning rate of circa 1.0 °/s was recorded, when a power of 7 MW was used in one propulsion unit and the all three bow thrusters produced their full thrust to the opposite direction.

6. CONCLUSIONS

a) The excellent powering performance of the selected podded propulsors with pulling propellers was confirmed in the speed trials of the Elation.
b) Pulling type podded propulsor generates good possibilities to reach superior cavitation pattern on the propeller blade and thus low level of pressure pulse excitation can be reached.
c) The cavitation pattern and the excitation level of the pulling type pod propeller are not sensitive to small short term-steering angles up to ±10°.
d) The turning ability of the Elation was underpredicted in the computer simulations based on the analysed results of the PMM model tests.
e) In the described case the turning performance of the passenger cruiser in general was remarkably better with pulling type podded propulsors than with conventional propulsion arrangement.

REFERENCE

1. R. Kurimo, A. Poustoshniy and E. Syrkin, Azipod Propulsion for Passanger Cruisers, NAV'97, Sorrento, March 18-21, 1997.

1998 Elsevier Science B.V.
Practical Design of Ships and Mobile Units
M.W.C. Oosterveld and S.G. Tan, editors.

AN ANALYSIS OF FULL SCALE TRIAL RESULTS THAT TAKES ACCOUNT OF NON-SCALED ENVIRONMENTAL CONDITIONS

R. ROCCHI [+]

Istituto Nazionale per Studi ed Esperienze di Architettura Navale,
Research Department,
Via di Vallerano 139 - 00128 Roma ITALY

ABSTRACT

This paper refers to the results of a regression analysis of the sea trials data of 16 Italian Navy twin-screwed ships conducted in order to determine equations that enable accurate predictions of the load factor (1+X). Non-scaled environmental factors such as wind, waves, water and air temperature, atmospheric pressure, days after last dry-docking (hull surface condition) have been considered among the independent variables. Polynomials for the prediction of (1+X) have been derived from a sample of eighty-two sea trials data of 16 naval vessels(displacement ranging from 1200 to 12200 t, length from 80 to 170 m, CBTQ or *speed-form* coefficient (*) from 1.6 to 4.6.). Tank tests where performed at INSEAN according to the *continental method* whereas sea trials data have been kindly released by the Italian Navy. The proposed regression equations allow the prediction of (1+X) with an acceptable degree of accuracy. The scale ratio, the Reynolds number of model, sea conditions, the atmospheric pressure and days after last dry-docking resulted to be independent variables highly related to the load factor (1+X) : = (averaged power at mean speed in actual sea trial conditions) / (ship delivered power predicted by model self propulsion tests). The regression equations proposed where derived by this author within an INSEAN research plan.

1. INTRODUCTION

The extrapolation of model experimental tank data represents a delicate task for every Ship Model Institution, when the at-sea delivered power for the real ship is to be given, for instance through the prediction of the load factor (1+X)'s value.

All the different component aspects of factor X were deeply considered from the very beginning of the model testing and an exhaustive *resumé* of the problems involved and the progress made where clearly reported at the "Ship Trials and Service Performance Analysis" International Symposium held at Newcastle Upon Tyne [1] (NECIES 1960).

Fifteen years later the Performance Committee reported to the ITTC'75 the results of an extensive international program of analysis of 833 sea trial tests, from more than 212 ships of quite different type, with the following conclusion:

"If, as a result of these efforts, common model and trial test techniques are adopted all over the world, the deviations in the results will undoubtedly decrease".

In 1990 the Powering and Performance Committee closed its own report to the 19th.ITTC as follows:
"Produce a standard method for the analysis of full scale trial results, taking account of the wind, waves, currents and shallow water effects. Perform post-trial model correlation tests where possible" [2].

At ITTC'96 Conference the Powering Performance Committee reported the synthesis of a survey made on the basis of 39 responses (16 from shipyards, 16 model basins and 7 universities) to a questionnaire prepared by the Committee [3].

From [3], page 411, we report the following:
"The results of the survey were discussed in order to attempt to produce a standard method to analyse the results of full scale trials. It was determined that this

---

+ The views expressed here in are the Author's opinions and not necessarily those of the INSEAN Research Department neither those of the Italian Navy.

(*) See Nomenclature at the end of the text.

task was of sufficient complexity that it should be spilt into two efforts: standardised conduct of trials, and standardised analysis. It was agreed that only the former task would be further addressed by the committee." "The SNAME Guide for Sea Trials 1989, and "Standardisation Speed Trial Analysis, 1993", prepared by the SR-208 Panel of the S.R.A.J, along with the BSRA Standard Method of Speed Trial Analysis, are additional references which can be used for the conduct and analysis of Speed/Powering Trials.". "In the Appendix I : An Updated Guide for Speed/Powering Trials [3] a guide is reported to outline a procedure for obtaining data on speed/powering trials so that the results may be of practical scientific value and be used in the development of model-ship correlation. This guide addresses trials that are specifically dedicated to obtaining speed/powering data under tightly controlled circumstances as opposed to commercial Builder's Trials conducted to fulfil contractual requirements".

From the above it is clear that much investigation is still needed and more reliable data must be collected. Anyone who is engaged with trial power and power service performance predictions knows how important the availability of previous reliable data is.

How elaborated and cumbersome the analysis of trial data maybe appears clearly from the above. Readers are kindly requested to refer to the exhaustive international literature, publications and specialised texts., many of which are among the 80 selected references in [3].

In closing this introduction it is worth noting what Prof. M.A: Abkowitz said in his contribution at the 19th ITTC [4].

"As is well known, the hydrodynamic coefficients involved in resistance and powering suffer from *scale effects* because of the inability in the model test to satisfy the viscous parameter of Reynolds' number. In order to predict a power-speed relationship for the ship, several model tests are carried out to measure the resistance coefficient(smooth hull)(CR), the wake fraction(w), the thrust deduction factor(t), and the propeller thrust(KT) and torque (KQ) coefficients in both open water and self-propelled conditions.

These five coefficients are all functions of Reynolds number and therefore suffer from *scale effects*. To these a sixth coefficient in the form of roughness factor (ΔCF) must be estimated for the ship from

hydrodynamic phenomena measured at low model test Reynolds' number. In the process of predicting ship power performance from model test, six different factors must be extrapolated from model tests or predicted from some hydrodynamic basis all of which suffer from *scale effects*:

$$\text{predicted power} = f\,[(CR + \Delta CF),\ w,\ t,\ KT,\ KQ]$$

When measured power from ship trials are compared to that predicted from model test, (1) if there is an error in the prediction, then one or more of the coefficients may have been wrongly predicted. The standard culprit in the past was the roughness factor(ΔCF) until negative roughness factors appeared when there was a significant change in ship size and fullness. In reality, the roughness factor has represented *a roughness in the extrapolation process* rather than specifically the ship roughness".

2. SEA TRIALS DATA

The analysed samples were obtained from the sea trials of five different types of naval vessels as follows: 1 was a Garibaldi Class ship, 1 an Audace Class ship, 3 sister vessels were Minerva Class, 4 sister ships were Lupo Class and 8 sister sips were Maestrale Class. Principal dimensional particulars are: length from 80 to 162.8 m; displacement from 1200 to 12100 t.; block coefficient (CB) from 0.42 to 0.49, (BT) from 3.26 to 4.23 and m (LV) from 7.33 to 8.27. All ships were twin propelled (CPP)..Sea trials tests were performed according to the requirements and specifications dictated by the Italian Navy.

Trials data comply with quite all the majority of the recommendations of the above referred ITTC '96 Guide.

During speed trials the following quantities were recorded:

- ship's draught, and main fluids consumption, for displacement determination at the beginning and at ending of each run,
- water temperature over the water depth in the trial area,
- torque and rotation rate on shafts, independent of on-board instruments,
- the ship's position by means of the Raydist,
- water depth,

- wind speed and direction by visual observations,
- air temperature,
- atmospheric pressure,
- sea conditions and direction by visual observation,
- fuel consumption,
- days after last dry-docking.

Trial data were corrected for sea currents following the classic method of 3 runs in opposite directions, whereas water depth of the trial area assured that in any case there was no shallow water inconvenience at all.

Power values were determined from fuel consumption and temperature of exhaust gases. Power values were checked with the values obtained from torque and rpm measurements performed separately according to Navy routine: torsionmeter and Kelso counter.

Only power corrections at constant speed, due to differences between trial displacement and corresponding ship model displacement have been made. In every case these corrections were applied even if the differences of displacement were very small. For the ship Grecale were reported some trial tests at a displacement value 1.3% higher than that at tank test conditions. Higher differences were registered during trial tests of ships Perseo and Sagittario. In all other tests the maximum difference resulted to be less than 1.0%.

The at-rest design trim conditions (for all ships established at even keel :$\Delta T=0$) were rigorously verified and in all cases, despite minimal variations as seen in the majority of tests, differences were remarkably small.

Samples of variables data were subjected to preliminary descriptive statistical analyses for the detection of possible influential cases and/or suspicious anomalous values. Samples data made-up from 40 trial sets of *speed-power* data collected from sister ships of Lupo Class, and samples of 33 analogous data obtained from Maestrale Class sister ships showed that some values of wind force and sea state were statistically anomalous. Those data were excluded from successive analyses.

3. MODEL TANK TESTS

All ship models were made in wood and equipped with all the same appendages as the ship at

trials. Turbulence was guaranteed by a standard row of pins. Extrapolation of towing data was conducted according to ITTC'57 methodology adopting a roughness allowance (ΔCF) = 0.0002. Appendages were extrapolated considering their surface added to that of the naked hull as an equivalent strip which length is equal to the overall submerged length.

Open water tests were conducted at propeller Reynold numbers higher than $5*10**5$ (min.: 5.23 and max.: 8.11) : propeller diameters ranging from 2.95 to 5.10 m.

Self-propulsion tests were performed, according to INSEAN standards, following the *continental method* and friction deduction values (FD) were determined following ITTC'57 methodology.

Resistance and self-propulsion experimental data we-re also extrapolated according to ITTC'78 methodology (slightly modified at INSEAN in 1983 for twin screw ships). The correlation found among the new load factor $(1+X)78$ and independent variables, remained those first obtained with analogous analyses made with the load factor previously determined according to ITTC'57 methodology $(1+X)$. This fact suggested that analyses of $(1+X)$ should be continued only with delivered power values extrapolated according to ITTC'57 method.

4. THE APPROACH TO THE PROBLEM

Analyses of variance and correlation techniques applied to reliable data are statistical tools that enable the detection of *determining factors*, which contribute to the variability of the quantity we assume as dependent. Multivariate regression analyses allow to determine quantitatively the linkage that exists among determining factors and the quantity we assumed as the dependent one. It is also well known that the success of any regression analysis depends largely on the suitability of the chosen independent variables assumed as representative quantities of the above said *determining factors*. In other words : the detection of useful independent variables is the crucial phase of any regression analysis.

The load factor $(1+X)$ represents a simple comparison criteria of corresponding data (measured power from ship trials and power predicted from model tests) that may be considered as a quantitative expression of the compendium of all physical

752

differences that exist between the ship at sea and its model in a tank. According to this point of view the ship-model correlation problem has been approached considering any variable suitable for the interpretation of the previously named *determining factors*, as the culprit of a part of the discrepancies that exist between the two very different *scenarios*: ship at sea and model in basin. In other words, this approach necessitates the inclusion of independent variables with which to take into account also *non-scaled environmental sea trial conditions*; the importance of which has been continuously confirmed by the majority of research institutions and towing tanks, as recently reported in [3],[5].

According to this author's approach it ought to be possible to predict the load factor value as a function of these physical differences:

$(1+X)=f[viscosity, environment, geometry, speed]$
where:

- *viscosity* is considered to include all the consequences of inability in model test to satisfy the Reynolds' number. Ship to model scale ratio and the Reynolds' number for the model have been considered as possible variables to take account of *hull scale effects*. Propeller scale effects are considered of minor importance: propeller Reynolds' number being high enough (paragraph 3);

- *environment* refers to trial area environmental conditions that are not scaled in the physical modelling of the ship. As a matter of fact in standard testing techniques only hull and propeller are scaled and speed is scaled according to Froude number identity. The time the ship's hull has been at sea should be an indirect measure of fouling and should give an interpretation of the degradation of the cleaned hull due to sea-water effects. The hull wetted surface may be considered a parameter that can concur with the days after last dry-docking. The sea state is introduced to account for the quantitative difference between supposed calm sea water(as assumed in the extrapolation of tank data) and actual sea-water conditions at trials. Wind strength and direction must be introduced as independent variables that take account of the action of the wind at trials (considered nil in the extrapolation of tank data). Atmospheric pressure and air temperature should be considered too for both are physically related to prime-motor performance.

- *geometry* should include all differences due to geometric discrepancies between model (hull, propeller, appendages, etc.) and ship. Suitable variables should be considered to account for the influence of hull roughness, welding beads protrusion, waviness of the hull, zinc anodes and sea water inlets and outlets. Also hull form parameters and coefficients like block coefficient(CB), length to volume(LV), beam to volume(BV), wetted surface to volume,(SV) must be considered. The volume being that at trials. The wetted surface should be that at trials defined by the wave profile at hull. Above-water ship dimensions must be considered for they related to wind effects.

- *speed* includes independent variables that may account for the attitude of the moving ship and its consequent effects. The Froude number may be a suitable variable. From [6] a *speed-form* independent variable CBTQ has been also introduced. This variable includes the block coefficient (CB) and speed to ship's length ratio.

5. THE REGRESSION EQUATIONS

As said before, the selection of variables was made simply considering that each variable should represent the aspect introduced in the multivariate analyses. Multivariate linear regression analyses have been performed following forward, backward and stepwise methods taking (1+X) as dependent variable against independent variables defined as above. From the available trial data the variables listed in the paragraph NOMENCLATURE could be derived.

A simple equation was determined from the sample after exclusion of anomalous data. This model has the same structure as the one proposed for twin-screw ships in 1975 [6]:

$(1+X) =1.214454 - 0.009775*CBTQ2$ (1)

Main statistics: SE = 0.058   R = 0.521   [% of ε<5% = 60%]
Applicability :   CBTQ from 1.76 to 4.52.

From the same data the introduction of variables that take account of wind, sea state, roughness, air pressure and temperature and hull form coefficients allowed us to derive the following regression equation:

$$(1+X) = SV * 0.414213 - TV *9.282907 +$$
$$- LV * 0.306374 + SCALA *0.005151 +$$
$$- CBTQ*0.050021 +2.591771 \qquad (2)$$

Main statistics : SE = 0.047   R = 0.768   [% of ε<5.5%=90%]
Applicability :   CBTQ = 1.76 to 4.52   CB = 0.42 to 0.49
                  SV   = 8.00 to 9.0   BT = 3.25 to 4.23
                  LOS   = 80.3 to 167.0 m   Δ = 1200 to 12000 t
                  SCALA = 16.0 to 24.0

For more accurate predictions of (1+X) the following equations are proposed:

a) *LOS < 100 m*

$$(1+X) = SV *0.354446 +TATM *0.004562 +$$
$$+GG*0.0008342 - RNM *0.008980 +$$
$$- CBTQ*0.369484 + 4.745663 \qquad (3)$$

Main statistics : SE = 0.035   R = 0.684   [% of ε<5.5%=100%]
Applicability :   CBTQ =1.76 to 3.35   CB = 0.46 to 0.49
                   SV   =8.40 to 8.70   BT = 3.25 to 3.38
                   LOS = 80.3 to 106.0 m   Δ = 1200 to 2500 t
                   SCALA = 13.5 to 16.7

b) *LOS > 100 m*

$$(1+X) = -TV *18.558559 +SCALA* 0.009474 +$$
$$+GG*0.002105 - HG * 0.003575 +$$
$$- MARE*0.048139 -CBTQ2*0.08090 +$$
$$+9.32555 \qquad (4)$$

Main statistics : SE = 0.021   R = 0.982   [% of ε<3.0%=100%]
Applicability :   CBTQ =2.32 to 4.52   CB = 0.42 to 0.48
                   SV   = 8.0 to 9.0   BT = 3.25 to 4.23
                   LOS = 114 to 167.0 m   Δ = 2800 to 12000 t
                   SCALA = 18.0 to 24.0

6. CONCLUSIONS - FUTURE WORK

The samples considered are being augmented with trials data of other types of twin-screwed ships : Coast-Guard planing-crafts and ferries and passengers ships.

The analyses that are on-the-way will allow to double the applicability of new equations : in terms of values of the independent variable CBTQ the range values will expand from 1.76-4.52 to .0.55-7.25.

It has been possible to verify, that the model-ship correlation may be approached taking into account some of the major physical differences that exist between ship at trial and its scaled model. In fact some non-scaled environmental conditions of trial tests, and new variables such as scale ratio,

model Reynolds number and main hull form coefficients related to the actual values of the displacement. This approach is not new: in [9], for instance, are reported the results of a tentative of its application to trial data of merchant ships, but the prediction model for (1+X) was derived by neural network techniques instead of regression ones.

Main statistics of proposed equations guarantee that the predictions of (1+X) will be accurate. They are valid for twin-screw naval vessels only. Their applicability is defined by range values specified at paragraph 5.

The analyses performed according to the approach here reported can be enhanced with the inclusion of several new variables to account for the following discrepancies between trial and tank conditions:

- trim, sinkage and actual wetted surface deduced from the profile of wave generated by the model at corresponding sea trial's speeds: Figure A gives an example of these differences,

- inlets and outlets; zinc anodes position and dimensions,

- power, speed and rpm to be derived from measurements continuously registered at trials, allowing more detailed and non-biased samples of data,

- model data to be obtained from post-trial model correlation tests.

-

The load factor (1+X) ought to be evaluated instant by instant of the sea trial course, i.e. relating instantaneous powers to the speeds taken at the same instants.
With this more detailed information it would be possible to obtain some equations that enable the designer to make more accurate predictions of the load factor.

The same approach duly applied to continuously recorded data during normal service of the ship should allow to determine regression equations for the prediction of the service performance of ships at varying conditions of hull surface, weather and sea conditions, wind, load, trim, etc.. Even more: such analyses will allow us to get feedback for the operation of the ship, the design process and the derivation of model-ship correlation coefficients [8].

754

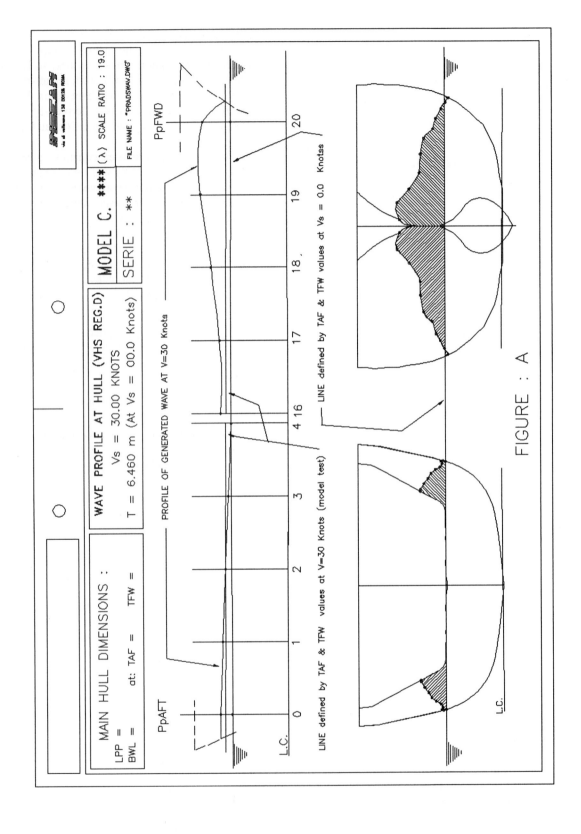

FIGURE : A

7. NOMENCLATURE

Computer compatible symbology adopted for standard and new independent variables is as follows:

- variables related to sea trials:

GG ....…........ days after last dry-docking
HG ....…….. atmospheric pressure (mmHg)
MARE ....…….. sea state
PDS ....….......... power, mean value (CV)
TATM ....…….................. air temperature (°C)

- variables related to the ship's geometry:

BWL ....…….......................breadth at waterline (m)
BT = ...…….......................... BWL / TMED
LOS ....…….............. length overall submerged (m)
LV ....…….................... LOS/VOL**1/3
CB = ...…….......…...VOL / LOS*TMED*BWL
SCALA ...……...................ship to model scale ratio
SUP ....:....... wetted surface at rest at TMED (m^2)
SV ....…….............................. SUP/ VOL**2/3
TMED .....……….....................mean draft (m)
TV ....…….......................... TMED /VOL**1/3
VOL..............…......... volume of displacement (m^3)

- variables related to model tests:

CBTQ = .CB * TMED * SQRT(V/(LOS)**0.5))
PDM = ship delivered power(CV) obtained from self-propulsion tests at corresponding sea trial speed

V ....…….............................speed (knots)
Δ ....…….......displacement of volume (t)

- variables related to statistics:
  SE ....….........standard error of estimates
  R ....…......multiple correlation coefficient
  ε ....…….....................error of estimate in %

- independent variable:…......... (1+X) = PDS / PDM

8. ACKNOWLEDGMENTS

Many members of the Italian Navy were involved in the performance of sea trial tests and in the acquisition of trial data used. Their contribution to this paper is invaluable, and it is gratefully acknowledged here.

REFERENCES

1. Ship Trials And Service Performance Analysis, Symposium, Newcastle Upon Tyne, April 1960 NECIES (1961).
2. 20th ITTC Powering Performance Committee Report, Proceedings, San Francisco, California (Sept. 1993).
3. 21st ITTC Powering- Performance Committee Report, (1996).
4. Abkowitz: Full Scale Measurements of the Resistance and Powering Coefficients and the Resulting Improvement in the Extrapolation Process from Model to Ship.19th ITTC, Report of the P.P.C (1990).
5. Hubregtse and G.G.J. Mennen: Speed Trial or Trial speed ?, PRADS'95, Seul, Korea,(1995).
6. Coppola : Statistic Investigation on the Towing Tank / Sea Trial Correlation Regarding the Shaft Power for Various Types of Displacement Ships, INSEAN Technical Report, Roma,(1974).
7. E.V.Telfer The Reconciliation of Model Data, Mile Results and Service Performance of Ships, NECIES, Transactions,(1961).
8. C.C.Schneiders: The Prediction of Ship Performance by Calculation or by Measurement? Paper to the PCC ITTC'90 Madrid (1990).
9. G.Lauro, A.Zini, R.Zunino: Neural Networks: A Modern Technology To Study Marine Problems.NAV'94Proceedings,Vol.2,Roma (1994). _____

Practical Design of Ships and Mobile Units
M.W.C. Oosterveld and S.G. Tan, editors.

An investigation into effective boss cap designs to eliminate propeller hub vortex cavitation

M. Atlar[a] and G. Patience[b]

[a] Department of Marine Technology,
University of Newcastle upon Tyne,
NE1 7RU, United Kingdom

[b] Stone Manganese Marine Ltd.,
Dock Road, Birkenhead, Merseyside,
L41 1DT, United Kingdom

The rudder of a containership behind its relatively heavily loaded propeller can be subjected to severe hub vortex cavitation that may cause undesirable vibration and noise and, in extreme cases, erosion on the rudder. This paper investigates the effectiveness of various boss caps to suppress the cavitating hub vortex. Five different boss caps were designed and tested at the Emerson Cavitation Tunnel of Newcastle University in "reverse pot condition". These were a standard cone shape, a cylindrical shape and three cylindrical boss caps with fins, trailing edge flaps and slots. The comparison of the cavitation characteristics for all the caps tested indicated the most favourable performance with the slot type cap with a considerable loss in the propeller efficiency requiring further investigation.

1. INTRODUCTION

1.1. Background

It is a well known phenomenon that abaft the fairing cap of a fast-running propeller, which is moving ahead rapidly through the water, a cavity of irregular cylindrical shape trailing behind the tip of the cap may develop. This cavity, which has the appearance of a large rope about to be completely unwound, is called a "hub vortex" because of its location.

The formation of the hub vortex can be associated with various mechanisms. With a conical shape of boss cap, the rotation of water set around the outside of the propeller hub, may form a vortex cavity due to inversely magnified values of the tangential velocities with contracting radius of rotation towards to tip of the cap [1]. Other mechanism is the coalescing of the several root vortices coming off the blades just outside the hub surface. This secondary flow is caused by the migrating flow from the pressure side to the suction side off the blades and hence forming a vortex filament shedding from each blade around the root region. In addition to the above two mechanisms, the viscous boundary layer generated by

the frictional drag over the flow surfaces upstream of the propeller (e.g. shaft bearing, shaft, appendages etc.) will generate another secondary flow which may also contribute to the development of the hub vortex [2].

The practical significance of hub vortex cavitation is, firstly, that waste energy is spent by the excessive swirl in the propeller slipstream near the propeller rotational axis. This excessive swirl will result in a pressure reduction across the downstream face of the slipstream relative to the ambient pressure and thus will create an undesirable drag force on the propeller. Furthermore the swirl may become so large that the energy imparted to the fluid near the hub will produce no axial thrust but will be dissipated by turbulent mixing. Secondly, it is a large cavity and not "solid" water, so that a rudder or control surface placed in line with the propeller-shaft axis loses part of the lift force that it is intended to produce. Thirdly, it is a source of undesirable vibration, noise and erosion on any object that may lie its path.

In some cases, it has been reported that, the lower parts of spade type rudders have disintegrated and broken-off. This loss of area did not affect the steering because the lower portion of the rudder, lying in the path of the strong hub vortex was not effective even when new [1].

1.2. A general review of the work on boss caps

The history of the work on effective boss cap design may be as old as the first observation of propeller cavitation by Sir Charles Parsons in his famous small cavitation tunnel in 1895. However, the oil crisis in the late 70's and early 80's was one reason for increased research which have resulted in successful designs used on merchant ships such as **P**ropeller **B**oss **C**ap **F**ins (PBCF) [3]. On the other hand the attempts to design naval propellers for utmost silence and least signature have been continuous since the beginning of this century resulting in various confidential concepts e.g. slotted or shrouded cones, which have been used by the British Admiralty.

Some activities in Japan

Recent activities on effective boss cap design have mainly focused on the PBCF which was proposed by a group of Japanese inventors as a means of increasing the efficiency of a ship screw propeller. A UK patent entitled " *A Screw Propeller Boss Cap with Fins* " was granted in 1988 [3]. According to the inventors the principle behind the device is that the hub vortex can be considerably weakened and hence the kinetic energy of the rotating flow around the boss can be recovered, by using an equal number of small fins associated with each propeller blade. These fins are attached to the boss cap within a limited range of inclination angles with respect to the geometric pitch angle at the root, as well as having a limitation on their maximum diameter relative to the propeller diameter. The inventors stated that the fins were not intended to generate a thrust themselves but for guiding the wake behind the boss cap to reduce the generation of the hub vortex. Because of this effect the hub vortex is diffused and hence the drag force induced by the vortex on a propeller blade plane is reduced. This results in a respectable improvement in the propeller efficiency without remarkable increase of the propeller torque. The general tendency is such that PBCF is more effective on propellers having higher pitch to diameter ratio which encourages the development of strong hub vortex. The patent included the results of 8 series of propeller tests in "reverse pot" condition in a

circulation channel to demonstrate the effectiveness of the device through some systematic investigation.

In [4], early work on PBCF involving flow visualisation, detailed model tests and the first full-scale measurements on the 44,979 GT PCC "Mercury Ace" were included. This reported a gain of 3-7% in propeller efficiency in the reverse pot condition, approx. 2% gain in propulsive efficiency based on the self-propulsion tests and finally a gain of 4% in power output on the actual vessel confirming the effectiveness of the device. Further investigations on PBCF were reported in [4,5] including 3-D LDV measurements in the model propeller slipstream and full scale measurements on 11 vessels to quantify the effect of scale. These investigations clarified that the PBCF have a rectification effect on eliminating the hub vortex and hence thrust increase on the propeller and also a torque reduction effect by receiving the hydrodynamic force on fins in the direction of propeller rotation (i.e. functioning like a turbine). The comparative analysis of the sea trial results of the 11 ships and their models indicated considerable scale effect between the model and actual measurements such that the efficiency gain in full scale could be two or three times that in the model scale. This was attributed to the laminar flow around the boss cap and possible laminar flow separation at the back of the fins resulting in different scale of forces. Finally a suitable scale effect factor, which can be determined from a large number of tests on models and actual ships in the same manner as roughness allowance "ΔC_f" in the resistance of ships, is recommended for taking into account the scale effect.

In more recent papers [6] various aspects of PBCF have been explored. It was found that the presence of a rudder significantly reduces the strength of the hub vortex. As a result the propeller efficiency gain can be reduced 10-30%. Self-propulsion tests to investigate the propulsive efficiency by PBCF indicated that the efficiency gain obtained from the tests were similar to the gain obtained from the "reverse pot" tests with rudder. The effectiveness of the PBCF was hardly affected by the presence of the hull at the operational draft condition suggesting that most of the efficiency gain was due to the propeller flow itself. The measurements of noise by a single hydrophone with and without the rudder in cavitating and non-cavitating conditions displayed favourable results for the PBCF.

The first numerical model to predict the performance of a propeller with PBCF as an analysis tool was reported in [7] and [8]. In this model the blades and fins are represented by lifting surfaces while the boss was represented by non-lifting panels. The numerical model showed that the PBCF generates a circulation opposite to that of the propeller blades, so that the strong tangential velocity near the boss centre is reduced and hence the propeller torque. The method has been found to be effective for performance prediction and flow analysis of a propeller with PBCF but the calculated thrust and torque did not agree well enough with the experimental data which was attributed to the complex flow around the PBCF.

In ref. [9] it was reported that 130 ships have been fitted with PBCF. At the last ITTC the propulsor committee reported that more than 215 ships have been fitted with PBCF.

Some activities in USA

In order to shed more light on the function of PBCF, in [2], detailed flow analysis and power measurements with PBCF were carried out. The theoretical analysis indicated two sources for saving energy: The first source was the recovery of the rotational (tangential) energy by the fins similar to the action of guide vanes used for energy saving devices. The second source was the recovery of energy associated with the secondary flow passing the propeller blade particularly around the root region.

The experimental analysis of the efficiency gain from a PBCF utilised the results of the LDV measurements in ARL Penstate water tunnel with a model propeller behind a hull including its rudder. This concluded that 2 % out of a total 6% efficiency gain was due to the thrust increase while the remaining 4% was associated with the torque decrease. It was envisaged that the total gain would be somewhat reduced due to the frictional drag of the fins. Indeed their thrust and torque measurements with the same model presented 5% gain in efficiency with the PBCF confirming the earlier predictions independently. It was also claimed that it was not apparent from the respective studies that the PBCF type device had to contain the same number of vanes as the number of the propeller blades to achieve the max efficiency gain.

Some activities in Germany

In [10,11], Potsdam Model Basin and Schottel reported on a joint innovative concept, which is known as **Hub Vortex Vane (HVV) propeller**. This concept includes a small vane propeller fixed to the tip of the cone shape hub cap within a limiting radial boundary where the tangential velocities due to hub vortex is greater than those of the propeller. In this range the small vane propeller diverts the high tangential velocities in the direction of the jet, thereby generating additional thrust. It is claimed that this mechanism is different from the mechanism for PBCF where the fins are located usually beyond this limiting radius. Another effect of the HVV is that the diverting torque of the vortex reduces the engine torque which results in saving in the drive power. In both references cavitation photographs of model propellers are included demonstrating the dramatic reduction in the hub vortex cavitation by the HVV. The number of vanes used appears to be greater than the number of propeller blades (e.g. 4 bladed propellers fitted with 6 bladed vane propeller).

In [10] was reported that on the successful full scale application of the HVV propeller behind a Schottel-Rudder propeller demonstrated about 3% higher efficiency compared to a conventional propeller.

Some activities in UK

Over the last decade various types of boss caps have been designed and tested in the Emerson Cavitation Tunnel at Newcastle University to investigate their effectiveness on propeller efficiency and cavitation performance as reported in [e.g. 12,13,14 and 15]. These investigations involved uniform flow tests with the same model propeller which is a relatively high pitch ratio of a right handed Meridian Series propeller with the following main particulars: D = 304.8 mm, P/D = 0.80, BAR = 0.65 and Z = 4. The propeller was tested in "reverse pot" position and no rudder arrangement was associated in any of these tests. Typical range of J values between 0.3 and 0.8 was selected at an uniform flow velocity of 3-4 m/s and at varying cavitation conditions.

Early tests reported in [12] explored the effect of differing outline shapes of boss caps. These included various types including the standard cone, cylindrical types with various divergent (diffuser type) and

760

convergent angles as well as the differing radius of corners and length to diameter ratios. None of these caps are associated with any fins or slots.

The experimental work in [13] involved the comparative tests of two identical cylindrical caps: one with fins like PBCF (see Fig 3) and other without the fins (see Fig 2). The caps had a 43 mm length and 49 mm diameter. Each fin fitted had a relative inclination angle of 54° and located mid way between the trailing edge of the propeller blades. The outline of the fins had parabolic shape with a maximum span of 20.5 mm, a chord of 36 mm and thickness of a 1mm. The comparison of the test results for various conditions presented an overall efficiency improvement of about 1% and slight delay in the inception of hub vortex cavitation in favour of the cap with fins.

In a later test series [14] 3 new caps were manufactured in addition to the above mentioned two caps and tested. The first cap was fitted with an "accelerating ring" attached to boss cap fins, which were equal to the number of propeller blades and located near the mid way of the cap length. The second cap was fitted with a "decelerating ring" and the third had boss cap fins with "end plates" attached to them. The comparison of the propeller efficiencies of the five caps presented the maximum value around at J = 0.6 for the PBCF like type by approx. 2% difference relative to the cylindrical one. In terms of the maximum efficiency, the second best cap was the "end plated" one by 1.8% relative to the cylindrical cap. However, the end plated cap presented a continuous gain relative to all the others over the entire operating range (i.e. J = 0.3 to 0.8). The decelerating ring type was the worst amongst the all caps tested, displaying a maximum of 5% efficiency loss relative to the cylindrical cap.

In a very recent investigation [15], to attempt to eliminate or reduce hub vortex cavitation effectively, two new caps have been designed and tested together with the earlier three caps. The two new caps were one with a cylindrical boss cap fitted with Trailing Edge Flaps "TEF" and the other with the cylindrical Boss cap with Slots "BoS". This paper summarises the theoretical and experimental details of this continuing research investigation. The comparative

results of the propeller performance in terms of the propeller efficiency and cavitation are presented and discussed with the all caps tested in the following sections of the paper.

2. PROPELLER AND BOSS CAP MODELS

2.1 Propeller

In order to represent a propeller, which would easily produce a strong hub vortex, a 4-bladed Meridian series model propeller was selected. The main particulars of the propeller are given in Table 1.

Table 1 Main particulars of model propeller tested

Model no	192
Diameter (D)	304.8 mm
Pitch to diameter ratio (P/D)	0.80
Blade Area Ratio (B.A.R.)	0.65
Number of Blade (Z)	4
Direction of rotation	Right hand
Material	Manganese bronze

2.2 Boss Caps

Five different boss cap types were considered for the investigation. These were: **Standard** cone shape; **Cylindrical** cap type; Cylindrical boss Cap with **Fins (CCF)**; cylindrical boss cap with **Trailing Edge Flaps (TEF)**; and cylindrical boss cap with **Boss Slots (BoS)**. Because of the fundamental and comparative nature of the investigation the main shape of the caps was restricted to the cylindrical shape, except for the standard cap, and the main parameters of the boss cap, which are the length and the diameter, were kept constant. The detailed sketches including the main dimensions of the caps are shown in Figs 1-5, while their photograph with the model propeller is included in Picture 1 at the end of the paper.

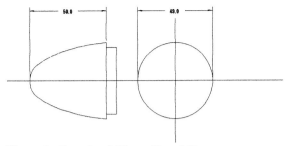

Figure 1. **Standard** (Cone Shape) Boss cap

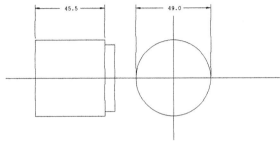

Figure 2. **Cylindrical** Boss Cap

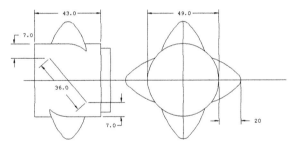

Figure 3. Cylindrical Boss Cap with **Fins (CCF)**

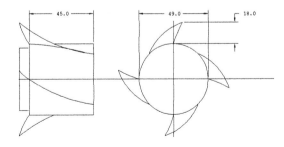

Figure 4. Cylindrical Boss Cap with **Trailing Edge Fins (TEF)**

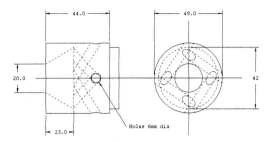

Figure 5. Cylindrical **Boss** Cap with Slots (**BoS**)

Although the fundamental nature of the investigation was one of the important effects in the selection of cylindrical shape, the ease of manufacture of the cylindrical cap was another reason. Moreover, the revolving **flow** particles around the boss caps, such as the standard one, would encourage the development of the hub vortex due to the increased rotational flow velocities caused by the converging boss cap. Therefore the use of the cylindrical boss cap would relatively discourage the development of the vortex.

Design of TEF

As stated earlier the first three of the boss caps were already available at the tunnel and the subject of the earlier investigations. These caps formed the basis for the two new caps, **TEF** and **BoS**.

The two new caps were based on the two independent theoretical approaches. The philosophy behind the TEF type relies on the reduction of the circulation around the hub region as this is one of the main contributions to the generation of the hub vortex. Therefore, a theoretical investigation was carried out to design an effective flap at the trailing edge of each blade to alter the circulation characteristics of the blade, particularly near the hub. This required the use of a systematic and accurate procedure to alter the circulation distribution over the blade by carefully taking into account the effect the boss cap fitted with various flaps of different size and pitch. Therefore a computer program, utilising a potential flow based surface panel method, was used for this investigation [16,17]. The method used in the software is based on the Morino Method and particularly suitable to the heavily loaded propellers due to the improved wake models, which are "Hydrodynamic Pitch Method" and "Streamline Method", employed. The former of these wake models uses the circumferentially averaged axial and tangential velocities in order to obtain the pitch of the wake. The latter uses the streamlines of the hub, propeller blades and their wake. From a prescribed wake geometry, so called initial wake, the streamlines are calculated for finding the final geometry of the wake.

For the discretisation of the propeller geometry in span wise the density of the radial stations were increased around the hub region as shown in Fig 6.

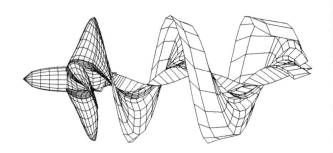

Figure 7. Panel arrangement of the propeller with TEF and its deformed wake after the second iteration at J=0.6.

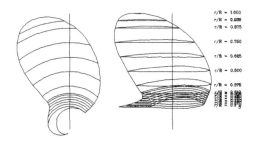

Figure 6. Expanded blade outline of the propeller with 0 degree flap angle.

In chord wise the cosine spacing rule was applied. The hub was considered as the axisymmetrical body on which the propeller blades were mounted. The trailing edge flaps were assumed as the natural extension of the blades and discretised with the blades. The total number of panels allocated for 1/4th of the hub and blades are 72 and 112 respectively. The wake surface was approximated by a prescribed helical surface with a pitch, which is assumed to be equal to the average value of the pitch of the onset flow and that of the blade. This surface is discretised by 560 panels in total for two complete rotations of the propeller blades. Fig 6 shows the expanded blade outline of the propeller with the TEF at zero degree of flap angle.

Based upon an advance coefficient of J = 0.6, which represents a typical operating point of a container ship, panel distribution of the propeller and its wake after the second iteration are shown in Fig 7.

The initial size of the flap was based upon 2D thin foil theory with no effect of the hub, number of blades and 3D flow to cancel the circulation at the hub, which was obtained from the panel code. This formed a basis for the final size and pitch of the flap which were devised through a systematic analysis of the circulation of the propeller and flaps in combination. Figure 8 shows a variation of the non-dimensional circulation Γ along the fractional radius r/R are given for the propeller with the standard cone and trailing edge flaps with varying pitch angles of flap. The non-dimensional expression for the circulation is related to the perturbation potentials by the following relationship:

$$\Gamma_i = \frac{\displaystyle\sum_{j=1}^{N_c}\left(\phi_{ij}^B - \phi_{ij}^F\right)}{\pi.D.V_I} \qquad i = 1, N_s$$

where

Γ_i Non-dimensional circulation

$\phi_{i,j}^{B,F}$ Perturbation potential on the face and back panels respectively

D   Diameter of propeller
V_I Inflow velocity
N_C Number of chord wise panels on the propeller surface.

N_S Number of span wise panels on the propeller surface.

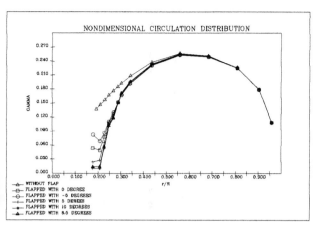

Figure 8. Non-dimensional circulation distribution along the blade without and with the trailing edge flaps at varying flap angles.

As shown in Figure 8 the introduction of the flap in line with the geometric pitch angle of the blade (i.e. zero flap angle relative to the blade) reduces the circulation at the root region. Increasing the flap angle systematically up to 10 degrees displays a systematic reduction in the circulation. Although a higher flap angle would present more reduction in the circulation it was decided to use the trailing edge flap with zero angle relative to the blade as this combination appeared to be modestly effective in reducing the circulation and expected to be causing less flow disturbance.

Design of BoS type cap

On the other hand, the BoS type cap is based on the idea of increasing the flow pressure in the core of the hub vortex by using slots. These slots would effectively direct the natural flow movement behind the blades to the vortex core where the pressure would be very low. In determining the size of the inlet and outlet of the slots as well as their slopes etc one would require a sophisticated CFD analysis. However this was beyond the scope of the project. Instead a heuristic approach was made to design the slots. A crude 2D flow analysis was carried out by using Navier-Stokes solver (FLUENT) neglecting the effect of rotary motion of the propeller and presence of the blades and hub. In this analysis the initial inlet and outlet velocities obtained from the panel code and effective slot size and slope was investigated [18].

All the caps were manufactured from manganese and bronze which was the same material as for the propeller.

3. OPEN WATER (REVERSE POT) TESTS

The tests were carried out in a uniform stream and at two values of constant water speed. The model propeller was mounted on the large K&R H33 dynamometer in " reverse pot " position to be able to observe the hub vortex being developed. In this position the propeller was working in the wake of the dynamometer housing and the direction of rotation was reversed. The distribution of the velocity field behind the dynamometer housing was not measured but is considered to have a very low value of wake fraction i.e. less than 0.05. During the tests the tunnel was open to atmosphere as well as being exposed to certain amount of vacuum to achieve low cavitation numbers over a range of revolutions to simulate realistic operating conditions to the full scale propeller . With each cap fitted to the model propeller the test conditions given in Table 2 were imposed in the cavitation tunnel in terms of the cavitation number σ and Reynolds number R_n.

Table 2 Test conditions

Water Speed	3 m/sec		4 m/sec	
Propeller rpm.	650	2000	650	2000
σ	23	23	12	12
$R_{n0.7}$	$0.95*10^6$	$0.965*10^6$	$2.62*10^6$	$2.67*10^6$

The results of the tests are given in terms of the standard coefficients described in Table 3 and plotted in Figures 9, 10 and 11 for η_0, K_T, and K_Q, respectively, for all caps in comparison. In these figures the values of J are corrected for tunnel wall interference using the Wood & Harris channel speed correction.

Table 3 Definition of propeller constants

Thrust coefficient, K_T	$T/(\rho n^2 D^4)$
Torque coefficient, K_Q	$Q/(\rho n^2 D^5)$
Advance coefficient, J	$V/(nD)$
Efficiency, η_0	$(2\Pi/J)(K_T/K_Q)$
Cavitation number, σ	$(p-e)/(0.5\rho V^2)$

3.1 Discussion of results

Although the main focus of the research was to investigate the cavitation performance of the caps, the comparison of the propeller efficiencies with the caps was equally important.

As shown in Fig 9 the maximum efficiency of the propeller with all caps occurs at around J=0.6 Overall, the best efficiency is obtained with the cylindrical cap. The worst efficiency appears to be obtained with the BoS type with a considerable margin (~10%) at the same J number compared to the cylindrical one. The remaining three caps provide comparable propeller efficiency values relative to each other although the efficiency with the TEF type is marginally better.

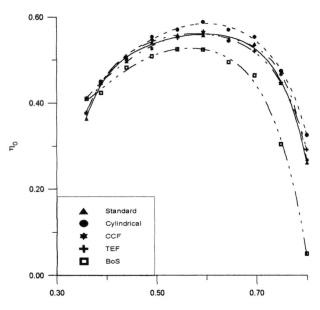

Figure 9   Comparison of propeller efficiency (η_o) with all caps tested.

Looking at the comparative thrust characteristics of the propeller with all the caps in Fig 10, while the cylindrical and TEF types present better thrust characteristics, the BoS type provides the poorest thrust performance compared to the others by a considerable margin. The margin is about 14% at J=0.6 relative to the thrust produced with the cylindrical cap. In contrast to its poor thrust performance, the BoS cap displays the best torque characteristics amongst the all with a relatively small margin compared to the cylindrical cap (~ 7%) at the

same J number as shown in Fig 11. The largest torque was produced with the standard boss cap.

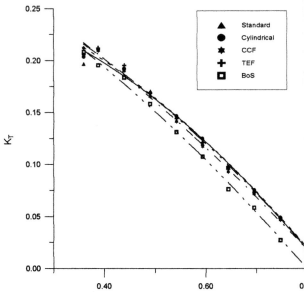

Figure 10 Comparison of measured propeller thrust coefficients (K_T) with all caps tested

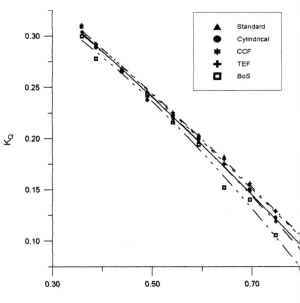

Figure 11 Comparison of measured propeller torque coefficients (K_Q) with all caps tested

As described earlier the BoS type cap is based on the idea of increasing the pressure in the vortex core and hence to suppress the development of the cavitation. On the other hand, as it will be seen in the following section, there is a clear visual evidence of a considerable improvement in the hub vortex cavitation with this cap. Therefore one would expect that the increase in the pressure across the vortex core should reduce the associated drag or thrust loss on the propeller and thus to improve the efficiency. However this appears to be not the case. Although the answer to this paradoxical situation may not be so easy without more detailed investigations, it may be associated with an unfavourable pressure drop at the inlets of the slots. In increasing the pressure in the core vortex there appears to be a considerable pressure drop at the inlets of the slots which are positioned at the trailing edges of the blades. This pressure drop seems unfavourably affecting the pressure increase at the face of the blades thus causing a thrust loss. Therefore the radial and axial positions of the inlets with respect to the blades may be crucial factors. This would require more tests supported by sophisticated CFD modelling and Laser Doppler Anemometry (LDA) measurements of the complex flow around the boss cap.

4. CAVITATION TESTS

In order to assess the cavitation performance with different caps, cavitation tests were carried out with each boss cap fitted to the model propeller. These tests involved the measurements of the inception points for the hub and tip cavitation as well as the observation of the nature of the cavitation developed.

During preliminary tests to study the nature of the hub vortex cavitation, two distinct types of cavitation phenomena were observed. Initially fluctuating flushes of cavitating vortices appeared at certain points on the hub and in its downstream wake. Then these vortices rolled up forming a steady hub vortex attaching to the hub. Therefore it was thought to be appropriate to define two set of inception points in association with the "fluctuating" and "steady" nature of the hub vortex cavitation.

In the determination of the inception points it was assumed that the cavitation would occur when the local pressure at the point concerned falls to a value at or below the saturated vapour pressure of the working fluid. The inception tests were carried out at a reduced tunnel static tunnel pressure with

uniform tunnel velocity of 4 m/s, which allowed a range of cavitation numbers between 2 to 12. During the tests the gas content of the water in % of air saturation was kept at 25 -30% by using DO_2 meter.

For all the boss caps, the comparative cavitation inception curves are given for the fluctuating hub vortex, steady hub vortex and the blade tip vortex in figs 12, 13 and 14 respectively. These figures are presented as cavitation inception number σ, which is based on the axial velocity vs. J.

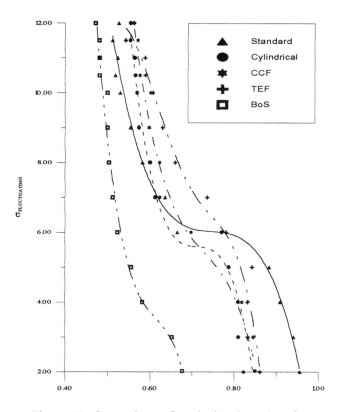

Figure 12 Comparison of cavitation inception for "Fluctuating Hub Vortex" with all boss caps tested

4.1 Discussion of results

From the hub vortex cavitation inception point of view, as shown in fig 12 and 13, the BoS type cap displays clear advantage over the rest of the caps both in the fluctuating and steady form. For the practical range of operations, i.e. low σ values, the second best appears to be the cylindrical cone. The poorest performance was observed with the standard cone and CCF type for the fluctuating and steady forms respectively. The TEF cap displays a mixed trend.

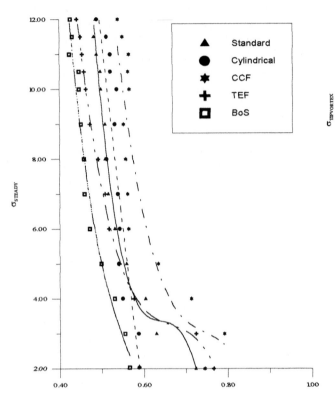

Figure 13 Comparison of cavitation inception for "Steady Hub Vortex" with all boss caps tested

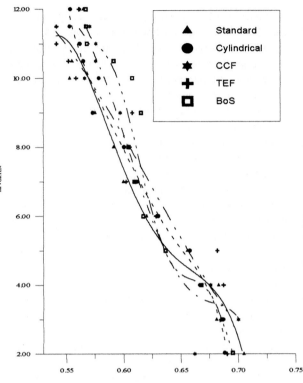

Figure 14 Comparison of cavitation inception for "Blade Tip Vortex" with all boss caps tested

As far as the tip vortex inception is concern, as shown in fig 14, for the low range of σ, the propeller with the BoS type cap still appears to be marginally advantageous displaying a slight delay in the onset of the tip vortex cavitation. In any case the differences in cavitation inception due to all the caps tested are marginal and changes depending upon the vacuum conditions applied. For example while at the atmospheric condition (i.e. high values of σ) the TEF type display a favourable inception characteristics this trend changes at low σ.

On the other hand, from the observations and analysis of the video films, it was noticed that the standard cap produced a distinctively thick cavitating core vortex in its developed form. The vortex is very

stable and glassy in appearance (see Picture 2). At the inception point the vortex appears as a thin stream of bubbles just behind the tip of the cone shape of cap and this quickly manifests itself as a fluctuating mass of misty cavitation which forms into a thin core trailing some way downstream. The increase in the propeller speed produces a stable core extending some distance downstream. The core increases to a maximum diameter of approximately 15mm for the model propeller and has the largest diameter of core cavity of all the boss caps tested. This stable hub vortex cavitation is strong and intense which is undoubtedly to cause an undesirable effect on the rudder. The tip cavitation originated some distance away from the trailing edge of each blade and eventually attached to it.

The cylindrical cap had quite different hub cavitation pattern to that of the Standard one. The cavitation began with a trace of cavitating bubbles just behind the centre of the cap and these bubbles travelled for a short distance downstream. As the propeller speed increased the cavitating bubbles travelled for a longer distance downstream and began to merge together to form a thin fluctuating core. At this point the core was quite unstable and appeared to have a small strength. A further increase in the rpm resulted in growth in the core length as well as the diameter and the vortex became very stable (see Picture 3). The core had the characteristics of a thick rope twisted about the end of the boss cap appearing to rotate in the same way as the direction of rotation of the propeller

The CCF type cap displayed the smallest diameter vortex of the four caps tested. Inception started in a similar fashion to the Standard cap where a trace of small cavitating bubbles formed behind the tip of the cap. These bubbles increased in size until they merged together and formed a long fluctuating core vortex. At the inception point the core was an intermittent long cavity extending well over three times the diameter of the propeller downstream. The core soon became stable and it was noticed that smaller vortex cavities intermittently rotate about the main core which were probably a result of vortices shed from the fins of the boss. Increasing the rate of revolutions produced a steady glass like core which did not increase in size (see Picture 5). As this core seemed to have the smallest diameter of those tested, it might be thought of as being the most effective at reducing possible cavitation erosion on rudder. However the stable nature of this core would have more detrimental effects than if the core was to be a fluctuating misty core.

The cavitation pattern of the TEF type was somewhat similar to that of the Cylindrical one. Hub cavitation inception occurred with a series of small cavitating bubbles streaming behind the boss for a short distance downstream. With increasing revolutions these bubbles coalesced to form a fluctuating misty core that increased in size. There was no solid core as such in the previous caps but a misty mass of cavitating fluid that fluctuated about the centreline of the boss. This fluctuation could be attributed to the asymmetrical fins positioned on the boss as a result of manufacturing inaccuracy. The cavitating core had a smaller diameter to that of the Cylindrical and Standard caps and the detrimental effect on the rudder would be less due to the elimination of the "solid" nature of the core. At high propeller speeds the cavitating core broke down extensively close to the boss whereas far downstream

a more stable core was noticed

The BoS type cap displayed the best cavitation characteristics in terms of a considerable delay in the inception of the hub cavitation and greatly reduced diameter of the cavitating core once it developed. The core cavity manifested itself as a fine stream of intermittent bubbles streaming behind the boss cap. The first signs of inception appeared approximately at a distance of two propeller diameter downstream of the end the boss and on increasing the revolutions this origin moved closer to the boss. However it never attached to the hub even at the highest propeller speed tested (see Picture 6). The hub cavity appeared at much higher propeller speed than for the blade tip vortices which was not the case with all the other boss caps. The cavity began as a stream of unconnected oscillating bubbles which never really merge together to form a solid core. At high propeller speed the core that was present comprised of a mist of unstable bubbles which started at a minimum of a boss cap length away from the cap. At no point through the range of advance coefficients did the cavitating core diameter grew greater than any other hub cavity and in fact the diameter was drastically reduced. The core was almost non-existent at advance coefficients where a cavitating hub vortex would be well established in all the other boss cap types.

5. CONCLUSIONS

The investigations on the hub vortex cavitation and efficiency performance of the five different types of boss caps, so far, indicated that:

- The best cavitation performance was obtained with the slot (BoS) type cap whilst the standard cap displayed the early inception and most erosive cavitation characteristics.
- The highest value of the propeller efficiency was obtained with the cylindrical cap while the slot type cap presented the lowest efficiency values of the all caps.
- There was no discernible advantages or disadvantages of the CCF and TEF type caps relative to each other
- Although the cylindrical cap appears to be the best compromise amongst the caps tested from the overall performance point of view, the slot type cap presented the superior cavitation

Picture 1. Photographs of boss caps tested

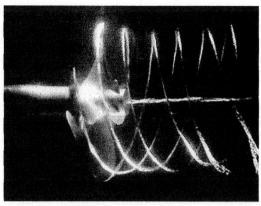

Picture 4. Fully developed hub vortex cavitation with CCF type cap at J=0.4 and σ=Atm.

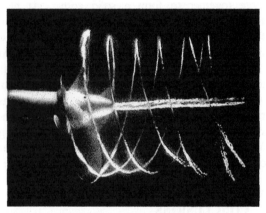

Picture 2. Fully developed hub vortex cavitation with Standard boss cap at J=0.4 and σ=Atm.

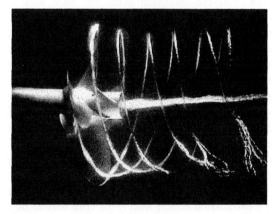

Picture 5. Fully developed hub vortex cavitation with TEF type cap at J=0.4 and σ=Atm.

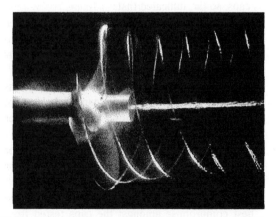

Picture 3. Fully developed hub vortex cavitation with Cylindrical boss cap at J=0.4 and σ=Atm.

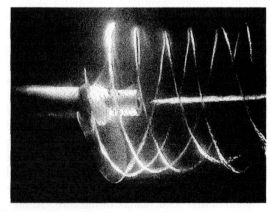

Picture 6. Fully developed hub vortex cavitation with BoS type cap at J=0.4 and σ=Atm.

performance. However the efficiency loss associated with this cap requires further research work supported by CFD analysis and LDA measurements of the flow around the cap.

The research work presented here has been continuing to quantify the effect of the rudder and noise as part of a student project during the writing up of this paper.

ACKNOWLEDGEMENTS

Along the course of the research work various members of the Cavitation Tunnel Group have contributed into the project. Particular efforts from ST Bianchi, Dr D Wang, Assist. Prof. AC Takinaci, Mr E Korkut and Ian Paterson are gratefully acknowledged.

REFERENCES

1. H.E. Saunders, Hydrodynamics in Ship Design, SNAME publications, Sec 23, Vol. 1 and Sec 47, Vol. 2, (1957)
2. W.S. Gearhart and M.W. McBride, "Performance Assessment of Propeller Boss Cap Fin Type Device", 22nd (ATTC), St John's, (1989)
3. M. Ogura, H. Koizuka, T. Takeshita, Y. Kohno, K. Ouchi and T. Shiotsu, "A Screw Propeller Boss Cap with Fins", UK Patent Application GB2194295A, (1988)
4. K. Ouchi, M. Ogura, Y. Kono, H. Orito, T. Shiotsu, M. Tamashima and H. Koizuka, " A Research and Development of PBCF (Propeller Boss Cap Fins) – Improvement of Flow from Propeller Boss", Journal of Society of Japan, Vol 163, (1988)
5. K. Ouchi, M. Tamashima , T. Kawasaki and H. Koizuka, " A Research and Development of PBCF (Propeller Boss Cap Fins) – 2nd Report: Study on Propeller Slipstream and Actual Ship Performance", Journal of Society of Japan, Vol 165, (1988)
6. K. Ouchi, M. T. Kawasaki, M. Tamashima and H. Koizuka, " Research and Development of PBCF (Propeller Boss Cap Fins) – Novel Energy Saving Device to Enhance Propeller Efficiency",
Naval Architecture and Ocean Engineering (Eds), Ship and Ocean Foundation of Japan (Pubs),Vol 163, (1992)
7. C.J. Yang, M. Tamashima, R.and R. Yamazaki, "Calculation of the Performance and Flow Field of a propeller with Boss Cap Fin – in Uniform flow", Trans. Of the West Japan Society of Naval Architects, No 81, (1991)
8. M Tamashima, C.J. Yang and K. Ouchi, "Calculation of the Performance of Propeller with Boss Cap Fins in Uniform Flow"" 2nd Intl. Symposium on Cavitation, Tokyo, (1994)
9. K. Ouchi, M. Tamashima and K. Arai, "Propeller Noise Reduction caused by PBCF", PRADS'92, Newcastle upon Tyne, (1992).
10. Potsdam Model Basin (SVA) Advertisement Brochure, "TVV and HVV Propellers - an Innovative Concept developed by SVA in Cooperation with SCHOTTEL"
11. SCHOTTEL Advertisement Brochure, "TVV and HVV Propeller, A Joint Develeopment of SCHOTTEL and SVA"
12. E.J. Glover and F.I. Leathard, "Tests with a series of fairing covers, SMM Project no RB/15/81", Emerson Cavitation Tunnel report no 4/85, May 1985.
13. E.J. Glover and F.I. Leathard, "Tests with a vaned cone", Emerson Cavitation Tunnel report no 1/89, February 1989.
14. E.J. Glover and G.H.G. Mitchell, "Further tests with Vaned Cones", Emerson Cavitation Tunnel Report no 1/90, June 1990.
15. M. Atlar, S.T. Bianchi and I. Paterson, "Experimental Investigation into Effective Boss Cap Design", Department of Marine Technology report No MT-1996-005, University of Newcastle, (1996)
16. A.C. Takinaci, "A Wake Rollup Model for Heavily Loaded Marine Propellers", Intl. Shipbuilding Progress, no 435 (1996)
17. A.C. Takinaci, "Hydrodynamic Analysis of a Propeller with Trailing Edge fFap by using a Surface Panel Method", Emerson Cavitation Tunnel Report No: 1/96, Department of Marine Technology, March 1996.
18. S.T. Bianchi "Effective Boss Cap design", MPhil Thesis (in preparation) Department of Marine Technology

Practical Design of Ships and Mobile Units
M.W.C. Oosterveld and S.G. Tan, editors.

LIUTO Development and Optimisation of the Propulsion System; Study, Design and Tests

G. Bertolo[a], A. Brighenti[b], S. Kaul[c] and R. Schulze[d]

[a] ACTV, Azienda Consorzio Trasporti Veneziano
San Marco 3880, Venezia, Italy

[b] Systems & Advanced Technologies Engineering S.r.l.,
San Marco 3911, Venezia, Italy

[c] SCHOTTEL Shipyard, JOSEF BECKER GmbH & Co. KG
D-56322 SPAY, Germany

[d] Propellerhydrodynamics , Schiffbau-Versuchsanstalt Potsdam GmbH
Marquardter Chaussee 100, D-14469 Potsdam, Germany

In fall 1996 ACTV, the two industrial companies SCHOTTEL WERFT and INTERMARINE and the three R&D institutions University of Naples (DIN), Maritime Research Institute Netherlands (MARIN) and Schiffbau-Versuchsanstalt Potsdam (SVA), with the financial support of the EC Brite-Euram programme, started an R&D project to develop and full scale test a new motor boat for public urban transports, to be used in water cities, such as primarily but not exclusively Venice.

LIUTO (Low Impact Urban Transport water Omnibus) is the name of the project, whose goal is to develop the prototype of Venice's 2000 M/b fleet with the following main aims and innovative features:

- achieve low hydrodynamic impact by wave and propeller washing generated in the navigation and the frequent manoeuvring, by means of an optimised hull and an innovative propeller designs;
- qualify and compare results with existing vessels by validated CFD numerical tools, model and full scale tests;
- test and apply composite materials, resisting heavy duty services, vandalism and environmental conditions, for the hull construction and the superstructure, to reduce maintenance costs and help achieving stability.

ACTV is responsible for co-ordinating the project and defining, as prime interested end user, the specification of the motor boat. This will also feature innovative technologies for propulsion energy generation (a hybrid diesel electric system with buffer batteries) and external and internal noise limitation which are the object of concurrent activities under ACTV own support. This paper after an overview of the vessel operational profile develops in greater detail the characteristics of the propulsion system.

1. OPERATIONAL PROFILE

The passenger transport system of the city of Venice and its lagoon relies on a fleet in excess of 110 vessels, among water busses (54 M/B) and motor crafts (59 M/S). They have steel hull and are powered by diesel engines. Their navigation pattern, compared to other water transports, features a wide variation of payload and displacement, varying speed limits, relatively high acceleration and deceleration performances, for timetable optimisation and safety in the very congested urban traffic.

Fixed axis propellers and manoeuvring by rudder cause a significant turbulence and jetting in areas near berths, canal turns and buildings foundations.

The LIUTO characteristics (Table 1) allow an efficient and environmentally friendly operation in a variety of lines, both across the city and between the central Venice and the nearby islands.

The optimisation of the vessel took into account particularly the most important service lines, involving the crossing of the central urban area along the Grand Canal (30-60 m wide) and the larger Canal of Giudecca by very frequent stops (one every 2-3 minutes) and prevailing transient power conditions.

Traffic is there intense and congested. The service must comply with the speed limits stated by the competent authorities to keep boats waves low, respectively the Borough of Venice for the canals in the historical centre of the city and the Harbour Authority, for the larger canals for the maritime navigation.

The LIUTO design and main test conditions thus cover the following basic speed requirements, in calm waters and no wind conditions:

V_1 urban speed   5.94   knots

V_2 max full load speed   10.0   knots

V_3 max half load speed   10.8   knots

		S. 80 Existing	E1 Existing	LIUTO New
Passengers capacity		219	208	234
Seats		83	72	100
Displ. (ls./fl.)	t	39/57	35/50	34/50
Length B.P.	m	21.0	20.9	24.7
Max Speed	kn	11.5	8.9	10.8
Constr. height	m	1.85	1.90	2.03
Driver cont. power	kW	147	60	90
Type of driver		Diesel	Electric	Hybrid

Table 1 - Comparison of the LIUTO characteristics with present ACTV M/bs

The full range of operational water depths goes down to 2 m in the shallowest areas of the canals, the most common falling between 3 and 10 m.

The manoeuvrability will also be improved with respect to present M/bs, presently characterised by 25 m turning radius, with a rudder angle of 42°.

The choice of a directional propeller goes with this target, as it allows continuous 360° thrust direction variation. The maintained possibility of reversing the propeller revolution, thanks to the electric drive provided by a hybrid energy system, allows maximum flexibility and possibility to

gradually shift from present pilots practice to the improved features offered by the directional thruster.

An important requirement of the propulsion and energy system is its capability to stop the vessel in a short distance both during regular service and when in emergency, e.g. to avoid collisions with crossing vessels. To limit the hydrodynamic impact under normal operation it should be avoided to exceed certain power limits, by a relatively "smooth" power demand attitude, within the nominal motor performances. Instead during occasional emergencies these power limits should be overcome allowing a stop within 2.5 x LOA from the speed of 10 kn and within 1 x LOA from the speed of 5.94 kn.

The power to the thruster motor is thus managed by a supervisory control system that is also in charge of the efficient and safe management of the whole hybrid energy system.

2. HYBRID PROPULSION ENERGY SYSTEM

The LIUTO energy system will be a *series type, hybrid diesel electric* system as shown in Figure 1, including the following main components:
- diesel engine, running at constant speed
- 3-phase synchronous electric generator
- 3-phase rectifier
- batteries stack (high charge/discharge rate lead-acid with gel type electrolyte)
- power management and battery charge control
- inverter unit
- 3-phase asynchronous electric motor
- Interface mechanical coupling with the propulsion system drive shaft

The system was studied in cooperation with the University of Naples Department of Naval Engineering [1,2] after detailed analysis of existing vessels performances and is now being designed and manufactured by ANSALDO, under ACTV contract.

The gen-set converts all the mechanical energy delivered by the diesel engine into AC electric energy. This is then rectified and supplied in parallel to the battery stack and the user functions, the main one being the propeller AC asynchronous motor, fed by the inverter. This latter operates at variable frequency and voltage, as shown in the control loop dotted in Figure 1.

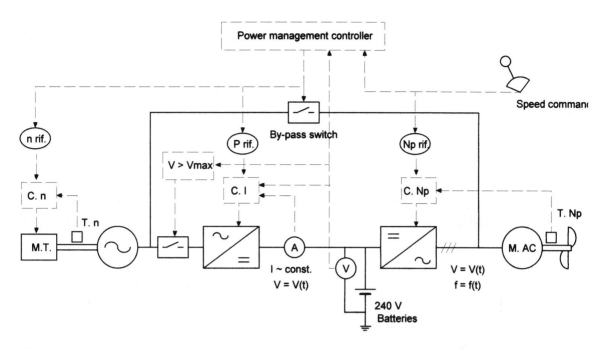

Figure 1 - Scheme of the hybrid system and its main control structure - Feedback to battery charger (T = transducer, C = controller). The main and the emergency AC motors are shown as a unique one.

The power generating set runs at either no power (idle or off) or at a single power level. The battery stack features mostly peak power supply and energy buffer, under very high but short charge/discharge power levels.

The system is twofold redundant, against failure of any of the static or rotating subsystems. If the inverter fails a by-pass switch allows the AC generator to directly drive an emergency, lower power, AC motor (not shown in Figure 1). The batteries are made by two stacks in parallel, that can individually be isolated, keeping the system operational, yet at lower power performance, if either fails. Should the engine, the generator or the rectifier fail instead, the energy capacity of the battery stack is largely sufficient to complete the mission and go to the repair yard, with the engine off.

3. PROPULSION SYSTEM DESIGN

3.1. Propulsion unit generally

SVA and SCHOTTEL-Werft were in charge and have developed an efficient propulsion system for LIUTO. Propellers with different diameters and optimum numbers of revolutions have been designed. The calculations showed that a propeller diameter of $D \approx 0.750$ m should be used.

Different propulsion systems are practicable for the water omnibus. Aspects of the propulsion systems conventional propeller, propeller made out of fibre reinforced material, ducted propeller, SCHOTTEL Twin Propeller and LINEAR-Jet have been discussed

Finally a new Z-drive was designed. An optimised, streamlined shape of the lower gearbox housing with low drag was realised. The drag of the housing is 3 - 5 % lower than at conventional shaped housings. The influence of the housing on the propellers was minimised. This means that the increase of thrust and torque coefficient and the decrease of the efficiency caused by the disturbed wake field due to the housing could be reduced distinctly. Compared with the state of the art it can be stated that there is no housing existing today with such outstanding stream properties. The use of spheroidal cast iron makes it possible to realise streamlined laminar profiles with an unconventional

774

form with an optimised relation of length to thickness. The blades camber and the angle of attack was optimised in model tests so that flow separation is completely prevented. Previous investigations about asymmetrical arrangement of the shaft showed that the spin recovery can be improved but the resistance of the housing increases in the same value. Therefore the symmetrical version was chosen.

Figure 2 - 3D-Drawing of the Lower Gear Box

3.2. Materials - Propellers and Foils / Fins

Propellers are available normally in CuNiAl or manganese-bronze and, for special fields of operation, even in stainless steel. For the LIUTO-project it is suitable to have parts with low weight due to the frequent speed reversals. In this case the use of unconventional materials (carbon fibre reinforced material) for the propellers and the foils/fins of the TWIN-Propeller was examined.

It was the first time that such kind of propeller was driven at a Z-drive with it's special operation conditions (oblique flow during steering, disturbed wake field due to the housing etc.)

The manufacture of this kind of propeller has shown a very high accuracy. The material's very good damping properties lead to operation with low noise and pressure vibrations. Another effect is the

positive influence to the elastic mass system. Because of the low rotational inertia it is possible to relinquish an additional elastic clutch at the cardan shaft. A propeller weight saving of approx. 65 % could be achieved compared to standard materials.

4. PROPULSION SYSTEM

4.1 The TWIN-Propeller

The TWIN-Propeller Technology, is characterised by two propellers on the same shaft with the same rotation direction and a guide arrangement between the propeller (Figs. 3,4,5).

Figure 3 - SCHOTTEL-Twin Propeller, Standard Version

This results in improvement of efficiency and noise emission (pressure fluctuations) for two main factors:

a. Power distribution at two propellers: a lower thrust load of each propeller reduces the impulse losses. The blade geometry of both low loaded propellers can thus be designed in a more efficient way (profile geometry, chord-length, thickness, camber etc.). Interference effects among the blades are by comparison lower than in a high loaded single propeller, usually designed with a higher number of blades. Furthermore this power distribution leads to low level of cavitation and pressure fluctuation, which cannot be achieved by another system.

b. Recovery of lost spin energy: lost spin energy is recovered and flow is directed to the rear propeller by using an integrated guide arrangement consisting of a specially formed housing and additional guide fins.

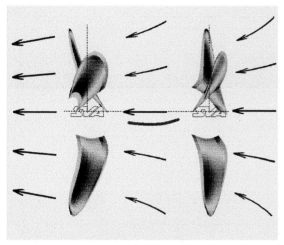

Figure 4 - Hydrodynamic model of a SCHOTTEL-Twin-Propeller

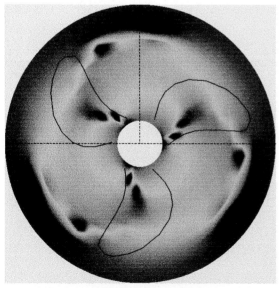

Figure 5 - Computed wake distribution in the position of the second propeller

4.2. Previous developments and LIUTO's step forward

SCHOTTEL has delivered or in order 26 STP-units in a power range from 80 kW to 1250 kW. For example some Double Ended Ferries for Norway, among which one is equipped with two STP 1010 (2 x 1250 kW). The STP-units enable an efficient operation at low noise and high degree of manoeuvrability. Extensive tests were done recently and confirmed the excellent performance of the units. The full scale test results were also very close to the model based predictions.

SCHOTTEL is using the Twin-Propeller Technology also in PoD-Propulsion. In co-operation with SIEMENS AG SCHOTTEL has developed the SSP. The SSP is a podded electric drive available in a power range from 5 to 30 MW. A new permanently excited electric motor from SIEMENS is integrated in the underwater gondola of the Rudderpropeller. Boths propellers are driven directly. The motor is the most efficient and smallest electric motor that is built today. Therefore a very slim streamlined gondola with a low resistance has been developed.

The combination with the TWIN-Propeller Technology leads to a convincing PoD-Propulsor.

The TWIN-Propeller Technology was a big step to improve the efficiency of high loaded rudder-propellers. Using the same technology with two plus two bladed propeller system is the consequent development for low loaded rudder-propellers.

LIUTO made it possible to examine such systems and to approve it in practice.

It was the first examination of 2-bladed TWIN-Propeller. The investigations include theoretical calculations, model tests in cavitation tank and full scale tests with a prototype at test pontoon of SCHOTTEL.

The safety against cavitation is given by the fact that each one of the two propellers is driven with lower load and the total propeller load of the LIUTO-vessel is very low too.

The result is an additional efficiency increase by using 2-bladed propellers was achieved. The mechanical problems of 2-bladed propeller systems - the torsional vibrations and the pressure vibrations - could be solved by a special blade geometry, a comfortable distance between propeller and housing and by the use of very light propeller blades made of carbon fibre reinforced material.

776

The procedure of calculation of 2-bladed TWIN-Propeller and the transformation of the model test results to full-scale values was verified.

Figure 6 - Double Ended Ferry, MRF-Norway

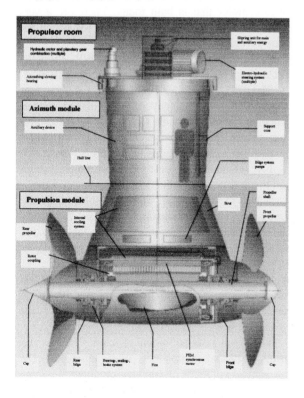

Figure 7 - SSP 7 TWIN-Propeller as PoD-Propulsor

Figure 8 - SSP 7 TWIN-Propeller as PoD-Propeller (test model in the SVA cavitation tunnel

Figure 9 - SCHOTTEL- STP 1010

777

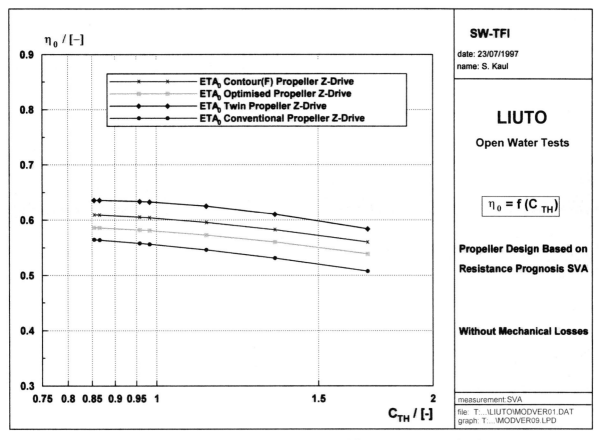

Figure 10 - Model Test Result, Comparison of the different Systems studied for LIUTO.

4.3. CONTOUR(F/S)-Propeller

The CONTOUR(S)-Propeller is a specially designed propeller. Carbon fibre reinforced composites enables a very slim profile geometry. This kind of profile geometry leads to a higher total efficiency of the propeller.

The combination of CONTOUR(S)-design and the TWIN-Propeller join the mechanical advantages of the light and highly accurate manufactured propeller with a new hydrodynamic technology. The result: propulsion system with highest efficiency for the LIUTO vessel.

The CONTOUR(F)-Propeller is the flexible type of CONTOUR-Propellers which deforms in operation in a defined way depending on the load. That leads to an optimum efficiency in a wide range of operation. For example, at overload conditions (stopping, acceleration) the pitch ratio is reduced

automatically. Therefore overload can be reduced and the engine operates at better conditions.

The theoretical pre-calculation led to an efficiency increase of 2 % at higher ship speed and up to 10 % at low speed (overload conditions).

There are still some difficulties of the fibre structure design to achieve the hydrodynamic parameters at each operation point.

Therefore for the LIUTO M/b prototype the 2 blade STP is being made by CONTOUR(S) type blades.

Further research work is required to control the deformation characteristic. A lot of companies all over the world have started the development of such systems. The interest in this technology has increased. Special installations are military and hydrographical ships or special requirements on environmental protection aspects for example.

PFFEDIT the routine PFFEDIT yields the propeller draw and represents the main data as a function of the radius r	
VTXPLOT yields the graph of K_T, $10K_Q$ and η_0 as a function of J or $C_{th,}$ the graph of the cavitation buckets	
DENPLOT represents the vortex strength distribution	
CAVPLOT represents the cavitation bucket chart and the cavitation behaviour of the blades in the wake (regions with critical pressure are marked)	
WAKPLOT represents the wake	
VELPLOT computation and representation of the velocity distribution around the propeller (in the propeller jet)	

Table 2 - Components of the numerical cavitation tunnel of the SVA-Potsdam GmbH

5. PROPELLER DESIGN BY SVA POTSDAM

5.1. Introduction

A modern design method for marine propellers is in general based on a collection of computer programs (Table 2) to calculate

- the propeller - hull interaction
- the propeller - machine interaction
- the propulsive performance which includes propeller efficiency
- the cavitation on blades and fluctuation pressure on ship hull
- the strength.

The usual strategy for the realisation of the design process consists in a iterative trial and error algorithm to

- increase the propeller efficiency
- decrease the cavitation on blades and the fluctuation pressure on ship hull.

This iteration process can be started with a conventional design method such as some series charts techniques.

The calculation for propellers can be accomplished by using the lifting line and the lifting surface theories under steady and unsteady conditions respectively.

Contrary to the trial and error strategy we determine the propeller geometry by an optimization technique.

5.2. Inverse methods for the design and optimisation of marine propellers

The strategy of Potsdam Ship Model Basin consists of the construction of a "Toolbox Propulsive Performance Optimisation"

- definition of a "weighting function" to determine the propeller efficiency (η), cavitation behaviour, etc.
- variation of the propeller geometry by an optimisation algorithm
- calculation of propellers by using series charts techniques lifting line methods lifting surface methods NSE-Solver

The geometry (G) of propellers can be described

r/R	c	P/D	rake	Xe	f/c	t/c
.200	479.71	1.10	.000	295.41	.0385	.118
.400	587.56	1.20	.000	352.51	.0274	.074
.600	641.93	1.30	.000	358.32	.0202	.047
.700	636.72	1.40	.000	335.21	.0171	.037
.900	486.39	1.30	.000	194.56	.0111	.022
1.000	20.00	1.20	.000	.000	.00000	.000

Table 3 – Propeller geometry definition
(c, P/D, rake, Xe, f/c and t/c are functions of r/R.)

by (e.g. 42) real numbers like in Table 3.

The geometry G of the "best propeller" in the sense of efficiency (η) is the solution of the following optimisation problem:

$$f(G) := \eta_0(G) ===> \max !$$

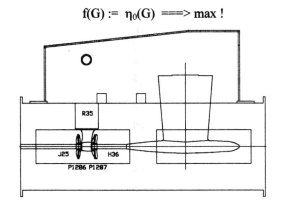

Figure 11 - SVA test facility (Kempf and Remmers) for the SCHOTTEL - TWIN - Propeller

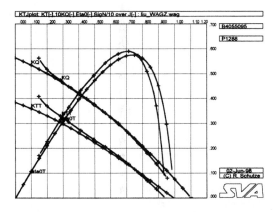

Figure 12 - Comparison of the open water behaviours of a single Wageningen propeller (continuous line) with the optimised SVA-propeller P1288 (dotted line), with the influence of the Z-drive.

6. MODEL AND FULL SCALE TESTS

6.1 Model Tests of the SCHOTTEL Twin Propeller (SVA design TP1286/1287)

The tests performed at SVA followed the steps listed herebelow:

- open water tests with the first and the second propellers of the twin pair with the dynamometer J 25

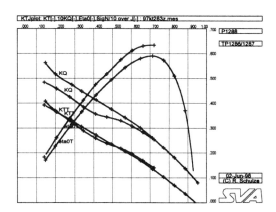

Figure 13 - Comparision of the single optimised pull-propeller P1288 (continuous line) and the Twin-propeller-system TP1286/1287 (dotted line), with the influence of the Z-drive.

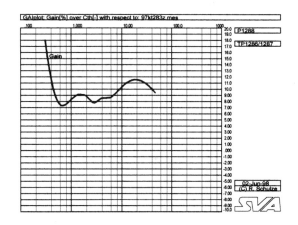

Figure 14 - Comparision of the single optimised pull-propeller P1288 and the Twin-propeller-system TP1286/DP1287/Z-drive, efficiency gain over C_{TH} [%] with respect to P1288.

- open water tests with the z-drive housing and the pull propeller (dynamometer J 25, balance R 35X). (control of the pitch ratio for the design point of the pull propeller)
- open water tests with the z-drive housing and the pull and push propeller (dynamometer J 25, H 36, balance R 35X). (control of the pitch ratio for the design point of the push propeller and the STP)
- open water and cavitation tests with a variation of the phase angle between both twin propellers (fixing of the phase angle)

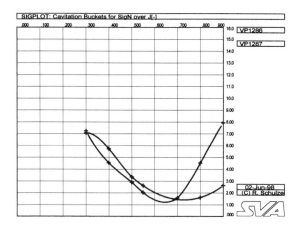

Figure 15 - Cavitation bucket of the Twin-propeller-system TP1286/1287. The cavitation bucket of the first propeller P1286 (continuous line) and of the second propeller P1287 (dotted line). Only the end of the suction/pressure side cavitation depending on the advance coefficient was drawn.

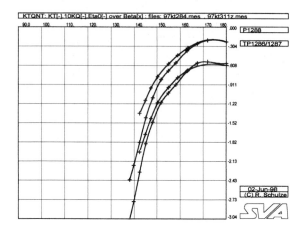

Figure 16 - Crash stop manoeuvre for the Schottel-Twin-Propeller TP1286/1287 in comparison with the optimised pull propeller P1288

- cavitation tests for all design points in homogeneous inflow
- open water tests for measuring the characteristic of the STP during the reversing process

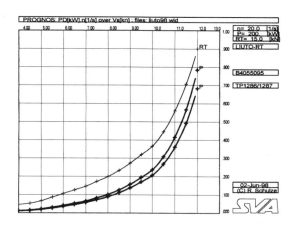

Figure 17 - Comparison of delivered powers in dependence of the water omnibus speed (kn) for the Schottel-Twin-Propeller and a classical Wageningen propeller, with Z - drive correction (RT: resistance curve. Continuous line: PD for the Wageningen propeller. Dotted line: PD for the Twin-Propeller).

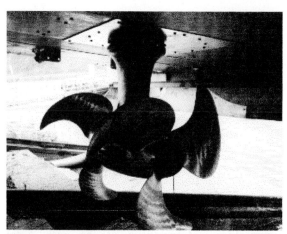

Figure 18 - Test arrangement of a SCHOTTEL-TWIN Propeller

6.2. Full Scale Tests

The test facilities at SCHOTTEL allow extensive examinations in full scale including the following tests like

- Torsional Vibrations
- Crash back
- Crash ahead
- Steering
- Load of the foils
- Bollard Pull

A careful comparison of full-scale tests and

model tests is an important step to show some scale effects especially for new developments. This gave the opportunity to verify model tests and computerised calculations. In the case of LIUTO full-scale tests confirmed the very good model tests.

After the parallel activities made in the project on the hull hydrodynamics optimisation and, in particular, after the resistance tests results on large scale ship model, it could be confirmed that more than 60% efficiency is achieved over a large range of LIUTO's operational speeds (Figure 21).

7. CRASH STOP PERFORMANCE OF THE PROPULSION SYSTEM

An important aspect of the performance of LIUTO's propulsion system is its breaking and acceleration performance, particularly the former as regards safety. As after the discussion of the operational profile under emergency conditions the stop distance should be within the limits of 2.5 or 1 times the LOA, i.e. 63 and 25 m, respectively when stopped from 10 and 5.94 kn respectively.

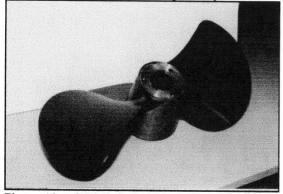

Figure 19 - 2-Bladed Optimised Propeller for the Twin-Propeller-System for LIUTO.

To verify this performance and the requirement on the electric motor, inverter and batteries dynamic simulations of the motor-mechanical transmission-propeller-ship dynamics were performed, based on the characteristics of the vessel and the STP propeller, derived from model tests. The dynamics account for the delay in motor shaft acceleration to reverse the propeller speed and the variation of the propeller working point (K_T & K_Q vs. J) and ship total resistance at each time.

The electric motor is supposed to be driven at constant negative torque during the whole breaking phase.

The results of the simulations and of a sensitivity analysis are summarised in Figure 22. With a torque overload factor of 1.4 both breaking requirements are satisfied, the overload lasting 24 and 16 s respectively. The peak motor power under these conditions is 132 kW.

The electric motor shall withstand these conditions only occasionally for the duration of the breaking phase without damage, starting from the thermal state corresponding to a typical mission power profile [Refs. 1, 2], without repetition of this stress. Under normal operating conditions, instead the electric motor shall be used within its nominal torque, yielding a breaking distance of 2.9 or 1.3 LOA, respectively from 10 and 5.94 kn, and a peak motor power of 76 kW.

Figure 20 - 5 bladed model of the CONTOUR (F) single rotor propeller (Carbon fibre reinforced material).

8. CONCLUSIONS

The LIUTO project is a step towards the achievement of low impact and efficient propulsion systems. It pursued by a broad view of applicable solutions and extensive modelling and tests this goal and it achieved it.

The most important results are:
- the efficiency of the propulsion system exceeding 62% for speeds going from 2 kn up to the maximum speed foreseen of 11 kn.
- the acceleration and crash-back performances, allowing a reduction of nearly 40% of the installed motor power with respect to present M/b's.

J, KT, KQ vs. speed at stationary conditions

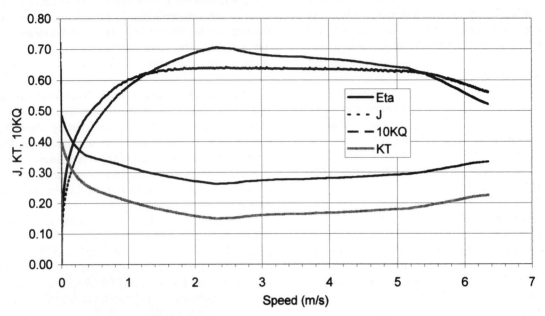

Figure 21 – J, KT, KQ and η curves vs. LIUTO speed.

STOP DISTANCE AND TIME FROM 10 or 5.94 kn

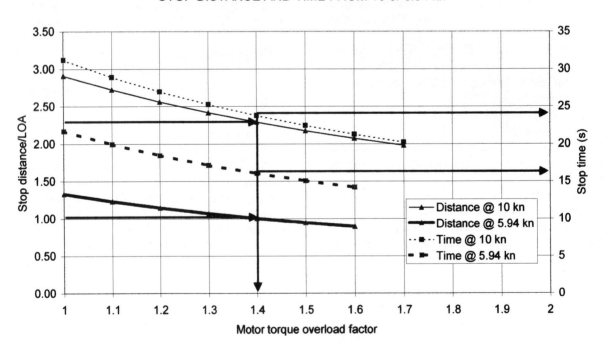

Figure 22 - Stop distance and duration - Sensitivity analysis vs. motor torque overload factor.

ACKNOWLEDGEMENTS

The LIUTO project partners wish to acknowledge and thank the EC DGXII for the support to this important project.

REFERENCES

[1] Balsamo F., Brighenti A., Landri G., Paciolla A., Quaranta F.: "The propulsion of public transport vessels in coastal and inner waters: working data acquisition, elaboration and study of alternative solutions", Proceedings of 1994 NAV Conference, Rome (Italy), 5-7 Oct. 1994

[2] Balsamo, F.; Brighenti, A.; Landri, G.; Paciolla, A.; Quaranta, F. " The propulsion of coastal and inland water transport vessels - Working data acquisition and preliminary design of innovative systems to reduce pollutant emissions 21st CONGRESS OF CIMAC, Interlaken - Switzerland, May 15-18, 1995

[3] Heinke, H.-J.; D.; Schulze, R. L.I.U.T.O. - Low Impact Urban Transport Omnibus; Propulsion Systems for the Omnibus, Report No. 2247, Schiffbau-Versuchsanstalt Potsdam, Jan. 1997

[4] Schulze, L.I.U.T.O.–Test–Report, Report No. 2299, Schiffbau-Versuchsanstalt Potsdam, Aug. 1997

[5] Bohm, M.; Heinke, H.-J.; Schmidt, DSchulze, R. Untersuchung zum SCHOTTEL- als Doppel-propeller Bericht Nr. 2204, Versuchsanstalt Potsdam GmbH, May 1996

[6] Propeller einmal anders Boote 12/93

[7] Bohm, M.; Heinke, H.- Hydrodynamische Untersuchungen Modellpropellern aus Bericht Nr. 2050, Schiffbau-Versuchsanstalt Potsdam GmbH, April 1994

[8] Propeller aus CFKProspekt der AIR Fertigung-Technologie GmbH

[9] Jonk, A.; Holtrop, Inboard noise in passenger ships propelled azimuthing thrustersFax-report, MARIN, Sept. 1995

[10] Contur-propeller, Contur(F)- Prospect AIR Fertigung-Technologie GmbHRostock 1996

[11] The SCHOTTEL Twin Propeller - an propulsion system, SCHOTTEL Information, 75 Years SCHOTTEL, 10th August 1996

[12] Jürgens, Der LINEAR-Jet - ein Propulsionssystem für flachgehende Wasserfahrzeuge, Fachkongreß ROTECH Rostock, June 1994

[13] Bohm, Entwicklung des Antriebssystems LINEAR-Jet Yachten Bericht Nr. 2023, Versuchsanstalt Potsdam GmbH, January 1997

[14] Schulze, RProjektstudie zur Entwicklung eines zur optimalen Gestaltung von Bericht 2030, Schiffbau-Versuchsanstalt Potsdam, 1994

[15] Schulze, R. Nachrechen- und Entwurfsverfahren für den Propellerentwurf,1. SVA-Forum „Numerischer Tank Potsdam 2.3.1995

[16] Schulze, R. Propellerumströmungs-und Festigkeits- berechnung mit VORTEX und ANSYS, Teil1: Berechnung der Propellerumströmung mit dem Wirbelgitterverfahren VORTEX(0), Bericht 2163, Schiffbau-Versuchsanstalt Potsdam, August 1995

[17] Schulze, R. VORTEX - Benutzerhandbuch für den numerischen Kavitationstunnel der SVA Bericht Nr. 2217, Schiffbau-Versuchsanstalt Potsdam, Juli 1996

[18] Schulze, R. Untersuchungenz ur Doppelpropelleranordnung VP4041/42 am Z-Antrieb mit Drallminimierung, Bericht 2166, Schiffbau-Versuchsanstalt Potsdam, August 1995

[19] Schulze, R. Entwurf einer Doppelpropelleranordnung für den Z-Antrieb der Experimentalplattform der SCHOTTEL-Werft Bericht 2172, Schiffbau- Versuchsanstalt Potsdam, Sept. 1995

[20] Bohm, M.; Heinke, H.-J.; Schmidt, D. Schulze, R. Untersuchungen zum SCHOTTEL-Ruderpropeller als Doppelpropeller, Bericht 2204, Schiffbau-Versuchsanstalt Potsdam, März 1996

A new concept of pushboat design

B. Bilen and M. Žerjal

Institute of Technical Sciences of the Serbian Academy of Sciences and Arts
Knez Mihailova 35/IV, P.O. Box 745, 11000 Belgrade, Yugoslavia

This paper briefly presents design of a new pushboat concept for European major inland waterway Rhine - Main - Danube. By studying model testing and trial records of a number of existing pushboats, very low propulsion efficiency is noticed as a basic weakness of conventional pushboats' propulsion system. A new pushboat concept featuring improved maneuver and better operating efficiency is proposed. The backbone of new design is an originally developed hybrid propulsion system with two main diesel engines driving three ducted propellers. In the complete absence of any rudders and appendages and with no stern tunnels, this new pushboat concept offers better propellers inflow conditions resulting in reduced level of propeller induced vibrations. By optimizing load distribution between a central and two steering propellers, an increased overall propulsive efficiency is achieved.

1. INTRODUCTION

The usage of pushing technology on European rivers Rhine and Danube started at the end of sixth decade of twentieth century. First pushing units were reconstructed existing river tugboats, then a rapid development of pushing technology began and the towing concept completely disappeared in a short time. The main characteristic of that development up to nowadays was an abrupt growth in barge capacity and pushed convoy size which had to be followed by a corresponding growth in installed propulsion power of pushboats. Thus, starting from the first barges having cargo capacity of about 600t, a capacity of more than 1800t has been reached with standardized barges EIIb being dominant on European rivers and channels now. The installed propulsive power on European pushboats was growing from about 2x500kW with the first units, up to 3x1330kW with the latest pushboats built in 1990. Therefore, the most powerful boats can push large convoys of six and more standard EIIb barges coupled, having total cargo capacity of 15000t to 18000t and even more.

Such a course of the pushing technology development was an exclusive product of shipowners' aspirations to increase the capacity of their transports as much as possible.

Therefore, it was quite natural that increasingly powerful pushboats were ordered. On the other hand, the growth in pushboats' power was followed by almost no any development of their propulsion - maneuvering system configuration.

Formerly, there were some attempts to point out to a theoretical inadequacy of such a development reduced to the growth in convoy capacity and propulsive power. However, those attempts never impressed European shipowners. Recently, some information have started to come from the USA reporting that development trend on the Mississippi river has been changing to a quite opposite in relation to the course described previously. Finally, the circumstances in European inland waterborne transport have completely changed as a consequence of the opening of Danube - Main channel creating a major European inland waterway from the North to the Black Sea. That way, the favorable conditions for restarting a discussion about a new concept of river pushing technology including a new concept of pushboat design have arisen.

2. CONVENTIONAL RIVER PUSHBOATS

In spite of an enormous growth in pushed convoys' size and pushboats' power, the basic

786

configuration of propulsion - maneuvering arrangement of pushboats remained unchanged during the whole developing period until nowadays. Within a conventional pushboat's propulsion and maneuvering system, screw propellers operate in extremely unfavorable conditions near water surface. Those propellers are heavily loaded, working in deformed hull-integrated nozzles with extremely small clearance between blade tip and hull bottom plate. The inflow of such a propeller is usually obstructed by a pair of flanking rudders and a strut, while one or two main rudders are situated in a propeller's wake. Moreover, the complete underwater part of the propulsion and maneuvering system is located in stern tunnels of emphasized shape, Fig. 1.

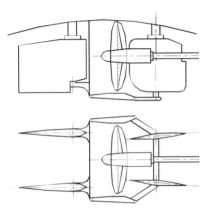

Figure 1. Conventional propulsive arrangement of a river pushboat

By analyzing data taken during towing tank testing, full-scale trial testing and regular service operation of different pushed convoys and pushboats, some bad operational characteristics were usually noticed. They can be directly connected to the configuration of conventional design pushboats' propulsion arrangement.

2.1 Propulsive efficiency

In described circumstances, propellers operate in a non-axysimetric and very non-uniform water inflow. The nozzle-generated thrust is reduced, while global and particularly local fluctuating cavitation is unavoidable. Most of conventional pushboats operating on Danube feature dramatically low propulsive efficiency of

about 30% to 40%. On the other hand, boat's crew experience very bad and health damaging hull vibrations generated by fluctuating propeller forces. Unusually low values of relative rotative efficiency η_R were determined for most of existing pushboats, laying in the range between 0.72 to 0.82 [1-3]. Except it's original definition, the η_R coefficient here includes the other phenomena making such a large difference between the model propeller open water conditions and cavitating conditions on the stern of a pushboat. Those circumstances also make many problems for a designer to choose or design an optimum propeller.

2.2. Speed drop in meandering navigation

Contrary to the sea borne navigation, river navigation is characterized with a frequent use of the rudders in overcoming numerous river curves. The Danube captains reported that it is very rear situation when more than 5 minutes of navigation pass without any small or big rudder deflection. Even with restricted rudder deflections, significant speed drops can be recorded during river navigation full-scale testing [4]. Except for a direct braking influence of deflected rudders as a consequence of generated axial parasite rudder force, the average service speed is also reduced due to yaw and lateral overshoot of maneuvering vessel. An estimation of speed reduction has been calculated for pushed convoys to lay in the range between 10% and 30%, depending on the rudder deflection and the convoy size.

2.3 Utilization of installed propulsive power

The operational efficiency of a pushed convoy can be defined as a product of cargo capacity and speed of navigation divided by propulsive power engaged. If the cargo capacity is presented by displacement, then the operational efficiency η equals the overall propulsive efficiency of the pushboat divided by specific resistance of the convoy:

$$\eta = \frac{\Delta[t] \cdot g[\frac{m}{s^2}] \cdot v[\frac{m}{s}]}{P_B[kW]} = \frac{\eta_S \cdot \eta_R \cdot \eta_O \cdot \eta_H}{(\frac{R_T}{\Delta})} \qquad (1)$$

According to resistance characteristics [5] of different convoy configurations, it can be shown that a large capacity convoy of six barges pushed by a powerful pushboat of more than 1600kW total propulsive power is not an optimum design solution when the operational efficiency is considered. Moreover, a powerful pushboat is very often engaged in off-design conditions either performing some harbor assistance or navigating on a waterway restricted in dimensions. For example, there is a large section of more than 700km on the European major inland waterway permitting the navigation of convoys consisting of a pushboat and just one or two barges EIIb in line. The navigation in such off-design conditions, using a powerful pushboat in operation with a small or unloaded convoy, results in a further dramatic drop in the operating efficiency.

3. HYBRID PUSHBOAT DESIGN

According to equation (1), it was noticed that the most influencing factors giving bad efficiency of conventional pushboats are low propeller open water efficiency η_O, low propeller relative rotative efficiency η_R and high values of specific resistance of the convoy R_T/Δ. Naturally, within attempting to increase the operating efficiency of pushed convoys, the greatest attention has been directed to the circumstances governing the values of R_T/Δ, η_O and η_R. Therefore, the operating efficiency of a pushed convoy can be increased by:
- choosing a nominal convoy configuration of lower specific resistance,
- decreasing the propellers' load and
- reconfiguring the propulsion - maneuvering arrangement and stern form of the pushboat.

After a careful analysis, the proposed three steps were implemented to an innovative pushboat design featuring a "hybrid", triple - screw propulsion and maneuvering system [6].

3.1 Determination of required propulsive power

The amount of propulsive power installed on a line navigation pushboat relates to the service demands, i.e. to the speed and cargo capacity requirements. Contrary to described practice of increasing a pushboat's propulsive power to

enable pushing of a large capacity nominal convoy regardless of its very poor resistance characteristics and corresponding high propeller load, a different approach was applied here. Namely, by disassembling a large convoy into two or three convoys pushed by two or three pushboats, the same cargo can be transported at a considerably greater operational efficiency. In that aspect, a convoy configuration consisting of two standard EIIb barges coupled in line is particularly interesting. That convoy configuration, marked as (1+1), with its large L/B ratio, has superior resistance characteristics, Fig. 2, showing the weakest sensitivity to changes in waterway depth and width [5]. Moreover, the convoy (1+1) is permitted to navigate on a great majority of European rivers and channels. That is why the convoy (1+1) is considered an optimum nominal convoy for a pushboat design, unless a shipowner's specific transport organization scheme requests something else.

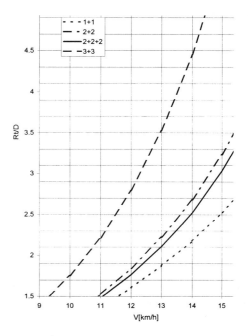

Figure 2. R_T/Δ vs. speed for different convoy configurations (d=2.5m, h_w=5m, b_w=160m)

Considering resistance characteristics of a (1+1) convoy at different draughts and water depths and taking into account low speed limit and extreme European river flow speeds, required propulsive power of the pushboat was

788

determined about 1000kW. That way, the propellers' load has been simultaneously reduced and higher values of the open water efficiency can be achieved too.

Such a low power pushboat design should also realize superior service efficiency as the possibility of inadequate usage of installed power in off-design regimes is reduced to a minimum.

Practically, a pushed convoy consisting of the pushboat and two standard barges in line is very similar in main dimensions to a self-propelled river ship pushing a barge, but surely represents a more flexible solution.

3.2 Propulsion system evaluation

The following step toward increased operating efficiency of pushed convoys was a reconfiguration of pushboat's conventional propulsive arrangement. Previously described conventional pushboat propeller arrangement strongly corrupts the propeller performance achieved in open water conditions, making the design optimization using open water series data an inadequate one. By simplifying the stern and propulsive arrangement and by minimizing the influence of hull, appendages and nozzle deformation, the conditions behind hull become very close to conditions existing in open water testing. That way, three goals can be achieved:

- Most of elements making a conventional pushboat propulsion arrangement inferior are removed giving no reason for drastic reduction of propeller performance in comparison to open water conditions;
- The propeller optimization based on open water results can be successfully applied in design process again (or at least in preliminary design) as the hull influence factors are going to be less important;
- The conditions are created the state-of-the-art numerical propeller models and modern propeller design and analysis techniques can be applied.

So, a triple screw arrangement has been proposed with one central main propeller and two side steering propellers - azimuth thrusters covering either propulsion and maneuvering function, Fig. 3.

As side mounted steering propellers experience almost no any inflow obstructions,

they can be considered operating in open water with η_R value being very close to 1.

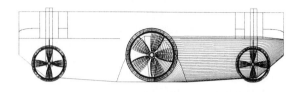

Figure 3. Stern view scheme of a hybrid boat

The central propeller has a dominant role in propulsive thrust production and a particular attention has been paid to the central propeller configuration. Above all, the intention was to make the central part of a pushboat stern configuration as much similar to a usual single screw ship arrangement as possible. By correlation with a depressurized towing tank testing [7-9], favorable values of relative rotative efficiency for the central propeller were estimated in the range from 0.97 to 1.02 for an expected range of cavitation numbers. Those values of η_R are valid for the central propeller arrangement with no stern tunnel, no rudders nor flanking rudders, a circular non-deformed nozzle and zero hull-duct clearance. In front of the central propeller, only a stream-shaped skeg is located carrying stern tube with shaft bearings.

Except improvement in η_R value, the reconfigured propulsion system features a reduced level of propeller induced vibrations due to more uniform propellers' inflow. The propeller load is additionally decreased by distributing the power between three instead of two propellers giving higher values of η_O and a reduced level of general cavitation danger.

3.3 A hybrid pushboat 950kW - 1200kW

According to proposed propulsion system modifications, a complete design of a new hybrid pushboat has been worked out. The boat is propelled by an originally developed hybrid propulsion system with two main diesel engines driving three ducted propellers. A central propeller is mechanically driven by a main diesel engine via reversible reduction gear and conventional propeller shaft. The other diesel engine drives two side propulsion - steering units via indirect hydraulic transmission system

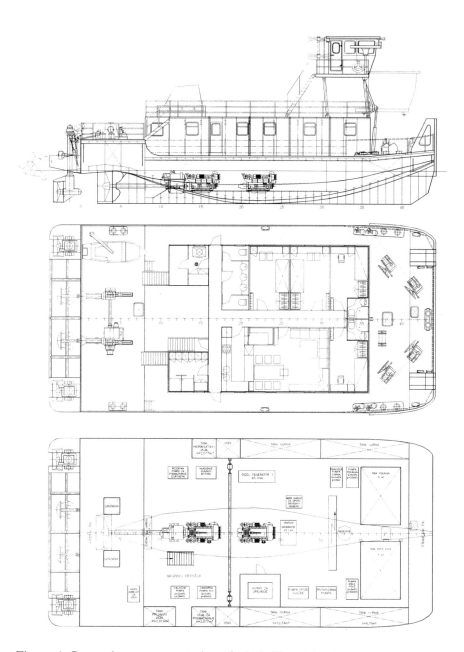

Figure 4. General arrangement plan of a hybrid pushboat

enabling independent reversing and continuous control of side propellers RPM. By optimizing load distribution between the central and two side propellers, unusually bad efficiency of side propellers' hydraulic transmission system can be overcompensated and increased total propulsive efficiency is achieved.

Main particulars of the pushboat are as follows:

- Length overall: $L_{OA}=24.2m$
- Breadth: $B=11.4m$
- Depth: $H=2.8m$
- Draught: $d=1.9m$
- Crew accommodation: 6 persons
- Propulsion diesel: $2 \times 600kW/1800min^{-1}$
- Central propeller: $D_C=1.85m$ + nozz. 19A
- Side propellers: $D_S=1.35m$ + nozz. 37
- Nominal prop. power: $P_B=960kW$

Due to small propeller immersion, and air suction danger, a compromise final solution for the central propeller configuration was found with a small stern tunnel, but no rudders nor flanking rudders, a circular non-integrated nozzle, zero hull-duct clearance and relative rotative efficiency estimated to $\eta_R = 0.95$.

A reconfigured stern form also enables the central propeller shaftline to be reduced to a short propeller shaft with no more than two shaft bearings in the stern tube. The consequences are reduced mechanical transmission losses and reduced shaftline bearings cost and maintenance expenses.

Main propulsion engines are two identical diesels situated in line, one behind the other. The rear diesel engine drives the central propeller, while the front engine on the flywheel side drives two variable displacement axial piston hydraulic pumps for side propellers drive. Nominal design propulsion power engaged is only 960kW, while the total installed power is 1200kW. That power difference is usually spent by a shaft electric generator, also driven by the front diesel engine. In emergency navigation cases, the side propulsion units in a full power regime can also use that spare power. Working pressure in the hydraulic propulsion system is between 140bar, during the nominal run, and 250bar at full engine power absorbed. A spare hydraulic pump driven by a PTO of the rear engine reduction gear can perform emergency navigation and maneuvering in case of the front engine failure.

In spite of very low transmission efficiency of about 75%, the hydraulic transmission system of side propellers' drive offers certain important advantages. For example, one diesel engine can drive two propellers featuring independent reversing, continuous propeller RPM and absorbed power control; capability of adjusting power absorption in off-design conditions; higher thrust in off-design conditions; flexible operation in relation to variable load pushboat services; diesel engine operation at constant RPM; reliable and durable system even capable of withstanding possible propeller blockage.

As an alternative to the diesel-hydraulic drive for side steering propellers, a solution with two separate mechanically driven azimuth thrusters is applicable too. That alternative solution requires three diesel engines to drive three propellers independently. The efficiency of such a mechanical hybrid system is greater as 25% hydraulic transmission losses are reduced to about 10% with mechanical Z-drive. On the other hand, the mechanical solution is more expensive, less flexible and more delicate in exploitation. The advantages of providing the propulsive power emergency reserve can not be obtained with the mechanical system. Contrary to the solution with hydraulic side propellers drive, the hybrid propulsion system with all three propellers mechanically driven can be designed for the total propulsive power exceeding 1500kW.

4. HYBRID VS. CONVENTIONAL DESIGN

A quantitative presentation of the differences between a hybrid and a conventional pushboat design can be given in the following several categories.

4.1 Propulsive efficiency

The propulsive efficiency of the hybrid pushboat design has been compared with a conventional pushboat twin-screw propulsion system of equal nominal power engaged (960kW) and with a conventional pushboat of 1500kW total propulsive power installed. Maximum propeller diameter was limited to 1.85m for all three units. The convoy's resistance has been calculated for calm water straight-line navigation conditions at the barges' draft of 2.5m. Corresponding propellers' performances were calculated using open water propeller series data [10]. With the optimum propellers chosen for nominal pushed convoy configurations, the maximum continuous performance is shown in Fig. 5, for three pushboats to be compared.

The main points could be summarized as follows. In a calm water straight line navigation,
- the hybrid pushboat operating with a nominal (1+1) EIIb barge train would be able to achieve 0.5km/h greater speed than a corresponding conventional pushboat at the same diesel engine power engaged;
- the hybrid pushboat propulsion system could be designed to perform 9% to 10% greater

overall propulsive efficiency than a corresponding conventional pushboat in the whole speed range;

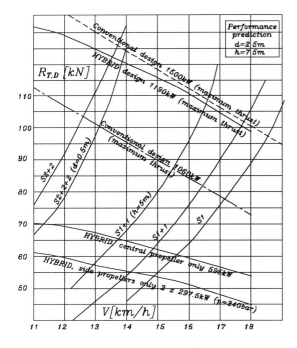

Figure 5. Maximum performance prediction

- in the speed range between 14km/h and 18km/h, the hybrid pushboat MCR (1190kW) performance is almost equal to the pushing performance of a twin-screw conventional pushboat of 1500kW installed propulsive power;
- the hybrid pushboat operating in side propellers only regime would be able to achieve a continual "economy" speed greater than V=13km/h with a nominal (1+1) convoy at d=2.5m draught in restricted water depth of h_w=5m.

4.2 Service efficiency

With the hybrid pushboat design, the service efficiency of a convoy is improved both by installing steering propellers instead of conventional rudders and by minimizing the appearance of navigation in off-design regimes.

Except better and easier low speed maneuvering, the steering propellers offer certain advantages over conventional rudders in the line navigation service too. Namely, the significant average speed drop in meandering

river navigation is dominantly influenced by the direct braking effect and propeller throttling effect of deflected rudders. With steering propellers, there is no propeller throttling, while the braking effect exists, but it is indirect and less influencing than with deflected rudders. Based on analysis of generated forces and maneuvering convoys, the average service speed drop on a given section of river waterway is estimated to be 25% to 30% lower by using steering propellers instead of conventional rudders. The exact amount of improvement depends on the convoy and the waterway configurations.

The influence of the other given improvement on the service efficiency of a pushboat is very hard to quantify. Having reduced propulsive power installed, the hybrid pushboat could be rarely found in the situation of pushing a convoy smaller than nominal one with a bad utilization of installed power. In such a situation, the hybrid pushboat could operate in the side propellers only regime with the rear diesel engine and the central propeller turned off. That way, the hybrid line navigation pushboat could be easily transformed into a typical harbor pushboat having enough power for different harbor operations or pushing unloaded barges. In an opposite case, when the hybrid pushboat is forced to push a convoy larger than nominal one, the spare emergency power could be efficiently used by hydraulically driven side propellers to perform that off-design operation.

4.3 Profitability and maintenance expenses

A local shipyard made an approximate comparative calculation of prices for the hybrid and the conventional pushboat of the same power. Their calculation showed some 8% lower price for the hybrid pushboat building, mainly as the consequence of described propulsion system simplifications.

Based on those price calculations, propulsion and service performance predictions, a complex profitability analysis has been worked out for a defined transport operation performed by the hybrid and the conventional pushboat of 900kW. The cargo transport consisted of 70% upstream and 30% downstream navigation, while the navigation with unloaded barges was excluded in

792

Table 1
Pushed convoy (1+1)

Type of the pushboat	Conventional	Hybrid
Pushboat power	900	900
Purchasing price pushboat + barges [$]	2720000	2420000
Cargo load capacity [t]	3400	3400
Service speed in flowing shallow (h_w=5m) water [km/h]	12.5	13.7
QP [$/h]	179	258

calculations. As the final result, quasiprofit (QP) has been calculated to be a sum of realized net profit and crew expenses per hour of line navigation, Tab. 1. The speed values shown in the Table have been calculated as average service speed in relation to river banks, taking into account the river flow speed and the speed drop generated by the usage of rudders or steering propellers. So, considerably higher value of QP could be realized by using the proposed hybrid pushboat concept thanks to the lower purchase price and higher service speed in comparison to the conventional one.

Moreover, some experts have estimated maintenance expenses of the hybrid pushboat to be approximately 20% lower than for a corresponding conventional one. That is a consequence of reduced number of propeller shafts and shaftline bearings, as well as missing the flanking rudders, main rudders, steering gears and struts with the hybrid concept. Next, the repair of steering propellers could be performed without taking the complete boat out of water, and finally, hydraulically driven propellers are more robust than the propellers driven by direct mechanical transmission.

REFERENCES

1. Model Tests with a Pushed Barge Train, Report No. 1070, Versuchsanstalt fur Binnenschiffbau E. V. Duisburg, June 1983.
2. Prototype Trials of Push Boat MG-40 "KARAĐORĐE", Report no. 4355-B, The Institute for Ship Hydrodynamics - Zagreb, March 1986.
3. Model Testing of Pushboat M-672 (Unit MP-14, 2x735HP), Report no. 2054-M, The Institute for Ship Hydrodynamics - Zagreb, 1972.
4. E. Schäle, Naturgroße Versuche mit Schubverbänden auf dem Rhein, Sonderdruck aus der Fachzeitschrift "Schiff und Hafen", Jahrgang 16 - Heft 11, November 1964.
5. J.L.J. Marchal, Y.-D. Shen, D. Kicheva, An Empirical Formula to Estimate the Resistance of a Convoy in a Restricted Waterway, Journal of Ship Research Vol. 40, No. 2, pp. 107-111, SNAME, June 1996.
6. B. Bilen, News in river pushing technology, Monograph, Institute of technical sciences of the Serbian academy of sciences and arts, Belgrade, 1997.
7. W. van Gent, J. van der Kooij, Influence of Hull Inclination and Hull-Duct Clearance on Performance, Cavitation and Hull Excitation of a Ducted Propeller - Part I, Monograph M4, The Netherlands Maritime Institute, April 1976.
8. J. van der Kooij, W. van den Berg, Influence of Hull Inclination and Hull-Duct Clearance on Performance, Cavitation and Hull Excitation of a Ducted Propeller - Part II, Monograph M15, The Netherlands Maritime Institute, May 1977.
9. J. van der Kooij, Personal communication, 1997.
10. Oosterveld M. W. C., van Oossanen P., Recent Developments in Marine Propeller Hydrodynamics, Publication no. 433 of the NSMB, International Jubilee Meeting in Occasion of the 40th Anniversary of the NSMB, August 30 - September 1, 1972.

Practical Design of Ships and Mobile Units
M.W.C. Oosterveld and S.G. Tan, editors.

On the practical computation of propulsion factors of ships

Do-Sung Kong[a], Young-Gi Kim[a] and Jae-Moon Lew[b]

[a]Shipbuilding & Plant Research Institute, Samsung Heavy Industries Co., Ltd.
103-6 Nunji-Dong, Yusong-Ku, Taejon, 305-600, Korea

[b]Department of Naval Architecture & Ocean Engineering, Chungnam National University
220 Kung-Dong, Yusong-Ku, Taejon, 305-600, Korea

A potential-based panel method together with vortex lattice lifting surface theory are used to predict the propulsion factors of a ship. Instead of considering all the interaction system at the same time, a stepwise iteration scheme is developed. Stern shear flows are represented by a system of radial and circumferential vortices from the nominal circumferential mean velocity and the induced velocity due to a propeller action. Axial variations of stream tube are computed by streamline tracing method until converged solutions of propeller performance are obtained. Rudder effect on effective wake is calculated by computing the interaction between propeller and rudder system. Propeller performances in open water condition and in behind condition with calculated effective wake distributions are computed to evaluate the effective wake of ships. Effective wake is obtained based on thrust identity principle. Thrust deduction factor is calculated by the sum of pressure differences between towing condition and self propulsion condition on the hull surface. Double body assumption and predicted effective wake distributions are used in the computations.

Numerical computations are made for three full ships and one container ship. Predicted self propulsion factors, effective wake and thrust deduction factor show good agreement with the experimental

1. INTRODUCTION

To improve the propulsive performance of a ship, it is essential to understand the interaction effects between stern flow and a propeller-rudder system. The interactions between hull and propeller can be calculated by solving Navier-Stokes equations with a propeller as body-force generator[1]. But the application of this results to practical design seems to be not achieved due to the complicated unsteady interaction, the limitation of computing machine capability and some uncertainty in turbulence structure of stern flow.

Present method is an alternative approach to avoid this difficulty by assuming the interactions with a propeller is in principle potential nature. Estimation of propulsion factors can be initiated from propeller side with specified nominal wake and known propeller and rudder geometry.

Huang & Gloves[2] computed effective wake of axi-symmetric body using the principle of energy conservation and Bleslin et al.[3] extended their results to non-uniform ship flow. Lee, C.S. et al.[4] calculated effective wake of axi-symmetric body which is defined as a form of triple integrals of induced velocities due to the vortices contained in the stern shear flow and used pre-determined vortex tube contraction shape. This method may be applied to actual propeller design.

In present study, as an extensions of Lee et al[4][5], stream line tracing method and iteration scheme are adopted to determine the contraction shape of vortex tube and to obtain the converged effective velocities along propeller radius. To obtain the effective wake of ships, propeller performance prediction with computed effective velocity distribution was compared with open water performance based on thrust identity.

The rudder effect on the effective wake was considered by the computed results of propeller-rudder interaction problem.

Thrust deduction factor as resistance auguments due to propeller action is evaluated by the pressure differences between in towing condition and in propeller operating condition on the hull surface.

Lifting surface theory using vortex lattice method

for propeller with computed effective wake distributions and double body approximation for hull are adopted to compute the induced velocities on hull surfaces and the boundary value problems of hull. Calculated results of three full form ships and one container ship show good agreement with experimental data obtained from the model test.

2. ITERATION METHOD TO COMPUTE EFFECTIVE WAKE

The velocity field around propeller and stern shear flow can be decomposed as following three components.

$$\underline{u}(x) = \underline{u}_0(x) + \underline{u}_p(x) + \underline{u}_s(x) \qquad (1)$$

where $\underline{u}(\underline{x})$, $\underline{u}_0(\underline{x})$, $\underline{u}_p(\underline{x})$ and $\underline{u}_s(\underline{x})$ are total velocity, on-coming velocity represented as nominal wake, induced velocity due to propeller action and flow interaction respectively.

According to I.T.T.C propulsor committee and Eq(1), effective wake can be defined as

$$\underline{u}_e(\underline{x}) = \underline{u}(\underline{x}) - \underline{u}_p(\underline{x}) = \underline{u}_0(\underline{x}) + \underline{u}_s(\underline{x}) \qquad (2)$$

From Lee et al.[4], Eq(2) can be expressed as follows if the fluid is incompressible and inviscid but filled with vorticity;

$$\underline{u}_e(\underline{x}) = \iiint w_q \times \nabla_p \left(\frac{-1}{4\pi r} \right) dV + \underline{q}_0(\underline{x}) \quad (3)$$

To solve Eq(3), the vorticity in the integrand, w_q, has to be known beforehand, which requires prior knowledge of nominal velocities and vortex tube deformation shape. According to Helmholtz' third vortex theorem, the strengths of the vortices before and after the elongation, are related each other by the associated lengths of the vortices as follows.

$$\frac{w'_\theta}{w_\theta} = \frac{r'}{r} \qquad (4)$$

To apply Eq(4) along vortex tubes, the induced velocities by discrete vortex ring and source/sink disc which represent a half infinite constant vortex cylinder are calculated by following discritized equation.

$$\vec{V} = \sum_{j=1}^{N_\Gamma} \Gamma_j \vec{V}_j + \sum_{j=1}^{N_\delta} \delta_j \vec{V}_j^\delta \qquad (5)$$

where N_Γ and N_δ are Nring$\times$Mring, $2\times$Mring respectively. Nring is the number of vortex ring along transition area represented by longitudinal vortex tube and Mring is the number of vortex ring along radius direction. Fig. 1 shows the flow chart of present iterative scheme. Fig. 2 depicts an example of nominal circumferential mean and computed effective velocity profiles. Converged streamlines for effective wake are shown in Fig. 3. The example of convergence history of effective velocities are shown in Table 1.

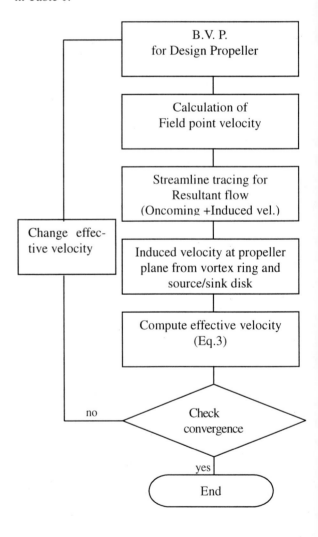

Fig. 1   Flow chart of iterative scheme

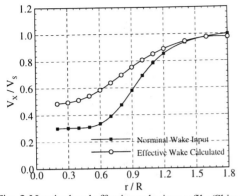

Fig. 2 Nominal and effective velocity profile (Ship A)

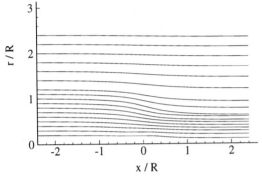

Fig 3 Converged streamlines for effective wake simulation

Table 1
Effective velocity convergence example (Ship A)

r/R	Nomi-nal veloc-ity	Effective velocity			
		1st iter.	2nd iter.	3rd iter.	4th iter.
0.2	0.301	0.528	0.477	0.488	0.484
0.3	0.304	0.538	0.485	0.496	0.494
0.4	0.305	0.557	0.502	0.513	0.511
0.5	0.310	0.588	0.531	0.544	0.541
0.6	0.335	0.630	0.574	0.587	0.584
0.7	0.390	0.679	0.628	0.639	0.637
0.8	0.474	0.731	0.688	0.697	0.695
0.9	0.580	0.781	0.748	0.755	0.753
1.0	0.690	0.827	0.803	0.808	0.807
Vol. Mean	0.436	0.680	0.635	0.644	0.642

2.2 Rudder effect on effective wake

It is well known that rudder increases effective wake due to additional thrust on a propeller from the interaction between propeller and rudder.

Most of the theoretical treatment on a propeller-rudder interaction problem used the infinite number of propeller blade theory and thick wing theory for rudder.[6][7]

In the present study, lifting surface theory for a propeller with actual geometry and potential based panel method for a rudder are used in the computations. The computation of interaction is made iteratively using computed effective velocity as input for a propeller until converged solutions are achieved.

Detailed description of the present method and the comparisons with other calculations[8] and experimental results[9] in uniform flow are described in reference[13].

2.2.1 Discritization of integral equation.

Discritized integral equation for flow field around a propeller can be described as below

$$\sum_{\substack{n=1 \\ m=1}}^{N_p M_p} \hat{n}_{ij} \cdot \vec{u}_{p\,ijnm} \Gamma_{nm} =$$

$$- \hat{n}_{ij} \cdot (\vec{U}_\infty + \vec{u}_{r\,ij}) - \sum_{\substack{n=1 \\ m=1}}^{N_p M_p} \hat{n}_{ij} \cdot \vec{v}_{p\,ijnm} \sigma_{nm}$$

(6)

where $\vec{u}_{p\,ijnm}$, $\vec{v}_{p\,ijnm}$ are induced velocity at control point **ij** due to vortex/source located at field point **nm** and induced velocity due to rudder. N_p and M_p denote the number of radial and circumferential panels of the propeller surface.

Discritized equation for flow field around rudder may be computed as;

$$\sum_{\substack{n=1 \\ m=1}}^{N_p M_p} \beta_{ijnm} \phi_{nm} + \sum \beta_{ijnm}^w (\phi_{N_p m} - \phi_{1m})$$

$$= - \sum_{\substack{n=1 \\ m=1}}^{N_p M_p} \alpha_{ijnm} (\hat{n}_{ij} \cdot (\vec{U}_\infty + \vec{u}_{pnm}))$$

(7)

$$\beta_{ijnm} = \int_{C_{nm}} \frac{\partial G_{ijnm}}{\partial n} \, ds \quad (8)$$

$$\beta_{ijnm} = \int_{C_{nm}} G_{ijnm} \, ds \quad (9)$$

where β_{ijnm}, α_{ijnm} are induced potential at control point **ij** due to unit dipole and source distributed on the panel **nm** and $\vec{u}_p$ is the induced velocity due to propeller.

2.2.2 Numerical scheme and examples

Greeley-Kerwin model[10] is used for the wake model. However nonlinearity of rudder wake and the distortion of propeller slip stream due to a rudder are not considered in the present study. Under the assumption that steady force equals the time mean average of unsteady forces, 10 point Gaussian quadrature is used for the evaluation of steady propeller and rudder force in propeller-rudder system. For rudder drag force, following equation is used.

$$C_{RD} = C_f + 0.234\alpha_F^2 \qquad (10)$$

where C_{RD}, C_f are local drag coefficient, local friction coefficient according to I.T.T.C 1957 friction line. α_F denotes an incidence angle for each panel of the rudder surface.

To obtain the steady interaction force for propeller and rudder, propeller positions are varied and induced velocities at the control points of rudder are obtained within one blade angle.

Table 2 shows the characteristics of rudders and propellers. Example of the convergence history of the present iterative scheme is shown in Table 3. Fig. 4 shows the schematic flow chart of the present method. The panel arrangement for propeller-rudder system with propeller wake is shown in Fig. 5.

Table 2
Characteristics of propellers and rudder

	A	B	C	D	E
Dia(Prop)	0.220	0.220	0.251	0.220	0.250
Pitch Ratio	0.709	0.709	0.704	0.710	0.978
Area Ratio	0.452	0.452	0.524	0.437	0.749
No. of Blade	4	4	4	4	5
Chord(Rudder)	0.210	0.207	0.264	0.198	0.197
Span(Rudder)	0.345	0.287	0.348	0.324	0.324
Section(NACA)	0018	0018	0020	0020	0018

Table 3
Convergence history of the iterative scheme (Ship A)

Iter. No	Kt (prop)	10Kq (prop)	Krx (rudder)
1st iter.	0.1893	0.2155	w/o rudder
2nd iter.	0.2002	0.2322	0.0022
3rd iter	0.2021	0.2338	0.0014
4th iter.	0.2023	0.2340	0.0014
5th iter.	0.2020	0.2337	0.0015

3. COMPUTATION OF THRUST DEDUCTION FACTOR

As an interaction problem between hull and propeller, thrust deduction factor is usually computed by viscous flow solver assuming a propeller as body force generator with infinite number of blades.

In the present study, thrust deduction factor is computed by potential theory with actual geometry. Computed effective wake distributions with radial variation are used to compute the induced velocity due to propeller action. For hull panel method with double body approximations is used and lifting surface theory with vortex lattice method is used for propeller.

Laplace equation is used as a governing equation. Total velocity and the flow tangency condition at each panel are described as follows.

$$\underline{V}_i^T = \underline{V}_i^s + \underline{u}_i^p + \underline{u}_i^h \qquad (11)$$

$$(\underline{V}_i^s + \underline{u}_i^p + \underline{u}_i^h) \cdot \hat{n}_i = 0 \qquad (12)$$

where $\underline{V}_i^s$, $\underline{u}_i^p$, $\underline{u}_i^h$ and $\hat{n}_i$ are ship speed, induced velocity due to propeller, induced velocity due to hull and normal vector at control point of the hull surface respectively.

To determine the source strength on the panel of hull at self propelled condition, Eq(12) can be converted as follows.

$$(\underline{V}_i^s + \underline{u}_i^p) \cdot \hat{n}_i + \sum_{j=1}^{N_p}(V_{ij}^s \cdot \hat{n}_i) \cdot \sigma_j = 0 \qquad (13)$$

where σ_j denotes source strength of each panel. For the comparisons with tank test results, thrust deduction factor is evaluated as follows.

$$t = \frac{T - R}{T} \qquad (14)$$

Eq(13) is used to compute T and R in Eq(14). However the induced velocities due to propeller, $\underline{u}_i^p$, are set to zero to compute the resistance R.

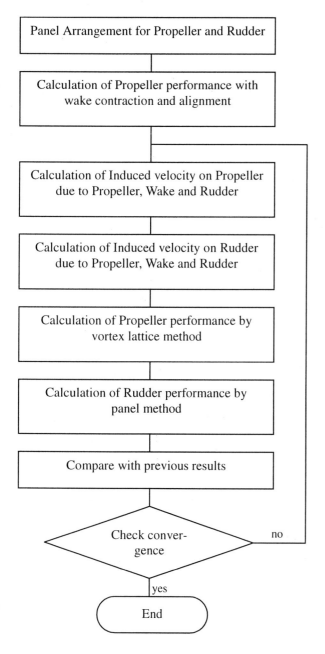

Fig. 4.   Schematic flow chart of the present method

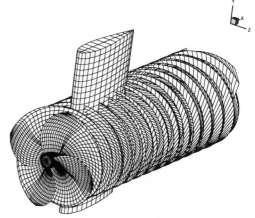

Fig 5.   Panel arrangement of the propeller with wake and rudder

4. RESULTS AND DISCUSSIONS

To obtain self propulsion factors of ships, a stepwise numerical computations were made systematically. Rudder-propeller system with nominal wake at self propelled condition and effective wake without rudder are simulated. Rudder effect is computed with simulated effective wake. To calculate the thrust deduction factor, field point velocity at the hull surface is computed with double body assumption using same propeller loading when the rudder is installed. To confirm the validity of the present method, three kinds of full form ship and one fine ship are selected in the computations. Ship A and ship B are same one, but loading conditions are different. Principal particulars of five conditions are shown in Table 4. Results of self-propulsion test for five cases are shown in Table 5 and the computed results are shown in Table 6. $W_{m(H)}$, $W_{m(R)}$ denote effective wake with hull only and effective wake increase due to rudder respectively.

In general, simulated results shows good agreement with test results but calculated effective wake is slightly higher than the experimental values. It may be caused by ignoring the unsteady term of stream tube contraction and uncertainty of extrapolation scheme for nominal wake upto no shear regions. For the contribution of rudder on effective wake seems to be reasonable compared to the results of Lee et al.[5] and Li et al.[8], however further experimental studies are required.

For thrust deduction factors, the computed results shows rather good agreement although simple double body model and potential theory is applied. The correlation among five hull forms are well re-produced except ship D for effective wake and ship C for thrust deduction factor.

Table 4
Principal Dimensions of Simulated Models

Ship	A	B	C	D	E
L/B	5.51	5.51	5.57	5.77	8.73
B/T	2.60	5.42	3.42	2.99	2.68
Cb	0.813	0.745	0.806	0.835	0.672
Load	full	ballast	full	full	Full

Table 5
Results of self-propulsion test

Ship	A	B	C	D	E
L(m)	7.25	7.25	8.53	7.55	8.48
V(m/s)	1.162	1.278	1.477	1.207	2.187
Wme	0.426	0.549	0.409	0.423	0.250
T	0.184	0.190	0.207	0.235	0.182

Table 6
Computed self-propulsion factors

Ship	A	B	C	D	E
Wm(H)	0.408	0.500	0.397	0.438	0.241
Wm(R)	0.037	0.033	0.032	0.043	0.027
Wme	0.445	0.533	0.429	0.481	0.268
t	0.183	0.180	0.240	0.224	0.177

5. CONCLUSIONS

An iterative computational method for propulsion factors from propeller side and rudder effect on effective wake is developed. Numerical computations are made for various type of ships to show the validity of the present method.

With specified nominal wake and actual propeller geometry, simulated effective wake and thrust deduction factor show relatively good agreement with experimental results.

A method to compute the propeller-rudder interaction in ship model is developed with simulated circumferential mean effective wake distribution along propeller radius. However experimental verifications are required for further study.

Unsteady terms should be included in the future and hybrid method using viscous solver to obtain

nominal wake field are recommended for the initial design work for reliable accuracy.

ACKNOWLEDGEMENT

This work is a part of a project supported by the Korean Science and Engineering Foundation, Contract 981-1013-232-2.

REFERENCES

[1] T. Kawamura and K. Mashimo et al., Finite-Volume Simulation of Self-Propelled Tanker Models, Proc. of the third Korea-Japan Joint Workshop On Hydrodynamics in Ship Design, Taejon(1996)
[2] T. T. Huang and N. C. Groves, Effective wake: Theory and Experiment, 13th Symp. on Naval Hydrodynamics, Tokyo(1980), pp651-673
[3] J.P. Breslin, J.E. Kerwin and C.A. Johnsson, Theoretical and Experimental Propeller Induced Hull Pressure Arising from Intermittent Blade Cavitation, Loading and Thickness, T. SNAME Vol. 90 (1980), pp 111-151
[4] C.S. Lee and J.T. Lee., Prediction of Effective Wake considering Propeller-Shear Flow Interaction., J. of Soc. of Naval Arch. of Korea, Vol. 27, No. 2(1990), pp1-12.
[5] C.S. Lee, Y.G. Kim and J.W. Ahn, Interaction between a propeller and the stern shear flow, Proc. of the first Korea-Japan Joint Workshop on Hydrodynamics in Ship Design, (1991) Seoul.
[6] F. Moriyama, On the effect of Rudder on Propulsive Performance of a Ship, J. of SNAJ, Vol. 150 (1981)
[7] M. Tamashima, C.J. Yang and R. Yamazaki, A Study of the Flow around a Rudder with Rudder Angle behind Propeller, T. of West-Japan Soc. of Naval Arch., No. 83(1992)
[8] D.Q, Li. and G. Dyne, Study of Propeller-Rudder Interaction based on a linear method. I.S.P. 42, No. 431(1995)
[9] E.J. Stierman, The influence of the Rudder on the Propulsive Performance on ships Part 1., I.S.P. 36, No. 407(1989)
[10] D.S. Greeley and J.E. Kerwin, Numerical method for Propeller design and Analysis in steady flow., T. SNAME. Vol. 90(1982)
[11] J.C. Suh, J.T. Lee and S.B. Suh, A bilinear source and doublet distribution over a planar panel and its application to surface panel method, 19th Symp. on

Naval Hydrodynamics, Seoul, Korea(1992)

[12] D.S. Kong. Y.G. Kim and J.M. Lew, On the Numerical Computation of Propulsion factors of ships, Proc. of Hull Form 96, Korea(1996)

[13] D.S. Kong, Y.G. Kim and J.M. Lew, A Study on the Propeller and Rudder interaction, Proc. of SNAK, Annual Meeting, Korea(1997)

1998 Elsevier Science B.V.
Practical Design of Ships and Mobile Units
M.W.C. Oosterveld and S.G. Tan, editors.

Model test results of a twin screw vessel with only one shaft line working.

Antonio Guerrero[a]

[a] El Pardo Model Basin (CEHIPAR)
Carretera de la Sierra S/N. El Pardo. 28048 MADRID (SPAIN)
Telephone + 34 (9)1 3760200 FAX + 34 (9)1 3760176
E-MAIL ceh.mail@cehipar.es

This work presents model test results with several alternatives that can be used. As blocking totally the shaft with the propeller at the design pitch, which allows the not driven shaft to turn freely or at a certain RPM regime or block the shaft with the propeller blades in feathering (infinite pitch) condition.

1. INTRODUCTION

Nowadays most of the Naval Vessel have a propulsion system with two shaft lines with controllable pitch propellers driven by a combination of Gas Turbines and Diesel Motors. Several configurations are possible, like CODOG, (Combined Diesel Or Gas), where Diesel Motors are used for propulsion at low speed operation and Gas Turbines are used for the high speed range, COGAG, (Combined Gas And Gas), using one Gas Turbine for each shaft line for low speed and two Gas Turbine by shaft line for high speed or COGOG, (Combined Gas Or Gas) using one Gas Turbine by shaft line for low speed and other different Gas Turbine by shaft line for high speed.

Gas Turbines have a very high specific fuel consumption when they work in the low power regime, (Figure 1). For medium range speeds it can be interesting, from a fuel saving point of view, to use only one shaft line moved by a Gas Turbine in the high power regime instead of two shaft lines moved with two Gas Turbines working in the low power regime. The problem is what to do with the propeller mounted in the shaft that does not work.

There are two alternatives: One is to block the propeller, at the design pitch, or in a feathering condition (infinite pitch). The other is to allow the not driven shaft to turn, freely or forcing it at a certain RPM regime with the help of an auxiliary electric or diesel motor.

In this paper model tests results of a Naval Vessel with a CODOG propulsion system with the different alternatives are presented. Also full scale results of a

DDG and a Frigate taken from reference 1 are used for comparisons with the tests.

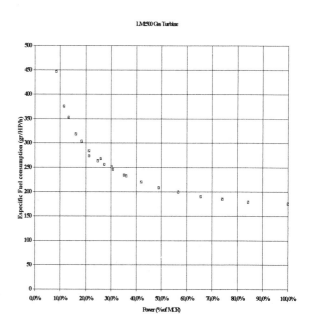

Figure 1. Specific Fuel Consumption of LM2500 Gas Turbine

2. EXPERIMENTAL PROCEDURE

Tests were performed with all the appendages corresponding to the shaft lines fitted to the model. Two movable rudders were fitted to the model, and oriented in angle of attack, to compensate the yaw torque induced by the differences in trust between shaft lines. The rudders angle were adjusted with the help of lateral force indicators placed forward and aft of the model, so the induced resistance coming for the rudders was taken into account. All tests were performed in propulsion condition. In the case of the free turning propeller tests, two different motors were used, one for each shaft-line. Test results were extrapolated to full scale with ITTC-78 method.

3. ALTERNATIVES

As has been mentioned in a ship with controllable pitch propellers there are several alternatives for the use of a single shat for propulsion. Tests results, comparisons with empirical methods and with results presented in reference 1 and technical advantages or disadvantages of the solution, are presented.

3.1 PROPELLER BLOCKED AT DESIGN PITCH

This solution is only feasible for very slow speeds, up to 6 or 7 knots, due to two different reasons. One is the very high resistance of the blocked propeller. The other reason is that the shaft-line brake should be dimensioned to absorb a very high torque coming from the propeller.

The drag coefficient of the propeller in this condition is defined by the expression:

$$C_D = \frac{D}{\frac{1}{2}\rho * A_D * Va^2} \qquad (1)$$

According with reference 2, it takes a value of 1,1.

Test results in this condition were performed in the speed range from 18 to 20 knots and confirm the expected value of C_D, a medium value of 1.102 was measured.

The increment in power to move the ship at this condition, the brake torque of the blocked shaft line, and the C_D value measured during the test are presented in Table 1.

3.2 PROPELLER BLOCKED IN FEATHERING CONDITION

The propulsion test in this condition was made with the port propeller adjusted in an infinite pitch condition at 0,80 r/R.

Ship Speed Knots	INCREASE IN POWER %	BRAKE TORQUE KN*M	C_D
18.0	215,2 %	464.1	1.128
19.0	259,1 %	523.7	1.083
20.0	260,1 %	572.3	1.099

Table 1. Port Propeller Blocked at design pitch

In this test, a value of C_D of 0.0921 was measured.

Using formulae from reference 3, a C_D value of 0.011 is obtained which is very small comparing with the experimental value. This difference is due to the formulation used in reference 3 valid only for symmetrical profiles and without angle of attack. In a propeller, none of the section profiles are symmetrical and most of them are not oriented to the incident flow.

In reference 1, a C_D value of 0.0533 is reported. This value was obtained from Free Water Propeller tests with the propeller blocked at different angles of attack, and was the optimum one. It is important to mention here that the propellers used in these type of ships, usually have a variable pitch distribution with strong variations from the root to the tip, so the optimum angle of attack, can differ from one propeller to the other. So open water or propulsion tests with different angles could be necessary to evaluate the optimum position of the feathered propeller.

The increment in the power necessary to move the ship in this condition, the brake torque on the blocked shaft line and the reduction of fuel consumption when the ship is propelled with Turbines of Gas and taking into account the specific fuel consumption of figure 1, are presented in Table 2.

Ship Speed Knots	INCREASE IN POWER %	BRAKE TORQUE KN*M	REDUCTION IN CONSUMPTION %
15.00	9.3 %	65.0	-29,3 %
17.00	12.6 %	86.2	-27,4 %
19.00	10.6 %	110.1	-24,1 %
21.00	14.2 %	135.8	-22,1 %
23.00	16.3 %	169.8	-20.9 %

Table 2. Test Results. Port Propeller Feathered

Table 3, taken from reference 1, represent similar results from Sea-Trials of a DDG.

Ship Speed Knots	INCREASE IN POWER %	REDUCTION IN CONSUMPTION %
16.00	25.8 %	-23.0 %
18.00	11.8 %	-20.2 %
19.00	12.9 %	-19.5 %
20.00	11.5 %	-18.4 %
21.00	14.0 %	-17.3 %
22.00	15.3 %	-15.3 %
23.00	14.7 %	-13.1 %
24.00	19.9 %	-10.9 %
25.00	20.5 %	-9.3 %
26.00	19.8 %	-8.6 %

Table 3. Sea-Trials Results. One Propeller Feathered (From reference 1)

It can be seen, from the comparison between tables 2 and 3, that the increase of power in the ship used in the tests was less than the one of the ship of reference 1, so the reduction in fuel consumption was more important.

As conclusion, this solution has two principal advantages over the former: The first is the great reduction in the brake torque needed to block the feathered propeller, as can be seen in tables 1 and 2. The second is the considerable reduction in resistance, comparing with the blocked propeller at design pitch. It allows the use of one shaft-line up to the high-speed range with a considerable reduction in fuel consumption.

The main relative disadvantage is the increased hub length necessary to allow the actuator cylinder to move the blades until the feather pitch.

3.3 PROPELLER FREE TURNING AT DESIGN PITCH

As it has been mentioned propulsion tests in this condition were made using different motors and dynamometers for each model shaft line. In one of

them, the motor was also used as a generator to simulate different shaft line losses. Obviously it is very difficult to estimate full scale shaft line losses, but this does not have influence in the conclusions of the test.

In figures 2, 3, 4 and 5 several results of this test are shown.

In figure 2 can be seen the effect of the change of the port shaft RPM (port propeller simulates the propeller free turning) in the thrust of this propeller. The diagram shows that, below a certain level of RPM, thrust is negative. It means an increment of the ship resistance and the necessary power in the starboard shaft line. As conclusion, if the port propeller will turn totally free the optimum RPM would be those than the thrust of the port propeller would be zero.

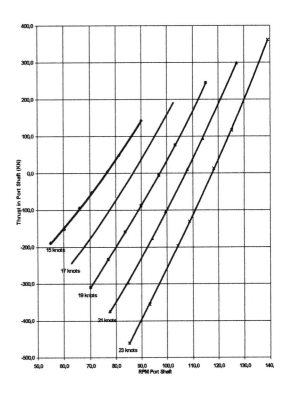

Figure 2. Thrust versus RPM (Port Shaft)

804

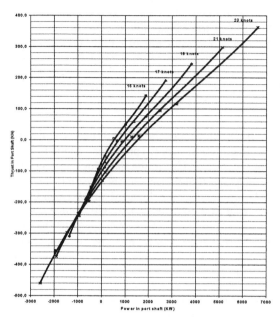

Figure 3. Thrust versus Power in Port Shaft.

When power is supplied to the port shaft line this increment reduces.

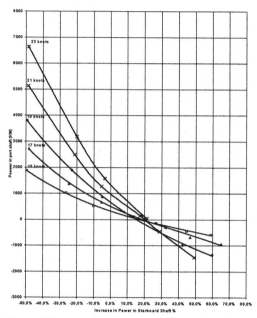

Figure 4. Increase in Power in Starboard Turbine of Gas versus Power in Port Shaft.

Figure 3 shows the power in the port propeller versus the thrust of this propeller. On this diagram negative power means than in the shaft line the power is absorbed, just as mechanical losses of the shaft line or by a generator connected to the shaft line. It can be seen that the negative thrust which incrementn the ship resistance is lower when the losses of the shaft line will be less. So it is necessary to reduce these losses to a minimum. Another conclusion of the test is that, if the zero thrust is required in the port propeller, it is necessary to supply power. This supply could be done in a CODOG propulsion system with the Diesel Motor. Another solution could be to include in the shaft line an electric motor or motor-generator to supply the required power. The electricity power could be generated with the other shaft line, increasing the power supplied by the Turbine of Gas, or by the Diesel-Generators of the ship.

Figure 4 represents the increasing power in the Turbine of Gas, starboard shaft, versus the power dissipated or supplied in the port shaft. For usual shaft line losses the increase in the Turbine of Gas power is about 20 % to 30 % of the power necessary to move the ship with two operating propellers.

In figure 5 the effect of the mechanical losses in the port shaft versus the fuel consumption is represented. In the figure is not included the increment in the fuel consumption due to the power supplied to the port shaft line, just with a Diesel or Electric Motors. The diagram shows that, with the normal values of the mechanical losses, reductions of fuel consumption of about 10 to 20 % can be achieved.

The main technical disadvantage of use of a free shaft line, are the lubrication problems due to the low regime of revolutions of the not driven shaft. If an auxiliary, Diesel or Electric Motor, is used, most of the lubrication problems would disappear due to the increased RPM of the shaft.

In figure 6 a similar case taken from reference 1 is represented. The figure shows the Sea-Trials results of two different ships, a DDG and a Frigate, in terms of increment of power and fuel consumption reduction. The DDG mounted a feathered propeller, and the Frigate, a free turning propeller. The increase in power in the Frigate is very important, indicating a very high level of mechanical losses in their shaft line. Nevertheless the reduction in consumption of this ship is considerable.

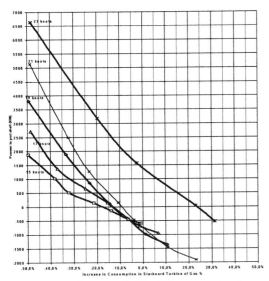

Figure 5. Increase in Fuel Consumption in Starboard Turbine of Gas versus Power in Port Shaft.

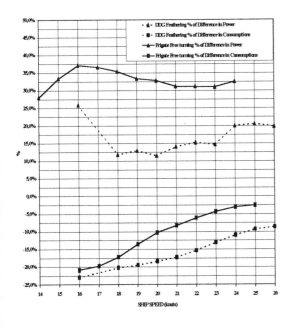

Figure 6. Comparison Free Turning and Feathering

4. CONCLUSIONS

Test results presented in this paper are made with models of a Naval Vessel. In these type of vessels, the CODAG or CODOG propulsion systems are very usual. The results can be helpful for design of new types of Commercial Vessels, as Fast Ferries, cargo vessels for Short Sea Shipping, etc, when the continuos demanding of more speed can force the use of Gas Turbines for their propulsion. In ballast or light load navigation, probably, the use of one shaft line driven by a Gas Turbine working at full power, and one of the solutions described here; a low power Diesel or Electric Motor driven the other shaft, or the use of a feathered propeller, could be enough to achieve the required speed and be an interesting solution, from a point of view of fuel consumption.

REFERENCES

1. Giuseppe Maria Bailo, Luigi Grossi and Stelio Vaccarezza, "Feathering Propellers: Full-Scale Propulsion and Cavitattion Performance", 1993 Meeting MWDDEA-N-I-73-4420, Washington, D.C. 18-20th May, 1993

2. Hoerner, S.F., "Fluid-Dynamic Drag", Second Edition. Published by the author. 1965.

3. Karl L. Kirkman and John W.Kloetzzli., "Scaling Problems of Model Appendages", 19th American Towing Tank Conference, University of Michigan, Ann Arbor, Michigan (July 1980) USA..

Practical Design of Ships and Mobile Units
M.W.C. Oosterveld and S.G. Tan, editors.

Design Studies of the Manoeuvring Performance of Rudder-Propeller Systems

A.F. Molland[a], S.R. Turnock[b] and J.E.T. Smithwick[c]

[a]Senior Lecturer, Department of Ship Science, University of Southampton, SO17 1BJ, United Kingdom.
[b]Lecturer, Department of Ship Science, University of Southampton, SO17 1BJ, United Kingdom.
[c]Research Engineer, Wolfson Unit M.T.I.A., University of Southampton, SO17 1BJ, United Kingdom.

Abstract: The fundamental basis for a rudder design methodology written as a computational tool for deriving forces and moments on a ship rudder is described. Rudder performance for a wide range of rudder geometries and positions downstream of a propeller or in the free-stream can be determined. The software uses the fundamental physical processes which control rudder and propeller interaction coupled with a large set of wind tunnel test data to produce accurate and workable results for design. Rather than curve fitting, the correction algorithm applies a change to the force on the rudder to the nearest known set of data. This ensures that the software experimental database is utilised fully. The rapid prediction of rudder forces, be they for manoeuvring force prediction or rudder design strength calculation, allows the designer to consider the effect on rudder performance of the principal design constraints.

1. INTRODUCTION

An extensive programme of experimental and theoretical research into the effect on manoeuvring of the stern arrangement of vessels and in particular the interaction between the ship hull, propeller and rudder has been on-going at the University of Southampton over a long period of time. The result of this work has been the development of a proven methodology which accounts for the physical basis of the interaction between the various stern components, References 1 and 2.

One of the results of the research has been the development of a large data-base. This relates the changes in propeller operating condition and the geometric description of the rudder-propeller system to the resultant performance of the rudder in terms of developed rudder forces and moments. The parametric data-base and underlying physical understanding have been coupled together in the form of a computer program which allows the designer easy access to practical rudder design information. This allows, for instance, the manoeuvring performance of a specific rudder-propeller system to be assessed and at the same time generate the resultant rudder stock torque (e.g. for steering gear sizing or rudder scantlings). The effect of design changes and trade-off studies can be

carried out in a rapid manner. Rudder performance is quantified for zero speed through to normal service speed and for rudder angles up to 70°. The software allows a complete stern arrangement, including multiple rudder-propeller systems, to be investigated.

The aim of this paper is to demonstrate the development and use of the software as a practical design tool through a series of relevant studies related to ship design. The paper describes the interpolation mechanism of the data-base based on a tree search algorithm which uses both the closest experimental data and, where necessary, parametric relationships based on the underlying physics. Three design case studies are included to demonstrate the use of the approach:

- an investigation into the choice between a skeg rudder, an all movable rudder and a high performance rudder, investigating the trade-off between reduced rudder area and increased manoeuvring side-force.

- the prediction of rudder torque for off-design conditions.

- the influence of rudder shape (aspect-ratio) related to the vertical position of the propeller.

The use of the approach provides ship designers with a much clearer understanding of the implications of design choices on the likely effective manoeuvring performance of rudder-propeller systems.

2. THE DATA BASE

2.1 Governing Physical Parameters

A ship designer is required to devise an overall stern arrangement which satisfies the requirements of the owner as regards speed and overall fuel consumption while ensuring the vessel is able to maintain its course and satisfy manoeuvring requirements at low and service speed. Prediction of rudder forces and moments has its use in detailed rudder design. In particular their use is important in establishing rudder scantlings, stock diameter and likely torques for the steering gear.

When a ship is moving ahead the flow passing through the propeller is accelerated and rotated. The swirl and acceleration induced in the flow by the propeller alters the speed and incidence of the flow arriving at a rudder aft of the propeller. This controls the forces and moments developed by the rudder. In addressing the rudder-propeller interaction problem it is necessary to identify the various independent parameters on which the rudder forces depend.

It is convenient to group the parameters which govern rudder-propeller interaction in four

categories. These groups can then be used to assess their affect on rudder performance. The four groups include flow variables, rudder geometric variables, propeller geometric variables and relative position and size of the rudder and propeller. The four groups may be represented as in Equation [1].

$$
\begin{pmatrix} L, \\ d, \\ M_X, \\ M_Y, \\ M_Z \end{pmatrix} = f \left\{ \begin{matrix} \left[J, Rn, \beta \right], \\ \left[\alpha, AR, \dfrac{t}{c} \right], \\ \left[\dfrac{P}{D} \right], \\ \left[\dfrac{X}{D} \right] \left[\dfrac{Y}{D} \right] \left[\dfrac{Z}{D} \right] \left[\dfrac{\lambda D}{S} \right] \end{matrix} \right\} \quad [1]
$$

The parameters governing rudder-propeller interaction are described in some detail in Reference 3.

2.2 The Experiments

The extensive experimental measurements and theoretical prediction of rudder forces provide a wide range of necessary data for use in predicting the manoeuvring performance. The extent of the rudder models tests are detailed in Table 1 and the propeller model particulars are detailed in Table 2. Details of the experimental method are given in Reference 4.

Table 1. Rudder Models Particulars

Designation	Type	Span (m)	Tip Chord (m)	Root Chord (m)	Tip Offset (m)	Aspect Ratio
Rudder 0	Skeg	1.0	0.593	0.741	0.148	1.5
Rudder 1	All Movable	1.0	0.593	0.741	0.148	1.5
Rudder 2	All Movable	1.0	0.667	0.667	0.0	1.5
Rudder 3	All Movable	1.2	0.667	0.667	0.0	1.8
Rudder 4	All Movable	1.3	0.667	0.667	0.0	1.95
Rudder 5	All Movable	1.0	0.800	0.800	0.0	1.25
Rudder 6	All Movable	1.0	0.556	0.556	0.0	1.8

All rudders have a NACA 0020 constant section with square tips.

Table 2. Propeller Model Particulars

Designation	Modified Wageningen B.4.40 Series
Range of revolutions (rpm)	0 to 3000
Number of blades	4
Diameter (m)	0.8
Boss Diameter (max)	0.2
Mean Pitch Ratio	0.95
Blade Area Ratio	0.4
Rake (degrees)	0
Blade thickness ratio t/D	0.050
Section shape	Based on Wageningen B series
Blade outline shape	Based on Wageningen but with reduced skew

Tests on the propeller model and various rudder models were conducted to provide a matrix of data for various parameters including: lateral separations, longitudinal separations, vertical separations, propeller thrust loadings and bollard pull condition.

3. SOFTWARE FUNDAMENTAL BASIS

3.1 Implementation

The software harnesses the experimental data with physical and empirical formulae to obtain forces and moments on a rudder operating downstream of a propeller. It was developed using the C^{++} programming language to provide a graphical user interface under a Windows based operating system.

The concept is based on a series of objects namely; a ship, its associated rudders and its associated propellers. By linking rudders and propellers an interaction set can be described. A set of operation conditions for all the components such as rudder angles, propeller rate of revolutions and ship speed can be defined to derive forces and moments for a specific condition or set of conditions.

The software has a specific data flow throughout. Data resources are from text input file and user input. The general data flow is indicated in Figure 1.

The basic data sources are as follows:
- Free stream rudder data (forces and moments, L, d, M_X, M_Y, M_Z)

- Open water propeller data (K_T against J for given P/D)
- Rudder and propeller interaction data (forces and moments for a particular set of parameters)

3.2 Data Correction

Data correction is based on theoretical predictions of performance. By curve fitting the various base data a correction can be applied to represent the required condition.

As corrections to the propeller data cannot be defined simply as a function of its diameter or pitch, propeller performance is described implicitly in the software and the interpolation is carried out to a base of non-dimensional thrust loading.

Rather than relying solely on curve fitting, the correction algorithm applies a change to the force on the rudder to the nearest known set of data. This ensures that the software experimental database is utilised fully. Moments are also corrected from an actual rudder to a required rudder

The applied principle is that the rudder can be split into an area controlled by the free stream flow and an area in way of the propeller in which the performance is controlled by the propeller thrust loading but related to the free stream rudder performance (References 3 and 5).

810

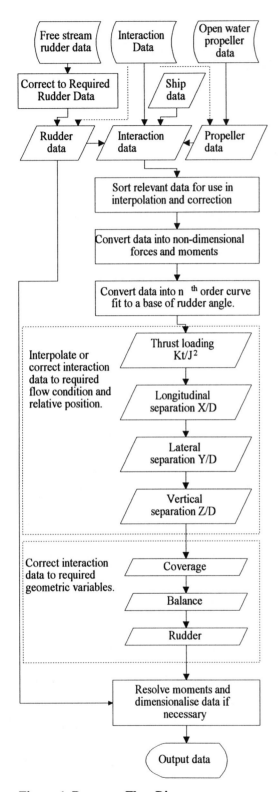

Figure 1. Program Flow Diagram

In order to derive the change in force or moment a correction is applied to the free stream properties of the rudder. The force or moment is calculated at the actual value of the parameter and then at the required value by scaling and/or applying an offset to the free stream rudder performance. The scaling value is proportional to the thrust loading. For low J (i.e. $K_T/J^2>8.0$) a different parameter must be used as the thrust loading K_T/J^2 tends to infinity. This parameter is based on the difference between the operational thrust coefficient K_T and the thrust coefficient at an advance ratio of J=0.

The angular offset for zero lift is proportional to the balance, b, which is defined in Equation [2]

$$b = 2\left(\frac{Y}{D}\right)\sqrt{0.25 - \left(\frac{Y}{D}\right)^2} + \left(\frac{A_1 - A_2}{D \times c}\right)$$ [2]

where A_1 is the area above the propeller axis and A_2 the area below the propeller axis all within the propeller race.

There is also a correction applied for the direction of the propeller rotation. The correction involves a complete reversal of the y-axis. This is represented by calculating forces or moments at the negative values of the required rudder angles. By applying a reversal to the signs of the forces and angles the forces are resolved back to the correct axis system and required angles.

3.3 Interpolation Tree

The interpolation tree is the fundamental method used in the software. A tree search is used to obtain results at the required geometrical and flow properties from a given rudder database by interpolation and, where necessary, correction. The interpolation tree method maximises the use of actual experimental data and thereby reduces the amount of correction required for each result.

Figure 2 describes the interpolation tree for a full database. Subscript U indicates a value above that required and L a value below that required.

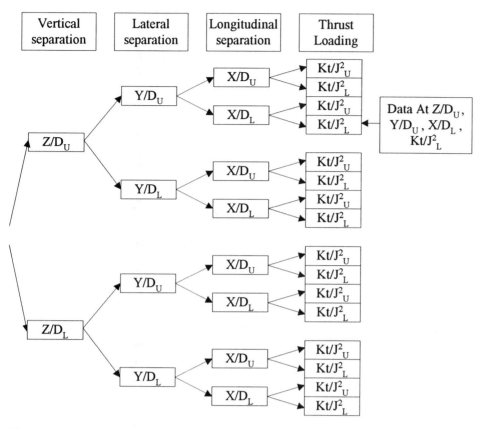

Figure 2. Interpolation Tree

The algorithm starts by searching the database and selecting the most densely branched tree. It then starts at K_T/J^2 and interpolates or corrects the data to the required thrust loading. It repeats this for all the parameters until the algorithm reduces to one data table at the required set of parameters. The allocated order of the parameters has been chosen according to the degree of variation of the parameters in the physical tests.

Where an upper and lower set of data are not found in the tree search, rather than interpolation, the software corrects to the required value. If the parameter is equal to that required it will use the data directly.

Corrections and interpolations are made to derive a correct set of forces and moments on the rudder from a user defined database. The choice of the

database is left to the user and it is important to apply the most valid and effective database to obtain accurate results.

4. CASE STUDIES

4.1 Rudder and Propeller Definition

A rudder in the context of the application is defined by its geometry, position, free stream performance and a set of operating conditions (i.e. angles).

Figure 3 indicates the various rudder particulars used to describe a rudder geometry.

The free stream rudder performance is selected in a variety of ways:
- 2D lift curve slope $(dC_l/d\alpha)$: This is the lift curve slope of the rudder section.

- 3D lift curve slope ($dC_L/d\alpha$) : This is the lift curve slope of the actual rudder.
- data table : actual performance data at discrete rudder angles.

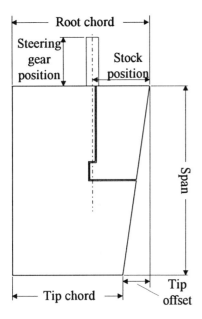

Figure 3. Rudder Geometry Definition

A propeller in the context of the application is defined by its geometry, position, open water performance and a set of operating conditions (i.e. rate of revolutions). Geometry is defined by a diameter and a direction of rotation.

Propeller open water performance characteristics can be defined in three forms:

- Fixed Thrust: The delivered thrust of the propeller for the specified operating point.

- A power curve fit of the form in Equation [3].

$$K_T = K_{TJ=0} - \left(\frac{K_{TJ=0}}{J^n_{K_T=0}} \right) \times J^n \quad [3]$$

where K_T at J=0, J at K_T=0 and the power n are user defined to match propeller data.

- data table: actual performance at discrete advance ratios.

4.2 Ship Definition

Ship definition is principally denoted by ship speed and wake fraction. In using the rudder design software Case study 1 and 3 use a typical ship operating at a speed ,V, of 14 knots and a propeller rate of revolutions of 120 rpm. This condition corresponds to an advance ratio ,J, of 0.497 using a wake fraction, w_T. of 0.32. Case study 2 has a range of operating conditions.

4.3 Case Study 1- Rudder Types

A study into the choice between a skeg rudder, all movable rudder and a high performance rudder, investigating the trade-off between reduced rudder area and increased manoeuvring side-force.

Three different rudder types have been used for the investigation of the effect of rudder type and these are detailed in Table 3. Table 4 details the propeller upstream of each of the rudders.

Table 3. Case 1, Rudder Particulars

Type	Span (m)	Tip Chord (m)	Root Chord (m)	Tip Offset (m)	Aspect Ratio	Stock Position (m)
All Movable	6.0	4.0	4.0	0.0	1.5	1.0
Skeg	6.0	3.5	4.5	1.0	1.5	1.0
High Lift	6.0	4.0	4.0	0.0	1.5	1.0

All rudders have a NACA 0020 constant section with square tips.

Table 4. Case 1 Propeller Particulars

Designation	Modified Wageningen B.4.40 Series
Range of revolutions (rpm)	100-200 rpm
Number of blades	4
Diameter (m)	5.0
Boss Diameter (max)	1.25
Mean Pitch Ratio P/D	0.95
Blade Area Ratio	0.4
Rake (degrees)	0
Blade thickness ratio t/D	0.050
Section shape	Based on Wageningen B series
Blade outline shape	Based on Wageningen but with reduced skew

The all movable and skeg rudders use data for designated experimental Rudders 2 and 0 (detailed in Table 1) respectively for their source free stream data whereas the high lift rudder has a 3D lift curve slope defined and a stall angle. Both the all-movable and high lift rudder use the designated Rudder 2 interaction data with the skeg rudder using designated Rudder 0 as the base interaction data set.

The relative positions of the rudders and propeller are detailed in Table 5.

Table 5. Case 1, Relative Positions

Type	X/D	Y/D	Z/D
All Movable	0.4	0.0	0.5
Skeg	0.4	0.0	0.5
High Lift	0.4	0.0	0.5

4.4 Case Study 2 - Off Design Conditions

For this particular study the all movable rudder with particulars shown in Table 3 is used with the propeller in Table 4. The relative positions of the rudder and propeller are the same as Case 1, indicated in Table 5. The all movable rudder uses data for designated Rudder 2 for its free stream data and interaction data set. The predictions of rudder torque are at several operating conditions as detailed in Table 6.

Table 6. Case 2 Operating Conditions

V (knots)	w_T	rpm	J	K_T
0.0	---	120.0	0.0	0.365
7.0	0.32	120.0	0.245	0.311
14.0	0.32	120.0	0.497	0.229
14.0	0.32	Free Stream		

4.5 Case Study 3 - Propeller Position and Rudder Particulars

The influence of rudder shape (aspect-ratio) related to the vertical position of the propeller.

Two different rudder types were used for the investigation of the effect of rudder particulars as detailed in Table 8. The propeller is the same as the propeller used in Case 1 and is detailed in Table 4.

The relative positions of the rudder and propeller are detailed in Table 7 and Figure 4 indicates the relative propeller coverage on all movable Rudders A and B.

Table 7. Case 3, Relative Positions

Type	X/D	Y/D	Z/D
All Movable A	0.4	0.0	0.7
All Movable B	0.4	0.0	0.7

814

Table 8. Case 3, Rudder Particulars

Type	Span (m)	Tip Chord (m)	Root Chord (m)	Tip Offset (m)	Aspect Ratio	Stock Position (m)
All Movable A	6.0	4.0	4.0	0.0	1.5	1.0
All Movable B	5.0	4.0	4.0	0.0	1.25	1.0

All rudders have a NACA 0020 constant section with square tips.

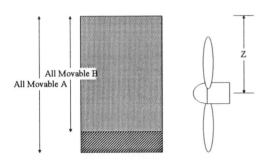

Figure 4. Case 3 Rudder Geometries

5. DISCUSSION - DESIGN IMPLICATIONS

5.1 Effect of Rudder Types (Case Study 1)

Figure 5 shows the rudder lift and drag curves for the defined operating condition. As expected the high lift rudder has an increased performance with respect to lift curve slope ($dC_L/d\alpha$) compared to that exhibited by the all movable and skeg rudder, but an increase in drag is experienced. There is a small offset in the zero lift angle of the skeg rudder and this is due to the sweepback of the skeg rudder due to the increased area of the rudder in the top of the propeller flow compared with the lower part of the propeller flow.

The design implication is that the high lift rudder area could be reduced somewhat to match the lift produced to that say of the all movable rudder, with consequent reduction in rudder drag. The reduction in rudder size could be attractive in terms of rudder layout and cost.

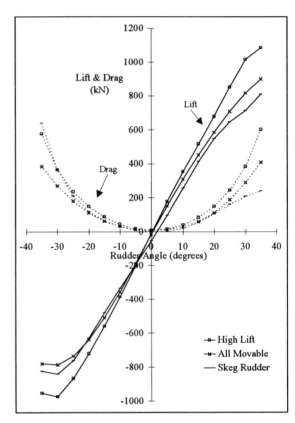

Figure 5. Case 1 Lift and Drag

5.2 Operating Conditions (Case Study 2)

Figure 6 shows the rudder torque at the rudder stock for the defined operating conditions. The highest torque exhibited is in the design condition. The results indicate a higher torque in the bollard pull condition (J=0) than that of an advance ratio of J=0.245. This is due to the thrust loading being higher for the bollard pull condition and the propeller flow compared to the free stream flow is more dominant.

The design implications are such that the highest torque is experienced at the design condition at 40 degrees rudder incidence, but at this condition a large helm angle may not often be applied. Therefore one should still be wary of the high torque exhibited by the bollard condition where a higher rudder angle may be used in this case.

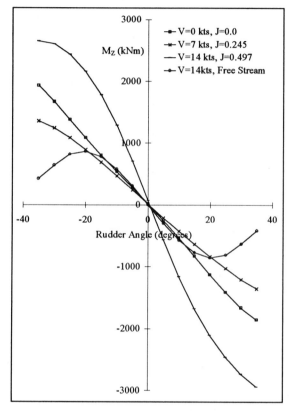

Figure 6. Case 3 Torque Moment

5.3 Propeller Position and Rudder Particulars Effects (Case Study 3)

Figure 7 shows the rudder lift and drag curves for the defined operating condition. The effect of having partial coverage from the propeller is to offset the lift curve slope such that the angle of attack at no lift is not at zero degrees. This effect is due to the asymmetry of the propeller race on to the rudder.

If the all movable Rudder B were chosen the design implications are such that there would have to be an incidence offset on the rudder. This will change for different propeller advance ratios and the design speed and matching propeller thrust would have to be used to calculate a suitable offset. The direction of the propeller rotation is also critical. If the propeller is rotating in the opposite direction to the results presented the incidence offset would have to be reversed.

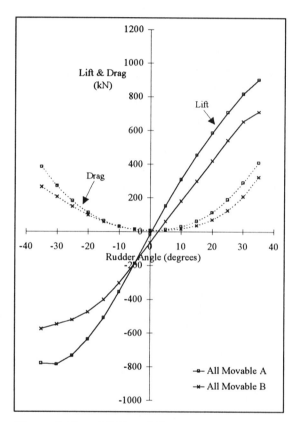

Figure 7. Case 3 Lift and Drag

6. CONCLUSIONS

The methodology allows parametric studies to be carried out and comparisons to be made between varying rudder and propeller types, and the development of rudder design and manoeuvring forces for a wide range of stern arrangements. The calculation process uses a specific database to interpolate and fit results by correcting experimental rudder and propeller data to the required rudder and propeller geometry and flow parameters.

The interpolation mechanism uses the most appropriate test data from the complete physical test data base and harnesses this, with the correction algorithms, to arrive at a solution for the required operating condition. The case studies presented have shown the rudder design methodology and software to be a practical design tool for manoeuvring considerations and for steering gear calculations.

Other design studies such as twin screw arrangements are easily implemented within the current framework. Effects such as lateral offset of rudders from the centreline of propellers can also be investigated providing useful design information.

ACKNOWLEDGMENTS

The work described in this paper covers part of the MOSES (Manoeuvring Of Ships and Estimation Schemes) research project funded by EPSRC and industry.

REFERENCES

1. A.F. Molland and S.R. Turnock, Developments in Modelling Ship Rudder-Propeller Interaction, CADMO '94.

2. A.F. Molland and S.R. Turnock, Some Effects of Rudder-Propeller-Hull Arrangements on Manoeuvring and Propulsion, The Sixth International Symposium on Practical Design of Ships and Mobile Units (PRADS), 17-22 September 1995.

3. A.F. Molland and S.R. Turnock, The Prediction of Ship Rudder Performance Characteristics in the Presence of a Propeller, MCMC '92

4. A.F. Molland and S.R. Turnock, Wind Tunnel Investigation of the Influence of Propeller Loading on Ship Rudder Performance, Transactions of The Royal Institution of Naval Architects, Vol 135, 1993.

5. A.F. Molland and S.R. Turnock, Prediction of Ship Rudder-Propeller Interaction at Low Speeds and in Four Quadrants of Operation, MCMC '94

NOMENCLATURE

A	Rudder Area (S.c)
AR	Aspect Ratio
c	Rudder Mean Chord
D	Propeller diameter
d	Rudder drag force, body (ship) axis
$dC_L/d\alpha$	Lift curve slope (the rate of change of lift with rudder angle)
L	Rudder lift force, normal to body (ship) axis
n	Propeller revs per sec
P/D	Propeller pitch ratio
J	Propeller advance coefficient (V/nD)
K_T	Propeller thrust coefficient ($T/\rho n^2 D^4$)
M	Moment
Rn	Reynolds Number ($\rho Vc/\mu$)
S	Rudder span
t	Rudder section thickness
T	Propeller Thrust
V	Ship Speed
w_T	Wake Fraction
X	Longitudinal distance, propeller plane to rudder leading edge in line with propeller axis
Y	Lateral distance between propeller axis and rudder stock
Z	Vertical distance between rudder root and propeller axis
α	Rudder Incidence
β	Propeller Inflow Angle (or Drift Angle)
λ	Proportion of D impinging on rudder
μ	Dynamic Viscosity
ρ	Mass Density

SHIP STRUCTURES AND MATERIALS

Practical Design of Ships and Mobile Units
M.W.C. Oosterveld and S.G. Tan, editors.

The Development of a Fatigue Centred Safety Strategy for Bulk Carriers

Braidwood, I. T., Buxton, I. L., Marshall, P.W., Clarke, D. and Zhu, Y.Z.

(University of Newcastle upon Tyne)

ABSTRACT

This paper reports the broad conclusions of a co-operative study into bulk carrier structural integrity. Strain measurements, at critical locations such as hatch corners and bracket frames, as well as those at less critical locations, were taken onboard the bulk carrier "British Steel". Making use of a finite element model of the ship, stresses at critical and non-critical locations were utilised to obtain stress predicting influence coefficients. This technique allowed the fatigue damage rates measured at the critical locations to be compared with those forecast from the stresses predicted by means of the influence coefficients. It is then possible to predict fatigue life at critical locations and institute corrective action.

1. INTRODUCTION

This paper describes the final part of the Bulk Carrier Structural Integrity Project, which was a joint undertaking by the University of Newcastle upon Tyne, British Steel, Lloyds Register of Shipping and Courtaulds Coatings (International Paints). The primary objective of the project was to provide technology to improve the structural integrity of bulk carriers, by addressing the problem areas of corrosion, physical damage and fatigue stress monitoring. It is these particular areas which are considered to be the primary cause of the continuing and unacceptable rate at which bulk carriers continue to be lost at sea, and which give rise to the requirements for a Fatigue Centred Safety Strategy for Bulk Carriers.

The project has been carried out over a three-year period and may be considered as a feasibility study to assess the use of an Influence Function (IF) approach to stress and fatigue life prediction. The basis of this method is to predict the stress and fatigue behaviour at several critical or inaccessible locations in the ship, such as the side shell frames, from similar data obtained at other non-critical locations such as the double bottom duct keel or the hopper tanks. Such a technique used in practice could lead to better planning of structural surveys and inspections, and most importantly to see if critical components are approaching their fatigue lives more rapidly than originally anticipated.

Primary contributions to the project were made by both British Steel and Lloyds Register, concerned with the areas of fatigue and structural integrity. Courtaulds Coatings endeavoured to find candidate coatings in their existing product range, which may give indications of high local stresses, from the early development of cracks in the coatings.

The subject ship which was built in 1984 was provided by British Steel, the aptly named 173,000 tonne deadweight *British Steel*, which brings iron ore to South Wales or Teesside from various ports in USA, Canada, Brazil and South Africa. The nine-hold ship was extensively fitted with strain gauges, which were monitored by an onboard data logging and recording system. The data collection and subsequent analysis has been considered previously by Braidwood et al (1), showing how the recorded stress data was used to compare with similar data derived from a finite element model. From this comparison a viable fatigue life prediction method based upon influence functions was developed. The technique of using influence functions to predict fatigue damage rates at critical locations, other than those where the measurements were taken, was described in an earlier paper by Braidwood et al (2).

The fatigue damage prediction techniques was extended to include aspects of fracture mechanics and structural integrity. An approach to survivability was considered, calculating many flooding scenarios and constructing a fault tree concerned with foundering, which as well as fracture mechanics also incorporated corrosion and crack propagation into a preliminary look at this important area.

2. THE CONCEPT OF INFLUENCE FUNCTIONS

The influence function approach is to derive and validate stress IFs, which allow global hull response measurements at a few locations (say 8-20) to be used to predict local fatigue damage at a large number of potential failure sites (numbering in the thousands). The basic premise is that local stress can be expressed as a simple linear combination of two global actions, using influence coefficients A and B, as follows:

$$\sigma_L = A\sigma_{G_A} + B\sigma_{G_B} \qquad (1)$$

where σ_L is local stress to be determined, σ_{GA} and σ_{GB} are global stresses of interest at global locations G_A and G_B, and A and B are influence functions or coefficients. A and B are, in general, functions of wave direction and frequency and the more general term "influence function" (IF) is used. The process was fully described by Braidwood et al (1, 2)

3. THE DEVELOPMENT OF INFLUENCE FUNCTIONS

3.1. Finite Element Model and Stress RAOs Developed by Lloyd's Register

For the purposes of this research project, Lloyd's Register performed advanced finite element analyses of *British Steel*. Figure 1 shows the structure of the finite element model. The analyses were undertaken with a quasi-dynamic procedure, to produce stress RAOs (response amplitude operators, or transfer functions) in the frequency domain.

3.2. Ship's Data Collected by British Steel

A strain gauge instrumentation system was set up onboard the *British Steel*. Data for 10 round trips, from December 1996 until December 1997, were collected and processed into stress statistics by British Steel, as described in Ref. 2.

3.3. Development of Influence Functions within the EDC

To investigate the influence function approach, and to provide theoretical and practical baseline data against which to test it, four fatigue analysis programs were developed at the University of Newcastle, which are:-

Fatigue Analysis Using RAOs (FURAO). This program uses the data supplied by Lloyds Register to evaluate short term (hourly) and long term (annual) fatigue damage rates at specified locations.

Fatigue Analysis using Influence Functions and RAOs (IFRAO). This program initially derives the influence functions from the stress RAOs, using the Lloyds Register finite element data from one local site and two global sites. The stress response spectra are then used to give RMS stresses and eventually forecast the short and long-term fatigue damage rates. These forecasts are for the inaccessible locations derived by means of the influence functions.

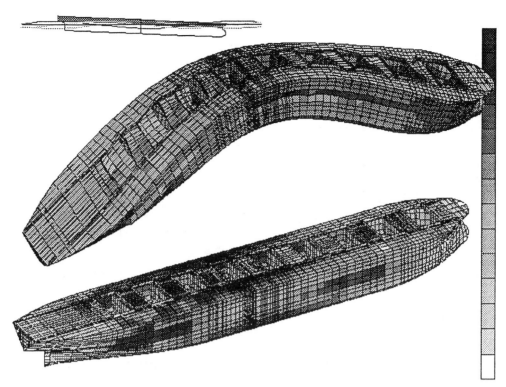

Fatigue Analysis Using Shipboard Data (FUSD). This program forecasts the short term and long term fatigue damage rates directly from the British Steel data measurements from the *British Steel*.

Fatigue Analysis using Influence Functions and Ship Data (IFSD). In this program the influence functions derived in the program IFRAO were used together with the ship data from *British Steel* to forecast the short and long-term damage rates at the inaccessible locations. This represents the practical implementation of the IF approach.

4. FATIGUE DAMAGE PREDICTION - RESULTS AND COMPARISONS

Fatigue lives predicted by FURAO, using expected seastates for the ship's routeing and Lloyds Register RAOs directly, are summarised in Table 1 for local damage sites of interest. A suitable S-N curve was selected from BS7608:1993 depending on the type of structural detail.

The hatch corner results appear to be consistent with the ship's service history of fatigue-induced repairs at these 36 locations (i.e. 9 holds × 4 corners) which have been replaced roughly every five years. There are roughly 1500 sideshell stiffener bracket toes on the ship, and their repairs during routine drydockings have been too numerous to report individually.

Comparisons between IFRAO and FURAO were used during program development, to test alternative strategies for finding "best fit" constant values of the influence coefficients A and B for each wave direction. For the hatch corner, the centre of the cross deck strip and a long base strain gauge outside the hatch opening were used as two global locations. Several options were investigated as to which global stress was to be extracted from the cross deck strip. Maximum principal stress (largest magnitude) with elimination of extreme values and removal of bias provided the best match of long term fatigue damage at the forward starboard 45° hatch corner location of hold NO. 6.

Table 1 Fatigue life predictions due to FURAO

Location	Fatigue life predictions
45° hatch corner No 6 hold	4.7 years
sideshell stiffener bracket toe	7.2 years
longitudinal deck stress	25.6 years
topside tank stiffener/ webframe	28.7 years

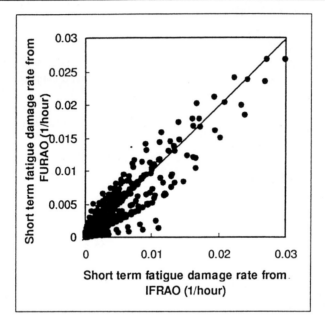

Figure 2 Comparison of the short term fatigue damage rates between FURAO and IFRAO for strain gauge 21 (hatch corner radius at 45°)

The resulting IFRAO influence coefficients, A and B as functions of relative wave direction only, were used for further theoretical comparisons, of which an example is given by the scatter plot in Figure 2, and for work on real ship's data using IFSD.

Predicting fatigue damage with IFs and measured ship data is the real challenge. Success is gauged by comparing local fatigue damage predicted by IFSD, using only global stress measurements, versus actual damage from FUSD, using measured local stresses.

The results analysed so far represent one loaded and ballasted trip across the North Atlantic during the winter and early spring of 1997. Some 100 data segments make up the deep sea portion of the trip. Eastbound, the ship is fully laden with iron ore; westbound, she is in ballast and at a shallower draft.

Fatigue results for the 45° hatch corner location have been examined. Initial results from using "best fit" IFs are disappointing, because the fatigue damage for the hatch corner for the round trip was overpredicted by a factor of 11, as compared with actual damage from FUSD. A detailed investigation into the procedure and the results of the analysis revealed that stress vs.

time traces at the two global gauges (one responding to longitudinal stress, one to cross deck maximum principal stress) show significant high frequency springing stresses at the ship's hull girder natural frequency. Thus stress cycle counts in the ship's data appear to be inflated by the many small springing cycles, but the variance and RMS stress were dominated by the larger wave frequency cycles. While rainflow counts (i.e. allocating the number of stress cycles into 'bins' of specified stress range) presumably sort this out correctly for FUSD, combining rainflow cycle rates, RMS stress, and Rayleigh statistics in IFSD would be inconsistent, producing a bias on the side of overpredicting fatigue damage. We therefore used the reported wave period to estimate the number of stress cycles, rather than the number from the rainflow counts.

Another cause for the over-prediction of fatigue damage lay in the fact that the influence functions were derived for ten fixed wave directions from IFRAO, while the wave directions reported by the ship's weather data files might take any value and therefore rarely matched any of the ten fixed directions. To avoid the influence of the mismatch of wave directions, the IFs were interpolated on actual wave directions used in IFSD.

Using wave frequency cycle counts and interpolating the coefficients for actual reported wave direction, the overprediction by IFSD was reduced, as can be seen from Figure 3. Short term fatigue damage comparison is given in Figure 4, for the loaded voyage, and Figure 5, for the ballasted voyage. Both show overprediction, i.e. damage rate from full scale prediction (FUSD) is much less than from IF prediction (IFSD).

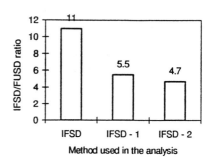

Figure 3 IFSD to FUSD ratio for long term fatigue damage rates, for the hatch corner

IFSD: *original method with IFs dependent on fixed wave directions and measured number of stress cycles.*

IFSD - 1:   *using wave frequencies to calculate number of stress cycles.*

IFSD - 2:   *using wave frequnicies to calculate number of stress cycles and IFs interpolated for actual wave directions.*

FUSD: *estimating fatigue life based on full scale measurements.*

Real seas are directionally spread, not unidirectional. For a steady wind, the wind generated seas have a typical spreading of 32° RMS (using a circular normal distribution), which corresponds to the standard cosine squared form. In direction sensitive offshore structures, accounting for spreading typically reduces the calculated fatigue damage by a factor of two. Applying 32° RMS spreading to the ship's data (interpolating between IFs for no spreading and IFs for equal energy from all directions) reduces the IFSD vs. FUSD damage overprediction from a factor of 4.7 fold down to 2.2-fold, corresponding to predicted local stresses being 26% too high (since fatigue damage rate is roughly proportional to $(stress)^{3.5}$ and $(1.26)^{3.5} = 2.25$). In the context of using messy real world data, this is not a

824

bad match, with modest bias on the safe side even being desirable.

Long term cumulative damage predictions, by the voyage or by the year, are what will be used in structural integrity strategies. Using the interpolated influence functions for reported wave directions and wave cycle counts, the following results as shown in Table 2 for IFSD were obtained, with the other methods to which it may be compared. The ship typically makes ten round voyages a year.

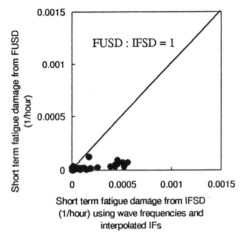

Figure 4 Comparison of short term fatigue damage for the loaded voyage

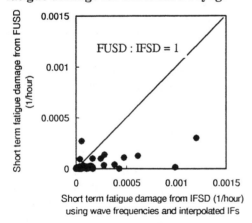

Figure 5 Comparison of short term fatigue damage for the ballasted voyage

5. FEASIBILITY OF THE INFLUENCE FUNCTION APPROACH

A principal objective of this project has been to derive and validate stress influence functions, which allow global hull response measurements at a few sparse locations to be used to predict local fatigue damage at a large number of potential failure sites.

The theoretical approaches indicated that, for the hatch corner, direction-dependent influence functions did very well, providing a good long term match without after-the-fact adjustments, as well as reasonable scatter in short term results (the plot of Figure 2). It should be remembered, however, that this is a theoretical test of the method, using the FE model predictions rather than the actual recorded stresses.

Demonstrating the influence method on real ship's data is the toughest challenge of all, and our real objective. Long term comparisons have been made of IFSD predicted damage vs. FUSD actual damage, for a full North Atlantic round trip. Initial results were disappointing. Further modifications to the influence function method - using spectral zero crossing frequencies, interpolating on wave direction, and considering directional spreading show potential for improving the prediction to 2.2 times the actual on damage rates. This is as good as it gets in fatigue work, as the coefficient of variation for laboratory S-N data is about this large.

Fatigue analysis software to test the influence function method has been successfully developed, applied, and tried with theoretical data. The influence function method shows promise, although further refinement is desirable.

Table 2 Fatigue Damage Rate Comparisons for Hatch Corner

Method used	Loaded damage rate	Ballast damage rate	Total damage rate	Fatigue Life (at D=1) in round voyages
FUSD	0.0049	0.0069	0.0118	85 with actual local stress
FURAO with real seas		0.0165		34 using RAOs directly
IFSD	0.0256	0.0298	0.0554	18 (no spreading) 38 (32° spreading)

6. A FATIGUE CENTRED SAFETY STRATEGY

With the influence function concept shown to be technically feasible in predicting fatigue life, the question arises as to its practical application to improve ship survivability. The objective is to identify the critical locations and monitor their fatigue life, by linking the theoretical IF models with actual ship measurements. The broad approach is outlined in Fig 6. The key steps are:

Preliminary Analysis of Selected Ship

1. Identify critical locations on board the specific bulk carrier, both from previous experience and from direct calculations, e.g. LR ShipRight. (for single hull ships, these are likely to be forward holds, including side shell frames and hatch corners)
2. Propose the (remote) global locations from which to derive IFs for (critical) local locations, for example long base strain gauges on deck, rosettes on cross deck strips and double bottom
3. From existing or modified FE model of the ship, derive RAOs for global and local locations
4. Select appropriate S-N curves for critical locations and actual thickness of structure

5. Run IFRAO for anticipated voyage pattern
6. Derive influence functions from (2) and (5), weighted as necessary to get best correlation for long term damage rates
7. Produce (theoretical) fatigue life predictions at local locations to confirm whether critical or not
8. Target the most important local and global locations
9. Prepare a 'survivability envelope' which assesses the probability of the ship surviving likely combinations of
   - Extent of flooding due to cracks in critical locations, especially where single hull
   - Deteriorated hull condition, due to corrosion, cracks, dents etc
   - Loading condition, cargo
   - Anticipated weather, i.e. Demand vs. Capability.

Service Shipboard Analysis

10. Install strain gauges at remote locations, calibrate and set up recording mechanism, including associated voyage and weather data
11. Analyse strain histories to give RMS stresses and correlations between the remote locations. Using IFs from (6), run IFSD to predict fatigue life at critical locations for anticipated

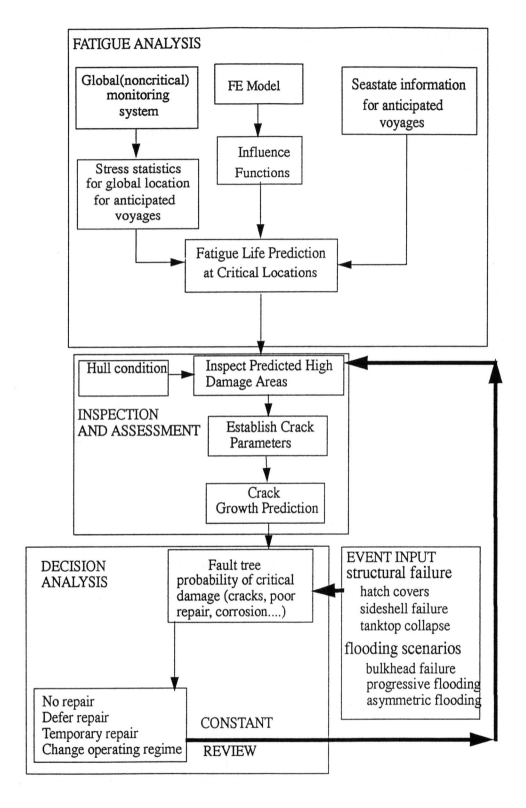

Figure 6 Overview of structural integrity and fatigue predictions for survivability

voyage pattern (initially done ashore, eventually on board ship)

12. For new ships, this will cumulate to show when elapsed time approaches predicted life in critical areas. For existing ships, this will also apply where new material has been fitted, e.g. replacement hatch corner inserts. For other areas, the part-expired fatigue life can be estimated on the basis of the ship's previous voyage history and 'hindcasting'

13. Estimate crack growth rate and corresponding compartments likely to be flooded

14. If resultant condition falls within critical area of survivability envelope, institute corrective action
   - Inspect, and repair if necessary
   - Operate vessel in a more survivable regime, e.g. better weather route, greater freeboard etc

It has to be recognised that a significant effort would be required to convert the approach into one for everyday use, depending on how much had already been done, e.g. existing shipboard instrumentation, suitable FE model etc. The authors believe that the potential is there for further development by interested parties.

REFERENCES

1. Braidwood, I. T., Buxton, I. L., White N. J., Zhu Z. Y., " A New Approach to Predicting the Fatigue Life for Bulk Carrier Designs", Sixth International Marine Design Conference, Newcastle upon Tyne, Vol. 1, pp 147-161, 23-25 June 1997.

2. Braidwood, I. T., Buxton, I. L., Hills, W., Marshall, P. W., Stevens, P., White N. J., Zhu, Z. Y., "Bulk Carrier Structural Integrity: Predicting Fatigue Life with Influence Functions", RINA Conference "Design and Operation of Bulk Carriers", London, 30 April 1998.

Acknowledgements

The paper draws on the Bulk Carrier Structural Integrity project supported by the Engineering and Physical Sciences Research Council (Marine Technology Directorate), Lloyds Register, British Steel and International Paints. The project integration and the fatigue analysis was carried out in the Engineering Design Centre at Newcastle University.

Practical Design of Ships and Mobile Units
M.W.C. Oosterveld and S.G. Tan, editors.

Single or Double Side Skin for Bulk Carriers ?

W. Fricke

Strength & Vibrations Dept., Germanischer Lloyd
Vorsetzen 32, D-20459 Hamburg, Germany

As a consequence of structural problems experienced on single side skin bulk carriers during the past few years, several measures have been internationally introduced in order to increase the structural integrity, among others the consideration of hold flooding in structural design, local reinforcements of side frames and protective coating of the side areas of all cargo holds. Still open is the question if a design with a double side skin is actually the better and more economic alternative. The vulnerable single side skin is replaced by the double skin, offering redundancy as well as protection of primary and secondary structural members from damage due to loading and unloading operations and from corrosion due to aggressive cargoes. In the paper, different designs of a 75,000 dwt panamax bulk carrier, suited for alternate hold loading conditions, are described. The design parameters such as steel weight, deadweight and cargo volume are evaluated and compared. The effect of the new requirements of IACS concerning hold flooding on steel weight is discussed in detail. Special care is taken with regard to detail design in view of the high fluctuating pressure loads on the side shell which caused structural failures in VLCCs some years ago. Finally, economical aspects of the different designs are compared, considering also discharge operation, cleaning as well as maintenance costs. In addition to the ship's safety, the main advantages of the double side skin are seen in reduced operation costs as well as in the reduced failure susceptibility and related costs.

1. INTRODUCTION

In the recent past, ageing bulk carriers have shown severe problems. Several total losses were recorded where structural problems may have been a factor (Adam et al., 1991). Frequently, water ingressed into the holds during heavy weather when carrying ore, followed by failure of the bulkheads and rapid sinking of the vessels.

In many cases, side frames, shell plating and bulkheads were severely corroded due to aggressive cargo, and deformed by rough discharge operations (Lehmann, 1990). Cracks occurring at the end connections of frames propagated into the side shell and caused leakage or even the detachment of larger structural areas from the side shell.

As a result of this development, more stringent measures were taken with respect to periodical surveys and their consequences

(Enhanced Survey Programme ESP), and repair proposals were elaborated (IACS, 1992). More recently, additional requirements with respect to the strength of longitudinal members, bulkheads and bottom structures in flooded conditions have been issued by the International Association of Classification Societies (IACS).

All measures mentioned have the intention of reinforcing the structure of conventional bulk carriers having a single side skin. The alternative measure for the standard bulk carrier characterised by hopper and wing tanks would be the arrangement of a double side skin. This would remove the actual cause of the problems mentioned, i. e. the exposure of primary structural members to the mechanical and corrosive attack by the cargo and by tools for discharge. The side boundary of the hold would be a smooth plate. Furthermore, redundancy of the watertight side skin is given, which is a logical conse-quence of the experience made.

In principle, a double hull construction is nothing new. It was prescribed for large tank-ers following the "Exxon Valdez" disaster to reduce the risk of oil spills after collisions and groundings. It has also been applied in OBO (ore-bulk-oil) carriers having a similar com-partmentation as conventional bulk carriers. For the latter, not the collision resistance is of main interest, but the protection of primary members against corrosion and mechanical damage. This means that the double hull can be relatively thin so that the loss of cargo volume compared to the conventional design is small.

Today's designers are faced with the ques-tion if a single or double side skin should be chosen for a new design of a standard bulk carrier, i. e. being characterised by hopper and wing tanks at the sides. It seems to be worthwhile to consider this question more deeply. If it is assumed that the ship's safety of both alternatives is the same - after intro-duction of the new strength requirements of IACS and good structural condition during service life, verified by ESP - how does the double side affect the cargo volume and steelweight and what is the effect on the eco-nomics of the ship ? In the following, these questions are discussed by the example of a typical 75,000 dwt panamax bulk carrier. Emphasis will be placed upon structural de-sign aspects of the side structure, including fatigue aspects which have shown to be gen-erally significant at the ship's sides.

The investigations described are based on a research project which has been performed by Germanischer Lloyd in cooperation with the German shipyard Flensburger Schiffbau-Gesellschaft (Fricke and Nagel, 1998).

2. DESIGN OF A 75,000 DWT PANAMAX BULK CARRIER

2.1 General Assumptions

A great number of bulk carriers of this size is characterised by a breadth of 32.24 m

(panamax) and an overall length of 225 m (max. length for the B-60 freeboard acc. to International Load Line Convention 1966 for one-compartment status). A typical geometry of the midship section of the conventional design with single side skin is shown in the upper part of Fig. 1. The double bottom height is 2 m and the inclination of the hopper plat-ing and wing tank bottom is 45° and 30° re-spectively. The bulkheads are corrugated and fitted with stools in the upper and lower part. The volume of seven cargo holds is approx. 86,000 m^3 and the deadweight of the ship 74,600 t, assuming a relatively high block coefficient ($C_B = 0.828$). Alternate hold load-ing with heavy cargo is permitted as well as two ballast conditions, "light" and "heavy" (the latter with ballast water in hold 4).

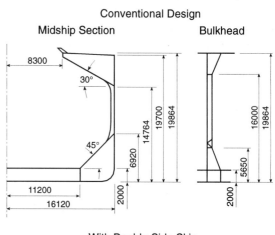

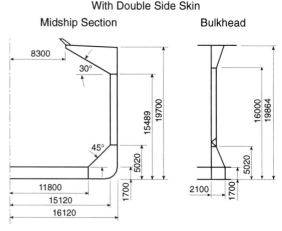

Figure 1. Hold Geometry of the Panamax Bulk Carrier Investigated

The geometry of a double side skin design is shown in the lower part of Fig. 1. In a first step, the width of the double side has been set to 1 m. The hopper area and double bottom height have been reduced in order to compensate partly for the loss in cargo volume. Reinforcements of the double bottom structure and associated weight increase are accepted in this connection. The space in the double sides is used for ballast water by arranging L-shaped tanks (i. e., side and bottom area).

2.2 Structural Alternatives of the Double Side

A number of alternatives exists when designing the double side, three of them have been considered in more detail in the project mentioned, cf. Fig. 2:

Alternative 1:
Longitudinals at Inner and Outer Skin

This is the usual method of stiffening panels in large ships. If transverse webs are arranged at a distance of about 2.6 m, longitudinals with a depth of 200 - 240 mm are required which means that the width of the double hull of 1 m as originally chosen is actually necessary. Two stringers are proposed to allow for easy performance of surveys.

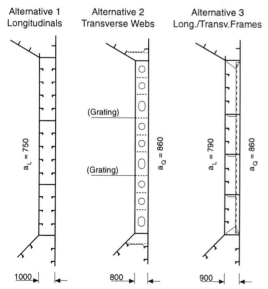

Figure 2. Structural Alternatives of the Double Side Investigated

Alternative 2:
Transverse Webs at Each Frame

The arrangement of transverse webs at each frame location (spacing chosen: a = 860 mm) enables the width of the double hull to be reduced to approx. 800 mm, because the space is not affected by longitudinals. However, sufficiently large access openings (min. 400 x 600 mm) and platforms (e. g. gratings) for surveys have to be provided. The structure with transverse webs is heavier than that with longitudinals (approx. 0.4 t/m + grating weight), because additional plates are arranged and because increased plate thicknesses will be required in the upper part of the inner and outer skin to satisfy buckling criteria.

Alternative 3:
Mixed Longitudinal and Transverse Framing

The selection of transverse framing at the sides meets the wishes of owners who expect from this a reduction of damages during berthing. The support of the frames is provided by stringers while the longitudinals at the inner skin are supported by transverse webs. The advantage compared to the structure with longitudinals only is more space in the double hull, which allows a width of approx. 900 mm to be realised. This alternative has nearly the same weight as alternative 2 due to the additional stringer and relatively thick side shell plating in the upper part regarding buckling strength.

3. DETERMINATION OF SCANTLINGS

3.1 Initial Design of Midship Section

The scantlings were determined according to the Construction Rules of Germanischer Lloyd (1997), using the design program POSEIDON (Cabos et al., 1997). The sizing of the components is performed automatically, taking into account local loads (hydrostatic and hydrodynamic pressure on bottom and sides, tank pressures, cargo and deck loads) as well as global bending moments and shear forces in the hull girder.

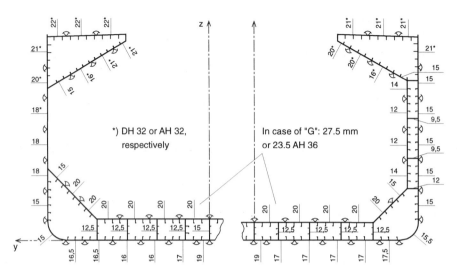

Figure 3. Midship Section of the Conventional Design (left) and of the Double Hull Alternative 1 (right)

Fig. 3 shows the resulting plate thicknesses for the conventional design and for the alternative 1 of the double hull design, assuming a spacing of 750 - 800 mm between the longitudinals and 2580 mm (3 x 860 mm) between transverse webs and floors. Higher-tensile steel HTS 32 is used for the upper flange of the hull girder, while mild steel is chosen for the bottom and sides. If the class notation "G" (strengthened for grabs) is requested, an increased inner bottom plate thickness as given in Fig. 3 is necessary.

The midship section complies with the minimum section modulus according to the Rules. The side shell thickness of the conventional design (18 mm) does not cover all shear force peaks so that further reinforcements will be necessary at the bulkheads. Also the reinforcements for flooded conditions are not yet included, which will be described later.

3.2 Finite Element Analysis of the Hold Area

The main purpose of the finite element analysis of a model extending over two cargo holds was to verify the strength of the double bottom of the double hull design which has a decreased height as mentioned before. The highest stresses in the bottom plating and girders occur in alternate hold loading conditions. Fig. 4 shows the deformations for such

a case. It turned out that the outer bottom of the empty holds requires a plate thickness of up to 23.5 mm in order to meet buckling criteria. Furthermore, the bottom girders have to be reinforced in the neighbourhood of the bulkheads where a plate thickness of up to 25 mm is necessary. This applies also to the conventional design, but to a lesser extent due to the larger double bottom height. The structural design of the corrugated bulkheads was also based on the finite element analysis.

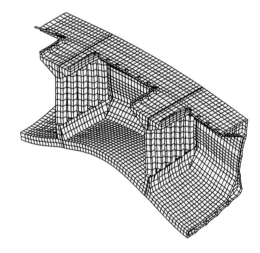

Figure 4. Finite Element Model of Holds with Deflections for Alternate Hold Loading (Deflections Magnified)

The reinforcements, which are summarised in the right part of Fig. 5, amount to 60 t for the double hull design, while 35 t additional steel is needed in the bottom and hopper area of the conventional design compared to the initial design of the midship section.

3.3 Analysis of Flooded Conditions for the Conventional Design

The new requirements of IACS for conventional bulk carriers regarding the strength in flooded condition were analysed in detail. Reinforcements compared to the initial design of the midship section and bulkheads concerned (a) the longitudinal strength and (b) the bulkheads.

The proof of the longitudinal strength acc. to IACS requires the analysis of a great number of combinations of loading conditions and holds flooded with water resulting in various distributions of possible bending moments and shear forces over the ship's length. Fig. 6 shows for example all shear force distributions which may occur in only two loading conditions ("altern. hold loading" and "heavy ballast"). Also included is the limit curve for the plate thickness chosen for the side shell

(18 mm). It becomes obvious that the single side skin has to be considerably reinforced at several bulkhead locations (up to 23.5 mm if higher-tensile steel is used). Furthermore, also the permissible bending moment is exceeded, resulting in an increased thickness of the deck plating at approx. half ship length.

The requirements of IACS concerning bulkhead strength resulted in this case only in minor reinforcements of the plating, i. e. 0.5 mm in general and 1.5 mm for the foremost corrugated bulkhead due to the increased pressure head.

All reinforcements are summarised in the left part of Fig. 5, resulting in an additional steelweight of 61 t for the conventional single side skin design. It is interesting to note that under consideration of these reinforcements, the steelweight of the conventional design is slightly higher than that of all alternatives of the double hull design. The reason for this unexpected result lies not only in the reinforcements mentioned, but also in relatively heavy frames especially in the hold 4 suited for ballast water. It should be mentioned in this connection that higher-tensile steel has not been assumed to the extent which was occasionally the case.

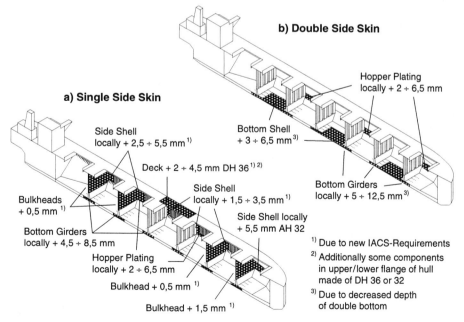

Figure 5. Reinforced Areas Compared to Initial Design of Midship Section

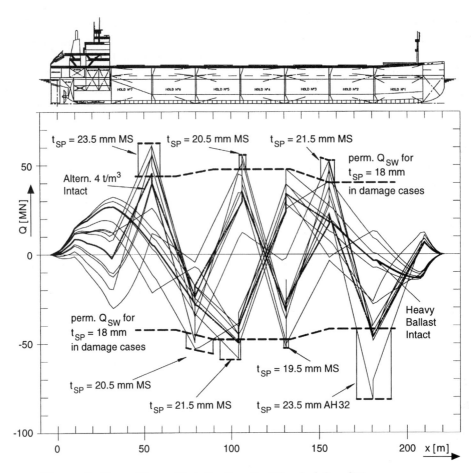

Figure 6. Shear Force Distribution for Flooded Conditions
of the Conventional Design and Resulting Side Shell Thickness t_{SP}

4. FATIGUE CONSIDERATIONS

4.1 General

It is known that fatigue cracks occurred in side frames of conventional single side skin bulk carriers, frequently in connection with severe corrosion and mechanical damages. As a consequence, end connections consisting of the more damage-tolerant "integral bracket" were proposed together with certain minimum scantlings. Since 1992, the Rules of Germanischer Lloyd have required 20 % higher section moduli for main frames of bulk carriers compared to usual requirements for other ships. This also compensates for the reduction of the fatigue life in corrosive environment (Lehmann et al., 1997). It is ex-

pected that these measures together with ESP and good maintenance will reduce fatigue problems with transverse frames in conventional bulk carriers to a tolerable limit.

Similar attention should be given to the double hull construction as well because it is known that the ship's sides belong to the structures which are particularly prone to fatigue.

4.2 Identification of Fatigue-Prone Areas in the Double Hull

The finite element model for double hull alternative 1 described in 3.2 was used also for fatigue analyses. A deterministic analysis was performed assuming unfavourable situations in waves in order to evaluate the highest expected stress range in various structural members (Germanischer Lloyd, 1998). Hull girder bending moments, dynamic pressures and acceleration components were assumed in accordance with the Rules of Germanischer Lloyd (1997).

Artificial load cases were generated by subtracting the loads in extreme wave situations (e. g. wave crest loads minus wave trough loads), in order to identify areas with high stress ranges. Fig. 7 shows the distribution of equivalent stresses in the side structure for such a load case, indicating a very high stress level in the lower part of the transverse webs (above the hopper area). Further evaluations showed that notable cyclic stresses are acting also close to the knuckled transition between the inner side skin and the hopper plating.

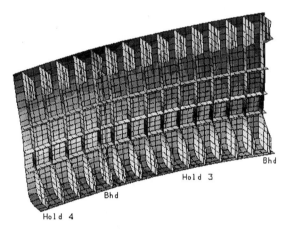

Figure 7. Side Structure of Double Hull (Alternative 1) with Equivalent Stress Ranges

The result is reasonable because the fluctuating pressure loads on the side shell are mainly transferred by the structural components mentioned. On the other hand, the evaluation of stresses in the side longitudinals showed that no remarkable stress increase occurs at the bulkheads - as it was observed in tanker structures (Yoneya et al., 1993, Fricke et al., 1995). The reason for this is seen in the high transverse flexibility of corrugated bulkheads which is comparable to that of the transverse webs.

4.3 Fatigue Assessment of Structural Details

Different structural details in the area with high cyclic stresses were analysed in detail. The first was the lightening hole 600/400 mm which is arranged in the lower part of the transverse web for reasons of fabrication and inspection. Due to shear forces, very high cyclic stresses are acting at the plate edge of the hole. A fatigue strength assessment according to Germanischer Lloyd (1997) showed that only an elliptical shape in connection with an increased web thickness (15 mm instead of 13 mm) is possible if the hole dimensions remain unchanged, see also Fig. 8. Alternative 3 requires a further increase of this plate thickness by 2 mm due to the reduced width of the double hull.

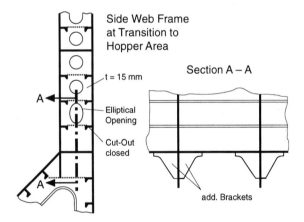

Figure 8. Measures in the Lower Part of the Double Hull (Alternative 1) Regarding Fatigue

Additional collar plates are necessary at the cut-outs for side shell longitudinals at least in the lower part of the transverse webs. The lowermost cut-out should be closed if the dimensions (e. g. for welding robots) are relatively large.

The knuckled transition was also investigated using a fine-mesh model. Extremely high stresses in a small area close to the transverse webs indicated that the effectiveness of the plating has to be increased which is, for example, possible by a vertical extension of the inner side skin using brackets as shown in the right part of Fig. 8.

These reinforcements of double hull alternatives 1 and 3 regarding fatigue result in an additional steelweight of approx. 7 t compared to the initial design of the midship section.

5. COMPARISON OF THE ALTERNATIVES AND CONCLUSIONS

The different designs can now be compared from their economical side, taking into account:
a) building costs
b) deadweight and cargo volume, resulting in different receipts
c) other costs and incomes during service life.

The building costs are generally higher for the double hull alternatives. The cost differences have been estimated within the project by the Flensburger Schiffbau-Gesellschaft, considering differences in steelweight, fabrication costs for the side construction as well as costs for presentation of the ballast tanks and holds (Fricke and Nagel, 1998).

The estimated differences in building costs given in the second column of Table 1 are relatively small in relation to the total building costs of a panamax bulk carrier.

The next two columns of Table 1 show the deadweight and cargo volume. The deadweight reflects the relatively small differences in steelweight. Concerning the cargo volume it is interesting to note that the loss in cargo volume by the double side of alternative 2 is fully compensated by the decreased double

bottom height and hopper area.

The differences in receipts, cf. last column of Table 1, have been estimated by evaluating typical voyages of panamax bulk carriers in the tramp trade. Here, alternative 2 turns out to be the best due to the highest cargo volume and the relatively small reduction in deadweight.

The following other costs and incomes during service life have been considered:
– A renewal of the cargo hold preservation is estimated to be necessary for the conventional single side skin bulk carrier after approx. 10 years, which will generally be carried out in a country with low wages, costing today about 250,000 $. The corresponding preservation for the double hull construction may take place after 15 years at a cost of about 100,000 $.
– Damages due to discharge of cargo with heavy grabs and special tools such as pneumatic hammers are mainly caused by the shore companies and thus will be normally paid by them accordingly. Therefore, these costs are not considered in the present analysis reflecting the owner's viewpoint although they affect the economics at least from a general viewpoint.
– The cleaning of holds is a very important factor in the charter market which is much easier in case of a double hull construction. The corresponding cost reduction has been estimated to be 12,500 $ per year. Furthermore, the risk of "off hire" is reduced, resulting in a better availability of the double hull designs in the market which will increase the annual receipts by, say, another 12,500 $.

Table 1

Differences in building costs, deadweight, cargo volume and related differences Δr in receipts

	Additional Building Costs for Double Hull Construction	Deadweight	Cargo Volume	Δr per year
Conventional Design	Reference	74,570 t	86,080 cub.m	Reference
Alternative 1	110,000 $	74,690 t	85,400 cub.m	- 15,400 $
Alternative 2	164,000 $	74,600 t	86,110 cub.m	+ 1,700 $
Alternative 3	146,000 $	74,590 t	85,750 cub.m	- 8,000 $

The double hull construction maintained in an appropriate way will certainly gain in future a better resale value than the conventional single side skin design which has been estimated to be about 200,000 $ after 15 years of operation.

It becomes clear that benefits can be expected during service life from a double hull bulk carrier although the initial investment costs are higher and the annual receipts may be smaller due to the reduced cargo volume. An economic analysis based on above figures has shown that - over the long term - the double hull designs are more profitable. The initial costs of investment are regained with an interest rate of more than 15 % (Fricke and Nagel, 1998). Most profitable has shown to be double hull alternative 2 due to its high cargo volume, despite of the relatively high building costs.

Finally may be stated that in total the double hull design is more economical. This holds true even if the new IACS requirements are applied also to this new type. Therefore, not only aspects of ship's safety as well as structural redundancy and robustness call for a fundamental change in the design of today's bulk carriers.

6. REFERENCES

Adam, J.C.; J.M. Ferguson and D.W. Robinson (1991): "Bulk Carriers - The Safety Issue." XII. Pan-American Congress of Naval Engineering, Maritime Transportation and Port Engineering, Buenos Aires.

Cabos, C.; W. Grafe and J. Schulte (1997): "New Computerized Approach in Classification." Proc. of ICASS'97, Yokohama.

Fricke, W.; M. Scharrer and H. von Selle (1995): "Integrated Fatigue Analysis of Tanker Structures." 6. Int. Symp. on Practical Design of Ships and Mobile Units (PRADS'95), Seoul.

Fricke, W. and R. Nagel (1998): "Technical and Economical Benefits from Double Side Skin Bulk Carriers." Int. Conf. on Design and Operation of Bulk Carriers (RINA), London.

Germanischer Lloyd (1997): "Rules for Classification and Construction, I - Ship Technology, Part 1 - Seagoing Ships, Chapter 1 - Hull Structures." Hamburg.

Germanischer Lloyd (1998): "Rules for Classification and Construction, V - Analysis Techniques, Chapter 2 - Guidelines for Fatigue Analyses of Ship Structures." Hamburg.

IACS (1994): "Bulk Carriers - Guidelines for Surveys, Assessment and Repair of Hull Structure." International Association of Classification Societies, London.

Lehmann, E. (1990): "Damages in Ship Structures (in German)." Trans. STG, Vol. 84, Springer-Verlag, Berlin.

Lehmann, E., M. Böckenhauer, W. Fricke and H.-J. Hansen (1997): "Structural Design Aspects of Bulk Carriers." Proc. of 2nd Int. Conf. on Maritime Technology ODRA 97, Szczecin.

Yoneya, T., A. Kumano, A. et al. (1993): "Hull Cracking of Very Large Ship Structures." Proc. of 5th Int. Conf. on Integrity of Offshore Structures, Glasgow.

Practical Design of Ships and Mobile Units
M.W.C. Oosterveld and S.G. Tan, editors.

Fatigue of Bulk Carrier Side Frame Structures

Anil K. Thayamballi and Zheng-Wei Zhao

American Bureau of Shipping
Two World Trade Center, 106 Floor
New York, NY 10048, USA

The loss of the side shell is a possible factor in some of the disappearances of bulk carriers at sea. One mechanism leading to such potential loss of side shell is the corrosion assisted fatigue of side frame structures, particularly at their end connections. In this paper, a study of fatigue effects in bulk carrier side frame structures located in the forward cargo hold is carried out, to judge the relative importance of various factors including corrosion. The results obtained illustrate the effects of various relevant factors such as the shape of end brackets, rate of corrosion, spacing of frames, and vessel size as they affect side shell fatigue. Based on the fatigue assessment, some conclusions relevant to design and maintenance of bulk carrier side framing are drawn. Recommendations are made for future studies.

Key Words: Bulk Carrier Side Shell Fatigue

1. INTRODUCTION

In the last decade, there have been several bulk carrier casualties including total losses. Over a thousand seafarers have lost their lives in the related incidents. Some of those vessels were lost under circumstances where the crew had no time for calling in a signal of distress, meaning that the loss occurred somewhat suddenly. The possible causes of such losses can be categorized into the three groups, namely (a) loss of reserve buoyancy or floatability, (b) hull girder collapse and (c) loss of stability, all initiated totally or partly by unintended water ingress into cargo holds.

In a significant number of the bulk carrier incidents reported, the vessels were carrying iron ore or coal, the former being one of the denser types of cargo, and the latter being one of the more corrosive, see for example BTCE (1993).

It has also been indicated that the age of most vessels concerned was over 15 years, and so significant defects related to corrosion and fatigue may have been present. Also, there was the possibility that part of the side shell forward could have been lost due to a combination of circumstances, e.g., excessive corrosion and fatigue cracking damage together with perhaps the shifting of solid cargo due to roll in rough weather. This could cause ingress of sea water into a cargo hold, particularly forward, and as a result, a vessel in the laden condition might loose significant reserve buoyancy, potentially leading to sinking particularly if the transverse bulkhead in the hold could not withstand the accidental flooding. Water ingress into the forward hold could also occur through failed hatch covers.

In this paper, a study of fatigue effects in bulk carrier side frame structures located in the forward cargo hold is carried out, to judge the relative importance of various factors including corrosion. The results obtained illustrate the effects of various relevant factors such as the shape of end brackets, rate of corrosion, spacing of frames, and vessel size as they affect side shell fatigue. Based on the fatigue assessment, some conclusions relevant to fatigue related design and maintenance of bulk carrier side frame structures are drawn. Recommendations are made for future studies.

2. REVIEW OF LOCAL STRUCTURAL DAMAGE

A brief look at local structural failures typical in bulk carriers that have survived is of some interest. While forensic examination of lost vessels is difficult, indirect information from damage experienced by such vessels may serve as an indirect indicator of the larger potential causes of vessel losses at sea.

In this regard, a study by Shama (1995) indicates where local structural failures occur in bulk carriers. It is seen from the study that damage to side frames is the most frequent, followed by damage to bottom floors and girders. Together, these three structural elements accounted for a third of all local structural damages in the vessels Shama considered. Also related to this, Ivanov (1993) reported the following types of side frame structural damage as being common:

- At bracket connections of the side frame to the wing and hopper tanks: fractures and potential detachment at brackets. The possible causes are said to be insufficient fatigue strength, intense corrosion, and deformations induced by rotation of the double bottom.
- Twisting of the frames, enhanced by mechanical damage and corrosion.

The detachment of hold frame ends forward, and the spread of that type of damage through a domino effect involving load shedding to other nearby corrosion / wear weakened structure has also been cited in the literature as an important factor in bulk carrier losses. Anecdotally, it has been said that in a number of vessels that survived to tell the tale, water ingress occurred in Hold 1, through failure of the side shell structure between the topside and hopper tanks, typically near one of the transverse bulkheads. Structurally, it is of interest that in conventional bulk carriers (unlike say, in tankers), each side shell frame does not necessarily form part of a continuous ring, i.e. the transverse frame spacing in the topside tanks can be different from that in the cargo hold, which can in turn be different from the frame spacing in the double bottom.

3. VESSELS SELECTED AND ANALYSIS LOCATION

We designed two hypothetical vessels for the purpose of comparative structural (fatigue) performance calculations- a 150,000 DWT Capesize vessel and a 67,000 DWT Panamax vessel. Both vessels were of conventional section, with single side skin between topside and hopper tanks. Generally, Capesize vessels are of DWT 80,000 or more, while Panamaxes are in the 50-80,000 DWT

Table 1
Principal Particulars of the Vessels

Item	Capesize	Panamax
Length, Overall	276.0 m	220 m
Length, Perpen.	264.0 m	215 m
Length, Scantlg	262.3 m	213 m
Breadth, Molded	43.5 m	32 m
Depth, Molded	24.0 m	18.3 m
Draft, Design	17.6 m	12 m
Speed, Design	14.1 kn	15.5 kn
Block Coefficient Design	0.84	0.855
Hold Aspect Ratio	0.55 Typical	0.78 Typical 0.68 Hold1
Side Frame End Brackets	Non-Integral	Integral
Deadweight tonnes	150,000	67,000
Material	All H36	MS(H32 Dk)

range. Of today's bulk carrier fleet of over 5000 vessels worldwide, about 20 % are Panamax and 10 % are Capesize.

The principal particulars of the vessels in the study are given in Table 1. All particulars shown are as built, i.e. they are not for the "net" vessel. The vessels are designed to the 1985 Rules. Hence their as built scantlings are different in critical areas from what would be required today under the SafeHull based ABS Rules (1997), but the selected vessels are perhaps more representative of the existing fleet. The Capesize is entirely of H36 steel, but in the Panamax the structure is mostly mild steel, except at the deck region where H32 is used.

It is of interest to note that the hold lengths in the Panamax were comparable to the Capesize, but the Panamax hold area is approximately 50 % less than that of the Capesize in case of Hold 1, and 25 % less in the case of the other holds. Any

postulated flooding of the Capesize Hold 1 is hence of more load consequence than flooding of the Panamax hold 1, at least in the case of these two specific vessels.

First, calculations of the envelope wave induced bending moments, shear forces, and external hydrodynamic and internal cargo inertial pressures were calculated for the two vessels, from hold 4 through hold 1 forward, using the load component formulae in Section 3 of the SafeHull Bulk Carrier Guide of 1995 (which is now part of the ABS Rules, see also Liu, et. al., 1995). The Capesize calculations were made for a draft of 18.7m and speed of 14.7kn, while the Panamax calculations used a draft of 13.3m and speed of 15.5kn.

The load calculations showed that the external pressure loads are larger toward the forward locations of the vessel in comparison to midship. From mid-hold 2 to mid-hold 1, the waterline pressure amplitudes varied from 18 to 23 t/m² for the Capesize, and 15 to 21 t/m² for the Panamax. The external dynamic pressures at the bilge are about 40% of the waterline values. The local draft used for the calculations was found to mainly affect the static and not dynamic pressures.

Based on the preliminary load study, side shell fatigue is checked in hold No.1, in the vicinity of the aft transverse bulkhead of the hold. Specifically the side frame lower connection details at the intersection of the side shell and the sloping plate of the lower hopper tank are of interest because that location is affected by wave profile changes in both the laden and the ballast conditions. It also happens to be a known location of side shell damage in some vessels. From a study of the loads involved, it was evident that the fatigue process at that location can be considered to be driven primarily by external dynamic pressure if one assumes that the ore in the laden condition is left as poured, although in our fatigue analysis we have included the additional effect of stresses due to secondary bending of the double bottom structure arising from the differential (internal minus external) double bottom pressure as well.

4. LOADS FOR FATIGUE ANALYSIS

To appreciate the origin of alternating loads affecting side shell fatigue, an idea of the location of the structural detail (i.e. the connection detail at the lower end of side frame) with respect to the laden and ballast waterlines is of interest. For the Capesize, the mean still water draft in an alternate hold loading ore laden condition is 17.2m (even keel). The typical draft in a heavy ballast condition is 10.9m (10.3m forward). In the case of the Panamax, the design draft is 12.2m, the scantling draft is 13m, a typical light ballast draft is 6.2m (4.5m forward), and a typical heavy ballast draft is 8.1m (7.6m fwd). These drafts, when plotted on a common figure, would show that for all practical purposes, the still waterline in the vessels hence varies roughly between the top and bottom connections of the side frame.

The comparative fatigue analysis that we undertake is of the simplified direct calculation type used in SafeHull. To calculate the fatigue damage for a structural detail by such a method, we need to know the following:

- Extreme stress range
- Weibull shape parameter

Using these the long term distribution of the stress range can be defined. Once that long term distribution has been defined, the fatigue damage may then be calculated using the appropriate S-N curve for the structural detail location being studied. In our study, we will pessimistically use a design (e.g., lower-bound) S-N curve rather than a curve representing mean or average fatigue capacity.

The extreme stress range in a nominal 20 year life is calculated from the corresponding extreme pressure ranges. The extreme pressure range and hence the extreme stress range corresponding to the laden and ballast cases are first calculated, and the larger of the two taken as the extreme stress range to be used in the fatigue analysis. In our case the larger stress was estimated to result from the laden pressure distribution, which is consistent with the locations of the still-water lines in the laden and heavy ballast cases and the fact that the laden and heavy ballast side shell extreme dynamic pressures are not vastly different. The pressure in the head seas case was assumed to be the 20 year extreme pressure, which adds some uncertainty to the stress range estimates in the sense that the other directions are not specifically considered. This could have been avoided, but the head seas assumption in severe waves was felt to be adequate for comparative purposes.

In the laden condition, the side pressure vertical distribution at the location of interest is idealized as uniform, with a value given by the average of the values occurring at the top and bottom of the frame ends (20.4, 10.2 for the Capesize, 18.4, 9.2 for the Panamax). This results in a uniform external pressure of 15.3 t/m² for the Capesize and 13.8 t/m² for the Panamax, compared to the waterline values of 20.4 and 18.4, i.e., roughly 75 % of the waterline values. The accompanying net dynamic pressures on the double bottom

Table 2
Fatigue Analysis Load Parameters

Item	Capesize	Panamax
Side Pressure (t/m²)	15.3	13.8
Intl. Pressure (t/m²)	3.3down	2.3down
Weibull Parameter	0.9	0.9
Cycles (20 years)	0.5E8	0.5E8

Table 3
Details of Side Structure Forward

Item	Capesize	Panamax
Frame Spacing (mm)	880	830
Plate Thickness(mm)	22.4	22
Stiffening	T Web	T Web
	500x12.5	350x11
(mm)	Flange	Flange
	170x25	150x19
Length (mm)	8000	6850
Material	H36	MS
Bracket Thickness	18mm	11mm

structure were also estimated and included in the fatigue calculations, but their final effect was small, meaning that subsequent fatigue estimates could have been considered to be driven mostly by the side shell pressure alone.

The Weibull shape parameter values defining the long term stress range distribution for the side shell hold frame are estimated from Section C2.2 of the Bulk Carrier Guide (now absorbed into the ABS Rules, 1997). These are 0.912 for the Capesize and 0.931 for the Panamax, so a Weibull shape parameter of 0.9 was eventually used in both vessels. The number of load cycles (zero to extreme amplitude) was nominally taken as 0.5×10^8 in 20 years, which should be adequate for comparative purposes

although specific values for pressure processes at specific locations may differ somewhat from the assumed value. The fatigue analysis load parameters are shown in Table 2. The side frame spacing is 880 mm for the Capesize and 830 mm for the Panamax, the plate thickness being about 22mm in either case. Figure 1 and Table 3 show the related side shell structural information.

5. FATIGUE ANALYSIS AND SENSITIVITY STUDY

As previously noted, all fatigue calculations are made for the bracket lower end of the frame where it joins the sloping plate of the hopper tank. The fatigue procedure used is essentially the same as that suggested in the SafeHull bulk carrier guide, except that three structural cases are considered: the gross structure (i.e., no corrosion in 20 years), a 10 % corroded structure (i.e., 10% corrosion in 20 years), and a 20 % corroded structure (i.e., 20 % corrosion in 20 years). The time varying nominal stress range response is calculated using the SafeHull simplified side frame response model. Calculations for the Capesize do not give credit to any bracket flange plate at the lower end, while the calculations for the Panamax do. This is because the particular Panamax considered employs an integral bracket end connection, while the particular Capesize does not, see Figure 1. The UK DEN in air E curve is used with an SCF of 2, which is a "default" assumption for illustrative purposes. In real cases, the actual SCF would be obtained by finite element analysis or other means. The present fatigue damage calculations are of course sensitive to such assumptions, which one must appropriately consider in interpreting the results.

844

Table 4
Fatigue Damage Estimates, E Curve and
SCF=2

Vessel	Corrosion	Stress Range (kg/mm²)	Miner sum
Capesize	gross	29.69	0.61
	10 %	32.99	0.88
	20 %	37.12	1.30
Panamax	gross	21.09	0.18
	10 %	23.43	0.26
	20 %	26.37	0.40

The resulting damage estimates (i.e. Miner sums) are shown in Table 4. Fatigue lives for a Miner sum of unity may be readily obtained. Using the same S-N curve and SCF, the sensitivity of fatigue damage estimates to decrease in side shell thickness, decrease in bracket thickness, and decrease in side frame scantlings were also studied. The results, shown in part in Table 5, tend to support the view that the side frame scantlings and the end bracket thickness affect the fatigue damage significantly, and that the damage is relatively insensitive to the thickness of the side shell plating. The following conclusions can be drawn from the 20 year baseline fatigue damage estimates and related sensitivity studies:

1) The side pressures calculated for the Capesize are about 10 % greater and the stiffener spacing is about 6 % more, meaning that the Capesize pressure force is about 15 % higher than for the Panamax. This, together with the added effect of the difference in unsupported span (e.g., 15 % larger in the case of the Capesize than the Panamax) gives rise to fixed end moments at the Capesize side

frame that are nearly 50% greater compared to the Panamax (71.81 ton-m versus 44.79 ton-m). For identical end details, the fatigue stress range is proportional to the fixed end moment, and the fatigue damage varies roughly as the stress cubed. Hence for identical end details one would expect the Capesize side frame fatigue damage to be more than three times that of the Panamax (for these particular vessels). In the particular vessels considered, the end details are of course not identical, which is reflected in the fatigue damage results of Table 4.

Table 5
Illustration of Fatigue Sensitivity

Case	Corrn.	Stress Range kg/mm²	Miner Sum
Capesize (3)			
16mm bracket,	Gross	33.30	0.90
22mm shell,	10 %	37.00	1.28
Tee Stiffener	20 %	41.63	1.89
16mm bracket,	Gross	33.50	0.92
18mm shell,	10 %	37.22	1.31
Tee Stiffener	20 %	41.87	1.93
16mm bracket,	Gross	51.37	3.69
22mm shell,	10 %	57.08	5.14
HP400x16 Stfr	20 %	64.21	7.39
Panamax (1)			
11mm bracket,	Gross	20.56	0.16
18mm shell,	10 %	22.85	0.24
Tee Stiffener	20 %	25.70	0.37

2) In the two vessels / locations considered, the side frame end connection details are not similar as was previously noted. The features of the end

bracket are quite important to fatigue performance. In general, an integral bracket with an effective flange plate (such as that in the Panamax) is far superior in terms of fatigue performance than the non-integral bracket (assumed for the Capesize). The calculation parameter critical to the difference is the effective flange used with the bracket.

3) Fatigue damage estimates for the Panamax considered are acceptable. For the Capesize considered, the fatigue damage is seen to be unacceptable for a corrosion level of roughly 10% and over.

6. CONCLUDING REMARKS

In summary, in the context of side shell fatigue in bulk carriers, the importance of corrosion and the strength of the end bracket connections should be emphasized. Levels of corrosion in way of bulk carrier side framing can be large in specific cases, particularly in Capesize vessels that tend to carry mostly coal and iron ore, see for example data in Akita (1984), although lesser values are reported by Paik, et al. (1998). In the particular vessels considered, the fatigue loading in the Capesize is about 1.5 time the Panamax, due to the accumulated effect of unsupported span, frame spacing, and local pressure differences between the two vessels. To what extent this observation would hold for Capesizes and Panamaxes in general needs to be studied.

Another aspect to be highlighted relates to accounting for corrosion effects in the fatigue calculation procedure itself. Because of the complex nature of local structural behavior, and the sensitivity of fatigue lives to stresses / corrosion, it would seem that fatigue damage calculation procedures need to

account for corrosion in a time dependent manner (much as we did in the above analysis). The question of how to define a "net" structure for design under such circumstances may warrant further study. Also, considering the relatively high corrosion rates possible, the strength of fillet welds at the side frame end connections may warrant further study as well, see for example Tsai, et al., (1980). In general, the side shell fatigue damage was found to be quite sensitive to the scantlings of the side frame and end brackets used and to levels of corrosion. Explicit fatigue design of the related hull girder details may thus need to be pursued.

ACKNOWLEDGMENTS

The authors wish to thank the American Bureau of Shipping for permission to publish this paper. The views expressed herein are, however, those of the authors and not necessarily those of the Bureau. The authors gratefully acknowledge the various comments received from J.F. Conlon, H.-H.Chen and Y.K.Chen, which have lead to improvements in this paper.

REFERENCES

ABS (1995). Guide for dynamic based design and evaluation of bulk carrier structures, American Bureau of Shipping, SafeHull Project, March 1995.

ABS (1997). Rules for building and classing steel vessels, American Bureau of Shipping.

Akita, Y. (1984). Reliability analysis in strength of ships (1st report), J. of the

Society of Naval Architects of Japan, Vol.155, pp.207-214 (in Japanese).

BTCE (1993). ``Structural Failure of Large Bulkships". Bureau of Transport and Communication Economics, Report 85. Australian Government Publishing Service, Canberra, December 1993.

Ivanov, L.D. (1993). Bulk carriers- Brief analysis of structural failures based on department data files, ABS R & D Department Internal Report, American Bureau of Shipping, New York, November.

Liu, D., Jan, H.Y., Chen, H.H., Scotto, F.J. and Akiyama, A. (1995). Recent development of design criteria for hull structures of bulk carriers, Proc. of the 6th International Symposium on Practical Design of Ships and Mobile Units (PRADS'95), Vol.2, Seoul, pp.2.883-2.897.

Paik, J.K., Kim, S.K. and Lee, S.K. (1998). A probabilistic corrosion rate estimation model for longitudinal strength members of bulk carriers, to appear in Ocean Engineering.

Shama, M.A. (1995). Ship casualties-Types, causes and environmental impacts, Alexandria Engineering Journal, University of Alexandria, Alexandria, Vol.34, No.2, April 1995.

Tsai, C.L. et al. (1980). ``Review of Fillet Weld Strength for Shipbuilding". Ship Structure Committee Report 296, February 1980.

Panamax

Capesize

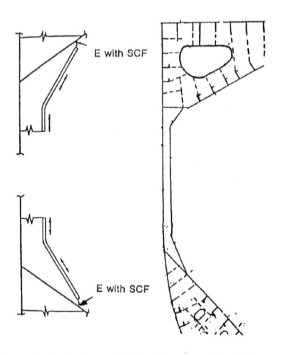
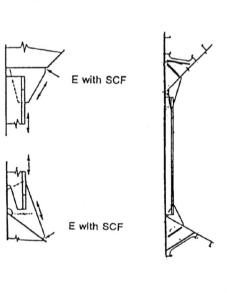

Figure 1   Side Shell Details in way of Bulkhead 1

Practical Design of Ships and Mobile Units
M.W.C. Oosterveld and S.G. Tan, editors.

FATIGUE LIFE PREDICTION FOR SHIP STRUCTURES"

J.H. Vink[a], M. Mukhopadhyay[b], B. Boon[a]

[a] Ship Structures Laboratory, Delft University of Technology
[b] On leave from Indian Institute of Technology, Kharagpur

ABSTRACT

The S-N method and the fracture mechanics approach are tools for predicting the fatigue life of ship structures. Fatigue life predictions according to the SN approach can be seen as an estimate of the crack initiation period.
In order to determine the total fatigue life of large constructions, including the crack growth stage up to a critical crack length, the fracture mechanics approach is an essential tool.
As cracks do originate in marine structures in arbitrary directions, a mixed mode crack growth model is to be adopted. In these situations the elastic T-term is an important parameter in view of crack growth direction. Results of assessing this T-term with help of the Finite Element Method will be presented.

1 INTRODUCTION

Due to the increased use of high strength steels in ship structures, in combination with optimised constructions and new details to further reduce steel weight and building cost, fatigue failure has gradually become a relevant design criterion, supplementary to the conventional yield strength and buckling criteria.

Realizing the importance of fatigue as a criterion to evaluate the adequacy of ship structural details, the methods for prediction of the fatigue life, taking into account both the crack initiation and crack growth stage, are becoming much of interest. The following methods for evaluation of the fatigue life will be briefly discussed in view of their principles, their possibilities and limitations:

- S-N concept. This is an accumulated damage analysis, which is independent of a visible crack. The damage parameter is defined as the ratio of number of stress cycles that have been experienced and number of stress cycles that can be sustained until failure, both using the same load spectrum. As it does not explicitly account for the crack growth stage, predictions according to the S-N approach are to be considered as giving the crack initiation life instead of the full fatigue life.
- Fracture Mechanics Approach. The singular stress field near a crack tip is characterized with help of the Stress Intensity Factors. They play an important role in predictions of the crack growth rate as well as

the crack growth path and the critical crack length. Using this method, a prediction of the crack growth life is possible, whereas it is less suited to estimate the crack initiation life.

However, for crack path predictions a third parameter, the so called elastic T-term, is equally important as the stress intensities. As follow up of previous crack growth simulations, additional macros have been written in the finite element code ANSYS to evaluate this T-term. Calculation results will be presented for simple geometries, and will be compared with theoretical results for the T-term as presented in literature.

2. S-N CONCEPT

2.1. S-N curve representations
In general, S-N lines can be represented as, see fig. 1:

$$\Delta\sigma = Max\left[\left(\frac{N_f}{A_1}\right)^{\frac{1}{-m_1}} ; \left(\frac{N_f}{A_2}\right)^{\frac{1}{-m_2}} ; \Delta\sigma_{co}\right] \quad (1)$$

The material constants, m_i and A_i are based upon the results of constant amplitude loading on small specimens. The stress range, $\Delta\sigma$, to be used in conjunction with the S-N approach is defined as the algebraic difference between the maximum and minimum stress in a cycle: $\Delta\sigma = \sigma_{max} - \sigma_{min}$
Design codes use "Detail Classes" for grouping detail geometries on a scale of fatigue vulnerability, see fig.

848

2 [1]. In case the proper Detail Class is selected for the subject detail, the probability of survival is said to be of the order of 97.7% (2 standard deviations below the mean line of the test results). However, as a class contains several different geometries, this figure can vary depending upon the location of the subject detail with respect to the middle of the relevant class.

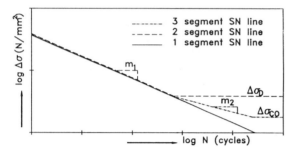

Fig. 1:   S-N curves with 1, 2 and 3 segments

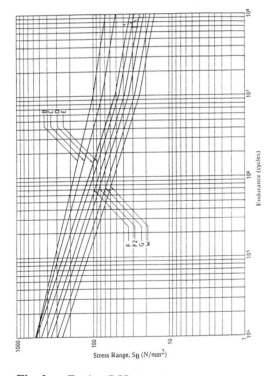

Fig. 2:   Design S-N curves

The choice between a one-, two or three segment S-N representation is dictated by the code, based upon constant or variable amplitude loading and environmental

conditions.

2.2. Fatigue life prediction

A fatigue life prediction for variable amplitude loading will be based upon the linear damage accumulation hypothesis according to Miner. The total damage due to a load spectrum, as e.g. the Weibull distribution, which contains n_i cycles at stress range $\Delta\sigma_i$, i=1..I, is:

$$D_{tot} = \sum_{i=1}^{I} d_i = \sum_{i=1}^{I} \frac{n_i}{N_{f,i}} \leq D_{failure} \qquad (2)$$

This damage accumulation concept implicitly uses a damage scaling where $D_{tot}=0$ means no damage and $D_{tot} \geq D_{failure}=1$ means detail failure.

2.3. Definition of "failure"

The failure criterion as normally used in conjunction with the S-N approach, $D_{tot}=1$, is based upon tests on relatively small specimens of standard geometry, see fig. 3 [2]. Different definitions of "failure" are in use, as: a "visible crack" or a "through thickness crack" [3], "loss of strength" of the specimen [3], "complete fracture" of the specimen [4], or "crack initiation" defined as 20% drop of tensile load [5].

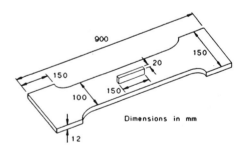

Fig. 3:   Class F fatigue specimen

It has to be realised that these failure criteria for small specimens are not capable to predict the total fatigue life of large structures as they do not reflect the residual fatigue life during the crack growth stage. In ship structures rather long cracks may be acceptable, depending upon stress range level, the actual place of the crack and the configuration, which means that a significant fraction of the total fatigue life may remain after the formation of a through the thickness crack. Determination of this rest period is of utmost importance upon the decision whether a crack needs urgent repair at the moment of its detection, or that it can be postponed until the first following survey.

3. FRACTURE MECHANICS APPROACH

3.1. Concept of the fracture mechanics approach

The fracture mechanics approach assumes a sharp initial crack of any length. In general a fatigue crack can propagate in three modes, depending on the relative orientation of loading near the crack tip, see fig. 4, which can be characterized with three different stress intensitiy factors, K_I, K_{II} and K_{III}. For relatively thin plates as used in marine structures, mode III (torsion) leads to a +/-ΔK_{II} component at both outer faces of the plate, which means that only 2 modes will be considered furtheron.

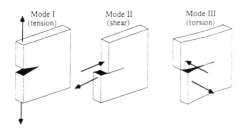

Fig. 4: Crack propagation modes

In view of mixed mode fatigue crack growth simulation, two aspects are important: crack growth rate and crack growth direction. A recent review of existing criteria for both these aspects is given in [6]

The main strength of the fracture mechanics approach is its ability to predict the number of cycles required to propagate from an initial small crack to a final crack size, and to calculate the critical crack size, in view of brittle fracture, using the toughness parameter K_c.

3.2. Crack growth rate simulation

The most well known model for crack growth is the Paris-Erdogan law for subcritical (ductile) fatigue crack growth:

$$\frac{da}{dN} = C(\Delta K_{eff})^{m_P} \qquad (3)$$

Here, $\Delta K_{eff} = K_{max} - K_{min}$, is the loading parameter (stress intensity factor range), while C and m_P are crack propagation constants of the material. For mixed mode loading ΔK_{eff} according to Tanaka [7] is used:

$$\Delta K_{eff} = (\Delta K_I^4 + 8\Delta k_{II}^4)^{0.25} \qquad (4)$$

The number of cycles for growth of a crack from an initial length, a_0, until a final length, a_1, follows from:

$$N = \int_0^N dn = \int_{a_0}^{a_1} \frac{da}{C(\Delta K_{eff})^{m_P}} \qquad (5)$$

The mixed mode stress intensities, K_I and K_{II}, as used in this paper are based upon the displacement extrapolation method, [8,9,10], which is inbuilt in the ANSYS software package. In order that a sufficiently accurate description of the stress field near the cracktip can be made, this method requires a ring of triangular quarterpoint elements around the cracktip, see fig. 5.

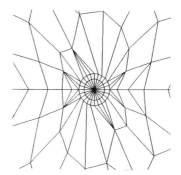

Fig. 5: Typical cracktip mesh

Whereas in the literature is suggested that this method is not very accurate, no evidence was found for it by comparisons with results based upon the stiffness derivative method [11], as programmed by the authors in ANSYS with help of macros [12]. The results of both methods compare equally well with theoretical results for simple configurations.

3.3. Crack direction simulation

For predicting the direction of the crack growth under mixed mode loading several criteria can be found in literature [6,13], as: direction where K_{II} vanishes, direction of maximum tangential stress, direction which gives maximum elastic energy release rate, direction of minimum strain energy density, direction with maximum K_I.

No uniform opinion exists on which of the above criteria is the best one to be used. For modest ratios of $S = K_{II}/K_I$ the above methods predict almost equal angles for the crack growth direction, and Cotterell and Rice [14] proved that the 4 first mentioned criteria for crack growth direction are equivalent up to the first order of the crack growth direction. This means that the predictions according to these three techniques will start to differ only for increasing values of S. From this it can be concluded that a crack in general will propagate in such a way that K_{II} will be negligible along the path. This was indeed the case in the simulations presented in [12], using the maximum tangential stress criterion proposed by Erdogan and Sih [15].

The maximum tangential stress criterion is well supported by experiments, and is rather simple to implement, because the crack growth direction, θ_T, is determined by S only:

$$\sin\theta_T = \frac{S - 3S\sqrt{1 + 8S^2}}{1 + 9S^2} \qquad (6)$$

Comparisons of predictions with this crack direction criterion and experimental results were presented in [12] for plates with inclined central cracks and uniaxial cyclic loading. The correlation between the theoretical and experimental crack paths was reasonable.

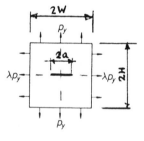

Fig. 6: Centrally notched plate under biaxial load

From literature [13,14,16,17,18,], there is evidence that the crack path prediction can be improved by inclusion of the constant stress component parallel to the crack, known as the elastic T-term. Cotterell [19] showed that cracks with a negative T-term (compression) are path-stable, and path-unstable in case of a positive T-term (tension). For a small central crack in a large

plate under uniaxial normal stress, σ_y, see fig. 6, it follows from theoretical solutions of the stress field in the region of the cracktip that $T = -\sigma_y$, so that such a crack will always be path-stable, except when a uniaxial parallel stress, $\sigma_x \geq \sigma_y$, is applied.

Cotterell and Rice [14] presented a crack direction criterion which is based upon a first order perturbation procedure for a kinked and slightly curved crack. It accounts for the T-term and the path is assumed to be one for which K_{II} vanishes. Using the parameters $\theta_0 = -2S$ and $\beta = 2\sqrt{2}T/K_I$, the crack extension, see fig. 7, for small values of $\beta^2 x$ is:

$$\lambda_{(x)} = \theta_0 x[1 + \frac{4T}{3K_I}\sqrt{\frac{2x}{\pi}} + 4\frac{T^2 x}{K_I^2}] \qquad (7)$$

Fig. 7: Crackpath increment in relation with the elastic T-term

With help of this result, the stability of a straight crack under mode I loading can be explained as follows: while normally $K_{II} = 0$, it is assumed that K_{II} differs slightly from zero due to inevitable imperfections. Hence, θ_0 is a random parameter, and it follows from eq. (7) that any imperfection will cause the crack path to deviate from the straight line when $T > 0$, whereas there will only be a local effect in case $T < 0$.

Sumi et al. [13,18,20] extended the work of Cotterell and Rice by considering also the $\sqrt{x}$ term of the Williams expansion. The resulting crack path extension is:

$$\lambda_{(x)} = \theta_0 x[1 + \frac{8T}{3K_I}\sqrt{\frac{2x}{\pi}} + \gamma_1 x]$$

where: (8)

$$\gamma_1 = 4\frac{T^2}{K_i^2} + \textit{additional corrections}$$

This result is very much the same as the one of Cotterell and Rice, except of a factor 2 in the second term, and additional corrections in the third term, which account for the stress redistribution due to the crack growth in a finite body. When these additional corrections are neglected, the third term is equal to the one of Cotterell and Rice too.

A further support for the equivalence of the different crack growth direction criteria follows from conclusion that the resulting angle from the tangential stress criterion of Erdogan and Sih, eq. (6), for small values of S reduces to $\theta_T = -2S$, which is equal to the initial angle θ_0 of the latter two criteria in case T=0.

3.4. Crack initiation simulation

In order that a prediction of the total fatigue life of a component loaded in fatigue can be based upon the fracture mechanics approach, the crack initiation period for the development of a small initial crack, length a_0, is an important parameter. In literature, three methods are found to account for the initiation period:

- S-N approach: As the standard S-N curves are based upon small specimen, they predict total fatigue lives which are close to the initiation period for a small (0.25 inch) or a through thickness crack.

- fracture mechanics approach: It assumes the existence of small imperfections (pre-existing microscopic cracks, notches at the surface, weld defects etc.) where cracks initiate from the very beginning. The initial crack length is to be based upon experience [21,22,23,24].

- local strain approach. This method uses a strain life relationship for the most highly stressed area, based upon the Coffin-Manson equation for plastic strain with a correction for the elastic strain according to Basquin, see [24]. It is mainly applicable for non welded details, and requires a summation of the strain history of each element ahead of the crack path.

As the phenomena involved with crack initiation in welded structures are dominated by stress concentrations due to random factors as weld defects, local weld geometry, misalignment etc. the crack initiation life shows a wide scatter which probably can be best represented with the S-N concept.

4. METHODS TO EVALUATE THE T-TERM

Based upon the above discussion of the crack growth direction, it is clear that the elastic T-term is a fracture mechanics parameter that cannot be excluded from crack growth predictions. For that purpose the methods to evaluate the T-term as found in literature are summarised, in order to make a proper choice on the

way it will be implemented in the macros already programmed for crack growth simulation using the finite element code ANSYS [12].

- A first group of methods evaluates the elastic T-term from the stress distribution along the crack flanks. Theoretically, the radial stresses at the crack flanks are zero for pure mode I, and of opposite sign on both crack flanks for pure mode II loading. This means that the non singular stress term can be determined by averageing the radial stress of both crack flanks at a number of radial positions. Extrapolation of these values to the crack tip position will yield the elastic T-term as constant part of the radial stress. When based upon results of the finite element method, the stress field as calculated is of course an approximation, especially near the non singular crack tip. Because of this the results are not very accurate. In order to improve the accuracy of their results, Larsson and Carlson [25] corrected the numerical stress field by subtracting the pure mode I and mode II stress fields as determined with the same finite element mesh.

- Another group of methods, as described by Cardew et. al. [26] and Kfouri [27], is based upon a theorem due to Eshelby, where T is avaluated with help of path independent integrals for two stress fields. The first stress field which is due to the external loading, F_0, results in an integral value $J(F_0)$, while the second integral value, $J(F_0+F_P)$, is based upon a stress field which is due to a combination of the loading F_0 plus a self equilibrating stress field due to a point load F_p at the crack tip and aligned with the crack flanks. As result of this theorem T can be calculated from:

$$\frac{T.F_P}{E} = J(F_0+F_P)-J(F_0) \tag{9}$$

- A third group of methods, as described by Swedlow et. al. [28] and Leevers and Radon [29], uses the displacements or load distribution along a boundary curve around the cracktip to solve the vector of coefficients of the Williams eigenfunctions.

5. IMPLEMENTATION

As an engineering solution, it was decided to evaluate the elastic T-term from the stress and/or displacement distribution along the crack flanks, because this is the most simple method to implement in a finite element code. In the way as implemented, for a number of radial positions the average of the top and bottom flank is used of either the radial stress or the radial displace-

852

ment. A linear least squares fit on the data points was used to find the relevant value at R=0 for evaluating the T-term in three ways, as follows:

- T-σ, using the radial distribution of the nodal stresses σ_r as supplied by the finite element code. At each nodal point location the average value of radial stress of top and bottomflank is determined. This method is the most logical one, but it is to be realised that these nodal stresses are average values of all elements which are connected to the relevant node. As a consequence, this value is meaningless at the cracktip node. Furthermore, the triangular cracktip elements with quarterpoint nodes have a very complicated strain distribution, resulting in stresses which are not very accurate at all for the first 2 or 3 nodes. Because of this, only stresses of the 3rd and subsequent nodes can be used for the least squares fit.

- T-m, based upon the local radial strain, $d(u_{ra})/dr$. Whereas the nodal stresses of the finite element code are averages for all elements attached to the node, and include also effects of tangential strains, this parameter simply evaluates locally the radial strain between 2 adjacent nodes along the crack flanks. The average of the radial stress of both crack flanks, $Ed(u_{ra})/dr$ is used at locations halfway between the nodes to evaluate T.

- T-u, using the radial distribution of Eu_{rca}/R: at each nodal point position the radial displacement, relative to the radial displacement of the cracktip node, is determined and averaged between top and bottom flank: u_{rca}. By considering u_{rca}/R the average strain over the full length between nodal point and cracktip node is used, instead of the local strain. As a consequence. Extreme strains further away from the cracktip node are filtered out in this method.

6. RESULTS

As a very first test, the elastic T-term was calculated for the geometry of a centrally cracked plate, fig. 6, for which analytical data are available [25,26,27,29].

A global impression of the results for plates with constant H/W=2 is presented in fig. 8, where the average of the three calculated results is compared with data of Leevers and Radon [29]. For this purpose, the biaxiality ratio $B_0=-(1+0.085.a/W)$, for loading ratio $\lambda=0$, as presented in their fig. 9, was used to calculate the biaxiality ratio B_λ for cases where $\lambda \neq 0$ as follows:

$$B_\lambda = \frac{T_\lambda \sqrt{\pi a}}{K_I} = \frac{(T_0 + \lambda \sigma_y)\sqrt{\pi a}}{\sigma_y Y \sqrt{\pi a}} = B_0 + \frac{\lambda}{Y} \quad (10)$$

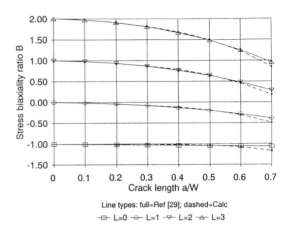

Line types: full=Ref [29]; dashed=Calc
L=0 L=1 L=2 L=3

Fig. 8: Comparison of calculated stress biaxiality ratio B with theoretical values

The shape parameter, Y, as used was a modified version of Fedderson's formula, see [30], error <0.1%. It is to be noted that the lines for B_1, B_2 and B_3 in fig. 8 differ from the lines in fig. 11 of [29], probably because they used another Y (Isada) to transform the B_0 line. From fig. 8 it can be seen that the avearge of the three calculated values, T-av, compares reasonably with the theoretical values. Having a closer look into the results for $\lambda=0$, see fig. 9, it appears that the three methods show an almost equivalent tendency of being slightly to high at a/W=0.4 and about 10% to low at a/W=0.7.

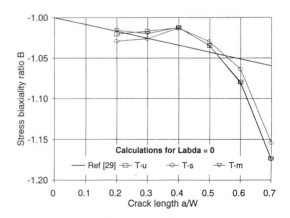

Fig. 9: Results for stress biaxiality ratio B_0 in case of uniaxial loading

As the error is probably attributable to the parameters which control the meshing, some further runs were made to investigate the influence of those parameters which are supposed to be the most important ones:
- radius of triangular cracktip mesh, delr, see fig. 5.
- crack growth increment, da.

The crack growth increment has two functions. It controls the global meshing in ANSYS, as a nodal point will be created at the crack flank at each da-interval. Furthermore, during crackgrowth simulation, da is the straight cracklength available for nodal points to be used as data points in the least squares fitting process. The ratio of da/delr controls the number of nodes which will be created along da.

Some general conclusions were drawn from these exercises:
- in most cases, T-m and T-u give almost equivalent results,
- T-s is more sensitive to additional parameters, as e.g. number of datapoint not used near cracktip, irregularities in mesh as generated by ANSYS just outside the cracktip rings etc.

A typical example of results for da=5 and delr=0.4 is presented in fig. 10, where T-u and T-m have an error <6%, while T-s is about 20% in error.

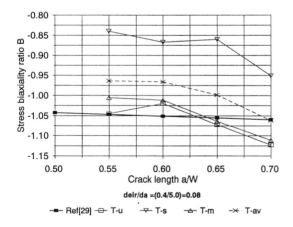

Fig. 10: Results for B due to improved values for parameters delr and da

From the results so far it follows that better tuning probably can be arrived at by finding optimal combinations of the parameters as mentioned. However, depending upon the accuracy that can be obtained, and a minimum degree of robustness regardless of the pa-

rameters, it may be necessary to enhance the method by inclusion of corrections for the stresses and displacements due to the singular fields near the crack tip.

7. CONCLUSIONS

It is stated that the fatigue life according to the S-N approach is merely a crack initiation period, whereas the fracture mechanics approach is capable to evaluate the rest life after crack initiation.

In order to improve crack growth predictions under mixed mode loading, the relevance of the elastic T-term is discussed.

Results are presented of numerical evaluation of the elastic T-term with help of the finite element method. The value of the elastic T-term has been calculated with help of the stress and displacement distribution on the crack flanks, using three methods. The results are acceptable in a global sense, but when looking into it more carefully, it appears that some further research is needed for finding a better tuning of the parameters which are used in the calculations.

REFERENCES

1. ABS: Guide for dynamic based design and evaluation of container carrier structures, 1996
2. Chalmers, D.W.: Design of ships' structures, Chapter 13, London 1993
3. Report of Committee V.1: Applied design - strength limit states formulations, 12 th ISSC, Vol 2, 1995
4. Lloyds Register: Fatigue design assessment procedure, LR-Ship Right, London 1996
5. Petinov, S., Yermolaeva, N.: Load-history sensitive cyclic curve concept in random load fatigue life predictions, Schiffstechnik, Band 40, p 107, 1993
6. Quin,J., Fatemi,A.: Mixed mode fatigue crack growth: a literature survey, Engineering Fracture Mechanics, Vol 55, No 6, 1996, p 969-990
7. Tanaka,K.: Fatigue crack propagation from a crack inclined to the cyclic tensile axis, Engineering Fracture Mechanics, Vol. 6 (1974), pp. 493-507
8. Banks-Sills,L.: Application of the finite element method to linear elastic fracture mechanics, Random Vibrations and Fract Mechanics, ASME

book: Appl. Mech. Review 100, 1991, p 447-461

9. Carpenter,W.C.: Extrapolation techniques for the determination of stress intensity factors, Engineering Fracture Mechanics, Vol 18(2), 1983, p 325-332

10. Lim,I.L., Johnston,I.W. Choi,S.K.: Comparison between various displacement based stress intensity factor computation techniques, International Journal of Fracture 58, 193-210,1992

11. Parks,D.M.: A stiffness derivative finite element technique for determination of crack tip stress intensity factors, Int Journ of Fracture, Vol 10 No 4 Dec 1974, p 487-502

12. Vink,J.H., Mukhopadhyay,M., Boon,B.: Numerical and experimental investigation of inclined cracks in ship plating, To be presented at OMAE june 1998

13. Sumi,Y., Chen Yang, Hayashi,S.: Morphological aspects of fatigue crack propagation, Part I: Computational Procedure, Intenational Journal of Fracture, Vol. 82-3 (1996), pp. 205-220

14. Cotterell,B., Rice,J.R.: Slightly curved or kinked cracks, Int. Journ. of Fracture, Vol. 16 (1980), pp. 155-169

15. Erdogan,F., Shih,G.C.: On the crack extension in plates under plane loading and transverse shear, Journ. of Basic Engineering, Vol. 85 (1963), pp. 519-527

16. Finnie,I., Saith,A.: A note on the angled crack problem and the directional stability of cracks, Int. Journ. of Fracture, Vol. 9 (1973), pp. 484-486

17. Leevers,P.S., Radon,J.C.: Culver,L.E., Fracture trajectories in a biaxially stressed plate, Journ. of the Mech. and Phys. of Solids, Vol. 24 (1976), pp. 381-395

18. Sumi,Y., Nemat-Nasser,S., Keer,L.M.: On crack branching and curving in a finite body, Int. Journ. of Fracture, Vol. 21 (1983),pp. 67-79

19. Cotterell,B.: Notes on the path and stability of cracks, Int. Journ. of Fract. Mech., Vol. 2 (1966), pp. 526-533

20. Sumi,Y., Nemat-Nasser,S., Keer,L.M.: On crack path stability in a finite body, Engineering Fracture Mechanics, Vol. 22 (1985), pp. 759-771

21. Stenseng,A.: Cracks and structural redundancy, Marine Technology, Vol 33, p 290-298, 1996

22. Sumi,Y.: Fatigue crack propagation and remaining life assesssment of ship structures, Advances in Fracture Research, Vol. 1: Failure analysis, Remaining Life Assessment, Life Extension and Repair, 1997, pp 63-76

23. DNV: Fatigue strength analysis for mobile offshore units, Class Note 30.3, 1984

24. Peeker,E.: Extended numerical modelling of fatigue behaviour, Thesis no 1617, Ecole Polytechnique Federale Lausanne, 1997

25. Larsson,G., Carlsson,A.J.: Influence of non-singular stress terms and specimen geometry on small scale yielding at crack tips in elastic-plastic material, Journ. of the Mech. and Phys. of Solids, Vol. 21 (1973), pp 263-277

26. Cardew,G.E., Goldthorpe,M.R., Howard,I.C., Kfouri,A.P.: On the elastic T-term, Fundamentals of Deformation and Fracture, Proceedings of the Eshelby Memorial Symposium, Ed: B.A. Bilby, K.J. Miller, J.R. Wills, Canbridge 1985, pp. 465-476

27. Kfouri,A.P.: Some evaluations of the elastic T-term using eshelby's method, Int. Journ. of Fracture, Vol. 30 (1986), pp. 301-315

28. Swedlow,J.L., Karabin,M.E., Maddux,G.E.: Cracktip stress analysis from field values of the displacements using complementary energy, Advances in Research on the Strength and Fracture of Materials, edited by Taplin, 4th Int. Conf. on Fracture 1977, Univ. of Waterloo, Canada 1977, pp 103-109

29. Leevers,P.S., Radon,J.C.: Inherent stress biaxiality in various fracture specimen geometries, Int. Journ. of Fracture, Vol. 19, (1982), pp. 311-325

30. Tada,H., Paris,P.C., Irwin,G.H.: The stress analysis of cracks handbook, Del Research Corp. 1985

Practical Design of Ships and Mobile Units
M.W.C. Oosterveld and S.G. Tan, editors.

Long term accumulation of fatigue damage in ship side structures

Are Johan Berstad and Carl Martin Larsen

Department of Marine Structures, Faculty of Marine Technology, Norwegian University of Science and Technology, N-7034 Trondheim-NTNU. Norway

Long term fatigue accumulation in the side structure of large ships is considered. The purpose is to improve fatigue life prediction of structural parts near the water-line. The crack growth is governed by time varying local stresses in the detail where the crack is developing. These stresses are influenced by global moments in the ship hull from waves and ship motions, and also variations of the local pressure on the side plates near the actual detail. External loads as well as internal pressure from a fluid cargo are considered. Long term fatigue damage is calculated by accumulating damage in each short term condition experienced by the ship.

A case study on a realistic ship operating on the west coast of North America is carried out. Average accumulated damage pr. year is calculated. Variation in accumulated damage caused by statistical variation in the wave environment is investigated, and also the relative contribution to fatigue accumulation from different sea areas along the route.

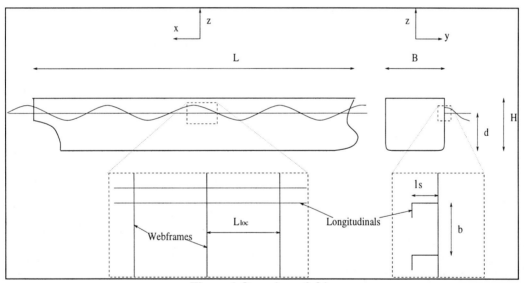

Figure 1 Overview of ship.

1   INTRODUCTION

Several ships have recently suffered from fatigue crack growth in the side longitudinals in the forepart and midship area close to the waterline (See Fig 1). This type of damage has occurred for oil tankers operating on the west coast of North America and on the east coast of South Africa. The crack growth is caused by time varying stresses in the detail where the crack is developing. These stresses will be influenced by global moments in the ship hull from waves and ship motions, and also variations of the local pressure on the side plates near the actual detail. External loads, as well as internal pressure from fluid cargo will depend on the actual sea state, ship draught, speed and ship orientation relative to wave propagation.

856

Several authors have addressed the problem. Witmer and Lewis (1994) consider fracture occurrence in oil tankers operating on the west coast of North America. They have also performed stress measurements. Sucharski (1995) gives an overview of fracture occurrence and discusses causes and consequences. Chen and Shin (1995) discuss what loads to take into account. Cramer et al. (1993) outline a procedure for calculating fatigue damage in side longitudinals taking into account local external pressure. Watanabe et al. (1995) propose a simplified calculation model for fast determination of long term fatigue statistics. Friis Hansen and Winterstein (1995) present a model that considers both local and global effects. Naito et al. (1995) have developed a method for calculating wave loads on the side shells when the ship is large relative to the wavelength. Berstad, Faltinsen and Larsen (1997) have established a simple calculation method to predict fatigue damage as a function of voyage parameters.

Scope of the present work is to predict long term fatigue accumulation, by accumulating damage in short term conditions using long term statistics for state parameters. Short term damage is calculated using the procedure presented by Berstad, Faltinsen and Larsen (1997).

This method is a time domain algorithm that calculates the stress response $\sigma(t)$ as the sum of the local bending response in the longitudinal from lateral pressure and axial stresses due to the horizontal and vertical global bending moments. The local bending moment in the longitudinal is due to the difference in external and internal pressure. A local beam analysis for the longitudinals considered is performed.

2   SHIP MOTIONS AND SEA LOADS

Wave induced ship motion and global wave loads in the linear frequency domain are calculated by strip theory (Salvesen, Tuck and Faltinsen 1970). The computer program VERES (Fathi 1996) is used. A linearized quadratic viscous roll damping term is introduced. The original theory did not find the pressure distribution. This is done in VERES by assuming the total velocity potential due to the presence of the ship satisfies a two-dimensional Laplace equation in the cross-sectional planes. The

free surface condition for the diffraction potential is the same as for the velocity potentials due to forced oscillations of the ship in six degrees of freedom.

A right-handed coordinate system (x,y,z) that follows the mean forward speed U of the ship and is fixed with respect to the mean position of the ship is used. The origin is in the plane of the undisturbed free surface. The z-axis is vertically upwards and goes through the centre of gravity of the ship. The mean forward speed of the ship appears as a flow along the positive x-axis. Let the translatory displacements in the x- , y- and z- directions with respect to the origin be η_1, η_2 and η_3 so that η_1 is the surge, η_2 is the sway and η_3 is the heave displacements. Furthermore, let the angular displacement of the rotational motion about the x-, y-, and z-axis be η_4, η_5 and η_6 respectively so that η_4 is the roll, η_5 is the pitch and η_6 is the yaw angle.

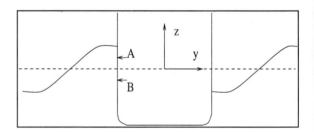

Figure 2 Local external fluid pressure.

2.1   Local external pressure

For a point A (see Figure 2) located above the mean free surface the time varying pressure, p is assumed to be "hydrostatic" relative to the instantaneous wave elevation if the point is submerged. If A is out of water p is zero, then as shown by Berstad, Faltinsen and Larsen (1997)

$$p = \max\left(-\rho\left(\frac{\partial}{\partial t} + U\frac{\partial}{\partial x}\right)\phi_t\big|_{z=0} - \rho g(\eta_3 - x\eta_5 + y\eta_4 + z_A), 0\right) \quad (1)$$

Here ϕ_t is the time-domain potential for the irregular sea accounting for both the incident waves as well as the disturbances caused by the oscillations of the vessel and the diffraction of the incident waves by the vessel, ρ is the mass density of the fluid, t is the time unit and (x,y,z_A) is the equilibrium coordinates of point A.

For a point B on the ship side with an equilibrium position below the mean free surface a Taylor series expansion of the potential ϕ_I around point B is performed. Keeping the linear terms, the following expression for the pressure is derived.

$$p = \max\left(-\rho\left(\frac{\partial}{\partial t} + U\frac{\partial}{\partial x}\right)\phi_I|_{z=B} - \rho g(\eta_3 - x\eta_5 + y\eta_4 + z_B),0\right) \quad (2)$$

Here (x,y,z_B) is the equilibrium coordinates of point B.

2.2 Internal local pressure

The flow in the tank is assumed to be two-dimensional in the y-z plane. The tank is full and the internal local pressure is calculated using the following expression from Berstad, Faltinsen and Larsen (1997):

$$p = -\rho((\ddot{\eta}_2' - z_m\ddot{\eta}_4)(y - y_0) + (\ddot{\eta}_3' + y_m\ddot{\eta}_4)(z - z_0) + \ddot{\eta}_4(f(y,z) - f(y_0,z_0))) \quad (3)$$

where $\dot{\eta}_2' = \eta_2 + \eta_6 x$ and $\dot{\eta}_3' = \eta_3 - \eta_5 x$. Dot means time derivative. y_m and z_m are the coordinates of the center of the tank, y_0 and z_0 are the coordinates of the opening of the tank, see Figure 3. ρ means the mass density of the fluid in the tank. The first and second term is caused by translation of the tank. The third term is due to forced roll oscillation around an axis through (y_m,z_m). Generally, the function $f(y,z)$ has to be found numerically, in the present study an analytical solution for a rectangular tank from Berstad, Faltinsen and Larsen (1997) is used. The velocity field caused by roll motion around tank center in a square tank is shown in Figure 3.

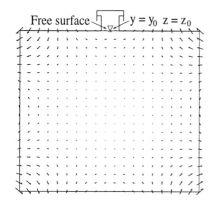

Figure 3 Full tank, with velocity field caused by roll motion around tank center.

3 STRESS ANALYSIS

A modified Pierson-Moskowitch (PM) wave spectrum is used to describe the stochastic long crested incoming waves in a short term sea state. The desired spectra for motion and global responses are then found by a frequency domain analysis as discussed. These spectra has been adjusted to take into account the forward speed of the vessel (Denis and Pierson 1953). The FFT-transform (Cooley et al. 1969), is used to generate consistent response time series which for the local external loads is adjusted for the effect of moving free surface. One should note that all phase angles have to be carefully considered in order to ensure consistency.

Having found consistent time series for all response components the total stress process is calculated as:

$$\sigma(t) = \sigma_V(t) + \sigma_H(t) + \sigma_{loc}(t) \quad (4)$$

Here $\sigma_V(t)$ is stress due to the vertical bending moment, $\sigma_H(t)$ is stress due to the horizontal bending moment, and $\sigma_{loc}(t)$ is stress due to the difference in pressure between the local external pressure and local internal pressure. The local stresses are calculated using a static beam model.

It has been shown by Berstad, Faltinsen and Larsen (1997) that all terms in Eq. (16) are of importance and should be considered in a fatigue analysis.

4 FATIGUE ACCUMULATION

Having calculated the stress response time history in the crack growth area the Palmgren-Miner (Miner 1945) hypothesis is applied for calculation of fatigue accumulation. For each short term condition, k accumulated damage D_k is found as

$$D_k = \sum \frac{n_i}{N_i} \quad (5)$$

Here n_i is the number of stress cycles for a given stress range, and N_i is the number of stress cycles before failure for the same stress range level. N_i is found from the relevant SN-curve. The stress cycles are identified from the response process according to the 'Rainflow' approach (Matsuishi and Endo, 1968) using a sufficient number of stress blocks and a response time series long enough to get acceptable accuracy. Accumulated damage D_k is a function of the actual sea state, load condition, speed and ship orientation relative to the wave propagation. Long term fatigue accumulation can be found by summing short term damages over conditions experienced by the ship.

$$D = \sum_k D_k \qquad (6)$$

When $D=1$ the crack is assumed to be critical, and the structural detail has failed.

5   CASE STUDY

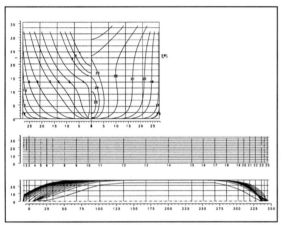

Figure 4 Body plan of ship

A large oil tanker is used as a case study. Experience has shown that ships of this type has suffered from damage in the side structure, in particular at the intersection between longitudinals and transverse web frames or bulkheads (see Figure 1). The ship is operating on the Trans Alaska Pipeline Service (TAPS) trade. The ship travels from Los Angeles to Valdez, Alaska in ballast condition, returning fully laden with crude oil. The ship spend a total of 18 days at sea on each round trip, 9 days in ballast condition and 9 days in laden condition. The ship makes 13 round trips in one year.

Wave statistics are found from Hogben et al (1986). The statistical distribution of significant wave height and spectral peak period is assumed to be constant in defined geographical areas. In each area wave statistics is given as scatter diagrams for 8 directions. Probability of occurrence for each direction is given. There are 5 area subdivisions between Los Angeles and Valdez, as shown in Figure 5. That is area 22,

14, 13, 7 and 6 counted as they appear on the ballast leg from Los Angeles to Valdez. The heading angle θ is zero if the vessel heads north. If the vessel heads west, θ = 90°. The vessel heading and percentage time spent in each area along the route is given in Table 1.

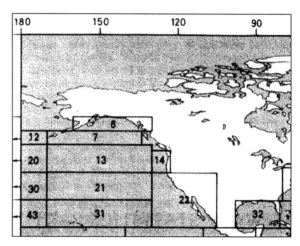

Figure 5 Area subdivisions. North America and Pacific ocean. From Hogben et al (1986).

Table 1 Time percentage in each area in Figure 5 and heading angle relative to north.

Area	% Time	θ, Laden	θ, Ballast
6	20 %	220°	40°
7	20 %	215°	35°
13	20 %	210°	30°
14	15 %	210°	30°
22	25 %	210°	30°

The body plan of the ship is shown in Figure 4, and main structural hull data is given in Table 2. In laden condition the ship has a draft of *22 m* and zero trim. In ballast the mid ship draft is *9.7 m* and the trim angle is *0.7°* giving a trim by the stern of *4,1 m*. The ship speed *U* is *12* knots both in ballast and laden condition. The center of gravity is located *163* and *167* meters from the forward perpendicular in laden and ballast condition respectively. The block coefficient is *0.84* in laden condition and *0.78* in ballast . The pitch and yaw radii of gyration are both *79.5 m* during the laden leg and *76.4 m* during the ballast leg. The roll radius of gyration is *18.7 m* and *24.2 m* for laden and ballast condition respectively. The transverse metacentric height is *8.6 m* in laden condition and *16 m*. In ballast. The ships displacement is *350,000* tonnes and *144,400* tonnes respectively. The longitudinal mass distribution is shown in Figure 6.

Table 2 Main structural data for the ship.

Length between perpendiculars	L_{pp}	335 m
Breadth at mid ship	B	55.4 m
Mid ship horizontal area moment of inertia	I_y	2831 m^4
Mid ship vertical area moment of inertia	I_z	1203 m^4
Mid ship vertical neutral axis (from keel)	z_{neu}	13.8 m
Spacing between web frames	L_{loc}	5 m
Vertical distance between side longitudinals	b	880 mm
Stress concentration factor for axial stress	SCF_{gl}	1.7
Stress concentration factor for bending stress	SCF_{loc}	2.2

The ship has totally *9* center tanks and twice as many wing tanks. In laden condition all tanks except center tank no. *2* is filled by crude oil with a mass density of *800 kg/m³*. The ship has single skin side and bottom structures. In the mid ship area the tanks are almost rectangular, with the wing tanks being *16.7* meters wide and *27.7* meters high. The tank opening is in the upper inner corner of the tank. The side of the ship is in the local analysis modeled by horizontal beams with fixed rotation at the frame or bulkhead intersection. The ship side form the plate flange of each beam with a cross section as shown in Figure 7. The pressure found at the vertical neutral axis of the beam is assumed to be uniformly distributed over the plate flange. The width of the flange has not been adjusted to account for the effective flange effect, and local bending stress in the top flange is not considered.

Data for the longitudinal stiffeners is given in Table 3. The stiffeners are numbered from *1* close to the deck, to *30* close to the keel. The stress concentration factor is found for point A in Figure 7 at the intersection between the longitudinal and the web frame. The stress concentration factors in Table 2 and SN-curve data for this case are found from Cramer et al. (1994) and Bøe and Tveit (1993). N_i is

found from Paris' law: $log N_i = log a - m \, log \, \Delta\sigma_i$, with *m=3* and *log a = 12.38*. $\Delta\sigma_i$ is the stress-range.

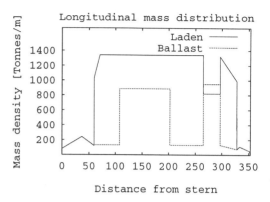

Figure 6 Longitudinal mass distribution

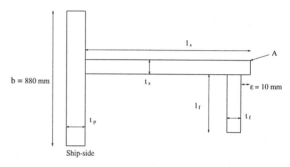

Figure 7 Longitudinal stiffener cross section

Table 3 Dimensions of side longitudinals. W_{loc} is the local sectional modulus for point A in Fig. 7.

Stiffener number	T_p (mm)	l_s (mm)	t_s (mm)	l_f (mm)	t_f (mm)	$W_{loc} \cdot 10^3$ m^3
1-3	24.74	350	28.42	-	-	1.1276
4	20.53	350	28.42	-	-	1.0947
5-8	20.53	410	12.1	150	18.95	1.6970
9-12	20.53	460	12.1	150	21.05	2.1099
13-15	20.53	510	12.1	150	23.16	2.5561
16-17	20.53	560	12.1	150	26.32	3.1365
18-19	21.05	560	12.1	150	26.32	3.1437
20	21.05	610	12.1	150	29.47	3.7684
21-23	21.48	610	12.1	150	29.47	3.7756
24	21.48	660	12.1	150	31.58	4.3711
25-27	22.63	660	12.1	150	31.58	4.3928
28-29	22.63	610	12.1	150	29.47	3.7938
30	22.63	760	17.89	150	31.58	5.8563

6   RESULTS AND DISCUSSION

Average accumulated damage during one year of operation on the route Los Angeles - Valdez return is calculated. The tank where the longitudinals are located is empty on the ballast leg from Los Angeles to Valdez, and laden with crude oil on the return.

Figure 8 shows mean accumulated damage pr. year in a mid ship cross section on the starboard side. In addition, damage accumulated during the laden and ballast legs respectively are shown.

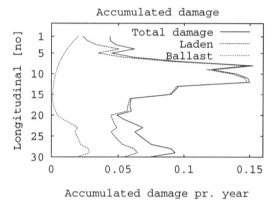

Figure 8 Mean accumulated damage pr. year. Starboard side. Mid ship cross section. Longitudinal no. 1 is located close to the deck.

In laden condition longitudinal no. 7 is located at the mean free surface. As seen from Figure 8 longitudinal no. 7 to 15 are the longitudinals accumulating most damage. As seen from this figure almost all fatigue damage in these longitudinals is accumulated during the laden leg from Valdez to Los Angeles.

Figure 9 shows mean accumulated damage pr. year on the port side.

As seen in Figure 8 and 9 longitudinals no. 7–15 accumulate more damage on the starboard side than the port side. In the considered area waves move predominantly in a westerly, north westerly direction. Therefore, most of the time the starboard side is the weather side when the ship is laden. On the weather side parts of the incident waves are reflected leading to larger local external pressure variations. As seen from figure 8 and 9 both the laden and the ballast leg are of importance for fatigue accumulation in the longitudinals located close to the keel. As seen from Figure 9 the port side longitudinals located close to the keel accumulate more damage during the ballast leg than during the laden leg.

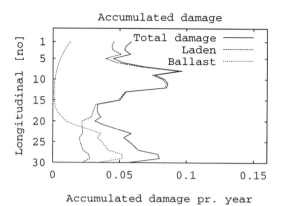

Figure 9 Mean accumulated damage pr. year. Port side. Mid ship cross section. Longitudinal no. 1 is located close to the deck.

Accumulated damage in a given time interval depend on what wave conditions the ship experience in that specific time interval. Statistical variation in wave environment is one source causing statistical variation in fatigue accumulation. This effect is investigated. Sets of sea state realizations during one journey are generated based on the probability of occurrence for short term parameters. A sea state is often considered to last for 3 hours. Consecutive sea states are correlated. Therefore it might be non-conservative to make realizations of fatigue accumulation using 3 hour long non-correlated sea states. Correlation of consecutive sea states can not be found from scatter diagrams. In the present study consecutive sea states are therefore assumed to be non-correlated. In order to investigate the importance of sea state correlation in an approximate way, fatigue damage accumulated during one year is calculated using 3 hours and 24 hours average sea state duration. It should be pointed out that the statistical variation investigated in the present analysis is only one of several components leading to statistical variation in fatigue damage results.

Figure 10 shows a histogram of accumulated damage pr. year in longitudinal no. 8 on the starboard side in a mid ship cross section. Each short term sea condition last for 24 hours. 10 000 realizations of one year operation are generated. Average accumulated damage pr. year, E(D) is found to be 0.1524 and the standard deviation, STD(D) is 0.0366.

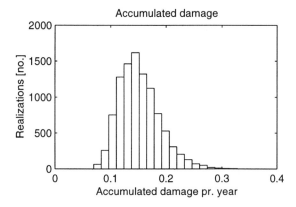

Figure 10 Accumulated damage pr. year. Longitudinal no. 8. Starboard. Mid ship cross section. 10000 realizations. E(D) = 0.1524. STD(D) = 0.0366. Each short term condition last for 24 hours.

Figure 11 shows a histogram of accumulated damage pr. year for the same detail and identical sea state statistics as in Figure 10. However, the average sea state duration is in this case 3 hours. Again 10 000 realizations are generated, and the results are now: E(D) = 0.1524 and STD(D) = 0.0129.

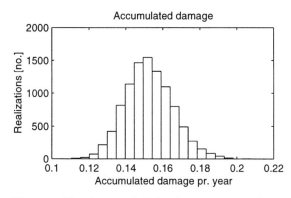

Figure 11 Accumulated damage pr. hour. Longitudinal no. 8. Starboard. Mid ship cross section. 10000 realizations. E(D) = 0.1524. STD(D) = 0.0129. Each short term condition last for 3 hours.

In Figure 10 the duration of each short term sea state is 8 times longer than in Figure 11. Hence $E(D_k)$ for

each short term condition k is 8 times larger and the variance is 64 times larger for this case. The number of short term conditions, k endured for the case in Figure 11 is 8 times larger than in Figure 10. Therefore the variance of accumulated damage during one year of operation is 8 times larger for the case in Figure 10 than in Figure 11, while E(D) of course is the same. From these figures it is seen that it can be non-conservative to assume 3 hour long non correlated sea states if the design is based on a specific level of probability of exceedance of fatigue life.

Because accumulated fatigue damage is a sum of k short term damages, the damage will become asymptotically Gauss distributed as k approach infinity. Since k is 8 times larger in Figure 11 than in Figure 10 the result in Figure 11 is seen to be more similar to a Gauss distribution.

It is 5 areas of different wave statistics (see Figure 5) on the trade between Los Angeles and Valdez. On a round trip from Los Angeles to Valdez each area is traversed twice, first on the ballast leg and then on the return laden leg. Relative contribution to fatigue accumulation from each area may now be investigated.

Figure 12 and 13 shows average accumulated damage pr. hour in each area (cf. Figure 5) on a return voyage Los Angeles – Valdez. The 5 area subdivisions are traversed twice. First on the ballast leg from Los Angeles to Valdez shown as the first five tics on the x- axis, then in opposite order of appearance the areas are traversed on the laden return.

Figure 12 shows damage pr. hour in longitudinal no. 8 located *0.88 m.* below the mean free surface in laden condition. As seen from this figure only the laden leg contributes to a significant amount of accumulated damage. On the laden leg it is seen that area no. 6, 7 and 13 in figure 5 contributes more to fatigue accumulation than area 14 and 22. Note that the adjacent areas 13 and 14 in Figure 5 have a very different contribution to fatigue accumulation. This indicates that the wave climate model may introduce inaccuracies due to discretization that must be considered in a reliability based approach to fatigue.

862

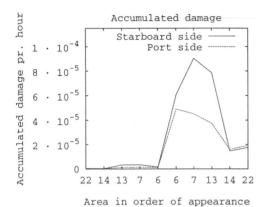

Figure 12 Mean accumulated damage pr. hour in each area (cf. Figure 5) on a return voyage Los Angeles - Valdez. Mid ship cross section. Longitudinal no. 8.

Figure 13 shows damage pr. hour in longitudinal no. 28. As seen from this figure both the laden leg and the ballast leg contributes to fatigue accumulation in this longitudinal. It is seen that on the port side the ballast leg contributes most to fatigue accumulation, while on the starboard side the largest contribution comes from the laden leg. Also in this longitudinal most damage is accumulated in area no. 6, 7 and 13.

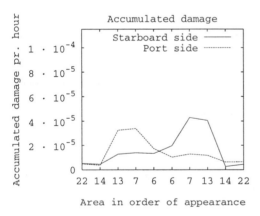

Figure 13 Mean accumulated damage pr. hour in each area (cf. Figure 5) on a return voyage Los Angeles - Valdez. Mid ship cross section. Longitudinal no. 28.

Results are discussed in view of uncertainties in a broad context. Model uncertainties and statistical uncertainties are considered. Uncertainties are subdivided into 5 main areas. 1. Wave climate. 2. Load model. 3. Structural model. 4. SN- curve data. and 5. Fatigue accumulation. Some potential model

error sources and statistical uncertainties in each of these areas are listed and discussed below. The following discussion is by no means exhaustive.

1. Wave climate. In the wave data found from Hogben et al (1986) there are model uncertainties concerning statistics of sea state parameters, directionality and the area subdivision. The importance of directionality has previously been studied by Berstad, Faltinsen and Larsen (1997), and a significant influence on fatigue damage was found. The wave statistics from Hogben et al (1986) is mainly based on observations from travelling ships in contrast to more reliable data based on direct wave measurements using buoys. The importance of using alternative wave statistics was studied by Bittner Gregersen et al (1993) by comparing results from ships operating in the North Sea. A scatter of ± 40% was found, and no clear tendency regarding which set of data that gave conservative results could be seen. There are also model uncertainties related to how each short term sea state is described as well as seasonal variations in wave climate and correlation of short term sea states. Ships tend to try to avoid bad weather. This effect is to some extent included in the data from Hogben et al (1986). How this effect is included may also introduce model errors. Statistical uncertainties may also be present in these parameters. In the present analysis only statistical uncertainty caused by the long term distribution of wave heading, significant wave height and mean zero crossing frequency is considered.

2. Loads. In each short term condition a linear 2-dimensional hydrodynamic boundary value problem is solved. There are model uncertainties regarding three dimensional effects, nonlinear second order effects, strong nonlinear effects and viscous roll damping effects. Guedes Soares and Trovao (1991) investigated the uncertainty in the transfer function from wave elevation to vertical bending moment. The results scattered between 80% to 118% of the mean value. They concluded that wave climate uncertainty is significantly larger than the uncertainty in the transfer function from wave elevation to vertical bending moment.

3. Structural model. The global response model is based on simple beam theory and might be improved by introducing a more correct representation of the stress distribution over the hull beam cross section. The local model assumes each longitudinal to be influenced from the local pressure on its own plate flange only. More accurate results may be obtained by using a model for local bending of longitudinals where a larger part of the ship side is included. Berstad and Larsen (1997) compared a beam model with a local FEM model that accounted for the lateral pressure variation in a more refined way. However this did not have a large influence on fatigue damage results in the case they studied. Relative deflections between transverse web frames and bulkheads, which is a model error source, was not included in their model. There are statistical uncertainties associated with the stress concentration factor caused by statistical variation in local geometry. The mass distribution along the length of the ship may also introduce statistical errors.

4. SN-curve. There are model uncertainties in the SN- curve data. It is also large statistical uncertainties. In Cramer et al (1993) the standard deviation in N_i is modeled as STD(log N_i) = 0.20 leading to an STD in fatigue accumulation of about 50% of the mean value.

5. Fatigue accumulation. There are both model and statistical uncertainties in the fatigue accumulation model based on the Palmgren - Miner approach and also in the 'Rainflow' cycle counting procedure.

In addition there are other uncertainties among them statistical uncertainties connected to the ships operation condition. Each potential error source do have a different correlation from one longitudinal to another, which should also be accounted for in a reliability based design and inspection scheme.

7   CONCLUSIONS

Long term fatigue accumulation in the side structure of a tanker has been investigated. A case study is presented.

The results show that fatigue accumulation in side longitudinals are strongly dependant on their vertical location. Results show that the longitudinals in a mid ship cross section located 16 – 22 meters above the keel accumulates more damage than other longitudinals. In these longitudinals more damage is accumulated on the starboard side than the port side. This show that wave directionality is important in a long term fatigue analysis.

The statistical variation of accumulated damage depending on the statistical variation of short term sea conditions is investigated and shown to be of importance. The length of each short term condition is of importance for the shape of the probability density function of accumulated damage, in addition to the standard deviation.

The results show that fatigue accumulation is strongly dependant on where the ship is operating. This indicates that errors in the wave environment model may be a significant error source.

Model and statistical uncertainties can have a large influence on fatigue accumulation. Some parameters having a large influence on accumulated damage are discussed. Future investigations should include systematic identifications of error sources and statistical uncertainties. A reliability based approach to design and inspection should include all uncertainties in a consistent manner, including uncertainties related to the inspection procedure.

REFERENCES

Berstad, A. J., O. M. Faltinsen and C. M. Larsen. (1997) "Fatigue crack growth in side longitudinals". In *NAV & HSMV*, Sorrento, Napoli, Italy. ATENA & CETENA.

Berstad, A. J. And C. M. Larsen. (1997) "Fatigue crack growth in the hull structure of high speed vessels". In *FAST' 97*. Sydney, Australia. Baird publications. 10 Oxford street, South Yarra Victoria 3141 Australia.

Bitner-Gregersen, E. M., E. H. Cramer and Robert Løseth. (1993) "Uncertainties of load characteristics and fatigue damage of ship structures" In *OMAE – Volume II*. ASME.

864

Bøe, Å. and O.Tveit (1993). "Study of stress concentrations in local details". Technical Report DNVC 93-0430, Det Norske Veritas.

Chen, Y. N. and Y. Shin (1995). "Consideration of loads for fatigue assessment of ship structures". In *Symposium and Workshop on The Prevention of Fracture in Ship Structure.* Washington DC. National Research Council.

Cooley, J. W., P. D. Lewis, and P. D. Welch (1969). "The fast Fourier transform and its applications". *IEEE Transactions on Education 12* (1), 27-34.

Cramer, E., S. Gran, G. Holtsmark, I. Lotsberg, R. Løseth, K Olaisen, and S. Valsgård (1994). "Fatigue assessment of ship structures". Technical Report DNVC 93-0432, Det Norske Veritas.

Cramer, E. H., R. Løseth, and E. Bitner-Gregersen (1993). "Fatigue in side shell longitudinals due to external wave pressure". In *OMAE*, pp. 267-272. ASME.

Denis, M. S. and W. J. Pierson (1953). "On the motions of ships in confused seas". *Transactions, SNAME 61*, 280-357.

Fathi, D. (1996). *VERES Version 2.0 - User's manual*. Marintek, 7034 Trondheim, Norway.

Friis Hansen, P. and S. R. Winterstein (1995). "Fatigue damage in the side shells of ships". *Marine Structures 8* (6), 631-655.

Guedes Soares, C. And M. F. S. Trovao (1991) "Influence of wave climate modeling on the long-term prediction of wave induced responses of ship structures". In *Dynamics of marine vehicles and structures in waves*. Elsevier Science Publishers.

Hogben, N., N. M. C. Dacunha and G. F. Olliver (1986). *Global Wave Statistics*, Brown Union Publ. London.

Matsuishi, M. And T. Endo (1968) "Fatigue of Metals Subjected to Varying Stress". *Japan society of Mechanical Engineers*, Fukuoka Japan.

Miner, M. A. (1945). "Cumulative damage in fatigue". *Journal of Applied Mechanics 12* (3), 159.

Naito, S., H. Kihara, and N. Nishimura (1995). "Wave loads acting on the side shells of large ships in very short wave length". *Journal of The Kansai Society of Naval Architects, Japan* (224), 95-104.

Salvesen, N., E. O. Tuck, and O. Faltinsen (1970). "Ship motions and sea loads". *Transactions, SNAME 78*, 250-287.

Sucharski, D. (1995). "Crude oil tanker structure fracturing. An operator's perspective". In *Symposium and workshop on The Prevention of Fracture in Ship Structure,* Washington DC. National Research Council.

Watanabe, E., S. Inue, K. Hashimoto, and K. S. H. Soeoka (1995). "Proposal of simplified fatigue design method for side longitudinals". *Journal of The Society of Naval Architects of Japan 177*.

Witmer, D. J. and Lewis J. W. (1994). "Operational and Scientific Hull Structure Monitoring on TAPS Trade Tankers". *SNAME Trans. 102,* 501-533.

Practical Design of Ships and Mobile Units
M.W.C. Oosterveld and S.G. Tan, editors.

865

Fatigue testing of large scale details of a large size aluminium surface effect ship

O.D. Dijkstra[a], A.W. Vredeveldt[b], G.T.M. Janssen[b], O. Ortmans[c].

[a] TNO Building and Construction Research
[b] TNO Centre for Mechanical Engineering
[c] Chantiers de l'Atlantique

The paper presents the results of two large scale fatigue tests on a detail of an aluminium surface effect ship. The overall dimensions of the specimens were: length 4.8 m, height 3.2 m and width 1.5 m (equal to the main frame spacing). The specimens consist of an aluminium structure of welded plates (thicknesses up to 20 mm).

The specimens were loaded with a constant amplitude loading, with an R ratio of 0,1. The nominal stress range at the weld detail was 18 N/mm^2. The stress concentration factor at the main frame location was approximately 5.

The comparison of the large scale results with the results of the small scale results leads to the conclusion that the lifetime of large scale tests (structures) is much longer. In large scale structures there is a longer stable crack grows period due to the redundancy in the structure.

1.   INTRODUCTION

Within the frame work of the Brite Euram project MATSTRUTSES, two fatigue tests were carried out on large scale specimens at the laboratory of TNO Building and Construction Research.

The MATSTRUTSES project is devoted to the structural design of a Surface Effect Ship (SES).

The specimens, welded structures made of aluminium plates, were fabricated by Chantiers de l'Atlantique. The specimens are a part of the midship cross-section.

A constant amplitude fatigue load was applied to the specimens until failure occurred.

During the tests strain readings were made and the development of the fatigue crack was monitored.

An evaluation of the test results was carried out.

2.   TEST SPECIMENS

The overall dimensions of the specimens are as follows: length 4800 mm, height 3203 mm and width 1500 mm. A web frame is located in the centre of the specimen (see figure 1). Two ordinary frames are located at each side of the main frame at a distance of 500 mm.

The detail where the crack will occur is the weld detail where the diagonal plate (thick 20 mm) is welded to the vertical plate (thick 20 mm). The detail is supported by a horizontal frame with a web of 10 mm thick. Note that the orientations here are given in the position used in the ship.

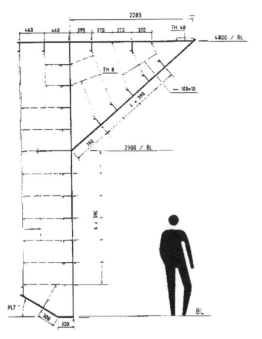

Figure 1   Specimen Main Frame view

The specification of the material used is as follows: plates - Al Mg4,5Mn (5083) or equivalent with minimum yield 220 N/mm2 (145 in welded areas) and stiffeners - ALMgSi1 - T6 (6082) or equivalent with minimum yield 250 N/mm2 (115 in welded areas).

3.   TEST RIG AND TEST PROCEDURE

The specimen is fixed in the TNO beam and column system with the platform deck in vertical position (Note that from here the orientations are given in the test position). The load was supplied at the main frame location at 180 mm from the BL and 2720 mm from the weld. The platform deck was fixed to three horizontal HE 320 M beams. One at 455 mm from the top, one near the continuing horizontal plate and one at the bottom flange position.

The specimen was fixed to the top beam by 6 strips (40 x 200 x 590) at a distance of 250 mm. Each strip was connected to the beam by two M30 bolts. The middle and the lower beam were not fixed to the specimen. So, only compressive loads can be transferred. The three beams were supported by columns at a distance of 2 meter.

The test set up and the most important detail are given in figure 2.

Figure 2   Test set up

After the specimen was fixed in the test rig a small load was applied to check strain gauge readings. After this check the specimen was loaded a few times up to 200 kN to get a linear behaviour and the strain ranges were measured.

Then the fatigue test was started with a constant amplitude fatigue load varying between 20 kN and 200 kN (load ratio $R = F_{min}/F_{max} = 0.1$).

During the fatigue test strain gauge readings were carried out at regular intervals and the development of the fatigue crack was monitored.

4.   FATIGUE TEST RESULTS

4.1   General

In both specimens the fatigue crack starts at the scallop near the welded detail at the main frame location. However this crack can be considered as a secondary crack as it does not grow to a large size.

The primary crack is the crack at the weld toe of the plate at the front side. This crack grows to a large size and when the plate was totally cracked over the whole width the specimen could not carry the fatigue load any more and the test was stopped.

The discovery of the first crack at the weld toe in the main plate occurred at about 20 to 25% of the total number of cycles at the end of test:

Specimen	First crack at weld toe	cracklength	End of test
1	125000	35	640138
2	184000	27	764275

The length of the first crack is such that it may be expected that the initiation phase to a small crack in the order of a few tenths of mm is negligible.
Further details of the crack developments will be given below.

4.2   Strain gauge readings at the beginning of the fatigue test

The calculated nominal strain range for a load increase from 20 to 200 kN (ΔF = 180 kN) at the weld detail is 252 micro strain.

The measured strain range at 28 mm from the weld toe across the specimen showed that at the main frame location there is a concentration factor (see figure 3). This is probably due to shear lag over the width of the specimen. The values of the strain gauge at the main frame, at 28 mm from the weld toe, can be considered as *local* nominal values. For the two specimens the measured values results in an average stress range of 35 N/mm² (approx. 2x nominal value).

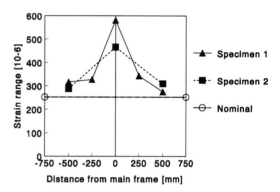

Figure 3   Strain distribution across the specimen (28 mm from weld toe)

Extrapolation of the strain ranges at the main frame location near the weld toe gives the hot spot strain range at the weld toe. The extrapolation can be done linear or quadratic (see figure 4). The hot spot strain and stress and the SNCF/SCF are as given in table 1. This results in an hot spot stress range of approximately 90 N/mm² (approx. 5x nominal value).

868

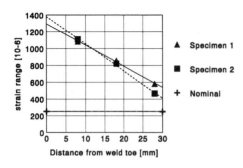

Figure 4   Strain range near weld toe at main frame (linear extrapolation)

Table 1   Hot spot strain / stress and SNCF / SCF

	Nominal values	Specimen 1	Specimen 2
Linear extrapolation			
Strain	252	1293	1381
Stress	17.6	90.5	96.7
SNCF-SCF	1.0	5.13	5.48
Quadratic extrapolation			
Strain	252	1233	1301
Stress	17.6	86.3	91.1
SNCF SCF	1.0	4.89	5.16

5.   Crack development

During the fatigue tests a number of cracks developed near the weld of the diagonal plate to the horizontal plate. At regular intervals the surface crack pattern was recorded.
The cracks are numbered from A to H (chronological for specimen 1). The front, rear, left and right side of the specimen are defined as given in the figures, where the location of the cracks are given (Figure 5 and 6).

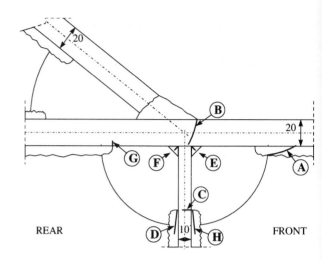

Figure 5   Crack location (main frame view)

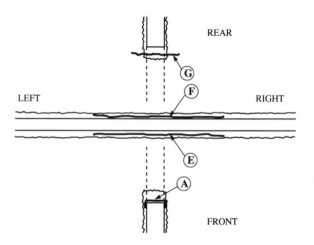

Figure 6   Crack location (lower side horizontal plate)

Some of the cracks stopped after a certain number of cycles. Table 2 gives the start and eventually stop of the cracks in the two specimens.

Table 2   Start and stop of the cracks

Crack	Specimen 1	Specimen 2
	Start	Start
A	70000	28700
B	125000	184000
C	347800	225450
D	353990	700000
E	415800	320000
F	415800	380000
G	550000	700000
H	-	700000

The cracks can further be described as follows.

Crack A

Crack A starts at the cope hole on the front side. This crack can be considered as a secondary crack as it does not grow to a large size. In specimen 2 the crack grow through the plate to the top side.

Crack B

Crack B starts at the weld toe of the main weld on the front side. This crack can be considered as the primary crack as it is growing into the horizontal plate. This plate is acting as the tension flange of the specimen. The crack starts near the centre of the plate and is growing in left and right direction. In the plate thickness direction the crack grows in the direction of the transverse stiffener and comes through the plate between the two fillet welds. From the fracture surface it can be concluded that the crack is through the plate at approximately 175000 cycles for specimen 1 and at 200000 cycles for specimen 2. The crack development can be found in detail in table 3 (specimen 1) and 4 (specimen 2) and figure 7.

Table 3 Development surface crack at weld toe (crack B) in specimen 1

Number of cycles	Crack tip location measured from centre line [mm]		Total crack length from left tip to right tip [mm]
	Tip left	Tip right	
125000	-14	36	50
150000	-85	101	186
175000	-96	110	206
200000	-128	126	254
225000	-154	144	298
250000	-174	153	327
275000	-193	166	359
300000	-202	178	381
347800	-237	197	434
400919	-275	223	498
450000	-325	248	573
500000	-380	276	656
550000	-448	314	762
600000	-548	335	883
640138	-750	410	1160

Table 4 Development surface crack at weld toe (crack B) in specimen 2

Number of cycles	Crack tip location measured from centre line [mm]		Total crack length from left tip to right tip [mm]
	Tip left	Tip right	
184000	-15	59	74
225450	-46	88	134
262000	-77	101	178
300000	-83	121	204
320000	-103	141	244
380000	-149	171	320
432000	-170	192	362
495000	-212	236	454
542000	-236	273	509
589000	-270	304	574
661000	-311	377	688
700000	-339	479	818
731000	-369	571	940
764275	-369	750	1119

870

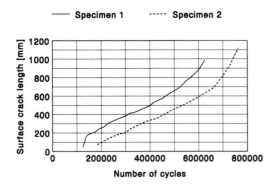

— Specimen 1   ----- Specimen 2

Figure 7   Development   crack   B
specimen 1 and 2

Crack C

Crack C starts at the cope hole in the transverse stiffener at the front side. This crack can be considered as a secondary crack as it does not grow to a large size.

Crack D

Crack D starts at the fillet weld of the transverse stiffener-mainframe connection at the rear side. This crack can be considered as a supporting connection for crack B.

Crack E

Crack E starts at the fillet weld of the transverse stiffener-plate connection at the front side. This crack can be considered as a supporting connection for crack B. It prevents crack B to become a real through crack.

Crack F

Crack F starts at the fillet weld of the transverse stiffener-plate connection at the rear. This crack can also be considered as a supporting connection for crack B. It prevents crack B to become a real through crack.

Crack G

Crack G starts at the cope hole on the rear side, growing in the plate. This crack can also be considered as a secondary crack as it does not grow to a large size.

Crack H (only in specimen 2)

Crack H starts at the fillet weld of the transverse stiffener-mainframe connection at the front side. This connection can be considered as a supporting connection for crack B.

6.   EVALUATION

The large scale tests are compared in a S_N diagram with small scale tests on strips with a longitudinal stiffener (TNO) [2] and butt welds (CDA [1]).

The two parameters used in an S-N graph are easy to determine for the small scale specimens, namely: for S the nominal stress range and for N the number of cycles to failure.

However for the large scale specimens these figures are more difficult to determine.

For the value of S the following numbers are available:

a)   nominal stress range
$\Delta\sigma_{nom}$ = 17.6 N/mm^2

b)   local nominal stress range
$\Delta\sigma_{l-nom}$ = 35 N/mm^2

c)   Hot Spot stress range
$\Delta\sigma_{HS}$ = 90 N/mm^2

For the number of cycles there are the following values (spec.1 / spec.2):

1) number of cycles to crack initiation
   N_i = 125000 / 184000

2) number of cycles to through crack
   N_{ThCr} = 175000 / 200000

3) number of cycles to end of test
   N_{EOT} = 640138 / 764275

Most fatigue codes and literature with regard to fatigue (specially for tubular joints in offshore structures) are using the Hot Spot stress range ($\Delta\sigma_{HS}$) and the number of cycles to through crack (N_{ThCr}) as the governing parameters. With this approach the most severe value for the stress is combined with a serious crack in the structure. Figure 8 shows that the large scale tests results with $\Delta\sigma_{HS}$ and N_{EOT} are much better than the small scale tests.

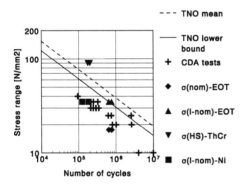

Figure 8 Comparison small scale (TNO and CDA) and large scale specimens (other dots) in an S-N graph

This better behaviour is probably due to the high value of the SCF of 5. Due to this high value the crack will initiate rapidly in the high stressed region and then growths into an area of lower stress, where it slows down or does not accelerate (see figure 7).

This means that using an S-N curve based on small scale test results in combination with the hot spot approach will result in conservative fatigue lifes for large scale structures.

In figure 8 other combinations of stress range and number of cycles are plotted.

This shows that using the nominal stress range ($\Delta\sigma_{nom}$) will lead to unsafe fatigue life estimates and is therefore not recommended.

Using the local nominal stress range ($\Delta\sigma_{l-nom}$) gives results which fall in the scatter band of the CDA tests. However, this can differ from one geometry to another and is therefore not recommended.

Further experimental fatigue research with large scale structures, supported by a refined stress analyses and fracture mechanics crack growth analyses are recommended.

7. CONCLUSIONS

Two fatigue test on large welded aluminium structures (4.8 x 3.2 x 1.5 meter) were carried out.

The failed detail is a plate welded under an angle with a full penetration weld to another plate (both plates 20 mm tick). The detail is supported by a 10 mm thick web fillet welded to the continuing plate.

A strain concentration factor of approximately 5 was present at the main frame location (in the centre of the specimen).

The calculated nominal stress (strain) range at the detail was 17.6 MPa (252 micro strain), while the measured hot spot stress (strain) range at the detail was 90 MPa (1300 micro strain).

The main test results are:

The specimen showed a long period of stable fatigue crack grows. This means that a crack in a real structure has a long safe life in which a crack can easily be found and repaired.

A comparison with available small scale tests showed that the use of the local nominal stress range (at 28 mm from weld toe) gives the best comparison. However, this can differ from one geometry to another.

Using S-N curves based on small scale specimens in combination with the hot spot approach will result in conservative fatigue lifes for large scale structures.

Further experimental fatigue research with large scale structures, supported by a refined stress analyses and fracture mechanics crack growth analyses are recommended.

REFERENCES

[1] O. Ortmans, Results of fatigue tests on aluminium alloy welded joints. Document TEC 42/055/001/19

[2] IJ.J. van Straalen, F. Soetens and O.D. Dijkstra, EUREKA 269 - Fatigue tests on longitudinal non-load carrying fillet welds. TNO report 94-CON-R1566, October 1994.

Practical Design of Ships and Mobile Units
M.W.C. Oosterveld and S.G. Tan, editors.

Fracture of a Stiffened Panel with Multiple Site Cracks under Lateral Pressure

Y. Sumi[a], Z. Bozic[b], H. Iyama[a], and Y. Kawamura[a]

[a] Department of Naval Architecture and Ocean Engineering, Yokohama National University,
Tokiwadai 79-5, Hodogaya-ku, Yokohama 240-8501, E-Mail:sumi@structlab.shp.ynu.ac.jp

[b] Faculty of Mechanical Engineering and Naval Architecture, University of Zagreb,
I. Lucica 5, 10000 Zagreb, Croatia, E-Mail: zeljko.bozic@fsb.hr
Formerly, Graduate Student, Yokohama National University

Considering the catastrophic failure of aged ship structures, damages caused by corrosion and fatigue are found to be the main reasons. In a ship structure a crack may initiate at a stress-concentrated region, where it extends by fatigue mechanism under moderate loading conditions. It may grow to a critical crack size, which leads to an instantaneous failure of the structure under an extreme loading condition. Sometimes cracks develop at several adjacent structural members creating multiple site damage (MSD), which may considerably decrease the structural integrity of ship structures. In the present paper, investigations are made for the fracture mechanisms of a stiffened panel, which is damaged by a single crack or multiple site cracks subjected to lateral pressure, by fracture experiments and the corresponding numerical simulation.

1. INTRODUCTION

This paper deals with experimental and numerical investigation of fracture of stiffened panels, damaged either by a single crack or by multiple site collinear cracks, under lateral pressure, where the experimental and numerical models are to a certain extent related to a part of a side shell structure of a ship exposed to wave pressure [1]. The modes of failures associated with fracture in stiffened panels such as crack propagation, crack curving, and crack arrest have been investigated experimentally by using stiffened and unstiffened panel specimens with various initial notches.

Finite element analyses, including both material and geometrical nonlinearity, were carried out to investigate fracture onset conditions under Mode I deformation. The elasto-plastic fracture mechanics concept (EPFM) was employed. The critical J-integral values associated with the onset of crack propagation were determined by using a centrally notched plate in tension in combining numerical results of the J-

integral values with the experimentally observed onset of crack propagation.

The critical pressure for the stiffened and unstiffened specimens subjected to lateral pressure, associated with the onset of stable crack propagation, was estimated based on the critical J-integral value. Estimated critical pressure agrees fairly well with experimental results for stiffened specimens, while the calculated critical pressure is slightly lower in the case of unstiffened specimens probably due to imperfect clamping conditions in the experiments.

2. TENSILE TESTS AND EXPERIMENTS OF NOTCHED TENSION SPECIMEN AND PRESSURIZED PANELS

2.1 Tensile test and notched plate test

The tension specimens and centrally notched plate specimens were tested in order to determine the mechanical properties and the material resistance to fracture so that the failure mechanism of the panel specimens subjected to lateral pressure can be estimated properly. The material of the specimens was aluminum alloy plate (JIS 5052 H32) of thickness, t=2mm. Proof and ultimate strengths are obtained by tensile tests as 182MPa and 231MPa, respectively.

Centrally notched plate specimens of thickness t=2mm were tested under slow crosshead speed in order to determine the material resistance at the onset of crack propagation. The video record provides information about crack propagation, by which the crack opening displacement associated with the onset of crack propagation was determined. According to the video record the stable crack propagation commenced at the maximum load, and the corresponding crack tip opening displacement was estimated to be within the range between 1.7 and 2mm. The centrally notched tension specimens failed by plastic collapse after stable crack growth of 5 - 7 mm.

2.2 Tests of panel specimens under lateral pressure

The unstiffened and stiffened specimens, notched either by a single crack or by an array of cracks, were tested. The specimens were instrumented with the strain gages, crack gages and the displacement (deflection) transducer as shown in Fig. 1. The specimens are shown in Figs. 2 and 3, and a list of specimens is given in Table 1.

Table 1   Specimens

Specimen	Material	Cracks	Stiffener
PPR-1a	5052 H32	1	no
PPR-1b	5052 H32	1	no
PPR-3	5052 H32	3	no
SPPR-1	5052P H112	1	3
SPPR-3	5052P H112	3	3

During the pressure loading the specimens initially undergo large elastic deformations, which are followed by considerable plastic deformations. In Fig. 4 measured deflections are plotted versus applied pressure. For specimens PPR-1a (see Fig.5) and PPR-1b, the measured maximum pressures and the corresponding deflections are almost equal, which verifies the reproducibility of the present experiments.

For specimen SPPR-1 the deflection is initially small because of the intact stiffeners located at the both sides of the specimen. The crack initially propagates in a stable manner followed by a rapid propagation at the maximum pressure, which induces the slight depressurization leading to the arrest of crack propagation. The pressure increases again and the crack propagates further in a stable manner until it is arrested temporarily at the intact stiffener. Finally, pressure increases again, where the crack sharply turns its direction so that it propagates parallel to the intact stiffener (see Fig.6).

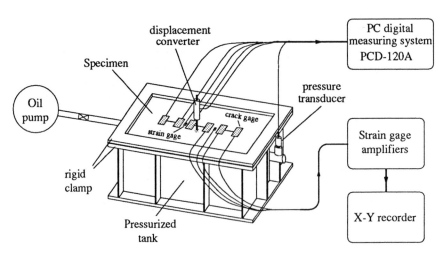

Figure 1.   Experimental setup.

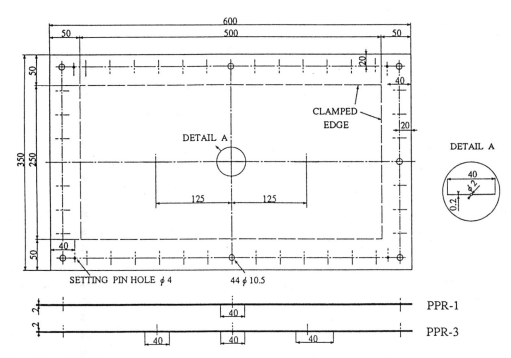

Figure 2. Unstiffened specimen.

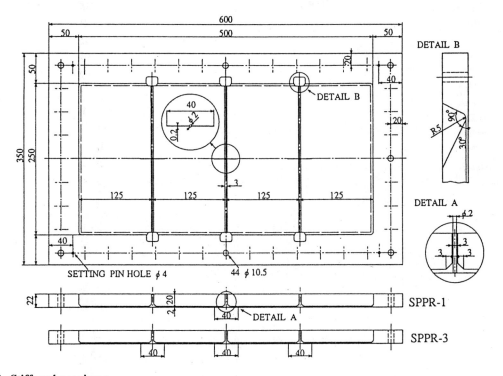

Figure 3. Stiffened specimen.

876

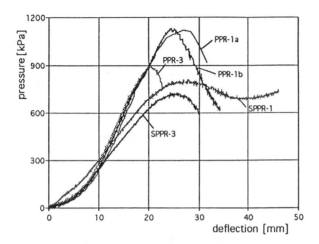

Figure 4. Experimental relation between pressure and deflection at the point, which locates at the center of the width of the specimen and 25mm away from the crack-line.

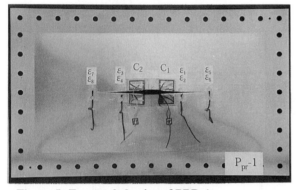

Figure 5. Fracture behavior of PPR-1a.

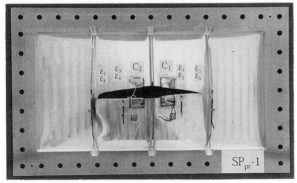

Figure 6. Fracture behavior of SPPR-1.

It was observed from video record of specimen PPR-1, that after stable fracture, rapid crack propagation commences, and it lasts until the driving force (oil pressure) is exhausted due to the depressurization. For specimen PPR-3, the cracks propagate in a stable manner in the beginning, and at the critical pressure, the cracks coalesce by instantaneous ductile fracture accompanied by fracture sound (see Fig.7). In broken ligaments one can distinct the stable and unstable fracture regions. Longer stable crack propagation is observed for the center crack in comparison with the side cracks because of its higher stress level.

Specimen SPPR-3 breaks initially in a stable fashion followed by an unstable ductile fracture, which leads to the coalescence of the all three cracks, basically similar to the fracture mechanism of specimen PPR-3 (see Fig.8). This specimen failed at the lowest pressure both due to the collinear multiple site cracks and the high stress concentration at the broken parts of the stiffeners.

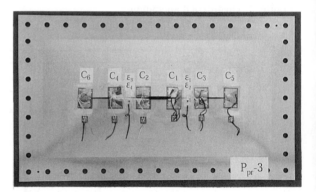

Figure 7. Fracture behavior of PPR-3.

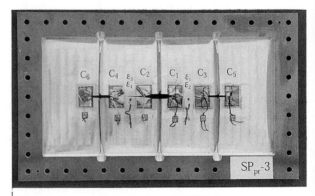

Figure 8. Fracture behavior of SPPR-3.

3. FINITE ELEMENT ANALYSES

Elasto-plastic, large deformation finite element analyses were carried out for all specimens. Isoparametric eight-node shell elements were used in modeling the panel specimens, and plane elements were used for the centrally notched tension specimen. The crack tip was meshed by triangular non-singular elements, which are degenerated from the quadrilateral elements having three nodes tied to one at the crack tip.

The radial length of crack tip elements was 1/20 of the half crack length, which is proper for the evaluation of J-integral based on the contour integral procedure. The plasticity behavior of the material was taken into account in the simulations by applying a multilinear isotropic hardening model, based on von Mises yield criterion coupled with an isotropic work hardening assumption. The stress-strain curve applied to FE simulation is modeled by using the data obtained from the tensile test.

In the case of panel specimens, load increments for lateral pressure were taken as 100kPa, and each load step has been divided into 200 substeps. Such a fine division was necessary because of the large-scale plastic deformation and large geometric non-linearity.

For centrally notched tension specimen, the prescribed nodal displacement were increased step-by-step, and the analysis was carried out until full scale plastic deformation was formed in the ligament. J-integral and crack tip opening displacement are examined as the fracture parameters. J-integrals were calculated by using ANSYS postprocessing routine [2].

3.1 Finite element analysis for centrally notched tension specimens

Calculated J-integral values with respect to applied tensile load and corresponding average axial stress in the net section are depicted in Fig. 9. At the maximum load, 216.27 MPa, the corresponding J-integral value equals to 360.6 MPa mm. After the maximum load the J-value still increases as strain energy increases with plastic deformation. At the maximum load the crack tip opening displacement equals to 1.84 mm.

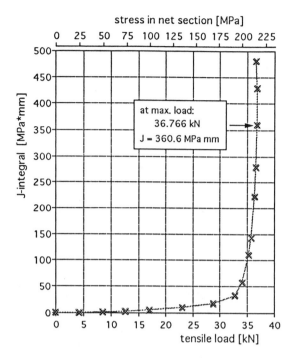

Figure 9. J-value of centrally notched specimen in tension.

3.2 Finite element analysis of stiffened and unstiffened panels under lateral pressure

Simulated deflections for pressurized specimens are illustrated in Fig. 10 with the experimental results. The simulated deflection curves agree fairly well in shape with experimental data, but the magnitudes of the simulated deflections are smaller than those measured in experiments. This could be explained by the possible pull-in condition along the clamped edges, which could have occurred in experiments. The edges of the specimen may slightly slip between the clamping frames due to the significant membrane force so that the membrane stiffness could be reduced, enabling larger deflections. Calculated deflection of stiffened panel specimen with a single crack, SPPR-1, is smaller than those of other specimens, due to the higher bending stiffness as observed in experiment.

Calculated J-values versus applied pressure for pressurized specimens are shown in Fig. 11. Because of higher stress concentration due to three broken stiffeners, the J-values of specimen SPPR-3 are much higher than those of specimen SPPR-1.

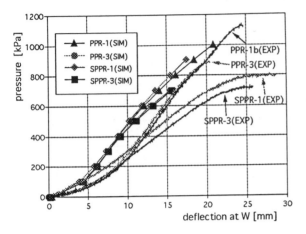

Figure 10. Calculated relation between pressure and deflection.

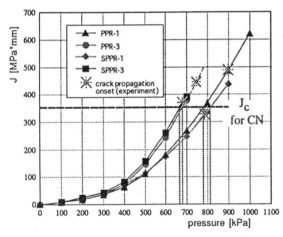

Figure 11. Relation between pressure and J-integral.

4. ESTIMATION OF CRITICAL PRESSURE

By combining numerical results of the J-integral and the load at the onset of crack propagation of the centrally notched specimen, one can estimate the critical values for J-integral as 360 MPa mm. It can be expected that for specimens with different geometry, having ligaments subjected primarily to tensile loading under large-scale yielding, the stable crack propagation commences at this critical points. In the case of panel specimens, stable crack propagation

started prior to maximum pressure. In Fig.11 a line for the critical value, J_c, is also plotted, and the intersection with the obtained J curves for each specimen gives the estimated pressure at the onset of stable crack growth. The estimated pressures are in fairly good agreement with the experimental results of the onsets of stable crack propagation.

5. CONCLUSIONS

Experimental and numerical investigations of single and multiple site cracks of stiffened and unstiffened panels exposed to lateral pressure have been investigated. In the case of stiffened panel under lateral pressure, intact stiffeners may act as crack arrester effectively, which may be contrasted to the fracture behavior of similar panel under tension[3].

Centrally notched tension plate specimens were used to determine the fundamental material resistance to crack propagation.

Finite element analyses including both large deformation and elasto-plastic material behavior were carried out. The critical J-integral and CTOD values, associated with the onset of stable crack propagation were determined for the centrally notched tension specimen. The critical pressures, at which stable crack propagation commences were estimated for the panel specimens, where the estimated critical pressures agree well with experimental results for stiffened panel specimens, while relatively lower critical pressures were calculated for unstiffened specimens.

REFERENCES

1. T. Yao, Y. Sumi, A. Murakami, K. Abe, and A. Kumano, Deterioration in Structural Strength of Aged Ships, Ship Structure Committee, Society of Naval Architects of Japan 45-104 (1995) (in Japanese).
2. Swanson Analysis System, Inc., ANSYS User's Manual, Revision 5.0 (1995).
3. Y. Sumi, Z. Bozic, H. Iyama, and Y. Kawamura, Journal of the Society of Naval Architects of Japan, 179 (1995) 407.

Practical Design of Ships and Mobile Units
M.W.C. Oosterveld and S.G. Tan, editors.

879

FATIGUE OF ALL STEEL SANDWICH PANELS - APPLICATIONS ON BULKHEADS AND DECKS OF A CRUISING SHIP

P. Kujala , K. Kotisalo and T. Kukkanen

Helsinki University of Technology, Ship Laboratory
PO Box 4100, 02015 TKK, Finland

Applications of all steel sandwich panels with corrugated cores as a longitudinal bulkhead and as a deck on a cruising ship are studied in the paper. The panels are produced by laser welding. Computer code developed to conduct spectral fatigue analysis for the ship girder is used to evaluate the fatigue life for the studied panels. The S-N curves needed in the analysis are obtained by fatigue testing of the most critical joints developed for the studied sandwich panels. Two panels are manufactured for the strength testing of longitudinal bulkhead panels enabling in testing of a total of 6 test pieces cut from these panels. The test series to study the fatigue characteristics of longitudinal joints on deck panels include in total 18 tests. The results of the strength tests are used to estimate the fatigue life of these panels when applied to a cruising ship built in the 80's.

1. INTRODUCTION

The demand for faster and lighter ships has increased the need for more efficient structures. Sandwich panels form one type of efficient structure enabling the application of steel, aluminium or composites in the construction. The present interest in steel sandwich structures has been awakened by the developments in laser welding technology enabling efficient production of these panels. The US Navy has studied the applications of laser welded corrugated core steel sandwich panels (LASCOR) since 1987 (Marsico et al., 1993). So far they have built five applications on Navy ships using stainless steel as material with plate thicknesses varying from 0.6 to 1.6 mm. These applications include bulkheads and decks in accommodation areas, deckhouses, deck edge elevator doors, and hangar bay division doors on conventional navy ships. The maximum weight savings on these applications have been reported to be 30 to 50 %. Meyer Werft has studied since 1993 applications of laser welded steel sandwich panels onboard cruising ships and also installed a number of test panels as bulkheads and staircase landings for passenger ships (Roland, 1996)

The studies related to the all steel sandwich panels at HUT/Ship Laboratory were initiated in 1987 (Tuhkuri, 1991, Tuhkuri, 1993). Tuhkuri studied the application of all steel sandwich panels as shell structures of an icebreaker. The application to the shell of an icebreaker was found to be problematic due to a high demand for the local strength of structures under ice loading.

As part of the Shipyard 2000 project in Finland, the applications of all steel sandwich panels as deck and bulkhead structures of a cruising ship were studied (Kujala et al., 1995). The studies included development of design methods, weight optimisation, ultimate strength testing under hydrostatic loading, fire and noise testing of the panels. The studies were continued as part of the Weld 2000 project, in which fatigue testing of the all steel sandwich panels used as longitudinal bulkheads were conducted (Kujala and Salminen, 1997).

The design methods for all steel sandwich panels were further developed in the Finnish national research project during the years 1996-1997. The project covered e.g. design and fatigue testing of longitudinal joints for sandwich panels planned to be used as deck structures on a cruising ship (Kotisalo, 1998). In addition local strength of sandwich panels was analysed (Kujala and Naar, 1998). In 1997, the first all steel sandwich test panels used as longitudinal and transverse bulkheads were installed on a cruising ship by Kvaerner Masa Yards, Helsinki New Shipyard (Nallikari, 1997).

In this paper, the fatigue testing of deck and bulkhead steel sandwich panels are described in more detail. Finally, the obtained S-N curves are used to evaluate the fatigue life for the tested panels when installed on a cruising ship navigating in the northern Atlantic and the Caribbean Sea.

2. SPECTRAL FATIGUE ANALYSIS PROCEDURE

2.1 General

An integrated program system has been developed for spectral fatigue analysis of ship structures (Kukkanen 1996). The procedure is based on linear theory and only low frequency loads are taken into account. The hull girder is assumed to behave as a rigid beam when evaluating the loads. The evaluation of fatigue life is based on standard well known approaches to determine long-term loading for ships and the application of Miner's rule. Special attention is given to the determination of the stress distribution on a cruising ship cross section with not fully effective superstructure and with large window openings (Fransman, 1989, Holopainen and Hakala, 1998). The theoretical background of the procedures is shortly described in the following.

2.2 Evaluation of long-term statistical distributions for ship response

Wave loads are calculated by a linear strip method and the added masses and damping coefficients are determined by Frank close-fit method. The linear theory is assumed to be sufficient in the fatigue analysis, because the most frequent cycles that occur during ship's life time are due to relatively small waves. Vertical and horizontal bending moments are determined together with the vertical shear force along the ship length.

When the wave-induced loads in regular waves and the responses of the structure are known, the next phase is to derive a combined response of the different load effects in irregular waves. Assuming linearity between excitation (waves) and response, the linear superposition principle can be used for deriving the responses in irregular waves. The calculation procedure is based on the linear spectral analysis in the frequency domain. The response spectrum in irregular waves is determined by using transfer functions of the response components per unit wave amplitude and modified Pierson-Moskowitz wave spectrum. The long-term distribution is obtained by summing up the different operational and environmental conditions during the ship's life-time.

2.3 Structural model of cross section

The cross-sectional properties for the studied ship are determined assuming the ship girder behaves as a thin-walled beam. For thin walled beams the shear stresses due to the bending are of equal importance to the normal stresses. In Finland a special program called SECPRO is developed to determine the cross sectional properties for calculation of shear and bending stress distributions on the ship cross-sections. Finite element approach is applied to solve the governing differential equations. Warping displacements in shear are first solved and thereafter the necessary cross-sectional properties and stress distributions are calculated. In Finland the finite element solution for the thin walled sections was first studied by Varsta (1974) based on the classical work of Love (1944) and the approach was further developed and computer coded by Hakala (1989). The finite element method provides a very suitable and versatile tool for determination of warping functions of thin-walled beams. There are no restrictions regarding the complexity of the cross section.

The recent development of the SECPRO program include extension of the code to be able to use orthotropic elements and to use multimaterial properties for the elements (Holopainen and Hakala, 1998). Also some analysis is conducted to study the shear lag effect on the cross section. The orthotropic material properties and shear lag effects are important when the stress distribution on the cross- section with large window openings is evaluated. The effect of windows is taken into account following the parametric studies conducted by Fransman (1989) to determine effective shear and bending stiffness of the structures with large window openings.

2.4 Prediction of fatigue damage

In this paper, a linear damage accumulation hypothesis is used and the fatigue strength characteristic of materials and structures are described by S-N curves (Hughes, 1988).

Linear damage accumulation hypothesis, sc. Miner's rule (Miner 1945) can be expressed as:

$$D = \sum_{i=1}^{M} \frac{n_i}{N_i} \qquad (1)$$

where n_i is the number of stress ranges that occur in a constant stress range S where N_i cycles cause a failure. The total number of the different stress blocks is M. Usually it is assumed that the damage occur on a structure when the fatigue damage D exceeds unity ($D > 1$).

The fatigue strength of the material and structure are described by the S-N curves

$$NS^m = C \qquad (2)$$

where N is the number of cycles to failure under a constant stress range S. m and C are fatigue strength exponent and fatigue strength coefficient, respectively.

Applying the Miner's rule together with an S-N curve, it is possible to derive an equation for the average value of the fatigue damage D expected after N_T cycles. The equation of the fatigue damage can be presented in the form

$$D = \frac{N_T}{C} E\left[S^m\right] \qquad (3)$$

where $E[S^m]$ is the expected value of S^m

$$E\left[S^m\right] = \int_0^\infty S^m f(S)dS \qquad (4)$$

and N_T is the total number of cycles in a ship's service time. $f(S)$ is a probability distribution of the stress ranges. The expected value of the stress range $E[S^m]$ is determined here by using a Rayleigh-distribution.

In the spectral analysis procedure the different operational and environmental conditions are taken into account by weighting the different conditions by their occurrence probabilities. The fatigue damage for a ship's service time is obtained by taking into account all conditions that the ship will encounter in her service time. All occurrence probabilities are assumed to be statistically independent. Once the fatigue damage D is calculated, the fatigue life T_f for the structure can be estimated from the relationship:

$$T_f = \frac{T_S}{D} \qquad (5)$$

where T_S is the design service time of the ship.

3. SHEAR FATIGUE TESTING OF LONGITUDINAL PANELS

3.1 Description of the tested structures and joints

The carbon steel used in this study is CR 280 produced by Rautaruukki, the measured yield strength of the steel is 300 MPa. Assuming a design pressure of 5 kPa for the deck above the longitudinal bulkhead, the vertical load carried by the bulkhead is related to the distance between the longitudinal bulkheads onboard the ship, e.g. with 5 m spacing between the bulkheads, the load carried by the bulkhead is 25 kN/m. The load increases when the number of decks above the bulkhead increases. The design against shear is done assuming 3 mm maximum relative deflection between two decks causing 110 MPa shear stresses on the longitudinal bulkhead.

Assuming the panel height is 50 mm, corrugation angle $\phi=60°$ and using the minimum plate thickness defined by steel manufacturers for thin steel plates of 0.5 mm, the load carrying capacity of the cross section under compressive loading can be shown to be about 100 kN/m (Kujala and Kotisalo, 1996). This means that the bulkhead with these minimum plate thicknesses can carry loads from 4 decks above it.

It is assumed that the bulkhead is constructed with standard size cassettes as shown in Figure 1. The maximum breadth of the panels is determined by the maximum breadth of the thin steel plates produced by the steel manufacturer. This breadth is 1250 mm which after corrugation gives a breadth of 675 mm. The edges of the panels are strengthened by 3 mm thick U-profiles so that the panels can be welded to surrounding structures and to each other.

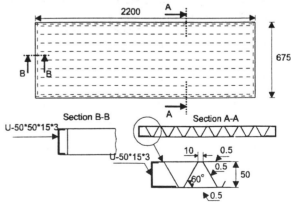

Figure 1. Configurations of the panel used as a longitudinal bulkhead.

The weight of the panels is 24 kg/m². At present these bulkheads are constructed typically from 4 mm corrugated plates, which have weight of 35 kg/m². This means that all steel sandwich panels can be 33 % lighter than the traditional structure.

3.2 Fatigue testing of the panels

The panels used in the test were manufactured by Laserplus Oy in Riihimäki (Kujala and Salminen, 1997). The tests aimed to simulate the fluctuating shear stresses on the bulkhead induced by relative displacement between decks while the ship is navigating in various wave conditions. The shear stresses are obtained through three point bending test arrangements, with the test piece length and height about equal so that the shear dominates on the test pieces, Figure 2 illustrates the test arrangements. The force is produced with a hydraulic cylinder and it is measured with a force gauge installed between test piece and hydraulic cylinder.

The test pieces are cut from the panels shown in Figure 1 so that the U-profiles at the edges are included in all test pieces. The length of the test specimen is 400 mm and height 185 mm. The U-profiles are included to simulate in the test the attachment between the deck and the all-steel sandwich longitudinal bulkhead. The U-profile is MIG welded to the end plate, which is then attached to the test rig. The force is introduced to the structure through the riveted and clued plates on the surfaces of the test pieces. The riveted and clued attachments are used to prevent the failure of the structure near the supports, where high stress concentrations can take place.

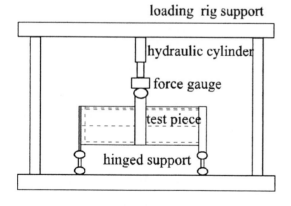

Figure 2. Layout of the test arrangements for the shear testing.

3.3 Obtained S-N curves

In altogether, 6 tests are carried out with the shear stress on the surface plates varying from 73.9 MPa to 99.4 MPa at maximum and with the minimum value of 2 MPa. Then the value of the stress ratio $R=\tau_{min}/\tau_{max}=0.03-0.02$.

The cracks initiated typically on the heat affected zone, develop first along the weld and then turn 45° due to the shear and propagate to the next weld. The test results are summarised in Figure 3. The shear stress is obtained assuming that the surface plates carry the total shear force:

$$\tau=F/(4t_f h) \tag{6}$$

where F is the force measured by the force gauge at midspan and h is height of the test panel and t_f is the thickness of the surface plate.

Based on the data given in Figure 3, the S-N curve parameters can be estimated and the following figures are obtained m=6.1 and C=2.31 10^{17} . The S-N curve is also shown in Figure 3.

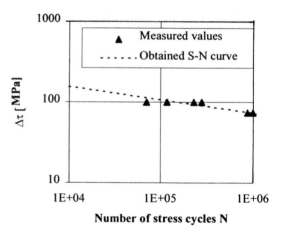

Figure 3. The measured shear fatigue strength values together with the obtained S-N curve.

4. FATIGUE TESTING OF THE LONGITUDINAL JOINTS

4.1 Description of the tested structures and joints

The deck panels for cruising ship panels are designed assuming a maximum longitudinal tensile stress of 170 MPa and a compressive stress of 70 MPa (Kotisalo, 1998). In addition a 5 kPa water pressure on deck is included as loading on the panel. The panel is assumed to be supported by webframes with a spacing of 2900 mm. Three types of material are used: Racold 320 cold formed steel, Polarit 725 and 757 stainless. The measured yield stress for the Racold is 405 MPa, for Polarit 725 the measured yield stress is 335 MPa and for Polarit 757 the measured yield stress is 312 MPa. The weight optimised height of the sandwich panel is 60 mm and plate thicknesses are 1 mm.

For deck structures the most critical joint is the longitudinal joint between the panels. Figure 4 illustrates the joint studied. C-profiles are attached at the end of the sandwich panel and these profiles are then welded by conventional welding onto the web frame. The plate thickness of the C-profile is 3 mm. This joint is critical because the attachment of the corrugated core to C-profile is practically difficult causing high stress concentrations on the surface plate when the longitudinal stresses from the core run over the joint. The weight of the panels is 33 kg/m^2, which is about 60 % of the weight for a typical deck structure with conventional stiffened plating.

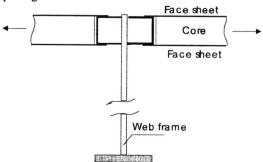

Figure 4. Illustration of the fatigue tested longitudinal joint.

4.2 Fatigue testing of the panels

The panels used in the tests are manufactured by Laserplus Oy in Riihimäki. The manufacturing is found to be problematic. The main reason for this is the inaccurate height of the "hand made" corrugated core. Due to the variation in the core height also the air gap between the core and surface plates vary causing problems in laser welding. This can cause bad welding quality at the end of the corrugated core weld on the most critical areas of the joint.

The tests aim to simulate the fluctuating longitudinal stresses on the deck panels. The length of the tests panels are 700 mm and breadth 230 mm. The studied joint is located at midspan. The force is produced with a hydraulic cylinder and it is measured with a force gauge installed between the test piece and hydraulic cylinder. The force is introduced to the structure through the riveted and clued plates on the surfaces of the test pieces. The panel breadth includes two longitudinal corrugations.

4.3 Obtained S-N curves

The tests include altogether 18 longitudinal fatigue tests for C-profile joints. This means that 6 tests are made with each material used. No big difference in fatigue behaviour is observed between the studied materials. Typically the fatigue cracks are observed to initiate at the end of the weld between the surface and core plates. This is the location where the core plate ends near the transverse weld of the C-profile and the surface plate.

Figure 5 summarises the obtained results. Based on the data given in Figure 5, the S-N curve parameters can be estimated and the following figures are obtained: m=3.07 and C=6.25 10^{10}. Figure 5 shows also the S-N curve for fatigue class 45, which is the class for a transverse butt weld on plate welded from one side only without backing bad.

The fatigue class of the studied joint is fairly low. The main reason for this is the problematic laser welding of the joints caused by the inaccuracies in the panel dimensions. In addition the lack of attachement of the corrugated core to the C-profile due to the manufacturing difficulties lowers the fatigue class of the joint.

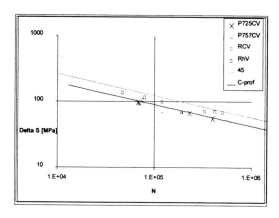

Figure 5. Obtained fatigue test results for the longitudinal joints.

5. APPLICATION OF THE TESTED PANELS ONBOARD PASSENGER CRUISE SHIP

5.1 Description of the ship

In this Chapter, fatigue strength of metal sandwich panels are studied by replacing normal bulkhead structures and decks by metal sandwich panels. A passenger cruise ship built in the 80's is used as an example ship. The main dimensions of the ship are presented below:

Length over all	L_{OA}	230.90 m
Waterline length	L_{WL}	196.50 m
Breadth	B	29.20 m
Draught	T	7.80 m
Height to 4th deck (main)	D_4	19.40 m
Height to 9th deck	D_9	32.95 m

5.2 Evaluation of long-term response

The approach described in section 2 is applied in the analysis, the calculations are described in more detail in the reference (Kujala et al., 1998). Occurrence probabilities of different sea states are described by Global Wave Statistics scatter diagrams (GWS 1986). Two different combinations of sea areas are considered in the investigations: the Caribbean Sea (70 %) and Atlantic (30 %), and the Caribbean Sea only (100 %).

In regular waves, calculations are carried out using wave frequencies from 0.2 to 2.0 rad/s in 0.05 rad/s intervals. Ship's heading angles between 0° to 180° in 22.5 degree intervals are used in calculations in regular waves. All heading angles are equally probable. In the fatigue analysis, a speed range 0 - 19 knots as a function of significant wave height is applied. The speed reduction correspond roughly to a speed loss in heavy seas due to added resistance in waves. Ship service time is assumed to be 20 years, where 15% is spent in a harbour.

Long-term prediction of responses are derived by weighting the cumulative distribution function of the short-term response peaks by the occurrence probabilities of different operational and environmental conditions during the assumed 20 year period. Cross sectional properties for the ship cross section are calculated by the SECPRO-

program. Orthotropic material properties are used in the structural model to take into account shear lag effects due to height of the superstructure and the large window openings.

S-N curves specified in section 3.3 for the longitudinal bulkhead and in section 4.3 for the longitudinal joint on the deck are used in the analysis.

5.3 Obtained fatigue life for the all steel sandwich as longitudinal bulkhead

Long-term prediction for vertical shear force along the ship length is presented in Figure 6. These predictions are given for two different sailing areas, where the ship is assumed to operate for 20 years. As can be seen from Figure 6, the maximum value for vertical shear force occurs at about $x/L \approx 0.3$. This location is used in this study when calculating the fatigue life of the steel sandwich panel. According to the DNV rules the nondimensional value for maximum wave shear force is 0.36, so the curves given in Figure 6 are somewhat above this value.

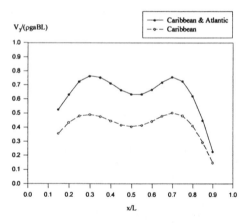

Figure 6. Long-term prediction for the vertical shear force along the ship length.

The metal sandwich panel is applied as a longitudinal bulkhead in three different decks. Effective thickness of the sandwich plate in the cross sectional model is $t = 2.26$ mm, when the plate thicknesses used in the sandwich panels are all 1 mm. Shear stress distributions per unit vertical shear force on the cross-section when the studied sandwich bulkhead is located between decks 4 and 5 is given in Figure 7.

Calculated fatigue damage and fatigue life are given in Table 1 for the assumed ship operating areas in the Caribbean Sea and Atlantic, and in the Atlantic. As can be seen from Table 1, the calculated fatigue life of the all steel sandwich panel is adequate in all locations used in this study.

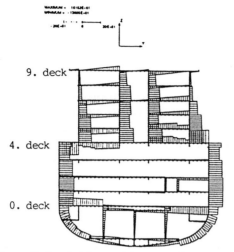

Figure 7. Sandwich panel between decks 4 and 5. Shear stress distribution for unit vertical force.

Table 1. Calculated fatigue damage and fatigue life for all steel sandwich bulkheads (Kujala et al., 1998).

	Caribbean Sea and Atlantic	Caribbean Sea
Longitudinal bulkhead	Fatigue life T_f [years]	Fatigue life T_f [years]
2. - 3. deck	$\sim 10^6$	$\sim 10^7$
4. - 5. deck	$\sim 10^4$	$\sim 10^5$
7. - 8. deck	86.3	589

Uncertainties exist in wave loads, and stress responses as well as in fatigue tests when determining S-N curve parameters. The S-N curve parameters are derived from a limited tests series using the mean values of the fatigue test results, that does not give any safety margin against fatigue. However the wave loads and statistical descriptions of operational conditions might be on the conservative side. Thus it can be concluded that at least 20 years service time for the all steel sandwich panel can be easily achieved.

5.4 Obtained fatigue life for the all steel sandwich as deck panel

Similarly to the section 5.3, the S-N curve given in Figure 5 is used to estimate the fatigue life of deck panels installed on the studied ship. In this case the spectrum fatigue analysis is applied by calculating the long term distribution of longitudinal stresses on the deck. Figure 8 shows the calculated long term distribution for the vertical bending moment and Figure 9 shows the bending stress on the cross-section with the bending moment value of 0.43 10^6 kNm (Holopainen and Hakala, 1998). When calculating the fatigue life, also the horizontal bending stresses are included, which are about 35 % of the vertical bending stress values.

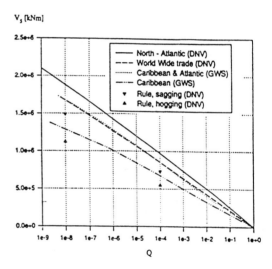

Figure 8. Long-term distribution for vertical bending moment. Also the long-term distributions according to DNV rules (1995) are given (Kujala et al., 1998).

Calculated fatigue damage and fatigue life are given in Table 2 for the assumed ship operating areas in the Caribbean Sea and Atlantic, and in the Atlantic (Kotisalo, 1998). As can be seen from Table 2, the calculated fatigue life of the all steel sandwich panel is adequate only on decks 4 and 5 when North-Atlantic and Caribbean Sea is used as the studied sea areas.

Kotisalo (1998) is studying also using finite element calculations the case when the corrugated core is attached to the C-profile by welding. He founds out that the stress level can then be about 15 % smaller than with tested joints. This means that

886

the deck panels can be applied as high as the 7. deck with a fatigue life longer than 20 years on the North Atlantic and Caribbean Sea as the studied sea areas. This indicates the existing possibilities for longitudinal joints after further development. The better accuracy of the corrugated core height can remarkably increase as such the fatigue life for the joints. Also, longitudinal joints can be further developed to increase the fatigue life.

ROYAL PRINCESS, MIDSHIP, MODEL 10, REDUCED WIHDOWS , 27.11.96 THO
NON-LINEAR STRESS DISTRIBUTION
MAXIMUM = .24112E+02
MINIMUM = -.20935E+02

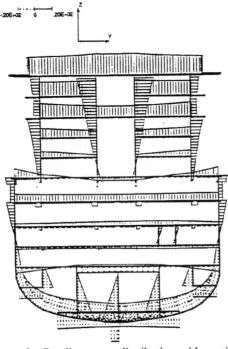

Figure 9. Bending stress distribution with vertical bending moment of $0.43 \cdot 10^6$ kNm (Holopainen and Hakala, 1998).

Table 2. Calculated fatigue damage and fatigue life for the all steel sandwich deck panels (Kotisalo, 1998).

	Caribbean Sea and Atlantic	Caribbean Sea
Deck	Fatigue life T_f [years]	Fatigue life T_f [years]
4. deck	85.7	127.1
5. deck	92.6	199.4
6. deck	17.0	33.6
8. deck	5.0	8.4

6. CONCLUSIONS

A spectral fatigue analysis of all steel, laser welded, sandwich panels applied as longitudinal bulkheads and deck panels onboard a cruising ship is presented. Spectral methods are applied to obtain the long term wave loads on varying sea conditions. Only wave induced hull girder primary stresses due to the vertical shear, horizontal and vertical bending moments have been taken into account. Miner's fatigue accumulation hypothesis has been used to obtain the fatigue life for the studied structures. The S-N curve data required in the analysis is obtained by conducting laboratory fatigue testing of the studied all steel sandwich panels.

An all-steel sandwich panel welded by laser is a new solution to obtain light and efficient ship structures. It has high potential for weight savings, but also require new approaches e.g. for detail design of the joints for attachment of these panels to each other and to surrounding structures. The fatigue tests summarised in this paper indicate that the panels can, at present, be used as longitudinal bulkheads on a cruising ship. Further development is, however, needed before these panels can be applied e.g. on the upper deck panels of a cruising ship. One of the main problems is accurate production of the corrugated core. Also the longitudinal joints between the panels need some further research and development work.

7. ACKNOWLEDGEMENTS

The work presented in this paper is based on the research conducted in a number of national research projects. The projects are financed by the Technology Development Center (TEKES) and the Finnish companies: Finnish Board of Navigation, KCI Konecranes International, Kvaerner Masa Yards, Outokumpu, Rautaruukki, VR Engineering. These financing supports are here gratefully acknowledged. Kvaerner Masa Yards is specially thanked for giving the detail design drawings of the cruising ship required in the analysis.

REFERENCES

Fransman, J. 1989. The influence of passenger ships superstructures on the response of the hull girder. Transaction of RINA 131. Pp 57-71.

GWS. 1986. Global wave statistics. Edited by Hogben N., Dacunha N. M. C. & Olliver G. F., British Maritime Technology Limited (BMT). Published by Unwin Brothers Limited. 661 p. ISBN 0 946653 38 0.

Hakala, M. K. 1989. Determination of ship cross-sectional properties using finite element method. PRADS'89, Varna, Bulgaria.

Holopainen, T and Hakala, M., 1998. Simplified methods for ship hierarchial structural design Paper presented at the Maritime Institute Symposium, 26.3.1998. Otaniemi, Finland.

Hughes, O., W., 1988. Ship structural design. A rationally-based, computer-aided optimisation approach. The Society of Naval Architects and Marine Engineers. Jersey City, New Jersey.

Kotisalo, K., 1998. Fatigue strength of the longitudinal joints of the corrugated core steel sandwich panels. Diploma thesis. Helsinki University of Technology. Faculty of Mechanical Engineering.(In Finnish).

Kujala, P., Metsä, A., Nallikari, M., 1995. All metal sandwich panels for ship applications. Shipyard 2000, Spin-off project. Helsinki University of Technology, Ship Laboratory, Report M-196. 65 p.

Kujala, P. and Kotisalo, K., 1996. Shear and bending fatigue testing of all steel sandwich structures. Weld 2000 project. Helsinki University of Technology, Ship Laboratory, Report M-210. 33 p.(In Finnish)

Kujala, P. and Salminen A., 1997. Fatigue strength testing of laser welded all steel sandwich panels for ships. 6th NOLAMP Conference. Luleå, Sweden. August 27-29.

Kujala, P. and Naar, H., 1998. Local strength analysis of all steel sandwich panels. Paper presented at the Maritime Institute Symposium, 26.3.1998. Otaniemi, Finland.

Kujala, P., Kukkanen, T. and Kotisalo, K., 1998. Fatigue of all metal sandwich panels. Application for cruise ship longitudinal bulkhead and decks. Helsinki University of Technology, Ship Laboratory, M series report (To be published).

Kukkanen, T. 1996. Spectral fatigue analysis for ship structures. Uncertainties in fatigue actions. Licentiate's Thesis, Helsinki University of Technology. Faculty of Mechanical Engineering. 101 p.

Love, A.E.H, 1944. A treatise of the mathematical theory of elasticity. 4th edition. New York, Dover Publications, 1944. 643 p.

Marsico, T.A. et al., 1993. Laser welding of lightweight structural steel panels. Proceedings of the Laser Materials Processing Conference. ICALEO'93. Orlando.

Miner, M. M. 1945. Cumulative damage in fatigue. Journal of Applied Mechanics, Vol. 12, No. 3, Trans. of the American Society of Mechanical Engineering (ASME). pp. A-159 - A-164.

Nallikari, M., 1997. Light weight structures of cruise ship. Weld 2000, Final Conference, 1.-2.10.1997. Nokia. (In Finnish).

Roland, F., 1996. Trends, problems and experience with laser welding in shipbuilding. IIW Shipbuilding Seminar, Odense, April 17-19.

Tuhkuri, J., 1991. Sandwich structures for ships under ice loading. Licentiate thesis. Helsinki University of Technology, Faculty of Mechanical Engineering.

Tuhkuri, J., 1993. Laboratory tests of ship structures under ice loading. Vols 1-3. Helsinki University of Technology, Ship Laboratory, Report M-166.

Varsta, P., 1974. On the application of the finite element method to torsion problems in ships. Helsinki University of Technology, Licentiate thesis, Espoo. 106 p. (In Finnish).

Practical Design of Ships and Mobile Units
M.W.C. Oosterveld and S.G. Tan, editors.

Enhanced Structural Connection between Longitudinal Stiffener and Transverse Web Frame

S.N. Kim[1], D.D. Lee[1], W.S. Kim[1], D.H. Kim[1], O.H. Kim[2], M.H. Hyun[2], U.N. Kim[2],
F.L.M. Violette[3], H.W. Chung[3]

[1]. Hull Initial Design Dep't, Ship Building Division, Hyundai Heavy Industries Co., Ltd.
   1 Cheonha-dong, Dong-ku, Ulsan, 682-792, KOREA
[2]. Hyundai Maritime Research Institute, Hyundai Heavy Industries Co., Ltd.
   1 Cheonha-dong, Dong-ku, Ulsan, 682-792, KOREA
[3]. TPDD, Marine Division, Lloyd's Register of Shipping
   100 Leadenhall street, London, EC3A 3BP, ENGLAND

To verify the structural soundness and safety of a new stiffening arrangement with circular type slot in a Double Hull VLCC structure, various comparative analyses between a conventional design and a new design have been carried out, such as the comparison of 1) the stress distribution around the connection, 2) the buckling strength of the transverse web frame, and 3) the fatigue strength through a full spectral analysis. In addition, comparative fatigue tests have been carried out with half scale structures.

The results of both analyses and experiments show improved performance of the newly developed structural design. By adopting the newly proposed arrangement of web stiffeners with the circular type slot, a decrease of the potential possibility of fatigue damage is expected, as well as easier fabrication work with reduced trouble-some quality control of boxing fillet welds at connections between longitudinal stiffeners and transverse web frame.

1. INTRODUCTION

The connections between longitudinal stiffeners and transverse web frames in ship structure are complicated and liable to fatigue damage. In particular, the cracks are the most likely to initiate at the connection between the web stiffener and the face plate of the longitudinal stiffener, due to higher stress concentration, as shown in Fig.1.1.

To reduce the possibility of the fatigue damage at the connections, we propose a new arrangement of the web stiffener, where the longitudinal stiffeners are supported by the transverse web frame directly eliminating the connections between the longitudinal stiffeners and the web stiffeners. Due to the removal of the web stiffener at the longitudinal stiffener connection, it is expected that the stresses around the slot of the web frame will increase. To smooth the stress distribution and to avoid the high stress concentrations around the slot, the circular type slot without a collar plate has been adopted as shown in Fig.1.2.

Various comparative analyses and fatigue tests have been carried out to verify the structural soundness and safety of the newly proposed stiffening arrangement.

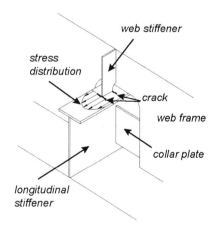

Fig.1.1 Connection between longitudinal stiffener and web frame

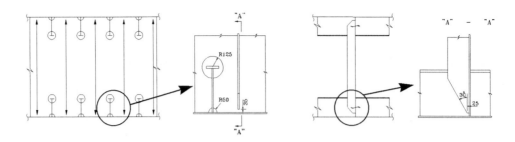

Fig.1.2 Newly proposed stiffening arrangement and slot

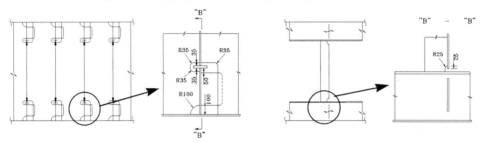

Fig.2.1 Conventional stiffening arrangement and slot

2. STRESS DISTRIBUTIONS AROUND CONNECTION

Using the finite element method, the maximum stress at the free edge around the slot has been compared between the conventional structure shown in Fig.2.1 and the newly stiffened structure with the circular slot shown in Fig.1.2.

The double bottom structure at mid tank region in way of center line of Double Hull (D/H) VLCC, where the static load by pressure is the most dominant, has been chosen to carry out the comparative stress evaluation. Fig.2.2 shows the 3-D global model of D/H VLCC and Fig.2.3 shows the detailed finite element model of the longitudinal stiffener connection for both the conventional design and the new design. Full loaded condition with the dynamic load induced by wave has been considered as shown in Fig.2.4[1].

Because the shape of connection is very different between the conventional structure and the new one, the comparison of the resulting stress value has been separated to two(2) locations, that is, the one is the intersection of the longitudinal stiffener and the web frame or the web stiffener and the other is around the slot.

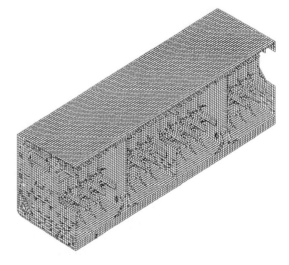

Length B.P.	: 320 m
Length Scant.	: 317.82 m
Breadth Mld.	: 58 m
Depth Mld.	: 31 m
Draught Mld.(design)	: 20.95 m
Draught Mld.(scant)	: 22.7 m

Fig.2.2 Finite element model of 3-D global analysis of D/H VLCC

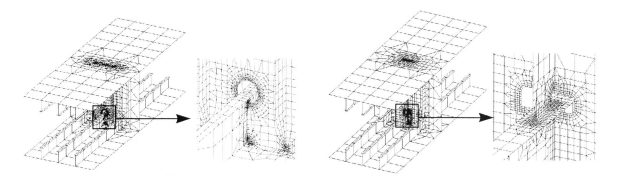

Fig.2.3 Detailed finite element model of longitudinal stiffener connection

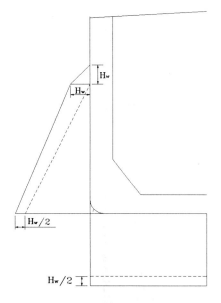

where, $H_w = 4.6 \times 10^{-2} \times L \times e^{-0.0044L}$, $L = Ship\ Length$

Fig.2.4 Hydrodynamic wave profile for comparison of stress distributions

Location	Stiffening Method		
	New Stiffening	Conventional Stiffening	
	Circular Type	Stiffener End	Slot Edge
(diagram)	257.6 370.4	734.0	295.2
(diagram)	217.4 348.6	696.6	268.3

Fig.2.5 Stress distribution and maximum stress

Fig.2.5 shows the stress distribution and the value of the maximum stress at each location. In the case of the intersection of the inner bottom longitudinal stiffener, where the stress level is higher than in the case of bottom longitudinal stiffener, the value of the maximum stress is 734 MPa for the conventional structure whilst it is 370 MPa for the new structure. The stress level along the slot is also lower for the new structure than the one for the conventional structure by about 15%.

3. BUCKLING STRENGTH ANALYSIS

Eigenvalue analyses using the finite element method have been carried out for the two different types of double bottom structure, the conventional one and the new one, to compare the buckling strength of the web frame.

Fig.3.1 shows the characteristic dimensions of the analyzed model, which is one floor space with four longitudinal stiffeners. Each of the three basic load types shown in Fig.3.2 have been applied individually and in combination (load 1 + load 2, load 1 + load 3, and load 2 + load 3). Fig.3.3 shows the FE models for buckling strength analysis and the models are composed of four node shell elements. Fig.3.4 and Fig.3.5 show the buckling mode shapes under each basic loading condition.

The interaction relationships of the elastic buckling strength under combined loadings are

892

shown in Fig.3.6 ~ 3.8. The non-dimensional terms of σ_m / σ_0 and τ_m / τ_0 normalized by design yield stress of ship structural mild steel ($\sigma_0 = 235$MPa and $\tau_0 = \sigma_0 / \sqrt{3}$) are introduced to compare the buckling strength, where σ_m and τ_m are the applied mean stresses under axial compression and shear force, respectively. For the double bottom structure subjected to bi-axial compression (load 1+load 2), the elastic buckling strength of the new structure is slightly higher than that of the conventional structure (see Fig.3.6).

It can be seen from Fig.3.8 and Fig.3.9 that the new structure is more effective than the conventional structure in buckling when axial compression is dominant, but the conventional structure shows slightly better performance than the new structure when shear is dominant.

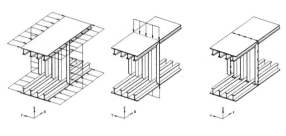

Load type1   Load type 2   Load type 3

Fig.3.2 Basic loading conditions of double bottom structure

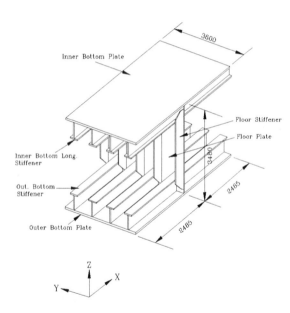

Floor Spacing   : 4,970
Long. Spacing   : 900
Inner/Outer Bottom Long. Stiffener
                      : 500×12+150×30 (T type)
Floor Stiffener   : 250×14
Floor & Collar Plate Thickness   : 14
Inner/Outer Bottom Plate Thickness : 20

Fig.3.1 Characteristic dimensions of the conventional and the new structural analysis model (unit:mm)

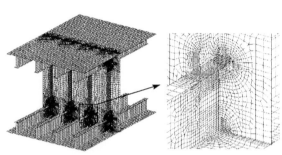

Conventional Structure

New Structure

Fig.3.3 FE model for buckling strength analysis

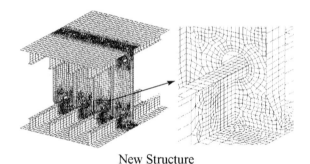

Load type 1   Load type 2   Load type 3

Fig.3.4 Buckling mode shapes under each basic loading condition for conventional structure

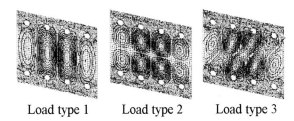

Load type 1   Load type 2   Load type 3

Fig.3.5 Buckling mode shapes under each basic loading condition for new structure

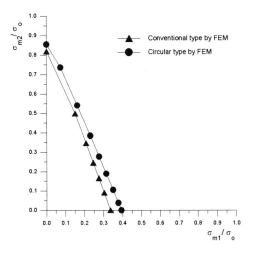

Fig.3.6 Interaction curve of elastic buckling strength under bi-axial compression (Load 1 + Load 2)

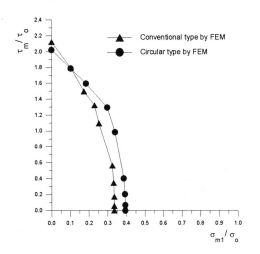

Fig.3.7 Interaction curve of elastic buckling strength under axial compression in Y-direction and shear force (Load 1 + Load 3)

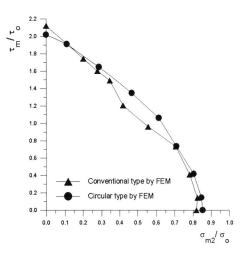

Fig.3.8 Interaction curve of elastic buckling strength under axial compression in Z-direction and shear force (Load 2 + Load 3)

4. COMPARATIVE FATIGUE STRENGTH ANALYSIS

Lloyd's Register's ShipRight Fatigue Design Assessment (FDA) Level 3 direct calculation procedure has been applied to compare the fatigue performance of the conventional design and the new design of the side longitudinal structure in the mid tank region in way of the loadwater line for the same D/H VLCC. The computational steps of the FDA Level 3 procedure are summarized in the flow chart shown in Fig.4.1.

The critical locations have been identified based on the ShipRight Structural Detail Design Guide [2] and the available service experience was used to select the finite elements to be considered as fatigue stress check points. These finite elements are adjacent to structural intersections, and along free edge around the slots and cutouts. The fatigue performance has been computed according to stress check point areas arranged along the structural intersections and slots as shown in Fig.4.2

The wave induced load and motion responses to unit amplitude, regular sinusoidal waves have been computed using the linear 2-D strip theory software LR2570, and the hydrodynamic pressure software LR Press [3]. The voyage simulation module of the ShipRight FDA Level 2 [4] software has been used

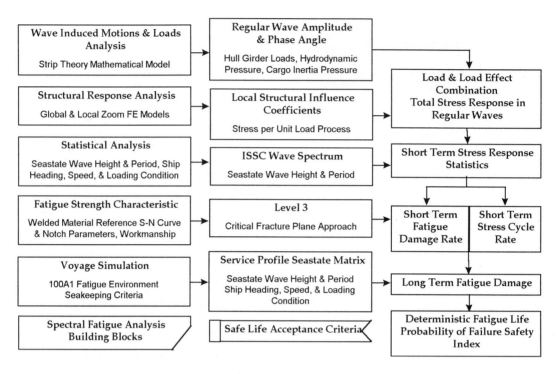

Fig.4.1 Flow Chart of Ship Right FDA Direct Calculation Procedure

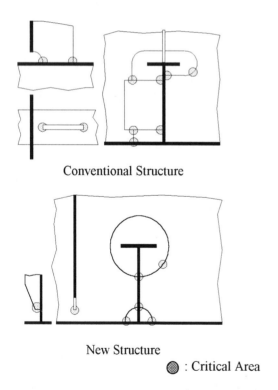

Conventional Structure

New Structure

⊚ : Critical Area

Fig.4.2 Critical areas for fatigue stress check points

to obtain the service profile matrix for the long term computation of the fatigue damage based upon the BMT Global Wave Statistics(GWS), and the 100A1 Fatigue Wave Environment for large crude oil tankers.

For a stress check point in way of a fillet weld, the hot spot welded material reference S-N Curve with a stress range of 75.6MPa at 10^7 stress cycles and a slope parameter of 3 has been used. For a stress check point in way of a free edge, the hot spot reference S-N curve with a stress range of 102.2MPa at 10^7 stress cycles and a slope parameter of 3.5 has been used. The Haibach correction has been applied to both S-N Curves at 10^7 stress cycles. The critical fracture plane approach, which allows the computation of the fatigue damage along several fracture planes in complex stress field situations, has been used to predict the fatigue damage at the stress check points.

It is observed from the comparative analysis of fatigue performance that the new structure performs significantly better than the conventional structure. The fatigue damage results for side longitudinals No. 50, 51 and 52, which are located at 2.5 m, 1.6 m and 0.7 m below the scantling draft respectively,

are summarized in Table 4.1, together with improve
-ment factors for the new design.

Table 4.1 Comparative summary of fatigue damage

Max Fd 97.5%	Conven -tional	New -weld	New -edge	Improvement factor
Long. 50	1.077	0.368	0.189	2.926
Long .51	2.210	0.347	0.378	5.847
Long. 52	1.758	0.156	0.380	4.626

5. COMPARATIVE FATIGUE TEST

Six models have been used in the fatigue test :
three are of the conventional type structure and the
remaining three models are of the new type one.
The model scale is about half of the actual bottom
structure of the D/H VLCC and the material is ship
structural mild steel. Fig.5.1 shows the details of
model. Models are loaded by two 25ton MTS
hydraulic actuators. One actuator is set on the face
plate of the middle longitudinal stiffener to apply
concentrated load in Z direction and the other is set
to fix the model in Z direction as shown in Fig.5.1.
The applied load wave form is sinusoidal with 7 ~
8Hz.

The inspection of initiation and propagation of
fatigue crack was carried out by the dye penetration
check method at every $3 \times 10^4 \sim 5 \times 10^4$ cycle and
strain monitoring. Table 5.1 shows the results of
fatigue strength tests. The fatigue strength is
defined as the number of stress cycles to crack
initiation and the crack initiation stage is defined
here as a visible crack with a length of about 15mm.

Table 5.1 Results of fatigue test

Model	Load Range (KN)	Load Ratio	Crack Initiation ($\times 10^4$cycle)
Conventional 1	140.2	0.1	10.0
Conventional 2	124.5	-1.0	15.0
Conventional 3	200.0	-1.0	3.0
New 1	139.4	0.1	25.0
New 2	50.7	-1.0	No crack
New 3	100.0	-1.0	225.0

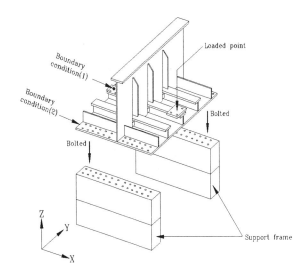

Model size (L×B×D)   : 200×2000×920 mm
Bottom plate thickness   : 10 mm
Bottom longitudinal   : 300×10 + 75×15 mm
Bottom longitudinal space : 400 mm
Floor plate thickness   : 10 mm
Floor stiffener   : 150×12 mm
Radius of scallop   : 50 mm
Welding leg length   : 6 mm

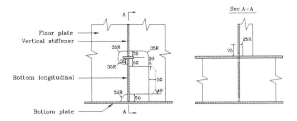

Conventional Structure

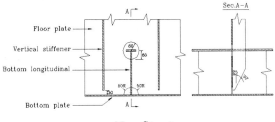

New Structure

Fig5.1 Details of model and boundary condition

The positions of crack initiation are shown in Fig.5.2. The first crack in the conventional type initiated at the weld toe of the face plate of longitudinal stiffener connected to the web stiffener. On the contrary, in the case of the new type, the first crack initiated at the weld toe of the slot intersected with the web of longitudinal stiffener. Fig.5.3 shows ΔP-Nc diagram which is the relation between the range of applied load range (ΔP) and the number of load cycles to crack initiation (Nc). It is shown in Fig.5.3 that the crack initiation life of the new type is far longer than that of the conventional one (more than five times at 100KN of ΔP)

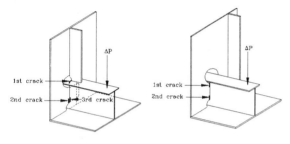

Conventional Structure   New Structure

Fig.5.2 Positions of crack initiation.

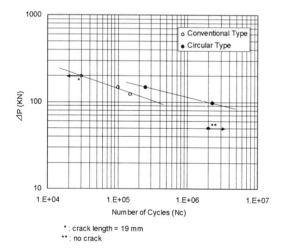

* : crack length = 19 mm
** : no crack

Fig.5.3 ΔP - Nc Diagram

6. CONCLUDING REMARK

To reduce the possibility of fatigue damage at the longitudinal stiffener / transverse web connection, a new arrangement of web stiffeners with the circular type slot has been proposed in this paper. To verify the structural soundness and safety of the newly developed design, various comparative analyses of 1) the stress distribution around the slot, 2) the buckling strength of transverse web frame, and 3) the fatigue strength assessment by a full spectral analysis have been carried out with a Double Hull VLCC structure. In addition, comparative fatigue tests have been carried out with structures at half scale.

The results of both analyses and experiments show improved performance for the newly developed structural design. By adopting the newly proposed arrangement of web stiffeners with the circular type slot, a decrease of the potential possibility of crack damage is expected, as well as easier fabrication work with reduced troublesome quality control of boxing fillet welds at connections between longitudinal stiffeners and the transverse web frame.

REFERENCES

1. Lloyd's Register, ShipRight Structural Design Assessment Procedure, Direct calculation- guidance notes,1996

2. Lloyd's Register, ShipRight Structural Detail Design Guide, Procedures Manual, 1996

3. Lloyd's Register, User Manual or LR2570 Program Suite - Calculation of Ship Responses in Regular Waves by Strip Theory, ASRD Report 97/21/RD

4. Lloyd's Register, ShipRight FDA Level 2 - Fatigue Damage Assessment - Software User Manual, 1996

Practical Design of Ships and Mobile Units
M.W.C. Oosterveld and S.G. Tan, editors.

STUDY ON FATIGUE DAMAGE ACCUMULATION PROCESS BY USING CRYSTALLINE FEM ANALYSIS

N. Osawa[a], Y. Tomita[a] and K. Hashimoto[a]

[a] Department of Naval Architecture and Ocean Engineering, Osaka University,
2-1, Yamadaoka, Suita, Osaka 565-0871 Japan

A crystalline FEM code which can analyze cyclic plastic deformation behavior during fatigue process is developed by proposing strain hardening rule which takes account of the Bauschinger effects. Using this code, the deformation behavior of f.c.c. single crystal under cyclic loading conditions is investigated. The code has been shown to successfully simulate the microscopic process of crack initiation, that is, the localization of plastic deformation and the generation of irreversible slip in the course of loading history.

1. INTRODUCTION

The microscopic view of the damage accumulation process (initiation of micro-crack) during cyclic loading can be summarized as follows: i) appearance of local plastic strain, ii) localization of these plastic strains and then generation of irreversible slip, iii) formation of persistent slip band (PSBs), iv) accumulation of irreversible slip within PSBs, v) formation of intrusion within PSBs, vi) growth of intrusion and then formation of a micro-crack.

Therefore, in order to explain the process of fatigue damage accumulation, it is necessary to examine the microscopic mechanisms of plastic strain localization and irreversible slip generation.

Crystalline FEM analysis is applicable to such investigation. Crystalline FEM theory first was established by Peirce et al.[1]. The authors [2-3] expanded this theory in order to analyze cyclic plastic deformation behavior during fatigue process, and developed a FEM code which can take into account the effect of lattice rotation and rate-dependency for cyclic deformation analysis. This code has been shown to successfully simulate localization of plastic deformation and the generation of irreversible slip in the course of loading history [3].

In this paper, using the crystalline FEM code, we calculate the non-uniform deformations of f.c.c. single crystal under cyclic loading conditions, and investigate the microscopic mechanisms of plastic strain localization and irreversible slip generation which are the process of fatigue damage accumulation.

2. BASIC EQUATIONS

2.1 Constitutive Equation

The constitutive equation used in our analysis is based on the rate-dependent crystalline plasticity theory developed by Peirce et al.[1]. It gives the relation between the Kirchhoff stress rate $\dot{\boldsymbol{\tau}}$, the shear strain rate on the slip system (a), $\dot{\gamma}^{(a)}$ and the total rates of stretching, $\mathbf{D}$ as

$$\dot{\boldsymbol{\tau}} = \mathbf{L} : \mathbf{D} - \mathbf{D}\boldsymbol{\tau} - \mathbf{D}\boldsymbol{\tau} - \sum_{(a)} \dot{\gamma}^{(a)} \mathbf{R}^{(a)} \tag{1}$$

$$; \mathbf{R}^{(a)} = \mathbf{L} : \mathbf{P}^{(a)} + \boldsymbol{\beta}^{(a)}$$

where

$$\mathbf{P}^{(a)} = \frac{1}{2}\left(\mathbf{s}^{*(a)} \otimes \mathbf{m}^{*(a)} + \mathbf{m}^{*(a)} \otimes \mathbf{s}^{*(a)}\right)$$

$$\mathbf{W}^{(a)} = \frac{1}{2}\left(\mathbf{s}^{*(a)} \otimes \mathbf{m}^{*(a)} - \mathbf{m}^{*(a)} \otimes \mathbf{s}^{*(a)}\right) \tag{2}$$

$$; \mathbf{s}^{*(a)} = \mathbf{F}^{*}\mathbf{s}^{(a)}, \quad \mathbf{m}^{*(a)} = \mathbf{F}^{*-T}\mathbf{m}^{(a)}$$

and

$$\boldsymbol{\beta}^{(a)} = \mathbf{W}^{(a)}\boldsymbol{\tau} - \boldsymbol{\tau}\mathbf{W}^{(a)} . \tag{3}$$

Here, $\mathbf{F}^{*}$ is the elastic deformation gradient which represents the elastic stretching and rigid rotation; $\mathbf{s}^{(a)}$ and $\mathbf{m}^{(a)}$ are the unit vector in the slip direction and the unit normal vector of the slip plane of slip system (a) in the undeformed configuration, and $\mathbf{L}$ is the elastic moduli tensor. From the constitutive equation, the

898

rate of the resolved shear stress (R.S.S.) of slip system (a), $\dot{\tau}^{(a)}$ is given by $\mathbf{D}$ and $\dot{\gamma}^{(a)}$ as

$$\dot{\tau}^{(a)} = \mathbf{R}^{(a)} : \left(\mathbf{D} - \sum_{(a)} \dot{\gamma}^{(a)} \mathbf{P}^{(a)} \right). \qquad (4)$$

2.2 Strain Hardening Rule

Peirce et al.[1] used the power law form expression of the shear rate $\dot{\gamma}^{(a)}$ as

$$\dot{\gamma}^{(a)} = \dot{a}^{(a)} \, \mathrm{sgn}\left(\frac{\tau^{(a)}}{g^{(a)}} \right) \left| \frac{\tau^{(a)}}{g^{(a)}} \right|^{\frac{1}{m}}; \quad \dot{g}^{(a)} = \sum_{(b)} h_{ab} \left| \dot{\gamma}^{(b)} \right|. \qquad (5)$$

This expression is not be applicable to the analysis of cyclic deformation behavior because it does not describe the Bauschinger effects. It is a well-known experimental result that dislocation density remains constant after cyclic hardening is saturated. This leads to the following hypotheses; multiplication does not occur during dislocation movements (a), and/or dislocations multiplied during slips disappear when the loading cycle is reversed (b).

Hypothesis Part (a) forces us to introduce back stress of slip systems (a), $\alpha^{(a)}$, in order to consider the Bauschinger effects. This is because strain hardening is mainly caused by the long range stress field which accelerates the dislocation movement in the opposite direction[2][3]. In this case, the strain hardening rule is described as

$$\dot{\gamma}^{(a)} = \dot{a}^{(a)} \, \mathrm{sgn}\left(\frac{\tau^{(a)} - \alpha^{(a)}}{g^{(a)}} \right) \left| \frac{\tau^{(a)} - \alpha^{(a)}}{g^{(a)}} \right|^{\frac{1}{m}}; \alpha^{(a)} = \sum_{(b)} h_{ab} \left| \dot{\gamma}^{(b)} \right|. \qquad (6)$$

We use the above rule in this paper.

The form of the latent hardening moduli h_{ab} is the same as the one used by Peirce et al.[1]

$$h_{ab} = qh + (1-q)h\delta_{ab} \qquad (7)$$

Here, the parameter q sets the level of latent hardening.

3. NUMERICAL METHOD

3.1 Finite Element Method

Boundary value problems in this theory can be solved using the finite element method. In the same manner as Pierce et al.[1], analysis is based on the Lagrangian formulation with the initial unstressed state taken as reference, and the convected coordinate formulation reviewed in Needleman [4] is adopted. Hereafter, the contravariant and covariant components of tensors or vectors on the deformed convected coordinates is abbreviated to 'cont. comp.'

and 'cov. comp.'. Similarly, the components on the reference Cartesian coordinates is abbreviated to 'ref. comp.'.

3.2 Time Integration Algorithm

When the residual force is not dispelled at any loading step, numerical solutions may lose the accuracy due to errors accumulated at each loading step. In order to calculate the deformation behavior during fatigue process over a great many loading steps, an iterative time integration algorithm is developed by the authors, by which the residual force is dispelled at every step.

The slip increment on slip system (a), $\Delta\gamma^{(a)}$, for the time increment Δt is calculated by a linear interpolation of shear rates at time t and $t+\Delta t$

$$\Delta\gamma^{(a)} / \Delta t = (1-\theta)\dot{\gamma}^{(a)}\big|_t + \theta\dot{\gamma}^{(a)}\big|_{t+\Delta t} \quad (0 \le \theta \le 1). \qquad (9)$$

$\dot{\gamma}^{(a)}\big|_{t+\Delta t}$ is approximated by the Taylor expansion of Eq. (6) to the first order in incremental quantities as

$$\dot{\gamma}\big|_{t+\Delta t} = \dot{\gamma}\big|_t + \frac{\partial\dot{\gamma}^{(a)}}{\partial\tau^{(a)}}\bigg|_t \Delta\tau^{(a)} + \frac{\partial\dot{\gamma}^{(a)}}{\partial\alpha^{(a)}}\bigg|_t \Delta\alpha^{(a)} \qquad (10)$$

The cov. comp. of $\mathbf{D}$; $\overline{D}_{ij}$ can be represented by the nodal displacement rate $\dot{U}^J$ as

$$\overline{D}_{ij} = \sum_J \overline{V}_{ij} U_J \qquad (11)$$

Eq. (6) and Eqs. (9),(4),(10),(11) lead to the relation between the slip increments and the nodal displacement increments

$$\Delta\gamma^{(a)} = \dot{f}^{(a)} + \overline{F}^{(a)ij} \sum_J \overline{V}_{ij}^J\big|_t \Delta U_J$$
$$; \dot{f}^{(a)} = \sum_{(b)} M_{ab} \dot{\gamma}^{(b)}\big|_t, \quad \overline{F}^{(a)ij} = \sum_{(b)} M_{ab} \overline{Q}^{(b)ij} \qquad (12)$$

where

$$\overline{Q}^{(a)ij} = \frac{\theta\dot{\gamma}^{(a)}\big|_t \Delta t}{m\left(\tau^{(a)}\big|_t - \alpha^{(a)}\big|_t\right)} \overline{R}^{(a)ij}\big|_t, \left[M_{ab}\right] = \left[N_{ab}\right]^{-1} \qquad (13)$$

$\overline{R}^{(a)ij}$ is the cont. comp. of $\mathbf{R}^{(a)}$ in Eq. (1). N_{ab} is given as

$$N_{ab} = \delta_{ab} + \frac{\theta\Delta t\dot{\gamma}^{(a)}\big|_t}{m}\left[\frac{1}{\tau^{(a)}\big|_t - \alpha^{(a)}\big|_t}\left\{\mathbf{R}^{(a)}\big|_t : \mathbf{P}^{(a)}\big|_t + \right. \right.$$
$$\left. \left. \mathrm{sgn}\left(\tau^{(a)}\big|_t - \alpha^{(a)}\big|_t\right)h_{ab}\,\mathrm{sgn}\left(\dot{\gamma}^{(b)}\big|_t\right)\right\}\right] \qquad (14)$$

Integrating the constitutive equation (1) with respect to time and substituting Eq. (12) for this integrated equation, a set of equations for the cont. comp. of Kirchhoff stress, $\overline{\tau}^{ij}$ and the nodal displacement U_J at t and $t+\Delta t$ is derived

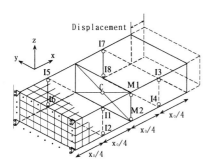

Figure 1. Finite element model

Table 1
Material properties

Young's modulus	58800(MPa)
Poisson's Ratio	0.3
$\dot{a}^{(a)}$ of Eq. (6)	0.002
m of Eq. (6)	0.005
h of Eq. (6)	30.40(MPa)
$g^{(a)}$ of Eq. (6)	58.8(MPa)
θ of Eq. (9)	1.0

$$\varphi^{ij} = \left(\overline{\tau}^{ij}\big|_{t+\Delta t} - \overline{\tau}^{ij}\big|_t\right) - \sum_J \overline{G}_J^{ij}\left(U_J\big|_{t+\Delta t} - U_J\big|_t\right)$$

$$+ \sum_{(a)} \dot{f}^{(a)} \overline{R}^{(a)ij}\big|_t \Delta t = 0$$

$$; \overline{G}_J^{ij} = \left(\overline{L}^{ijkl}\big|_t - \overline{g}^{ik}\big|_t \overline{\tau}^{lj}\big|_t - \overline{g}^{jl}\big|_t \overline{\tau}^{ik}\big|_t - \sum_{(a)} \overline{R}^{(a)ij}\big|_t \overline{F}^{(a)ij}\right) \quad (15)$$

$$\cdot \frac{1}{2}\left(F_{mk}\big|_t B_{ml}^J + F_{ml}\big|_t B_{mk}^J\right) , \ u_{i,j} = \sum_J B_{ij} U_J$$

where $\overline{L}^{ijkl}$ and $\overline{g}^{ij}$ are the cont. comp. of the elastic moduli and the metric tensor, and F_{ij} and u_i are the ref. comp. of the deformation gradient and the displacement vector.

The principle of virtual work at $t + \Delta t$ leads to

$$\Psi_J = \int_V \overline{\tau}^{ik}\big|_{t+\Delta t}\left(\delta_{jk} + \sum_L B_{jk}^L U_L\big|_{t+\Delta t}\right) B_{ji}^J dV - F_J\big|_{t+\Delta t} = 0$$

$$; F_J\big|_{t+\Delta t} = \int_{S_t} T^i\big|_{t+\Delta t} N_j^i dS \quad (16)$$

Here, T_i is the ref. comp. of the nominal traction of surface. V and S are the volume and surface of the body in the reference configuration.

Taking the first 'guess' of $\overline{\tau}^{ij}\big|_{t+\Delta t}$ and $U_J\big|_{t+\Delta t}$ as of $\overline{\tau}^{ij}\big|_{t+\Delta t} = \overline{\tau}^{ij}\big|_t$ and $U_J\big|_{t+\Delta t} = U_J\big|_t$, two sets of successive iterations for $\overline{\tau}^{ij}\big|_{t+\Delta t}$ and $U_J\big|_{t+\Delta t}$ can be derived by using the Newton-Raphson procedure for Eqs. (15) and (16)

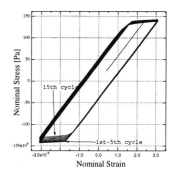

Figure 2. Nominal stress-strain curves for A3100

$$\delta U_J\big|_{t+\Delta t} = \sum_L \left([S_{JL}]^{-1}\right)_{JL} \left(-\Psi_L + F_L^v\right)$$

$$\delta \overline{\tau}^{ij}\big|_{t+\Delta t} = -\varphi^{ij} + \sum_J \overline{G}_J^{ij} \delta U_J\big|_{t+\Delta t} \quad (17)$$

where,

$$S_{JL} = \int_V \left(\overline{\tau}^{ik}\big|_{t+\Delta t} B_{jk}^L B_{ji}^J + F_{kj}\big|_{t+\Delta t} B_{ki}^J \overline{G}_L^{ij}\right) dV$$

$$F_J^v = \int_V F_{kj}\big|_{t+\Delta t} B_{ki}^J \varphi^{ij} dV \quad (18)$$

$[S_{JL}]$ in Eq. (18) corresponds to the stiffness matrix.

4. RESULTS AND DISCUSSION

4.1. Model Perspectives

In this paper, cyclic plastic deformation behavior of a rectangular parallelepiped f.c.c. crystal, shown in Figure 1, is analyzed. In Figure 1, coordinates x, y, z are the longitudinal, horizontal, and vertical edges of the crystal. The ratio of initial length, x_0, y_0, z_0, is $24 : 8 : 4$. The finite element model employed here is composed of $24 \times 8 \times 4$ 8-node isoparametric solid elements. Because of initial imperfections, the width and height of the crystal are reduced about 1% at $x_0/5$ from the crystal's ends.

The elasticity of the crystal is taken as isotropic. Material constants which characterize the elastic deformation and plastic shear deformation are listed in Table 1.

The rate of the end displacement is controlled during cyclic loading while both ends remain shear free. End displacement is controlled so that the nominal strain rate is $\pm 10^{-4}(1/s)$. The loading is fully-reversed tensile-compressive cyclic loading. We call the period from the beginning of the loading to the first maximum tensile loading "the 0th cycle", and the period from the i th maximum tensile loading

900

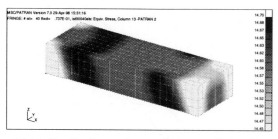

(a) $\bar{\sigma}$ distribution at the maximum tensile loading point of the 1st cycle

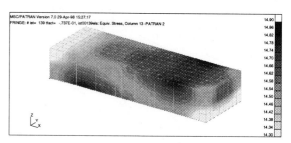

(b) $\bar{\sigma}$ distribution at the maximum compressive loading point of the 1st cycle

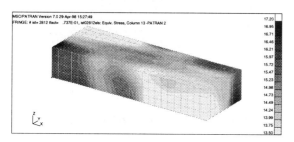

(c) $\bar{\sigma}$ distribution at the maximum tensile loading point of the last cycle

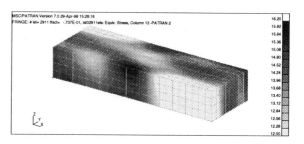

(d) $\bar{\sigma}$ distribution at the maximum compressive loading point of the last cycle

Figure 3. Distribution of equivalent stress $\bar{\sigma}$

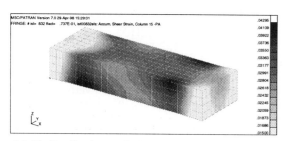

(a) Γ distribution at the maximum tensile loading point of the 5th cycle

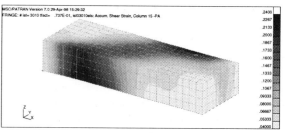

(b) Γ distribution at the maximum tensile loading point of the last cycle

Figure 4. Distribution of accumulated sum of slips Γ

to the $i+1$ th maximum tensile loading "the i th cycle". The loading direction is set to $[\sqrt{1/6}, \sqrt{1/6}, \sqrt{1/3}]$. The nominal strain amplitude is 2500 (represented to A2500), 2800 (A2800), 3100×10^{-6} (A3100). The number of loading cycles is 15.

4.2. Numerical Results

Figure 2 shows the computed nominal stress-strain curve for A3100. The nominal stress-strain curves of each loading cycle retrace almost the same course for the first few cycles. After that, they differ because the

yield points during the tensile and compressive loadings diminish as the loading cycle proceeds. The deviation of stress-strain curve from the initial stable loop arises at the 5th and 9th cycles for A3100 and A2800, but it does not arise until the last cycle for A2500.

The symbols $I1$, ..., $I8$, $M1$, $M2$ and C stand for the points located near the initial imperfections and on the midsection shown in Figure 1. For A3100, stress centers on each imperfection, and the stress distribution during tensile / compressive loading varies for each cycle from the beginning of the

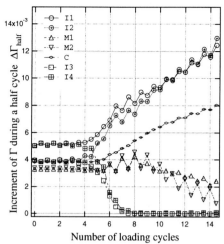

Figure 5. Time history of $\Delta\Gamma_{half}$ at various points in the crystal for A3100.

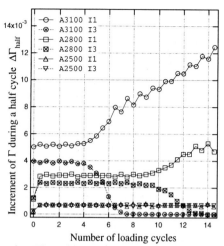

Figure 6. Time history of $\Delta\Gamma_{half}$ at $I1$ and $I3$ for A3100, A2800 and A2500.

loading to the 5th cycle. The values of equivalent stress $\bar{\sigma}$ tend to be larger at $I2$, $I3$, $I6$ and $I7$ during tensile loading, and at $I1$, $I4$, $I5$ and $I8$ during compressive loading. Though $\bar{\sigma}$ distribution is almost symmetric with respect to the xz plane which passes through the center of the model. After the 6th cycle, this symmetry is lost, and the difference in stress distribution during tensile / compressive loading becomes larger. Figure 3 shows the $\bar{\sigma}$ distribution at the maximum compressive and tensile loading points of the 1st and the last cycle for A3100. The appearance of $\bar{\sigma}$ distribution is similar in the other two cases. However, when the load amplitude is smaller, the change in distribution pattern arises later.

The accumulated sum of slips Γ defined by Eq. (19) is employed as the parameter, which indicates the amount of plastic deformation.

$$\Gamma = \int_{s=0}^{t} \sum_{a=1,12} \left| \dot{\gamma}^{(a)} \right| ds \qquad (19)$$

Figure 4 shows the distribution of Γ at the maximum tensile loading points of the 5th cycle and the last cycle for A3100. For A3100, Γ centers on $I1$, $I4$, $I5$ and $I8$ until the 5th cycle; after that, a concentration of Γ occurs at $I1$, $I2$, $I7$ and $I8$. The appearance of Γ distribution is similar in the other cases. However, when the load amplitude is smaller, the concentration of Γ is weaker.

$\Delta\Gamma_{half}$ at each measuring point is defined by the increment of Γ during a half cycle (the period from the maximum tensile / compressive loading to the maximum compressive / tensile loading). The degree

of irreversible slip generation and plastic deformation localization can be judged by the time history of $\Delta\Gamma_{half}$. $\Delta\Gamma_{half}$ remains constant when reversible slip occurs, while it changes when irreversible slip occurs. $\Delta\Gamma_{half}$ increases when plastic deformation concentrates, while $\Delta\Gamma_{half}$ decreases when it declines. Figure 5 shows the time histories of $\Delta\Gamma_{half}$ at various points in the model for A3100. In this figure, the ordinate represents the increment of Γ during the tensile loading of the i th cycle when the abscissa equals $i - 1/2$, and during compressive loading when the abscissa equals i. From the figure, one can see that reversible slip occurs and the amount of plastic deformation during each cycle is almost constant throughout the entire crystal until the 5th cycle. After that, plastic deformation increases rapidly with irreversible slip at $I1$, $I2$ and C, while it decreases at $I3$, $I4$, $M1$ and $M2$.

From the above results, it is clear that the same deformation behavior is repeated for several cycles from the beginning of the loading throughout the whole crystal, but after that, deformation behavior varies at every cycle when plastic deformation localization and irreversible slip occur. This change can be detected by the time history of $\Delta\Gamma_{half}$ and / or the deviation of stress-strain curve from the initial stable loop. Figure 6 shows the time history of $\Delta\Gamma_{half}$ at $I1$ and $I3$ for A3100, A2800 and A2500. Plastic deformation localization and irreversible slip generation start early and advance rapidly when the loading amplitude becomes larger.

4.3. Mechanisms of Plastic Deformation Localization and Irreversible Slip Generation

In each case studied in this paper, the D4 slip system ($\bar{1}$11)[101] and the C5 slip system (1$\bar{1}$1)[011] work from the beginning of loading until a certain cycle throughout the entire crystal. Figure 7 and 8 show the normalized values of $\bar{\sigma}$ and the absolute values of R.S.S $\left|\tau^{(a)}\right|$ at $I1,I3,M2$ and C for A3100, which are normalized by the values at the maximum tensile loading point of the 0th cycle. It is also apparent from these figures is that $\bar{\sigma}$ and $\left|\tau^{(a)}\right|$ during tensile / compressive loading are different from each other. Figures 9 and 10 show the normalized back stresses at $I1,I3,M2$ and C for A3100, which are normalized by the flow stress $g^{(a)}$. Figures 9 shows that $\alpha^{(a)}$ on D4 at the maximum tensile / compressive

loading point deviates as the loading cycle proceeds until finally the maximum 'drive force' $\left|\tau^{(a)} - \alpha^{(a)}\right|$ cannot exceed the flow stress $g^{(a)}$ at $I3$. At that instant, the D4 slip system stops working. These phenomena can be explained as follows. The initial imperfections cause non-uniform residual plastic deformation. This makes stress distribution and slip deformation during tensile / compressive loading different from each other. This, in turn, biases the back stresses of slip systems and causes lattice rotation. The deviation of back stress causes the critical resolved stress (C.R.S.S.) to vary, and the lattice rotation does the same for R.S.S. (That is, it causes geometrical softening / hardening).

After the D4 slip system ceases to operate at $I3$, deformation behavior changes rapidly. Figures 7 to 10 show the following: $\bar{\sigma}$ increases at $I1$ while it decreases at $I3$ (Figure 7); during compressive loading $\left|\tau^{(a)}\right|$ increases on C5 at $I1$ and on D4 at C while it decreases at all other points (Figure 8); notable geometrical hardening occurs on D4 at $M2$ and $I1$ while notable geometrical softening occurs on C5 at $I1$ and D4 at C (Figure 7 and 8); slip systems D4 and C5 continue to be active and $\alpha^{(a)}$ continues to be biased at $I1,M2$ and C while no slip system operates at $I3$ (Figures 8 and 10). With these changes, the D1 slip system ($\bar{1}$11)[110] comes into play during compressive loading at $I1$. It can be considered that these rapid changes are triggered by the sudden decrease of slip deformation at $I3$ caused by the cessation of the D4 slip system. These changes also happen for A2800 but proceed slowly, and do not happen until the last cycle for A2500.

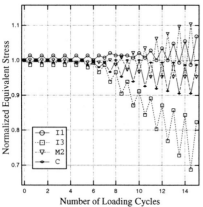

Figure 7. Time history of normalized equivalent stress

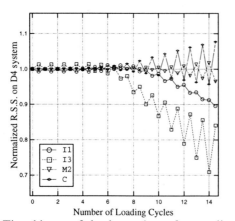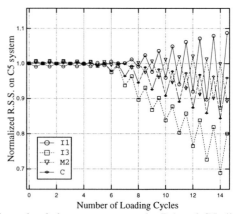

Figure 8. Time history of absolute values of normalized resolved shear stresses on the D4 and C5 slip systems

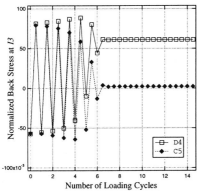

Figure 9. Time history of normalized back stress on the D4 and C5 slip systems at *I3*

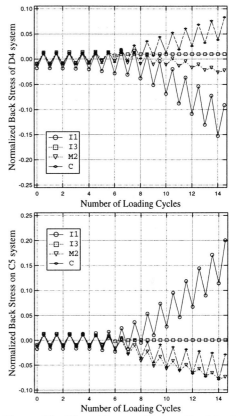

Figure 10. Time history of normalized back stresses on the D4 and C5 slip systems

The following can be concluded from the above results, obtained by employing the hardening rule Eq. (6). i) Plastic deformation localization and irreversible slip generation can be simulated. ii) The causes of these phenomena are the re-distribution of stress, the deviation of back stress and lattice rotation, all of which are induced by the difference of slip deformation during tensile / compressive loading. iii) These changes proceed slowly from the beginning of loading until a certain cycle, and suddenly accelerate when one of the active slip systems ceases to operate. Therefore, a linear relationship between the advance of these changes and the number of loading cycles does not exist. iv) When the load amplitude is small, the onset of acceleration is delayed because the difference of slip deformation during tensile / compressive loading is small.

Because plastic deformation localization and irreversible slip generation are part of the process of fatigue crack initiation, the FEM code used in this paper may successfully simulate the fatigue damage accumulation process.

REFERENCES

1. D. Peirce, R.J. Asaro, and A. Needleman, Acta Metall. Vol.31, No.12 (1983) pp.1951-1976

2. Y. Tomita, K. Hashimoto, N. Osawa and T. Ozaki, Proc. of 7th ISOPE, Vol. IV(1997) pp.664-671.

3. Y. Tomita, K. Hashimoto, N. Osawa and K. Hirose, Proc. of 8th ISOPE, Vol. IV(1998) now printing.

4. A. Needleman, Brown University Report MRL E-134 (1981)

Fatigue Damage in the Expansion Joints of ss ROTTERDAM.

Stapel, H.W. [a], Vredeveldt, A.W. [b], Journee, J.M.J. [c] and Koning, W. de [a]

[a] Retired from the Rotterdam Dockyard Company

[b] Netherlands Organisation for Applied Scientific Research, TNO, Delft

[c] Delft University of Technology

After 38 years of satisfactory service the Holland-America line sold her flagship "ss ROTTERDAM". Although promised to Lloyds, studies and tests made during construction on the expansion joints were never published. This PRADS '98 conference is a good occasion to report on these now. Moreover, the behaviour is placed in the light of an fatigue assessment in retrospect.

1. INTRODUCTION

In 1959 the Rotterdam Dockyard Company delivered the 200 m passenger ship ss ROTTERDAM to the Holland-America Line, HAL [1]. Until last year the vessel was operated by the HAL. She now sails under the name REMBRANDT for Premier Cruises.

The vessel is one of the first fully welded passenger ships.

At the time of building of the ss ROTTERDAM, most passenger ships had expansion joints. However much research was carried out on incorporating the superstructures in the strength of the hull girder [2, 3, 4, 5, 6]. A full-scale test was already made in 1913 by building the CALGERIAN with and her sister ship ALSATION without joints [7, 8].

The yard decided to fit four expansion joints, based on five arguments:

1. At the time the scantlings had to be decided on, the arrangement of the superstructure and its contribution to the strength was not known.

2. The risk of cracks in the super-structure at uncontrolled spots was considered to be larger than at the joints.

3. The expansion joints fitted on the ss NIEUW AMSTERDAM, built by the same yard in 1938, gave satisfactory results, although cracks did develop.

4. The construction of this ship was not a routine job for the yard; therefore a proven design was favoured.

5. There was no financial pressure to save building costs by leaving out the joints.

After commissioning in 1959, the vessel sailed on the North Atlantic service in the summer and made cruises in winter time. In April 1963 a crack was reported at the joints on frame 138/139. After repair and modification no cracks were reported since. The ship's logbooks are still available. From these it was possible to draw up a history of sea states and headings to which the vessel had been subjected until the crack occurred. With the current computational tools a fatigue damage analysis was carried out on the expansion joint. The results of this research are reported in this article.

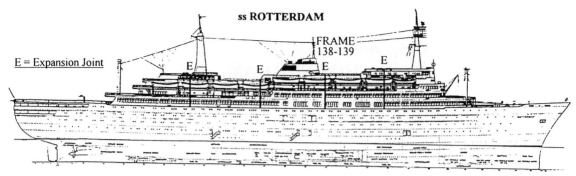

Figure 1. Side view of ss ROTTERDAM.

2. THE SHIP

Figure 1 shows a side view of the ship and figure 2 gives a general impression of the midship section. For comparison the cross section of the new ROTTERDAM-1997, built in Italy, is shown as well. The 1959 ship has the promenade deck (P) as strength deck. The plating of the 1997 ship on deck 3 and above is of high tensile steel, while the superstructure up to deck 8 forms part of the hull girder. It is remarkable to see that plate thicknesses have been reduced considerably over the years. The ROTTERDAM-1959 is fitted with transverse frames, while the ROTTERDAM-1997 is fitted with high tensile longitudinal frames. Owners extra's are indicated in brackets. Both ships are classified by Lloyds Register of Shipping. The sectional modulus of ROTTERDAM-1997 at deck 8 (34.80 m above base) is almost equal to the section modulus of ROTTERDAM-1959 at the promenade deck P (21.96 m above base).

Figure 3 shows a detail of the expansion joint as applied just above the promenade deck. Initially bolt holes were present in the edge strengthening bar. At the repair and modification in 1963 stud bolts were welded on the bar.

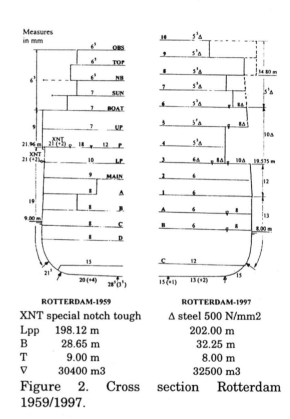

	ROTTERDAM-1959	ROTTERDAM-1997
	XNT special notch tough	Δ steel 500 N/mm2
Lpp	198.12 m	202.00 m
B	28.65 m	32.25 m
T	9.00 m	8.00 m
∇	30400 m3	32500 m3

Figure 2. Cross section Rotterdam 1959/1997.

3. APPLIED ANALYSIS

A fatigue calculation has been carried out in retrospect. The procedure as described below has been followed.

1. All logbooks from the maiden voyage up to the first reported crack at the expansion joint were analysed.

907

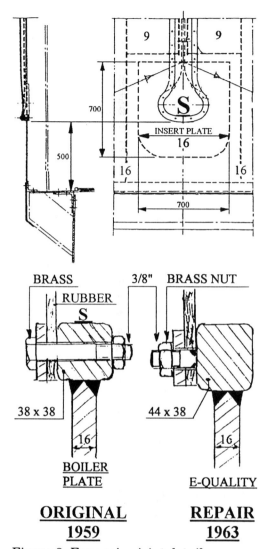

Figure 3. Expansion joint detail

BRASS   3/8"   BRASS NUT
RUBBER
S

38 x 38   44 x 38

16   16

BOILER
PLATE
E-QUALITY

**ORIGINAL
1959**   **REPAIR
1963**

2. The ship's behaviour in seaway was calculated by applying a general purpose ship motions prediction computer program. This analysis yielded a set of vertical en horizontal hull bending moment frequency response functions (FRF) for the cross section at frame 138/139.
3. This set of FRF's was used to calculate the stress spectrum in the promenade deck at the side during each watch of 4 hours. The environmental conditions and the ship speed were assumed to be constant during each watch.
4. From each stress spectrum the cumulative probability distribution of the stress range was calculated by assuming a Rayleigh distribution.

5. For each spectrum the number of zero up-crossings was determined.
6. The number of cycles at 23 stress ranges was calculated
7. The number of cycles for the considered stress ranges for each watch were finally added.
8. Next, the number of cycles per stress range were divided by the "required" number of cycles up to damage for the given stress range.
9. Finally the individual damage ratios were added up, yielding the cumulative damage D.

Items 8 and 9 describe the Palmgren Miner approach for calculating fatigue damage in a structural detail [10].

4. BEHAVIOUR IN SEAWAY

To calculate the behaviour of the ship in the experienced wave conditions until cracking of the expansion joints, the computer code SEAWAY of the Delft University of Technology [9] has been used. This program calculates the loads and motions of ships in waves in the frequency domain by the linear strip theory method. Depending on the shape of each cross section, a 10 parameter close-fit conformal mapping method or Frank's pulsating source method is used to calculate the 2-D potential coefficients.

After taking into account the forward speed effect, the coefficients of the equations of motion in the frequency domain are obtained by a longitudinal integration of the 2-D values. The presence of bilge keels and fin stabilisers has been taken into account. Bretschneider wave energy spectra are used to obtain statistical data on motions and loads in irregular waves.

The ship's under water hull form is given in figure 4

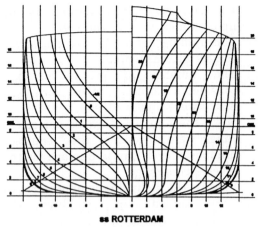

ss ROTTERDAM

Figure 4. Body plan ss ROTTERDAM

The distribution of the mass along the ship length is shown in figure 5.

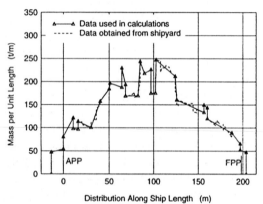

Figure 5. Mass distribution SS ROTTERDAM

Further input for the hydromechanical calculations were:

Draught (average)		8.86 m
Metacentric height		1.25 m
Radii of inertia	k_{xx}	11.45 m
	k_{yy}	51.80 m

The data from the logbooks were taken per sea watch, i.e. 4 hours. The time span investigated starts in September 1959 and continues until April 1963. In total 4634 observations were recorded. Figure 6 shows the exposure of the vessel to sea states during the considered period.

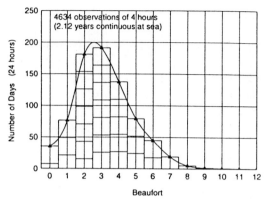

Figure 6. Exposure of vessel to sea states from 1959 till 1963.

It can be seen that the vessel has not often been subjected to very heavy weather. For each period a Bretschneider sea spectrum is assumed based on the observed sea state. From this spectrum and the heading of the ship, the horizontal and vertical bending moment in the ship's hull in way of the expansion joint were calculated. Because of the closed cross section, torsion could be ignored. Next, they were devided by the respective section moduli and added, thus yielding a spectrum of longitudinal stresses in the promenade deck at the side. As illustration figure 7 is included to give an impression of stresses in the promenade deck.

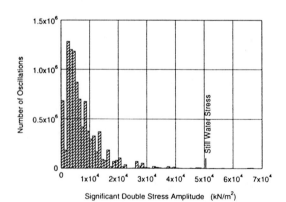

Figure 7. Stress levels versus number of stresses in the promenade deck.

The area properties $m_{0\sigma}$ and $m_{2\sigma}$ of the stress spectra were used as input for the fatigue analysis.

5. STRESS ASSESSMENT

The cracked expansion joint is situated at frame 138-139 (0.55 Lpp from APP), 500 mm above the promenade deck. Due to its geometry, the stress at the bottom of the expansion joint will be larger than the stress level in the promenade deck.

Prediction during design.

During the design of the ship a Stress Concentration Factor $SCF = 3.5$ has been estimated based on analytical conside-rations.

Measurements during launching.

During the launch of the vessel 1958 the Ship Structures Laboratory of the Delft University of Technology, DUT, and the Netherlands Organisation of Applied Scientific Research, TNO, carried out strain measurements on several spots in the vessel [12]. The bottom of the expansion joint and the promenade deck were included. Stresses during launching were 40% of the stresses calculated for the design wave (see figure 8).

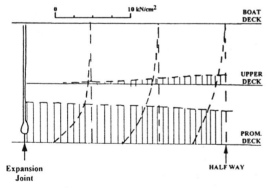

Figure 8. Stresses in promenade deck and superstructure side.

The major conclusion of the measurements was that there was only a slight increase in the stresses in the strength deck below the expansion joints. A stress concentration factor $SCF = 4.4$ was found in the bottom of the joint (point S in figure 3), which was higher than expected. It was realised that here the fracture strength of the steel 41 would be surpassed, as Lloyds set for this ship a maximum allowable hogging stress in the strength deck of 125 MPa.

The section modulus at the strength deck, including owwners extra's was 20 % higher than required by Lloyds. Therefore a crack might occur after sufficient heavy loading. This was not considered to be a risk for the hull, as it would not lead to any major rise in the stress in the topsides. Both the stringer plate and the sheer strake are of special notch tough (XNT) steel. The riveted deck stringer angle and the angle bar connecting the side plates, work as crack arrestors while spreading the forces over the length.

An attempt was made to correlate the measured stresses during launching with the stresses from the launching calculations. Two differences were found:

1. The measured maximum sagging stress at the point of uplift as measured was smaller than calculated.
2. The hogging stress measured while the ship was fully afloat proved to be smaller than calculated.

The first difference is mainly due to the effect of the presence of breaking shields and maybe also due to a difference between effective and calculated section modulus. The second difference is due to a hogging stress while the ship was still at her berth. At this position strain gauges were set to zero. This hogging was probably due to the weight distribution over the flexible berth and stresses caused during welding of the hull.

Recent finite element calculations.

An attempt has been made to estimate a stress concentration factor, SCF, based on both a coarse mesh and a fine mesh finite element calculation.

For this purpose the environment of the expansion joint has been modelled with plate elements capable of describing membrane stresses. The analysis is limited to in plane deformations only. The lower edge of the model is at promenade deck level.

910

The left hand side of the model and the lower edge are subjected to an imposed horizontal displacement equivalent with a strain of 238 microstrain, i.e. a stress level of 50 MPa. The lower edge is restrained in vertical direction. The right hand side edge of the model, between bottom of the joint and the promenade deck is restrained in horizontal direction. The upper edge is subjected to an imposed displacement and rotation, taken from the strength analysis carried out by the yard. Figure 9 shows a contour plot of the calculated stresses. The stress increase between lower edge (deck level) and the bottom of the joint was similar in both cases: $SCF = 2.7$.

Review of SCF's

A brief review of the obtained stress concentration factors at the bottom of the expansion joint is given in table 1.

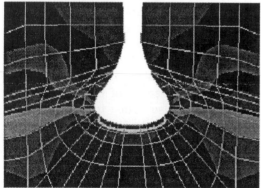

Figure 9. Stress contour in way of expansion joint (frame 137-138), coarse mesh.

Table 1 Assessed stress concentration factors

Method of assessment	SCF
Prediction during design	3.5
Measured during launching	4.4
Recent FE calculation	2.7

These factors, determined at the centre line of the expansion joint, do not include the effect of the presence of bolt holes.

It is noted that the SCF's do not match very well. The effect of mesh size was checked but proved in this case negligible. The effect of the stress built up between

the expansion joint (Figure 8) proved to contribute substantially to the stress increase in the joint.

In case of the presence of bolt holes an additional SCF must be applied of 3, see figure 10 obtained from ref. [10].

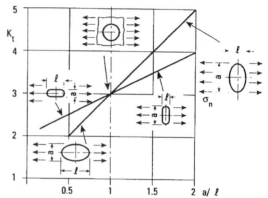

Figure 10. SCF for cut outs from [10].

Therefore the actual SCF to be used in a fatigue assessment should lie between 5.1 and 13.2!

6. FATIGUE ASSESSMENT

For the considered life period of the vessel the number of cycles at 23 stress levels ranging from 1 Mpa to 1000 Mpa has been determined. For this purpose cumulative probability density functions were assumed, based on the stress spectra as determined for each watch.

A Rayleigh distribution is assumed which could be characterised by the area $m_{0\sigma}$ of the stress spectra. The number of cycles is calculated by dividing the 4-hour watch period in seconds, by the average zero up-crossing period (based $m_{0\sigma}$ and $m_{2\sigma}$). The number of cycles for the considered stress ranges for each watch were finally added, yielding a final set of pairs with stress range and number of cycles. Next, the number of cycles per stress range were divided by the "required" number of cycles up to damage for the given stress range. Thus damage ratios per stress range were obtained. Finally the individual damage ratios were added up yielding the damage ratio D. A value larger than 1.0 implies fatigue damage and D lower than 1.0 implies no damage.

Without applying any *SCF* the result is shown in figure 11. This diagram is valid for the stresses in the promenade deck. From [11] an S-N curve (curve D) has been taken, considered to be valid for the structural detail under consideration.

This curve shows the number of stress cycles "required" to obtain fatigue damage at any stress level.

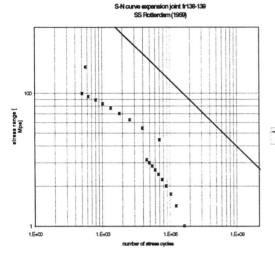

Figure 11. Stress level versus number of stress cycles (S-N-curve).

The fatigue assessment on the bottom of the expansion joint. was carried out with four different *SCF* values. Table 2 shows the results.

Table 2 Fatigue assessment results.

SCF from	SCF	D
Analytical considerations x 3	10.5	2.65
Measured (launching)	4.4	0.19
Measurement x 3	13.2	5.27
FE-calculations x 3	8.1	1.22

Note that a fatigue crack occurs when *D* is larger than 1.0. A crack is to be expected much earlier than the considered period, when the *SCF*'s from analytical considerations or measurements are applied. When using the SCF from the FE analysis a fatigue life is found which is nearer to the actual reported life. When the *SCF* increase due to the presence of the bolt hole is discarded, no damage is expected. It is interesting to note that the effect of the presence of a bolt hole is decisive. Survey reports of Lloyds were scrutinised from 1959 till 1997.

The joints did not give any problems after the modification in 1963.

The fact that the crack developed in the first years and none afterwards will have had three probable reasons:

1. The built in welding stresses were releaved by heavy loading of the hull.
2. The detail of the joint was improved deleting bolt holes.
3. The ship was taken out of the Trans-Atlantic service in 1969 where after she was mainly cruising in good weather areas.

7. DEVELOPMENTS

Nowadays all passenger ships are built without expansion joints. On some, high tensile steel is used. Apparently classification societies are satisfied with the performance of the ships as the surveyors don't report cracks. The fact that cruise ships predominantly sail in fine weather areas will have its influence in this matter. With the enormous growth of the market cruising will become world wide. Notwithstanding weather routing, the ships will have to sail to their destination and may face heavy weather close to port. Recent examples are QUEEN ELIZABETH 2 in September 1995 and the ROTTERDAM in April 1997 close to the U.S. East coast, where both ships suffered damage. This is not too serious as long as only bulwarks and front bulkheads are involved. Plastic deformations and cracks in the hull girder must be avoided by careful analysis of critical spots including fatigue assessments. One should bear in mind that high tensile steels do not have any higher resistance to fatigue than mild steel.

Full scale measurements on the cruise ship ROYAL PRINCESS built in Finland give an good picture of the contribution of the superstructure to the strength of the hull [13]. Further reference is made to interesting papers by Mr. M.J. Gudmunsen [14], Mr. Violette and Mr. Shenoi [15].

912

8. CONCLUSIONS

The assessment of a stress concentration factor *SCF*, in the bottom of the expansion joint, based on finite element calculations shows a difference with an earlier analytical assessment and strain measurements during the launch of the vessel.

The effect of the bolt holes in the strengthening bars in the expansion joints prove to be paramount with respect to fatigue damage.

9. ACKNOWLEDGEMENTS

The authors greatly acknowledge the Holland America Line and Premier Cruises for their permission to publish on their ships and Fincantieri for an impression of the plate thicknesses of the ROTTERDAM-1997. Morover gratitude is expressed for the cooperation of Lloyds Register of Shipping on accessing their files on the ss ROTTERDAM. Mr. Piet de Heer (DUT) is thanked for his help in recording the data from the ships journals. Finally we express our gratitude to Mr. Wouter Pastoor (DUT) for his guidance and checking on the fatigue assessment.

REFERENCES

1. "D.S.S.Rotterdam - Holland Amerika Lijn", Extra Edition of "Schip en Werf" - September 1959 (in Dutch).
2. Chapman, J.C.: "The Interaction between a Ship's Hull and a Long Superstructure", Trans. R.I.N.A. 1957.
3. Johnson A.J.: " Stresses in Deckhouses and Superstructures", Trans. R.I.N.A. 1957
4. Caldwell, J.B.: " The Effect of Superstructures on the longitudinal Strength of Ships",Trans. R.I.N.A. 1957.
5. Vasta, J.: " Structural Tests on the S.S. President Wilson" , Trans S.N.A.M.E. 1949.
6. Vasta J.: " Full Scale Ship Structural Tests", Trans S.N.A.M.E. 1958
7. Foster King, J.: " On large Deckhouses", Trans. I.N.A. 1913.
8. Montgomery, J.: " The Scantlings of Light Superstructures", Trans. I.N.A. 1915.
9. Journee J.M.J. (1992), "SEAWAY – DELFT, User manual of release 4.00". Thechnical Report 910, Delft University of Technology, Ship Hydromechanics Laboratory, The Netherlands.
10. Fricke Dr. W., Petershagen Prof. Dr. H., Paetzhold Dr. H. (1997), "Fatigue Strength of Ship Structures, Part I: Basic Principles". GL-Technology Number 1/97, Germanischer Lloyd, Hamburg.
11. Classification notes Note No. 30.2., "Fatigue strength analysis for mobile offshore units", Det Norske Veritas, august 1984.
12. "Spanningsmetingen gedurende de Afloop van het S.S.ROTTERDAM", laboratorium voor Scheepsconstructies Rapport no 58 - 30.10.1950, University Delft (in Dutch).
13. Fransman, J.W.: "The Influence of Passenger Ship Superstructures on the Response of the Hullgirder" Trans. R.I.N.A. 1988.
14. Gudmunsen, M.J.: " Some Aspects of Modern Cruiseship Structural Design", Lloyd Register of Shipping- May 1996.
15. Violette F.L.M. Shenoi R.A., "On the fatigue performance prediction of ship structural details", RINA paper No. 4, spring meeting 1998.

Practical Design of Ships and Mobile Units
M.W.C. Oosterveld and S.G. Tan, editors.

A development of Technical Database for Hull Structures

-A technical information system and database for design, construction, and inspection/maintenance of hull structures "HullExpert"-

Hirohiko Emi, Toshiyuki shigemi and Hiroshi Ochi

Research center, Nippon kaiji Kyokai, 1-8-3 Ohnodai Midori-ku Chiba, Japan

1. Introduction

Feedback based on past experience and accurate records is essential for the proper design, construction, operation, inspection and maintenance of ships. With this in mind, the Society is currently developing a database for hull structures known as "HullExpert". This database includes not only data compiled from an extensive base of experience and records of actual performance but also all information concerning practical and technical aspects of hull structures. The general concept behind the database system is illustrated in **Figure 1**.This paper presents an outline description of the database system.

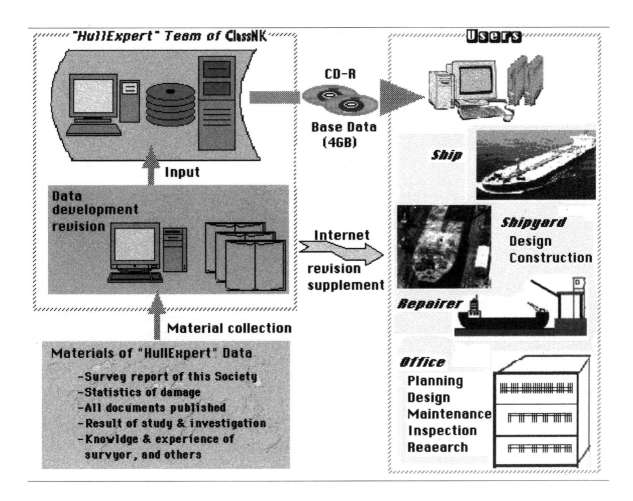

Fig.1 Concept of HullExpert

2. Specification

The overall specifications of the database are shown in Table 1. Information comprising the database is organized into several categories of output for easy understanding. A strong point of the system is the fact that users can search and output any item related to the problem at hand from the wealth of data stored in the separate portions of the database.

It is assumed that users of the database will be those who have an interest in hull structures, such as ship designers, shipbuilders, superintendents acting for shipowners, repairers, inspectors, surveyors, students, researchers and the like. Given this broad range of needs, the database will require a computer system with a sufficiently large capacity capable of storing and processing the nine or ten CD-ROMs worth of data which will be provided at the first stage. The database is organized into three parts, the first having a further five sub-parts, as follows:

I. Technical information regarding hull structures (Catalog of members/location)

   I.1 Single hull VLCCs and large oil tankers

Table 1 Specification of the "HullExpert"

Hardware	All types of personal computer with the O.S. of Windows 95 / N.T.4.0 in English, or Japanese CD-R Recommend: 4GB or more harddesk for copy of all the data supplied by nine or more CD-Roms.	In order that all data are jointedly searched and outputted, a big capacity of the harddesk is necessary because the information will be about 4GB.
Software	Application soft "HullExpert" developed on the base "Visual Works". No other application is necessary to operate the "HullExpert"	
Picture size	800 x 1200 dots with 256 color	Monitor's display with 1600 x 1200 dots can look a picture without scroll
Language	in English, and in Japanese translation, separate release	

Frame work	Technical information regarding hull structures
	(Members/location catalog for ship types)
	  -Single hull VLCC and large oil tanker
	  -Double hull VLCC and large oil tanker
	  -Bulk carrier
	  -Container ship
	  -General cargo ship (Cargo ship with aft engine and one or two tiers decks)
	
	Damage data of hull structure(Part 1,2,....)
	
	Inspection and maintenance of hull structure (Part 1 and 2)
		The underlines are the first release in 1999, and
		
		no underline is the second release in 2000 or 2001

I.2   Double hull VLCCs and large oil tankers
I.3   Bulk carriers
I.4   Container ships
I.5   General cargo ships with aft engine and one or two decks (Multi-cargo ships, lumber carriers, refrigerated cargo ships, heavy cargo ships, chip carriers, etc.)

II. Hull damage data (Part 1, as well as subsequent Parts 2, 3, etc.)

III. Inspection and maintenance of hull structures

An example of the total application of such a large and separate database is shown in **Figure 2**.

3. Technical information regarding hull structures

(Catalog of members/location of hull structures)

This part of the database is divided into five sub-parts or sub-sections corresponding to ship type, as shown above. One catalog is prepared for each member and location. Each sub-section contains a total of about sixty or more catalogs.

Each catalog consists of a cover page with search items and principal information, as well as three areas of detailed information consisting of:

(1) Details and causes of failure,
(2) Maintenance, inspection and repair/ reinforcement, and
(3) Precaution in design and construction.

Each catalog contains detailed information regarding members and location. Necessary information related to a given member or location can be quickly accessed and outputted. **Figure 3** shows an example of the process followed to output desired data.

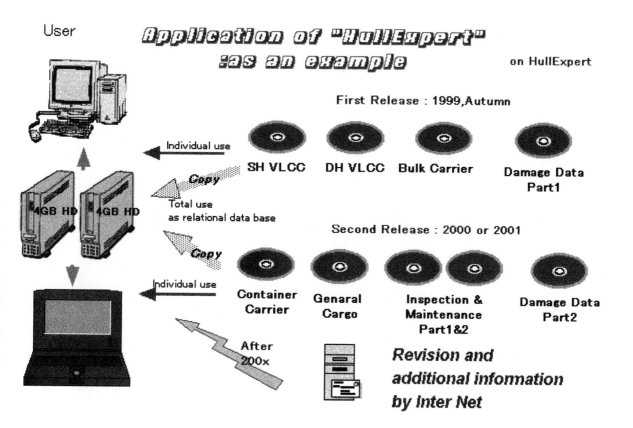

Fig.2 Application of HullExpert

916

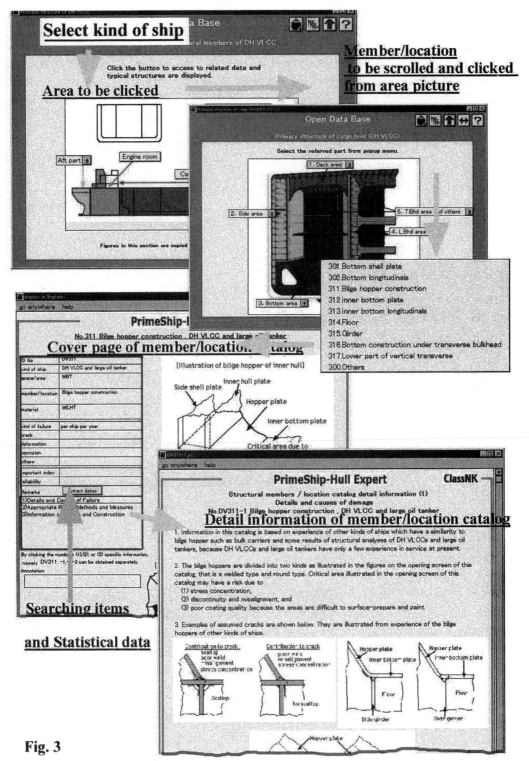

Fig. 3

Process to output of information
from "HullExpert" Members/location catalog

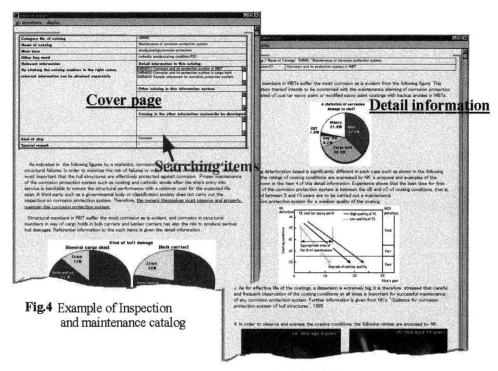

Fig.4 Example of Inspection
and maintenance catalog

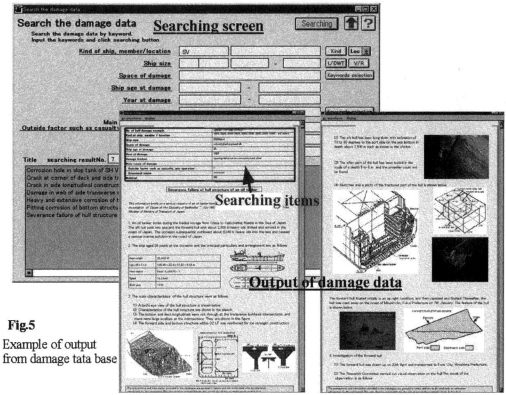

Fig.5

Example of output
from damage tata base

4. Inspection and maintenance of hull structures

This section will be supplied by two CD-ROMs and consists of the following chapters:
- SM100 Survey
- SM200 Inspection
- SM300 Repair
- SM400 Temporary repair, rescue and detour of damaged hull
- SM500 Maintenance of hull
- SM600 Information from NK
- SM700 Information from all the world

This section contains all data and information concerning the inspection, survey and maintenance of hull structures. An example of the output of the inspection and maintenance catalog is shown in **Figure 4**.

5. Hull damage data

Examples of hull damage as well as the repair work done or other countermeasures taken are contained in this portion of the database. Ships covered by the database include the following types:

SV: SH VLCCs and large oil tankers
DV: DH VLCCs and large oil tankers
BL: Bulk carriers
CN: Container ships
General cargo ship (ship with aft engine and one, two or three decks

    MC : Multi-purpose cargo ship
    RF : Refrigerated cargo carrier (Reefer)
    LC : Lumber carrier
    CH : Chip carrier
    HC : Heavy cargo ship
    PT : Product tanker
    CT : Chemical tanker

PP   : Pressurized LPG tanker
RP   : Refrigerated LPG tanker
VC   : Car carrier
RO   : Roll on roll off ship
OS   : Other ship

Searches for relevant hull damage data can be conducted using the following parameters:
- Kind of ship
    - Members/location
    - Space or area of damage
    - Ship age at damage
    - Year at damage
    - Damage feature
    - Main cause of damage
    - Material

This portion of the database contains about 500 examples of damage. At the first stage of this project due for completion in the year 2000 or 2001, about 1000 cases of damage are to be included in Part 1 and Part 2. Damage data will then be updated and supplemented each year.

An example of damage data is shown in **Figure 5**.

6. Conclusion

About half of the parts of the database (members/location catalog of single and double hull VLCCs and large oil tankers, bulk carriers and hull damage data Part 1) will be released next year, while the remaining portions are due to be released in 2000 or 2001. In order to design, construct, operate and maintain ships at optimum cost while ensuring high levels of safety, it is desirable to provide proper feedback on the valuable experiences and various records of related work done up to the present. The authors hope that the database will prove helpful in accomplishing these aims.

Practical Design of Ships and Mobile Units
M.W.C. Oosterveld and S.G. Tan, editors.

Prediction of propeller cavitation noise on board ships

C.A.F. de Jong and M.J.A.M. de Regt

TNO Institute of Applied Physics (TNO-TPD)
P.O. Box 155, 2600 AD Delft, The Netherlands

The different approaches that are available for predicting the on board noise levels caused by cavitating propellers are reviewed. The applicability and reliability of both numerical methods for a deterministic description of low-frequency tonal noise at the propeller blade rate frequencies and of semi-empirical and statistical methods for broadband noise at higher frequencies are discussed.

1. INTRODUCTION

Propeller cavitation noise often contributes considerably to the noise levels measured in the accommodations and work spaces on board of ships. Because noise control has become an important parameter in ship design, the ability to perform noise level predictions in an early stage is essential for the designer. It enables to foresee if noise requirements can be met, or which design modifications or counter measures are needed.

The state-of-the-art in ship acoustic knowledge has been the topic of two international symposia that were held in the Netherlands in 1976 [1] and 1986 [2]. Several important studies have been presented. In Europe, the Nordic countries have performed a major ship acoustic research programme with an emphasis on the development of theoretical prediction methods [3], while in the Netherlands MARIN and TNO-TPD have developed experimental procedures and semi-empirical prediction formulas [4,5,6,7]. Nevertheless, an accurate prediction of on board noise levels due to propeller cavitation appears to be a difficult task, especially in the low to mid-frequency range (say 40 to 400 Hz).

This paper reviews the applicability of different types of prediction methods and presents the results of some additional investigations that have been carried out at TNO-TPD for the former 'Foundation for the Co-ordination of Maritime Research in the Netherlands' (CMO) in the years between 1988 and 1996.

2. PROPELLER CAVITATION NOISE

Propeller cavitation occurs wherever the local pressure in the water flow around the propeller becomes lower than the vapour pressure. The growth and implosion of gas bubbles causes strong pressure fluctuations in the water. These pressure fluctuations induce unsteady forces on the hull, which excite vibrations that propagate through the ship structure and cause the radiation of noise into the accommodation and work spaces.

The frequency spectrum of the pressure fluctuations due to propeller cavitation (e.g. fig.1), exhibits tonals at harmonics of the 'blade rate' (the shaft revolution rate times the number of blades), that are mainly caused by unsteady blade loading and unsteady sheet cavitation at the propeller blades, and broadband noise, with a broad peak at the 'bubble frequency' (typically at 5 to 10 times the blade rate), above which the noise spectrum is determined by the statistics of expansion and implosion of individual cavitation bubbles.

The low frequency tonals are mainly of importance because they excite global ship vibrations, while the acoustic noise level in the accommodation and work spaces is predominated by the broadband spectrum. In the intermediate frequency range, harmonics of the blade rate may contribute to the noise level, especially when they coincide with local resonances in the ship structure.

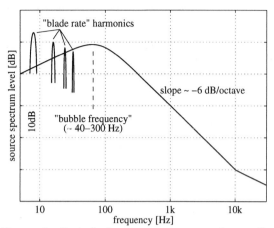

Figure 1. Typical frequency spectrum of propeller cavitation induced pressure fluctuations, after [8].

A first step in the development of prediction methods for propeller cavitation noise on board is to model the process of noise transmission as a series of transfer functions [6]:

$$p_{acc} = \frac{p_{acc}}{v_{hull}} \cdot \frac{v_{hull}}{p_{hull}} \cdot \frac{p_{hull}}{p_i} \cdot \frac{p_i}{U_{cav}} \cdot U_{cav} \qquad (1)$$

where:
U_{cav} = cavitation source strength [m³/s],
p_i = incident sound pressure at the hull [Pa],
p_{hull} = total sound pressure at the hull [Pa],
v_{hull} = vibration velocity of the hull [m/s],
p_{acc} = sound pressure on board [Pa].

For noise prediction purposes, all parameters are analysed in 1/3-octave frequency bands and averaged over multiple measurement points.

3. CAVITATION SOURCE STRENGTH

In the model of eq.(1) the unsteady propeller cavitation is usually described as an equivalent acoustic monopole source, that is located near the top of the propeller blade in upward position [5,6,7]. The actual shape of the cavitation volume is assumed to be unimportant for the pressure field, which is true for the farfield radiation of an acoustic monopole that is small compared to the wavelength in water. The size of the propeller cavitation volume, however, is of the same order of magnitude as its distance to the ship hull, so that the pressure distribution at the hull is rather determined by its

hydraulic nearfield than by acoustic radiation. The validity of the single monopole model is therefore questionable. For the same reason, Nilsson's [3,10] approach using a 'plane wave' model is unrealistic.

A more accurate source description might be to model the cavitation as a spatial array of monopoles. The motion of the cavitation volume with the blades can thus be taken into account via the phase relation between these monopoles. A tentative study of this approach [11] has shown that the single monopole model leads to an overestimation of the hull pressure in comparison with the array model. However, the array model requires detailed information of the dynamics of the cavitation volumina, that can only be acquired from hydrodynamic calculations, see section 3.1.

If one accepts the monopole description, for lack of a more accurate description, its source strength U_{cav} can be estimated on the basis of measurements on real ships or scale models, or of analytical or semi-empirical calculation methods.

3.1 Theoretical models

Hydrodynamic calculations can be used to determine the unsteady sheet cavitation volume on the blades of the rotating propeller in the non-uniform wake field of the ship as a function of time, see e.g. [3,7]. These calculations are generally only valid at the blade passing frequency and its first few harmonics. The corresponding frequency range is more important for vibration control in ships than for noise in the accommodation spaces. The maximum frequency is limited by the number of discrete time steps in the hydrodynamic calculations. Attempts to extend the frequency range by interpolation between these steps [3] are not based on the actual physical processes involved and therefore not necessarily correct.

3.2 Semi-empirical calculation methods

The broadband noise spectrum that is due to the dynamic behaviour of individual cavitation bubbles, (fig.1) is highly statistical. The corresponding source levels can be predicted from semi-empirical calculation methods that are based on experimental data in combination with an assessment of the relevant correlation parameters. For example, Brown [8] states that an upper bound of the volume velocity source strength above the bubble frequency is proportional with the number of blades B, the propeller diameter D [m], its revolution rate N [Hz] and the frequency f [Hz]:

$$U_{cav} \sim BD^4 N^3 f^{-3} \qquad (2)$$

Equation (2) does not take into consideration the effects of propeller geometry and wake distribution nor of the ship structure above the propeller. More elaborate semi-empirical models, like the model by Wind and De Bruijn [5], are therefore considered to be more reliable.

The available semi-empirical models for the cavitation source strength are all indirectly derived from measurements of pressure pulsation or hull vibration near the propeller. These measurement data have to be somehow corrected for effects of the presence of the flexible ship structure near the source. An elegant solution for this correction is the independent measurement of the transfer function between the monopole source strength and the measured response, via a reciprocal technique that has been described in [5]. In this method the validity of the monopole description is somewhat enhanced by averaging the transfer functions over multiple positions in the source region and in frequency bands.

3.3 Incident hull pressure

The incident hull pressure p_i is defined [4,6] as the sound pressure that would be found at the position of the hull for the equivalent monopole source with volume velocity U_{cav} in free field conditions. The corresponding transfer function of the source strength to the incident pressure at the hull surface is defined as:

$$\frac{p_i}{U_{cav}} = \frac{\rho f U_{cav}}{2r} \qquad (3)$$

where ρ [kg/m^3] is the water density, f [Hz] the frequency and r [m] the distance between the source position and the hull surface.

4. HULL PRESSURE RATIO AND SURFACE IMPEDANCE

The "hull pressure ratio" p_{hull}/p_i gives the ratio of the actually measured pressure fluctuations at the hull and the incident pressure as defined above. The actual hull pressure distribution will be strongly influenced by the dynamic interaction between the water and the flexible hull structure. Because prediction methods for noise transmission through the ship structure use a hull velocity level as source descriptor, the hull pressure ratio is multiplied by a "hull surface impedance" p_{hull}/v_{hull} that describes the response of the water loaded ship structure.

In the semi-empirical approach described in [6], these hull pressure related transfer functions are averaged over space (several hull positions) and/or frequency (one-third octave bands). The resulting hull surface impedance spectra show a stiffness behaviour at low and a mass behaviour at higher frequencies. The corresponding stiffness and mass parameters may be estimated on the basis of the geometry and material properties of the hull plates. Nilsson [3,9] has developed a numerical method to calculate the hull surface impedance for simply supported hull plates, in case of plane wave excitation, which leads to comparable results. At very low frequencies, e.g. below 40 Hz, the propeller cavitation excites global vibrations of the entire aft body of the ship rather than hull plates. This global behaviour can be calculated using the Finite Element Method [9].

The following practical general conclusions can be drawn from the theoretical and experimental studies [6,9] of hull pressure ratio and surface impedance:

- Sound pressure or vibration measurements at the hull plates above the propeller are strongly influenced by the dynamic response of the ship hull, hence they do not give undisturbed information of the propeller cavitation source strength.
- Increasing the stiffness of the hull plate fields, e.g. by increasing the plate thickness or decreasing the distance between the stiffeners, will reduce the excitation of the hull by propeller cavitation in the frequency range below the first plate resonance.
- If hull plate resonances occur in the frequency range where tonal noise predominates the propeller source strength, the level of excitation of the hull is strongly dependent on the ratio of the excitation and resonance frequencies.

The latter conclusion indicates that it might be worthwhile to be able to predict the local response of the hull plates above the propeller. As a first step, a computer program has been developed to calculate the response of a rectangular plate with simply supported edges in a rigid baffle and with water loading on one side, that is driven by a point source in the water [11]. The equations of motion are solved via a truncated series expansion in the in-vacuo eigenmodes of the plate, following [12].

Calculations have been performed for a typical 8 mm thick steel plate of 3 by 0.6 . The first three in-vacuo modes of this plate occur at 57, 75 and 110 Hz. The low frequency velocity response of the centre of the water loaded plate, driven by a monopole source at 1 m above the plate's centre is shown in fig.2.

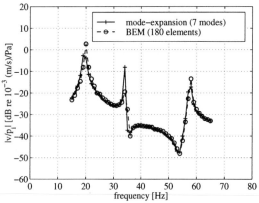

Figure 2: the velocity at the centre of a simply supported, rectangular, water loaded plate, excited by a monopole that induces an incident hull pressure p_i.

The lowest resonance frequency shifts down from 57 Hz to 20 Hz under influence of the fluid loading (note that the simplified formula for the calculation of the natural frequency of submerged plates, e.g. [13], is not very accurate: it predicts 28 Hz).

Fig.2 also shows the results obtained from numerical calculations with a coupled Finite-Element/ Boundary-Element Method [14]. The results of both calculation methods agree satisfactorily. This agreement gives confidence in the application of FEM/BEM, which makes it possible to study more realistic geometries.

In fig. 3 the response of the single simply supported plate of fig. 2 is compared with the response of the same plate when mounted between two identical plates, connected via beam stiffeners. This comparison demonstrates the influence of the actual boundary conditions on the plate response.

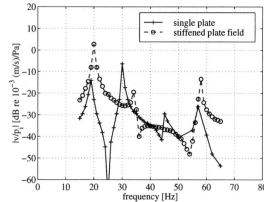

Figure 3: the velocity at the centre of a simply supported plate (fig. 2) and the same plate between adjacent plates (see fig. 4), calculated using FEM/BEM.

Fig. 4 shows the displacement of the stiffened plate at the two lowest natural frequencies. The lowest resonance for the three plate geometry is a global resonance of the complete stiffened plate. The lowest mode of the centre plate shifts from 19 Hz for the simply supported plate to 30 Hz, under influence of the additional rotatory stiffness at the edges that is caused by the presence of the beams and adjacent plates.

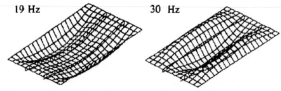

Figure 4: mode shape at the two lowest natural frequencies of the three connected plates (FEM/BEM result), total dimensions: $3 \times 1.8 \ m^2$.

The large effects of the actual boundary conditions on the plate response at low frequencies that are observed in these simple examples indicate that deterministic methods to calculate the dynamic response of the local structure above the propeller require a very accurate and detailed description of geometry and material properties to produce a reliable response prediction. In practice, the available information will seldom be sufficiently accurate to avoid the coincidence of local resonances and blade rate tonals by design. Nevertheless, deterministic calculation methods can be useful to investigate the relative effects of design parameters, like plate thickness, material, stiffener distance and geometry, etc. on the hull surface impedance.

5 EQUIVALENT VELOCITY LEVEL

The velocity level of the hull plates above the propeller does not necessarily give a good description of the excitation of the aft ship, because these levels can strongly depend upon local dissipation in the structure. Therefore some prediction models [4,15] use a source description in which the excitation by the propeller is described as an equivalent structure-borne noise source, positioned at tank top. The source strength is then given in terms of an "equivalent velocity level", that is indirectly calculated from a number of measured velocity and sound pressure levels on board.

6. TRANSMISSION THROUGH THE SHIP

Once the velocity source level due to propeller cavitation has been established, it can be balanced against the source levels of other major noise sources, like for example the propulsion engines. The sound pressure levels on board follow from the transfer functions p_{acc}/v_{hull}. These transfer functions describe the structure-borne sound transmission through the ship and the radiation of sound from floor, ceiling and walls into the accommodation and work spaces. The radiation is determined from the average velocity level of the six bounding surfaces and their radiation efficiency, see e.g. [16].

Three major approaches are being used to determine the structure-borne noise transmission through the ship structure [17]:

1. semi-empirical models,
2. wave guide models,
3. Statistical Energy Analysis (SEA).

The semi-empirical models give the velocity level difference between two positions in the ship as a function of the number of decks and transverse frames that are crossed [4]. These models are based on the statistics of measurements on a series of existing ships. They can give good results for similar ships, but fail when new concepts are introduced in the design.

The wave guide models [3,9,18] give a deterministic description of the structure-borne noise transmission through a vertical section of the ship structure between two transverse frames. The corresponding noise transmission across frames is estimated on the basis of semi-empirical models.

SEA [19] is an efficient method for the prediction of structure-borne sound transmission, that has been under development since the early sixties. In an SEA model, the ship is divided into subsystems, which exchange acoustical energy via resonant modes. The energy distribution in the ship is calculated by solving the linear system of energy balance equations that describe the internal losses in and coupling losses between these subsystems. The advantage of the SEA method is that it is based on an analytical description of the parameters of the subsystems, so that the effects of design parameters like material properties, plate thickness, stiffener distance, etc., are explicitly taken into account. Several computer codes for SEA (*AutoSEA, SEAM, SEADS*, etc.) have become commercially available in recent years and effort has been put in facilitating the pre-processing of models for large, complex ship structures [20].

The results of a comparison of SEA calculations with full scale measurements on three ships [20] indicate that a typical accuracy of 3 to 6 dB in the prediction of the A-weighted sound pressure levels is achievable. The accuracy of the SEA calculations is critically dependent on the source level estimation.

In SEA the source is described in terms of a vibro-acoustic input power into a subsystem. The input power P into the ship caused by propeller cavitation is not only dependent on the average velocity $\langle v^2 \rangle$ of the hull, but also on the real part of the complex impedance Z (the ratio of driving force and velocity response) of the aft ship:

$$P = \tfrac{1}{2} \operatorname{Re}\{Z\} < v^2 > \qquad (4)$$

As stated in section 3, the aft ship is driven by a pressure distribution. Therefore, attempts to estimate the impedance on the basis of measured or calculated impedance values for point force excitation of the aft ship [20,21] will not necessarily lead to reliable results. This is illustrated by the results of FEM/BEM calculations on the simply supported, water loaded plate as described in section 4. In figure 5 the impedance for point force excitation at the centre of this plate is compared with the point impedance for excitation by a monopole at 1 m above the plate's centre. In this case the driving force due to monopole excitation has been defined as the actually measured point pressure at the centre position multiplied by the area of the plate. Although one might argue that the hull surface impedance

924

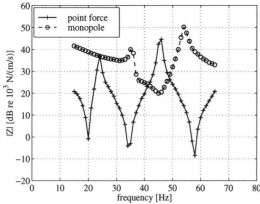

Figure 5: the impedance Z for point force excitation at the centre of a rectangular plate (fig.2), compared with the hull impedance (Ap_{hull}/v_{hull}) in case of excitation by a monopole, where A is the area of the plate (1.8 m^2).

should be described on the basis of the average velocity and the pressure distribution integrated over the plate area, fig. 5 indicates that there is no direct relationship between the two impedance spectra.

A correct estimation procedure for the input power into the ship structure due to the propeller induced pressure fluctuation distribution at the aft ship hull might be developed in analogy with the "effective mobility" concept for describing the input power due to the multi-point force distribution at the foundation of large engines [22].

7. CONCLUSION

Prediction methods for propeller cavitation noise on board include models that describe the source mechanism and models for the transmission of noise through the ship structure. Semi-empirical models, that are based on full scale measurement data on existing ships, have proved to be very useful in an early design stage, to decide whether noise reducing measures will be necessary. However, these methods have their limitations. Usually they are not very reliable in the low frequency range where propeller tonals determine the noise levels. And they can only be used reliably when the design of ship and propeller does not seriously deviate from the ships on which the measurements have been taken. If major design modifications are considered, or if noise problems are foreseen at low frequencies, it becomes worthwhile to apply analytical or numerical

models, in combination with (scale model) experiments wherever that is possible.

The most natural extension of the statistical, semi-empirical methods for the prediction of noise transmission through the ship is the method of Statistical Energy Analysis (SEA), that takes into account design parameters like material properties, plate thickness, frame distance, etc., without requiring a too detailed description. The main difficulty associated with the application of SEA for propeller induced noise is that a correct description of the excitation mechanism remains to be developed.

In the low frequency range, one needs deterministic models to describe the cavitation source strength, the response of the aft ship and the noise transmission through the ship. These require detailed information of the design parameters.

The applicability and reliability of the available analytical and numerical methods, that are reviewed in this paper, is strongly dependent on how detailed and how accurate the input data can be given, but also on the limitations in the accuracy of the different models, due to the specific approximations that have been taken. Some examples of these limitations are discussed in this paper and possibilities for further development are indicated.

REFERENCES

1. *Proceedings of the International Symposium on Shipboard Acoustics* 1976, Amsterdam: Elsevier Scientific Publishing Company
2. *Proceedings of the 2nd International Symposium on Shipboard Acoustics* 1986, Dordrecht: Martinus Nijhoff Publishers
3. NILSSON A.C. & N.P. TYVAND 1981 *Noise sources in ships I: Propellers.* Stockholm: Nordforsk,
4. JANSSEN, J.H. & J. BUITEN 1973 On acoustical designing in naval architecture. *Proc. Internoise'73*, Copenhagen, 349-356
5. DE BRUIJN, A., W.H. MOELKER & F.G.J. ABSIL 1986 Prediction method for the acoustic source strength of propeller cavitation. In [2], 1-19
6. JANSSEN, J.H. & W.H. MOELKER 1986 Some experiments on the transmission of propeller cavitation noise into the ship's structure. In [2], 103-119

7. VAN DER KOOIJ, J. 1986 Experimental and analytical aspects of propeller induced pressure fluctuations. In [2], 43-62

8. BROWN, N.A. 1976 Cavitation noise problems and solutions. In [1], 21-38

9. NILSSON, A.C. 1980 Propeller induced hull plate vibrations. *Journal of Sound and Vibration* **69**(4), 539-557

10. BRUGGEMAN, J.C., C.A.F. DE JONG & F.G.P. VAN DER KNAAP 1991 Numerical calculations of hull pressures caused by propeller cavitation. Report TPD-SA-RPT-91-0050 (in Dutch), TNO Institute of Applied Physics, Delft

11. BRUGGEMAN, J.C. & C.A.F. DE JONG 1993 Sound radiated by a ship's hull plate. *Proc. Internoise'93*, Leuven, Belgium

12. LOMAS, N.S. & S.I. HAYEK 1977 Vibration and acoustic radiation of elastically supported rectangular plates. *Journal of Sound and Vibration* **52**(1), 1-25

13. VERITEC, Høvik, Norway 1985 *Vibration control in ships.*

14. *SYSNOISE 4.4a.* LMS NUMERICAL TECHNOLOGIES NV, Leuven, Belgium

15. HECKL, M. 1988 Entwicklung von Methoden zur Schallpegelprognose für den Unterkunfts- und Maschinenraumbereich von Schiffen. Forschungszentrum des Deutschen Schiffbaus, Bericht Nr.193

16. CREMER, L., M. HECKL & E.E. UNGAR 1988 *Structure-borne sound.* New York: Springer Verlag

17. NILSSON, A.C. 1984 A method for the prediction of noise and velocity levels in ship constructions. *Journal of Sound and Vibration* **94**(3), 411-429

18. HYNNÄ, P. 1986 A literature survey concerning propeller as noise source and prediction methods of structure-borne noise in ships. In [2], 233-243

19. LYON, R.H. & R.H. DEJONG 1995 *Theory and application of Statistical Energy Analysis.* Boston: Butterworth-Heinemann

20. HYNNÄ, P., P. KLINGE & J. VUOKSINEN 1995 Prediction of structure-borne sound transmission in large welded ship structures using Statistical Energy Analysis. *Journal of Sound and Vibration* **180**(4), 583-607

21. IRIE, Y. 1981 Structure-borne sound power induced by propeller. *Proc. Internoise'81*, 687-692

22. PETERSSON, B.A.T. & J. PLUNT 1982 On effective mobilities in the prediction of structure-borne sound transmission between a source structure and a receiving structure. *Journal of Sound and Vibration* **82**(4), 517-540

© 1998 Elsevier Science B.V. All rights reserved.
Practical Design of Ships and Mobile Units
M.W.C. Oosterveld and S.G. Tan, editors.

Computation of Structure-borne Noise Propagation in Ship Structures using Noise-FEM

C. Cabos, J. Jokat

Strength & Vibrations Dept., Germanischer Lloyd
Vorsetzen 32, D-20459 Hamburg, Germany

The Noise Finite Element Method (Noise-FEM) allows the computation of structure-borne sound intensity flux through complex ship structures. In case finite-element models for global strength and vibration calculations are available, they can be further used for Noise-FEM analyses. The method can be employed to optimize local structural elements along the main structure-borne sound transmission paths, thereby increasing the structure-borne sound insulation index in the frequency ranges of interest. Applied Noise-FEM results are shown and discussed. By comparing computed and measured structure-borne sound levels, the accuracy is plotted against the octave mid-band frequency, ranging from 31.5 Hz to 8,000 Hz.

1. INTRODUCTION

The **Noise Finite Element Method** (Noise-FEM) developed by Germanischer Lloyd in a BMBF (Federal Ministry of Education, Science, Research and Technology) project makes it possible to forecast the propagation of structure-borne noise in complex ship structures. The aim of the development of Noise-FEM (NFEM) was to predict the propagation of structure-borne noise on the basis of the geometry data of existing finite element models created mainly for strength and vibration computations.

To check the reliability of Noise-FEM, extensive measurements on real ship structures were performed in addition to structure-borne, water-borne and air-borne noise measurements carried out on a ship model structure. The procedure and accuracy of the Noise-FEM results are discussed below, taking some applications as examples.

2. COMPUTATION OF STRUCTURE-BORNE NOISE PROPAGATION

Basically, this is a vibration or wave propagation problem which, in the case of harmonic excitation, leads mainly to the Helmholtz equation. If the finite element method is used to solve this equation, it can be assumed as a rule of thumb that, in the discretization process, about ten elements are needed per wavelength. In the case of a container ship of length 250 m in the range of frequencies around 1000 Hz, this leads to models with over 10^6 degrees of freedom. Because predictions for the mean propagation of structure-borne noise are usually required in a particular frequency band (e.g. an octave), the above-mentioned vibration computations would have to be repeated for a very large number of frequencies. On the other hand, precisely the fact that information averaged over a frequency band is required makes it appear unnecessary to perform a large number of narrowband computations with a highly detailed model.

The problems mentioned above led to the development of statistical methods for computing the average propagation of energy. Thus, for example, the energy exchange between discrete subsystems performing coupled oscillations is investigated in the *Statistical Energy Analysis* (SEA) [4]. One difficulty in the application of SEA in shipbuilding is that a special 3D model of the ship has to be created, tailored to suit the method. An FE model already in existence cannot be used for a computation with the SEA.

In the derivation of the SEA, it is assumed that there is
• a uniform distribution of the energy per mode of a subsystem.

Furthermore, it is assumed that there is
• weak coupling of the subsystems, and
• a large number of modes in the frequency band for each subsystem.

It is precisely the last two requirements mentioned that impede the use of the finite element method: convergence of an FE approach can be achieved only if finer subdivision of the model does not lead to poorer results than the coarse model. For this reason, in the Noise-FEM the basic equations of the SEA are used, with appropriate corrections.

In the search for a way of using FE models for the computation of structure-borne noise propagation, one possibility that presents itself consists of an analogy to the propagation of heat:

There, the vibration problem is not solved in the microscopic, molecular domain: instead, the solution to the macroscopic energy propagation problem is found. A key role is assigned to the definition of the temperature T. It describes the energy content per vibrating subsystem (mode) and determines the direction of energy flux. In this way, the simple law $I = -\Lambda\nabla T$ is obtained, where I is the intensity and Λ stands for the thermal conductivity.

The propagation of structure-borne noise can be described by means of the same equation if, analogously, the *sound temperature T* is defined as the mean energy per mode of a subsystem that is capable of vibrating. The greatest difficulty in the application of the above equation consists in determining the *sound conductivities* Λ. If these are known, then - as in the case of heat propagation too - it is comparatively simple to use a power balance in order to derive a parabolic propagation-equation, which can be solved with the aid of the method of finite elements, for example. The result is the sound temperature distribution and hence the energy distribution based on the given power fed into the system.

To derive the conductivity relation, methods for the computation of structure-borne noise propagation that are known from the literature are used as the starting point. These methods include the **Acoustical Waveguide Model**, the **Power Flow Finite Element Analysis** (PFFEA) and the **SEA**. They serve as the basis of the **Noise-FEM** developed for the computation of structure-borne noise propagation in complex structures. A sketch illustrating this is given in Fig. 1.

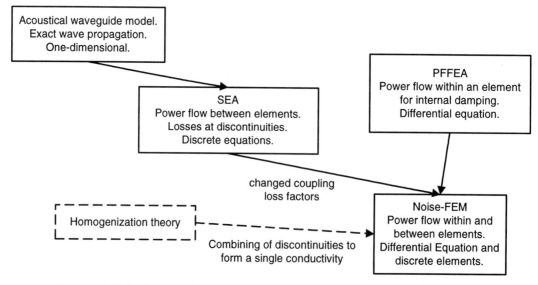

Figure 1. Relationship between Noise-FEM and the other methods used

3. OVERVIEW OF NOISE-FEM

In a ship structure, the main differences in structure-borne noise velocity level (vibration velocity on solid bodies in the audible frequency range) do not occur within homogeneous plates or beams. There, the propagation losses caused by damping are the most important factor. Instead, these differences occur at discontinuities such as stiffeners or angles between plates, i.e. they are caused by structure-borne noise attenuation. In order to model both of the effects mentioned, Noise-FEM proceeds as follows:

Damping within homogeneous plates and beams can be computed with the aid of PFFEA [5]. This theory leads to a propagation equation of the desired type, separated depending on wave type. It can be largely modelled by means of finite elements, as used for heat propagation computations. The thermal conductivities Λ needed for plates are derived in [1]. For example, in the case of uniform plates the relation $\Lambda = \dfrac{\Delta\omega}{\pi\eta}$ is obtained, where $\Delta\omega$ represents the width of the frequency band being considered and η is the loss factor of the material. The drawing-up of a power balance for the closed section of a plate leads to the elliptic equation for the steady-state sound temperature distribution:

$$-\nabla\cdot(\Lambda\nabla T) + C\omega_0\Delta\omega\eta T = P_0''. \tag{1}$$

Here ω_0 is the mean frequency in the frequency band being considered and C is the *mode density*, i.e. the number of eigenfrequencies per plate area and width of the frequency band. The symbol P_0'' denotes the power introduced per area.

The **attenuation** effects at a discontinuity are covered separately for each edge adjoining several finite elements. One example might be two plates meeting one another at an angle. For this case, the SEA predicts a net power flow of magnitude

$$P_{12} = -\omega_0\eta_{12}N_1(T_2 - T_1). \tag{2}$$

The "coupling loss factor" η_{12} depends on the material and geometry of the coupled plates; N_1 is the number of modes in subsystem 1. Reliable computation of the coupling loss factors is important for the accuracy of prediction of the structure-borne noise propagation. If the coupling is weak and the mode density is high, η can be determined with the aid of detailed calculations; for this purpose, the acoustical waveguide method [2], for example, is used. If these conditions are not fulfilled, then correction factors must be introduced [4].

At a plate junction a wave of any type on one plate can in general cause power flow of any type on any other plate connected to the junction. Therefore, a coupling matrix occurs which, if there are n elements meeting at an edge, has $3n$ rows and columns (corresponding to the three wave types: bending waves, longitudinal waves and shear waves).

In a finite element model of a ship, it is not standard practice at present to model all stiffeners. For this reason, the Noise-FEM offers the possibility of "smearing" the effects of several stiffeners on a plate to form a single homogeneous conductivity. For this purpose, the propagation equation of the plate and the coupling relations of the discontinuities are jointly **homogenized** by means of the methods described in [3]. The main difficulty here lies in correctly taking account of the different wave types.

Thus, for Noise-FEM, there are mainly two **element types** that are used:
- An element which describes the couplings along an edge - it has $2n$ nodes if there are n elements meeting each other at the edge.
- A 3- or 4-node surface element which models a possibly orthotropic sound energy conductivity. The orthotropy depends on the direction in which the homogenized stiffeners run. The conductivity of an element without any stiffeners depends only on the effects of internal damping.

4   APPLICATION OF NOISE-FEM

4.1 Partially Equipped Deckhouse

Extensive measurements of structure-borne sound were performed on a partially equipped deckhouse of a container ship (type ECOBOX 42). The deckhouse was situated at the fabrication site provided for it on the premises of the shipyard Flensburger Schiffbau-Gesellschaft (FSG), Germany. The principal dimensions of the deckhouse were:

- Length   approx. 15 m
- Width   approx. 26 m
- Height   approx. 20 m
- Mass   approx. 800 t

The structure was excited at the rear transverse wall of the deckhouse with the aid of a shaker (an electrodynamic vibration-exciter) in the frequency range from 40 Hz to 6300 Hz with third-octave band noise ("white" noise). The response of the structure was simultaneously measured with a piezoelectric force sensor and several accelerometers; the measurement signals were stored on a DAT cassette via measurement cables and a DAT recorder, and were then evaluated in Germanischer Lloyd's measurement laboratory.

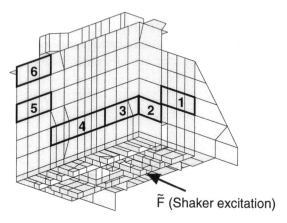

$\tilde{F}$ (Shaker excitation)

Figure 2. Finite element model used for Noise-FEM analysis (FSG deckshouse)

The mesh of the deckhouse FE model was chosen to be fine enough to ensure that every longitudinal and transverse wall could be replicated. Built-up web frames and bulb profiles were not modelled as plane elements. Information about their dimensions and

positions is contained in a special file which the Noise-FEM needs when computing the sound propagation.

Fig. 2 shows the finite element model of the FSG deckhouse. The results of the marked areas are compared as an example for the distinctions between measurement and computation in Fig. 3. Altogether, 550 different structure-borne noise measurement points were distributed over approximately 41 wall and deck surfaces of the deckhouse and measured (each surface having an area of about 20 to 50 square metres). At least nine and at the most twelve structure-borne sound measurement points were provided per "surface". The structure-borne noise velocity levels per "surface" were averaged on the basis of energy, and the corresponding results from the FEM computation were compared. The items compared consisted of the structure-borne noise level differences (third-octave mid-band frequency, re $v_0 = 5*10^{-8}$ m/s) between the velocity level (vibration velocity) at the excitation point (reference point) and the various measurement points at the side, interior and deck surfaces of the deckhouse (Fig. 2).

From the comparison (Fig. 3), it can be seen that good agreement exists between measurement and computation in the frequency range between 160 Hz and 4000 Hz. The evaluation of the measurement results revealed that, starting at the excitation point (in the lower region of the rear wall of the deckhouse), the structure-borne noise produced by the excitation (point excitation) was propagated mainly via the decks into the side walls and front wall. These principal structure-borne noise propagation paths were reproduced by the simulation computations (Noise-FEM). The comparison for measurement area 1 (Fig. 3) shows larger deviations. The measured structure-borne noise difference in velocity level is greater than the predicted level difference. This is due to a large swimming pool trunk construction which exists in reality at the level of the 2nd superstructure-deck but was not incorporated into the FE model. For this reason, in the Noise-FEM computation more structure-borne noise energy can distribute itself into the deck concerned and thus into the port side wall.

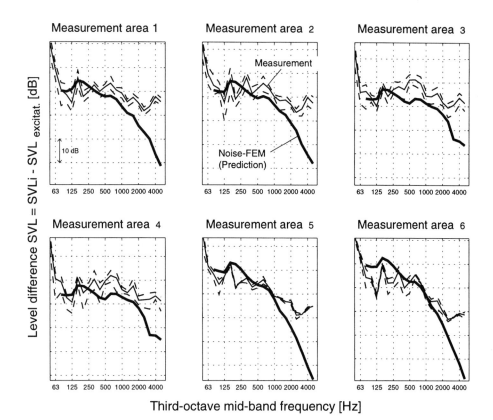

Figure 3. Measured and predicted structure-borne noise velocity levels (SVL) for the FSG deckshouse

Below 160 Hz, larger deviations generally occur. This is mainly due to the method itself, because in this frequency range the mode density per finite element is lower.

Below 63 Hz, it was not possible to excite the whole structure adequately by means of the selected measurement set-up.

4.2 Aft Body of a Container Ship

During a cargo-carrying voyage on the container ship "Trade Sol" (FSG nb. 685, type ECOBOX 42), extensive measurements of structure-borne and air-borne sound were performed in the operational state 90% MCR and in harbour operation. The velocity levels and the mechanical input-impedance (frequency-dependent vibration resistance of the structure) were determined by measurement at several points on the shell plating above the propeller, on the engine foundation and below the resilient mountings

of the auxiliary diesel unit. In the operational state 90% MCR, a large number of structure-borne noise acceleration measurements were performed on the exterior surfaces of the deckhouse, and airborne noise measurements were carried out in all accommodation rooms. The aim of these measurements was to be able to estimate the structure-borne noise power fed into the ship's structure by the relevant structure-borne noise sources, and to ascertain the structure-borne noise distribution in the exterior walls of the deckhouse.

4.2.1 Finite Element Model for the Noise-FEM Computation

The finite element model for the container ship mentioned was created for strength examinations and global vibration investigations (Fig. 4). In order to be able to use these geometry data for the Noise-FEM computations, minor changes had to be made in the model geometry and a file containing the dimensions and positions of the stiffeners had to be generated. The changes in the model geometry mainly involved the regions in which the structure-borne noise excitation had to take place in the model. The coarse modelling level and the simplifications in the modelling that are usual for global vibration models were retained.

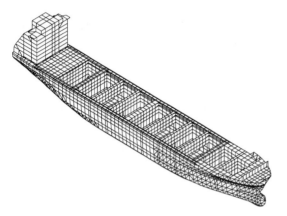

Figure 4. Finite element model used for global strength and vibration analysis

The length of the model used for the Noise-FEM computation was restricted to a region ranging to about 20 m ahead of the forward engine-room bulkhead. The effect of this reduction on the results of the propagation computation is negligible, because in this region the structure-borne noise level during the operational state 90% MCR was almost too low to measure.

Noise-FEM predicts the propagation of the structure-borne noise power, and at the location of the structure-borne noise source concerned it needs to know the structure-borne noise power being fed in. With the aid of the mean mechanical input impedance Z_i determined by measurement and the energy-averaged structure-borne noise velocity $v_{\text{eff},i}$ for the source i, the structure-borne noise power $P_{K,i}$ being fed in was estimated by means of the relation $P_{Ki} = v_{\text{eff},i}^2 \cdot \text{Re}(Z_i)$ and was used as the excitation for the Noise-FEM model.

4.2.2 Comparison of the Results

As an example, Fig. 5 shows the 3D presentation of the bending wave velocity distribution in the aft body of the ship for the octave mid-band frequency 125 Hz.

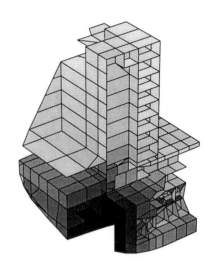

Figure 5. Noise-FEM results for the octave mid-band frequency 125 Hz in dB

The longitudinal section of the ship's aft body can be seen with the excitation on the engine foundation, the shell plating above the propeller, and the foundation of the auxiliary diesel unit. Whereas the velocity levels in the region of the sources exhibit larger differences or level reductions, the levels in the deckhouse are fairly evenly distributed. The levels lie in the range between 55 dB and 90 dB, corresponding to the grey scale.

In Fig. 6, the energy-averaged velocity levels from the measurement and from the Noise-FEM computation are compared for all exterior surfaces of the deckhouse between the 5th and 6th superstructure-decks.

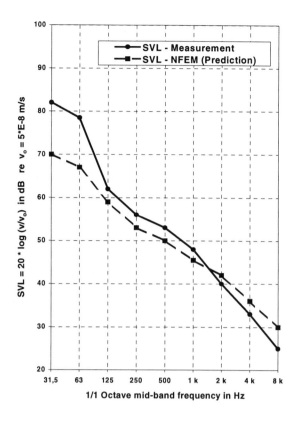

Figure 6. Measured and predicted structure borne noise velocity levels (SVL) for all walls between the 5th and 6th superstructure decks

In the frequency range between 125 Hz and 2000 Hz, the octave levels of the structure-borne noise prediction exhibit relatively good agreement. Above 2000 Hz, the predicted levels are overestimated. The cause must be looked for in the assumption of the loss factors, because a mean loss factor of 0.007 (63 Hz) to 0.0025 (8000 Hz) was assumed for all surfaces of the deckhouse. This loss factor does not take account of special acoustic insulation measures (e.g. effective cavity-insulation, fire protection insulation etc.) such as were present in reality to some extent. In the frequency range below the 125 Hz octave, a larger deviation from the measured values is evident. The levels of the Noise-FEM prediction are underestimated. The causes must be looked for in the lower mode-density per finite element and in the ignoring of structure-borne noise sources

situated in the upper deckhouse (rigidly mounted combined exhaust-gas boiler and exhaust-gas silencer in the engine casing, air-borne noise excitation of the deckhouse rear wall and of the deckhouse top deck by the exhaust-gas noise). Because the levels below the 125 Hz octave are mainly due to the propeller excitation, the excitation zone of the shell plating above the propeller - especially that of the longitudinal wave component - likewise cannot be contained to an adequate extent in the prediction model. Possible sources of error are currently being investigated.

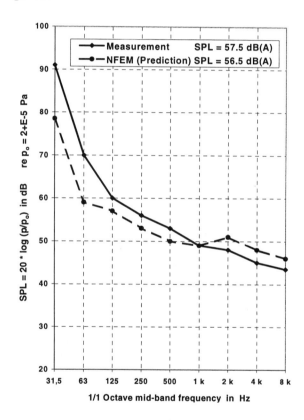

Figure 7. Measured and predicted average airborne sound pressure levels (SPL) for all cabins on the 5th superstructure deck

In Fig. 7, a prediction of the air-borne noise levels on the 5th superstructure-deck was performed on the basis of the predicted structure-borne noise levels and of an empirically determined transfer function between

934

the mean structure-borne noise level in the exterior walls of the deckhouse and the mean air-borne noise level in acoustically similar cabins. Although the dB(A) values between measurement and prediction differ by only 1 dB(A) (see Fig. 7). The frequency-dependent level curve exhibits relatively small differences in the lower frequency ranges (Fig. 7, 31.5 Hz to 125 Hz).

REFERENCES

[1]   O.M. Bouthier, R. J. Bernhard. *Models of space-averaged energetics of plates* AIAA Journal, 30(3):616-623, 1992.

[2]   L. Cremer, M. Heckl: *Körperschall.* *Physikalische Grundlagen und technische Anwendungen*. 2nd Edition,. Springer-Verlag, Berlin, 1996.

[3]   V.V. Jikov, S.M. Kozlov, and O.A. Oleinik. *Homogenization of Differential Operators and Integral Functionals*. Springer-Verlag, Berlin, 1994.

[4]   R. H. Lyon, R.G. DeJong: *Theory and Application of Statistical Energy Analysis*. Butterworth-Heinemann, Boston 1995.

[5]   D.J. Nefske, S.H. Sung. *Power flow finite element analysis of dynamic systems: Basic theory and application to beams*. Transactions of the ASME, Journal of Vibration, Acoustics, Stress, and Reliability in Design, 111:94-100, January 1989.

Practical Design of Ships and Mobile Units
M.W.C. Oosterveld and S.G. Tan, editors.

The acoustic source strength of waterjet installations

K.N.H. Looijmans[a], R. Parchen[a], and H. Hasenpflug[b]

[a]TNO Institute of Applied Physics (TNO-TPD)
P.O. Box 155, 2600 AD Delft, The Netherlands

[b]Royal Netherlands Navy
Directorate of Material, Naval Design Department
P.O. Box 20702, 2500 ES The Hague, The Netherlands

This paper describes the main noise producing mechanisms of waterjets. A prediction model for the acoustic source strength of waterjet installations under non-cavitating conditions will be discussed. Furthermore, a laboratory experiment on the acoustic source strength of waterjet pumps and on installation effects is described. The model predictions are compared with experimental results.

1. INTRODUCTION

With the advent of high-power waterjet installations, waterjet propulsion becomes increasingly interesting for applications such as ferries and naval applications. Owing to the high

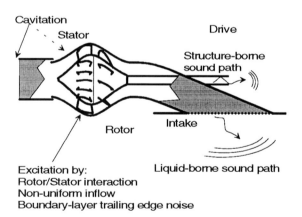

Figure 1: Schematic drawing of a waterjet installation with hydrodynamic noise sources and transmission paths to underwater noise.

cavitation inception speed, waterjets may be up to 10 dB less noisy than conventional propeller propulsion at high speeds. However, depending on details in the installation, the character of the noise may contain

undesirable features, such as having strong tonal components.

Figure 1 shows a schematic drawing of a waterjet installation in which its most important noise sources and transmission paths to underwater noise are indicated. The waterjet installation consists of a pump of mixed flow type mounted in the waterjet channel with the impeller and stator as its main parts. Important noise sources are: 1) cavitation near the intake and in the waterjet pump which may generate both broadband and tonal noise, 2) wake noise caused by a non-uniform time-averaged inflow of the impeller resulting in tonal noise, 3) tonal noise due to rotor-stator interaction, 4) broadband noise caused by turbulence in the impeller inflow, and 5) broadband noise due to excitation of the impeller and stator blades by boundary-layer turbulence.

Two sound transmission paths are sketched in figure 1, viz. the liquid-borne transmission path and the structure-borne transmission path. The liquid-borne sound path radiates underwater noise through the intake opening. The structure-borne sound path transfers the excitation forces on the impeller and waterjet duct wall via the shaft, gearbox, and partitions in the construction to the ship hull. Subsequently, the ship hull radiates underwater noise.

The next section gives an overview in more detail of the noise sources that will contribute to underwater radiated noise. Models that describe the acoustic source strength of the source mechanisms are discussed. The source models that concern the acoustic source strength at non-cavitating operation of the waterjet have been implemented in a computer program to predict underwater noise of waterjet propulsion systems. The model can be used to study the effect of waterjet parameters, such as geometry, installation, and operation conditions on radiation of underwater noise. Results of computations are presented.

2. NOISE SOURCE MECHANISMS AND PREDICTION MODEL

2.1 Cavitation

Cavitation can be a source of strong broadband as well as tonal noise. In a waterjet propulsion system different types of cavitation can be distinguished, resulting in different types of noise radiation as well. Cavitation in the waterjet channel occurs as a consequence of pressure variations in the flow. When the pressure drops below the vapour pressure, cavitation volumes arise, creating individual vapour bubbles (bubble cavitation) or vapour volumes that may cover a large part of an impeller blade (sheet cavitation). Bubble cavitation can take place near the impeller and stator blades and near the waterjet intake (inlet lip cavitation).
Cavitation noise occurs due to the changes in cavitation volume. Therefore, it is a monopole noise source. Inlet lip cavitation and impeller cavitation cause broadband noise in the frequency range from approximately 100 Hz to above 100 kHz, due to the growth and rapid collapse of individual bubbles. Moreover, bubble cavitation and sheet cavitation may give rise to tonal noise related to the impeller blade frequency, as changes in cavitation volume are induced by periodic pressure variations in the neighbourhood of the impeller blades. Brown [1] has formulated a relation to describe the high-frequency broadband spectrum of propeller cavitation.

2.2 Non-uniform inflow

Velocity variations in the inflow of the impeller blade result in fluctuating lift forces on the impeller, which act as a hydrodynamic source of dipole noise. The impeller inflow variations originate from different mechanisms: 1) non-uniform time-averaged inflow, 2) rotor-stator interaction, and 3) inflow

turbulence noise. In this section we will concentrate on non-uniform time-averaged inflow, while rotor-stator interaction and inflow turbulence are dealt with in the next sections.

Non-uniform time-averaged inflow in the impeller plane of the waterjet is caused by intake of the hull boundary layer, and by the geometry of the waterjet channel. As a consequence of the non-uniform time-averaged inflow, the impeller blades experience a periodic variation of inflow velocity. Disturbances in amplitude and direction of the inflow velocity cause variations of lift force on an impeller blade. The time dependent periodic lift force variations radiate tonal noise with frequency of the tones equal to the blade rate and its harmonics.

Lift force variations can be considered to be of a dipole source type. The sound power W of such a dipole source in a pipe system is proportional to the fourth power of the Mach number $W \propto M^4$, with the Mach number M being the ratio between effective inflow velocity and velocity of sound. As a consequence of this strong dependence on Mach number, most part of the noise is radiated at the impeller tips, where the effective inflow velocity is largest. If it is assumed that the effective inflow velocity is proportional to the distance to the impeller shaft, the outer 10 percent of the blade contributes to over 50 percent of the radiated sound power.

The fluctuating forces on the blades of the waterjet caused by a non-uniform time-averaged inflow can be described by the quasi-steady lift force F on a body:

$$F = C_L \frac{1}{2} \rho U_{eff}^2 cs, \qquad (1)$$

where C_L is the lift coefficient, ρ (kg/m^3) the density of the fluid, U_{eff} (m/s) the effective inflow velocity of a blade section, c (m) the blade chord, and s (m) the length of the blade section in spanwise direction. The strength of the acoustic source is obtained by combining the expression for the lift force with the Green's function of the waterjet ducting.

Figure 2 shows the dependency of the acoustic source strength of the blade rate tonal on the rotational speed of the impeller for a small waterjet with a six bladed impeller. The blade rate tonal is generated by the non-uniform inflow of the impeller

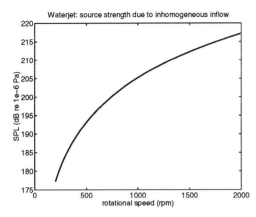

Figure 2: Source strength due to non-uniform inflow caused by a curvature in the inlet channel, as a function of the rotational speed of the impeller.

caused by the curvature in the waterjet duct. The strong dependency of the source strength on rotational speed is the result of the 4th order relation between Mach number and acoustic source power.

Figure 3 shows an example of a prediction of acoustic source strength caused by non-uniform inflow for a representative velocity profile in the impeller plane. The figure illustrates that besides the

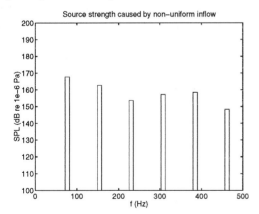

Figure 3: Source strength of blade rate and harmonics, due to non-uniform inflow.

blade rate tonal at 77 Hz a number of related harmonics is present in the source strength spectrum.

2.3 Rotor-stator interaction

Interaction between rotor and stator of the waterjet induces the third type of non-homogeneous inflow. Rotor-stator interaction can be separated into two different processes:

1) Non-viscous interaction as a consequence of the forces that rotor and stator blades exert on the fluid and therefore on each other. This interaction force is inversily proportional to the distance squared between the blades. Non-viscous rotor-stator interaction is only important for small axial separation between rotor and stator, which is the case in waterjets.

The model for the acoustic source strength of non-viscous rotor-stator interaction is based upon the occurrence of circulation around the rotor and stator blades. The circulation exert forces on rotor and stator blades:

$$F \propto \rho s \frac{\Gamma_{rotor} \Gamma_{stator}}{\delta}, \qquad (2)$$

where Γ (m²/s) is the circulation around a blade, and δ (m) the distance between the respective rotor and stator blade. Since the distance between rotor and stator blades varies with time as a consequence of the impeller rotation, a harmonically fluctuating

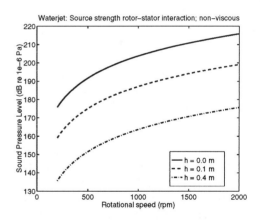

Figure 4: Source strength of non-viscous rotor-stator interaction as a function of rotor speed and distance between rotor and stator.

force exists, which radiates tonal noise of dipole type. For an ideal rotor and stator (equally spaced blades, uniform impeller inflow) the frequency of the tone is equal to the blade frequency multiplied by the number of stator blades. For non-ideal rotor and stator, harmonics of the shaft frequency arise. Figure 4 shows the source strength of non-viscous rotor-stator interaction as a function of rotational impeller speed and distance between rotor and stator.

2) Viscous rotor-stator interaction. The impingement of impeller blade wakes on the stator induces a time varying inflow on the stator blades. Since the depth of the wake decreases relatively slow with distance, viscous rotor-stator interaction remains important even at large distance between rotor and stator. Most important frequencies of the tones generated by rotor-stator interaction are the blade rate frequency and its harmonics. However, non-uniformities in the flow, turbulence in the rotor blade wakes, and irregularities in rotor and stator give rise to stator frequencies (shaft frequency times number of stator blades, and harmonics) and all harmonics of the shaft frequency.

Viscous rotor-stator interaction is modelled by combining a model for the rotor blade wakes with the model for quasi-steady lift forces on the stator blades. In figure 5 an illustration is given of the source strength of the viscous rotor-stator interaction where turbulence in the wakes and an irregular spacing of the rotor blades has been included in the calculation. All harmonics of the shaft frequency appear in the spectrum.

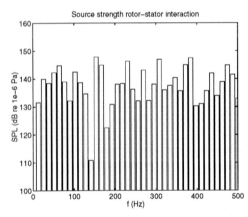

Figure 5: Example of the source strength of viscous rotor-stator interaction.

2.4 Inflow turbulence noise

Turbulence in the waterjet channel is a third type of non-homogeneous inflow of impeller blades. The Reynolds number based on the diameter of the channel is of the order $Re=O(10^6)$, indicating a turbulent flow in the waterjet channel. Inflow turbulence causes noise of both broadband and tonal character. If the size of the turbulent eddies in the flow direction is that large that the impeller blades cut the eddies several times, the noise will exhibit humps in the frequency spectrum at the blade frequency and its harmonics. Especially, if the flow is accelerated in, e.g., a contraction, stretching of the eddies can occur so that the size of the eddies in the direction of the flow is increased. This results in noise of tonal character that can hardly be distinguished from the tonal noise caused by non-uniform time-averaged inflow of the impeller. On the other hand, if the turbulent eddies are small, broadband noise is generated with a frequency spectrum that depends on the turbulence spectrum of the flow.

In our prediction model, we applied the model of Amiet [2] for inflow turbulence noise. Amiet developed a prediction model for acoustic radiation of an airfoil in a turbulent stream for the case that the blade chord is larger than a quarter of the wavelength of the radiated noise, and for small variations of the incident flow velocity. In this case the response of a blade section can be developed from a combination of two solutions for semi-infinite flat plates with a sharp upstream edge and a sharp downstream edge respectively. If the blade chord is smaller than a quarter of the wavelength, Amiet proposes a correction factor based on the Sears function. The sound pressure level in 1/3rd octave bands for non-compact sources, for a blade section of width s, is given by:

$$SPL_{1/3}^{hf} \propto 10\log_{10}\left(\frac{sL}{r^2} M^5 \frac{\overline{u^2}}{U_{eff}^2} \frac{\hat{K}_x^3}{\left(1+\hat{K}_x^2\right)^{7/3}}\right), \quad (3)$$

where:

$SPL_{1/3}^{hf}$	Sound pressure level.
s	blade segment width (m),
U_{eff}	effective incident velocity (m/s),
r	distance source-observer [m],
$\overline{u^2}$	turbulence intensity (m²/s²),
L	integral length scale of turbulence (m),
M	Mach number, M = U/c$_0$,
$\hat{K}_x^2$	normalised wave number of turbulence,
r	distance to the source (m),

The expression

$$\frac{\overline{u^2}}{U^2}\frac{\hat{K}_x^2}{\left(1+\hat{K}_x^2\right)^{7/3}} \qquad (4)$$

represents the spectrum of turbulent velocity fluctuations. In the Amiet model, turbulence is characterised by two parameters: the turbulence intensity $\overline{u^2}$, and an integral length scale L. Expression (4), which determines the spectrum of the inflow turbulence noise, is obtained assuming that the spectrum of the turbulence in the inflow can be described by a von Kármán spectrum.

Figure 6 shows a calculation of the source strength of inflow turbulence noise with estimated turbulence parameters. Note that the assumption of isotropic turbulence, which is connected with the von Kármán spectrum, is probably not correct for the waterjet channel owing to stretching of the turbulent eddies in contractions. Therefore, a more accurate prediction of inflow turbulence noise can be expected when a more accurate estimate of the turbulence spectrum is applied, including this phenomenon.

2.5 Boundary-layer trailing edge noise

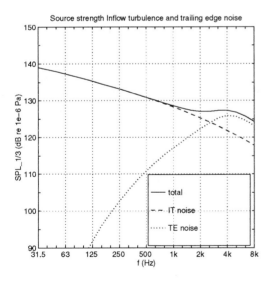

Figure 6 Source strength in 1/3-octave bands of inflow turbulence noise and trailing edge noise.

Turbulence produces noise. In free space the sound power W, that is generated by the turbulent vortices, shows a dependence on flow Mach number to the power eight: $W \propto M^8$. If scattering objects are in the vicinity of the vortices this proportionality changes. Hence, turbulent vortices in a turbulent boundary layer of an impeller or stator blade will radiate noise much more efficiently. In present prediction models the scaling of the trailing-edge noise is based on an analysis of Ffowcs Williams and Hall [3]. From their analysis it follows that the sound power of the trailing-edge noise is proportional to the Mach number to the power five: $W \propto M^5$.

Prediction models for trailing edge noise have been developed by, e.g., Grosveld [4], Brooks et al. [5], Lowson [6] en Howe [7]. All these models are based on the analysis of Ffowcs Williams and Hall, and assume, with only some small exceptions, the following relation for the sound power:

$$W \propto \frac{4\rho U_{eff}^5 \frac{\overline{u^2}}{U^2}\delta s}{\pi c_0^3} \qquad (5)$$

where δ [m] denotes the boundary-layer thickness.

The difference between the various prediction models mainly consists in the expression that is used for the spectrum shape of the trailing-edge noise which, in all cases, was derived from experimental data.

Brooks et al. [5] give an extensive correlation for the noise spectrum as a function of Reynolds number, angle of attack, boundary-layer displacement thickness, etc., on the basis of experiments on airfoil sections in a wind tunnel.

The spectrum of the trailing-edge noise is significantly determined by the boundary-layer structure. The boundary-layer characteristics, in their turn, are strongly dependent on the angle of attack of the flow, and the profile geometry. The only model mentioned that accounts for these correlations in detail is the model of Brooks et al. [5]. For this reason we assume this model as most appropriate to describe the boundary-layer trailing-edge noise mechanism. A calculation of the source strength of boundary-layer trailing-edge noise with the model of Brooks et al. [5] is also given in Fig. 6. The trailing edge noise spectrum shows a maximum near the 4 kHz 1/3-octave band which is determined by the

Strouhal number based on the boundary-layer displacement thickness and effective inflow velocity.

2.6 Liquid-borne transmission path in the waterjet channel

The waterjet pump is installed in a duct with a length of the order of the wavelength of the noise. Therefore, the acoustic source is situated in a reverberant sound field. Consequently, the sound transfer from dipole source in the pump to underwater noise is strongly dependent on frequency and duct geometry.

Figure 7 shows a principle sketch of the acoustic model for the liquid-borne transmission path in the waterjet channel. The pump is situated in a straight pipe at a distance L1 from the outlet opening. The intake opening at a distance L2 from the pump is modelled as a radiation impedance of flanged open end. For the frequency range of interest, only plane waves propagate in the waterjet duct. The sound transfer over the pump is modelled with a transfer matrix, in which the pump is incorporated as a two port system with added mass effects and compressibility of the pump housing included.

Figure 8 shows the result of a calculation of the transfer function of source strength p' to underwater radiated noise.

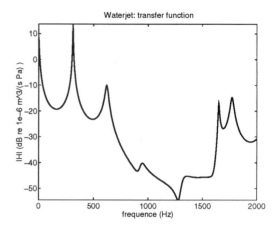

Figure 8: Transfer function of the liquid-borne transmission path of the dipole source strength p' to the volume flux at the intake opening.

schematically given in Fig. 9. In the suction and discharge piping of the waterjet pump pressure transducers and executors were installed for the acoustic measurements.

Figure 10 shows a typical noise spectrum measured with the two-microphone method of the in-duct sound pressure generated by the waterjet pump. The spectrum shows much of the properties that are predicted with the models discussed in the previous paragraphs. First, the spectrum exhibits large peaks that correspond with tones at the harmonics of the blade frequency of 100 Hz, i.e., 200 Hz, 300 Hz and 400 Hz. These tones are attributed to non-uniform time-averaged inflow. Intermediate tones are harmonics of the shaft frequency of 16.7 Hz and of the stator frequency (shaft frequency times number of stator blades). These harmonics are attributed to rotor-stator interaction. Furthermore, the level of the

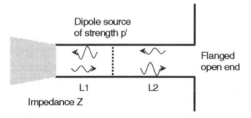

Figure 7: Sketch of the acoustic model of the liquid-borne transmission path.

3. EXPERIMENTS

In co-operation with LIPS Jets b.v. acoustic source strength measurements were performed on a waterjet pump in a closed water tunnel facility at NEL. A two-source location method in combination with a two-microphone method was employed to measure the transfer matrix and the acoustic source strength of the waterjet pump. The test set-up is

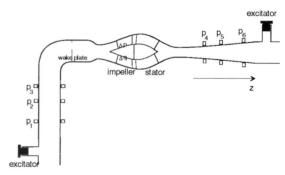

Figure 9: Laboratory test set-up for acoustic source strength measurements on a waterjet pump.

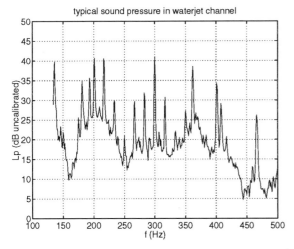

Figure 10: Typical in-duct noise spectrum measured on a waterjet pump in a laboratory test set-up.

broadband spectrum varies around 130 dB re 1.10^{-6} Pa which roughly corresponds to the level of the broadband source strength estimated for inflow turbulence noise and trailing edge noise. This broadband level exhibits qualitatively the peaks and troughs that are predicted by the transfer function. It must be noted here that the geometry and length of the ducts in the laboratory experiment differ from the duct geometry and length of the waterjet system used to calculate the transfer function of Fig. 7. This explains why the large peak at 300 Hz in Fig. 7 can not be found in the measured sound spectrum of Fig. 10. We expect that a much better quantitative agreement will be obtained when more accurate input parameters such as duct geometry and turbulence spectrum are used.

4. CONCLUSIONS

We have developed a prediction model for the underwater noise of waterjets radiated through the intake. In this paper we have concentrated on developing models for tonal and broadband noise at non-cavitating operation of waterjets. The following noise sources are included in the model: inhomogeneous inflow of the waterjet pump caused by non-uniform time-averaged inflow, rotor-stator interaction, and inflow turbulence, and noise production by turbulent boundary-layer trailing-edge noise. The model predicts the occurrence of tones at harmonics of the blade frequency due to non-uniform time-averaged inflow, and harmonics of the shaft frequency caused by rotor-stator interaction. The models for inflow turbulence noise and trailing edge noise describe the source strength of broadband noise.

The liquid-borne transmission path from the acoustic sources in the pump to underwater noise is modelled by a transfer matrix method. The calculation shows that the transfer function exhibits sharp peaks and troughs determined by the geometry of the duct.

The predictions of the model were compared to laboratory measurements of the acoustic source strength of a waterjet pump in a water tunnel facility. The results show qualitative agreement, which can be improved by a more accurate description of the inflow turbulence spectrum and the waterjet geometry.

ACKNOWLEDGEMENT

The possibility to perform acoustic source strength measurements on a waterjet pump of LIPS Jets b.v. in a laboratory set-up is gratefully acknowledged.

REFERENCES

1. Brown, N.A., Cavitation noise problems and solutions. *Proc. ISSA'76 (1976)*, Elsevier Scientific, Amsterdam, 21-38.
2. Amiet, R.K., Acoustic radiation from an airfoil in a turbulent stream, J. Sound Vib., 41 (1975), 407-420.
3. Grosveld, F.W., Prediction of broadband noise from horizontal axis wind turbines, J. Propulsion, 1 (1985), 292
4. Ffowcs Williams, J.E., Hall, L.H., Aerodynamic sound generation by turbulent flow in the vicinity of a scattering half-plane, J. Fluid Mech., 40 (1970), 657
5. Brooks, T.F., Pope, D.S., Marcolini, M.A., Airfoil self noise and prediction, NASA Reference Publication 1218, (1989)
6. Lowson, M.V., Assessment and prediction of wind turbine noise, Department of Trade and Industry, ETSU W/13/00284/REP, (1993)
7. Howe, M.S., Noise produced by a sawtooth trailing edge, JASA, 90 (1991)

Practical Design of Ships and Mobile Units
M.W.C. Oosterveld and S.G. Tan, editors.

VISCOELASTIC PASSIVE DAMPING TECHNOLOGY ON SHIP'S VIBRATION AND NOISE CONTROL

Wei-Hui Wang, Rong-Juin Shyu and Jiang-Ren Chang

Department of Naval Architecture
National Taiwan Ocean University
Keelung 202, Taiwan, R.O.C.

Vibration and noise control of structures are vital to many high performance ships. This article first outline the most important physical aspect of structure-borne sound, including the generation, transfer and control of such phenomena. Also, the mechanical four pole system is used to introduce the effect of damping mechanism into the structural model. This article also outlines and summarizes most important aspects of the viscoelastic passive damping technologies associated with vibration and noise control onboard ship. Besides the fundamentals of damping material properties, vibrations of discrete damped systems and fundamentals of vibration control technique have been discussed, approach concentrated on the vibrational behaviour of ship's structures with constrained viscoelastic layers is undertaken. Finite element model for beams, plates like a ship's structure were constructed using PATRAN. Modelling code was then generated for MSC/NASTRAN to compute the driving point impedance. Experiments were also carried out in an engine room structure. A shaker driven by a power amplifier with band-limited white noise signal was used as the input excitation, the driving force signal was measured by an impedance head. Accelerometers were used to pick up the acceleration signals. Input and response signal were analyzed by a spectrum analyzer to estimate the driving point and transfer impedances. The comparisons of the impedances from FEM modelling show good agreement with the experimental data. Also it is clear that the constrained damping layer is good for controlling the resonant vibration response for ship's stiffened plate.

1. INTRODUCTION

Structure-borne noise and vibration power transmitted to a sound carrying structure form a source predominantly via a number of contact points. Noise and vibrations propagating in the structure possibly will cause sensitive equipment to vibrate or undesired noise radiation. In principle, this may be avoided by countermeasures at the source, in transmission paths, or at radiation surfaces. It is, of course, preferable to handle the problem at the source and thereby avoid more extensive and expensive measures in the propagation paths.

Viscous damper in a dynamics system provides a mechanism for energy dissipation and results in amplitude reduction at resonance. Problems caused by resonance include noise, fatigue, performance, and human discomfort. The utilization of damping treatment techniques can be a relatively simple solution to resonant vibration problem. It can be incorporated in design level or can be used after the facts with minimal system alternation and relatively inexpensive alternative which means the low material cost , low application cost and no maintenance cost.

2. LOGIC OF STRUCTURE-BORNE SOUND REDUCTION AND VIBRATION CONTROL

The low noise design problem for a simple engine room onboard a vessel is schematically shown in Figure 1. The primary noise generating mechanism is the main engine that will inject vibrations into the structure via the mounts and foundation. These vibrations are transmitted throughout the structure and may eventually radiate to the surroundings.

To reduce the transmitted power, viscoelastic material such as rubber is used inevitably in a mount. Owing to the high damping ratio and loss factor, the rubber has a very good performance for reducing

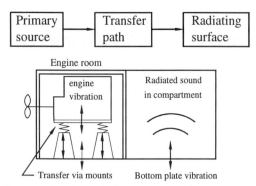

Figure 1. Schematic of machinery noise analysis for low noise design

vibration and structure-borne noise from amechanical vibrating source. The vibration reduction behaviour of the resilient mounts of an engine may be expressed in terms of the input/output transfer functions.

Figure 2 represents the analysis models which can be used to model the structure-borne sound transfer between the primary sources and the radiating surface. For a uni-directional simple harmonic excitation force F_1, the engine structure vibrates with a simple harmonic velocity V_1 at the driving point. Structural waves propagate from the excitation point to the adjacent bottom plate, where

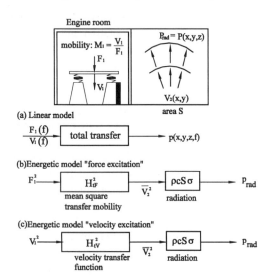

(a) Linear model

(b)Energetic model "force excitation"

(c)Energetic model "velocity excitation"

Figure 2. Schematic for the analysis of structure-borne noise transfer

bending waves generate a velocity field $V_2(x,y)$. This vibration field radiates a sound field $p(x,y,z)$.

The linear equations which relate the radiated sound pressure with the force and the velocity at the driving point are:

$$p(x, y, z, f) = H_{tF} F_1(f);\qquad(1)$$

$$p(x, y, z, f) = H_{tV} V_1(f)\qquad(2)$$

H_{tF} and H_{tV} are so-called frequency response functions, describing the sound transfer for force and velocity excitation respectively.

The factors which determine the sound power transfer and will be influenced by design are seen from the following equations:

$$\frac{P_{rad}}{F_1^2} = (H_{tF}^2 S\sigma)\rho c = (M_{11}^2 H_{tV}^2 S\sigma)\rho c;\qquad(3)$$

$$\frac{P_{rad}}{V_1^2} = (H_{tV}^2 S\sigma)\rho c\qquad(4)$$

where the driving point mobility M_{11} is defined by :

$$M_{11} = \frac{V_1}{F_1}\qquad(5)$$

If the nature of the excitation is such that for different engine design F_1 is unaffected , equation (3) implies that sound reduction is obtained by: decrease of driving point mobility; reduction of radiating surface area; decrease of velocity transfer function; decrease of radiation efficiency.

If the nature of the excitation is such that for different engine designs V_1 is unaffected, equation (4) implies that the sound reduction is obtained by: reduction of radiating surface area; decrease of velocity transfer function; decrease of radiation efficiency.

Design measures to decrease S significantly, have very limited applications. However, measures to decrease M_{11} and H_{tV}^2 are very important and will be discussed in this paper.

3. MOBILITY REDUCTION OF STRUCTURE-BORNE SOUND TRANSFER

Not only for beams, but also for structure like plates, cylinders, etc., the asymptotic approximation that is obtained by averaging over a frequency is closely equal to the mobility of the corresponding infinitely extended system [1] [2]. Such infinite system mobilities are often applied in machinery acoustics calculations. The driving point mobilities for infinite beams and plates are given by Cremer et.al., [1]:

$$\hat{M}_{b\infty} = \frac{0.19(1-j)}{\rho S \sqrt{C_{Lb}} \, hf} \; ; \tag{6}$$

$$|M|_{P\infty} = \frac{0.125}{\sqrt{B'm''}} \approx \frac{0.453}{\rho_p c_{Lp} h^2} \approx \frac{5.41}{\omega \rho_p \lambda_B^2} \tag{7}$$

where m''; mass per unit area; B': bending stiffness per unit width; ρ, ρ_p: density of beam/plate material; c_{LP}, c_{Lb}: longitudinal wave speed in plate/beam ; h: thickness of plate / beam ; λ_B : bending wavelength; j:unit imaginary number.

The velocity transfer function H_{tV}^2 depends strongly on the size and nature of a structure. For noise reduction purposes the main interest is in the natural frequency range. Here again the discussion will be limited to the elementary case of a finite plate. In principle H_{tV}^2 can be calculated with the aid of an eigenfunction model and finite element method. Again it can be said that if in the frequency range of interest a large number of eigenfrequecies is involved the calculation become time consuming. Moreover, from the viewpoint of the designer, it is important to look at smoothed average data and to understand the relation between design parameters and these data.

The derivation of H_{tV}^2 uses the equality of injected and dissipated power

$$P_{in} = P_{diss} \tag{8}$$

The power injected by a point source into the plate at a single frequency is given by :

$$P_{in} = \frac{1}{2} \mathrm{Re}\{\hat{F}_1 \, \hat{V}_1^*\} = F_{1,rms}^2 \, \mathrm{Re}\{\hat{M}_1\} = V_{1,rms}^2 \, \mathrm{Re}\{\frac{1}{\hat{M}_1}\} \tag{9}$$

For averaging over frequencies $\hat{M}_1$ may be replaced by $M_{p\infty}$. Using equations (7) and (9) this leads to :

$$\hat{P}_{in} = \frac{0.453 F_{1,\Delta f}^2}{\rho_p c_{LP} h^2} = 2.2 \rho_p c_{LP} h^2 V_{1,\Delta f}^2 \tag{10}$$

The power dissipation from the plate is caused by material damping, by energy transportation across the plate boundaries into the support and by sound radiation from the plate. A well known parameter which includes all these damping mechanisms is the (apparent) loss factor η. For each frequency or frequency band it may be defined as:

$$\eta(f) = \frac{P_{diss}}{\omega m'' S \overline{V_2^2}} = \frac{\text{energy dissipation per vibration period}}{2\pi \times \text{mechanical energy (reversible)}} \tag{11}$$

In equation (11) it is assumed that the total mechanical energy (i.e. the sum of kinetic and potential energy) is twice the kinetic energy, i.e.

$$E = m'' S \overline{V_2^2} \tag{12}$$

Therefore, the dissipated power may be written as :

$$\hat{P}_{diss} = \eta \omega m'' S \overline{V_2^2} \tag{13}$$

Using equations (8), (10) and (13) one obtains

$$H_{tV}^2(f) = \frac{0.35 c_{LP} h}{f S \eta(f)} \tag{14}$$

Figure 3 shows the effect on H_{tV}^2 of different loss factor. It is seen that at low frequencies $H_{tV}^2 \gg 1$. The frequency at which $H_{tV}^2 = 1$, decreases with increasing η and decreasing plate thickness.

The squared transfer mobility H_{tF}^2 in the eigenfrequency range follows from equations (8), (9), (10) and (13) :

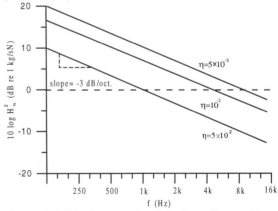

Figure 3. Velocity transfer function of a steel late (480mm×340mm×5mm) calculated according to equation (14).

$$H_{tF}^2 = \frac{\overline{V_{2,\Delta f}^2}}{F_{1,\Delta f}^2} = \frac{M_{P\infty}}{\eta \omega m''S} \approx \frac{2.85}{\eta f \rho_P^2 c_{LP} h^3 S} \qquad (15)$$

Comparing equations (14) and (15) reveals that the influence of damping upon H_{tV}^2 and H_{tV}^2 is the same, but the influence of plate thickness variation is quite different. Therefore, it is of vital importance for a designer to have appropriate knowledge on the source (or component) which drives the structure.

4. MECHANICAL FOUR-POLE PARAMETERS METHOD IN SYSTEM MOBILITY VALUATION

Assume a mechanical system, in which the harmonic force F_1 and the harmonic velocity V_1 were applied to the left end (or input end) of the system and the force F_2 and the velocity V_2 were induced on the right end (or output end). The relations between V_1, F_1 and V_2, F_2 are[3]:

$$\begin{Bmatrix} V_1 \\ F_1 \end{Bmatrix} = \begin{bmatrix} \alpha_1 & \alpha_{12} \\ \alpha_{21} & \alpha_{22} \end{bmatrix} \begin{Bmatrix} V_2 \\ F_2 \end{Bmatrix} \qquad (16)$$

where α_{11}, α_{12}, α_{21} and α_{22} are called the four-pole parameters[4], and

$$\alpha_{11} = \frac{M_{11}}{M_{12}}, \quad \alpha_{12} = \frac{M_{12}M_{21} - M_{11}M_{22}}{M_{21}},$$

$$\alpha_{21} = \frac{1}{M_{12}}, \quad \alpha_{22} = -\frac{M_{22}}{M_{12}}, \qquad (17)$$

and M_{ij} (i=1,2 and j=1,2) are the mobility functions.

If the two ends of the subsystems a,b are 1,2 and 3,4 respectively, as shown in Figure 4, in which F represents force and V represents velocity. When the subsystems link together, then $V_3 = V_2$ and $F_3 = -F_2$, equation(16) becomes

$$\begin{Bmatrix} V_1 \\ F_1 \end{Bmatrix} = \begin{bmatrix} \alpha_{11} & \alpha_{12} \\ \alpha_{21} & \alpha_{22} \end{bmatrix} \begin{Bmatrix} V_2 \\ F_2 \end{Bmatrix} = \begin{bmatrix} \alpha_{11} & \alpha_{12} \\ \alpha_{21} & \alpha_{22} \end{bmatrix} \begin{bmatrix} \alpha_{33} & \alpha_{34} \\ -\alpha_{43} & -\alpha_{44} \end{bmatrix} \begin{Bmatrix} V_4 \\ F_4 \end{Bmatrix}$$

$$= \begin{bmatrix} \alpha'_{11} & \alpha'_{14} \\ \alpha'_{41} & \alpha'_{44} \end{bmatrix} \begin{Bmatrix} V_4 \\ F_4 \end{Bmatrix} \qquad (18)$$

It requires that $F_4 = 0$ since 4 is the receiving point and 1 is the driving point . Thus ,

$$\frac{1}{\alpha'_{41}} = M_{41} = \frac{V_4}{F_1} \qquad (19)$$

From equation (18)

$$\alpha'_{41} = \alpha_{21}\alpha_{33} - \alpha_{22}\alpha_{43} = \frac{M_{22} + M_{33}}{M_{12}M_{34}},$$

or $\quad M_{41} = \frac{M_{12} + M_{34}}{M_{22}M_{33}}. \qquad (20)$

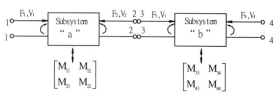

Figure 4. Mobility of integrated system combined by mobility matrices of the subsystems

From equation (19), it can be seen that the transfer mobility of the combined system (M_{41}) is not simply the combination of the transfer mobilities (M_{12} and M_{34}) of the two subsystems. The linking point mobilities (M_{22} and M_{33}) of the subsystems play a predominant role for determining the total transfer mobility . As a damping material is introduced into the input end of the subsystem b, such as Figure 5(b) which represents the improved system of Figure 5(a), then the mobility M_{33} increases significantly. It follows from equation (19) that the total transfer mobility M_{41} is reduced. In other words, to reduce the total transfer mobility, it requires the introduction of a larger mobility than the original one. This is the basic principle for a good structure-borne noise attenuation result that has to properly design the structure on both sides of a resilient mount with small mobility than that of the mount itself.

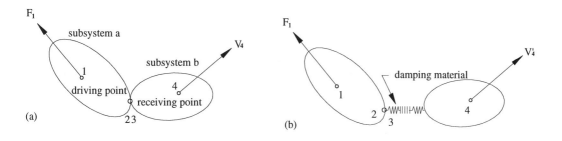

Figure 5. System before and after installation of isolator damping material

The other way of reducing structure-borne sound is the insertion of damping material between point 3 and 4. This is equivalent to the application of a damping material on the surface of the foundation girder of an engine. The type of damping material used is the constrained damping layer. That is, a damping layer sandwiched between the foundation girder and a steel-like constrained plate. This material is good for providing vibration and noise attenuation in higher frequency ranges. Coupled with resilient mount which provided good vibration attenuation characteristics in low frequency, these two viscoelastic passive damping technology provide a broader frequency range solution to the vibration problems.

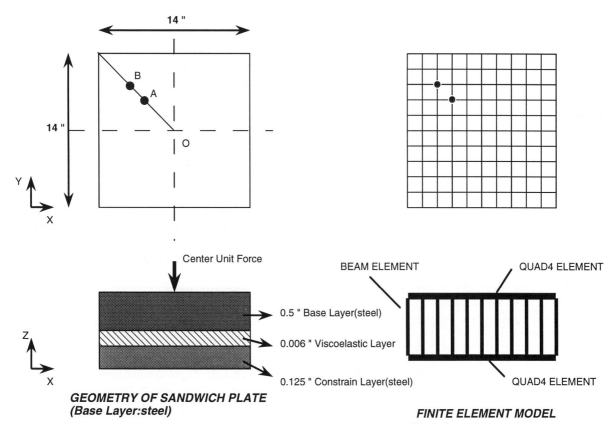

Figure 6. Finite element model of the driving point mobility analysis of a $14'' \times 14''$ sandwich steel plate

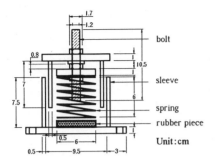

(a)Rubber piece specimen in a sleeve type resilient mount

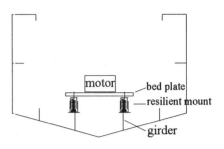

(b)Motor-resilient mount-girder system

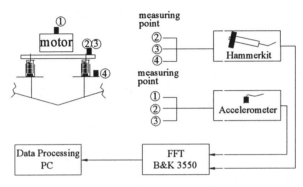

(c) Mobility measuring points and instrumentation

Figure 8. Experiment set-up

5. FEM MODELING OF CONSTRAINED DAMPING LAYER

To be more flexibly modeling damping layer, a FEM procedure proposed by Lu [5] was used to simulate the dynamic behavior of the damped structure. As usual, in FEM model, the base structure and the constraint plate were modeled as quadrilateral plate elements. But the damping material was modeled as beam element that can withstand extension and shear. A numerical experiment was performed on a $14'' \times 14''$ steel plate with damping layer. The FEM model is shown in Figure 6, and the result compare with an analytical solution[6] is shown in Figure 7. It is clear that the FEM model shows good approximation with the exact solution.

6. EXPERIMENTS

The experiments on the damping technology were carried out on a ship model. The dimension of this model is approximately a 3m long and 1.5m wide box girder. There are two longitudinal girders on the ship bottom which served as the engine foundation. An electric motor with a bed plate was used to simulate the actual mounting. The

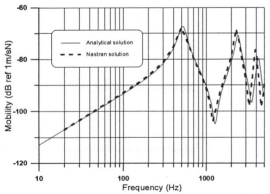

Figure 7. Comparison of driving point mobility by FEA and analytical solution

experimental set up and the resilient mount are shown in Figure 8. The types of rubber used in the test are summarized in Table 1. Experiment was also preformed on one of the longitudinal girder to obtain its mobility. The mobility is used to correct the FEM prediction model for later use in the prediction after damping treatment.

Table 1. Properties of rubber specimens

Type No.	Material	Specific Weight	Diameter (mm)	Thickness (mm)
A	Butyl	1.251	60	10
B	Butyl	1.251	60	5
C	Silicone	1.315	60	10
D	Silicone	1.315	60	5
E	Styrene Butadiene Rubber (SBR)	1.151	60	5.2
F	Neoprene	1.223	60	12.1

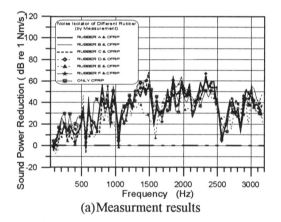

(a)Measurment results

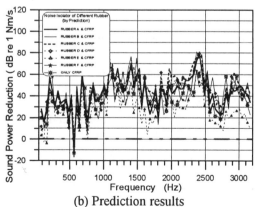

(b) Prediction results

Figure 9. Sound power reduction by insertion of resilient mount with different rubbers

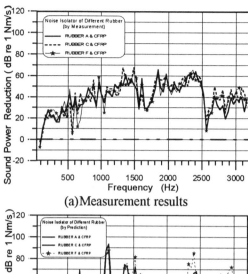

(a)Measurement results

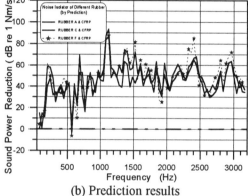

(b) Prediction results

Figure 10. Sound power reduction by insertion of resilient mount and sandwich beam with damping layer

7. RESULTS AND DISCUSSION

Consider the motor-resilient mount-girder system shown in Figure 8(b). Comparing to the case of the motor bed plate linked to girder directly the sound power reduction of structure-borne noise transmitted from motor to the girder by insertion of resilient mounts and insertion of resilient mounts and sandwich beam with damping layer simultaneously are shown in Figures 9 and 10. The results by measurement and prediction show good agreement. The results reveal that all the four kinds of rubber have an effectiveness to attenuate the structure-borne noise level 20~30 dB in the frequency range of 800Hz to 3,200Hz. While the sandwich beam with damping layer has the capability to reduce 5-10dB structure-borne noise transmission in the same frequency range.

950

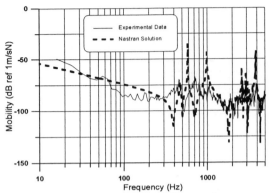

Figure 11 Driving point mobility at the center of face plate

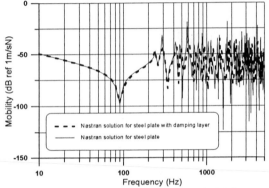

Figure 12 Driving point mobility at 1/6 W from one side of the face plate (W:width of face plate)

Figure 11 shows both the analytical and experimental driving point mobilities at the midpoint of the longitudinal girder, the measuring position is at the center of the face plate. The purpose of this is to check the validity of the FEM model. Figure 12 shows the analytical results from FEM predictions with and without damping layer, the point being computed is at 1/6 of the width from one side of the face plate. The result shows a 5-10 dB reduction of vibration at higher frequency ranges.

8. CONCLUSIONS

From the works and results in this paper, it is concluded that:

1. The prediction model for the transfer mobility function of structure-borne noise from the exciting source to a receiving structure shows good agreement with measurements.

2. The sound power reduction using resilient mounts with four types of rubber material was shown to attenuate the structure-borne noise level of about 20-30 dB in the frequency range of 800 Hz to 3,200Hz.

3. Damping layer treatment can reduce substantial amount of vibrational energy, usually, 5-10 dB in the magnitude of the mobility. This can be an effective way for reducing the energy transfer from structural vibration into airborne sound.

REFERENCES

1. L. Cremer, M. Heckl and E. E. Ungar, "Structure-Borne Sound", 2nd edition, Springer-Verlag, Berlin etc., (1988).

2. E. Skudrzyd, "Understanding the Behaviour of Complex Vibrators", Acoustica, Vol.64, pp.123-147, (1987).

3. W.H. Wang, R. Sutton and B. Dobson, "Vibration Reduction Behaviour Simulation of Resilient Mount by Series Double Ends Transfer Mobility", The 5th Conference on the Chinese Society of Sound and Vibration, Hsin-Chu, ROC, pp.507-524, (1997).

4. J.C. Snowdown, " Mechanical Four-Pole Parameters and Their Application", Journal of Sound and Vibration, Vol. 15(3), pp.307-323, (1971).

5. J.W. Killian and Y.P. Lu, "A Finite Element Modeling Approximation for Damping Material Used in Constraint Damped Structures.", Journal of Sound and Vibration. 97(2), pp.352-354, (1984).

6. Y.P. Lu, J.C. Clemens and A.J. Roscoe, "Vibration of Composite Plate Structures Consisting of a Constrained Layer Damping Sandwich with Viscoelastic Core.", Journal of Sound and Vibration, 158(3), pp.552-558, (1992).

Practical Design of Ships and Mobile Units
M.W.C. Oosterveld and S.G. Tan, editors.

Dynamic Loads on Fast Ferry Hull Structures Induced by the Engine-Propeller System

D. Boote[a], A.Carcaterra[b], P.G.Esposito[b], M. Figari[a]

[a] Dipartimento di Ingegneria Navale e Tecnologie Marine, Università di Genova, Italy

[b] INSEAN, *Istituto Nazionale Studi ed Esperienze di Architettura Navale*, Roma, Italy

ABSTRACT

In this paper the problem of the vibrations induced on the hull by the propulsion system is considered. Although the problem has been widely studied in the technical literature, some new aspects concerning the propulsion system model are here investigated. In particular the engine is modelled taking into account its exact crankshaft dynamics, leading to variable inertia effects and parametric excitation. The pulsating propeller torque in the hull wake is studied as well, by solving the flow about the propeller by a boundary element technique. The derived model allows an accurate calculation of the reaction forces transmitted by the engine-propeller system to the ship structures; in this way the characteristic time history of the vibrational motion of important parts of the ships can be predicted. The attention is focused on a typical fast passenger ship, the finite element model of which is developed. Preliminary test results are shown and the performed numerical simulations suggest that the global propeller-engine-structure model could be a useful tool in the frame of the optimization of the vibrational ship design.

1. INTRODUCTION

Some important aspects of the marine engine induced vibrations deserve further developments, especially in view of the requirements concerning weight reduction, large stress values and vibrational comfort of new high speed marine vehicles [1-5]. This imposes a careful investigation on dynamic problems related to the shafting system design and its effects on the hull vibrations.

The exciting forces acting on the ship's structure are recovered by the dynamic analysis of the engine and the propeller. Several approximations are generally introduced. The standard design technique reduces the crankshaft mechanism to a system of equivalent flywheels connected by torsional springs (a classical approach to this problem is found in [6]). The propeller is also modelled as an equivalent flywheel taking into account, by empirical formulas, the added mass and damping effects. Moreover the hydrodynamic torque on the propeller is considered constant and equal to the average engine torque. However several authors have reconsidered the problem in the light of a different point of view,

where some traditionally accepted simplifications are removed [7-8]. In a recent paper [9] a model of the propeller-engine interaction has been developed by considering both the previous mentioned effects. The crankshaft mechanism is modelled taking into account the variable inertia effects related to the complex piston-rod motion and the propeller-engine interaction is described through an integral-differential mathematical model. The mathematical formulation includes several additional effects related to the operating conditions of the engine such as misfiring, order of ignition and engine power loading. The obtained results show several differences with respect to the prediction of the standard design technique.

In this paper a development of the previous work is presented, analysing the effects of the vibrations induced by the engine-propeller system on the hull structure. The time history of the propulsion system excitation provides the forces acting on the ship structures and allows to predict its dynamic response.

As a case study a fast ferry is considered equipped with four stroke diesel engines and propellers. The dynamic response of the ship is investigated by a finite element model of the hull structure, taking into account the effects of the added mass as well. The exciting forces are applied to the hull in correspondence of engine foundations. The amplitude of vibrations is investigated in several points of the passengers deck and wheelhouse, where the comfort requirements and the reliability of electronic instruments are particularly important.

It should be pointed out that this kind of units are characterised by high specific power and severe weight restrictions. The presence of spourious spectral components due to the variable inertia effects and a precise estimate of the periodic torque on the propeller are important elements in predicting fatigue phenomena in shafts and engine foundations that should be considered very carefully. Moreover this is the first step to correctly describe some important vibration and noise sources in the frame of the acoustic design of ships.

2. SHIP'S PROPULSION SYSTEM: A GENERAL MODEL

In the following sections the theoretical models providing the forces acting on the propulsion system, the propeller torque and the engine gas pressure, are briefly presented. General non-linear equations for the crankshaft dynamics are derived as well. The model is simplified by recognizing the weakly non-linear nature of the problem and a parametric differential system is approached. The computation of the reaction forces transmitted by the engine to the foundation is finally analysed.

2.1 Diesel engine thermodynamic model

A numerical lumped parameters simulator of a four stroke turbo-charged diesel engine (direct injection) has been developed.

The principal assumptions of the model are listed in the following.

- The engine is considered as a set of control volumes (*i.e.* inlet manifold, cylinder, exhaust manifold) in which time-dependent pressures and temperatures are modelled by standard thermodynamic equations. Uniform conditions in the control volumes are assumed for each crankshaft angle.

- Inlet and exhaust valve motion are modelled by standard laws.
- Flow between control volumes is described by standard gas-dynamic equations. Choking phenomena are also taken into account.
- Heat release in the cylinder, heat transfer and friction losses are modelled with the help of empirical formulas. In particular the code uses a combustion heat release curve (Wiebe) in which pre-mixed and diffusive phases in combustion are considered.
- Turbo-charger performances are quasi-statically modelled by standard maps.

Successful comparisons with real devices data have been made to test the flexibility of the code particularly for cycle pressures, torque and shaft power.

2.2 Propeller dynamics: theoretical formulation

The propeller rotates in a non uniform flow due to the perturbation induced by the hull wake. Therefore each blade experiences, during its periodic rotation, lift and drag fluctuations and a pulsating torque is generated and transmitted to the engine.

The non uniform flow about the propeller is described by $\mathbf{V}_w(X,Y,Z)$ that, in the propeller rotating frame, becomes $\mathbf{V}_I(x,y,z,t)$. Towing tank model tests provided the inflow velocity field information.

The perturbation induced by the propeller is assumed to be described by a velocity potential ϕ which satisfies the Laplace equation.

Under these assumptions the total velocity in the propeller co-ordinate system can be written as

$$\mathbf{V}(x,y,z,t) = \mathbf{V}_I(x,y,z,t) + \nabla\phi(x,y,z,t)$$

We can define a boundary surface S made by the blade surface S_B, the hub surface S_H and the wake surface S_W and an outward normal vector $\mathbf{n}_Q$. The perturbation potential can be written in terms of an integral equation by applying Green's identity at point $P(x,y,z,t)$ on the boundary surface:

$$2\pi\phi(P,t) = \iint\limits_{S} \phi(Q,t)\frac{\partial}{\partial n_{Q}}\left(\frac{1}{R(P,Q)}\right)dS +$$

$$-\iint\limits_{S} \frac{\partial\phi(Q,t)}{\partial n_{Q}}\frac{1}{R(P,Q)}dS$$

where $R(P,Q)$ is the distance between the point $P(x,y,z,t)$ and the point $Q(x',y',z',t)$.

The considered unsteady integral formulation for the propeller and the wake surfaces, leads to a boundary element method (BEM), solved in terms of the perturbation potential. The pressure field on each blade is determined by the Bernoulli's equation and the associated propeller torque is finally recovered by integration.

2.3 General equations of the engine-propeller motion.

In this section the basic formulation for the crankshaft-propeller dynamic equation, given in [9], is resumed. The reference configuration is shown in fig.1. The system is composed by a crankshaft mechanism, forced by the gas pressure p generated during the combustion stroke in each cylinder and by the hydrodynamic propeller torque.

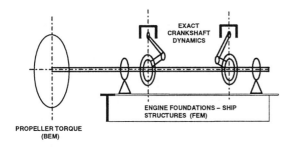

Figure 1. Propeller-engine-structures coupling

The power balance equation of the system states: $\dot{W}_{e} = d/dt(T+U)$, where W_{e} is the work done by the external forces, *i.e.* the pressure of the gas trapped in the cylinders and the propeller torque; T and U are the kinetic and the potential energy respectively. The external power provided to the system is:

$$\dot{W}_{e} = \sum_{i=1}^{N} Ap^{i}(x_{B}^{i})\dot{x}_{B}^{i} + m_{p}\dot{\theta}_{p}$$

where A is the piston surface, x_{B}^{i} the i-*th* piston displacement, m_{P}, θ_{P} the propeller torque and rotation, respectively. The kinetic energy T, recovered by using the kinematic analysis, and the potential energy U are:

$$T = \sum_{i=1}^{N}\frac{1}{2}J_{T}(\theta_{c}^{i})\dot{\theta}_{c}^{i\,2} + \frac{1}{2}J_{p}\dot{\theta}_{p}^{2}$$

$$U = \sum_{i=1}^{N}\frac{1}{2}k_{s}^{i}\left(\theta_{c}^{i+1} - \theta_{c}^{i}\right)^{2} + \frac{1}{2}k_{t}\left(\theta_{p} - \theta_{c}^{N}\right)^{2}$$

where θ_{c}^{i} is the i-*th* crankshaft rotation, J_{p} the propeller moments of inertia, k_{s}^{i} and k_{t} stiffness coefficients. The function $J_{T}(\theta_{c}^{i})$, given in [9], takes into account the inertial effects related to the motion of each crankshaft-rod-piston mechanism of the engine. Substituting the previous expressions into the power balance and developing the derivatives with respect to t, a set of non-linear differential equations in terms of the unknown functions $\theta_{c}^{i}(t),\theta_{P}(t)$ is obtained.

Let be ω_{c} the constant design speed of revolution of the engine; the following expansions can be used: $\theta_{c} = \omega_{c}t + \varepsilon\tilde{\theta}_{c}; \theta_{P} = \omega_{c}t + \varepsilon\tilde{\theta}_{P}$, where $\varepsilon\tilde{\theta}_{c}^{i}, \varepsilon\tilde{\theta}_{P}$ are small deviations of the crankshaft and propeller rotations with respect to the uniform motion of revolution. In fact the engine and the propeller torque are both pulsating and therefore an irregular motion of the system and elastic vibrations of the shaft are produced. However the amplitude of these oscillations are small in comparisons with the average motion ($\omega_{c}t$) due to the high stiffness and inertia of the shafting system, this last purposely increased by the flywheel. The assumed expressions for the $\theta_{c}^{i}, \theta_{P}$ angles allows a Taylor series expansion of all the terms in the equations of motion, up to the first order with respect to ε.

The expression of the propeller torque, for small perturbation on θ_{P}, has been already analysed in [9]. Introducing the last expressions in the non-linear general equations, one finally obtains:

$$
\begin{cases}
J'_T \ddot{\tilde{\theta}}^i_c + \left[\dfrac{\partial^2 J'_T}{\partial \theta^{i2}_c} \omega^2_c - \dfrac{\partial}{\partial \theta^i_c}\left(Ap^i F^i_1 \right) + k^i_s + k^{i+1}_s \right]\tilde{\theta}^i_c + \\[2mm]
+ \dfrac{1}{2}\dfrac{\partial J'_T}{\partial \theta^i_c}\omega_c \dot{\tilde{\theta}}^i_c - k^{i+1}_s \tilde{\theta}^i_c - k^i_s \tilde{\theta}^{i-1}_c = Ap^i F^i_1 - \dfrac{1}{2}\dfrac{\partial J'_T}{\partial \theta^i_c}\omega^2_c \\[2mm]
J_P \ddot{\tilde{\theta}}_P - \displaystyle\int_o^t g(t-\tau)\dot{\tilde{\theta}}_P\, d\tau + k_t\left(\tilde{\theta}_P - \tilde{\theta}^N_c \right) = m_{P0}
\end{cases}
$$

where F_1 is a known function of $\tilde{\theta}^i_c$ [9] and the $\varepsilon\tilde{\theta}^i_c, \varepsilon\tilde{\theta}_P$ variables are replaced by $\tilde{\theta}^i_c, \tilde{\theta}_P$ for the sake of simplicity.

The first are linear differential equations with time dependent periodic coefficients. In fact they are now only dependent on t through the linear argument $\omega_c t$. The second is instead an integral-differential equation, where the convolution integral, characterised by the kernel $g(t)$, takes into account the wake memory. m_{P0} is the pulsating torque related to the design speed of revolution of the propeller.

2.4 Exciting forces on the structures

The previously developed analysis provides the time history of any crankshaft angle $\theta^{(i)}_c(t)$. Therefore the law of motion of the three bodies system crankshaft-rod-piston, is completely known for each cylinder.

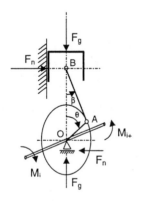

Figure 2. Engine reaction forces analysis

Let us determine the contact forces between them and the reaction forces acting on the crankshaft foundation and the cylinder wall. Let us consider the equilibrium of the system (fig. 2).

The general statement of the problem involves eight unknowns, corresponding to eight dynamic equilibrium equations. In fact the following unknown forces (seven scalar components) must be determined: $\mathbf{R}_O$, $\mathbf{R}_A$, $\mathbf{R}_B$, F_n associated to the points O, A, B and to the cylinder wall – piston contact respectively. $\theta^{(i)}_c(t)$ should be the eighth (kinematics) unknown. On the piston surface acts the known gas pressure force F_g, and to the crankshaft portion, associated to the i-th cylinder, known elastic torques M_i, M_{i+1} are transmitted by the neighbour shaft segments.

The eight unknowns $\mathbf{R}_O$, $\mathbf{R}_A$, $\mathbf{R}_B$, F_n and $\theta^i_c(t)$ are related through eight equilibrium equations of the considered three bodies system. In fact three equilibrium equations can be stated for the rod and the related portion of the crankshaft, respectively and two equations can be written for the piston motion. For each cylinder they assume the form:

$$ \mathbf{A}(\theta)\cdot\mathbf{R} = \mathbf{F}_i\,(\theta,\dot{\theta},\ddot{\theta}) + \mathbf{F}_e $$

where $\mathbf{R} = \begin{Bmatrix} \mathbf{R}_O \\ \mathbf{R}_A \\ \mathbf{R}_B \\ F_n \end{Bmatrix}$, $\mathbf{F}_e$ includes the external known forces F_g, M_i and M_{i+1}, $\mathbf{F}_i(\theta,\dot{\theta},\ddot{\theta})$ is the inertial forces vector and $\mathbf{A}$ is a 8x7 matrix whose terms are trigonometric functions of the crankshaft angle.

The previous equation can be solved only in terms of $\mathbf{R}$, being the kinematic variable $\theta^i_c(t)$ already determined by the solution of the general equations given in the previous section. Therefore a suitable elimination of one row in the equilibrium equation is allowed, reducing the matrix $\mathbf{A}$ to a 7x7 size.

The force acting on the engine foundation in correspondence of a single cylinder is given by (fig. 2):

$$ \mathbf{R}_f = \mathbf{R}_O + \begin{Bmatrix} -F_n \\ F_g \end{Bmatrix} $$

and the torque reaction, due to the wall-piston force F_n, is $M_r = F_n \cdot \overline{OB}$ (fig. 2).

The developed model does not introduce simplifications in the analysis of the inertial forces involved in the equilibrium equations. This implies that the forces produced by the reciprocating piston and rod motion are exactly computed up to an arbitrary high order. Moreover an effect related to the elastic crankshaft deformation is accounted for in both $\mathbf{A}$ and $\mathbf{F}_i$ terms.

3. SHIP FINITE ELEMENT MODEL

In order to test the procedure on a real case, a medium size fast passenger ship has been chosen as case study. The vessel is characterised by an overall length of 50 meters, a beam of 8.5 meters, a displacement of about 170 tons and a service speed, in full load condition, of 32 kn. The hull has a longitudinal structure with transverse reinforced frames 1000 mm spaced, a double bottom and a main deck, where two orders of superstructures are installed on. Both hull and superstructures are built in 5083 H321 and H111 light alloy.

The unit is equipped with two turbo-charged diesel engines of about 2300 kW coupled with two fixed pitch propellers with five blades and a diameter of about 950 mm.

A finite elements dynamic analysis has been carried out in order to determine the natural frequencies of the ship. For this study the f.e.m. code MAESTRO [10] has been adopted. The program allows to perform static and dynamic analysis in the linear domain for structures built up of reinforced shells, like ships are. The software contains a powerful pre-processor module, specifically realised to easily reproduce ship structures, and a post-processor module that allows to display the results in terms of displacements, stresses and adequacy parameters of the structure. The fluid added mass is automatically calculated on the base of the wetted hull surface, defined through the draft of the ship.

The numerical model of the ship have been obtained using the available elements of the MAESTRO library. Ordinary frames and girders and secondary stiffeners have been modelled together with platings by a special element called "strake". This element is a combination of a quadrilateral shell element, that includes the bending stiffness of the stiffeners in the longitudinal or transverse direction, and 3 dimensional beam elements, to represent ordinary frames and girders. Pillars have been modelled by rods. Transverse bulkheads have been modelled by membrane elements and beams.

The whole structure has been divided into blocks, called "modules", on the basis of the hull forms, bulkheads position and changes in the structural organisation. Every block have been schematised as close as possible to the real structure and completed with outfitting and equipment weights. The modules have been grouped into three substructures relative to the after, central and fore parts of the vessel. The resulting three substructures are represented in fig.3.

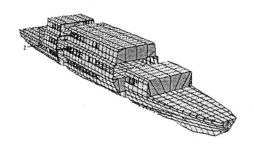

Figure 3. Finite elements model of the ship

The global model of the vessel has been generated assembling the three substructures. The complete model has then been used to compute the modal and the eigenvalues Φ and Λ matrices, respectively. The first four resulting natural frequencies of the vessel are listed below:

First mode: two nodes - vertical 11.6 Hz
Second mode: three nodes - vertical 13.2 Hz
Third mode: local – super structures 15.2 Hz
Fourth mode: local – main deck 16.4 Hz

As an example the first two mode shapes are shown in Fig. 4.

4. SHIP STRUCTURAL RESPONSE TO ENGINE EXCITATION

Several simulations have been carried out on the numerical model of the ship assumed as case study.
A typical eight cylinder configuration, 2300 kW shaft-power, has been considered with a speed of revolution of about 750 rev/min. All the characteristic engine parameters are referred to an actual sample.

The numerical simulation consists of five basic steps. First the external forces on the propulsion system, namely the cylinder gas pressure and the

956

pulsating propeller torque, are provided. The gas pressure in each cylinder is computed by simulating the gas evolution along the engine cycle. In this phase particular effects related to the misfiring and cylinders ignition order can also be considered.

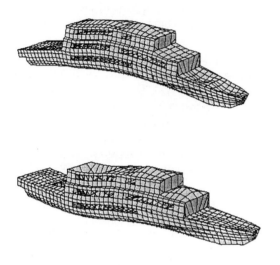

Figure 4. Modal shapes of the two first natural frequencies: (upper) two nodes deformed pattern; (lower) three nodes deformed pattern.

The second step is related to the evaluation of the pulsating torque on the propeller. The flow about the blades is studied by the integral formulation associated to the Laplace equation leading to a boundary element solution technique. The propeller inflow is instead determined by hull wake experiments performed in towing tank.

Once the mentioned external force are determined, the third step, concerning the numerical solution of the equation of motion of the crankshaft, is finally performed. The non-conventional model for the coupled crankshaft-propeller dynamics has been used where the exact dynamics definitely replaces the simple equivalent flywheels hypothesis. In this way the complex phenomena of the parametric excitation, related to the variable inertia effects, is taken into account.

The fourth numerical calculation is related to the evaluation of the forces acting on the crankshaft foundation, that can be determined when the motion

of all engine components is known from the dynamic analysis performed in the third phase. The analysis provides the harmonic components $\mathbf{F}_m$ of the engine reaction forces. These, transmitted to the ship structures, allow to predict the structural response, once the FEM analysis of the system has been performed, i.e. when the eigenvalues and the modal matrices are known.

The general displacements response $\mathbf{w}(t)$ of the ship's structure to the propulsion system's excitation is therefore computed by the following modal expansion:

$$\mathbf{w}(t) = \sum_{m=1}^{M} \Phi (\Lambda - \omega_m^2 \mathbf{I})^{-1} \Phi^{\mathrm{T}} \ ^{\mathrm{T}} \mathbf{F}_m e^{j\omega_m t}$$

where:
Φ = modal matrix;
Λ = eigenvalues matrix;
$\mathbf{I}$ = unit matrix;
$\mathbf{F}_m$ = m-*th* harmonic component of the exciting force;
ω_m = m-*th* angular frequency of the exciting force.

In Fig. 5 the spectrum of the engine reaction torque is given in the frequency range 0-200 Hz. The characteristic peaks related to the harmonic components of the engine torque are evident.

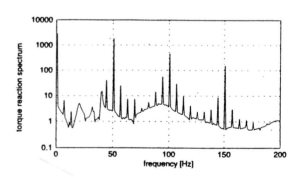

Figure 5. Spectrum of the reaction torque

It must be noticed that in this analysis the propulsion system model is able to account for a richer frequency content of the reaction forces with respect to a traditional analysis. In fact the kinematics of the crankshaft considers the contributions of the elastic perturbation on each crankshaft angle that affects the motion of the rod-piston system. Moreover these perturbations account for the variable inertia effects and the reaction forces are performed by considering the exact dynamics of the problem, as shown in section 2.4.

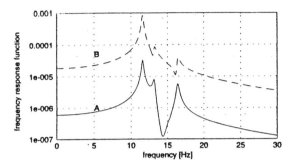

Figure 6. A: FRF between engine foundation and main deck. B: FRF between engine foundation and 2° super-structure

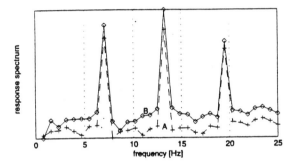

Figure 7. Spectrum of the response. A: main deck point, B: 2° super-structure point

The modal analysis of the ship's structures allows to determine the frequency response function (FRF) between two arbitrary points of the ship. In Fig. 6 two FRFs are considered: the curves represent the FRF between one of the point of the engine foundation and a point of the main deck floor, curve A , and a point of the second super-structure, curve B. In Fig. 7 the response spectrum corresponding to the mentioned locations are reported.

5. CONCLUSIONS

The attention of the present work is focused on the use of a non-conventional ship's propulsion model developed in [9] in order to predict some important structural vibrations induced by the engine-propeller system on the drive train and the ship's structure. In particular the model accounts for the variable inertia effects and the pulsating propeller torque. First these effects play a crucial role in a correct prediction of torsional vibrations as shown in [9]. On the other hand once an 'exact' dynamic model of the engine-propulsion system is available, an accurate prediction of the reaction forces transmitted to the ship's structure is possible.

The modelling of the propulsion system as an excitation source is only the first step of a more complicated process. In fact two main tasks are considered for future developments.

The first is related to fatigue phenomena associated to the vibrations transmitted by the engine to the structure that should be carefully considered.

The second is related to the acoustic ship's design. In this case the prediction of the noise and vibration level on the ship is a challenge for the whole acoustic scientific community being the problem, at the actual state of the art, far from an effective solution.

6. ACKNOWLEDGENENTS

This work has been supported by Ministero dei Trasporti e della Navigazione in the frame of the INSEAN Research Program 1997-99.

REFERENCES

[1] A. C. Nilsson, 'A method for the prediction of noise and velocity levels in ship constructions', Journal of Sound and Vibration, 1984, 94(4).

[2] A. C. Nilsson, 'Attenuation of structureborne sound in superstructures on ships', Journal of Sound and Vibration, 1977, 55, pp.71-91.

[3] K. Iino, I Honda, 'Total noise prediction system for a passenger cruise ship', Proc. of the 5-th Int. Symp. on Practical Design of Ships and Mobile Units, PRADS'92, vol.1,pp. 1648-1661.

[4] J. Plunt, J Odegaard (editors), 'Noise sources in ships, II: diesel engines', Miljovardsserien Publication, 1981, Nordforsk.

[5] P. Hynna, P. Klinge, J. Vuoksinen,' Prediction of structure-borne sound transmission in large welded ship structures using statistical energy analysis', Journal of

958

Sound and Vibration, 1995, 180(4), pp. 583-607.

[6] W. Ker Wilson, 'Practical Solution of Torsional Vibration Problems', Chapman & Hall, London, 1940.

[7] M.S. Pasricha, W.D. Carnegie, 'Formulation of the Equations of Dynamic Motion Including the Effects of Variable Inertia on the Torsional Vibrations in Reciprocating Engines, Part I', Journal of Sound and Vibration, 1979, 66(2).

[8] M.S. Pasricha, W.D. Carnegie, 'Diesel Crankshaft Failures in Marine Industry - A Variable Inertia Aspect', Journal of Sound and Vibration, 78(3).

[9] A. Carcaterra, P.G. Esposito, A. Petritoli, 'A Model of the Propeller-Engine Interaction for the Dynamic Response of the Ship Propulsion System', AIMETA '97, Siena, Italy.

[10] O. Hughes, "MAESTRO V.7.0 – User and Application Manuals", Maryland, 1995.

1998 Elsevier Science B.V.
Practical Design of Ships and Mobile Units
M.W.C. Oosterveld and S.G. Tan, editors.

Minimum Plate Thickness in High-Speed Craft

P. Terndrup Pedersen and Shengming Zhang

Department of Naval Architecture and Offshore Engineering,
Technical University of Denmark, DK-2800 Lyngby, Denmark

The minimum plate thickness requirements specified by the classification societies for high-speed craft are supposed to ensure adequate resistance to impact loads such as collision with floating objects and objects falling on the deck. The paper presents analytical methods of describing such impact phenomena and proposes performance requirements instead of thickness requirements for hull panels in high-speed craft made of different building materials.

1. INTRODUCTION

High-speed craft is highly weight-sensitive structures. Therefore, novel building materials with high strength-to-weight ratios such as aluminium, fibre-reinforced plastic (FRP) single-skin plates or sandwich panels are widely used. In order to provide sufficient strength to withstand local impact loads, all the major classification societies have recently established empirical rule requirements of the minimum panel thickness for these new hull materials. As usual, deviations from these minimum requirements may be accepted, provided equivalent load resistance can be documented. But due to lack of analysis procedures, such documentation can at present only be obtained by physical experiments.

In the first part of the present paper, the impact energy to be dissipated by the panel is studied. An analytical expression is derived for ships colliding with floating objects in the sea with different impact angles, taking into account hydrodynamic effects by application of simple momentum and energy considerations.

Secondly, the critical energy causing rupture is studied for panels made of steel, aluminium and FRP single-skin plates. Again simplified procedures are derived on the basis of maximum strain criteria. The derived penetration strength results are compared to some previously published experimental results.

Finally, the derived analysis procedure is used to analyse the minimum thickness requirements made by one of the classification societies and to convert these existing empirical requirements into critical impact energies or minimum floating object masses which just rupture the panel at a given ship speed.

2. BASIC FORMULAS

2.1 Energy to be dissipated by the ship structure

Fig. 1 shows a ship, with the forward speed V, colliding with a motionless floating object. We assume that the mass M of the object is small compared to the mass of the ship.

The floating object may slide away after impacting the ship panel if the collision is light, but in the case of rupture, no sliding will occur at the impact point. In the latter case, we have derived the following formulas for the energy to be dissipated by the shell plating in the perpendicular and the parallel direction of the shell plating:

$$E_\xi = \frac{1}{2}MV^2 \cdot \frac{\sin^2 \beta}{D_{a\xi} + \mu \cdot D_{a\eta}}$$

$$\left.\begin{array}{c} \end{array}\right\} \quad (1)$$

$$E_\eta = \frac{1}{2}MV^2 \cdot \frac{\cos^2 \beta}{\dfrac{1}{\mu}K_{a\xi} + K_{a\eta}}$$

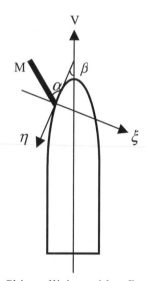

Fig. 1. Ship collision with a floating object.

where

$$D_{a\xi} = \frac{1}{1+m_{ax}}\sin^2\alpha + \frac{1}{1+m_{ay}}\cos^2\alpha + \frac{3}{1+j_a}\cos^2\alpha$$

$$D_{a\eta} = (\frac{1}{1+m_{ax}} - \frac{1}{1+m_{ay}} - \frac{3}{1+j_a})\sin\alpha\cos\alpha$$

$$K_{a\xi} = (\frac{1}{1+m_{ax}} - \frac{1}{1+m_{ay}} - \frac{3}{1+j_a})\sin\alpha\cos\alpha$$

$$K_{a\eta} = \frac{1}{1+m_{ax}}\cos^2\alpha + \frac{1}{1+m_{ay}}\sin^2\alpha + \frac{3}{1+j_a}\sin^2\alpha$$

and

$$\mu = \frac{D_{a\xi}\cos\beta - K_{a\xi}\sin\beta}{K_{a\eta}\sin\beta - D_{a\eta}\cos\beta}$$

and m_{ax} is the added mass coefficient for the surge motion of the object, m_{ay} is the added mass coefficient for the sway motion of the object, j_a is the added mass coefficient of moment for the rotation around the centre of gravity, α is the impact angle between the object and the shell plating, β is the shell face angle between the shell plating and the ship sailing direction.

2.2 Critical rupture energy

Let us consider a sharp object impacting the centre of a shell plate bounded by longitudinal stiffeners and transverse frames. The stiffener spacing is $2b$, the frame spacing is $2a$, t is the thickness of the plate, A is the area of the plate, σ_0 is the flow stress of the material. When the maximum strain is equal to the critical strain ε_c, the plate is assumed to rupture. Then we have derived the following relationship ($1 \le \frac{a}{b} \le 2$):

$$\frac{E_\xi}{\dfrac{1}{15}(1+(\frac{a}{b})^2)} + \frac{E_\eta}{\dfrac{1}{8}(1+\dfrac{2}{21}\dfrac{a}{b})} = \frac{2}{\sqrt{3}}\sigma_0 t \varepsilon_c A \quad (2)$$

Substituting Eq. (1) into Eq. (2), we get

$$\frac{1}{2}MV^2 = \frac{\dfrac{2}{\sqrt{3}}\varepsilon_c\sigma_0 tA}{\dfrac{\sin^2\beta}{\dfrac{1}{15}(1+(\frac{a}{b})^2)(D_{a\xi}+\mu D_{a\eta})} + \dfrac{\cos^2\beta}{(\dfrac{1}{8}+\dfrac{2}{21}\dfrac{a}{b})(\dfrac{1}{\mu}K_{a\xi}+K_{a\eta})}} \quad (3)$$

When the ship velocity V is known, the mass M of the floating object which just ruptures the shell plating can be determined from Eq. (3). The critical rupture energy can be expressed as:

$$E_c = \frac{1}{2}MV^2\{\frac{\sin^2\beta}{D_{a\xi}+\mu D_{a\eta}}+\frac{\cos^2\beta}{\frac{1}{\mu}K_{a\xi}+K_{a\eta}}\} \quad (4)$$

For a dropped sharp object impacting the centre of a clamped panel in air (see Fig. 2), the critical rupture energy is

$$E_c = \frac{\frac{2}{\sqrt{3}}\varepsilon_c\sigma_0 tA}{\frac{\sin^2\alpha}{\frac{1}{15}(1+(\frac{a}{b})^2)}+\frac{\cos^2\alpha}{\frac{1}{8}(\frac{2}{8}+\frac{2}{21}\frac{a}{b})}} \quad (5)$$

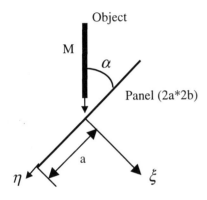

Fig. 2. A dropped object impacts a panel.

Similarly, for a dropped cylinder impacting perpendicularly to the centre of a clamped circular plate in air (see Fig. 3), the critical energy is

$$E_c = \frac{\pi}{\sqrt{3}}\sigma_0 t(\frac{1}{3}\varepsilon_c(a-a_0)(a+3a_0)+t\cdot a_0\sqrt{2\varepsilon_c}) \quad (6)$$

Here a is the radius of the circular plate, a_0 is the radius of the cylinder.

Fig. 3. A cylinder impacts the centre of a clamped circular plate.

3. COMPARISON WITH TESTS

3.1 Comparison with DNV test [1]

DNV [1] has carried out tests where a dropped cylinder impacts various plates. The radius of the cylinder is 40 mm, the dimensions of the clamped plate are $500*500\,mm^2$. The cylinder impacts the centre of the plate at an angle $\alpha = 35°$. Since the impact is oblique, it can be assumed that the impactor is a sharp object for the initial impact. The flow stress of the aluminium material is $\sigma_0 = 260 MPa$ and the critical strain is $\varepsilon_c = 10\%$. A comparison of the present results with the experimental results in [1] is shown in Table 1. It is seen that the agreement is reasonable.

Table 1. Comparison of critical energy.

Thickness (mm)	Present (kJ)	DNV test (kJ)
2	2.72	1.86
3	4.08	4.33
4	5.44	>5.87

3.2 Comparison with test by Wen and Jones

An experimental investigation is reported by Wen and Jones [7], where fully clamped circular plates are struck by blunt projectiles travelling at relatively low velocities. A comparison with the present calculation, using Eq. (6) and the test results [7], is shown in Table 2. Again the agreement is reasonable. In the present calculation, the critical rupture strain is calculated from $\varepsilon_c = 0.10\cdot(\varepsilon_f/0.32)$ in Ref. [8], where ε_f is the tensile-strain ductility.

Table 2. Comparison with test results [7].

Parameters (mm), (N/mm^2)				Energy (Nm)	
t	a	a_0	σ_0	Present	Test [7]
2	25.4	2.975	303.9	35.0	39.2
4	50.8	5.95	372.2	319.6	310.0
6	76.2	8.925	331.6	917.1	1004.7
8	101.6	11.9	327.4	2708.9	2437.6

4. THE INFLUNCE OF IMPACT ANGLE

4.1 A dropped object impacting a plate at various angles α in air

Here we consider a dropped object impacting the centre of a plate, such as the deck, at various angles α, to examine the influence of the impact angle α on the critical energy. The parameters of the plate are: $500*500\,mm^2$, $t = 3mm$, $\sigma_0 = 260\,MPa$ and $\varepsilon_c = 10\%$.

Fig. 4 shows the critical energy with various impact angles. It can be seen that the perpendicular impact requires the least energy to rupture the plate.

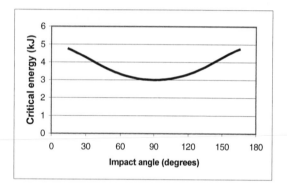

Fig. 4. Critical energy versus different impact angles α (dropped object).

4.2 Ship collision with a floating object in water

Now, let us consider a high-speed ship with the velocity V=20 m/s colliding with a floating slender object, as shown in Fig. 1. The purpose is to examine which impact angle is the most dangerous. For the chosen example, the dimensions of the shell plating are $500*500\,mm^2$, the thickness is t=3mm, $\sigma_0 = 260\,MPa$, $\varepsilon = 10\%$. The calculation results of the critical object mass vs. different impact angles for the cases $\beta = 35°$, $50°$ and $70°$ are shown in Fig. 5. The results show that

when the impact angle α equals the shell face angle β, the object mass which just ruptures the shell plating is the minimum. This means that when the slender object is orientated in the ship sailing direction, it most easily ruptures the shell plating.

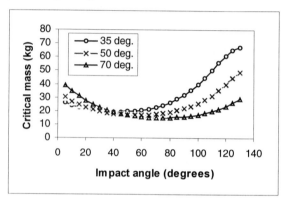

Fig. 5. Critical object mass which just ruptures the shell plate (shell face angles: $\beta = 35°$, $50°$ and $70°$).

5. SIMPLIFIED FORMULAS

5.1 Sharp object

From the present analyses, we have seen that the easiest way to rupture the shell plating is when the floating object is orientated in the craft sailing direction. In this case, the expression of the critical object mass which just ruptures the shell plating for a moving craft with the speed V can be simplified to

$$M_c = \frac{\frac{4}{\sqrt{3}}\varepsilon_c\sigma_0 t \cdot A}{[\frac{\sin^2\beta}{\frac{1}{15}(1+\lambda^2)} + \frac{\cos^2\beta}{(\frac{1}{8}+\frac{2}{21}\lambda)}](1+m_{ax})V^2} \qquad (7)$$

where s is stiffener spacing, λ is aspect ratio of the plate, $A = s^2\lambda$ is plate area.

The critical rupture energy is

$$E_c = \frac{1}{2} M_c (1 + m_{ax}) V^2 \qquad (8)$$

5.2 Object with line contact s_0

Let us, as shown in Fig. 6, consider an object impacting a panel with the line contact s_0. The floating object lies in the ship sailing direction. The critical object mass which just ruptures the shell plating can be expressed as

$$M_c = \frac{\frac{4}{\sqrt{3}} \varepsilon_c \sigma_0 t [s \lambda (s - s_0)]}{\left\{ \frac{\sin^2 \beta}{\left[\frac{1}{15} + \frac{1}{15} \frac{(s\lambda)^2}{(s-s_0)^2} + \frac{1}{3} \frac{s_0}{(s-s_0)} \right]} + \frac{\cos^2 \beta}{\left[\frac{1}{8} + \frac{2}{21} \frac{s\lambda}{s-s_0} + \frac{1}{2} \frac{s_0}{s-s_0} \right]} \right\} (1 + m_{ax}) V^2}$$

(9)

and the critical energy can be calculated from

$$E_c = \frac{1}{2} M_c (1 + m_{ax}) V^2 .$$

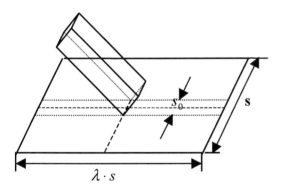

Fig. 6. Object impacting a panel with a line contact.

5.3 Example

Fig. 7 shows a comparison of the critical energy of a ship collision with a sharp object ($s_0 = 0$) and blunt objects. The ship panel thickness is based on DNV rules, see Eq.(10), $\sigma_y = 220 MPa$, $t_0 = 4.0 mm$, $k = 0.03$, $s = 300 mm$, $\lambda = 1.5$. From the results it is seen, as expected, that the sharp object ruptures the plate at less energy than blunt objects.

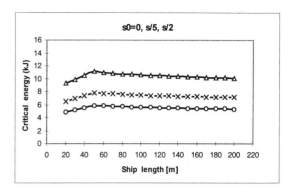

Fig. 7. Ship collision with a sharp object and blunt objects (V=20 m/s).

6. APPLICATIONS TO HIGH-SPEED CRAFT

6.1 Aluminium craft

The minimum thickness requirement of the DNV rules [6] for aluminium high-speed craft is

$$t = \frac{t_0 + kL}{\sqrt{\sigma_y / 240}} \frac{s}{S_R}, \qquad (mm) \qquad (10)$$

σ_y is the yield stress (MPa),

s is the actual stiffener spacing (m),

$S_R = \frac{2(100 + L)}{1000}$ (m), $0.5 \le \frac{s}{S_R} \le 1.0$,

L is the ship length (m),

$k = 0.03$ for the shell plate.

Here we shall use the present formula to convert the minimum thickness requirement into the critical object mass which just ruptures the shell plating for a given craft speed.

Fig. 8 shows the present results of the critical object mass vs. craft length, when the ship velocity is V=20 m/s and the shell face angles are $\beta = 35°$ and $70°$ respectively. The plate aspect ratio is $\lambda = 1.5$.

From the figure it is seen that when the shell face angle increases, the shell plating is ruptured more easily. According to the present results, a 25 kg-mass object may rupture the hull of a high-speed craft moving at a speed of 20m/s, the corresponding critical energy is about 5 kJ.

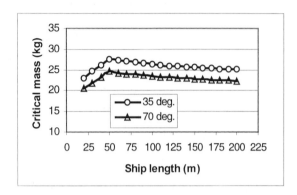

Fig. 8. Critical object mass vs. various ship lengths (the speed V=20 m/s, $\beta = 35°, 70°$).

6.2 FRP single -skin craft

Here we shall use the formula to predict the critical energy of a FRP single-skin craft.

Before we do this, we shall firstly make a comparison with FRP single-skin plate test. The tests were clamped square plates, with the dimensions $500*500\,mm^2$, struck by a dropped cylinder with the diameter $s_0 = 80mm$. The impact angle is $\alpha = 35°$. The material ultimate stress is $\sigma_u = 180MPa$, the critical strain is $\varepsilon_c = 3.5\%$ [5]. The comparison of the critical energy by the present method and DNV [1] test results is shown in Table 3. Good agreement is found.

Table 3. Critical energy making a hole.

Thickness (mm)	Present (kJ)	DNV test (kJ)
2.9	1.28	0.99
8.1	3.57	3.40
11.2	4.92	5.87
13.1	5.76	>5.87

Now we convert the thickness requirement into the critical energy. The thickness requirement of the DNV rules for FRP single skin craft is [6]:

$$t = \frac{t_0 + kL}{\sqrt{\dfrac{\sigma_u}{160}}}, \quad (mm) \qquad (11)$$

σ_u is the ultimate tensile stress (MPa), L is the ship length (m).

The stiffener spacing is $s = 300mm$, the aspect ratio is $\lambda = 1.5$, $\sigma_u = 180MPa$, $\varepsilon_c = 3.5\%$, $t_0 = 5mm$, $k = 0.09$ and the ship velocity is V = 20m/s. The present calculation results for the critical energy are shown in Fig. 9 when the shell face angles are $\beta = 35°$ and $70°$ respectively.

From the results it is seen that the critical energy increase with increasing ship length.

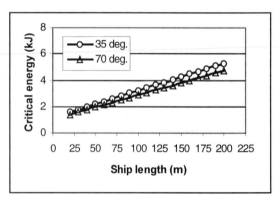

Fig. 9. The critical energy of FRP single-skin craft vs. ship length.

7. CONCLUSIONS

A study on the impact strength of high-speed craft colliding with floating objects and dropped objects impacting plates has been presented. The comparisons show an acceptable agreement between the present results and experimental results. The existing minimum thickness requirement of aluminium craft and FRP single-skin craft is converted into critical impact energy and critical object mass. The major conclusions are summarised as follows:

(1) The impact strength is determined by the following parameters:
- Thickness, size and aspect ratio of the shell plating
- Impact location and angle
- Yield stress and critical strain of the plates

(2) For a dropped object impacting a panel in air (such as the deck plate), the perpendicular impact requires the least energy to rupture the panel.

(3) For a high-speed craft colliding with floating slender objects, the shell plating is most easily ruptured when the slender object is orientated in the sailing direction of the ship.

(4) Simple formulas have been presented from which the critical energy and the critical object mass can be determined.

REFERENCES

1. O. Aamlid, Oblique Impact Testing of Aluminium and Composite Panels, Det Norske Veritas, Report No. 95-2042, Oslo, Norway, 1995.

2. O. Aamlid and G. A. Antonsen, Oblique Impact Testing of Single Skin, Aramid Fibre Reinforced Plastic Panels, Det Norske Veritas, Report No. 97-2000, Oslo, Norway, 1997.

3. N. Jones, Dynamic Plastic Behaviour of Ship and Ocean Structures, The Royal Institution of Naval Architects, 1997.

4. P. Terndrup Pedersen and Shengming Zhang, The Mechanics of Ship Impacts with Bridges, Int. Symposium Advances in Bridge Aerodynamics, Ship Collisions & Maintenance, Lyngby, Denmark, Balkema Press, May, 1998.

5. M. Hildebrand, The Effect of Raw-Material Related Parameters on the Impact Strength of Sandwich Boat-Laminates, VTT Report No.211, Finland, 1994.

6. Rules for Classification of High-Speed Light Craft, DNV, Norway, 1993.

7. M. Wen and N. Jones, Experimental Investigation of the Scaling Laws for Metal Plates Struck by Large Masses, Int. J. of Impact Engng., Vol.13, No.3, 1993.

8. J. McDermott, R. Kline, E. Jones, N. Maniar and W. Chiang, Tanker Structural Analysis for Minor Collisions, SNAME Transactions, 1974.

X-joints in Composite Sandwich Panels

A.W. Vredeveldt[a], G.T.M. Janssen[a]

[a] TNO-CMC, Netherlands Organisation for Applied Scientific Research, Delft.

The small structural weight of fast large ships such as fast mono hulls or catamaran type of ships is of extreme importance to their success. One possible light weight structural solution is the sandwich panel with fibre reinforced laminates and a balsa, honeycomb or foam core. A severe obstacle for applying such panels is the design and manufacturing of adequate joints between panels. This is especially true for X-type of joints. This article shows how this type of joint can be analysed in a manner which can be suitable for design purposes. It is shown how delamination can be treated by applying a well established fracture mechanics technique. Some comparison is made with a structural test on a full scale X-joint specimen. Fatigue analysis is not referred to.

1. INTRODUCTION

The ambition to build large fast ships, particularly for ferry services, causes a need for light ship structures. Because of this need high tensile steels and non conventional shipbuilding materials have become of interest to designers and builders. In the framework of the Brite Euram project MATSTRUTSES a large Surface Effect Ship, SES, has been designed in both aluminium and composite. Manufacturing large composite ship structures requires a suitable method is required for joining sandwich panels. Suitable in this respect means the technology to manufacture and the possibility to predict the strength.

A typical connection which can be identified in the structural design of a large SES vessel is the connection of the tunnel corner the tank deck and the vertical inner walls (tunnel inside hull and tank bulkhead) of the hulls (see Figure 1). The SES structure features an inclined part of the tunnel corner. Therefore the above mentioned connection has to be of the X-type.

This article contains some results of an analysis calculation and a static strength test on the X-joint between the tunnel corner the tank deck and the vertical inner walls of the hulls.

It is possible to find solutions for joining sandwich panels by applying metallic inserts. However they are deliberately left out of this project, to make way for an investigation into an all-composite solution.

This paper reports on part of a research effort which includes prediction of loads, fatigue behaviour and fire resistance as well. This research is carried out in the framework of Brite Euram and in that capacity sponsored by the European Union. The partners in this project are: CETENA, Chantier de l'Atlantique, Danyard, ETH Zurich, DnV, Marin, Hexel, NTUA and TNO.

2. LOADS

Figure 1 shows the cross section of the vessel under consideration. The area of interest is indicated by a circle.

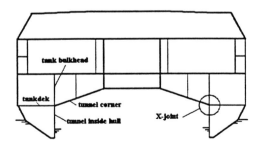

Figure 1 Schematic cross section SES vessel

Table 1 shows all relevant loads that could be identified to be acting on an x-joint.

968

The table also indicates the origin of such loads. Only the situation at mid span between transverse webs is described.

Table 1. Loads on X-joint

Description	Caused by	Remarks
Long. membrane stress	Hull bending	Global load
Transverse and vertical membrane stress	Hull splitting moment and vessels weight	Global transverse load
In plane shear	Hull long. and transverse torsion and shear	Global load
Lateral shear	Water/air pressure	Local load
Transverse bending	Water/air pressure	Local load

From a loads point of view there are two operational modes;
- on cushion,
- survival.

For the ultimate strength assessment the survival mode is paramount.

The complexity of the loads on the X-joint is rather large. When the location at an intersection with webs is considered, the situation becomes even more complex. Moreover the behaviour of composite materials is so much different from steel that conventional strength assessment concepts are not adequate. This is especially true for the delamination failure mode being a mode which does appear and cannot be analysed with the conventional stress concept. Because of this situation it was decided to consider only one aspect of the complex loading. Since it is an import global load for a SES, the hull splitting moment was chosen for further analysis. Moreover this load may cause delamination Figure 2 shows this loading case. The load in the tunnel corner, Ftc, originates from the hull splitting moment. The load in the tank bulkhead, Fb, makes equilibrium with the

vertical component of Ftc and the vertical force Ftih. The force in the tunnel inner hull Ftih is mainly due to buoyancy. The force in the tank deck Fpd is there to make equilibrium with the horizontal component of Ftc

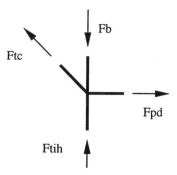

Figure 2 X-joint loading.

The actual load level in the tunnel corner has been chosen at an adequate unit level. However the results such as stresses and strains are assumed to vary linearly with the loads, the energy release rate (see section 4) varies quadratic with the loads. Therefore calculated results can be scaled to match actual load levels. Vertical forces due to buoyancy have been neglected.

3. STRENGTH ASSESSMENT

A proper strength assessment requires knowledge of the failure modes and an insight in how the loads are distributed in the joint. Therefore the stiffness distribution in the X-joint and the various failure modes must be analysed.

Stiffness
Material stiffness is not as straight forward a parameter as it is with steel or aluminium. In principal the effective stiffness of a panel depends on a combination of stiffness properties of fibres and matrix material. In plane stiffness depends on the considered direction. In the analysis reported here, moduli have been used as shown in Table 2.

Table 2 Elasticity values

Material	Moduli, Poison	Values [MPa], []
Laminate	longitudinal	52400, 0.229
"	transverse	52400, 0.229
"	lateral	10800, 0.307
Core	-	92, 0.3

Failure modes

In general five failure modes can be recognised. Table 3. shows a review.

Table 3 Considered failure modes

Failure mode	Description
laminate fracture	in plane crack occurs mainly due to fracture of reinforcement
laminate buckling	denting due to in plane compression
laminate-core debonding	bonding between laminate and core fractures
fracture/crush in core	cracks in core material and crumbling of core
delamination	bonding between two laminates fails

Laminate fracture

For the design stages concept design and contract design, it is considered adequate to assess laminate fracture with a simple stress criterion. For the considered material, i.e. carbon fibre reinforced epoxy composite. Some maximum values are given in Table 4. Please note the different strengths for the different directions. The values are based on a ply by ply laminate strength analysis, not included in this paper.

Table 4 Allowable stress levels

Stress type	Allowable value [MPa]
longitudinal	150
vertical/transverse	110
lateral	20

For allowable stresses DnV applies a reduction factor of 0.3 on the fracture strength of laminates applied below the water line.For the purpose of analysing the X-joints by applying a FE calculation method these figures were simply taken as reference. No effort has been made to actually analyse the laminates in terms of their ply by ply build up. In the de production design stage a ply by ply analysis should be carried out. It is interesting to note that some literature [1] shows how design charts can be of use at this stage. These charts shows strength/stiffness properties as function of fibre content and fibre direction.

Laminate buckling

Laminate buckling can be assessed by applying formulas for Euler buckling of elastically mounted plating.

For this purpose the formula as suggested in the tentative DnV rules for high speed and light craft (ref [2]) is used;

$$\sigma_{cr} = \frac{1}{2}\sqrt[3]{EE_cG_c} \qquad (1)$$

with : σ_{cr} Critical buckling stress in laminate [MPa]

E Young's modulus in compression direction [MPa]

Ec Young's modulus in core [MPa]

Gc Shear modulus in core [MPa]

With E = 52400 MPa, E_c = 92 MPa, G_c = 35 MPa and a reduction factor of 0.3, for laminates used in ships side or bottom, an allowable buckling stress is found of

$$\sigma_{cr} = 82 \text{ MPa.}$$

Core failure

Core fracture can be assessed with a simple stress criterion. For core shear fracture also a simple stress criterion can be used. Table 5. lists some allowable values.

Table 5. Strength properties core

Failure type	Mpa
Tension strength	0.9
Shear strength	0.4

Maximum strains will occur in the core material which is easier to scan in a post processor. Therefore while assessing failure, it is easier to take strain as a criterion rather than stress, moreover a von Mises criterion is applied. Simply applying the stress strain relation an allowable stress of 0.9 MPa yields an allowable strain value of $\varepsilon_{crit} = 0.01$.

Delamination

The material between two panel laminates coming together consists of either resin only or resin+glue+resin. In any case the connection is basically a glued connection. It is widely accepted that glue connections are only adequate in shear loading, allowable tension stresses are very low. Unfortunately X-joints of panels, applying fillet laminates, do have glue type connections in them which may subjected to both shear and tension.

This failure mode cannot be analysed by adopting conventional stress criteria which are adequate for metallic materials. The next section in this paper is dedicated to this failure mode.

4. DELAMINATION

Delamination cannot be analysed with conventional stress analysis. The concept of stress energy release rate is introduced instead.

Energy release rate

The usual way of connecting sandwich panels which are situated in different planes, e.g. bulkhead and deck, is applying connecting laminates, so called fillets. Figure 3 shows such a connection.

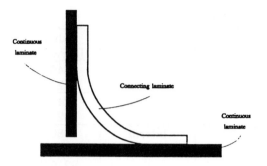

Figure 3 Typical fillet connection between laminates

One particular failure mode which can be recognised is delamination between fillet(connecting) laminate and sandwich laminates. These delaminations can be caused by overloads in either shear or tension. With respect to shear it seems reasonable to take shear stress as a criterion for judging the interlaminar strength. With respect to tension the stress criterion is not adequate. When peeling forces are present it is obvious that at the contact line between sandwich laminates and fillet laminate a stress concentration is to be anticipated. From a strict theoretical point of view this stress concentration is infinite.

In a paper by Kildegard ref. [3], an attempt is reported to apply the concept of Stress Energy Release Rate, as has been developed for fracture mechanics.

The method has been adopted in a modified form, for the analysis of the X-joints.

To qualify the strength situation at a crack tip the stress energy release rate G, can be used. This energy release rate equals the elastic energy W_{in} minus the external potential P_{ex} when the crack area increases with one unit area A, while displacements at the outer boundary are kept constant.

$$G = -\frac{\delta W_{in}}{\delta A} + \frac{\delta P_{ex}}{\delta A} \qquad [2]$$

The contributions of external potential energy thermal energy etc. are assumed zero.

In case of a two dimensional analysis the increase in cracked area can be related to the increase in crack length a If W_{in} is the elastic energy and t is the thickness of a plate, then the energy release rate is defined as

$$G = -\frac{1}{t}\frac{\delta W_{in}}{\delta a} \qquad [3]$$

at constant displacements. In finite element analysis the elastic energy W can be calculated from pre- and post multiplication of the stiffness matrix K by the displacement vector u

$$W_{in} = \frac{1}{2}u^T K u \qquad [4]$$

Since the displacements are kept constant, the derivative in the right hand side of the equation [2] can be calculated from

$$\frac{\delta Win}{\delta a} = \frac{1}{2}u^T\frac{\delta K}{\delta a}u \qquad [5]$$

DCB Test

In the paper by Kildegaard [3], results are reported on a Double Cantilever Beam (DCB) Tests, which yield experimental stress energy release rates. This section shows results of Fe calculations on these tests. A strip with a length of 350 mm, a width of 50 mm and a thickness of 10 mm is subjected to a peeling force at one end. At this end an initial crack of 50 mm is present. The load is gradually increased up to a critical load where first crack propagation occurs. Crack progress, critical load, displacements of the loaded end and crack length are measured. After the crack growth has stopped the load is again gradually increased until the next critical load where crack progress restarts. Thus at several crack lengths, the critical load, the displacement of the loaded end and the crack growth can be determined. From these data it is possible to calculate a stress energy release rate G_{ic} .at each crack length;

$$G_{ic} = \frac{3}{B}P^2\frac{a}{EI} \qquad [6]$$

with :
G_{ic}	Stress energy release rate [J/m2]	
B	width of the specimen [m]	
P	Load [N]	
a	crack length [m]	
E	Young's modulus [N/m2]	
I	Moment of inertia of strips adjacent to crack [m4]	

These stress energy release rates are reported in the paper. With a Young's modulus of 11000 MPa, also given in the paper, the matching critical loads could be determined by applying the above formula. These loads have been used as input for FE calculations simulating the DCB test. The initial crack of 50 mm is included in the FE model. The FE code DIANA [4], which has been used, has an option for calculating the stress energy release rate for a given load and crack length. For six different crack lengths a FE calculation was carried out with a fixed external load. The stress energy release rates as calculated by the FE code are corrected by multiplying with the square of the ratio between actual critical load as derived from the paper and the load as used in the FE calculation (see previous formula).

The results of this exercise are shown in Figure 4, including the G-values as reported in the paper by Kildegaard.

972

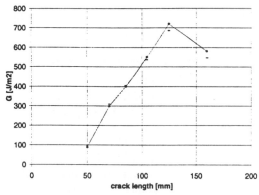

Figure 4 Stress energy release rates

It is important to note that a constant G-level of 600 [Nm/m2] exists at crack lengths larger than 120 mm. This value can be used for design purposes.
Figure 5 shows both the undeformed and deformed model at a crack length of 50 mm.

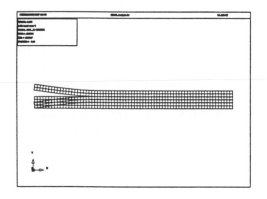

Figure 5 Deformed geometry of DCB at crack of 50 mm.

It seems that the applied FE calculation method is adequate to determine the stress energy release rate for a given structure, initial crack length and loading. The shown example refers to glass fibre reinforced polyester. The X-joint under consideration is manufactured from carbon fibre and an epoxy resin. For this material DCB tests have been carried out as well.

5. FE ANALYSIS

A finite element analysis has been carried out to assess the strength situation of the X-joint. The FE model is used to calculate strain and stress distributions in both laminates and core. Moreover at critical locations a stress energy release rate is calculated for a given assumed crack length.
The choice of this locations is based on vector plots of the principal stresses. At locations where these stresses are lateral to laminate interfaces a critical situation is assumed.

Figure 6 shows a picture of the actual X-joint mounted in its test rig.

Figure 6 Cross section X-joint

Laminates and core material can be recognised (see also fig 8).
In this analysis orthotropic properties have been assumed for the laminates. The core material is described with isotropic properties.

Boundary conditions
Many valid considerations can be given with respect to the choice of proper boundary conditions. Basically they are output from overall strength and stiffness calculations on the ships hulls. At the start of the analysis on X-joints the results of these overall calculations were still subject to discussions and further analysis. However from simple considerations some, more or less obvious, intuitive assumptions could be made with respect to the boundary conditions. They can be described as given in Table 6.

Table 6 Boundary conditions

Panel	restrained in
Tunnel inside hull	both transverse en vertical direction
Platform deck	Transverse direction
Tank bulkhead deep	vertical direction

It is clear that these assumptions are arbitrary. However it can be argued that the actual assumptions will not affect the main goal of the investigation i.e. show that the strength of composite ship structures can be assessed in a rational manner.

Loads

As stated in section 4 the analysis has been restricted to a two dimensional case in the ships transverse plane. Therefore the only relevant load case which remains is the one as stated in section 2. The actual load which was chosen for the analysis was 2000 kN, resulting in a stress of 100 MPa in the laminates where the loads are applied.

FE model

Figure 7 shows the applied FE mesh. Both the deformed and undeformed geometry are shown. Please note that no effort has been made to carry out any mesh refinement because stress concentrations are not considered particularly important. Adequacy of the structure with respect to delamination strength is assessed by applying the parameter of stress energy release rates rather than stresses.

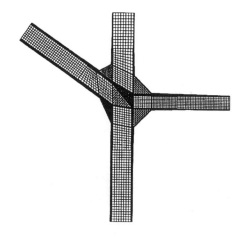

Figure 7 Finite element mesh of X-Joint.

6. RESULTS

From the FE calculations it was concluded that the first failure mode is delamination. Other failure modes, laminate buckling and fracture occur at much higher loads. In this paper only delamination results are given.

The area of first delamination could be found by searching for the highest principal stress perpendicular to a laminate. The location was found by studying the vector plots of these stresses. Principal stresses perpendicular to the laminate were found in the lower right hand fillet laminate, at the lower part. The strength situation at this spot was further investigated according the method described in 4.

Some experimental data is available from DCB tests on carbon fibre epoxy composite laminates at crack-lengths starting from 50 mm upwards. Average G-values are found between 600 and 800 J/m2. For the calculation of a G-value for the X-joint a crack length of 52 mm was modelled. With a load of 100 kN a G-value was found of 168 Nm/m2. Baring in mind the quadratic relation between load and G, failure is expected at 204 kN.

974

From two static failure tests fracture loads were found of 125 and 160 kN. Figure 9 shows the fracture during one of these tests.

Figure 9 Delamination of X-joint at 125 kN

The calculated fracture load is higher than measured. On the other hand between two fracture tests a difference in cracking load was found by a factor of 1.28.

It should be noted that the reported cracking does not mean a total failure of the joint.

8. CONCLUSIONS

Applying the concept of stress energy release, as developed for fracture mechanics, seems adequate for the assessment of delamination strength of laminates.

Determining allowable stress energy release rates G requires further investigation.

9. ACKNOWLEDGEMENTS
Mr. Oostvogels, from TNO, is kindly thanked for his efforts on performing the finite element calculations.

REFERENCES

1. Guy, D., Materieaux composite, Edition Hermes. ISBN 2-86601-268-2

2. Tentative rules for the classification of high speed and light craft, Det Norske Veritas, January 1991.

3. Kildegaard C,. Experimental and numerical fracture mechanical studies of FRF-sandwich T-joints in maritime constructions.

4. Diana User Manual, Release 6.1, TNO Building and Construction Research, Delft 1996.

Practical Design of Ships and Mobile Units
M.W.C. Oosterveld and S.G. Tan, editors.

An Energy-Based Approach to Determine Critical Defect Sizes in FRP Ship Structures.

H. J. Phillips [a] & R. A. Shenoi [b]

[a] Hamworthy Marine Technology Ltd., Poole, Dorset, UK
[b] Department of Ship Science, University of Southampton, Southampton, SO17 1BJ,UK

Abstract

Fibre reinforced plastic materials are finding increased usage in high performance ships and offshore structures. One of the major issues affecting designers, shipbuilders and operators is the long term performance of such ships. From a safety and commercial viewpoint, it is essential to determine the significance of defects when they arise. The main defect in FRP structures is delamination, when one ply in a plate separates from another in a locality. This paper seeks to address the defect tolerance of such delaminations in ship structures. Typically, this covers delaminations within flat plating areas, in top hat stiffener configurations and in tee joints. An energy based approach, using the concept of J-integral and/or G, is advocated. The fracture mechanics background to this is explained. The problems are solved using FEA approaches. Typical results in a ship structural context will be presented and discussed.

1. INTRODUCTION

Fibre reinforced plastic (FRP) materials are finding increased usage in a variety of marine vehicles and structures. The major advantage of these materials is their high strength-to-weight ratio. The drawback is that FRP, especially when the reinforcement phase is E-glass, has a relatively low specific stiffness. This implies the need for apt measures to ensure that structural stability and stiffness are maintained. This can be done in three ways, through using a sandwich topology, having a monocoque single skin structure or adopting a single skin top hat stiffened configuration. The latter is widely used in many advanced applications such as minehunters and high speed racing craft. Although this structural configuration has been used in many craft over the years and initial

strength/stiffness/stability have been assessed adequately, the question of in-service maintenance and repair still needs to be addressed.

FRP is strongest when the response is in-plane of any given load. In large structures such as ships and offshore platforms, there is a need for joints within the structure. Many joints represent a connection between two orthogonal sets of plating. This leads to the question of assessing the load transfer between two panels in an out-of-plane direction. Herein lies another problem issue with FRP. The joints in both the top hat and tee connections are of a generic nature. They are formed by placing laminated strips of reinforcement cloth on the external sides of a joint. The resulting gap between the cloth and plates is filled with an appropriate resin. The plating in this case is usually made from glass reinforced isophtahalic polyester resin. The fillet material usually has a high strain-to-failure value in order to give the joint some flexibility.

There is a growing body of literature on the structural assessment of such generic joints. This paper focuses attention on defects that arise in service. The most common of these are in the form of "whitening" of the structure. This happens because of delaminations, where one ply within the laminate has separated from another. Delaminations weaken the structure both from a viewpoint of affecting effective thickness of plating and therefore stability of a panel and from a perspective of inducing a crack-like phenomenum which leads to local stress concentrations in the structure. Both can have dangerous consequences if the delamination is located in critical areas. One such critical area with delamination problems is at the roots of tee joints or top hat connections.

This paper reports on a detailed study (1) of the

effects of such defects on the damage tolerance of the structure. After a brief outline of the fracture mechanics concepts, modelling of the tee and top hat connections are discussed. The discussion centres around the manner in which the results of the fracture studies can be translated into guidelines for repair strategies of shipowners.

2. FRACTURE MECHANICS CONCEPTS

The condition for a delamination induced crack to propagate can be explained in terms of the strain energy release rate, G, whose general form is (2):

$$G = \frac{1}{B} \frac{dU}{da} \qquad (1)$$

where U is the strain energy stored in the body, B is the material thickness and a is the crack length dimension. It can be shown that for plane strain, the above can be written as:

$$G = \frac{\pi (1-v^2)}{E} \sigma^2 a \qquad (2)$$

where E is the Young's modulus of the material, σ is the field stress and υ is the Poisson's ratio. Using the relationship:

$$K_I = \sigma \sqrt{\pi a} \qquad (3)$$

with K_I being the stress intensity factor, or SIF, for the opening mode, and substituting it in eqn. (2), the strain energy release rate for the corresponding mode is:

$$G_I = \frac{K_I^2}{E} (1-v^2) \qquad (4)$$

A similar expression is valid for the pure shearing mode II. For the tearing mode III, the strain energy release rate is given by:

$$G_{III} = (1+v) \frac{K_{III}^2}{E} \qquad (5)$$

Very often, especially in the case of laminated FRP materials it is not possible to ascribe one single mode. Rather the failure is a result of mixed mode consequences. The strain energy release rate in such a case, is:

$$G = (K_I^2 + K_{II}^2) \frac{(\kappa+1)(1+v)}{4E}$$

$$+ K_{III}^2 \frac{(1+v)}{2E} \qquad (6)$$

where:

$$\kappa = 3 - 4v \qquad (7)$$

If the values of G are known and found to be less than critical values for the material then it is likely that the crack will not extend under the given load and boundary conditions and hence the defect associated with the crack or delamination can be safely tolerated. The non-linear equivalent to G is known as the J-integral (2).

3. APPLICATION TO JOINTED STRUCTURES

The first task in assessing crack damage tolerance is to identify the potential zones of weakness and most importantly, those most susceptible to delaminations. Following this initial assessment, additional fracture models must be generated which contain cracks. This is best achieved through a detailed Finite Element Analysis (FEA) of a typical stiffener or tee joint. A series of FE models have been generated in two-dimensions using linear quadrilateral elements which possess extra displacement shapes to improve bending performance. The software package which has been used throughout is ANSYS.

3.1 Top Hat Stiffeners

In the case of the stiffeners, the overlaminated regions were constructed of 12 elements through the thickness with one element representing a layer of woven roving material. The flange, being of lesser interest, was constructed as a single element through the thickness. The material properties used in the modelling of the top hat stiffeners are given in Table 1.

Material	Location	Property	Value
Polyester/ Woven Roving Glass	Stiffener, Flange and Overlaminate	Ex	13060 MPa
		Ey	7770 MPa
		vxy	0.25
Urethane Acrylate	Fillet	Ex	1500 MPa
		Ey	1500 MPa
		vxy	0.25
Core Material		Ex	10^{-6} MPa
		Gxy	10^{-6} MPa
		vxy	0.25

Table 1. Material Properties used in Top Hat Finite Element Models.

Validation of the FEA model was achieved by comparing its load-deflection characteristics against those generated through laboratory testing of physical specimens (3). Three loading configurations were chosen: (a) Three-point bend; (b) reverse bend; (c) pull-off. These are shown schematically in Figure 1. The load-deflection curves pertaining to the FEA models have been compared with those of the physical specimens. The results for the three-point bend situation exhibit the best agreement.

Having validated the FEA models, stress distributions in the overlaminate (through-thickness and in-plane), flange (through-thickness and in-plane) and fillet (principal) were investigated. The regions where the maximum stresses occurred were noted in each case in order to yield a table of damage sites. These are shown in Table 2. The first load corresponds to delaminations in the curved part of the overlaminate for the three-point bending,

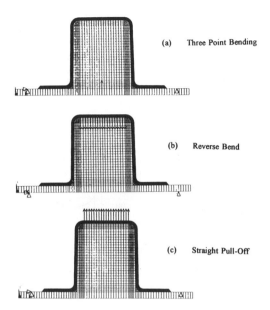

(a) Three Point Bending

(b) Reverse Bend

(c) Straight Pull-Off

Figure 1. Diagrams to show the Three Loading Configurations.

fillet cracking for the reverse bend and cracks at the flange/fillet interface for the pull-off. The second load level corresponds to the failure load of the Top-Hat.

From the diagrams in Table 2 it can be noted that the critical regions where delaminations are most likely to occur are in the curved region of the overlaminate and in the flange. The delaminations are only seen to arise in the three point bending and pull-off loading configurations. It is these two cases which been investigated for fracture purposes.

Having identified the potential damage sites, a finite element fracture analysis (4) was undertaken so as to carry out a sensitivity study on the effect of (i) crack depth and (ii) crack length on the strain energy release rate, G. These were intended to represent possible delaminations in the overlaminate of the top-hat. The FEA model used for the strength analysis was modified for the fracture studies in the area of the intended cracks. Elements in the crack region were six-noded triangular elements with midside nodes at the quarter point and quadratic displacement shapes, so as to be able to cope with stress singularities at the crack tip. To be certain of element compatibility between the four-noded quadrilateral elements and the six-noded

977

978

Test	Load (KN)	Fillet Value	Location	Overlaminate In-plane Value	Location	Overlaminate Through-thickness Value	Location	Flange In-plane Value	Location	Flange Through-thickness Value	Location
Three point bend	13.5	14.81		55.08		75.95		170.52		5.03	
Three point bend	16.5	18.09		67.33		92.83		208.42		6.15	
Reverse Bend	5.0	4.80		7.63		21.31		34.56		0.86	
Reverse Bend	14.0	13.44		21.36		59.67		96.77		2.41	
Pull-Off	5.5	6.16		21.77		38.03		34.69		2.18	
Pull-off	7.0	7.84		27.71		48.4		44.15		2.77	

Table 2. Locations of Maximum Stresses in the Top Hat Stiffener Model.

triangular crack elements, it was important that the crack elements which are connected to the interface between the two element types had their midside nodes removed along the appropriate edge. Gap elements with a very high compressive stiffness and zero tensile stiffness were placed between the two crack faces to prevent cross-over. The modelling assumed plane strain conditions. Values for G were yielded in each case and then compared with a critical value, G_C of 0.54 N/mm(5).

If the calculated value of G is greater than G_C then the crack may reasonably be assumed to propagate under the given load conditions. The parameters varied in the sensitivity study, crack depth and length, are shown schematically in Figure 2.

In the case of the depth variation, the 8 mm crack is horizontal and its depth varies from 2 mm below the outer surface to an 8 mm depth. The total overlaminate thickness is 12 mm. In the case of the length variation, each crack is 6 mm deep and the location of the crack tip furthest away from the fillet is kept constant. i.e the length is increased around the radius of the overlaminate.

Results under three-point bend and pull-off loads each of 10 kN are shown in Figure 3(a) and 3(b) respectively.

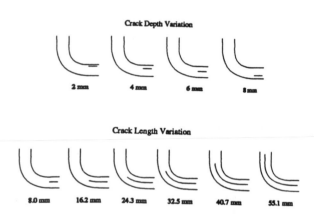

Figure 2. Parameters Varied in the Fracture Study of Top Hat Stiffeners.

3.2 Tee Joints

In the case of the tee joint specimens, the finite element model results were compared with those from a series of tests carried out by Shenoi et.al.(6). The specimens comprised a 560 mm flange and a 260 mm web cut into 100 mm wide sections. Connection was achieved with pure urethane acrylate resin, which was injected between the plates on both sides of the web to form the fillet. The web and flange were constructed from 17 plies

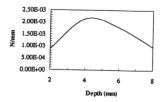

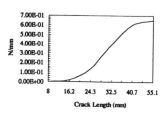

Figure 3(a). Results of the Fracture Study of Top Hat Stiffeners - Three Point Bending.

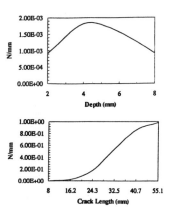

Figure 3(b). Results of the Fracture Study of Top Hat Stiffeners - Pull Off.

of E-Glass Woven roving (WR)set in polyester resin. The construction of the boundary angles over the 35 mm radiused fillet comprised 2 × 600g Chopped Strand Mat (CSM), 1 × 780g WR, 2 × CSM, 1 × WR, 2 × CSM, 1 × WR, 1 × CSM, 2 × WR set in a 50-50 mix of polyester and urethane acrylate resins.

The specimens were subjected to a displacement-controlled pull-off of the web at a 45^0 angle to the flange. The specimens were restrained by rigid clamps on the flange close to the fillet on both sides of the web. The joint failed initially by delamination within the curved part of the boundary angle on the side in tension. This delamination occurred within the third or sixth ply of the overlaminate with whitening being noticed at loads from about 6 kN. With application of further load, it was found that the joint had sustained only a small reduction in strength and stiffness. Final failure occurred when the remaining boundary angle plies delaminated (with the delamination spreading to the flat part of the boundary angle attached to the flange) coupled with failure of the fillet.

Initially, an FE model was analysed so as to represent the tested specimens under a 45^0 pull-off load. The material properties used in the finite element models are shown in Table 3. A comparison of its load-deflection curve can be made with the experimental curve for Sample B, shown in Figure 4, and shows very good agreement. In addition, at a load of 6 kN which was when delaminations became visible in the tested specimens, the maximum through-thickness stress in the overlaminate is equal to 9 MPa which is greater than the interlaminar tensile strength of the material of 7 MPa (7). The maximum through-thickness stress occurs in the inner regions of the overlaminate, indicating that delaminations are

Material	Location	Property	Value
Polyester/ Woven Roving Glass	Web, Flange and Overlaminate	Ex	13060 MPa
		Ey	7770 MPa
		vxy	0.25
Urethane Acrylate	Fillet	Ex	1500 MPa
		Ey	1500 MPa
		vxy	0.25
Polyester/CSM	Overlaminate	Ex	6890 MPa
		Ey	7770 MPa
		vxy	0.25

Table 3. Material Properties used in Tee Joint Finite Element Models.

980

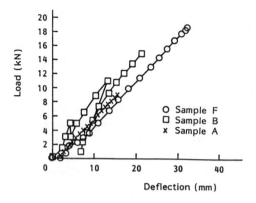

Figure 4. Experimental Load-Deflection Curve.

likely to occur in these regions which is in complete agreement with the tested specimens. Having verified the FE model, a fracture analysis can now be carried out. FE analyses were carried out for a tee joint with a series of cracks in two different locations (8) and the values of both G and J were calculated since it was inticipated that the behaviour may show non-linear characteristics.

A typical plot of the finite element model used in the analyses is shown in Figure 5. Four-noded quadrilateral elements were used for the whole model except for the region containing the crack. Six-noded triangular elements were used for the region containing the crack. As in the case of the stiffener FE models, it was also deemed necessary to insert gap elements along the crack line to ensure that the crack did not close thus truly representing the joint and crack behaviour. Plane strain conditions were again assumed to prevail in all cases.

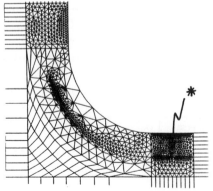

Figure 5. Crack Region of Tee Joint Finite Element Model.

So as to simulate the experimental test programme, the FE models were also subjected to a 45^0 pull-off load with flanges being clamped. The applied load was 10 kN. This represents a feasible maximum load to be experienced by the structure. Thus any crack which remains stable at this load may be deemed to be safe and the structure to be damage tolerant.

The effect of crack depth was investigated for a single, horizontal crack of 10 mm in the overlaminate above the flange. Three depths were analysed for a crack at 2, 6 and 8 plies deep from the overlaminate surface. For the cracks in the overlaminate above the flange the J-integral was calculated for the crack tip nearest the joint radius for each depth. The results are shown in Figure 6.

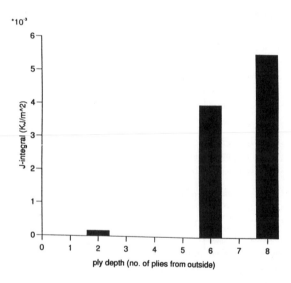

Figure 6. Effect of Crack Depth on G Values.

The effect of the radiused crack length was also investigated. The right-hand crack tip (marked * in Figure 5) was kept at the same location in each case. A total of four radiused crack lengths were analysed. For the cracks in the curved part of the overlaminate, the J-integral values were calculated at both crack tips for each of the 4 crack lengths. The results are plotted in Figure 7. In addition, the values of G have also been calculated for 4 different crack lengths and compared with the J values.

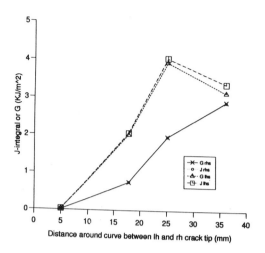

Figure 7. Effect of Crack Length on G and J Values.

4. DISCUSSION

4.1 Top Hat Stiffeners

The results of the strength analysis give reasonable correlation with the experimental behaviour. The models predicted the regions which were subjected to the highest stresses and hence which sections of the top-hat stiffeners are likely to fail under a given set of boundary conditions. The results from the modelling of the three-point bending tests was particularly encouraging since it not only correctly predicted the regions under the highest levels of stress but also gave comparable actual values. For example, the top-hat ultimately failed in the flange at a load of 16.5 kN. The assumed in-plane failure stress of the flange plate material is 210 MPa and the stress predicted by the finite element modelling is 208 MPa.

In the case of the crack depth variation for the three-point bending model, it can be noted that all the values of G are well below the critical value and thus inferring that none of these cracks are likely to propagate under these loading conditions. The gradual increase in G as a depth of 4 mm is approached and then a gradual decrease indicates that cracks at depths of about 4 mm are more likely to propagate. It must be pointed out however, that the values of G near the surface may in fact be

influenced by the fact that the element mesh becomes more coarse near the surface as the crack becomes shallower. This feature needs further investigation.

In the case of the crack length variation a definite trend is achieved. For a critical value of 0.54 N/mm (6), it can be shown that cracks over a length of about 37 mm are likely to propagate.

Similar trends are obtained for the pull-off model except that critical crack length is about 30 mm.

4.2 Tee Joints

It can be noted that cracks nearest the surface i.e at a depth of 2 plies give lower values of the J-integral than those deeper within the overlaminate. This implies that for this loading and boundary condition, surface cracks are less likely to propagate than deep cracks. It is also interesting that even the highest value of J-integral (0.0056 kJ/m^2) is a lot less than the critical value of strain energy release rate of 0.54 kJ/m^2 for such materials (6).

This implies that for a load of 10 kN, straight cracks in the overlaminate above the flange are not of importance in a damage tolerance context. This also images the experimental results where at a load level of 10 kN only delamination in the curved part of the overlaminate had occurred.

It is clear that the highest values of the J-integral are obtained for the cracks which extend into the region of overlaminate curvature. The lowest values are for the cracks which are almost horizontal. This is directly comparable with the previous results for horizontal cracks in the overlaminate above the flange. Noting that the critical value of the J-integral is of the order of 0.5 kJ/m^2 then it is likely that cracks whose curved length is greater than 16 mm are likely to propagate under these loading and boundary conditions based upon the values at the right hand tip. The equivalent critical length based upon the results at the left hand tip is 8 mm.

5. CLOSURE

The operational defects in FRP ships are most likely to be in the form of delaminations. These can occur within plate panels or at the roots of tee joints and top hat stiffeners. Such defects have the potential to cause serious structural problems either in the form of instability-related effects or in the form of crack extensions. The former are relatively easy to assess and, in the main, are not problematical because of the comparatively small in-plane compressive strains and short panel lengths. The latter are more problematical in that if they occur at the root of out-of-plane joints, which are in themselves a zone of weakness, thay can have serious consequences with regard to structural integrity.

The kay question to resolve in the operational context therefore is one of assessing the possibility of unstable delamination (crack) growth. One approach to this assessment of damage tolerance is to compare strain energy release rates associated with a given delamination and (local) loading condition with the fracture toughness of the material in question. This paper has outlined an approach for such treatment. Using this approach, a shipowner should be able to evolve a data bank of fault location, delamination length, loading situation and strain energy release rates. From this and a full characterisation of the materials' fracture toughness, it should be possible to deduce a repair strategy for the vessel class.

6. ACKNOWLEDGEMENTS

The work outlined in this paper was funded by EPSRC/MOD (DERA). The authors wish to acknowledge the help and guidance from colleagues at DERA, Lt. Cdr. Gray, Prof. Sumpter, Mr. Court, Mr. Elliot, Mr. Lay, Mr. Swift and Mr. Trask whose input was invaluable.

7. REFERENCES

(1) Phillips, H. J. "Assessment of Damage Tolerance Levels in FRP Ship Structures", PhD Thesis, Department of Ship Science, University of Southampton, January 1997.

(2) Parker, A. P. "The Mechanics of Fracture and Fatigue", E&F Spon Ltd., London, 1981.

(3) Elliot, D. M. "Mechanical Testing of Composite Joints - Interim Report." DRA Report, April 1994.

(4) Phillips, H. J., Shenoi, R. A. & Moss, C. E. "Damage Mechanics of Top Hat Stiffeners used in FRP Ship Construction", Marine Structures (Submitted for Publication).

(5) Court,R. : "Fracture Toughness of Woven Glass Reinforced Composites", DRA Dunfermline, December 1994.

(6) Shenoi,R.A., Read,P.J.C.L. & Hawkins,G.L. : "Fatigue Failure Mechanisms in Fibre Reinforced Plastic Laminated Tee Joints", International Journal of Fatigue, Vol 17, No. 6, 1995, pp. 415-426.

(7) Bird,J. & Allan, R.C.. The determination of the Interlaminar Tensile Strength of Ship Type Laminates. Proc. 7 th International Conference on Experimental Stress Analysis, Haifa, Israel, 1982, pp. 91-104.

(8) Phillips, H. J. & Shenoi, R. A. "Damage Tolerance of Laminated Tee Joints in FRP Structures". Composites Part A (Accepted for Publication, to appear).

OFFSHORE ENGINEERING

OFFSHORE ENGINEERING

Verification of FPSO Structural Integrity

R. Potthurst and K. Mitchell

Offshore Floating Units Group, Industry Division, Lloyds Register of Shipping

This paper is intended to provide an overview of FPSO structural integrity aspects based on experience gained from recent approvals in relation to the new UK verification requirements. Main hull strength and fatigue design issues are addressed. Related FPSO structural design aspects which require consideration in order to ensure long term structural integrity are also discussed.

1 INTRODUCTION

Since 1991 there has been rapid growth in deployment of FPSO installations for field development. This trend is set to continue, especially in harsh environments, with new projects demanding tight design, approval and construction schedules. This places increasing technical challenges on designers, operators and regulatory bodies to ensure that satisfactory standards are met. With the majority of vessels intended for long term service without dry-docking this particularly applies to hull strength and fatigue life.

In UK waters the increased use of FPSOs is occurring at the same time as key changes in the regulatory regime. New verification regulations have replaced the former prescriptive certification scheme and these place clear accountability for safety with the owner or operator. This change of requirements to a goal setting approach allows flexibility to select optimum solutions whilst at the same time permit a more robust and formalised approach to the management of safety and hazards throughout the life-cycle of the installation.

This paper outlines some recent experience of verification schemes concentrating on structural issues and how class rules can provide an appropriate standard for structural requirements of an FPSO. Hull structural strength and fatigue life are discussed together with important contributory factors.

2 REGULATIONS

Following the recommendations made after the Piper Alpha incident in 1988 the previous requirements for UK certification as given in SI (1974) No.289 have been revoked. Regulatory responsibility for offshore safety has been transferred from the Department of Energy to the Health and Safety Executive and new Safety Case Regulations, UK SI (1992) No.2885, introduced.

To support the provisions of the Safety Case Regulations a number of additional regulations have also been introduced which give more specific requirements. Included among these is SI (1996) No.913: Design and Construction Regulations (DCR). The DCR is concerned with the integrity of offshore installations, the safety of wells, the workplace environment offshore, and the suitability by verification of safety critical elements of an offshore installation.

3 OUTLINE OF VERIFICATION SCHEME REQUIREMENTS

The new regulations require a verification scheme to be set up for each installation. Under the scheme Safety Critical Elements (SCE) of an offshore installation are to be verified as suitable by independent and competent persons. The scheme for each installation will be the responsibility of the duty holder who will appoint the independent competent body in the capacity as verifier.

To assist duty holders to fulfil their obligations in respect of implementing a scheme, the HSE has published Guidance (1), which summarises the matters to be included in a scheme, and how the work performed within the scheme relates to other regulation requirements.

As part of the new regulations the duty holder is responsible for setting performance standards. These may be set at several levels. High level safety goals and performance standards for complete parts or systems will normally be part of the Safety Case and will be reviewed and accepted by the statutory authority. Lower level performance standards may be full performance standards or system specifications, industry codes or recommended practices. The verification body clearly needs to be familiar with the high level performance standards as these provide the link between the Safety Case and the verification scheme.

For owners/duty holders of FPSO installations which operate possibly at a single fixed location within UK waters for their entire operating life, classification is not a statutory requirement, therefore a duty holder is unlikely to adopt such a voluntary provision unless there is a demonstrable case that adopting classification will contribute more than another method, toward achieving compliance within a verification scheme.

LR's recent experience (2) with duty holders of FPSO's indicates that the classification requirements in relation to the vessel hull, essential marine systems and moorings, embraced within LR Rules are providing an adequate basis to establish a performance standard for these elements within a scheme. The class system also has the merit of well established survey procedures, a reporting system to record the findings of ongoing examinations, and a diary system to ensure implementation of corrective actions, all of which is consistent with the overall objectives of the verification scheme.

Structural design verification is intended to ensure that the design of structural elements identified as being safety critical and included within the Verification Scheme Matrix has addressed all appropriate loadings and load combinations and that resultant deflections and stresses are within allowable values and the design provides adequate fatigue and wear resistance.

High level components of the FPSO structure which would be expected to be considered as SCE include:
- Hull
- Mooring
- Bearings
- Topsides
- Subsea
- Accommodation block
- Helideck
- Lifting equipment

A definitive list for a specific project would be developed on a continuous consultation basis and kept under review as the project develops and the Safety Case is completed.

4 STRUCTURAL ARRANGEMENT
4.1 General
General requirements regarding location and separation of spaces, layout and arrangement of primary structural components, scantlings and structural details are contained in class rules. Overall subdivision of the hull should take full account of strength and stability requirements and minimise consequence of damage, pollution risk and loss of the unit in the event of damage. Additional subdivision of the hull may be required to account for ballast water needed to control hull stresses and for the storage of process-related liquids.

Account should be taken of the interaction between structural strength and stability. Particular consideration should be given to tank dimensions with respect to tank inspection requirements and sloshing/free surface effects for partially filled tanks. Intact and damage stability should comply with applicable national authority requirements.

Self-propelled units should meet the Load Line requirements defined in the 1966 International Convention of Load Lines (ICLL). Units which do not engage in international voyages except for transfers between fabrication sites and the installation voyage to the designated site, should have marks which indicate the maximum permissible draught as calculated under the terms of the ICLL Rules.

The International Maritime Organisation (IMO) and ICLL rules, or applicable flag state /

national administration requirements, should be followed with respect to weathertightness and watertightness of decks, superstructures, deckhouses, doors, vents, etc.

The effects and consequences of accidental damage to the hull should be addressed. The effect of the extent of damage from penetration or flooding of one or more compartments as outlined in class rules and guidelines, national authority and flag state regulations should be considered in terms of stability, strength and impact on the environment.

4.2   Deckhouse Superstructure

Living quarters, lifeboats and other means of evacuation should be located in non-hazardous areas and be protected and separated from production, storage and wellhead areas. As a minimum the arrangement and separation of living quarters, storage tanks, machinery rooms, etc. should be in accordance with the International Convention for the Safety of Life at Sea 1974 (as amended) Regulations. Deckhouse superstructures may be located forward or aft of the oil storage tanks. However, where superstructure is located forward of the cargo tank area, arrangements should provide a suitable level of separation and protection.

4.3   Topside Facilities

The location of the process facility deck and structural arrangements should comply with class requirements and acceptable national codes and standards regarding dangerous zones or divisions and provision of adequate access. Areas and compartments of units are defined as 'Dangerous Zones' according to their proximity to equipment, pipes or tanks containing certain flammable liquids and whether these fluids are at temperatures approaching or exceeding their flashpoints.

On oil storage units, the main dangerous zone extends over the cargo tank area up to a height of between 2.4m and 3.0m above the main deck. Dangerous zones also exist around tank vent outlets and any other areas connected with the loading or discharge of cargo. To avoid any technical contravention of the dangerous zone concept on many FPSO units the process

plant is accommodated on a suitable deck structure constructed at a height of approximately 3m above the cargo/upper deck.

5   HULL - GLOBAL STRENGTH

Both longitudinal and transverse hull girder strength require evaluation, and will be expected to comply with class rule requirements for floating production installations. These indicate that, as a minimum, the structure should be designed in accordance with the full seagoing requirements of class rules for ships. In addition it can be anticipated that dynamic loads to which the hull may be subjected will vary from those associated with seagoing trading ships and will require assessment. This assessment may be based on results of model testing or by suitable direct calculation methods to predict the actual wave loads on the hull at the service location, taking into account site specific service related factors including relevant non-linear effects.

All modes of operation should be investigated using a full range of realistic loading conditions. All anticipated pre-service and in-service conditions require evaluation to determine the most unfavourable design cases. For units intended for multi-field developments the most onerous still water and site specific environmental loadings should be considered.

A unit may be kept on station by various methods, including several different types of station-keeping systems such as internal and submerged turret systems, external turret, CALM buoy, fixed spread mooring and dynamic positioning. Each mooring system configuration will impose loads into the hull structure which are characteristic to that system.

The effects of all life-cycle loadings should be evaluated to ensure that all relevant design loadings are considered. These would include construction, transportation, installation and decommissioning, on-site intact and on-site accidental loadings.

If it is intended to dry-dock the unit the bottom structure should be suitably strengthened to withstand the loadings involved.

An on-board approved loading instrument should be installed to monitor still water bending moments and shear forces and ensure they are maintained within the approved permissible levels.

6 FINITE ELEMENT ANALYSIS
6.1 General
Structural assessment is generally based on direct calculation methods, although results of model testing can also provide design data in certain circumstances. Finite element analysis of the hull structure and other major structural components will normally be required. All relevant loadings should be adequately accounted for. The complexity of mathematical models of the structure, together with the associated computer element types used should be sufficiently representative of all the parts of the primary structure to enable accurate stress distribution to be obtained. A combination of global and local analytical models may be necessary, particularly when the global model does not fully account for local load effects or does not contain sufficient detail to ascertain response to the level required. Account should be taken of boundary condition effects on the structural evaluation. Particular attention should be given to structural evaluation in way of critical interfaces and abrupt changes of section such as the mooring structure and its integration within the hull.

Finite Element structural analysis of the following would typically be required:
- representative portion of the hull and containment system;
- mooring structure and integration of the mooring system;
- topside facilities support arrangements;
- PAU's etc;
- turret deckhouse, helideck, crane pedestal, flare stack etc;
- other major discontinuities and identified safety critical structure etc.

7 STEEL GRADES
Steel should be manufactured and tested in accordance with class rules or other acceptable standard. As a minimum, steel grades should comply with the class rules for ships.

Structural components may be grouped into categories based on applied loading and stress levels, critical load transfer points, stress concentrations and consequence of failure, and the steel grade selected accordingly.

Material where the principle loads or welding stresses are perpendicular to the plate thickness should have suitable through thickness properties.

8 FATIGUE
8.1 General
Fatigue life evaluation is one of the most important design considerations for FPSO structures, particularly for vessels intended to remain on site for an extended period without scheduled dry docking. Experience from vessels in service has shown that ship type structures used in an FPSO role may be prone to fatigue damage particularly in way of primary hull connections. Since fatigue cracks can be possible points of initiation for structural fractures or costly repair it is essential that fatigue design is given careful consideration.

For new build hulls the assessment will confirm the suitability of critical structural details for site specific design criteria. For conversions assessment will indicate critical areas of the structure requiring upgrading and confirm that the proposed scheme of modifications is acceptable.

Structural fatigue life evaluation together with experience gained from existing project survey records can also provide a valuable means to target inspections towards those areas demonstrated to be most at risk.

8.2 Analysis
The extent of the analysis will be dependent on the mode and area of operation. The analysis should be performed using the long-term prediction of environment for the site and account for directionality. For conversions due allowance should be made of previous operational history of the vessel. Deterministic or spectral analysis methods may be used. The analysis should address the primary hull

structure, mooring structure and other primary structure subject to significant dynamic loading e.g. flaretower, and account should be taken of all important sources of cyclic loading.

Fatigue life assessment may be based on the Miner's summation method using appropriate stress ranges, which are to include stress concentration effects where these have not already been accounted for within the joint classification. Data relating to previous service may be based on a study of vessel records, or where these are not readily available, on an appropriate simplified representation of trading history.

A representative range of loading conditions should be included . It is generally acceptable to consider three loading conditions, typically: ballast (or light load) condition, 50% loaded, and the fully loaded condition, with appropriate amount of time at each condition. An appropriate range of wave approach directions and wave energy spreading should be considered.

Both the global and the local dynamic stress components should be considered. The global stress component results from the effects of hull wave bending moments. The local stress component results from the effects of the localised fluid pressure. The method of combining these stresses for the fatigue damage calculation will depend on the location of the structural detail.

The minimum design fatigue life for structural elements should not be less than the intended field life, but, in general should not be less than 20 years. The cumulative damage ratio for individual components should take account of the degree of redundancy and accessibility of the structure and also the consequence of failure.

8.3   Structural details

An important consideration related to fatigue performance in-service is the quality of detail design and construction. It is apparent that in many cases premature fracture can be traced to inadequate detail design.

Fatigue strength is seriously reduced by the introduction of a stress raiser such as a notch or hole. Since actual structural elements invariably contain stress raisers like fillet welds, end brackets, cut-outs etc, it is not surprising to find that fatigue cracks in structural parts usually start at such geometrical irregularities. An effective way of minimising fatigue cracking is by the reduction of avoidable stress raisers through careful design and the prevention of accidental stress raisers by careful fabrication together with attention to the relative stiffness of structural members to avoid secondary local restraint effects.

9   CORROSION PROTECTION

It is recognised that an important factor regarding in-service strength and fatigue performance is the degree of protection against corrosion afforded to the structure. The combination of significant local cyclic stresses combined with corrosion effects has led to a large number of structural fatigue problems, particularly on tankers and VLCC's. The most severe corrosion has occurred in ballast tanks and slop tanks.

To mitigate against the effects of corrosion all structural steelwork is required to be suitably protected. In general suitable protection systems may include coatings, metallic claddings, cathodic protection, corrosion allowances or other acceptable method. Combinations of methods may be used. Consideration should be paid to the design life and the maintainability of the surfaces in the design of protective systems.

10   GREEN SEAS

Recent experience indicates that very sizeable green water loading can occur on FPSO's operating in harsh environments and account of this should be taken in the design. Significant amounts of green water will have an impact on the vessel deck structural design, accommodation superstructure, equipment design and layout and may induce vibrations in the hull.

At present green water loadings are difficult to predict and it is recommended that provisions be made during model testing for suitable measurements to determine design pressures for local structural design.

MARIN are currently involved in a Joint Industry Project on green water loading, and this should provide valuable data for design purposes.

Measures which have been used to mitigate the effect of green water include bow shape design, flare, breakwaters and other protective structure such as turret housings. Adequate drainage arrangements must be provided.

11 SLAMMING

Slamming occurs when wave impact causes a dynamic impulsive pressure load which can lead to vibration of the hull and structural damage. The most important location to be considered with respect to slamming is the forward bottom structure, hence slamming forces should be taken into account for both hull and turret design. However other locations on the hull which may be subject to effects of wave impacts include: stern structure, bow flare and bow side, masts, crane posts, piping etc.

Every effort should be made to minimise the possibility of slamming by careful selection of bow design and drafts linked to site criteria. The effects of slamming on the structure should be considered in design particularly with regard to local strength aspects and also possible enhancement of global hull girder bending moments and shear loadings.

Although class rules contain minimum requirements relating to draft, pressure, extent and scantlings, it is recommended that model testing can again provide suitable data on pressure loading for structural design.

12 SLOSHING

Sloshing is defined as a dynamic magnification of internal pressures acting on the boundaries and internal structure of cargo tanks etc. to a level greater than that obtained from static considerations alone. Sloshing occurs if the natural periods of the fluid and the vessel are close to each other.

Trading tankers are normally operated so that cargo tanks are typically either full or empty. For a floating production unit in general no restrictions are imposed in partial loading of cargo tanks, and tanks may be only part full for long periods. Because of this a sloshing analysis would be expected to be carried out.

The acceptability of cargo tanks structural components such as transverse and longitudinal bulkheads, wash bulkheads and deck plating should be demonstrated. Pressures obtained from the sloshing analysis can be compared with ultimate strength formulations for the plating and stiffeners.

For new build vessels the scantlings of some plating and stiffening may be required to be increased, in excess of class rule minimum scantlings, to sustain resultant high pressures. Other options such as placing restrictions on filling ranges or imposing limitations on GM and draught for particular fill levels may be available depending on vessel operation considerations. Another solution is to designate certain tanks for partial filling, and increase the scantlings of these tanks only.

Sloshing is an important consideration for tanker conversions and in some cases the fitting of swash bulkheads or other baffling devices may be necessary together with some limited operational restrictions.

13 MODEL TESTING

Model testing will generally be necessary as the total FPSO configuration is usually too complex to be reliably analysed by numerical methods alone. It therefore forms an essential stage in the design process and should be capable of providing loads and motions for all vital components. Most installations will need project-specific model testing with supporting dynamic analysis which will be carried out to establish the riser and mooring configuration, vessel responses and sizing of anchor lines.

The specification for the testing programme will include tests which allow calibration and measurements of static, regular and irregular values.

Observation of model testing is now greatly enhanced by the use of video cameras. These can be located close to the model and at strategic positions which allow important interactions to be viewed remotely without impeding model motions or affecting the wind and wave conditions.

For shallow and medium water depth locations, initial design of the mooring system can be based on quasi-static studies and the application of dynamic amplification factors based on data from previous similar cases. For such systems, that method can provide sufficiently accurate results to make final model testing only confirmatory. Mooring systems in deep water locations however, should not be considered using this method alone as there will generally not be a comprehensive database from similar designs or locations to reliably derive the basic characteristics of the system by quasi-static methods. In addition, dynamic effects will be more significant in deep water, and assumptions on the relationship between static and dynamic tensions which may be valid for moderate water depths, will not be so for deep water.

A suitable study is recommended in order to estimate appropriate combinations of wave, wind and current in extreme conditions.

The test programme will be expected to provide a range of key data and help to ensure that critical loadings and motion responses will be determined. This will enable verification of the current design or identify required alterations.

14 MOTIONS AND ACCELERATIONS

Hull motions and accelerations are of prime importance in establishing design criteria for topside structure. All six degrees of freedom can provide contributions, however heave, pitch and roll effects generally dominate. It is important to recognise that maximum accelerations depend upon both motion and period. Model test results can confirm validity of results obtained from analytical methods.

Account should be taken of all relevant factors which may have a significant influence on vessel motions and accelerations.

Components of motions should be combined in a rational manner using appropriate analytical or model test data. In the absence of such data class rules can be applied.

15 MOORING STRUCTURE

The diameter of the turret is decided by a variety of factors which include, size of FPSO; number and size of risers; number and size of mooring lines, and access arrangements.

Structural analysis for the integration of the mooring system within the structure of the unit should be carried out in accordance with project and class requirements. Structural strength should be evaluated considering all relevant, realistic loading conditions and combinations. In particular load combinations due to the following should be accounted for in the design:
- mooring and riser system loads;
- overall hull bending moments and shear forces;
- internal and external pressure loads, covering the intended range of draughts and loading conditions, including non-symmetric cases as applicable.

Design calculations should take into account the following:
- strength;
- fatigue;
- one line failed condition;
- accidental loadings;
- maximum design loadings;
- motion load spectra.

Particular attention should be given to the design of critical interfaces. Continuity of primary longitudinal structural elements is to be maintained as far as practicable in way of the turret opening. Reductions in hull section modulus should be kept to a minimum and compensation fitted where necessary.

16 TOPSIDE STRUCTURE

The following items would normally be identified as safety critical and included under this general heading:
- process deck and support structure;
- skid mounted structures (identified as SCE);
- flare stack;
- crane pedestals;
- helideck.

The design of skid structures, should take account of loadings due to bending of the hull, and vessel hull / skid structure interaction. The main support members of skid structures are in general regarded as primary structure. Support plating and stiffening is regarded as secondary structure.

Helideck structural design and arrangements should be verified. The helideck arrangement should be developed paying particular attention to the following:

- access to the helideck;
- helicopter type and size specified;
- markings, lighting, obstruction free zones, windsock, floodlights etc. to comply with the relevant requirements.

Requirements for markings are contained in the ICAO Annex 14, or in national regulations (e.g. UK CAA requirements CAP437).

Structural assessment can cover various loading phases including loadout, transportation, lift, in-place.

Structural analysis and design should be carried out and take into account appropriate loading combinations due to the following:

- gravity loading;
- dynamic loadings, due to hull motions and accelerations;
- hull deflection;
- wind loading, including vortex shedding;
- snow and ice loading;
- fatigue loading;
- live loads.

In addition, loading cases accounting for static angles simulating the vessel inclination in damaged conditions should be considered.

As outlined previously, material selection for topside structures is to be based on a consideration of applied loading, structural arrangement, location, design temperature and consequence of failure in accordance with class rules.

17 LOCAL STRENGTH AND STRUCTURAL DETAILS

Design for local strength and structural details including connections, continuity and alignment, tolerances, closing arrangement, vents, welding etc. should comply with class rules and requirements of the flag state, if applicable. Consideration should be given to the following:

- proportions of built-up members;
- the design of structural details against the harmful effects of stress concentrations and notches;
- details of the ends and intersections of members and associated brackets;
- shape and location of air, drainage, and lightening holes;
- shape and reinforcement of slots and cut-outs for internals;
- elimination or closing of weld scallops in way of butts, "softening" bracket toes, reducing abrupt changes of section or structural discontinuities;
- proportions and thickness of structural members to reduce fatigue damage due to engine, propeller or wave-induced cyclic stresses;
- the thickness of internal structure in locations susceptible to excessive corrosion;
- the structure supporting the components of the mooring system such as fairleads, winches, etc. (should be designed to withstand, as a minimum, the stresses corresponding to a mooring line loaded to its breaking strength).;
- scantlings necessary to maintain strength in way of large openings;
- the strength of the unit in the transit condition. (For a turret-moored unit or a unit with a moonpool well, the plating of the well should be suitably stiffened to prevent damage in transit).

18 ACCIDENTAL LOADS
18.1 General

As part of the verification process the effects of accidental loadings on the design of structural elements needs to be addressed. Such loadings include:

- collision and vessel impact, due to supply boat or shuttle tanker;
- dropped objects, from deck cranes;

- fire and explosion, (this would be dealt with in the Safety Case, however structural effects require evaluation).

18.2 Collision

Collision should be considered for all elements of the unit which may be impacted by sideways, bow or stern collision. The vertical extent of the collision zone should be based on the depth and draught of attending vessels and the relative motion between the attending vessels and the unit. Minimum impact energy levels and loading combinations are outlined in class rules and national authority requirements.

18.3 Dropped objects

The accidental impact loads caused by dropped objects from cranes should be considered when the arrangements are such that the failure of a vital structural member could result in the local collapse of the structure. Critical areas for dropped objects shall be determined on the basis of the actual movement of crane loads over the unit. Structural damage resulting from dropped objects should be evaluated taking into account the nature and size of the likely loads to be lifted during operations.

18.4 Blast

It should be demonstrated that the fire insulation remains effective for the duration of the fire/blast scenario. The scenario should be defined; e.g. fire followed by blast followed by fire; or blast followed by fire etc. Blast resistance requirements shall be addressed concurrently with fire resistance requirements, taking into account probability of occurrence, blast safety evaluation, layout and area of importance, venting system, missile damage and access to escape etc. The resistance to fire after blast should also be addressed.

18.5 Strength criteria

The resistance of the structure may be based on the ultimate strength and estimated in terms of collapse pressure, maximum deformation, maximum ductility ratio, energy absorption or brittle failure, whichever dominates.

Calculations are required to demonstrate the integrity of the system in accordance with relevant requirements.

REFERENCES

1) A guide to the installation verification and miscellaneous aspects of amendments by the Offshore Installation and Wells (Design and Construction, etc.) Regulations 1996 to the Offshore Installation (Safety Case) Regulations 1992.

2) P C Campbell and H J Sewell. Experience with verification schemes under new UK offshore regulations. LRTA internal paper 1997.

3) LR Classification Rules for Floating Production, Storage and Offloading Installations.

4) LR ShipRight procedures manuals.

5) Draft ISO/TC67/SC7/WG5 - panel 2: structural design.

6) J R Lane and M C Jones. Performance standards and written schemes - recent experience from the UK offshore industry. LRTA internal paper 1996.

7) JIP - FPSO integrity, press bulletin February 1997.

8) A G Gavin. Double hull tankers. LRTA internal paper 1995.

9) J MacGregor and J-Loup Isnard. Developments in monohull production vessels for North Sea and Atlantic applications, FPSO world congress, Norway 1996.

10) B Buchner. Green water and FPSOs - a new challenge, FPSO world congress, Norway 1996.

11) IACS, bulk carriers - guidelines for surveys, assessment and repair of hull structure, 1994.

Practical Design of Ships and Mobile Units
M.W.C. Oosterveld and S.G. Tan, editors.

Integrated Motion, Load and Structural Analysis for Offshore Structures

Yung Shin[1], Craig Lee[1] and D.E. Jones[1]

[1] American Bureau of Shipping , Two World Trade Center 106th Fl., New York, USA 10048

This paper presents the recent development of dynamic load based analysis procedure and computer system for analyzing various types of offshore structures.   This fully integrated analysis method incorporates the all aspects of motion characteristics, wave loading and structural response.   Two levels of analysis sophistication are provided starting from simple space frame analysis to full finite element analysis. The space frame analysis is first used to provide global responses for initial screening to identify the critical load conditions which will be analyzed via a full FE analysis.

This space frame analysis utilizing Morrison's equations considers all the wave headings and a range of wave frequencies. Platform motions, mooring and wave loads are calculated and checked for dynamic equilibrium before they are applied in structural analysis. Finally structural responses are checked for the three failure modes of yielding, buckling, and fatigue. Based on the simplified Phase A analysis for all conditions, a few selected critical conditions will be re-examined by Phase B detailed analysis. This Phase B analysis is based on a diffraction theory utilizing a 3-D panel and Green's function approach.   The wave diffraction effect due to large diameter columns and pontoons, and the hydrodynamic interaction between these large members are fully taken into account.   Complete pressure distribution on the structure of a 8 column semi-submersible is calculated and used for FEM analysis.   Comparison of theoretical calculations with model test and other theoretical methods are presented.   The same analysis approach can also be applied to other types of offshore structures such as tension leg platforms and spars.

1.   INTRODUCTION

The offshore industry is moving into deeper water operation than ever before and as such conventional design practices may not be adequate. Thus the limitations of present practice based on classification rules needs to be fully examined using a First principles based design approach.

For this purpose Offshore Structure Analysis System (OSAS) was developed to help ABS engineers in reviewing the designs of compliant offshore structures such as semi-submersibles, TLPs, spars, etc. by performing direct calculation based on the First Principles rather than conventional classification rules. This system can be used for an integrated analysis for determining motion, mooring loads, structural loads and the critical load sets needed for the Finite Element Analysis (FEA) of the offshore structures. The general procedure and approach for the hydrodynamic-structural analyses are discussed.

1.1 General analysis procedure

The analysis procedure consists of two phases, Phase A and B (see Figure 1). Phase A is a comprehensive screening step using a fully integrated hydrodynamics, structural analysis, and mooring analysis package for column-stabilized offshore platforms moored by catenaries, or by tension legs. and risers utilizing a simplified space frame model. It consists of a quasi-linear load and motion module, a linear finite element module and a spectral analysis module. The user will be able to accomplish a full range of evaluation for platform motions, mooring loads and the development of hydrodynamic loads to assess the structural response/adequacy via structural code checks for strength and fatigue based on extreme combined stress and fatigue assessment.   The objective of Phase A analysis is to determine critical load cases required for more detailed FEM analysis in Phase B.   Two different methods are employed in determining critical load cases: stress based and load based approaches.   Phase B is an enhanced hydrodynamic-structural analysis package for column-stabilized offshore platforms of FEM configuration using the design wave approach. It consists of a three dimensional diffraction load analysis using 3-D panels, a FEM analysis, and interface to generate boundary conditions for substructures and a post-processing module. The user will be able to accomplish a further detail structural analysis for the critical design load cases obtained from Phase A analysis and a fine mesh FEM model. The objective of Phase B analysis is to evaluate the strength of structural details at critical locations.

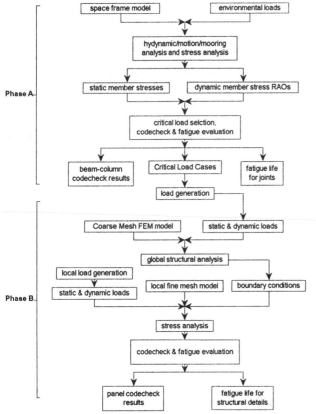

Figure 1. Analysis Flow of OSAS system

The detail of Phase A and B analysis procedure is discussed in the following sections. In addition, an example analysis for a modern 8 column semi-submersible (displacement of approx. 21,000 tons) using this methodology is presented in the paper.

2. SIMPLIFIED ANALYSIS BY A SPACE FRAME MODEL (PHASE A)

2.1 Modeling of a space frame

Based on the structural scantlings and weight distribution, a space-frame model (see figure 2) for a 8 column semi-submersible was constructed. This model, mainly consisting of beam elements, will give a good representation of force flow. The properties of each structural element, stiffness and mass, represent the gross properties of the corresponding section. The junctions between column/pontoon, column/bracing, column/deck and deck/bracing are modeled as rigid elements, since these parts are considerably stiffer than other parts. Cut-outs, other openings and local structural details are not taken into account in the model, and will be considered in Phase B.

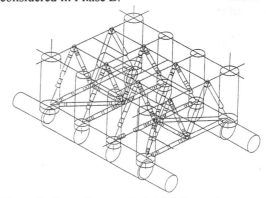

Figure 2. Space Frame Model for Phase A

The hydrodynamic properties of the platform are represented as a space frame array of slender members having specified dimensions and coefficients of added mass and drag with fluid forces calculated using Morrison's equation. Member cross sections are assumed to be circular or elliptical. Other sectional shapes can be analyzed by defining equivalent diameter and corresponding hydrodynamic coefficients.

Whole Year *(For All Nine Weather Ships)*									(227,497 Observations)
		Wave Period (sec)							
		5 s	7 s	9 s	11 s	13 s	15 s	17 s	Sum over All Periods
	0 - 0.75	20.91	11.79	4.57	2.24	0.47	0.06	0.00	0.60	40.64
	0.75 - 1.75	72.78	131.08	63.08	17.26	2.39	0.33	0.11	0.77	287.80
W	1.75 - 2.75	21.24	126.41	118.31	30.24	3.68	0.47	0.09	0.56	301.00
a	2.75 - 3.75	3.28	49.60	92.69	32.99	5.46	0.68	0.12	0.27	185.09
v	3.75 - 4.75	0.53	16.19	44.36	22.28	4.79	1.14	0.08	0.29	89.66
e	4.75 - 5.75	0.12	4.34	17.30	12.89	3.13	0.56	0.13	0.04	38.51
	5.75 - 6.75	0.07	2.90	9.90	8.86	3.03	0.59	0.08	0.03	25.46
H	6.75 - 7.75	0.03	1.39	4.47	5.22	1.93	0.38	0.04	0.04	13.50
e	7.75 - 8.75	0.00	1.09	2.55	3.92	1.98	0.50	0.03	0.02	10.09
i	8.75 - 9.75	0.00	0.54	1.36	2.26	1.54	0.68	0.20	0.04	6.62
g	9.75 - 10.75	0.01	0.01	0.10	0.11	0.10	0.05	0.02	0.00	0.40
h	10.75 - 11.75	0.00	0.00	0.03	0.08	0.17	0.06		0.00	0.34
t	11.75 - 12.75		0.05	0.00	0.14	0.22	0.06	0.01		0.48
	12.75 - 13.75		0.02		0.07	0.09	0.03		0.01	0.22
(m)	13.75 - 14.75				0.02	0.06	0.02	0.00	0.01	0.11
	14.75 - 15.75	0.00	0.02	0.00	0.01	0.01	0.02	0.01	0.04	0.08
Sum over All Heights		118.97	345.43	358.72	138.59	29.05	5.63	0.92	2.69	1000.00

Figure 3. Typical Wave Scatter Diagram

2.2 Wave data and condition

The sea environment is described by a collection of sea states with a probability of occurrence associated with each sea state which represent the long-term experience of the platform at the operation site or along the transit route. This information can be obtained from many sources: measured, visual, and hindcast wave data. Hindcast data derived from GSOWM (Caruso et al., 1998) provides approximately 6000 grids covering entire northern and southern hemispheres. The wave data is mostly represented in the form of joint probability of two major variables, namely, the significant wave height and a characteristic period (e.g., average period or zero-crossing period), commonly known as the wave scatter diagram. A typical wave scatter diagram is shown in Figure 3. Each of the individual sea states and the associated platform response is assumed to represent a stationary random process.

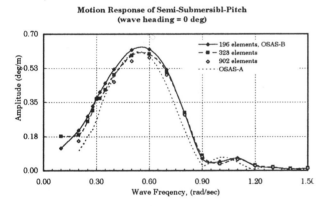

Figure 4. Motion RAOs by OSAS

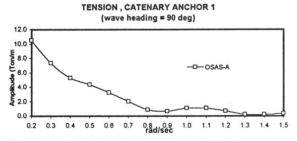

Figure 5. Mooring line force RAO by OSAS

2.3 Motion and load analysis

Hydrodynamic load and motion analyses are performed to generate motion RAOs (or transfer function) for the appropriate frequency range and wave directions. Various mooring systems such as catenaries, slanted and vertical taut cables, and risers can be included in the motion and load analysis. Motion RAOs and the mooring line tension RAO are shown in Figures 4 and 5, respectively. Motion from OSAS has been validated with model test data for a 6 column generic semi-submersible (ABS, 1988) and shown in Figure 6.

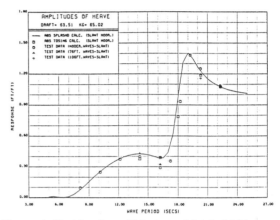

Figure 6. Verification of Motion with Model Test

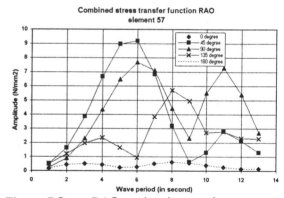

Figure 7 Stress RAO at a bracing member

2.4 Structural analysis for a space frame

All the forces due to wave action on the member, inertial force of distributed mass and concentrated mass, and mooring line forces are calculated and applied to a space frame structure model. Dynamic equilibrium should be checked before carrying out structural analysis. Through a linear elastic stress analysis, the member static stresses and stress RAOs can be calculated. Typical member stress RAOs are shown in Figure 7. Further, the most probable extreme responses and member stresses of the structure can be computed by incorporating the stress RAOs and the scatter diagram via a spectral analysis for every member of the space frame model.

2.5 Evaluation of fatigue, yield and instability

Using the calculated extreme stress for every node, code checks for the yielding failure mode and the instability (restricted to beam-column) failure mode can be performed. The most probable extreme values of the combined stress at nodes are also utilized in the prediction of the worst cases. The first criterion used to select worst case scenarios is based on stresses and their variation with respect to all frequencies and wave headings. Similarly a simpler screening method based on critical design load rather than the stress was also developed and included in the system. These two different approaches are to be discussed in the later section. In addition to code checks and worst case scenario determination, fatigue life for selected nodes can also be evaluated at this Phase A utilizing closed form integration of the cumulative damage of the short term probability density function with specified S-N data (Chen and Mavrakis, 1988). The long term stress distribution parameter, the Weibull shape parameter, can then be calculated from the spectral analysis results. The value of the parameter at each node will be applied to local fine mesh FEA results to evaluate the fatigue lives of structural details at the corresponding location in phase B analysis.

3. CRITICAL DESIGN LOAD SELECTION

An alternative but simpler and effective method to determine the critical load cases needed for more detailed FEM analysis based solely on the load rather than member stress for a twin hull semi-submersible can be used. A similar approach has been successfully applied to the structural analysis for ocean going ships (Liu, et al. 1992) and twin hull vessels such as SWATH and Catamarans.

3.1 Dominant Loading Parameters

Dominant Loading Parameters (DLPs) are considered here as the influential parameters that can be used to establish the critical loading conditions for the strength assessment. The term DLP herein refers to internal load effects, a global motion of a unit, or a local dynamic response such as inertial effect on deck, etc.

The instantaneous response of the unit can be evaluated by each one of several Dominant Load Parameters. For the strength assessment of semi-submersibles, the following representative DLPs have been identified as being necessary to establish the critical loadings on the structure:

Prying-Squeezing Moment: Major joints and braces of the unit are affected by prying and squeezing forces acting on the pontoons. The critical condition happens in a beam sea with its wave length approximately equal to twice the unit breadth. The response normally causes the maximum axial forces in the transverse bracing members and the maximum bending actions on columns and deck structures.

Pitching Torsional Moment: The torsional moment is maximized in quartering seas with a wave length approximately equal to the diagonal length of the unit. Exact wave condition will be determined from the seakeeping analysis described in the next sections. The response normally yields the maximum axial forces in the diagonal horizontal and vertical bracing members.

Yaw Splitting Moment: The splitting moment is maximized in quartering seas with a wave length approximately equal to the diagonal length of the unit. The response normally yields the maximum axial forces in the horizontal bracing members.

Vertical Bending Moment: The vertical bending moment is maximized in head seas when a wave length is slightly larger than the unit length. The response causes a maximum bending action on pontoons.

Shear Force between Pontoons: The longitudinal shear force between pontoons is maximized in quartering seas when the wave length is approximately equal to 1.5 times than the diagonal length of the unit. The response usually produces the critical design loading case for all horizontal bracing members. The response induces opposite longitudinal displacements for each pontoon, thus also yielding bending moments on pontoons and transverse members.

Vertical shear forces along the deck centerline cut and midship section, horizontal shear forces at column-deck and column-pontoon connections should also be investigated.

Accelerations: The longitudinal acceleration on deck is maximized in head seas. The response may introduce maximum longitudinal racking due to the inertial effect on deck structures, maximum shear and bending effects on columns, and maximum axial actions on pontoons and longitudinal diagonal members as well. Similarly accelerations in other directions should also be investigated.

For each DLP stated above, a set of sea conditions is selected to facilitate the determination of the representative wave conditions corresponding to its critical load cases for Phase B. This process is carried out through a seakeeping and load analyses which are described below.

3.2 Critical internal loads

Among the DLPs defined above, internal loads or load effects are discussed in this section. These loads can be calculated by introducing an imaginary cut plane through the platform. Three cutting planes, i.e. parallel to x-y, y-z, z-x planes, respectively, can be placed at any location. The DLPs are calculated first in term of the RAO to be used in the long-term prediction to determine the dominant heading and the wave frequency which produces the maximum response in the wave heading for each DLP of interest . Example of those internal loads (3 global forces and moments) at the centerline cut are shown in Figure 8.

3.3 Extreme Value Predictions

The short term and long-term extreme values are to be calculated for each DLP of interest using its frequency transfer functions and certain wave data. The long-term extreme values refer to the most probable extreme values at a probability level corresponding to a specified service life (fifty or a hundred year) of the offshore platform.

The most probable extreme values by Ochi's theory (1978) is employed. The wave data are presented in a form of scatter diagram. H-family or Walden data representing the wave statistics for the North Atlantic Ocean are used for ocean going ships with unrestricted service. For offshore structures, site specific and route specific for transit wave data can be used. Various spreading functions are available and in general the cosine squared function may be used to simulate the short crestedness of ocean waves. A equal probability occurrence of each wave heading is assumed but preferred heading probability can be applied for site specific operation. The extreme value of each DLP is used

to determine the design wave system for each critical load case.

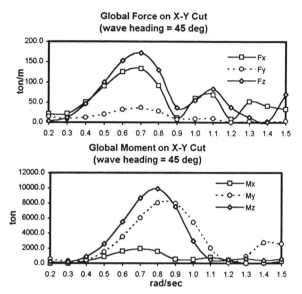

Figure 8. Critical Internal Loads

3.4 Design Wave System

For each critical load case determined either by the maximum DLP or by the maximum stress, a design wave is to be characterized by its amplitude, length, heading, and crest position referenced to the unit. For each structural load case corresponding to the maximum value of each DLP, an equivalent wave system is determined. The instantaneous loads on the unit structure resulting from the equivalent system is applied to the unit for the structural analysis for each maximum DLP.

The wave amplitude of the equivalent wave system is calculated by dividing the extreme value of the DLP under consideration by its maximum transfer function in the dominant heading. The corresponding frequency and phase are used to compute the wave length and crest position, respectively, of the equivalent wave system (Liu et al. 1992).

4. 3-D DIFFRACTION AND FEM ANALYSIS (PHASE B)

Based on the critical load analysis from Phase A, a 3-D panel based diffraction analysis can be carried out to generate the pressure distribution for a FEM analysis. Two steps are employed for the structural FEM analysis. Step 1 is a global structural analysis followed by step 2 with a local finer mesh analysis. In this two-step analysis, several selected design waves, which are determined in Phase A analysis, are first applied to the coarse mesh model for global behavior. Then, using sub-structural analysis concepts, the results of the coarse mesh model analysis are further carried into a local fine mesh model analysis for the evaluation of stresses of the structural details. The details for these two steps are described below.

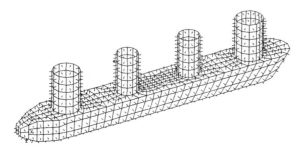

Figure 9. Hydrodynamic model

4.1   3-D Diffraction Analysis for Pressure

For the critical conditions from Phase A analysis, 3-D panel based diffraction analysis is performed using a 3-D hydro-model shown in Fig. 9. Only half of the structure is needed because of symmetry. The diffraction effect due to large diameter columns and pontoons and interaction between these large members can be fully taken into account. First rigid body motion of the offshore structure is calculated to determine motion dependent dynamic and quasi-static components of pressure which are added to the pressure due to incident wave and diffracted wave to get the total pressure. Complete pressure distribution on the wetted surface are calculated and applied to the panels of FEM model.

Rigid body motions by OSAS compared with other recognized programs of AQWA and MIT-WAMIT and its comparison are shown in Fig 10. Dynamic pressures for a number of panels are also compared with MIT-WAMIT in Fig. 11, since the latter provides only dynamic component without quasi-static motion induced component. The total pressures that the structures feel and therefore needed for FEM analysis are compared with AQWA in Figure 12. These pressures are practically identical for all the 1804 hydrodynamic wetted panels.

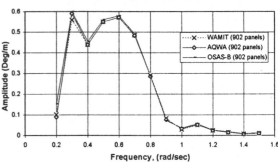

Figure 10 Verification of OSAS Diffraction Module

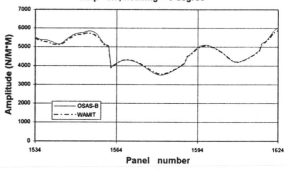

Figure 11 Comparison of Hydrodynamic pressure between OSAS and WAMIT

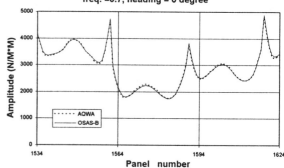

Figure 12 Comparison of total pressure between OSAS and AQWA

The same mooring systems used in the Phase A space frame model can also be applied for 3-D diffraction based motion analysis. The pressure distribution on the example semi-submersible is also shown in gray scale in Fig. 13.

Since the hydrostatic and hydrodynamic forces are applied on the structural surface under water, only the wetted surface of the structure is required for loading. Load and motion analyses can be performed with the hydro-model and known mass properties for each of the design waves identified in phase A analysis.

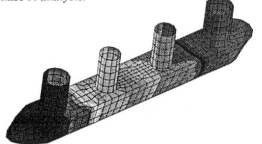

Fig. 13 Pressure distribution for a semi-submersible

4.2 Load Interface to FEM analysis
In order to speed up the analysis process, the model loading, boundary condition definitions, and post-processing of the results have been automated. For the global structural analysis, the external wave loads from the hydrodynamic analysis are distributed to the wetted surface of the structure through an interface program. The inertia loads are obtained from motion analysis and are used for the structural analysis.

4.3 Global Structural Analysis
Similar to Phase A, an coarse mesh FEM model (see figure 14) of the global structure, including the hull, bracing and deck, is created based on the principal dimensions and scantlings of the primary structural members in accordance with the construction drawings. For this analysis level, all primary load bearing structures such as shell plates, bulkheads, flats toward the overall stiffness of the structure are modeled:
- bending plate elements are used to model all flat plates and shells,
- deep girders and circumferential rings are modeled with beam elements,
- secondary stiffeners are lumped (based on transverse web frame spacing in pontoons) and attached to the plate element, which may represent several panels.

Normally, the structural model consists of many components such as pontoons, column, deck and bracings, etc. Each component is built separately, and the integrated model is generated by

assembling these components. Typically, the coarse model of a semi-submersible consists of approximately 3,000 nodes and 7,000 elements made up of 4,000 plate elements and 3,000 beam elements.

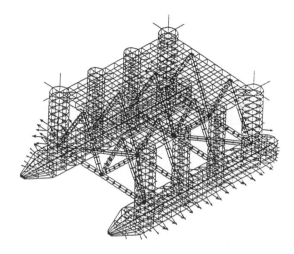

Figure 14. Loaded FEM model

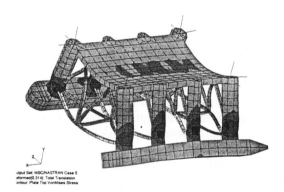

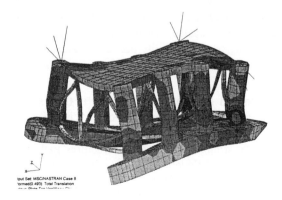

Fig. 15 Structural responses of global coarse mesh

After loading the structural model, a linear finite element analysis is performed. A loaded FEM model is shown in Figure 14. The results of this analysis provide the necessary information for the subsequent local fine mesh model analyses. The strengths of the above mentioned primary structural members against yielding and buckling can be evaluated in this step. Representative structural responses for the maximum prying moment and torsional moment are shown in Figure 15.

4.4 Local Fine Mesh Model Analysis

Following the global structural analysis, fine mesh model analyses for selected local structures are performed. Analysis procedure for local structure is shown in Fig. 16. The selection of the local structures is made as warranted by the higher stress distributions and the lower yielding and buckling strengths of the global structure. Typically these local structures are the joints which represent the connections between the columns, the pontoons and the bracing members.

Detailed fine mesh models are created for these local structures. Fig. 16 shows an example of a simplified model for a structural node detail. A more elaborate model should be used at this stage of analysis. Each of these individual models primarily consists of plate elements with more detail than the coarse mesh model. Typically, the webs and flanges of girders and stiffeners are explicitly modeled with plate elements. In less critical areas, flanges are modeled with beam elements.

The fine mesh models are loaded with local loads which consist of hydrostatic and the hydrodynamic pressure, inertia loads, and other externally applied loads. These load sets are consistent with those used in the coarse mesh model analysis, and thereby ensure traceability between various design load conditions and overall force and moment balance.

This loaded local fine mesh model is then solved using a finite element analysis solver. The resulting stresses are used to evaluate the yielding and buckling strengths of the local structures including the girders, stiffeners, brackets, connections, and penetrations. The API bulletins 2U and 2V can be employed for panel checking. The fatigue life of a structural detail can be determined using a stress range, the Weibull shape parameter calculated in Phase A analysis, and specified S-N data.

5. CONCLUSIONS

A rational and efficient multi-level analysis procedure (Phase A and B) and computer program system OSAS have been developed. Using this method and the OSAS system, ABS engineers are able to efficiently perform consistent analysis needed to determine motion, mooring, and wave loads, used in the structural and fatigue analyses for offshore compliant structures moored with various systems.

The analysis methods and computer system OSAS have been verified with model tests and other recognized theoretical methods.

It is important to note that to identify the critical cases required for detailed FEM analysis, two unique methods (namely design load based approach and member stress approach) have been developed. This approach is applicable to all types of offshore structures such as conventional semi-submersible, TLPs and Spars.

In conclusion advancements in computational methods and computer hardware make it possible to apply state-of-the-art First-principles based analysis method to the design analysis of offshore structures.

ACKNOWLEDGEMENT

The authors wish to express their appreciation and gratitude to the management of the American Bureau of Shipping especially Dr. D. Liu for his encouragement and support for the OSAS project. Special thanks should be given to Professor J.R. Paulling for his consultation throughout the project. The authors are also grateful to Drs. Y.N. Chen, C.T. Zhao, and S. Tang for their assistance in preparing this paper.

REFERENCES

American Bureau of Shipping, "Comparison of Model Test Experimental Results and analytical Predictions," Technical Report RD-87009, May 1987

American Bureau of Shipping, "OSAS User's Manual", ABS R&D Report, April 1997

Bulletin on Design of Flat Plate structures (Bul. 2V), American Petroleum Institute, May 1987.

Bulletin on Stability Design of Cylindrical Shells (Bul. 2U), American Petroleum Institute, May 1987.

Caruso, N. Johnson, M., Grove, T.(1998), "A Structural Strength and Fatigue Assessment Methodology Applicable to Conversion-Type FPSO Designs", to be presented at OTC 1998 Paper No. 8772.

Chen, Y. N. and Mavrakis, S.A.(1988), "Closed Form spectral Fatigue Analysis for Compliant Offshore Structures," Journal of Ship Research, Vol. 32, No. 4, December 1988.

Liu, D., Chen, Y.N., Shin, Y.S., and Chen, P.C.(1980), "Integrated Computational Procedure for Hydrodynamic Loads and Structural Response of a Tension Leg Platform," ASME Annual meeting.

Liu, D., Spencer, J., Itoh, T., Kawachi, S., and Shigematsu, K.,(1992), "Dynamic Load Approach in Tanker Design", SNAME Transaction 1992.

Ochi, M.K., "Wave Statistics for the Design of Ships and Ocean Structures," Transaction, SNAME, Vol. 86, 1978

Paulling, J.R., Hong, Y.S., Stiansen, S.G. and Chen, H.H. (1978), "Structural Loads on Twin-Hull Semi-Submersible Platform", OTC paper No. 3246

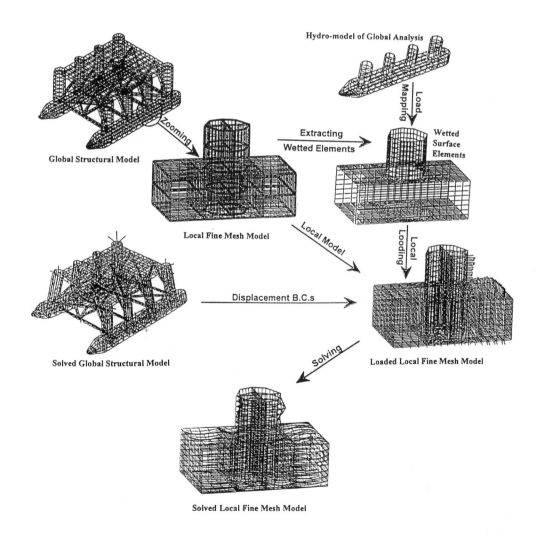

Figure 16 Flow diagram of local FEM analysis

Practical Design of Ships and Mobile Units
M.W.C. Oosterveld and S.G. Tan, editors.

Wave Drift Forces and Responses in Storm Waves

C.T. Stansberg[a], R. Yttervik[a] and F.G. Nielsen[b]

[a]Norwegian Marine Technology Research Institute A/S (MARINTEK)
P.O.Box 4125 Valentinlyst, N-7002 Trondheim, Norway

[b]Norsk Hydro Research Centre,
N-5020 Bergen, Norway

Slow-drift surge and pitch motion records from model tests with a moored semisubmersible in irregular waves are analysed. They are compared to numerical time series reconstructions, based on conventional potential theory models as well as on empirical drift coefficients. In high sea states, large discrepancies are observed between measurements and potential theory, while empirical reconstructions compare well. For surge, they are explained by viscous drift forces in the splash zone. Viscous force and response simulations confirm this. For pitch, they may be explained by off-diagonal variations in the quadratic transfer function matrix.

1. INTRODUCTION

Slow-drift vessel motion is an important issue in the analysis of floating offshore production systems. It is often a critical parameter which has a large influence on e.g. extreme mooring line tensions. Normally the slow-drift excitation is theoretically modelled by drift excitation coefficients from potential theory, see e.g. Faltinsen (1990). In moderate sea states, this approach works fine for most cases. In severe weather conditions, however, scaled model tests have shown that there is a sea-state dependent increase in empirically estimated surge drift coefficients for semisubmersible platforms. Examples are presented in e.g. Stansberg (1994) and Dev & Pinkster (1997). As described in the latter reference, this effect can at least partly be explained by viscous drift forces in the splash zone.

In the following, we study this problem further through a set of new model test results on a semisubmersible. They are analysed by numerical simulations of the model tests, using the measured wave elevation as input. Theoretical as well as empirical excitation models will be used. In addition to surge motions, low-frequency pitch measurements are analysed as well, and the validity of Newman's (1974) approximation will be discussed. We also present simulations including viscous forces on the hull.

2. MODEL TESTS WITH SEMISUBMERSIBLE

Model tests in scale 1:55 were carried out in the 50m x 80m Ocean Basin at MARINTEK. A catenary moored floating production platform was tested in various irregular wave, current and wind conditions. The system was modelled with complete mooring and riser systems on 335m water depth. The platform was a 4-column semisubmersible with a displacement of 55000 tons and with a square ring pontoon. The natural surge and pitch periods of the moored platform was 124s and 47s, respectively. All tests were done with the platform in 0 deg heading. Recordings made during the tests included wave elevation, all 6 DOF platform motions, mooring line tensions and air-gap under the deck. The duration of the records corresponds to 3 hours in full scale.

Here we study selected results from tests in 3 different longcrested wave conditions, zero heading:

Run 145: Hs= 5.0m Tp=10.0s (Current/Wind=0)
Run 143: Hs= 8.0m Tp=12.0s   "
Run 144: Hs=14.7m Tp=14.5s   "

The latter sea state corresponds to a steep 100-year North Sea condition. From the large amount of data obtained, we shall here look mainly into the slow-drift surge and pitch platform motions. They are obtained from the motion measurements by a digital low-pass filter at 25 seconds.

3. MEASURED SLOW-DRIFT AND NUMERICAL RECONSTRUCTIONS

3.1. Numerical Reconstruction Procedure

Here we shall consider the slow-drift (LF) platform motions only, by numerical simulation reconstruction of recorded time series. We use a coupled analysis simulation program, see Ormberg & Larsen (1997), where the dynamics of the subsea mooring and riser systems are fully integrated in the time domain with the large-volume floater motions. It is a combination of the independent programs RIFLEX (1995) and SIMO (1996). The subsea slender body hydrodynamics is given by the RIFLEX part. The top end motion results from the hydrodynamic forces on the floater, given by the SIMO part, and from the time-varying mooring and riser loads. Although the LF floater motion reconstruction is the main topic here, the simulations include complete modelling of wave-frequency (WF) and LF motions and mooring line forces.

The irregular wave measurements from the pre-calibration of the actual sea states are used as input. The WF platform motions in 6 DOF are modelled with WAMIT (Newman & Sclavounos, 1988). The wave-induced slow-drift motions are simulated in two alternative ways: A) WAMIT drift excitation + user-defined damping, and B) Empirical drift excitation coefficients + user-defined damping. Thus a major goal of the present work is to compare the different LF reconstructions against the measurements. The present WAMIT drift coefficients represent the diagonal of the quadratic transfer function matrix (QTF), i.e. Newman's approximation is invoked.

In addition to the "user-defined" damping, there is an implicit damping contribution arising from the mooring and riser system due to the coupled simulation. This includes e.g. the mooring-line damping effect described in Huse (1991). The "user-defined" input accounts for the other contributions such as wave drift damping, hull drag and friction, and can be specified in various ways. Here we model the surge wave drift damping explicitly, by linear differentiation of the drift coefficients with respect to the relative velocity. Surge hull drag from the relative velocity is modelled by a quadratic current force coefficient. Other contributions are put in an additional linearized term, tuned to make the total response spectrum match the measurements.

For pitch we have modelled all the "user-defined" damping input by a tuned linearized term.

The empirical drift coefficients are estimated from the measured LF motion records. A recently developed version of cross-bi-spectral analysis has been applied, see Stansberg (1997). Since only the motion, and not the excitation, has been measured directly, iterations are included in the estimation. Simplified excitation and response simulations are made, based upon drift estimates and tuning of linearized system parameters (damping + period) where response spectra are matched to the measured. Full QTF's can in principle be estimated this way, but here we have chosen to use Newman's approximation. Thus we use drift coefficients, only. However, the estimated coefficients are "equivalent" diagonal values primarily based on off-diagonal QTF values at a difference-frequency corresponding to the natural frequency of the motion response, where the main signal energy is located. Therefore the empirical estimates can include off-diagonal effects. This is normally most relevant for responses with short natural periods, like heave and pitch.

3.2. Results with Comparison

The empirical surge and pitch drift coefficients from the 3 different sea states are compared to the WAMIT predictions in Figure 1. The frequency ranges of the estimates depend on the actual wave spectra (the peak frequencies are indicated on the figure). The empirical surge coefficients are generally higher than predicted by second-order potential theory. The relative discrepancy is increasing with decreasing wave frequency. In addition, the highest sea state seems to show slightly higher drift coefficients for long waves, although that trend is not very pronounced. Thus the discrepancy will have the smallest effects in the lowest sea state. For the highest sea state, however, the effects will be quite significant. The difference between WAMIT and measurements at low frequencies is most likely due to viscous drift forces in the splash zone, as concluded by Dev & Pinkster (1997). This will be discussed in a later section.

Also the empirical pitch moment coefficients are higher then the WAMIT results, except from a region around 0.1Hz. This may to some extent be due to off-diagonal variations in the excitation QTF, i.e. a full second-order potential-flow effect. As indicated earlier, these are not included in the

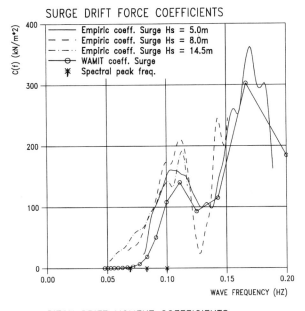

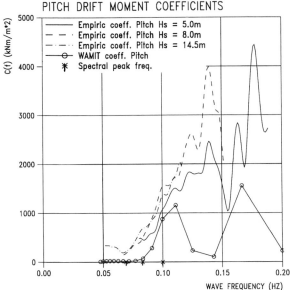

Figure 1. WAMIT and empirical drift coefficients.

considerable quasi-static contribution from coupling to surge, through the catenary mooring system. This contribution is a disturbing element in the wave drift study, and was empirically estimated from the test results and subtracted in the time domain from the signal before the cross-bi-spectral analysis.

Low-frequency surge and pitch record samples from the numerical reconstructions of run 143 (Hs=8m) are compared to the measurements in Figure 2. The corresponding power spectra are shown in Figure 3 (center plots), where also the spectra for the other two runs are included (upper and lower plots). Corresponding statistics is presented in Table 1. The figures and the table confirm the observation from Figure 1: WAMIT and the empirical reconstruction match the experiments quite well in low and short waves, while in higher sea states, a significant underprediction is seen in the WAMIT results. In the 100-year sea state, the surge measurements and the predictions differ as much as by a factor 2.5 – 3. With the empirical drift coefficients, however, the plots and table show that the reconstructions match the measurements quite well, for all the sea states.

The influence from the slow-drift motion on the extreme mooring line tensions is illustrated in Figure 4. With WAMIT-based slow-drift, the predicted extreme is about 1000 kN lower than measured, while the empirical version is much closer. We also observe that the coupled analysis program reconstructs the mooring line dynamics very well.

The numerical pitch record reconstructions are more complicated. This is partly because the LF motion consists of two contributions, seen as two peaks in the spectra in Figure 4. There is one peak at the 124

Table 1. Slow-Drift Statistics

Sea State [m]	Type	Surge [m] Mean	St.Dev.	Pitch [deg] Mean	St.Dev.
	Measur	1.54	1.67	0.26	0.52
Hs=5	WAMIT	1.36	1.35	0.18	0.30
	Empiric.	1.64	1.55	0.15	0.44.
	Measur	2.89	2.39	0.37	0.76
Hs=8	WAMIT	1.90	1.57	0.22	0.42
	Empric.	3.08	2.46	0.27	0.72
	Measur.	5.63	4.02	0.79	1.18
Hs=14	WAMIT	2.37	1.55	0.24	0.78
	Empiric.	5.58	3.58	0.51	1.10 .

present WAMIT coefficients, while the empirical coefficient estimates will be influenced by them. In fact, the cross-bi-spectral analysis of the pitch measurements did indicate such variations, although the present application makes use of "equivalent" diagonal values only, as discussed in section 3.1.

The pitch estimation process was more complex than for surge, since the LF pitch motion also includes a

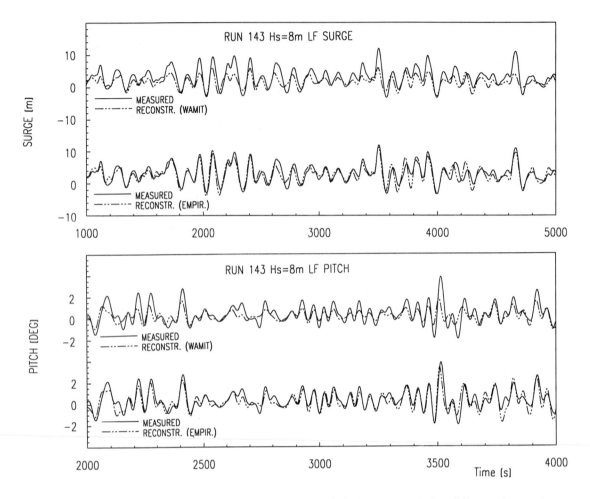

Figure 2. Measured and reconstructed LF surge (upper) and pitch (lower). Notice different time scales.

sec surge period due to quasi-static coupling via the catenary system, and one peak at the natural pitch period of 47 sec. At a given frequency, the two parts act in opposite directions. Another complicating factor is the possible off-diagonal variation in the QTF. But as a rough conclusion from the spectra, we conclude that WAMIT underestimates by a factor of 2 in all 3 sea states. The spectra also show that the empirical reconstructions compare reasonably well with the measurements for the two lowest sea states, while the 100-year sea state is more troublesome. The time series samples in Figure 3 show that for the 12 sec sea state, this is the case also in the time domain.

Linearized damping coefficients have been estimated from the response spectra, and are

presented in Table 2. The different contributions have been found by successively "switching off" the mooring line coupling, the wave drift damping and the current force coefficient, and tuning the linearized term to give the measured spectrum. In addition, the linearized relative damping found directly from tuning of the spectrum in a linear system model (done in connection with the cross-bi-spectral analysis), is shown. A significant increase in the various damping contributions is observed with increasing sea state, except from the linear one. The wave drift damping model is simplified, with somewhat uncertain estimates. It probably includes some viscous hull drag, reflecting the anticipated viscous drift force. Care should be taken in detailed interpretation of linearized damping values, since the nonlinear effects may disturb the picture.

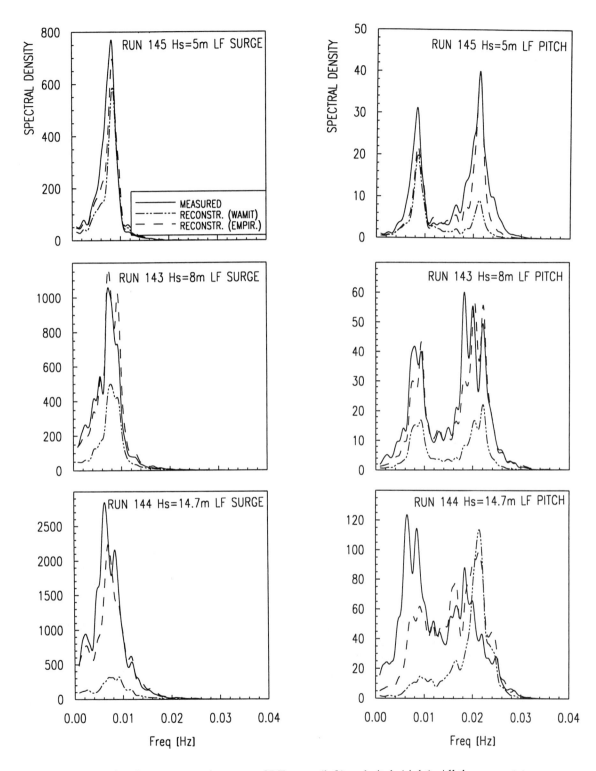

Figure 3. Measured and reconstructed spectra of LF surge (left) and pitch (right). All three sea states.

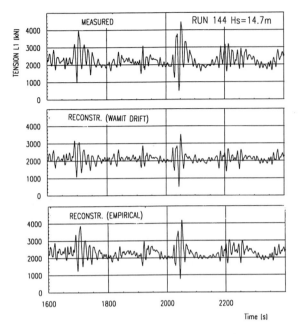

Figure 4. Measured and reconstructed tensions

Table 2. Linearized Damping Estimates

Sea State [m]	Surge [kN/(m/s)]					Pitch [GNm/(rad/s)]		
	ML	WD	HD	AL	Rel %	ML	AL	Rel %
5	600	250	200	300	14	1.2	0.2	8
8	1050	400	450	300	24	2.1	0.9	16
14	1000	1100	700	300	38	4.5	2.5	18

(ML=Mooring line, WD=Wave drift, HD=Hull Drag AL=Added linearized, Rel=Relative damping level)

4. VISCOUS DRIFT FORCES

In the previous section, it was suggested that the deviations observed between WAMIT and empirical surge drift coefficients are caused by splash zone viscous drift forces on the hull. The implementation of this into our coupled analysis simulation software is to be completed. Here we present results from simulations with another program, SWIM, see e.g. Kim et.al. (1997). For pure potential-flow problems, the latter reference documents good comparison with WAMIT. It therefore makes sense to compare these results with those of the previous section.

In the present simulations viscous drag forces are included by a Morison equation approach. The restoring forces from the mooring system are linear, without the fully integrated dynamical coupling to the floater as in the previous section.. The full force QTF is included, while for the wave drift damping only the diagonal is included. We use the pre-calibrated irregular wave record as input.

Invoking Morison's equation, the viscous drag forces may be modelled in different ways. Here we use the "total relative velocity" approach. Thus the drag force on a fully submerged segment of a vertical member may be written as:

$$dF = \tfrac{1}{2}\,\rho D\, C_D\,(U + u)\,|U + u|\,dz \qquad (1)$$

Here D=diameter of member, C_D=drag coefficient, U=local horizontal slow-drift velocity, and u=horizontal wave frequency velocity (member + wave particle velocity). In the splash zone, we must account for the variation in the wetted area, see e.g. Faltinsen (1990). This gives an extra viscous load term related to the water line:

$$F_{wl} = \tfrac{1}{2}\,\rho D\, C_D\,\eta\,(U + u)\,|U + u|_{z=0} \qquad (2)$$

Here η is the instantaneous relative wave elevation at the member. Eq. (2) gives a force that is cubic in the wave elevation. It also gives a mean force:

$$F_{mean} = 2/3\,\rho D\, C_D\,\omega^2\eta_a^3 \qquad (3)$$

Here ω=circular frequency, η_a =wave amplitude. The viscous drag forces also have a significant impact on the damping of the slow-drift motions.

A proper choice of drag coefficients for real geometries is always difficult, in particular in the case of combined WF and LF motions. We assume that the flow characteristics are dominated by the WF flow and pick the drag coefficients for characteristic KC numbers as given by e.g. Faltinsen (1990). $C_D = 1.5$ has been chosen for the columns, which are square with rounded corners. For the pontoons, $C_D = 2.5$ and 2.0 has been chosen in the vertical and horizontal direction, respectively. In these simulations, no additional linearized damping terms have been added to tune the measurements.

Results from the simulations are presented and compared to the measurements in Figures 5 – 6 and Table 3. Here we consider low-frequency surge from the test with Hs=8m, only. In the table, Case A

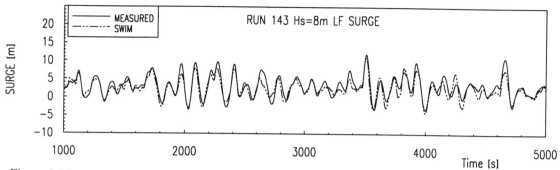

Figure 5. Measured and reconstructed (SWIM) LF surge.

is the base case with forces modelled as described above. The present results are not very far from the measurements: 11% too low mean value, and 5% too low standard deviation. This is clearly higher than the pure WAMIT case in Table 2 which gave 40% too low mean and standard deviation, while it is still a little lower than the empirical reconstruction. Thus for LF surge, the including of viscous hull forces explains most of the deviations observed in the previous section. We also see this in the LF spectra of Figure 5, and in the time series samples in Figure 6, if we compare them to the corresponding figures in section 3.2. The fact that the SWIM-based spectrum looks similar to the measured one, without the mooring line damping included in the modelling, indicates that the simulated hull drag damping may

in reality be too high. Thus the C_D-values chosen for the pontoons may perhaps be too high.

In order to further study the viscous effects in the SWIM simulations, we have run a case with $C_D=0$ (Case B in Table 3). Here we see that the mean value drops to the same level as in the WAMIT case in Table 2, which confirms that the two programs compare well. The standard deviation in Case B is too high since all drag damping has been removed.

The mean drift force due to potential effects is quadratic in the wave elevation. The mean viscous forces on the submerged hull is also primarily a quadratic effect, while the splash zone forces are cubic. Thus if we divide the incident wave amplitude by a factor k and multiply the computed drift offset by k^2, we should obtain the same result as in Case A except from the contribution from force terms which are higher than quadratic. In Case C we have used k=10. We observe that for the mean offset, we obtain almost the same as with $C_D=0$, i.e. the viscous effects on the mean force is dominated by effects higher than quadratic. The splash zone is thus very important to the mean force in high waves, as concluded earlier by Dev and Pinkster (1997). This result is further confirmed by simulation Case D,

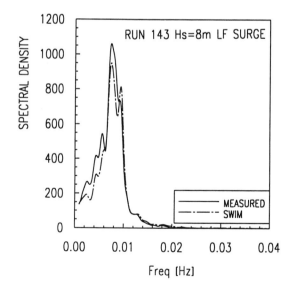

Figure 6. Measured and reconstructed surge spectra

Table 3. Slow-Drift Statistics Surge (SWIM).

Case	Mean [m]	St.Dev. [m]
Model test 143	2.89	2.39
A. Base Case.	2.58	2.26
B. $C_D=0$	1.90	5.39
C. As A, but H_s/10 Resp*100	1.87	4.81
D. $C_{Dc}=0$	1.88	2.56

where the drag coefficient on the cylinders is set to zero while the drag coefficient on the pontoons are remained the same as in A). The mean offset in D) is almost identical to the offset in B) and C). The standard deviation, however, is reduced to nearly the same level as in A). This means that with the present choice of drag coefficients, the viscous damping from the pontoons is significant, but there is also a contribution from the columns: The viscous excitation on the columns is accompanied by a corresponding damping force keeping the standard deviation more or less unchanged.

5. CONCLUDING REMARKS

A study of measurements from model tests with a catenary moored semisubmersible has shown that in large irregular waves, there are significantly higher wave drift forces and moments than predicted by standard potential theory. It is also shown that this can be empirically compensated for by drift force coefficients estimated from the experiments by cross-bi-spectral analysis. Numerical reconstructions of surge and pitch motions as well as of mooring forces then agree well with the measurements, except from pitch in the largest waves. For surge, the deviations from potential theory are explained by viscous drag forces in the splash zone on the columns. This has been confirmed by simulations including hull drag forces. Our study has also shown that the total damping increases strongly with the sea state. The largest contribution comes from mooring line damping, but wave drift and hull drag damping also contribute. These tests were run in waves without current. In current we expect relatively more hull drag damping.

For the pitch drift moments, the observed deviations from standard potential theory may be partly due to viscous effects, but parts of it may also be explained by off-diagonal variations in the excitation QTF matrix. As for surge, the low-frequency damping increases strongly with the sea state, with the mooring line damping as the largest constribution.

ACKNOWLEDGEMENTS

This work has been carried out as a part of the Verideep project, financed by Norsk Hydro and the Royal Norwegian Research Council.

REFERENCES

Faltinsen, O.M. (1990), *Sea Loads on Ships and Offshore Structures*, Cambridge Univ. Pr., Cambridge, U.K.

Huse, E. (1991), "New Developments in Prediction of Mooring Line Damping", *OTC Paper No. 6593, Offshore Technology Conf.*, Houston, TX, U.S.A.

Kim, S., Sclavounos, P.D. and Nielsen, F.G. (1997), "Slow-Drift Responses of Moored Platforms", *Proc., Vol. 2, 8th BOSS Conf.*, Delft, The Netherlands, pp.161-176.

Newman, J.N. (1974), "Second Order, Slowly Varying Forces on Vessels in Irregular Waves", *Proc., Int. Symp. On Dynamics of Marine Vehicles and Structures in Waves*, London, U.K.

Newman, J.N. and Sclavounos, P.D. (1988), "The Computation of Wave Loads on Large Volume Structures", *Proc., 5th BOSS Conf.*, Trondheim, Norway.

Ormberg, H. and Larsen, K. (1997), "Coupled Analysis of Floater Motion and Mooring Dynamics for a Turret Moored Tanker", *Proc., Vol. 2, 8th BOSS Conf.*, Delft, The Netherlands, pp. 469-483.

Dev, A.K. and Pinkster, J.A. (1997), "Viscous Mean and Low Frequency Drift Forces on Semi-Submersibles", *Proc., Vol. 2, 8th BOSS Conf.*, Delft, The Netherlands, pp. 351-365.

RIFLEX User Documentation (1995), SINTEF, Trondheim, Norway.

SIMO User Documentation (1996), MARINTEK, Trondheim, Norway.

Stansberg, C.T. (1994), "Low-Frequency Excitation and Damping Characteristics of a Moored Semisubmersible in Irregular Waves – Estimation from Model Test Data", *Proc., Vol. 2, 7th BOSS Conf.*, Boston, Mass, USA.

Stansberg, C.T. (1997), "Linear and Nonlinear System Identification in Model Testing", *Proc., Int'l Symp. on Nonlinear Design Aspects in Physical Model Testing*, OTRC, Texas A&M Univ., U.S.A.

Practical Design of Ships and Mobile Units
M.W.C. Oosterveld and S.G. Tan, editors.

Practical Method for Mooring Systems Optimum Design

Oscar Brito Augusto[a], Carlos Alberto Nunes Dias[a] and Ronaldo Rosa Rossi[b]

[a]*Naval Architecture and Marine Engineering Dept, University of Sao Paulo*
Av. Prof. Mello Moraes, 2231, São Paulo Brazil

[b]*Cenpes, Petrobras*
Cidade Universitaria Qd 7-Ilha do Fundao
Rio de Janeiro Brazil

Abstract

A practical, rationally based method is presented for the automated optimum semi-submersible mooring design. The method required the development of a powerful program for the non-linear analysis of catenary based systems and an optimisation algorithm for the solution of a large, highly constrained, non-linear redesign problem. In the pre-design phase the platform heading and the mooring pattern are searched. In the second phase, based on the depths of each line, the diameters to be used in each line segment are defined and finally, in the third phase, the optimum segments lengths, anchor points and anchors type are search to optimise the anchoring system total cost or weight. In the present version only three different segments in each line are permitted, with systems hanging up to sixteen mooring lines. The lines should not have buoys or clamp weights. Slackening leeward lines, when necessary and permitted as operational measure, is performed automatically. All these developments have been incorporated into a DOS based user-friendly program called EXMOOR (EXpert system for deep water MOORing design). The efficiency and robustness of the method are illustrated by using it to determine the optimum design of the mooring lines of two semi-submersible platforms already in operation for the Brazilian Oil Company PETROBRAS. With the same conditions used in the original design the program EXMOOR achieved the optimum mooring lines with an expressive material cost saving.

1. Introduction

The dimensioning of a mooring system requires a big effort of the designer to realise the non linear structural mechanics behaviour of the moored vessel and mooring lines, therefore only continuing practice can give him the necessary expertise. As a result it is possible to deduce the relevance of the expert to get a good engineering solution for the mooring problem, with all the constraints being satisfied.

However considering the limitations of the human brain the expert knowledge can not be spread in a short time, a young expert does not rise up easily and, as consequence, the knowledge maybe lost.

In an integrated computational system, the EXMOOR program joins some of the calculating tools developed by NDAI[1], allowing the optimum design of mooring systems in a quick and automated way that keeps the knowledge in computer media.

2. The design of mooring systems

We will not discuss the advantages and disadvantages of a particular station keeping system. At this point a station keeping system with mooring lines was adopted and the designer should find the better line spreading and composition to fulfill the design requirements. In general as design condition the station keeping system should withstand to:

i. Environmental loads that act upon the system,

ii. Loads resulting from operational conditions coming from connected risers,

iii. Damaged conditions were one or more lines are broken,

with tension in lines, offsets of the anchored system and other operational constraints being under design limits as suggest by rules or by previous experience.

The design is accomplished through a series of design iterations, in which a trial design based in similar ones is chosen, analyzed, by a computer analysis tool and then modified by the designer after the examination of the numerical results. The system modified is reanalyzed, the results examined and if they are not adequate the system is again modified. This procedure is followed until a satisfactory design is achieved. Since the designer's judgment and intuition are influencing factors in the redesign process, and since only a small number of design iterations is practical, there is no guarantee that this process will involve a design of minimum cost or weight or any other merit for all design conditions. Furthermore, such a process is very inefficient, and there is clearly a need for an automated one where the human designer is replaced by the computer.

Under this point of view we developed a computer program to perform the optimum mooring system design, that is, a design for minimum weight or minimum cost which satisfies all design requirements.

For sake of simplicity we have decided to perform optimization in conventional mooring systems with lines composed up to three segments without clump weights or buoys. So as design variables we selected the bottom chain segment length, the cable segment length, the anchor and the anchor point in each mooring line. The chain segment near the surface has a fixed length since it is required for operational reasons only. Choosing the anchor point position the line composition and pre-tension is optimized taking in account all load cases including those corresponding to the damaged conditions.

3. Basic hypothesis for the synthesis model

The moored system will be investigated under the well-known quasi-static determinist approach consecrated in the technical literature. These basics are:

i. The mooring lines are treated as static with appropriate safety factors to take care any dynamic amplification originate by their eventual vibrations.

ii. Wind and currents are considered constants over the moored system. So they affect the mean drift offset only, without generating system oscillations.

iii. The wave effects are divided in mean drift (static), dynamic of first order (dynamic in the wave frequency) and dynamic of second order (slow drift).

iv. The system stiffness is disregarded in calculating the first order offsets caused by the waves. They are not disregarded in calculating the slow drift offsets. So the dynamic behaviour of mooring and moored system must be evaluated.

Being safe wind, waves and currents act simultaneously in a given direction. All possible directions are investigated.

The design of a mooring system should at least comply with general rules fixed by classification societies[2] or recommendations[3] and meet requirements such minimum cost and maximum efficiency.

4. Basics aspects of the search algorithm

The design variables search algorithm used in the EXMOOR system is based on optimisation procedures. Optimisation is purely a mathematical process and therefore it can and should be a fully automated process. That is, in the mooring system design process, the optimisation step should be simply a "black box" which performs a specific task. It accepts, as input, an objective function and a set of constraints, and it returns, as output, the specific optimal values of the design variables, that is, the values that maximise the objective while satisfying all of the constrains. Since there is only one optimum point in the design space the optimisation task is straightforward and unambiguous. The only requirement is that the method must find the optimum rapidly and efficiently. The particular manner in which it does is not important and any optimisation method that meets the requirement can be used in the design process.

The method we used in this work is based on a fully non-linear constrained search originally proposed by Box[4]. The algorithm is simple and uses the same idea developed by Nelder and Mead[5] with

1015

extended simplex technique. A group of points containing the search variables, all satisfying interior constraints, i.e., inequality constraints are randomly chosen. The point whose objective function is the worst is substituted by another chosen by reflection from the worst point through the centroid of the remaining points and whose the objective function is better. These points are called vertices and they form the complex. This procedure proceeds until all vertices are "close" together.

As many others constrained algorithms that proposed by Box needs a trial set of feasible variables which satisfies all internal and external constraints. Since this may be a hard task for the designer, we modified it to begin with any set of variables that need not satisfy the constraints. We split the algorithm in two phases. In the first phase we make a penalty search trying to find a feasible region for the searching variables. Delimited this region we proceed with the second phase performing a search satisfying all the constraints, as does the original Box algorithm.

For the penalty search we define an effective function considering the objective, weights or costs, and including a penalty term, that is, a term proportional to the sum of constraint violations scaled by a penalty parameter. If no feasible region is encountered at around the minimum found by the algorithm the penalty factor is modified and a new search is performed. Again if no feasible region is encountered the preceding procedure is repeated. Fletcher[6] refers this method as Sequential Unconstrained Minimisation Technique (SUMT). The penalty factor is arbitrarily modified in power of ten. If in two trying of minimum search the algorithm finds the same unfeasible point then the problem, as defined by the user, has, potentially, no solution.

At this stage the program decides slackening lines in leeward in order to reduce the system offsets and consequently constraint violations. The search is performed again and if no feasible region is found the problem definitely has no solution.

As constraints to the mooring problem we adopted the following objectives:

i. Line tensions should be less than limit tensions for each load condition;
ii. System offsets should be less than admissible offsets for each load condition;

iii. The cable segment in the line, if it exists, should not touch the sea floor in selected load conditions;
iv. Complementing the constraint above, a specified length of the chain segment in the chain-cable interface should ever be suspended.
v. Drag anchors should not be solicited with uplift forces;
vi. Complementing the constraint above, a specified length of the chain segment in the chain-anchor connection should ever be lied on the sea floor;
vii. The anchor should lie in a defined anchoring zone;
viii. The line touch down point should be inside the anchoring zone;

The conditions i, ii and v must be figured out as safety measures for the system. The specified lengths in conditions iv and vi can be interpreted as margins for some final adjustments in the line segments after the mathematical search finished. Conditions vii and viii take into account the sub sea layout avoiding those designs interfering with pipelines or seabed unfavourable conditions.

Any violation in any of the constraints above is treated as penalty term to the effective objective function, attributing a poor objective function value for infeasible points and aim the search toward the dominion were all constraints can be satisfied. If this region exists it is located and the search is changed to the vectors whose constraint properties are satisfied. In those cases where this region does not exist, the algorithm will find the minimum penalty solution and, if permitted, it will try to find solutions slackening leeward lines. For the lines slackened the constraints iii and iv are not verified and they are not also for damaged conditions although they are for pre-damaged conditions.

5. The load conditions

EXMOOR can process up to four load conditions with different allowances for tension in lines and offsets for the platform. Environmental forces, which come from waves, wind and currents and act in several directions, compose each load condition. With the load condition four are define two additional damaged conditions with different allowances for line tensions and vessel offsets. In the first damaged condition the most tensioned line in pre-damaged condition is supposed to fail carrying the vessel to

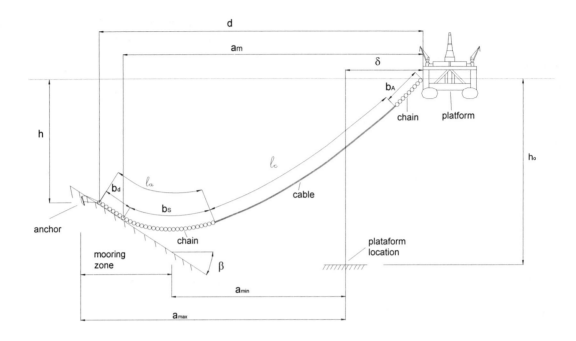

Figure 1 – Physical and decision variables

maximise its offset. In the second, the second most tensioned line fails attempting to maximise force in the most tensioned line.

6. The design variables and the explicit constraints

The searching variables for each line as mentioned before are the cable length, the chain length, the anchor weight and the anchoring radius, as defined in figure 1.

The bounds on each variable must be defined carefully otherwise by virtue of once the process is started it will choose the variables guided only by its objective and sometimes a problem with no physical meaning can arise.

7. The problem formulation

For the penalty search it is defined and effective objective function that considers the merit function and includes a penalty term, generally a term proportional to the sum of constraint violations scaled by a penalty parameter as

$$\phi(\mathbf{X}, r_k) = f(\mathbf{X}) + r_k \, p(\mathbf{X}) \qquad (1)$$

where,

$\mathbf{X}$ is the vector containing the problem variables;

$f(\mathbf{X}$ is the merit function such as cost or weight;

$p(\mathbf{X})$ is the penalty function that takes into account the accumulated penalties for all not satisfied constraints;

r_k is the penalty factor in the kth search sequence;

All functions are normalised relative to maximum possible value in any variable.

Hence the function $p(\mathbf{X})$ is defined as

$$p(\mathbf{X}) = \sum_{l}^{N_{LC}} \left\{ \sum_{i}^{N_L} \sum_{c}^{N_{CO}} g_{lic}(\mathbf{X}) + \overline{g}_l(\mathbf{X}) \right\} \qquad (2)$$

where,

N_{LC} is the number of load conditions

N_L number of lines

N_{CO} number of line constraints

l is the load condition number

i is the line identification number

c is the identification number of each constraint as defined in next sub-items.

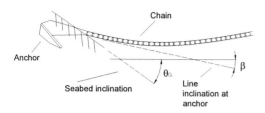

Figure 2 – Anchor with uplift force

7.1 Drag anchor being pulled uplift

An anchor is being pulled uplift if the line force at the anchor point is not parallel to sea floor. The constraint is

$$g_{li1}(\mathbf{X}) = \tan(\theta_a - \beta) \tag{3}$$

where,

β is the sea bed inclination at the anchor point as shown in Figure 2;

θ_a is the line inclination in the bottom segment near the anchor;

7.2 The minimum seabed dead line not satisfied

As an extra feature we permit a minimum dead line length to be kept near the anchor and if this is not satisfied the penalty is computed as

$$g_{li2}(\mathbf{X}) = \max[0, \frac{(l_{dc} - b_d)}{l_{am}}] \tag{4}$$

where,

b_d is the calculated dead line lying on the seabed.

l_{dc} the minimum required dead length specified in the design

7.3 The cable segment touching the sea floor or the minimum length of chain segment on the seabed to be always suspended

To provide a safety measure for prevent cable segment touch the seabed its possible to choose a

minimum extension of the bottom chain segment to stay always. This constraint is stated mathematically as

$$g_{li3}(\mathbf{X}) = \max[0, \frac{(l_{sc} - b_s)}{l_c}] \tag{5}$$

where,

b_s is the length suspended of the bottom segment (near position)

l_{sc} is the minimum required suspended chain length specified in the design

l_c is the length of the cable segment;

7.4 Line tension exceed limit tension

To prevent any overrun in the admissible traction on each line segment it is defined the constraint

$$g_{li4}(\mathbf{X}) = \max[0, (\frac{T_{max}}{T_{ad}} - 1)] \tag{6}$$

where T_{max}/T_{ad} is the maximum ratio between maximum acting traction per admissible traction for each segment in the line.

7.5 Anchor limit force exceeded

Any uplift forces and also any traction beyond its holding power should not solicit the anchor. This constraint is stated as

$$g_{li5}(\mathbf{X}) = \max[0, \frac{(f_{max} - f_{ad})}{f_{ad}}] \tag{7}$$

where,

f_{max} maximum line tension in the anchor point

f_{ad} anchor limit force

7.6 Anchor point lying outside the anchoring zone

Some locations can be overpopulated with equipment and may have restrictions on the anchoring zone. Defining

a_{max} far limit of the anchoring zone for the ith line;

a_{min} near limit of the anchoring zone for the ith line;

d_o anchoring radius, the horizontal distance taken from the fairlead to the anchoring point when the platform is on its initial position;

it can be stated the constraint for the anchor drop point

$$g_{li6}(\mathbf{X}) = \max\left[0, \frac{d_0 - a_{max}}{a_{max} - a_{min}}\right] + \max\left[0, \frac{a_{min} - d_0}{a_{max} - a_{min}}\right]$$

(8)

7.7 Touch down point outside the anchoring zone

Similarly the before mentioned constraint, the line touch down point should be inside the same anchoring zone. The constraint can be stated as

$$g_{li7}(\mathbf{X}) = \max\left[0, \frac{a_m - a_{max}}{a_{max} - a_{min}}\right] + \max\left[0, \frac{a_{min} - a_m}{a_{max} - a_{min}}\right]$$

(9)

where,

a_m is the horizontal touch down point distance taken from fairlead when the platform is on its equilibrium position;

7.8 Offsets exceeding limits

In general, when designing anchoring system the total vessel offset must be limited. As a versatility we have permitted limits total offset as well as on each of its components: the static that comes from currents, winds and mean drift from wave forces and the dynamics, from second order wave excitation. The constraint is stated as:

$$g_{li8}(\mathbf{X}) = \max\left[0, \frac{\delta_{max} - \delta_{adm}}{\delta_{adm}}\right]_{static} +$$

$$\max\left[0, \frac{\delta_{max} - \delta_{adm}}{\delta_{adm}}\right]_{dynamic} + (10)$$

$$\max\left[0, \frac{\delta_{max} - \delta_{adml}}{\delta_{adm}}\right]_{total} +$$

where,

δ_{max} is the maximum offset of specific component

δ_{adm} is admissible offset for the specific component.

7.9 Generals

As shown each constraint can be stated as a mathematical relationship able to compute a penalty value when this constraint is not satisfied, being positive when the vector analysed is outside the feasible sub domain and zero inside. When the search is not restricted to the feasible domain the program search by a penalised objective function. When a minimum is unfeasible the penalty parameter is automatically modified and a new search is performed. This procedure is repeated until program reaches the feasible region. If this occurs the search is switched to those solutions which satisfy the constraint and the objective function is replaced by the original $f(\mathbf{X})$ and at least one minimum solution will be find. If the problem initial conditions result in no feasible region the program proceeds slackening lines in leeward trying do reduce the offsets and consequently other associated problems. Thereafter if no feasible region is found the problem has no solution for the chain and/or cables diameters initially selected.

8. Practical Results of EXMOOR use's

As a final validation test the developed computational tool was used to calculate the mooring lines of two semi-submersible platform, which were in operation already, and had their mooring lines designed by the conventional approach using DMOOR program as a design tool. Both platform operate in a 600 meters deep oil field and were designed for two load conditions, operational and survival modes. The lines spread azimuths and line segments composition were fixed to the original design Additionally, if the design limits are exceeded a 300 hundred meters of line are permitted to slack in leeward load directions. As a measure of safety was imposed for all lines a 150 meters of chain permanently lying on the se bead. The constraints are the platform maximum offset, the maximum tension in the lines, anchoring zone unrestricted, the cable segment in lines not slacked can not touch the sea bed and no uplift force in the anchor was permitted. In a conventional approach design, based on trial and error analysis an experimented designer found the results shown in tables T1 and T2.

The same conditions were inputted in the EXMOOR system and the results, after few hours of time processing, are shown in the same tables.

Table T1 – Comparison of Exmoor Results and the original design for PXX Anchoring System

Mooring Line		Top Segment (m)		Bottom Segment (m)		Cable Segment (m)		Anchor Radius (m)		Pre-Tension (KN)		Anchor Tract (KN)	
N	Azim	Conv.	Exmoor	Conv.	Exmoor	Conv.	Exmoor	Conv.	Exmoor	Conv.	Exmoor	Conv.	Exmoor
1	24°	200	200	1250	895	600	927	1850	1855	1179	1126		1675
2	42°	200	200	1300	899	600	927	1898	1854	1190	1169		1696
3	60°	200	200	1300	677	600	927	1895	1603	1202	906		1041
4	109°	200	200	1350	982	600	927	1931	1922	1210	1129		1652
5	132°	200	200	1300	1009	600	927	1895	1954	1202	1128		1732
6	155°	200	200	1300	820	600	927	1928	1763	1422	993		1476
7	205°	200	200	1200	966	600	927	1831	1950	1386	1296		2014
8	228°	200	200	1200	759	600	927	2259	1740	1134	985		1520
9	251°	200	200	1200	950	600	927	1856	1947	1119	1111		2304
10	300°	200	200	1300	1183	600	927	1958	2193	1105	1204		3386
11	318°	200	200	1250	1127	600	927	1904	2125	1119	1126		2941
12	336°	200	200	1200	1119	600	927	1849	2117	1127	1325		2828
	max	200	200	1350	1183	600	927	2359	2193	1422	1325	3750	3386

Type	φ76mm K4		φ76 mm ORQ/K4		φ96mm 6x37 IWRC	
Kgf/m	126.2		126.2		38.7	
US$/kg	2.50		2.50		3.00	
Tons	303	303	2215	1437	279	431
US$x10⁶	0.757	0.757	5.537	3.592	0.836	1.291

Type	Bruce TS	
Tons	23.0	21.0
US$/Kg	3.00	3.00
Tons	276	252
US$x10⁶	0.828	0.756

Totals	Conv.	Exmoor
Tons	3072	2119
US$x10⁶	7.958	6.396

Savings	
%	31
US$x10⁶	1.562

PXX: 610 m

Reprint with permition from reference 8

The solution find by the system is more economical than those achieved by the conventional approach but the main benefit feature of the EXMOOR process is the automation. The system do the tedious part design task freeing the designer for other creative assignment.

9. Conclusions

As we can see the EXMOOR system has proven to be a powerful tool helping to design anchoring systems, but some improvements yet can be done.

EXMOOR performs a complete quasi-static analysis of the mooring lines. An adequate line dynamics suitable for design proposals should be searched and added to the system.

To increase its flexibility the nature and which design variables must be selected is another front to worked. Since EXMOOR uses a pure mathematical search algorithm to find the best design including line diameters, platform headings, line angle and number of lines, segments, clump weights and buoys in each line as additional parameters to be found is a matter of growing the time consumed in the search only.

Table T1 – Comparison of Exmoor Results and the original design for PXIII Anchoring System

Mooring Line		Top Segment (m)		Bottom Segment (m)		Cable Segment (m)		Anchor Radius (m)		Pre-Tension (KN)		Anchor Tract (KN)	
N	Azim	Conv.	Exmoor	Conv.	Exmoor	Conv.	Exmoor	Conv.	Exmoor	Conv.	Exmoor	Conv.	Exmoor
1	35°	-	-	750	657	1200	1399	1843	1926	478	651		1419
2	55°	-	-	750	597	1200	1399	1839	1862	442	590		1062
3	110°	-	-	750	660	1200	1399	1854	1936	638	783		1323
4	140°	-	-	750	675	1200	1399	1847	1944	527	656		1301
5	220°	-	-	750	675	1200	1399	1847	1944	527	656		1301
6	250°	-	-	750	660	1200	1399	1854	1936	638	783		1323
7	305	-	-	750	597	1200	1399	1839	1862	442	590		1062
8	325°	-	-	750	657	1200	1399	1843	1926	478	651		1419
	max	-	-	750	675	1200	1399	1854	1944	638	783	1726	1419

Type	-	φ76 mm ORQ	φ70mm 6x36 IWRC		Type	Bruce TS	
Kgf/m	-	132.2	20.4		Tons	9.0	8.0
US$/kg	-	2.50	3.00		US$/Kg	3.00	3.00

Tons	-	-	793.2	684.5	195.8	228.3		Tons	72	64
US$x10⁶	-	-	1.983	1.711	0.588	0.685		US$x10⁶	0.216	0.192

Totals	Conv.	Exmoor		Savings	
Tons	1061	977		%	8
PXIII: 600m					
US$x10⁶	2.787	2.588		US$x10⁶	0.199

Reprint with permition from reference 8

The actual version of EXMOOR runs in PC Compatible computers and may spend a large amount of time in the search. The inclusion of the above mentioned items may require a version to run in more powerful machines.

The platform heading and line spread is being actually optimised for one load condition with the idea developed by Lie and Fylling[7]. Including them as variables is a matter of a cost/benefit relationship. The increase in the computer time must compensate the savings achieved.[8]

Bibliography

[1] NDAI Noble Denton Associated Inc, 1986, *DMOOR Deterministic Mooring Analysis Program Suite.*

[2] DnV, *Rules for the Design Construction and Inspection of Offshore Structures.*

[3] API, *Analysis of Spread Mooring Systems for Floating Drilling Units*, Recommended Practice 2P (RP2P).

[4] Box, M. J., 1965, *A New Method of Constrained Optimisation and Comparison with other*

Methods, Computer Journal, Vol. 8, No. 1, April, pp. 42-52.

[5] Nelder, J. A., Mead, R., 1965, *A Simplex Method for Function Minimisation,* Computer Journal, Vol. 7, No. 4, Jan., pp. 308-313.

[6] Fletcher, R., *Practical Methods of Optimisation, Constrained Optimisation,* Vol. 2. John Wiley & Sons.

[7] Fylling, I.J. and Lie, H., *Mooring System Design Aspects of Environmental Loading and Mooring System Optimisation Potential,* First OMAE Speciality Symposium on Offshore and Arctic Frontiers, New Orleans, 1986.

[8] Nunes Dias, C. A., Augusto, O. B., Rossi, R. R. , *Expert System for Anchor System Design,* Petrobras/University of São Paulo Project, 1994, final report (in portuguese).

Practical Design of Ships and Mobile Units
M.W.C. Oosterveld and S.G. Tan, editors.

A Practical Design and Dynamic Characteristics of a Deep Sea Mooring System

H.S. Shin, J.W. Cho, I.K. Park

Maritime Research Institute, Hyundai Heavy Industries Co., Ltd.
1 Cheonhadong Dong-ku, Ulsan, KOREA

This paper presents a practical design for a deep sea mooring system with the dynamic characteristics of it. Catenary mooring system is widely used for position-keeping of floating offshore structures in the moderate water depth less than about 600m. In deeper water, the dynamic positioning system or other mooring systems such as tendon for TLP are generally adopted. Through the challenges for the deep sea mooring, many valuable engineering tools and new materials of light and strong mooring lines are developed. Deeper water more than 1,000 m is challenged by the turret moored FPSO.
The analysis of line dynamics is recommended to perform in the design of the mooring system. In order to find tension transfer function at the line top, calculations based on a simplified method and model tests are performed in the present paper. Forced oscillation tests in the surge mode are performed to find tension transfer functions at the line top and anchor point. In the case of a simple catenary mooring line, good agreements are found between calculations and experiments. Comparisons are performed between quasi-static analysis and dynamic analysis.

1. INTRODUCTION

Deep sea and floating structure are two important key words in recent developments of offshore drilling and production structures. Recently offshore drilling entered the deep sea deeper than 2,000 m. The water depth of offshore oil production reached almost 1,000 m. For example, several FPSOs are going to be installed in the Marlim field at Campos Basin of Brazilian offshore, where the water depth is close to 1,000 m (Park et al, 1998). The turret mooring systems in these cases are composed of only chain and wire. A semisubmersible drilling rig is now under construction in Hyundai Heavy Industries which will be installed in the Gulf of Mexico site deeper than 2,000 m. The position keeping is the most important function in the floating structure to keep the excursion boundary of floating structure.

Owing to the development of strong wire rope, offshore structure can be moored in the deep sea with or without assistance of DP system. In the mild sea such as Gulf of Mexico, DP assistance is normally unnecessary in keeping the vessel within the watch circle.

Practical mooring design and analysis procedures are well described in API(1997). If the mooring analysis is performed by the quasi-static analysis, the water depth is not so much important factor. It is because the catenary equation is simply applied and the equation gives acceptably accurate results in viewpoint of static analysis. In fact, however, the dynamic effects are very important in the case of deep water mooring since they govern the dynamic behaviors of less stiff mooring line in deep water. Dynamic mooring analyses need to be verified by model tests. The verification of the dynamic analysis methods and subsequent results for deep sea mooring system has been hampered by the lack of experimental data. Scaling the water depth in the model basin, geometry and elasticity of the mooring line, the surface model and environmental conditions simultaneously is almost impossible.

Kwan and Bruen(1991) emphasized the dynamic analysis of the mooring lines showing the comparisons of dynamic analysis to quasi-static analysis. The ratios at peak wave frequency tensions are particularly large. The poor estimation of quasi-static analysis will give a poor estimation of wave frequency tensions. This causes worse effect on fatigue analysis, since the fatigue life is governed by wave frequency tensions.

The present paper summarizes 1) practical design procedure for mooring systems and presents 2) a simplified dynamic analysis and 3) model tests for verification.

2. PRACTICAL DESIGN AND ANALYSIS PROCEDURE

The purpose of station keeping is to keep the floating structure within excursion limit against environmental forces. The excursion limit is determined by the riser design criteria in the case of oil production or the distance to the other structure at vicinity in the case of offshore working vessel. The types of station keeping system for floating structures are the single point mooring, the spread mooring, and the dynamic positioning. The single point mooring is widely used for the ship type vessel and the spread mooring is taken for working barges and semi submersible type vessel. The dynamic positioning is used independently or as the assistance for the mooring system. The type of station keeping is dependent on mooring duration; permanent or temporary mooring.

Generally the mooring analysis is performed by a quasi-static analysis method. In the quasi-static mooring analysis, performance evaluation of mooring systems is made based on the static excursion-load(tension or force) relation. The excursion-load relation is well defined by the catenary equation for conventional offshore wire ropes and chains. If clump weights and spring buoys are used for better mooring capability, some modifications are required in the catenary equation program.

If load-excursion relations are set up for designed mooring system, static forces (wind, current and mean wave drift forces) on and motions of floating structure are calculated. Motions include the 6-DOF first order wave frequency motions and the 3-DOF second order low frequency motions. Additionally low frequency motions due to wind forces fluctuating at low frequencies can be added. However, caution is demanded as there are still uncertainties in the estimation of low frequency wind forces.

Static forces due to wind and current are estimated through empirical data base or model tests. Static wave drift forces and motions are estimated generally by the 3-D panel method. Recently the high order panel method is well used for higher accuracy and time efficiency.

In the quasi-static mooring analysis, at first, static excursion due to environmental forces is estimated. Then tensions due to wave motions are estimated and checked against breaking strength of specific mooring line. Pretensions are important parameters in viewpoints of excursion limit and maximum line tensions. Optimum pretensions are found to satisfy excursion limit and not to exceed specified maximum allowable line tensions. The design pretension is generally determined such that summation of the tension due to limit excursion and the pretension equals to the maximum allowable line tension. For the on board operators the characteristics of the mooring system for various pretensions are tabulated and contained in the operating manual.

The larger values of motions from the summation of the maximum high frequency motions and the significant low frequency motions or the summation of the significant high frequency motions and the maximum low frequency motions are selected as the limit excursions. Extreme values of motions are obtained in frequency domain analysis or time domain analysis. Methods for predicting the low frequency motions are still under development. Especially there exist uncertainties in the estimation of viscous damping, wave drift damping and mooring system damping. Quadratic transfer functions for the time history of low frequency wave exciting forces are the other sources of uncertainties. They are not so accurately calculated by the conventional 3-D panel method, particularly in high frequency region. Newman's(1974) approximation method to obtain the low frequency wave exciting forces is practically well used.

It is required by the rule that dynamic analysis should be performed in deeper water than 450 m(DNV, 1997). Larsen(1990) suggests a simplified method to calculate dynamic tensions due to upper motions of mooring lines. Geometrical configuration of the mooring line in dynamic behavior is assumed to be similar to that of quasi-static behavior, which is determined by the top end motions. In the present paper, dynamic tensions are surveyed through the simplified dynamic calculation and model tests. Details are described below.

3. DYNAMIC CHARACTERISTICS OF MOORING LINE

3.1 Simplified Dynamic Analysis

There exist nonlinearities to have influence on line dynamics; nonlinear elastic behavior of the line, geometrical nonlinearity, nonlinear fluid loading, nonlinear interaction between the line and sea floor. Time domain analysis is widely used to consider fully these nonlinearities. At every time step every term need to be calculated and hence the computations are complex and time consuming. On the other hand, the frequency domain analyses should be linear to satisfy the superposition principle. Linearizations are required and iterative schemes are well used. Generally frequency domain analyses are accepted as to give reasonable accuracy and also practically very convenient.

Instead of complicated and time consuming FEM analysis or lumped mass analysis for line dynamics, simplified dynamic analysis was introduced by Larsen(1990). The key point of this simplified dynamic analysis method is in the presentation of dynamic tensions in the form of transfer function. Assuming that the shape of the dynamic motion is equal to the change in the static line geometry, it will be the difference in the static line configuration caused by a upper point motion. The amplitude of the dynamic motion of the mooring line is found from the equation of motion which constitutes moment equilibrium for the touch down point. Forces to cause moment at the touch down point consist of drag forces, inertia forces and line tensions at the line top. Through a series of simple numerical manipulations, transfer functions are obtained as the ratios of tension to motion. Spectral analyses using tension transfer functions give extreme tensions and fatigue life of the mooring line. This dynamic method is validated by the model tests

described below.

3.2 Model Test

The purpose of model test is to verify indirectly how accurately the analysis method and analysis results show the actual phenomena. In some cases, to conduct model tests is more difficult than proto type experiment. The dynamic behavior of deep sea mooring line is a typical example, since there is no model basin to conform to very deep sea. Hence the verification of analysis has been hampered by the lack of model test results.

As model tests of deep sea mooring systems are very difficult to perform due to many limitations such as scaling the water depth and line elasticity, simplified model tests are frequently conducted. Linear springs are typically well used instead of actual nonlinear catenary mooring lines. If the test purpose is to investigate wave frequency motions of moored vessel, this simplification can be accepted. When, however, now frequency motions and line dynamics are to be investigated, the mooring system should be modeled as it is.

It is assumed in the present study that a tanker type FPSO model is moored in deep sea by two catenary mooring lines. First of all, forced oscillation tests for a single line were performed in the calm water to find dynamic behaviors of the line. And then, motion tests for moored model in regular and irregular waves. The ratios of dynamic tensions to forced oscillation amplitudes are investigated as transfer functions and compared with simplified analysis results. The test basin is a deep towing tank (210 m long, 14 m wide and 6 m deep). The catenary mooring line is composed of chain and wire rope as Fig. 1. The weight of the chain and wire rope in water are 28.1 g/m and 4.5 g/m respectively. The diameters of them are 1.4 mm and 1.0 mm respectively. A weight is installed at bottom as an anchor point. Line tensions at the anchor point and the line top are measured using load cells. In the case of the forced oscillation test, the upper load cell is the 3 component type. In the other cases only tensions are measured with the tension load cell.

Forced oscillation tests are performed for a single mooring line. Motion tests in regular and irregular waves are performed for the model moored by two mooring lines. The mooring lines are deployed in the forward and backward directions as Fig. 2. Fairleaders are installed for smooth connection of the load cell and mooring lines. With this, only tension components are measured by the load cell.

The tanker model has 0.121 m of draft, 2.0 m of length, 0.365 m of breadth and 71.69 kg of displacement. Imagining a bow turreted FPSO, the mooring lines are connected to the bow of the model. From weather-vaning characteristics of turret mooring, only the head sea condition is considered.

Motions are measured by an optical measuring system. Tensions are measured by a 3 component load cell and tension type load cells.

3.3   Results and Discussions

Fig. 3 shows time histories of the horizontal forces measured by the upper load cell in forced oscillation tests. It is found that tensions are not sinusoidal against sinusoidal horizontal oscillations. It can be caused by the nonlinear effects of mooring line dynamics. In order to measure the pure line tensions, another experimental apparatus is required to connect the mooring line and the load cell smoothly.

Fig. 4 shows transfer functions of tensions against forced surge oscillation. Pretension is set as 3.235 N by adjusting the distance between the model and the anchoring point. Differences between calculations and model tests look to be resulted from different oscillation modes. In the calculations, the transfer function is calculated as the tension change against the unit tangential motion amplitude at line top end. In the model test, however, line top end is oscillated horizontally. Horizontal motions have tangential component and also lateral component about the line direction at the line top. The lateral component of oscillation will give additional dynamic tensions. As a further validation study, it is required to perform model tests where forced motions are oscillated only in the tangential direction.

Fig. 5 shows time histories of the tensions of two mooring lines at top end and anchor points obtained from the regular wave test of frequency 0.913 Hz. The tension signal measured by tension load cells

are smooth and close to sinusoidal compared to the forced oscillation test of Fig.3. The tensions at the anchor points are not periodic and especially the tension at the forward anchor point is very large compared to the top tension. Further investigation is needed.

Tension transfer functions are obtained from time histories of tensions as Fig. 6. It is found that the tension transfer functions of forward line and backward line show different frequency characteristics. The tension transfer function of forward line shows one peak. However, the tension transfer function of backward line shows two peaks along the increase of the frequency. Both decreases in high frequencies, where elastic deformation is dominant. The different frequency characteristics of two mooring lines are caused by the different mean tensions due to the mean offset. Since the mean offset is caused by the mean wave drift forces, tensions of forward line must be large and so be the tension transfer function. However, model test results are reversed. Further experimental studies are required. Comparing Fig. 6 with Fig. 4, it is found that the tension transfer functions from regular wave tests are similar to those of forced oscillation tests except in the high frequency region.

Fig. 7 shows time histories of irregular wave tests. The significant wave height is 0.0475 m and the zero crossing period is 0.913 sec, which conform to 5.7 m and 10.0 sec respectively considering the scale ratio 120. It is found that low frequency surge motions are dominant factors to determine the line tensions. Hence the investigation about the damping for low frequency motions is very important.

4.   CONCLUSIONS

The present paper summarizes the general procedures of mooring system design and analysis, and model tests. A simplified method for mooring dynamic analysis is validated through forced oscillation tests and wave tests for a deep sea mooring system. Transfer functions of the dynamic tensions against top end motions are compared between calculations and model tests. Model test results look to show much larger values than calculations in the figure. However, considering different oscillation mode at the line top end between model tests and calculations, two results can be regarded as close. Simple modeling of the mooring system for the deep sea condition could also cause differences. As further comparison studies, model tests of the same oscillation mode with the analysis are required. Through comprehensive model tests, some dynamic characteristics of mooring lines are understood and required further research works are drawn.

REFERENCES

API(1997), 'Recommended Practice for Design and Analysis of Stationkeeping Systems for Floating Structures', API RP 2SK

DNV(1997), 'Rules for Classification of Mobile Offshore Units', Part 6, Ch 2, POSPOOR

Kwan C.T. and Bruen F.J.(1991), 'Mooring line Dynamics: Comparison of Time Domain. Frequency Domain and Quasi-static Analysis', OTC 6657

Larsen K and Sandvik P.C.(1990), 'Efficient Methods for the Calculation of Dynamic Mooring Line Tension', 1st European Offshore Mechanics Symposium

Newman J.N.(1974), 'Second-order Slowly varying Forces on Vessels in Irregular Waves', International Symposium on Marine Vehicles and Structures in Waves, pp193-197

Park I.K., Jang Y.S., Shin H.S. and Yang Y.T.(1998), 'Conceptual Design and Analyses of Deep-Sea FPSO Converted from VLCC', OTC 8809

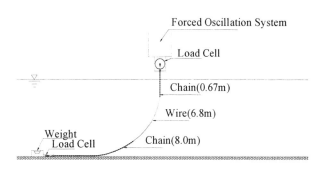

Fig. 1 Set up of forced oscillation test

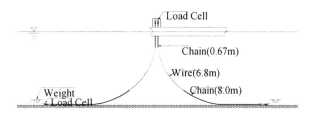

Fig. 2 Set up of a moored model

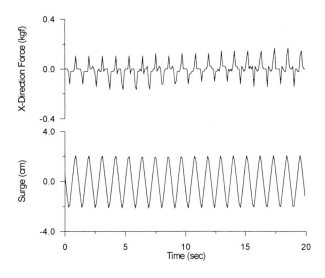

Fig. 3 Time histories of tension from forced oscillation tests

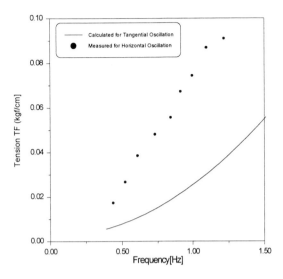

Fig. 4 Tension transfer functions from forced oscillation tests and simplified analysis

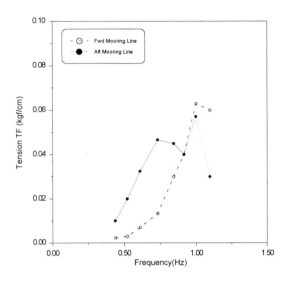

Fig. 6 Tension transfer functions from regular wave tests

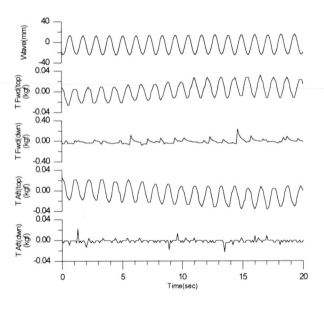

Fig. 5 Time histories of regular wave tests

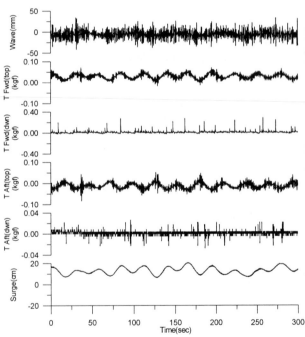

Fig. 7 Time histories of irregular wave tests

Analysis of Dynamic Response of a Moored Tanker and Mooring Lines in a Single Point Mooring System

Yojiro Wada and Yoichi Yamaguchi

Ship & Ocean Engineering Laboratory

Nagasaki Research & Development Center, Mitsubishi Heavy Industries, Ltd.

5-717-1 Fukahori-machi, Nagasaki 851-0392, JAPAN

ABSTRACT

Dynamic response of a moored tanker and mooring lines in a Turret Mooring system and a Catenary Anchor Leg Mooring system, in waves, wind and current was investigated theoretically and experimentally. Theoretical analysis is based on the time domain simulation of coupled equations of motions, which include slewing motion of the tanker by maneuvering equations, dynamic response of mooring lines by 3-D lumped mass method, and wave frequency mooring buoy motion in case of CALM system.

Model experiments were performed to obtain the data for the verification of the simulation program and for the reference in designing a system. Simulated results on the dynamic response of the tanker and the mooring lines showed good agreement with measured results. It was verified that the simulation program could be applied in the design of various SPM systems.

1. INTRODUCTION

Recently, it has been becoming a trend in the development of oil and gas fields to shift into the marginal and deeper field. Therefore, vessel-type storage and offloading system with processing equipment comes before the footlights instead of usual fixed platform, because it saves initial cost compared with fixed platform.

In these developments, Single Point Mooring (SPM) system plays a great important role as a effective station-keeping system of a floating vessel. It has economical advantage and offers safety operation due to the minimization of environmental loads on the vessel by weather vane effect.

However, It has been increasing for a SPM system to be operated in severer conditions, in which large motions of the vessel and mooring lines may be induced and those effects can not be neglected. So, there is a possibility that design loads of the mooring lines has to be increased because of such dynamic response of vessels and mooring lines as in severe seas.

Usually, in designing a system, a quasi-static analysis is performed for evaluating mooring load on the lines, and the effects of dynamic response are taken into account only through the use of a relatively conservative safety factor.

Therefore, to make a system design more reliable and rational, it is getting of great importance to evaluate the effects of dynamic response exactly in the completion of the design. For this purpose, it is indispensable to develop a dynamic analysis method considering vessel and line dynamics.

This paper presents a simulation method for evaluating the characteristics of dynamic response of a moored tanker, mooring lines and a mooring buoy in such a Single Point Mooring system as Turret system or CALM system. The coupled equations of dynamic behavior of mooring lines and low frequency slewing motions of a tanker, or wave frequency buoy motions are solved in the time domain.

Model experiments were performed for both the Turret system and the CALM system. Motions of a tanker, dynamic load on the mooring lines were measured in the experiments. Calculated results agreed with measured results fairly well. It was shown that the present method could be applied in the evaluation of dynamic characteristics of a single point mooring system.

2. THEORETICAL ANALYSIS OF DYNAMIC RESPONSE

2.1 Tanker Motion

Motion of a moored tanker is divided into two components. One is the low frequency slewing motion and the other is the wave frequency motion. It was assumed in the present method that the wave frequency motion are independent of the low frequency slewing motion, and those motions can be treated separately. Wave frequency motion was not taken into account.

The low frequency slewing motions are taken into account based on the manoeuvring equations in the horizontal plane. These are defined in the longitudinal, the lateral and the rotational direction, as follows.

$$(M + m_x)\dot{u} - (M + m_y)v \cdot r = F_{XH} + F_{XT} + F_{XE}$$
$$(M + m_y)\dot{v} + (M + m_x)u \cdot r = F_{YH} + F_{YT} + F_{YE} \quad (1)$$
$$(I_{zz} + J_{zz})\dot{r} = F_{NH} + F_{NT} + F_{NE}$$

in which,

M, I_{zz} : mass and mass moment of inertia of the tanker

m_x, m_y, J_{zz} : added mass and added mass moment of inertia of the tanker

u, v, r : longitudinal, lateral and angular velocity

F_{XH}, F_{YH}, F_{NH} : components of hydrodynamic force acting on the submerged hull of the tanker

F_{XT}, F_{YT}, F_{NT} : components of restoring force due to hawser tension

F_{XE}, F_{YE}, F_{NE} : components of the force due to external disturbances

The hydrodynamic forces on the submerged hull, including current force, are modeled by the manoeuvring derivatives using our database [1]. External force components are composed of steady wind force and low frequency wave drifting force. Wind force is calculated by Isherwood's method [2], and wave drifting force is by Pinkster's method [3] (direct double summation of the transfer function of wave drifting force) to the relative wind and wave direction in each time step. The wave drift damping force is not taken into account, because the lateral motions due to slewing are mainly focused in this paper, and the effect of wave drift damping is small in the lateral motions [4].

2.2 Wave Frequency Motions of a Mooring Buoy

Wave frequency mooring buoy motions are important in the survival condition of CALM system, in which sea conditions are much severer than the operating condition. In those cases, larger amplitude buoy motions may cause fatal damage on the mooring lines.

The equations of wave frequency buoy motions are given by the following equations, which are same as the equations used in the estimation of ship motion in waves.

$$\sum_{k=1}^{6}\left[\left(M_{jk}+A_{jk}\right)\ddot{\eta}_k+B_{jk}\cdot\dot{\eta}_k+D_{jk}\cdot\dot{\eta}_k|\dot{\eta}_k|+C_{jk}\cdot\eta_k\right]=F_{Tj}+F_{Ej} \quad (2)$$

$$(j=1\sim 6)$$

in which,

η_j : buoy motion vector

M_{jk} : mass matrix

A_{jk} : added mass matrix

B_{jk} : wave damping coefficient matrix

D_{jk} : non-linear damping coefficient matrix

C_{jk} : restoring coefficient matrix

F_{Tj} : force vector due to the mooring tensions

F_{Ej} : force vector due to the external disturbances

The exciting force due to external disturbances is composed of not only the first order wave force, but also the steady components due to the second order wave drifting force and current force. These equations are solved in time domain, being coupled with the motions of mooring lines.

2.3 Motions of Mooring Lines

Motions of mooring lines are modeled by 3-D Lumped Mass method[5]. The motion equations of each segment are given as follows.

$$\begin{bmatrix} _k(I_{11})_j & _k(I_{12})_j & _k(I_{13})_j \\ _k(I_{21})_j & _k(I_{22})_j & _k(I_{23})_j \\ _k(I_{31})_j & _k(I_{32})_j & _k(I_{33})_j \end{bmatrix}\begin{bmatrix} _k\ddot{x}_j \\ _k\ddot{y}_j \\ _k\ddot{z}_j \end{bmatrix}=\begin{bmatrix} _k(F_X)_j \\ _k(F_Y)_j \\ _k(F_Z)_j \end{bmatrix} \quad (3)$$

$$(k=1\sim m,\ j=2\sim N_k)$$

in which,

m : number of mooring lines

N_k : number of segments of the k-th line

$[_k(I_{in})_j]$: inertia matrix of segments

$_k\ddot{x}_j, _k\ddot{y}_j, _k\ddot{z}_j$: accelerations of segments

$_k(F_X)_j, _k(F_Y)_j, _k(F_Z)_j$: external forces acting on the segments

The external force is composed of tension between segments, current force, weight of segment and reaction force from seabed. Current force is originated from relative velocity including both motion of segment and wave particle.

The segments are imposed following boundary condition on the continuity of the line.

$$\left(_kx_j-_kx_{j-1}\right)^2+\left(_ky_j-_ky_{j-1}\right)^2+\left(_kz_j-_kz_{j-1}\right)^2=_ks_{0j}^2\left(1+\frac{_kT_{j-1}}{_kA_j\cdot_kE_j}\right) \quad (4)$$

Furthermore, the position of the connecting segments to the mooring buoy is defined in each time step, taking the results on the coupled motion with tanker or mooring buoy into account.

3. MODEL EXPERIMENTS

3.1 Outline of Model Experiments

Model experiments on both the Turret mooring system and the CALM system were performed to investigate the system response, as tanker motion and mooring loads, for the verification of simulation program and for the reference to the design of various SPM systems.

Table 1 Principal Particulars of the Tanker

	Turret	CALM
Length(pp)	255.7	328.1
Breadth [m]	42.3	54.2
Depth [m]	22.0	28.2
Condition	Full Load	
Draft [m]	14.6	22.5
GM$_T$ [m]	6.2	4.7

The principal particulars of the tanker are given in Table 1. A same tanker model was used in both experiments. Scale ratio of the model in Turret system is 1/50, and that is 1/64 in CALM system.

3.2 Experiments on the Turret system

The arrangement of the Turret system is given in Figure 1. Water depth is 160 m in full scale. A turret is mounted internally in the fore part of the tanker. The particulars of mooring chains are given in Table 2. In case of the turret system, experiments were performed only in regular and irregular head waves. Height of regular wave is 5 m for various wave periods, and significant wave height of irregular waves is 10 m for peak period of 14 sec. Pierson-Moskowitz type spectra were used in generating irregular waves.

Measured items in the experiments were the horizontal displacements at the turret point on the tanker, the turret load and the chain tensions at the attaching point to the turret. Measurements were executed for about two hours in full scale.

Optical tracking system was used for the measurement of displacements and mooring load were measured by ring-shaped force transducer incorporated in the line.

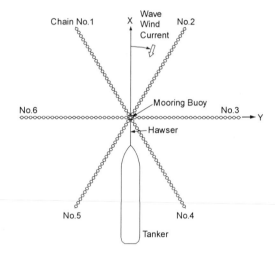

Fig. 2 Arrangement of the CALM System

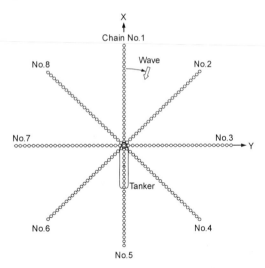

Fig. 1 Arrangement of the Turret Mooring System

Table 3 Principal Particulars of the Buoy

Dia. (outside) [m]	12.0
(inside) [m]	4.0
Depth [m]	5.0
Draft [m]	3.0
GM [m]	1.0

Table 2 Principal Particulars of Mooring Chains

Diameter [mm]	114
Length [m]	700
Submerged Weight [kg$_f$/m]	248
Longitudinal Stiffness [ton$_f$]	116,000
Number of Lines	8

Table 4 Principal Particulars of Mooring Chains

Diameter [mm]	92
Length [m]	310
Submerged Weight [kg$_f$/m]	168
Longitudinal Stiffness [ton$_f$]	76,000
Number of Lines	6

3.3 Experiments on the CALM system

The arrangement of the CALM system is given in Figure 2. It is composed of a tanker, hawser at the tanker bow, mooring buoy and six mooring chains. Water depth is 33 m in full scale. The principal particulars of mooring buoy and mooring chains are given in Table 3 and Table 4.

In the CALM system, responses in survival condition (without tanker) were also measured in addition to operating condition, because there is a possibility that mooring loads on the lines may be larger in survival condition. Larger model of mooring buoy and chains, having scale ratio of 1/39, was used in survival condition to raise the accuracy of the measurements.

Measurements were performed under the composite external disturbances due to current, wind and waves in operating condition, and due to current and waves in case of survival condition. Velocity of wind and current is 35 m/s and 0.3 m/s, respectively. Significant wave height is 5 m and peak period is 10 sec.

Horizontal motions of the tanker (surge, sway, yaw), horizontal motions of the mooring buoy (surge, sway), and tension of the hawser and chains were measured. Measurements were executed for about two hours also in CALM system.

4. EXPERIMENTAL VERIFICATION

Response of a tanker and mooring lines in a Turret Mooring system is discussed at first.

The response amplitudes of turret load and chain tension of No.1 and No.2 in regular waves of various periods are compared with measured results in Figure 3 to 5. These results were obtained from Fourier analysis of calculated and measured time

traces. External disturbance is only due to waves (no current and wind). Wave direction is heading waves and wave height is 5 m in full scale. Wave frequency tanker motions were not taken into account in the calculation, and surge damping coefficient was determined experimentally by free surging test.

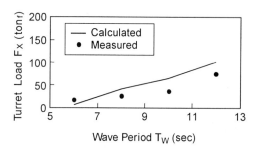

Fig.3 Response Amplitude in Regular Head Waves (Turret Load)

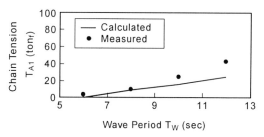

Fig.4 Response Amplitude in Regular Head Waves (Chain Tension No.1)

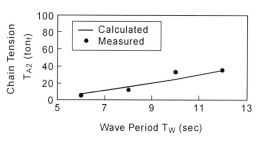

Fig.5 Response Amplitude in Regular Head Waves (Chain Tension No.2)

1034

It is shown that the amplitudes of fluctuations become large as wave period becomes long.

Same calculation in irregular head waves of significant height 10 m, peak period 14 sec was performed. Spectrum of measured wave was used in the calculation to generate irregular waves. Wave conditions are statistically equivalent in the calculation and the experiment. As an example, calculated and measured time traces of turret surging, tension fluctuation of chain No.1 and No.2 during 30 minutes are shown in Figure 6. The maximum amplitude of tension fluctuation from these time traces is compared in Figure 7.

As shown in these comparisons in regular and irregular waves, calculated results coincide fairly well with measured results.

Responses in composite external disturbances were analyzed by the calculation. Calculated results on the tanker motion and the chain tension were shown in Figure 8. The external disturbances are composed of waves, wind and current. Significant wave height is 10 m and peak period is 14 sec. Velocity of wind and current is 50 m/s and 1.8 m/s, respectively. Direction of disturbances is −30 degrees. Slewing motion of the tanker occurs and double amplitude of yaw fluctuation is about 40 degrees. Chain tension fluctuates according to the tanker motions.

As an example of the analysis on the CALM system, surging of mooring buoy and tension fluctuations in survival condition are shown in Figure 9 in comparison with measured results. Current velocity is 0.3 m/s and significant wave height is 5 m and peak period is 10 sec from direction of −30 degrees. It is said that calculated results give a good estimation.

As for the responses in operating condition, same analysis is possible considering the coupled motion of tanker, mooring buoy and mooring lines. But, it consumed very much time in performing calculations. In these cases, studies on the simplified models are indispensable for practical use. So, it was decided in this paper not to discuss the applicability of the complicated models.

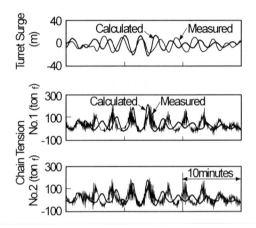

Fig.6 Time Traces for Responses in Irregular Head Waves

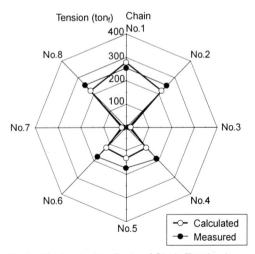

Fig. 7 Maximum Amplitude of Chain Tension in Irregular Head Waves

5. CONCLUSIONS

A simulation program to evaluate the dynamic response of a moored tanker and mooring lines in a Single Point Mooring system was presented.

According to the analysis on a Turret Mooring system and a Catenary Anchor Leg Mooring system, calculated results on the dynamic response of the tanker and the mooring lines showed good agreement with measured results. Compared with the usual quasi-static method, the rational evaluation becomes possible and the accuracy of the estimation can be raised by the direct simulation as in the present study, which takes into account the dynamic effects of coupled motion between the tanker and the mooring lines. Therefore, this kind of analysis method seems powerful in the design of SPM systems, though it should be improved to analyze more complicated cases as in operating condition of CALM system for practical use.

REFERENCES

[1] E.Kobayashi, "A Simulation Study on Ship Manoeuvrability at Low Speeds", Mitsubishi Technical Bulletin No.180 (1988)

[2] R.M.Isherwood, "Wind Resistance of Merchant Ships", The Royal Institution of Naval Architects (1972)

[3] J.A.Pinkster, "Low Frequency Phenomena Associated with Vessels Moored at Sea", Society of Petroleum Engineers, No.4837 (1974)

[4] J.E.W.Wichers, "A Simulation Model For A Single Point Moored Tanker", PhD Thesis, Delft University of Technology (1988)

[5] T.Nakajima, "On The Dynamic Analysis of Multi Component Mooring Lines", The 14th Annual Offshore Technology Conference (1987)

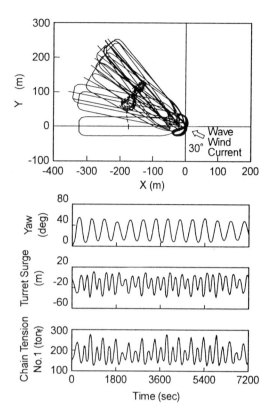

Fig. 8   Calculated Results on the Slowing Motion

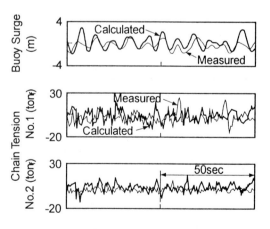

Fig.9   Time Traces for Responses in Survival Condition (CALM System)

© 1998 Elsevier Science B.V. All rights reserved.
Practical Design of Ships and Mobile Units
M.W.C. Oosterveld and S.G. Tan, editors.

Wave Drift Forces of a Very Large Flexible Floating Structure

H. Maeda[a], T. Ikoma[a] and K. Masuda[b]

[a] Institute of Industrial Science, University of Tokyo
 7-22-1 Roppongi, Minato-ku, Tokyo 106, Japan

[b] College of Science & Technology, Nihon University
 7-24-1 Narashinodai, Funabashi-shi 274, Chiba, Japan

The authors consider a pontoon type flexible floating airport as the very large floating structure. In order to design this kind of structure, we have to investigate several key points in advance. Among those key points, estimation of the wave drift forces is very important in order to design of the mooring system. In this paper, the authors consider the square type flexible flat plate with shallow draft, exactly speaking, zero draft as the floating airport. The authors develop the computer code of the slowly varying wave drift force for an elastic floating structure with shallow draft based on the so called Near Field Method which is validated by both the Far Field Method and the corresponding Experiments.

The authors conclude that even though the water depth is shallower, there is no big difference among the steady wave drift forces in each water depth, while the rigidity of the structure is smaller, more flexible, then, the steady wave drift forces or mooring forces are smaller.

Finally, they showed the flexibility of a floating structure decreases not only steady wave drift forces but also slowly varying wave drift forces under the numerical time domain simulation of a very large floating structure with the multiple dolphins mooring system.

1. INTRODUCTION

The mooring system is one of the most important items for designing a very large floating airport. Therefore, the design tool to analyze the wave drift force is necessary. However, the characteristics of the wave drift force on a very large floating airport that behaves elastically in ocean waves are known little. There have been little discussions of the slowly varying wave drift force on a very large floating structure, which oscillates around a long-period resonance.

There are two methods for estimation of wave drift forces. One of them is called the far field theory or Maruo's theory [6], which is based on the momentum theory. Another one is called the near field theory, which is a method such as integration of the wave pressure on the wetted structure surface. The near field theory developed first by Pinkster [7], and then Matsui [8] and Masuda [9] etc. have used its theory to estimate the slowly varying wave drift forces including the second order potential. However, even though those two methods have been often applied to structures that are rigid, they have been used little for estimation of the wave drift forces on the very large and elastic structure. The authors have showed how to apply the far field theory to a structure that behaved elastically in regular waves [4], while the slowly varying wave drift force on a elastic structure in irregular waves are investigated little. So, there are little discussion about the mooring system of a very large floating airport.

In this paper, the theoretical methods to analyze the wave drift force on very shallow draft floating structures, which have elastic motions in waves, are shown. The slowly varying wave drift forces in two-component waves are calculated by the near field theory. The present method that is the near field theory is validated by comparison with the experimental results and computational results by the far field theory. Furthermore, the effects due to the water depth and the elasticity of a structure to the drift forces are investigated. And, fender reaction forces of the dolphin mooring system on the elastic structure that have various bending stiffness are analyzed by time domain simulations, and are compared varying the bending stiffness.

2. THEORY

It is assumed that the fluid is perfect fluid. The analysis of the hydrodynamic forces and wave excitations is based on the pressure distribution method because the draft of a pontoon type floating structure is very small. The coordinate system is the right hand Cartesian and the z-axis is positive upward. In addition, velocity potentials Φ, pressures P and vertical displacement η of a free surface or the structure are expressed by perturbation expansion using a perturbation parameter ε. The time-dependent terms are decomposed into the regular exponential terms, furthermore, the following equations can be available:

$$\varepsilon \Phi^{(1)} = \mathrm{Re}\left[\sum_{i=1}^{N} i\omega_i a_i \phi_i^{(1)} e^{-i\omega t} \right] \tag{1}$$

$$\varepsilon P^{(1)} = \mathrm{Re}\left[\sum_{i=1}^{N} -\rho g a_i p_i^{(1)} e^{-i\omega t} \right] \tag{2}$$

$$\varepsilon \eta^{(1)} = \mathrm{Re}\left[\sum_{i=1}^{N} a_i \eta_i^{(1)} e^{-i\omega t} \right] \tag{3}$$

$$\varepsilon^2 \Phi^{(2)} = \mathrm{Re}\left[\sum_{i=1}^{N} \sum_{j=1}^{N} \left(i\omega^+ a_i a_j \phi_{ij}^+ e^{-i\omega^+ t} \right. \right.$$
$$\left. \left. + i\omega^- a_i a_j^* \phi_{ij}^- e^{-i\omega^- t} \right) \right] \tag{4}$$

$$\varepsilon^2 P^{(2)} = \mathrm{Re}\left[\sum_{i=1}^{N} \sum_{j=1}^{N} \left(-\rho g a_i a_j p_{ij}^+ e^{-i\omega^+ t} \right. \right.$$
$$\left. \left. -\rho g a_i a_j^* p_{ij}^- e^{-i\omega^- t} \right) \right] \tag{5}$$

$$\varepsilon^2 \eta^{(2)} = \mathrm{Re}\left[\sum_{i=1}^{N} \sum_{j=1}^{N} \left(a_i a_j \eta_{ij}^+ e^{-i\omega^+ t} \right. \right.$$
$$\left. \left. + a_i a_j^* \eta_{ij}^- e^{-i\omega^- t} \right) \right] \tag{6}$$

where, subscripts i and j mean the component wave, but i that is not subscript means an imaginary unit. Superscripts $+$ represents the second order sum-frequency component, $-$ the second order difference-frequency component and $*$ a conjugate complex. a_i is a complex amplitude of an incident wave, ρ is a fluid density, g is the acceleration of gravity, and ω_i is a circular frequency. $\omega^+ = \omega_i + \omega_j$ and $\omega^- = \omega_i - \omega_j$. In case of the pressure distribution method, a first-order velocity potential $\phi_{ir}^{(1)}$ of r-th radiation modes or diffraction mode is obtained as follows:

$$\phi_{ir}^{(1)}(x,y) = -\iint_{S_H} p_{ir}^{(1)}(x',y') \cdot G \, dx' dy', \tag{7}$$

where, G is the Green's function [1] and S_H means an area of a floating body wetted surface.

2.1. Far Field Theory

The steady wave drift forces in regular waves are given by the momentum theory, i.e. the far field theory as follows.

In this paper, H_{li} is defined as the Kochin function and can be expressed as follows:

$$H_{li}(k_i, \alpha_i) = \iint_{S_H} p_{li}^{(1)}(x', y') e^{-ik_i(x'\cos\alpha_i + y'\sin\alpha_i)} dS_H . \tag{8}$$

where, α is a wave direction angle, K is a wave number in deep water and k is a wave number in shallow water. Using this Kochin function, steady wave drift forces of surge and sway modes in regular waves are given as follows:

$$F_{xii}^- = \rho g a_i^2 \frac{K_i \cdot k_i}{4\pi} \bar{k}_i \int_0^{2\pi} |A(k_i, \theta)|^2 (\cos\alpha_i - \cos\theta) d\theta, \tag{9}$$

$$F_{yii}^- = \rho g a_i^2 \frac{K_i \cdot k_i}{4\pi} \bar{k}_i \int_0^{2\pi} |A(k_i, \theta)|^2 (\sin\alpha_i - \sin\theta) d\theta. \tag{10}$$

Where,

$$\bar{k} = \begin{cases} \dfrac{K}{2} & \text{(in deep water)} \\ \dfrac{h\cosh^2 kh}{2kh + 2\sinh kh} & \text{(in shallow water)} \end{cases},$$

and, A is the Kochin function.

$$A(k_i, \theta) = H_D(k_i, \theta) + \sum_{r=1}^{M} q_{ri}^{(1)} H_r(k_i, \theta). \tag{11}$$

Here, $q_{ri}^{(1)}$ is the first-order principal coordinate of r-th rigid and elastic mode. H_D is the Kochin function corresponds to the diffraction wave. Considering the second term in the right side of eq.(11), wave drift forces can be calculated even though the floating structure behaves an elastic motion.

2.2. Near Field Theory

Next, the authors show the theory of the slowly varying wave drift force. The far field theory is the effective method to estimate the wave drift forces in regular waves. However, this method is not suitable for the analysis of the slowly varying wave drift forces in irregular waves.

In this paper, the pressure integral method, i.e. the near field theory is applied to calculate the slowly varying wave drift forces in two-component waves, i.e.; second-order terms of the wave exciting forces are integrated on the body surface. When the shallow draft theory is applied, there is no side wall of the structure to integrate the pressure, while the wave run-up term and second-order terms due to the product of the first-order elastic motions must be considered. In this section, the authors show the theoretical method based on the near field theory for the wave drift forces on the structure with a shallow draft which oscillates elastically.

The following definition of the wave excitation $\vec{F}$ and a normal vector $\vec{N}$ up to the second-order level can be established by a perturbation expansion:

$$\vec{F} = \vec{F}^{(0)} + \varepsilon\vec{F}^{(1)} + \varepsilon^2\vec{F}^{(2)} + O(\varepsilon^3), \quad (12)$$

$$\vec{N} = \vec{N}^{(0)} + \varepsilon\vec{N}^{(1)} + \varepsilon^2\vec{N}^{(2)} + O(\varepsilon^3). \quad (13)$$

The second-order wave drift force is obtained as

$$\vec{F}^{(2)} = \iint_{S_H} \left(P^{(0)}\cdot\vec{N}^{(2)} + P^{(1)}\cdot\vec{N}^{(1)} + P^{(2)}\cdot\vec{N}^{(0)}\right)dS_H, \quad (14)$$
$$+ \int_C\int_{\zeta^{(1)}}^{\eta^{(1)}} P^{(0)}\cdot\vec{N}^{(0)}dz^{(1)}dC$$

where, $\eta^{(1)}$ is the first-order displacement of water surface, and $\zeta^{(1)}$ is the first-order vertical displacement of the floating body at side end. And, an integral along C means a line integral. Here, normal vectors and the second-order wave excitation are defined as follows, respectively, by decomposing the time-dependent terms:

$$\varepsilon^2\vec{F}^{(2)} = \mathrm{Re}\left[\sum_{i=1}^{N}\sum_{j=1}^{N}\left\{\vec{F}_{ij}^{+}e^{-i\omega^+t} + \vec{F}_{ij}^{-}e^{-i\omega^-t}\right\}\right], \quad (15)$$

$$\varepsilon\vec{N}^{(1)} = \mathrm{Re}\left[\sum_{i=1}^{N} a_i\vec{N}_i^{(1)}e^{-i\omega_i t}\right], \quad (16)$$

$$\varepsilon^2\vec{N}^{(2)} = \mathrm{Re}\left[\sum_{i=1}^{N}\sum_{j=1}^{N}\left\{a_ia_j\vec{N}_{ij}^{+}e^{-i\omega^+t} + a_ia_j^*\vec{N}_{ij}^{-}e^{-i\omega^-t}\right\}\right]. \quad (17)$$

The zero-th order pressure is zero because of the shallow draft assumption. A direction cosine of x and y component is appeared because the floating body is deformed elastically. So, the normal vector term of the first-order must be considered. After all, eq.(14) can be changed to the following equation:

$$\vec{F}^{(2)} = \iint_{S_H} P^{(1)}\cdot\vec{N}^{(1)}dS_H + \int_C\int_{\zeta^{(1)}}^{\eta^{(1)}} P^{(1)}\cdot\vec{N}^{(0)}dz^{(1)}dC. \quad (18)$$

Moreover, the first-order pressure is given by

$$p_i^{(1)} = p_{ei}^{(1)} + \sum_{r=1}^{M} q_{ri}^{(1)}p_{r\omega_i}, \quad (19)$$

where, $p_{ei}^{(1)}$ is the sum of the first-order incident and diffraction wave pressure, and $p_{r\omega i}$ is the radiation wave pressure which is a linear pressure related to an arbitrary angular frequency ω_i.

On the wave run-up term which is the second term in eq.(18) can be rewritten using the velocity potential as follows:

$$\int_C\int_{\zeta^{(1)}}^{\eta^{(1)}} P^{(1)}\cdot\vec{N}^{(0)}dz^{(1)}dC$$
$$= \int_C\int_{\zeta^{(1)}}^{\eta^{(1)}}\left(-\rho\Phi_t^{(1)} - \rho g z^{(1)}\right)\vec{N}^{(0)}dz^{(1)}dC \quad (20)$$

After all, eq.(20) is changed to

$$\int_C\int_{\zeta^{(1)}}^{\eta^{(1)}} P^{(1)}\cdot\vec{N}^{(0)}dz^{(1)}dC = \frac{\rho g}{2}\int_C\xi^{(1)^2}\vec{N}^{(0)}dC, \quad (21)$$

where,

$$\xi^{(1)} = \eta^{(1)} - \zeta^{(1)}. \quad (22)$$

The vertical displacement of the free surface $\eta_i^{(1)}$ and the vertical displacement of the floating body end $\zeta_i^{(1)}$ are given as follows:

$$\eta_i^{(1)} = \eta_{Di}^{(1)} + \eta_{Ii}^{(1)} + \sum_{r=i}^{M} q_{ri}^{(1)}\eta_{Ri}^{(1)}, \quad (23)$$

$$\zeta_i^{(1)} = \sum_{r=i}^{M} q_{ri}^{(1)}\eta_{ri}^{(1)}, \quad (24)$$

where, $\eta_{ri}^{(1)}$ is the principal mode functions of the r-th motion mode, furthermore,

$$\left.\begin{array}{l}\eta_{Ii(x,y)}^{(1)} = e^{-ik_i(x\cos\alpha_i + y\sin\alpha_i)}\\[4pt]\eta_{Di(x,y)}^{(1)} = K_i\iint_{S_H} p_{ei}^{(1)}(x',y')\cdot G(x,y,0,x',y',0)dS_H\\[4pt]\eta_{Ri(x,y)}^{(1)} = K_i\iint_{S_H} p_{ri}^{(1)}(x',y')\cdot G(x,y,0,x',y',0)dS_H\end{array}\right\} \quad (25)$$

In eq.(25), subscripts I means incident wave component, D stands for diffraction component, and R stands for radiation component of r-th mode motion. Here, $\zeta_i^{(1)}$ only has to be calculated at the same coordinate as $\eta_i^{(1)}$.

Thus, the wave drift force is given as:

$$\vec{F}_{ij}^{-} = \frac{1}{2}\left(\vec{f}_{ij}^{-} + \vec{f}_{ji}^{-*}\right), \quad (26)$$

where,

$$\vec{f}_{ij}^{-} = -\frac{\rho g a_i a_j^*}{2}\iint_{S_H} p_i^{(1)}\cdot\vec{N}_i^{(1)*}dS_H - \frac{\rho g a_i a_j^*}{4}\int_C\xi_i^{(1)}\xi_j^{(1)*}dC. \quad (27)$$

The wave drift force calculated by eqs.(26), (27) is the case of a regular wave condition if subscript i equals to j, and only a real part exists. If i is not equal to j, the slowly varying wave drift force in a two component wave is calculated by eqs.(26) and (27). In this paper, the drift force of the surge mode is expressed as F_{xij}^{-}, and that of the sway mode F_{yij}^{-}.

2.3. Validation of the Near Field Theory

The authors validated the above theoretical method by the comparison of computation results and experimental results. The experiments were carried out at water depth of 2.7m and a wave height of incident waves is 3.0cm in the model basin of the University of Tokyo. The wave direction angle was 0 degree, i.e. the beam sea condition. The experimental model was an elastic plate made of polyurethane foam (Young's modulus; E=1.5 × 10^7kgf/m^2). As the dimensions of the model, length is 3.97m, width is 0.97m, thickness is 4cm, and the draft is about 3mm. The comparison of the near field theory and the far field theory, and the comparison with the corresponding experimental results are shown in Fig.1, where L is length of the model, B width of the model and λ_i the incident wavelength. The theoretical method for the elastic motions was based on the shallow draft assumption [10]. In Fig.2, the two results of the steady wave drift force on the model calculated by the near and far field theory are compared.

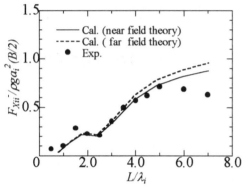

Fig.1 Comparison of experimental and computational results of steady wave drift forces of surge mode in head sea condition

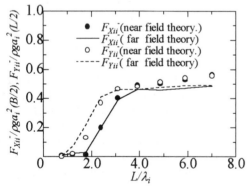

Fig.2 Comparison of far and near field computational results of steady wave drift forces of surge and sway mode in oblique wave condition (30 degrees)

The calculated value agrees well with the experimental results, and there is no big difference between the results of the near field theory and far fields theory. The authors think that a difference

between the both computational results is caused by accuracy at high frequency. It is confirmed that the near field theory and the far field theory are useful.

3. RESULTS AND DISCUSSION

3.1. Effects of water depth

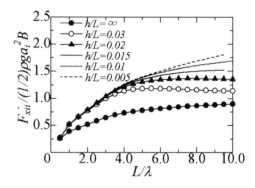

Fig. 3 Wave drift forces of surge mode on fixed model in various water depth (α=0deg.)

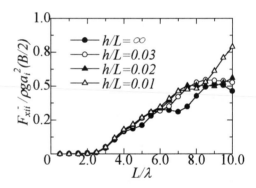

Fig.4 Wave drift forces on floating model in various water depth (EI/B^5=0.618, W/LB^2=20.0, α=0deg.)

In this section, the authors investigate the effects of the water depth on the steady wave drift force in the head sea regular waves by numerical calculations using the far field theory. The generic model has L/B=4. Figure 3 shows the results of the wave drift forces on the fixed model, i.e., the model does not oscillate, in head sea waves. The wave drift forces are larger in the cases of the shallower water depth. Figure 4 shows the steady wave drift forces when the model has elastic motions. We can know easily that the effects of water depth on the steady wave drift force are very small on the contrary of the fixed structure case.

3.2. Effects of bending stiffness

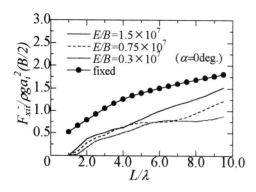

Fig.5 Wave drift forces of surge mode with various bending stiffness (h/L=0.005)

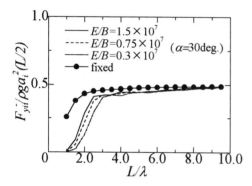

Fig.6 Wave drift forces of sway mode with various bending stiffness (h/L=0.005)

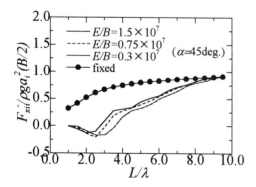

Fig.7 Wave drift forces of surge mode with various bending stiffness (h/L=0.005)

It can be supposed that the bending stiffness has a linear effect on the characteristics of the wave drift force. Then, the authors illustrated the calculation results of the steady wave drift forces. Figures 5 shows the wave drift forces in head sea regular

waves, in the water depth h/λ=0.005 with various bending stiffness. These results show the tendency such as the wave drift forces are made small by the smaller bending stiffness. Figures 6~8 show the wave drift forces of surge and sway mode in regular waves that have the wave direction angle of 30 degrees and 45 degrees. The same tendency as in the head sea condition was obtained.

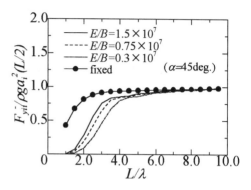

Fig.8 Wave drift forces of sway mode with various bending stiffness (h/L=0.005)

3.3. Time domain simulations

The authors investigated the difference between the mooring force on elastic structures and that of rigid structures in irregular waves. The time domain simulation was done which considered the slowly varying wave drift forces and the nonlinearity of the mooring system. The radiation forces are expressed by the convolution integral based on the memory effect function. The effects of the elasticity are included to the second-order wave excitations, which are the slowly varying wave drift forces.

The authors used the following calculation model. The length is 1000.0m, the width is 200.0m and the draft is 3.0m. However, the shallow draft assumption was applied to the computation of the hydrodynamic forces. Water depth is 30.0m, and the fender height of the dolphins is 3.0m. And, Young's modulus is used two cases: hard model type is 4.32 $\times$ 10^8[kgf/m^2] and soft model type is 4.32 $\times$ 10^7[kgf/m^2].

Figure 9 shows the coordinate system and the dolphin mooring system. The condition of the incident wave is as follows; the significant wave height is 4.0m and the mean wave period is 8.0 second. And, the wave direction angle is 30 degrees. The numerical simulation is calculated 100 times for each bending stiffness condition.

Examples of the simulation results of the fender reaction forces are illustrated in Fig.10. We can find that the fender reaction force is smaller if the bending stuffiness is smaller. The average of extreme, the maximum value, 1/3 significant value $F_{1/3}$ and 1/10 significant value $F_{1/10}$ of the fender reaction force on the No.10 fender are shown in Table 1. When the results of the elastic model are compared with those of the rigid model, we can find that the fender reaction forces of the elastic model are clearly 10%~20% smaller than the results of the rigid model.

The authors illustrate the probability of exceedance on the fender deformation to understand the effect of elasticity in Fig.11. Solid lines mean a fitting line by Weibull distribution in Fig.11. We find that we have to consider the elasticity of the very large floating structure when we estimate the expectation of the fender deformation according to Fig.11.

Table 1 Results of statistical analysis on fender reaction force of No.10 fender

	Rigid model	Elastic model	
		$E=4.32\times10^{8}$	$E=4.32\times10^{7}$
Average(m)	2.601×10^{5}	2.332×10^{5}	1.655×10^{5}
Max.(m)	5.552×10^{5}	5.012×10^{5}	4.682×10^{5}
$F_{1/10}$ (m)	5.155×10^{5}	4.588×10^{5}	3.768×10^{5}
$F_{1/3}$ (m)	4.379×10^{5}	3.943×10^{5}	2.899×10^{5}

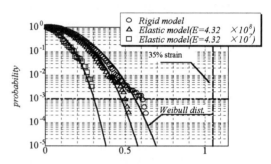

Fig.11 Exceedance probability of No.10 fender deformation

4. CONCLUSION

In this paper, the authors show the analytical method for the estimation of the wave drift force on a very large flexible-floating airport. Then, the characteristics of the wave drift force in various water depths and with various bending stiffness are investigated. Furthermore, the authors show the time domain simulation results of the fender reaction forces of a dolphin mooring system. After all, the authors obtained the following conclusions.

1) The present near field theory to analyze the slowly varying wave drift force on a very large and shallow draft floating structure is shown. In addition, the usefulness of this theory is confirmed.
2) In case of a fixed structure, the steady wave drift force becomes bigger if a water depth gets shallower. However, in case of an elastic structure, the effects due to a water depth on the wave drift force is smaller.
3) If the bending stiffness is made smaller, a magnitude of the wave drift forces can be reduced.
4) The mooring force in irregular waves on a very large floating structure that has an elastic motion is 10~20% smaller than that of a rigid

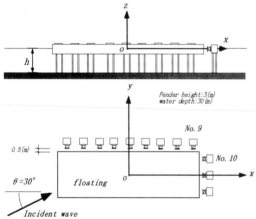

Fig.9 Coordinate system and dolphin mooring system

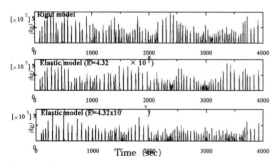

Fig.10 Time series of fender reaction forces on No.10 fender with various bending stiffness

floating structure. If the structure has a smaller bending stiffness, its tendency becomes more remarkable.

5) We must consider the effects of the elasticity on the estimation of the wave drift force of a very large floating airport.

REFERENCE

1. Yamashita, S.," Motions and Hydrodynamic Pressures of a Box-shaped Floating Structure of Shallow Draft in Regular Waves –Comparisons between Calculations by the Use of Pulsating Pressure Distributions and experiments–," Journal of the Society of Naval Architects of Japan, Vol.146, (1979), (in Japanese)
2. Maeda, H., Masuda, K., Miyajima, S. and Ikoma, T., "Hydroelastic Responses of Pontoon Type Very Large Floating Offshore Structure," Proceeding of the 15th International Conference on OMAE, Vol.I, ASME, pp.407-414,(1996)
3. Ikoma, T., Masuda, K., Maeda, H., Miyajima, S. and Shimizu, K., "The Estimations of Hydrodynamic Force on A Very Large Floating Body by The Pressure Distribution Method," TECHNO-OCEAN '96 INTERNATIONAL SYMPOSIUM, PROCEEDINGS Vol.II, pp.639-644,(1996), (in Japanese)
4. Maeda, H., Masuda, K., Miyajima, S. and Ikoma, T., "Hydroelastic Responses of Pontoon Type Very Large Floating Offshore Structures (The 2nd report, The effect of the water depth and the drift forces)," Journal of the Society of Naval Architects of Japan, Vol.180, pp.365-371, (1996), (in Japanese)
5. Kashiwagi, M., "A Calculation Method of Hydrodynamic Forces on a Shallow-Draft and Very Large Floating Structure," TECHNO-OCEAN '96 INTERNATIONAL SYMPOSIUM, PROCEEDINGS Vol.II, pp.627-639, (1996)
6. Maruo, H., "The drift of a body floating in waves," Journal of Ship Research, Vol.4, No.3, pp.1-10,(1960)
7. Pinkster, J.A., "Low-frequency second order wave exciting force on floating structures," Netherlands Ship Model Basin, Publication No.650, (1980)
8. Matsui, T., "Computation of slowly varying second order hydrodynamic forces on floating structures in irregular waves," Proceedings of the 7th International Conference on OMAE, ASME, (1988)
9. Masuda, K. and Osawa, H., "PREDICTION OF MOTIONS AND TENSION IN MOORING LINES ON SLACK-MOORED FLOATING OCEANIC ARCHITENTURAL BUILDING I COASTAL ZONES," Proceedings of the 14th International Conference on OMAE, Vol.I-A, ASME, pp483-489,(1995)
10. Nagata, S., Yoshida, H., Fujita, T. and Isshiki, H., "THE ANALYSIS OF THE WAVE-INDUCED RESPONSES OF AN ELASTIC FLOATING PLATE," Proceeding of the 16th international conference on OMAE, Vol.VI, ASME, pp.163-178, (1997)
11. Kashiwagi, M. and Furukawa, C., "A MODE-EXPANSION METHOD FOR PREDICTING HYDROELASTIC BEHAVIOE OF A SHALLOW-DRAFT VLFS," Proceeding of the 16th international conference on OMAE, Vol.VI, ASME, pp.179-186, (1997)

1998 Elsevier Science B.V.
Practical Design of Ships and Mobile Units
M.W.C. Oosterveld and S.G. Tan, editors.

Numerical and experimental study on attitude control of a large floating offshore structure by pneumatic actuator

Tsugukiyo Hirayama[a], Ning Ma[a] and Yasuhiro Saito[b]

[a]Department of Naval Architecture and Ocean Engineering, Yokohama National University
Tokiwadai 79-5, Hodogaya-ku, Yokohama, Japan

[b]Kawasaki Heavy Industries Co. Ltd.*
1-1 Higashikawasaki-cho 3, Chuo-ku, Kobe, Japan

An attitude control system using pneumatic actuators for very large floating structure (VLFS) was developed and verifications were conducted by applying it to scale model test. The preliminary investigations into the characteristics of actuator, static recovery of attitude and active attitude control in waves were done numerically and experimentally. Numerical simulations on hydroelastic response were carried out based on hydroelasticity theory by simplifying the model as an elastic beam and the results were compared with experiments. This paper introduces the results of model tests and numerical simulations. Based on the results, the applicability of the proposed pneumatic type attitude control for VLFS has been discussed. The influence of bending rigidity, control method, actuator arrangement, etc. on response and its control are investigated.

1.INTRODUCTION

The environmental loads on and response of very large floating structure (VLFS) have attracted many interests of research recently. As it is well known, VLFS exhibits significant elastic deformations in waves comparing to its rigid motions. It could behave like a beam in head sea or a plate in quartering sea. These deflections could affect unexpectedly the operability and safety of structure, especially when usie it as an airport. Therefore it becomes necessary to suppress these deflections from view point of practical use. However, up to now, there are very few studies concerning the active attitude control of VLFS. For the aim of long durability, we have proposed a semi-submersible type VLFS which consists of multiple removable units for floating airport [1,2]. Hence it is essential to shorten the downtime due to the unit replacement and to ensure the safety. In order to meet with the requirement of the limitation of attitude fluctuation of floating airport, enhancement of bending rigidity or

hydrodynamic damping will be the practical answer. But on the other hand, the possibility of attitude control should be studied alternatively.

As for the active control of the motion of offshore structure, some recent studies can be found and categorized to mechanical type [3] and pneumatic type [4,5]. Air pressure utilization brings many advantages, such as very few pollution effects upon the environment because of its cleanness and relatively low cost to build and to maintain the apparatus. Here we considered the introduction of pneumatic actuator and carried out investigations in model scale firstly. Hirayama et al [6] have developed a pneumatic control device for floating structure and applied it to a VLFS recently [7]. In this system, "air columns" (bottom open tanks) are attached to the bottom of column footings and air pressure inside are controlled.

The principal objective of this paper is to present the new results of investigations on the characteristics of actuator, the effectiveness of passive and active control. To

---

*Formerly Graduate School of Yokohama National University

suppress vertical motions and deflections of a semi-submersible type VLFS, the actuators were installed to column footings uniformly with special considerations of its elastic deformation modes. The air pressure inside air columns was measured simultaneously. Numerical investigations into hydroelastic response are also made and the results were compared with the experiments in regular waves of various periods. In the simulations, two types of control, i.e. PD control and optimal control (multiple variables control) were adopted. The influences of control method, number and location of actuator are studied through simulations by changing parameters. Further investigations into the adaptability with respect to wave frequency of the experimental apparatus were performed considering the compressibility of air, response characteristics of air valves. Finally, the applicability of proposed pneumatic type active control are discussed.

2.NUMERICAL SIMULATIONS

The prediction of the response of a VLFS requires a special analysis technique, because its flexible deformation modes exist in addition to rigid motions. Hence, the hydrodynamic force and elastic deformation must be determined simultaneously. We have developed a numerical method based on the linear hydroelasticity theory by using the mode superposition method [2,8]. The hydrodynamic forces are evaluated in three dimensions by using sink source method and the structural analysis is executed by using modal analysis based on FEM. From the comparison in Table 1, the analyzed natural periods show very good coincidences with experimental ones.

In this study, the model was simply represented as an elastic uniform beam supported by an elastic foundation as shown in Figure 1 schematically. A numerical method based on mode superposition for time domain analysis was developed. Assuming linear elastic deformation, the motion equation of beam can be expressed as follows.

$$\rho A \ddot{z} + \frac{\eta EI}{\omega}\frac{\partial^4 \dot{z}}{\partial x^4} + EI\frac{\partial^4 z}{\partial x^4} + kz = q(x,t) \quad (1)$$

where ρ is the mass density, A is the

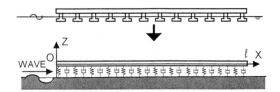

Figure 1. Beam model for semi submersible type VLFS and coordinate system.

sectional area, η is the structural damping coefficient, EI is the bending rigidity, k is the spring coefficient of foundation and q(x,t) represents external load acting on beam. The overdots denote differentiation with respect to time.

According to the principle of mode super-position, the deflection z can be represented as eq. (2).

$$z(x,t) = \sum_{r=1}^{\infty} Z_r(x) \cdot p_r(t) \quad (2)$$

where Z_r is the natural mode, p_r is the principal coordinate of mode. As for the dry mode of beam, following analytical mode function for free ends condition are used.

$$Z_r(x) = \{\cos K_r x + \cosh K_r x - \\ \alpha_r(\sin K_r x + \sinh K_r x)\} \quad (3)$$

$$\alpha_r = (\cosh K_r l - \cos K_r l)/(\sinh K_r l - \sin K_r l) \quad (4)$$

in which l is the beam length and K_r is the eigenvalue of r-th mode and is expressed as follows.

$$K_r = \begin{cases} 4.73/l & (r=1) \\ (2r+1)\pi/(2l) & (r=2,3,4...) \end{cases} \quad (5)$$

By substituting eq.(3) into eq.(1), and applying the orthogonality of principal modes,

Table 1. Principal dimensions & natural periods

	Model(1/256)	Prototype
Length(m)	7.1	1818
Breadth(m)	1.7	435
Depth(m)	0.36	92
Draft(m)	0.115	29
Displacement(ton)	0.267	$4.49*10^6$
KG(m)	0.27	68
EI(Nm²)	992	$1.1*10^{15}$
Natural periods	exp.(cal.)	cal.
Heave(sec)	1.72(1.54)	24.6
Pitch(sec)	1.84(1.76)	28.2
Roll(sec)	1.62(1.68)	26.9
1st Bending(sec)	1.18(1.16)	18.6
2nd Bending(sec)	0.61(0.59)	9.4
3rd Bending(sec)	(0.32(0.32)	5.1

following equation of motion with respect to principal coordinate is obtained.

$$M_r \ddot{p}_r(t) + C_r \dot{p}_r(t) + K_r p_r(t) = Q_r(t) \qquad (6)$$

where M_r, C_r, K_r are the generalized mass, damping and stiffness of mode respectively. $Q_r(t)$ is generalized external force.

For a pneumatic controlled structure floating in waves, the effect of pressured air trapped in air column could be take into accounts by considering its pneumatic stiffness and varying pressure forces [5], therefore the motion equation can be expressed as follows.

$$(M_r + a)\ddot{p}_r + (C_r + c)\dot{p}_r + (K_r + k_s + k_p)p_r = Q_r(t) \qquad (7)$$

$$Q_r(t) = \sum_{n=1}^{N} \left(F_n^w(t) + F_n^p(t) \right) \cdot Z_r(x_n) \qquad (8)$$

where a and c are the added mass and wave making damping of mode, k_s and k_p represent the hydrostatic and pneumatic stiffness respectively. The time varying pneumatic stiffness is ignored here since it is of small order. F_n^w is the wave exciting force and F_n^p is the pneumatic pressure force (control force) with different phase to wave exciting force. N is the number of columns. The pneumatic stiffness (k_i) is a function of mean mass(m_i) and height(h_i) of trapped air in air column and can be determined by following equation.

$$k_i = \frac{nRTm_i}{h_i} \qquad (9)$$

where R is the air constant and T is the absolute temperature, n is index of expansion.

In the calculations, added mass and wave damping evaluated from 3-dimensional hydroelasticity analysis were used at certain fixed frequency, i.e. 1st bending mode.

Number of modes to be aggregated is taken up to 5. Equations (7), (8) are solved numerically using Newmark-β method in time domain.

Two control methods, PD control and optimal control were adopted for simulations, but for experiments, only PD control was adopted. The feed back gains for PD-control are obtained by trial and error and optimization of gains for optimal control are determined analytically [9].

In PD-control, the flow rate of air mass ($\dot{m}_i$) associated with i-th air column has been controlled. Generally, $\dot{m}_i$ is given by the equation of the state vector X.

$$\dot{m}_i = f(\mathbf{X}, \dot{\mathbf{X}}) \qquad (10)$$

$$\mathbf{X} = \left\{ \mathbf{Z}^T, \mathbf{m}^T \right\} \qquad (11)$$

where Z is the displacement vector, m is the vector of air mass containing in air column. Assuming linear control, the equation of state can be rewritten as,

$$\dot{m}_i = \mathbf{g}_{Pi}^T \mathbf{X}(t) + \mathbf{g}_{Di}^T \dot{\mathbf{X}}(t) \qquad (12)$$

where g_{Pi}, g_{Di} are gain vectors of P-D control associated with i-th air column, which are assumed to be constant and were determined through preparatory tests [9].

3. MODEL EXPERIMENTS

Experiments using semi-submersible type large model were carried out in the towing tank of Yokohama National University. The model is supposed as a medium scale airport which consists of 36 removable floating units. Each unit is compromised of 4 column footings arrayed squarely. For the sake of simplicity, the elasticity of structure is modeled by connecting the units with elastic beams transversely and longitudinally at deck.

The model and measurements of deflections and bending moments are shown in Figure 2. Hence, Zn represents the vertical displacement of point-'n'. The deflections were detected by

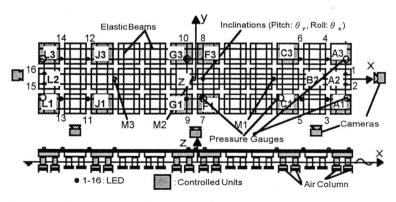

Figure 2. Experimental model and set-up.

using a non-touch type optical sensing system as shown in the figure as LED and Cameras. The model were pneumatically controlled at 12 units by controlling air mass trapped inside the underwater air column (bottom open air tank). The pressure gauges were mounted to topside of air column to measure the variation of air pressure. A linear spring mooring system was adopted, which prevent only the horizontal motions. The principal dimensions of model, measured and calculated natural periods are shown in Table 1. The calculated natural periods are obtained from wet mode analysis [8].

Two air compressors (max. initial pressure: 9.9 kgf/cm², capacity: 65 liters)were used to provide pressured air flow to fore and rear part of model respectively. Four air columns of one unit are connected by tubes, so that one unit is controlled individually. The description of control of one unit is shown in Figure 3. It is desirable to control the mass of air flow into and from the air column directly, however it not realizable due to the simplicity of experimental apparatus. Therefore in this system, the on/off of electromagnetic valves (open time of valve) between compressor and air columns are intended to control. This was achieved by a set of electromagnetic valves and a hand made piston, so the air masses inside air columns are controlled consequently and results in an additional buoyancy of supporting column. The control provides three conditions to each unit named "in", "hold", "out", where "hold" means there is neither flow-in nor flow-out from air column. The calculation of control command is performed on a personal computer based on traditional P-D control algorithm. Applying this control to all units individually, the different vertical control forces at units can be

generated, thus the motions of different modes can be controlled.

Air mass flow has strong non-linearity [6] with respect to the valve open length and results in poor adaptability of control especially at initial stage of flow-in. To overcome this, we chose initial pressure and the control gains carefully through a number of tests. Prior to control experiments in waves, the frequency characteristics of actuator were investigated through forced oscillations. Then two kinds of experiments, static attitude recovery and dynamic control in waves, were conducted. The later was mainly carried out in long regular waves.

4. RESULTS AND DISCUSSIONS

In order to achieve an effective control, it is essential to understand the characteristics of wave-induced motion and deflection of VLFS. The response functions of deflections in Z-direction at Z2 (fore) and Z7 (center) are shown in Figure 4. The predictions were done by the method aforementioned, and the experimental results were obtained from tests in transient water waves. The calculations show fairly good agreements with measured values including resonance at natural frequencies of elastic deformation modes. However, the response amplitudes and the phases due to long waves are much simpler than those for short waves, this implies that the controls at these frequencies is relatively easier to achieve.

In the following sections, the results of verifications of actuator characteristics, passive and active control are shown. Since it is impractical to scale atmospheric pressure, the tests were done within available

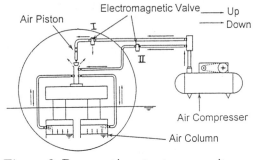

Figure 3. Pneumatic actuator on unit.

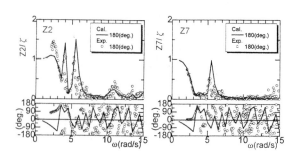

Figure 4. Transfer functions of deflection.

apparatus. All data and dimensions are in model scale units.

4.1. Characterics of Pneumatic Actuator

To obtain the frequency characteristics of control system, forced oscillation tests in still water were carried out. As the feedback signal, an external transient signal was made and inputted to controller. The signal for each air column can be different in phase with each other so that the desired oscillation mode can be easily obtained. As an example, the time histories of heave mode forced oscillation is shown in Figure 5. It can be recognized the valves repeat on/off actions which correspond to the external signal, vertical displacement at fore and mid show reasonable response due to the excited heave. Figure 6 shows the transfer functions of vertical displacement (z4, fore) and air pressure (PG7, fore). Air pressure, which corresponds to actuator force, was obtained satisfactorily at frequencies lower than 1 rad/sec, and becomes lower when frequency increases. The phase lag

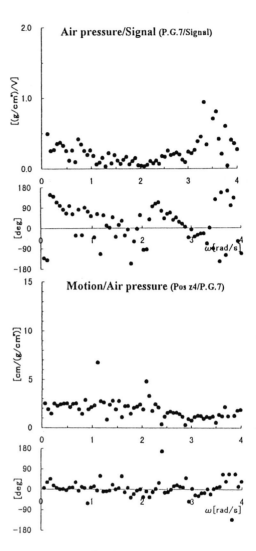

Figure 6. Transfer functions of motion and air pressure (frequency characteristics).

between motion and air pressure is considerably small at these frequencies. Accordingly it can be concluded that motion is able to be excited precisely at relatively low frequencies, if the air mass flow is sufficiently enough. The effect of compressibility of air on actuator force, i.e. the resultant phase lag seems to be negligibly small at tested frequencies. The forced oscillations of pitch and bending mode have also been carried out, the similar results have been obtained and we omitted them here.

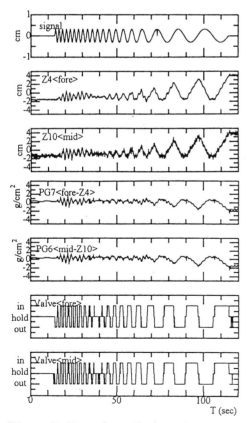

Figure 5. Forced oscillation of heave.

4.2. Effects of air cushion

Passive type suppression of motions and deflections was conducted by using the pre-pressured air trapped inside air columns. The air cushions were formed by blowing air to air columns and holding the valves. Figure 7 shows an example of measured deflections, air pressure for cushion and no cushion conditions in long regular waves. Although the incident waves for two experiments are slightly different, it can be observed that the deflections was suppressed with cushion and

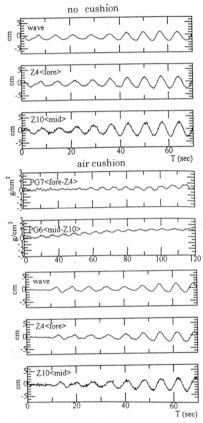

Figure 7. Experiments in long regular waves (T=7 sec) with and w/o air cushion.

air pressure is out of phase with heave (the time axes are different in figure). The simulated deflections at this wave frequency for no control, cushion and active control (designated as: 'No Ctrl', 'Cushn', 'Ctrl') are shown in Figure 8. 'Ctrl' represents the case where the actuators are supposed to be fully idealized. From the results, 18% reduction of deflections was achieved by cushion. Figure 9

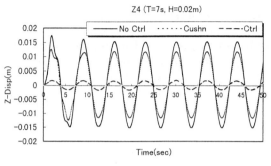

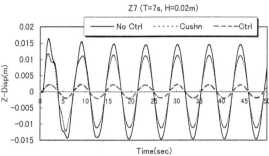

Figure 8. Simulated deflections in regular waves (T=7 sec).

shows the measured deflections (non-dimensioned by wave amplitude ζ_a) with and without air cushion for wave period T=7sec. The results are obtained from the harmonic analyses of measured time histories shown in Figure 7. At the wave frequency, deflection was reduced to 85% at bow by cushion.

The air cushion has two effects on motion suppression, i.e. its pneumatic stiffness and wave attenuation effect. Both of them are associated with air compressibility. Air stiffness exerts a force on the structure which is out of phase with the inertia force. Thus, vertical motions can be compensated

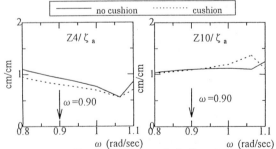

Figure 9. Suppression of deflection by air cushion (experimental) in regular waves.

consequently due to this force cancellation. The wave attenuation causes reduction of wave heights especially for short waves and reduces wave exciting force directly. Although the tests were limited to long waves and with partial air compartments, a considerable effect by using air cushion was obtained.

4.3. Effects of active control

Active controls at frequencies ranging from 4 to 7 rad/s where the structure has resonant deflections is desirable, however due to the shortage of available air flow in experiments, we abandoned to control at these frequencies. As expected, because of the compressibility of air, the system has a limited adaptability at high frequencies, thus we restricted investigations to relatively long waves.

Calculated deflections at bow of two structures, present structure and another one which bending rigidity was reduced to 1/10 (designated as 'EI/10') in regular waves (T=7sec) with and without control are shown in Figure 10. Simulations are all based on PD control, results of two different actuator forces, i.e. idealized actuator and actual

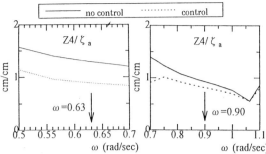

Figure 11. Suppression of deflection by control (exp.) in regular waves (T=7, 10 sec).

actuator force using measured air pressure including its phase (designated as 'Ideal' and 'Act.F') are compared. The reduction of deflection amplitude by actual actuator, referenced to ideal actuator, is approximately 55% and 75% respectively. Figure 11 shows the transfer functions by harmonic analyses of experiments in regular waves. Reduction of deflection amplitude, approximately 31% and 14% were obtained for wave period T=10 and 7 seconds respectively. It was observed that the dependency on frequency is remarkable, i.e. deflection due to longer wave is easier to control.

Further simulations were conducted with focusing on the influence of arrangement and number of actuators, method of control etc. An example of comparison of two controls assuming idealized actuators is shown in Figure 12. At the actuator points, both PD and optimal control gives good results and for the other points, the optimal control becomes superior. It is also found that PD control requires excess force than optimal control.

The relationship between motion

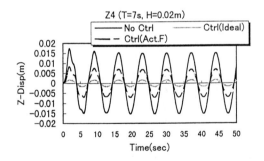

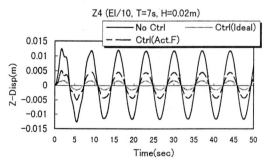

Figure 10. Simulated deflections in regular waves (T=7 sec) with and w/o controls.

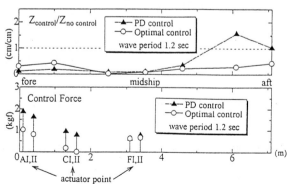

Figure 12. Deflection amplitude and control force of different controls.

1052

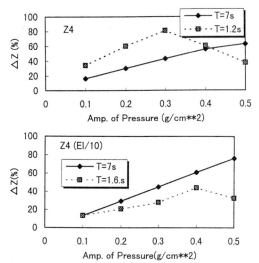

Figure 13. Ratio of decrement vs. amplitude of air pressure.

decrement ratio and varying air pressure amplitude for active control is calculated and shown in Figure 13. The shorter wave period corresponds to the 2-nodes bending mode natural frequency respectively. It is confirmed that the necessary actuator force is proportional to the motion decrement and the lower EI structure is easier to suppress for long wave period, but for short wave period this is not true, therefore further investigations on feedback gains and optimization are needed.

5. CONCLUSIONS

The proposed control system was originally conceived for the purpose of attitude recovery supposing removing of unit in replacement condition for a VLFS, but the possibilities of active motion suppression in waves, especially for long waves are derived.

Through the examinations by experiments and numerical simulations for long regular waves, the usefulness and effectiveness of the proposed system has been confirmed. The potential of the system exists considerably and the applicability to shorter waves can be enhanced through further improvements of apparatus and control algorithm. The application of the proposed system to a prototype VLFS will be realizable with special considerations of scale effect of air.

ACKNOWLEDGMENTS

The authors wish to thank the staffs and students of our laboratory for their assists in experiments. Particularly we appreciate K. Miyakawa and T. Takayama for their efforts in preparation and manufacture of the apparatus for experiment.

REFERENCES

1. Hirayama, T. and Ma, N. : Response Characteristics of a Long Life Type Floating Airport in Waves (1st report), *Journal of SNAJ*, Vol. 177, (1995).
2. Hirayama, T., Ma, N. and Nishio, F.O. : Response Characteristics of a Long Life Type Floating Airport in Waves (2nd report), *Journal of SNAJ*, Vol. 178, (1995).
3. Yamashita, S. : An Experimental Investigation of Active Control of Floating Body, *11th Ocean Engng. Symp. of SNAJ*, (1992).
4. Numata, T., Kudo, K. and Sao, K. : Study on Motion Suppression System for Marine Structure, *J. of SNAJ*, Vol. 157, (1985).
5. Patel, M.H. and Witz, J.A. : On Active Control of Marine Vehicles with Pneumatic Compliances, *Intl. Shipbuilding Progress*, Vol. 34, No. 395, (1987).
6. Hirayama, T., Takayama, T., Miyakawa, K. and Ma, N. : Development of Active Attitude Control System for Floating Structure by Using Air Pressure, *J. of Kansai SNAJ*, No. 226, (1996).
7. Hirayama, T. and Ma, N. : Dynamic Response of a Very Large Floating Structure with Active Pneumatic Control, *7th Intl. Offshore & Polar Engng. Conf.*, Vol. 1, (1997).
8. Ma, N. and Hirayama, T. : Hydroelastic Response of Two Types of Very Large Floating Structures, *16th Intl. Conf. Offshore Mechanics & Arctic Engng.*, Vol. VI, (1997).
9. Hirayama, T., Ma, N. and Saito, Y. : Attitude Control of a Large Floating Structure by Pneumatic Actuator (1st report), *Journal of SNAJ*, Vol. 182, (1997).

Practical Design of Ships and Mobile Units
M.W.C. Oosterveld and S.G. Tan, editors.

Simulation study on oceanophysical environment around a large floating offshore structure moored in Tokyo Bay

M.Fujino, K.Seino, M.Hasebe and D.Kitazawa

Department of Environmental and Ocean Engineering, Graduate School of Engineering
University of Tokyo
7-3-1 Hongo, Bunkyo-ku, Tokyo 113-8656, Japan

To create an artificial land space for sustaining sound development of human activity, a large floating offshore structure is now expected to be built in Japan. To realizing such an offshore structure, it is indispensable to examine various expected impacts, due to construction of the structure, on the marine environment. For an assessment in the oceanophysical aspect, numerical simulation by means of the so-called multilevel model is one of powerful tools. In this paper, some results are illustrated to show usefulness of the numerical simulation, and then an example of examining environmental effects of an imaginary huge floating offshore structure (HFOS) moored in Tokyo Bay is briefly presented.

1. INTRODUCTION

For sustaining sound development of human activity in line with rapid growth of the world population, the need of land space is vastly increasing. As one of the solutions to this problem, large floating offshore structure (hereafter called LFOS) is expected to be built and emplaced in the sea. To realizing such an LFOS, it is indispensable to examine comprehensively various expected impacts, due to construction of the LFOS, on the marine environment around that structure. For example, an LFOS may cause unnegligible change of flow field around the structure, hamper the intrusion of solar radiation into the sea water, and prevent precipitation and evaporation on the water surface. These phenomena may change adversely the marine environment in the vicinity of the LFOS. Therefore, it is extremely important to assess carefully the environmental impacts from a wide range of physical, biological, and chemical viewpoints. For an assessment in the physical aspect, numerical simulation of the oceanophysical environment is one of the useful tools; variation of flow field around a reclaimed airport or an imaginary "huge floating offshore structure (HFOS)" was examined by numerical simulation based on the so-called multilevel model [1-5].

This paper is intended to show several examples of numerical simulation concerning the oceanophysical environment around an existing LFOS, called Mega-Float Structure, moored in Tokyo Bay, and an imaginary HFOS as well.

2. NUMERICAL MODEL

2.1. Basic equations

Cartesian coordinate system in which the origin is placed at the mean sea level, the x axis points east, the y axis points north, and the z axis is positive upward is introduced to describe the governing equations. Let u,v and w be velocity components in the direction of the x,y and z axis, respectively. Then the equations of fluid motions and continuity are as follows[6]:

$$\frac{Du}{Dt} = -\frac{1}{\rho_0}\frac{\partial p}{\partial x} + fv + A_M \nabla_H^2 u + \frac{\partial}{\partial z}\left(K_M \frac{\partial u}{\partial z}\right) \quad (1)$$

$$\frac{Dv}{Dt} = -\frac{1}{\rho_0}\frac{\partial p}{\partial y} - fu + A_M \nabla_H^2 v + \frac{\partial}{\partial z}\left(K_M \frac{\partial v}{\partial z}\right) \quad (2)$$

$$0 = -\frac{1}{\rho}\frac{\partial p}{\partial z} - g \quad (3)$$

$$\frac{\partial u}{\partial x} + \frac{\partial v}{\partial y} + \frac{\partial w}{\partial z} = 0 \tag{4}$$

In these equations, t is time, p the pressure, f the Coriolis parameter, ρ the density of sea water, ρ_0 a constant reference density, g the acceleration of gravity, A_M the horizontal eddy viscosity coefficient, K_M the vertical eddy viscosity coefficient, and ∇^2_H the two-dimensional Laplacian defined in the horizontal xy plane. Similarly, variation of water temperature T and salinity S can be written

$$\frac{DT}{Dt} = A_C \nabla^2_H T + \frac{1}{\delta}\frac{\partial}{\partial z}\left(K_C \frac{\partial T}{\partial z}\right) \tag{5}$$

$$\frac{DS}{Dt} = A_C \nabla^2_H S + \frac{1}{\delta}\frac{\partial}{\partial z}\left(K_C \frac{\partial S}{\partial z}\right) - \frac{RS}{V_R} \tag{6}$$

where A_C and K_C are horizontal and vertical eddy diffusivity coefficient, respectively, and R is river inflow per unit time that flows into the volume V_R, which is taken to be the volume of the uppermost level mesh immediately adjacent to the river mouth. The parameter δ in Eqs. (5) and (6) is defined by

$$\delta = \begin{cases} 0 & \partial\rho/\partial z > 0 \\ 1 & \partial\rho/\partial z \leq 0 \end{cases} \tag{7}$$

This means that when $\partial\rho/\partial z$ is positive, water mass in upper and lower layers are mixed instantaneously. The water density is assumed to be a function of temperature(°C) and salinity (psu);

$$\rho = 1028.14 - 0.0735T - 0.00469T^2$$
$$+ (0.802 - 0.002T)(S - 35.0) \tag{8}$$

When the sea water stratifies, the density gradients suppresses the vertical turbulent transportation of momentum, heat and salinity. To consider this effect, the following formulae are used for evaluating vertical eddy viscosity and diffusivity coefficients.

$$\frac{K_M}{K_{MO}} = (1 + 5.2R_i)^{-1} \tag{9}$$

$$\frac{K_C}{K_{CO}} = \left(1 + \frac{10}{3}R_i\right)^{-1.5} \tag{10}$$

where K_{MO} and K_{CO} mean the K_M and K_C values under homogenous condition, and R_i is the Richardson Number defined by

$$R_i = -\frac{g \cdot \partial\rho/\partial z}{\rho\left((\partial u/\partial z)^2 + (\partial v/\partial z)^2\right)} \tag{11}$$

These equations are transformed to finite-difference equations by means of a multilevel model with staggered mesh in the horizontal direction [6].

2.2. Boundary conditions

At the sea bottom and at the coastline, the no-slip condition is applied, and the heat and salinity fluxes through those boundaries are assumed to be zero. At the open boundary, which is treated as the non-reflecting boundary[7] the sea level and both temperature and salinity are fixed to those of the outer sea, and gradients of current velocity is assumed to be zero.

The boundary conditions at the sea bottom $(z = -h)$ are as follows:

$$u\frac{\partial h}{\partial x} + v\frac{\partial h}{\partial y} + w = 0 \tag{12}$$

$$K_M \frac{\partial u}{\partial z} = \frac{\tau_{xB}}{\rho_0} \tag{13}$$

$$K_M \frac{\partial v}{\partial z} = \frac{\tau_{yB}}{\rho_0} \tag{14}$$

where $\vec{\tau}_B = (\tau_{xB}, \tau_{yB})$ denotes the bottom friction, and is given by

$$\tau_{xB} = \gamma^2 \rho_0 u\sqrt{u^2 + v^2} \tag{15}$$

$$\tau_{yB} = \gamma^2 \rho_0 v\sqrt{u^2 + v^2} \tag{16}$$

In these equations, γ^2 is the drag coefficient of sea bottom.

The boundary conditions at the sea surface $(z = \varsigma)$ are as follows:

$$p = p_a \tag{17}$$

$$-\frac{\partial\varsigma}{\partial t} - u\frac{\partial\varsigma}{\partial x} - v\frac{\partial\varsigma}{\partial y} + w = 0 \tag{18}$$

$$K_M \frac{\partial u}{\partial z} = \frac{\tau_{xW}}{\rho_0} \tag{19}$$

$$K_M \frac{\partial v}{\partial z} = \frac{\tau_{yW}}{\rho_0} \tag{20}$$

$$-K\frac{\partial T}{\partial z} = \frac{Q_T}{\rho_0 C_p} \tag{21}$$

$$-K_C\frac{\partial S}{\partial z} = \frac{Q_S}{\rho_0} \tag{22}$$

where p_a is the atmospheric pressure at the sea surface, $\vec{\tau}_W$ the wind stress vector, Q_T downward heat flux per unit time per unit area, Q_S downward salinity flux, and C_p specific heat of sea water. $\vec{\tau}_W$, Q_T and Q_S are given by

$$\vec{\tau}_W = \rho_a C_d \vec{W}\left|\vec{W}\right| \tag{23}$$

$$Q_T = Q_r - Q_b - Q_e - Q_c \tag{24}$$

$$Q_S = S\left(E_{vap} - P_r\right) \tag{25}$$

where ρ_a is the air density, C_d the drag coefficient at the sea surface, $\vec{W}$ the wind velocity vector, Q_r global solar radiation, Q_b long wave radiation from the sea, Q_e latent heat transport by evaporation, Q_c sensible heat transport by convection or conduction, E_{vap} evaporation rate at the sea surface, and P_r precipitation rate.

Each of Q_r, Q_b, Q_e, Q_c and E_{vap} is calculated by the so-called bulk formulae of transport[8,9] described in terms of water temperature, atmospheric data such as wind speed, air temperature, vapor pressure and amount of cloudness, and various bulk transport coefficients. However, those bulk formulae are not shown here to avoid verbose description.

2.3. Variable computational grid size

In order to investigate the effects of a floating offshore structure (FOS) on the oceanophysical environment of the surrounding sea water, it is extensively important to get detailed knowledge on the spatial variation of current speed, water temperature and salinity around the FOS. For this purpose, grid sizes in an area adjacent to the FOS should be much finer than those far from the FOS. In the present numerical simulation, three kinds of square grid with different side length are used ; they are 180m(Rank 3), 540m(Rank 2), and 1620m(Rank 1). The finest grids of 180m are used to cover a region adjacent to the FOS.

The values of horizontal eddy viscosity coefficient A_M and horizontal eddy diffusion coefficient A_C should be varied depending on the grid size; in the following numerical simulation, the values of A_M and A_C are varied according to the Richardson's 4/3 power law[10].

3. NUMERICAL RESULTS AND DISCUSSIONS

3.1. Tidal current at the mooring site of Mege-Float Structure

The construction of Mega-Float Structure (300m in length, 60m in breadth, 2m in depth, and 0.5m in draft) was completed in July 1996, and is still now being moored off Oppma, Yokosuka, in Tokyo Bay in order to obtain various data for the purpose of verifying the practicality of the LFOS. In Fig.1 shown are the mooring site of Mega-Float Structure in Tokyo Bay as well as the detailed topography in a region close to that structure. The mean water depth at the mooring site is about 8m.

Continuous measurement of current velocity by means of ADCP (Acoustic Doppler Current Profiler) was carried out at two positions, A and B, of the structure as shown in Fig.1. Besides, in situ measurement of the current field in a narrow area adjacent to the structure was executed several times before and after the completion of the Structure. According to the observation, the southward current is dominant in the ebb tide, while in the flood tide the dominant current is northward, but the southward current prevails over the northward one.

In the present numerical simulation, periodic sea level change due to the M_2 tide, solar radiation and inflow from the major five rivers flowing into Tokyo Bay are taken into consideration. The initial values of water temperature and salinity are estimated from the oceanographical observation data, which were provided by the Japan Oceanographycal Data Center. The atmospheric temperature was assumed to be constant, that is to say, fixed to the mean temperature of August in

Tokyo, and the river inflows from the major rivers were also fixed to the average values and the water temperature was fixed to 25.7°C. Numerical values of various parameters used in the simulation are summarized in Table 1.

In Fig. 2, the simulation results of tidal current in the ebb tide, in the region adjacent to Mega-Float Structure, are compared with the observation. Agreement between the simulation and the observation seems satisfactory. Fig.3 depicts the hodographs of tidal current measured by ADCP at two positions shown in Fig.1. This figure indicates that the north-south component (v) is more dominant compared with the east-west component (u), in particular, at the position of ADCP-2.

Similarly, Fig.4 depicts the hodographs of computed tidal current at three different depth, to say, at z=−1.5m,−4.5m and −7.5m, together with the hodograph of the mean tidal current averaged in the vertical direction. Comparing this figure with the previous one, Fig.3, it can be said the east-west component is underestimated, but the general characteristic

Table 1　Numerical values of principal parameters used in the simulation

f	8.42×10^{-5}　$1/s$
ρ_0	1025.0　kg/m^3
ρ_a	1.226　kg/m^3
γ^2	0.0026
C_d	0.0015
A_M	50 m^2/s for $1000m$ $grid$
A_C	10 m^2/s for $1000m$ $grid$
K_{MO}	0.001　m^2/s
K_{CO}	0.001　m^2/s
Δt	$1.0s \sim 5.0s$

features, that the current speed at ADCP-2 surpasses that at ADCP-1 and that the residual current speed at ADCP-2 is southward and amounts to about $0.08ms^{-1}$, are predicted fairly well by numerical simulation.

3.2. Oceanophysical effects of an HFOS on the surrounding sea water

In order to examine to what extent the marine environment is affected by the emplacement of

(a)　Mooring site in Tokyo Bay

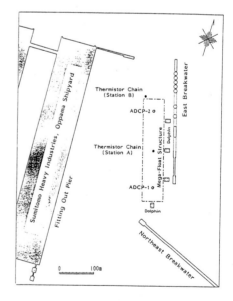

(b)　Topography of mooring site

Figure 1.　Mooring site of Mega-Float Structure and its topography

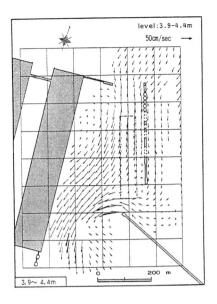

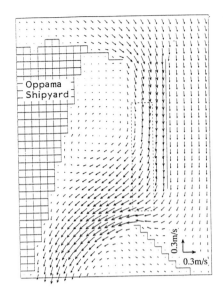

(a) observed   (b) computed

Figure 2.   Comparison of observed and computed tidal current in the ebb tide around
Mega-Float Structure

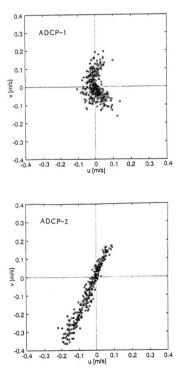

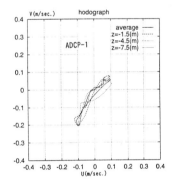

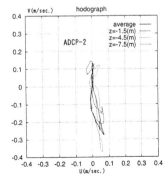

Figure 3.   Scattering diagram of observed
tidal current (July 1996)

Figure 4.   Hodograph of computed tidal
current

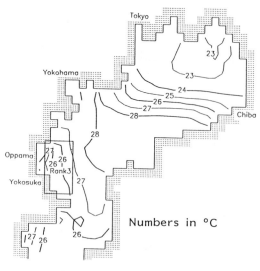
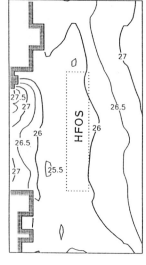

(a) Tokyo Bay (Rank 1)　　　　　　(b) Mooring site of HFOS (Rank 3)

Figure 5.　Temperature distribution in the ebb tide when HFOS is assumed to exist

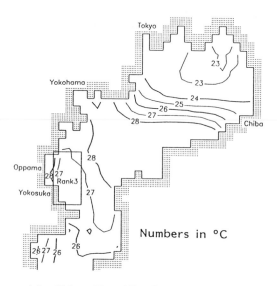
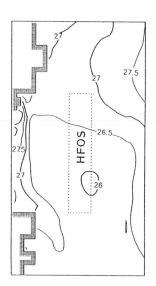

(a) Tokyo Bay (Rank 1)　　　　　　(b) Mooring site of HFOS (Rank 3)

Figure 6.　Temperature distribution in the ebb tide when HFOS does not exist

an HFOS in a bay, an imaginary HFOS (5040m in length, 1080m in breadth and 1.0m in draft) was assumed to be put in place about 600m apart from the nearest coast, off Oppama, Yokosuka, in Tokyo Bay. And the numerical simulation, with respects to current speed, water temperature and salinity, was executed under the same computational conditions as those in the previous section 3.1. In the following, a part of numerical results is presented.

Fig.5 shows the temperature distribution, in the uppermost level ($z = -1.5m$), (a) in the

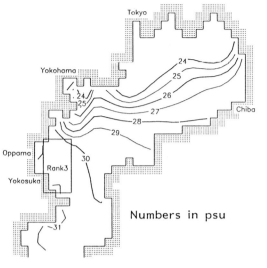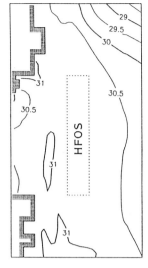

(a) Tokyo Bay (Rank 1)   (b) Mooring site of HFOS (Rank 3)

Figure 7. Salinity distribution in the ebb tide when HFOS is assumed to exist

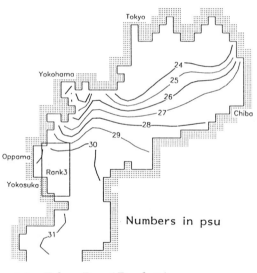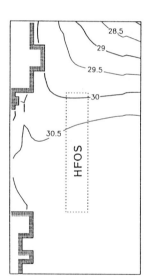

(a) Tokyo Bay (Rank 1)   (b) Mooring site of HFOS (Rank 3)

Figure 8. Salinity distribution in the ebb tide when HFOS does not exist

whole region of Tokyo Bay (Rank 1) and (b) in the vicinity of the HFOS (Rank 3), at the instance of 352,000sec (about 4 days) after the commencement of computation for the case where the HFOS is assumed to exist; at that instance, the tidal current attained almost periodic state although the water temperature and salinity still increase gradually with time. Fig.6 shows similar results for the case where the HFOS does not exist.

Figs.7 and 8 depict the horizontal salinity distribution. From these figures, it can be said that the effects of the HFOS on temperature and salinity of the surrounding sea water are recognized in a sea region adjacent to the HFOS, but in an area far from the HFOS, for example, in the middle part and in the head of Tokyo Bay, the HFOS dose not definitely bring about difference in the distribution of temperature and salinity. Further the results of numerical simulation indicate that the characteristic features, stated above for the case of the ebb tide, regarding the oceanophysical effects of HFOS on the surrounding sea water remain true during other phases of a tidal period, for instance, in the flood tide.

Here, it should be noted that further investigation in other aspects, for instance, in the chemical and/or biological aspects is necessary for comprehensive environmental assessment.

4.CONCLUDING REMARKS

The principal purpose of this paper is not to discuss the extent of environmental effects, caused by an HFOS, on the surrounding sea water, but to show the practicality of applying numerical simulation method to the oceanophysical assessment of the marine environment in the vicinity of the HFOS. For this object, the so-called multilevel model describing the time variation of tidal current, water temperature and salinity was used to compare the difference in the spatial distribution of water temperature and salinity between two cases where an HFOS is assumed, and is not assumed, to be emplaced in Tokyo Bay. Comparison of simulation results indicates that the spatial distribution of temperature and salinity in the vicinity of the mooring site of the HFOS differs definitly in these two cases, but the difference itself in the

temperature and salinity is very small. In a region apart from the mooring site, on the contrary, the spatial distribution of temperature and salinity is almost the same even in the two cases. As far as the results of several sample simulations presented in this paper are concerned, it seems that the numerical simulation is a useful tool to assess comprehensively the effects of an HFOS, for instance, on the oceanophysical environment.

REFERENCES

1. S.Tabeta and Y. Inoue, Proc. International Conference on Technology for Marine Environment Preservation (1995) 505.
2. Y.Inoue, M.Arai, S.Tabeta, K.Nakazawa, X.Zhang and Y.Takai, Proc.International Offshore and Polar Engineering Conference (1995) 406.
3. C.H.Hu and Y.Kyozuka, Trans. The West-Japan Soc. Nav.Archit.,No.91(1996)51.
4. M.Fujino, H.Kagemoto and K.Hamada, Jour. The Soc.Nav.Archit. Japan, Vol.180 (1996) 393.
5. Y.Kyozuka, C.H.Hu, H.Hasemi and A.Hikai, Jour. The Soc. Nav.Archit.Japan, Vol.181 (1997) 151
6. S.Tabeta and M.Fujino, Jour.Marine Science and Technology, Vol.1, No.2 (1996) 94.
7. M.Fujino, H.Kagemoto, S.Tabeta and T.Hamada, Jour.The Soc. Nav.Archit.Japan, Vol.175 (1994) 161.
8. T.Yanagi, Coastal Oceanography, Koseisha-Koseikaku Co.Ltd.,1989.
9. J.Kondo, Meteorology in Water Environment, Asakura Publishing Co. Ltd., 1994
10. M.Fujino, K.Seino and M.Hasebe, Jour.The Soc.Nav.Archit. Japan, Vol.182 (1997) 97.

Practical Design of Ships and Mobile Units
M.W.C. Oosterveld and S.G. Tan, editors.

Downtime Minimization by Optimum Design of Offshore Structures

G.F. Clauss and L. Birk

Institute of Naval Architecture and Ocean Engineering, Berlin University of Technology
SG 17, Salzufer 17-19, 10587 Berlin, Germany

Hydrodynamic shape optimization is one of the key tools for the development of new system concepts in offshore engineering. This paper presents a fully automated numerical procedure for optimum adjustment of shapes to environmental conditions. Nonlinear programming algorithms vary the design parameters to find a minimum of the objective function within a few iterations. The resulting hull shapes are characterized by minimized wave loads and motions. Spectral analysis combined with standard wave spectra and wave scatter diagrams is used to minimize expected downtime of offshore structures. The paper presents the optimization of a semisubmersible and of large buoys to illustrate the efficiency of the proposed procedure.

1. INTRODUCTION

Floating offshore structures frequently operate in rough and hostile sea states. Wave induced motions interrupt normal service, if limitations of the motion compensation are reached. Thus design improvement minimizing weather induced downtime yields an economic advantage. Currently best seakeeping behavior is obtained by interactive and heuristic design variations or by series of expensive model tests. In this paper a fully automated numerical procedure is described, which achieves an optimum adjustment of the shapes of offshore structures to environmental conditions.

Relevant criteria are necessary to distinguish between bad and good designs. In order to minimize downtime the system response caused by the irregular wave climate has to be analyzed. For intervals ranging from one to three hours the statistical parameters of irregular seas do not show much variations, i.e. the process is stationary [3]. The description of these short-term sea states is commonly based on design spectra, representing the frequency dependent energy distribution of waves. Pierson and St. Denis [19] published the first application of spectral analysis to ship motions, which was the basis of many future developments [5].

Chou [6] presents a remarkable paper proposing an analytical procedure for the optimum design of offshore structures, based on spectral analysis. To evaluate the seakeeping behavior the variance of the response spectrum is calculated. The minimization problem is solved by means of calculus of variations. An optimum displacement distribution along the vertical axis is developed. Due to the analytical formula used for the evaluation of wave forces and motions [17] the geometry is restricted to hydrodynamically transparent bodies of revolution with vertical axis. Chou derives optimal shapes for a buoy and an ocean platform supported by four columns applying ISSC spectra.

The same restriction with respect to geometry applies to the work of Saito [21]. The heave exciting force of a buoy is minimized for selected wave frequencies. A semisubmersible is optimized with respect to heave motion in a design sea state applying the objective function proposed by Chou. Forces are calculated on the basis of a slender body assumption. Viscous effects are neglected.

Akagi and Ito [1] optimize the heave motion of a hydrodynamic transparent semisubmersible using a quadratic programming technique. The diameter of the hull as well as diameter and longitudinal position of the columns are varied to minimize the variance of the heave response spectrum. The short-term environmental conditions are defined

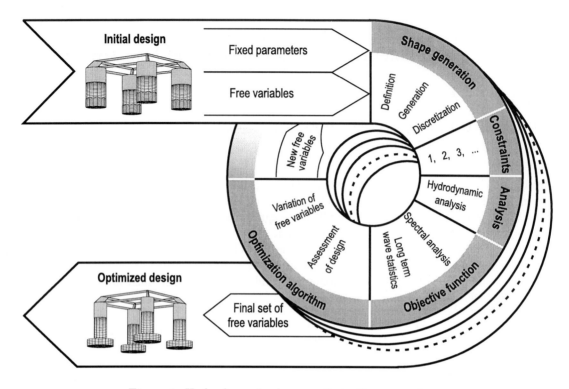

Figure 1: Hydrodynamic shape optimization procedure.

by an ISSC spectrum. As a result optimum basic dimensions for various configurations of semisubmersibles are presented. In a following paper Akagi et al. [2] apply a multi-objective optimization combining displacement, variable load and heave motion into a single objective function.

Clauss and Birk [7] extend the application of spectral analysis to the optimization of arbitrarily shaped offshore structures. The optimization procedure comprises an automated form generation procedure and applies a panel program to solve the three-dimensional wave-body interaction problem. Applications to different types of fixed and floating offshore structures are presented in subsequent papers [9, 11]. Recently long-term statistics of sea states have been introduced to comprise all sea states occurring during the lifetime of a structure [4].

This paper focuses at the minimization of downtime due to excessive heave motions in harsh sea states. The following Section 2 presents a short overall description of the shape optimization procedure. Section 3 reviews aspects of spectral analysis and Section 4 extends the system as-

sessment to the prediction of downtime probability based on long-term wave statistics at a specific location. Applications of the shape optimization method are presented for a semisubmersible and large buoys, which have been recently considered for the development of smaller fields.

2. HYDRODYNAMIC OPTIMIZATION

Fig. 1 illustrates the shape optimization process. The user selects the objective function and provides the parameters p and the start vector of free variables $\underline{x}$. Using this set of data the hydrodynamic shape optimization is started by creating and discretizing the initial design. Its shape is stored on disk for the subsequent hydrodynamic analysis. Note that the shape generation process is completely automated. The design is checked against the set of constraints before entering the time consuming stage necessary to evaluate wavebody interaction.

Constraints are necessary for the design optimization to ensure that all technical and economical boundary conditions are satisfied. In particular, simple subroutines and results associated to the preliminary design analysis which may follow from

- hydrostatic stability,

- strength and fatigue, or

- fabrication, installation and maintenance cost.

According to the dependency between free variables and the objective function a mathematical model is defined for every constraint.

The subsequent hydrodynamic analysis is based on the 3D-diffraction-radiation program WAMIT (Wave Analysis Massachusetts Institute of Technology) [18]. Viscous effects are not considered. The results comprise coefficients of added mass and potential damping along with response amplitude operators of wave exciting forces and motions. This is the basis for the evaluation of the objective function by spectral analysis (Section 3) and downtime prediction (Section 4) using a probabilistic theory of wave loads. Depending on the changes of the objective function, the optimization algorithm decides how to alter the set of free variables. After processing of the initial design the loop of shape generation, check of constraints, hydrodynamic analysis and assessment of designs is repeated until a minimum of the objective function is obtained.

3. SPECTRAL ANALYSIS

In spectral analysis irregular seas are interpreted as a random superposition of a great number of harmonic waves of different amplitudes ζ_{ai} and frequencies ω_i. The wave crests are assumed to be of infinite length, i.e. the water elevation depends on time t and location x [14]:

$$\zeta(x,t) = \lim_{N_w \to \infty} \sum_{i=1}^{N_w} \zeta_{ai} \cos(k_i x - \omega_i t + \epsilon_i) \quad (1)$$

The random variable is the phase shift ϵ_i between the component waves.

Each component wave i contributes an amount of energy to the seaway proportional to its squared wave amplitude ζ_{ai}:

$$S_\zeta(\omega_i)\, d\omega = \frac{1}{2} \cdot \zeta_{ai}^2. \quad (2)$$

The spectral density S_ζ represents the energy distribution as a function of circular frequency ω. Integration of the term $S_\zeta \cdot \omega^i$ yields the ith-order moment $m_{i\zeta}$ of the spectrum:

$$m_{i\zeta} = \int_0^\infty S_\zeta(\omega)\omega^i d\omega. \quad (3)$$

Note that $m_{0\zeta}$ corresponds to the total spectral energy and equals the variance σ_ζ^2 of the random process [16].

Various standard spectra have been defined for design purposes. In this paper all computations are based on Pierson-Moskowitz (PM) spectra $S_\zeta(\omega)$ [13]:

$$S_\zeta = 4\pi^3 \frac{H_s^2}{T_0^4 \, \omega^5} e^{\left(-\frac{16\pi^3}{T_0^4 \omega^4}\right)}. \quad (4)$$

Two characteristic parameters define the shape of the energy distribution. The first parameter is the significant wave height H_s. It is defined as the average of the one third highest waves, and is related to the energy density spectrum by

$$H_s = 4 \sqrt{m_{0\zeta}}. \quad (5)$$

The second parameter is the mean zero-up-crossing period T_0, describing the average duration between two up-crossings of the mean water level:

$$T_0 = 2\pi \sqrt{\frac{m_{0\zeta}}{m_{2\zeta}}}. \quad (6)$$

The selection of T_0 which determines the range of relevant wave frequencies characterizing a certain sea state strongly affects dynamic responses of offshore structures.

In analogy to the water surface elevation (1) the response of the structure is also described by superposition of harmonic components. Corresponding to the wave spectrum $S_\zeta(\omega)$ of the seaway the response spectrum $S_s(\omega)$ represents the energy distribution of the output signal. $S_s(\omega)$ follows from the product of the squared magnitude of the transfer function and the wave spectrum [12]:

$$S_s(\omega) = |H_{s\zeta}(\omega)|^2 S_\zeta(\omega). \quad (7)$$

The corresponding significant double response amplitude $(2s_a)_s$ follows from this response spectrum in analogy to (5):

$$(2s_a)_s = 4 \cdot \sqrt{\int_0^\infty S_s(\omega)\, d\omega} = 4\sqrt{m_{0s}}. \quad (8)$$

As the total spectral energy implies the factor H_s^2 we usually apply the normalized form of (8)

$$\frac{(2s_a)_s}{H_s} = \frac{4\sqrt{m_{0s}}}{H_s}. \quad (9)$$

These normalized significant double amplitudes of motions, forces, etc. characterize the behavior of offshore structures in stationary sea states. Optimization processes based on this design criterion are presented in [9, 11]. Of course, any algebraic combination of different significant double response amplitudes can also be used as a measure of merit, e.g. the combination of heave and pitch [10]. The increase in computational effort is negligible, because the time consuming calculation of the velocity potential is performed only once.

The numerical solution and the implementation of (3) - (9) is straightforward. Instead of an infinite number of wave frequencies a finite number N_w - large enough to represent the spectrum correctly - is chosen, with numerical integration of (3) using the trapezoidal rule.

4. DOWNTIME PREDICTION

For downtime prediction of offshore operations limiting values of significant double amplitudes of forces and motions $(2s_a)_{s,limit}$ are introduced and the objective function (8) defined above is extended to include long-term wave statistics.

Tab. 1 shows a wave scatter diagram of the region *Haltenbanken* located West of Trondheim [15]. The respective number of observations r_{ij} represent the joint probability $q_{ij} = r_{ij}/r$ of a stationary sea state characterized by a zero-up-crossing period T_{0j} and a significant wave height H_{si}. All observations sum up to $r = \sum_i \sum_j r_{ij}$.

The duration T_{ij} of all occurrences of an individual sea state (T_{0j}, H_{si}) is given by

$$T_{ij} = q_{ij} \cdot T_T. \quad (10)$$

T_T is the total time considered, e.g. one year.

Evaluating (3) and (8) for all upper interval boundaries of the zero-up-crossing period T_0 yields the function

$$\frac{(2s_a)_s}{H_s} = f(T_0). \quad (11)$$

If operational requirements limit the significant double amplitude of motion to $(2s_a)_{s,limit}$ the highest acceptable significant wave height at a specified zero-up-crossing period T_0 follows from

$$H_{s,limit}(T_0) = (2s_a)_{s,limit} \cdot \left(\frac{(2s_a)_s}{H_s}\right)^{-1}. \quad (12)$$

Because the probabilities of significant wave heights and periods in the wave scatter diagram (Tab. 1) are assigned to discrete segments, computed values of $H_{s,limit}(T_0)$ are approximated by round off values. If these values $H_{s,limit}(T_0)$ are introduced into the wave scatter diagram, sea states are assigned to a feasible $(H_{si} < H_{s,limit}(T_0))$ and an infeasible $(H_{si} > H_{s,limit}(T_0))$ domain as illustrated by the white and gray areas in Tab. 1. Simple summation of all probabilities in the infeasible domain, $\sum \sum q_{dj}$ yields the expected probability of downtime P_d.

$$P_d = \sum_j \sum_d q_{dj} = \frac{\sum_j \sum_d r_{dj}}{r} \quad (13)$$

Index d ranges over all H_s-classes, where $H_s(T_0)$ is higher than $H_{s,limit}(T_0)$. The design criterion (13) is applied to the optimization of a semisubmersible and large buoys.

5. APPLICATIONS

Based on 100-year design spectra the authors have optimized various types of offshore structures, like GBS, TLP, caisson and twin-hull semisubmersibles and large buoys [7–11]. As compared to existing structures these new concepts proved to be surprisingly good. The following applications expand the optimization process by considering wave scatter diagrams in the design of structures with minimum downtime.

Table 1
Assigning sea states to a feasible and an infeasible domain. The total number of observations is 16834 (location *Haltenbanken*, West of Trondheim, Norway).

Sign. wave height H_s [m]	Zero-up-crossing period T_0 [s]											
	0–4	4–5	5–6	6–7	7–8	8–9	9–10	10–11	11–12	12–13	13–14	14–15
11.5 – 12.0	0	0	0	0	0	0	0	1	0	0	0	0
11.0 – 11.5	0	0	0	0	0	0	0	1	0	0	0	0
10.5 – 11.0	0	*$H_{s,\mathrm{limit}}=10.70$m*	0	0	0	0	*$T_0=8$s*	0	0	0	0	0
10.0 – 10.5	0	0	0	0	0	0	0	0	0	0	0	0
9.5 – 10.0	0	0	0	0	0	0	2	3	1	0	0	0
9.0 – 9.5	0	0	0	0	0	0	10	4	1	0	0	0
8.5 – 9.0	0	0	0	0	0	0	13	5	*infeasible sea states*	0	0	0
8.0 – 8.5	0	0	0	0	0	6	22	3	0	0	0	0
7.5 – 8.0	0	0	0	0	0	19	27	5	0	0	0	0
7.0 – 7.5	0	0	0	0	3	30	14	1	0	0	0	0
6.5 – 7.0	0	*$H_{s,\mathrm{limit}}=6.70$m*	0	0	8	61	16	*$T_0=9$s*	0	0	0	0
6.0 – 6.5	0	0	0	0	31	91	7	0	0	0	0	0
5.5 – 6.0	0	0	0	0	117	91	14	1	0	0	0	0
5.0 – 5.5	0	0	0	7	194	75	17	0	1	0	0	0
4.5 – 5.0	0	0	0	79	256	86	16	1	0	0	0	0
4.0 – 4.5	0	0	*feasible*		271	75	28	2	0	0	0	0
3.5 – 4.0	0	0	*sea states*		263	101	9	0	0	0	0	0
3.0 – 3.5	0	0	187	514	293	78	16	0	0	0	0	1
2.5 – 3.0	0	16	640	642	304	97	15	2	0	0	0	0
2.0 – 2.5	0	310	1010	744	283	72	18	3	0	0	0	0
1.5 – 2.0	83	1151	110	*$H_{s,\mathrm{limit}}=1.74$m*		89	7	0	*$T_0=10$s*	0	0	0
1.0 – 1.5	725	1600	875			7	3	0	0	0	0	0
0.5 – 1.0	729	755	328	98	*$H_{s,\mathrm{limit}}=0.65$m*	0	0	0	*$T_0=11$s*	0	0	0
0.0 – 0.5	22	18	9	0	0	0	0	0	0	0	0	0

5.1. Semisubmersible

Following the procedure developed in Section 4 a twin-hull semisubmersible with minimum downtime is designed. The description of the wave climate is based on the *Haltenbanken* wave scatter diagram (Tab. 1) and Pierson-Moskowitz spectra. The expected downtime probabilities P_d (13) are calculated assuming an upper limit of the significant double amplitude of heave motion $(2s_{3a})_{limit} = 1.0$ m.

The initial design is a semisubmersible with circular columns and two pontoons capped by hemispheres (Fig. 2). The column spacing is 70 m in longitudinal and in transverse direction. Column spacing as well as the column diameter of 15 m and the displacement of $\forall = 60,000$ m^3 were fixed throughout the optimization process. The set of free variables includes the ratio of displacement $x_1 = V_{column}/\forall$, the pontoon diameter midships, the diameter of the hemispherical end cap and the longitudinal center of volume of a half pontoon. Variation of the displacement ratio x_1 entails draught variations, as the column diameter is held constant.

Fig. 2 presents the optimization results: the broken line separates feasible and infeasible sea states of the initial design, whereas the solid line gives the corresponding dividing line of the opti-

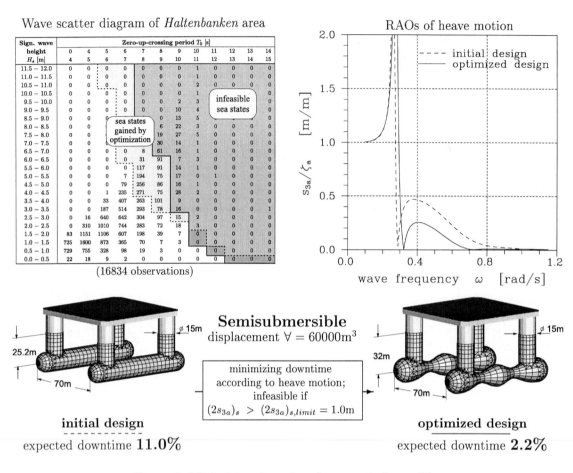

Figure 2: Minimizing downtime for a semisubmersible.

mized semisubmersible. All sea states indicated by the area shaded in light gray become feasible due to optimization. The expected downtime is decreased from 11.0 % to 2.2 % accordingly. The associated RAOs of heave motion (Fig. 2) illustrate the improvement of seakeeping qualities in the range of relevant wave frequencies. The displacement reduction of the pontoons yields an increased draught which in turn results in a substantially lower heave exciting force. Due to a decreased heave added mass a minor increase of the heave resonance frequency is acceptable as it is sufficiently small.

5.2. Large Buoys

Large buoys have recently gained attention for the development of marginal fields [20]. At our next demonstration the proposed concept of a conical shaped buoy (displacement $\forall = 173670\text{m}^3$) is introduced as an initial design,

not considering effects of moon pool and damping devices. As shown in Fig. 3 this *Maritime Tentech AS* design proves to be hydrodynamically excellent as the expected downtime is very low (2.3%). However, concentrating on the hydrodynamic design aspect only a further reduction of downtime is obtained by reducing lower and upper diameters of the conical tank accompanied by a draught increase. The associated heave RAOs illustrate that wave exciting forces are significantly reduced during the optimization process.

The previous example demonstrates that it is very convenient to start with a sophisticated initial structure. However, even if the initial design is rather primitive, the optimization procedure proves to be very efficient. Starting with a straight circular cylinder with the same draught and displacement as before, Fig. 4 shows that the variation of four free variables (vertical location of the center of buoyancy, diameter at the water-

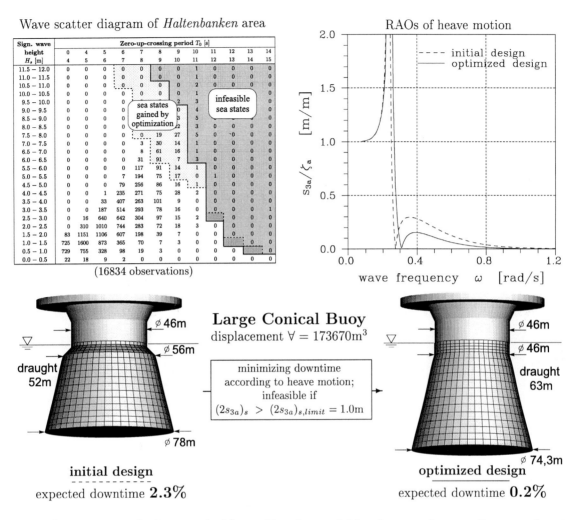

Figure 3: Optimization of a large conical buoy. Two free variables: lower and upper diameter of conical tank.

line, draught of the body and diameter at the basis) yields an optimized buoy with very low downtime. Note that the set of geometric constraints includes vertical side walls in the waterline, a minimum cross section and an edge angle of more than 90° at the bottom.

As shown in Fig. 5 the optimization process increases heave added mass which in turn reduces the heave resonance frequency. In addition, the heave exciting force is decreased at low frequencies as the inertia forces on the upper and on the lower part of the bulgy hull are compensating, canceling each other at $\omega_c = 0.32\text{rad/s}$ (Fig. 5). Summarizing, the optimization procedure successfully adapts the transfer functions of

added mass and exciting force to the energy distribution of the design sea state. As a result the expected downtime is reduced from 19.3% to 0.9%.

6. CONCLUSIONS

The described procedure combines numerical tools of CAD, CFD, and probabilistic theory of wave loads to evaluate and optimize the seakeeping behavior in an automated process. The applications presented above demonstrate the potential of hydrodynamic shape optimization. Significant reduction of expected downtime is achieved in all test cases. Although many other objectives

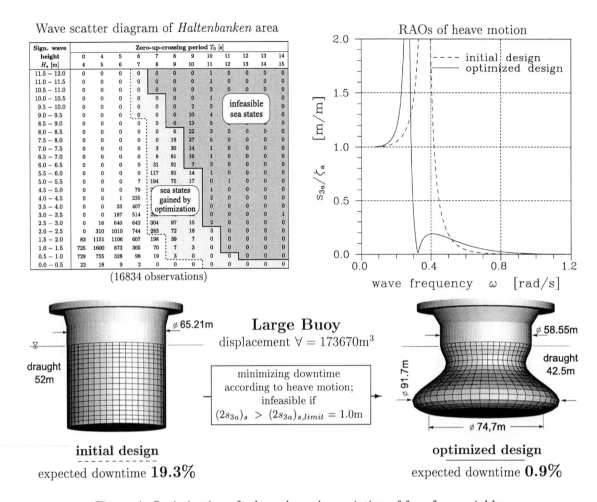

Figure 4: Optimization of a large buoy by variation of four free variables.

may also have an impact on the design of offshore structures, the hydrodynamic shape optimization is one of the key tools for the development of new system concepts in short time.

REFERENCES

1. S. Akagi and K. Ito. Optimal design of semisubmersible form by minimizing its motion in random seas. *Jour. of Mechanisms, Transmissions, and Automation in Design, ASME Trans.*, 106:23–30, 1984.

2. S. Akagi, K. Ito, and R. Yokoyama. Optimal design of semisubmersible's form based on systems analysis. *Jour. of Mechanisms, Transmissions, and Automation in Design, ASME Trans.*, 106:524–530, 1984.

3. N.D.P. Barltrop and A.J. Adams. *Dynamics of Fixed Marine Structures*. Butterworth-Heinemann Ltd, Oxford, 3rd edition, 1991.

4. L. Birk. *Hydrodynamic Shape Optimization of Offshore Structures*. PhD thesis, Technische Universität Berlin (D83), Berlin, Germany, 1998. To be published.

5. R.E.D. Bishop and W.G. Price. *Probabilistic Theory of Ship Dynamics*. Chapman & Hall, London; Wiley, New York, 1974.

6. F.S. Chou. A minimization scheme for the motions and forces of an ocean platform in random seas. *SNAME Trans.*, 85:32–50, 1977.

7. G.F. Clauss and L. Birk. Minimizing forces and motions of large offshore structures. In *Conf. on Offshore Mechanics and Arctic Engineering (OMAE '93)*, pages 321–329, Glas-

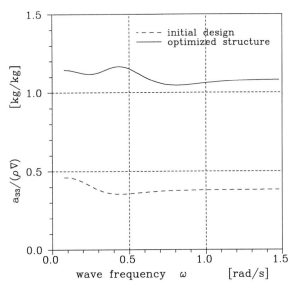 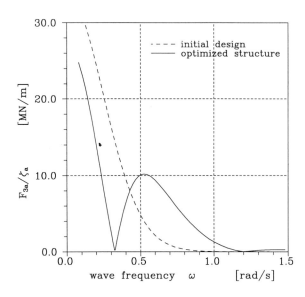

Figure 5: Heave added mass a_{33} and heave exciting force amplitude F_{3a}/ζ_a of initial and optimized buoy shown in Fig. 4.

gow, UK, 1993.

8. G.F. Clauss and L. Birk. Hydrodynamic optimization of large offshore structures. In *Int. Marine Design Conf. (IMDC '94)*, pages 595–608, Delft, The Netherlands, 1994.

9. G.F. Clauss and L. Birk. Optimizing the motion behaviour of offshore structures. In *Int. Conf. on Behaviour of Offshore Structures (BOSS '94)*, pages 665–684, Cambridge (MA), USA, 1994.

10. G.F. Clauss and L. Birk. Design optimization of large offshore structures by nonlinear programming. In *IX Int. Symp. on Offshore Eng., BRASIL OFFSHORE '95*, Rio de Janeiro, Brasil, 1995.

11. G.F. Clauss and L. Birk. Hydrodynamic shape optimization of large offshore structures. *Applied Ocean Research*, 18(4):157–171, 1996.

12. G.F. Clauss, E. Lehmann, and C. Östergaard. *Offshore Structures*, Volume 1: Conceptual Design and Hydrodynamics. Springer Verlag London, 1992.

13. G.F. Clauss, E. Lehmann, and C. Östergaard. *Offshore Structures*, Volume 2: Strength and Safety for Structural Design. Springer Verlag London, 1994.

14. O.M. Faltinsen. *Sea loads on ships and off-shore structures*. Ocean Technology. Cambridge University Press, 1990.

15. J. Mathisen and E. Bitner-Gregerson. Joint distributions for significant wave height and wave zero-upcrossing period. *Applied Ocean Research*, 12(2):93–103, 1990.

16. D.E. Newland. *Random Vibration and spectral analysis*. Longman Inc., New York, 1975.

17. J.N. Newman. The motion of a SPAR buoy in regular waves. Techn. Report 1499, David Taylor Model Basin, 1963.

18. J.N. Newman and P.D. Sclavounos. The computation of wave loads on large offshore structures. In *Int. Conf. on Behaviour of Offshore Structures (BOSS '88)*, pages 605–622, Trondheim, Norway, 1988.

19. W.J. Pierson and M. St. Denis. On the motion of ships in confused seas. In *SNAME Trans.*, volume 61, pages 280–354, 1953.

20. S. Popov. Tech Trends: New platform concept makes small field development economic. *Hart's Petroleum Engineer Int.*, page 19, 1997.

21. K. Saito. An optimization method for the motions and forces of an ocean structure in waves. Techn. Report 91/3, Inst. of Naval Arch. and Ocean Eng. Technische Universität Berlin, Berlin, Germany, 1991.

Optimisation of DP Stationkeeping for New Generation Early Production Drillships

Albert B. Aalbers[1] and Richard P. Michel[2]

[1] Maritime Research Institute Netherlands
P.O. Box 28, 6700 AA Wageningen, The Netherlands

[2] Bennett & Associates
1140 St. Charles Avenue, New Orleans LA 70130, U.S.A.

Abstract

In July, 1996 a contract was signed between Astilleros y Talleres del Noroeste (ASTANO) and Transocean Offshore Inc. for the construction of a 5th generation drillship. The vessel named "Discoverer Enterprise" combines a unique dual drilling activity with a modern hull form to produce a drilling unit capable of world wide operations in water depths up to 3000 meters.

The drillship incorporates dual activity drilling operations (capable of drilling in 3000m water depth), has the ability to support a variable deck load of 20,000 tons, is fully dynamically positioned (DP), has storage for 120,000 barrels of crude and can transit at 15 knots.

In close cooperation with AMOCO and Transocean Offshore Inc. of Houston an extensive model test program was carried out by MARIN for optimisation and approval of the DP system for this FPSO/ deep water drill ship. The main aspects were:

- Optimisation of DP control for the six azimuthing thrusters;
- Measurement of motions DP capability in limit design condition;
- Determination of limit sea state for turning the vessel around against the weather.

The tests were carried out in the Wave and Current Basin of MARIN, using a closed loop DP control system to steer the thrusters.

The paper outlines the philosophy with respect to the DP control strategy and presents selected results of the DP tests for optimisation of heading control and positioning capability.

1. INTRODUCTION

The naval architectural design of the vessel hull presented many challenges due to the geometry requirements, essential performance characteristics, and mode of construction. The above considerations led to the following vessel dimensions:

- Length overall ....254.4 m
- Length b.p.p. ....240.0 m
- Breadth moulded ....38.0 m
- Depth moulded ....19.0 m
- Design draft ....12.0 m

A general arrangement is shown in Figure 1. [1].

Hull form

The need to accommodate six huge thrusters (5000 kW each), three forward and three aft, was also of paramount importance in the hull shape definition, by avoiding large trunks protruding the bottom which would increase the ship resistance. Also, the maintenance of the thrusters was facilitated as much as possible.

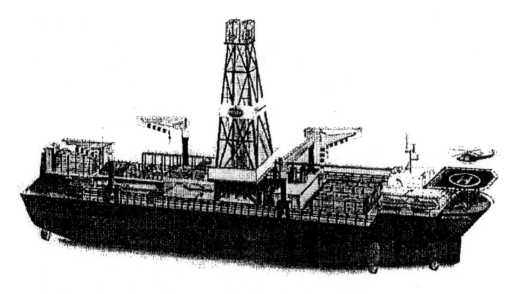

Figure 1. The Discoverer Enterprise Drilling Unit.

MARIN was committed to perform an assessment on the hydrodynamic performance of the vessel speeds for different drafts, based on their experience in multithruster propulsion. The prevention of slamming occurrence either forward or aft was also considered of a main concern, and the hull forms were adapted accordingly. Also the greenwater occurrence was specially considered in the design, moreover when the vessel can be facing the waves by stern due to the dual activity of the drilling operation.

Bilge keels of suitable width and length were fitted to reduce rolling which is critical for the drilling operation as well as for the design of the huge derrick.

Structural Design

The vessel is designed in accordance with DNV Rules (MODU) for a 20 years fatigue life at the specified area of operation (Gulf of Mexico, West of Africa and summer North Sea).

More information on vessel functions and the selection process of the main hull dimensions is given by Lopéz-Cortijo Garcia and Michel [1].

Therefore, the vessel structure complies with the loading imposed by the specified 50 years return period storm (Hurricane condition in the G.O.M.).

Furthermore, the vessel main structure as well as the substructure and derrick have been verified to comply with an ULS condition associated to a 50 years North Sea winter storm by the beam.

2. DP SYSTEM AND OPTIMIZATION

The unit is assigned the DNV class notation DYNPOS AUTR. One of the thrusters in the machinery spaces forward and aft, is segregated from the other two, in order to provide an enhanced integrity against fire or flooding scenarios.

The number and size of thrusters installed proved to be sufficient for operation of the vessel under the specified maximum environmental conditions mentioned earlier. This was verified during the model tests, as will be described later.

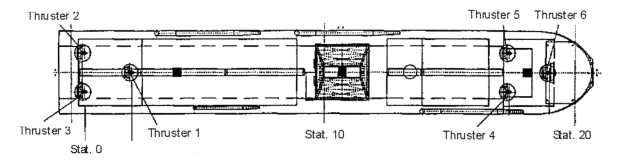

Figure 2. Thruster lay-out on the model

The thruster system consists of six variable speed azimuthing thrusters (three forward and three aft), of 5000kW each. The units can be dismounted in site for repair or maintenance, without the need of external means.

Adequate control systems, reference systems and environmental sensors are provided in accordance with the rules and regulations.

2.1. DP allocation strategy

The Discoverer Enterprise thruster layout is shown in Figure 2. The thrusters are azimuthing and RPM controlled. They cannot change the direction of rotation of the propeller. The nominal thrust capacity is 874 kN. The thrusters are grouped in a triangle with the forward unit on the centre line of the ship.

For this thruster layout a thrust allocation had to be designed which would result in a high effectiveness for heading control as well as for surge and - to a lesser extent - sway control. The following considerations played a role in the design:

- Combined heading and sway control puts a high thrust requirement on the forward thrusters. Heading control is very important because loss of heading will immediately lead to a sharp increase of the transverse environmental load and sideways drift-off.
- As long as the thrusters are (nominally) able to generate the required forces, the positioning is governed by the PID feedback control coefficients. However, if more force is required (saturation), the priorities in the allocation determine the positioning (i.c. the status and trend of position error).

- The relatively small distance between the three azimuthing thrusters at the bow as well as between those at the stern, implies situations of thruster-thruster interactions with severe thrust degradation.
- The azimuthing of the thrusters allows to orient the thrust vector T in such way that a significant transverse force can be generated without loosing much force in longitudinal direction: e.g. at a thruster azimuth of 45 degr., 0.7 of T acts transverse and 0.7 of T acts longitudinal.
- Wind feed forward would have to be applied, because the wind load on the vessel is a significant part of the total environmental force.

2.2. DP control

DP model testing requires a real time automatic closed loop feedback system. as shown in Figure 3 [2]. In the model basin the motions are measured and an Extended Kalman filter is used to calculate a low frequency position estimate (eq. 3) from the inputs (eq. 1) and (eq. 2), being the mathematical model prediction of the LF ship motions and the measured total motions respectively:

$$\hat{x}_{k+1} = A_k\,\hat{x}_k + B.u(k) + G.W_k \qquad (1)$$
$$y_k = C\,x_k + V_k \qquad (2)$$
$$\hat{x}_{k+1} = A_k\,\hat{x}_k + B.u(k) + K_k.(y_k - C.\hat{x}_k) \qquad (3)$$

in which k is the time step, A_k is the transition matrix basically derived from Newton's 2nd Law, u(k) is the thruster action, W and V represent the noise in the mathematical model and in the measurement respectively, while B, G and C are matrices to match the dimensions of the various contributions to the true state vector $\hat{x}_k$.

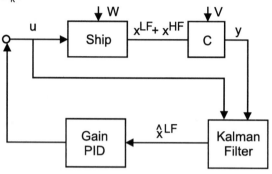

Figure 3. Control loop for DP model testing

The noise vectors W and V determine the Kalman Gain K_k (eq. 3) in such a way that for low measurement noise V a high gain results. Then, the Kalman position estimate will closely follow the measurements. The opposite may also occur: a high measurement noise will result in low Kalman Gain and the position estimate will rely more on the mathematical prediction.

The drift velocity and the position errors (i.e. the difference between the required position and the estimated low frequency position) is input for the PID controller. This yields the required horizontal forces and yaw moment to keep on station:

$$F_x^{req} = -P_x . \Delta \hat{x} - D_x . d/dt\,(\hat{x}) - I_x . \int \hat{x}.dt \qquad (4)$$

An optimum thrust allocation distributes the required forces and moment over the available thrusters.

Optimisation

Prior to the design tests, the allocation and the control settings were optimised in 'wind only' and in 'wind plus waves'. The optimisation concerns the KALMAN gains for surge, sway and yaw, and the PID control coefficients. The control settings had to generate stable positioning and effectively use of the limited possibilities of the thrusters. The following aspects played a role:

- Since the thrusters could not reverse RPM, too frequent changes in sign of the required force had to be avoided. The time to turn 180 degrees is 15 seconds, which is rather long and may lead to destabilisation if the thrust requirement is building up after the change of sign.
- The heading control would be very important, especially in those conditions in which the DP is taxed to the limits.

In table below, the range of control parameter variation and the effect on the positioning is shown.

DP try-outs in 50-year storm (case A, B and C) and in Hurricane (case D) with the heading set-point into the waves.

Case	Px	Dx	Py	Dy	Pø	Dø	K1	K2	K3	Positioning
A	150	2500	250	3000	1.E6	1.E6	5.E-4	5.E-4	5.E-7	reasonable
B	150	2500	150	3000	1.E6	6.E6	5.E-4	5.E-4	5.E-7	good
C	150	4000	150	3000	1.E6	6.E6	5.E-4	5.E-4	5.E-8	easy
D	150	4000	100	2000	4.E6	6.E6	5.E-4	5.E-4	5.E-8	hovering capabilities

Case A showed inaccurate positioning because the heading control was insufficient. By reducing the sway restoring (which tends to oppose heading control) and by increasing the yaw damping, a significant improvement was obtained in Case B. This allowed better Kalman gains for yaw, applied in Case C. Furthermore, the surge damping in Case C was enhanced to improve positioning accuracy. On this optimised control the test series in the

50 Year storm condition were carried out.

Case D shows the coefficients used for the 50-year Hurricane test, in which hovering capability was investigated. Comparing the control settings for the 50-year storm and the Hurricane conditions, it appears that the hovering Hurricane condition requires low sway restoring control coefficient and an increased heading restoring coefficient. Note that for the heading control the restoring may be put at a very high level compared to the damping. This "limit condition" strategy results in a vessel positioning response with maximised heading control for saturated thrusters.

DP accuracy as function of sea state and water depth

Figure 4 shows the general dependency of positioning accuracy on of the sea state. The main conclusion from it is that the positioning is quite good until frequent thruster saturation starts to occur. Systematic variation of PID control parameters in closed loop model tests has shown that the optimum control coefficients provide good positioning accuracy as well as economic power consumption. The optimum coefficients, however, depend slightly on the sea state (= environmental forces) as long as thruster saturation is not too frequent.

Near the limit sea state, it may show to be beneficial to adapt the control coefficients for heading and sway as discussed before.

The dependency on water depth is marginal: it plays a role through the environmental forces (e.g. drift forces may increase due to shallow water effects), but this aspect applies only for water depths below 100 m. Since the most economic power utilisation practically coincides with the optimum of positioning accuracy, little can be gained from 'loosening' the control coefficients if the positioning requirements are low.

3. MODEL TESTS

Model test scope

The purpose of the tests was to study the DP capability of the vessel in stand-by and limit operational conditions. Moreover, its hovering capability in Hurricane conditions was investigated. The results of the model tests will provide design information and will be used to evaluate of the vessel workability. The tests were carried out in a number of sea states with different orientations of wind, wave and current which apply to typical Gulf of Mexico sea conditions, see the table on the next page

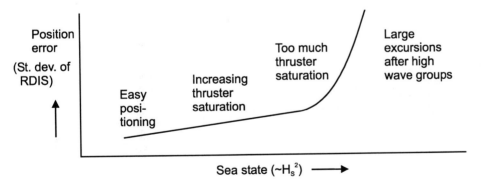

Figure 4. Position error in various sea states

Environmental conditions for DP test programme

	Pierson-Moskowitz spectrum type				Wind		Current	
	Sign.height (m)	Peak period (s)	Dir. (deg)	Duration (hour)	Speed (m/s)	Dir. (deg)	Speed (m/s)	Dir. (deg)
50-year Hurricane	12.00	14.00	200	3.0	42.2	200	1.16	180
50-year storm: Parallel	7.30	12.00	200	3.0	24.2	200	0.46	180
50-year storm: Transverse	7.30	12.00	270	3.0	24.2	240	0.46	180
1-year storm	4.60	9.50	180	0.5	19.9	180	0.37	180

The model test program consisted among others of the following test series:

- Stationary tests in calm water;
- Full DP tests;
- Turning tests;
- Squall tests, see [1].

3.1. Stationary tests in calm water

The initial part of the test program concerns stationary tests to obtain specific information on thrust degradation and to calibrate the wind forces on the vessel for different headings. To this purpose, the model was hold stationary between light weight rods, with force transducers in it to measure Fx, Fy and Mz in the horizontal plane.

Thrust degradation effects

All thrusters work on the principle of accelerating water. So, there is a suction side flow and a jet flow. The suction flow is characterised by relatively low flow velocity over a wide area, while the jet flow is high speed and concentrated in a relatively small cross section. Furthermore, the jet may induce other flow patterns, depending on the local hull form and the intensity and direction of the jet.

Most of the thruster-hull interactions are based on the forces induced on the hull by the suction flow (the Coanda effect, see Ref. [4]), the jet and the jet-induced flow. These forces may be of viscous origin, of potential pressure origin or they may be caused by blockage effects. An example of the last effect is the force caused when the jet hits a skeg, bilge keel or shaft.

Thruster-hull interaction is important, and depends on the type of thruster and its location on the vessel, the intensity of the jet and current. A review of mechanisms and effects is given in the table below.

Thruster type	Location on ship	Mechanism	See Ref.
Tunnel	Transverse through hull, at bow or stern	Suction flow and jet induced flow interacts with hull	[3]
Azimuthing/fixed	Under keel, at bow or stern	Jet flow friction and jet induced flow inter action on hull	[4]
Fixed waterjet	Inlet under keel, jet outlet at sides of ship, under or above water	Jet induced flow interacts with hull if outlet under water	[5]
Rotating waterjet	Inlet and outlet in one rotatable unit under keel	Jet flow friction and jet induced flow inter action on hull	-
Main propeller(s)	At the ship stern	Suction flow interaction on hull in forward thrust mode, strong jet flow interaction in astern thrust mode	[3]

For the present vessel some indication of the magnitude of thrust degradation could be obtained from the static thruster performance tests.

Measured thrust losses

The open water thrust/RPM curves are compared with the built-in thrust/RPM curves, for two centreline thrusters and two port side thrusters (Nos. 1 and 4 and Nos. 2 and 5 respectively). Due to symmetry, the thrusters Nos. 3 and 6 were not investigated. For each of the investigated thrusters, the degradation due to thruster-hull interaction differs due to their unique position. Additional thrust loss is found due to thruster-thruster interaction.

In principle, the accelerated fluid in the thruster jet causes the forces which lead to the observed thrust degradation. Some results are shown in Figures 5 and 6.

Figure 5 shows that the port side aft thruster, No. 2, suffers little degradation for conditions in which the jet flows aft or to port. The proximity of the thruster to the side makes the jet flow separate from the hull at the side, so that negligible Coanda effect is found. For the jet flowing to starboard, large degradation results from the interference with starboard thruster and Coanda effect.

Due to the long area of the hull-jet flow interference, the thrust degradation of the bow thrusters (Figure 6) when giving forward thrust is significant. It is mostly caused by frictional effects.

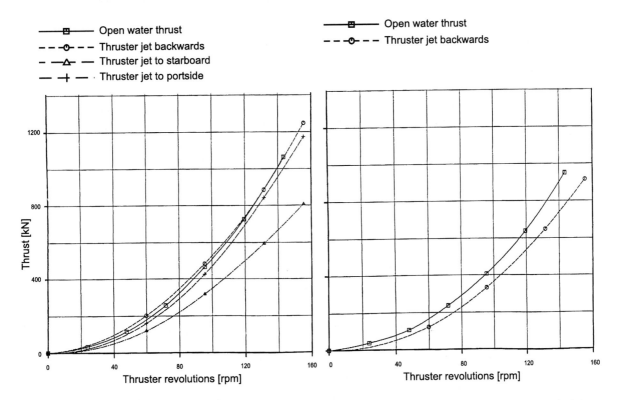

Figure 5.   Thrust revolutions characteristics of port side aft thruster (2)

Figure 6.   Thrust revolutions characteristics of centre fore thruster (4)

3.2. Full DP tests

With respect to the present testing, the use of a closed loop DP system at model scale, steering 6 individual azimuthing thrusters is a novel feature and will be described below.

Instrumentation

In order to properly investigate the DP system of the vessel, six azimuthing thruster models were manufactured at the MARIN

1078

workshops. Azimuth and propellers RPM are driven by servo motors and controlled by a computer which could be fitted in the model. The DP control program RUNSIM controls the thrusters on basis of a Kalman filter and a PID controller. "Soll" value of thruster RPM and azimuth are communicated through RS232 to the computer in the model.

The program RUNSIM allows real time display of position estimate and thrust allocation. Another important feature is the possibility to modify control settings and position set points during the process, allowing control optimisation.

The thruster system consisted of six model thrusters matching the full-scale characteristics of thrust, azimuthing speed, and RPM control speed. The thrusters were controlled in three mains groups, viz. group one contained the three aft thrusters, group two contained the two lateral forward thrusters, and group three consisted of only the forward most center line thruster. While each group was controlled independently, all thrusters within one group were given the same commands with regard to RPM and heading. Differences within groups would occur as the result of the applying "forbidden" sectors in the azimuth etc.

Wave and Current basin

The DP tests were performed in the Wave and Current Basin at MARIN, which corresponds to a full-scale water depth of 140 meters. Wind forces were generated using a series of wind fans. Current was generated by means of water pumps. Waves were generated by flap-type wave generators.

DP tests in limit design condition

The criteria for investigation of the thruster system was to maintain station and heading in the- 50-year storm cases, and to maintain heading only in the 50-years Hurricane condition with an acceptable level of 100% thruster utilisation (saturation). Further, a "one thruster down" condition was run in the 50-year storm case.

The 50-years storm tests were run in two

sets, "parallel" with wind, wave and current coming from the same direction, and "transverse" where there is separation between the headings of wind, wave and current. The vessel was also tested with the stern into the environment. The results of the tests in parallel condition showed that the vessel could maintain heading and station, with reasonable thruster saturation (approximately 25%) within a vessel heading angle of approximately +/-20° off the bow (see Figure 7). The ability to maintain heading was easier in the transverse weather condition, as the heading set point was between the wind and wave, reducing the yaw moment on the vessel. The window for a stern heading was not tested in these tests, but is expected to be slightly less than the bow-on case due to higher wave drift forces on the blunt stern.

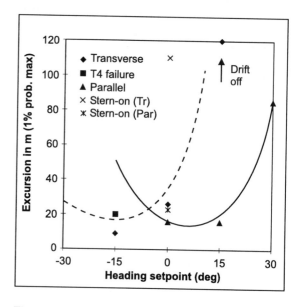

Figure 7. The relation between heading set-point and positioning accuracy

In the "one thruster down" case, the thruster made inactive was the foreward center line thruster. This thruster was selected, as it was the most critical for heading control. Further, the DP allocation logic was not modified for this thruster being out of service. These parameters were considered as a worst scenario. Even in this case, the thruster system was able to maintain heading and station, though the allowable heading window was less than for the intact case, as shown in Figure 8.

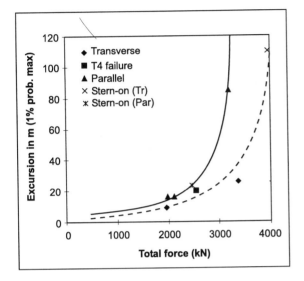

Figure 8. The relation between mean thrust and positioning accuracy

The 50-year Hurricane condition test was performed twice; once with the nominal thrust capacity, and a second time with the thrust increased 15%. The second case was run as the thruster system has an allowable overload capacity built into it, and the effect on heading recovery time was to be studied. In both cases, the thruster system demonstrated the capability to hover in the environment, whereas the 15% overload capability visibly improved heading control.

3.3. Turning tests

The Discoverer Enterprise is a dual activity drillship. This dual activity allows well operations to be performed simultaneously from the two rotary tables located on the vessel centreline. The possibility exists, then, that the vessel may have two well strings connected to the seafloor at any given time. In this scenario, the vessel does not have the liberty to maintain

vessel heading into the environment at all times. The possibility exists, then, for the requirement to turn the vessel 180° to re-align the vessel bow into the weather. To test the thruster system's capability to maintain station and heading in the 1-year storm operating case, a series of tests were performed.

The tests consisted of maintaining station and heading for 30 minutes (full scale), then rotating the vessel 30° more, etc., until the vessel had rotated 180°. The thruster system was able to maintain station and heading throughout the series of tests, with the beam-on case resulting in the highest percentage of thruster saturation, Figure 9. It was concluded from the combined rotation steps that a total turn of the vessel would be accomplished in about 20 minutes.

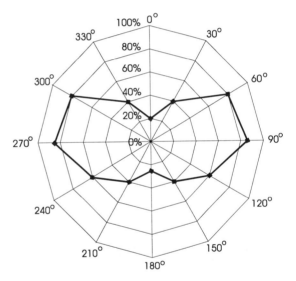

Figure 9. One-year storm - mean thruster utilization

4. CONCLUSIONS

The main purpose of the model tests described in this paper was to optimise and evaluate the DP positioning capability for an FPSO/Drillship designed for Gulf of Mexico (GoM) conditions.
- DP system optimisation in terms of optimising thruster layout, allocation logic and PID control could efficiently be carried

out with help of model tests utilising a closed loop DP system and modelling of all thrusters.

- In the limit operational conditions, corresponding to a 50-year storm in the GoM (non-Hurricane) a heading set-point window of about 40 degr. (20 degr. each side of the optimum heading) is feasible for optimum PID control coefficients. Stern-to-wave positioning also is possible, though at a somewhat smaller set-point window because at this orientation more thrust is required.
- In the 1-year storm condition, the positioning is possible at almost any heading: for beam and aft-of-beam wave directions high thruster utilisation was measured with significant saturation and only 'reasonable' positioning accuracy. In such a condition it is possible to make a 180 degrees turn of the vessel in about 20 minutes.
- In the survival conditions, corresponding to a 50-year Hurricane in the GoM, hovering is expected to be feasible.

5. REFERENCES

1. Lopéz-Cortijo Garcia, J. and Michel R.P.: "Naval Architectural Design of a 5th Generation Drillship", 9th Deep Offshore Technology, The Hague, The Netherlands, 1997.

2. Aalbers, A.B. and Merchant A.A.: "The Hydrodynamic Model Testing for Closed Loop DP Assisted Mooring", Offshore Technology Conference, Houston, U.S.A., 1996.

3. Nienhuis, U.: "Analysis of Thruster Effectivity for Dynamic Positioning and Low Speed Manoeuvring" PhD Thesis, Delft University of Technology, The Netherlands, 1992.

4. Nienhuis, U.: Propulsive Aspects of Dynamically Positioned Semi-submersibles", Conference on Stationing and Stability of Semi-submersibles, Strathclyde University, Glasgow, United Kingdom, 1986.

5. Brix, J.: "Querstrahl Steure", Forschungszentrum der Deutsche Schiffbau, Bericht No. 80, 1978.

1998 Elsevier Science B.V.
Practical Design of Ships and Mobile Units
M.W.C. Oosterveld and S.G. Tan, editors.

Mathematical description of Green function for radiation problem of floating structures in waves

Y.Y.Wang, K.Qian and D.Z.Wang

Department of Naval Architecture and Ocean Engineering

Dalian University of Technology

2 Linggong Rd. Dalian 116024, China

Based on the boundary value conditions of Green function and the analysis of define and particular solutions a generating function of Green function for floating structure with forward speed at finite water depth in waves is deduced and can be applied to all of cases such as deep water and zero-speed problems. Also relative numerical procedure is discussed including oscillation and increasing amplitude for integrand of Green function at special computational area.

1. INTRODUCTION

The radiation problem for floating structures such as ships and platforms is one of hydrodynamic problems in the field of naval architecture and ocean engineering and can be applied to predict motion and load encountered by ship and floating body in waves. The Green function theory or the so-called source-sink distribution approach with the boundary element method is one of efficient numerical procedures in the engineering practice.

For ship and floating structure they are working in the sea-way with infinite field at the horizontal plan and with finite field on the vertical plan and in the floating or the navigating state with zero or forward speed. Therefore the Green function can be expressed into the following different cases for the practical application as Tab. 1. The basic formula is G_1 for the case with finite water depth and with forward speed and it can be deduced with the degenerate transformation into G_2, G_3 and G_4 for all of other cases respectively.

To be numerical procedure a treatment of oscillation with high frequency and of increasing

amplitude at a certain computational area for integrand of Green function are introduced by using variable replacement approach respectively.

Table 1
Green function for different cases

Green Function	with finite water depth	with infinite water depth
with forward speed	G_1	G_2
with zero speed	G_3	G_4

Some calculating examples are given including the comparison with experimental and published data and it can be seen that the mathematical model for the Green function at different boundary condition is efficient. The calculation of radiation problem with three dimensional method for floating structures should be developed in order to replace the strip approach.

1082

2. DEDUCTION FOR GREEN FUNCTION

2.1 Definite conditions

It well known that the continuity [L] in the fluid domain, the kinematic and dynamic conditions [F] at the free surface, the condition of no flow through the sea bed [B] on sea bottom and the radiation condition [R] in the far way for the radiation problem should be satisfied for incompressible and inviscid fluid and for irrotational motion as Eq. 1. All of these conditions consist of the radiation problem of the fluid flow for floating structures in waves.

2.2 Particular solutions

It is assumed that the particular solution for Green function can be expressed as Eq. 2 [1] in which the relation between radius r_1, r_1', r_2, r_2' and corresponding coordinates is listed as Eq. 3 [2]. The nomenclature for all of variable symbols is the same with reference [3].

Calling Fourier transformation for $1/r$ and considering z<0, ζ<0, 2h>0, and z>ζ (on the free surface) the following relations can be deduced as Eq. 4 to be the particular solution partly in Eq. 2.

2.3 Definite solutions

To obtain the definite solution for Green function the Eq. 2 is substituted into [F] condition in Eq. 1 and the relation formula can be written as Eq. 5. Now it is assumed again about G* as Eq. 6 and substituting it into the right of Eq. 5 then the function A(m,θ) can be solved as Eq. 7. It can be found that the function of A(m,θ) to be an integral kernel for G* is composed with effect of 3 terms on flow field parameters, i.e. water depth, free surface form and forward speed. Further deduction for Eq. 7 the simplification formal can be obtained as Eq. 8.

Then the Green function of radiation problem for floating structures at finite water depth with forward speed can be written as Eq. 9 in which the first term is the result for floating structures at finite water depth with zero speed and the second term is considering the effect of forward speed especially. In Eq. 9

$v = \dfrac{\omega^2}{g} = kth(kh)$. It should be pointed that Eq. 9 is the general solution for the radiation problem in fact and corresponding equation for other cases of the radiation problem can be rededuced from Eq. 9 with particular conditions, i.e. the so-called degenerate basic solution in mathematically.

3. REGENERATE BASIC SOLUTIONS

3.1 The case of deep water with forward speed

Now it is just considered that the water depth, h in Eq. 9 should be taken an infinite value, or more than half wave length and $\omega^2 = kgth(mh) = kg$. Then it can be found that $G^*_1 = 0$, and G^*_2 is regenerated into Eq. 10. Therefore Eq. 11 is the result for the case of deep water with forward speed corresponding regenerate condition from Eq. 2 in which term of $(\dfrac{1}{r_2} - \dfrac{1}{r_2'})$ is disappearance and Green function is defined as $G = \dfrac{1}{r_1} - \dfrac{1}{r_1'} + G^*$.

3.2 The case of deep water with zero speed

For this case taking $h = \infty$ and $U = 0$ in Eq. 11, then Eq.s 12 and 13 can be deduced and corresponding regenerate condition is the same with section 3.1.

3.3 The case of finite water depth with zero speed

Taking $U = 0$ Eq. 9 can be regenerated into Eq. 14 with corresponding condition in Eq. 2. On another occasion, it is assumed that $G = \dfrac{1}{r_1} + \dfrac{1}{r_2} + G^{**}$, i.e. $G^{**} = G^* - \dfrac{1}{r_1'} - \dfrac{1}{r_2'}$, then A_0 and G^{**} can be written as Eq.s 15 and 16 to be a general form.

4. SUPPLEMENT FOR GREEN FUNCTION

4.1 On integrand of Green function

It is found that dreadful oscillation and increasing

$$[L] \quad \nabla^2 G(P,Q) = \delta(P-Q)$$

$$[F] \quad -\nu g G + 2iU\sqrt{\nu g}\,\frac{\partial G}{\partial x} + U^2\,\frac{\partial^2 G}{\partial x^2} + g\,\frac{\partial G}{\partial z} = 0$$

$$[B] \quad \left.\frac{\partial G}{\partial z}\right|_{z=-h} = 0 \tag{1}$$

$$[R] \quad suitable \ \ radiation \ \ condition$$

$$G = \frac{1}{r_1} - \frac{1}{r_1'} + \frac{1}{r_2} - \frac{1}{r_2'} + G^* \tag{2}$$

$$\begin{matrix} r_1 \\ r_1' \end{matrix} = \left[(x-\xi)^2 + (y-\eta)^2 + (z\mp\varsigma)^2\right]^{\frac{1}{2}}$$

$$\begin{matrix} r_2 \\ r_2' \end{matrix} = \left[(x-\xi)^2 + (y-\eta)^2 + (z+2h\pm\varsigma)^2\right]^{\frac{1}{2}} \tag{3}$$

$$\frac{1}{r_1} - \frac{1}{r_1'} = \frac{1}{2\pi}\int_0^\infty\int_{-\pi}^\pi e^{im[(x-\xi)\cos\theta+(y-\eta)\sin\theta]}\left[e^{-m(z-\varsigma)} - e^{m(z+\varsigma)}\right]d\theta dm$$

$$\frac{1}{r_2} - \frac{1}{r_2'} = \frac{1}{2\pi}\int_0^\infty\int_{-\pi}^\pi e^{im[(x-\xi)\cos\theta+(y-\eta)\sin\theta]}\left[e^{-m(z+2h+\varsigma)} - e^{m(z+2h-\varsigma)}\right]d\theta dm \tag{4}$$

$$-\frac{1}{2\pi}\int_0^\infty\int_{-\pi}^\pi\left\{2mge^{m\varsigma} - 2\left[\left(\sqrt{\nu g} + mU\cos\theta\right)^2 + mg\right]e^{-2mh}sh(m\varsigma)\right\}e^{im[(x-\xi)\cos\theta+(y-\eta)\sin\theta]}d\theta dm$$

$$= \nu g G^* - 2iU\sqrt{\nu g}\,\frac{\partial G^*}{\partial x} - U^2\,\frac{\partial^2 G^*}{\partial x^2} - g\,\frac{\partial G^*}{\partial z} \tag{5}$$

$$G^* = \frac{1}{2\pi}\int_0^\infty\int_{-\pi}^\pi A(m,\theta)ch[m(z+h)]\,e^{im[(x-\xi)\cos\theta+(y-\eta)\sin\theta]}d\theta dm \tag{6}$$

$$A(m,\theta) = \frac{-2mge^{-mh}ch[m(\varsigma+h)] - 2mge^{-mh}e^{m\varsigma}sh(mh) + 2\left(\sqrt{\nu g}+mU\cos\theta\right)^2 e^{-2mh}sh(m\varsigma)}{\left[\left(\sqrt{\nu g}+mU\cos\theta\right)^2 - mgth(mh)\right]ch(mh)} \tag{7}$$

$$A(m,\theta) = \frac{2e^{-2mh}sh(m\varsigma)}{ch(mh)} - \frac{2mge^{-mh}ch[m(\varsigma+h)]}{\left[\left(\sqrt{\nu g}+mU\cos\theta\right)^2 - mgth(mh)\right]ch(mh)}\left[1+th(mh)\right] \tag{8}$$

$$G^* = G_1^* + G_2^* = \frac{1}{2\pi}\int_0^\infty\int_{-\pi}^\pi \frac{2e^{-2mh}sh(m\varsigma)\,ch[m(z+h)]}{ch(mh)}e^{im[(x-\xi)\cos\theta+(y-\eta)\sin\theta]}d\theta dm$$

$$-\frac{1}{2\pi}\int_0^\infty\int_{-\pi}^\pi \frac{2mge^{-mh}ch[m(\varsigma+h)]\,ch[m(z+h)]}{ch(mh)\left[\left(\sqrt{\nu g}+mU\cos\theta\right)^2 - mgth(mh)\right]}\left[1+th(mh)\right]\,e^{im[(x-\xi)\cos\theta+(y-\eta)\sin\theta]}d\theta dm \tag{9}$$

$$G_2^* = \frac{1}{2\pi}\int_0^\infty\int_{-\pi}^\pi \frac{2mg}{mg - (\omega + mU\cos\theta)^2}e^{im[(x-\xi)\cos\theta+(y-\eta)\sin\theta]}d\theta dm \tag{10}$$

1084

$$G^* = \frac{1}{2\pi} \int_0^\infty \int_{-\pi}^\pi \frac{2mg}{mg-(\omega+mU\cos\theta)^2} e^{im[(x-\xi)\cos\theta+(y-\eta)\sin\theta]} d\theta dm \tag{11}$$

$$G^* = \frac{1}{2\pi} \int_0^\infty \int_{-\pi}^\pi \frac{2mg}{mg-\omega^2} e^{im[(x-\xi)\cos\theta+(y-\eta)\sin\theta]} d\theta dm \tag{12}$$

$$G^* = \frac{1}{2\pi} \int_0^\infty \int_{-\pi}^\pi \frac{2m}{m-k} e^{im[(x-\xi)\cos\theta+(y-\eta)\sin\theta]} d\theta dm \tag{13}$$

$$G^* = G^*_1 + G^*_2 = \frac{1}{2\pi} \int_0^\infty \int_{-\pi}^\pi \frac{2e^{-2mh}sh(m\varsigma)ch[m(z+h)]}{ch(mh)} e^{im[(x-\xi)\cos\theta+(y-\eta)\sin\theta]} d\theta dm$$

$$+ \frac{1}{2\pi} \int_0^\infty \int_{-\pi}^\pi \frac{2mge^{-mh}ch[m(\varsigma+h)]\ ch[m(z+h)]}{msh(mh)-\nu ch(mh)}[1+th(mh)]\ e^{im[(x-\xi)\cos\theta+(y-\eta)\sin\theta]} d\theta dm \tag{14}$$

$$A_0 = \frac{2(m+\nu)\ e^{-mh}ch[m(\varsigma+h)]}{msh(mh)-\nu ch(mh)} \tag{15}$$

$$G^{**} = \frac{1}{2\pi} \int_0^\infty \int_{-\pi}^\pi \frac{2(m+\nu)e^{-mh}ch[m(\varsigma+h)]\ ch[m(z+h)]}{msh(mh)-\nu ch(mh)} e^{im[(x-\xi)\cos\theta+(y-\eta)\sin\theta]} d\theta dm \tag{16}$$

$$I = \int_{\sqrt2}^\infty dt \frac{1}{\sqrt{1+\frac{4\tau}{t}}} \cdot \frac{-1}{t^2\sqrt{1-\frac{1}{t^2}}} \cdot \frac{t^2}{2}(1+\frac{2\tau}{t}+\sqrt{1+\frac{4\tau}{t}})e^{\frac{t^2}{2}(1+\frac{2\tau}{t}+\sqrt{1+\frac{4\tau}{t}})\omega}$$

$$= \int_{\sqrt2}^\infty dt \frac{-1}{\sqrt{1+\frac{4\tau}{t}}\sqrt{1-\frac{1}{t^2}}} \cdot (1+\frac{2\tau}{t}+\sqrt{1+\frac{4\tau}{t}})e^{\frac{t^2}{2}(1+\frac{2\tau}{t}+\sqrt{1+\frac{4\tau}{t}})\omega} \le \int_{\sqrt2}^\infty dt| \bullet | \tag{17}$$

$$\le \int_{\sqrt2}^\infty dt \frac{1}{\sqrt{1-\frac{1}{t^2}}} \cdot (1+\frac{2\tau}{t}+\sqrt{1+\frac{4\tau}{t}})e^{\frac{t^2}{2}(1+\frac{2\tau}{t}+\sqrt{1+\frac{4\tau}{t}})Z}$$

$$\lim_{t\to\infty}\left[\frac{1}{\sqrt{1-\frac{1}{t^2}}} \cdot (1+\frac{2\tau}{t}+\sqrt{1+\frac{4\tau}{t}})e^{\frac{t^2}{2}(1+\frac{2\tau}{t}+\sqrt{1+\frac{4\tau}{t}})Z}\right]\Big/\frac{1}{t^2} = \lim_{t\to\infty} t^2 e^{t^2 z} \overset{\because z<0}{=} 0 \tag{18}$$

$$E(M) = \int_M^\infty dt \frac{1}{\sqrt{1+\frac{4\tau}{t}}} \cdot \frac{-1}{t^2\sqrt{1-\frac{1}{t^2}}} \cdot \frac{t^2}{2}(1+\frac{2\tau}{t}+\sqrt{1+\frac{4\tau}{t}})e^{\frac{t^2}{2}(1+\frac{2\tau}{t}+\sqrt{1+\frac{4\tau}{t}})\omega}$$

$$\le \int_M^\infty dt| \bullet | \le \int_M^\infty dt \frac{1}{2\sqrt{1+\frac{4\tau}{M}}\sqrt{1-\frac{1}{M^2}}} \cdot (1+\frac{2\tau}{M}+\sqrt{1+\frac{4\tau}{M}})e^{\frac{t^2}{2}(1+\frac{2\tau}{t}+\sqrt{1+\frac{4\tau}{t}})Z} \le A\int_M^\infty \frac{1}{2}e^{\frac{t^2}{2}z} dt \tag{19}$$

$$\le A\int_M^\infty \frac{1}{2}e^{\frac{Mt}{2}z} dt$$

$$= \frac{A}{Mz}e^{\frac{Mt}{2}z}\Big|_M^\infty = -\frac{A}{Mz}e^{\frac{1}{2}M^2 z}$$

amplitude appear at $\theta = \pm\dfrac{\pi}{2}$ and $z \to 0$ for the integrand of Green function and integral accuracy is limited by integral step size. To solve this difficulty in the computation an effective technique is that the upper limit of integration, $\dfrac{\pi}{2}$ is mapped into infinite by using the variable substitution, such as $t = scs\theta$ or $t = tg\theta$ [4]. In order to make an identification the following function is selected to be example:

$$I = \int_{\pi/4}^{\pi/2} \frac{k_i e^{k_i\omega}}{\sqrt{1+4\tau\cos\theta}}\,d\theta, \quad \text{in which} \quad i=1,\ 2;$$

$$k_1, k_2 = \frac{1}{2\cos^2\theta}(1+2\tau\cos\theta \pm \sqrt{1+4\tau\cos\theta});\quad \text{and}$$

$$\tau = \frac{U\omega_e}{g}, \quad \omega = Z + i(x\cos\theta + y\sin\theta), \quad Z<0.\ \text{Now}$$

taking $i=1, t=\dfrac{1}{\cos\theta}$, then "I" can be transformed into the Eq. 17 in which $|\bullet|$ is the absolute value of integrant for the above formula. From the last formula it can be obtained as Eq. 18. Based on Cauchy criterion for infinite improper integral it is well known that the integral $\int_a^{+\infty} f(x)dx$ is convergence absolutely when

$$\lim_{x\to\infty}\frac{|f(x)|}{\dfrac{1}{x^P}} = \lim_{x\to\infty} x^P|f(x)| = K, \text{ with } 0<K<+\infty;$$

and P>1.

4.2 On accuracy of Green function

The truncation error of " I " is important part for total error of Green function and the accuracy of Green function can be measured by the analysis of the

$$A = \frac{1+\dfrac{2\tau}{M}+\sqrt{1+\dfrac{4\tau}{M}}}{\sqrt{1+\dfrac{4\tau}{M}}\sqrt{1-\dfrac{1}{M^2}}} \tag{20}$$

truncation error of "I". Let assuming point M which is far away from the origin, then the expectation of truncation error of " I " at M, E(M) can be written as

Eq. 19. In this equation, A is a finite value as expressing from Eq. 20. It can be seen that A will trend to 1 when M trends infinite. It can be found that the expectation of error at M is decreasing speedy when the distance from origin to M is increasing and is also depended on the distance of Z, i.e. the sum of vertical coordinates of source and field points. When both source and field points are close to the free surface the expectation of error at E is decreasing slowly.

5. COMPUTATIONAL EXAMPLES

A submerged slender spheroid and a floating semi-submersible model have been computed. Considering both bodies have two planes of symmetry only one-quarter of the underwater surface should be discretized. The non-dimensional response of exciting force for heave motion acting on a underway submerged spheroid is shown in Fig. 1 where 48 plane quadrilateral elements are used to discretize one-quarter of the body- surface. It can be

$$|F_3|/(\rho g\varsigma_a BL)$$

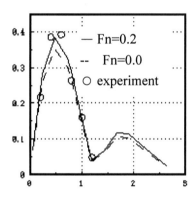

Figure 1. Wave exciting force of spheroid for heave in head wave ($\chi=180$ deg.)

found that forward speed of body hardly affects on the wave exciting force in the heading waves which has been confirmed by the model experiments. The response functions with frequency domain of linear force of heave, roll and pitch motion for the semi-submersible model given by 17th ITTC are listed at Fig.s 2, 3 and 4 respectively and corresponding results under condition of Fn=0 are also drawn in each figure provided by Wang and Li's work [5]. It should be pointed that the different of both results have to be proved by model tests in future.

$$|F_3|/(\rho g V \varsigma_a / L)$$

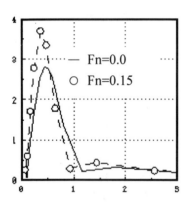

Figure 2. Wave exciting force of heave for semi-submersible in oblique wave (χ=135 deg.)

$$|F_4|/(\rho g V \varsigma_a)$$

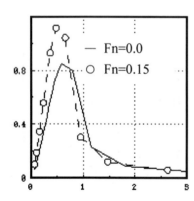

Figure 3. Wave exciting force of pitch for semi-submersible in oblique wave (χ=135 deg.)

6. CONCLUDING REMARKS

1) Eq. 9 can be seen a generating function of Green function for floating structures with forward speed at finite water depth in waves.

2) A suitable variable substitution can be used to eliminate the effect of dreadful oscillation and increasing amplitude at special computing area on truncation error to ensure computation accuracy of Green function.

$$|F_5|/(\rho g V \varsigma_a)$$

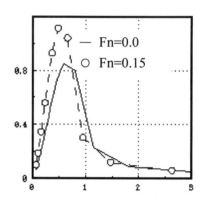

Figure 4. Wave exciting force of pitch for semi-submersible in oblique wave (χ=135 deg.)

REFERENCES

1. J.V.Wehausen and E.V.Laitone. Surface Waves. Handbuch der Physik, Springer-Verlag, Berlin, Vol.9(1960), 446-778.
2. M.Abramowtiz and I.A.Stegun. Handbook of Mathematical Functions with Formulas, Graphs, and Mathematical Tables. Government Printing Office, Washington, D.C.(1964).
3. M.Takagi and M.Ganno. A Calculation of Finite Depth Effect on Ship Motion in Waves. JSNME(Japan), No.22(1968),10-17.
4. H.Iwashita and M.Ohkusu. The Green Function Method for Ship Motions at Forward Speed. Schiffstechnik, Bd. 39(1992), 3-21.
5. Y.Wang. and F.Lie. Motion Calculation in Waves for 3-dimensional Floating Structures. Proc.The Special Offshore Symp. China(1994), 489-502.

Author Index